Student Solutions Manual

Calculus
Early Transcendental Functions

SEVENTH EDITION

Ron Larson
The Pennsylvania State University, The Behrend College

Bruce Edwards
University of Florida

Prepared by

Ron Larson
The Pennsylvania State University, The Behrend College

Bruce Edwards
University of Florida

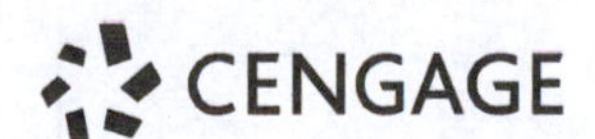

Australia • Brazil • Mexico • Singapore • United Kingdom • United States

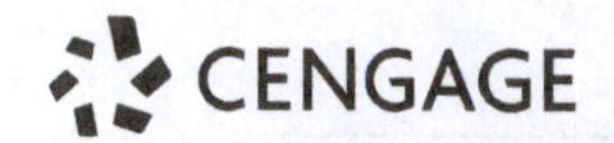

ISBN: 978-1-337-55256-1

Cengage Learning
20 Channel Center Street
Boston, MA 02210
USA

Cengage Learning is a leading provider of customized learning solutions with office locations around the globe, including Singapore, the United Kingdom, Australia, Mexico, Brazil, and Japan. Locate your local office at: **www.cengage.com/global**.

Cengage Learning products are represented in Canada by Nelson Education, Ltd.

To learn more about Cengage Learning Solutions, visit **www.cengage.com**.

Purchase any of our products at your local college store or at our preferred online store **www.cengagebrain.com**.

Printed in the United States of America
Print Number: 01 Print Year: 2018

CONTENTS

CHAPTER 1
Preparation for Calculus

CHAPTER 1
Preparation for Calculus

Section 1.1 Graphs and Models

1. To find the x-intercepts of the graph of an equation, let y be zero and solve the equation for x. To find the y-intercepts of the graph of an equation, let x be zero and solve the equation for y.

3. $y = -\frac{3}{2}x + 3$

   x-intercept: $(2, 0)$

   y-intercept: $(0, 3)$

   Matches graph (b).

4. $y = \sqrt{9 - x^2}$

   x-intercepts: $(-3, 0), (3, 0)$

   y-intercept: $(0, 3)$

   Matches graph (d).

5. $y = 3 - x^2$

   x-intercepts: $(\sqrt{3}, 0), (-\sqrt{3}, 0)$

   y-intercept: $(0, 3)$

   Matches graph (a).

6. $y = x^3 - x$

   x-intercepts: $(0, 0), (-1, 0), (1, 0)$

   y-intercept: $(0, 0)$

   Matches graph (c).

7. $y = \frac{1}{2}x + 2$

x	-4	-2	0	2	4
y	0	1	2	3	4

9. $y = 4 - x^2$

x	-3	-2	0	2	3
y	-5	0	4	0	-5

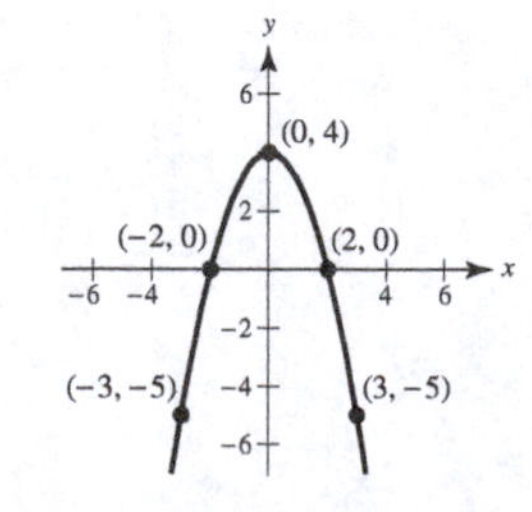

11. $y = |x + 1|$

x	-4	-3	-2	-1	0	1	2
y	3	2	1	0	1	2	3

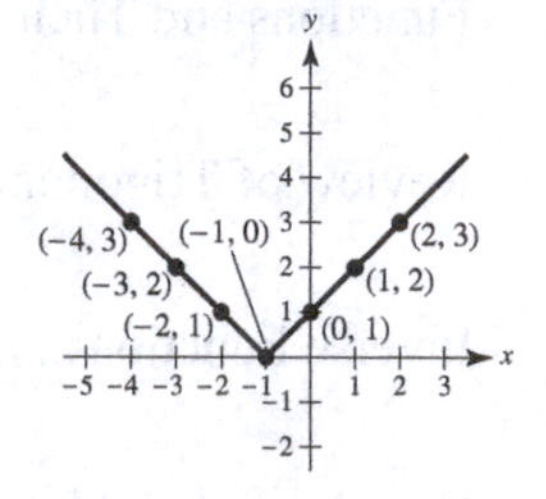

13. $y = \sqrt{x} - 6$

x	0	1	4	9	16
y	-6	-5	-4	-3	-2

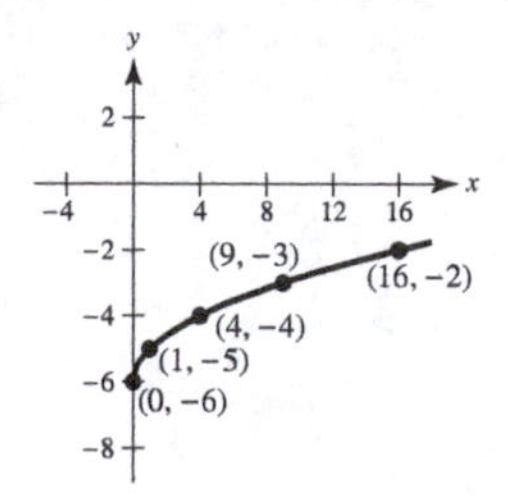

15. $y = \dfrac{3}{x}$

x	-3	-2	-1	0	1	2	3
y	-1	$-\frac{3}{2}$	-3	Undef.	3	$\frac{3}{2}$	1

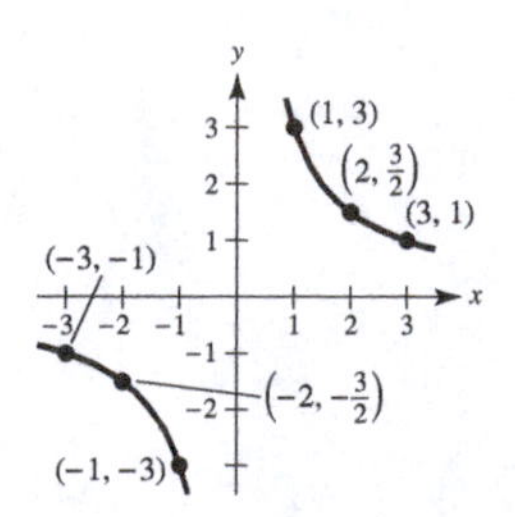

17. $y = \sqrt{5 - x}$

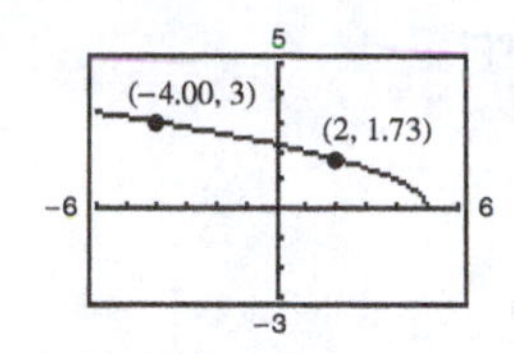

(a) $(2, y) = (2, 1.73)$ $\left(y = \sqrt{5 - 2} = \sqrt{3} \approx 1.73\right)$

(b) $(x, 3) = (-4, 3)$ $\left(3 = \sqrt{5 - (-4)}\right)$

19. $y = 2x - 5$

y-intercept: $y = 2(0) - 5 = -5$; $(0, -5)$

x-intercept: $0 = 2x - 5$

$5 = 2x$

$x = \frac{5}{2}$; $\left(\frac{5}{2}, 0\right)$

21. $y = x^2 + x - 2$

y-intercept: $y = 0^2 + 0 - 2$

$y = -2$; $(0, -2)$

x-intercepts: $0 = x^2 + x - 2$

$0 = (x + 2)(x - 1)$

$x = -2, 1$; $(-2, 0)$, $(1, 0)$

23. $y = x\sqrt{16 - x^2}$

y-intercept: $y = 0\sqrt{16 - 0^2} = 0$; $(0, 0)$

x-intercepts: $0 = x\sqrt{16 - x^2}$

$0 = x\sqrt{(4 - x)(4 + x)}$

$x = 0, 4, -4$; $(0, 0)$, $(4, 0)$, $(-4, 0)$

25. $y = \dfrac{2 - \sqrt{x}}{5x + 1}$

y-intercept: $y = \dfrac{2 - \sqrt{0}}{5(0) + 1} = 2$; $(0, 2)$

x-intercept: $0 = \dfrac{2 - \sqrt{x}}{5x + 1}$

$0 = 2 - \sqrt{x}$

$x = 4$; $(4, 0)$

27. $x^2y - x^2 + 4y = 0$

y-intercept: $0^2(y) - 0^2 + 4y = 0$

$y = 0$; $(0, 0)$

x-intercept: $x^2(0) - x^2 + 4(0) = 0$

$x = 0$; $(0, 0)$

29. Symmetric with respect to the y-axis because $y = (-x)^2 - 6 = x^2 - 6$.

31. Symmetric with respect to the x-axis because $(-y)^2 = y^2 = x^3 - 8x$.

33. Symmetric with respect to the origin because $(-x)(-y) = xy = 4$.

35. $y = 4 - \sqrt{x + 3}$

No symmetry with respect to either axis or the origin.

37. Symmetric with respect to the origin because

$$-y = \frac{-x}{(-x)^2 + 1}$$

$$y = \frac{x}{x^2 + 1}.$$

39. $y = \left|x^3 + x\right|$ is symmetric with respect to the y-axis because $y = \left|(-x)^3 + (-x)\right| = \left|-(x^3 + x)\right| = \left|x^3 + x\right|$.

41. $y = 2 - 3x$

$y = 2 - 3(0) = 2$, y-intercept

$0 = 2 - 3(x) \Rightarrow 3x = 2 \Rightarrow x = \frac{2}{3}$, x-intercept

Intercepts: $(0, 2)$, $\left(\frac{2}{3}, 0\right)$

Symmetry: none

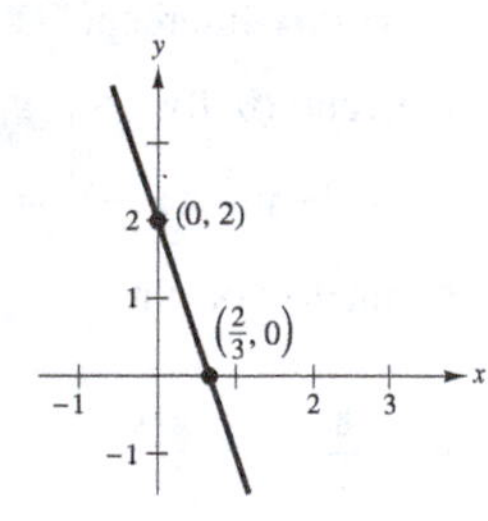

43. $y = 9 - x^2$

$y = 9 - (0)^2 = 9$, y-intercept

$0 = 9 - x^2 \Rightarrow x^2 = 9 \Rightarrow x = \pm 3$, x-intercepts

Intercepts: $(0, 9)$, $(3, 0)$, $(-3, 0)$

$y = 9 - (-x)^2 = 9 - x^2$

Symmetry: y-axis

45. $y = x^3 + 2$

$y = 0^3 + 2 = 2$, y-intercept

$0 = x^3 + 2 \Rightarrow x^3 = -2 \Rightarrow x = -\sqrt[3]{2}$, x-intercept

Intercepts: $\left(-\sqrt[3]{2}, 0\right)$, $(0, 2)$

Symmetry: none

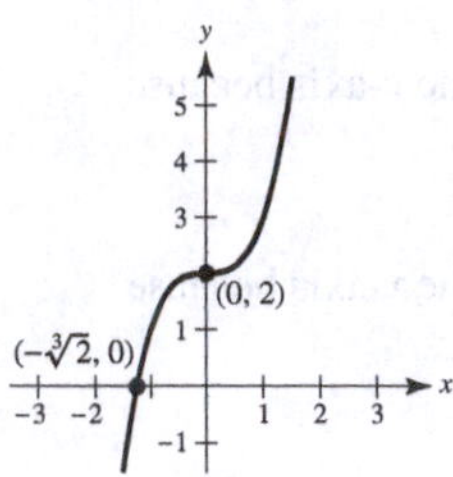

47. $y = x\sqrt{x + 5}$

$y = 0\sqrt{0 + 5} = 0$, y-intercept

$x\sqrt{x + 5} = 0 \Rightarrow x = 0, -5$, x-intercepts

Intercepts: $(0, 0)$, $(-5, 0)$

Symmetry: none

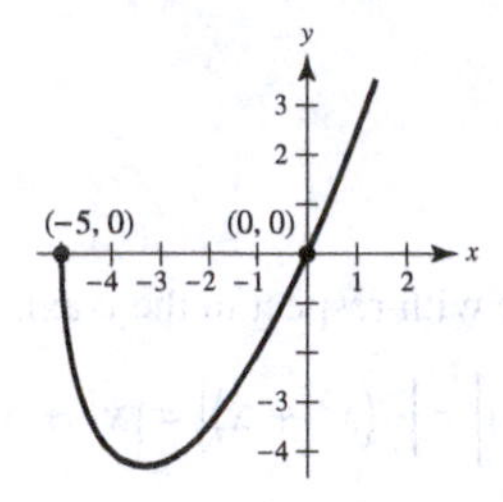

49. $x = y^3$

$y^3 = 0 \Rightarrow y = 0$, y-intercept

$x = 0$, x-intercept

Intercept: $(0, 0)$

$-x = (-y)^3 \Rightarrow -x = -y^3$

Symmetry: origin

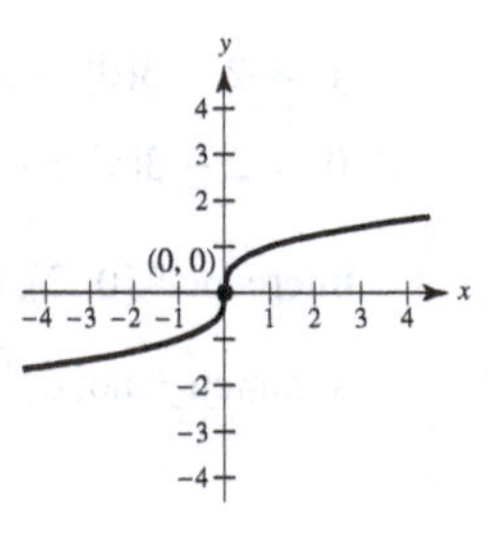

51. $y = \dfrac{8}{x}$

$y = \dfrac{8}{0} \Rightarrow$ Undefined $\Rightarrow$ no y-intercept

$\dfrac{8}{x} = 0 \Rightarrow$ No solution $\Rightarrow$ no x-intercept

Intercepts: none

$-y = \dfrac{8}{-x} \Rightarrow y = \dfrac{8}{x}$

Symmetry: origin

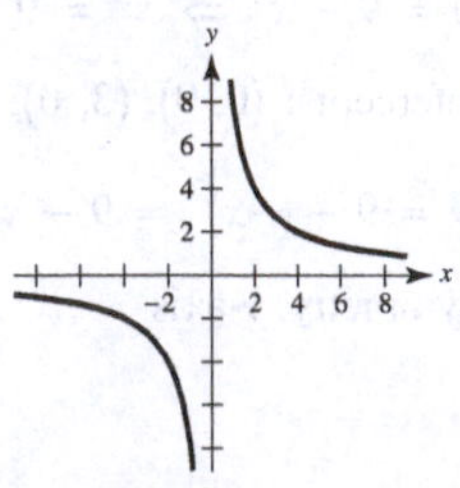

53. $y = 6 - |x|$

$y = 6 - |0| = 6$, y-intercept

$6 - |x| = 0$

$6 = |x|$

$x = \pm 6$, x-intercepts

Intercepts: $(0, 6)$, $(-6, 0)$, $(6, 0)$

$y = 6 - |-x| = 6 - |x|$

Symmetry: y-axis

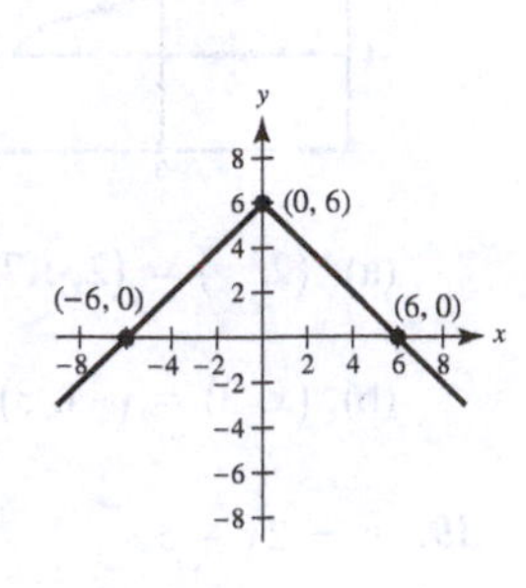

55. $3y^2 - x = 9$

$3y^2 = x + 9$

$y^2 = \frac{1}{3}x + 3$

$y = \pm\sqrt{\frac{1}{3}x + 3}$

$y = \pm\sqrt{0 + 3} = \pm\sqrt{3}$, y-intercepts

$\pm\sqrt{\frac{1}{3}x + 3} = 0$

$\frac{1}{3}x + 3 = 0$

$x = -9$, x-intercept

Intercepts:

$\left(0, \sqrt{3}\right)$, $\left(0, -\sqrt{3}\right)$, $(-9, 0)$

$3(-y)^2 - x = 3y^2 - x = 9$

Symmetry: x-axis

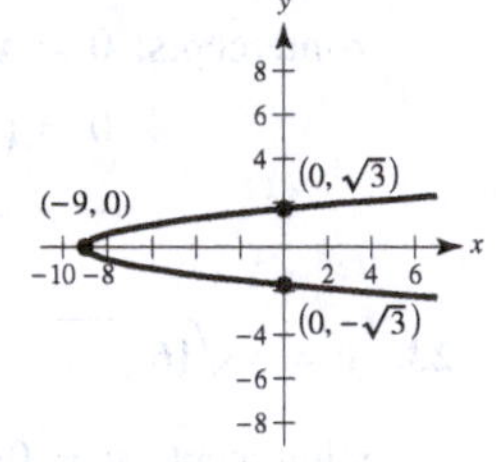

57. $x + y = 8 \Rightarrow y = 8 - x$

$4x - y = 7 \Rightarrow y = 4x - 7$

$8 - x = 4x - 7$

$15 = 5x$

$3 = x$

The corresponding y-value is $y = 5$.

Point of intersection: $(3, 5)$

59. $x^2 + y = 15 \Rightarrow y = -x^2 + 15$

$-3x + y = 11 \Rightarrow y = 3x + 11$

$-x^2 + 15 = 3x + 11$

$0 = x^2 + 3x - 4$

$0 = (x + 4)(x - 1)$

$x = -4, 1$

The corresponding y-values are $y = -1$ (for $x = -4$) and $y = 14$ (for $x = 1$).

Points of intersection: $(-4, -1)$, $(1, 14)$

61. $x^2 + y^2 = 5 \Rightarrow y^2 = 5 - x^2$

$x - y = 1 \Rightarrow y = x - 1$

$5 - x^2 = (x - 1)^2$

$5 - x^2 = x^2 - 2x + 1$

$0 = 2x^2 - 2x - 4 = 2(x + 1)(x - 2)$

$x = -1 \text{ or } x = 2$

The corresponding y-values are $y = -2$ (for $x = -1$) and $y = 1$ (for $x = 2$).

Points of intersection: $(-1, -2), (2, 1)$

63. $y = x^3 - 2x^2 + x - 1$

$y = -x^2 + 3x - 1$

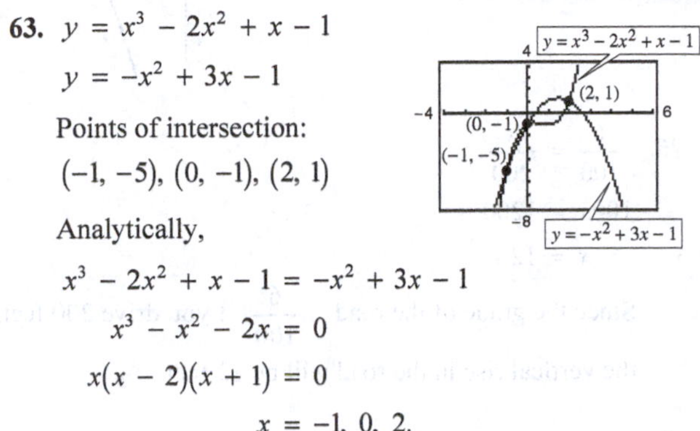

Points of intersection:

$(-1, -5), (0, -1), (2, 1)$

Analytically,

$x^3 - 2x^2 + x - 1 = -x^2 + 3x - 1$

$x^3 - x^2 - 2x = 0$

$x(x - 2)(x + 1) = 0$

$x = -1, 0, 2.$

65. $y = \sqrt{x + 6}$

$y = \sqrt{-x^2 - 4x}$

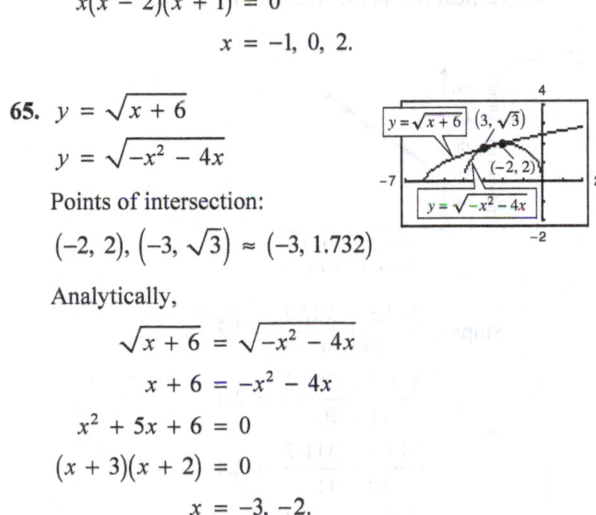

Points of intersection:

$(-2, 2), (-3, \sqrt{3}) \approx (-3, 1.732)$

Analytically,

$\sqrt{x + 6} = \sqrt{-x^2 - 4x}$

$x + 6 = -x^2 - 4x$

$x^2 + 5x + 6 = 0$

$(x + 3)(x + 2) = 0$

$x = -3, -2.$

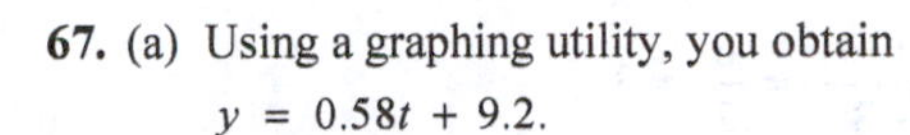

67. (a) Using a graphing utility, you obtain

$y = 0.58t + 9.2.$

(b)

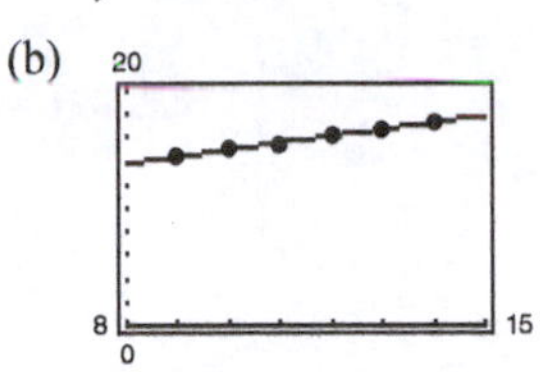

The model is a good fit for the data.

(c) For 2024, $t = 24$:

$y = 0.58(24) + 9.2 \approx 23.1$

The GDP in 2024 will be approximately \$23.1 trillion.

69.

$C = R$

$2.04x + 5600 = 3.29x$

$5600 = 3.29x - 2.04x$

$5600 = 1.25x$

$x = \dfrac{5600}{1.25} = 4480$

To break even, 4480 units must be sold.

71. Answers may vary. *Sample answer*:

$y = \left(x + \frac{3}{2}\right)(x - 4)\left(x - \frac{5}{2}\right)$ has intercepts at

$x = -\frac{3}{2}$, $x = 4$, and $x = \frac{5}{2}$.

73. Yes. Assume that the graph has x-axis and origin symmetry. If (x, y) is on the graph, so is $(x, -y)$ by x-axis symmetry. Because $(x, -y)$ is on the graph, then so is $(-x, -(-y)) = (-x, y)$ by origin symmetry. Therefore, the graph is symmetric with respect to the y-axis. The argument is similar for y-axis and origin symmetry.

75. False. x-axis symmetry means that if $(-4, -5)$ is on the graph, then $(-4, 5)$ is also on the graph. For example, $(4, -5)$ is not on the graph of $x = y^2 - 29$, whereas $(-4, -5)$ is on the graph.

77. True. The x-intercepts are $\left(\dfrac{-b \pm \sqrt{b^2 - 4ac}}{2a}, 0\right)$.

Section 1.2 Linear Models and Rates of Change

1. In the form $y = mx + b$, m is the slope and b is the y-intercept.

3. $m = 2$

5. $m = -1$

7. $m = \dfrac{2 - (-4)}{5 - 3} = \dfrac{6}{2} = 3$

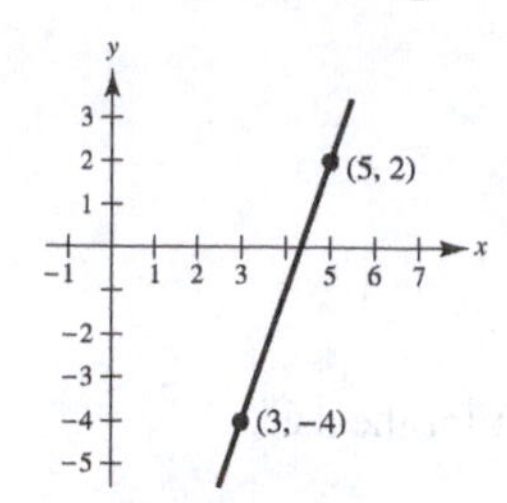

9. $m = \dfrac{1 - 6}{4 - 4} = \dfrac{-5}{0}$, undefined.

The line is vertical.

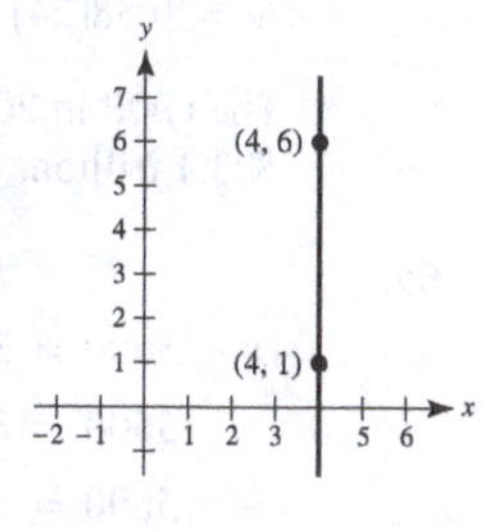

11. $m = \dfrac{\frac{2}{3} - \frac{1}{6}}{-\frac{1}{2} - \left(-\frac{3}{4}\right)} = \dfrac{\frac{1}{2}}{\frac{1}{4}} = 2$

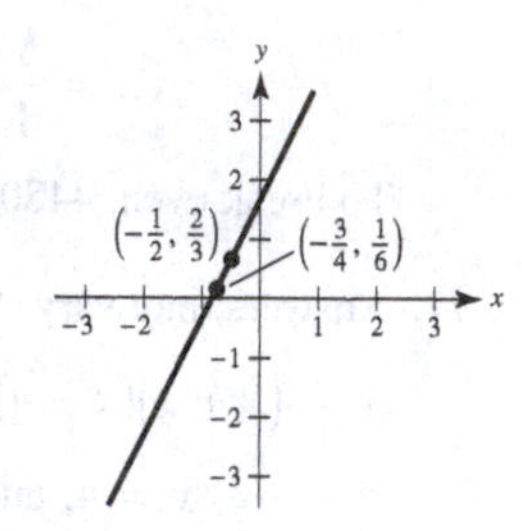

13.

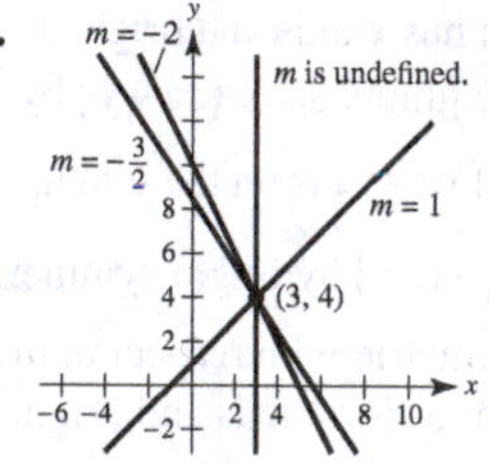

15. Because the slope is 0, the line is horizontal and its equation is $y = 2$. Therefore, three additional points are (0, 2), (1, 2), (5, 2).

17. The equation of this line is

$$y - 7 = -3(x - 1)$$
$$y = -3x + 10.$$

Therefore, three additional points are (0, 10), (2, 4), and (3, 1).

19.

$$y = \tfrac{3}{4}x + 3$$
$$4y = 3x + 12$$
$$0 = 3x - 4y + 12$$

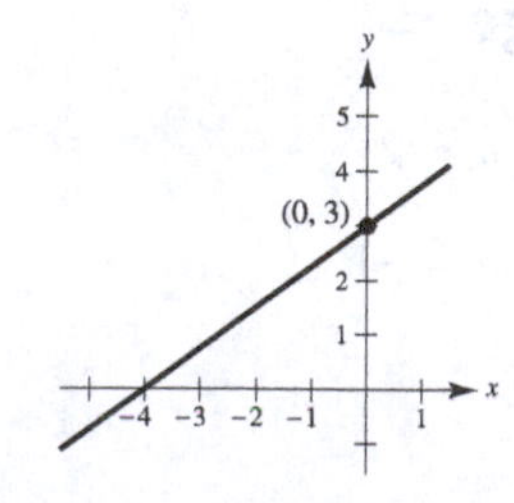

21. Because the slope is undefined, the line is vertical and its equation is $x = 1$.

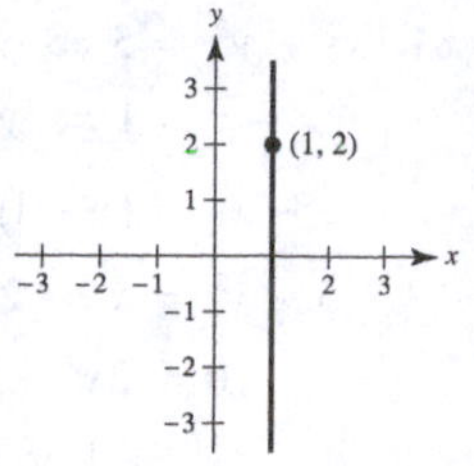

23.

$$y + 2 = 3(x - 3)$$
$$y + 2 = 3x - 9$$
$$y = 3x - 11$$
$$0 = 3x - y - 11$$

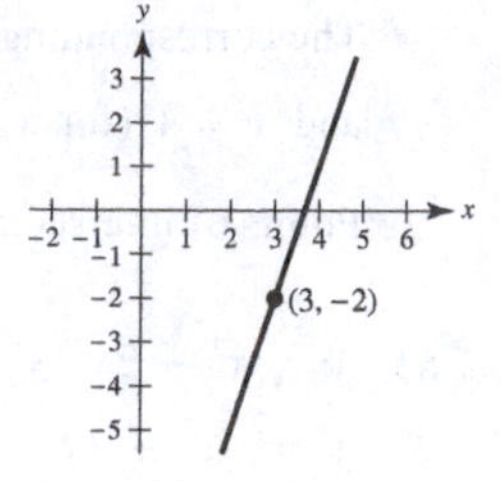

25.

$$\frac{6}{100} = \frac{x}{200}$$
$$100x = 1200$$
$$x = 12$$

Since the grade of the road is $\frac{6}{100}$, if you drive 200 feet, the vertical rise in the road will be 12 feet.

27. (a)

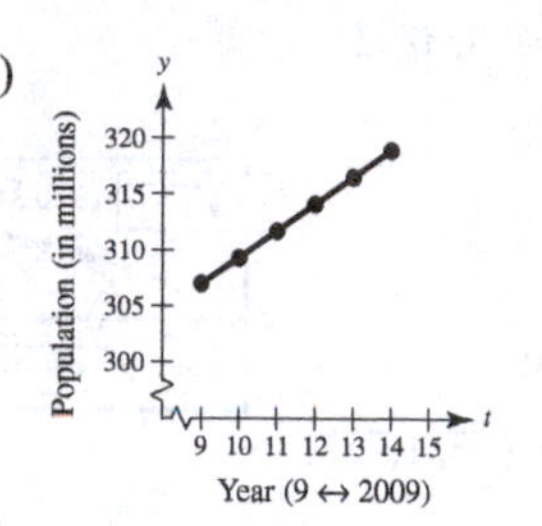

Slopes: $\dfrac{309.3 - 307.0}{10 - 9} = 2.3$

$\dfrac{311.7 - 309.3}{11 - 10} = 2.4$

$\dfrac{314.1 - 311.7}{12 - 11} = 2.4$

$\dfrac{316.5 - 314.1}{13 - 12} = 2.4$

$\dfrac{318.9 - 316.5}{14 - 13} = 2.4$

The population increased least rapidly from 2009 to 2010.

(b) $\dfrac{318.9 - 307.0}{14 - 9} = 2.38$ million people per year

(c) For 2025, $t = 25$:

$$\frac{P - 307.0}{25 - 9} = 2.38 \Rightarrow P = 2.38(16) + 307.0 \approx 345.1$$

The population of the United States in 2025 will be about 345.1 million people.

29. $y = 4x - 3$

The slope is $m = 4$ and the y-intercept is $(0, -3)$.

31. $5x + y = 20$

$y = -5x + 20$

The slope is $m = -5$ and the y-intercept is $(0, 20)$.

33. $x = 4$

The line is vertical. Therefore, the slope is undefined and there is no y-intercept.

35. $y = -3$

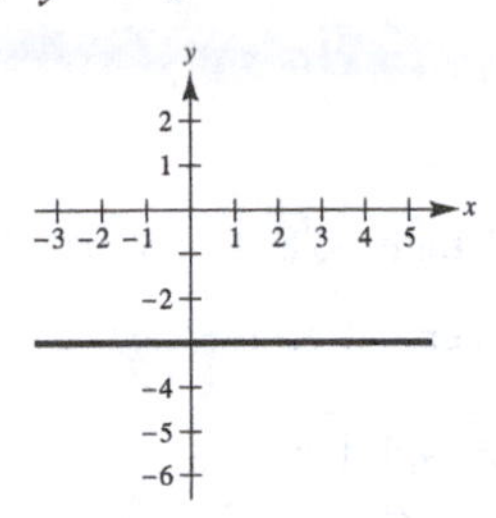

37. $y = -2x + 1$

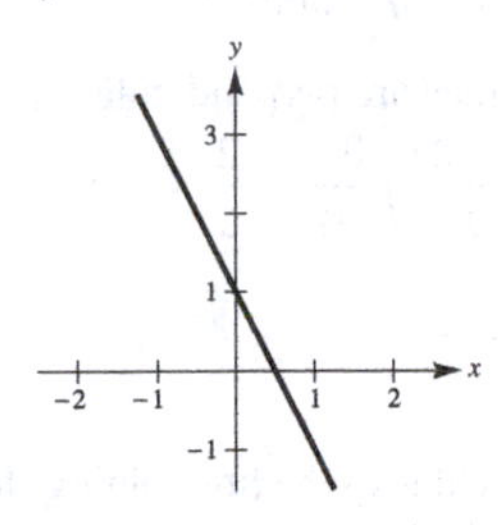

39. $y - 2 = \frac{3}{2}(x - 1)$

$y = \frac{3}{2}x + \frac{1}{2}$

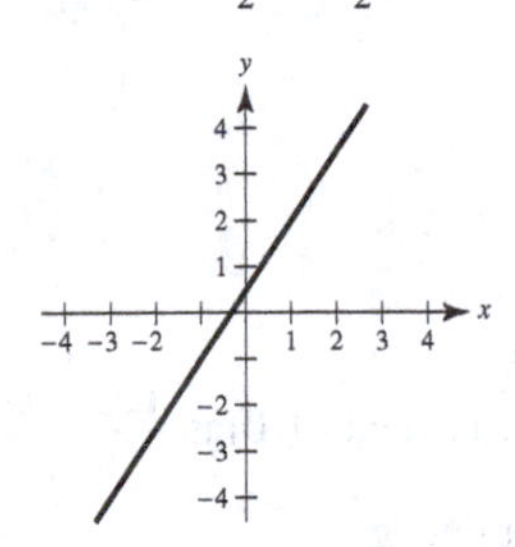

41. $3x - 3y + 1 = 0$

$3y = 3x + 1$

$y = x + \frac{1}{3}$

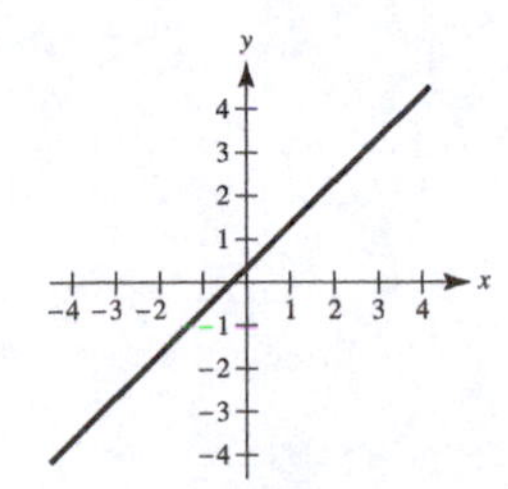

43. $m = \dfrac{-5 - 3}{0 - 4} = \dfrac{-8}{-4} = 2$

$y - (-5) = 2(x - 0)$

$y + 5 = 2x$

$0 = 2x - y - 5$

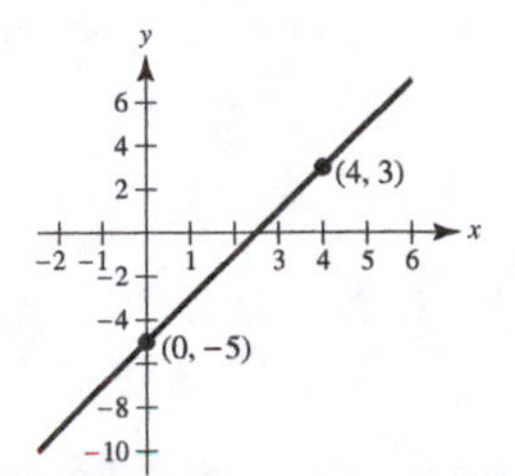

45. $m = \dfrac{8 - 0}{2 - 5} = -\dfrac{8}{3}$

$y - 0 = -\frac{8}{3}(x - 5)$

$y = -\frac{8}{3}x + \frac{40}{3}$

$8x + 3y - 40 = 0$

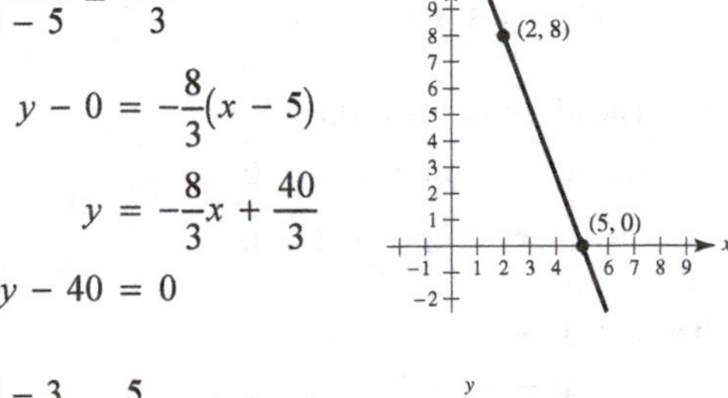

47. 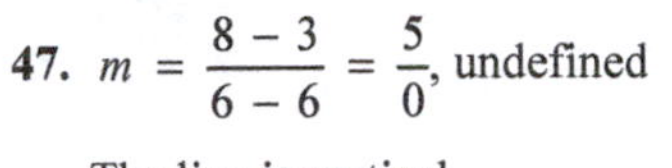 $m = \dfrac{8 - 3}{6 - 6} = \dfrac{5}{0}$, undefined

The line is vertical.

$x = 6$ or $x - 6 = 0$

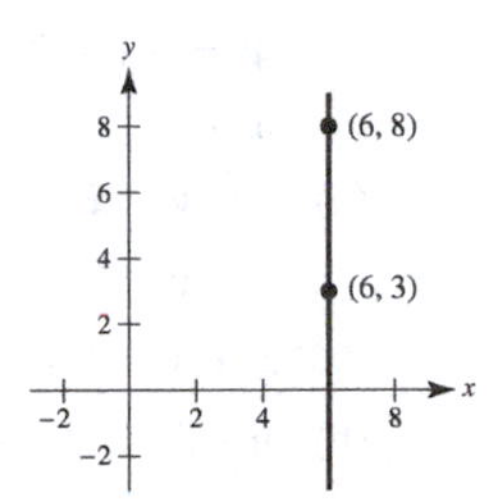

49. $m = \dfrac{1 - 1}{5 - 3} = 0$

The line is horizontal.

$y = 1$ or $y - 1 = 0$

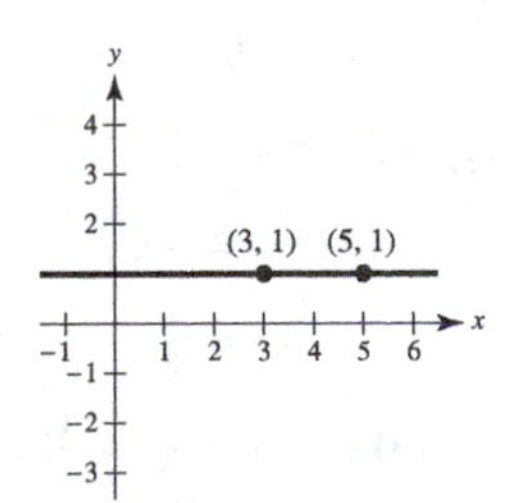

51. The slope is $\dfrac{1 - b}{3 - 0} = \dfrac{1 - b}{3}$.

The y-intercept is $(0, b)$. Hence,

$$y = mx + b = \left(\frac{1 - b}{3}\right)x + b.$$

53. $\dfrac{x}{2} + \dfrac{y}{3} = 1$

$3x + 2y - 6 = 0$

55. $\frac{x}{2a} + \frac{y}{a} = 1$

$\frac{9}{2a} + \frac{-2}{a} = 1$

$\frac{9-4}{2a} = 1$

$5 = 2a$

$a = \frac{5}{2}$

$\frac{x}{2\left(\frac{5}{2}\right)} + \frac{y}{\left(\frac{5}{2}\right)} = 1$

$\frac{x}{5} + \frac{2y}{5} = 1$

$x + 2y = 5$

$x + 2y - 5 = 0$

57. The given line is vertical.

(a) $x = -7$, or $x + 7 = 0$

(b) $y = -2$, or $y + 2 = 0$

59. $x + y = 7$

$y = -x + 7$

$m = -1$

(a) $y - 2 = -1(x + 3)$

$y - 2 = -x - 3$

$x + y + 1 = 0$

(b) $y - 2 = 1(x + 3)$

$y - 2 = x + 3$

$0 = x - y + 5$

61. $5x - 3y = 0$

$y = \frac{5}{3}x$

$m = \frac{5}{3}$

(a) $y - \frac{7}{8} = \frac{5}{3}\left(x - \frac{3}{4}\right)$

$24y - 21 = 40x - 30$

$0 = 40x - 24y - 9$

(b) $y - \frac{7}{8} = -\frac{3}{5}\left(x - \frac{3}{4}\right)$

$40y - 35 = -24x + 18$

$24x + 40y - 53 = 0$

63. The slope is 250.

$V = 1850$ when $t = 6$.

$V = 250(t - 6) + 1850$

$= 250t + 250$

65. $m_1 = \frac{1-0}{-2-(-1)} = -1$

$m_2 = \frac{-2-0}{2-(-1)} = -\frac{2}{3}$

$m_1 \neq m_2$

The points are not collinear.

67.

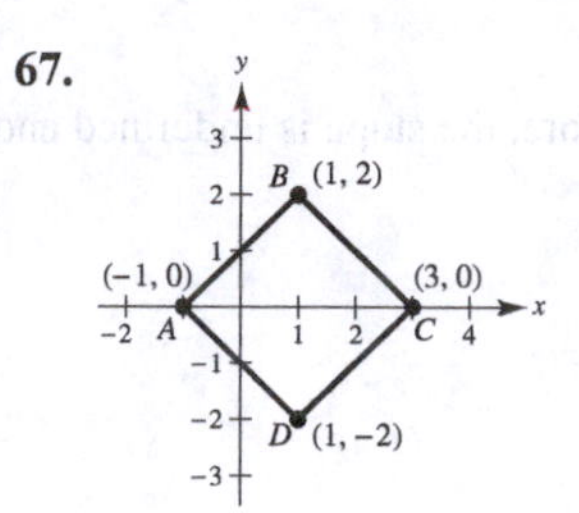

The four sides are of equal length: $\sqrt{8} = 2\sqrt{2}$.

For example, the length of segment AB is

$\sqrt{(1-(-1))^2 + (2-0)^2} = \sqrt{4+4}$

$= \sqrt{8}$

$= 2\sqrt{2}$ units.

Furthermore, the adjacent sides are perpendicular because the slope of $\overline{AB}$ is $\frac{2-0}{1-(-1)} = \frac{2}{2} = 1$, whereas the slope of $\overline{BC}$ is $\frac{2-0}{1-3} = -1$.

69. The tangent line is perpendicular to the line joining the point (5, 12) and the center (0, 0).

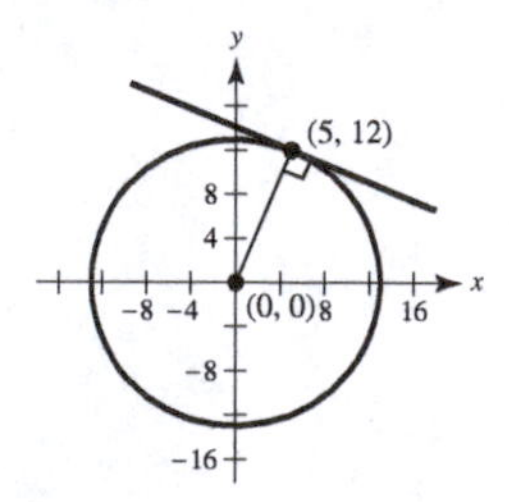

Slope of the line joining (5, 12) and (0, 0) is $\frac{12}{5}$.

The equation of the tangent line is

$y - 12 = \frac{-5}{12}(x - 5)$

$y = \frac{-5}{12}x + \frac{169}{12}$

$5x + 12y - 169 = 0$.

71. (a) The slope of the segment joining (b, c) and $(a, 0)$ is $\dfrac{c}{(b-a)}$. The slope of the perpendicular bisector of this segment is $\dfrac{(a-b)}{c}$. The midpoint of this segment is $\left(\dfrac{a+b}{2}, \dfrac{c}{2}\right)$.

So, the equation of the perpendicular bisector to this segment is

$$y - \frac{c}{2} = \frac{a-b}{c}\left(x - \frac{a+b}{2}\right).$$

Similarly, the equation of the perpendicular bisector of the segment joining $(-a, 0)$ and $(a, 0)$ is

$$y - \frac{c}{2} = \frac{a-b}{-c}\left(x - \frac{b-a}{2}\right).$$

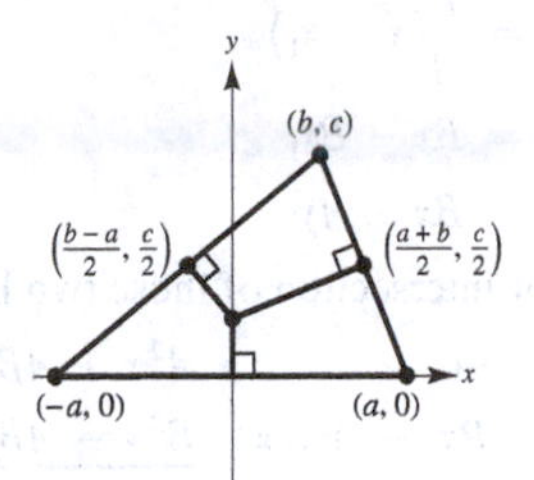

Equating the right-hand sides of each equation, you obtain $x = 0$.

Letting $x = 0$ in either equation yields the point of intersection:

$$y = \frac{c}{2} + \frac{a-b}{c}\left(0 - \frac{a+b}{2}\right) = \frac{c^2}{2c} + \frac{b^2 - a^2}{2c} = \frac{c^2 + b^2 - a^2}{2c}.$$

The point of intersection is $\left(0, \dfrac{-a^2 + b^2 + c^2}{2c}\right)$.

(b) The equations of the medians are:

$$y = \frac{c}{b}x$$

$$y = \frac{c/2}{\left(\dfrac{b-a}{2}\right) - a}(x - a) = \frac{c}{b - 3a}(x - a)$$

$$y = \frac{c/2}{\left(\dfrac{a+b}{2}\right) + a}(x + a) = \frac{c}{3a + b}(x + a).$$

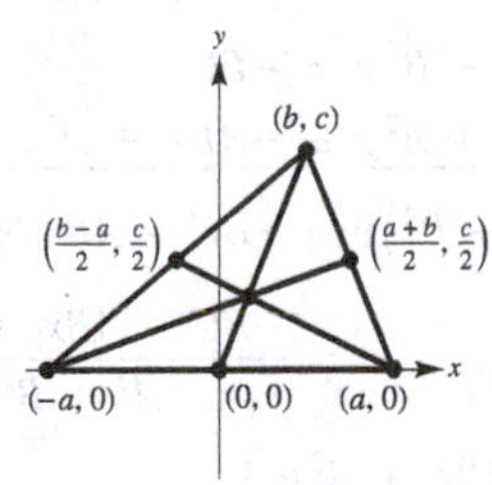

Solving these equation simultaneously for (x, y), you obtain the point of intersection $\left(\dfrac{b}{3}, \dfrac{c}{3}\right)$.

73. Find the equation of the line through the points (0, 32) and (100, 212).

$$m = \tfrac{180}{100} = \tfrac{9}{5}$$

$$F - 32 = \tfrac{9}{5}(C - 0)$$

$$F = \tfrac{9}{5}C + 32$$

or

$$C = \tfrac{1}{9}(5F - 160)$$

$$5F - 9C - 160 = 0$$

For $F = 72°$, $C \approx 22.2°$.

75. (a) Two points are (50, 780) and (47, 825).

The slope is

$$m = \frac{825 - 780}{47 - 50} = \frac{45}{-3} = -15.$$

$$p - 780 = -15(x - 50)$$

$$p = -15x + 750 + 780 = -15x + 1530$$

or

$$x = \frac{1}{15}(1530 - p)$$

(b)

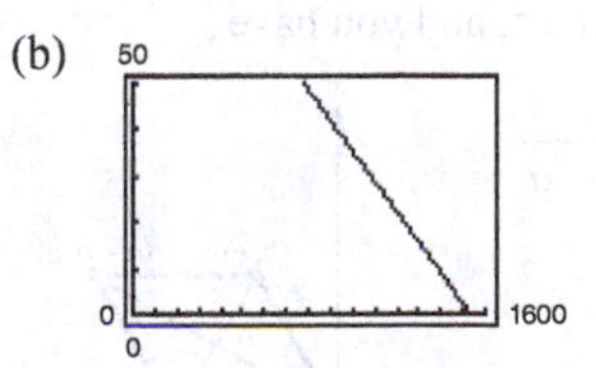

If $p = 855$, then $x = 45$ units.

(c) If $p = 795$, then $x = \dfrac{1}{15}(1530 - 795) = 49$ units

77. If $A = 0$, then $By + C = 0$ is the horizontal line $y = -C/B$. The distance to (x_1, y_1) is

$$d = \left|y_1 - \left(\frac{-C}{B}\right)\right| = \frac{|By_1 + C|}{|B|} = \frac{|Ax_1 + By_1 + C|}{\sqrt{A^2 + B^2}}.$$

If $B = 0$, then $Ax + C = 0$ is the vertical line $x = -C/A$. The distance to (x_1, y_1) is

$$d = \left|x_1 - \left(\frac{-C}{A}\right)\right| = \frac{|Ax_1 + C|}{|A|} = \frac{|Ax_1 + By_1 + C|}{\sqrt{A^2 + B^2}}.$$

(Note that A and B cannot both be zero.) The slope of the line $Ax + By + C = 0$ is $-A/B$.

The equation of the line through (x_1, y_1) perpendicular to $Ax + By + C = 0$ is:

$$y - y_1 = \frac{B}{A}(x - x_1)$$
$$Ay - Ay_1 = Bx - Bx_1$$
$$Bx_1 - Ay_1 = Bx - Ay$$

The point of intersection of these two lines is:

$$Ax + By = -C \quad \Rightarrow A^2x + ABy = -AC \quad (1)$$
$$Bx - Ay = Bx_1 - Ay_1 \Rightarrow \underline{B^2x - ABy = B^2x_1 - ABy_1} \quad (2)$$
$$(A^2 + B^2)x = -AC + B^2x_1 - ABy_1 \quad \text{(By adding equations (1) and (2))}$$
$$x = \frac{-AC + B^2x_1 - ABy_1}{A^2 + B^2}$$

$$Ax + By = -C \quad \Rightarrow ABx + B^2y = -BC \quad (3)$$
$$Bx - Ay = Bx_1 - Ay_1 \Rightarrow \underline{-ABx + A^2y = -ABx_1 + A^2y_1} \quad (4)$$
$$(A^2 + B^2)y = -BC - ABx_1 + A^2y_1 \quad \text{(By adding equations (3) and (4))}$$
$$y = \frac{-BC - ABx_1 + A^2y_1}{A^2 + B^2}$$

$$\left(\frac{-AC + B^2x_1 - ABy_1}{A^2 + B^2}, \frac{-BC - ABx_1 + A^2y_1}{A^2 + B^2}\right) \text{ point of intersection}$$

The distance between (x_1, y_1) and this point gives you the distance between (x_1, y_1) and the line $Ax + By + C = 0$.

79. $x - y - 2 = 0 \Rightarrow d = \dfrac{|1(-2) + (-1)(1) - 2|}{\sqrt{1^2 + 1^2}} = \dfrac{5}{\sqrt{2}} = \dfrac{5\sqrt{2}}{2}$

81. For simplicity, let the vertices of the rhombus be (0, 0), $(a, 0)$, (b, c), and $(a + b, c)$, as shown in the figure.

The slopes of the diagonals are then $m_1 = \dfrac{c}{a + b}$ and $m_2 = \dfrac{c}{b - a}$. Because the sides of the rhombus are equal, $a^2 = b^2 + c^2$, and you have

$$m_1m_2 = \frac{c}{a + b} \cdot \frac{c}{b - a} = \frac{c^2}{b^2 - a^2} = \frac{c^2}{-c^2} = -1.$$

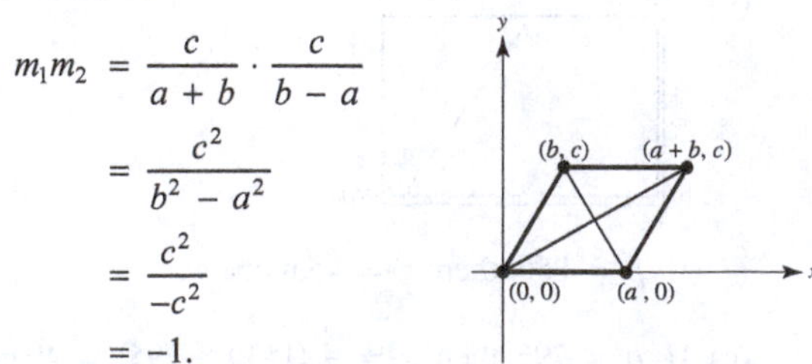

Therefore, the diagonals are perpendicular.

83. Consider the figure below in which the four points are collinear. Because the triangles are similar, the result immediately follows.

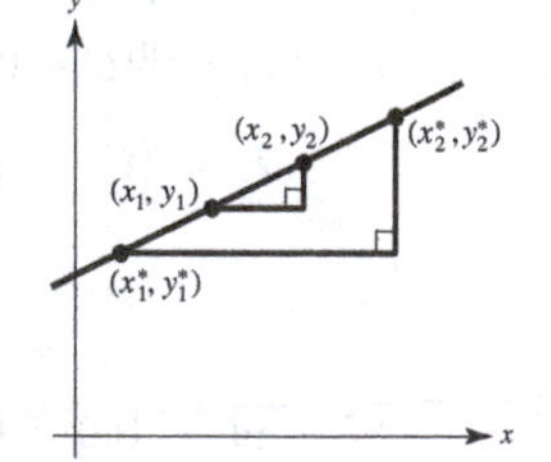

$$\frac{y_2^* - y_1^*}{x_2^* - x_1^*} = \frac{y_2 - y_1}{x_2 - x_1}$$

85. True.

$$ax + by = c_1 \Rightarrow y = -\frac{a}{b}x + \frac{c_1}{b} \Rightarrow m_1 = -\frac{a}{b}$$

$$bx - ay = c_2 \Rightarrow y = \frac{b}{a}x - \frac{c_2}{a} \Rightarrow m_2 = \frac{b}{a}$$

$$m_2 = -\frac{1}{m_1}$$

Section 1.3 Functions and Their Graphs

1. A relation between two sets X and Y is a set of ordered pairs of the form (x, y), where x is a member of X and y is at member of Y.

A function from X to Y is a relation between X and Y that has the property that any two ordered pairs with the same x-value also have the same y-value.

3. The three basic types are vertical shifts, horizontal shifts, and reflections.

5. $f(x) = 3x - 2$

(a) $f(0) = 3(0) - 2 = -2$

(b) $f(5) = 3(5) - 2 = 13$

(c) $f(b) = 3(b) - 2 = 3b - 2$

(d) $f(x - 1) = 3(x - 1) - 2 = 3x - 5$

7. (a) $f(-2) = \sqrt{(-2)^2 + 4} = \sqrt{4 + 4} = \sqrt{8} = 2\sqrt{2}$

(b) $f(3) = \sqrt{3^2 + 4} = \sqrt{9 + 4} = \sqrt{13}$

(c) $f(2) = \sqrt{2^2 + 4} = \sqrt{4 + 4} = \sqrt{8} = 2\sqrt{2}$

(d) $f(x + bx) = \sqrt{(x + bx)^2 + 4}$

$= \sqrt{x^2 + 2bx^2 + b^2x^2 + 4}$

9. (a) $g(0) = 5 - 0^2 = 5$

(b) $g(\sqrt{5}) = 5 - (\sqrt{5})^2 = 5 - 5 = 0$

(c) $g(-2) = 5 - (-2)^2 = 5 - 4 = 1$

(d) $g(t - 1) = 5 - (t - 1)^2 = 5 - (t^2 - 2t + 1)$

$= 4 + 2t - t^2$

11. $$\frac{f(x + \Delta x) - f(x)}{\Delta x} = \frac{(x + \Delta x)^3 - x^3}{\Delta x} = \frac{x^3 + 3x^2\Delta x + 3x^2(\Delta x)^2 + (\Delta x)^3 - x^3}{\Delta x} = 3x^2 + 3x\Delta x + (\Delta x)^2, \ \Delta x \neq 0$$

13. $f(x) = 4x^2$

Domain: $(-\infty, \infty)$

Range: $[0, \infty)$

15. $f(x) = x^3$

Domain: $(-\infty, \infty)$

Range: $(-\infty, \infty)$

17. $g(x) = \sqrt{6x}$

Domain: $6x \geq 0$

$x \geq 0 \Rightarrow [0, \infty)$

Range: $[0, \infty)$

19. $f(x) = \sqrt{16 - x^2}$

$16 - x^2 \geq 0 \Rightarrow x^2 \leq 16$

Domain: $[-4, 4]$

Range: $[0, 4]$

Note: $y = \sqrt{16 - x^2}$ is a semicircle of radius 4.

21. $f(x) = \dfrac{3}{x}$

Domain: all $x \neq 0 \Rightarrow (-\infty, 0) \cup (0, \infty)$

Range: $(-\infty, 0) \cup (0, \infty)$

23. $f(x) = \sqrt{x} + \sqrt{1 - x}$

$x \geq 0$ and $1 - x \geq 0$

$x \geq 0$ and $x \leq 1$

Domain: $0 \leq x \leq 1 \Rightarrow [0, 1]$

25. $f(x) = \dfrac{1}{|x + 3|}$

$|x + 3| \neq 0$

$x + 3 \neq 0$

Domain: all $x \neq -3$

Domain: $(-\infty, -3) \cup (-3, \infty)$

27. $f(x) = \begin{cases} 2x + 1, & x < 0 \\ 2x + 2, & x \geq 0 \end{cases}$

(a) $f(-1) = 2(-1) + 1 = -1$

(b) $f(0) = 2(0) + 2 = 2$

(c) $f(2) = 2(2) + 2 = 6$

(d) $f(t^2 + 1) = 2(t^2 + 1) + 2 = 2t^2 + 4$

(**Note:** $t^2 + 1 \geq 0$ for all t.)

Domain: $(-\infty, \infty)$

Range: $(-\infty, 1) \cup [2, \infty)$

29. $f(x) = \begin{cases} |x| + 1, & x < 1 \\ -x + 1, & x \ge 1 \end{cases}$

(a) $f(-3) = |-3| + 1 = 4$

(b) $f(1) = -1 + 1 = 0$

(c) $f(3) = -3 + 1 = -2$

(d) $f(b^2 + 1) = -(b^2 + 1) + 1 = -b^2$

Domain: $(-\infty, \infty)$

Range: $(-\infty, 0] \cup [1, \infty)$

31. $f(x) = 4 - x$

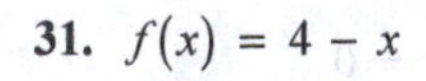

Domain: $(-\infty, \infty)$

Range: $(-\infty, \infty)$

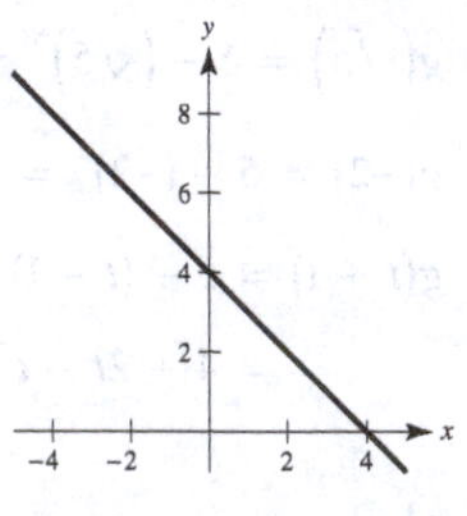

33. $g(x) = \dfrac{1}{x^2 + 2}$

Domain: $(-\infty, \infty)$

Range: $\left(0, \frac{1}{2}\right]$

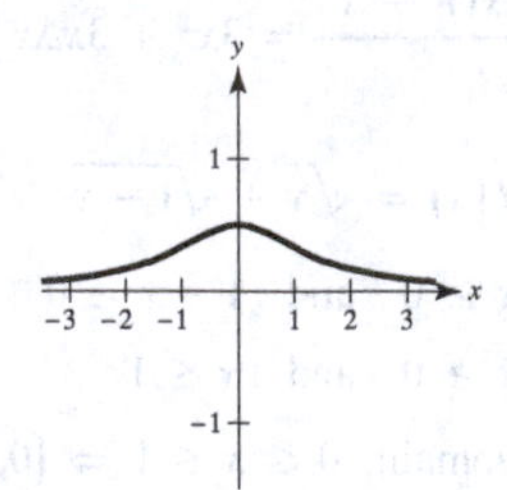

35. $h(x) = \sqrt{x - 6}$

Domain: $x - 6 \ge 0$

$x \ge 6 \Rightarrow [6, \infty)$

Range: $[0, \infty)$

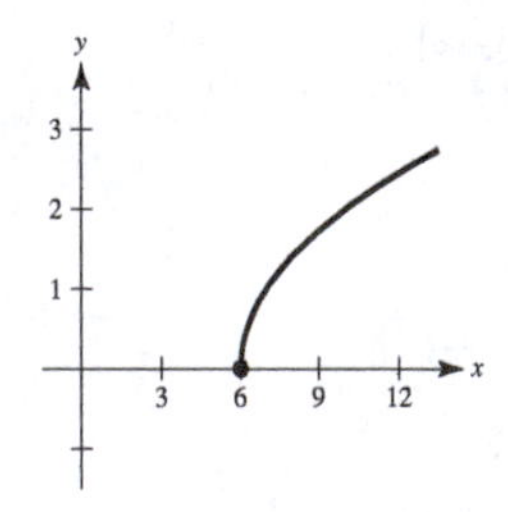

37. $f(x) = \sqrt{9 - x^2}$

Domain: $[-3, 3]$

Range: $[0, 3]$

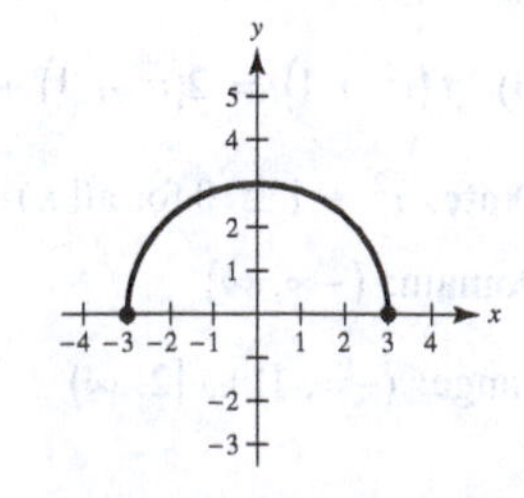

39. $x - y^2 = 0 \Rightarrow y = \pm\sqrt{x}$

y is not a function of x. Some vertical lines intersect the graph twice.

41. y is a function of x. Vertical lines intersect the graph at most once.

43. $x^2 + y^2 = 16 \Rightarrow y = \pm\sqrt{16 - x^2}$

y is not a function of x because there are two values of y for some x.

45. $y^2 = x^2 - 1 \Rightarrow y = \pm\sqrt{x^2 - 1}$

y is not a function of x because there are two values of y for some x.

47. The transformation is a horizontal shift two units to the right of the function $f(x) = \sqrt{x}$.

Shifted function: $y = \sqrt{x - 2}$

49. The transformation is a horizontal shift 2 units to the right and a vertical shift 1 unit downward of the function $f(x) = x^2$.

Shifted function: $y = (x - 2)^2 - 1$

51. $y = f(x + 5)$ is a horizontal shift 5 units to the left. Matches d.

52. $y = f(x) - 5$ is a vertical shift 5 units downward. Matches b.

53. $y = -f(-x) - 2$ is a reflection in the y-axis, a reflection in the x-axis, and a vertical shift downward 2 units. Matches c.

54. $y = -f(x - 4)$ is a horizontal shift 4 units to the right, followed by a reflection in the x-axis. Matches a.

55. $y = f(x + 6) + 2$ is a horizontal shift to the left 6 units, and a vertical shift upward 2 units. Matches e.

56. $y = f(x - 1) + 3$ is a horizontal shift to the right 1 unit, and a vertical shift upward 3 units. Matches g.

57. (a) The graph is shifted 3 units to the left.

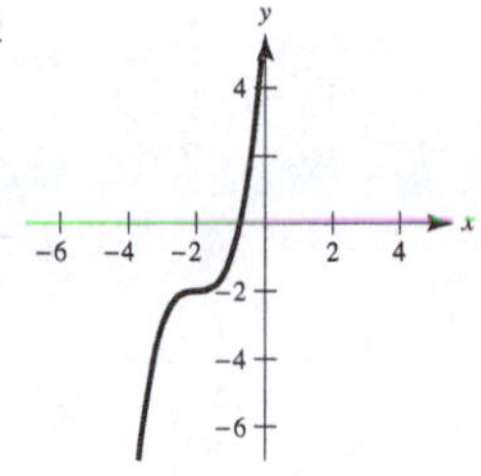

(b) The graph is shifted 1 unit to the right.

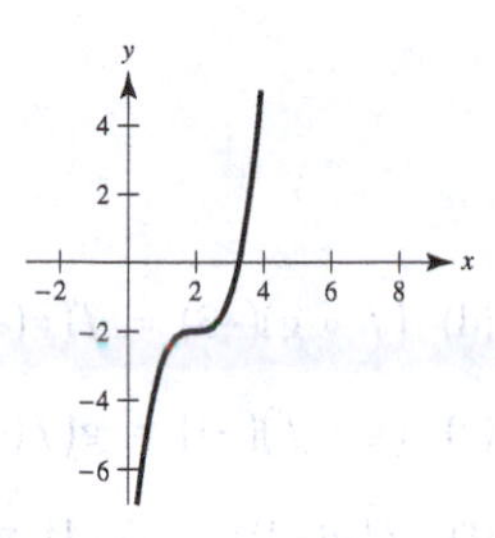

(c) The graph is shifted 2 units upward.

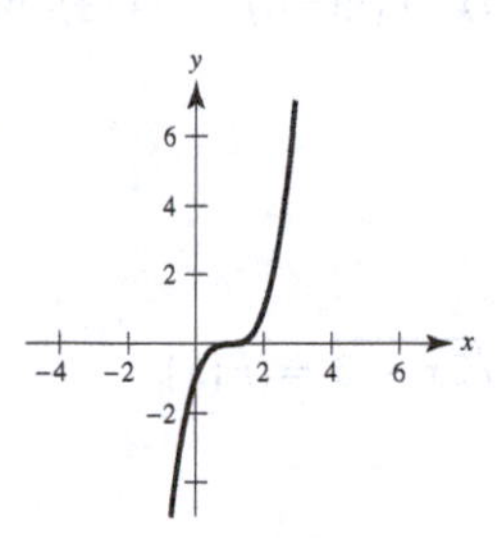

(d) The graph is shifted 4 units downward.

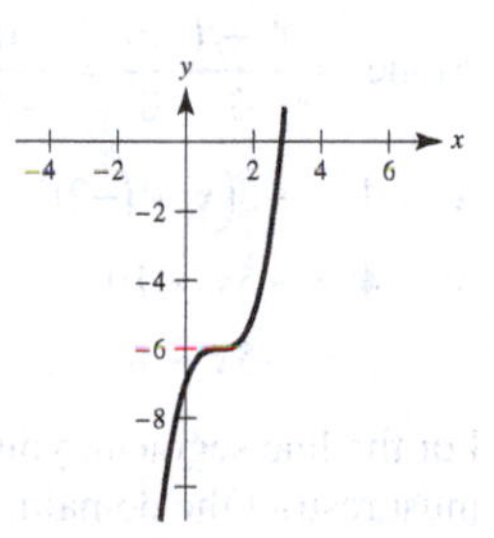

(e) The graph is stretched vertically by a factor of 3.

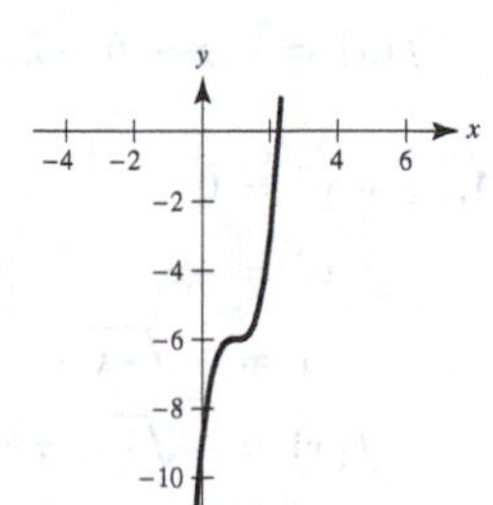

(f) The graph is stretched vertically by a factor of $\frac{1}{4}$.

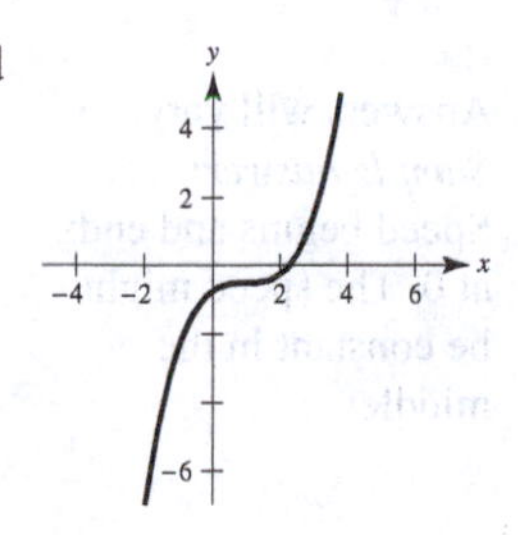

(g) The graph is a reflection in the x-axis.

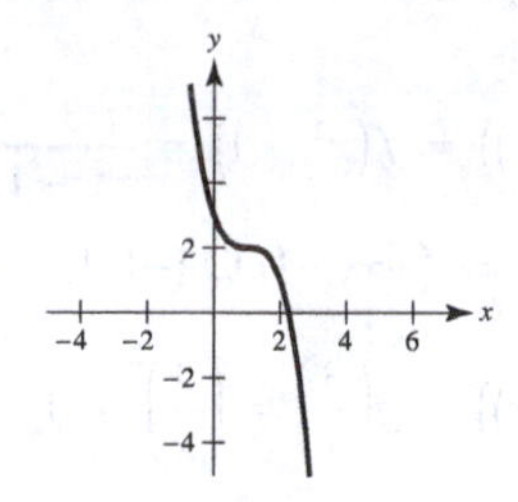

(h) The graph is a reflection about the origin.

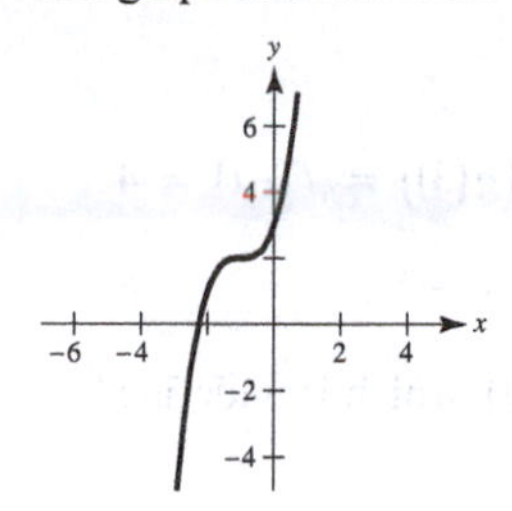

59. $f(x) = 2x - 5,\ g(x) = 4 - 3x$

(a) $f(x) + g(x) = (2x - 5) + (4 - 3x) = -x - 1$

(b) $f(x) - g(x) = (2x - 5) - (4 - 3x) = 5x - 9$

(c) $f(x) \cdot g(x) = (2x - 5)(4 - 3x)$

$= -6x^2 + 8x + 15x - 20$

$= -6x^2 + 23x - 20$

(d) $\dfrac{f(x)}{g(x)} = \dfrac{2x - 5}{4 - 3x}$

61. (a) $f(g(1)) = f(0) = 0$

(b) $g(f(1)) = g(1) = 0$

(c) $g(f(0)) = g(0) = -1$

(d) $f(g(-4)) = f(15) = \sqrt{15}$

(e) $f(g(x)) = f(x^2 - 1) = \sqrt{x^2 - 1}$

(f) $g(f(x)) = g(\sqrt{x}) = (\sqrt{x})^2 - 1 = x - 1,\ (x \geq 0)$

63. $f(x) = x^2,\ g(x) = \sqrt{x}$

$(f \circ g)(x) = f(g(x)) = f(\sqrt{x}) = (\sqrt{x})^2 = x,\ x \geq 0$

Domain: $[0, \infty)$

$(g \circ f)(x) = g(f(x)) = g(x^2) = \sqrt{x^2} = |x|$

Domain: $(-\infty, \infty)$

No. Their domains are different. $(f \circ g) = (g \circ f)$ for $x \geq 0$.

65. $f(x) = \frac{3}{x},\ g(x) = x^2 - 1$

$(f \circ g)(x) = f(g(x)) = f(x^2 - 1) = \frac{3}{x^2 - 1}$

Domain: all $x \neq \pm 1 \Rightarrow (-\infty, -1) \cup (-1, 1) \cup (1, \infty)$

$(g \circ f)(x) = g(f(x)) = g\left(\frac{3}{x}\right) = \left(\frac{3}{x}\right)^2 - 1 = \frac{9}{x^2} - 1 = \frac{9 - x^2}{x^2}$

Domain: all $x \neq 0 \Rightarrow (-\infty, 0) \cup (0, \infty)$

$f \circ g \neq g \circ f$

67. (a) $(f \circ g)(3) = f(g(3)) = f(-1) = 4$

(b) $g(f(2)) = g(1) = -2$

(c) $g(f(5)) = g(-5)$, which is undefined

(d) $(f \circ g)(-3) = f(g(-3)) = f(-2) = 3$

(e) $(g \circ f)(-1) = g(f(-1)) = g(4) = 2$

(f) $f(g(-1)) = f(-4)$, which is undefined

69. $F(x) = \sqrt{2x - 2}$

Let $h(x) = 2x,\ g(x) = x - 2$ and $f(x) = \sqrt{x}$.

Then $(f \circ g \circ h)(x) = f(g(2x)) = f((2x) - 2) = \sqrt{(2x) - 2} = \sqrt{2x - 2} = F(x)$.

(Other answers possible.)

71. (a) If f is even, then $\left(\frac{3}{2}, 4\right)$ is on the graph.

(b) If f is odd, then $\left(\frac{3}{2}, -4\right)$ is on the graph.

73. f is even because the graph is symmetric about the y-axis. g is neither even nor odd. h is odd because the graph is symmetric about the origin.

75. $f(x) = x^2(4 - x^2)$

$f(-x) = (-x)^2\left(4 - (-x)^2\right) = x^2(4 - x^2) = f(x)$

f is even.

$f(x) = x^2(4 - x^2) = 0$

$x^2(2 - x)(2 + x) = 0$

Zeros: $x = 0, -2, 2$

77. $f(x) = 2\sqrt[6]{x}$

The domain of f is $x \geq 0$ and the range is $y \geq 0$. Hence, the function is neither even nor odd. The only zero is $x = 0$.

79. Slope $= \frac{4 - (-6)}{-2 - 0} = \frac{10}{-2} = -5$

$y - 4 = -5(x - (-2))$

$y - 4 = -5x - 10$

$y = -5x - 6$

For the line segment, you must restrict the domain.

$f(x) = -5x - 6,\ -2 \leq x \leq 0$

81. $x + y^2 = 0$

$y^2 = -x$

$y = -\sqrt{-x}$

$f(x) = -\sqrt{-x},\ x \leq 0$

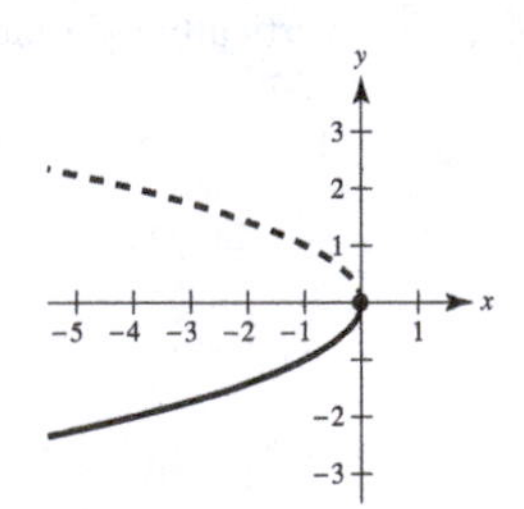

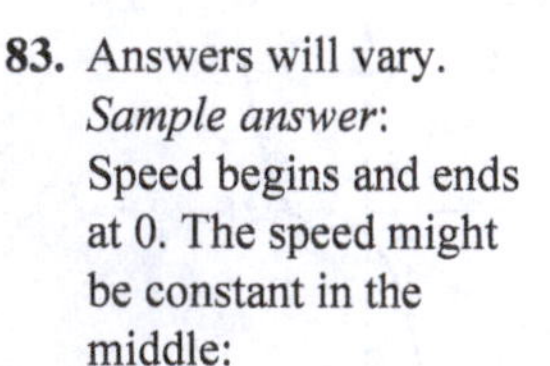

83. Answers will vary.
Sample answer:
Speed begins and ends at 0. The speed might be constant in the middle:

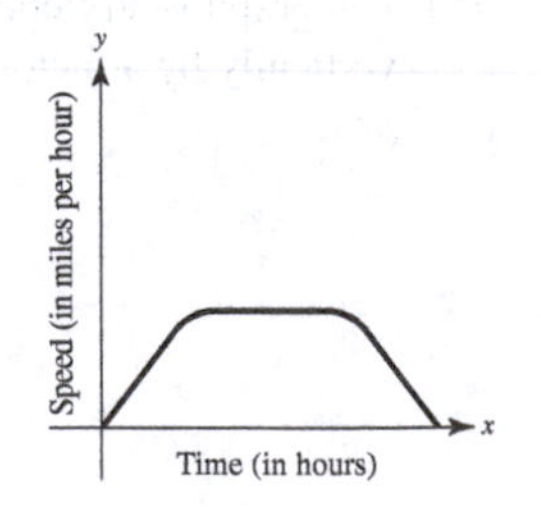

85. Answers will vary.

Sample answer:

Distance begins at 0, then the graph has a sharp turn after a few minutes and goes back to 0. Then the graph goes back upward steeply.

87. $y = \sqrt{c - x^2}$

$y^2 = c - x^2$

$x^2 + y^2 = c$, a circle.

For the domain to be $[-5, 5]$, $c = 25$.

89. No. If a horizontal line intersects the graph more than once, then there is more than one x-value corresponding to the same y-value.

91. No. For example, $y = x^3 + x + 2$ is not odd since $f(-x) \neq -f(x)$.

93. (a) $T(4) = 16°$, $T(15) \approx 23°$

(b) If $H(t) = T(t - 1)$, then the changes in temperature will occur 1 hour later.

(c) If $H(t) = T(t) - 1$, then the overall temperature would be 1 degree lower.

95. (a)

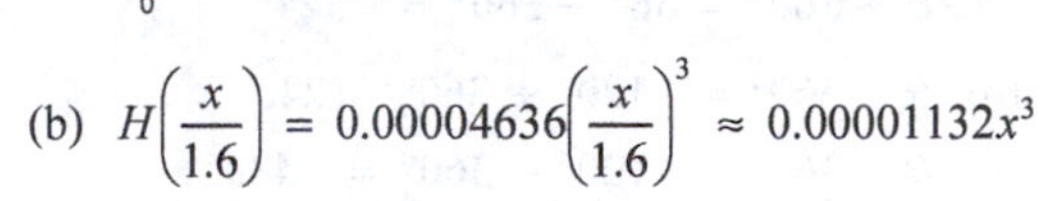

(b) $H\left(\frac{x}{1.6}\right) = 0.00004636\left(\frac{x}{1.6}\right)^3 \approx 0.00001132x^3$

97. $f(-x) = a_{2n+1}(-x)^{2n+1} + \cdots + a_3(-x)^3 + a_1(-x) = -\left[a_{2n+1}x^{2n+1} + \cdots + a_3x^3 + a_1x\right] = -f(x)$

Odd

99. Let $F(x) = f(x)g(x)$ where f and g are even. Then $F(-x) = f(-x)g(-x) = f(x)g(x) = F(x)$.

So, $F(x)$ is even. Let $F(x) = f(x)g(x)$ where f and g are odd. Then

$F(-x) = f(-x)g(-x) = [-f(x)][-g(x)] = f(x)g(x) = F(x)$.

So, $F(x)$ is even.

101. By equating slopes,

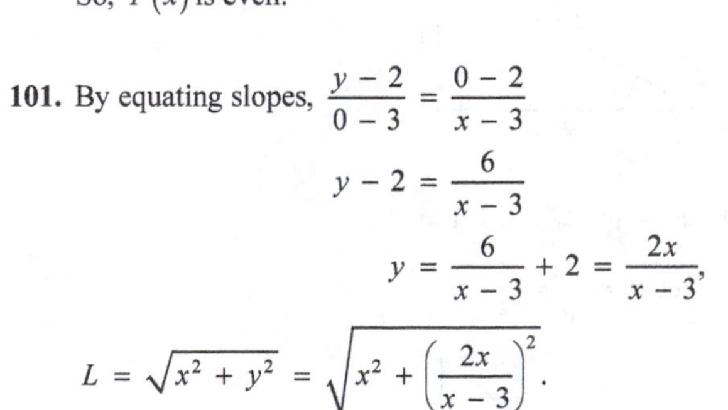

$$\frac{y - 2}{0 - 3} = \frac{0 - 2}{x - 3}$$

$$y - 2 = \frac{6}{x - 3}$$

$$y = \frac{6}{x - 3} + 2 = \frac{2x}{x - 3},$$

$$L = \sqrt{x^2 + y^2} = \sqrt{x^2 + \left(\frac{2x}{x - 3}\right)^2}.$$

103. False. If $f(x) = x^2$, then $f(-3) = f(3) = 9$, but $-3 \neq 3$.

105. True. The function is even.

107. False. The constant function $f(x) = 0$ has symmetry with respect to the x-axis.

109. First consider the portion of R in the first quadrant: $x \geq 0$, $0 \leq y \leq 1$ and $x - y \leq 1$; shown below.

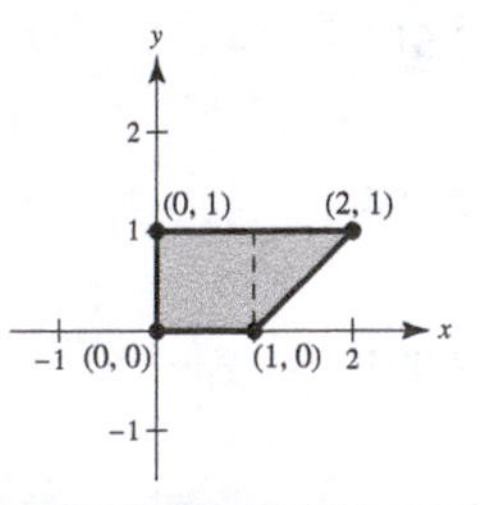

The area of this region is $1 + \frac{1}{2} = \frac{3}{2}$.

By symmetry, you obtain the entire region R:

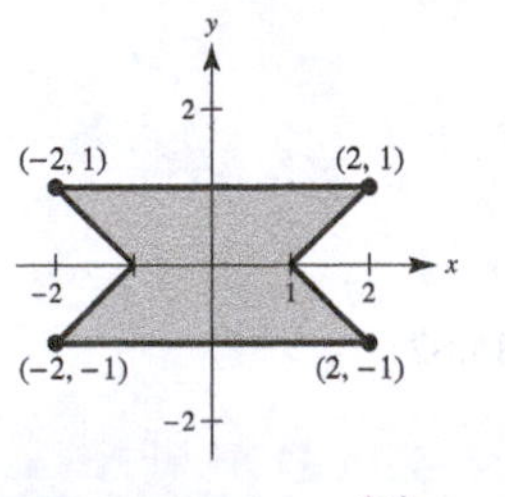

The area of R is $4\left(\frac{3}{2}\right) = 6$.

Section 1.4 Review of Trigonometric Functions

1. In general, if θ is any angle measured in degrees, then the angle $\theta + n(360°)$, n a nonzero integer, is coterminal with θ.

3. $\sin\theta = \dfrac{\text{opp}}{\text{hyp}} = \dfrac{7}{25}$

$\cos\theta = \dfrac{\text{adj}}{\text{hyp}} = \dfrac{24}{25}$

$\tan\theta = \dfrac{\text{opp}}{\text{adj}} = \dfrac{7}{24}$

5. (a) $\theta + 360° = 36° + 360° = 396°$

$\theta - 360° = 36° - 360° = -324°$

(b) $\theta + 360° = -120° + 360° = 240°$

$\theta - 360° = -120° - 360° = -480°$

7. (a) $\theta + 2\pi = \dfrac{\pi}{9} + 2\pi = \dfrac{19\pi}{9}$

$\theta - 2\pi = \dfrac{\pi}{9} - 2\pi = -\dfrac{17\pi}{9}$

(b) $\theta + 2\pi = \dfrac{4\pi}{3} + 2\pi = \dfrac{10\pi}{3}$

$\theta - 2\pi = \dfrac{4\pi}{3} - 2\pi = -\dfrac{2\pi}{3}$

9. (a) $30°\left(\dfrac{\pi}{180°}\right) = \dfrac{\pi}{6} \approx 0.524$

(b) $150°\left(\dfrac{\pi}{180°}\right) = \dfrac{5\pi}{6} \approx 2.618$

(c) $315°\left(\dfrac{\pi}{180°}\right) = \dfrac{7\pi}{4} \approx 5.498$

(d) $120°\left(\dfrac{\pi}{180°}\right) = \dfrac{2\pi}{3} \approx 2.094$

11. (a) $\dfrac{3\pi}{2}\left(\dfrac{180°}{\pi}\right) = 270°$

(b) $\dfrac{7\pi}{6}\left(\dfrac{180°}{\pi}\right) = 210°$

(c) $-\dfrac{7\pi}{12}\left(\dfrac{180°}{\pi}\right) = -105°$

(d) $-2.367\left(\dfrac{180°}{\pi}\right) \approx -135.62°$

13.

r	8 ft	15 in.	85 cm	24 in.	$\dfrac{12{,}963}{\pi}$ mi
s	12 ft	24 in.	63.75π cm	96 in.	8642 mi
θ	1.5	1.6	$\dfrac{3\pi}{4}$	4	$\dfrac{2\pi}{3}$

15. (a) $x = 3, y = 4, r = 5$

$\sin\theta = \frac{4}{5}$ $\csc\theta = \frac{5}{4}$

$\cos\theta = \frac{3}{5}$ $\sec\theta = \frac{5}{3}$

$\tan\theta = \frac{4}{3}$ $\cot\theta = \frac{3}{4}$

(b) $x = -12, y = -5, r = 13$

$\sin\theta = -\frac{5}{13}$ $\csc\theta = -\frac{13}{5}$

$\cos\theta = -\frac{12}{13}$ $\sec\theta = -\frac{13}{12}$

$\tan\theta = \frac{5}{12}$ $\cot\theta = \frac{12}{5}$

17. $x^2 + 1^2 = 2^2 \Rightarrow x = \sqrt{3}$

$\cos\theta = \dfrac{x}{2} = \dfrac{\sqrt{3}}{2}$

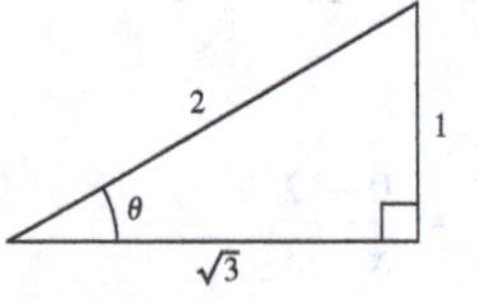

19. $4^2 + y^2 = 5^2 \Rightarrow y = 3$

$\cot\theta = \dfrac{4}{y} = \dfrac{4}{3}$

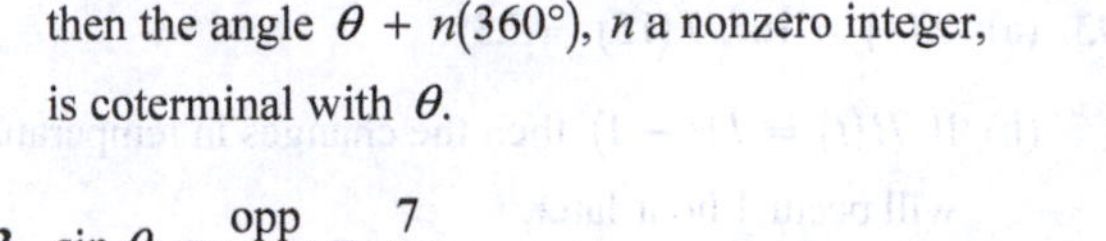

21. (a) $\sin 60° = \frac{\sqrt{3}}{2}$

$\cos 60° = \frac{1}{2}$

$\tan 60° = \sqrt{3}$

(b) $\sin 120° = \sin 60° = \frac{\sqrt{3}}{2}$

$\cos 120° = -\cos 60° = -\frac{1}{2}$

$\tan 120° = -\tan 60° = -\sqrt{3}$

(c) $\sin \frac{\pi}{4} = \frac{\sqrt{2}}{2}$

$\cos \frac{\pi}{4} = \frac{\sqrt{2}}{2}$

$\tan \frac{\pi}{4} = 1$

(d) $\sin \frac{5\pi}{4} - \sin \frac{\pi}{4} = -\frac{\sqrt{2}}{2}$

$\cos \frac{5\pi}{4} = \cos \frac{\pi}{4} = -\frac{\sqrt{2}}{2}$

$\tan \frac{5\pi}{4} = \tan \frac{\pi}{4} = 1$

23. (a) $\sin 225° = -\sin 45° = -\frac{\sqrt{2}}{2}$

$\cos 225° = -\cos 45° = -\frac{\sqrt{2}}{2}$

$\tan 225° = \tan 45° = 1$

(b) $\sin(-225°) = \sin 45° = \frac{\sqrt{2}}{2}$

$\cos(-225°) = -\cos 45° = -\frac{\sqrt{2}}{2}$

$\tan(-225°) = -\tan 45° = -1$

(c) $\sin\frac{5\pi}{3} = -\sin\frac{\pi}{3} = -\frac{\sqrt{3}}{2}$

$\cos\frac{5\pi}{3} = \cos\frac{\pi}{3} = \frac{1}{2}$

$\tan\frac{5\pi}{3} = -\tan\frac{\pi}{3} = -\sqrt{3}$

(d) $\sin\frac{11\pi}{6} = -\sin\frac{\pi}{6} = -\frac{1}{2}$

$\cos\frac{11\pi}{6} = \cos\frac{\pi}{6} = \frac{\sqrt{3}}{2}$

$\tan\frac{11\pi}{6} = -\tan\frac{\pi}{6} = -\frac{\sqrt{3}}{3}$

25. (a) $\sin 10° \approx 0.1736$

(b) $\csc 10° \approx 5.759$

27. (a) $\tan\frac{\pi}{9} \approx 0.3640$

(b) $\tan\frac{10\pi}{9} \approx 0.3640$

29. (a) $\sin \theta < 0 \Rightarrow \theta$ is in Quadrant III or IV.

$\cos \theta < 0 \Rightarrow \theta$ is in Quadrant II or III.

$\sin \theta < 0$ **and** $\cos \theta < 0 \Rightarrow \theta$ is in Quadrant III.

(b) $\sec \theta > 0 \Rightarrow \theta$ is in Quadrant I or IV.

$\cot \theta < 0 \Rightarrow \theta$ is in Quadrant II or IV.

$\sec \theta > 0$ **and** $\cot \theta < 0 \Rightarrow \theta$ is in Quadrant IV.

31. (a) $\cos \theta = \frac{\sqrt{2}}{2}$

$\theta = \frac{\pi}{4}, \frac{7\pi}{4}$

(b) $\cos \theta = -\frac{\sqrt{2}}{2}$

$\theta = \frac{3\pi}{4}, \frac{5\pi}{4}$

33. (a) $\tan \theta = 1$

$\theta = \frac{\pi}{4}, \frac{5\pi}{4}$

(b) $\cot \theta = -\sqrt{3}$

$\theta = \frac{5\pi}{6}, \frac{11\pi}{6}$

35. $2 \sin^2 \theta = 1$

$\sin \theta = \pm\frac{\sqrt{2}}{2}$

$\theta = \frac{\pi}{4}, \frac{3\pi}{4}, \frac{5\pi}{4}, \frac{7\pi}{4}$

37. $\tan^2 \theta - \tan \theta = 0$

$\tan \theta(\tan \theta - 1) = 0$

$\tan \theta = 0$ $\qquad$ $\tan \theta = 1$

$\theta = 0, \pi, 2\pi$ $\qquad$ $\theta = \frac{\pi}{4}, \frac{5\pi}{4}$

39. $\sec \theta \csc \theta - 2 \csc \theta = 0$

$\csc \theta(\sec \theta - 2) = 0$

($\csc \theta \neq 0$ for any value of θ)

$\sec \theta = 2$

$\theta = \frac{\pi}{3}, \frac{5\pi}{3}$

41. $\cos^2 \theta + \sin \theta = 1$

$1 - \sin^2 \theta + \sin \theta = 1$

$\sin^2 \theta - \sin \theta = 0$

$\sin \theta(\sin \theta - 1) = 0$

$\sin \theta = 0 \qquad \sin \theta = 1$

$\theta = 0, \pi, 2\pi \qquad \theta = \dfrac{\pi}{2}$

43. $(275 \text{ ft/sec})(60 \text{ sec}) = 16{,}500 \text{ feet}$

$$\sin 18° = \frac{a}{16{,}500}$$

$$a = 16{,}500 \sin 18° \approx 5099 \text{ feet}$$

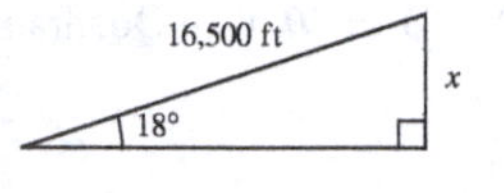

45. $y = 2 \sin 2x$

$\text{Period} = \dfrac{2\pi}{2} = \pi$

$\text{Amplitude} = |2| = 2$

47. $y = -3 \sin 4\pi x$

$\text{Period} = \dfrac{2\pi}{4\pi} = \dfrac{1}{2}$

$\text{Amplitude} = |-3| = 3$

49. $y = 5 \tan 2x$

$\text{Period} = \dfrac{\pi}{2}$

51. $y = \sec 5x$

$\text{Period} = \dfrac{2\pi}{5}$

53. (a) $f(x) = c \sin x$; changing c changes the amplitude.

When $c = -2$: $f(x) = -2 \sin x$.

When $c = -1$: $f(x) = -\sin x$.

When $c = 1$: $f(x) = \sin x$.

When $c = 2$: $f(x) = 2 \sin x$.

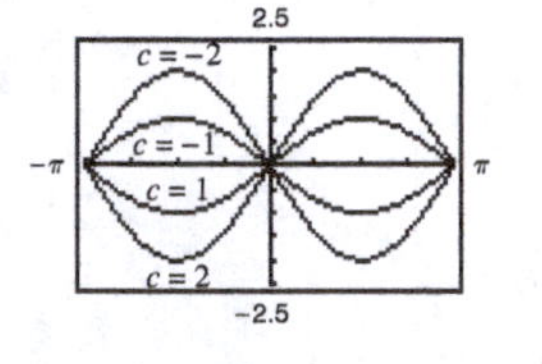

(b) $f(x) = \cos(cx)$; changing c changes the period.

When $c = -2$: $f(x) = \cos(-2x) = \cos 2x$.

When $c = -1$: $f(x) = \cos(-x) = \cos x$.

When $c = 1$: $f(x) = \cos x$.

When $c = 2$: $f(x) = \cos 2x$.

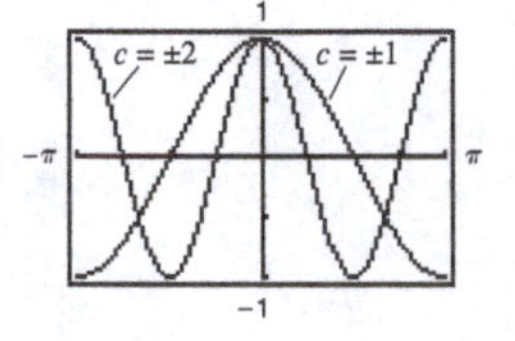

(c) $f(x) = \cos(\pi x - c)$; changing c causes a horizontal shift.

When $c = -2$: $f(x) = \cos(\pi x + 2)$.

When $c = -1$: $f(x) = \cos(\pi x + 1)$.

When $c = 1$: $f(x) = \cos(\pi x - 1)$.

When $c = 2$: $f(x) = \cos(\pi x - 2)$.

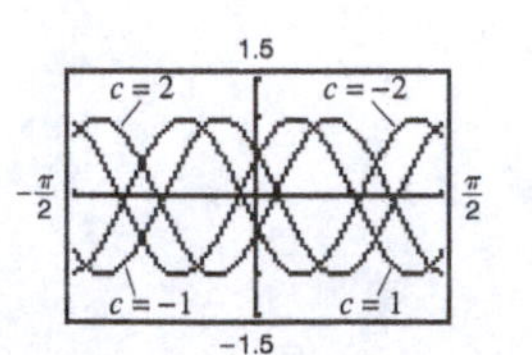

55. $y = \sin \dfrac{x}{2}$

Period: 4π

Amplitude: 1

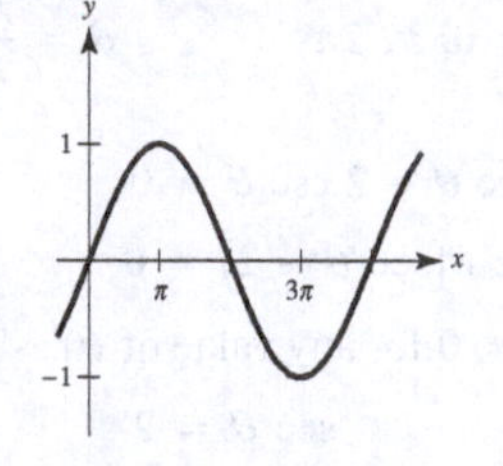

57. $y = -\sin \dfrac{2\pi x}{3}$

Period: 3

Amplitude: 1

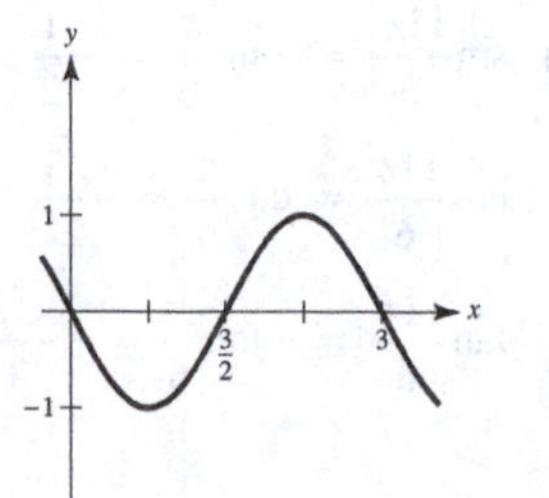

59. $y = \csc\dfrac{x}{2}$

Period: 4π

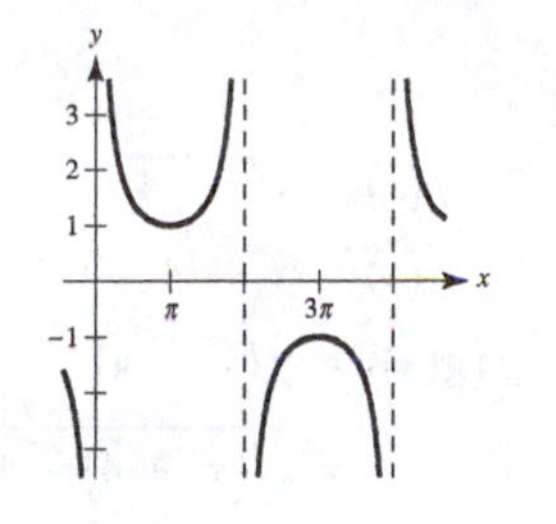

61. $y = 2\sec 2x$

Period: π

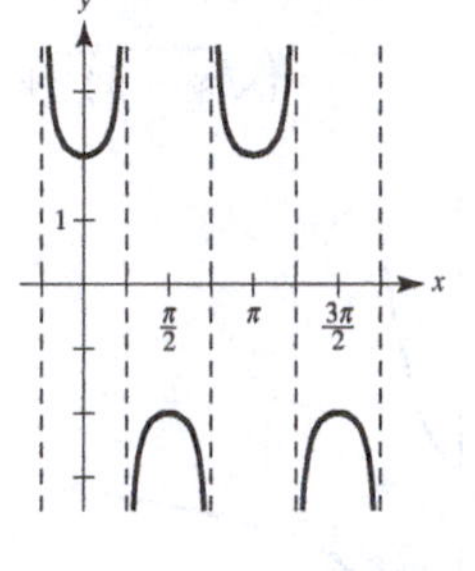

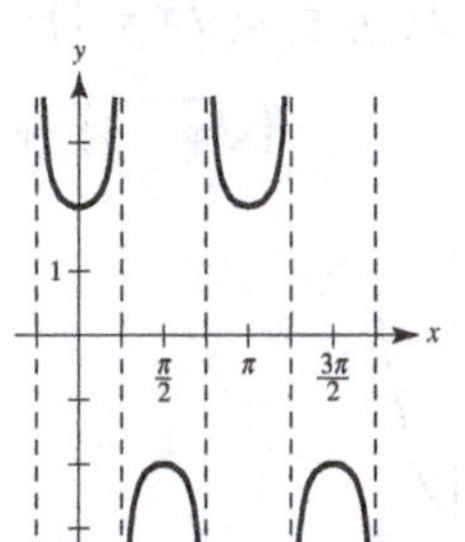

63. $y = \sin(x + \pi)$

Period: 2π

Amplitude: 1

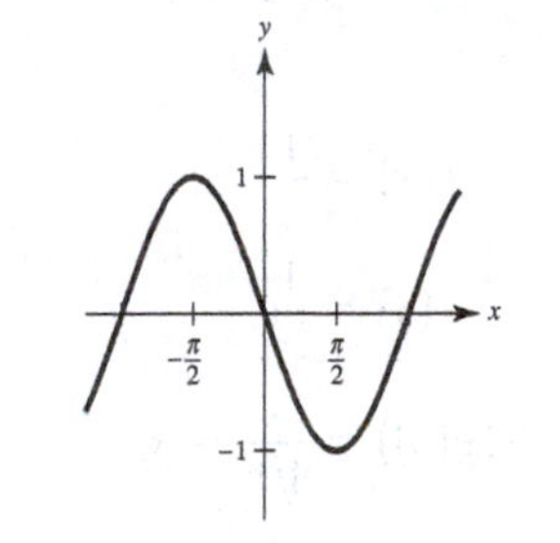

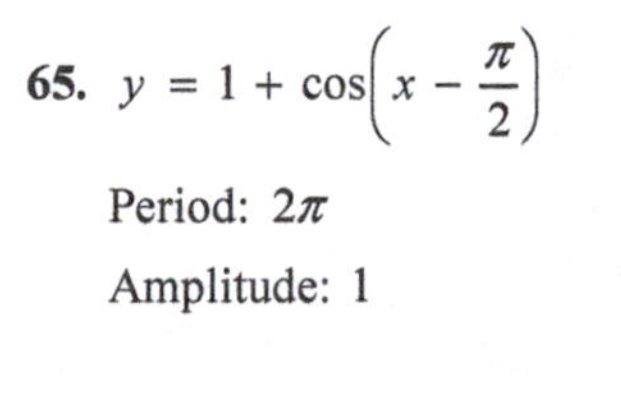

65. $y = 1 + \cos\left(x - \dfrac{\pi}{2}\right)$

Period: 2π

Amplitude: 1

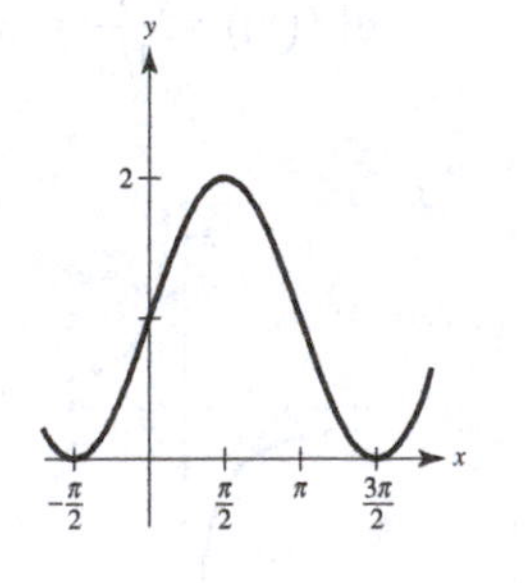

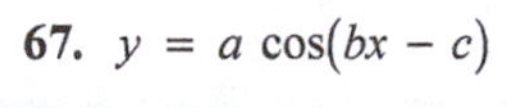

67. $y = a\cos(bx - c)$

From the graph, we see that the amplitude is 3, the period is 4π, and the horizontal shift is π. Thus,

$$a = 3$$

$$\frac{2\pi}{b} = 4\pi \Rightarrow b = \frac{1}{2}$$

$$\frac{c}{b} = \pi \Rightarrow c = \frac{\pi}{2}.$$

Therefore, $y = 3\cos[(1/2)x - (\pi/2)]$.

69. Yes. Use the right-triangle definitions of the trigonometric functions.

71. The range of the cosine function is $[-1, 1]$. The range of the secant function is $(-\infty, -1] \cup [1, \infty)$.

73. $f(x) = \sin x$

$g(x) = |\sin x|$

$h(x) = \sin|x|$

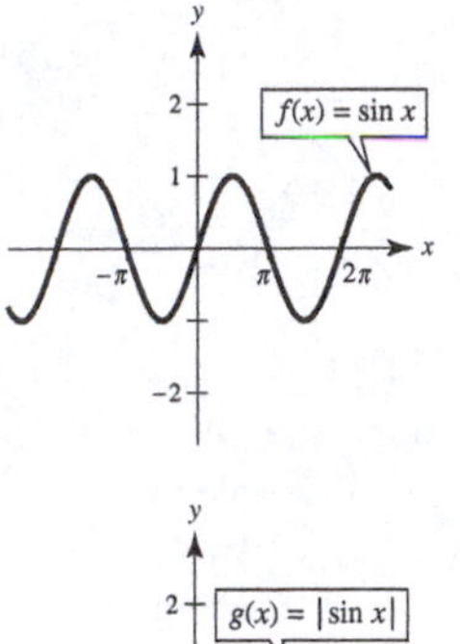

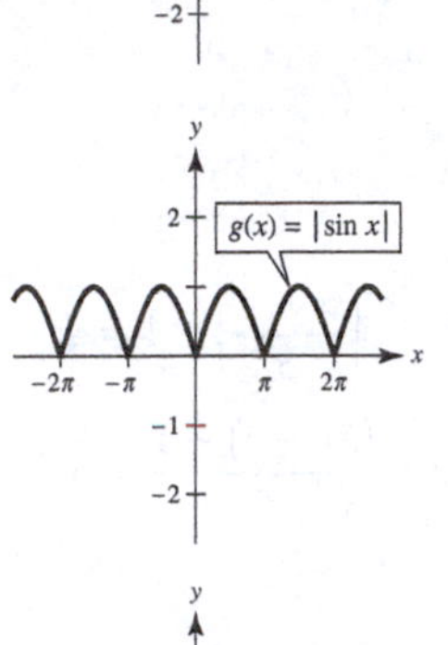

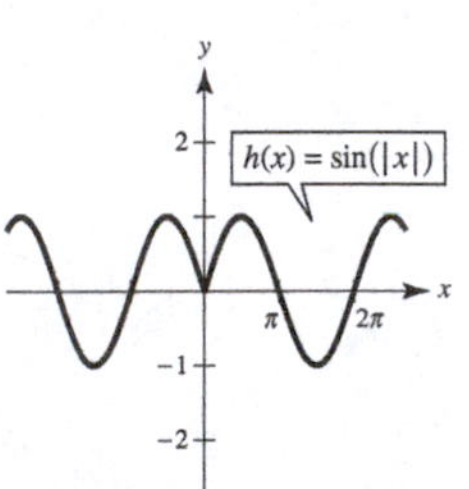

The graph of $|f(x)|$ will reflect any parts of the graph of $f(x)$ below the x-axis about the y-axis.

The graph of $f(|x|)$ will reflect the part of the graph of $f(x)$ to the right of the y-axis about the y-axis.

75. $S = 58.3 + 32.5\cos\dfrac{\pi t}{6}$

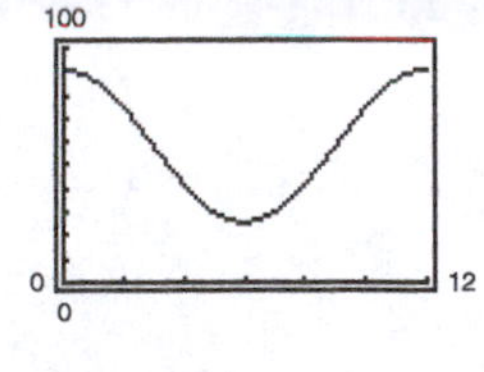

Sales exceed 75,000 during the months of January, November, and December.

77. False. 4π radians (not 4 radians) corresponds to two complete revolutions from the initial side to the terminal side of an angle.

79. False. The amplitude of the function $y = \frac{1}{2}\sin 2x$ is one-half the amplitude of the function $y = \sin x$.

Section 1.5 Inverse Functions

1. The graphs of f and f^{-1} are mirror images with respect to the line $y = x$.

3. $\arccos x$ is the angle θ whose cosine is x, where $0 \le \theta \le \pi$.

5. Matches (c)

6. Matches (b)

7. Matches (a)

8. Matches (d)

9. (a)

$$f(x) = 5x + 1$$
$$g(x) = \frac{x - 1}{5}$$
$$f(g(x)) = f\left(\frac{x-1}{5}\right) = 5\left(\frac{x-1}{5}\right) + 1 = x$$
$$g(f(x)) = g(5x + 1) = \frac{(5x + 1) - 1}{5} = x$$

(b)

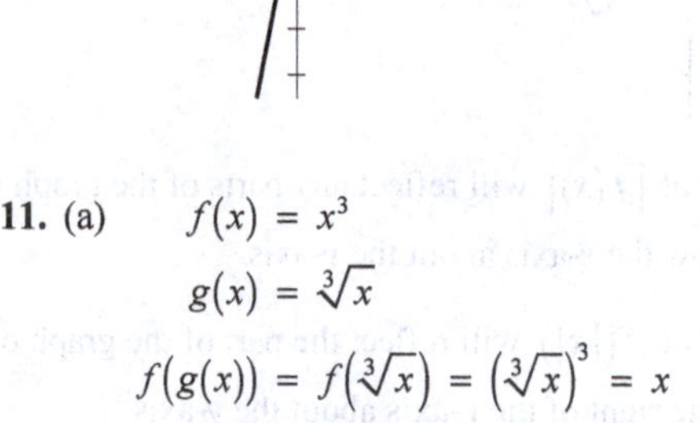

11. (a)

$$f(x) = x^3$$
$$g(x) = \sqrt[3]{x}$$
$$f(g(x)) = f(\sqrt[3]{x}) = (\sqrt[3]{x})^3 = x$$
$$g(f(x)) = g(x^3) = \sqrt[3]{x^3} = x$$

(b)

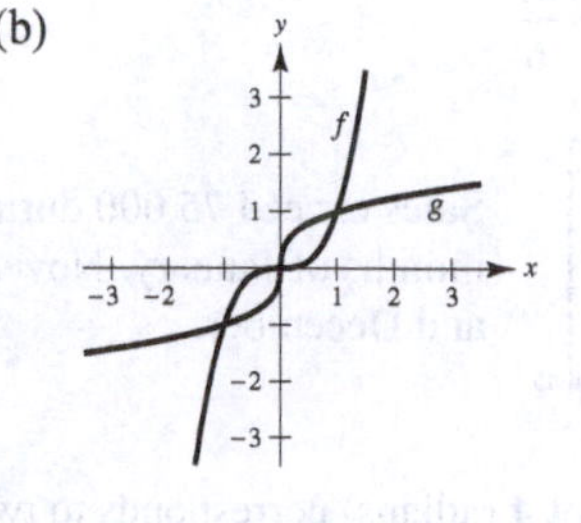

13. (a)

$$f(x) = \sqrt{x - 4}$$
$$g(x) = x^2 + 4, \quad x \ge 0$$
$$f(g(x)) = f(x^2 + 4) = \sqrt{(x^2 + 4) - 4} = \sqrt{x^2} = x$$
$$g(f(x)) = g(\sqrt{x - 4}) = (\sqrt{x - 4})^2 + 4 = x - 4 + 4 = x$$

(b)

15. (a)

$$f(x) = \frac{1}{x}$$
$$g(x) = \frac{1}{x}$$
$$f(g(x)) = \frac{1}{1/x} = x$$
$$g(f(x)) = \frac{1}{1/x} = x$$

(b)

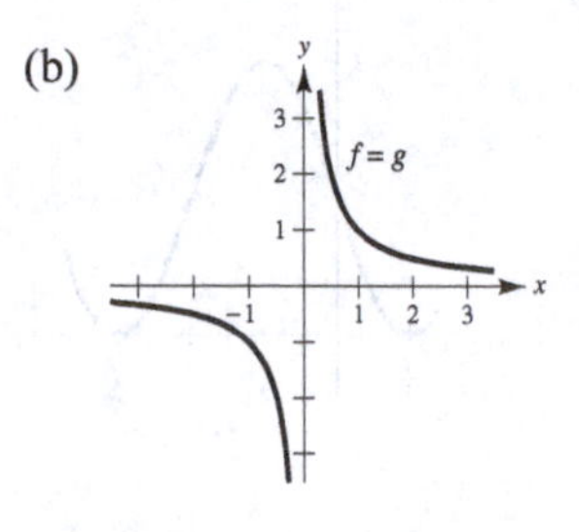

17. $f(\theta) = \sin\theta$

Not one-to-one; does not have an inverse

19. $f(x) = 2 - x - x^3$

One-to-one; has an inverse

21. $f(x) = \dfrac{1}{3x + 1}$

One-to-one; has an inverse

23. $f(x) = \tan 2\pi x$

Not one-to-one; does not have an inverse

25. $h(s) = \dfrac{1}{s-2} - 3$

One-to-one; has an inverse

27. $g(t) = \dfrac{1}{\sqrt{t^2+1}}$

Not one-to-one; does not have an inverse

29. $g(x) = (x+5)^3$

One-to-one; has an inverse

31. (a) $f(x) = 2x - 3 = y$

$$x = \frac{y+3}{2}$$

$$y = \frac{x+3}{2}$$

$$f^{-1}(x) = \frac{x+3}{2}$$

(b)

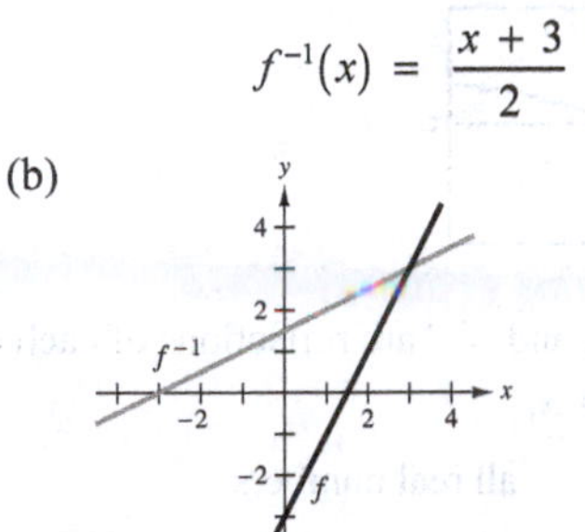
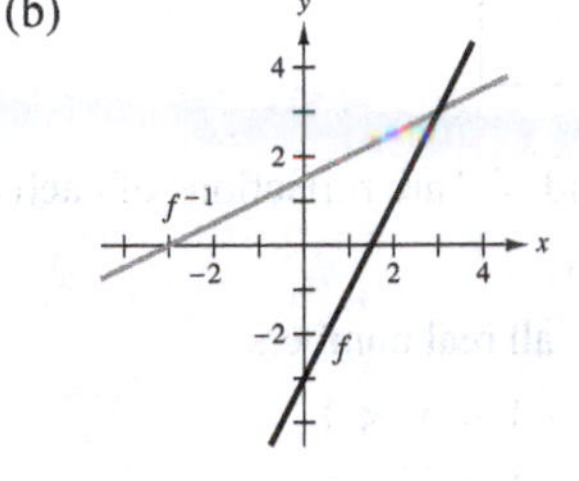

(c) The graphs of f and f^{-1} are reflections of each other in the line $y = x$.

(d) Domain of f: all real numbers
Range of f: all real numbers
Domain of f^{-1}: all real numbers
Range of f^{-1}: all real numbers

33. (a) $f(x) = x^5 = y$

$$x = \sqrt[5]{y}$$

$$y = \sqrt[5]{x}$$

$$f^{-1}(x) = \sqrt[5]{x} = x^{1/5}$$

(b)

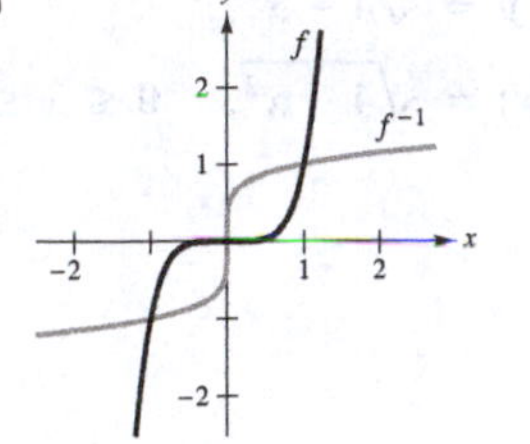
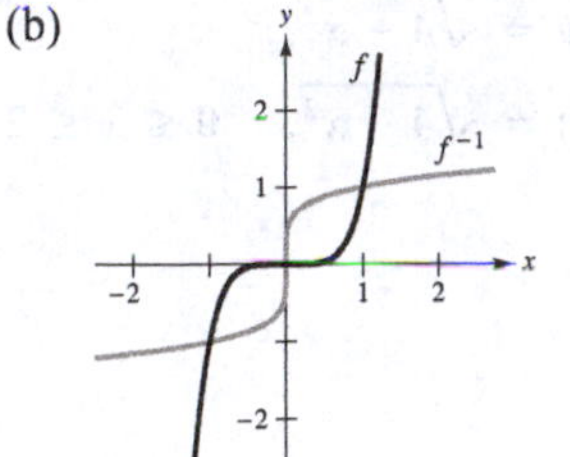

(c) The graphs of f and f^{-1} are reflections of each other in the line $y = x$.

(d) Domain of f: all real numbers
Range of f: all real numbers
Domain of f^{-1}: all real numbers
Range of f^{-1}: all real numbers

35. (a) $f(x) = \sqrt{x} = y$

$$x = y^2$$

$$y = x^2$$

$$f^{-1}(x) = x^2, \quad x \ge 0$$

(b)

(c) The graphs of f and f^{-1} are reflections of each other in the line $y = x$.

(d) Domain of f: $x \ge 0$
Range of f: $y \ge 0$
Domain of f^{-1}: $x \ge 0$
Range of f^{-1}: $y \ge 0$

37. (a) $f(x) = \sqrt{4 - x^2} = y, \quad 0 \le x \le 2$

$$4 - x^2 = y^2$$
$$x^2 = 4 - y^2$$
$$x = \sqrt{4 - y^2}$$
$$y = \sqrt{4 - x^2}$$
$$f^{-1}(x) = \sqrt{4 - x^2}, \quad 0 \le x \le 2$$

(b)

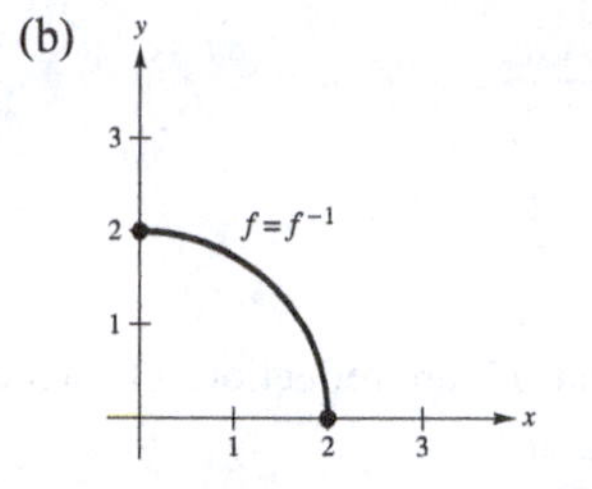

(c) The graphs of f and f^{-1} are reflections of each other in the line $y = x$. In fact, the graphs are identical.

(d) Domain of f: $0 \le x \le 2$
Range of f: $0 \le y \le 2$
Domain of f^{-1}: $0 \le x \le 2$
Range of f^{-1}: $0 \le y \le 2$

39. (a) $f(x) = \sqrt[3]{x - 1} = y$

$$x - 1 = y^3$$
$$x = y^3 + 1$$
$$y = x^3 + 1$$
$$f^{-1}(x) = x^3 + 1$$

(b)

(c) The graphs of f and f^{-1} are reflections of each other in the line $y = x$.

(d) Domain of f: all real numbers
Range of f: all real numbers
Domain of f^{-1}: all real numbers
Range of f^{-1}: all real numbers

41. (a) $f(x) = x^{2/3} = y, \quad x \ge 0$

$$x = y^{3/2}$$
$$y = x^{3/2}$$
$$f^{-1}(x) = x^{3/2}, \quad x \ge 0$$

(b)

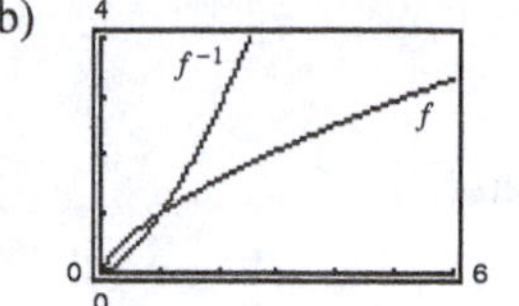

(c) The graphs of f and f^{-1} are reflections of each other in the line $y = x$.

(d) Domain of f: $x \ge 0$
Range of f: $y \ge 0$
Domain of f^{-1}: $x \ge 0$
Range of f^{-1}: $y \ge 0$

43. (a) $f(x) = \dfrac{x}{\sqrt{x^2 + 7}} = y$

$$x = y\sqrt{x^2 + 7}$$
$$x^2 = y^2(x^2 + 7) = y^2x^2 + 7y^2$$
$$x^2(1 - y^2) = 7y^2$$
$$x = \frac{\sqrt{7}y}{\sqrt{1 - y^2}}$$
$$y = \frac{\sqrt{7}x}{\sqrt{1 - x^2}}$$
$$f^{-1}(x) = \frac{\sqrt{7}x}{\sqrt{1 - x^2}}, \quad -1 < x < 1$$

(b)

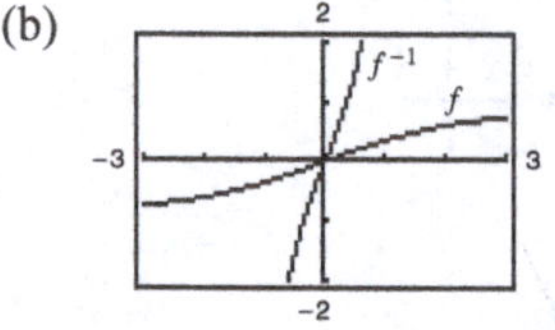

(c) The graphs of f and f^{-1} are reflections of each other in the line $y = x$.

(d) Domain of f: all real numbers
Range of f: $-1 < y < 1$
Domain of f^{-1}: $-1 < x < 1$
Range of f^{-1}: all real numbers

45.

x	0	1	2	4
$f(x)$	1	2	3	4

x	1	2	3	4
$f^{-1}(x)$	0	1	2	4

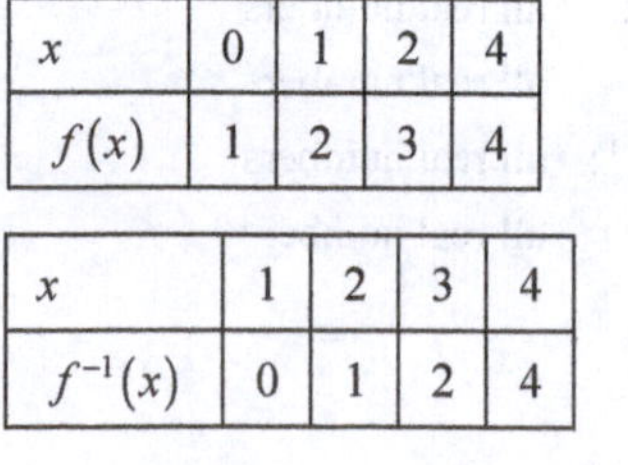

47. (a) Let x be the number of pounds of the commodity costing \$1.25 per pound. Because there are 50 pounds total, the amount of the second commodity is $50-x$. The total cost is

$$f(x) = y = 1.25x + 2.75(50 - x)$$
$$= -1.5x + 137.5, 0 \le x \le 50$$

(b)
$$y = -1.5x + 137.5$$
$$1.5x = 137.5 - y$$
$$x = \frac{(137.5 - y)}{1.5}$$
$$y = f^{-1}(x) = \frac{2}{3}(137.5 - x)$$

x represents the total cost and y represents the number of pounds of the less expensive commodity.

(c) The range of f is $[62.5, 137.5]$, so the domain of f^{-1} is the same. $50(1.25) = 62.5$ gives the total cost when purchasing 50 pounds of the less expensive commodity, and $50(2.75) = 137.5$ gives the total cost when purchasing 50 pounds of the more expensive commodity.

(d) If $x = 73$, then $f^{-1}(73) = 43$ pounds.

49. $f(x) = \sqrt{x - 2},\ x \ge 2$

f is one-to-one; has an inverse.

$$y = \sqrt{x - 2},\ x \ge 2,\ y \ge 0$$
$$y^2 = x - 2$$
$$x = y^2 + 2$$
$$f^{-1}(x) = x^2 + 2,\ x \ge 0$$

51. $f(x) = -3$

Not one-to-one; does not have an inverse.

53. $f(x) = ax + b$

f is one-to-one; has an inverse.

$$ax + b = y$$
$$x = \frac{y - b}{a}$$
$$y = \frac{x - b}{a}$$
$$f^{-1}(x) = \frac{x - b}{a},\quad a \ne 0$$

55. $f(x) = (x - 4)^2$ on $[4, \infty)$

f passes the Horizontal Line Test on $[4, \infty)$, so it is one-to-one.

57. $f(x) = \dfrac{4}{x^2}$ on $(0, \infty)$

f passes the Horizontal Line Test on $(0, \infty)$, so it is one-to-one.

59. $f(x) = \cos x$ on $[0, \pi]$

f passes the Horizontal Line Test on $[0, \pi]$, so it is one-to-one.

61. $f(x) = (x - 3)^2$ is one-to-one for $x \ge 3$.

$$(x - 3)^2 = y$$
$$x - 3 = \sqrt{y}$$
$$x = \sqrt{y} + 3$$
$$y = \sqrt{x} + 3$$
$$f^{-1}(x) = \sqrt{x} + 3,\quad x \ge 0$$

(Answer is not unique.)

63. (a) $f(x) = (x + 5)^2$

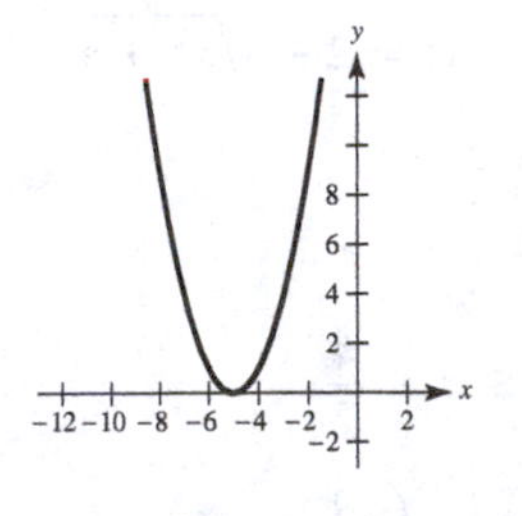

(b) f is one-to-one on $[-5, \infty)$. (Note that f is also one-to-one on $(-\infty, -5]$.)

(c)
$$f(x) = (x + 5)^2 = y,\qquad x \ge -5$$
$$x + 5 = \sqrt{y}$$
$$x = \sqrt{y} - 5$$
$$y = \sqrt{x} - 5$$
$$f^{-1}(x) = \sqrt{x} - 5$$

(d) Domain of f^{-1}: $x \ge 0$

65. (a) $f(x) = \sqrt{x^2 - 4x}$

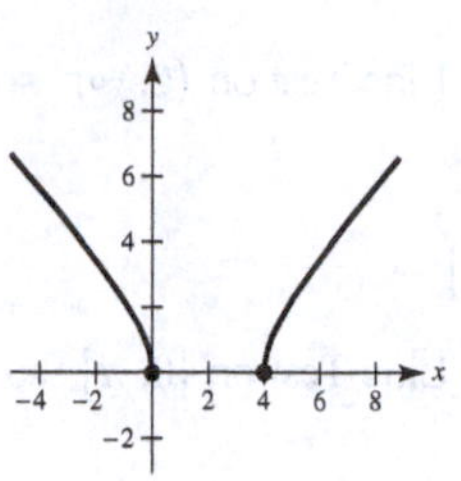

(b) f is one-to-one on $[4, \infty)$. (Note that f is also one-to-one on $(-\infty, 0]$.)

(c)
$$\begin{aligned} f(x) = \sqrt{x^2 - 4x} &= y,\ x \ge 4 \\ x^2 - 4x &= y^2 \\ x^2 - 4x + 4 &= y^2 + 4 \\ (x-2)^2 &= y^2 + 4 \\ x - 2 &= \sqrt{y^2 + 4} \\ x &= 2 + \sqrt{y^2 + 4} \\ y &= 2 + \sqrt{x^2 + 4} \\ f^{-1}(x) &= 2 + \sqrt{x^2 + 4} \end{aligned}$$

(d) Domain of $f^{-1}: x \ge 0$

67. (a) $f(x) = 3 \cos x$

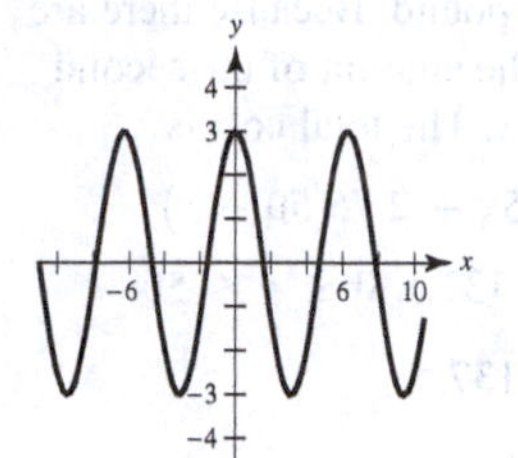

(b) f is one-to-one on $[0, \pi]$. (other answers possible)

(c)
$$\begin{aligned} f(x) = 3 \cos x &= y \\ \cos x &= \frac{y}{3} \\ x &= \arccos\left(\frac{y}{3}\right) \\ y &= \arccos\left(\frac{x}{3}\right) \\ f^{-1}(x) &= \arccos\left(\frac{x}{3}\right) \end{aligned}$$

(d) Domain of $f^{-1}: -3 \le x \le 3$

69. $f(x) = x^3 + 2x - 1$

$f(1) = 2 = a \Rightarrow f^{-1}(2) = 1$

71. $f(x) = 5 \sin x,\ -\frac{\pi}{2} \le x \le \frac{\pi}{2}$

$$f\left(-\frac{\pi}{6}\right) = 5 \sin\left(-\frac{\pi}{6}\right) = 5\left(-\frac{1}{2}\right) = -\frac{5}{2} = a \Rightarrow f^{-1}\left(-\frac{5}{2}\right) = -\frac{\pi}{6}$$

73. $f(x) = x^3 - \frac{4}{x}$

$f(2) = 6 = a \Rightarrow f^{-1}(6) = 2$

In Exercises 75–77, use the following.

$f(x) = \frac{1}{8}x - 3$ **and** $g(x) = x^3$

$f^{-1}(x) = 8(x + 3)$ **and** $g^{-1}(x) = \sqrt[3]{x}$

75. $(f^{-1} \circ g^{-1})(1) = f^{-1}(g^{-1}(1)) = f^{-1}(1) = 32$

77. $(f^{-1} \circ f^{-1})(-2) = f^{-1}(f^{-1}(-2)) = f^{-1}(8) = 88$

In Exercises 79–81, use the following.

$f(x) = x + 4$ **and** $g(x) = 2 - x^3$

$f^{-1}(x) = x - 4$ **and** $g^{-1}(x) = \sqrt[3]{2 - x}$

79.
$$\begin{aligned} (g^{-1} \circ f^{-1})(x) &= g^{-1}(f^{-1}(x)) \\ &= g^{-1}(x - 4) \\ &= \sqrt[3]{2 - (x - 4)} \\ &= \sqrt[3]{6 - x} \end{aligned}$$

81.
$$\begin{aligned} (f \circ g)(x) &= f(g(x)) \\ &= f(2 - x^3) \\ &= (2 - x^3) + 4 \\ &= 6 - x^3 \end{aligned}$$

So, $(f \circ g)^{-1}(x) = \sqrt[3]{6 - x}$.

Note: $(f \circ g)^{-1} = g^{-1} \circ f^{-1}$

83. (a) f is one-to-one because it passes the Horizontal Line Test.

(b) The domain of f^{-1} is the range of f: $[-2, 2]$.

(c) $f^{-1}(2) = -4$ because $f(-4) = 2$.

85.

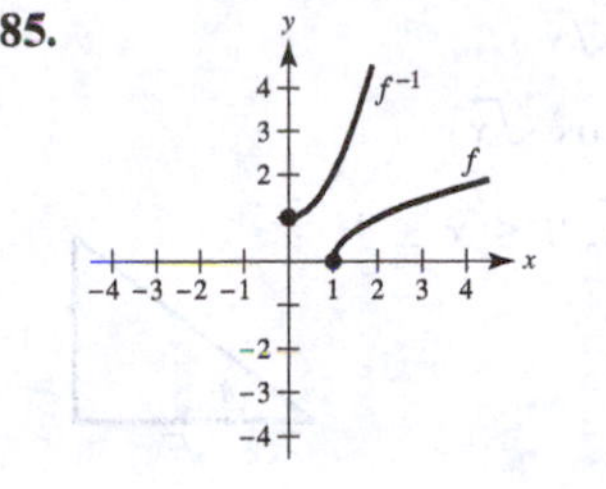

87. $y = \arcsin x$

(a)

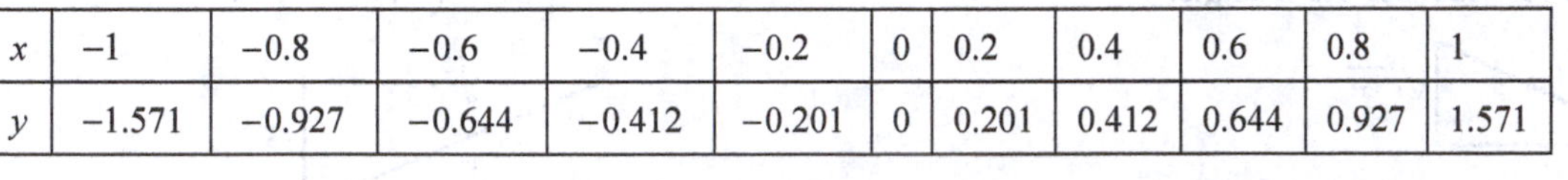

x	-1	-0.8	-0.6	-0.4	-0.2	0	0.2	0.4	0.6	0.8	1
y	-1.571	-0.927	-0.644	-0.412	-0.201	0	0.201	0.412	0.644	0.927	1.571

(b)

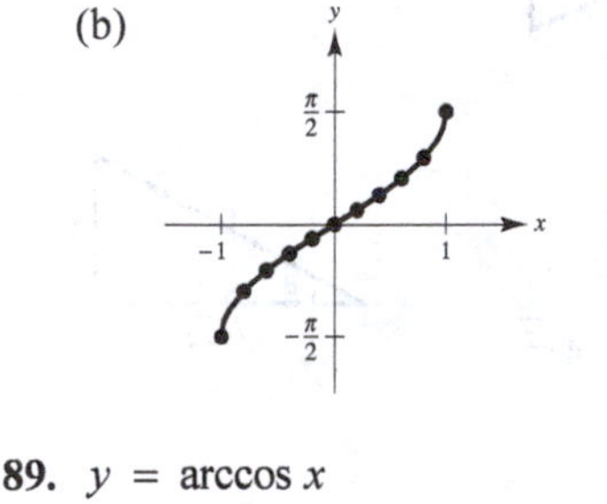

(c)

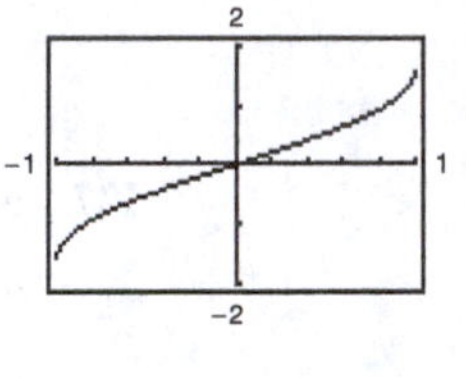

(d) Symmetric about origin: $\arcsin(-x) = -\arcsin x$

Intercept: $(0, 0)$

89. $y = \arccos x$

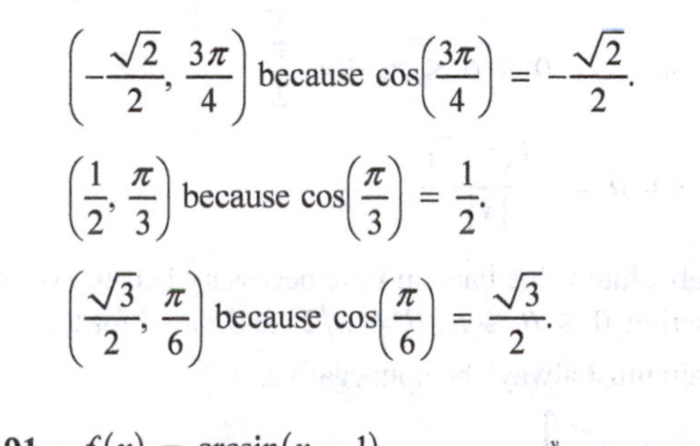

$\left(-\frac{\sqrt{2}}{2}, \frac{3\pi}{4}\right)$ because $\cos\left(\frac{3\pi}{4}\right) = -\frac{\sqrt{2}}{2}$.

$\left(\frac{1}{2}, \frac{\pi}{3}\right)$ because $\cos\left(\frac{\pi}{3}\right) = \frac{1}{2}$.

$\left(\frac{\sqrt{3}}{2}, \frac{\pi}{6}\right)$ because $\cos\left(\frac{\pi}{6}\right) = \frac{\sqrt{3}}{2}$.

91. $f(x) = \arcsin(x - 1)$

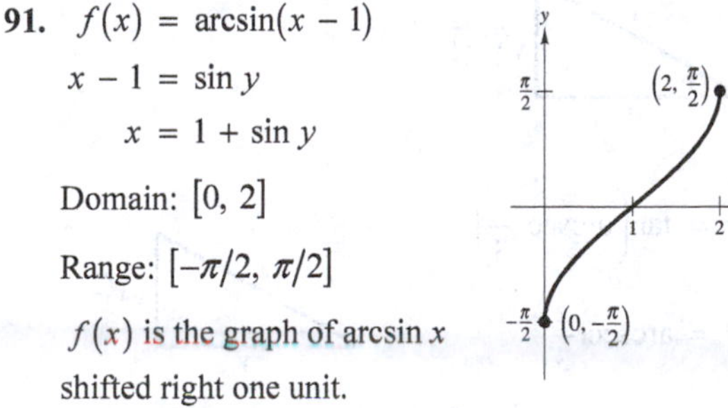

$x - 1 = \sin y$

$x = 1 + \sin y$

Domain: $[0, 2]$

Range: $[-\pi/2, \pi/2]$

$f(x)$ is the graph of $\arcsin x$ shifted right one unit.

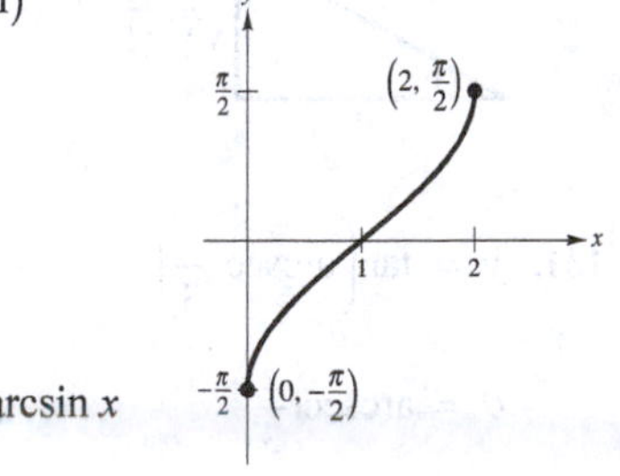

93. $f(x) = \arctan x + \frac{\pi}{2}$

$x = \tan\left(y - \frac{\pi}{2}\right)$

Domain: $(-\infty, \infty)$

Range: $[0, \pi]$

$f(x)$ is the graph of $\arctan x$ shifted $\pi/2$ unit upward.

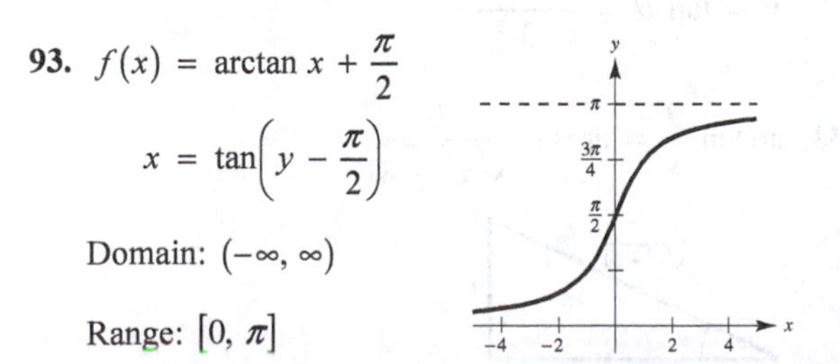

95. $\arcsin \frac{1}{2} = \frac{\pi}{6}$

97. $\arccos \frac{1}{2} = \frac{\pi}{3}$

99. $\arctan \frac{\sqrt{3}}{3} = \frac{\pi}{6}$

101. $\operatorname{arccsc}\left(-\sqrt{2}\right) = -\frac{\pi}{4}$

103. $\arccos(0.051) \approx 1.52$

105. $\operatorname{arcsec}(1.269) = \arccos\left(\frac{1}{1.269}\right) \approx 0.66$

107. $\cos\left[\arccos(-0.1)\right] = -0.1$

109. No. Graphically, adding a constant shift the graph vertically.

111. The trigonometric functions are not one-to-one. So, their domains must be restricted to define the inverse trigonometric functions.

113. $\arcsin(3x - \pi) = \frac{1}{2}$

$3x - \pi = \sin\left(\frac{1}{2}\right)$

$x = \frac{1}{3}\left[\sin\left(\frac{1}{2}\right) + \pi\right] \approx 1.207$

115. $\arcsin \sqrt{2x} = \arccos \sqrt{x}$

$$\sqrt{2x} = \sin(\arccos \sqrt{x})$$

$$\sqrt{2x} = \sqrt{1 - x},\ 0 \le x \le 1$$

$$2x = 1 - x$$

$$3x = 1$$

$$x = \tfrac{1}{3}$$

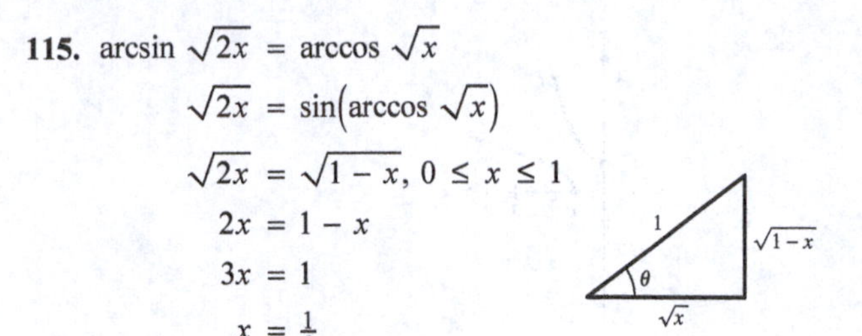

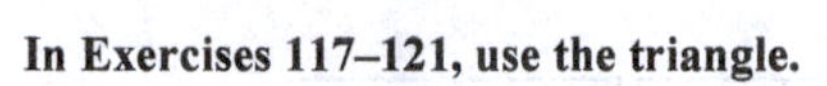

In Exercises 117–121, use the triangle.

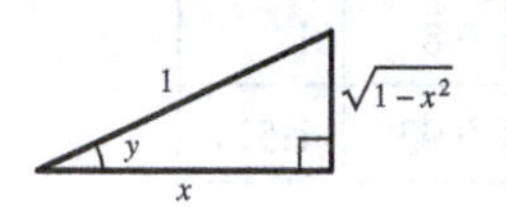

117. $y = \arccos x$

$\cos y = x$

119. $\tan y = \dfrac{\sqrt{1 - x^2}}{x}$

121. $\sec y = \dfrac{1}{x}$

123. (a) $\sin\left(\arctan\dfrac{3}{4}\right) = \dfrac{3}{5}$

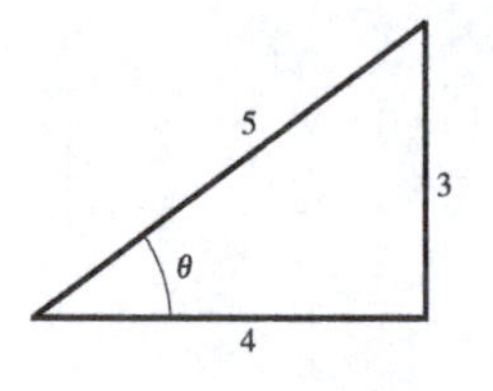

(b) $\sec\left(\arcsin\dfrac{4}{5}\right) = \dfrac{5}{3}$

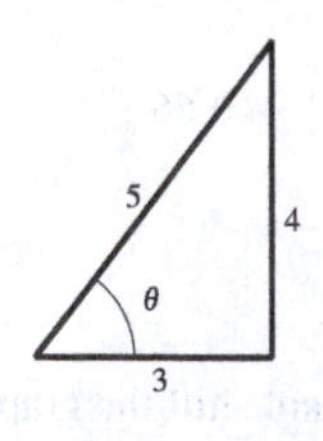

125. (a) $\cot\left[\arcsin\left(-\dfrac{1}{2}\right)\right] = \cot\left(-\dfrac{\pi}{6}\right) = -\sqrt{3}$

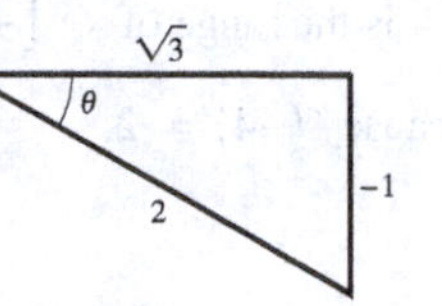

(b) $\csc\left[\arctan\left(-\dfrac{5}{12}\right)\right] = -\dfrac{13}{5}$

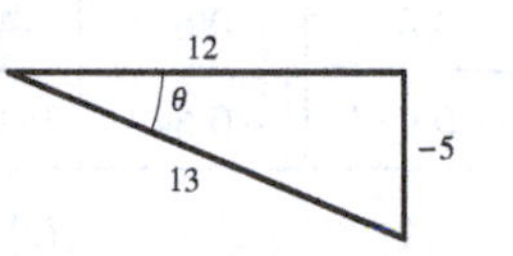

127. $y = \cos(\arcsin 2x)$

$$\theta = \arcsin 2x$$

$$y = \cos\theta = \sqrt{1 - 4x^2}$$

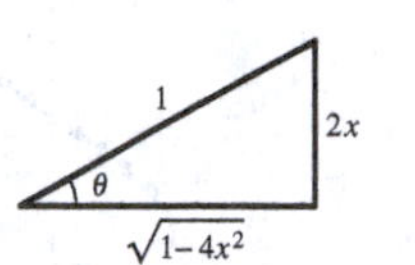

129. $y = \sin(\operatorname{arcsec} x)$

$$\theta = \operatorname{arcsec} x,\ 0 \le \theta \le \pi,\ \theta \ne \frac{\pi}{2}$$

$$y = \sin\theta = \frac{\sqrt{x^2 - 1}}{|x|}$$

The absolute value bars on x are necessary because of the restriction $0 \le \theta \le \pi$, $\theta \ne \pi/2$, and $\sin\theta$ for this domain must always be nonnegative.

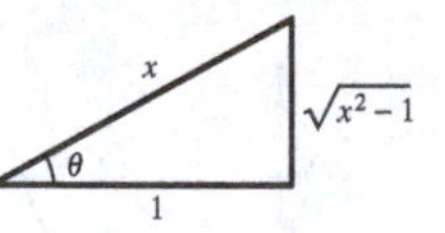

131. $y = \tan\left(\operatorname{arcsec}\dfrac{x}{3}\right)$

$$\theta = \operatorname{arcsec}\frac{x}{3}$$

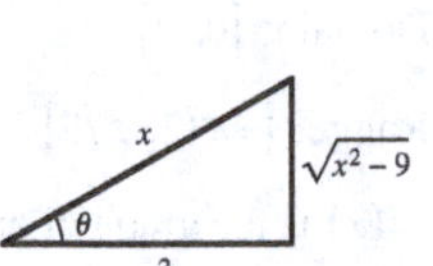

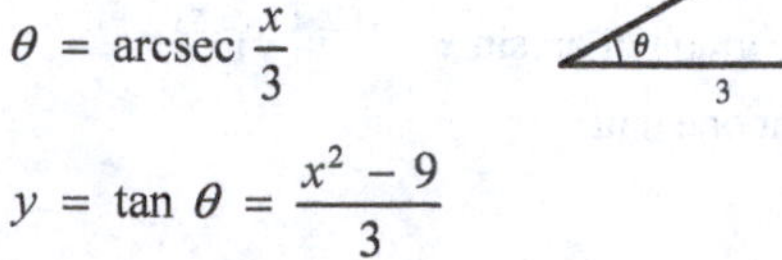

$$y = \tan\theta = \frac{x^2 - 9}{3}$$

133. $\arctan\dfrac{9}{x} = \arcsin\dfrac{9}{\sqrt{x^2 + 81}}$

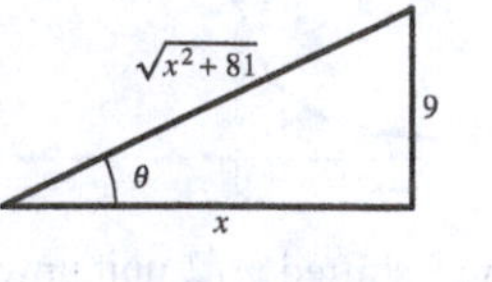

135. (a) $\operatorname{arccsc} x = \arcsin \frac{1}{x}, |x| \ge 1$

Let $y = \operatorname{arccsc} x$.

Then for $-\frac{\pi}{2} \le y < 0$ and $0 < y \le \frac{\pi}{2}$,

$\csc y = x \Rightarrow \sin y = \frac{1}{x}$.

So, $y = \arcsin\left(\frac{1}{x}\right)$. Therefore,

$\operatorname{arccsc} x = \arcsin\left(\frac{1}{x}\right)$.

(b) $\arctan x + \arctan\frac{1}{x} = \frac{\pi}{2}, x > 0$

Let $y = \arctan x + \arctan(1/x)$.

$$\text{Then } \tan y = \frac{\tan(\arctan x) + \tan[\arctan(1/x)]}{1 - \tan(\arctan x)\tan[\arctan(1/x)]}$$

$$= \frac{x + (1/x)}{1 - x(1/x)}$$

$$= \frac{x + (1/x)}{0} \text{ (which is undefined).}$$

So, $y = \pi/2$. Therefore,

$\arctan x + \arctan(1/x) = \pi/2$.

137. (a) $\operatorname{arccot} x = y$ if and only if $\cot y = x$, $0 < y < \pi$.

For $x > 0$, $\cot y > 0$ and $0 < y < \frac{\pi}{2}$.

So, $\tan y = \frac{1}{x} > 0$ and $y = \arctan\left(\frac{1}{x}\right)$.

For $x = 0$, $\operatorname{arccot}(0) = \frac{\pi}{2}$.

For $x < 0$, $\cot y < 0$ and $\frac{\pi}{2} < y < \pi$.

So, $\tan y = \frac{1}{x} < 0$ and $\arctan\left(\frac{1}{x}\right) < 0$.

Therefore, you need to add π to get

$y = \pi + \arctan\left(\frac{1}{x}\right)$.

(b) $y = \operatorname{arcsec} x$ if and only if $\sec y = x, |x| \ge 1$,

$0 \le y \le \pi, y \ne \frac{\pi}{2}$.

So, $\cos y = \frac{1}{x}$ and $y = \arccos\left(\frac{1}{x}\right)$.

(c) $y = \operatorname{arccsc} x$ if and only if $\csc y = x$, $|x| \ge 1$,

$-\frac{\pi}{2} \le y \le \frac{\pi}{2}, y \ne 0$.

So, $\sin y = \frac{1}{x}$ and $y = \arcsin\left(\frac{1}{x}\right)$.

139. False. Let $f(x) = x^2$.

141. False

$$\arcsin^2 0 + \arccos^2 0 = 0 + \frac{\pi^2}{2} \ne 1$$

143. True

145. Let f and g be one-to-one functions.

Let $(f \circ g)(x) = y$, then $x = (f \circ g)^{-1}(y)$. Also:

$$(f \circ g)(x) = y$$
$$f(g(x)) = y$$
$$g(x) = f^{-1}(y)$$
$$x = g^{-1}(f^{-1}(y))$$
$$x = (g^{-1} \circ f^{-1}(y))$$

So, $(f \circ g)^{-1}(y) = (g^{-1} \circ f^{-1})(y)$ and $(f \circ g)^{-1} = g^{-1} \circ f^{-1}$.

147. Let $y = \sin^{-1} x$. Then $\sin y = x$ and $\cos(\sin^{-1} x) = \cos(y) = \sqrt{1 - x^2}$, as indicated in the figure.

149. $$\tan(\arctan x + \arctan y) = \frac{\tan(\arctan x + \arctan y)}{1 - \tan(\arctan x)\tan(\arctan y)} = \frac{x + y}{1 - xy}, xy \ne 1$$

So, $\arctan x + \arctan y = \arctan\left(\frac{x + y}{1 - xy}\right), xy \ne 1$.

Let $x = \frac{1}{2}$ and $y = \frac{1}{3}$.

$$\arctan\left(\frac{1}{2}\right) + \arctan\left(\frac{1}{3}\right) = \arctan \frac{\frac{1}{2} + \frac{1}{3}}{1 - \left(\frac{1}{2} \cdot \frac{1}{3}\right)} = \arctan \frac{\frac{5}{6}}{1 - \frac{1}{6}} = \arctan \frac{\frac{5}{6}}{\frac{5}{6}} = \arctan 1 = \frac{\pi}{4}$$

151. $f(x) = kx + \sin x$

For $k \geq 1$, f is one-to-one, and for $k \leq -1$, f is one-to-one. Therefore, f has an inverse for $k \geq 1$ and $k \leq 1$.

153. f is one-to-one if $f(x_1) = f(x_2)$ implies $x_1 = x_2$. So assume

$$\begin{aligned} f(x_1) &= f(x_2) \\ \frac{ax_1 + b}{cx_1 + d} &= \frac{ax_2 + b}{cx_2 + d} \\ acx_1x_2 + adx_1 + bcx_2 + bd &= acx_1x_2 + adx_2 + bcx_1 + bd \\ adx_1 + bcx_2 &= adx_2 + bcx_1 \\ (ad - bc)x_1 &= (ad - bc)x_2. \end{aligned}$$

So, $x_1 = x_2$ if $ad - bc \neq 0$. To find f^{-1}, solve for x as follows.

$$\begin{aligned} y &= \frac{ax + b}{cx + d} \\ ycx + yd &= ax + b \\ (yc - a)x &= b - yd \\ x &= \frac{b - yd}{yc - a} \\ f^{-1}(x) &= \frac{b - dx}{cx - a} \end{aligned}$$

Section 1.6 Exponential and Logarithmic Functions

1. $f(x) = e^x$. Domain is $(-\infty, \infty)$ and range is $(0, \infty)$. f is continuous, increasing, one-to-one, and concave upwards on its entire domain.

$\lim_{x\to-\infty} e^x = 0$ and $\lim_{x\to\infty} e^x = \infty$

3. The functions $f(x) = e^x$ and $g(x) = \ln x$ are inverses of each other. So, $\ln e^x = g(f(x)) = x$.

5. (a) $25^{3/2} = 5^3 = 125$

(b) $81^{1/2} = 9$

(c) $3^{-2} = \frac{1}{3^2} = \frac{1}{9}$

(d) $27^{-1/3} = \frac{1}{27^{1/3}} = \frac{1}{3}$

7. (a) $(5^2)(5^3) = 5^{2+3} = 5^5 = 3125$

(b) $(5^2)(5^{-3}) = 5^{2-3} = 5^{-1} = \frac{1}{5}$

(c) $\frac{5^3}{25^2} = \frac{5^3}{5^4} = \frac{1}{5}$

(d) $\left(\frac{1}{4}\right)^2 2^6 - \frac{2^6}{2^4} = 2^2 = 4$

9. (a) $e^2(e^4) = e^6$

(b) $(e^3)^4 = e^{12}$

(c) $(e^3)^{-2} = e^{-6} = \frac{1}{e^6}$

(d) $\left(\frac{e^{-6}}{e^{-2}}\right)^2 = \left(\frac{e^2}{e^6}\right)^2 = \left(\frac{1}{e^4}\right)^2 = \frac{1}{e^8}$

11. $3^x = 81 \Rightarrow x = 4$

13. $6^{x-2} = 36 \Rightarrow x - 2 = 2 \Rightarrow x = 4$

15. $\left(\frac{1}{2}\right)^x = 32 \Rightarrow 2^{-x} = 32 \Rightarrow -x = 5 \Rightarrow x = -5$

17. $\left(\frac{1}{3}\right)^{x-1} = 27 \Rightarrow 3^{1-x} = 27 \Rightarrow 1 - x = 3 \Rightarrow x = -2$

19. $4^3 = (x + 2)^3 \Rightarrow 4 = x + 2 \Rightarrow x = 2$

21. $x^{3/4} = 8 \Rightarrow x = 8^{4/3} = 2^4 = 16$

23. $e^x = e^{2x+1} \Rightarrow x = 2x + 1 \Rightarrow x = -1$

25. $e^{-2x} = e^5 \Rightarrow -2x = 5 \Rightarrow x = -\frac{5}{2}$

27. $y = 4^x$

x	-1	0	1	2
y	$\frac{1}{4}$	1	4	16

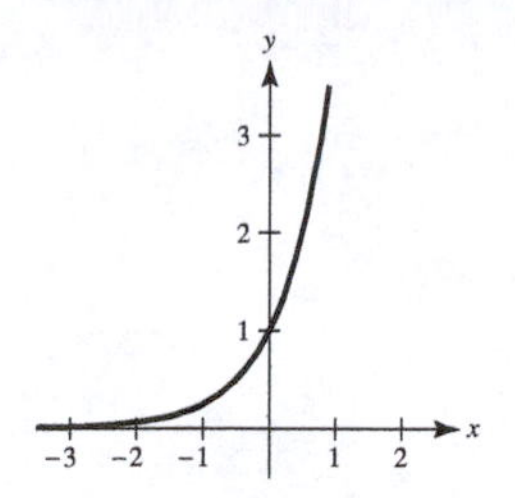

29. $y = \left(\frac{1}{3}\right)^x = 3^{-x}$

x	-2	-1	0	1	2
y	9	3	1	$\frac{1}{3}$	$\frac{1}{9}$

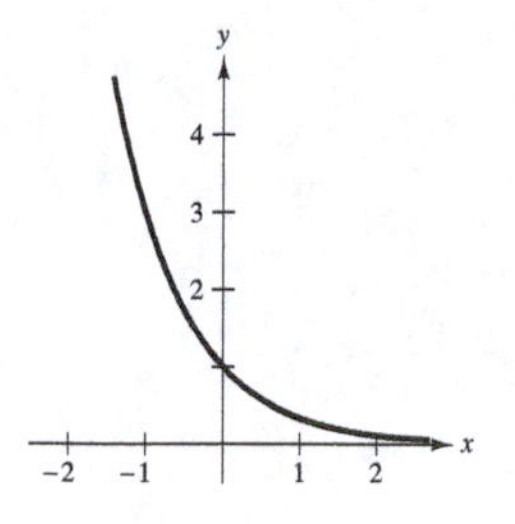

31. $f(x) = 3^{-x^2}$

x	0	± 1	± 2
y	1	$\frac{1}{3}$	0.0123

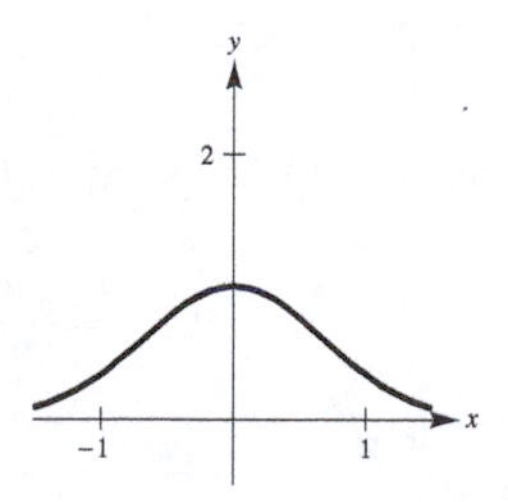

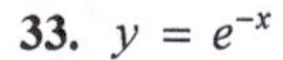

33. $y = e^{-x}$

x	-1	0	1
y	e	1	$\frac{1}{e}$

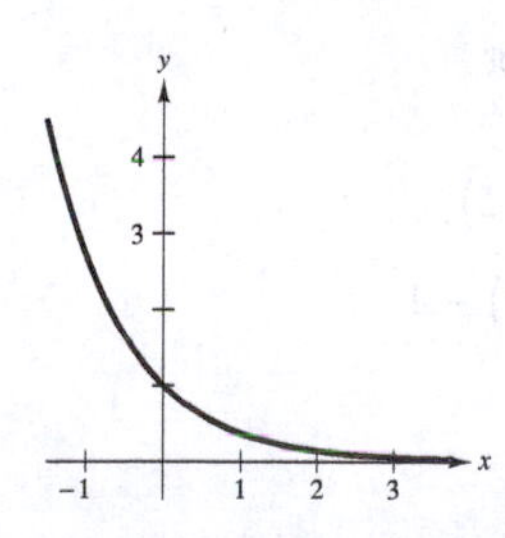

35. $y = e^x + 2$

x	-2	-1	0	1	2
y	$\frac{1}{e^2} + 2$	$\frac{1}{e} + 2$	3	$e + 2$	$e^2 + 2$

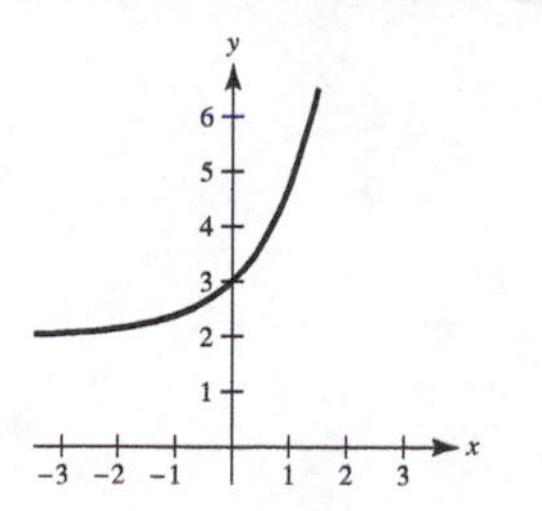

37. $h(x) = e^{x-2}$

x	0	1	2	3	4
y	e^{-2}	e^{-1}	1	e	e^2

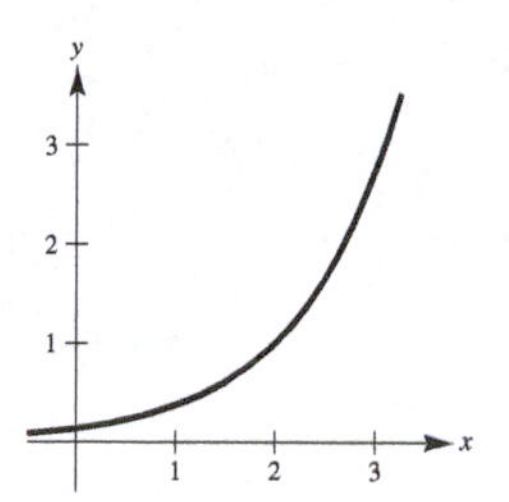

39. $y = e^{-x^2}$

x	-1	-0.5	0	0.5	1
y	$\frac{1}{e}$	$\frac{1}{e^{1/4}}$	1	$\frac{1}{e^{1/4}}$	$\frac{1}{e}$

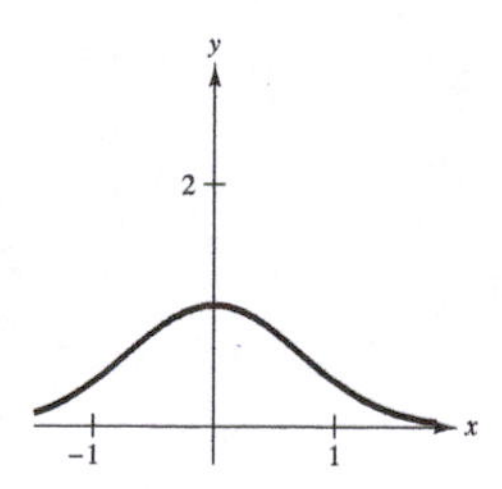

41. $f(x) = \dfrac{1}{3 + e^x}$

Because $e^x > 0$, $3 + e^x > 0$.

Domain: all real numbers

43. $f(x) = \sqrt{1 - 4^x}$

$1 - 4^x \geq 0 \Rightarrow 4^x \leq 1 \Rightarrow x \ln 4 \leq \ln 1 = 0$

Domain: $x \leq 0$

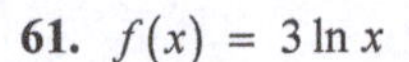

45. $f(x) = \sin e^{-x}$

Domain: all real numbers

47. $y = Ce^{ax}$

Graph rises from left to right

Matches (c)

48. $y = Ce^{-ax}$

Reflection in the y-axis

Matches (d)

49. $y = C(1 - e^{-ax})$

Vertical shift C units

Reflection in both the x- and y-axes

Matches (a)

50. $y = \dfrac{C}{1 + e^{-ax}}$

Matches (b)

51. $y = Ca^x$

$(0, 2): 2 = Ca^0 = C$

$(3, 54): 54 = 2a^3$

$27 = a^3$

$3 = a$

$y = 2(3^x)$

53. $e^0 = 1$

$\ln 1 = 0$

55. $\ln 4.15 = 1.4231\ldots$

$e^{1.4231\ldots} = 4.15$

57. $f(x) = \ln x + 1$

Vertical shift 1 unit upward

Matches (b)

58. $f(x) = -\ln x$

Reflection in the x-axis

Matches (d)

59. $f(x) = \ln(x - 1)$

Horizontal shift 1 unit to the right

Matches (a)

60. $f(x) = -\ln(-x)$

Reflection in the y-axis and the x-axis

Matches (c)

61. $f(x) = 3 \ln x$

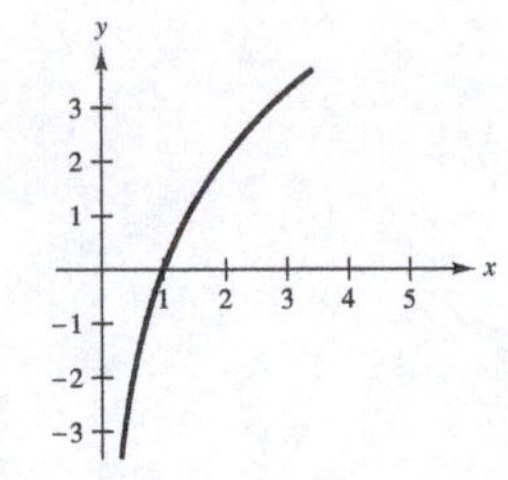

Domain: $x > 0$

63. $f(x) = \ln 2x$

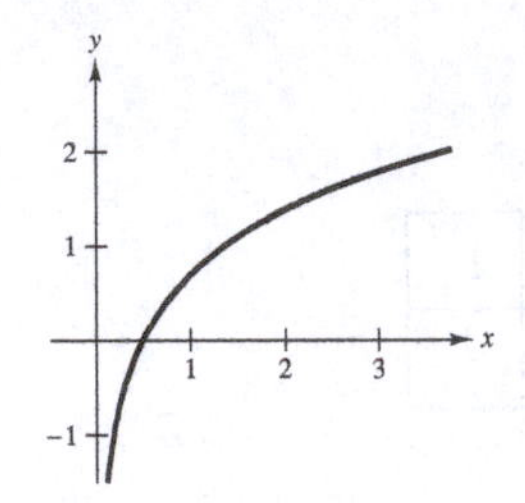

Domain: $x > 0$

65. $f(x) = \ln(x - 3)$

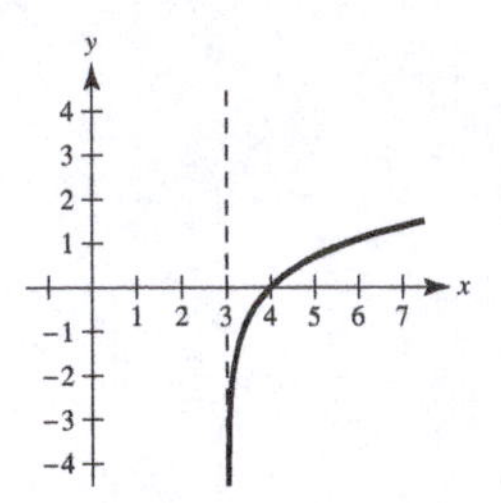

Domain: $x > 3$

67. $f(x) = -\ln(x + 2)$

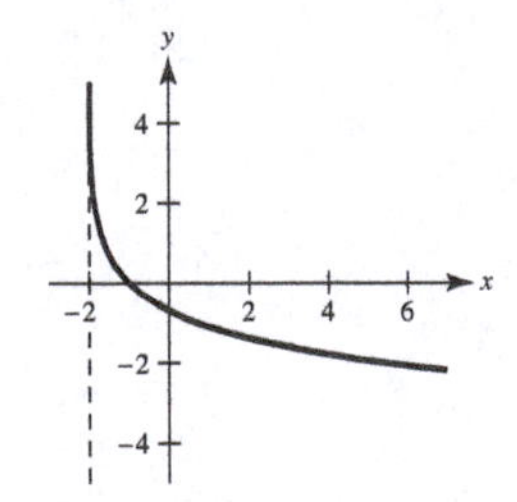

Domain: $x > -2$

69. 8 units upward: $e^x + 8$

Reflected in x-axis: $-(e^x + 8)$

$y = -(e^x + 8) = -e^x - 8$

71. 5 units to the right: $\ln(x - 5)$

1 unit downward: $\ln(x - 5) - 1$

$y = \ln(x - 5) - 1$

73. $f(x) = e^{2x}$

$g(x) = \ln\sqrt{x} = \dfrac{1}{2}\ln x$

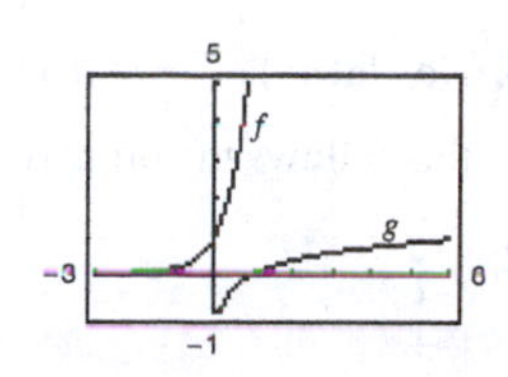

75. $f(x) = e^x - 1$

$g(x) = \ln(x + 1)$

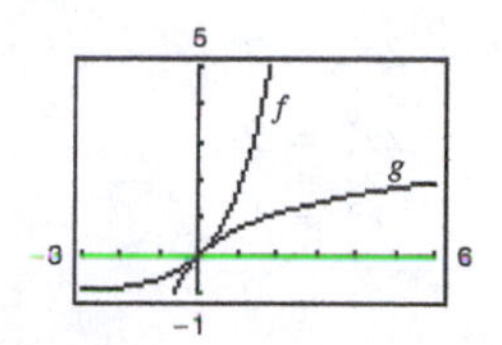

77. (a)

$$\begin{aligned} y &= e^{4x-1} \\ \ln y &= 4x - 1 \\ \ln y + 1 &= 4x \\ x &= \tfrac{1}{4}(\ln y + 1) \\ f^{-1}(x) &= \tfrac{1}{4}(\ln x + 1) \end{aligned}$$

(b)

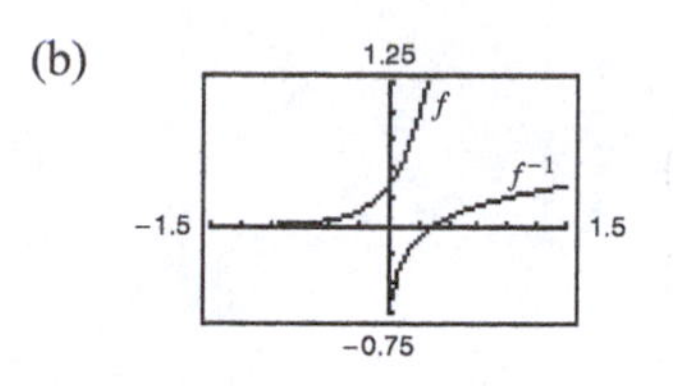

(c) $f^{-1}(f(x)) = f^{-1}(e^{4x-1}) = \frac{1}{4}(\ln e^{4x-1} + 1) = \frac{1}{4}(4x - 1 + 1) = x$

$f(f^{-1}(x)) = f\left(\frac{1}{4}(\ln x + 1)\right) = e^{(\ln x + 1) - 1} = e^{\ln x} = x$

79. (a)

$$\begin{aligned} y &= 2\ln(x - 1) \\ \frac{y}{2} &= \ln(x - 1) \\ e^{y/2} &= x - 1 \\ x &= 1 + e^{y/2} \\ f^{-1}(x) &= 1 + e^{x/2} \end{aligned}$$

(b)

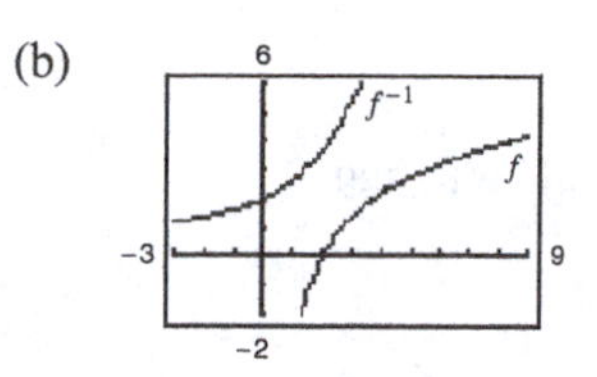

(c) $f^{-1}(f(x)) = f^{-1}(2\ln(x - 1)) = 1 + e^{\ln(x-1)} = 1 + x - 1 = x$

$f(f^{-1}(x)) = f(1 + x^{x/2}) = 2\ln\left[(1 + e^{x/2}) - 1\right] = 2\left(\dfrac{x}{2}\right) = x$

81. $\ln e^{x^2} = x^2$

83. $e^{\ln(5x+2)} = 5x + 2$

85. $-1 + \ln e^{2x} = -1 + 2x$

87. (a) $\ln 6 = \ln 2 + \ln 3 \approx 1.7917$

(b) $\ln\frac{2}{3} = \ln 2 - \ln 3 \approx -0.4055$

(c) $\ln 81 = 4\ln 3 \approx 4.3944$

(d) $\ln\sqrt{3} = \frac{1}{2}\ln 3 \approx 0.5493$

89. $\ln\dfrac{x}{4} = \ln x - \ln 4$

91. $\ln\dfrac{xy}{z} = \ln x + \ln y - \ln z$

93. $\ln\left(x\sqrt{x^2 + 5}\right) = \ln x + \ln(x^2 + 5)^{1/2}$

$= \ln x + \frac{1}{2}\ln(x^2 + 5)$

95. $\ln\sqrt{\dfrac{x - 1}{x}} = \ln\left(\dfrac{x - 1}{x}\right)^{1/2} = \dfrac{1}{2}\ln\left(\dfrac{x - 1}{x}\right)$

$= \dfrac{1}{2}\left[\ln(x - 1) - \ln x\right]$

$= \dfrac{1}{2}\ln(x - 1) - \dfrac{1}{2}\ln x$

97. $\ln 3e^2 = \ln 3 + 2\ln e = 2 + \ln 3$

99. $\ln x + \ln 7 = \ln(x \cdot 7) = \ln(7x)$

101. $\ln(x - 2) - \ln(x + 2) = \ln\dfrac{x - 2}{x + 2}$

103. $\dfrac{1}{3}\left[2\ln(x + 3) + \ln x - \ln(x^2 - 1)\right] = \dfrac{1}{3}\ln\dfrac{x(x + 3)^2}{x^2 - 1}$

$= \ln\sqrt[3]{\dfrac{x(x + 3)^2}{x^2 - 1}}$

105. $2\ln 3 - \dfrac{1}{2}\ln(x^2 + 1) = \ln 9 - \ln\sqrt{x^2 + 1} = \ln\dfrac{9}{\sqrt{x^2 + 1}}$

107. $e^x = 12$

$x = \ln 12 \approx 2.485$

109. $9 - 2e^x = 7$

$$2e^x = 2$$
$$e^x = 1$$
$$x = 0$$

111. $50e^{-x} = 30$

$$e^{-x} = \frac{3}{5}$$
$$-x = \ln\left(\frac{3}{5}\right)$$
$$x = \ln\left(\frac{5}{3}\right)$$
$$\approx 0.511$$

113. $\ln x = 2$

$$x = e^2 \approx 7.389$$

115. $\ln(x - 3) = 2$

$$x - 3 = e^2$$
$$x = 3 + e^2 \approx 10.389$$

117. $\ln\sqrt{x + 2} = 1$

$$\sqrt{x + 2} = e^1 = e$$
$$x + 2 = e^2$$
$$x = e^2 - 2 \approx 5.389$$

119. $e^{2x+1} > 3$

$$2x + 1 > \ln 3$$
$$2x > \ln 3 - 1$$
$$x > \tfrac{1}{2}[\ln 3 - 1]$$

121. $-2 < \ln x < 0$

$$e^{-2} < x < e^0 = 1$$
$$\frac{1}{e^2} < x < 1$$

123. (a)

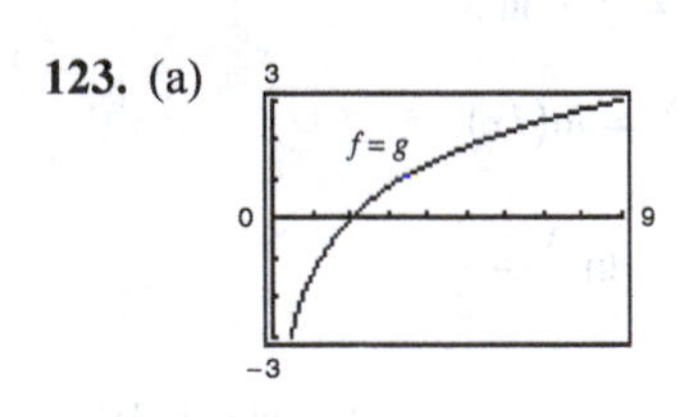

(b) $f(x) = \ln\left(\dfrac{x^2}{4}\right),\ x > 0$

$$= \ln x^2 - \ln 4$$
$$= 2\ln x - \ln 4$$
$$= g(x)$$

125. No; $\ln(a^b) = b\ln a$ is only true when $a > 0$ because this follows the properties of logarithms.

127.

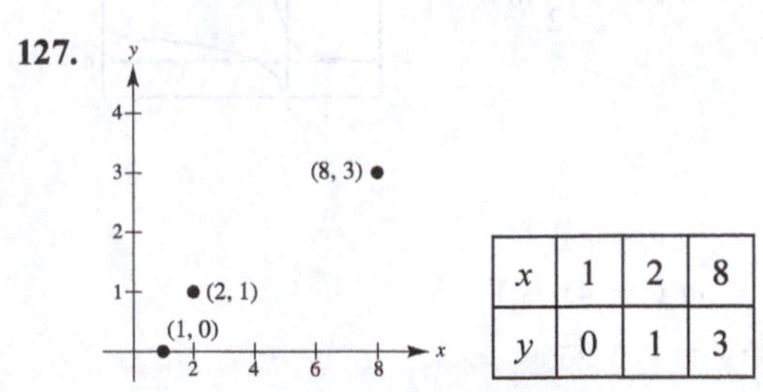

x	1	2	8
y	0	1	3

(a) y is a logarithmic function of x: True; $y = \dfrac{\ln x}{\ln 2}$

(b) y is an exponential function of x: False

(c) x is an exponential function of y: True; $2^y = x$

(d) y is a linear function of x: False

129. $\beta = \dfrac{10}{\ln 10}\ln\left(\dfrac{I}{10^{-16}}\right)$

$$= \frac{10}{\ln 10}\left[\ln I - \ln 10^{-16}\right]$$
$$= \frac{10}{\ln 10}[\ln I + 16\ln 10]$$
$$= \frac{10}{\ln 10}\ln I + 160$$
$$= 10\log_{10} I + 160$$

131.

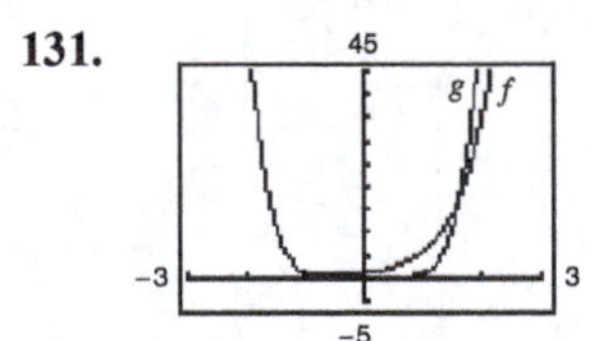

The graphs intersect three times: $(-0.7899, 0.2429)$, $(1.6242, 18.3615)$ and $(6, 46{,}656)$.

The function $f(x) = 6^x$ grows more rapidly.

133. $f(x) = \ln\left(x + \sqrt{x^2 + 1}\right)$

(a)

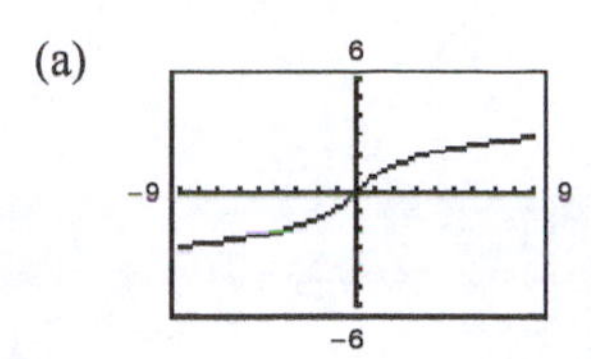

Domain: $-\infty < x < \infty$

(b) $f(-x) = \ln\left(-x + \sqrt{x^2 + 1}\right)$

$$= \ln\left[\frac{\left(-x + \sqrt{x^2 + 1}\right)\left(-x - \sqrt{x^2 + 1}\right)}{\left(-x - \sqrt{x^2 + 1}\right)}\right]$$

$$= \ln\left[\frac{\left(x^2\sqrt{x^2 + 1}\right)}{\left(-x - \sqrt{x^2 + 1}\right)}\right]$$

$$= \ln\left[\frac{1}{\left(x + \sqrt{x^2 + 1}\right)}\right]$$

$$= -\ln\left(x + \sqrt{x^2 + 1}\right) = -f(x)$$

(c)

$$y = \ln\left(x + \sqrt{x^2 + 1}\right)$$

$$e^y = x + \sqrt{x^2 + 1}$$

$$\left(e^y - x\right)^2 = x^2 + 1$$

$$2xe^y = e^{2y} - 1$$

$$x = \frac{e^{2y} - 1}{2e^x}$$

135. $n = 12$

$12! = 12 \cdot 11 \cdot 10 \cdots 3 \cdot 2 \cdot 1 = 479{,}001{,}600$

Stirlings Formula:

$$12! \approx \left(\frac{12}{e}\right)^{12}\sqrt{2\pi(12)} \approx 475{,}687{,}487$$

137. Let $m = \ln x$ and $n = \ln y$. Then $x = e^m$ and $y = e^n$.

$$\frac{x}{y} = \frac{e^m}{e^n} = e^{m-n}$$

$$\ln\left(\frac{x}{y}\right) = \ln\left(e^{m-n}\right)$$

$$\ln\left(\frac{x}{y}\right) = (m - n)\ln e$$

$$\ln\left(\frac{x}{y}\right) = m - n$$

$$\ln\left(\frac{x}{y}\right) = \ln x - \ln y,\ x > 0,\ y > 0$$

Review Exercises for Chapter 1

1. $y = 5x - 8$

$x = 0$: $y = 5(0) - 8 = -8 \Rightarrow (0, -8)$, y-intercept

$y = 0$: $0 = 5x - 8 \Rightarrow x = \frac{8}{5} \Rightarrow \left(\frac{8}{5}, 0\right)$, x-intercept

3. $y = \dfrac{x - 3}{x - 4}$

$x = 0$: $y = \dfrac{0 - 3}{0 - 4} = \dfrac{3}{4} \Rightarrow \left(0, \dfrac{3}{4}\right)$, y-intercept

$y = 0$: $0 = \dfrac{x - 3}{x - 4} \Rightarrow x = 3 \Rightarrow (3, 0)$, x-intercept

5. $y = x^2 + 4x$ does not have symmetry with respect to either axis or the origin.

7. Symmetric with respect to both axes and the origin because:

$y^2 = \left(-x^2\right) - 5$	$\left(-y\right)^2 = x^2 - 5$	$\left(-y\right)^2 = \left(-x\right)^2 - 5$
$y^2 = x^2 - 5$	$y^2 = x^2 - 5$	$y^2 = x^2 - 5$

9. $y = -\frac{1}{2}x + 3$

y-intercept: $y = -\frac{1}{2}(0) + 3 = 3$

$(0, 3)$

x-intercept: $-\frac{1}{2}x + 3 = 0$

$$-\frac{1}{2}x = -3$$

$$x = 6$$

$(6, 0)$

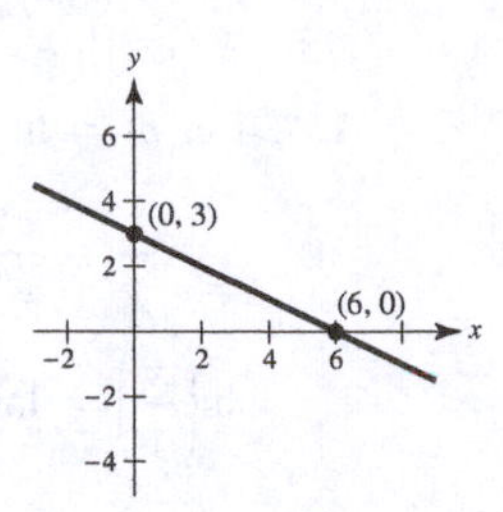

Symmetry: none

11. $y = 9x - x^3$

$9x - x^3 = x(9 - x^2) = x(3 - x)(3 + x) = 0 \Rightarrow x = 0, 3, -3$

Intercepts: $(0, 0), (3, 0), (-3, 0)$

Symmetric with respect to the origin because

$f(-x) = 9(-x) - (-x)^3 = -9x + x^3 = -(9x - x^3) = -f(x).$

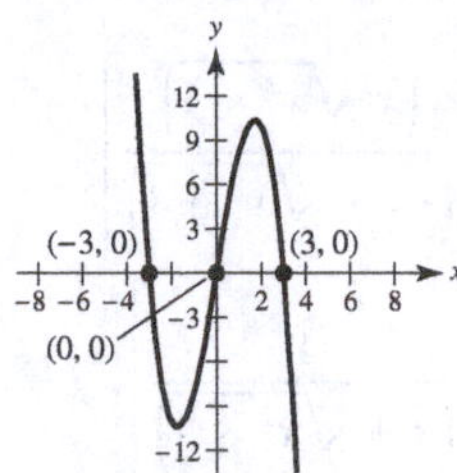

13. $y = 2\sqrt{4 - x}$

y-intercept: $y = 2\sqrt{4 - 0} = 2\sqrt{4} = 4$

$(0, 4)$

x-intercept: $2\sqrt{4 - x} = 0$

$$\sqrt{4 - x} = 0$$

$$4 - x = 0$$

$$x = 4$$

$(4, 0)$

Symmetry: none

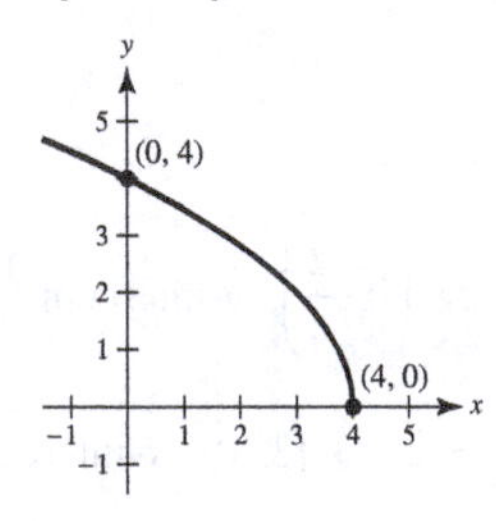

15. $5x + 3y = -1 \Rightarrow y = \frac{1}{3}(-5x - 1)$

$x - y = -5 \Rightarrow y = x + 5$

$\frac{1}{3}(-5x - 1) = x + 5$

$-5x - 1 = 3x + 15$

$-16 = 8x$

$-2 = x$

For $x = -2$, $y = x + 5 = -2 + 5 = 3$.

Point of intersection is: $(-2, 3)$

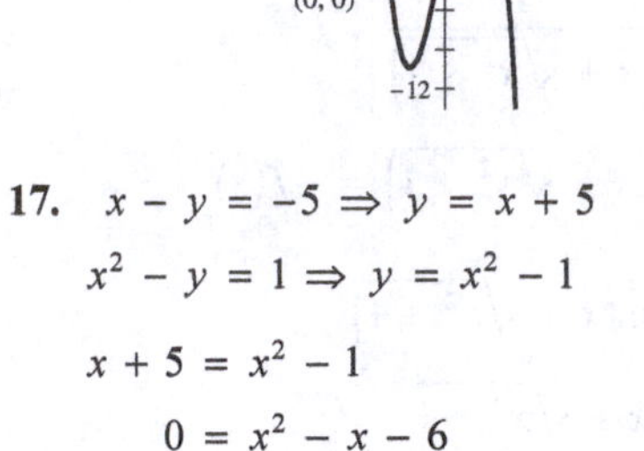

17. $x - y = -5 \Rightarrow y = x + 5$

$x^2 - y = 1 \Rightarrow y = x^2 - 1$

$x + 5 = x^2 - 1$

$0 = x^2 - x - 6$

$0 = (x - 3)(x + 2)$

$x = 3$ or $x = -2$

For $x = 3$, $y = 3 + 5 = 8$.

For $x = -2$, $y = -2 + 5 = 3$.

Points of intersection: $(3, 8), (-2, 3)$

19.

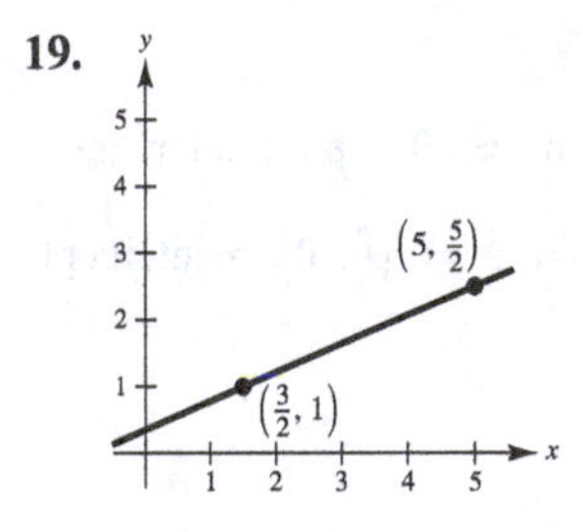

$$\text{Slope} = \frac{\left(\frac{5}{2}\right) - 1}{5 - \left(\frac{3}{2}\right)} = \frac{\frac{3}{2}}{\frac{7}{2}} = \frac{3}{7}$$

21. $y - (-5) = \frac{7}{4}(x - 3)$

$y + 5 = \frac{7}{4}x - \frac{21}{4}$

$4y + 20 = 7x - 21$

$0 = 7x - 4y - 41$

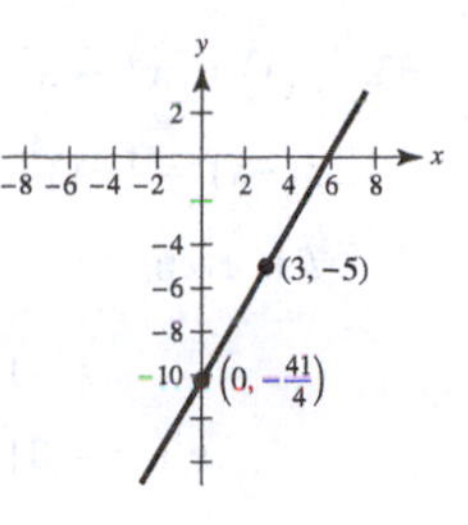

23. $y - 0 = -\frac{2}{3}(x - (-3))$

$y = -\frac{2}{3}x - 2$

$2x + 3y + 6 = 0$

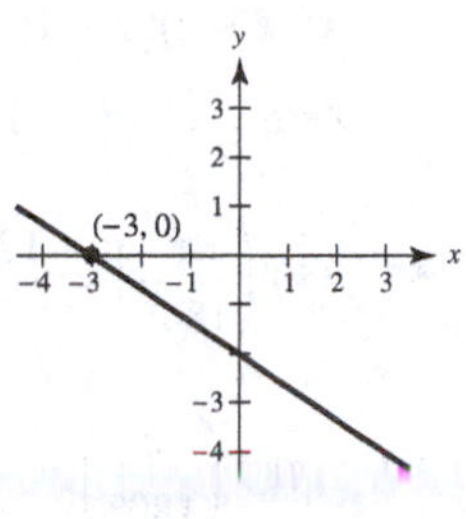

25. $y - 3x = 5$

$y = 3x + 5$

Slope: $m = 3$

y-intercept: $(0, 5)$

27. $y = 6$

Slope: 0

y-intercept: $(0, 6)$

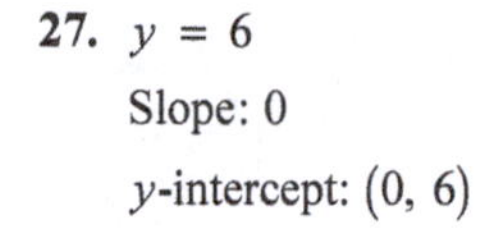

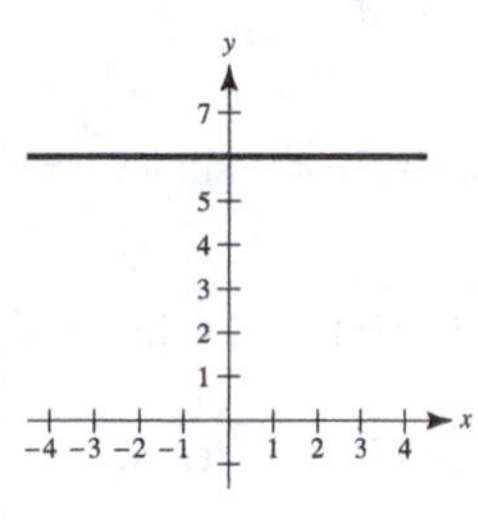

29. $y = 4x - 2$

Slope: 4

y-intercept: $(0, -2)$

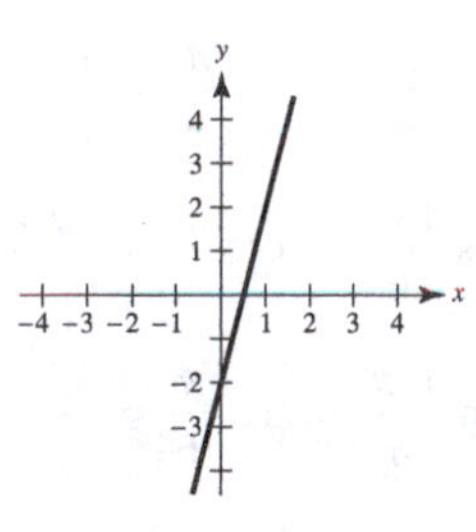

31. $m = \dfrac{2 - 0}{8 - 0} = \dfrac{1}{4}$

$y - 0 = \frac{1}{4}(x - 0)$

$y = \frac{1}{4}x$

$4y - x = 0$

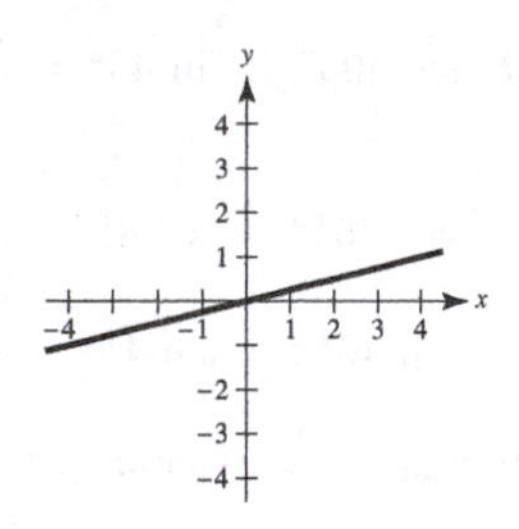

33. (a) $y - 5 = \frac{7}{16}(x + 3)$

$16y - 80 = 7x + 21$

$0 = 7x - 16y + 101$

(b) $5x - 3y = 3$ has slope $\frac{5}{3}$.

$y - 5 = \frac{5}{3}(x + 3)$

$3y - 15 = 5x + 15$

$0 = 5x - 3y + 30$

(c) $3x + 4y = 8$

$4y = -3x + 8$

$y = \dfrac{-3}{4}x + 2$

Perpendicular line has slope $\dfrac{4}{3}$.

$y - 5 = \dfrac{4}{3}(x - (-3))$

$3y - 15 = 4x + 12$

$4x - 3y + 27 = 0$ or $y = \dfrac{4}{3}x + 9$

(d) Slope is undefined so the line is vertical.

$x = -3$

$x + 3 = 0$

35. The slope is -850.

$V = -850t + 12{,}500.$

$V(3) = -850(3) + 12{,}500 = \$9950$

37. $f(x) = 5x + 4$

(a) $f(0) = 5(0) + 4 = 4$

(b) $f(5) = 5(5) + 4 = 29$

(c) $f(-3) = 5(-3) + 4 = -11$

(d) $f(t + 1) = 5(t + 1) + 4 = 5t + 9$

39. $f(x) = 4x^2$

$$\frac{f(x + \Delta x) - f(x)}{\Delta x} = \frac{4(x + \Delta x)^2 - 4x^2}{\Delta x}$$

$$= \frac{4\left(x^2 + 2x\Delta x + (\Delta x)^2\right) - 4x^2}{\Delta x}$$

$$= \frac{4x^2 + 8x\Delta x + 4(\Delta x)^2 - 4x^2}{\Delta x}$$

$$= \frac{8x\Delta x + 4(\Delta x)^2}{\Delta x}$$

$$= 8x + 4\Delta x, \quad \Delta x \neq 0$$

41. $f(x) = x^2 + 3$

Domain: $(-\infty, \infty)$

Range: $[3, \infty)$

43. $f(x) = \dfrac{4}{2x - 1}$

Domain: $\left(-\infty, \frac{1}{2}\right) \cup \left(\frac{1}{2}, \infty\right)$

Range: $(-\infty, 0) \cup (0, \infty)$

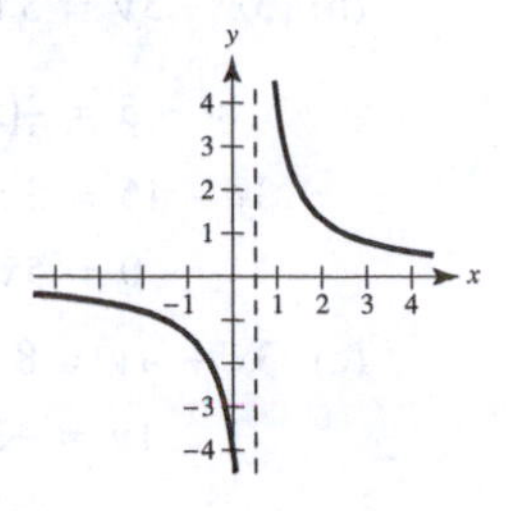

45. $x + y^2 = 2 \Rightarrow y = \pm\sqrt{2 - x}$

y is not a function of x.
Some vertical lines intersect the graph more than once.

47. $xy + x^3 - 2y = 0$

$(x - 2)y = -x^3$

$y = \dfrac{-x^3}{x - 2}$

y is a function of x.

49. $f(x) = x^3 - 3x^2$

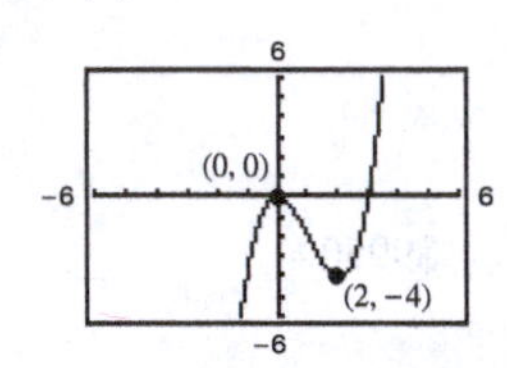

(a) The graph of g is obtained from f by a vertical shift down 1 unit, followed by a reflection in the x-axis:

$g(x) = -\left[f(x) - 1\right] = -x^3 + 3x^2 + 1$

(b) The graph of g is obtained from f by a vertical shift upwards of 1 and a horizontal shift of 2 to the right.

$g(x) = f(x - 2) + 1 = (x - 2)^3 - 3(x - 2)^2 + 1$

51. $f(x) = 3x + 1$, $g(x) = -x$

$(f \circ g)(x) = f(g(x)) = f(-x) = -3x + 1$

Domain: $(-\infty, \infty)$

$(g \circ f)(x) = g(f(x))$
$= g(3x + 1)$
$= -(3x + 1)$
$= -3x - 1$

Domain: $(-\infty, \infty)$

$f \circ g \neq g \circ f$

53. $f(x) = x^4 - x^2$

$f(-x) = (-x)^4 - (-x)^2 = x^4 - x^2 = f(x)$

f is even.

$f(x) = x^4 - x^2 = 0$

$x^2(x^2 - 1) = 0$

$x^2(x + 1)(x - 1) = 0$

Zeros: $x = 0, -1, 1$

55. $340°\left(\dfrac{\pi}{180°}\right) = \dfrac{17\pi}{9} \approx 5.934$

57. $-480°\left(\dfrac{\pi}{180°}\right) = -\dfrac{8\pi}{3} \approx -8.378$

59. $\dfrac{\pi}{6}\left(\dfrac{180°}{\pi}\right) = 30°$

61. $-\dfrac{2\pi}{3}\left(\dfrac{180°}{\pi}\right) = -120°$

63. $\sin(-45°) = -\sin 45° = -\dfrac{\sqrt{2}}{2}$

$\cos(-45°) = \cos 45° = \dfrac{\sqrt{2}}{2}$

$\tan(-45°) = -\tan 45° = -1$

65. $\sin \dfrac{13\pi}{6} = \sin \dfrac{\pi}{6} = \dfrac{1}{2}$

$\cos \dfrac{13\pi}{6} = \cos \dfrac{\pi}{6} = \dfrac{\sqrt{3}}{2}$

$\tan \dfrac{13\pi}{6} = \tan \dfrac{\pi}{6} = \dfrac{\sqrt{3}}{3}$

67. $\sin 405° = \sin 45° = \dfrac{\sqrt{2}}{2}$

$\cos 405° = \cos 45° = \dfrac{\sqrt{2}}{2}$

$\tan 405° = \tan 45° = 1$

69. $\tan 33° \approx 0.6494$

71. $\sec \dfrac{12\pi}{5} = \dfrac{1}{\cos(12\pi/5)} \approx 3.2361$

73. $\sin\left(-\dfrac{\pi}{9}\right) \approx -0.3420$

75. $2\cos\theta + 1 = 0$

$$\cos\theta = -\frac{1}{2}$$

$$\theta = \frac{2\pi}{3}, \frac{4\pi}{3}$$

77. $2\sin^2\theta + 3\sin\theta + 1 = 0$

$$(2\sin\theta - 1)(\sin\theta + 1) = 0$$

$$\sin\theta = -\frac{1}{2} \text{ or } \sin\theta = -1$$

$$\theta = \frac{7\pi}{6}, \frac{11\pi}{6} \text{ or } \theta = \frac{3\pi}{2}$$

79. $\sec^2\theta - \sec\theta - 2 = 0$

$$(\sec\theta - 2)(\sec\theta + 1) = 0$$

$$\sec\theta = 2 \text{ or } \sec\theta = -1$$

$$\cos\theta = \frac{1}{2} \text{ or } \cos\theta = -1$$

$$\theta = \frac{\pi}{3}, \frac{5\pi}{3} \text{ or } \theta = \pi$$

81. $y = 9\cos x$

Period: 2π

Amplitude: 9

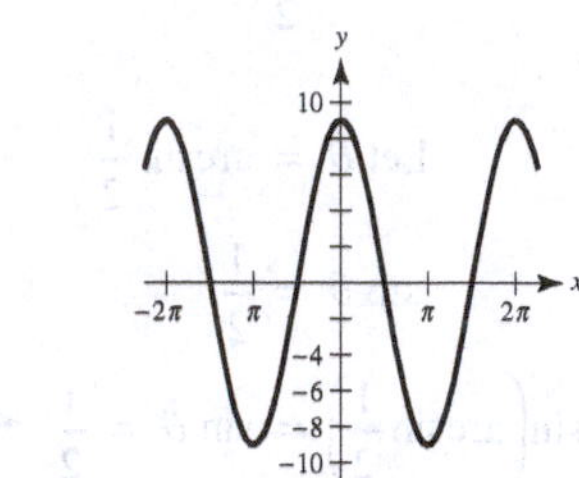

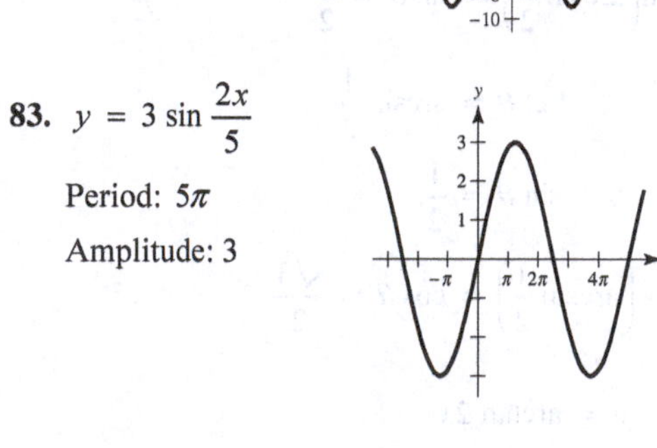

83. $y = 3\sin\dfrac{2x}{5}$

Period: 5π

Amplitude: 3

85. $y = \frac{1}{3}\tan x$

Period: π

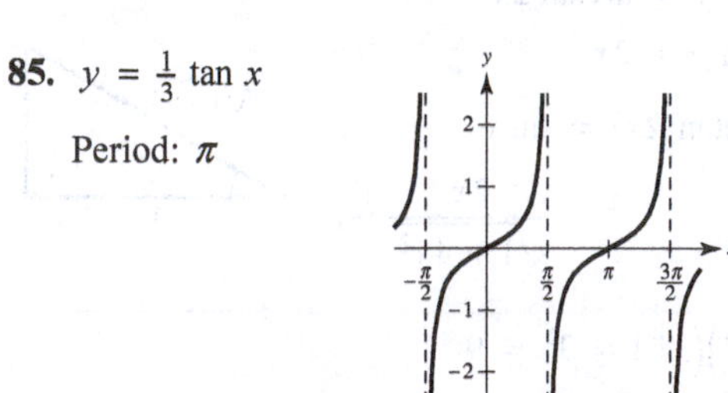

87. $y = -\sec 2\pi x$

Period: 1

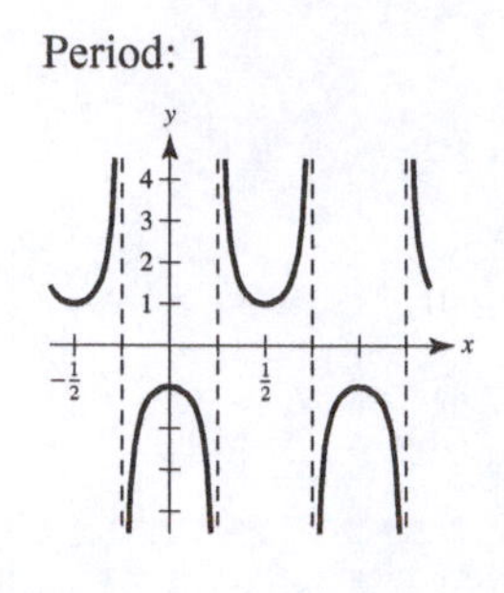

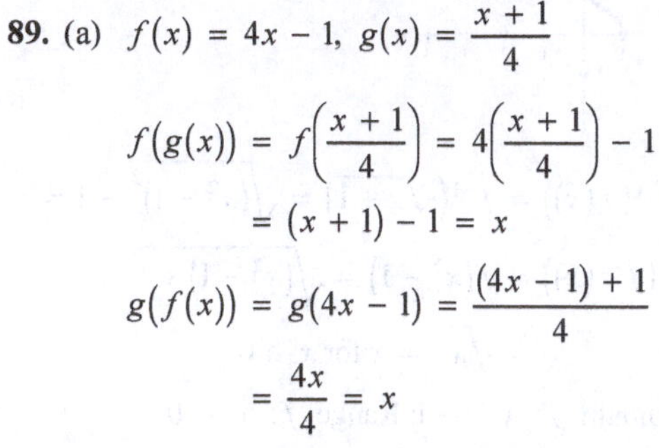

89. (a) $f(x) = 4x - 1,\ g(x) = \dfrac{x+1}{4}$

$$f(g(x)) = f\left(\frac{x+1}{4}\right) = 4\left(\frac{x+1}{4}\right) - 1$$

$$= (x+1) - 1 = x$$

$$g(f(x)) = g(4x - 1) = \frac{(4x - 1) + 1}{4}$$

$$= \frac{4x}{4} = x$$

(b)

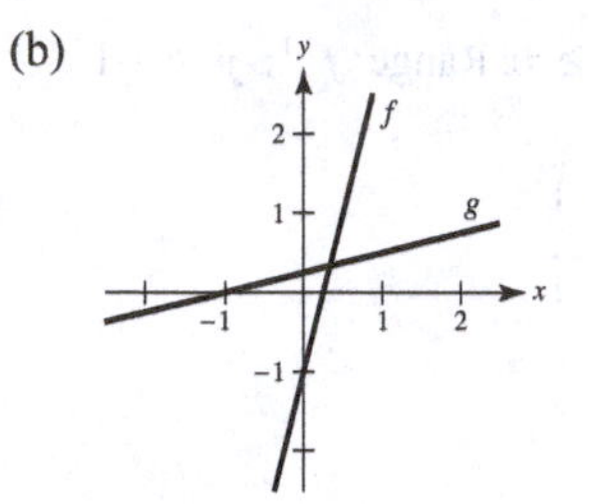

91. $f(x) = x^2 + 2x + 1$

Not one-to-one; does not have an inverse.

93. $f(x) = \csc 3\pi x$

Not one-to-one; does not have an inverse.

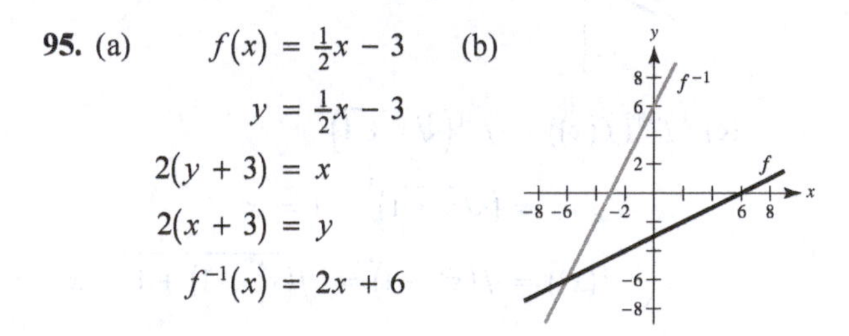

95. (a)

$$f(x) = \tfrac{1}{2}x - 3$$

$$y = \tfrac{1}{2}x - 3$$

$$2(y + 3) = x$$

$$2(x + 3) = y$$

$$f^{-1}(x) = 2x + 6$$

(b)

(c) $f^{-1}(f(x)) = f^{-1}\left(\frac{1}{2}x - 3\right) = 2\left(\frac{1}{2}x - 3\right) + 6 = x$

$f(f^{-1}(x)) = f(2x + 6) = \frac{1}{2}(2x + 6) - 3 = x$

(d) Domain f: all real numbers; Range f: all real numbers
Domain f^{-1}: all real numbers; Range f^{-1}: all real numbers

97. (a)
$$f(x) = \sqrt{x+1}$$
$$y = \sqrt{x+1}$$
$$y^2 - 1 = x$$
$$x^2 - 1 = y$$
$$f^{-1}(x) = x^2 - 1, \quad x \geq 0$$

(b)

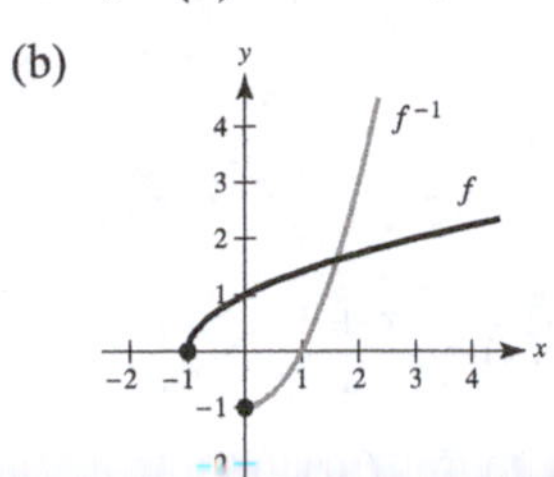

(c) $f^{-1}(f(x)) = f^{-1}(\sqrt{x+1}) = \sqrt{(x^2-1)^2} - 1 = x$

$f(f^{-1}(x)) = f(x^2 - 1) = \sqrt{(x^2-1)+1}$

$= \sqrt{x^2} = x$ for $x \geq 0$.

(d) Domain f: $x \geq -1$; Range f: $y \geq 0$

Domain f^{-1}: $x \geq 0$; Range f^{-1}: $y \geq -1$

99. (a)
$$f(x) = \sqrt[3]{x+1}$$
$$y = \sqrt[3]{x+1}$$
$$y^3 - 1 = x$$
$$x^3 - 1 = y$$
$$f^{-1}(x) = x^3 - 1$$

(b)

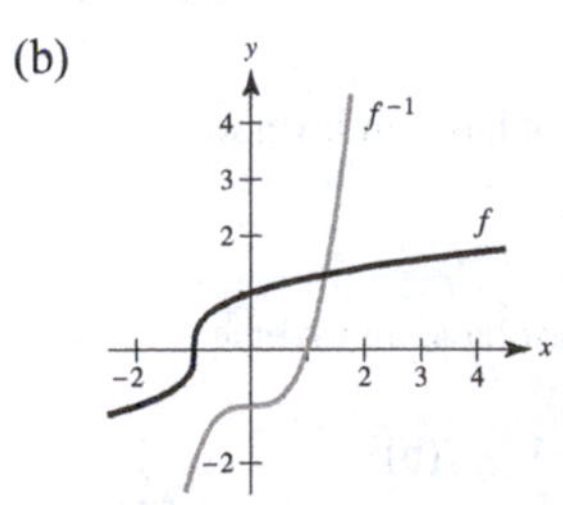

(c) $f^{-1}(f(x)) = f^{-1}(\sqrt[3]{x+1})$

$= (\sqrt[3]{x+1})^3 - 1 = x$

$f(f^{-1}(x)) = f(x^3 - 1) = \sqrt[3]{(x^3-1)+1} = x$

(d) Domain f: all real numbers; Range f: all real numbers

Domain f^{-1}: all real numbers; Range f^{-1}: all real numbers

101. $f(x) = \sin 3\pi x$ is not one-to-one and an inverse does not exist.

103. $f(x) = x^4 + 2x^2 + 2, \; x \geq 0$

$= (x^2 + 1)^2 + 1, \; x \geq 0$

By the Horizontal Line Test, f is one-to-one on $[0, \infty)$.

105. $f(x) = 2 \arctan(x + 3)$

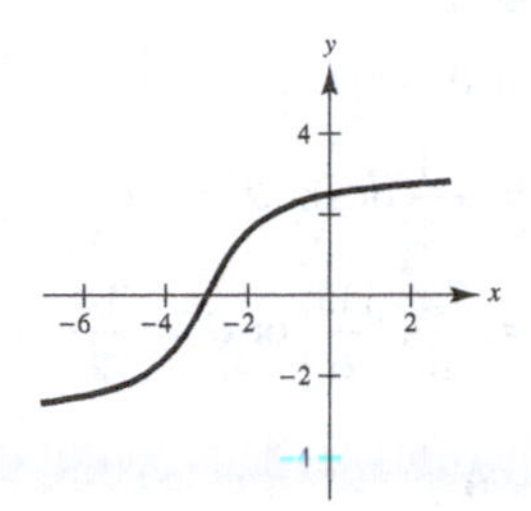

107. $\arctan 1 = \dfrac{\pi}{4}$ because $\tan \dfrac{\pi}{4} = 1$.

109.
$$\arccos(2x + 1) = 2$$
$$\cos 2 = 2x + 1$$
$$x = \frac{1}{2}(\cos 2 - 1) \approx -0.708$$

111. (a)
$$\text{Let } \theta = \arcsin \frac{1}{2}$$
$$\sin \theta = \frac{1}{2}.$$
$$\sin\left(\arcsin \frac{1}{2}\right) = \sin \theta = \frac{1}{2}$$

(b)
$$\text{Let } \theta = \arcsin \frac{1}{2}$$
$$\sin \theta = \frac{1}{2}.$$
$$\cos\left(\arcsin \frac{1}{2}\right) = \cos \theta = \frac{\sqrt{3}}{2}$$

113. Let $y = \arctan 2x$
$$\tan y = 2x.$$
$$\sin(\arctan 2x) = \sin y$$
$$= \frac{2x}{\sqrt{1 + 4x^2}}$$

115. (a) $(3^3)(3^{-1}) = 3^2 = 9$

(b) $(3^2)^4 = 3^8 = 6561$

(c) $\dfrac{3^4}{3^{-2}} = 3^6 = 729$

(d) $\left(\dfrac{1}{3}\right)^2 9^2 = \dfrac{1}{9}(9^2) = 9$

117. $3^{x/2} = 81 = 3^4$

$$\frac{x}{2} = 4$$
$$x = 8$$

119. $f = e^{-x/2}$

121. $f(x) = e^x$ matches (d).

The graph rises from left to right and the domain is all real x.

123. $f(x) = \ln(x + 1) + 1$ matches (c).

The graph is a vertical shift upward 1 unit and a horizontal shift to the left 1 unit of $f(x) = \ln x$ and the domain is $x > -1$.

125. $f(x) = \ln x + 3$

Vertical shift three units upward.

Domain: $x > 0$

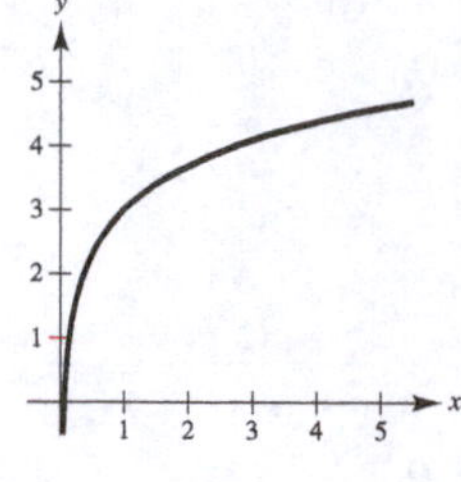

127. $$\ln\sqrt[5]{\frac{4x^2 - 1}{4x^2 + 1}} = \frac{1}{5}\ln\frac{(2x-1)(2x+1)}{4x^2+1}$$
$$= \frac{1}{5}\left[\ln(2x-1) + \ln(2x+1) - \ln(4x^2+1)\right]$$

129. $$\ln 3 + \frac{1}{3}\ln(4 - x^2) - \ln x = \ln 3 + \ln\sqrt[3]{4 - x^2} - \ln x$$
$$= \ln\left(\frac{3\sqrt[3]{4 - x^2}}{x}\right)$$

131. $$-4 + 3e^{-2x} = 6$$
$$3e^{-2x} = 10$$
$$e^{-2x} = \tfrac{10}{3}$$
$$\ln e^{-2x} = \ln\tfrac{10}{3}$$
$$-2x = \ln\tfrac{10}{3}$$
$$x = -\tfrac{1}{2}\ln\tfrac{10}{3} \approx -0.602$$

Problem Solving for Chapter 1

1. (a) $$x^2 - 6x + y^2 - 8y = 0$$
$$(x^2 - 6x + 9) + (y^2 - 8y + 16) = 9 + 16$$
$$(x - 3)^2 + (y - 4)^2 = 25$$

Center: (3, 4); Radius: 5

(b) Slope of line from (0, 0) to (3, 4) is $\frac{4}{3}$.

Slope of tangent line is $-\frac{3}{4}$. So,

$y - 0 = -\frac{3}{4}(x - 0) \Rightarrow y = -\frac{3}{4}x$, Tangent line

(c) Slope of line from (6, 0) to (3, 4) is $\frac{4 - 0}{3 - 6} = -\frac{4}{3}$.

Slope of tangent line is $\frac{3}{4}$. So,

$y - 0 = \frac{3}{4}(x - 6) \Rightarrow y = \frac{3}{4}x - \frac{9}{2}$, Tangent line

(d) $$-\frac{3}{4}x = \frac{3}{4}x - \frac{9}{2}$$
$$\frac{3}{2}x = \frac{9}{2}$$
$$x = 3$$

Intersection: $\left(3, -\frac{9}{4}\right)$

3. $H(x) = \begin{cases} 1, & x \geq 0 \\ 0, & x < 0 \end{cases}$

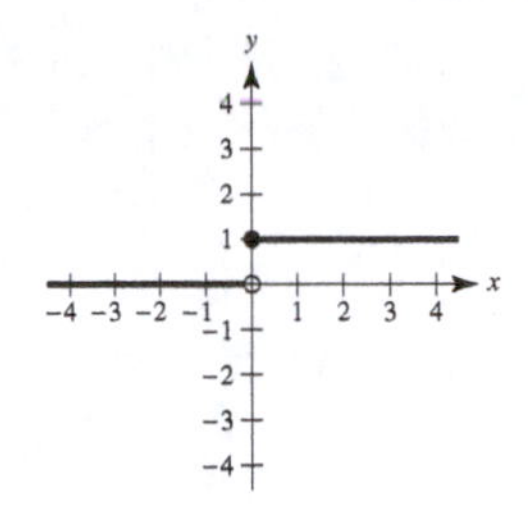

(a) $H(x) - 2 = \begin{cases} -1, & x \geq 0 \\ -2, & x < 0 \end{cases}$

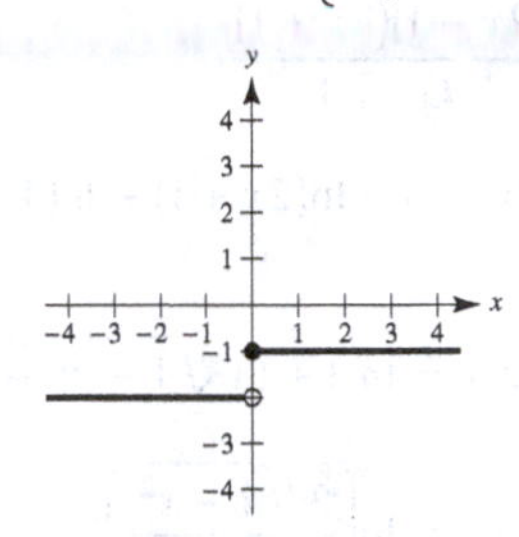

(b) $H(x - 2) = \begin{cases} 1, & x \geq 2 \\ 0, & x < 2 \end{cases}$

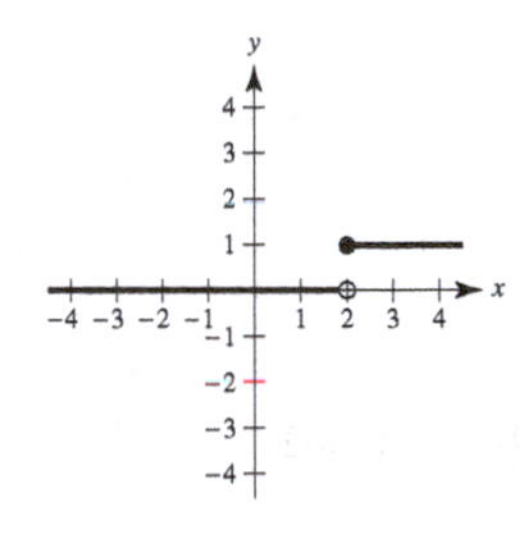

(c) $-H(x) = \begin{cases} -1, & x \geq 0 \\ 0, & x < 0 \end{cases}$

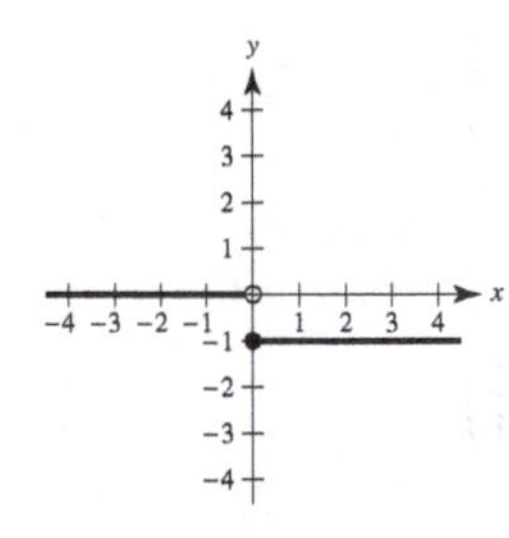

(d) $H(-x) = \begin{cases} 1, & x \leq 0 \\ 0, & x > 0 \end{cases}$

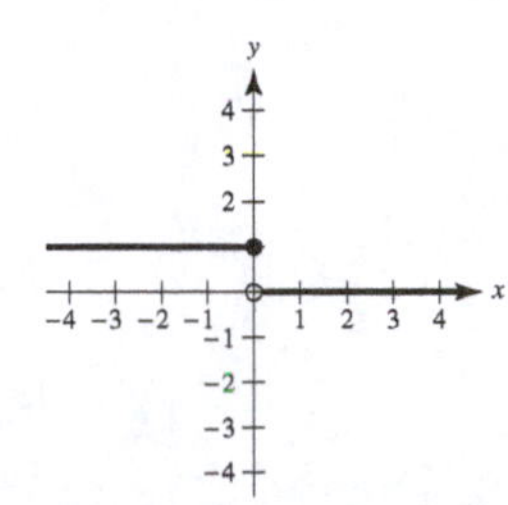

(e) $\frac{1}{2}H(x) = \begin{cases} \frac{1}{2}, & x \geq 0 \\ 0, & x < 0 \end{cases}$

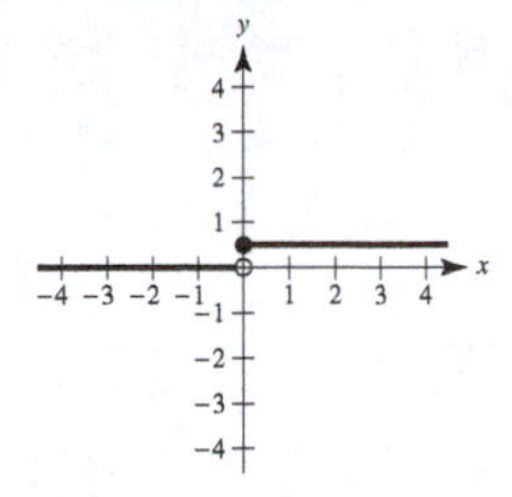

(f) $-H(x - 2) + 2 = \begin{cases} 1, & x \geq 2 \\ 2, & x < 2 \end{cases}$

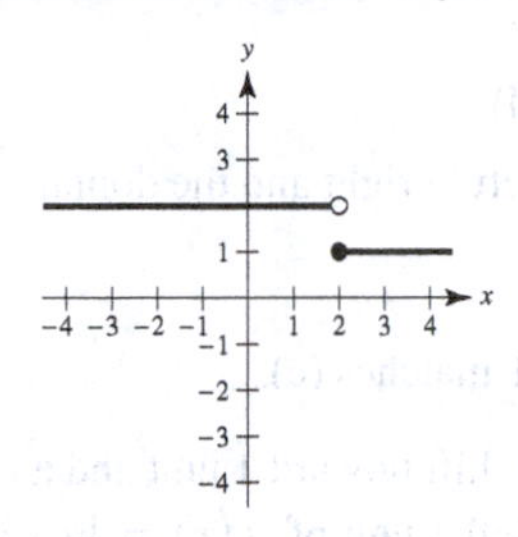

5. (a) $x + 2y = 100 \Rightarrow y = \frac{100 - x}{2}$

$A(x) = xy = x\left(\frac{100 - x}{2}\right) = -\frac{x^2}{2} + 50x$

Domain: $0 < x < 100$ or $(0, 100)$

(b)

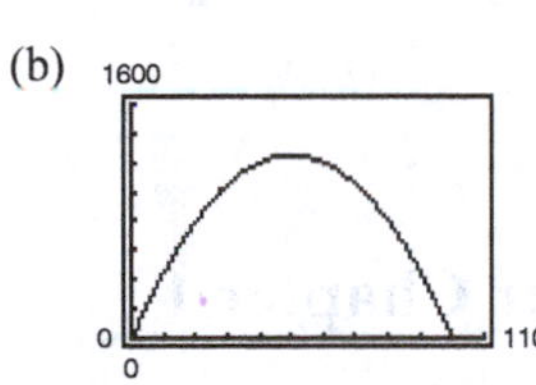

Maximum of 1250 m^2 at $x = 50$ m, $y = 25$ m.

(c) $A(x) = -\frac{1}{2}(x^2 - 100x)$

$= -\frac{1}{2}(x^2 - 100x + 2500) + 1250$

$= -\frac{1}{2}(x - 50)^2 + 1250$

$A(50) = 1250$ m^2 is the maximum.

$x = 50$ m, $y = 25$ m

7. The length of the trip in the water is $\sqrt{2^2 + x^2}$, and the length of the trip over land is $\sqrt{1 + (3 - x)^2}$.

So, the total time is $T = \frac{\sqrt{4 + x^2}}{2} + \frac{\sqrt{1 + (3 - x)^2}}{4}$ hours.

9. (a) Slope $= \dfrac{9-4}{3-2} = 5$. Slope of tangent line is less than 5.

(b) Slope $= \dfrac{4-1}{2-1} = 3$. Slope of tangent line is greater than 3.

(c) Slope $= \dfrac{4.41-4}{2.1-2} = 4.1$. Slope of tangent line is less than 4.1.

(d) Slope $= \dfrac{f(2+h)-f(2)}{(2+h)-2} = \dfrac{(2+h)^2-4}{h} = \dfrac{4h+h^2}{h} = 4+h,\ h \neq 0$

(e) Letting h get closer and closer to 0, the slope approaches 4. So, the slope at (2, 4) is 4.

11. $f(x) = y = \dfrac{1}{1-x}$

(a) Domain: all $x \neq 1$ or $(-\infty, 1) \cup (1, \infty)$

Range: all $y \neq 0$ or $(-\infty, 0) \cup (0, \infty)$

(b) $f(f(x)) = f\left(\dfrac{1}{1-x}\right) = \dfrac{1}{1-\left(\dfrac{1}{1-x}\right)} = \dfrac{1}{\dfrac{1-x-1}{1-x}} = \dfrac{1-x}{-x} = \dfrac{x-1}{x}$

Domain: all $x \neq 0, 1$ or $(-\infty, 0) \cup (0, 1) \cup (1, \infty)$

(c) $f(f(f(x))) = f\left(\dfrac{x-1}{x}\right) = \dfrac{1}{1-\left(\dfrac{x-1}{x}\right)} = \dfrac{1}{\dfrac{1}{x}} = x$

Domain: all $x \neq 0, 1$ or $(-\infty, 0) \cup (0, 1) \cup (1, \infty)$

(d) The graph is not a line. It has holes at (0, 0) and (1, 1).

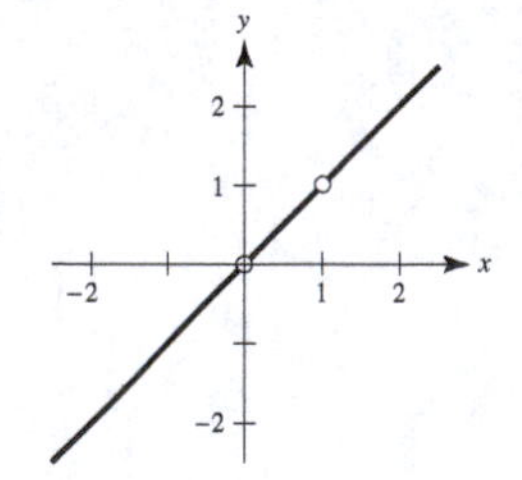

13. (a)

$$\frac{I}{x^2} = \frac{2I}{(x-3)^2}$$

$$x^2 - 6x + 9 = 2x^2$$

$$x^2 + 6x - 9 = 0$$

$$x = \frac{-6 \pm \sqrt{36+36}}{2} = -3 \pm \sqrt{18} \approx 1.2426,\ -7.2426$$

(b)

$$\frac{I}{x^2+y^2} = \frac{2I}{(x-3)^2+y^2}$$

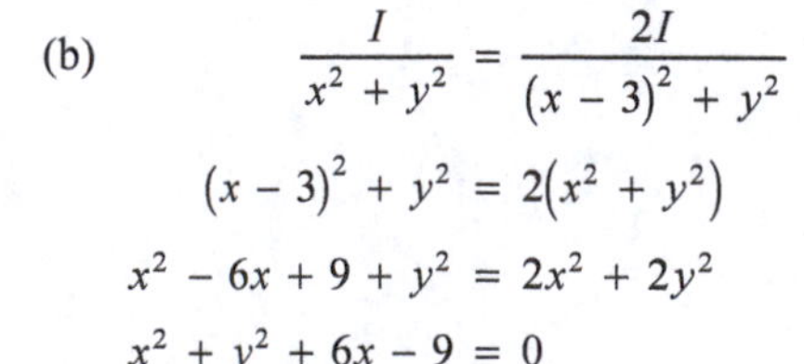

$$(x-3)^2 + y^2 = 2(x^2+y^2)$$

$$x^2 - 6x + 9 + y^2 = 2x^2 + 2y^2$$

$$x^2 + y^2 + 6x - 9 = 0$$

$$(x+3)^2 + y^2 = 18$$

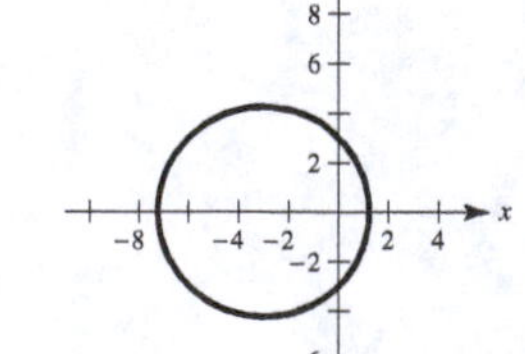

Circle of radius $\sqrt{18}$ and center $(-3, 0)$.

15.

$$
\begin{aligned}
d_1 d_2 &= 1\\
\left[(x+1)^2 + y^2\right]\left[(x-1)^2 + y^2\right] &= 1\\
(x+1)^2(x-1)^2 + y^2\left[(x+1)^2 + (x-1)^2\right] + y^4 &= 1\\
\left(x^2 - 1\right)^2 + y^2\left[2x^2 + 2\right] + y^4 &= 1\\
x^4 - 2x^2 + 1 + 2x^2y^2 + 2y^2 + y^4 &= 1\\
\left(x^4 + 2x^2y^2 + y^4\right) - 2x^2 + 2y^2 &= 0\\
\left(x^2 + y^2\right)^2 &= 2\left(x^2 - y^2\right)
\end{aligned}
$$

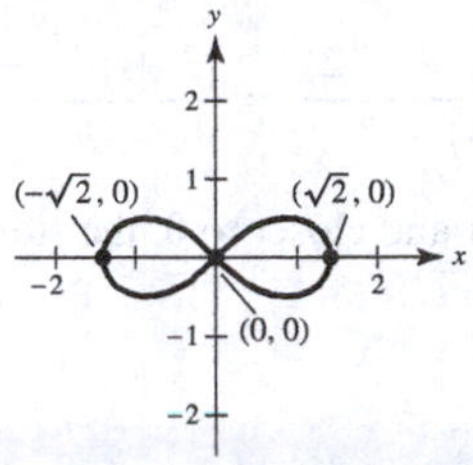

Let $y = 0$. Then $x^4 = 2x^2 \Rightarrow x = 0$ or $x^2 = 2$.

So, $(0, 0)$, $\left(\sqrt{2}, 0\right)$ and $\left(-\sqrt{2}, 0\right)$ are on the curve.

CHAPTER 2
Limits and Their Properties

CHAPTER 2
Limits and Their Properties

Section 2.1 A Preview of Calculus

1. Calculus is the mathematics of change. Precalculus is more static. Answers will vary. *Sample answer:*

Precalculus: Area of a rectangle
Calculus: Area under a curve

Precalculus: Work done by a constant force
Calculus: Work done by a variable force

Precalculus: Center of a rectangle
Calculus: Centroid of a region

3. Precalculus: $(20 \text{ ft/sec})(15 \text{ sec}) = 300 \text{ ft}$

5. Calculus required: Slope of the tangent line at $x = 2$ is the rate of change, and equals about 0.16.

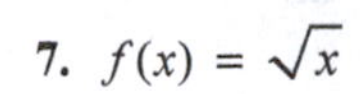

7. $f(x) = \sqrt{x}$

(a)

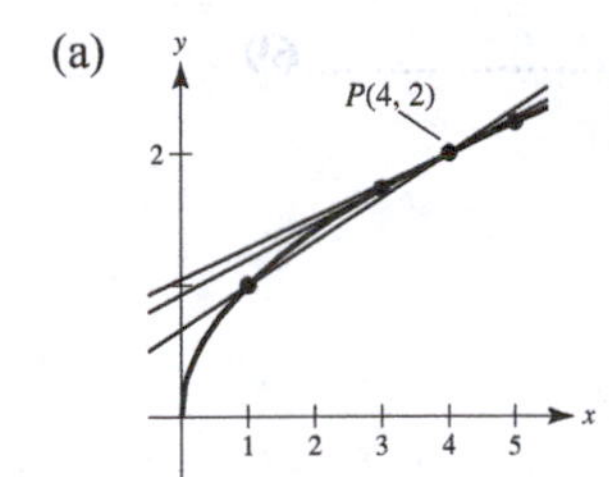

(b) $\text{slope} = m = \dfrac{\sqrt{x} - 2}{x - 4} = \dfrac{\sqrt{x} - 2}{\left(\sqrt{x} + 2\right)\left(\sqrt{x} - 2\right)}$

$= \dfrac{1}{\sqrt{x} + 2}, x \neq 4$

$x = 1: m = \dfrac{1}{\sqrt{1} + 2} = \dfrac{1}{3}$

$x = 3: m = \dfrac{1}{\sqrt{3} + 2} \approx 0.2679$

$x = 5: m = \dfrac{1}{\sqrt{5} + 2} \approx 0.2361$

(c) At $P(4, 2)$ the slope is $\dfrac{1}{\sqrt{4} + 2} = \dfrac{1}{4} = 0.25.$

You can improve your approximation of the slope at $x = 4$ by considering x-values very close to 4.

9. (a) $\text{Area} \approx 5 + \frac{5}{2} + \frac{5}{3} + \frac{5}{4} \approx 10.417$

$\text{Area} \approx \frac{1}{2}\left(5 + \frac{5}{1.5} + \frac{5}{2} + \frac{5}{2.5} + \frac{5}{3} + \frac{5}{3.5} + \frac{5}{4} + \frac{5}{4.5}\right)$

≈ 9.145

(b) You could improve the approximation by using more rectangles.

11. (a) $D_1 = \sqrt{(5 - 1)^2 + (1 - 5)^2} = \sqrt{16 + 16} \approx 5.66$

(b) $D_2 = \sqrt{1 + \left(\frac{5}{2}\right)^2} + \sqrt{1 + \left(\frac{5}{2} - \frac{5}{3}\right)^2} + \sqrt{1 + \left(\frac{5}{3} - \frac{5}{4}\right)^2} + \sqrt{1 + \left(\frac{5}{4} - 1\right)^2} \approx 2.693 + 1.302 + 1.083 + 1.031 \approx 6.11$

(c) Increase the number of line segments.

Section 2.2 Finding Limits Graphically and Numerically

1. As the graph of the function approaches 8 on the horizontal axis, the graph approaches 25 on the vertical axis.

3.

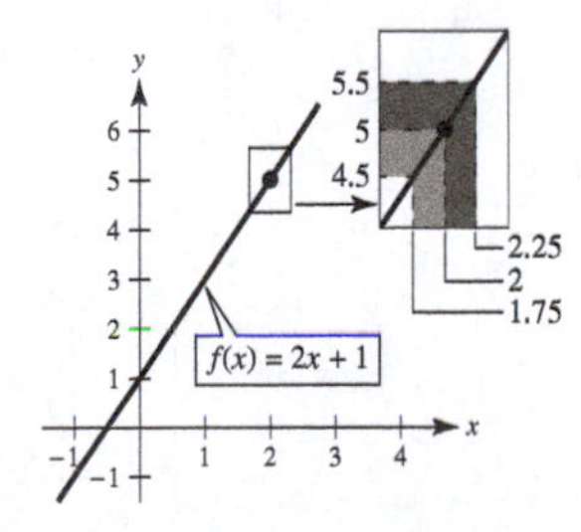

5.

x	3.9	3.99	3.999	4	4.001	4.01	4.1
$f(x)$	0.3448	0.3344	0.3334	?	0.3332	0.3322	0.3226

$\lim_{x \to 4} \dfrac{x - 4}{x^2 - 5x - 4} \approx 0.3333 \qquad \left(\text{Actual limit is } \dfrac{1}{3}.\right)$

7.

x	–0.1	–0.01	–0.001	0.001	0.01	0.1
$f(x)$	0.9983	0.99998	1.0000	1.0000	0.99998	0.9983

$\lim_{x \to 0} \frac{\sin x}{x} \approx 1.0000$ (Actual limit is 1.) (Make sure you use radian mode.)

9.

x	–0.1	–0.01	–0.001	0.001	0.01	0.1
$f(x)$	0.9516	0.9950	0.9995	1.0005	1.0050	1.0517

$\lim_{x \to 0} \frac{e^x - 1}{x} \approx 1.0000$ (Actual limit is 1.)

11.

x	0.9	0.99	0.999	1.001	1.01	1.1
$f(x)$	0.2564	0.2506	0.2501	0.2499	0.2494	0.2439

$\lim_{x \to 1} \frac{x - 2}{x^2 + x - 6} \approx 0.2500$ $\left(\text{Actual limit is } \frac{1}{4}.\right)$

13.

x	0.9	0.99	0.999	1.001	1.01	1.1
$f(x)$	0.7340	0.6733	0.6673	0.6660	0.6600	0.6015

$\lim_{x \to 1} \frac{x^4 - 1}{x^6 - 1} \approx 0.6666$ $\left(\text{Actual limit is } \frac{2}{3}.\right)$

15.

x	–6.1	–6.01	–6.001	–6	–5.999	–5.99	–5.9
$f(x)$	–0.1248	–0.1250	–0.1250	?	–0.1250	–0.1250	–0.1252

$\lim_{x \to -6} \frac{\sqrt{10 - x} - 4}{x + 6} \approx -0.1250$ $\left(\text{Actual limit is } -\frac{1}{8}.\right)$

17.

x	–0.1	–0.01	–0.001	0.001	0.01	0.1
$f(x)$	1.9867	1.9999	2.0000	2.0000	1.9999	1.9867

$\lim_{x \to 0} \frac{\sin 2x}{x} \approx 2.0000$ (Actual limit is 2.) (Make sure you use radian mode.)

19.

x	1.9	1.99	1.999	2.001	2.01	2.1
$f(x)$	0.5129	0.5013	0.5001	0.4999	0.4988	0.4879

$\lim_{x \to 2} \frac{\ln x - \ln 2}{x - 2} \approx 0.5000$ $\left(\text{Actual limit is } \frac{1}{2}.\right)$

21.

x	–0.1	–0.01	–0.001	0.001	0.01	0.1
$f(x)$	3.99982	4	4	0	0	0.00018

$\lim_{x \to 0} \frac{4}{1 + e^{1/x}}$ does not exist.

23. $\lim_{x\to 3}(4 - x) = 1$

25. $\lim_{x\to 2} f(x) = \lim_{x\to 2}(4 - x) = 2$

27. $\lim_{x\to 2}\dfrac{|x-2|}{x-2}$ does not exist.

For values of x to the left of 2, $\dfrac{|x-2|}{(x-2)} = -1$, whereas

for values of x to the right of 2, $\dfrac{|x-2|}{(x-2)} = 1$.

29. $\lim_{x\to 0}\cos(1/x)$ does not exist because the function oscillates between -1 and 1 as x approaches 0.

31. (a) $f(1)$ exists. The black dot at $(1, 2)$ indicates that $f(1) = 2$.

(b) $\lim_{x\to 1} f(x)$ does not exist. As x approaches 1 from the left, $f(x)$ approaches 3.5, whereas as x approaches 1 from the right, $f(x)$ approaches 1.

(c) $f(4)$ does not exist. The hollow circle at $(4, 2)$ indicates that f is not defined at 4.

(d) $\lim_{x\to 4} f(x)$ exists. As x approaches 4, $f(x)$ approaches 2: $\lim_{x\to 4} f(x) = 2$.

33.

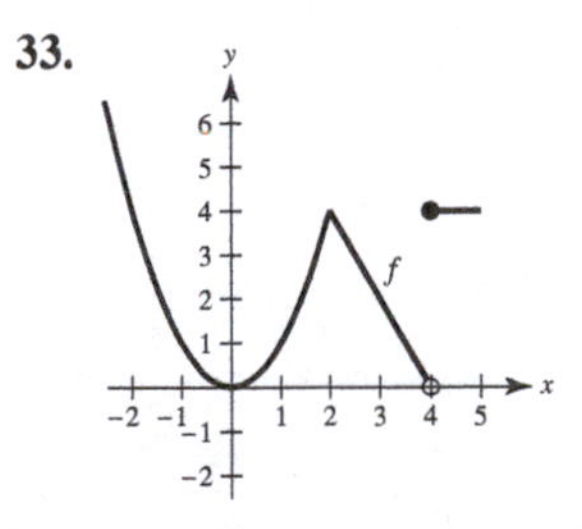

$\lim_{x\to c} f(x)$ exists for all values of $c \neq 4$.

35. One possible answer is

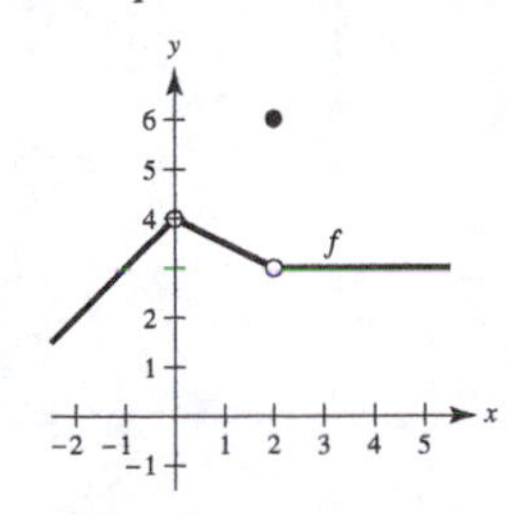

37. You need $|f(x) - 3| = |(x + 1) - 3| = |x - 2| < 0.4$.

So, take $\delta = 0.4$. If $0 < |x - 2| < 0.4$,

then $|x - 2| = |(x + 1) - 3| = |f(x) - 3| < 0.4$,

as desired.

39. You need to find δ such that $0 < |x - 1| < \delta$ implies $|f(x) - 1| = \left|\dfrac{1}{x} - 1\right| < 0.1$. That is,

$$-0.1 < \frac{1}{x} - 1 < 0.1$$
$$1 - 0.1 < \frac{1}{x} < 1 + 0.1$$
$$\frac{9}{10} < \frac{1}{x} < \frac{11}{10}$$
$$\frac{10}{9} > x > \frac{10}{11}$$
$$\frac{10}{9} - 1 > x - 1 > \frac{10}{11} - 1$$
$$\frac{1}{9} > x - 1 > -\frac{1}{11}.$$

So take $\delta = \dfrac{1}{11}$. Then $0 < |x - 1| < \delta$ implies

$$-\frac{1}{11} < x - 1 < \frac{1}{11}$$
$$-\frac{1}{11} < x - 1 < \frac{1}{9}.$$

Using the first series of equivalent inequalities, you obtain

$$|f(x) - 1| = \left|\frac{1}{x} - 1\right| < 0.1.$$

41. $\lim_{x\to 2}(3x + 2) = 3(2) + 2 = 8 = L$

(a) $|(3x + 2) - 8| < 0.01$

$|3x - 6| < 0.01$

$3|x - 2| < 0.01$

$0 < |x - 2| < \frac{0.01}{3} \approx 0.0033 = \delta$

So, if $0 < |x - 2| < \delta = \frac{0.01}{3}$, you have

$3|x - 2| < 0.01$

$|3x - 6| < 0.01$

$|(3x + 2) - 8| < 0.01$

$|f(x) - L| < 0.01.$

(b) $|(3x + 2) - 8| < 0.005$

$|3x - 6| < 0.005$

$3|x - 2| < 0.005$

$0 < |x - 2| < \dfrac{0.005}{3} \approx 0.00167 = \delta$

Finally, as in part (a), if $0 < |x - 2| < \dfrac{0.005}{3}$, you have $|(3x + 2) - 8| < 0.005$.

43. $\lim_{x \to 2} (x^2 - 3) = 2^2 - 3 = 1 = L$

(a) $$\left|(x^2 - 3) - 1\right| < 0.01$$
$$\left|x^2 - 4\right| < 0.01$$
$$\left|(x + 2)(x - 2)\right| < 0.01$$
$$|x + 2||x - 2| < 0.01$$
$$|x - 2| < \frac{0.01}{|x + 2|}$$

If you assume $1 < x < 3$, then $\delta \approx 0.01/5 = 0.002$.

So, if $0 < |x - 2| < \delta \approx 0.002$, you have

$$|x - 2| < 0.002 = \frac{1}{5}(0.01) < \frac{1}{|x + 2|}(0.01)$$
$$|x + 2||x - 2| < 0.01$$
$$\left|x^2 - 4\right| < 0.01$$
$$\left|(x^2 - 3) - 1\right| < 0.01$$
$$|f(x) - L| < 0.01.$$

(b) $$\left|(x^2 - 3) - 1\right| < 0.005$$
$$\left|x^2 - 4\right| < 0.005$$
$$\left|(x + 2)(x - 2)\right| < 0.005$$
$$|x + 2||x - 2| < 0.005$$
$$|x - 2| < \frac{0.005}{|x + 2|}$$

If you assume $1 < x < 3$, then

$$\delta = \frac{0.005}{5} = 0.001.$$

Finally, as in part (a), if $0 < |x - 2| < 0.001$, you have $\left|(x^2 - 3) - 1\right| < 0.005$.

45. $\lim_{x \to 4} (x^2 - x) = 16 - 4 = 12 = L$

(a) $$\left|(x^2 - x) - 12\right| < 0.01$$
$$\left|(x - 4)(x + 3)\right| < 0.01$$
$$|x - 4||x + 3| < 0.01$$
$$|x - 4| < \frac{0.01}{|x + 3|}$$

If you assume $3 < x < 5$, then

$$\delta = \frac{0.01}{8} = 0.00125.$$

So, if $0 < |x - 4| < \frac{0.01}{8}$, you have

$$|x - 4| < \frac{0.01}{|x + 3|}$$
$$|x - 4||x + 3| < 0.01$$
$$\left|x^2 - x - 12\right| < 0.01$$
$$\left|(x^2 - x) - 12\right| < 0.01$$
$$|f(x) - L| < 0.01$$

(b) $$\left|(x^2 - x) - 12\right| < 0.005$$
$$\left|(x - 4)(x + 3)\right| < 0.005$$
$$|x - 4||x + 3| < 0.005$$
$$|x - 4| < \frac{0.005}{|x + 3|}$$

If you assume $3 < x < 5$, then

$$\delta = \frac{0.005}{8} = 0.000625.$$

Finally, as in part (a), if $0 < |x - 4| < \frac{0.005}{8}$, you have $\left|(x^2 - x) - 12\right| < 0.005$.

47. $\lim_{x \to 4} (x + 2) = 4 + 2 = 6$

Given $\varepsilon > 0$:

$$\left|(x + 2) - 6\right| < \varepsilon$$
$$|x - 4| < \varepsilon = \delta$$

So, let $\delta = \varepsilon$. So, if $0 < |x - 4| < \delta = \varepsilon$, you have

$$|x - 4| < \varepsilon$$
$$\left|(x + 2) - 6\right| < \varepsilon$$
$$|f(x) - L| < \varepsilon.$$

49. $\lim_{x\to -4}\left(\frac{1}{2}x - 1\right) = \frac{1}{2}(-4) - 1 = -3$

Given $\varepsilon > 0$:

$$\left|\left(\tfrac{1}{2}x - 1\right) - (-3)\right| < \varepsilon$$
$$\left|\tfrac{1}{2}x + 2\right| < \varepsilon$$
$$\tfrac{1}{2}\left|x - (-4)\right| < \varepsilon$$
$$\left|x - (-4)\right| < 2\varepsilon$$

So, let $\delta = 2\varepsilon$.

So, if $0 < \left|x - (-4)\right| < \delta = 2\varepsilon$, you have

$$\left|x - (-4)\right| < 2\varepsilon$$
$$\left|\tfrac{1}{2}x + 2\right| < \varepsilon$$
$$\left|\left(\tfrac{1}{2}x - 1\right) + 3\right| < \varepsilon$$
$$\left|f(x) - L\right| < \varepsilon.$$

51. $\lim_{x\to 6} 3 = 3$

Given $\varepsilon > 0$:

$$\left|3 - 3\right| < \varepsilon$$
$$0 < \varepsilon$$

So, any $\delta > 0$ will work.

So, for any $\delta > 0$, you have

$$\left|3 - 3\right| < \varepsilon$$
$$\left|f(x) - L\right| < \varepsilon.$$

53. $\lim_{x\to 0} \sqrt[3]{x} = 0$

Given $\varepsilon > 0$: $\left|\sqrt[3]{x} - 0\right| < \varepsilon$

$$\left|\sqrt[3]{x}\right| < \varepsilon$$
$$\left|x\right| < \varepsilon^3 = \delta$$

So, let $\delta = \varepsilon^3$.

So, for $0|x - 0|\delta = \varepsilon^3$, you have

$$\left|x\right| < \varepsilon^3$$
$$\left|\sqrt[3]{x}\right| < \varepsilon$$
$$\left|\sqrt[3]{x} - 0\right| < \varepsilon$$
$$\left|f(x) - L\right| < \varepsilon.$$

55. $\lim_{x\to -5}\left|x - 5\right| = \left|(-5) - 5\right| = \left|-10\right| = 10$

Given $\varepsilon > 0$: $\left|\left|x - 5\right| - 10\right| < \varepsilon$

$$\left|-(x - 5) - 10\right| < \varepsilon \quad (x - 5 < 0)$$
$$\left|-x - 5\right| < \varepsilon$$
$$\left|x - (-5)\right| < \varepsilon$$

So, let $\delta = \varepsilon$.

So for $\left|x - (-5)\right| < \delta = \varepsilon$, you have

$$\left|-(x + 5)\right| < \varepsilon$$
$$\left|-(x - 5) - 10\right| < \varepsilon$$
$$\left|\left|x - 5\right| - 10\right| < \varepsilon \quad \text{(because } x - 5 < 0\text{)}$$
$$\left|f(x) - L\right| < \varepsilon.$$

57. $\lim_{x\to 1}\left(x^2 + 1\right) = 1^2 + 1 = 2$

Given $\varepsilon > 0$: $\left|\left(x^2 + 1\right) - 2\right| < \varepsilon$

$$\left|x^2 - 1\right| < \varepsilon$$
$$\left|(x + 1)(x - 1)\right| < \varepsilon$$
$$\left|x - 1\right| < \frac{\varepsilon}{\left|x + 1\right|}$$

If you assume $0 < x < 2$, then $\delta = \varepsilon/3$.

So for $0 < \left|x - 1\right| < \delta = \dfrac{\varepsilon}{3}$, you have

$$\left|x - 1\right| < \frac{1}{3}\varepsilon < \frac{1}{\left|x + 1\right|}\varepsilon$$
$$\left|x^2 - 1\right| < \varepsilon$$
$$\left|\left(x^2 + 1\right) - 2\right| < \varepsilon$$
$$\left|f(x) - 2\right| < \varepsilon.$$

59. $\lim_{x\to \pi} f(x) = \lim_{x\to \pi} 4 = 4$

61. $f(x) = \dfrac{\sqrt{x + 5} - 3}{x - 4}$

$$\lim_{x\to 4} f(x) = \frac{1}{6}$$

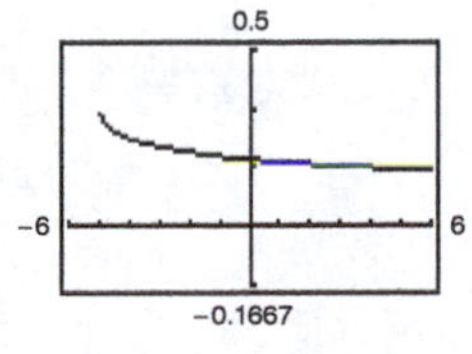

The domain is $[-5, 4) \cup (4, \infty)$. The graphing utility does not show the hole at $\left(4, \frac{1}{6}\right)$.

63. $C(t) = 9.99 - 0.79[\![1 - t]\!],\ t > 0$

(a) $C(10.75) = 9.99 - 0.79[\![1 - 10.75]\!]$

$= 9.99 - 0.79(-10)$

$= \$17.89$

$C(10.75)$ represents the cost of a 10-minute, 45-second call.

(b)

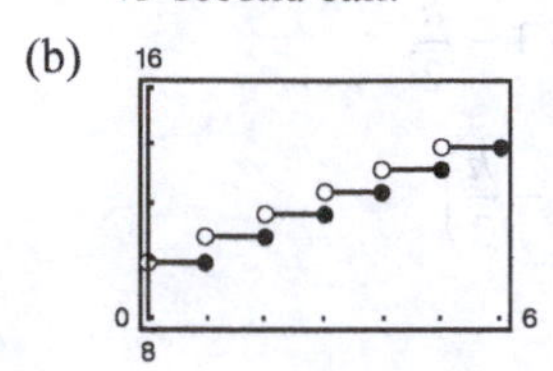

(c) The limit does not exist because the limits from the left and right are not equal.

65. Choosing a smaller positive value of δ will still satisfy the inequality $|f(x) - L| < \varepsilon$.

67. No. The fact that $f(2) = 4$ has no bearing on the existence of the limit of $f(x)$ as x approaches 2.

69. (a) $C = 2\pi r$

$$r = \frac{C}{2\pi} = \frac{6}{2\pi} = \frac{3}{\pi} \approx 0.9549 \text{ cm}$$

(b) When $C = 5.5$: $r = \dfrac{5.5}{2\pi} \approx 0.87535$ cm

When $C = 6.5$: $r = \dfrac{6.5}{2\pi} \approx 1.03451$ cm

So $0.87535 < r < 1.03451$.

(c) $\lim\limits_{x \to 3/\pi} (2\pi r) = 6;\ \varepsilon = 0.5;\ \delta \approx 0.0796$

71. $f(x) = (1 + x)^{1/x}$

$$\lim_{x \to 0} (1 + x)^{1/x} = e \approx 2.71828$$

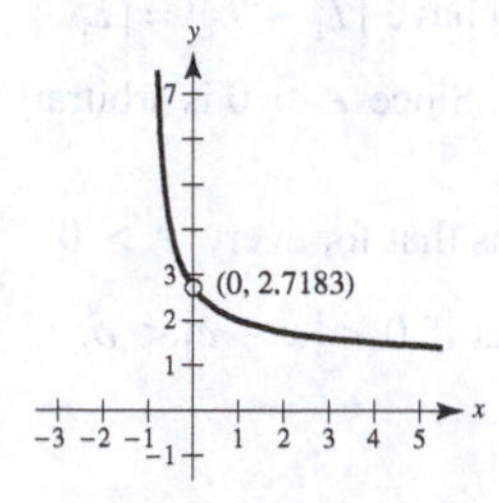

x	$f(x)$
–0.1	2.867972
–0.01	2.731999
–0.001	2.719642
–0.0001	2.718418
–0.00001	2.718295
–0.000001	2.718283

x	$f(x)$
0.1	2.593742
0.01	2.704814
0.001	2.716942
0.0001	2.718146
0.00001	2.718268
0.000001	2.718280

73.

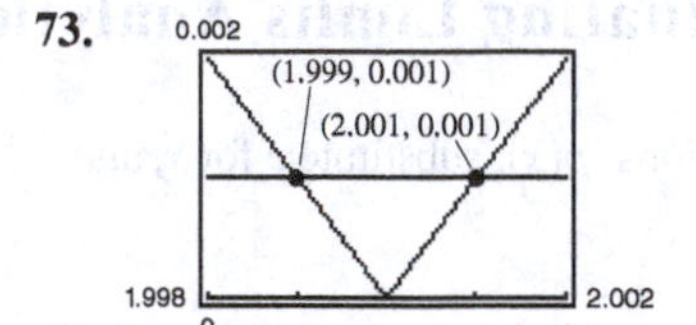

Using the zoom and trace feature, $\delta = 0.001$. So $(2 - \delta, 2 + \delta) = (1.999, 2.001)$.

Note: $\dfrac{x^2 - 4}{x - 2} = x + 2$ for $x \neq 2$.

75. False. The existence or nonexistence of $f(x)$ at $x = c$ has no bearing on the existence of the limit of $f(x)$ as $x \to c$.

77. False. Let

$$f(x) = \begin{cases} x - 4, & x \neq 2 \\ 0, & x = 2 \end{cases}$$

$f(2) = 0$

$$\lim_{x \to 2} f(x) = \lim_{x \to 2} (x - 4) = 2 \neq 0$$

79. $f(x) = \sqrt{x}$

$\lim\limits_{x \to 0.25} \sqrt{x} = 0.5$ is true.

As x approaches $0.25 = \frac{1}{4}$ from either side, $f(x) = \sqrt{x}$ approaches $\frac{1}{2} = 0.5$.

81. Using a graphing utility, you see that

$$\lim_{x \to 0} \frac{\sin x}{x} = 1$$

$$\lim_{x \to 0} \frac{\sin 2x}{x} = 2, \text{ etc.}$$

So, $\lim\limits_{x \to 0} \dfrac{\sin nx}{x} = n$.

83. If $\lim_{x\to c} f(x) = L_1$ and $\lim_{x\to c} f(x) = L_2$, then for every $\varepsilon > 0$, there exists $\delta_1 > 0$ and $\delta_2 > 0$ such that $|x - c| < \delta_1 \Rightarrow |f(x) - L_1| < \varepsilon$ and $|x - c| < \delta_2 \Rightarrow |f(x) - L_2| < \varepsilon$. Let δ equal the smaller of δ_1 and δ_2. Then for $|x - c| < \delta$, you have $|L_1 - L_2| = |L_1 - f(x) + f(x) - L_2| \le |L_1 - f(x)| + |f(x) - L_2| < \varepsilon + \varepsilon$. Therefore, $|L_1 - L_2| < 2\varepsilon$. Since $\varepsilon > 0$ is arbitrary, it follows that $L_1 = L_2$.

85. $\lim_{x\to c}\left[f(x) - L\right] = 0$ means that for every $\varepsilon > 0$ there exists $\delta > 0$ such that if $0 < |x - c| < \delta$,

then

$|(f(x) - L) - 0| < \varepsilon$.

This means the same as $|f(x) - L| < \varepsilon$ when $0 < |x - c| < \delta$.

So, $\lim_{x\to c} f(x) = L$.

87. The radius OP has a length equal to the altitude z of the triangle plus $\frac{h}{2}$. So, $z = 1 - \frac{h}{2}$.

$$\text{Area triangle} = \frac{1}{2}b\left(1 - \frac{h}{2}\right)$$

$$\text{Area rectangle} = bh$$

Because these are equal,

$$\frac{1}{2}b\left(1 - \frac{h}{2}\right) = bh$$

$$1 - \frac{h}{2} = 2h$$

$$\frac{5}{2}h = 1$$

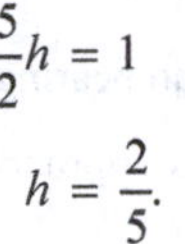

$$h = \frac{2}{5}.$$

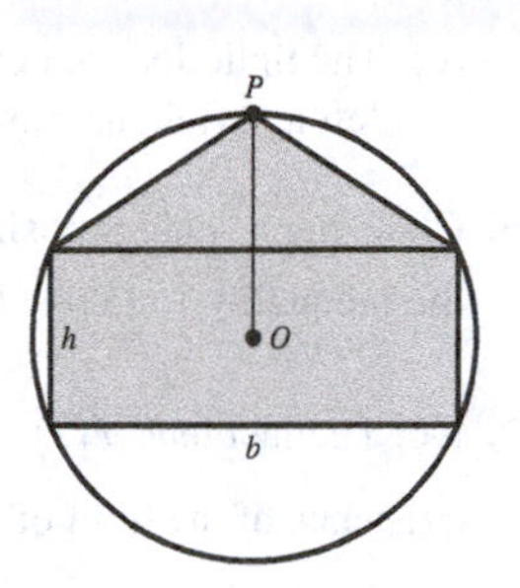

Section 2.3 Evaluating Limits Analytically

1. For polynomial functions $p(x)$, substitute c for x, and simplify.

3. If a function f is squeezed between two functions h and g, $h(x) \le f(x) \le g(x)$, and h and g have the same limit L as $x \to c$, then $\lim_{x\to c} f(x)$ exists and equals L

5. $\lim_{x\to -3}(2x + 5) = 2(-3) + 5 = -1$

7. $\lim_{x\to -3}(x^2 + 3x) = (-3)^2 + 3(-3) = 9 - 9 = 0$

9. $\lim_{x\to 3}\sqrt{x + 8} = \sqrt{3 + 8} = \sqrt{11}$

11. $\lim_{x\to -4}(1 - x)^3 = \left[1 - (-4)\right]^3 = 5^3 = 125$

13. $\lim_{x\to 2}\frac{3}{2x + 1} = \frac{3}{2(2) + 1} = \frac{3}{5}$

15. $\lim_{x\to 1}\frac{x}{x^2 + 4} = \frac{1}{1^2 + 4} = \frac{1}{5}$

17. $\lim_{x\to 7}\frac{3x}{\sqrt{x + 2}} = \frac{3(7)}{\sqrt{7 + 2}} = \frac{21}{3} = 7$

19. (a) $\lim_{x\to 1} f(x) = 5 - 1 = 4$

(b) $\lim_{x\to 4} g(x) = 4^3 = 64$

(c) $\lim_{x\to 1} g(f(x)) = g(f(1)) = g(4) = 64$

21. (a) $\lim_{x\to 1} f(x) = 4 - 1 = 3$

(b) $\lim_{x\to 3} g(x) = \sqrt{3 + 1} = 2$

(c) $\lim_{x\to 1} g(f(x)) = g(3) = 2$

23. $\lim_{x\to \pi/2}\sin x = \sin\frac{\pi}{2} = 1$

25. $\lim_{x\to 1}\cos\frac{\pi x}{3} = \cos\frac{\pi}{3} = \frac{1}{2}$

27. $\lim_{x\to 0}\sec 2x = \sec 0 = 1$

29. $\lim_{x\to 5\pi/6}\sin x = \sin\frac{5\pi}{6} = \frac{1}{2}$

31. $\lim_{x\to 3}\tan\left(\frac{\pi x}{4}\right) = \tan\frac{3\pi}{4} = -1$

33. $\lim_{x\to 0} e^x\cos 2x = e^0\cos 0 = 1$

35. $\lim_{x\to 1}\left(\ln 3x + e^x\right) = \ln 3 + e$

37. $\lim_{x\to c} f(x) = \frac{2}{5}, \lim_{x\to c} g(x) = 2$

(a) $\lim_{x\to c}\left[5g(x)\right] = 5\lim_{x\to c} g(x) = 5(2) = 10$

(b) $\lim_{x\to c}\left[f(x) + g(x)\right] = \lim_{x\to c} f(x) + \lim_{x\to c} g(x) = \frac{2}{5} + 2 = \frac{12}{5}$

(c) $\lim_{x\to c}\left[f(x)g(x)\right] = \left[\lim_{x\to c} f(x)\right]\left[\lim_{x\to c} g(x)\right] = \frac{2}{5}(2) = \frac{4}{5}$

(d) $\lim_{x\to c}\frac{f(x)}{g(x)} = \frac{\lim_{x\to c} f(x)}{\lim_{x\to c} g(x)} = \frac{2/5}{2} = \frac{1}{5}$

39. $\lim_{x\to c} f(x) = 16$

(a) $\lim_{x\to c}\left[f(x)\right]^2 = \left[\lim_{x\to c} f(x)\right]^2 = (16)^2 = 256$

(b) $\lim_{x\to c}\sqrt{f(x)} = \sqrt{\lim_{x\to c} f(x)} = \sqrt{16} = 4$

(c) $\lim_{x\to c}\left[3f(x)\right] = 3\left[\lim_{x\to c} f(x)\right] = 3(16) = 48$

(d) $\lim_{x\to c}\left[f(x)\right]^{3/2} = \left[\lim_{x\to c} f(x)\right]^{3/2} = (16)^{3/2} = 64$

41. $f(x) = \frac{x^2 - 1}{x + 1} = \frac{(x + 1)(x - 1)}{x + 1}$ and $g(x) = x - 1$ agree except at $x = -1$.

$\lim_{x\to -1} f(x) = \lim_{x\to -1} g(x) = \lim_{x\to -1}(x - 1) = -1 - 1 = -2$

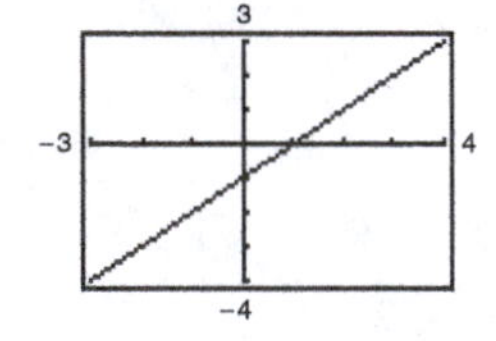

43. $f(x) = \frac{x^3 - 8}{x - 2}$ and $g(x) = x^2 + 2x + 4$ agree except at $x = 2$.

$\lim_{x\to 2} f(x) = \lim_{x\to 2} g(x) = \lim_{x\to 2}\left(x^2 + 2x + 4\right) = 2^2 + 2(2) + 4 = 12$

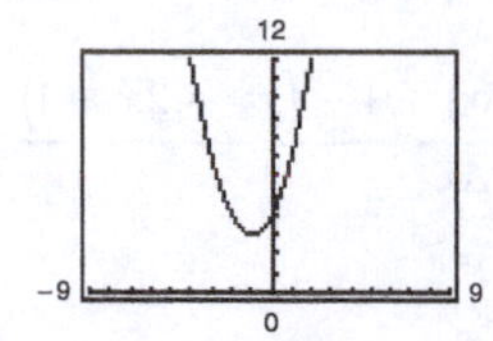

45. $f(x) = \frac{(x + 4)\ln(x + 6)}{x^2 - 16}$ and $g(x) = \frac{\ln(x + 6)}{x - 4}$ agree except at $x = -4$.

$\lim_{x\to -4} f(x) = \lim_{x\to -4} g(x) = \frac{\ln 2}{-8} \approx -0.0866$

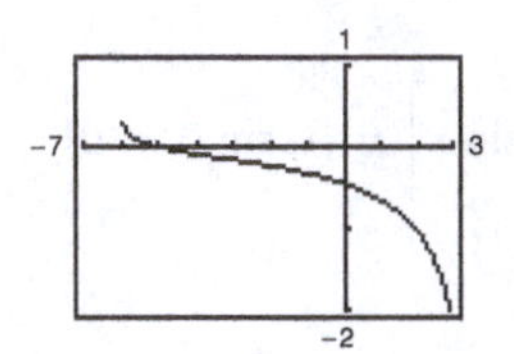

47. $\lim_{x\to 0}\frac{x}{x^2 - x} = \lim_{x\to 0}\frac{x}{x(x - 1)} = \lim_{x\to 0}\frac{1}{x - 1} = \frac{1}{0 - 1} = -1$

49. $\lim_{x\to 4}\frac{x - 4}{x^2 - 16} = \lim_{x\to 4}\frac{x - 4}{(x + 4)(x - 4)} = \lim_{x\to 4}\frac{1}{x + 4} = \frac{1}{4 + 4} = \frac{1}{8}$

51. $\lim_{x\to -3}\frac{x^2 + x - 6}{x^2 - 9} = \lim_{x\to -3}\frac{(x + 3)(x - 2)}{(x + 3)(x - 3)} = \lim_{x\to -3}\frac{x - 2}{x - 3} = \frac{-3 - 2}{-3 - 3} = \frac{-5}{-6} = \frac{5}{6}$

53. $\lim_{x\to 4}\frac{\sqrt{x + 5} - 3}{x - 4} = \lim_{x\to 4}\frac{\sqrt{x + 5} - 3}{x - 4}\cdot\frac{\sqrt{x + 5} + 3}{\sqrt{x + 5} + 3}$

$= \lim_{x\to 4}\frac{(x + 5) - 9}{(x - 4)\left(\sqrt{x + 5} + 3\right)}$

$= \lim_{x\to 4}\frac{1}{\sqrt{x + 5} + 3} = \frac{1}{\sqrt{9} + 3} = \frac{1}{6}$

55. $\lim_{x\to 0}\frac{\sqrt{x + 5} - \sqrt{5}}{x} = \lim_{x\to 0}\frac{\sqrt{x + 5} - \sqrt{5}}{x}\cdot\frac{\sqrt{x + 5} + \sqrt{5}}{\sqrt{x + 5} + \sqrt{5}}$

$= \lim_{x\to 0}\frac{(x + 5) - 5}{x\left(\sqrt{x + 5} + \sqrt{5}\right)} = \lim_{x\to 0}\frac{1}{\sqrt{x + 5} + \sqrt{5}} = \frac{1}{\sqrt{5} + \sqrt{5}} = \frac{1}{2\sqrt{5}} = \frac{\sqrt{5}}{10}$

57. $\lim_{x\to 0} \dfrac{\dfrac{1}{3+x} - \dfrac{1}{3}}{x} = \lim_{x\to 0} \dfrac{3-(3+x)}{(3+x)3(x)} = \lim_{x\to 0} \dfrac{-x}{(3+x)(3)(x)} = \lim_{x\to 0} \dfrac{-1}{(3+x)3} = \dfrac{-1}{(3)3} = -\dfrac{1}{9}$

59. $\lim_{\Delta x\to 0} \dfrac{2(x+\Delta x) - 2x}{\Delta x} = \lim_{\Delta x\to 0} \dfrac{2x + 2\Delta x - 2x}{\Delta x} = \lim_{\Delta x\to 0} \dfrac{2\Delta x}{\Delta x} = \lim_{\Delta x\to 0} 2 = 2$

61. $\lim_{\Delta x\to 0} \dfrac{(x+\Delta x)^2 - 2(x+\Delta x) + 1 - (x^2 - 2x + 1)}{\Delta x} = \lim_{\Delta x\to 0} \dfrac{x^2 + 2x\Delta x + (\Delta x)^2 - 2x - 2\Delta x + 1 - x^2 + 2x - 1}{\Delta x}$

$= \lim_{\Delta x\to 0} (2x + \Delta x - 2) = 2x - 2$

63. $\lim_{x\to 0} \dfrac{\sin x}{5x} = \lim_{x\to 0} \left[\left(\dfrac{\sin x}{x}\right)\left(\dfrac{1}{5}\right)\right] = (1)\left(\dfrac{1}{5}\right) = \dfrac{1}{5}$

65. $\lim_{x\to 0} \dfrac{(\sin x)(1-\cos x)}{x^2} = \lim_{x\to 0} \left[\dfrac{\sin x}{x} \cdot \dfrac{1-\cos x}{x}\right]$

$= (1)(0) = 0$

67. $\lim_{x\to 0} \dfrac{\sin^2 x}{x} = \lim_{x\to 0} \left[\dfrac{\sin x}{x} \sin x\right] = (1)\sin 0 = 0$

69. $\lim_{h\to 0} \dfrac{(1-\cos h)^2}{h} = \lim_{h\to 0} \left[\dfrac{1-\cos h}{h}(1-\cos h)\right]$

$= (0)(0) = 0$

71. $\lim_{x\to 0} \dfrac{6 - 6\cos x}{3} = \dfrac{6 - 6\cos 0}{3} = \dfrac{6-6}{3} = 0$

73. $\lim_{x\to 0} \dfrac{1 - e^{-x}}{e^x - 1} = \lim_{x\to 0} \dfrac{1 - e^{-x}}{e^x - 1} \cdot \dfrac{e^{-x}}{e^{-x}} = \lim_{x\to 0} \dfrac{(1 - e^{-x})e^{-x}}{1 - e^{-x}}$

$= \lim_{x\to 0} e^{-x} = 1$

75. $\lim_{t\to 0} \dfrac{\sin 3t}{2t} = \lim_{t\to 0} \left(\dfrac{\sin 3t}{3t}\right)\left(\dfrac{3}{2}\right) = (1)\left(\dfrac{3}{2}\right) = \dfrac{3}{2}$

77. $f(x) = \dfrac{\sqrt{x+2} - \sqrt{2}}{x}$

x	–0.1	–0.01	–0.001	0	0.001	0.01	0.1
$f(x)$	0.358	0.354	0.354	?	0.354	0.353	0.349

It appears that the limit is 0.354.

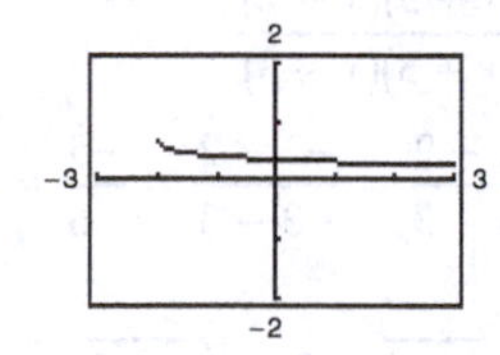

The graph has a hole at $x = 0$.

Analytically, $\lim_{x\to 0} \dfrac{\sqrt{x+2} - \sqrt{2}}{x} = \lim_{x\to 0} \dfrac{\sqrt{x+2} - \sqrt{2}}{x} \cdot \dfrac{\sqrt{x+2} + \sqrt{2}}{\sqrt{x+2} + \sqrt{2}}$

$= \lim_{x\to 0} \dfrac{x + 2 - 2}{x\left(\sqrt{x+2} + \sqrt{2}\right)} = \lim_{x\to 0} \dfrac{1}{\sqrt{x+2} + \sqrt{2}} = \dfrac{1}{2\sqrt{2}} = \dfrac{\sqrt{2}}{4} \approx 0.354.$

79. $f(x) = \dfrac{\dfrac{1}{2+x} - \dfrac{1}{2}}{x}$

x	–0.1	–0.01	–0.001	0	0.001	0.01	0.1
$f(x)$	–0.263	–0.251	–0.250	?	–0.250	–0.249	–0.238

It appears that the limit is –0.250.

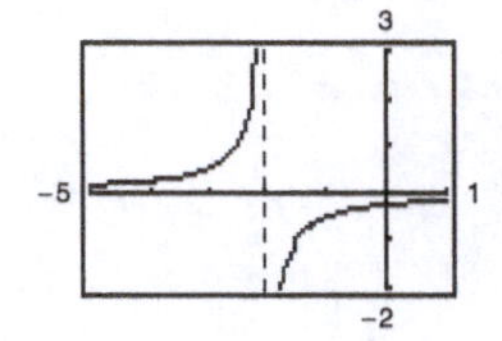

The graph has a hole at $x = 0$.

Analytically, $\lim_{x\to 0} \dfrac{\dfrac{1}{2+x} - \dfrac{1}{2}}{x} = \lim_{x\to 0} \dfrac{2-(2+x)}{2(2+x)} \cdot \dfrac{1}{x} = \lim_{x\to 0} \dfrac{-x}{2(2+x)} \cdot \dfrac{1}{x} = \lim_{x\to 0} \dfrac{-1}{2(2+x)} = -\dfrac{1}{4}.$

81. $f(t) = \dfrac{\sin 3t}{t}$

t	–0.1	–0.01	–0.001	0	0.001	0.01	0.1
$f(t)$	2.96	2.9996	3	?	3	2.9996	2.96

It appears that the limit is 3.

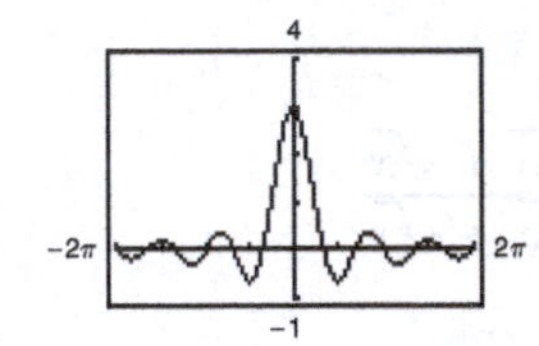

The graph has a hole at $t = 0$.

Analytically, $\lim_{t\to 0} \dfrac{\sin 3t}{t} = \lim_{t\to 0} 3\left(\dfrac{\sin 3t}{3t}\right) = 3(1) = 3.$

83. $f(x) = \dfrac{\sin x^2}{x}$

x	–0.1	–0.01	–0.001	0	0.001	0.01	0.1
$f(x)$	–0.099998	–0.01	–0.001	?	0.001	0.01	0.099998

It appears that the limit is 0.

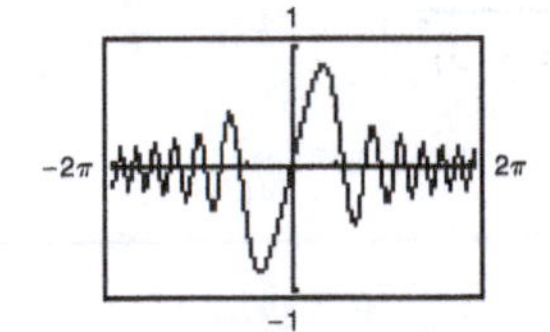

The graph has a hole at $x = 0$.

Analytically, $\lim_{x\to 0} \dfrac{\sin x^2}{x} = \lim_{x\to 0} x\left(\dfrac{\sin x^2}{x^2}\right) = 0(1) = 0.$

85. $f(x) = \dfrac{\ln x}{x - 1}$

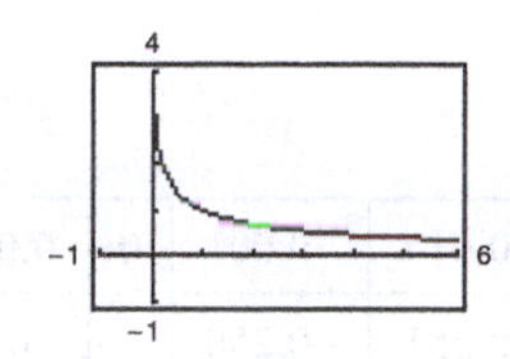

x	0.5	0.9	0.99	1.01	1.1	1.5
$f(x)$	1.3863	1.0536	1.0050	0.9950	0.9531	0.8109

It appears that the limit is 1.

Analytically, $\lim_{x\to 1} \dfrac{\ln x}{x-1} = \lim_{x\to 1}\left(\dfrac{1}{x-1}\right)\ln x = \lim_{x\to 1} \ln x^{1/(x-1)}$

Let $y = x - 1$, then $x = y + 1$ and $y \to 0$ as $x \to 1$.

So, $\lim_{x\to 1} \ln x^{1/(x-1)} = \lim_{y\to 0} \ln(y+1)^{1/y} = \ln e = 1$.

87. $f(x) = 3x - 2$

$$\lim_{\Delta x\to 0} \frac{f(x+\Delta x) - f(x)}{\Delta x} = \lim_{\Delta x\to 0} \frac{3(x+\Delta x) - 2 - (3x-2)}{\Delta x} = \lim_{\Delta x\to 0} \frac{3x + 3\Delta x - 2 - 3x + 2}{\Delta x} = \lim_{\Delta x\to 0} \frac{3\Delta x}{\Delta x} = 3$$

89. $f(x) = x^2 - 4x$

$$\lim_{\Delta x\to 0} \frac{f(x+\Delta x) - f(x)}{\Delta x} = \lim_{\Delta x\to 0} \frac{(x+\Delta x)^2 - 4(x+\Delta x) - (x^2 - 4x)}{\Delta x} = \lim_{\Delta x\to 0} \frac{x^2 + 2x\Delta x + \Delta x^2 - 4x - 4\Delta x - x^2 + 4x}{\Delta x}$$

$$= \lim_{\Delta x\to 0} \frac{\Delta x(2x + \Delta x - 4)}{\Delta x} = \lim_{\Delta x\to 0} (2x + \Delta x - 4) = 2x - 4$$

91. $f(x) = 2\sqrt{x}$

$$\lim_{\Delta x\to 0} \frac{f(x+\Delta x) - f(x)}{\Delta x} = \lim_{\Delta x\to 0} \frac{2\sqrt{x+\Delta x} - 2\sqrt{x}}{\Delta x} = \lim_{\Delta x\to 0} \frac{2\left(\sqrt{x+\Delta x} - \sqrt{x}\right)}{\Delta x} \cdot \frac{\sqrt{x+\Delta x} + \sqrt{x}}{\sqrt{x+\Delta x} + \sqrt{x}}$$

$$= \lim_{\Delta x\to 0} \frac{2(x + \Delta x - x)}{\Delta x\left(\sqrt{x+\Delta x} + \sqrt{x}\right)} = \lim_{\Delta x\to 0} \frac{2\Delta x}{\Delta x\left(\sqrt{x+\Delta x} + \sqrt{x}\right)}$$

$$= \lim_{\Delta x\to 0} \frac{2}{\sqrt{x+\Delta x} + \sqrt{x}} = \frac{2}{2\sqrt{x}} = \frac{1}{\sqrt{x}} = x^{-1/2}$$

93. $f(x) = \dfrac{1}{x+3}$

$$\lim_{\Delta x\to 0} \frac{f(x+\Delta x) - f(x)}{\Delta x} = \lim_{\Delta x\to 0} \frac{\dfrac{1}{x+\Delta x+3} - \dfrac{1}{x+3}}{\Delta x} = \lim_{\Delta x\to 0} \frac{x + 3 - (x + \Delta x + 3)}{(x + \Delta x + 3)(x+3)} \cdot \frac{1}{\Delta x}$$

$$= \lim_{\Delta x\to 0} \frac{-\Delta x}{(x + \Delta x + 3)(x+3)\Delta x} = \lim_{\Delta x\to 0} \frac{-1}{(x + \Delta x + 3)(x+3)} = \frac{-1}{(x+3)^2}$$

95. $\lim_{x\to 0}\left(4 - x^2\right) \le \lim_{x\to 0} f(x) \le \lim_{x\to 0}\left(3 + x^2\right)$

$4 \le \lim_{x\to 0} f(x) \le 4$

Therefore, $\lim_{x\to 0} f(x) = 4$.

97. $f(x) = |x|\sin x$

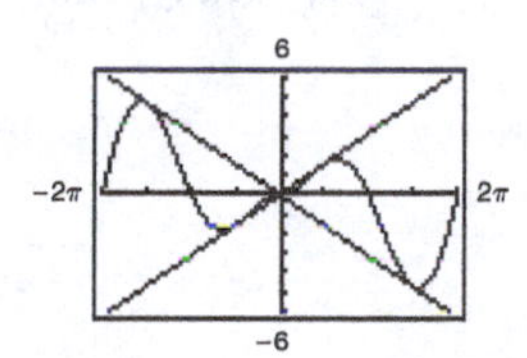

$\lim_{x\to 0} |x|\sin x = 0$

99. $f(x) = x \sin \frac{1}{x}$

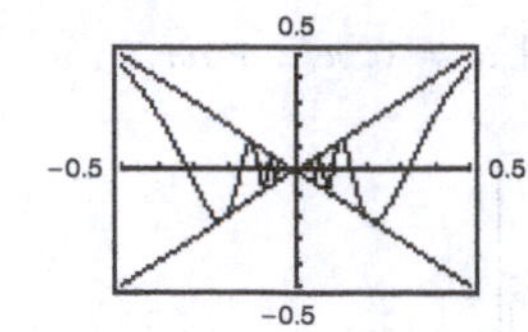

$$\lim_{x\to 0}\left(x \sin \frac{1}{x}\right) = 0$$

101. (a) Two functions f and g agree at all but one point (on an open interval) if $f(x) = g(x)$ for all x in the interval except for $x = c$, where c is in the interval.

(b) $f(x) = \dfrac{x^2 - 1}{x - 1} = \dfrac{(x + 1)(x - 1)}{x - 1}$ and $g(x) = x + 1$ agree at all points except $x = 1$.

(Other answers possible.)

103. $f(x) = x,\ g(x) = \sin x,\ h(x) = \dfrac{\sin x}{x}$

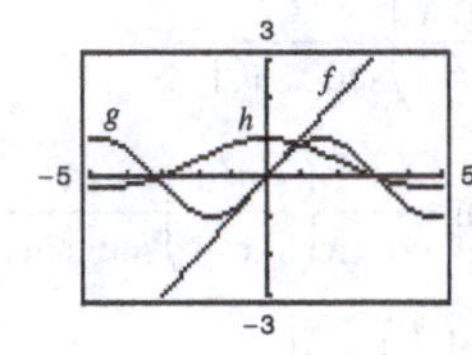

When the x-values are "close to" 0 the magnitude of f is approximately equal to the magnitude of g. So, $|g|/|f| \approx 1$ when x is "close to" 0.

105. $s(t) = -16t^2 + 500$

$$\begin{aligned}\lim_{t\to 2}\frac{s(2) - s(t)}{2 - t} &= \lim_{t\to 2}\frac{-16(2)^2 + 500 - \left(-16t^2 + 500\right)}{2 - t}\\ &= \lim_{t\to 2}\frac{436 + 16t^2 - 500}{2 - t}\\ &= \lim_{t\to 2}\frac{16\left(t^2 - 4\right)}{2 - t}\\ &= \lim_{t\to 2}\frac{16(t - 2)(t + 2)}{2 - t}\\ &= \lim_{t\to 2} -16(t + 2) = -64 \text{ ft/sec}\end{aligned}$$

The paint can is falling at about 64 feet/second.

107. $s(t) = -4.9t^2 + 200$

$$\begin{aligned}\lim_{t\to 3}\frac{s(3) - s(t)}{3 - t} &= \lim_{t\to 3}\frac{-4.9(3)^2 + 200 - \left(-4.9t^2 + 200\right)}{3 - t}\\ &= \lim_{t\to 3}\frac{4.9\left(t^2 - 9\right)}{3 - t}\\ &= \lim_{t\to 3}\frac{4.9(t - 3)(t + 3)}{3 - t}\\ &= \lim_{t\to 3}\left[-4.9(t + 3)\right]\\ &= -29.4 \text{ m/sec}\end{aligned}$$

The object is falling about 29.4 m/sec.

109. Let $f(x) = 1/x$ and $g(x) = -1/x$. $\lim_{x\to 0} f(x)$ and $\lim_{x\to 0} g(x)$ do not exist. However,

$$\lim_{x\to 0}\left[f(x) + g(x)\right] = \lim_{x\to 0}\left[\frac{1}{x} + \left(-\frac{1}{x}\right)\right] = \lim_{x\to 0}[0] = 0$$

and therefore does not exist.

111. Given $f(x) = b$, show that for every $\varepsilon > 0$ there exists a $\delta > 0$ such that $|f(x) - b| < \varepsilon$ whenever $|x - c| < \delta$. Because $|f(x) - b| = |b - b| = 0 < \varepsilon$ for every $\varepsilon > 0$, any value of $\delta > 0$ will work.

113. If $b = 0$, the property is true because both sides are equal to 0. If $b \neq 0$, let $\varepsilon > 0$ be given. Because $\lim_{x\to c} f(x) = L$, there exists $\delta > 0$ such that

$|f(x) - L| < \varepsilon/|b|$ whenever $0 < |x - c| < \delta$.

So, whenever $0 < |x - c| < \delta$, we have

$$|b||f(x) - L| < \varepsilon \quad\text{or}\quad |bf(x) - bL| < \varepsilon$$

which implies that $\lim_{x\to c}\left[bf(x)\right] = bL$.

115.

$$\begin{aligned}-M|f(x)| &\le f(x)g(x) \le M|f(x)|\\ \lim_{x\to c}\left(-M|f(x)|\right) &\le \lim_{x\to c}\left[f(x)g(x)\right] \le \lim_{x\to c}\left(M|f(x)|\right)\\ -M(0) &\le \lim_{x\to c}\left[f(x)g(x)\right] \le M(0)\\ 0 &\le \lim_{x\to c}\left[f(x)g(x)\right] \le 0\end{aligned}$$

Therefore, $\lim_{x\to c}\left[f(x)g(x)\right] = 0$.

117. Let

$$f(x) = \begin{cases} 4, & \text{if } x \ge 0\\ -4, & \text{if } x < 0\end{cases}$$

$\lim_{x\to 0}|f(x)| = \lim_{x\to 0} 4 = 4.$

$\lim_{x\to 0} f(x)$ does not exist because for $x < 0$, $f(x) = -4$ and for $x \ge 0$, $f(x) = 4$.

119. The limit does not exist because the function approaches 1 from the right side of 0 and approaches −1 from the left side of 0.

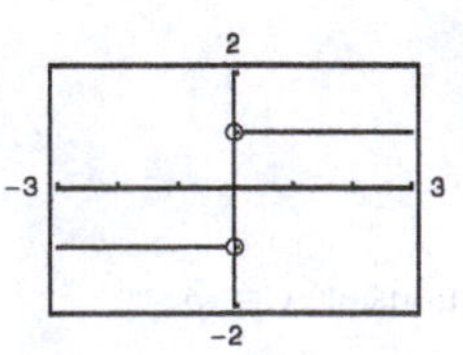

121. True.

123. False. The limit does not exist because $f(x)$ approaches 3 from the left side of 2 and approaches 0 from the right side of 2.

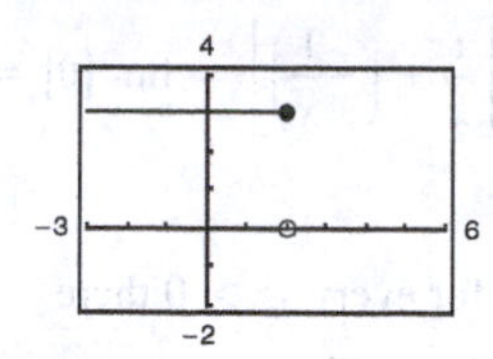

125. $\lim_{x\to 0}\frac{1-\cos x}{x} = \lim_{x\to 0}\frac{1-\cos x}{x}\cdot\frac{1+\cos x}{1+\cos x}$

$= \lim_{x\to 0}\frac{1-\cos^2 x}{x(1+\cos x)} = \lim_{x\to 0}\frac{\sin^2 x}{x(1+\cos x)}$

$= \lim_{x\to 0}\frac{\sin x}{x}\cdot\frac{\sin x}{1+\cos x}$

$= \left[\lim_{x\to 0}\frac{\sin x}{x}\right]\left[\lim_{x\to 0}\frac{\sin x}{1+\cos x}\right]$

$= (1)(0) = 0$

127. $f(x) = \frac{\sec x - 1}{x^2}$

(a) The domain of f is all $x \neq 0, \pi/2 + n\pi$.

(b)

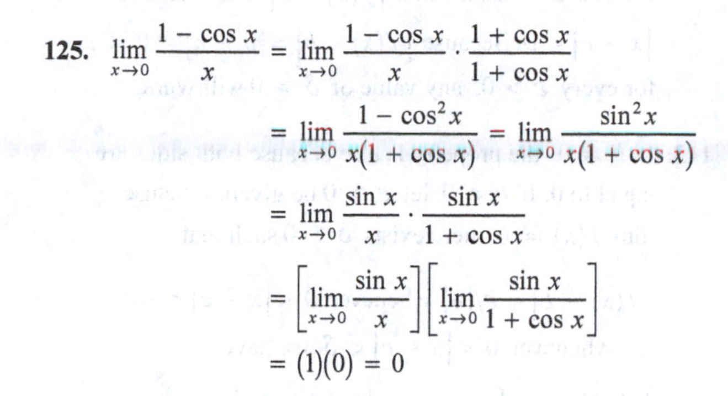

The domain is not obvious. The hole at $x = 0$ is not apparent.

(c) $\lim_{x\to 0} f(x) = \frac{1}{2}$

(d) $\frac{\sec x - 1}{x^2} = \frac{\sec x - 1}{x^2}\cdot\frac{\sec x + 1}{\sec x + 1}$

$= \frac{\sec^2 x - 1}{x^2(\sec x + 1)} = \frac{\tan^2 x}{x^2(\sec x + 1)}$

$= \frac{1}{\cos^2 x}\left(\frac{\sin^2 x}{x^2}\right)\frac{1}{\sec x + 1}$

So, $\lim_{x\to 0}\frac{\sec x - 1}{x^2} = \lim_{x\to 0}\frac{1}{\cos^2 x}\left(\frac{\sin^2 x}{x^2}\right)\frac{1}{\sec x + 1}$

$= 1(1)\left(\frac{1}{2}\right) = \frac{1}{2}.$

Section 2.4 Continuity and One-Sided Limits

1. A function f is continuous at a point c if there is no interruption of the graph at c.

3. The limit exists because the limit from the left and the limit from the right and equivalent.

5. (a) $\lim_{x\to 4^+} f(x) = 3$

(b) $\lim_{x\to 4^-} f(x) = 3$

(c) $\lim_{x\to 4} f(x) = 3$

The function is continuous at $x = 4$ and is continuous on $(-\infty, \infty)$.

7. (a) $\lim_{x\to 3^+} f(x) = 0$

(b) $\lim_{x\to 3^-} f(x) = 0$

(c) $\lim_{x\to 3} f(x) = 0$

The function is NOT continuous at $x = 3$.

9. (a) $\lim_{x\to 2^+} f(x) = -3$

(b) $\lim_{x\to 2^-} f(x) = 3$

(c) $\lim_{x\to 2} f(x)$ does not exist

The function is NOT continuous at $x = 2$.

11. $\lim_{x\to 8^+}\frac{1}{x+8} = \frac{1}{8+8} = \frac{1}{16}$

13. $\lim_{x\to 5^+}\frac{x-5}{x^2-25} = \lim_{x\to 5^+}\frac{x-5}{(x+5)(x-5)}$

$= \lim_{x\to 5^+}\frac{1}{x+5} = \frac{1}{10}$

15. $\lim_{x\to -3^-}\frac{x}{\sqrt{x^2-9}}$ does not exist because $\frac{x}{\sqrt{x^2-9}}$ decreases without bound as $x \to -3^-$.

17. $\lim_{x\to 0^-}\frac{|x|}{x} = \lim_{x\to 0^-}\frac{-x}{x} = -1$

19. $\lim_{\Delta x \to 0^-} \dfrac{\frac{1}{x + \Delta x} - \frac{1}{x}}{\Delta x} = \lim_{\Delta x \to 0^-} \dfrac{x - (x + \Delta x)}{x(x + \Delta x)} \cdot \dfrac{1}{\Delta x}$

$= \lim_{\Delta x \to 0^-} \dfrac{-\Delta x}{x(x + \Delta x)} \cdot \dfrac{1}{\Delta x}$

$= \lim_{\Delta x \to 0^-} \dfrac{-1}{x(x + \Delta x)}$

$= \dfrac{-1}{x(x + 0)} = -\dfrac{1}{x^2}$

21. $\lim_{x \to 3^-} f(x) = \lim_{x \to 3^-} \dfrac{x + 2}{2} = \dfrac{5}{2}$

23. $\lim_{x \to \pi} \cot x$ does not exist because

$\lim_{x \to \pi^+} \cot x$ and $\lim_{x \to \pi^-} \cot x$ do not exist.

25. $\lim_{x \to 4^-} (5[\![x]\!] - 7) = 5(3) - 7 = 8$

$([\![x]\!] = 3$ for $3 \le x < 4)$

27. $\lim_{x \to -1} \left(\left[\!\!\left[\dfrac{x}{3} \right]\!\!\right] + 3 \right) = \left[\!\!\left[-\dfrac{1}{3} \right]\!\!\right] + 3 = -1 + 3 = 2$

29. $\lim_{x \to 3^+} \ln(x - 3) = \ln 0$

does not exist.

31. $\lim_{x \to 2^-} \ln[x^2(3 - x)] = \ln[4(1)] = \ln 4$

33. $f(x) = \dfrac{1}{x^2 - 4}$

has discontinuities at $x = -2$ and $x = 2$ because $f(-2)$ and $f(2)$ are not defined.

35. $f(x) = \dfrac{[\![x]\!]}{2} + x$

has discontinuities at each integer k because $\lim_{x \to k^-} f(x) \ne \lim_{x \to k^+} f(x)$.

37. $g(x) = \sqrt{49 - x^2}$ is continuous on $[-7, 7]$.

39. $\lim_{x \to 0^-} f(x) = 3 = \lim_{x \to 0^+} f(x)$. f is continuous on $[-1, 4]$.

41. $f(x) = \dfrac{4}{x - 6}$ has a nonremovable discontinuity at $x = 6$ because $\lim_{x \to 6} f(x)$ does not exist.

43. $f(x) = 3x - \cos x$ is continuous for all real x.

45. $f(x) = \dfrac{x}{x^2 - x}$ is not continuous at $x = 0, 1$.

Because $\dfrac{x}{x^2 - x} = \dfrac{1}{x - 1}$ for $x \ne 0$, $x = 0$ is a removable discontinuity, whereas $x = 1$ is a nonremovable discontinuity.

47. $f(x) = \dfrac{x + 2}{x^2 - 3x - 10} = \dfrac{x + 2}{(x + 2)(x - 5)}$

has a nonremovable discontinuity at $x = 5$ because $\lim_{x \to 5} f(x)$ does not exist, and has a removable discontinuity at $x = -2$ because

$\lim_{x \to -2} f(x) = \lim_{x \to -2} \dfrac{1}{x - 5} = -\dfrac{1}{7}$.

49. $f(x) = \dfrac{|x + 7|}{x + 7}$

has a nonremovable discontinuity at $x = -7$ because $\lim_{x \to -7} f(x)$ does not exist.

51. $f(x) = \begin{cases} \dfrac{x}{2} + 1, & x \le 2 \\ 3 - x, & x > 2 \end{cases}$

has a **possible** discontinuity at $x = 2$.

1. $f(2) = \dfrac{2}{2} + 1 = 2$

2. $\left.\begin{array}{l} \lim_{x \to 2^-} f(x) = \lim_{x \to 2^-} \left(\dfrac{x}{2} + 1 \right) = 2 \\ \lim_{x \to 2^+} f(x) = \lim_{x \to 2^+} (3 - x) = 1 \end{array}\right\} \lim_{x \to 2} f(x)$ does not exist.

Therefore, f has a nonremovable discontinuity at $x = 2$.

53. $f(x) = \begin{cases} \tan \dfrac{\pi x}{4}, & |x| < 1 \\ x, & |x| \ge 1 \end{cases}$

$= \begin{cases} \tan \dfrac{\pi x}{4}, & -1 < x < 1 \\ x, & x \le -1 \text{ or } x \ge 1 \end{cases}$

has **possible** discontinuities at $x = -1$, $x = 1$.

1. $f(-1) = -1$ $\qquad f(1) = 1$
2. $\lim\limits_{x \to -1} f(x) = -1$ $\qquad \lim\limits_{x \to 1} f(x) = 1$
3. $f(-1) = \lim\limits_{x \to -1} f(x)$ $\qquad f(1) = \lim\limits_{x \to 1} f(x)$

f is continuous at $x = \pm 1$, therefore, f is continuous for all real x.

55. $f(x) = \begin{cases} \ln(x + 1), & x \ge 0 \\ 1 - x^2, & x < 0 \end{cases}$

has a **possible** discontinuity at $x = 0$.

1. $f(0) = \ln(0 + 1) = \ln 1 = 0$
2. $\left.\begin{array}{l} \lim\limits_{x \to 0^-} f(x) = 1 - 0 = 1 \\ \lim\limits_{x \to 0^+} f(x) = 0 \end{array}\right\} \lim\limits_{x \to 0} f(x)$ does not exist.

So, f has a nonremovable discontinuity at $x = 0$.

57. $f(x) = \csc 2x$ has nonremovable discontinuities at integer multiples of $\pi/2$.

59. $f(x) = [\![x - 8]\!]$ has nonremovable discontinuities at each integer k.

61. $f(2) = 8$

Find a so that $\lim\limits_{x \to 2^+} ax^2 = 8 \Rightarrow a = \frac{8}{2^2} = 2$.

63. $\lim\limits_{x \to a} g(x) = \lim\limits_{x \to a} \dfrac{x^2 - a^2}{x - a} = \lim\limits_{x \to a} (x + a) = 2a$

Find a such $2a = 8 \Rightarrow a = 4$.

65. $f(1) = \arctan(1 - 1) + 2 = 2$

Find a such that $\lim\limits_{x \to 1^-} \left(ae^{x-1} + 3\right) = 2$

$ae^{1-1} + 3 = 2$

$a + 3 = 2$

$a = -1.$

67. $f(g(x)) = \dfrac{1}{(x^2 + 5) - 6} = \dfrac{1}{x^2 - 1}$

Nonremovable discontinuities at $x = \pm 1$

69. $f(g(x)) = \tan \dfrac{x}{2}$

Not continuous at $x = \pm\pi, \pm 3\pi, \pm 5\pi, \ldots$ Continuous on the open intervals $\ldots, (-3\pi, -\pi), (-\pi, \pi), (\pi, 3\pi), \ldots$

71. $y = [\![x]\!] - x$

Nonremovable discontinuity at each integer

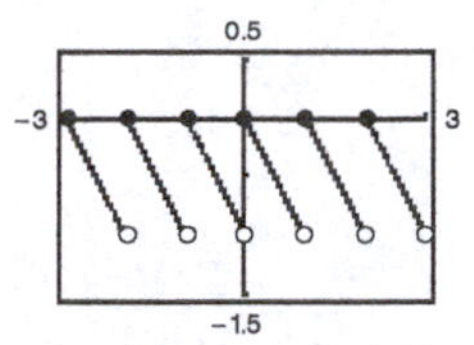

73. $g(x) = \begin{cases} x^2 - 3x, & x > 4 \\ 2x - 5, & x \le 4 \end{cases}$

Nonremovable discontinuity at $x = 4$

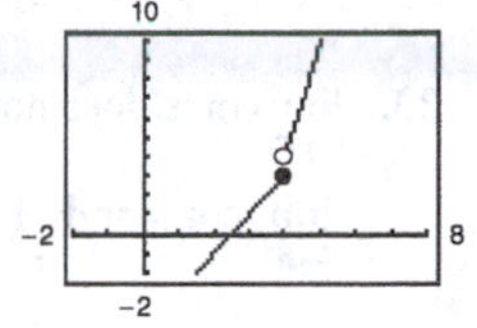

75. $f(x) = \dfrac{x}{x^2 + x + 2}$

Continuous on $(-\infty, \infty)$

77. $f(x) = 3 - \sqrt{x}$

Continuous on $[0, \infty)$

79. $f(x) = \sec \dfrac{\pi x}{4}$

Continuous on:

$\ldots, (-6, -2), (-2, 2), (2, 6), (6, 10), \ldots$

81. $f(x) = \begin{cases} \dfrac{x^2 - 1}{x - 1}, & x \ne 1 \\ 2, & x = 1 \end{cases}$

Since $\lim\limits_{x \to 1} f(x) = \lim\limits_{x \to 1} \dfrac{x^2 - 1}{x - 1} = \lim\limits_{x \to 1} \dfrac{(x - 1)(x + 1)}{x - 1}$

$= \lim\limits_{x \to 1} (x + 1) = 2,$

f is continuous on $(-\infty, \infty)$.

83. $f(x) = \frac{1}{12}x^4 - x^3 + 4$ is continuous on the interval $[1, 2]$. $f(1) = \frac{37}{12}$ and $f(2) = -\frac{8}{3}$. By the Intermediate Value Theorem, there exists a number c in $[1, 2]$ such that $f(c) = 0$.

85. $f(x) = x^2 - 2 - \cos x$ is continuous on $[0, \pi]$. $f(0) = -3$ and $f(\pi) = \pi^2 - 1 \approx 8.87 > 0$. By the Intermediate Value Theorem, $f(c) = 0$ for at least one value of c between 0 and π.

87. $h(x) = -2e^{-x/2} \cos 2x$ is continuous on the interval $\left[0, \frac{\pi}{2}\right]$. $h(0) = -2 < 0$ and $h\left(\frac{\pi}{2}\right) \approx 0.91 > 0$.

By the Intermediate Value Theorem, there exists a number c in $\left[0, \frac{\pi}{2}\right]$ such that $h(c) = 0$.

89. Consider the intervals $[1, 3]$ and $[3, 5]$ for

$f(x) = (x - 3)^2 = 2$.

$f(1) = 2 > 0$ and $f(3) = -2 < 0$, so f has at least one zero in $[1, 3]$.

$f(3) = -2 < 0$ and $f(5) = 2 > 0$, so f has at least one zero in $[3, 5]$.

So, f has at least two zeros in $[1, 5]$.

91. $f(x) = x^3 + x - 1$

$f(x)$ is continuous on $[0, 1]$.

$f(0) = -1$ and $f(1) = 1$

By the Intermediate Value Theorem, $f(c) = 0$ for at least one value of c between 0 and 1. Using a graphing utility to zoom in on the graph of $f(x)$, you find that $x \approx 0.68$. Using the *root* feature, you find that $x \approx 0.6823$.

93. $f(x) = \sqrt{x^2 + 17x + 19} - 6$

f is continuous on $[0, 1]$.

$f(0) = \sqrt{19} - 6 \approx -1.64 < 0$

$f(1) = \sqrt{37} - 6 \approx 0.08 > 0$

By the Intermediate Value Theorem, $f(c) = 0$ for at least one value of c between 0 and 1. Using a graphing utility to zoom in on the graph of $f(x)$, you find that $x \approx 0.95$. Using the *root* feature, you find that $x \approx 0.9472$.

95. $g(t) = 2 \cos t - 3t$

g is continuous on $[0, 1]$.

$g(0) = 2 > 0$ and $g(1) \approx -1.9 < 0$.

By the Intermediate Value Theorem, $g(c) = 0$ for at least one value of c between 0 and 1. Using a graphing utility to zoom in on the graph of $g(t)$, you find that $t \approx 0.56$. Using the *root* feature, you find that $t \approx 0.5636$.

97. $f(x) = x + e^x - 3$

f is continuous on $[0, 1]$.

$f(0) = e^0 - 3 = -2 < 0$ and

$f(1) = 1 + e - 3 = e - 2 > 0$.

By the Intermediate Value Theorem, $f(c) = 0$ for at least one value of c between 0 and 1. Using a graphing utility to zoom in on the graph of $f(x)$, you find that $x \approx 0.79$. Using the *root* feature, you find that $x \approx 0.7921$.

99. $f(x) = x^2 + x - 1$

f is continuous on $[0, 5]$.

$f(0) = -1$ and $f(5) = 29$

$-1 < 11 < 29$

The Intermediate Value Theorem applies.

$$x^2 + x - 1 = 11$$
$$x^2 + x - 12 = 0$$
$$(x + 4)(x - 3) = 0$$
$$x = -4 \text{ or } x = 3$$

$c = 3$ ($x = -4$ is not in the interval.)

So, $f(3) = 11$.

101. $f(x) = \sqrt{x + 7} - 2$

f is continuous on $[0, 5]$.

$f(0) = \sqrt{7} - 2 \approx 0.6458 < 1$

$f(5) = \sqrt{12} - 2 \approx 1.4641 > 1$

The Intermediate Value Theorem applies.

$$\sqrt{x + 7} - 2 = 1$$
$$\sqrt{x + 7} = 3$$
$$x + 7 = 9$$
$$x = 2$$
$$c = 2$$

So, $f(2) = 1$.

103. $f(x) = \dfrac{x - x^3}{x - 4}$

f is continuous on $[1, 3]$. The nonremovable discontinuity, $x = 4$, lies outside the interval.

$f(1) = \dfrac{1 - 1}{1 - 4} = 0 < 3$

$f(3) = 24 > 3$

The Intermediate Value Theorem applies.

$$\frac{x - x^3}{x - 4} = 3$$
$$x - x^3 = 3x - 12$$
$$x^3 + 2x - 12 = 0$$
$$(x - 2)(x^2 + 2x + 6) = 0$$
$$x = 2$$

$(x^2 + 2x + 6$ has no real solution.$)$

$$c = 2$$

So, $f(2) = 3$.

105. Answers will vary. *Sample answer:*

$f(x) = \dfrac{1}{(x - a)(x - b)}$

107. If f and g are continuous for all real x, then so is $f + g$ (Theorem 1.11, part 2). However, f/g might not be continuous if $g(x) = 0$. For example, let $f(x) = x$ and $g(x) = x^2 - 1$. Then f and g are continuous for all real x, but f/g is not continuous at $x = \pm 1$.

109. True

1. $f(c) = L$ is defined.
2. $\lim_{x \to c} f(x) = L$ exists.
3. $f(c) = \lim_{x \to c} f(x)$

All of the conditions for continuity are met.

111. False. $f(x) = \cos x$ has two zeros in $[0, 2\pi]$. However, $f(0)$ and $f(2\pi)$ have the same sign.

113. False. A rational function can be written as $P(x)/Q(x)$ where P and Q are polynomials of degree m and n, respectively. It can have, at most, n discontinuities.

115. The functions agree for integer values of x:

$$\left.\begin{aligned} g(x) &= 3 - [\![-x]\!] = 3 - (-x) = 3 + x \\ f(x) &= 3 + [\![x]\!] = 3 + x \end{aligned}\right\} \text{for } x \text{ an integer}$$

However, for non-integer values of x, the functions differ by 1.

$f(x) = 3 + [\![x]\!] = g(x) - 1 = 2 - [\![-x]\!]$.

For example,

$f\left(\frac{1}{2}\right) = 3 + 0 = 3$, $g\left(\frac{1}{2}\right) = 3 - (-1) = 4$.

117. $C(t) = 10 - 7.5[\![1 - t]\!]$, $t > 0$

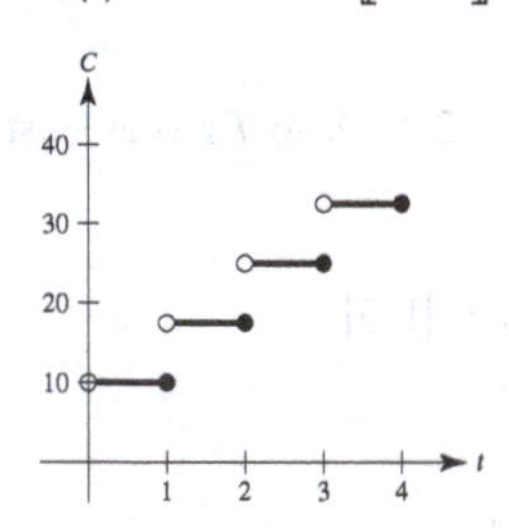

There is a nonremovable discontinuity at every integer value of t, or gigabyte.

119. Let $s(t)$ be the position function for the run up to the campsite. $s(0) = 0$ ($t = 0$ corresponds to 8:00 A.M., $s(20) = k$ (distance to campsite)). Let $r(t)$ be the position function for the run back down the mountain: $r(0) = k$, $r(10) = 0$. Let $f(t) = s(t) - r(t)$.

When $t = 0$ (8:00 A.M.),

$f(0) = s(0) - r(0) = 0 - k < 0$.

When $t = 10$ (8:00 A.M.), $f(10) = s(10) - r(10) > 0$.

Because $f(0) < 0$ and $f(10) > 0$, then there must be a value t in the interval $[0, 10]$ such that $f(t) = 0$. If $f(t) = 0$, then $s(t) - r(t) = 0$, which gives us $s(t) = r(t)$. Therefore, at some time t, where $0 \le t \le 10$, the position functions for the run up and the run down are equal.

121. Suppose there exists x_1 in $[a, b]$ such that $f(x_1) > 0$ and there exists x_2 in $[a, b]$ such that $f(x_2) < 0$. Then by the Intermediate Value Theorem, $f(x)$ must equal zero for some value of x in $[x_1, x_2]$ (or $[x_2, x_1]$ if $x_2 < x_1$). So, f would have a zero in $[a, b]$, which is a contradiction. Therefore, $f(x) > 0$ for all x in $[a, b]$ or $f(x) < 0$ for all x in $[a, b]$.

123. If $x = 0$, then $f(0) = 0$ and $\lim_{x\to 0} f(x) = 0$. So, f is continuous at $x = 0$.

If $x \neq 0$, then $\lim_{t\to x} f(t) = 0$ for x rational, whereas $\lim_{t\to x} f(t) = \lim_{t\to x} kt = kx \neq 0$ for x irrational. So, f is not continuous for all $x \neq 0$.

125. (a)

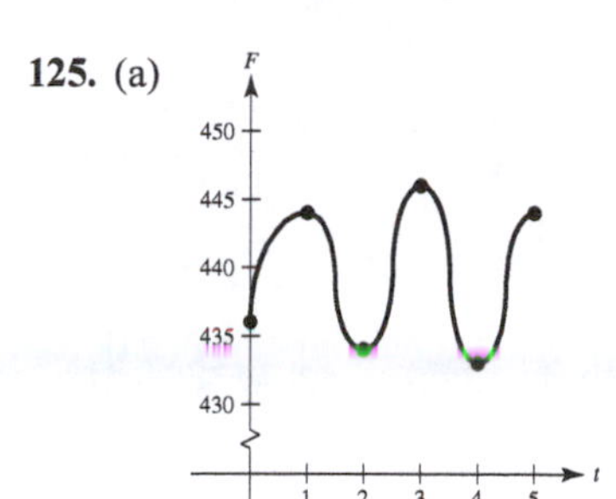

(b) No. The frequency is oscillating.

127. $f(x) = \begin{cases} 1 - x^2, & x \le c \\ x, & x > c \end{cases}$

f is continuous for $x < c$ and for $x > c$. At $x = c$, you need $1 - c^2 = c$. Solving $c^2 + c - 1$, you obtain

$$c = \frac{-1 \pm \sqrt{1+4}}{2} = \frac{-1 \pm \sqrt{5}}{2}.$$

129. $f(x) = \dfrac{\sqrt{x + c^2} - c}{x}, c > 0$

Domain: $x + c^2 \ge 0 \Rightarrow x \ge -c^2$ and $x \neq 0, \left[-c^2, 0\right) \cup (0, \infty)$

$$\lim_{x\to 0} \frac{\sqrt{x+c^2} - c}{x} = \lim_{x\to 0} \frac{\sqrt{x+c^2} - c}{x} \cdot \frac{\sqrt{x+c^2} + c}{\sqrt{x+c^2} + c}$$

$$= \lim_{x\to 0} \frac{(x + c^2) - c^2}{x\left[\sqrt{x+c^2} + c\right]}$$

$$= \lim_{x\to 0} \frac{1}{\sqrt{x+c^2} + c} = \frac{1}{2c}$$

Define $f(0) = 1/(2c)$ to make f continuous at $x = 0$.

131. $h(x) = x[\![x]\!]$

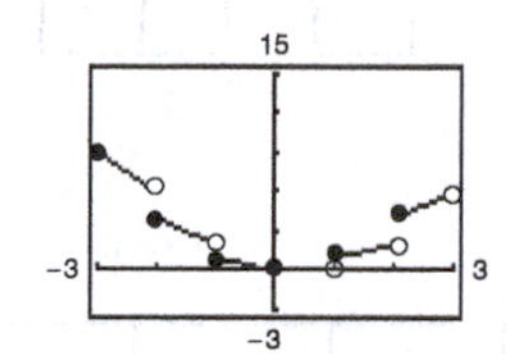

h has nonremovable discontinuities at $x = \pm 1, \pm 2, \pm 3, \ldots$.

133. The statement is true.

If $y \ge 0$ and $y \le 1$, then $y(y - 1) \le 0 \le x^2$, as desired. So assume $y > 1$. There are now two cases.

Case 1:

If $x \le y - \frac{1}{2}$, then $2x + 1 \le 2y$ and

$$\begin{aligned} y(y-1) &= y(y+1) - 2y \\ &\le (x+1)^2 - 2y \\ &= x^2 + 2x + 1 - 2y \\ &\le x^2 + 2y - 2y \\ &= x^2 \end{aligned}$$

Case 2:

If $x \ge y - \frac{1}{2}$

$$\begin{aligned} x^2 &\ge \left(y - \tfrac{1}{2}\right)^2 \\ &= y^2 - y + \tfrac{1}{4} \\ &> y^2 - y \\ &= y(y-1) \end{aligned}$$

In both cases, $y(y - 1) \le x^2$.

Section 2.5 Infinite Limits

1. A limit in which $f(x)$ increases or decreases without bound as x approaches c is called an infinite limit. ∞ is not a number. Rather, the symbol

$\lim_{x\to c} f(x) = \infty$

Says how the limit fails to exist.

3. $\displaystyle\lim_{x\to -2^+} 2\left|\frac{x}{x^2 - 4}\right| = \infty$

$\displaystyle\lim_{x\to -2^-} 2\left|\frac{x}{x^2 - 4}\right| = \infty$

5. $\displaystyle\lim_{x\to -2^+} \tan\frac{\pi x}{4} = -\infty$

$\displaystyle\lim_{x\to -2^-} \tan\frac{\pi x}{4} = \infty$

7. $f(x) = \dfrac{1}{x - 4}$

As x approaches 4 from the left, $x - 4$ is a small negative number. So, $\lim_{x \to 4^-} f(x) = -\infty$

As x approaches 4 from the right, $x - 4$ is a small positive number. So, $\lim_{x \to 4^+} f(x) = \infty$

9. $f(x) = \dfrac{1}{(x - 4)^2}$

As x approaches 4 from the left or right, $(x - 4)^2$ is a small positive number. So,

$\lim_{x \to 4^+} f(x) = \lim_{x \to 4^-} f(x) = \infty.$

11. $f(x) = \dfrac{1}{x^2 - 9}$

x	-3.5	-3.1	-3.01	-3.001	-2.999	-2.99	-2.9	-2.5
$f(x)$	0.308	1.639	16.64	166.6	-166.7	-16.69	-1.695	-0.364

$\lim_{x \to -3^-} f(x) = \infty$

$\lim_{x \to -3^+} f(x) = -\infty$

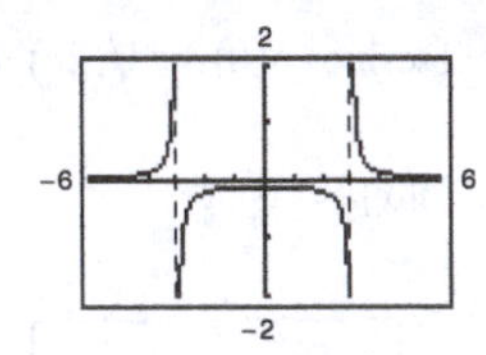

13. $f(x) = \dfrac{x^2}{x^2 - 9}$

x	-3.5	-3.1	-3.01	-3.001	-2.999	-2.99	-2.9	-2.5
$f(x)$	3.769	15.75	150.8	1501	-1499	-149.3	-14.25	-2.273

$\lim_{x \to -3^-} f(x) = \infty$

$\lim_{x \to -3^+} f(x) = -\infty$

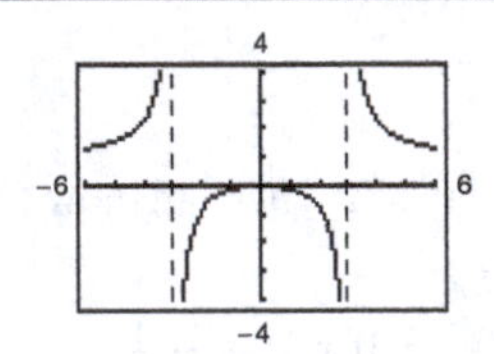

15. $f(x) = \cot \dfrac{\pi x}{3}$

x	-3.5	-3.1	-3.01	-3.001	-2.999	-2.99	-2.9	-2.5
$f(x)$	-1.7321	-9.514	-95.49	-954.9	954.9	95.49	9.514	1.7321

$\lim_{x \to -3^-} f(x) = -\infty$

$\lim_{x \to -3^+} f(x) = \infty$

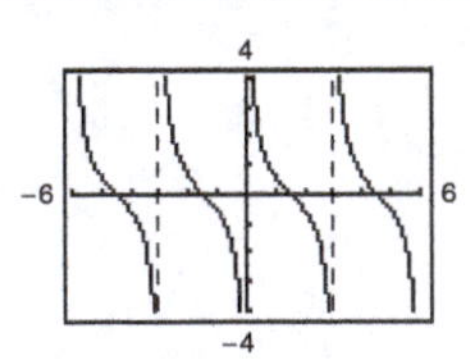

17. $f(x) = \dfrac{x^2}{x^2 - 4} = \dfrac{x^2}{(x + 2)(x - 2)}$

$\lim_{x \to -2^-} \dfrac{x^2}{x^2 - 4} = \infty$ and $\lim_{x \to -2^+} \dfrac{x^2}{x^2 - 4} = -\infty$

Therefore, $x = -2$ is a vertical asymptote.

$\lim_{x \to 2^-} \dfrac{x^2}{x^2 - 4} = -\infty$ and $\lim_{x \to 2^+} \dfrac{x^2}{x^2 - 4} = \infty$

Therefore, $x = 2$ is a vertical asymptote.

19. $f(x) = \dfrac{3}{x^2 + x - 2} = \dfrac{3}{(x + 2)(x - 1)}$

$\lim_{x \to -2^-} \dfrac{3}{x^2 + x - 2} = \infty$ and $\lim_{x \to -2^+} \dfrac{3}{x^2 + x - 2} = -\infty$

Therefore, $x = -2$ is a vertical asymptote.

$\lim_{x \to 1^-} \dfrac{3}{x^2 + x - 2} = -\infty$ and $\lim_{x \to 1^+} \dfrac{3}{x^2 + x - 2} = \infty$

Therefore, $x = 1$ is a vertical asymptote.

21. $f(x) = \dfrac{4(x^2 + x - 6)}{x(x^3 - 2x^2 - 9x + 18)} = \dfrac{4(x + 3)(x - 2)}{x(x - 2)(x^2 - 9)}$

$= \dfrac{4}{x(x - 3)},\ x \neq -3, 2$

$\lim_{x \to 0^-} f(x) = \infty$ and $\lim_{x \to 0^+} f(x) = -\infty$

Therefore, $x = 0$ is a vertical asymptote.

$\lim_{x \to 3^-} f(x) = -\infty$ and $\lim_{x \to 3^+} f(x) = \infty$

Therefore, $x = 3$ is a vertical asymptote.

$\lim_{x \to 2} f(x) = \dfrac{4}{2(2 - 3)} = -2$

and

$\lim_{x \to -3} f(x) = \dfrac{4}{-3(-3 - 3)} = \dfrac{2}{9}$

Therefore, the graph has holes at $x = 2$ and $x = -3$.

23. $f(x) = \dfrac{e^{-2x}}{x - 1}$

$\lim_{x \to 1^-} f(x) = -\infty$ and $\lim_{x \to 1^+} = \infty$

Therefore, $x = 1$ is a vertical asymptote.

25. $h(t) = \dfrac{\ln(t^2 + 1)}{t + 2}$

$\lim_{t \to -2^-} h(t) = -\infty$ and $\lim_{t \to -2^+} = \infty$

Therefore, $t = -2$ is a vertical asymptote.

27. $f(x) = \dfrac{1}{e^x - 1}$

$\lim_{x \to 0^-} f(x) = -\infty$ and $\lim_{x \to 0^+} f(x) = \infty$

Therefore, $x = 0$ is a vertical asymptote.

29. $f(x) = \csc \pi x = \dfrac{1}{\sin \pi x}$

Let n be any integer.

$\lim_{x \to n} f(x) = -\infty$ or ∞

Therefore, the graph has vertical asymptotes at $x = n$.

31. $s(t) = \dfrac{t}{\sin t}$

$\sin t = 0$ for $t = n\pi$, where n is an integer.

$\lim_{t \to n\pi} s(t) = \infty$ or $-\infty$ (for $n \neq 0$)

Therefore, the graph has vertical asymptotes at $t = n\pi$, for $n \neq 0$.

$\lim_{t \to 0} s(t) = 1$

Therefore, the graph has a hole at $t = 0$.

33. $\lim_{x \to -1} \dfrac{x^2 - 1}{x + 1} = \lim_{x \to -1} (x - 1) = -2$

Removable discontinuity at $x = -1$

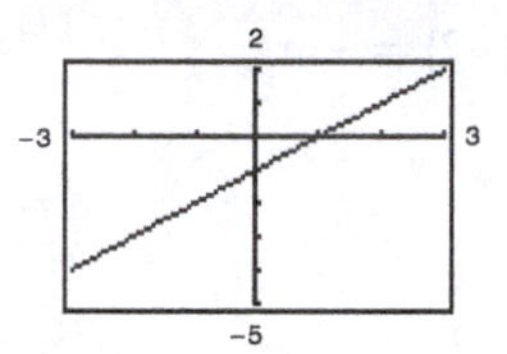

35. $\lim_{x \to -1^-} \dfrac{\cos(x^2 - 1)}{x + 1} = -\infty$

$\lim_{x \to -1^+} \dfrac{\cos(x^2 - 1)}{x + 1} = \infty$

Vertical asymptote at $x = 1$

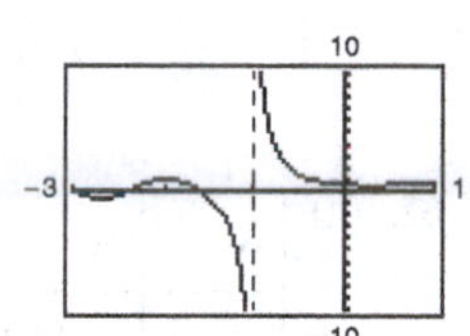

37. $\lim_{x \to 2^+} \dfrac{x}{x - 2} = \infty$

39. $\lim_{x \to -3^-} \dfrac{x + 3}{(x^2 + x - 6)} = \lim_{x \to -3^-} \dfrac{x + 3}{(x + 3)(x - 2)}$

$= \lim_{x \to -3^-} \dfrac{1}{x - 2} = -\dfrac{1}{5}$

41. $\lim_{x \to 0^-} \left(1 + \dfrac{1}{x}\right) = -\infty$

43. $\lim_{x \to -4^-} \left(x^2 + \dfrac{2}{x + 4}\right) = -\infty$

45. $\lim_{x \to 0^+} \left(\sin x + \dfrac{1}{x}\right) = \infty$

47. $\lim_{x \to 8^-} \dfrac{e^x}{(x - 8)^3} = -\infty$

49. $\lim_{x \to (\pi/2)^-} \ln|\cos x| = \ln\left|\cos \dfrac{\pi}{2}\right| = \ln 0 = -\infty$

51. $\lim_{x \to (1/2)^-} x \sec \pi x = \lim_{x \to (1/2)^-} \dfrac{x}{\cos \pi x} = \infty$

53. $f(x) = \dfrac{x^2 + x + 1}{x^3 - 1}$

$= \dfrac{x^2 + x + 1}{(x - 1)(x^2 + x + 1)}$

$\lim_{x \to 1^+} f(x) = \lim_{x \to 1^+} \dfrac{1}{x - 1} = \infty$

55. $\lim_{x\to c} f(x) = \infty$ and $\lim_{x\to c} g(x) = -2$

(a) $\lim_{x\to c}\left[f(x) + g(x)\right] = \infty - 2 = \infty$

(b) $\lim_{x\to c}\left[f(x)g(x)\right] = \infty(-2) = -\infty$

(c) $\lim_{x\to c} \dfrac{g(x)}{f(x)} = \dfrac{-2}{\infty} = 0$

57. One answer is

$$f(x) = \frac{x-3}{(x-6)(x+2)} = \frac{x-3}{x^2-4x-12}.$$

59.

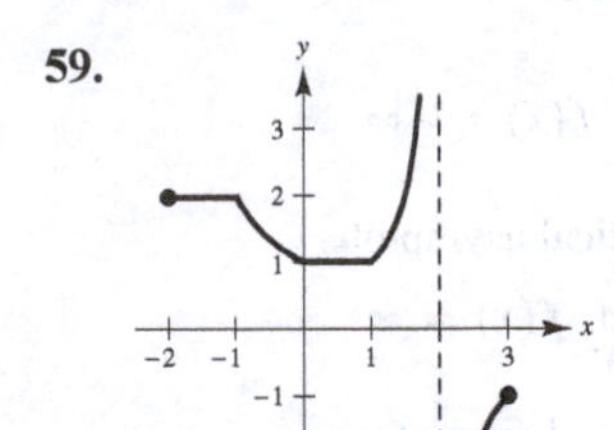

61. (a)

x	1	0.5	0.2	0.1	0.01	0.001	0.0001
$f(x)$	0.1585	0.0411	0.0067	0.0017	≈ 0	≈ 0	≈ 0

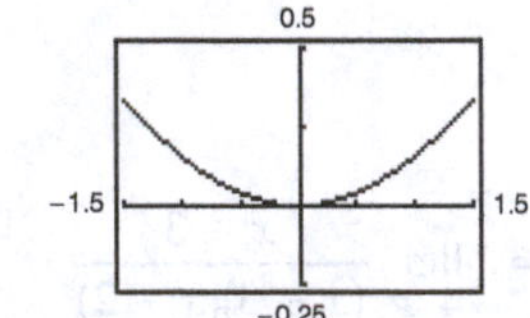

$$\lim_{x\to 0^+} \frac{x - \sin x}{x} = 0$$

(b)

x	1	0.5	0.2	0.1	0.01	0.001	0.0001
$f(x)$	0.1585	0.0823	0.0333	0.0167	0.0017	≈ 0	≈ 0

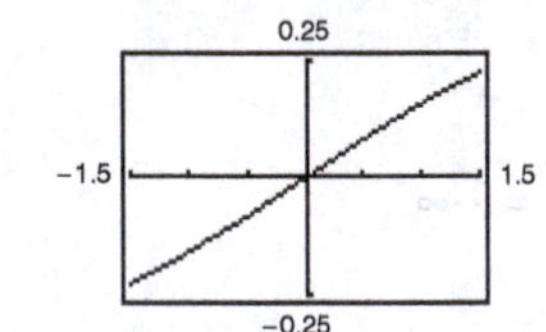

$$\lim_{x\to 0^+} \frac{x - \sin x}{x^2} = 0$$

(c)

x	1	0.5	0.2	0.1	0.01	0.001	0.0001
$f(x)$	0.1585	0.1646	0.1663	0.1666	0.1667	0.1667	0.1667

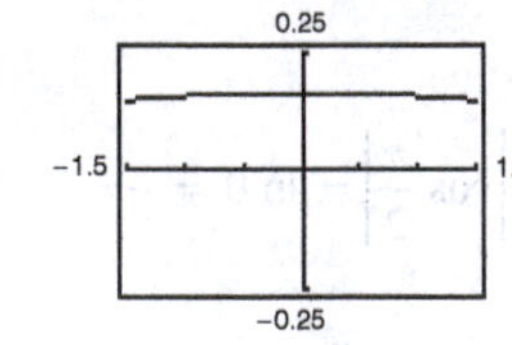

$$\lim_{x\to 0^+} \frac{x - \sin x}{x^3} = 0.1667 = 1/6$$

(d)

x	1	0.5	0.2	0.1	0.01	0.001	0.0001
$f(x)$	0.1585	0.3292	0.8317	1.6658	16.67	166.7	1667.0

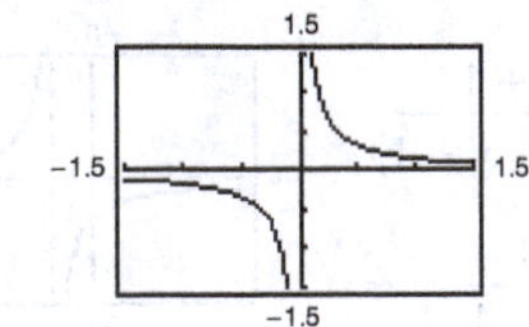

$\lim_{x\to 0^+} \dfrac{x - \sin x}{x^4} = \infty$ or $n > 3$, $\lim_{x\to 0^+} \dfrac{x - \sin x}{x^n} = \infty.$

63. (a) $r = \dfrac{2(7)}{\sqrt{625 - 49}} = \dfrac{7}{12}$ ft/sec

(b) $r = \dfrac{2(15)}{\sqrt{625 - 225}} = \dfrac{3}{2}$ ft/sec

(c) $\displaystyle\lim_{x \to 25^-} \frac{2x}{\sqrt{625 - x^2}} = \infty$

65. (a) $A = \dfrac{1}{2}bh - \dfrac{1}{2}r^2\theta$

$= \dfrac{1}{2}(10)(10 \tan \theta) - \dfrac{1}{2}(10)^2 \theta$

$= 50 \tan \theta - 50\theta$

Domain: $\left(0, \dfrac{\pi}{2}\right)$

(b)

θ	0.3	0.6	0.9	1.2	1.5
$f(\theta)$	0.47	4.21	18.0	68.6	630.1

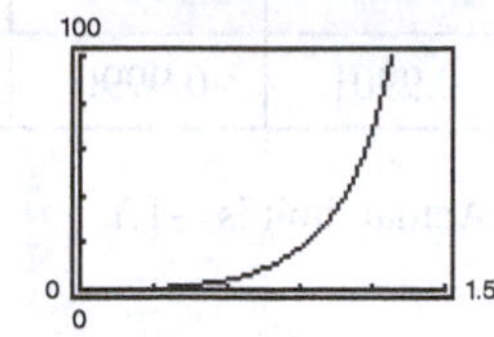

(c) $\displaystyle\lim_{\theta \to \pi/2^-} A = \infty$

67. True. The function is undefined at a vertical asymptote.

69. False. The graphs of $y = \tan x$, $y = \cot x$, $y = \sec x$ and $y = \csc x$ have vertical asymptotes.

71. Let $f(x) = \dfrac{1}{x^2}$ and $g(x) = \dfrac{1}{x^4}$, and $c = 0$.

$\displaystyle\lim_{x \to 0} \frac{1}{x^2} = \infty$ and $\displaystyle\lim_{x \to 0} \frac{1}{x^4} = \infty$, but $\displaystyle\lim_{x \to 0}\left(\frac{1}{x^2} - \frac{1}{x^4}\right) = \lim_{x \to 0}\left(\frac{x^2 - 1}{x^4}\right) = -\infty \neq 0.$

73. Given $\displaystyle\lim_{x \to c} f(x) = \infty$, let $g(x) = 1$. Then $\displaystyle\lim_{x \to c} \frac{g(x)}{f(x)} = 0$ by Theorem 1.15.

75. $f(x) = \dfrac{1}{x - 3}$ is defined for all $x > 3$. Let $M > 0$ be given. You need $\delta > 0$ such that $f(x) = \dfrac{1}{x - 3} > M$ whenever $3 < x < 3 + \delta$. Equivalently, $x - 3 < \dfrac{1}{M}$ whenever $|x - 3| < \delta, x > 3$. So take $\delta = \dfrac{1}{M}$. Then for $x > 3$ and $|x - 3| < \delta, \dfrac{1}{x - 3} > \dfrac{1}{8} = M$ and so $f(x) > M$. Thus, $\displaystyle\lim_{x \to 3^+} \frac{1}{x - 3} = \infty$.

77. $f(x) = \dfrac{3}{8 - x}$ is defined for all $x > 8$. Let $N < 0$ be given. You need $\delta > 0$ such that $f(x) = \dfrac{3}{8 - x} < N$ whenever $8 < x < 8 + \delta$. Equivalently, $\dfrac{8 - x}{3} > \dfrac{1}{N}$ whenever $|x - 8| < \delta, x > 8$. Equivalently, $|8 - x| < \dfrac{-3}{N}$ whenever $|x - 8| < \delta, x > 8$. So, let $\delta = \dfrac{-3}{N}$. Note that $\delta > 0$ because $N < 0$. Finally, for $|x - 8| < \delta$ and $x > 8$,

$\dfrac{1}{|x - 8|} > \dfrac{1}{\delta} = \dfrac{N}{-3}, \dfrac{-3}{|x - 8|} < N$, and $\dfrac{3}{8 - x} < N$. Thus, $\displaystyle\lim_{x \to 8^+} f(x) = -\infty$.

Review Exercises for Chapter 2

1. Calculus required. Using a graphing utility, you can estimate the length to be 8.3. Or, the length is slightly longer than the distance between the two points, approximately 8.25.

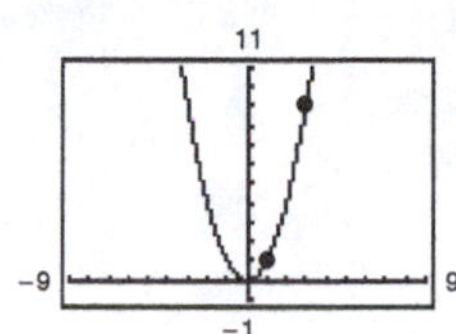

3. $f(x) = \dfrac{x-3}{x^2 - 7x + 12}$

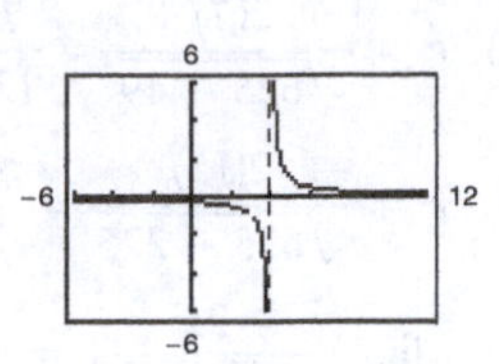

x	2.9	2.99	2.999	3	3.001	3.01	3.1
$f(x)$	−0.9091	−0.9901	−0.9990	?	−1.0010	−1.0101	−1.1111

$\lim_{x\to 3} f(x) \approx -1.0000$ (Actual limit is -1.)

5. $h(x) = \left[\!\left[-\dfrac{x}{2}\right]\!\right] + x^2$

(a) The limit does not exist at $x = 2$. The function approaches 3 from the left side of 2, but it approaches 2 from the right side of 2.

(b) $\lim_{x\to 1} h(x) = \left[\!\left[-\dfrac{1}{2}\right]\!\right] + x^2 = -1 + 1 = 0$

7. $\lim_{x\to 1}(x + 4) = 1 + 4 = 5$

Let $\varepsilon > 0$ be given. Choose $\delta = \varepsilon$. Then for $0 < |x - 1| < \delta = \varepsilon$, you have

$$|x - 1| < \varepsilon$$
$$|(x + 4) - 5| < \varepsilon$$
$$|f(x) - L| < \varepsilon.$$

9. $\lim_{x\to 2}(1 - x^2) = 1 - 2^2 = -3$

Let $\varepsilon > 0$ be given. You need

$$|1 - x^2 - (-3)| < \varepsilon \Rightarrow |x^2 - 4| = |x - 2||x + 2| < \varepsilon$$
$$\Rightarrow |x - 2| < \frac{1}{|x + 2|}\varepsilon$$

Assuming $1 < x < 3$, you can choose $\delta = \dfrac{\varepsilon}{5}$.

So, for $0 < |x - 2| < \delta = \dfrac{\varepsilon}{5}$, you have

$$|x - 2| < \frac{\varepsilon}{5} < \frac{\varepsilon}{|x + 2|}$$
$$|x - 2||x + 2| < \varepsilon$$
$$|x^2 - 4| < \varepsilon$$
$$|4 - x^2| < \varepsilon$$
$$|(1 - x^2) - (-3)| < \varepsilon$$
$$|f(x) - L| < \varepsilon.$$

11. $\lim_{x\to -6} x^2 = (-6)^2 = 36$

13. $\lim_{x\to 27}\left(\sqrt[3]{x} - 1\right)^4 = \left(\sqrt[3]{27} - 1\right)^4 = (3 - 1)^4 = 2^4 = 16$

15. $\lim_{x\to 4} \dfrac{4}{x - 1} = \dfrac{4}{4 - 1} = \dfrac{4}{3}$

17.
$$\lim_{x\to -3} \frac{2x^2 + 11x + 15}{x + 3} = \lim_{x\to -3} \frac{(2x + 5)(x + 3)}{x + 3} = \lim_{x\to -3}(2x + 5) = 2(-3) + 5 = -1$$

19.
$$\lim_{x\to 4} \frac{\sqrt{x - 3} - 1}{x - 4} = \lim_{x\to 4} \frac{\sqrt{x - 3} - 1}{x - 4} \cdot \frac{\sqrt{x - 3} + 1}{\sqrt{x - 3} + 1} = \lim_{x\to 4} \frac{(x - 3) - 1}{(x - 4)\left(\sqrt{x - 3} + 1\right)} = \lim_{x\to 4} \frac{1}{\sqrt{x - 3} + 1} = \frac{1}{2}$$

21.
$$\lim_{x\to 0} \frac{[1/(x + 1)] - 1}{x} = \lim_{x\to 0} \frac{1 - (x + 1)}{x(x + 1)} = \lim_{x\to 0} \frac{-1}{x + 1} = -1$$

23. $\lim_{x\to 0} \dfrac{1 - \cos x}{\sin x} = \lim_{x\to 0}\left(\dfrac{x}{\sin x}\right)\left(\dfrac{1 - \cos x}{x}\right) = (1)(0) = 0$

25. $\lim_{x\to 1} e^{x-1} \sin \dfrac{\pi x}{2} = e^0 \sin \dfrac{\pi}{2} = 1$

27.
$$\lim_{\Delta x\to 0} \frac{\sin[(\pi/6) + \Delta x] - (1/2)}{\Delta x} = \lim_{\Delta x\to 0} \frac{\sin(\pi/6)\cos \Delta x + \cos(\pi/6)\sin \Delta x - (1/2)}{\Delta x}$$
$$= \lim_{\Delta x\to 0} \frac{1}{2} \cdot \frac{(\cos \Delta x - 1)}{\Delta x} + \lim_{\Delta x\to 0} \frac{\sqrt{3}}{2} \cdot \frac{\sin \Delta x}{\Delta x} = 0 + \frac{\sqrt{3}}{2}(1) = \frac{\sqrt{3}}{2}$$

29. $\lim_{x\to c}\left[f(x)g(x)\right] = \left[\lim_{x\to c} f(x)\right]\left[\lim_{x\to c} g(x)\right]$
$= (-6)\left(\frac{1}{2}\right) = -3$

31. $\lim_{x\to c}\left[f(x) + 2g(x)\right] = \lim_{x\to c} f(x) + 2\lim_{x\to c} g(x)$
$= -6 + 2\left(\frac{1}{2}\right) = -5$

33. $f(x) = \dfrac{\sqrt{2x+9}-3}{x}$

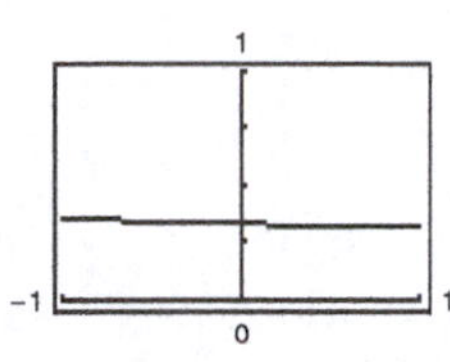

The limit appears to be $\frac{1}{3}$.

x	-0.01	-0.001	0	0.001	0.01
$f(x)$	0.3335	0.3333	?	0.3333	0.331

$\lim_{x\to 0} f(x) \approx 0.3333$

$$\lim_{x\to 0}\frac{\sqrt{2x+9}-3}{x}\cdot\frac{\sqrt{2x+9}+3}{\sqrt{2x+9}+3} = \lim_{x\to 0}\frac{(2x+9)-9}{x\left[\sqrt{2x+9}+3\right]} = \lim_{x\to 0}\frac{2}{\sqrt{2x+9}+3} = \frac{2}{\sqrt{9}+3} = \frac{1}{3}$$

35. $f(x) = \dfrac{x^3+729}{x+9}$

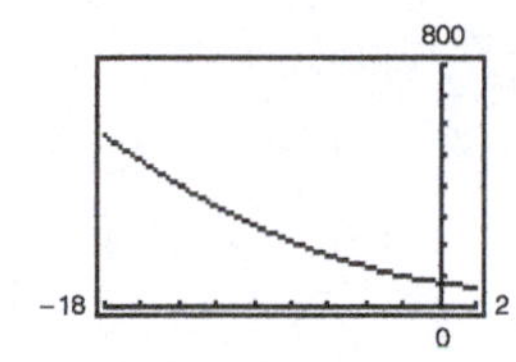

The limit appears to be 243.

x	-9.1	-9.01	-9.001	-9	-8.999	-8.99	-8.9
$f(x)$	245.7100	243.2701	243.0270	?	242.9730	242.7301	240.310

$$\lim_{x\to -9}\frac{x^3+729}{x+9} \approx 243.00$$

$$\lim_{x\to -9}\frac{x^3+729}{x+9} = \lim_{x\to -9}\frac{(x+9)(x^2-9x+81)}{x+9} = \lim_{x\to -9}(x^2-9x+81) = 81+81+81 = 243$$

37. $v = \lim_{t\to 4}\dfrac{s(4)-s(t)}{4-t}$
$= \lim_{t\to 4}\dfrac{\left[-4.9(16)+250\right]-\left[-4.9t^2+250\right]}{4-t}$
$= \lim_{t\to 4}\dfrac{4.9(t^2-16)}{4-t}$
$= \lim_{t\to 4}\dfrac{4.9(t-4)(t+4)}{4-t}$
$= \lim_{t\to 4}\left[-4.9(t+4)\right] = -39.2$ m/sec

The object is falling at about 39.2 m/sec.

39. $\lim_{x\to 3^+}\dfrac{1}{x+3} = \dfrac{1}{3+3} = \dfrac{1}{6}$

41. $\lim_{x\to 25^+}\dfrac{\sqrt{x}-5}{x-25} = \lim_{x\to 25^+}\dfrac{\sqrt{x}-5}{(\sqrt{x}+5)(\sqrt{x}-5)}$
$= \lim_{x\to 25^+}\dfrac{1}{\sqrt{x}+5}$
$= \dfrac{1}{\sqrt{25}+5} = \dfrac{1}{5+5} = \dfrac{1}{10}$

43. $\lim_{x\to 2} f(x) = 0$

45. $\lim_{t\to 1} h(t)$ does not exist because $\lim_{t\to 1^-} h(t) = 1 + 1 = 2$ and $\lim_{t\to 1^+} h(t) = \frac{1}{2}(1+1) = 1$.

47. $\lim_{x\to 2^-}\left(2[\![x]\!]+1\right) = 2(1) + 1 = 3$

49. $\lim_{x\to 2^-} \frac{x^2 - 4}{|x - 2|} = \lim_{x\to 2^-} \frac{(x - 2)(x + 2)}{2 - x}$

$= \lim_{x\to 2^-} -(x + 2) = -(2 + 2) = -4$

51. The function $g(x) = \sqrt{8 - x^3}$ is continuous on $[-2, 2]$ because $8 - x^3 \ge 0$ on $[-2, 2]$.

53. $f(x) = x^4 - 81x$ is continuous for all real x.

55. $f(x) = \frac{4}{x - 5}$ has a nonremovable discontinuity at $x = 5$ because $\lim_{x\to 5} f(x)$ does not exist.

57. $f(x) = \frac{x}{x^3 - x} = \frac{x}{x(x^2 - 1)} = \frac{1}{(x - 1)(x + 1)}, x \ne 0$

has nonremovable discontinuities at $x = \pm 1$ because $\lim_{x\to -1} f(x)$ and $\lim_{x\to 1} f(x)$ do not exist,

and has a removable discontinuity at $x = 0$ because

$\lim_{x\to 0} f(x) = \lim_{x\to 0} \frac{1}{(x - 1)(x + 1)} = -1.$

59. $f(2) = 5$

Find c so that $\lim_{x\to 2^+} (cx + 6) = 5.$

$c(2) + 6 = 5$

$2c = -1$

$c = -\frac{1}{2}$

61. $f(x) = -3x^2 + 7$

Continuous on $(-\infty, \infty)$

63. $f(x) = \sqrt{x} + \cos x$ is continuous on $[0, \infty)$.

65. $g(x) = 2e^{[\![x]\!]/4}$ is continuous on all intervals $(n, n + 1)$, where n is an integer. g has nonremovable discontinuities at each n.

67. $f(x) = \frac{3x^2 - x - 2}{x - 1} = \frac{(3x + 2)(x - 1)}{x - 1}$

$\lim_{x\to 1} f(x) = \lim_{x\to 1} (3x + 2) = 5$

Removable discontinuity at $x = 1$

Continuous on $(-\infty, 1) \cup (1, \infty)$

69. $f(x) = 2x^3 - 3$

f is continuous on $[1, 2]$. $f(1) = -1 < 0$ and $f(2) = 13 > 0$. Therefore by the Intermediate Value Theorem, there is at least one value c in $(1, 2)$ such that $2c^3 - 3 = 0$.

71. $f(x) = x^2 + 5x - 4$

f is continuous on $[-1, 2]$.

$f(-1) = (-1)^2 + 5(-1) - 4 = -8 < 2$

$f(2) = 2^2 + 5(2) - 4 = 10 > 2$

The Intermediate Value Theorem applies.

$x^2 + 5x - 4 = 2$

$x^2 + 5x - 6 = 0$

$(x + 6)(x - 1) = 0$

$x = 1$ ($x = -6$ lies outside the interval.)

$c = 1$

So, $f(1) = 2$.

73. $\lim_{x\to 6^-} \frac{1}{x - 6} = -\infty$

$\lim_{x\to 6^+} \frac{1}{x - 6} = \infty$

75. $f(x) = \frac{3}{x}$

$\lim_{x\to 0^-} \frac{3}{x} = -\infty$

$\lim_{x\to 0^+} \frac{3}{x} = \infty$

Therefore, $x = 0$ is a vertical asymptote.

77. $f(x) = \frac{x^3}{x^2 - 9} = \frac{x^3}{(x + 3)(x - 3)}$

$\lim_{x\to -3^-} \frac{x^3}{x^2 - 9} = -\infty$ and $\lim_{x\to -3^+} \frac{x^3}{x^2 - 9} = \infty$

Therefore, $x = -3$ is a vertical asymptote.

$\lim_{x\to 3^-} \frac{x^3}{x^2 - 9} = -\infty$ and $\lim_{x\to 3^+} \frac{x^3}{x^2 - 9} = \infty$

Therefore, $x = 3$ is a vertical asymptote.

79. $f(x) = \sec \dfrac{\pi x}{2} = \dfrac{1}{\cos \dfrac{\pi x}{2}}$

$\cos \dfrac{\pi x}{2} = 0$ when $x = \pm 1, \pm 3, \ldots$

Therefore, the graph has vertical asymptotes at $x = 2n + 1$, where n is an integer.

81. $g(x) = \ln(25 - x^2) = \ln[(5 + x)(5 - x)]$

$\lim_{x \to 5} \ln(25 - x^2) = 0$

$\lim_{x \to -5} \ln(25 - x^2) = 0$

Therefore, the graph has holes at $x = \pm 5$. The graph does not have any vertical asymptotes.

83. $\lim_{x \to 1^-} \dfrac{x^2 + 2x + 1}{x - 1} = -\infty$

85. $\lim_{x \to -1^+} \dfrac{x + 1}{x^3 + 1} = \lim_{x \to -1^+} \dfrac{1}{x^2 - x + 1} = \dfrac{1}{3}$

87. $\lim_{x \to 0^+} \left(x - \dfrac{1}{x^3}\right) = -\infty$

89. $\lim_{x \to 0^+} \dfrac{\sin 4x}{5x} = \lim_{x \to 0^+} \left[\dfrac{4}{5}\left(\dfrac{\sin 4x}{4x}\right)\right] = \dfrac{4}{5}$

91. $\lim_{x \to 0^+} \dfrac{\csc 2x}{x} = \lim_{x \to 0^+} \dfrac{1}{x \sin 2x} = \infty$

93. $\lim_{x \to 0^+} \ln(\sin x) = -\infty$

95. $C = \dfrac{80{,}000p}{100 - p}, \ 0 \le p < 0$

(a) $C(50) = \dfrac{80{,}000(50)}{100 - 50} = \$80{,}000$

(b) $C(90) = \dfrac{80{,}000(90)}{100 - 90} = \$720{,}000$

(c) $\lim_{p \to 100^-} C(p) = \infty$

It would be financially impossible to remove 100% of the pollutants.

Problem Solving for Chapter 2

1. (a) $\text{Perimeter } \Delta PAO = \sqrt{x^2 + (y - 1)^2} + \sqrt{x^2 + y^2} + 1$

$= \sqrt{x^2 + (x^2 - 1)^2} + \sqrt{x^2 + x^4} + 1$

$\text{Perimeter } \Delta PBO = \sqrt{(x - 1)^2 + y^2} + \sqrt{x^2 + y^2} + 1$

$= \sqrt{(x - 1)^2 + x^4} + \sqrt{x^2 + x^4} + 1$

(b) $r(x) = \dfrac{\sqrt{x^2 + (x^2 - 1)^2} + \sqrt{x^2 + x^4} + 1}{\sqrt{(x - 1)^2 + x^4} + \sqrt{x^2 + x^4} + 1}$

x	4	2	1	0.1	0.01
Perimeter ΔPAO	33.02	9.08	3.41	2.10	2.01
Perimeter ΔPBO	33.77	9.60	3.41	2.00	2.00
$r(x)$	0.98	0.95	1	1.05	1.005

(c) $\lim_{x \to 0^+} r(x) = \dfrac{1 + 0 + 1}{1 + 0 + 1} = \dfrac{2}{2} = 1$

3. (a) There are 6 triangles, each with a central angle of $60° = \pi/3$. So,

$$\text{Area hexagon} = 6\left[\frac{1}{2}bh\right] = 6\left[\frac{1}{2}(1)\sin\frac{\pi}{3}\right]$$
$$= \frac{3\sqrt{3}}{2} \approx 2.598.$$

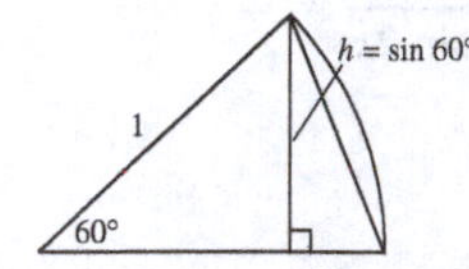

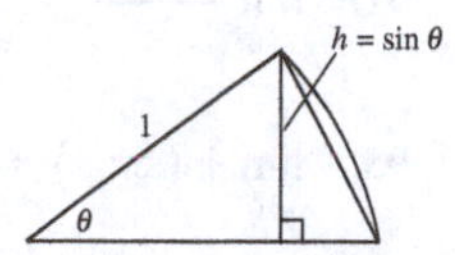

$$\text{Error} = \text{Area (Circle)} - \text{Area (Hexagon)}$$
$$= \pi - \frac{3\sqrt{3}}{2} \approx 0.5435$$

(b) There are n triangles, each with central angle of $\theta = 2\pi/n$. So,

$$A_n = n\left[\frac{1}{2}bh\right] = n\left[\frac{1}{2}(1)\sin\frac{2\pi}{n}\right] = \frac{n\sin(2\pi/n)}{2}.$$

(c)

n	6	12	24	48	96
A_n	2.598	3	3.106	3.133	3.139

As n gets larger and larger, $2\pi/n$ approaches 0. Letting $x = 2\pi/n$,

$$A_n = \frac{\sin(2\pi/n)}{2/n} = \frac{\sin(2\pi/n)}{(2\pi/n)}\pi = \frac{\sin x}{x}\pi$$

which approaches $(1)\pi = \pi$, which is the area of the circle.

5. (a) Slope $= -\dfrac{12}{5}$

(b) Slope of tangent line is $\dfrac{5}{12}$.

$$y + 12 = \frac{5}{12}(x - 5)$$
$$y = \frac{5}{12}x - \frac{169}{12} \quad \text{Tangent line}$$

(c) $Q = (x, y) = \left(x, -\sqrt{169 - x^2}\right)$

$$m_x = \frac{-\sqrt{169 - x^2} + 12}{x - 5}$$

(d)
$$\lim_{x\to5} m_x = \lim_{x\to5} \frac{12 - \sqrt{169 - x^2}}{x - 5} \cdot \frac{12 + \sqrt{169 - x^2}}{12 + \sqrt{169 - x^2}}$$
$$= \lim_{x\to5} \frac{144 - \left(169 - x^2\right)}{(x - 5)\left(12 + \sqrt{169 - x^2}\right)}$$
$$= \lim_{x\to5} \frac{x^2 - 25}{(x - 5)\left(12 + \sqrt{169 - x^2}\right)}$$
$$= \lim_{x\to5} \frac{(x + 5)}{12 + \sqrt{169 - x^2}} = \frac{10}{12 + 12} = \frac{5}{12}$$

This is the same slope as part (b).

7. (a)
$$3 + x^{1/3} \geq 0$$
$$x^{1/3} \geq -3$$
$$x \geq -27$$

Domain: $x \geq -27,\ x \neq 1$ or $[-27, 1) \cup (1, \infty)$

(b)

(c) $$\lim_{x\to-27^+} f(x) = \frac{\sqrt{3 + (-27)^{1/3}} - 2}{-27 - 1} = \frac{-2}{-28} = \frac{1}{14} \approx 0.0714$$

(d)
$$\lim_{x\to1} f(x) = \lim_{x\to1} \frac{\sqrt{3 + x^{1/3}} - 2}{x - 1} \cdot \frac{\sqrt{3 + x^{1/3}} + 2}{\sqrt{3 + x^{1/3}} + 2} = \lim_{x\to1} \frac{3 + x^{1/3} - 4}{(x - 1)\left(\sqrt{3 + x^{1/3}} + 2\right)}$$
$$= \lim_{x\to1} \frac{x^{1/3} - 1}{\left(x^{1/3} - 1\right)\left(x^{2/3} + x^{1/3} + 1\right)\left(\sqrt{3 + x^{1/3}} + 2\right)} = \lim_{x\to1} \frac{1}{\left(x^{2/3} + x^{1/3} + 1\right)\left(\sqrt{3 + x^{1/3}} + 2\right)}$$
$$= \frac{1}{(1 + 1 + 1)(2 + 2)} = \frac{1}{12}$$

9. (a) $\lim_{x\to 2} f(x) = 3$: g_1, g_4

(b) f continuous at 2: g_1

(c) $\lim_{x\to 2^-} f(x) = 3$: g_1, g_3, g_4

11.

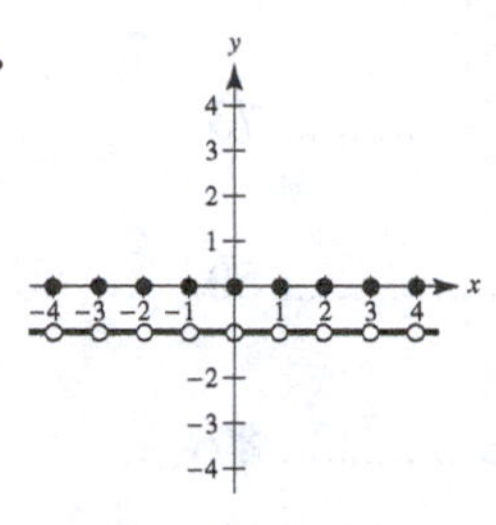

(a) $f(1) = [\![1]\!] + [\![-1]\!] = 1 + (-1) = 0$

$f(0) = 0$

$f\left(\frac{1}{2}\right) = 0 + (-1) = -1$

$f(-2.7) = -3 + 2 = -1$

(b) $\lim_{x\to 1^-} f(x) = -1$

$\lim_{x\to 1^+} f(x) = -1$

$\lim_{x\to 1/2} f(x) = -1$

(c) f is continuous for all real numbers except $x = 0, \pm 1, \pm 2, \pm 3, \ldots$

13. (a)

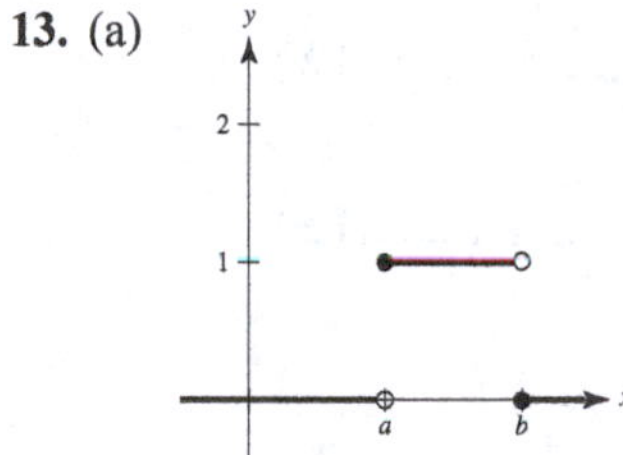

(b) (i) $\lim_{x\to a^+} P_{a,b}(x) = 1$

(ii) $\lim_{x\to a^-} P_{a,b}(x) = 0$

(iii) $\lim_{x\to b^+} P_{a,b}(x) = 0$

(iv) $\lim_{x\to b^-} P_{a,b}(x) = 1$

(c) $P_{a,b}$ is continuous for all positive real numbers except $x = a, b$.

(d) The area under the graph of U, and above the x-axis, is 1.

CHAPTER 3
Differentiation

CHAPTER 3
Differentiation

Section 3.1 The Derivative and the Tangent Line Problem

1. The problem of finding the tangent line at a point P is essentially finding the slope of the tangent line at point P. To do so for a function f, if f is defined on an open interval containing c, and if the limit

$$\lim_{\Delta x \to 0} \frac{\Delta y}{\Delta x} = \lim_{\Delta x \to 0} \frac{f(c + \Delta x) - f(c)}{\Delta x} = m$$

exists, then the line passing through the point $P(c, f(c))$ with slope m is the tangent line to the graph of f at the point P.

3. The limit used to define the slope of a tangent line is also used to define differentiation. The key is to rewrite the difference quotient so that Δx does not occur as a factor of the denominator.

5. At (x_1, y_1), slope $= 0$.

At (x_2, y_2), slope $= \frac{5}{2}$.

7. (a)–(c)

(d) $$\begin{aligned} y &= \frac{f(4) - f(1)}{4 - 1}(x - 1) + f(1) \\ &= \frac{3}{3}(x - 1) + 2 \\ &= 1(x - 1) + 2 \\ &= x + 1 \end{aligned}$$

9. $f(x) = 3 - 5x$ is a line. Slope $= -5$

11. $$\begin{aligned} \text{Slope at } (2, 5) &= \lim_{\Delta x \to 0} \frac{f(2 + \Delta x) - f(2)}{\Delta x} \\ &= \lim_{\Delta x \to 0} \frac{2(2 + \Delta x)^2 - 3 - \left[2(2)^2 - 3\right]}{\Delta x} \\ &= \lim_{\Delta x \to 0} \frac{2\left[4 + 4\Delta x + (\Delta x)^2\right] - 3 - (5)}{\Delta x} \\ &= \lim_{\Delta x \to 0} \frac{8\Delta x + 2(\Delta x)^2}{\Delta x} \\ &= \lim_{\Delta x \to 0} (8 + 2\Delta x) = 8 \end{aligned}$$

13. $$\begin{aligned} \text{Slope at } (0, 0) &= \lim_{\Delta t \to 0} \frac{f(0 + \Delta t) - f(0)}{\Delta t} \\ &= \lim_{\Delta t \to 0} \frac{3(\Delta t) - (\Delta t)^2 - 0}{\Delta t} \\ &= \lim_{\Delta t \to 0} (3 - \Delta t) = 3 \end{aligned}$$

15. $f(x) = 7$

$$\begin{aligned} f'(x) &= \lim_{\Delta x \to 0} \frac{f(x + \Delta x) - f(x)}{\Delta x} \\ &= \lim_{\Delta x \to 0} \frac{7 - 7}{\Delta x} \\ &= \lim_{\Delta x \to 0} 0 = 0 \end{aligned}$$

17. $f(x) = -5x$

$$\begin{aligned} f'(x) &= \lim_{\Delta x \to 0} \frac{f(x + \Delta x) - f(x)}{\Delta x} \\ &= \lim_{\Delta x \to 0} \frac{-5(x + \Delta x) - (-5x)}{\Delta x} \\ &= \lim_{\Delta x \to 0} \frac{-5x - 5\Delta x + 5x}{\Delta x} \\ &= \lim_{\Delta x \to 0} \frac{-5\Delta x}{\Delta x} \\ &= \lim_{\Delta x \to 0} (-5) = -5 \end{aligned}$$

19. $h(s) = 3 + \frac{2}{3}s$

$$h'(s) = \lim_{\Delta s \to 0} \frac{h(s + \Delta s) - h(s)}{\Delta s} = \lim_{\Delta s \to 0} \frac{3 + \frac{2}{3}(s + \Delta s) - \left(3 + \frac{2}{3}s\right)}{\Delta s} = \lim_{\Delta s \to 0} \frac{3 + \frac{2}{3}s + \frac{2}{3}\Delta s - 3 - \frac{2}{3}s}{\Delta s} = \lim_{\Delta s \to 0} \frac{\frac{2}{3}\Delta s}{\Delta s} = \frac{2}{3}$$

21. $f(x) = x^2 + x - 3$

$$\begin{aligned} f'(x) &= \lim_{\Delta x \to 0} \frac{f(x + \Delta x) - f(x)}{\Delta x} \\ &= \lim_{\Delta x \to 0} \frac{(x + \Delta x)^2 + (x + \Delta x) - 3 - (x^2 + x - 3)}{\Delta x} \\ &= \lim_{\Delta x \to 0} \frac{x^2 + 2x(\Delta x) + (\Delta x)^2 + x + \Delta x - 3 - x^2 - x + 3}{\Delta x} \\ &= \lim_{\Delta x \to 0} \frac{2x(\Delta x) + (\Delta x)^2 + \Delta x}{\Delta x} \\ &= \lim_{\Delta x \to 0} (2x + \Delta x + 1) = 2x + 1 \end{aligned}$$

23. $f(x) = x^3 - 12x$

$$\begin{aligned} f'(x) &= \lim_{\Delta x \to 0} \frac{f(x + \Delta x) - f(x)}{\Delta x} \\ &= \lim_{\Delta x \to 0} \frac{\left[(x + \Delta x)^3 - 12(x + \Delta x)\right] - \left[x^3 - 12x\right]}{\Delta x} \\ &= \lim_{\Delta x \to 0} \frac{x^3 + 3x^2\Delta x + 3x(\Delta x)^2 + (\Delta x)^3 - 12x - 12\,\Delta x - x^3 + 12x}{\Delta x} \\ &= \lim_{\Delta x \to 0} \frac{3x^2\Delta x + 3x(\Delta x)^2 + (\Delta x)^3 - 12\,\Delta x}{\Delta x} \\ &= \lim_{\Delta x \to 0} \left(3x^2 + 3x\,\Delta x + (\Delta x)^2 - 12\right) = 3x^2 - 12 \end{aligned}$$

25. $f(x) = \dfrac{1}{x - 1}$

$$\begin{aligned} f'(x) &= \lim_{\Delta x \to 0} \frac{f(x + \Delta x) - f(x)}{\Delta x} \\ &= \lim_{\Delta x \to 0} \frac{\dfrac{1}{x + \Delta x - 1} - \dfrac{1}{x - 1}}{\Delta x} \\ &= \lim_{\Delta x \to 0} \frac{(x - 1) - (x + \Delta x - 1)}{\Delta x(x + \Delta x - 1)(x - 1)} \\ &= \lim_{\Delta x \to 0} \frac{-\Delta x}{\Delta x(x + \Delta x - 1)(x - 1)} \\ &= \lim_{\Delta x \to 0} \frac{-1}{(x + \Delta x - 1)(x - 1)} = -\frac{1}{(x - 1)^2} \end{aligned}$$

27. $f(x) = \sqrt{x + 4}$

$$\begin{aligned} f'(x) &= \lim_{\Delta x \to 0} \frac{f(x + \Delta x) - f(x)}{\Delta x} \\ &= \lim_{\Delta x \to 0} \frac{\sqrt{x + \Delta x + 4} - \sqrt{x + 4}}{\Delta x} \cdot \left(\frac{\sqrt{x + \Delta x + 4} + \sqrt{x + 4}}{\sqrt{x + \Delta x + 4} + \sqrt{x + 4}}\right) \\ &= \lim_{\Delta x \to 0} \frac{(x + \Delta x + 4) - (x + 4)}{\Delta x\left[\sqrt{x + \Delta x + 4} + \sqrt{x + 4}\right]} \\ &= \lim_{\Delta x \to 0} \frac{1}{\sqrt{x + \Delta x + 4} + \sqrt{x + 4}} = \frac{1}{\sqrt{x + 4} + \sqrt{x + 4}} = \frac{1}{2\sqrt{x + 4}} \end{aligned}$$

29. (a) $f(x) = x^2 + 3$

$$f'(x) = \lim_{\Delta x \to 0} \frac{f(x + \Delta x) - f(x)}{\Delta x}$$

$$= \lim_{\Delta x \to 0} \frac{\left[(x + \Delta x)^2 + 3\right] - (x^2 + 3)}{\Delta x}$$

$$= \lim_{\Delta x \to 0} \frac{x^2 + 2x\Delta x + (\Delta x)^2 + 3 - x^2 - 3}{\Delta x}$$

$$= \lim_{\Delta x \to 0} \frac{2x\Delta x + (\Delta x)^2}{\Delta x} = \lim_{\Delta x \to 0} (2x + \Delta x) = 2x$$

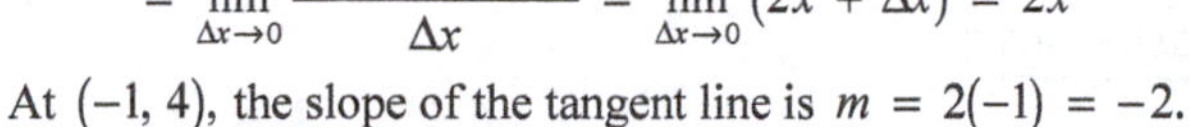

At $(-1, 4)$, the slope of the tangent line is $m = 2(-1) = -2$.

The equation of the tangent line is

$y - 4 = -2(x + 1)$

$y - 4 = -2x - 2$

$y = -2x + 2$

(b)

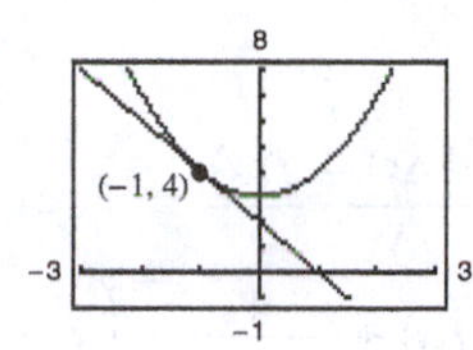

(c) Graphing utility confirms $\dfrac{dy}{dx} = -2$ at $(-1, 4)$.

31. (a) $f(x) = x^3$

$$f'(x) = \lim_{\Delta x \to 0} \frac{f(x + \Delta x) - f(x)}{\Delta x} = \lim_{\Delta x \to 0} \frac{(x + \Delta x)^3 - x^3}{\Delta x}$$

$$= \lim_{\Delta x \to 0} \frac{x^3 + 3x^2\Delta x + 3x(\Delta x)^2 + (\Delta x)^3 - x^3}{\Delta x}$$

$$= \lim_{\Delta x \to 0} \frac{3x^2\Delta x + 3x(\Delta x)^2 + (\Delta x)^3}{\Delta x}$$

$$= \lim_{\Delta x \to 0} \left(3x^2 + 3x\,\Delta x + (\Delta x)^2\right) = 3x^2$$

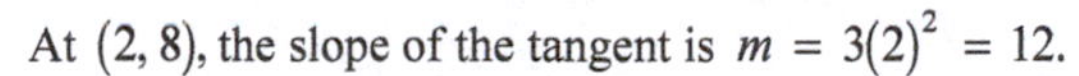

At $(2, 8)$, the slope of the tangent is $m = 3(2)^2 = 12$.

The equation of the tangent line is

$y - 8 = 12(x - 2)$

$y - 8 = 12x - 24$

$y = 12x - 16.$

(b)

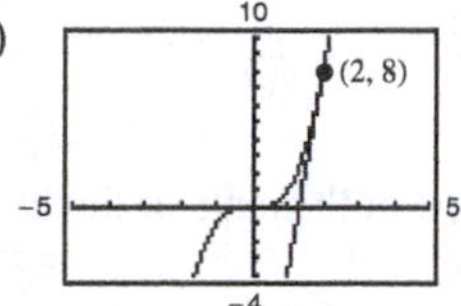

(c) Graphing utility confirms $\dfrac{dy}{dx} = 12$ at $(2, 8)$.

33. (a) $f(x) = \sqrt{x}$

$$f'(x) = \lim_{\Delta x \to 0} \frac{f(x + \Delta x) - f(x)}{\Delta x}$$

$$= \lim_{\Delta x \to 0} \frac{\sqrt{x + \Delta x} - \sqrt{x}}{\Delta x} \cdot \frac{\sqrt{x + \Delta x} + \sqrt{x}}{\sqrt{x + \Delta x} + \sqrt{x}}$$

$$= \lim_{\Delta x \to 0} \frac{(x + \Delta x) - x}{\Delta x\left(\sqrt{x + \Delta x} + \sqrt{x}\right)}$$

$$= \lim_{\Delta x \to 0} \frac{1}{\sqrt{x + \Delta x} + \sqrt{x}} = \frac{1}{2\sqrt{x}}$$

At $(1, 1)$, the slope of the tangent line is $m = \dfrac{1}{2\sqrt{1}} = \dfrac{1}{2}$.

The equation of the tangent line is

$y - 1 = \dfrac{1}{2}(x - 1)$

$y - 1 = \dfrac{1}{2}x - \dfrac{1}{2}$

$y = \dfrac{1}{2}x + \dfrac{1}{2}.$

(b)

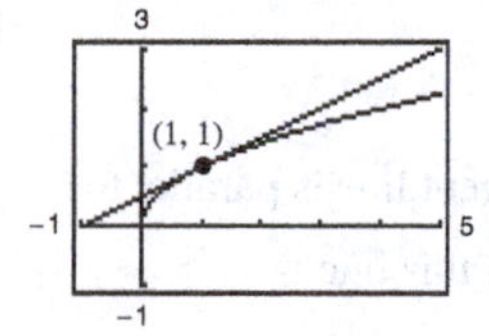

(c) Graphing utility confirms $\dfrac{dy}{dx} = \dfrac{1}{2}$ at $(1, 1)$.

35. (a) $f(x) = x + \frac{4}{x}$

$$f'(x) = \lim_{\Delta x \to 0} \frac{f(x + \Delta x) - f(x)}{\Delta x}$$

$$= \lim_{\Delta x \to 0} \frac{(x + \Delta x) + \frac{4}{x + \Delta x} - \left(x + \frac{4}{x}\right)}{\Delta x}$$

$$= \lim_{\Delta x \to 0} \frac{x(x + \Delta x)(x + \Delta x) + 4x - x^2(x + \Delta x) - 4(x + \Delta x)}{x(\Delta x)(x + \Delta x)}$$

$$= \lim_{\Delta x \to 0} \frac{x^3 + 2x^2(\Delta x) + x(\Delta x)^2 - x^3 - x^2(\Delta x) - 4(\Delta x)}{x(\Delta x)(x + \Delta x)}$$

$$= \lim_{\Delta x \to 0} \frac{x^2(\Delta x) + x(\Delta x)^2 - 4(\Delta x)}{x(\Delta x)(x + \Delta x)}$$

$$= \lim_{\Delta x \to 0} \frac{x^2 + x(\Delta x) - 4}{x(x + \Delta x)}$$

$$= \frac{x^2 - 4}{x^2} = 1 - \frac{4}{x^2}$$

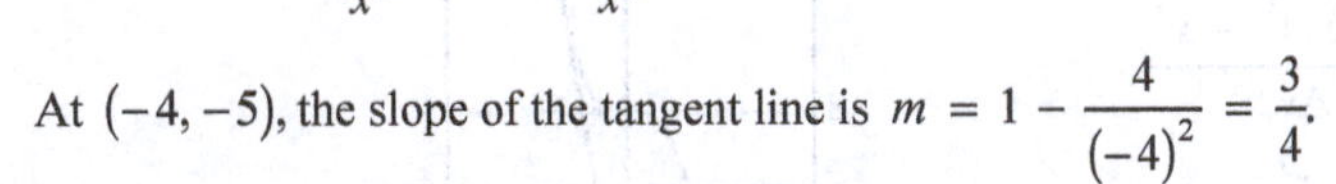

At $(-4, -5)$, the slope of the tangent line is $m = 1 - \frac{4}{(-4)^2} = \frac{3}{4}$.

The equation of the tangent line is

$$y + 5 = \frac{3}{4}(x + 4)$$

$$y + 5 = \frac{3}{4}x + 3$$

$$y = \frac{3}{4}x - 2.$$

(b)

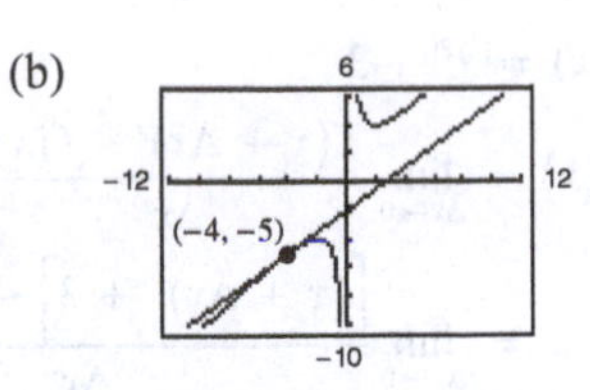

(c) Graphing utility confirms $\frac{dy}{dx} = \frac{3}{4}$ at $(-4, -5)$.

37. Using the limit definition of a derivative, $f'(x) = -\frac{1}{2}x$.

Because the slope of the given line is -1, you have

$$-\frac{1}{2}x = -1$$

$$x = 2.$$

At the point $(2, -1)$, the tangent line is parallel to $x + y = 0$. The equation of this line is

$$y - (-1) = -1(x - 2)$$

$$y = -x + 1.$$

39. From Exercise 31 we know that $f'(x) = 3x^2$.

Because the slope of the given line is 3, you have

$$3x^2 = 3$$

$$x = \pm 1.$$

Therefore, at the points $(1, 1)$ and $(-1, -1)$ the tangent lines are parallel to $3x - y + 1 = 0$.

These lines have equations

$$y - 1 = 3(x - 1) \text{ and } y + 1 = 3(x + 1)$$

$$y = 3x - 2 \qquad y = 3x + 2.$$

41. Using the limit definition of derivative,

$$f'(x) = \frac{-1}{2x\sqrt{x}}.$$

Because the slope of the given line is $-\frac{1}{2}$, you have

$$-\frac{1}{2x\sqrt{x}} = -\frac{1}{2}$$

$$x = 1.$$

Therefore, at the point $(1, 1)$ the tangent line is parallel to $x + 2y - 6 = 0$. The equation of this line is

$$y - 1 = -\frac{1}{2}(x - 1)$$

$$y - 1 = -\frac{1}{2}x + \frac{1}{2}$$

$$y = -\frac{1}{2}x + \frac{3}{2}.$$

43. The slope of the graph of f is 1 for all x-values.

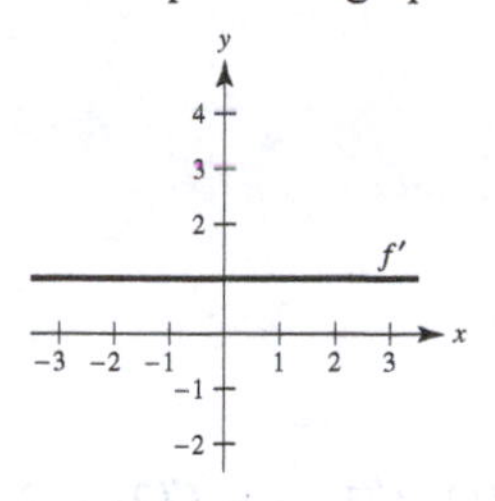

45. The slope of the graph of f is negative for $x < 4$, positive for $x > 4$, and 0 at $x = 4$.

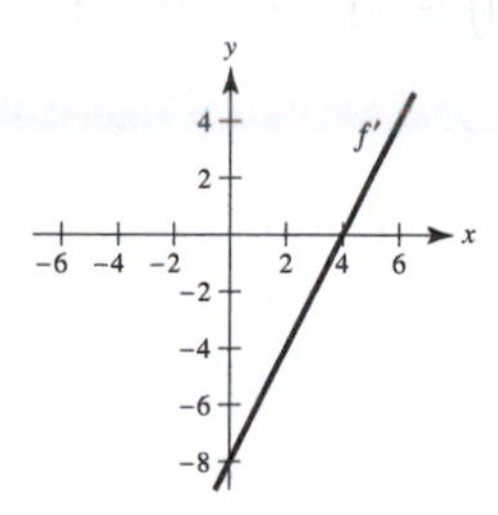

47. The slope of the graph of f is negative for $x < 0$ and positive for $x > 0$. The slope is undefined at $x = 0$.

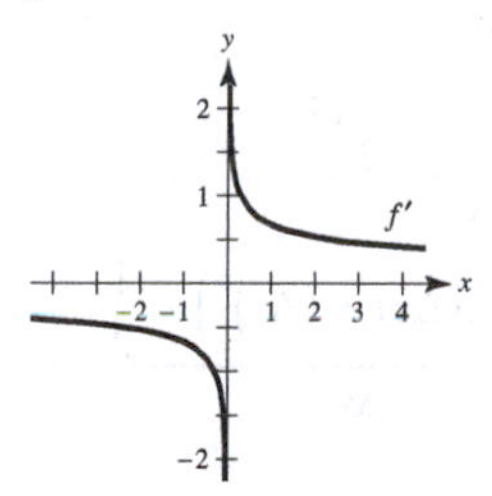

49. Answers will vary.

Sample answer: $y = -x$

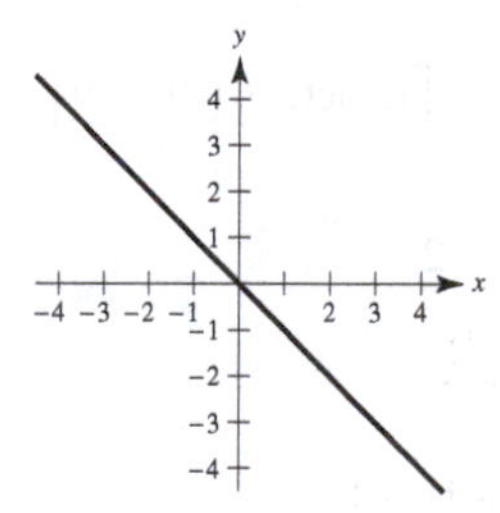

The derivative of $y = -x$ is $y' = -1$. So, the derivative is always negative.

51. No. For example, the domain of $f(x) = \sqrt{x}$ is $x \geq 0$, whereas the domain of $f'(x) = \dfrac{1}{2\sqrt{x}}$ is $x > 0$.

53. $g(4) = 5$ because the tangent line passes through $(4, 5)$.

$$g'(4) = \frac{5 - 0}{4 - 7} = -\frac{5}{3}$$

55. $f(x) = 5 - 3x$ and $c = 1$

57. $f(x) = -x^2$ and $c = 6$

59. $f(0) = 2$ and $f'(x) = -3, -\infty < x < \infty$

$f(x) = -3x + 2$

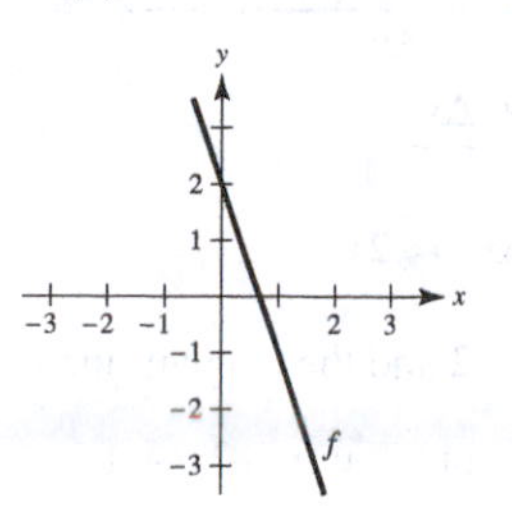

61. Let (x_0, y_0) be a point of tangency on the graph of f. By the limit definition for the derivative, $f'(x) = 4 - 2x$. The slope of the line through $(2, 5)$ and (x_0, y_0) equals the derivative of f at x_0:

$$\frac{5 - y_0}{2 - x_0} = 4 - 2x_0$$

$$5 - y_0 = (2 - x_0)(4 - 2x_0)$$

$$5 - (4x_0 - x_0^2) = 8 - 8x_0 + 2x_0^2$$

$$0 = x_0^2 - 4x_0 + 3$$

$$0 = (x_0 - 1)(x_0 - 3) \Rightarrow x_0 = 1, 3$$

Therefore, the points of tangency are $(1, 3)$ and $(3, 3)$, and the corresponding slopes are 2 and –2. The equations of the tangent lines are:

$y - 5 = 2(x - 2)$ $\quad$ $y - 5 = -2(x - 2)$

$y = 2x + 1$ $\quad$ $y = -2x + 9$

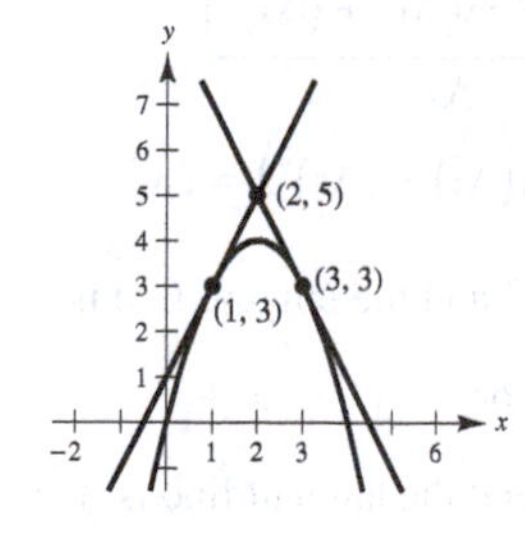

63. (a) $f(x) = x^2$

$$f'(x) = \lim_{\Delta x \to 0} \frac{f(x + \Delta x) - f(x)}{\Delta x}$$
$$= \lim_{\Delta x \to 0} \frac{(x + \Delta x)^2 - x^2}{\Delta x}$$
$$= \lim_{\Delta x \to 0} \frac{x^2 + 2x(\Delta x) + (\Delta x)^2 - x^2}{\Delta x}$$
$$= \lim_{\Delta x \to 0} \frac{\Delta x(2x + \Delta x)}{\Delta x}$$
$$= \lim_{\Delta x \to 0} (2x + \Delta x) = 2x$$

At $x = -1$, $f'(-1) = -2$ and the tangent line is

$y - 1 = -2(x + 1)$ or $y = -2x - 1$.

At $x = 0$, $f'(0) = 0$ and the tangent line is $y = 0$.

At $x = 1$, $f'(1) = 2$ and the tangent line is

$y = 2x - 1$.

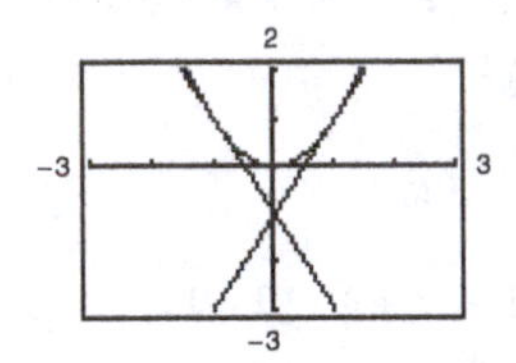

For this function, the slopes of the tangent lines are always distinct for different values of x.

(b) $$g'(x) = \lim_{\Delta x \to 0} \frac{g(x + \Delta x) - g(x)}{\Delta x}$$
$$= \lim_{\Delta x \to 0} \frac{(x + \Delta x)^3 - x^3}{\Delta x}$$
$$= \lim_{\Delta x \to 0} \frac{x^3 + 3x^2(\Delta x) + 3x(\Delta x)^2 + (\Delta x)^3 - x^3}{\Delta x}$$
$$= \lim_{\Delta x \to 0} \frac{\Delta x\left(3x^2 + 3x(\Delta x) + (\Delta x)^2\right)}{\Delta x}$$
$$= \lim_{\Delta x \to 0} \left(3x^2 + 3x(\Delta x) + (\Delta x)^2\right) = 3x^2$$

At $x = -1$, $g'(-1) = 3$ and the tangent line is

$y + 1 = 3(x + 1)$ or $y = 3x + 2$.

At $x = 0$, $g'(0) = 0$ and the tangent line is $y = 0$.

At $x = 1$, $g'(1) = 3$ and the tangent line is

$y - 1 = 3(x - 1)$ or $y = 3x - 2$.

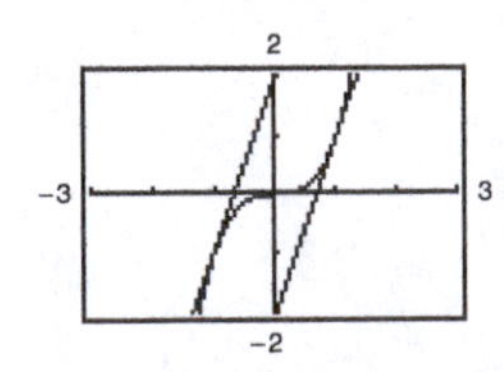

For this function, the slopes of the tangent lines are sometimes the same.

65. $f(x) = \frac{1}{2}x^2$

(a)

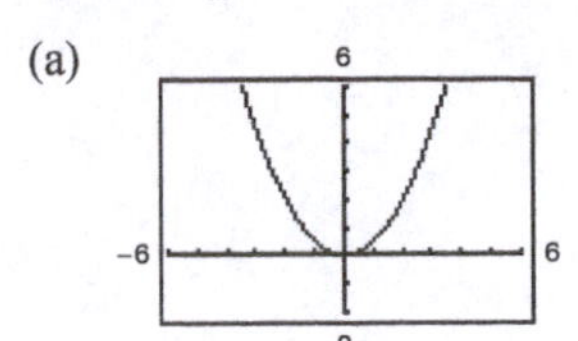

$f'(0) = 0$, $f'(1/2) = 1/2$, $f'(1) = 1$, $f'(2) = 2$

(b) By symmetry:

$f'(-1/2) = -1/2$, $f'(-1) = -1$, $f'(-2) = -2$

(c)

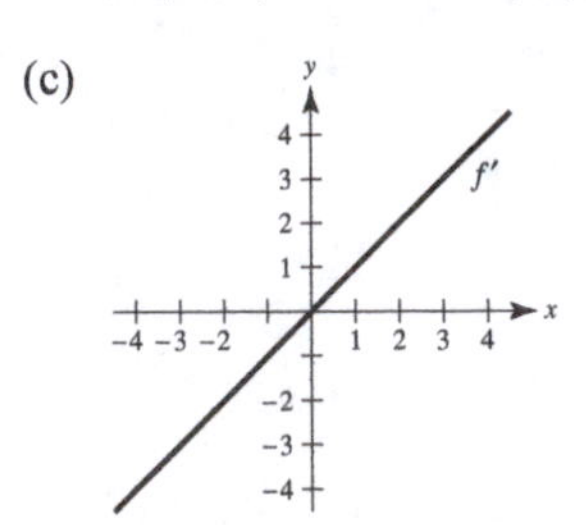

(d) $$f'(x) = \lim_{\Delta x \to 0} \frac{f(x + \Delta x) - f(x)}{\Delta x}$$
$$= \lim_{\Delta x \to 0} \frac{\frac{1}{2}(x + \Delta x)^2 - \frac{1}{2}x^2}{\Delta x}$$
$$= \lim_{\Delta x \to 0} \frac{\frac{1}{2}\left(x^2 + 2x(\Delta x) + (\Delta x)^2\right) - \frac{1}{2}x^2}{\Delta x}$$
$$= \lim_{\Delta x \to 0} \left(x + \frac{\Delta x}{2}\right) = x$$

67. $f(2) = 2(4 - 2) = 4$, $f(2.1) = 2.1(4 - 2.1) = 3.99$

$$f'(2) \approx \frac{3.99 - 4}{2.1 - 2} = -0.1 \quad [\text{Exact: } f'(2) = 0]$$

69. $f(x) = x^3 + 2x^2 + 1$, $c = -2$

$$f'(-2) = \lim_{x \to -2} \frac{f(x) - f(-2)}{x + 2}$$
$$= \lim_{x \to -2} \frac{(x^3 + 2x^2 + 1) - 1}{x + 2}$$
$$= \lim_{x \to -2} \frac{x^2(x + 2)}{x + 2} = \lim_{x \to -2} x^2 = 4$$

71. $g(x) = \sqrt{|x|}, c = 0$

$$g'(0) = \lim_{x\to 0}\frac{g(x) - g(0)}{x - 0} = \lim_{x\to 0}\frac{\sqrt{|x|}}{x}.$$ Does not exist.

As $x \to 0^-$, $\frac{\sqrt{|x|}}{x} = \frac{-1}{\sqrt{|x|}} \to -\infty$.

As $x \to 0^+$, $\frac{\sqrt{|x|}}{x} = \frac{1}{\sqrt{x}} \to \infty$.

Therefore $g(x)$ is not differentiable at $x = 0$.

73. $f(x) = (x - 6)^{2/3}, c = 6$

$$f'(6) = \lim_{x\to 6}\frac{f(x) - f(6)}{x - 6}$$
$$= \lim_{x\to 6}\frac{(x - 6)^{2/3} - 0}{x - 6} = \lim_{x\to 6}\frac{1}{(x - 6)^{1/3}}.$$

Does not exist.

Therefore $f(x)$ is not differentiable at $x = 6$.

75. $h(x) = |x + 7|, c = -7$

$$h'(-7) = \lim_{x\to -7}\frac{h(x) - h(-7)}{x - (-7)}$$
$$= \lim_{x\to -7}\frac{|x + 7| - 0}{x + 7} = \lim_{x\to -7}\frac{|x + 7|}{x + 7}.$$

Does not exist.

Therefore $h(x)$ is not differentiable at $x = -7$.

77. $f(x)$ is differentiable everywhere except at $x = -4$. (Sharp turn in the graph)

79. $f(x)$ is differentiable on the interval $(-1, \infty)$. (At $x = -1$ the tangent line is vertical.)

81. $f(x) = |x - 5|$ is differentiable everywhere except at $x = 5$. There is a sharp corner at $x = 5$.

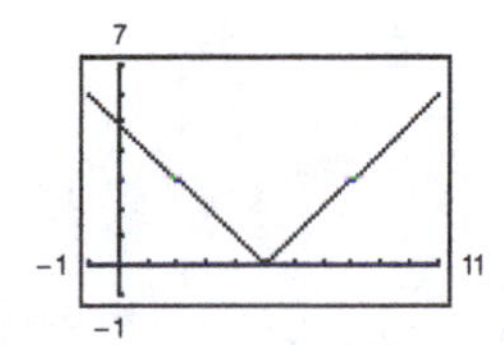

83. $f(x) = x^{2/5}$ is differentiable for all $x \neq 0$. There is a sharp corner at $x = 0$.

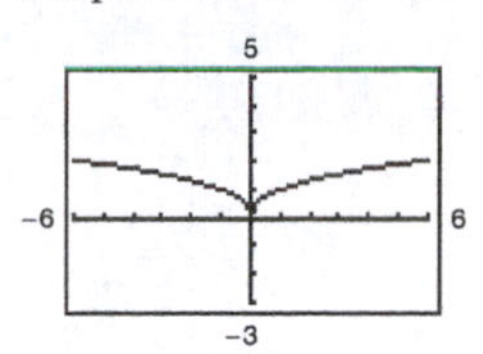

85. $f(x) = |x - 1|$

The derivative from the left is

$$\lim_{x\to 1^-}\frac{f(x) - f(1)}{x - 1} = \lim_{x\to 1^-}\frac{|x - 1| - 0}{x - 1} = -1.$$

The derivative from the right is

$$\lim_{x\to 1^+}\frac{f(x) - f(1)}{x - 1} = \lim_{x\to 1^+}\frac{|x - 1| - 0}{x - 1} = 1.$$

The one-sided limits are not equal. Therefore, f is not differentiable at $x = 1$.

87. $f(x) = \begin{cases}(x - 1)^3, & x \le 1\\ (x - 1)^2, & x > 1\end{cases}$

The derivative from the left is

$$\lim_{x\to 1^-}\frac{f(x) - f(1)}{x - 1} = \lim_{x\to 1^-}\frac{(x - 1)^3 - 0}{x - 1}$$
$$= \lim_{x\to 1^-}(x - 1)^2 = 0.$$

The derivative from the right is

$$\lim_{x\to 1^+}\frac{f(x) - f(1)}{x - 1} = \lim_{x\to 1^+}\frac{(x - 1)^2 - 0}{x - 1}$$
$$= \lim_{x\to 1^+}(x - 1) = 0.$$

The one-sided limits are equal. Therefore, f is differentiable at $x = 1$. $(f'(1) = 0)$

89. Note that f is continuous at $x = 2$.

$$f(x) = \begin{cases}x^2 + 1, & x \le 2\\ 4x - 3, & x > 2\end{cases}$$

The derivative from the left is

$$\lim_{x\to 2^-}\frac{f(x) - f(2)}{x - 2} = \lim_{x\to 2^-}\frac{(x^2 + 1) - 5}{x - 2}$$
$$= \lim_{x\to 2^-}(x + 2) = 4.$$

The derivative from the right is

$$\lim_{x\to 2^+}\frac{f(x) - f(2)}{x - 2} = \lim_{x\to 2^+}\frac{(4x - 3) - 5}{x - 2} = \lim_{x\to 2^+} 4 = 4.$$

The one-sided limits are equal. Therefore, f is differentiable at $x = 2$. $(f'(2) = 4)$

91.

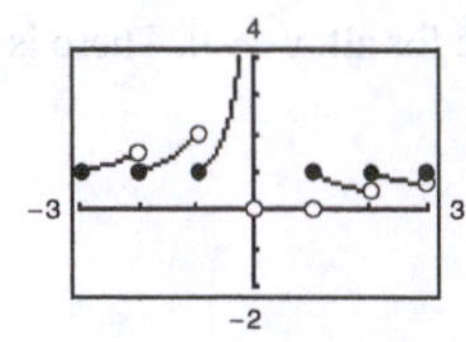

Let $g(x) = \dfrac{[\![x]\!]}{x}$.

For $f(x) = [\![x]\!]$, $\displaystyle\lim_{x\to 0^-} \frac{f(x) - f(0)}{x - 0} = \lim_{x\to 0^-} \frac{[\![x]\!] - 0}{x} = \lim_{x\to 0^-} \frac{[\![x]\!]}{x} = \lim_{x\to 0^-} [\![x]\!] \cdot \lim_{x\to 0^-} \frac{1}{x} = -1 \cdot \lim_{x\to 0^-} \frac{1}{x} = \lim_{x\to 0^-} \frac{-1}{x} = \infty.$

On the other hand, $\displaystyle\lim_{x\to 0^+} \frac{f(x) - f(0)}{x - 0} = \lim_{x\to 0^+} \frac{[\![x]\!] - 0}{x} = \lim_{x\to 0^+} \frac{[\![x]\!]}{x} = \lim_{x\to 0^+} [\![x]\!] \cdot \lim_{x\to 0^+} \frac{1}{x} = 0 \cdot \lim_{x\to 0^+} \frac{1}{x} = 0.$

So, f is not differentiable at $x = 0$ because $\displaystyle\lim_{x\to 0} \frac{f(x) - f(0)}{x - 0}$ does not exist. f is differentiable for all $x \neq n$, n an integer.

93. False. The slope is $\displaystyle\lim_{\Delta x\to 0} \frac{f(2 + \Delta x) - f(2)}{\Delta x}$.

95. False. If the derivative from the left of a point does not equal the derivative from the right of a point, then the derivative does not exist at that point. For example, if $f(x) = |x|$, then the derivative from the left at $x = 0$ is -1 and the derivative from the right at $x = 0$ is 1. At $x = 0$, the derivative does not exist.

97. $f(x) = \begin{cases} x \sin(1/x), & x \neq 0 \\ 0, & x = 0 \end{cases}$

Using the Squeeze Theorem, you have $-|x| \le x \sin(1/x) \le |x|$, $x \neq 0$. So, $\displaystyle\lim_{x\to 0} x \sin(1/x) = 0 = f(0)$ and f is continuous at $x = 0$. Using the alternative form of the derivative, you have

$$\lim_{x\to 0} \frac{f(x) - f(0)}{x - 0} = \lim_{x\to 0} \frac{x \sin(1/x) - 0}{x - 0} = \lim_{x\to 0}\left(\sin \frac{1}{x}\right).$$

Because this limit does not exist ($\sin(1/x)$ oscillates between -1 and 1), the function is not differentiable at $x = 0$.

$g(x) = \begin{cases} x^2 \sin(1/x), & x \neq 0 \\ 0, & x = 0 \end{cases}$

Using the Squeeze Theorem again, you have $-x^2 \le x^2 \sin(1/x) \le x^2$, $x \neq 0$. So, $\displaystyle\lim_{x\to 0} x^2 \sin(1/x) = 0 = g(0)$ and g is continuous at $x = 0$. Using the alternative form of the derivative again, you have

$$\lim_{x\to 0} \frac{g(x) - g(0)}{x - 0} = \lim_{x\to 0} \frac{x^2 \sin(1/x) - 0}{x - 0} = \lim_{x\to 0} x \sin \frac{1}{x} = 0.$$

Therefore, g is differentiable at $x = 0$, $g'(0) = 0$.

Section 3.2 Basic Differentiation Rules and Rates of Change

1. To find the derivative of $f(x) = cx^n$, multiply n by c, and reduce the power of x by 1.

$f'(x) = ncx^{n-1}$

3. For any constant c, $f(x) = ce^x$ is equal to its derivative.

5. (a) $y = x^{1/2}$

$y' = \frac{1}{2}x^{-1/2}$

$y'(1) = \frac{1}{2}$

(b) $y = x^3$

$y' = 3x^2$

$y'(1) = 3$

7. $y = 12$
$y' = 0$

9. $y = x^7$
$y' = 7x^6$

11. $y = \dfrac{1}{x^5} = x^{-5}$
$y' = -5x^{-6} = -\dfrac{5}{x^6}$

13. $f(x) = \sqrt[9]{x} = x^{1/9}$
$f'(x) = \dfrac{1}{9}x^{-8/9} = \dfrac{1}{9x^{8/9}}$

15. $f(x) = x + 11$
$f'(x) = 1$

17. $f(t) = -3t^2 + 2t - 6$
$f'(t) = -6t + 2$

19. $s(t) = t^3 + 5t^2 - 3t + 8$
$s'(t) = 3t^2 + 10t - 3$

21. $y = \dfrac{\pi}{2}\sin\theta$
$y' = \dfrac{\pi}{2}\cos\theta$

23. $y = x^2 - \frac{1}{2}\cos x$
$y' = 2x + \frac{1}{2}\sin x$

25. $y = \frac{1}{2}e^x - 3\sin x$
$y' = \frac{1}{2}e^x - 3\cos x$

	Function	Rewrite	Differentiate	Simplify
27.	$y = \dfrac{2}{7x^4}$	$y = \dfrac{2}{7}x^{-4}$	$y' = -\dfrac{8}{7}x^{-5}$	$y' = -\dfrac{8}{7x^5}$
29.	$y = \dfrac{6}{(5x)^3}$	$y = \dfrac{6}{125}x^{-3}$	$y' = -\dfrac{18}{125}x^{-4}$	$y' = -\dfrac{18}{125x^4}$

31. $f(x) = \dfrac{8}{x^2} = 8x^{-2}, (2, 2)$
$f'(x) = -16x^{-3} = -\dfrac{16}{x^3}$
$f'(2) = -2$

33. $f(x) = -\frac{1}{2} + \frac{7}{5}x^3, \left(0, -\frac{1}{2}\right)$
$f'(x) = \frac{21}{5}x^2$
$f'(0) = 0$

35. $y = (4x + 1)^2, (0, 1)$
$= 16x^2 + 8x + 1$
$y' = 32x + 8$
$y'(0) = 32(0) + 8 = 8$

37. $f(\theta) = 4\sin\theta - \theta, (0, 0)$
$f'(\theta) = 4\cos\theta - 1$
$f'(0) = 4(1) - 1 = 3$

39. $f(t) = \frac{3}{4}e^t, \left(0, \frac{3}{4}\right)$
$f'(t) = \frac{3}{4}e^t$
$f(0) = \frac{3}{4}e^0 = \frac{3}{4}$

41. $f(x) = x^2 + 5 - 3x^{-2}$
$f'(x) = 2x + 6x^{-3} = 2x + \dfrac{6}{x^3}$

43. $g(t) = t^2 - \dfrac{4}{t^3} = t^2 - 4t^{-3}$
$g'(t) = 2t + 12t^{-4} = 2t + \dfrac{12}{t^4}$

45. $f(x) = \dfrac{x^3 - 3x^2 + 4}{x^2} = x - 3 + 4x^{-2}$
$f'(x) = 1 - \dfrac{8}{x^3} = \dfrac{x^3 - 8}{x^3}$

47. $g(t) = \dfrac{3t^2 + 4t - 8}{t^{3/2}} = 3t^{1/2} + 4t^{-1/2} - 8t^{-3/2}$
$g'(t) = \dfrac{3}{2}t^{-1/2} - 2t^{-3/2} + 12t^{-5/2}$
$= \dfrac{3t^2 - 4t + 24}{2t^{5/2}}$

49. $y = x(x^2 + 1) = x^3 + x$
$y' = 3x^2 + 1$

51. $f(x) = \sqrt{x} - 6\sqrt[3]{x} = x^{1/2} - 6x^{1/3}$

$f'(x) = \frac{1}{2}x^{-1/2} - 2x^{-2/3} = \frac{1}{2\sqrt{x}} - \frac{2}{x^{2/3}}$

53. $f(x) = 6\sqrt{x} + 5\cos x = 6x^{1/2} + 5\cos x$

$f'(x) = 3x^{-1/2} - 5\sin x = \frac{3}{\sqrt{x}} - 5\sin x$

55. $y = \frac{1}{(3x)^{-2}} - 5\cos x$

$= (3x)^2 - 5\cos x = 9x^2 - 5\cos x$

$y' = 18x + 5\sin x$

57. $f(x) = x^{-2} - 2e^x$

$f'(x) = -2x^{-3} - 2e^x = \frac{-2}{x^3} - 2e^x$

59. (a) $f(x) = -2x^4 + 5x^2 - 3$

$f'(x) = -8x^3 + 10x$

At $(1, 0)$: $f'(1) = -8(1)^3 + 10(1) = 2$

Tangent line: $y - 0 = 2(x - 1)$

$y = 2x - 2$

(b) and (c)

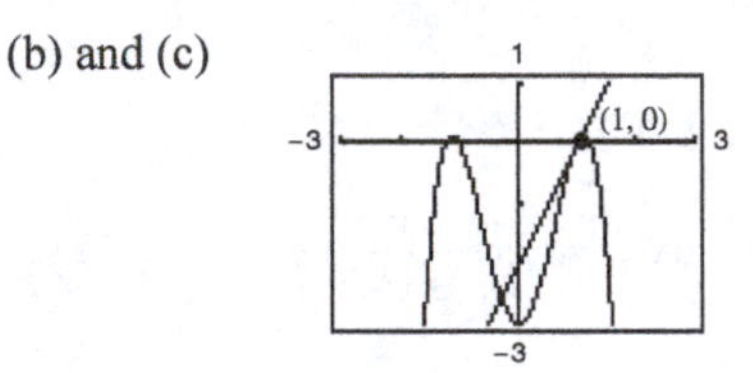

61. (a) $g(x) = x + e^x$

$g'(x) = 1 + e^x$

At $(0, 1)$: $g'(0) = 1 + 1 = 2$

Tangent line: $y - 1 = 2(x - 0)$

$y = 2x + 1$

(b) and (c)

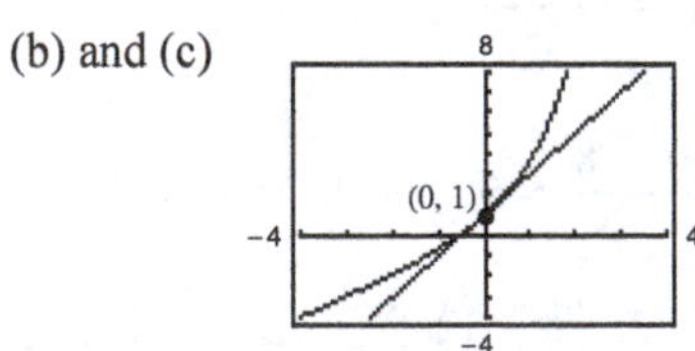

63. $y = x^4 - 2x^2 + 3$

$y' = 4x^3 - 4x = 4x(x^2 - 1) = 4x(x - 1)(x + 1)$

$y' = 0 \Rightarrow x = 0, \pm 1$

Horizontal tangents: $(0, 3)$, $(1, 2)$, $(-1, 2)$

65. $y = \frac{1}{x^2} = x^{-2}$

$y' = -2x^{-3} = -\frac{2}{x^3}$ cannot equal zero.

Therefore, there are no horizontal tangents.

67. $y = -4x + e^x$

$y' = -4 + e^x = 0$

$e^x = 4$

$x = \ln 4$

Horizontal tangent: $(\ln 4, -4\ln 4 + 4)$

69. $y = x + \sin x$, $0 \le x < 2\pi$

$y' = 1 + \cos x = 0$

$\cos x = -1 \Rightarrow x = \pi$

At $x = \pi$: $y = \pi$

Horizontal tangent: (π, π)

71. $f(x) = k - x^2$, $y = -6x + 1$

$f'(x) = -2x$ and slope of tangent line is $m = -6$.

$f'(x) = -6$

$-2x = -6$

$x = 3$

$y = -6(3) + 1 = -17$

$-17 = k - 3^2$

$8 = k$

73. $f(x) = \frac{k}{x}$, $y = -\frac{3}{4}x + 3$

$f'(x) = -\frac{k}{x^2}$ and slope of tangent line is $m = -\frac{3}{4}$.

$f'(x) = -\frac{3}{4}$

$-\frac{k}{x^2} = -\frac{3}{4}$

$x^2 = \frac{4k}{3}$

$x = \sqrt{\frac{4k}{3}}$

$y = -\frac{3}{4}x + 3 = -\frac{3}{4}\sqrt{\frac{4k}{3}} + 3$

$-\frac{3}{4}\sqrt{\frac{4k}{3}} + 3 = k\sqrt{\frac{3}{4k}}$

$k = 3$

75. $g(x) = f(x) + 6 \Rightarrow g'(x) = f'(x)$

77.

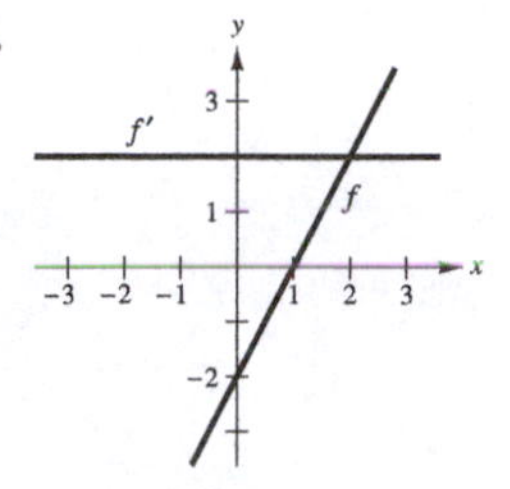

If f is linear then its derivative is a constant function.

$$f(x) = ax + b$$
$$f'(x) = a$$

79. The graph of a function f such that $f' > 0$ for all x and the rate of change of the function is decreasing (i.e., as x increases, f' decreases) would, in general, look like the graph below.

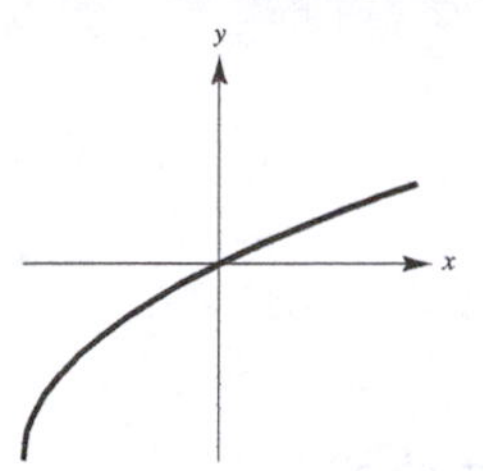

81. Let (x_1, y_1) and (x_2, y_2) be the points of tangency on $y = x^2$ and $y = -x^2 + 6x - 5$, respectively.

The derivatives of these functions are:

$$y' = 2x \Rightarrow m = 2x_1 \text{ and } y' = -2x + 6 \Rightarrow m = -2x_2 + 6$$
$$m = 2x_1 = -2x_2 + 6$$
$$x_1 = -x_2 + 3$$

Because $y_1 = x_1^2$ and $y_2 = -x_2^2 + 6x_2 - 5$:

$$m = \frac{y_2 - y_1}{x_2 - x_1} = \frac{\left(-x_2^2 + 6x_2 - 5\right) - \left(x_1^2\right)}{x_2 - x_1} = -2x_2 + 6$$

$$\frac{\left(-x_2^2 + 6x_2 - 5\right) - \left(-x_2 + 3\right)^2}{x_2 - \left(-x_2 + 3\right)} = -2x_2 + 6$$

$$\left(-x_2^2 + 6x_2 - 5\right) - \left(x_2^2 - 6x_2 + 9\right) = \left(-2x_2 + 6\right)\left(2x_2 - 3\right)$$
$$-2x_2^2 + 12x_2 - 14 = -4x_2^2 + 18x_2 - 18$$
$$2x_2^2 - 6x_2 + 4 = 0$$
$$2(x_2 - 2)(x_2 - 1) = 0$$
$$x_2 = 1 \text{ or } 2$$

$x_2 = 1 \Rightarrow y_2 = 0, x_1 = 2$ and $y_1 = 4$

So, the tangent line through $(1, 0)$ and $(2, 4)$ is

$$y - 0 = \left(\frac{4 - 0}{2 - 1}\right)(x - 1) \Rightarrow y = 4x - 4.$$

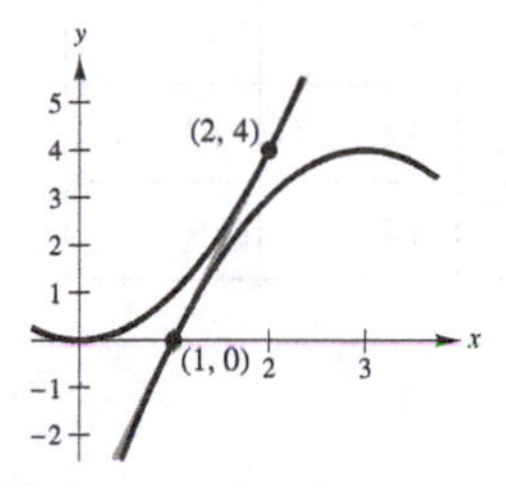

So, the tangent line through $(2, 3)$ and $(1, 1)$ is

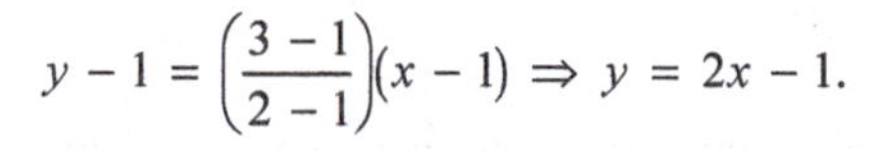

$$y - 1 = \left(\frac{3 - 1}{2 - 1}\right)(x - 1) \Rightarrow y = 2x - 1.$$

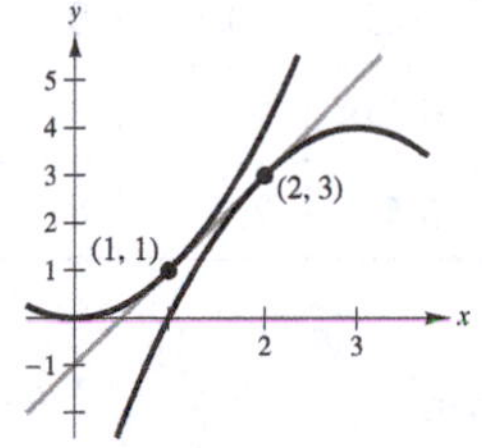

$x_2 = 2 \Rightarrow y_2 = 3, x_1 = 1$ and $y_1 = 1$

83. $f(x) = 3x + \sin x + 2$

$f'(x) = 3 + \cos x$

Because $|\cos x| \le 1$, $f'(x) \ne 0$ for all x and f does not have a horizontal tangent line.

85. $f(x) = \sqrt{x}, (-4, 0)$

$$f'(x) = \frac{1}{2}x^{-1/2} = \frac{1}{2\sqrt{x}}$$

$$\frac{1}{2\sqrt{x}} = \frac{0 - y}{-4 - x}$$

$$4 + x = 2\sqrt{x}y$$

$$4 + x = 2\sqrt{x}\sqrt{x}$$

$$4 + x = 2x$$

$$x = 4, y = 2$$

The point $(4, 2)$ is on the graph of f.

Tangent line: $y - 2 = \frac{0 - 2}{-4 - 4}(x - 4)$

$$4y - 8 = x - 4$$

$$0 = x - 4y + 4$$

87. (a) One possible secant is between $(3.9, 7.7019)$ and $(4, 8)$:

$$y - 8 = \frac{8 - 7.7019}{4 - 3.9}(x - 4)$$

$$y - 8 = 2.981(x - 4)$$

$$y = S(x) = 2.981x - 3.924$$

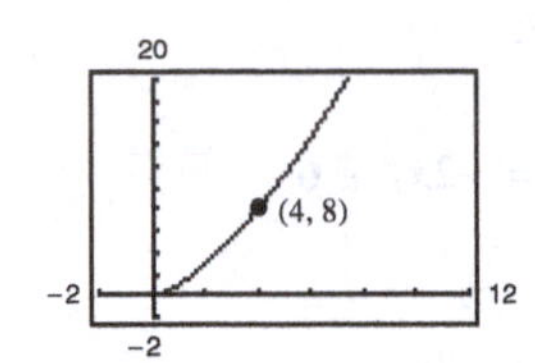

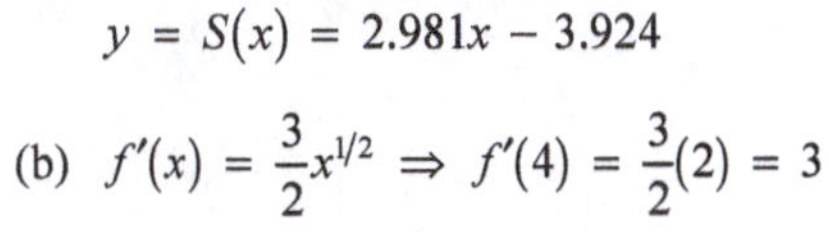

(b) $f'(x) = \frac{3}{2}x^{1/2} \Rightarrow f'(4) = \frac{3}{2}(2) = 3$

$T(x) = 3(x - 4) + 8 = 3x - 4$

The slope (and equation) of the secant line approaches that of the tangent line at $(4, 8)$ as you choose points closer and closer to $(4, 8)$.

(c) As you move further away from $(4, 8)$, the accuracy of the approximation T gets worse.

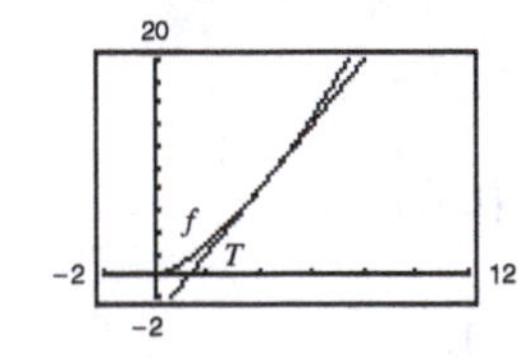

(d)

Δx	−3	−2	−1	−0.5	−0.1	0	0.1	0.5	1	2	3
$f(4 + \Delta x)$	1	2.828	5.196	6.548	7.702	8	8.302	9.546	11.180	14.697	18.520
$T(4 + \Delta x)$	−1	2	5	6.5	7.7	8	8.3	9.5	11	14	17

89. False. Let $f(x) = x$ and $g(x) = x + 1$. Then $f'(x) = g'(x) = x$, but $f(x) \neq g(x)$.

91. False. If $y = \pi^2$, then $dy/dx = 0$. (π^2 is a constant.)

93. False. If $f(x) = 0$, then $f'(x) = 0$ by the Constant Rule.

95. $f(t) = 3t + 5, \quad [1, 2]$

$f'(t) = 3$. So, $f'(1) = f'(2) = 3$.

Instantaneous rate of change is the constant 3.

Average rate of change:

$$\frac{f(2) - f(1)}{2 - 1} = \frac{11 - 8}{1} = 3$$

97. $f(x) = -\frac{1}{x}, \quad [1, 2]$

$f'(x) = \frac{1}{x^2}$

Instantaneous rate of change:

$(1, -1) \Rightarrow f'(1) = 1$

$\left(2, -\frac{1}{2}\right) \Rightarrow f'(2) = \frac{1}{4}$

Average rate of change:

$$\frac{f(2) - f(1)}{2 - 1} = \frac{(-1/2) - (-1)}{2 - 1} = \frac{1}{2}$$

99. $g(x) = x^2 + e^x, \quad [0, 1]$

$g'(x) = 2x + e^x$

Instantaneous rate of change:

$(0, 1): g'(0) = 1$

$(1, 1 + e): g'(1) = 2 + e \approx 4.718$

Average rate of change:

$$\frac{g(1) - g(0)}{1 - 0} = \frac{(1 + e) - (1)}{1} = e \approx 2.718$$

101. (a) $s(t) = -16t^2 + 1362$

$v(t) = -32t$

(b) $\frac{s(2) - s(1)}{2 - 1} = 1298 - 1346 = -48$ ft/sec

(c) $v(t) = s'(t) = -32t$

When $t = 1$: $v(1) = -32$ ft/sec

When $t = 2$: $v(2) = -64$ ft/sec

(d) $-16t^2 + 1362 = 0$

$$t^2 = \frac{1362}{16} \Rightarrow t = \frac{\sqrt{1362}}{4} \approx 9.226 \text{ sec}$$

(e) $v\left(\frac{\sqrt{1362}}{4}\right) = -32\left(\frac{\sqrt{1362}}{4}\right)$

$= -8\sqrt{1362} \approx -295.242$ ft/sec

103. $s(t) = -4.9t^2 + v_0 t + s_0$

$= -4.9t^2 + 120t$

$v(t) = -9.8t + 120$

$v(5) = -9.8(5) + 120 = 71$ m/sec

$v(10) = -9.8(10) + 120 = 22$ m/sec

105. From $(0, 0)$ to $(4, 2)$, $s(t) = \frac{1}{2}t \Rightarrow v(t) = \frac{1}{2}$ mi/min.

$v(t) = \frac{1}{2}(60) = 30$ mi/h for $0 < t < 4$

Similarly, $v(t) = 0$ for $4 < t < 6$. Finally, from $(6, 2)$ to $(10, 6)$,

$s(t) = t - 4 \Rightarrow v(t) = 1$ mi/min. $= 60$ mi/h.

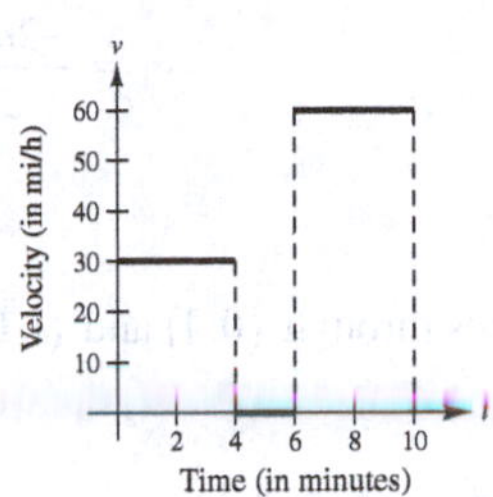

(The velocity has been converted to miles per hour.)

107. $V = s^3, \frac{dV}{ds} = 3s^2$

When $s = 6$ cm, $\frac{dV}{ds} = 108 \text{ cm}^3$ per cm change in s.

109. (a) Using a graphing utility,

$R(v) = 0.417v - 0.02.$

(b) Using a graphing utility,

$B(v) = 0.0056v^2 + 0.001v + 0.04.$

(c) $T(v) = R(v) + B(v) = 0.0056v^2 + 0.418v + 0.02$

(d)

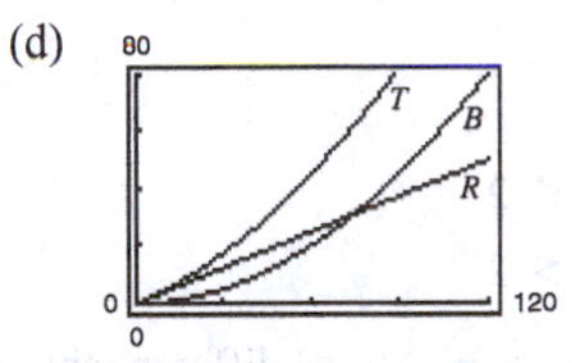

(e) $\frac{dT}{dv} = 0.0112v + 0.418$

For $v = 40$, $T'(40) \approx 0.866$

For $v = 80$, $T'(80) \approx 1.314$

For $v = 100$, $T'(100) \approx 1.538$

(f) For increasing speeds, the total stopping distance increases.

111. $s(t) = -\frac{1}{2}at^2 + c$ and $s'(t) = -at$

Average velocity: $\dfrac{s(t_0 + \Delta t) - s(t_0 - \Delta t)}{(t_0 + \Delta t) - (t_0 - \Delta t)} = \dfrac{\left[-(1/2)a(t_0 + \Delta t)^2 + c\right] - \left[-(1/2)a(t_0 - \Delta t)^2 + c\right]}{2\Delta t}$

$$= \frac{-(1/2)a\left(t_0^2 + 2t_0\Delta t + (\Delta t)^2\right) + (1/2)a\left(t_0^2 - 2t_0\Delta t + (\Delta t)^2\right)}{2\,\Delta t}$$

$$= \frac{-2at_0\,\Delta t}{2\,\Delta t} = -at_0 = s'(t_0) \quad \text{instantaneous velocity at } t = t_0$$

113. $y = ax^2 + bx + c$

Because the parabola passes through $(0, 1)$ and $(1, 0)$, you have:

$(0, 1)$: $1 = a(0)^2 + b(0) + c \Rightarrow c = 1$

$(1, 0)$: $0 = a(1)^2 + b(1) + 1 \Rightarrow b = -a - 1$

So, $y = ax^2 + (-a - 1)x + 1$. From the tangent line $y = x - 1$, you know that the derivative is 1 at the point $(1, 0)$.

$$y' = 2ax + (-a - 1)$$
$$1 = 2a(1) + (-a - 1)$$
$$1 = a - 1$$
$$a = 2$$
$$b = -a - 1 = -3$$

Therefore, $y = 2x^2 - 3x + 1$.

115. $y = x^3 - 9x$

$y' = 3x^2 - 9$

Tangent lines through $(1, -9)$:

$$y + 9 = (3x^2 - 9)(x - 1)$$
$$(x^3 - 9x) + 9 = 3x^3 - 3x^2 - 9x + 9$$
$$0 = 2x^3 - 3x^2 = x^2(2x - 3)$$
$$x = 0 \text{ or } x = \tfrac{3}{2}$$

The points of tangency are $(0, 0)$ and $\left(\frac{3}{2}, -\frac{81}{8}\right)$. At $(0, 0)$, the slope is $y'(0) = -9$. At $\left(\frac{3}{2}, -\frac{81}{8}\right)$, the slope is $y'\left(\frac{3}{2}\right) = -\frac{9}{4}$.

Tangent Lines:

$y - 0 = -9(x - 0)$ and $y + \frac{81}{8} = -\frac{9}{4}\left(x - \frac{3}{2}\right)$

$y = -9x$ $\qquad y = -\frac{9}{4}x - \frac{27}{4}$

$9x + y = 0$ $\qquad 9x + 4y + 27 = 0$

117. $f(x) = \begin{cases} ax^3, & x \le 2 \\ x^2 + b, & x > 2 \end{cases}$

f must be continuous at $x = 2$ to be differentiable at $x = 2$.

$$\left.\begin{aligned} \lim_{x\to 2^-} f(x) &= \lim_{x\to 2^-} ax^3 = 8a \\ \lim_{x\to 2^+} f(x) &= \lim_{x\to 2^+} (x^2 + b) = 4 + b \end{aligned}\right\} \quad \begin{aligned} 8a &= 4 + b \\ 8a - 4 &= b \end{aligned}$$

$$f'(x) = \begin{cases} 3ax^2, & x < 2 \\ 2x, & x > 2 \end{cases}$$

For f to be differentiable at $x = 2$, the left derivative must equal the right derivative.

$$3a(2)^2 = 2(2)$$
$$12a = 4$$
$$a = \tfrac{1}{3}$$
$$b = 8a - 4 = -\tfrac{4}{3}$$

119. $f_1(x) = |\sin x|$ is differentiable for all $x \neq n\pi$, n an integer.

$f_2(x) = \sin|x|$ is differentiable for all $x \neq 0$.

You can verify this by graphing f_1 and f_2 and observing the locations of the sharp turns.

121. You are given $f : R \to R$ satisfying

$(*) f'(x) = \dfrac{f(x+n) - f(x)}{n}$ for all real numbers x and all positive integers n.

You claim that $f(x) = mx + b,\ m, b \in R$.

For this case, $f'(x) = m = \dfrac{[m(x+n)+b] - [mx+b]}{n} = m$.

Furthermore, these are the only solutions:

Note first that $f'(x+1) = \dfrac{f(x+2) - f(x+1)}{1}$, and $f'(x) = f(x+1) - f(x)$.

From (*) you have $2f'(x) = f(x+2) - f(x) = [f(x+2) - f(x+1)] + [f(x+1) - f(x)] = f'(x+1) + f'(x)$.

Thus, $f'(x) = f'(x+1)$.

Let $g(x) = f(x+1) - f(x)$.

Let $m = g(0) = f(1) - f(0)$.

Let $b = f(0)$.

Then $g'(x) = f'(x+1) - f'(x) = 0$

$g(x) = \text{constant} = g(0) = m$

$f'(x) = f(x+1) - f(x) = g(x) = m \Rightarrow f(x) = mx + b$.

Section 3.3 Product and Quotient Rules and Higher-Order Derivatives

1. To find the derivative of the product of two differentiable functions f and g, multiply the first function f by the derivative of the second function g, and then add the second function g times the derivative of the first function f.

3. $\dfrac{d}{dx}\tan x = \sec^2 x$

$\dfrac{d}{dx}\cot x = -\csc^2 x$

$\dfrac{d}{dx}\sec x = \sec x \tan x$

$\dfrac{d}{dx}\csc x = -\csc x \cot x$

5. $g(x) = (2x-3)(1-5x)$

$$g'(x) = (2x-3)(-5) + (1-5x)(2)$$
$$= -10x + 15 + 2 - 10x$$
$$= -20x + 17$$

7. $h(t) = \sqrt{t}(1-t^2) = t^{1/2}(1-t^2)$

$$h'(t) = t^{1/2}(-2t) + (1-t^2)\frac{1}{2}t^{-1/2}$$
$$= -2t^{3/2} + \frac{1}{2t^{1/2}} - \frac{1}{2}t^{3/2}$$
$$= -\frac{5}{2}t^{3/2} + \frac{1}{2t^{1/2}}$$
$$= \frac{1-5t^2}{2t^{1/2}} = \frac{1-5t^2}{2\sqrt{t}}$$

9. $f(x) = e^x \cos x$

$$f'(x) = e^x(-\sin x) + e^x \cos x$$
$$= e^x(\cos x - \sin x)$$

11. $f(x) = \dfrac{x}{x-5}$

$$f'(x) = \frac{(x-5)(1) - x(1)}{(x-5)^2} = \frac{x-5-x}{(x-5)^2} = -\frac{5}{(x-5)^2}$$

13. $h(x) = \dfrac{\sqrt{x}}{x^3 + 1} = \dfrac{x^{1/2}}{x^3 + 1}$

$h'(x) = \dfrac{(x^3 + 1)\frac{1}{2}x^{-1/2} - x^{1/2}(3x^2)}{(x^3 + 1)^2} = \dfrac{x^3 + 1 - 6x^3}{2x^{1/2}(x^3 + 1)^2} = \dfrac{1 - 5x^3}{2\sqrt{x}(x^3 + 1)^2}$

15. $g(x) = \dfrac{\sin x}{e^x}$

$g'(x) = \dfrac{e^x \cos x - \sin x(e^x)}{(e^x)^2} = \dfrac{\cos x - \sin x}{e^x}$

17. $f(x) = (x^3 + 4x)(3x^2 + 2x - 5)$

$f'(x) = (x^3 + 4x)(6x + 2) + (3x^2 + 2x - 5)(3x^2 + 4)$

$= 6x^4 + 24x^2 + 2x^3 + 8x + 9x^4 + 6x^3 - 15x^2 + 12x^2 + 8x - 20$

$= 15x^4 + 8x^3 + 21x^2 + 16x - 20$

$f'(0) = -20$

19. $f(x) = \dfrac{x^2 - 4}{x - 3}$

$f'(x) = \dfrac{(x - 3)(2x) - (x^2 - 4)(1)}{(x - 3)^2} = \dfrac{2x^2 - 6x - x^2 + 4}{(x - 3)^2} = \dfrac{x^2 - 6x + 4}{(x - 3)^2}$

$f'(1) = \dfrac{1 - 6 + 4}{(1 - 3)^2} = -\dfrac{1}{4}$

21. $f(x) = x \cos x$

$f'(x) = (x)(-\sin x) + (\cos x)(1) = \cos x - x \sin x$

$f'\left(\dfrac{\pi}{4}\right) = \dfrac{\sqrt{2}}{2} - \dfrac{\pi}{4}\left(\dfrac{\sqrt{2}}{2}\right) = \dfrac{\sqrt{2}}{8}(4 - \pi)$

23. $f(x) = e^x \sin x$

$f'(x) = e^x \cos x + e^x \sin x$

$= e^x (\cos x + \sin x)$

$f'(0) = 1$

	Function	Rewrite	Differentiate	Simplify
25.	$y = \dfrac{x^3 + 6x}{3}$	$y = \dfrac{1}{3}x^3 + 2x$	$y' = \dfrac{1}{3}(3x^2) + 2$	$y' = x^2 + 2$
27.	$y = \dfrac{6}{7x^2}$	$y = \dfrac{6}{7}x^{-2}$	$y' = -\dfrac{12}{7}x^{-3}$	$y' = -\dfrac{12}{7x^3}$
29.	$y = \dfrac{4x^{3/2}}{x}$	$y = 4x^{1/2}, x > 0$	$y' = 2x^{-1/2}$	$y' = \dfrac{2}{\sqrt{x}}, x > 0$

31. $f(x) = \dfrac{4 - 3x - x^2}{x^2 - 1}$

$f'(x) = \dfrac{(x^2 - 1)(-3 - 2x) - (4 - 3x - x^2)(2x)}{(x^2 - 1)^2} = \dfrac{-3x^2 + 3 - 2x^3 + 2x - 8x + 6x^2 + 2x^3}{(x^2 - 1)^2}$

$= \dfrac{3x^2 - 6x + 3}{(x^2 - 1)^2} = \dfrac{3(x^2 - 2x + 1)}{(x^2 - 1)^2} = \dfrac{3(x - 1)^2}{(x - 1)^2(x + 1)^2} = \dfrac{3}{(x + 1)^2}, x \neq 1$

33. $f(x) = x\left(1 - \frac{4}{x+3}\right) = x - \frac{4x}{x+3}$

$f'(x) = 1 - \frac{(x+3)4 - 4x(1)}{(x+3)^2}$

$= \frac{(x^2 + 6x + 9) - 12}{(x+3)^2}$

$= \frac{x^2 + 6x - 3}{(x+3)^2}$

35. $f(x) = \frac{3x - 1}{\sqrt{x}} = 3x^{1/2} - x^{-1/2}$

$f'(x) = \frac{3}{2}x^{-1/2} + \frac{1}{2}x^{-3/2} = \frac{3x+1}{2x^{3/2}}$

Alternate solution:

$f(x) = \frac{3x-1}{\sqrt{x}} = \frac{3x-1}{x^{1/2}}$

$f'(x) = \frac{x^{1/2}(3) - (3x-1)\left(\frac{1}{2}\right)\left(x^{-1/2}\right)}{x}$

$= \frac{\frac{1}{2}x^{-1/2}(3x+1)}{x}$

$= \frac{3x+1}{2x^{3/2}}$

37. $f(x) = \frac{2 - (1/x)}{x - 3} = \frac{2x-1}{x(x-3)} = \frac{2x-1}{x^2 - 3x}$

$f'(x) = \frac{(x^2 - 3x)2 - (2x-1)(2x-3)}{(x^2 - 3x)^2}$

$= \frac{2x^2 - 6x - 4x^2 + 8x - 3}{(x^2 - 3x)^2}$

$= \frac{-2x^2 + 2x - 3}{(x^2 - 3x)^2} = -\frac{2x^2 - 2x + 3}{x^2(x-3)^2}$

39. $g(s) = s^3\left(5 - \frac{s}{s+2}\right) = 5s^3 - \frac{s^4}{s+2}$

$g'(s) = 15s^2 - \frac{(s+2)4s^3 - s^4(1)}{(s+2)^2}$

$= 15s^2 - \frac{3s^4 + 8s^3}{(s+2)^2}$

$= \frac{15s^2(s+2)^2 - (3s^4 + 8s^3)}{(s+2)^2}$

$= \frac{15s^2(s^2 + 4s + 4) - 3s^4 - 8s^3}{(s+2)^2}$

$= \frac{12s^4 + 52s^3 + 60s^2}{(s+2)^2}$

$= \frac{4s^2(3s^2 + 13s + 15)}{(s+2)^2}$

41. $f(x) = (2x^3 + 5x)(x-3)(x+2)$

$f'(x) = (6x^2 + 5)(x-3)(x+2) + (2x^3 + 5x)(1)(x+2) + (2x^3 + 5x)(x-3)(1)$

$= (6x^2 + 5)(x^2 - x - 6) + (2x^3 + 5x)(x+2) + (2x^3 + 5x)(x-3)$

$= (6x^4 + 5x^2 - 6x^3 - 5x - 36x^2 - 30) + (2x^4 + 4x^3 + 5x^2 + 10x) + (2x^4 + 5x^2 - 6x^3 - 15x)$

$= 10x^4 - 8x^3 - 21x^2 - 10x - 30$

Note: You could simplify first: $f(x) = (2x^3 + 5x)(x^2 - x - 6)$

43. $f(t) = t^2 \sin t$

$f'(t) = t^2 \cos t + 2t \sin t = t(t \cos t + 2 \sin t)$

45. $f(t) = \frac{\cos t}{t}$

$f'(t) = \frac{-t \sin t - \cos t}{t^2} = -\frac{t \sin t + \cos t}{t^2}$

47. $f(x) = -e^x + \tan x$

$f'(x) = -e^x + \sec^2 x$

49. $g(t) = \sqrt[4]{t} + 6 \csc t = t^{1/4} + 6 \csc t$

$g'(t) = \frac{1}{4}t^{-3/4} - 6 \csc t \cot t$

$= \frac{1}{4t^{3/4}} - 6 \csc t \cot t$

51. $y = \dfrac{3(1 - \sin x)}{2\cos x} = \dfrac{3 - 3\sin x}{2\cos x}$

$y' = \dfrac{(-3\cos x)(2\cos x) - (3 - 3\sin x)(-2\sin x)}{(2\cos x)^2}$

$= \dfrac{-6\cos^2 x + 6\sin x - 6\sin^2 x}{4\cos^2 x}$

$= \dfrac{3}{2}\left(-1 + \tan x \sec x - \tan^2 x\right)$

$= \dfrac{3}{2}\sec x(\tan x - \sec x)$

53. $y = -\csc x - \sin x$

$y' = \csc x \cot x - \cos x$

$= \dfrac{\cos x}{\sin^2 x} - \cos x$

$= \cos x\left(\csc^2 x - 1\right)$

$= \cos x \cot^2 x$

55. $f(x) = x^2 \tan x$

$f'(x) = x^2 \sec^2 x + 2x \tan x = x\left(x \sec^2 x + 2\tan x\right)$

57. $y = 2x \sin x + x^2 e^x$

$y' = 2x(\cos x) + 2\sin x + x^2 e^x + 2xe^x$

$= 2x\cos x + 2\sin x + xe^x(x + 2)$

59. $y = \dfrac{e^x}{4\sqrt{x}}$

$y' = \dfrac{4\sqrt{x}e^x - e^x\left(4/2\sqrt{x}\right)}{\left(4\sqrt{x}\right)^2} = \dfrac{e^x\left[4\sqrt{x} - \left(2/\sqrt{x}\right)\right]}{16x} = \dfrac{e^x(4x - 2)}{16x^{3/2}} = \dfrac{e^x(2x - 1)}{8x^{3/2}}$

61. $g(x) = \left(\dfrac{x + 1}{x + 2}\right)(2x - 5)$

$g'(x) = \left(\dfrac{x + 1}{x + 2}\right)(2) + (2x - 5)\left[\dfrac{(x + 2)(1) - (x + 1)(1)}{(x + 2)^2}\right] = \dfrac{2x^2 + 8x - 1}{(x + 2)^2}$

(Form of answer may vary.)

63. $y = \dfrac{1 + \csc x}{1 - \csc x}$

$y' = \dfrac{(1 - \csc x)(-\csc x \cot x) - (1 + \csc x)(\csc x \cot x)}{(1 - \csc x)^2} = \dfrac{-2\csc x \cot x}{(1 - \csc x)^2}$

$y'\left(\dfrac{\pi}{6}\right) = \dfrac{-2(2)\left(\sqrt{3}\right)}{(1 - 2)^2} = -4\sqrt{3}$

65. $h(t) = \dfrac{\sec t}{t}$

$h'(t) = \dfrac{t(\sec t \tan t) - (\sec t)(1)}{t^2} = \dfrac{\sec t(t \tan t - 1)}{t^2}$

$h'(\pi) = \dfrac{\sec \pi(\pi \tan \pi - 1)}{\pi^2} = \dfrac{1}{\pi^2}$

67. (a) $f(x) = \left(x^3 + 4x - 1\right)(x - 2), \quad (1, -4)$

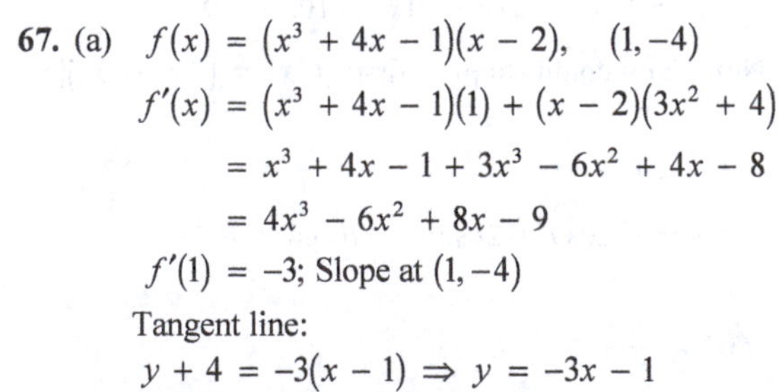

$f'(x) = \left(x^3 + 4x - 1\right)(1) + (x - 2)\left(3x^2 + 4\right)$

$= x^3 + 4x - 1 + 3x^3 - 6x^2 + 4x - 8$

$= 4x^3 - 6x^2 + 8x - 9$

$f'(1) = -3$; Slope at $(1, -4)$

Tangent line:

$y + 4 = -3(x - 1) \Rightarrow y = -3x - 1$

(b)

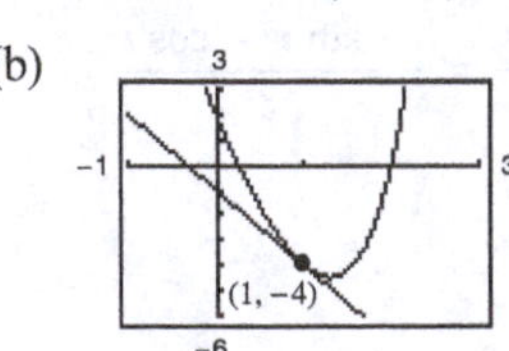

(c) Graphing utility confirms $\dfrac{dy}{dx} = -3$ at $(1, -4)$.

69. (a) $f(x) = \dfrac{x}{x+4}, \quad (-5, 5)$

$$f'(x) = \frac{(x+4)(1) - x(1)}{(x+4)^2} = \frac{4}{(x+4)^2}$$

$$f'(-5) = \frac{4}{(-5+4)^2} = 4; \quad \text{Slope at } (-5, 5)$$

Tangent line: $y - 5 = 4(x + 5) \Rightarrow y = 4x + 25$

(b)

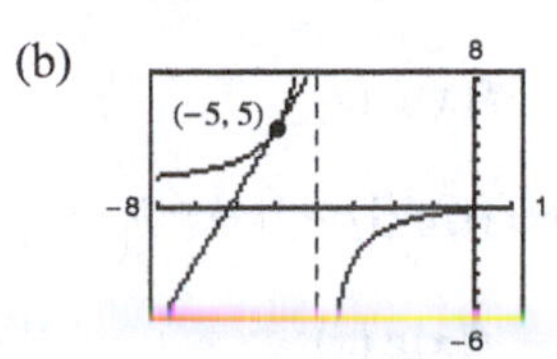

(c) Graphing utility confirms $\dfrac{dy}{dx} = 4$ at $(-5, 5)$.

71. (a) $f(x) = \tan x, \quad \left(\dfrac{\pi}{4}, 1\right)$

$$f'(x) = \sec^2 x$$

$$f'\left(\frac{\pi}{4}\right) = 2; \quad \text{Slope at } \left(\frac{\pi}{4}, 1\right)$$

Tangent line:

$$y - 1 = 2\left(x - \frac{\pi}{4}\right)$$

$$y - 1 = 2x - \frac{\pi}{2}$$

$$4x - 2y - \pi + 2 = 0$$

(b)

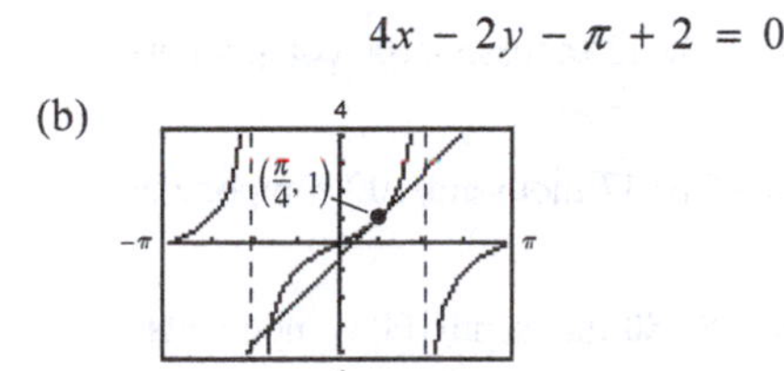

(c) Graphing utility confirms $\dfrac{dy}{dx} = 2$ at $\left(\dfrac{\pi}{4}, 1\right)$.

73. (a) $f(x) = (x - 1)e^x, \quad (1, 0)$

$$f'(x) = (x-1)e^x + e^x = xe^x$$

$$f'(1) = e$$

Tangent line: $y - 0 = e(x - 1)$

$$y = e(x - 1)$$

(b)

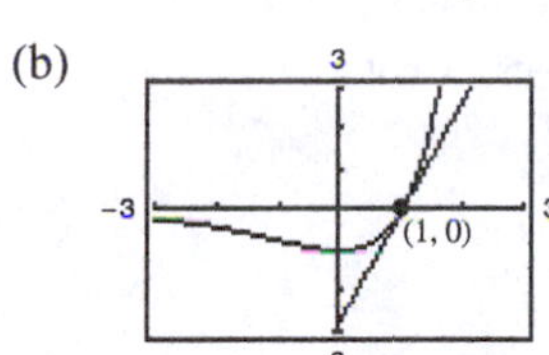

(c) Graphing utility confirms $\dfrac{dy}{dx} = e$ at $(1, 0)$.

75. $f(x) = \dfrac{8}{x^2 + 4}; \quad (2, 1)$

$$f'(x) = \frac{(x^2+4)(0) - 8(2x)}{(x^2+4)^2} = \frac{-16x}{(x^2+4)^2}$$

$$f'(2) = \frac{-16(2)}{(4+4)^2} = -\frac{1}{2}$$

$$y - 1 = -\frac{1}{2}(x - 2)$$

$$y = -\frac{1}{2}x + 2$$

$$2y + x - 4 = 0$$

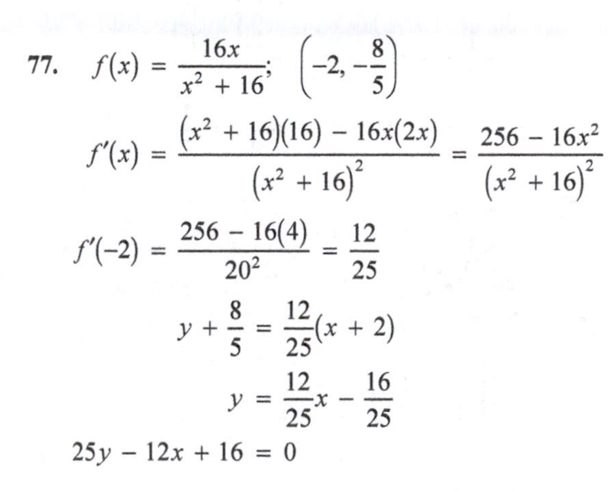

77. $f(x) = \dfrac{16x}{x^2 + 16}; \quad \left(-2, -\dfrac{8}{5}\right)$

$$f'(x) = \frac{(x^2+16)(16) - 16x(2x)}{(x^2+16)^2} = \frac{256 - 16x^2}{(x^2+16)^2}$$

$$f'(-2) = \frac{256 - 16(4)}{20^2} = \frac{12}{25}$$

$$y + \frac{8}{5} = \frac{12}{25}(x + 2)$$

$$y = \frac{12}{25}x - \frac{16}{25}$$

$$25y - 12x + 16 = 0$$

79. $f(x) = \dfrac{x^2}{x - 1}$

$$f'(x) = \frac{(x-1)(2x) - x^2(1)}{(x-1)^2}$$

$$= \frac{x^2 - 2x}{(x-1)^2} = \frac{x(x-2)}{(x-1)^2}$$

$f'(x) = 0$ when $x = 0$ or $x = 2$

Horizontal tangents are at $(0, 0)$ and $(2, 4)$.

81. $g(x) = \dfrac{8(x - 2)}{e^x}$

$$g'(x) = \frac{e^x(8) - 8(x-2)e^x}{e^{2x}} = \frac{24 - 8x}{e^x}$$

$g'(x) = 0$ when $x = 3$.

Horizontal tangent is at $(3, 8e^{-3})$.

83. $f(x) = \dfrac{x+1}{x-1}$

$f'(x) = \dfrac{(x-1)-(x+1)}{(x-1)^2} = \dfrac{-2}{(x-1)^2}$

$2y + x = 6 \Rightarrow y = -\dfrac{1}{2}x + 3$; Slope: $-\dfrac{1}{2}$

$\dfrac{-2}{(x-1)^2} = -\dfrac{1}{2}$

$(x-1)^2 = 4$

$x - 1 = \pm 2$

$x = -1, 3;\ f(-1) = 0,\ f(3) = 2$

$y - 0 = -\dfrac{1}{2}(x+1) \Rightarrow y = -\dfrac{1}{2}x - \dfrac{1}{2}$

$y - 2 = -\dfrac{1}{2}(x-3) \Rightarrow y = -\dfrac{1}{2}x + \dfrac{7}{2}$

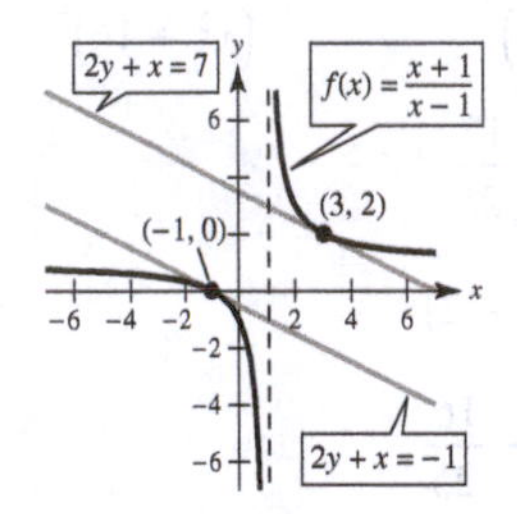

85. $f'(x) = \dfrac{(x+2)3 - 3x(1)}{(x+2)^2} = \dfrac{6}{(x+2)^2}$

$g'(x) = \dfrac{(x+2)5 - (5x+4)(1)}{(x+2)^2} = \dfrac{6}{(x+2)^2}$

$g(x) = \dfrac{5x+4}{(x+2)} = \dfrac{3x}{(x+2)} + \dfrac{2x+4}{(x+2)} = f(x) + 2$

f and g differ by a constant.

87. (a) $p'(x) = f'(x)g(x) + f(x)g'(x)$

$p'(1) = f'(1)g(1) + f(1)g'(1) = 1(4) + 6\left(-\dfrac{1}{2}\right) = 1$

(b) $q'(x) = \dfrac{g(x)f'(x) - f(x)g'(x)}{g(x)^2}$

$q'(4) = \dfrac{3(-1) - 7(0)}{3^2} = -\dfrac{1}{3}$

89. Area $= A(t) = (6t+5)\sqrt{t} = 6t^{3/2} + 5t^{1/2}$

$A'(t) = 9t^{1/2} + \dfrac{5}{2}t^{-1/2} = \dfrac{18t+5}{2\sqrt{t}}$ cm²/sec

91. $C = 100\left(\dfrac{200}{x^2} + \dfrac{x}{x+30}\right),\quad 1 \le x$

$\dfrac{dC}{dx} = 100\left(-\dfrac{400}{x^3} + \dfrac{30}{(x+30)^2}\right)$

(a) When $x = 10$: $\dfrac{dC}{dx} = -\$38.13$ thousand/100 components

(b) When $x = 15$: $\dfrac{dC}{dx} = -\$10.37$ thousand/100 components

(c) When $x = 20$: $\dfrac{dC}{dx} = -\$3.80$ thousand/100 components

As the order size increases, the cost per item decreases.

93. (a) $\sec x = \dfrac{1}{\cos x}$

$$\frac{d}{dx}[\sec x] = \frac{d}{dx}\left[\frac{1}{\cos x}\right] = \frac{(\cos x)(0) - (1)(-\sin x)}{(\cos x)^2} = \frac{\sin x}{\cos x \cos x} = \frac{1}{\cos x}\cdot\frac{\sin x}{\cos x} = \sec x \tan x$$

(b) $\csc x = \dfrac{1}{\sin x}$

$$\frac{d}{dx}[\csc x] = \frac{d}{dx}\left[\frac{1}{\sin x}\right] = \frac{(\sin x)(0) - (1)(\cos x)}{(\sin x)^2} = -\frac{\cos x}{\sin x \sin x} = -\frac{1}{\sin x}\cdot\frac{\cos x}{\sin x} = -\csc x \cot x$$

(c) $\cot x = \dfrac{\cos x}{\sin x}$

$$\frac{d}{dx}[\cot x] = \frac{d}{dx}\left[\frac{\cos x}{\sin x}\right] = \frac{\sin x(-\sin x) - (\cos x)(\cos x)}{(\sin x)^2} = -\frac{\sin^2 x + \cos^2 x}{\sin^2 x} = -\frac{1}{\sin^2 x} = -\csc^2 x$$

95. (a) $h(t) = 101.7t + 1593$

$p(t) = 2.1t + 287$

(b)

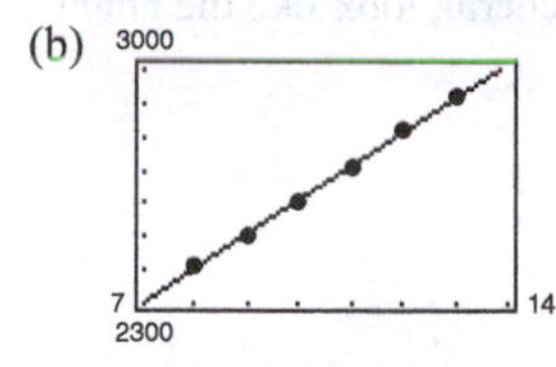

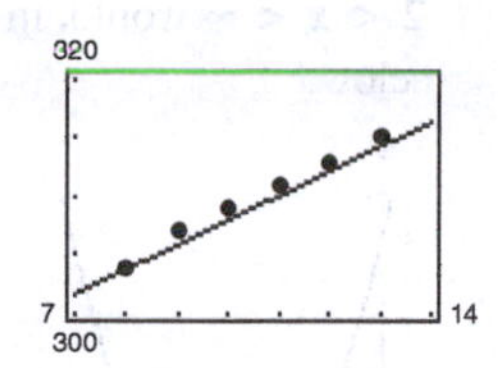

(c) $A = \dfrac{101.7t + 1593}{2.1t + 287}$

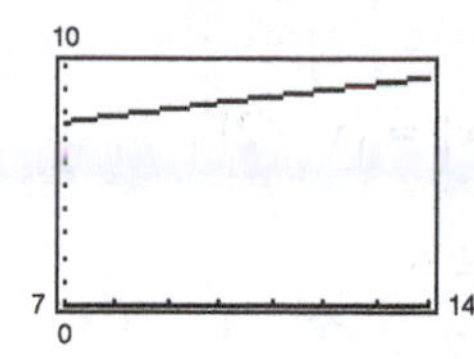

A represents the average health care expenditures per person (in thousands of dollars).

(d) $A'(t) \approx \dfrac{25{,}842.6}{4.41t^2 + 1205.4t + 82{,}369}$

$A'(t)$ represents the rate of change of the average health care expenditures per person for the given year *t*.

97. $f(x) = x^2 + 7x - 4$

$f'(x) = 2x + 7$

$f''(x) = 2$

99. $f(x) = 4x^{3/2}$

$f'(x) = 6x^{1/2}$

$f''(x) = 3x^{-1/2} = \dfrac{3}{\sqrt{x}}$

101. $f(x) = \dfrac{x}{x - 1}$

$f'(x) = \dfrac{(x - 1)(1) - x(1)}{(x - 1)^2} = \dfrac{-1}{(x - 1)^2}$

$f''(x) = \dfrac{2}{(x - 1)^3}$

103. $f(x) = x \sin x$

$f'(x) = x \cos x + \sin x$

$f''(x) = x(-\sin x) + \cos x + \cos x$

$= -x \sin x + 2 \cos x$

105. $f(x) = \csc x$

$f'(x) = -\csc x \cot x$

$f''(x) = -\csc x(-\csc^2 x) - \cot x(-\csc x \cot x)$

$= \csc^3 x + \cot^2 x \csc x$

107. $g(x) = \dfrac{e^x}{x}$

$g'(x) = \dfrac{xe^x - e^x}{x^2}$

$g''(x) = \dfrac{x^2(xe^x + e^x - e^x) - 2x(xe^x - e^x)}{x^4}$

$= \dfrac{e^x}{x^3}(x^2 - 2x + 2)$

109. $f'(x) = x^3 - x^{2/5}$

$f''(x) = 3x^2 - \dfrac{2}{5}x^{-3/5}$

$f'''(x) = 6x + \dfrac{6}{25}x^{-8/5} = 6x + \dfrac{6}{25x^{8/5}}$

111. $f''(x) = -\sin x$

$f^{(3)}(x) = -\cos x$

$f^{(4)}(x) = \sin x$

$f^{(5)}(x) = \cos x$

$f^{(6)}(x) = -\sin x$

$f^{(7)}(x) = -\cos x$

$f^{(8)}(x) = \sin x$

113. $f(x) = 2g(x) + h(x)$

$f'(x) = 2g'(x) + h'(x)$

$f'(2) = 2g'(2) + h'(2)$

$= 2(-2) + 4$

$= 0$

115. $f(x) = \dfrac{g(x)}{h(x)}$

$f'(x) = \dfrac{h(x)g'(x) - g(x)h'(x)}{[h(x)]^2}$

$f'(2) = \dfrac{h(2)g'(2) - g(2)h'(2)}{[h(2)]^2}$

$= \dfrac{(-1)(-2) - (3)(4)}{(-1)^2}$

$= -10$

117. Polynomials of degree $n - 1$ (or lower) satisfy $f^{(n)}(x) = 0$. The derivative of a polynomial of the 0th degree (a constant) is 0.

119.

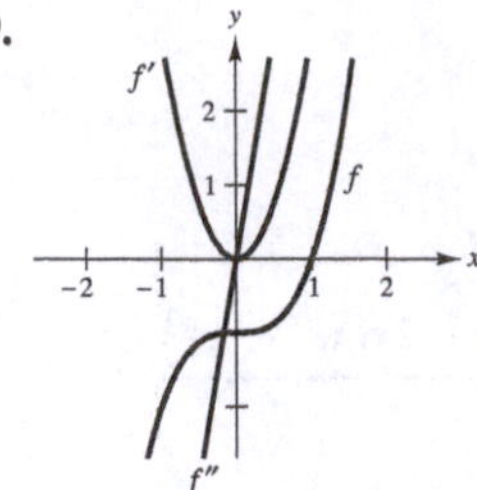

It appears that f is cubic, so f' would be quadratic and f'' would be linear.

121.

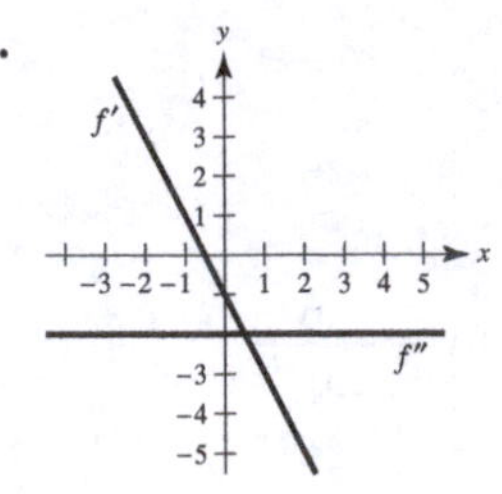

123. The graph of a differentiable function f such that $f(2) = 0$, $f' < 0$ for $-\infty < x < 2$, and $f' > 0$ for $2 < x < \infty$ would, in general, look like the graph below.

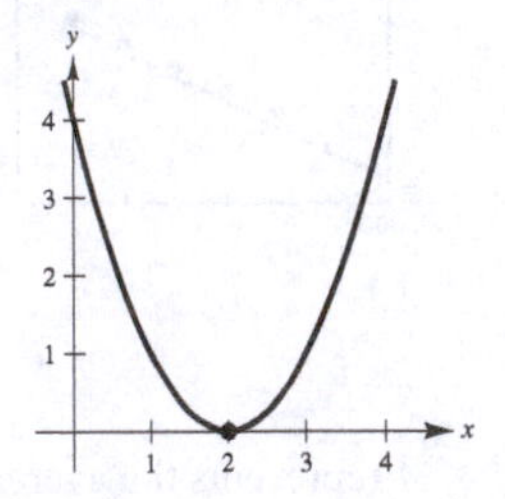

One such function is $f(x) = (x - 2)^2$.

125. $v(t) = 36 - t^2, 0 \le t \le 6$

$a(t) = v'(t) = -2t$

$v(3) = 27$ m/sec

$a(3) = -6$ m/sec^2

The speed of the object is decreasing.

127. $s(t) = -8.25t^2 + 66t$

$v(t) = s'(t) = -16.50t + 66$

$a(t) = v'(t) = -16.50$

t(sec)	0	1	2	3	4
$s(t)$ (ft)	0	57.75	99	123.75	132
$v(t) = s'(t)$ (ft / sec)	66	49.5	33	16.5	0
$a(t) = v'(t)$ (ft / sec^2)	−16.5	−16.5	−16.5	−16.5	−16.5

Average velocity on:

$[0, 1]$ is $\dfrac{57.75 - 0}{1 - 0} = 57.75$

$[1, 2]$ is $\dfrac{99 - 57.75}{2 - 1} = 41.25$

$[2, 3]$ is $\dfrac{123.75 - 99}{3 - 2} = 24.75$

$[3, 4]$ is $\dfrac{132 - 123.75}{4 - 3} = 8.25$

129. $f(x) = x^n$

$f^{(n)}(x) = n(n - 1)(n - 2)\cdots(2)(1) = n!$

Note: $n! = n(n - 1)\cdots 3 \cdot 2 \cdot 1$ (read "n factorial")

131. $f(x) = g(x)h(x)$

(a) $$f'(x) = g(x)h'(x) + h(x)g'(x)$$
$$f''(x) = g(x)h''(x) + g'(x)h'(x) + h(x)g''(x) + h'(x)g'(x)$$
$$= g(x)h''(x) + 2g'(x)h'(x) + h(x)g''(x)$$
$$f'''(x) = g(x)h'''(x) + g'(x)h''(x) + 2g'(x)h''(x) + 2g''(x)h'(x) + h(x)g'''(x) + h'(x)g''(x)$$
$$= g(x)h'''(x) + 3g'(x)h''(x) + 3g''(x)h'(x) + g'''(x)h(x)$$
$$f^{(4)}(x) = g(x)h^{(4)}(x) + g'(x)h'''(x) + 3g'(x)h'''(x) + 3g''(x)h''(x) + 3g''(x)h''(x) + 3g'''(x)h'(x) + g'''(x)h'(x) + g^{(4)}(x)h(x)$$
$$= g(x)h^{(4)}(x) + 4g'(x)h'''(x) + 6g''(x)h''(x) + 4g'''(x)h'(x) + g^{(4)}(x)h(x)$$

(b) $$f^{(n)}(x) = g(x)h^{(n)}(x) + \frac{n(n-1)(n-2)\cdots(2)(1)}{1\left[(n-1)(n-2)\cdots(2)(1)\right]}g'(x)h^{(n-1)}(x) + \frac{n(n-1)(n-2)\cdots(2)(1)}{(2)(1)\left[(n-2)(n-3)\cdots(2)(1)\right]}g''(x)h^{(n-2)}(x)$$
$$+ \frac{n(n-1)(n-2)\cdots(2)(1)}{(3)(2)(1)\left[(n-3)(n-4)\cdots(2)(1)\right]}g'''(x)h^{(n-3)}(x) + \cdots$$
$$+ \frac{n(n-1)(n-2)\cdots(2)(1)}{\left[(n-1)(n-2)\cdots(2)(1)\right](1)}g^{(n-1)}(x)h'(x) + g^{(n)}(x)h(x)$$
$$= g(x)h^{(n)}(x) + \frac{n!}{1!(n-1)!}g'(x)h^{(n-1)}(x) + \frac{n!}{2!(n-2)!}g''(x)h^{(n-2)}(x) + \cdots$$
$$+ \frac{n!}{(n-1)!1!}g^{(n-1)}(x)h'(x) + g^{(n)}(x)h(x)$$

Note: $n! = n(n-1)\cdots3\cdot2\cdot1$ (read "n factorial")

133. $f(x) = x^n \sin x$

$f'(x) = x^n \cos x + nx^{n-1}\sin x$

When $n = 1$: $f'(x) = x\cos x + \sin x$

When $n = 2$: $f'(x) = x^2\cos x + 2\sin x$

When $n = 3$: $f'(x) = x^3\cos x + 3x^2\sin x$

When $n = 4$: $f'(x) = x^4\cos x + 4x^3\sin x$

For general n, $f'(x) = x^n\cos x + nx^{n-1}\sin x$.

135. $y = \frac{1}{x}, y' = -\frac{1}{x^2}, y'' = \frac{2}{x^3}$

$$x^3y'' + 2x^2y' = x^3\left[\frac{2}{x^3}\right] + 2x^2\left[-\frac{1}{x^2}\right] = 2 - 2 = 0$$

137. $$y = 2\sin x + 3$$
$$y' = 2\cos x$$
$$y'' = -2\sin x$$
$$y'' + y = -2\sin x + (2\sin x + 3) = 3$$

139. False. If $y = f(x)g(x)$, then

$$\frac{dy}{dx} = f(x)g'(x) + g(x)f'(x).$$

141. True

$$h'(c) = f(c)g'(c) + g(c)f'(c)$$
$$= f(c)(0) + g(c)(0)$$
$$= 0$$

143. True

145. $$\frac{d}{dx}\left[f(x)g(x)h(x)\right] = \frac{d}{dx}\left[(f(x)g(x))h(x)\right]$$
$$= \frac{d}{dx}\left[f(x)g(x)\right]h(x) + f(x)g(x)h'(x)$$
$$= \left[f(x)g'(x) + g(x)f'(x)\right]h(x) + f(x)g(x)h'(x)$$
$$= f'(x)g(x)h(x) + f(x)g'(x)h(x) + f(x)g(x)h'(x)$$

Section 3.4 The Chain Rule

1. To find the derivative of the composition of two differentiable functions, take the derivative of the outer function and keep the inner function the same. Then multiply this by the derivative of the inner function.

$$[f(g(x))]' = f'(g(x))\,g'(x)$$

3. Because $d/dx = u'/u$ for $\ln u$, you get $d/dx = 2/2x$ for $\ln 2x$ and $d/dx = 3/3x$ for $\ln 3x$. So, both derivatives simplify to $1/x$.

	$y = f(g(x))$	$u = g(x)$	$y = f(u)$
5.	$y = (6x - 5)^4$	$u = 6x - 5$	$y = u^4$
7.	$y = \dfrac{1}{3x + 5}$	$u = 3x + 5$	$y = \dfrac{1}{u}$
9.	$y = \csc^3 x$	$u = \csc x$	$y = u^3$
11.	$y = e^{-2x}$	$u = -2x$	$y = e^u$

13. $y = (2x - 7)^3$

$$y' = 3(2x - 7)^2(2) = 6(2x - 7)^2$$

15. $g(x) = 3(4 - 9x)^{5/6}$

$$g'(x) = 3\left(\frac{5}{6}\right)(4 - 9x)^{-1/6}(-9)$$
$$= \frac{-45}{2}(4 - 9x)^{-1/6}$$
$$= -\frac{45}{2(4 - 9x)^{1/6}}$$

17. $h(s) = -2\sqrt{5s^2 + 3} = -2(5s^2 + 3)^{1/2}$

$$h'(s) = -2\left(\frac{1}{2}\right)(5s^2 + 3)^{-1/2}(10s)$$
$$= \frac{-10s}{(5s^2 + 3)^{1/2}} = -\frac{10s}{\sqrt{5s^2 + 3}}$$

19. $y = (x - 2)^{-1}$

$$y' = -1(x - 2)^{-2}(1) = \frac{-1}{(x - 2)^2}$$

21. $g(s) = \dfrac{6}{(s^3 - 2)^3} = 6(s^3 - 2)^{-3}$

$$g'(s) = 6(-3)(s^3 - 2)^{-4}(3s^2) = -\frac{54s^2}{(s^3 - 2)^4}$$

23. $y = \dfrac{1}{\sqrt{3x + 5}} = (3x + 5)^{-1/2}$

$$y' = -\frac{1}{2}(3x + 5)^{-3/2}(3)$$
$$= \frac{-3}{2(3x + 5)^{3/2}}$$
$$= -\frac{3}{2\sqrt{(3x + 5)^3}}$$

25. $f(x) = x(2x - 5)^3$

$$f'(x) = x(3)(2x - 5)^2(2) + (2x - 5)^3(1)$$
$$= (2x - 5)^2[6x + (2x - 5)]$$
$$= (2x - 5)^2(8x - 5)$$

27. $y = \dfrac{x}{\sqrt{x^2 + 1}} = \dfrac{x}{(x^2 + 1)^{1/2}}$

$$y' = \frac{(x^2 + 1)^{1/2}(1) - x\left(\frac{1}{2}\right)(x^2 + 1)^{-1/2}(2x)}{\left[(x^2 + 1)^{1/2}\right]^2}$$
$$= \frac{(x^2 + 1)^{1/2} - x^2(x^2 + 1)^{-1/2}}{x^2 + 1}$$
$$= \frac{(x^2 + 1)^{-1/2}\left[x^2 + 1 - x^2\right]}{x^2 + 1}$$
$$= \frac{1}{(x^2 + 1)^{3/2}} = \frac{1}{\sqrt{(x^2 + 1)^3}}$$

29. $g(x) = \left(\dfrac{x + 5}{x^2 + 2}\right)^2$

$$g'(x) = 2\left(\frac{x + 5}{x^2 + 2}\right)\left(\frac{(x^2 + 2) - (x + 5)(2x)}{(x^2 + 2)^2}\right)$$
$$= \frac{2(x + 5)(2 - 10x - x^2)}{(x^2 + 2)^3}$$
$$= \frac{-2(x + 5)(x^2 + 10x - 2)}{(x^2 + 2)^3}$$

31. $f(x) = \left((x^2 + 3)^5 + x\right)^2$

$f'(x) = 2\left((x^2 + 3)^5 + x\right)\left(5(x^2 + 3)^4(2x) + 1\right)$

$= 2\left[10x(x^2 + 3)^9 + (x^2 + 3)^5 + 10x^2(x^2 + 3)^4 + x\right] = 20x(x^2 + 3)^9 + 2(x^2 + 3)^5 + 20x^2(x^2 + 3)^4 + 2x$

33. $y = \cos 4x$

$\dfrac{dy}{dx} = -4 \sin 4x$

35. $g(x) = 5 \tan 3x$

$g'(x) = 15 \sec^2 3x$

37. $y = \sin(\pi x)^2 = \sin(\pi^2 x^2)$

$y' = \cos(\pi^2 x^2)\left[2\pi^2 x\right] = 2\pi^2 x \cos(\pi^2 x^2)$

$= 2\pi^2 x \cos(\pi x)^2$

39. $h(x) = \sin 2x \cos 2x$

$h'(x) = \sin 2x(-2 \sin 2x) + \cos 2x(2 \cos 2x)$

$= 2 \cos^2 2x - 2 \sin^2 2x$

$= 2 \cos 4x$

Alternate solution: $h(x) = \frac{1}{2} \sin 4x$

$h'(x) = \frac{1}{2} \cos 4x(4) = 2 \cos 4x$

41. $f(x) = \dfrac{\cot x}{\sin x} = \dfrac{\cos x}{\sin^2 x}$

$f'(x) = \dfrac{\sin^2 x(-\sin x) - \cos x(2 \sin x \cos x)}{\sin^4 x}$

$= \dfrac{-\sin^2 x - 2\cos^2 x}{\sin^3 x} = \dfrac{-1 - \cos^2 x}{\sin^3 x}$

43. $f(\theta) = \frac{1}{4} \sin^2 2\theta = \frac{1}{4}(\sin 2\theta)^2$

$f'(\theta) = 2\left(\frac{1}{4}\right)(\sin 2\theta)(\cos 2\theta)(2)$

$= \sin 2\theta \cos 2\theta = \frac{1}{2} \sin 4\theta$

45. $f(t) = 3 \sec(\pi t - 1)^2$

$f''(t) = 3 \sec(\pi t - 1)^2 \tan(\pi t - 1)^2 (2)(\pi t - 1)(\pi)$

$= 6\pi(\pi t - 1) \sec(\pi t - 1)^2 \tan(\pi t - 1)^2$

47. $y = \sin(3x^2 + \cos x)$

$y' = \cos(3x^2 + \cos x)(6x - \sin x)$

49. $y = \sin\sqrt{\cot 3\pi x} = \sin(\cot 3\pi x)^{1/2}$

$y' = \cos(\cot 3\pi x)^{1/2}\left[\frac{1}{2}(\cot 3\pi x)^{-1/2}(-\csc^2 3\pi x)(3\pi)\right]$

$= -\dfrac{3\pi \cos\left(\sqrt{\cot 3\pi x}\right)\csc^2(3\pi x)}{2\sqrt{\cot 3\pi x}}$

51. $y = \dfrac{\sqrt{x} + 1}{x^2 + 1}$

$y' = \dfrac{1 - 3x^2 - 4x^{3/2}}{2\sqrt{x}(x^2 + 1)^2}$

The zero of y' corresponds to the point on the graph of y where the tangent line is horizontal.

53. $y = \sqrt{\dfrac{x + 1}{x}}$

$y' = -\dfrac{\sqrt{(x + 1)/x}}{2x(x + 1)}$

y' has no zeros.

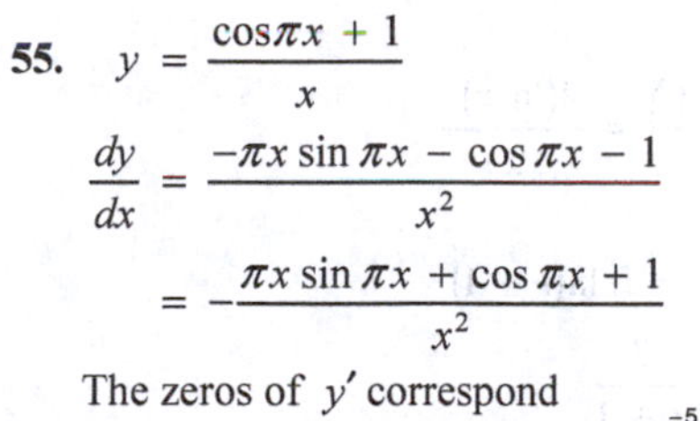

55. $y = \dfrac{\cos \pi x + 1}{x}$

$\dfrac{dy}{dx} = \dfrac{-\pi x \sin \pi x - \cos \pi x - 1}{x^2}$

$= -\dfrac{\pi x \sin \pi x + \cos \pi x + 1}{x^2}$

The zeros of y' correspond to the points on the graph of y where the tangent lines are horizontal.

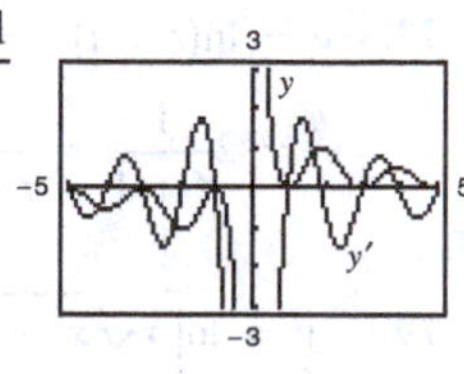

57. $y = e^{5x}$

$y' = 5e^{5x}$

59. $y = e^{\sqrt{x}}$

$\dfrac{dy}{dx} = \dfrac{e^{\sqrt{x}}}{2\sqrt{x}}$

61. $g(t) = (e^{-t} + e^t)^3$

$g'(t) = 3(e^{-t} + e^t)^2(e^t - e^{-t})$

63. $y = x^2e^x - 2xe^x + 2e^x = e^x(x^2 - 2x + 2)$

$\frac{dy}{dx} = e^x(2x - 2) + e^x(x^2 - 2x + 2) = x^2e^x$

65. $y = \frac{2}{e^x + e^{-x}} = 2(e^x + e^{-x})^{-1}$

$\frac{dy}{dx} = -2(e^x + e^{-x})^{-2}(e^x - e^{-x}) = \frac{-2(e^x - e^{-x})}{(e^x + e^{-x})^2}$

67. $y = \frac{e^x + 1}{e^x - 1}$

$y' = \frac{(e^x - 1)e^x - (e^x + 1)e^x}{(e^x - 1)^2} = \frac{-2e^x}{(e^x - 1)^2}$

69. $y = e^x(\sin x + \cos x)$

$\frac{dy}{dx} = e^x(\cos x - \sin x) + (\sin x + \cos x)(e^x)$

$= e^x(2\cos x) = 2e^x \cos x$

71. $g(x) = e^{\csc x}$

$g'(x) = e^{\csc x}[-\csc x \cot x] = -e^{\csc x}\csc x \cot x$

73. $f(x) = \ln(x^2 + 3)$

$f'(x) = \frac{1}{x^2 + 3}(2x) = \frac{2x}{x^2 + 3}$

75. $y = (\ln x)^4$

$\frac{dy}{dx} = 4(\ln x)^3\left(\frac{1}{x}\right) = \frac{4(\ln x)^3}{x}$

77. $y = \ln(t + 1)^2 = 2\ln(t + 1)$

$y' = 2\frac{1}{t + 1} = \frac{2}{t + 1}$

79. $y = \ln\left[x\sqrt{x^2 - 1}\right] = \ln x + \frac{1}{2}\ln(x^2 - 1)$

$\frac{dy}{dx} = \frac{1}{x} + \frac{1}{2}\left(\frac{2x}{x^2 - 1}\right) = \frac{2x^2 - 1}{x(x^2 - 1)}$

81. $f(x) = \ln\frac{x}{x^2 + 1} = \ln x - \ln(x^2 + 1)$

$f'(x) = \frac{1}{x} - \frac{2x}{x^2 + 1} = \frac{1 - x^2}{x(x^2 + 1)}$

83. $g(t) = \frac{\ln t}{t^2}$

$g'(t) = \frac{t^2(1/t) - 2t\ln t}{t^4} = \frac{1 - 2\ln t}{t^3}$

85. $y = \ln(\ln x^2)$

$\frac{dy}{dx} = \frac{1}{\ln x^2}\frac{d}{dx}(\ln x^2) = \frac{(2x/x^2)}{\ln x^2} = \frac{2}{x\ln x^2} = \frac{1}{x\ln x}$

87. $y = \ln\sqrt{\frac{x + 1}{x - 1}} = \frac{1}{2}[\ln(x + 1) - \ln(x - 1)]$

$\frac{dy}{dx} = \frac{1}{2}\left[\frac{1}{x + 1} - \frac{1}{x - 1}\right] = \frac{1}{1 - x^2}$

89. $y = \ln|\sin x|$

$\frac{dy}{dx} = \frac{\cos x}{\sin x} = \cot x$

91. $y = \ln\left|\frac{\cos x}{\cos x - 1}\right| = \ln|\cos x| - \ln|\cos x - 1|$

$\frac{dy}{dx} = \frac{-\sin x}{\cos x} - \frac{-\sin x}{\cos x - 1} = -\tan x + \frac{\sin x}{\cos x - 1}$

93. $f(x) = \ln(1 + e^{-3x})$

$f'(x) = \frac{1}{1 + e^{-3x}}(-3e^{-3x}) = -\frac{3}{e^{3x} + 1}$

95. $y = \sin 3x$

$y' = 3\cos 3x$

$y'(0) = 3$

3 cycles in $[0, 2\pi]$

97. $y = e^{3x}$

$y' = 3e^{3x}$

At $(0, 1)$, $y' = 3$.

99. $y = \ln x^3 = 3\ln x$

$y' = \frac{3}{x}$

At $(1, 0)$, $y' = 3$.

101. $y = \sqrt{x^2 + 8x} = (x^2 + 8x)^{1/2}$, $(1, 3)$

$y' = \frac{1}{2}(x^2 + 8x)^{-1/2}(2x + 8)$

$= \frac{2(x + 4)}{2(x^2 + 8x)^{1/2}}$

$= \frac{x + 4}{\sqrt{x^2 + 8x}}$

$y'(1) = \frac{1 + 4}{\sqrt{1^2 + 8(1)}} = \frac{5}{\sqrt{9}} = \frac{5}{3}$

103. $f(x) = 5(x^3 - 2)^{-1}, \quad \left(-2, -\frac{1}{2}\right)$

$f'(x) = -5(x^3 - 2)^{-2}(3x^2) = \dfrac{-15x^2}{(x^3 - 2)^2}$

$f'(-2) = -\dfrac{60}{100} = -\dfrac{3}{5}$

105. $y = \dfrac{4}{(x + 2)^2} = 4(x + 2)^{-2}, (0, 1)$

$y' = -8(x + 2)^{-3} = \dfrac{-8}{(x + 2)^3}$

$y'(0) = \dfrac{-8}{8} = -1$

107. $y = 26 - \sec^3 4x, \quad (0, 25)$

$y' = -3\sec^2 4x \sec 4x \tan 4x \cdot 4$

$= -12\sec^3 4x \tan 4x$

$y'(0) = 0$

109. (a) $y = (4x^3 + 3)^2, \quad (-1, 1)$

$y' = 2(4x^3 + 3)(12x^2) = 24x^2(4x^3 + 3)$

$y'(-1) = -24$

Tangent line:

$y - 1 = -24(x + 1) \Rightarrow 24x + y + 23 = 0$

(b)

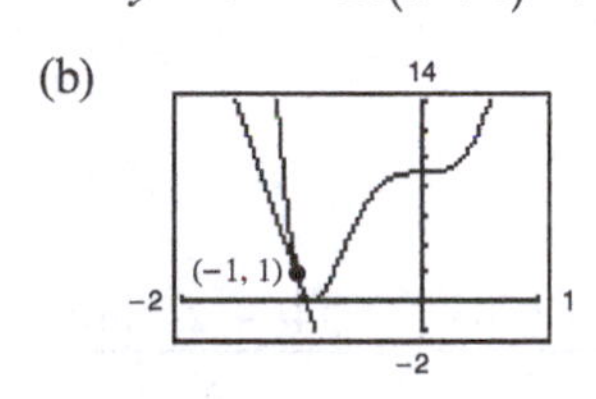

111. (a) $f(x) = \sin 8x, \quad (\pi, 0)$

$f'(x) = 8\cos 8x$

$f'(\pi) = 8$

Tangent line: $y = 8(x - \pi) = 8x - 8\pi$

(b)

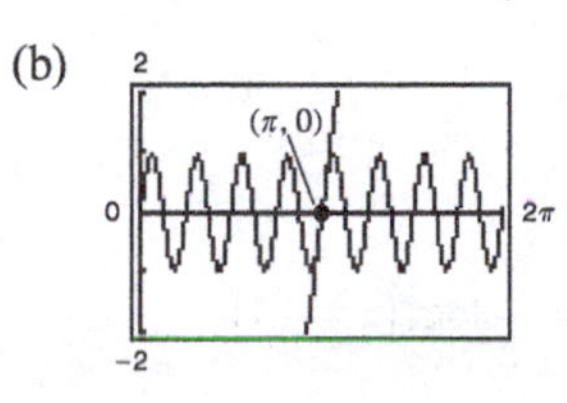

113. (a) $f(x) = \tan^2 x, \quad \left(\frac{\pi}{4}, 1\right)$

$f'(x) = 2\tan x \sec^2 x$

$f'\left(\frac{\pi}{4}\right) = 2(1)(2) = 4$

Tangent line:

$y - 1 = 4\left(x - \frac{\pi}{4}\right) \Rightarrow 4x - y + (1 - \pi) = 0$

(b)

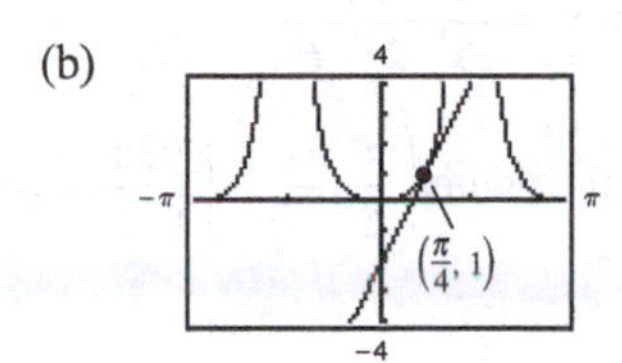

115. (a) $y = 4 - x^2 - \ln\left(\frac{1}{2}x + 1\right), \quad (0, 4)$

$\dfrac{dy}{dx} = -2x - \dfrac{1}{(1/2)x + 1}\left(\dfrac{1}{2}\right) = -2x - \dfrac{1}{x + 2}$

When $x = 0$, $\dfrac{dy}{dx} = -\dfrac{1}{2}$.

Tangent line: $y - 4 = -\frac{1}{2}(x - 0)$

$y = -\frac{1}{2}x + 4$

(b)

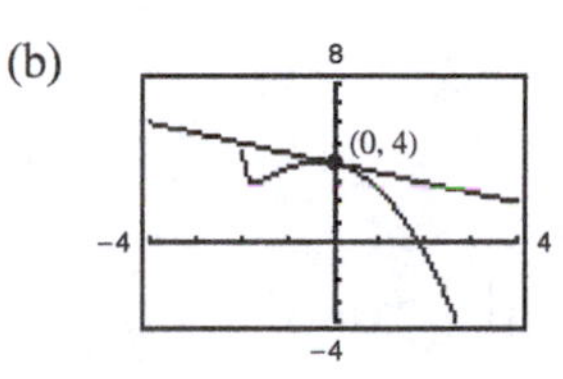

117. $f(x) = \sqrt{25 - x^2} = (25 - x^2)^{1/2}, \quad (3, 4)$

$f'(x) = \frac{1}{2}(25 - x^2)(-2x) = \dfrac{-x}{\sqrt{25 - x^2}}$

$f'(3) = -\dfrac{3}{4}$

Tangent line:

$y - 4 = -\frac{3}{4}(x - 3) \Rightarrow 3x + 4y - 25 = 0$

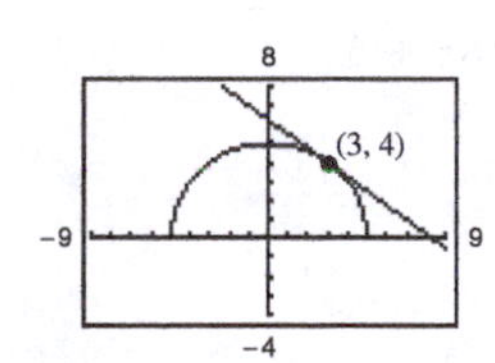

119.
$$f(x) = 2\cos x + \sin 2x, \quad 0 < x < 2\pi$$
$$f'(x) = -2\sin x + 2\cos 2x = -2\sin x + 2 - 4\sin^2 x = 0$$
$$2\sin^2 x + \sin x - 1 = 0$$
$$(\sin x + 1)(2\sin x - 1) = 0$$
$$\sin x = -1 \Rightarrow x = \frac{3\pi}{2}$$
$$\sin x = \frac{1}{2} \Rightarrow x = \frac{\pi}{6}, \frac{5\pi}{6}$$

Horizontal tangents at $x = \frac{\pi}{6}, \frac{3\pi}{2}, \frac{5\pi}{6}$

Horizontal tangent at the points $\left(\frac{\pi}{6}, \frac{3\sqrt{3}}{2}\right)$, $\left(\frac{3\pi}{2}, 0\right)$, and $\left(\frac{5\pi}{6}, -\frac{3\sqrt{3}}{2}\right)$

121.
$$f(x) = 5(2 - 7x)^4$$
$$f'(x) = 20(2 - 7x)^3(-7) = -140(2 - 7x)^3$$
$$f''(x) = -420(2 - 7x)^2(-7) = 2940(2 - 7x)^2$$

123.
$$f(x) = \frac{1}{11x - 6} = (11x - 6)^{-1}$$
$$f'(x) = -(11x - 6)^{-2}(11)$$
$$f''(x) = -22(11x - 6)^{-3}(11)$$
$$= 242(11x - 6)^{-3}$$
$$= \frac{242}{(11x - 6)^3}$$

125.
$$f(x) = \sin x^2$$
$$f'(x) = 2x\cos x^2$$
$$f''(x) = 2x\left[2x\left(-\sin x^2\right)\right] + 2\cos x^2$$
$$= 2\left(\cos x^2 - 2x^2 \sin x^2\right)$$

127.
$$f(x) = (3 + 2x)e^{-3x}$$
$$f'(x) = (3 + 2x)\left(-3e^{-3x}\right) + 2e^{-3x} = (-7 - 6x)e^{-3x}$$
$$f''(x) = (-7 - 6x)\left(-3e^{-3x}\right) - 6e^{-3x} = 3(6x + 5)e^{-3x}$$

129.
$$h(x) = \tfrac{1}{9}(3x + 1)^3, \quad \left(1, \tfrac{64}{9}\right)$$
$$h'(x) = \tfrac{1}{9}3(3x + 1)^2(3) = (3x + 1)^2$$
$$h''(x) = 2(3x + 1)(3) = 18x + 6$$
$$h''(1) = 24$$

131.
$$f(x) = \cos x^2, \quad (0, 1)$$
$$f'(x) = -\sin\left(x^2\right)(2x) = -2x\sin\left(x^2\right)$$
$$f''(x) = -2x\cos\left(x^2\right)(2x) - 2\sin\left(x^2\right)$$
$$= -4x^2\cos\left(x^2\right) - 2\sin\left(x^2\right)$$
$$f''(0) = 0$$

133.
$$f(x) = 4^x$$
$$f'(x) = (\ln 4)4^x$$

135.
$$g(t) = t^2 2^t$$
$$g'(t) = t^2(\ln 2)2^t + (2t)2^t = t2^t(t\ln 2 + 2)$$
$$= 2^t t(2 + t\ln 2)$$

137.
$$f(t) = \frac{-2t^2}{8^t}$$
$$f'(t) = \frac{8^t(-4t) + 2t^2(\ln 8)8^t}{8^{2t}} = \frac{-4t + 2t^2\ln 8}{8^t}$$

139.
$$h(\theta) = 2^{-\theta}\cos\pi\theta$$
$$h'(\theta) = 2^{-\theta}(-\pi\sin\pi\theta) - (\ln 2)2^{-\theta}\cos\pi\theta$$
$$= -2^{-\theta}\left[(\ln 2)\cos\pi\theta + \pi\sin\pi\theta\right]$$

141.
$$y = \log_3 x$$
$$\frac{dy}{dx} = \frac{1}{x\ln 3}$$

143.
$$y = \log_5\sqrt{x^2 - 1} = \frac{1}{2}\log_5\left(x^2 - 1\right)$$
$$\frac{dy}{dx} = \frac{1}{2}\cdot\frac{2x}{\left(x^2 - 1\right)\ln 5} = \frac{x}{\left(x^2 - 1\right)\ln 5}$$

145.
$$f(x) = \log_2\frac{x^2}{x - 1} = 2\log_2 x - \log_2(x - 1)$$
$$f'(x) = \frac{2}{x\ln 2} - \frac{1}{(x - 1)\ln 2} = \frac{x - 2}{(\ln 2)x(x - 1)}$$

147. $g(t) = \dfrac{10\log_4 t}{t} = \dfrac{10}{\ln 4}\left(\dfrac{\ln t}{t}\right)$

$g'(t) = \dfrac{10}{\ln 4}\left[\dfrac{t(1/t) - \ln t}{t^2}\right] = \dfrac{10}{t^2 \ln 4}[1 - \ln t] = \dfrac{5}{t^2 \ln 2}(1 - \ln t)$

149.

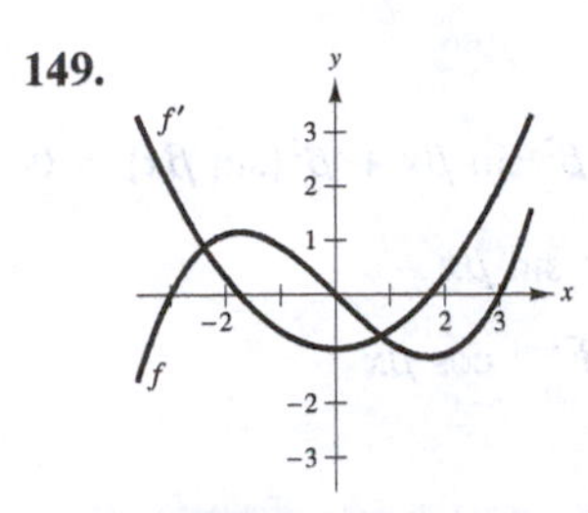

The zeros of f' correspond to the points where the graph of f has horizontal tangents.

151. (a) $g(x) = f(3x)$

$g'(x) = f'(3x)(3) \Rightarrow g'(x) = 3f'(3x)$

The rate of change of g is three times as fast as the rate of change of f.

(b) $g(x) = f(x^2)$

$g'(x) = f'(x^2)(2x) \Rightarrow g'(x) = 2x\, f'(x^2)$

The rate of change of g is $2x$ times as fast as the rate of change of f.

153. (a) $g(x) = f(x) - 2 \Rightarrow g'(x) = f'(x)$

(b) $h(x) = 2f(x) \Rightarrow h'(x) = 2f'(x)$

(c) $r(x) = f(-3x) \Rightarrow r'(x) = f'(-3x)(-3) = -3f'(-3x)$

So, you need to know $f'(-3x)$.

$r'(0) = -3f'(0) = (-3)\left(-\frac{1}{3}\right) = 1$

$r'(-1) = -3f'(3) = (-3)(-4) = 12$

(d) $s(x) = f(x + 2) \Rightarrow s'(x) = f'(x + 2)$

So, you need to know $f'(x + 2)$.

$s'(-2) = f'(0) = -\frac{1}{3}$, etc.

x	−2	−1	0	1	2	3
$f'(x)$	4	$\frac{2}{3}$	$-\frac{1}{3}$	−1	−2	−4
$g'(x)$	4	$\frac{2}{3}$	$-\frac{1}{3}$	−1	−2	−4
$h'(x)$	8	$\frac{4}{3}$	$-\frac{2}{3}$	−2	−4	−8
$r'(x)$		12	1			
$s'(x)$	$-\frac{1}{3}$	−1	−2	−4		

155. (a) $h(x) = f(g(x)),\ g(1) = 4,\ g'(1) = -\frac{1}{2},\ f'(4) = -1$

$h'(x) = f'(g(x))g'(x)$

$h'(1) = f'(g(1))g'(1) = f'(4)g'(1) = (-1)\left(-\frac{1}{2}\right) = \frac{1}{2}$

(b) $s(x) = g(f(x)),\ f(5) = 6,\ f'(5) = -1,\ g'(6)$ does not exist.

$s'(x) = g'(f(x))\, f'(x)$

$s'(5) = g'(f(5))f'(5) = g'(6)(-1)$

$s'(5)$ does not exist because g is not differentiable at 6.

157. $\theta = 0.2\cos 8t$

The maximum angular displacement is $\theta = 0.2$ (because $-1 \le \cos 8t \le 1$).

$\dfrac{d\theta}{dt} = 0.2[-8\sin 8t] = -1.6\sin 8t$

When $t = 3$, $d\theta/dt = -1.6\sin 24 \approx 1.4489$ rad/sec.

159. (a) $F = 132{,}400(331 - v)^{-1}$

$F' = (-1)(132{,}400)(331 - v)^{-2}(-1) = \dfrac{132{,}400}{(331 - v)^2}$

When $v = 30$, $F' \approx 1.461$.

(b) $F = 132{,}400(331 + v)^{-1}$

$F' = (-1)(132{,}400)(331 + v)^{-2}(-1) = \dfrac{-132{,}400}{(331 + v)^2}$

When $v = 30$, $F' \approx -1.016$.

161. $C(t) = P(1.05)^t$

(a) $C(10) = 29.95(1.05)^{10} \approx \$48.79$

(b) $\dfrac{dC}{dt} = P\ln(1.05)(1.05)^t$

When $t = 1$, $\dfrac{dC}{dt} \approx 0.051P$.

When $t = 8$, $\dfrac{dC}{dt} \approx 0.072P$.

(c) $\dfrac{dC}{dt} = \ln(1.05)\left[P(1.05)^t\right] = \ln(1.05)C(t)$

The constant of proportionality is $\ln 1.05$.

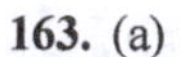

163. (a)

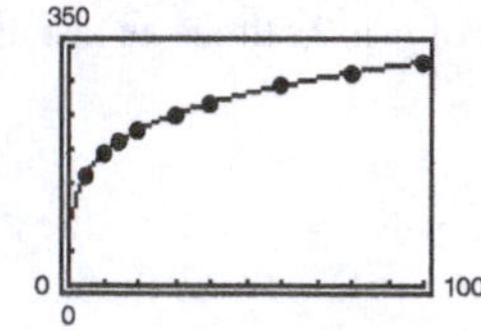

(b) $T'(p) = \dfrac{34.96}{p} + \dfrac{3.955}{\sqrt{p}}$

$T'(10) \approx 4.75$ deg/ lb / in.^{2}

$T'(70) \approx 0.97$ deg/ lb / in.^{2}

(c)

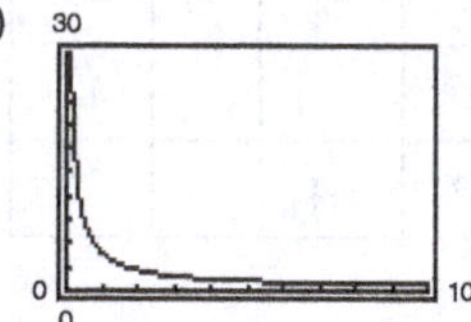

$\lim_{p\to\infty} T'(p) = 0$

Answers will vary. *Sample answer*: As the pounds per square inch approach infinity, the temperature will not change.

165. $N = 400\left[1 - \dfrac{3}{\left(t^2 + 2\right)^2}\right] = 400 - 1200\left(t^2 + 2\right)^{-2}$

$N'(t) = 2400\left(t^2 + 2\right)^{-3}(2t) = \dfrac{4800t}{\left(t^2 + 2\right)^3}$

(a) $N'(0) = 0$ bacteria/day

(b) $N'(1) = \dfrac{4800(1)}{(1 + 2)^3} = \dfrac{4800}{27} \approx 177.8$ bacteria/day

(c) $N'(2) = \dfrac{4800(2)}{(4 + 2)^3} = \dfrac{9600}{216} \approx 44.4$ bacteria/day

(d) $N'(3) = \dfrac{4800(3)}{(9 + 2)^3} = \dfrac{14{,}400}{1331} \approx 10.8$ bacteria/day

(e) $N'(4) = \dfrac{4800(4)}{(16 + 2)^3} = \dfrac{19{,}200}{5832} \approx 3.3$ bacteria/day

(f) The rate of change of the population is decreasing as $t \to \infty$.

167. $f(x) = \sin \beta x$

(a) $f'(x) = \beta \cos \beta x$

$f''(x) = -\beta^2 \sin \beta x$

$f'''(x) = -\beta^3 \cos \beta x$

$f^{(4)} = \beta^4 \sin \beta x$

(b) $f''(x) + \beta^2 f(x) = -\beta^2 \sin \beta x + \beta^2(\sin \beta x) = 0$

(c) $f^{(2k)}(x) = (-1)^k \beta^{2k} \sin \beta x$

$f^{(2k-1)}(x) = (-1)^{k+1} \beta^{2k-1} \cos \beta x$

169. (a) $r'(x) = f'(g(x))g'(x)$

$r'(1) = f'(g(1))g'(1)$

Note that $g(1) = 4$ and $f'(4) = \dfrac{5 - 0}{6 - 2} = \dfrac{5}{4}$.

Also, $g'(1) = 0$. So, $r'(1) = 0$.

(b) $s'(x) = g'(f(x))f'(x)$

$s'(4) = g'(f(4))f'(4)$

Note that $f(4) = \dfrac{5}{2}$, $g'\left(\dfrac{5}{2}\right) = \dfrac{6 - 4}{6 - 2} = \dfrac{1}{2}$ and

$f'(4) = \dfrac{5}{4}$. So, $s'(4) = \dfrac{1}{2}\left(\dfrac{5}{4}\right) = \dfrac{5}{8}$.

171. (a) If $f(-x) = -f(x)$, then

$\dfrac{d}{dx}[f(-x)] = \dfrac{d}{dx}[-f(x)]$

$f'(-x)(-1) = -f'(x)$

$f'(-x) = f'(x)$.

So, $f'(x)$ is even.

(b) If $f(-x) = f(x)$, then

$\dfrac{d}{dx}[f(-x)] = \dfrac{d}{dx}[f(x)]$

$f'(-x)(-1) = f'(x)$

$f'(-x) = -f'(x)$.

So, f' is odd.

173. $g(x) = |3x - 5|$

$g'(x) = 3\left(\dfrac{3x - 5}{|3x - 5|}\right), \quad x \neq \dfrac{5}{3}$

175. $h(x) = |x|\cos x$

$h'(x) = -|x|\sin x + \dfrac{x}{|x|}\cos x, \quad x \neq 0$

177. (a)

$$f(x) = \tan x \qquad f(\pi/4) = 1$$
$$f'(x) = \sec^2 x \qquad f'(\pi/4) = 2$$
$$f''(x) = 2\sec^2 x \tan x \qquad f''(\pi/4) = 4$$
$$P_1(x) = 2(x - \pi/4) + 1$$
$$P_2(x) = \frac{1}{2}(4)(x - \pi/4)^2 + 2(x - \pi/4) + 1$$
$$= 2(x - \pi/4)^2 + 2(x - \pi/4) + 1$$

(b)

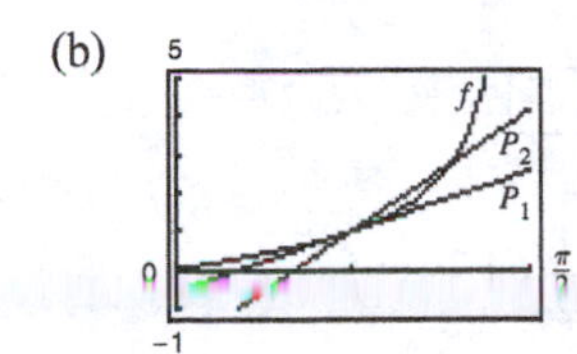

(c) P_2 is a better approximation than P_1.

(d) The accuracy worsens as you move away from $x = \pi/4$.

179. (a)

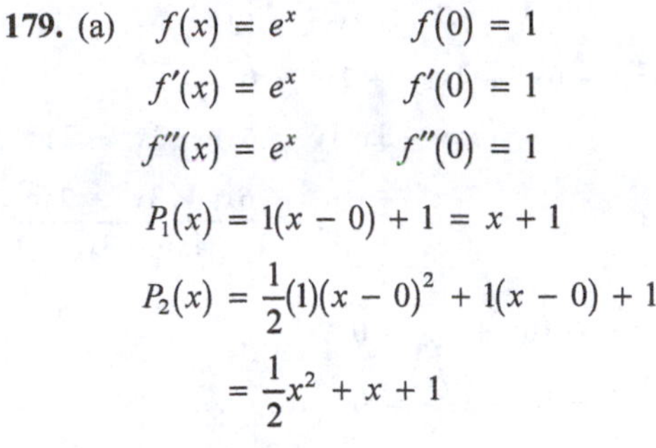

$$f(x) = e^x \qquad f(0) = 1$$
$$f'(x) = e^x \qquad f'(0) = 1$$
$$f''(x) = e^x \qquad f''(0) = 1$$
$$P_1(x) = 1(x - 0) + 1 = x + 1$$
$$P_2(x) = \frac{1}{2}(1)(x - 0)^2 + 1(x - 0) + 1$$
$$= \frac{1}{2}x^2 + x + 1$$

(b)

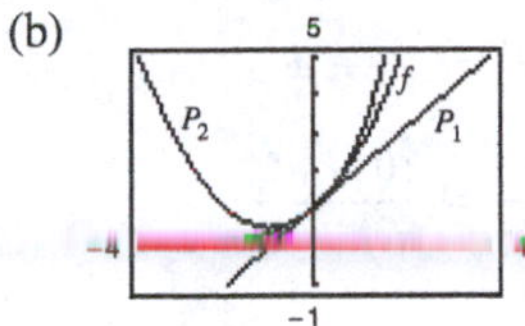

(c) P_2 is a better approximation than P_1.

(d) The accuracy worsens as you move away from $x = 0$.

181. True

183. True

185.

$$f(x) = a_1 \sin x + a_2 \sin 2x + \cdots + a_n \sin nx$$
$$f'(x) = a_1 \cos x + 2a_2 \cos 2x + \cdots + na_n \cos nx$$
$$f'(0) = a_1 + 2a_2 + \cdots + na_n$$
$$|a_1 + 2a_2 + \cdots + na_n| = |f'(0)| = \lim_{x\to 0}\left|\frac{f(x) - f(0)}{x - 0}\right| = \lim_{x\to 0}\left|\frac{f(x)}{\sin x}\right| \cdot \left|\frac{\sin x}{x}\right| = \lim_{x\to 0}\left|\frac{f(x)}{\sin x}\right| \le 1$$

Section 3.5 Implicit Differentiation

1. Answers will vary. *Sample answer:* In the explicit form of a function, the variable is explicitly written as a function of x. In an implicit equation, the function is only implied by an equation. An example of an implicit function is $x^2 + xy = 5$. In explicit form it would be $y = (5 - x^2)/x$.

3. If y is an implicit function of x, then to compute y', you differentiate the equation with respect to x. For example, if $xy^2 = 1$, then $y^2 + 2xyy' = 0$. Here, the derivative of y^2 is $2yy'$.

5.

$$x^2 + y^2 = 9$$
$$2x + 2yy' = 0$$
$$y' = -\frac{x}{y}$$

7.

$$x^5 + y^5 = 16$$
$$5x^4 + 5y^4y' = 0$$
$$5y^4y' = -5x^4$$
$$y' = -\frac{x^4}{y^4}$$

9.

$$x^3 - xy + y^2 = 7$$
$$3x^2 - xy' - y + 2yy' = 0$$
$$(2y - x)y' = y - 3x^2$$
$$y' = \frac{y - 3x^2}{2y - x}$$

11.

$$x^3y^3 - y - x = 0$$
$$3x^3y^2y' + 3x^2y^3 - y' - 1 = 0$$
$$(3x^3y^2 - 1)y' = 1 - 3x^2y^3$$
$$y' = \frac{1 - 3x^2y^3}{3x^3y^2 - 1}$$

13.
$$x^3 - 3x^2y + 2xy^2 = 12$$
$$3x^2 - 3x^2y' - 6xy + 4xyy' + 2y^2 = 0$$
$$(4xy - 3x^2)y' = 6xy - 3x^2 - 2y^2$$
$$y' = \frac{6xy - 3x^2 - 2y^2}{4xy - 3x^2}$$

15.
$$xe^y - 10x + 3y = 0$$
$$xe^y\frac{dy}{dx} + e^y - 10 + 3\frac{dy}{dx} = 0$$
$$\frac{dy}{dx}(xe^y + 3) = 10 - e^y$$
$$\frac{dy}{dx} = \frac{10 - e^y}{xe^y + 3}$$

17.
$$\sin x + 2\cos 2y = 1$$
$$\cos x - 4(\sin 2y)y' = 0$$
$$y' = \frac{\cos x}{4\sin 2y}$$

19.
$$\csc x = x(1 + \tan y)$$
$$-\csc x\cot x = (1 + \tan y) + x(\sec^2 y)y'$$
$$y' = -\frac{\csc x\cot x + 1 + \tan y}{x\sec^2 y}$$

21.
$$y = \sin xy$$
$$y' = [xy' + y]\cos(xy)$$
$$y' - x\cos(xy)y' = y\cos(xy)$$
$$y' = \frac{y\cos(xy)}{1 - x\cos(xy)}$$

23.
$$x^2 - 3\ln y + y^2 = 10$$
$$2x - \frac{3}{y}\frac{dy}{dx} + 2y\frac{dy}{dx} = 0$$
$$2x = \frac{dy}{dx}\left(\frac{3}{y} - 2y\right)$$
$$\frac{dy}{dx} = \frac{2x}{(3/y) - 2y} = \frac{2xy}{3 - 2y^2}$$

25.
$$4x^3 + \ln y^2 + 2y = 2x$$
$$12x^2 + \frac{2}{y}y' + 2y' = 2$$
$$\left(\frac{2}{y} + 2\right)y' = 2 - 12x^2$$
$$y' = \frac{2 - 12x^2}{2/y + 2}$$
$$y' = \frac{y - 6yx^2}{1 + y} = \frac{y(1 - 6x^2)}{1 + y}$$

27. (a)
$$x^2 + y^2 = 64$$
$$y^2 = 64 - x^2$$
$$y = \pm\sqrt{64 - x^2}$$

(b)

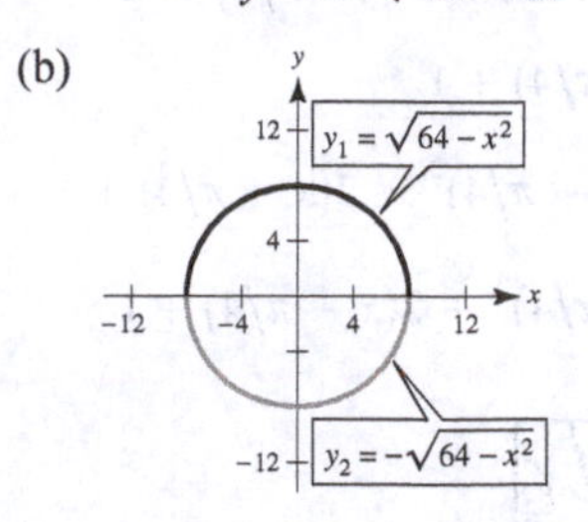

(c) Explicitly:
$$\frac{dy}{dx} = \pm\frac{1}{2}(64 - x^2)^{-1/2}(-2x)$$
$$= \frac{\mp x}{\sqrt{64 - x^2}}$$
$$= \frac{-x}{\pm\sqrt{64 - x^2}}$$
$$= -\frac{x}{y}$$

(d) Implicitly:
$$2x + 2yy' = 0$$
$$y' = -\frac{x}{y}$$

29. (a)
$$16y^2 - x^2 = 16$$
$$16y^2 = x^2 + 16$$
$$y^2 = \frac{x^2}{16} + 1 = \frac{x^2 + 16}{16}$$
$$y = \frac{\pm\sqrt{x^2 + 16}}{4}$$

(b)

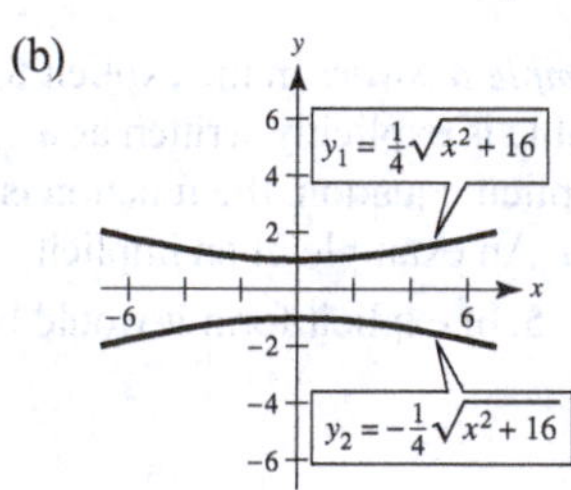

(c) Explicitly:
$$\frac{dy}{dx} = \frac{\pm\frac{1}{2}(x^2 + 16)^{-1/2}(-2x)}{4}$$
$$= \frac{\pm x}{4\sqrt{x^2 + 16}} = \frac{\pm x}{4(\pm 4y)} = \frac{x}{16y}$$

(d) Implicitly:
$$16y^2 - x^2 = 16$$
$$32yy' - 2x = 0$$
$$32yy' = 2x$$
$$y' = \frac{2x}{32y} = \frac{x}{16y}$$

31.
$$xy = 6$$
$$xy' + y(1) = 0$$
$$xy' = -y$$
$$y' = -\frac{y}{x}$$

At $(-6, -1)$: $y' = -\frac{1}{6}$

33.
$$y^2 = \frac{x^2 - 49}{x^2 + 49}$$
$$2yy' = \frac{(x^2 + 49)(2x) - (x^2 - 49)(2x)}{(x^2 + 49)^2}$$
$$2yy' = \frac{196x}{(x^2 + 49)^2}$$
$$y' = \frac{98x}{y(x^2 + 49)^2}$$

At $(7, 0)$: y' is undefined.

35.
$$\tan(x + y) = x$$
$$(1 + y')\sec^2(x + y) = 1$$
$$y' = \frac{1 - \sec^2(x + y)}{\sec^2(x + y)}$$
$$= \frac{-\tan^2(x + y)}{\tan^2(x + y) + 1}$$
$$= -\sin^2(x + y)$$
$$= -\frac{x^2}{x^2 + 1}$$

At $(0, 0)$: $y' = 0$

37.
$$3e^{xy} - x = 0$$
$$3e^{xy}[xy' + y] - 1 = 0$$
$$3e^{xy}xy' = 1 - 3ye^{xy}$$
$$y' = \frac{1 - 3ye^{xy}}{3xe^{xy}}$$

At $(3, 0)$: $y' = \frac{1}{9}$

39.
$$(x^2 + 4)y = 8$$
$$(x^2 + 4)y' + y(2x) = 0$$
$$y' = \frac{-2xy}{x^2 + 4}$$
$$= \frac{-2x[8/(x^2 + 4)]}{x^2 + 4}$$
$$= \frac{-16x}{(x^2 + 4)^2}$$

At $(2, 1)$: $y' = \frac{-32}{64} = -\frac{1}{2}$

(Or, you could just solve for y: $y = \frac{8}{x^2 + 4}$)

41.
$$(x^2 + y^2)^2 = 4x^2y$$
$$2(x^2 + y^2)(2x + 2yy') = 4x^2y' + y(8x)$$
$$4x^3 + 4x^2yy' + 4xy^2 + 4y^3y' = 4x^2y' + 8xy$$
$$4x^2yy' + 4y^3y' - 4x^2y' = 8xy - 4x^3 - 4xy^2$$
$$4y'(x^2y + y^3 - x^2) = 4(2xy - x^3 - xy^2)$$
$$y' = \frac{2xy - x^3 - xy^2}{x^2y + y^3 - x^2}$$

At $(1, 1)$: $y' = 0$

43. $(y - 3)^2 = 4(x - 5), \quad (6, 1)$
$$2(y - 3)y' = 4$$
$$y' = \frac{2}{y - 3}$$

At $(6, 1)$: $y' = \frac{2}{1 - 3} = -1$

Tangent line: $y - 1 = -1(x - 6)$
$$y = -x + 7$$

45. $x^2y^2 - 9x^2 - 4y^2 = 0, \ (-4, 2\sqrt{3})$
$$x^2 2yy' + 2xy^2 - 18x - 8yy' = 0$$
$$y' = \frac{18x - 2xy^2}{2x^2y - 8y}$$

At $(-4, 2\sqrt{3})$: $y' = \frac{18(-4) - 2(-4)(12)}{2(16)(2\sqrt{3}) - 16\sqrt{3}}$
$$= \frac{24}{48\sqrt{3}} = \frac{1}{2\sqrt{3}} = \frac{\sqrt{3}}{6}$$

Tangent line: $y - 2\sqrt{3} = \frac{\sqrt{3}}{6}(x + 4)$
$$y = \frac{\sqrt{3}}{6}x + \frac{8}{3}\sqrt{3}$$

47. $3(x^2 + y^2)^2 = 100(x^2 - y^2), \quad (4, 2)$

$6(x^2 + y^2)(2x + 2yy') = 100(2x - 2yy')$

At (4, 2):

$6(16 + 4)(8 + 4y') = 100(8 - 4y')$

$960 + 480y' = 800 - 400y'$

$880y' = -160$

$y' = -\frac{2}{11}$

Tangent line: $y - 2 = -\frac{2}{11}(x - 4)$

$11y + 2x - 30 = 0$

$y = -\frac{2}{11}x + \frac{30}{11}$

49. $4xy = 9, \quad \left(1, \frac{9}{4}\right)$

$4xy' + 4y = 0$

$xy' = -y$

$y' = \frac{-y}{x}$

At $\left(1, \frac{9}{4}\right)$, $y' = \frac{-9/4}{1} = \frac{-9}{4}$

Tangent line: $y - \frac{9}{4} = \frac{-9}{4}(x - 1)$

$4y - 9 = -9x + 9$

$4y + 9x = 18$

$y = \frac{-9}{4}x + \frac{9}{2}$

51. $x + y - 1 = \ln(x^2 + y^2), \quad (1, 0)$

$1 + y' = \frac{2x + 2yy'}{x^2 + y^2}$

$x^2 + y^2 + (x^2 + y^2)y' = 2x + 2yy'$

At (1, 0): $1 + y' = 2$

$y' = 1$

Tangent line: $y = x - 1$

53. Answers will vary. *Sample answers*:

$xy = 2 \Rightarrow y = \frac{2}{x}$

$yx^2 + x = 2 \Rightarrow y = \frac{2 - x}{x^2}$

$x^2 + y^2 = 4$

$xy + y^2 = 2$

55. (a) $\frac{x^2}{2} + \frac{y^2}{8} = 1, \quad (1, 2)$

$x + \frac{yy'}{4} = 0$

$y' = -\frac{4x}{y}$

At (1, 2): $y' = -2$

Tangent line: $y - 2 = -2(x - 1)$

$y = -2x + 4$

(b) $\frac{x^2}{a^2} + \frac{y^2}{b^2} = 1 \Rightarrow \frac{2x}{a^2} + \frac{2yy'}{b^2} = 0 \Rightarrow y' = \frac{-b^2x}{a^2y}$

$y - y_0 = \frac{-b^2x_0}{a^2y_0}(x - x_0)$, Tangent line at (x_0, y_0)

$\frac{y_0y}{b^2} - \frac{y_0^2}{b^2} = \frac{-x_0x}{a^2} + \frac{x_0^2}{a^2}$

Because $\frac{x_0^2}{a^2} + \frac{y_0^2}{b^2} = 1$, you have $\frac{y_0y}{b^2} + \frac{x_0x}{a^2} = 1$.

Note: From part (a),

$\frac{1(x)}{2} + \frac{2(y)}{8} = 1 \Rightarrow \frac{1}{4}y = -\frac{1}{2}x + 1 \Rightarrow y = -2x + 4$, Tangent line.

57. $\tan y = x$

$y' \sec^2 y = 1$

$y' = \frac{1}{\sec^2 y} = \cos^2 y, -\frac{\pi}{2} < y < \frac{\pi}{2}$

$\sec^2 y = 1 + \tan^2 y = 1 + x^2$

$y' = \frac{1}{1 + x^2}$

59. $x^2y - 4x = 5$

$x^2y' + 2xy - 4 = 0$

$y' = \frac{4 - 2xy}{x^2}$

$x^2y'' + 2xy' + 2xy' + 2y = 0$

$x^2y'' + 4x\left[\frac{4 - 2xy}{x^2}\right] + 2y = 0$

$x^4y'' + 4x(4 - 2xy) + 2x^2y = 0$

$x^4y'' + 16x - 8x^2y + 2x^2y = 0$

$x^4y'' = 6x^2y - 16x$

$y'' = \frac{6xy - 16}{x^3}$

61.

$$7xy + \sin x = 2$$
$$7xy' + 7y + \cos x = 0$$
$$y' = \frac{-7y - \cos x}{7x}$$
$$7xy'' + 7y' + 7y' - \sin x = 0$$
$$7xy'' = \sin x - 14y' = \sin x - 14\left(\frac{-7y - \cos x}{7x}\right)$$
$$7xy'' = \sin x + \frac{14y + 2\cos x}{x}$$
$$y'' = \frac{\sin x}{7x} + \frac{14y + 2\cos x}{7x^2}$$
$$y'' = \frac{x\sin x + 14y + 2\cos x}{7x^2}$$

63. $x^2 + y^2 = 25$

$$2x + 2yy' = 0$$
$$y' = \frac{-x}{y}$$

At $(4, 3)$:

Tangent line:

$$y - 3 = \frac{-4}{3}(x - 4) \Rightarrow 4x + 3y - 25 = 0$$

Normal line: $y - 3 = \dfrac{3}{4}(x - 4) \Rightarrow 3x - 4y = 0$

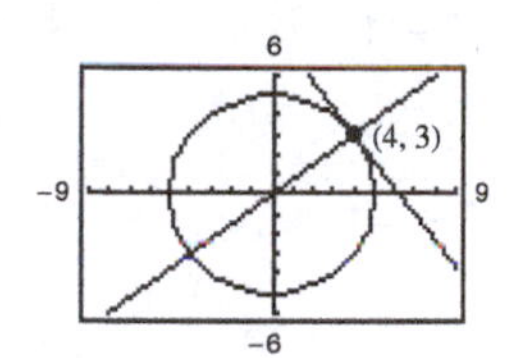

At $(-3, 4)$:

Tangent line:

$$y - 4 = \frac{3}{4}(x + 3) \Rightarrow 3x - 4y + 25 = 0$$

Normal line: $y - 4 = \dfrac{-4}{3}(x + 3) \Rightarrow 4x + 3y = 0$

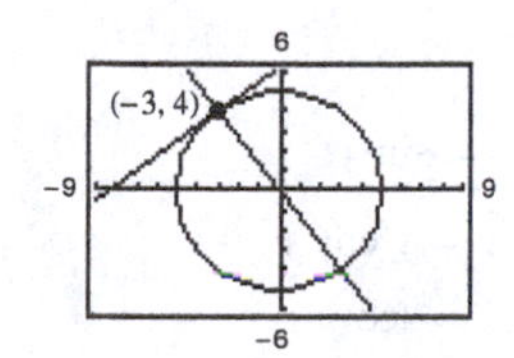

65. $x^2 + y^2 = r^2$

$$2x + 2yy' = 0$$
$$y' = \frac{-x}{y} = \text{slope of tangent line}$$
$$\frac{y}{x} = \text{slope of normal line}$$

Let (x_0, y_0) be a point on the circle. If $x_0 = 0$, then the tangent line is horizontal, the normal line is vertical and, hence, passes through the origin. If $x_0 \neq 0$, then the equation of the normal line is

$$y - y_0 = \frac{y_0}{x_0}(x - x_0)$$
$$y = \frac{y_0}{x_0}x$$

which passes through the origin.

67. $25x^2 + 16y^2 + 200x - 160y + 400 = 0$

$$50x + 32yy' + 200 - 160y' = 0$$
$$y' = \frac{200 + 50x}{160 - 32y}$$

Horizontal tangents occur when $x = -4$:

$$25(16) + 16y^2 + 200(-4) - 160y + 400 = 0$$
$$y(y - 10) = 0 \Rightarrow y = 0, 10$$

Horizontal tangents: $(-4, 0)$, $(-4, 10)$

Vertical tangents occur when $y = 5$:

$$25x^2 + 400 + 200x - 800 + 400 = 0$$
$$25x(x + 8) = 0 \Rightarrow x = 0, -8$$

Vertical tangents:

$(0, 5)$, $(-8, 5)$

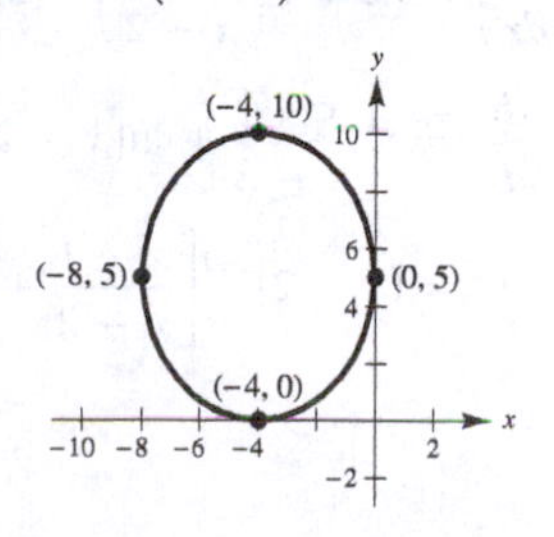

69. $y = x\sqrt{x^2 + 1}$

$\ln y = \ln x + \frac{1}{2}\ln(x^2 + 1)$

$\frac{1}{y}\left(\frac{dy}{dx}\right) = \frac{1}{x} + \frac{x}{x^2 + 1}$

$\frac{dy}{dx} = y\left[\frac{2x^2 + 1}{x(x^2 + 1)}\right] = \frac{2x^2 + 1}{\sqrt{x^2 + 1}}$

71. $y = \frac{x^2\sqrt{3x - 2}}{(x + 1)^2}$

$\ln y = 2\ln x + \frac{1}{2}\ln(3x - 2) - 2\ln(x + 1)$

$\frac{1}{y}\left(\frac{dy}{dx}\right) = \frac{2}{x} + \frac{3}{2(3x - 2)} - \frac{2}{x + 1}$

$\frac{dy}{dx} = y\left[\frac{3x^2 + 15x - 8}{2x(3x - 2)(x + 1)}\right]$

$= \frac{3x^3 + 15x^2 - 8x}{2(x + 1)^3\sqrt{3x - 2}}$

73. $y = \frac{x(x - 1)^{3/2}}{\sqrt{x + 1}}$

$\ln y = \ln x + \frac{3}{2}\ln(x - 1) - \frac{1}{2}\ln(x + 1)$

$\frac{1}{y}\left(\frac{dy}{dx}\right) = \frac{1}{x} + \frac{3}{2}\left(\frac{1}{x - 1}\right) - \frac{1}{2}\left(\frac{1}{x + 1}\right)$

$\frac{dy}{dx} = \frac{y}{2}\left[\frac{2}{x} + \frac{3}{x - 1} - \frac{1}{x + 1}\right]$

$= \frac{y}{2}\left[\frac{4x^2 + 4x - 2}{x(x^2 - 1)}\right] = \frac{(2x^2 + 2x - 1)\sqrt{x - 1}}{(x + 1)^{3/2}}$

75. $y = x^{2/x}$

$\ln y = \frac{2}{x}\ln x$

$\frac{1}{y}\left(\frac{dy}{dx}\right) = \frac{2}{x}\left(\frac{1}{x}\right) + \ln x\left(-\frac{2}{x^2}\right) = \frac{2}{x^2}(1 - \ln x)$

$\frac{dy}{dx} = \frac{2y}{x^2}(1 - \ln x) = 2x^{(2/x)-2}(1 - \ln x)$

77. $y = (x - 2)^{x+1}$

$\ln y = (x + 1)\ln(x - 2)$

$\frac{1}{y}\left(\frac{dy}{dx}\right) = (x + 1)\left(\frac{1}{x - 2}\right) + \ln(x - 2)$

$\frac{dy}{dx} = y\left[\frac{x + 1}{x - 2} + \ln(x - 2)\right]$

$= (x - 2)^{x+1}\left[\frac{x + 1}{x - 2} + \ln(x - 2)\right]$

79. $y = x^{\ln x}, \quad x > 0$

$\ln y = \ln x^{\ln x} = (\ln x)(\ln x) = (\ln x)^2$

$\frac{y'}{y} = 2\ln x(1/x)$

$y' = \frac{2y\ln x}{x} = \frac{2x^{\ln x}\cdot\ln x}{x}$

81. Find the points of intersection by letting $y^2 = 4x$ in the equation $2x^2 + y^2 = 6$.

$2x^2 + 4x = 6$ and $(x + 3)(x - 1) = 0$

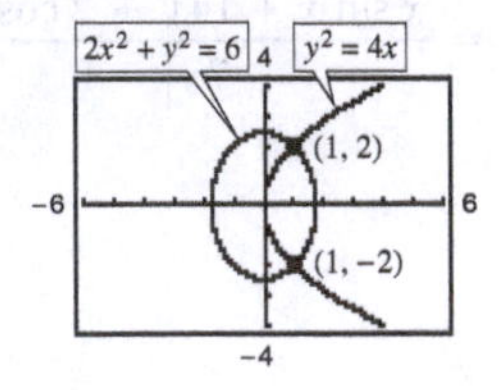

The curves intersect at $(1, \pm 2)$.

Ellipse:	*Parabola*:
$4x + 2yy' = 0$	$2yy' = 4$
$y' = -\frac{2x}{y}$	$y' = \frac{2}{y}$

At $(1, 2)$, the slopes are:

$y' = -1$ $\qquad$ $y' = 1$

At $(1, -2)$, the slopes are:

$y' = 1$ $\qquad$ $y' = -1$

Tangents are perpendicular.

83. $y = -x$ and $x = \sin y$

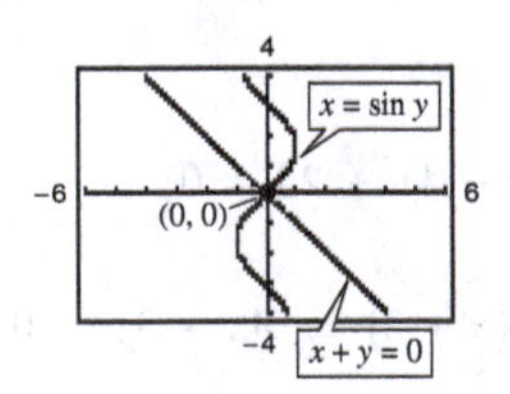

Point of intersection: $(0, 0)$

$y = -x$:	$x = \sin y$:
$y' = -1$	$1 = y'\cos y$
	$y' = \sec y$

At $(0, 0)$, the slopes are:

$y' = -1$ $\qquad$ $y' = 1$

Tangents are perpendicular.

85.

$$xy = C \qquad x^2 - y^2 = K$$
$$xy' + y = 0 \qquad 2x - 2yy' = 0$$
$$y' = -\frac{y}{x} \qquad y' = \frac{x}{y}$$

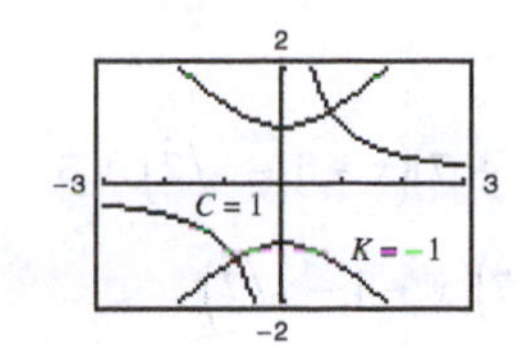

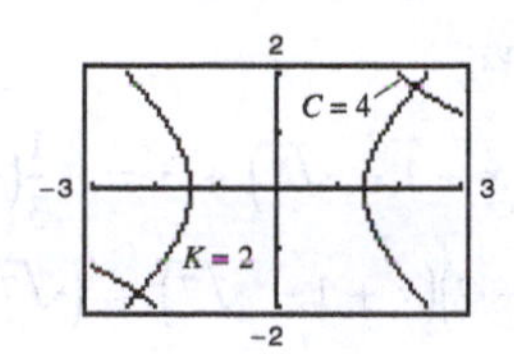

At any point of intersection (x, y) the product of the slopes is $(-y/x)(x/y) = -1$. The curves are orthogonal.

87. (a) True

(b) False. $\frac{d}{dy}\cos(y^2) = -2y\sin(y^2)$. (c) False. $\frac{d}{dx}\cos(y^2) = -2yy'\sin(y^2)$.

89. (a)

$$x^4 = 4(4x^2 - y^2)$$
$$4y^2 = 16x^2 - x^4$$
$$y^2 = 4x^2 - \frac{1}{4}x^4$$
$$y = \pm\sqrt{4x^2 - \frac{1}{4}x^4}$$

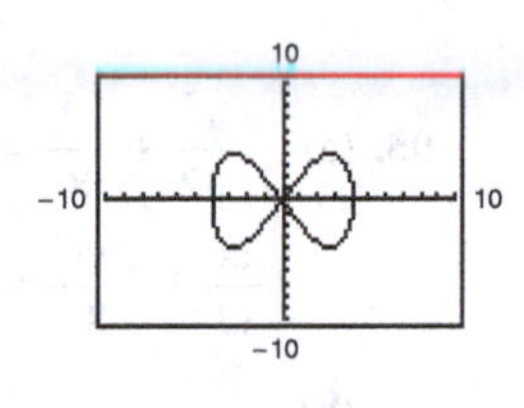

(b)

$$y = 3 \Rightarrow 9 = 4x^2 - \frac{1}{4}x^4$$
$$36 = 16x^2 - x^4$$
$$x^4 - 16x^2 + 36 = 0$$
$$x^2 = \frac{16 \pm \sqrt{256 - 144}}{2} = 8 \pm \sqrt{28}$$

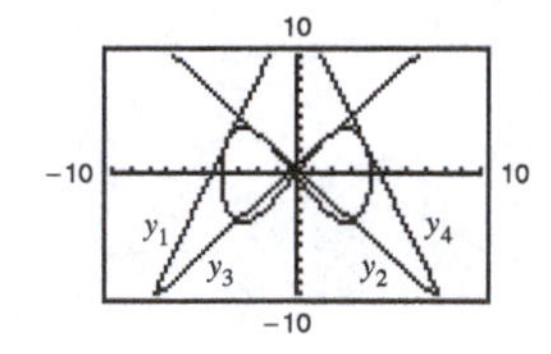

Note that $x^2 = 8 \pm \sqrt{28} = 8 \pm 2\sqrt{7} = \left(1 \pm \sqrt{7}\right)^2$. So, there are four values of x:

$-1 - \sqrt{7}, 1 - \sqrt{7}, -1 + \sqrt{7}, 1 + \sqrt{7}$

To find the slope, $2yy' = 8x - x^3 \Rightarrow y' = \dfrac{x(8 - x^2)}{2(3)}$.

For $x = -1 - \sqrt{7}$, $y' = \frac{1}{3}(\sqrt{7} + 7)$, and the line is

$$y_1 = \frac{1}{3}(\sqrt{7} + 7)(x + 1 + \sqrt{7}) + 3 = \frac{1}{3}\left[(\sqrt{7} + 7)x + 8\sqrt{7} + 23\right].$$

For $x = 1 - \sqrt{7}$, $y' = \frac{1}{3}(\sqrt{7} - 7)$, and the line is

$$y_2 = \frac{1}{3}(\sqrt{7} - 7)(x - 1 + \sqrt{7}) + 3 = \frac{1}{3}\left[(\sqrt{7} - 7)x + 23 - 8\sqrt{7}\right].$$

For $x = -1 + \sqrt{7}$, $y' = -\frac{1}{3}(\sqrt{7} - 7)$, and the line is

$$y_3 = -\frac{1}{3}(\sqrt{7} - 7)(x + 1 - \sqrt{7}) + 3 = -\frac{1}{3}\left[(\sqrt{7} - 7)x - (23 - 8\sqrt{7})\right].$$

For $x = 1 + \sqrt{7}$, $y' = -\frac{1}{3}(\sqrt{7} + 7)$, and the line is

$$y_4 = -\frac{1}{3}(\sqrt{7} + 7)(x - 1 - \sqrt{7}) + 3 = -\frac{1}{3}\left[(\sqrt{7} + 7)x - (8\sqrt{7} + 23)\right].$$

(c) Equating y_3 and y_4:

$$-\frac{1}{3}(\sqrt{7}-7)(x+1-\sqrt{7})+3 = -\frac{1}{3}(\sqrt{7}+7)(x-1-\sqrt{7})+3$$
$$(\sqrt{7}-7)(x+1-\sqrt{7}) = (\sqrt{7}+7)(x-1-\sqrt{7})$$
$$\sqrt{7}x+\sqrt{7}-7-7x-7+7\sqrt{7} = \sqrt{7}x-\sqrt{7}-7+7x-7-7\sqrt{7}$$
$$16\sqrt{7} = 14x$$
$$x = \frac{8\sqrt{7}}{7}$$

If $x = \dfrac{8\sqrt{7}}{7}$, then $y = 5$ and the lines intersect at $\left(\dfrac{8\sqrt{7}}{7}, 5\right)$.

91. $x^2 + y^2 = 100$, slope $= \dfrac{3}{4}$

$$2x + 2yy' = 0$$
$$y' = -\frac{x}{y} = \frac{3}{4} \Rightarrow y = -\frac{4}{3}x$$
$$x^2 + \left(\frac{16}{9}x^2\right) = 100$$
$$\frac{25}{9}x^2 = 100$$
$$x = \pm 6$$

Points: $(6, -8)$ and $(-6, 8)$

93. $\dfrac{x^2}{4} + \dfrac{y^2}{9} = 1, \quad (4, 0)$

$$\frac{2x}{4} + \frac{2yy'}{9} = 0$$
$$y' = \frac{-9x}{4y}$$
$$\frac{-9x}{4y} = \frac{y-0}{x-4}$$
$$-9x(x-4) = 4y^2$$

But, $9x^2 + 4y^2 = 36 \Rightarrow 4y^2 = 36 - 9x^2$. So,

$-9x^2 + 36x = 4y^2 = 36 - 9x^2 \Rightarrow x = 1.$

Points on ellipse: $\left(1, \pm\dfrac{3}{2}\sqrt{3}\right)$

At $\left(1, \dfrac{3}{2}\sqrt{3}\right)$: $y' = \dfrac{-9x}{4y} = \dfrac{-9}{4\left[(3/2)\sqrt{3}\right]} = -\dfrac{\sqrt{3}}{2}$

At $\left(1, -\dfrac{3}{2}\sqrt{3}\right)$: $y' = \dfrac{\sqrt{3}}{2}$

Tangent lines: $y = -\dfrac{\sqrt{3}}{2}(x-4) = -\dfrac{\sqrt{3}}{2}x + 2\sqrt{3}$

$$y = \frac{\sqrt{3}}{2}(x-4) = \frac{\sqrt{3}}{2}x - 2\sqrt{3}$$

95. (a) $\dfrac{x^2}{32} + \dfrac{y^2}{8} = 1$

$$\frac{2x}{32} + \frac{2yy'}{8} = 0 \Rightarrow y' = \frac{-x}{4y}$$

(b)

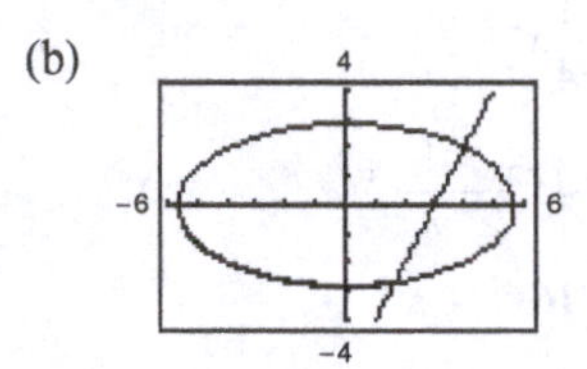

At $(4, 2)$: $y' = \dfrac{-4}{4(2)} = -\dfrac{1}{2}$

Slope of normal line is 2.

$$y - 2 = 2(x - 4)$$
$$y = 2x - 6$$

(c) $\dfrac{x^2}{32} + \dfrac{(2x-6)^2}{8} = 1$

$$x^2 + 4(4x^2 - 24x + 36) = 32$$
$$17x^2 - 96x + 112 = 0$$
$$(17x - 28)(x - 4) = 0 \Rightarrow x = 4, \frac{28}{17}$$

Second point: $\left(\dfrac{28}{17}, -\dfrac{46}{17}\right)$

Section 3.6 Derivatives of Inverse Functions

1. Because you know that f^{-1} exists and that $y_1 = f(x_1)$ by Theorem 3.17, then $\left(f^{-1}\right)'(y_1) = \dfrac{1}{f'(x_1)}$, provided that $f'(x_1) \neq 0$.

3. $f(x) = 5 - 2x^3, \quad a = 7$

$f'(x) = -6x^2$

f is continuous and differentiable on $(-\infty, \infty)$. Therefore, f has an inverse.

$$f(-1) = 7 \Rightarrow f^{-1}(7) = -1$$

$$\left(f^{-1}\right)'(7) = \frac{1}{f'\left(f^{-1}(7)\right)} = \frac{1}{f'(-1)} = \frac{1}{-6(-1)^2} = \frac{-1}{6}$$

5. $f(x) = x^3 - 1, \quad a = 26$

$f'(x) = 3x^2$

f is continuous and differentiable on $(-\infty, \infty)$. Therefore f has an inverse.

$$f(3) = 26 \Rightarrow f^{-1}(26) = 3$$

$$\left(f^{-1}\right)'(26) = \frac{1}{f'\left(f^{-1}(26)\right)} = \frac{1}{f'(3)} = \frac{1}{3(3^2)} = \frac{1}{27}$$

7. $f(x) = \sin x, \quad a = 1/2, -\dfrac{\pi}{2} \leq x \leq \dfrac{\pi}{2}$

$f'(x) = \cos x > 0 \text{ on } \left(-\dfrac{\pi}{2}, \dfrac{\pi}{2}\right)$

f is continuous and differentiable on $\left[-\dfrac{\pi}{2}, \dfrac{\pi}{2}\right]$. Therefore f has an inverse.

$$f\left(\frac{\pi}{6}\right) = \sin\frac{\pi}{6} = \frac{1}{2} \Rightarrow f^{-1}\left(\frac{1}{2}\right) = \frac{\pi}{6}$$

$$\left(f^{-1}\right)'\left(\frac{1}{2}\right) = \frac{1}{f'\left(f^{-1}\left(\frac{1}{2}\right)\right)}$$

$$= \frac{1}{f'\left(\frac{\pi}{6}\right)} = \frac{1}{\cos\left(\frac{\pi}{6}\right)}$$

$$= \frac{2}{\sqrt{3}}$$

$$= \frac{2\sqrt{3}}{3}$$

9. $f(x) = \dfrac{x + 6}{x - 2}, \quad x > 0, a = 3$

$$f'(x) = \frac{(x - 2)(1) - (x + 6)(1)}{(x - 2)^2}$$

$$= \frac{-8}{(x - 2)^2} < 0 \text{ on } (2, \infty)$$

f is continuous and differentiable on $(2, \infty)$. Therefore f has an inverse.

$$f(6) = 3 \Rightarrow f^{-1}(3) = 6$$

$$\left(f^{-1}\right)'(3) = \frac{1}{f'\left(f^{-1}(3)\right)} = \frac{1}{f'(6)} = \frac{1}{-8/(6 - 2)^2} = -2$$

11. $f(x) = x^3 - \dfrac{4}{x}, \quad a = 6, x > 0$

$$f'(x) = 3x^2 + \frac{4}{x^2} > 0$$

f is continuous and differentiable on $(0, \infty)$. Therefore f has an inverse.

$$f(2) = 6 \Rightarrow f^{-1}(6) = 2$$

$$\left(f^{-1}\right)'(6) = \frac{1}{f'\left(f^{-1}(6)\right)} = \frac{1}{f'(2)} = \frac{1}{3(2^2) + 4/2^2} = \frac{1}{13}$$

13. $f(x) = x^3, \quad \left(\dfrac{1}{2}, \dfrac{1}{8}\right)$

$f'(x) = 3x^2$

$f'\left(\dfrac{1}{2}\right) = \dfrac{3}{4}$

$$f^{-1}(x) = \sqrt[3]{x}, \quad \left(\frac{1}{8}, \frac{1}{2}\right)$$

$$\left(f^{-1}\right)'(x) = \frac{1}{3\sqrt[3]{x}}$$

$$\left(f^{-1}\right)'\left(\frac{1}{8}\right) = \frac{4}{3}$$

15. $f(x) = \sqrt{x - 4}, \quad (5, 1)$

$$f'(x) = \frac{1}{2\sqrt{x - 4}}$$

$$f'(5) = \frac{1}{2}$$

$$f^{-1}(x) = x^2 + 4, \quad (1, 5)$$

$$\left(f^{-1}\right)'(x) = 2x$$

$$\left(f^{-1}\right)'(1) = 2$$

17. $f(x) = \arcsin(x - 1)$

$$f'(x) = \frac{1}{\sqrt{1 - (x - 1)^2}} = \frac{1}{\sqrt{2x - x^2}}$$

19. $g(x) = 3 \arccos \dfrac{x}{2}$

$$g'(x) = \frac{-3(1/2)}{\sqrt{1 - (x^2/4)}} = \frac{-3}{\sqrt{4 - x^2}}$$

21. $f(x) = \arctan(e^x)$

$$f'(x) = \frac{1}{1 + (e^x)^2} e^x = \frac{e^x}{1 + e^{2x}}$$

23. $g(x) = \dfrac{\arcsin 3x}{x}$

$$g'(x) = \frac{x\left(3/\sqrt{1 - 9x^2}\right) - \arcsin 3x}{x^2} = \frac{3x - \sqrt{1 - 9x^2} \arcsin 3x}{x^2\sqrt{1 - 9x^2}}$$

25. $g(x) = e^{2x} \arcsin x$

$$g'(x) = e^{2x} \frac{1}{\sqrt{1 - x^2}} + 2e^{2x} \arcsin x = e^{2x}\left[2 \arcsin x + \frac{1}{\sqrt{1 - x^2}}\right]$$

27. $h(x) = \operatorname{arccot} 6x$

$$h'(x) = \frac{-6}{1 + 36x^2}$$

29. $h(t) = \sin(\arccos t) = \sqrt{1 - t^2}$

$$h'(t) = \frac{1}{2}(1 - t^2)^{-1/2}(-2t) = \frac{-t}{\sqrt{1 - t^2}}$$

31. $y = x \arccos x - 2\sqrt{x}$

$$y' = x\left(\frac{-1}{\sqrt{1 - x^2}}\right) + \arccos x - \frac{1}{\sqrt{x}} = \arccos x - \frac{x}{\sqrt{1 - x^2}} - \frac{1}{\sqrt{x}}$$

33. $y = \ln \dfrac{x + 1}{x - 1} + \arctan x$

$$= \ln(x + 1) - \ln(x - 1) + \arctan x$$

$$y' = \frac{1}{x + 1} - \frac{1}{x - 1} + \frac{1}{x^2 + 1} = \frac{-2}{x^2 - 1} + \frac{1}{x^2 + 1} = \frac{x^2 + 3}{1 - x^4}$$

35. $g(t) = \tan(\arcsin t) = \dfrac{t}{\sqrt{1 - t^2}}$

$$g'(t) = \frac{\sqrt{1 - t^2} - t\left(-t/\sqrt{1 - t^2}\right)}{1 - t^2} = \frac{1}{(1 - t^2)^{3/2}}$$

37. $y = x \arcsin x + \sqrt{1 - x^2}$

$$\frac{dy}{dx} = x\left(\frac{1}{\sqrt{1 - x^2}}\right) + \arcsin x - \frac{x}{\sqrt{1 - x^2}} = \arcsin x$$

39. $y = 8 \arcsin \dfrac{x}{4} - \dfrac{x\sqrt{16 - x^2}}{2}$

$$y' = 2\frac{1}{\sqrt{1 - (x/4)^2}} - \frac{\sqrt{16 - x^2}}{2} - \frac{x}{4}(16 - x^2)^{-1/2}(-2x)$$

$$= \frac{8}{\sqrt{16 - x^2}} - \frac{\sqrt{16 - x^2}}{2} + \frac{x^2}{2\sqrt{16 - x^2}} = \frac{16 - (16 - x^2) + x^2}{2\sqrt{16 - x^2}} = \frac{x^2}{\sqrt{16 - x^2}}$$

41. $y = \arctan x + \dfrac{x}{1 + x^2}$

$$y' = \frac{1}{1 + x^2} + \frac{(1 + x^2) - x(2x)}{(1 + x^2)^2} = \frac{(1 + x^2) + (1 - x^2)}{(1 + x^2)^2} = \frac{2}{(1 + x^2)^2}$$

43. $y = 2\arcsin x, \quad \left(\frac{1}{2}, \frac{\pi}{3}\right)$

$$y' = \frac{2}{\sqrt{1-x^2}}$$

At $\left(\frac{1}{2}, \frac{\pi}{3}\right)$, $y' = \dfrac{2}{\sqrt{1-(1/4)}} = \dfrac{4}{\sqrt{3}}$.

Tangent line: $y - \dfrac{\pi}{3} = \dfrac{4}{\sqrt{3}}\left(x - \dfrac{1}{2}\right)$

$$y = \frac{4}{\sqrt{3}}x + \frac{\pi}{3} - \frac{2}{\sqrt{3}}$$

$$y = \frac{4\sqrt{3}}{3}x + \frac{\pi}{3} - \frac{2\sqrt{3}}{3}$$

45. $y = \arctan\left(\frac{x}{2}\right), \quad \left(2, \frac{\pi}{4}\right)$

$$y' = \frac{1}{1+(x^2/4)}\left(\frac{1}{2}\right) = \frac{2}{4+x^2}$$

At $\left(2, \frac{\pi}{4}\right)$, $y' = \dfrac{2}{4+4} = \dfrac{1}{4}$.

Tangent line: $y - \dfrac{\pi}{4} = \dfrac{1}{4}(x-2)$

$$y = \frac{1}{4}x + \frac{\pi}{4} - \frac{1}{2}$$

47. $y = 4x\arccos(x-1), \quad (1, 2\pi)$

$$y' = 4x\frac{-1}{\sqrt{1-(x-1)^2}} + 4\arccos(x-1)$$

At $(1, 2\pi)$, $y' = -4 + 2\pi$.

Tangent line: $y - 2\pi = (2\pi - 4)(x-1)$

$$y = (2\pi - 4)x + 4$$

49. (a)–(c) $f(x) = \arccos(x^2)$

$$f'(x) = \frac{-1}{\sqrt{1-x^4}}(2x) = \frac{-2x}{\sqrt{1-x^4}}$$

$$f'(0) = 0$$

$$y - \frac{\pi}{2} = 0(x-0)$$

$y = \dfrac{\pi}{2}$, tangent line

51. (a)–(c) $f(x) = \arcsin 3x$

$$f'(x) = \frac{1}{\sqrt{1-(3x)^2}}(3) = \frac{3}{\sqrt{1-9x^2}}$$

$$f'(\sqrt{2}/6) = \frac{3}{\sqrt{1-9(1/18)}} = \frac{3}{\sqrt{1/2}} = 3\sqrt{2}$$

$$y - \frac{\pi}{4} = 3\sqrt{2}\left(x - \sqrt{2}/6\right)$$

$y = 3\sqrt{2}x + \dfrac{\pi}{4} - 1$, Tangent line

53. $f(x) = \arccos x$

$f'(x) = \dfrac{-1}{\sqrt{1-x^2}} = -2$ when $x = \pm\dfrac{\sqrt{3}}{2}$.

When $x = \sqrt{3}/2$, $f(\sqrt{3}/2) = \pi/6$.

When $x = -\sqrt{3}/2$, $f(-\sqrt{3}/2) = 5\pi/6$.

Tangent lines:

$$y - \frac{\pi}{6} = -2\left(x - \frac{\sqrt{3}}{2}\right) \Rightarrow y = -2x + \left(\frac{\pi}{6} + \sqrt{3}\right)$$

$$y - \frac{5\pi}{6} = -2\left(x + \frac{\sqrt{3}}{2}\right) \Rightarrow y = -2x + \left(\frac{5\pi}{6} - \sqrt{3}\right)$$

55. $f(x) = \arctan x, \; a = 0$

$$f(0) = 0$$

$$f'(x) = \frac{1}{1+x^2}, \quad f'(0) = 1$$

$$f''(x) = \frac{-2x}{(1+x^2)^2}, \quad f''(0) = 0$$

$$P_1(x) = f(0) + f'(0)x = x$$

$$P_2(x) = f(0) + f'(0)x + \frac{1}{2}f''(0)x^2 = x$$

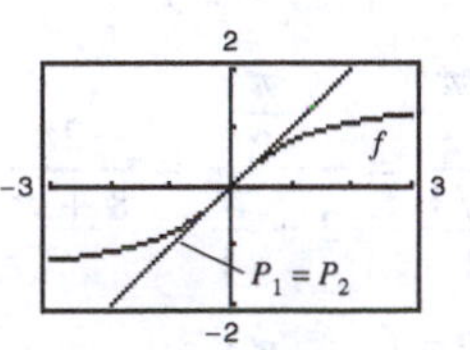

57. $f(x) = \arcsin x,\ a = \dfrac{1}{2}$

$f'(x) = \dfrac{1}{\sqrt{1 - x^2}}$

$f''(x) = \dfrac{x}{(1 - x^2)^{3/2}}$

$P_1(x) = f\left(\dfrac{1}{2}\right) + f'\left(\dfrac{1}{2}\right)\left(x - \dfrac{1}{2}\right) = \dfrac{\pi}{6} + \dfrac{2\sqrt{3}}{3}\left(x - \dfrac{1}{2}\right)$

$P_2(x) = f\left(\dfrac{1}{2}\right) + f'\left(\dfrac{1}{2}\right)\left(x - \dfrac{1}{2}\right) + \dfrac{1}{2}f''\left(\dfrac{1}{2}\right)\left(x - \dfrac{1}{2}\right)^2$

$= \dfrac{\pi}{6} + \dfrac{2\sqrt{3}}{3}\left(x - \dfrac{1}{2}\right) + \dfrac{2\sqrt{3}}{9}\left(x - \dfrac{1}{2}\right)^2$

59. $x = y^3 - 7y^2 + 2$

$1 = 3y^2\dfrac{dy}{dx} - 14y\dfrac{dy}{dx}$

$\dfrac{dy}{dx} = \dfrac{1}{3y^2 - 14y}$

At $(-4, 1)$: $\dfrac{dy}{dx} = \dfrac{1}{3 - 14} = \dfrac{-1}{11}$

61. $x \arctan x = e^y$

$x\dfrac{1}{1 + x^2} + \arctan x = e^y \cdot \dfrac{dy}{dx}$

At $\left(1, \ln\dfrac{\pi}{4}\right)$: $\dfrac{1}{2} + \dfrac{\pi}{4} = \dfrac{\pi}{4}\dfrac{dy}{dx}$

$\dfrac{dy}{dx} = \dfrac{\pi + 2}{\pi}$

63. $x^2 + x \arctan y = y - 1, \quad \left(-\dfrac{\pi}{4}, 1\right)$

$2x + \arctan y + \dfrac{x}{1 + y^2}y' = y'$

$\left(1 - \dfrac{x}{1 + y^2}\right)y' = 2x + \arctan y$

$y' = \dfrac{2x + \arctan y}{1 - \dfrac{x}{1 + y^2}}$

At $\left(-\dfrac{\pi}{4}, 1\right)$: $y' = \dfrac{-\dfrac{\pi}{2} + \dfrac{\pi}{4}}{1 - \dfrac{-\pi/4}{2}} = \dfrac{-\dfrac{\pi}{2}}{2 + \dfrac{\pi}{4}} = \dfrac{-2\pi}{8 + \pi}$

Tangent line: $y - 1 = \dfrac{-2\pi}{8 + \pi}\left(x + \dfrac{\pi}{4}\right)$

$y = \dfrac{-2\pi}{8 + \pi}x + 1 - \dfrac{\pi^2}{16 + 2\pi}$

65. $\arcsin x + \arcsin y = \dfrac{\pi}{2}, \quad \left(\dfrac{\sqrt{2}}{2}, \dfrac{\sqrt{2}}{2}\right)$

$\dfrac{1}{\sqrt{1 - x^2}} + \dfrac{1}{\sqrt{1 - y^2}}y' = 0$

$\dfrac{1}{\sqrt{1 - y^2}}y' = \dfrac{-1}{\sqrt{1 - x^2}}$

At $\left(\dfrac{\sqrt{2}}{2}, \dfrac{\sqrt{2}}{2}\right)$: $y' = -1$

Tangent line: $y - \dfrac{\sqrt{2}}{2} = -1\left(x - \dfrac{\sqrt{2}}{2}\right)$

$y = -x + \sqrt{2}$

67. $f(x) = \sec x, \quad 0 \le x < \dfrac{\pi}{2}, \pi \le x < \dfrac{3\pi}{2}$

(a) $y = \operatorname{arcsec} x, \quad x \le -1 \quad$ or $\quad x \ge 1$

$0 \le y < \dfrac{\pi}{2} \quad$ or $\quad \pi \le y < \dfrac{3\pi}{2}$

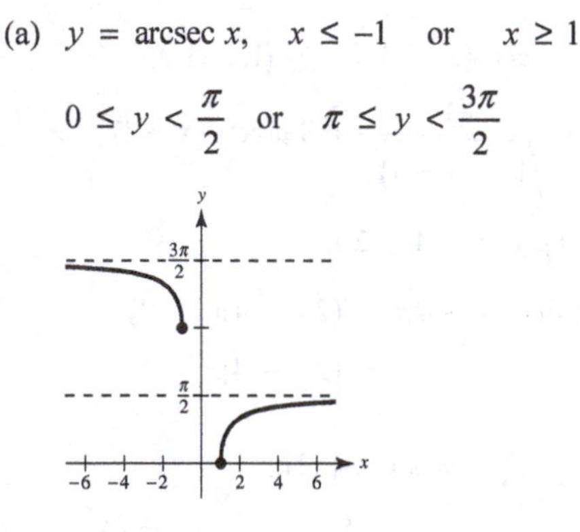

(b) $y = \operatorname{arcsec} x$

$x = \sec y$

$1 = \sec y \tan y \cdot y'$

$y' = \dfrac{1}{\sec y \tan y} = \dfrac{1}{x\sqrt{x^2 - 1}}$

$\tan^2 y + 1 = \sec^2 y$

$\tan y = \pm\sqrt{\sec^2 y - 1}$

On $0 \le y < \pi/2$ and $\pi \le y < 3\pi/2$, $\tan y \ge 0$.

69. The two tangent lines intersect at the line $y = x$ because the tangent lines have reciprocal slopes.

71. Because the slope of f at $(1, 3)$ is $m = 2$, the slope of f^{-1} at $(3, 1)$ is $1/2$.

73. (a) $\cot\theta = \dfrac{x}{5}$

$$\theta = \operatorname{arccot}\left(\frac{x}{5}\right)$$

(b) $\dfrac{d\theta}{dt} = \dfrac{-1/5}{1 + (x/5)^2}\dfrac{dx}{dt} = \dfrac{-5}{x^2 + 25}\dfrac{dx}{dt}$

If $\dfrac{dx}{dt} = -400$ and $x = 10$, $\dfrac{d\theta}{dt} = 16$ rad/h.

If $\dfrac{dx}{dt} = -400$ and $x = 3$, $\dfrac{d\theta}{dt} \approx 58.824$ rad/h.

75. (a) $h(t) = -16t^2 + 256$

$-16t^2 + 256 = 0$ when $t = 4$ sec

h, θ, 500

(b) $\tan\theta = \dfrac{h}{500} = \dfrac{-16t^2 + 256}{500}$

$$\theta = \arctan\left[\frac{16}{500}(-t^2 + 16)\right]$$

$$\frac{d\theta}{dt} = \frac{-8t/125}{1 + \left[(4/125)(-t^2 + 16)\right]^2} = \frac{-1000t}{15{,}625 + 16(16 - t^2)^2}$$

When $t = 1$, $d\theta/dt \approx -0.0520$ rad/sec.

When $t = 2$, $d\theta/dt \approx -0.1116$ rad/sec.

77. 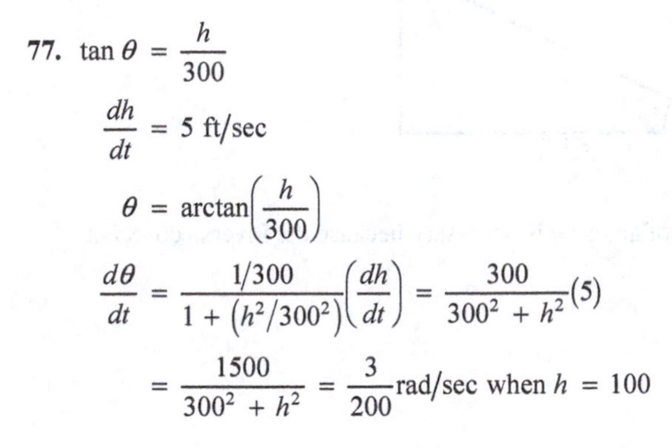

$\tan\theta = \dfrac{h}{300}$

$\dfrac{dh}{dt} = 5$ ft/sec

$\theta = \arctan\left(\dfrac{h}{300}\right)$

$$\frac{d\theta}{dt} = \frac{1/300}{1 + (h^2/300^2)}\left(\frac{dh}{dt}\right) = \frac{300}{300^2 + h^2}(5)$$

$$= \frac{1500}{300^2 + h^2} = \frac{3}{200}\text{ rad/sec when } h = 100$$

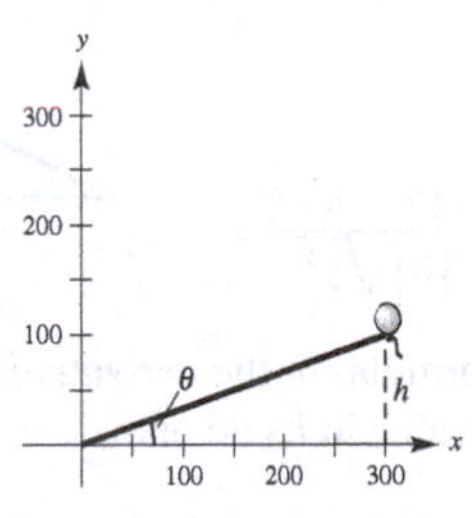

79. True

$$\frac{d}{dx}[\operatorname{arcsec} u] = \frac{u'}{|u|\sqrt{u^2 - 1}}$$

$$\frac{d}{dx}[\operatorname{arccsc} u] = \frac{-u'}{|u|\sqrt{u^2 - 1}}$$

81. True

$$\frac{d}{dx}[\arctan(\tan x)] = \frac{\sec^2 x}{1 + \tan^2 x} = \frac{\sec^2 x}{\sec^2 x} = 1$$

83. (a) Let $y = \arccos u$. Then

$$\cos y = u$$

$$-\sin y \frac{dy}{dx} = u'$$

$$\frac{dy}{dx} = -\frac{u'}{\sin y} = -\frac{u'}{\sqrt{1-u^2}}.$$

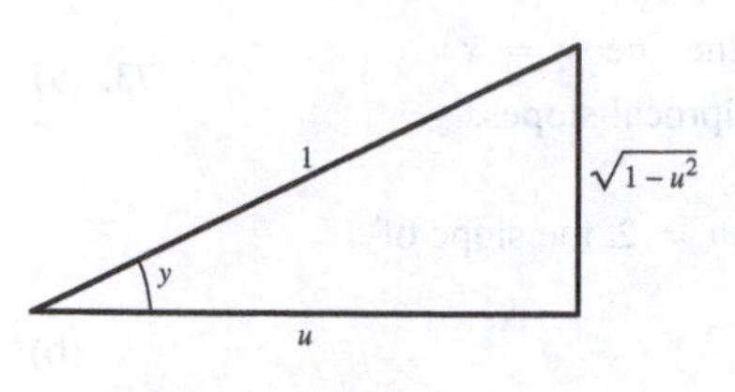

(b) Let $y = \arctan u$. Then

$$\tan y = u$$

$$\sec^2 y \frac{dy}{dx} = u'$$

$$\frac{dy}{dx} = \frac{u'}{\sec^2 y} = \frac{u'}{1-u^2}.$$

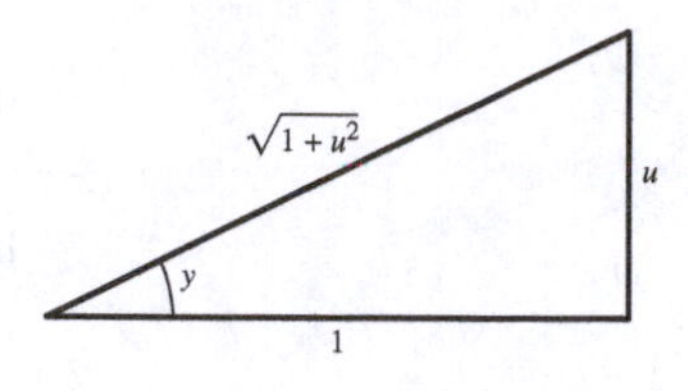

(c) Let $y = \operatorname{arcsec} u$. Then

$$\sec y = u$$

$$\sec y \tan y \frac{dy}{dx} = u'$$

$$\frac{dy}{dx} = \frac{u'}{\sec y \tan y} = \frac{u'}{|u|\sqrt{u^2-1}}.$$

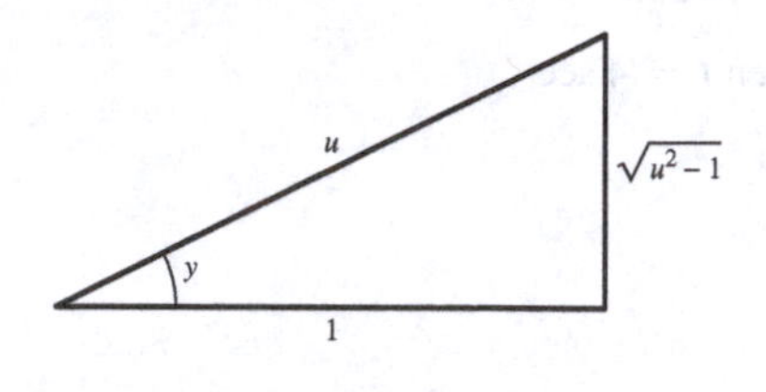

Note: The absolute value sign in the formula for the derivative of arcsec u is necessary because the inverse secant function has a positive slope at every value in its domain.

(d) Let $y = \operatorname{arccot} u$. Then

$$\cot y = u$$

$$-\csc^2 y \frac{dy}{dx} = u'$$

$$\frac{dy}{dx} = \frac{u'}{-\csc^2 y} = -\frac{u'}{1+u^2}.$$

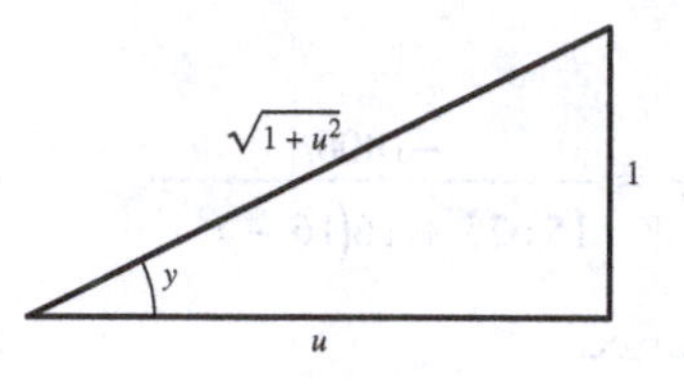

(e) Let $y = \operatorname{arccsc} u$. Then

$$\csc y = u$$

$$-\csc y \cot y \frac{dy}{dx} = u'$$

$$\frac{dy}{dx} = \frac{u'}{-\csc y \cot y} = -\frac{u'}{|u|\sqrt{u^2-1}}.$$

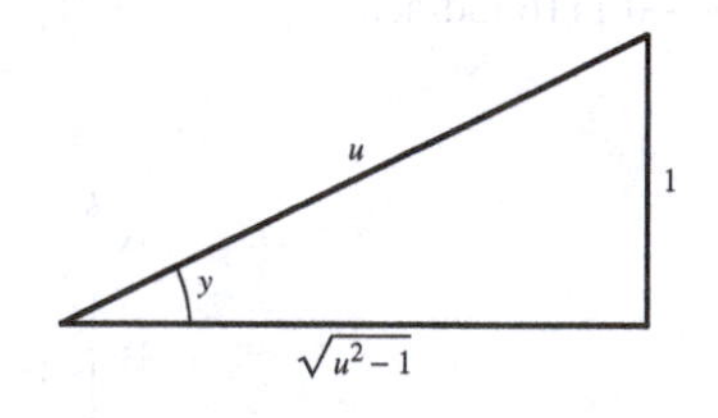

Note: The absolute value sign in the formula for the derivative of arccsc u is necessary because the inverse cosecant function has a negative slope at every value in its domain.

85. Let $\theta = \arctan\left(\frac{x}{\sqrt{1-x^2}}\right), \quad -1 < x < 1$

$$\tan \theta = \frac{x}{\sqrt{1-x^2}}$$

$$\sin \theta = \frac{x}{1} = x$$

$$\arcsin x = \theta.$$

So, $\arcsin x = \arctan\left(\frac{x}{\sqrt{1-x^2}}\right)$ for $-1 < x < 1$.

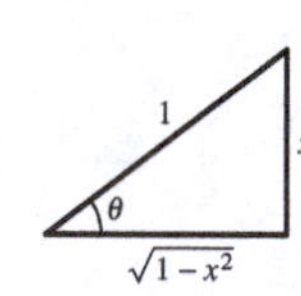

Section 3.7 Related Rates

1. A related-rate equation is an equation that relates the rates of change of various quantities.

3. $y = \sqrt{x}$

$$\frac{dy}{dt} = \left(\frac{1}{2\sqrt{x}}\right)\frac{dx}{dt}$$

$$\frac{dx}{dt} = 2\sqrt{x}\,\frac{dy}{dt}$$

(a) When $x = 4$ and $dx/dt = 3$:

$$\frac{dy}{dt} = \frac{1}{2\sqrt{4}}(3) = \frac{3}{4}$$

(b) When $x = 25$ and $dy/dt = 2$:

$$\frac{dx}{dt} = 2\sqrt{25}(2) = 20$$

5. $xy = 4$

$$x\frac{dy}{dt} + y\frac{dx}{dt} = 0$$

$$\frac{dy}{dt} = \left(-\frac{y}{x}\right)\frac{dx}{dt}$$

$$\frac{dx}{dt} = \left(-\frac{x}{y}\right)\frac{dy}{dt}$$

(a) When $x = 8$, $y = 1/2$, and $dx/dt = 10$:

$$\frac{dy}{dt} = -\frac{1/2}{8}(10) = -\frac{5}{8}$$

(b) When $x = 1$, $y = 4$, and $dy/dt = -6$:

$$\frac{dx}{dt} = -\frac{1}{4}(-6) = \frac{3}{2}$$

7. $y = 2x^2 + 1$

$$\frac{dx}{dt} = 2$$

$$\frac{dy}{dt} = 4x\frac{dx}{dt}$$

(a) When $x = -1$:

$$\frac{dy}{dt} = 4(-1)(2) = -8 \text{ cm/sec}$$

(b) When $x = 0$:

$$\frac{dy}{dt} = 4(0)(2) = 0 \text{ cm/sec}$$

(c) When $x = 1$:

$$\frac{dy}{dt} = 4(1)(2) = 8 \text{ cm/sec}$$

9. $y = \tan x, \dfrac{dx}{dt} = 3$

$$\frac{dy}{dt} = \sec^2 x \cdot \frac{dx}{dt} = \sec^2 x(3) = 3\sec^2 x$$

(a) When $x = -\dfrac{\pi}{3}$:

$$\frac{dy}{dt} = 3\sec^2\left(-\frac{\pi}{3}\right) = 3(2)^2 = 12 \text{ ft/sec}$$

(b) When $x = -\dfrac{\pi}{4}$:

$$\frac{dy}{dt} = 3\sec^2\left(-\frac{\pi}{4}\right) = 3\left(\sqrt{2}\right)^2 = 6 \text{ ft/sec}$$

(c) When $x = 0$:

$$\frac{dy}{dt} = 3\sec^2(0) = 3 \text{ ft/sec}$$

11. $A = \pi r^2$

$$\frac{dr}{dt} = 4$$

$$\frac{dA}{dt} = 2\pi r\frac{dr}{dt}$$

When $r = 37$, $\dfrac{dA}{dt} = 2\pi(37)(4) = 296\pi \text{ cm}^2/\text{min}$.

13. $V = \dfrac{4}{3}\pi r^3$

$$\frac{dr}{dt} = 3$$

$$\frac{dV}{dt} = 4\pi r^2\frac{dr}{dt}$$

(a) When $r = 9$,

$$\frac{dV}{dt} = 4\pi(9)^2(3) = 972\pi \text{ in.}^3/\text{min.}$$

When $r = 36$,

$$\frac{dV}{dt} = 4\pi(36)^2(3) = 15{,}552\pi \text{ in.}^3/\text{min.}$$

(b) If dr/dt is constant, dV/dt is proportional to r^2.

15. $V = x^3$

$$\frac{dx}{dt} = 6$$

$$\frac{dV}{dt} = 3x^2\frac{dx}{dt}$$

(a) When $x = 2$,

$$\frac{dV}{dt} = 3(2)^2(6) = 72 \text{ cm}^3/\text{sec}.$$

(b) When $x = 10$,

$$\frac{dV}{dt} = 3(10)^2(6) = 1800 \text{ cm}^3/\text{sec}.$$

17. $V = \frac{1}{3}\pi r^2 h = \frac{1}{3}\pi\left(\frac{9}{4}h^2\right)h$ [because $2r = 3h$]

$$= \frac{3\pi}{4}h^3$$

$$\frac{dV}{dt} = 10$$

$$\frac{dV}{dt} = \frac{9\pi}{4}h^2\frac{dh}{dt} \Rightarrow \frac{dh}{dt} = \frac{4(dV/dt)}{9\pi h^2}$$

When $h = 15$, $\frac{dh}{dt} = \frac{4(10)}{9\pi(15)^2} = \frac{8}{405\pi}$ ft/min.

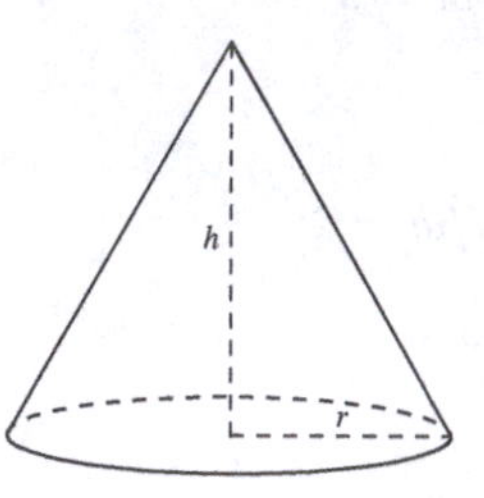

19.

(a) Total volume of pool $= \frac{1}{2}(2)(12)(6) + (1)(6)(12) = 144 \text{ m}^3$

Volume of 1 m of water $= \frac{1}{2}(1)(6)(6) = 18 \text{ m}^3$ (see similar triangle diagram)

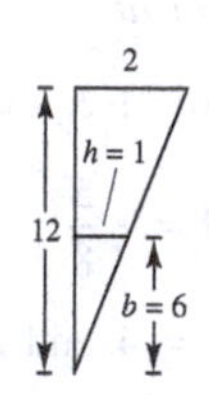

% pool filled $= \frac{18}{144}(100\%) = 12.5\%$

(b) Because for $0 \le h \le 2$, $b = 6h$, you have

$$V = \frac{1}{2}bh(6) = 3bh = 3(6h)h = 18h^2$$

$$\frac{dV}{dt} = 36h\frac{dh}{dt} = \frac{1}{4} \Rightarrow \frac{dh}{dt} = \frac{1}{144h} = \frac{1}{144(1)} = \frac{1}{144} \text{ m/min}.$$

21. $x^2 + y^2 = 25^2$

$$2x\frac{dx}{dt} + 2y\frac{dy}{dt} = 0$$

$$\frac{dy}{dt} = \frac{-x}{y}\cdot\frac{dx}{dt} = \frac{-2x}{y} \quad \text{because } \frac{dx}{dt} = 2.$$

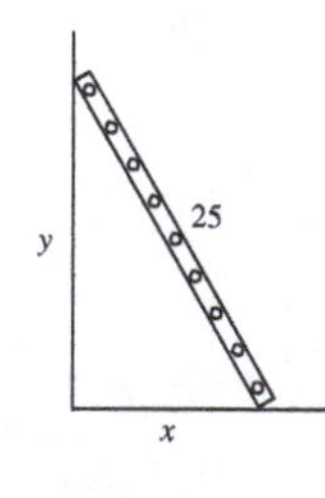

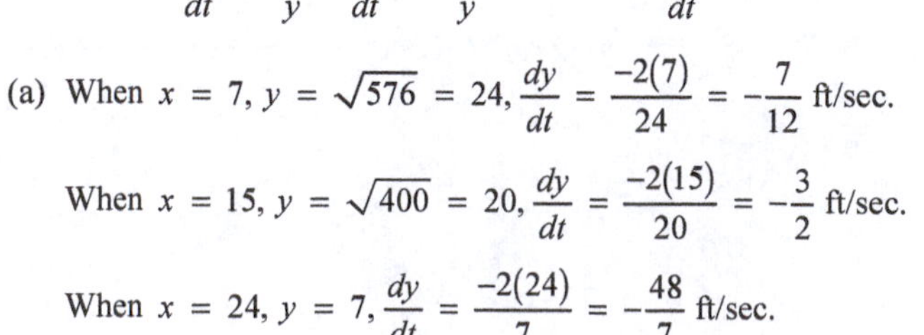

(a) When $x = 7$, $y = \sqrt{576} = 24$, $\frac{dy}{dt} = \frac{-2(7)}{24} = -\frac{7}{12}$ ft/sec.

When $x = 15$, $y = \sqrt{400} = 20$, $\frac{dy}{dt} = \frac{-2(15)}{20} = -\frac{3}{2}$ ft/sec.

When $x = 24$, $y = 7$, $\frac{dy}{dt} = \frac{-2(24)}{7} = -\frac{48}{7}$ ft/sec.

(b) $A = \frac{1}{2}xy$

$$\frac{dA}{dt} = \frac{1}{2}\left(x\frac{dy}{dt} + y\frac{dx}{dt}\right)$$

From part (a) you have $x = 7$, $y = 24$, $\frac{dx}{dt} = 2$, and $\frac{dy}{dt} = -\frac{7}{12}$.

So, $\frac{dA}{dt} = \frac{1}{2}\left[7\left(-\frac{7}{12}\right) + 24(2)\right] = \frac{527}{24}$ ft²/sec.

(c) $\tan\theta = \frac{x}{y}$

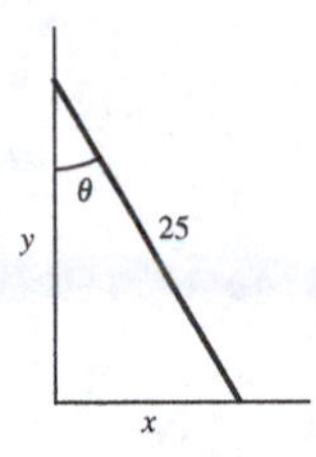

$$\sec^2\theta\frac{d\theta}{dt} = \frac{1}{y}\cdot\frac{dx}{dt} - \frac{x}{y^2}\cdot\frac{dy}{dt}$$

$$\frac{d\theta}{dt} = \cos^2\theta\left[\frac{1}{y}\cdot\frac{dx}{dt} - \frac{x}{y^2}\cdot\frac{dy}{dt}\right]$$

Using $x = 7$, $y = 24$, $\frac{dx}{dt} = 2$, $\frac{dy}{dt} = -\frac{7}{12}$ and $\cos\theta = \frac{24}{25}$, you have

$$\frac{d\theta}{dt} = \left(\frac{24}{25}\right)^2\left[\frac{1}{24}(2) - \frac{7}{(24)^2}\left(-\frac{7}{12}\right)\right] = \frac{1}{12} \text{ rad/sec.}$$

23. When $y = 6$, $x = \sqrt{12^2 - 6^2} = 6\sqrt{3}$, and $s = \sqrt{x^2 + (12 - y)^2} = \sqrt{108 + 36} = 12$.

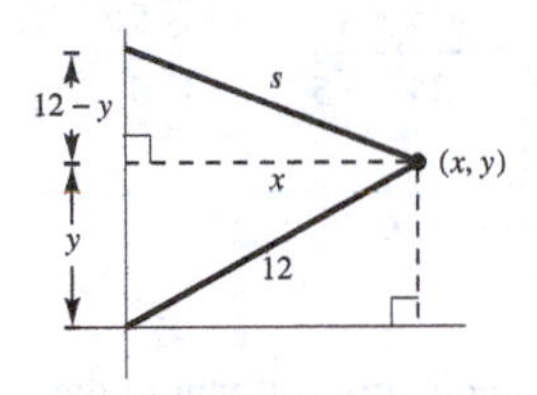

$$x^2 + (12 - y)^2 = s^2$$

$$2x\frac{dx}{dt} + 2(12 - y)(-1)\frac{dy}{dt} = 2s\frac{ds}{dt}$$

$$x\frac{dx}{dt} + (y - 12)\frac{dy}{dt} = s\frac{ds}{dt}$$

Also, $x^2 + y^2 = 12^2$.

$$2x\frac{dx}{dt} + 2y\frac{dy}{dt} = 0 \Rightarrow \frac{dy}{dt} = \frac{-x}{y}\frac{dx}{dt}$$

So, $x\frac{dx}{dt} + (y - 12)\left(\frac{-x}{y}\frac{dx}{dt}\right) = s\frac{ds}{dt}$.

$$\frac{dx}{dt}\left[x - x + \frac{12x}{y}\right] = s\frac{ds}{dt} \Rightarrow \frac{dx}{dt} = \frac{sy}{12x}\cdot\frac{ds}{dt} = \frac{(12)(6)}{(12)(6\sqrt{3})}(-0.2) = \frac{-1}{5\sqrt{3}} = \frac{-\sqrt{3}}{15} \text{ m/sec (horizontal)}$$

$$\frac{dy}{dt} = \frac{-x}{y}\frac{dx}{dt} = \frac{-6\sqrt{3}}{6}\cdot\frac{\left(-\sqrt{3}\right)}{15} = \frac{1}{5} \text{ m/sec (vertical)}$$

25. (a) $s^2 = x^2 + y^2$

$$\frac{dx}{dt} = -450$$

$$\frac{dy}{dt} = -600$$

$$2s\frac{ds}{dt} = 2x\frac{dx}{dt} + 2y\frac{dy}{dt}$$

$$\frac{ds}{dt} = \frac{x(dx/dt) + y(dy/dt)}{s}$$

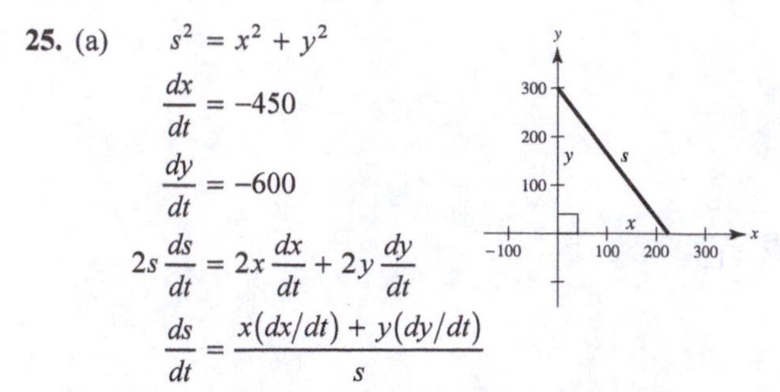

When $x = 225$ and $y = 300$, $s = 375$ and

$$\frac{ds}{dt} = \frac{225(-450) + 300(-600)}{375} = -750 \text{ mi/h.}$$

(b) $t = \dfrac{375}{750} = \dfrac{1}{2}$ h $= 30$ min

27. $s^2 = 90^2 + x^2$

$$x = 20$$

$$\frac{dx}{dt} = -25$$

$$2s\frac{ds}{dt} = 2x\frac{dx}{dt} \Rightarrow \frac{ds}{dt} = \frac{x}{s}\cdot\frac{dx}{dt}$$

When $x = 20$, $s = \sqrt{90^2 + 20^2} = 10\sqrt{85}$,

$$\frac{ds}{dt} = \frac{20}{10\sqrt{85}}(-25) = \frac{-50}{\sqrt{85}} \approx -5.42 \text{ ft/sec.}$$

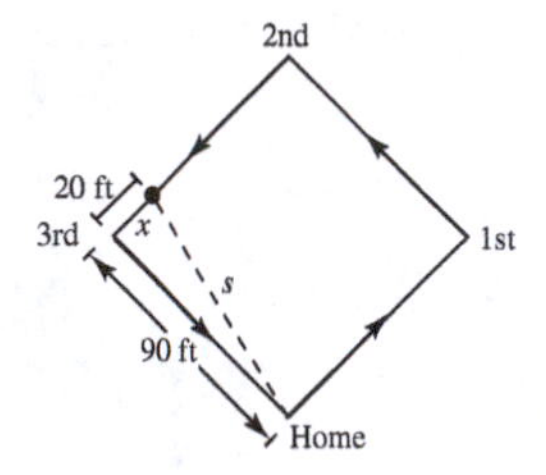

29. (a) $\dfrac{15}{6} = \dfrac{y}{y - x} \Rightarrow 15y - 15x = 6y$

$$y = \frac{5}{3}x$$

$$\frac{dx}{dt} = 5$$

$$\frac{dy}{dt} = \frac{5}{3}\cdot\frac{dx}{dt} = \frac{5}{3}(5) = \frac{25}{3} \text{ ft/sec}$$

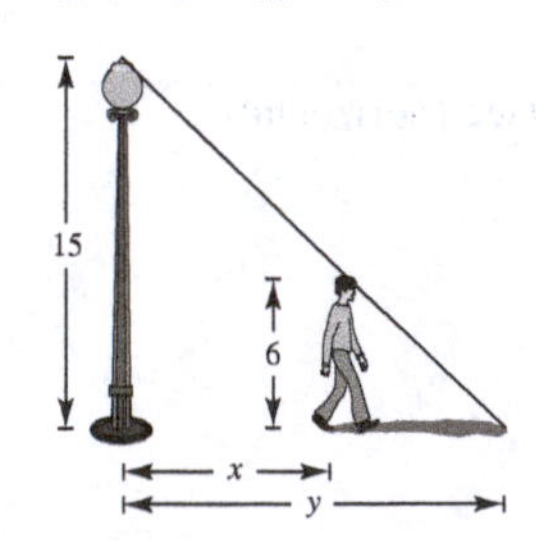

(b) $\dfrac{d(y - x)}{dt} = \dfrac{dy}{dt} - \dfrac{dx}{dt} = \dfrac{25}{3} - 5 = \dfrac{10}{3}$ ft/sec

31. $x(t) = \dfrac{1}{2}\sin\dfrac{\pi t}{6}$, $x^2 + y^2 = 1$

(a) Period: $\dfrac{2\pi}{\pi/6} = 12$ seconds

(b) When $x = \dfrac{1}{2}$, $y = \sqrt{1^2 - \left(\dfrac{1}{2}\right)^2} = \dfrac{\sqrt{3}}{2}$ m.

Lowest point: $\left(0, \dfrac{\sqrt{3}}{2}\right)$

(c) When $x = \dfrac{1}{4}$,

$$y = \sqrt{1 - \left(\frac{1}{4}\right)^2} = \frac{\sqrt{15}}{4} \text{ and } t = 1:$$

$$\frac{dx}{dt} = \frac{1}{2}\left(\frac{\pi}{6}\right)\cos\frac{\pi t}{6} = \frac{\pi}{12}\cos\frac{\pi t}{6}$$

$$x^2 + y^2 = 1$$

$$2x\frac{dx}{dt} + 2y\frac{dy}{dt} = 0 \Rightarrow \frac{dy}{dt} = \frac{-x}{y}\frac{dx}{dt}$$

So, $\dfrac{dy}{dt} = -\dfrac{1/4}{\sqrt{15}/4}\cdot\dfrac{\pi}{12}\cos\left(\dfrac{\pi}{6}\right)$

$$= \frac{-\pi}{\sqrt{15}}\left(\frac{1}{12}\right)\frac{\sqrt{3}}{2} = \frac{-\pi}{24}\frac{1}{\sqrt{5}} = \frac{-\sqrt{5}\pi}{120}.$$

Speed $= \left|\dfrac{-\sqrt{5}\pi}{120}\right| = \dfrac{\sqrt{5}\pi}{120}$ m/sec

33. Because the evaporation rate is proportional to the surface area, $dV/dt = k(4\pi r^2)$. However, because $V = (4/3)\pi r^3$, you have $\dfrac{dV}{dt} = 4\pi r^2\dfrac{dr}{dt}$. Therefore,

$$k(4\pi r^2) = 4\pi r^2\frac{dr}{dt} \Rightarrow k = \frac{dr}{dt}.$$

35. (a) $dy/dt = 3(dx/dt)$ means that y changes three times as fast as x changes.

(b) y changes slowly when $x \approx 0$ or $x \approx L$. y changes more rapidly when x is near the middle of the interval.

37. $\dfrac{1}{R} = \dfrac{1}{R_1} + \dfrac{1}{R_2}$

$\dfrac{dR_1}{dt} = 1$

$\dfrac{dR_2}{dt} = 1.5$

$\dfrac{1}{R^2} \cdot \dfrac{dR}{dt} = \dfrac{1}{R_1^2} \cdot \dfrac{dR_1}{dt} + \dfrac{1}{R_2^2} \cdot \dfrac{dR_2}{dt}$

When $R_1 = 50$ and $R_2 = 75$:

$R = 30$

$\dfrac{dR}{dt} = (30)^2\left[\dfrac{1}{(50)^2}(1) + \dfrac{1}{(75)^2}(1.5)\right] = 0.6$ ohm/sec

39. $\sin 18° = \dfrac{x}{y}$

$0 = -\dfrac{x}{y^2} \cdot \dfrac{dy}{dt} + \dfrac{1}{y} \cdot \dfrac{dx}{dt}$

$\dfrac{dx}{dt} = \dfrac{x}{y} \cdot \dfrac{dy}{dt} = (\sin 18°)(275) \approx 84.9797$ mi/hr

41. $\sin\theta = \dfrac{10}{x}$

$\dfrac{dx}{dt} = (-1)$ ft/sec

$\cos\theta\left(\dfrac{d\theta}{dt}\right) = \dfrac{-10}{x^2} \cdot \dfrac{dx}{dt}$

$\dfrac{d\theta}{dt} = \dfrac{-10}{x^2}\dfrac{dx}{dt}(\sec\theta)$

$= \dfrac{-10}{25^2}(-1)\dfrac{25}{\sqrt{25^2 - 10^2}}$

$= \dfrac{10}{25}\dfrac{1}{5\sqrt{21}} = \dfrac{2}{25\sqrt{21}}$

$= \dfrac{2\sqrt{21}}{525} \approx 0.017$ rad/sec

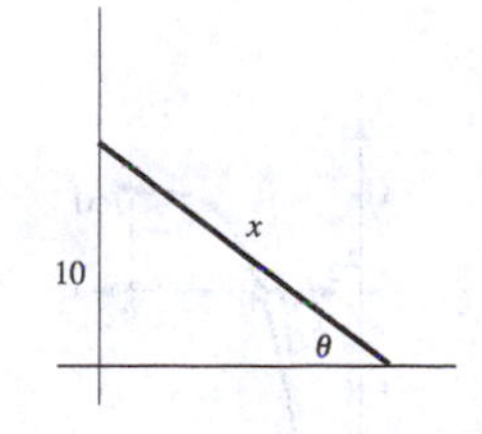

43. $\tan\theta = \dfrac{x}{50}$

$\dfrac{d\theta}{dt} = 30(2\pi) = 60\pi$ rad/min $= \pi$ rad/sec

$\sec^2\theta\left(\dfrac{d\theta}{dt}\right) = \dfrac{1}{50}\left(\dfrac{dx}{dt}\right)$

$\dfrac{dx}{dt} = 50\sec^2\theta\left(\dfrac{d\theta}{dt}\right)$

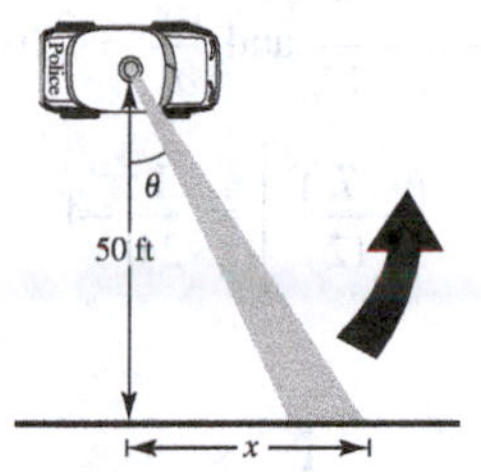

(a) When $\theta = 30°$, $\dfrac{dx}{dt} = \dfrac{200\pi}{3}$ ft/sec.

(b) When $\theta = 60°$, $\dfrac{dx}{dt} = 200\pi$ ft/sec.

(c) When $\theta = 70°$, $\dfrac{dx}{dt} \approx 427.43\pi$ ft/sec.

45. (a) $A = (\text{base})(\text{height}) = 2xe^{-x^2/2}$

(b) $\dfrac{dA}{dt} = \left[2x\left(-xe^{-x^2/2}\right) + 2e^{-x^2/2}\right]\dfrac{dx}{dt}$

$= \left(-2x^2 + 2\right)e^{-x^2/2}\dfrac{dx}{dt}$

For $x = 2$ and $\dfrac{dx}{dt} = 4$,

$\dfrac{dA}{dt} = -6e^{-2}(4) = \dfrac{-24}{e^2} \approx -3.25\text{ cm}^2/\text{min}.$

47. $H = \dfrac{4347}{400{,}000{,}000}e^{369{,}444/(50t+19{,}793)}$

(a) $t = 65° \Rightarrow H \approx 99.79\%$

$t = 80° \Rightarrow H \approx 60.20\%$

(b) $H' = H \cdot \left(\dfrac{-369{,}444(50)}{(50t + 19{,}793)^2}\right)t'$

At $t = 75$ and $t' = 2$, $H' \approx -4.7\%/h$.

49. $x^2 + y^2 = 25$; acceleration of the top of the ladder $= \dfrac{d^2y}{dt^2}$

First derivative:

$$2x\frac{dx}{dt} + 2y\frac{dy}{dt} = 0$$

$$x\frac{dx}{dt} + y\frac{dy}{dt} = 0$$

Second derivative:

$$x\frac{d^2x}{dt^2} + \frac{dx}{dt}\cdot\frac{dx}{dt} + y\frac{d^2y}{dt^2} + \frac{dy}{dt}\cdot\frac{dy}{dt} = 0$$

$$\frac{d^2y}{dt^2} = \left(\frac{1}{y}\right)\left[-x\frac{d^2x}{dt^2} - \left(\frac{dx}{dt}\right)^2 - \left(\frac{dy}{dt}\right)^2\right]$$

When $x = 7$, $y = 24$, $\dfrac{dy}{dt} = -\dfrac{7}{12}$, and $\dfrac{dx}{dt} = 2$ (see Exercise 25). Because $\dfrac{dx}{dt}$ is constant, $\dfrac{d^2x}{dt^2} = 0$.

$$\frac{d^2y}{dt^2} = \frac{1}{24}\left[-7(0) - (2)^2 - \left(-\frac{7}{12}\right)^2\right] = \frac{1}{24}\left[-4 - \frac{49}{144}\right] = \frac{1}{24}\left[-\frac{625}{144}\right] \approx -0.1808 \text{ ft/sec}^2$$

51. $y(t) = -4.9t^2 + 20$

$$\frac{dy}{dt} = -9.8t$$

$$y(1) = -4.9 + 20 = 15.1$$

$$y'(1) = -9.8$$

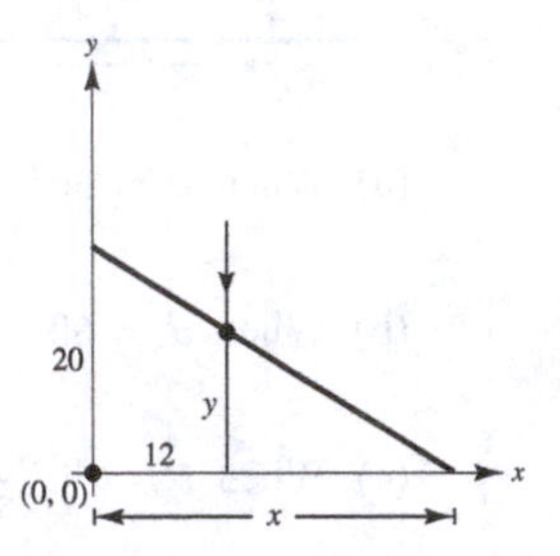

By similar triangles:

$$\frac{20}{x} = \frac{y}{x - 12}$$

$$20x - 240 = xy$$

When $y = 15.1$: $20x - 240 = x(15.1)$

$$(20 - 15.1)x = 240$$

$$x = \frac{240}{4.9}$$

$$20x - 240 = xy$$

$$20\frac{dx}{dt} = x\frac{dy}{dt} + y\frac{dx}{dt}$$

$$\frac{dx}{dt} = \frac{x}{20 - y}\frac{dy}{dt}$$

At $t = 1$, $\dfrac{dx}{dt} = \dfrac{240/4.9}{20 - 15.1}(-9.8) \approx -97.96$ m/sec.

Section 3.8 Newton's Method

1. Answers will vary. *Sample answer:* If f is a function continuous on $[a, b]$ and differentiable on (a, b), where $c \in [a, b]$ and $f(c) = 0$, then Newton's Method uses tangent lines to approximate c. First, estimate an initial x_1 close to c. (See graph.) Then determine x_2 using $x_2 = x_1 - \dfrac{f(x_1)}{f'(x_1)}$. Calculate a third estimate using $x_3 = x_2 - \dfrac{f(x_2)}{f'(x_2)}$. Continue this process until $|x_n - x_{n+1}|$ is within the desired accuracy, and let x_{n+1} be the final approximation of c.

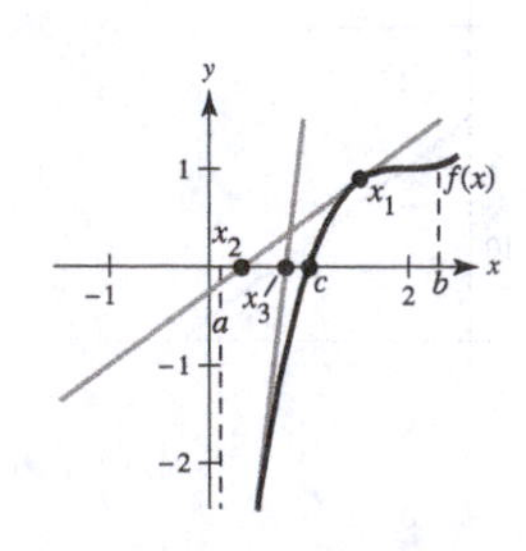

In the solutions for Exercises 3–5, the values in the tables have been rounded for convenience. Because a calculator and a computer program calculate internally using more digits than they display, you may produce slightly different values from those shown in the tables.

3. $f(x) = x^2 - 5$

$f'(x) = 2x$

$x_1 = 2$

n	x_n	$f(x_n)$	$f'(x_n)$	$\frac{f(x_n)}{f'(x_n)}$	$x_n - \frac{f(x_n)}{f'(x_n)}$
1	2	–1	4	–0.25	2.25
2	2.25	0.0625	4.5	0.0139	2.2361

5. $f(x) = \cos x$

$f'(x) = -\sin x$

$x_1 = 1.6$

n	x_n	$f(x_n)$	$f'(x_n)$	$\frac{f(x_n)}{f'(x_n)}$	$x_n - \frac{f(x_n)}{f'(x_n)}$
1	1.6000	–0.0292	–0.9996	0.0292	1.5708
2	1.5708	0.0000	–1.0000	0.0000	1.5708

7. $f(x) = x^3 + 4$

$f'(x) = 3x^2$

$x_1 = -2$

n	x_n	$f(x_n)$	$f'(x_n)$	$\frac{f(x_n)}{f'(x_n)}$	$x_n - \frac{f(x_n)}{f'(x_n)}$
1	–2.0000	–4.0000	12.0000	–0.3333	–1.6667
2	–1.6667	–0.6296	8.3333	–0.0756	–1.5911
3	–1.5911	–0.0281	7.5949	–0.0037	–1.5874
4	–1.5874	–0.0000	7.5596	0.0000	–1.5874

Approximation of the zero of f is –1.587.

9. $f(x) = x^3 + x - 1$

$f'(x) = 3x^2 + 1$

$x_1 = 0.5$

n	x_n	$f(x_n)$	$f'(x_n)$	$\frac{f(x_n)}{f'(x_n)}$	$x_n - \frac{f(x_n)}{f'(x_n)}$
1	0.5000	–0.3750	1.7500	–0.2143	0.7143
2	0.7143	0.0788	2.5307	0.0311	0.6832
3	0.6832	0.0021	2.4003	0.0009	0.6823

Approximation of the zero of f is 0.682.

11. $f(x) = 5\sqrt{x - 1} - 2x$

$f'(x) = \frac{5}{2\sqrt{x - 1}} - 2$

From the graph you see that there are two zeros. Begin with $x_1 = 1.2$.

n	x_n	$f(x_n)$	$f'(x_n)$	$\frac{f(x_n)}{f'(x_n)}$	$x_n - \frac{f(x_n)}{f'(x_n)}$
1	1.2000	–0.1639	3.5902	–0.0457	1.2457
2	1.2457	–0.0131	3.0440	–0.0043	1.2500
3	1.2500	–0.0001	3.0003	–0.0003	1.2500

Approximation of the zero of f is 1.250.

Now use $x_1 = 5.0$.

n	x_n	$f(x_n)$	$f'(x_n)$	$\frac{f(x_n)}{f'(x_n)}$	$x_n - \frac{f(x_n)}{f'(x_n)}$
1	5.0000	0.0000	–0.7500	0.0000	5.0000

Approximation of the zero of f is 5.0000.

13. $f(x) = x - e^{-x}$

$f'(x) = 1 + e^{-x}$

$x_1 = 0.5$

n	x_n	$f(x_n)$	$f'(x_n)$	$\frac{f(x_n)}{f'(x_n)}$	$x_n - \frac{f(x_n)}{f'(x_n)}$
1	0.5	–0.1065	1.6065	–0.0663	0.5663
2	0.5663	0.0013	1.5676	0.0008	0.5671
3	0.5671	0.0001	1.5672	–0.0000	0.5671

Approximation of the zero of f is 0.567.

15. $f(x) = x^3 - 3.9x^2 + 4.79x - 1.881$

$f'(x) = 3x^2 - 7.8x + 4.79$

$x_1 = 0.5$

n	x_n	$f(x_n)$	$f'(x_n)$	$\frac{f(x_n)}{f'(x_n)}$	$x_n - \frac{f(x_n)}{f'(x_n)}$
1	0.5000	–0.3360	1.6400	–0.2049	0.7049
2	0.7049	–0.0921	0.7824	–0.1177	0.8226
3	0.8226	–0.0231	0.4037	–0.0573	0.8799
4	0.8799	–0.0045	0.2495	–0.0181	0.8980
5	0.8980	–0.0004	0.2048	–0.0020	0.9000
6	0.9000	0.0000	0.2000	0.0000	0.9000

Approximation of the zero of f is 0.900.

$x_1 = 1.1$

n	x_n	$f(x_n)$	$f'(x_n)$	$\frac{f(x_n)}{f'(x_n)}$	$x_n - \frac{f(x_n)}{f'(x_n)}$
1	1.1	0.0000	–0.1600	–0.0000	1.1000

Approximation of the zero of f is 1.100.

$x_1 = 1.9$

n	x_n	$f(x_n)$	$f'(x_n)$	$\frac{f(x_n)}{f'(x_n)}$	$x_n - \frac{f(x_n)}{f'(x_n)}$
1	1.9	0.0000	0.8000	0.0000	1.9000

Approximation of the zero of f is 1.900.

17. $f(x) = 1 - x + \sin x$

$f'(x) = -1 + \cos x$

$x_1 = 2.0$

n	x_n	$f(x_n)$	$f'(x_n)$	$\frac{f(x_n)}{f'(x_n)}$	$x_n - \frac{f(x_n)}{f'(x_n)}$
1	2.0000	–0.0907	–1.4161	0.0640	1.9360
2	1.9360	–0.0019	–1.3571	0.0014	1.9346
3	1.9346	0.0000	–1.3558	0.0000	1.9346

Approximation of the zero of f is $x \approx 1.935$.

19. $h(x) = f(x) - g(x) = 2x + 1 - \sqrt{x + 4}$

$h'(x) = 2 - \dfrac{1}{2\sqrt{x + 4}}$

n	x_n	$h(x_n)$	$h'(x_n)$	$\dfrac{h(x_n)}{h'(x_n)}$	$x_n - \dfrac{h(x_n)}{h'(x_n)}$
1	0.6000	0.0552	1.7669	0.0313	0.5687
2	0.5687	0.0000	1.7661	0.0000	0.5687

Point of intersection of the graphs of f and g occurs when $x \approx 0.569$.

21. $h(x) = f(x) - g(x) = x - \tan x$

$h'(x) = 1 - \sec^2 x$

n	x_n	$h(x_n)$	$h'(x_n)$	$\dfrac{h(x_n)}{h'(x_n)}$	$x_n - \dfrac{h(x_n)}{h'(x_n)}$
1	4.5000	–0.1373	–21.5048	0.0064	4.4936
2	4.4936	–0.0039	–20.2271	0.0002	4.4934

Point of intersection of the graphs of f and g occurs when $x \approx 4.493$.

Note: $f(x) = x$ and $g(x) = \tan x$ intersect infinitely often.

23. $f(x) = x^3 - 3x^2 + 3$

(a)

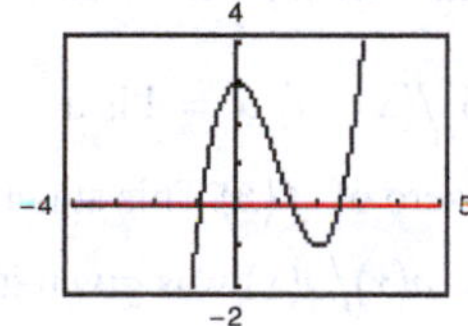

(b) $f'(x) = 3x^2 - 6x$

$x_1 = 1$

$x_2 = x_1 - \dfrac{f(x_1)}{f'(x_1)} \approx 1.3333$

Continuing, you obtain the zero 1.3473.

(c) $x_1 = \dfrac{1}{4}$

$x_2 = x_1 - \dfrac{f(x_1)}{f'(x_1)} \approx 2.4048$

Continuing, you obtain the zero 2.5321.

(d)

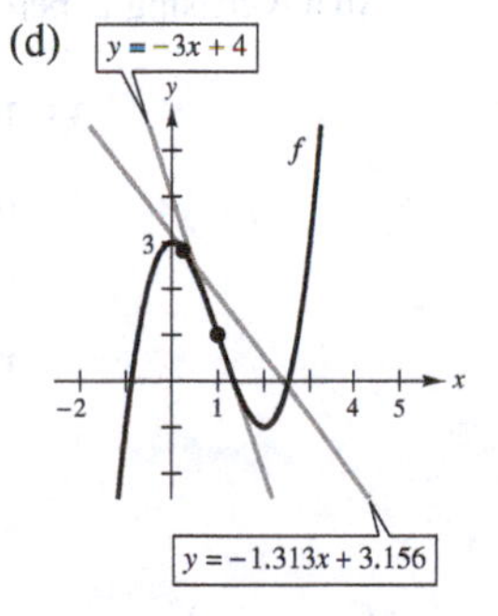

If the initial estimate $x = x_1$, is not sufficiently close to the desired zero of a function, then the x-intercept of the corresponding tangent line to the function may approximate a second zero of the function.

25. $y = 2x^3 - 6x^2 + 6x - 1 = f(x)$

$y' = 6x^2 - 12x + 6 = f'(x)$

$x_1 = 1$

$f'(x) = 0$; therefore, the method fails.

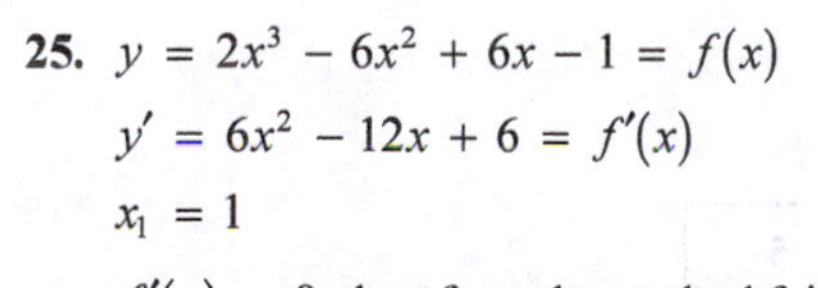

n	x_n	$f(x_n)$	$f'(x_n)$
1	1	1	0

27. Let $g(x) = f(x) - x = \cos x - x$

$g'(x) = -\sin x - 1.$

n	x_n	$g(x_n)$	$g'(x_n)$	$\frac{g(x_n)}{g'(x_n)}$	$x_n - \frac{g(x_n)}{g'(x_n)}$
1	1.0000	–0.4597	–1.8415	0.2496	0.7504
2	0.7504	–0.0190	–1.6819	0.0113	0.7391
3	0.7391	0.0000	–1.6736	0.0000	0.7391

The fixed point is approximately 0.74.

29. Let $g(x) = e^{x/10} - x$

$g'(x) = \frac{1}{10}e^{x/10} - 1.$

n	x_n	$g(x_n)$	$g'(x_n)$	$\frac{g(x_n)}{g'(x_n)}$	$x_n - \frac{g(x_n)}{g'(x_n)}$
1	1.0	0.1052	–0.8895	–0.1182	1.1182
2	1.1182	0.0001	–0.8882	–0.0001	1.1183

The fixed point is approximately 1.12.

31. Let $f(x) = -76x^3 + 4830x^2 - 2{,}820{,}000.$

$f'(x) = -228x^2 + 9660x.$

From the graph, choose $x_1 = 38.$

n	x_n	$f(x_n)$	$f'(x_n)$	$\frac{f(x_n)}{f'(x_n)}$	$x_n - \frac{f(x_n)}{f'(x_n)}$
1	38.0	–15,752	37,848	–0.4162	38.4162
2	38.4162	–699.2765	34,617.0834	–0.0202	38.4364
3	38.4364	28.3848	34,458.2634	0.0008	38.4356

An advertising expense of about \$384,364 will yield a profit of \$2,500,000.

33. The future iterations will all be x_1 because

$$x_{n+1} = x_n - \frac{f(x_n)}{f'(x_n)} = x_n, n = 1, 2, 3, \ldots.$$

35. No. Let $f(x) = (x^2 - 1)/(x - 1)$. $x = 1$ is a discontinuity. It is not a zero of $f(x)$. This statement would be true if $f(x) = p(x)/q(x)$ was given in **reduced** form.

37. (a) $f(x) = x^2 - a, a > 0$

$f'(x) = 2x$

$$x_{n+1} = x_n - \frac{f(x_n)}{f'(x_n)} = x_n - \frac{x_n^2 - a}{2x_n} = \frac{1}{2}\left(x_n + \frac{a}{x_n}\right)$$

(b) $\sqrt{5}$: $x_{n+1} = \frac{1}{2}\left(x_n + \frac{5}{x_n}\right), x_1 = 2$

n	1	2	3	4
x_n	2	2.25	2.2361	2.2361

For example, given $x_1 = 2$,

$$x_2 = \frac{1}{2}\left(2 + \frac{5}{2}\right) = \frac{9}{4} = 2.25.$$

$\sqrt{5} \approx 2.236$

$\sqrt{7}$: $x_{n+1} = \frac{1}{2}\left(x_n + \frac{7}{x_n}\right), x_1 = 2$

n	1	2	3	4	5
x_n	2	2.75	2.6477	2.6458	2.6458

$\sqrt{7} \approx 2.646$

39. $f(x) = \frac{1}{x} - a = 0$

$f'(x) = -\frac{1}{x^2}$

$x_{n+1} = x_n - \frac{(1/x_n) - a}{-1/x_n^2} = x_n + x_n^2\left(\frac{1}{x_n} - a\right) = x_n + x_n - x_n^2 a = 2x_n - x_n^2 a = x_n(2 - ax_n)$

41. True

43. True

45. $f(x) = -\sin x$

$f'(x) = -\cos x$

Let $(x_0, y_1) = (x_0, -\sin(x_0))$ be a point on the graph of f. If (x_0, y_0) is a point of tangency, then

$-\cos(x_0) = \frac{y_0 - 0}{x_0 - 0} = \frac{y_0}{x_0} = \frac{-\sin(x_0)}{x_0}.$

So, $x_0 = \tan(x_0)$.

$x_0 \approx 4.4934$

Slope $= -\cos(x_0) \approx 0.217$

You can verify this answer by graphing $y_1 = -\sin x$ and the tangent line $y_2 = 0.217x$.

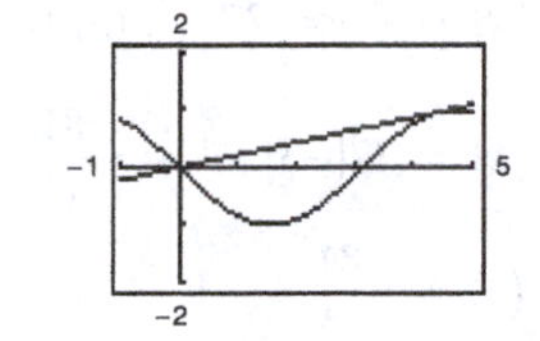

Review Exercises for Chapter 3

1. $f(x) = 12$

$f'(x) = \lim_{\Delta x \to 0} \frac{f(x + \Delta x) - f(x)}{\Delta x} = \lim_{\Delta x \to 0} \frac{12 - 12}{\Delta x} = \lim_{\Delta x \to 0} \frac{0}{\Delta x} = 0$

3. $f(x) = x^3 - 2x + 1$

$$\begin{aligned} f'(x) &= \lim_{\Delta x \to 0} \frac{f(x + \Delta x) - f(x)}{\Delta x} \\ &= \lim_{\Delta x \to 0} \frac{\left[(x + \Delta x)^3 - 2(x + \Delta x) + 1\right] - \left[x^3 - 2x + 1\right]}{\Delta x} \\ &= \lim_{\Delta x \to 0} \frac{x^3 + 3x^2(\Delta x) + 3x(\Delta x)^2 + (\Delta x)^3 - 2x - 2(\Delta x) + 1 - x^3 + 2x - 1}{\Delta x} \\ &= \lim_{\Delta x \to 0} \frac{3x^2(\Delta x) + 3x(\Delta x)^2 + (\Delta x)^3 - 2(\Delta x)}{\Delta x} \\ &= \lim_{\Delta x \to 0} \left[\frac{3x^2 + 3x(\Delta x) + (\Delta x)^2 - 2}{1}\right] = 3x^2 - 2 \end{aligned}$$

5. $g(x) = 2x^2 - 3x, c = 2$

$g'(2) = \lim_{x \to 2} \frac{g(x) - g(2)}{x - 2} = \lim_{x \to 2} \frac{(2x^2 - 3x) - 2}{x - 2} = \lim_{x \to 2} \frac{(x - 2)(2x + 1)}{x - 2} = \lim_{x \to 2} (2x + 1) = 2(2) + 1 = 5$

7. f is differentiable for all $x \neq 3$.

9. $y = 25$

$y' = 0$

11. $f(x) = x^3 - 11x^2$

$f'(x) = 3x^2 - 22x$

13. $h(x) = 6\sqrt{x} + 3\sqrt[3]{x} = 6x^{1/2} + 3x^{1/3}$

$h'(x) = 3x^{-1/2} + x^{-2/3} = \dfrac{3}{\sqrt{x}} + \dfrac{1}{\sqrt[3]{x^2}}$

15. $g(t) = \dfrac{2}{3}t^{-2}$

$g'(t) = \dfrac{-4}{3}t^{-3} = -\dfrac{4}{3t^3}$

17. $f(\theta) = 4\theta - 5\sin\theta$

$f'(\theta) = 4 - 5\cos\theta$

19. $f(t) = 3\cos t - 4e^t$

$f'(t) = -3\sin t - 4e^t$

21. $f(x) = \dfrac{27}{x^3} = 27x^{-3}, (3, 1)$

$f'(x) = 27(-3)x^{-4} = -\dfrac{81}{x^4}$

$f'(3) = -\dfrac{81}{3^4} = -1$

23. $f(x) = 4x^5 + 3x - \sin x, (0, 0)$

$f'(x) = 20x^4 + 3 - \cos x$

$f'(0) = 3 - 1 = 2$

25. $F = 200\sqrt{T}$

$F'(t) = \dfrac{100}{\sqrt{T}}$

(a) When $T = 4$, $F'(4) = 50$ vibrations/sec/lb.

(b) When $T = 9$, $F'(9) = 33\frac{1}{3}$ vibrations/sec/lb.

27. $s(t) = -16t^2 + v_0t + s_0;\ s_0 = 600, v_0 = -30$

(a) $s(t) = -16t^2 - 30t + 600$

$s'(t) = v(t) = -32t - 30$

(b) Average velocity $= \dfrac{s(3) - s(1)}{3 - 1} = \dfrac{366 - 554}{2}$

$= -94$ ft/sec

(c) $v(1) = -32(1) - 30 = -62$ ft/sec

$v(3) = -32(3) - 30 = -126$ ft/sec

(d) $s(t) = 0 = -16t^2 - 30t + 600$

Using a graphing utility or the Quadratic Formula, $t \approx 5.258$ seconds.

(e) When $t \approx 5.258$, $v(t) \approx -32(5.258) - 30 \approx -198.3$ ft/sec.

29. $f(x) = (5x^2 + 8)(x^2 - 4x - 6)$

$f'(x) = (5x^2 + 8)(2x - 4) + (x^2 - 4x - 6)(10x)$

$= 10x^3 + 16x - 20x^2 - 32 + 10x^3 - 40x^2 - 60x$

$= 20x^3 - 60x^2 - 44x - 32$

$= 4(5x^3 - 15x^2 - 11x - 8)$

31. $f(x) = (9x - 1)\sin x$

$f'(x) = (9x - 1)\cos x + 9\sin x$

$= 9x\cos x - \cos x + 9\sin x$

33. $f(x) = \dfrac{x^2 + x - 1}{x^2 - 1}$

$f'(x) = \dfrac{(x^2 - 1)(2x + 1) - (x^2 + x - 1)(2x)}{(x^2 - 1)^2}$

$= \dfrac{-(x^2 + 1)}{(x^2 - 1)^2}$

35. $y = \dfrac{x^4}{\cos x}$

$y' = \dfrac{(\cos x)4x^3 - x^4(-\sin x)}{\cos^2 x}$

$= \dfrac{4x^3\cos x + x^4\sin x}{\cos^2 x}$

37. $y = 3x^2\sec x$

$y' = 3x^2\sec x\tan x + 6x\sec x$

39. $y = x\cos x - \sin x$

$y' = -x\sin x + \cos x - \cos x = -x\sin x$

41. $y = 4xe^x - \cot x$

$y' = 4xe^x + 4e^x + \csc^2 x$

43. $f(x) = (x + 2)(x^2 + 5), (-1, 6)$

$f'(x) = (x + 2)(2x) + (x^2 + 5)(1)$

$= 2x^2 + 4x + x^2 + 5 = 3x^2 + 4x + 5$

$f'(-1) = 3 - 4 + 5 = 4$

Tangent line: $y - 6 = 4(x + 1)$

$y = 4x + 10$

45. $f(x) = \dfrac{x+1}{x-1}, \left(\dfrac{1}{2}, -3\right)$

$f'(x) = \dfrac{(x-1)-(x+1)}{(x-1)^2} = \dfrac{-2}{(x-1)^2}$

$f'\left(\dfrac{1}{2}\right) = \dfrac{-2}{(1/4)} = -8$

Tangent line: $y + 3 = -8\left(x - \dfrac{1}{2}\right)$

$y = -8x + 1$

47. $g(t) = -8t^3 - 5t + 12$

$g'(t) = -24t^2 - 5$

$g''(t) = -48t$

49. $f(x) = 15x^{5/2}$

$f'(x) = \frac{75}{2}x^{3/2}$

$f''(x) = \frac{225}{4}x^{1/2} = \frac{225}{4}\sqrt{x}$

51. $f(\theta) = 3\tan\theta$

$f'(\theta) = 3\sec^2\theta$

$f''(\theta) = 6\sec\theta(\sec\theta\tan\theta)$

$= 6\sec^2\theta\tan\theta$

53. $g(x) = 4\cot x$

$g'(x) = -4\csc^2 x$

$g''(x) = -8\csc x(-\csc x\cot x)$

$= 8\csc^2 x\cot x$

55. $v(t) = 20 - t^2, 0 \le t \le 6$

$a(t) = v'(t) = -2t$

$v(3) = 20 - 3^2 = 11$ m/sec

$a(3) = -2(3) = -6$ m/sec^2

57. $y = (7x+3)^4$

$y' = 4(7x+3)^3(7) = 28(7x+3)^3$

59. $y = \dfrac{1}{(x^2+5)^3} = (x^2+5)^{-3}$

$y' = -3(x^2+5)^{-4}(2x) = -\dfrac{6x}{(x^2+5)^4}$

61. $y = 5\cos(9x+1)$

$y' = -5\sin(9x+1)(9) = -45\sin(9x+1)$

63. $y = \dfrac{x}{2} - \dfrac{\sin 2x}{4}$

$y' = \dfrac{1}{2} - \dfrac{1}{4}\cos 2x(2) = \dfrac{1}{2}(1 - \cos 2x) = \sin^2 x$

65. $y = x(6x+1)^5$

$y' = x\,5(6x+1)^4(6) + (6x+1)^5(1)$

$= 30x(6x+1)^4 + (6x+1)^5$

$= (6x+1)^4(30x + 6x + 1)$

$= (6x+1)^4(36x+1)$

67. $f(x) = \left(\dfrac{x}{\sqrt{x+5}}\right)^3$

$f'(x) = 3\left(\dfrac{x}{\sqrt{x+5}}\right)^2\left[\dfrac{(x+5)^{1/2}(1) - x\left(\frac{1}{2}\right)(x+5)^{-1/2}}{x+5}\right]$

$= \dfrac{3x^2}{x+5}\left[\dfrac{2(x+5)-x}{2(x+5)^{3/2}}\right]$

$= \dfrac{3x^2(x+10)}{2(x+5)^{5/2}}$

69. $h(z) = e^{-z^2/2}$

$h'(z) = -ze^{-z^2/2}$

71. $g(t) = t^2e^{t/4}$

$g'(t) = \frac{1}{4}t^2e^{t/4} + 2te^{t/4}$

$= \frac{1}{4}te^{t/4}[t+8]$

73. $y = \sqrt{e^{2x} + e^{-2x}} = \left(e^{2x} + e^{-2x}\right)^{1/2}$

$y' = \dfrac{1}{2}\left(e^{2x} + e^{-2x}\right)^{-1/2}\left(2e^{2x} - 2e^{-2x}\right) = \dfrac{e^{2x} - e^{-2x}}{\sqrt{e^{2x} + e^{-2x}}}$

75. $g(x) = \ln\sqrt{x} = \dfrac{1}{2}\ln x$

$g'(x) = \dfrac{1}{2x}$

77. $f(x) = x\sqrt{\ln x}$

$f'(x) = \left(\dfrac{x}{2}\right)(\ln x)^{-1/2}\left(\dfrac{1}{x}\right) + \sqrt{\ln x}$

$= \dfrac{1}{2\sqrt{\ln x}} + \sqrt{\ln x}$

$= \dfrac{1 + 2\ln x}{2\sqrt{\ln x}}$

79. $g(x) = \ln\left(x^3\sqrt{4x+1}\right) = 3\ln x + \frac{1}{2}\ln(4x+1)$

$$g'(x) = \frac{3}{x} + \frac{1}{2}\cdot\frac{4}{4x+1}$$
$$= \frac{3}{x} + \frac{2}{4x+1}$$
$$= \frac{3(4x+1)+2x}{x(4x+1)}$$
$$= \frac{14x+3}{4x^2+x}$$

81. $f(x) = \ln\frac{x+4}{x(x+5)}$

$$= \ln(x+4) - \ln x - \ln(x+5)$$
$$f'(x) = \frac{1}{x+4} - \frac{1}{x} - \frac{1}{x+5}$$
$$= -\frac{x^2+8x+20}{x^3+9x^2+20x}$$

83. $f(x) = \sqrt{1-x^3}$, $(-2, 3)$

$$f'(x) = \frac{1}{2}\left(1-x^3\right)^{-1/2}\left(-3x^2\right) = \frac{-3x^2}{2\sqrt{1-x^3}}$$
$$f'(-2) = \frac{-12}{2(3)} = -2$$

85. $f(x) = \frac{x+8}{(3x+1)^{1/2}}$, $(0, 8)$

$$f'(x) = \frac{(3x+1)^{1/2}(1) - (x+8)\left(\frac{1}{2}\right)(3x+1)^{-1/2}(3)}{3x+1}$$
$$f'(0) = \frac{1-4(3)}{1} = -11$$

87. $y = \frac{1}{2}\csc 2x$, $\left(\frac{\pi}{4}, \frac{1}{2}\right)$

$$y' = -\csc 2x \cot 2x$$
$$y'\left(\frac{\pi}{4}\right) = 0$$

89. $y = (8x+5)^3$

$$y' = 3(8x+5)^2(8) = 24(8x+5)^2$$
$$y'' = 24(2)(8x+5)(8) = 384(8x+5)$$

91. $f(x) = \cot x$

$$f'(x) = -\csc^2 x$$
$$f''(x) = -2\csc x(-\csc x \cdot \cot x) = 2\csc^2 x \cot x$$

93. $T = \frac{700}{t^2+4t+10}$

$$T = 700\left(t^2+4t+10\right)^{-1}$$
$$T' = \frac{-1400(t+2)}{\left(t^2+4t+10\right)^2}$$

(a) When $t = 1$,

$$T' = \frac{-1400(1+2)}{(1+4+10)^2} \approx -18.667 \text{ deg/h}.$$

(b) When $t = 3$,

$$T' = \frac{-1400(3+2)}{(9+12+10)^2} \approx -7.284 \text{ deg/h}.$$

(c) When $t = 5$,

$$T' = \frac{-1400(5+2)}{(25+20+10)^2} \approx -3.240 \text{ deg/h}.$$

(d) When $t = 10$,

$$T' = \frac{-1400(10+2)}{(100+40+10)^2} \approx -0.747 \text{ deg/h}.$$

95. (a) You get an error message because $\ln h$ does not exist for $h = 0$.

(b) Reversing the data, you obtain $h = 0.8627 - 6.4474 \ln p$.

(c)

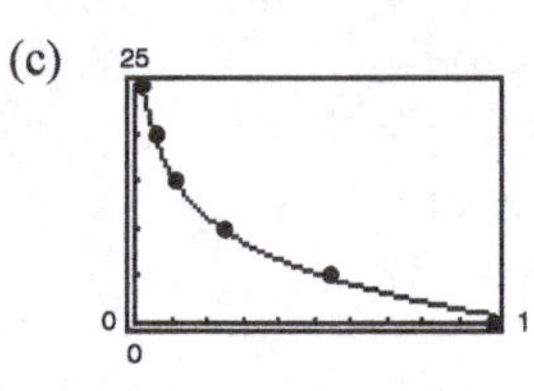

(d) If $p = 0.75$, $h \approx 2.72$ km.

(e) If $h = 13$ km, $p \approx 0.15$ atmosphere.

(f) $h = 0.8627 - 6.4474 \ln p$

$$1 = -6.4474\frac{1}{p}\frac{dp}{dh} \quad \text{(implicit differentiation)}$$
$$\frac{dp}{dh} = \frac{p}{-6.4474}$$

For $h = 5$,

$$p = 0.5264 \text{ and } \frac{dp}{dh} = -0.0816 \text{ atm/km}$$

For $h = 20$,

$$p = 0.0514 \text{ and } \frac{dp}{dh} = -0.0080 \text{ atm/km}$$

As the altitude increases, the rate of change of pressure decreases.

97. $x^2 + y^2 = 64$

$$2x + 2yy' = 0$$
$$2yy' = -2x$$
$$y' = -\frac{x}{y}$$

99. $x^3y - xy^3 = 4$

$$x^3y' + 3x^2y - x3y^2y' - y^3 = 0$$
$$x^3y' - 3xy^2y' = y^3 - 3x^2y$$
$$y'(x^3 - 3xy^2) = y^3 - 3x^2y$$
$$y' = \frac{y^3 - 3x^2y}{x^3 - 3xy^2}$$
$$y' = \frac{y(y^2 - 3x^2)}{x(x^2 - 3y^2)}$$

101. $x \sin y = y \cos x$

$$(x \cos y)y' + \sin y = -y \sin x + y' \cos x$$
$$y'(x \cos y - \cos x) = -y \sin x - \sin y$$
$$y' = \frac{y \sin x + \sin y}{\cos x - x \cos y}$$

103. $x^2 + y^2 = 10$

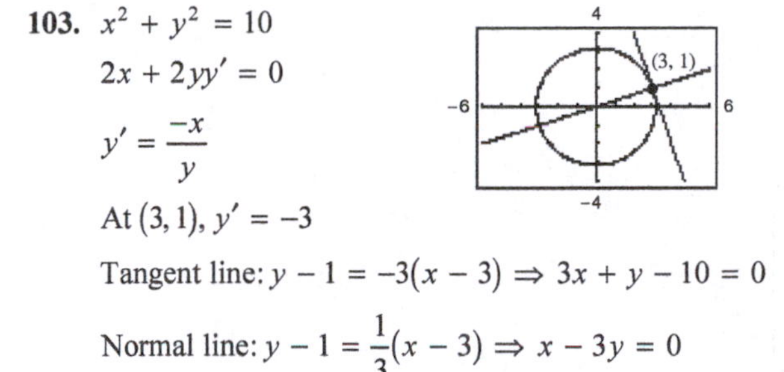

$$2x + 2yy' = 0$$
$$y' = \frac{-x}{y}$$

At $(3, 1)$, $y' = -3$

Tangent line: $y - 1 = -3(x - 3) \Rightarrow 3x + y - 10 = 0$

Normal line: $y - 1 = \frac{1}{3}(x - 3) \Rightarrow x - 3y = 0$

105. $y \ln x + y^2 = 0, \quad (e, -1)$

$$y' \ln x + \frac{y}{x} + 2yy' = 0$$
$$y'(\ln x + 2y) = \frac{-y}{x}$$
$$y' = \frac{-y}{x(\ln x + 2y)}$$

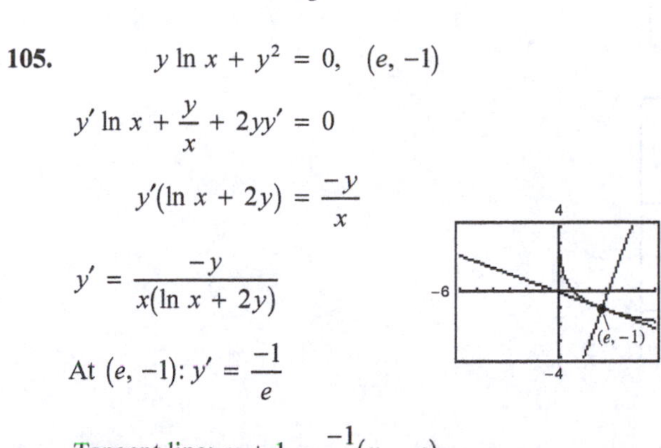

At $(e, -1)$: $y' = \dfrac{-1}{e}$

Tangent line: $y + 1 = \dfrac{-1}{e}(x - e)$

$$y = \frac{-1}{e}x$$

Normal line: $y + 1 = e(x - e)$

$$y = ex - e^2 - 1$$

107. $y = \dfrac{x\sqrt{x^2 + 1}}{x + 4}$

$$\ln y = \ln x + \frac{1}{2}\ln(x^2 + 1) - \ln(x + 4)$$
$$\frac{y'}{y} = \frac{1}{x} + \frac{x}{x^2 + 1} - \frac{1}{x + 4}$$
$$y' = \frac{x\sqrt{x^2 + 1}}{x + 4}\left(\frac{1}{x} + \frac{x}{x^2 + 1} - \frac{1}{x + 4}\right)$$
$$= \frac{x^3 + 8x^2 + 4}{(x + 4)^2\sqrt{x^2 + 1}}$$

109. $f(x) = x^3 + 2, \quad a = -1$

$$f'(x) = 3x^2 > 0$$

f is continuous and differentiable on $(-\infty, \infty)$. Therefore f has an inverse.

$$f(-3^{1/3}) = -1 \Rightarrow f^{-1}(-1) = -3^{1/3}$$
$$f'(-3^{1/3}) = 3^{2/3}$$
$$(f^{-1})'(-1) = \frac{1}{f'(f^{-1}(-1))} = \frac{1}{f'(-3^{1/3})} = \frac{1}{3(3^{2/3})} = \frac{1}{3^{5/3}}$$

111. $f(x) = \tan x, \quad a = \dfrac{\sqrt{3}}{3}, -\dfrac{\pi}{4} \le x \le \dfrac{\pi}{4}$

$$f'(x) = \sec^2 x > 0 \text{ on } \left(-\frac{\pi}{4}, \frac{\pi}{4}\right)$$

f is continuous and differentiable on $\left[-\dfrac{\pi}{4}, \dfrac{\pi}{4}\right]$. Therefore f has an inverse.

$$f\left(\frac{\pi}{6}\right) = \frac{\sqrt{3}}{3} \Rightarrow f^{-1}\left(\frac{\sqrt{3}}{3}\right) = \frac{\pi}{6}$$
$$f'\left(\frac{\pi}{6}\right) = \frac{4}{3}$$
$$(f^{-1})'\left(\frac{\sqrt{3}}{3}\right) = \frac{1}{f'\left(f^{-1}\left(\frac{\sqrt{3}}{3}\right)\right)} = \frac{1}{f'\left(\frac{\pi}{6}\right)} = \frac{1}{\left(\frac{4}{3}\right)} = \frac{3}{4}$$

113. $y = \sin(\arctan 2x) = \dfrac{2x}{\sqrt{4x^2 + 1}}$

$$y' = \frac{(4x^2 + 1)^{1/2}(2) - (2x)\left(\frac{1}{2}\right)(4x^2 + 1)^{-1/2}(8x)}{4x^2 + 1}$$
$$= \frac{2(4x^2 + 1) - 8x^2}{(4x^2 + 1)^{3/2}}$$
$$= \frac{2}{(4x^2 + 1)^{3/2}}$$

115. $y = x \operatorname{arcsec} x$

$$y' = \frac{x}{|x|\sqrt{x^2 - 1}} + \operatorname{arcsec} x$$

117. $y = x(\arcsin x)^2 - 2x + 2\sqrt{1 - x^2} \arcsin x$

$$y' = \frac{2x \arcsin x}{\sqrt{1 - x^2}} + (\arcsin x)^2 - 2 + \frac{2\sqrt{1 - x^2}}{\sqrt{1 - x^2}} - \frac{2x}{\sqrt{1 - x^2}} \arcsin x = (\arcsin x)^2$$

119. $y = \sqrt{x}$

$$\frac{dy}{dt} = 2 \text{ units/sec}$$

$$\frac{dy}{dt} = \frac{1}{2\sqrt{x}}\frac{dx}{dt} \Rightarrow \frac{dx}{dt} = 2\sqrt{x}\frac{dy}{dt} = 4\sqrt{x}$$

(a) When $x = \frac{1}{2}$, $\frac{dx}{dt} = 2\sqrt{2}$ units/sec.

(b) When $x = 1$, $\frac{dx}{dt} = 4$ units/sec.

(c) When $x = 4$, $\frac{dx}{dt} = 8$ units/sec.

121. $\tan\theta = x$

$$\frac{d\theta}{dt} = 3(2\pi) \text{ rad/min}$$

$$\sec^2\theta\left(\frac{d\theta}{dt}\right) = \frac{dx}{dt}$$

$$\frac{dx}{dt} = (\tan^2\theta + 1)(6\pi)$$

$$= 6\pi(x^2 + 1)$$

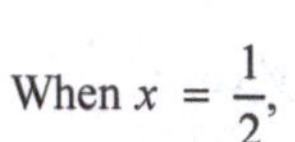

When $x = \frac{1}{2}$,

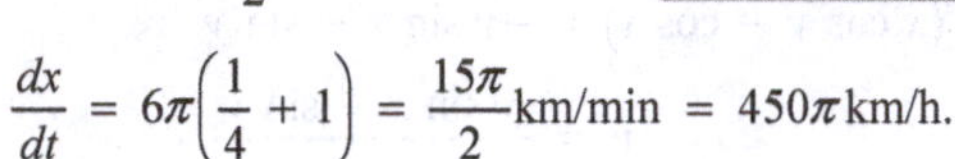

$$\frac{dx}{dt} = 6\pi\left(\frac{1}{4} + 1\right) = \frac{15\pi}{2} \text{km/min} = 450\pi \text{km/h}.$$

123. $f(x) = x^3 - 3x - 1$

From the graph you can see that $f(x)$ has three real zeros.

$f'(x) = 3x^2 - 3$

n	x_n	$f(x_n)$	$f'(x_n)$	$\frac{f(x_n)}{f'(x_n)}$	$x_n - \frac{f(x_n)}{f'(x_n)}$
1	−1.5000	0.1250	3.7500	0.0333	−1.5333
2	−1.5333	−0.0049	4.0530	−0.0012	−1.5321

n	x_n	$f(x_n)$	$f'(x_n)$	$\frac{f(x_n)}{f'(x_n)}$	$x_n - \frac{f(x_n)}{f'(x_n)}$
1	−0.5000	0.3750	−2.2500	−0.1667	−0.3333
2	−0.3333	−0.0371	−2.6667	0.0139	−0.3472
3	−0.3472	−0.0003	−2.6384	0.0001	−0.3473

n	x_n	$f(x_n)$	$f'(x_n)$	$\frac{f(x_n)}{f'(x_n)}$	$x_n - \frac{f(x_n)}{f'(x_n)}$
1	1.9000	0.1590	7.8300	0.0203	1.8797
2	1.8797	0.0024	7.5998	0.0003	1.8794

The three real zeros of $f(x)$ are $x \approx -1.532$, $x \approx -0.347$, and $x \approx 1.879$.

125. $g(x) = xe^x - 4$

$g'(x) = (x + 1)e^x$

From the graph, there is one zero near 1.

n	x_n	$g(x_n)$	$g'(x_n)$	$\dfrac{g(x_n)}{g'(x_n)}$	$x_n - \dfrac{g(x_n)}{g'(x_n)}$
1	1.0	−1.2817	5.4366	−0.2358	1.2358
2	1.2358	0.2525	7.6937	0.0328	1.2030
3	1.2030	0.0059	7.3359	0.0008	1.2022

To three decimal places, $x = 1.202$.

127. $f(x) = x^4 + x^3 - 3x^2 + 2$

From the graph you can see that $f(x)$ has two real zeros.

$f'(x) = 4x^3 + 3x^2 - 6x$

n	x_n	$f(x_n)$	$f'(x_n)$	$\dfrac{f(x_n)}{f'(x_n)}$	$x_n - \dfrac{f(x_n)}{f'(x_n)}$
1	−2.0	−2.0	−8.0	0.25	−2.25
2	−2.25	1.0508	−16.875	−0.0623	−2.1877
3	−2.1877	0.0776	−14.3973	−0.0054	−2.1823
4	−2.1823	0.0004	−14.3911	−0.00003	−2.1873

n	x_n	$f(x_n)$	$f'(x_n)$	$\dfrac{f(x_n)}{f'(x_n)}$	$x_n - \dfrac{f(x_n)}{f'(x_n)}$
1	−1.0	−1.0	5.0	−0.2	−0.8
2	−0.8	−0.0224	4.6720	−0.0048	−0.7952
3	−0.7952	−0.00001	4.6569	−0.0000	−0.7952

The two zeros of $f(x)$ are $x \approx -2.1823$ and $x \approx -0.7952$.

129. $h(x) = -x - \ln x$

$h'(x) = -1 - \dfrac{1}{x}$

From the graph you can see that there is one point of intersection near $x_1 = 0.5$.

n	x_n	$h(x_n)$	$h'(x_n)$	$\dfrac{h(x_n)}{h'(x_n)}$	$x_n - \dfrac{h(x_n)}{h'(x_n)}$
1	0.5	−0.1931	3.0000	−0.0644	0.5644
2	0.5644	−0.0076	2.7718	−0.0027	0.5671
3	0.5671	0.0001	2.7634	−0.0000	0.5671

Point of intersection of the graph of f and y occurs when $x \approx 0.567$.

Problem Solving for Chapter 3

1. (a) $x^2 + (y - r)^2 = r^2$, Circle

$x^2 = y$, Parabola

Substituting:

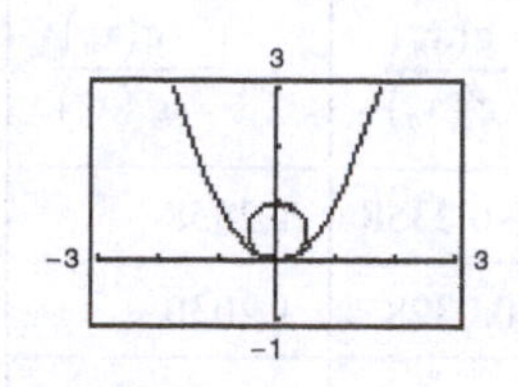

$$(y - r)^2 = r^2 - y$$
$$y^2 - 2ry + r^2 = r^2 - y$$
$$y^2 - 2ry + y = 0$$
$$y(y - 2r + 1) = 0$$

Because you want only one solution, let $1 - 2r = 0 \Rightarrow r = \frac{1}{2}$. Graph $y = x^2$ and $x^2 + \left(y - \frac{1}{2}\right)^2 = \frac{1}{4}$.

(b) Let (x, y) be a point of tangency:

$x^2 + (y - b)^2 = 1 \Rightarrow 2x + 2(y - b)y' = 0 \Rightarrow y' = \dfrac{x}{b - y}$, Circle

$y = x^2 \Rightarrow y' = 2x$, Parabola

Equating:

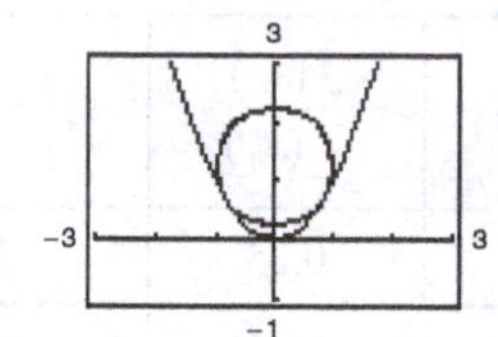

$$2x = \frac{x}{b - y}$$
$$2(b - y) = 1$$
$$b - y = \frac{1}{2} \Rightarrow b = y + \frac{1}{2}$$

Also, $x^2 + (y - b)^2 = 1$ and $y = x^2$ imply:

$$y + (y - b)^2 = 1 \Rightarrow y + \left[y - \left(y + \frac{1}{2}\right)\right]^2 = 1 \Rightarrow y + \frac{1}{4} = 1 \Rightarrow y = \frac{3}{4} \text{ and } b = \frac{5}{4}$$

Center: $\left(0, \frac{5}{4}\right)$

Graph $y = x^2$ and $x^2 + \left(y - \frac{5}{4}\right)^2 = 1$.

3. Let $p(x) = Ax^3 + Bx^2 + Cx + D$

$p'(x) = 3Ax^2 + 2Bx + C$.

At $(1, 1)$:

$A + B + C + D = 1$ Equation 1

$3A + 2B + C = 14$ Equation 2

At $(-1, -3)$:

$-A + B - C + D = -3$ Equation 3

$3A - 2B + C = -2$ Equation 4

Adding Equations 1 and 3: $2B + 2D = -2 \Rightarrow D = \frac{1}{2}(-2 - 2B) = -5$.

Subtracting Equations 1 and 3: $2A + 2C = 4$

Adding Equations 2 and 4: $6A + 2C = 12$

Subtracting Equations 2 and 4: $4B = 16$

So, $B = 4$ and $D = \frac{1}{2}(-2 - 2B) = -5$. Subtracting $2A + 2C = 4$ and $6A + 2C = 12$, you obtain $4A = 8 \Rightarrow A = 2$. Finally, $C = \frac{1}{2}(4 - 2A) = 0$. So, $p(x) = 2x^3 + 4x^2 - 5$.

5. (a) $y = x^2, y' = 2x$, Slope $= 4$ at $(2, 4)$

Tangent line: $y - 4 = 4(x - 2)$

$$y = 4x - 4$$

(b) Slope of normal line: $-\frac{1}{4}$

Normal line: $y - 4 = -\frac{1}{4}(x - 2)$

$$y = -\frac{1}{4}x + \frac{9}{2}$$

$$y = -\frac{1}{4}x + \frac{9}{2} = x^2 \Rightarrow 4x^2 + x - 18 = 0 \Rightarrow (4x + 9)(x - 2) = 0$$

$x = 2, -\frac{9}{4}$

Second intersection point: $\left(-\frac{9}{4}, \frac{81}{16}\right)$

(c) Tangent line: $y = 0$

Normal line: $x = 0$

(d) Let (a, a^2), $a \neq 0$, be a point on the parabola $y = x^2$. Tangent line at (a, a^2) is $y = 2a(x - a) + a^2$.

Normal line at (a, a^2) is $y = -(1/2a)(x - a) + a^2$. To find points of intersection, solve:

$$x^2 = -\frac{1}{2a}(x - a) + a^2$$

$$x^2 + \frac{1}{2a}x = a^2 + \frac{1}{2}$$

$$x^2 + \frac{1}{2a}x + \frac{1}{16a^2} = a^2 + \frac{1}{2} + \frac{1}{16a^2}$$

$$\left(x + \frac{1}{4a}\right)^2 = \left(a + \frac{1}{4a}\right)^2$$

$$x + \frac{1}{4a} = \pm\left(a + \frac{1}{4a}\right)$$

$$x + \frac{1}{4a} = a + \frac{1}{4a} \Rightarrow x = a \quad \text{(Point of tangency)}$$

$$x + \frac{1}{4a} = -\left(a + \frac{1}{4a}\right) \Rightarrow x = -a - \frac{1}{2a} = -\frac{2a^2 + 1}{2a}$$

The normal line intersects a second time at $x = -\frac{2a^2 + 1}{2a}$.

7. (a)

$$x^4 = a^2x^2 - a^2y^2$$

$$a^2y^2 = a^2x^2 - x^4$$

$$y = \frac{\pm\sqrt{a^2x^2 - x^4}}{a}$$

Graph: $y_1 = \dfrac{\sqrt{a^2x^2 - x^4}}{a}$

and $y_2 = -\dfrac{\sqrt{a^2x^2 - x^4}}{a}$.

(b)

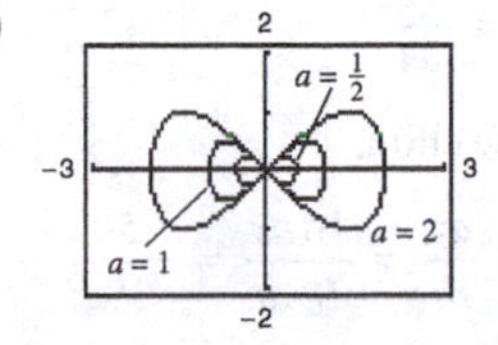

$(\pm a, 0)$ are the x-intercepts, along with $(0, 0)$.

(c) Differentiating implicitly:

$$4x^3 = 2a^2x - 2a^2yy'$$

$$y' = \frac{2a^2x - 4x^3}{2a^2y} = \frac{x(a^2 - 2x^2)}{a^2y} = 0 \Rightarrow 2x^2 = a^2 \Rightarrow x = \frac{\pm a}{\sqrt{2}}$$

$$\left(\frac{a^2}{2}\right)^2 = a^2\left(\frac{a^2}{2}\right) - a^2y^2$$

$$\frac{a^4}{4} = \frac{a^4}{2} - a^2y^2$$

$$a^2y^2 = \frac{a^4}{4}$$

$$y^2 = \frac{a^2}{4}$$

$$y = \pm\frac{a}{2}$$

Four points: $\left(\frac{a}{\sqrt{2}}, \frac{a}{2}\right), \left(\frac{a}{\sqrt{2}}, -\frac{a}{2}\right), \left(-\frac{a}{\sqrt{2}}, \frac{a}{2}\right), \left(-\frac{a}{\sqrt{2}}, -\frac{a}{2}\right)$

9. (a)

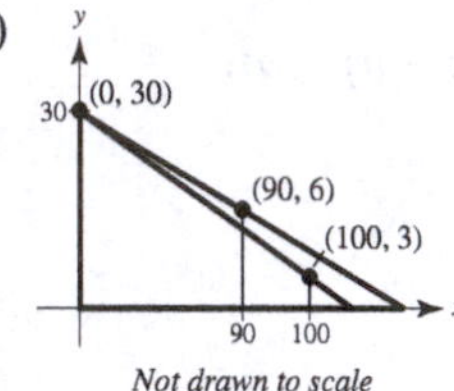

Not drawn to scale

Line determined by $(0, 30)$ and $(90, 6)$:

$$y - 30 = \frac{30 - 6}{0 - 90}(x - 0) = -\frac{24}{90}x = -\frac{4}{15}x \Rightarrow y = -\frac{4}{15}x + 30$$

When $x = 100$: $y = -\frac{4}{15}(100) + 30 = \frac{10}{3} > 3$

As you can see from the figure, the shadow determined by the man extends beyond the shadow determined by the child.

(b)

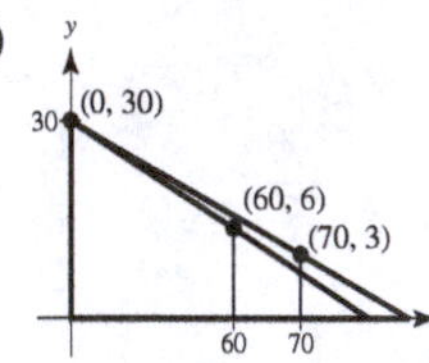

Not drawn to scale

Line determined by $(0, 30)$ and $(60, 6)$:

$$y - 30 = \frac{30 - 6}{0 - 60}(x - 0) = -\frac{2}{5}x \Rightarrow y = -\frac{2}{5}x + 30$$

When $x = 70$: $y = -\frac{2}{5}(70) + 30 = 2 < 3$

As you can see from the figure, the shadow determined by the child extends beyond the shadow determined by the man.

(c) Need $(0, 30), (d, 6), (d + 10, 3)$ collinear.

$$\frac{30 - 6}{0 - d} = \frac{6 - 3}{d - (d + 10)} \Rightarrow \frac{24}{d} = \frac{3}{10} \Rightarrow d = 80 \text{ feet}$$

(d) Let y be the distance from the base of the street light to the tip of the shadow. You know that $dx/dt = -5$.

For $x > 80$, the shadow is determined by the man.

$$\frac{y}{30} = \frac{y - x}{6} \Rightarrow y = \frac{5}{4}x \text{ and } \frac{dy}{dt} = \frac{5}{4}\frac{dx}{dt} = \frac{-25}{4}$$

For $x < 80$, the shadow is determined by the child.

$$\frac{y}{30} = \frac{y - x - 10}{3} \Rightarrow y = \frac{10}{9}x + \frac{100}{9} \text{ and } \frac{dy}{dt} = \frac{10}{9}\frac{dx}{dt} = -\frac{50}{9}$$

Therefore:

$$\frac{dy}{dt} = \begin{cases} -\frac{25}{4}, & x > 80 \\ -\frac{50}{9}, & 0 < x < 80 \end{cases}$$

dy/dt is not continuous at $x = 80$.

ALTERNATE SOLUTION for parts (a) and (b):

(a) As before, the line determined by the man's shadow is

$$y_m = -\frac{4}{15}x + 30$$

The line determined by the child's shadow is obtained by finding the line through $(0, 30)$ and $(100, 3)$:

$$y - 30 = \frac{30 - 3}{0 - 100}(x - 0) \Rightarrow y_c = -\frac{27}{100}x + 30$$

By setting $y_m = y_c = 0$, you can determine how far the shadows extend:

Man: $y_m = 0 \Rightarrow \frac{4}{15}x = 30 \Rightarrow x = 112.5 = 112\frac{1}{2}$

Child: $y_c = 0 \Rightarrow \frac{27}{100}x = 30 \Rightarrow x = 111.\overline{11} = 111\frac{1}{9}$

The man's shadow is $112\frac{1}{2} - 111\frac{1}{9} = 1\frac{7}{18}$ ft beyond the child's shadow.

(b) As before, the line determined by the man's shadow is

$$y_m = -\frac{2}{5}x + 30$$

For the child's shadow, $y - 30 = \frac{30 - 3}{0 - 70}(x - 0) \Rightarrow y_c = -\frac{27}{70}x + 30$

Man: $y_m = 0 \Rightarrow \frac{2}{5}x = 30 \Rightarrow x = 75$

Child: $y_c = 0 \Rightarrow \frac{27}{70}x = 30 \Rightarrow x = \frac{700}{9} = 77\frac{7}{9}$

So the child's shadow is $77\frac{7}{9} - 75 = 2\frac{7}{9}$ ft beyond the man's shadow.

11. (a) $v(t) = -\frac{27}{5}t + 27$ ft/sec

$a(t) = -\frac{27}{5}$ ft/sec^2

(b) $v(t) = -\frac{27}{5}t + 27 = 0 \Rightarrow \frac{27}{5}t = 27 \Rightarrow t = 5$ seconds

$s(5) = -\frac{27}{10}(5)^2 + 27(5) + 6 = 73.5$ feet

(c) The acceleration due to gravity on Earth is greater in magnitude than that on the moon.

13. $L'(x) = \lim_{\Delta x \to 0} \frac{L(x + \Delta x) - L(x)}{\Delta x} = \lim_{\Delta x \to 0} \frac{L(x) + L(\Delta x) - L(x)}{\Delta x} = \lim_{\Delta x \to 0} \frac{L(\Delta x)}{\Delta x}$

Also, $L'(0) = \lim_{\Delta x \to 0} \frac{L(\Delta x) - L(0)}{\Delta x}$. But, $L(0) = 0$ because $L(0) = L(0 + 0) = L(0) + L(0) \Rightarrow L(0) = 0$.

So, $L'(x) = L'(0)$ for all x. The graph of L is a line through the origin of slope $L'(0)$.

15. $j(t) = a'(t)$

(a) $j(t)$ is the rate of change of acceleration.

(b) $s(t) = -8.25t^2 + 66t$

$v(t) = -16.5t + 66$

$a(t) = -16.5$

$a'(t) = j(t) = 0$

The acceleration is constant, so $j(t) = 0$.

(c) a is position.

b is acceleration.

c is jerk.

d is velocity.

CHAPTER 4
Applications of Differentiation

CHAPTER 4
Applications of Differentiation

Section 4.1 Extrema on an Interval

1. The Extreme Value Theorem states that if f is continuous on a closed interval $[a, b]$, then f has both a minimum and a maximum on the interval.

3. 

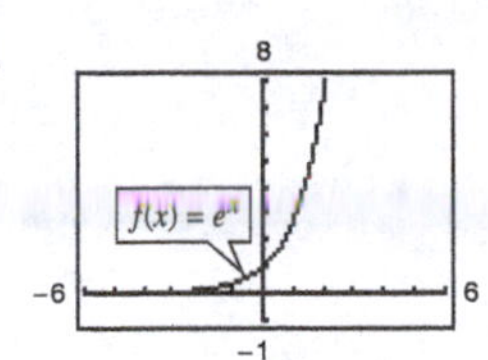

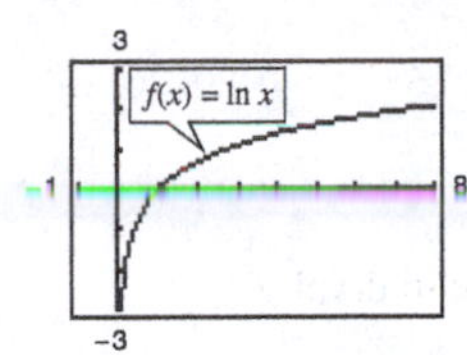

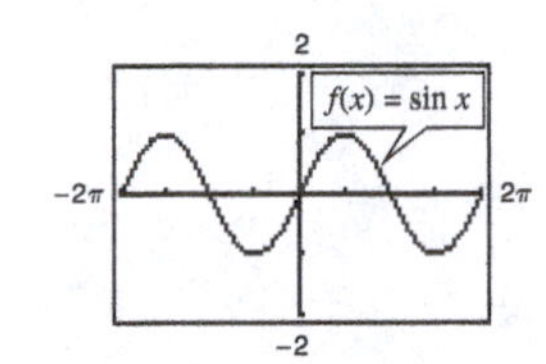

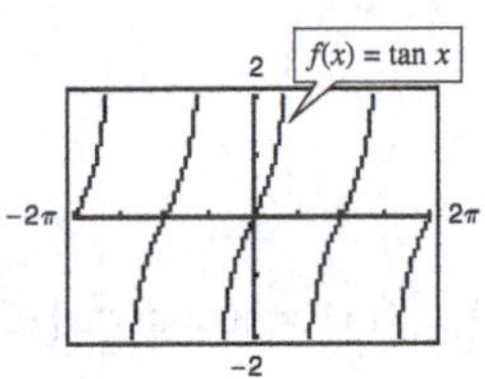

$f(x) = \sin x$; From the graph, you can see that f is defined and $f'(x) = 0$ for $f(x) = \sin x$ when $x = \pi/2 + n\pi$ (n is an integer).

5. $f(x) = \dfrac{x^2}{x^2 + 4}$

$f'(x) = \dfrac{(x^2 + 4)(2x) - (x^2)(2x)}{(x^2 + 4)^2} = \dfrac{8x}{(x^2 + 4)^2}$

$f'(0) = 0$

7. $f(x) = x + \dfrac{4}{x^2} = x + 4x^{-2}$

$f'(x) = 1 - 8x^{-3} = 1 - \dfrac{8}{x^3}$

$f'(2) = 0$

9. $f(x) = (x + 2)^{2/3}$

$f'(x) = \frac{2}{3}(x + 2)^{-1/3}$

$f'(-2)$ is undefined.

11. Critical number: $x = 2$

$x = 2$: absolute maximum (and relative maximum)

13. Critical numbers: $x = 1, 2, 3$

$x = 1, 3$: absolute maxima (and relative maxima)

$x = 2$: absolute minimum (and relative minimum)

15. $f(x) = 4x^2 - 6x$

$f'(x) = 8x - 6 = 2(4x - 3)$

Critical number: $\frac{3}{4}$

17. $g(t) = t\sqrt{4 - t}, \ t < 3$

$g'(t) = t\left[\frac{1}{2}(4 - t)^{-1/2}(-1)\right] + (4 - t)^{1/2}$

$= \frac{1}{2}(4 - t)^{-1/2}\left[-t + 2(4 - t)\right] = \dfrac{8 - 3t}{2\sqrt{4 - t}}$

Critical number: $t = \dfrac{8}{3}$

19. $h(x) = \sin^2 x + \cos x, \quad 0 < x < 2\pi$

$h'(x) = 2 \sin x \cos x - \sin x = \sin x(2 \cos x - 1)$

Critical numbers in $(0, 2\pi)$: $x = \dfrac{\pi}{3}, \pi, \dfrac{5\pi}{3}$

21. $f(t) = te^{-2t}$

$f'(t) = e^{-2t} - 2te^{-2t} = e^{-2t}(1 - 2t)$

Critical number: $t = \frac{1}{2}$

23. $f(x) = x^2 \log_2(x^2 + 1) = x^2 \dfrac{\ln(x^2 + 1)}{\ln 2}$

$f'(x) = 2x\dfrac{\ln(x^2 + 1)}{\ln 2} + x^2\dfrac{2x}{(\ln 2)(x^2 + 1)}$

$= \dfrac{2x}{\ln 2}\left[\ln(x^2 + 1) + \dfrac{x^2}{x^2 + 1}\right]$

Critical number: $x = 0$

25. $f(x) = 3 - x, \quad [-1, 2]$

$f'(x) = -1 \Rightarrow$ no critical numbers

Left endpoint: $(-1, 4)$ Maximum

Right endpoint: $(2, 1)$ Minimum

27. $h(x) = 5 - 2x^2, [-3, 1]$

$h'(x) = -4x$

Critical number: $x = 0$

Left endpoint: $(-3, -13)$ Minimum

Critical number: $(0, 5)$ Maximum

Right endpoint: $(1, 3)$

29. $f(x) = x^3 - \frac{3}{2}x^2, [-1, 2]$

$f'(x) = 3x^2 - 3x = 3x(x - 1)$

Left endpoint: $\left(-1, -\frac{5}{2}\right)$ Minimum

Right endpoint: $(2, 2)$ Maximum

Critical number: $(0, 0)$

Critical number: $\left(1, -\frac{1}{2}\right)$

31. $f(x) = 3x^{2/3} - 2x, \quad [-1, 1]$

$f'(x) = 2x^{-1/3} - 2 = \dfrac{2\left(1 - \sqrt[3]{x}\right)}{\sqrt[3]{x}}$

Left endpoint: $(-1, 5)$ Maximum

Critical number: $(0, 0)$ Minimum

Right endpoint: $(1, 1)$

33. $g(x) = \dfrac{6x^2}{x - 2}, [-2, 1]$

$$g'(x) = \frac{(x - 2)(12x) - 6x^2(1)}{(x - 2)^2} = \frac{6x^2 - 24x}{(x - 2)^2} = \frac{6x(x - 4)}{(x - 2)^2}$$

Critical number: $x = 0$ ($x = 2$ and $x = 4$ are outside the interval.)

Left endpoint: $(-2, -6)$ Mimimum

Right endpoint: $(1, -6)$ Minimum

Critical number: $(0, 0)$ Maximum

35. $y = 3 - |t - 3|, \quad [-1, 5]$

For $t < 3, y = 3 + (t - 3) = t$

and $y' = 1 \neq 0$ on $[-1, 3)$

For $t > 3, y = 3 - (t - 3) = 6 - t$

and $y' = -1 \neq 0$ on $(3, 5]$

So, $t = 3$ is the only critical number.

Left endpoint: $(-1, -1)$ Minimum

Right endpoint: $(5, 1)$

Critical number: $(3, 3)$ Maximum

37. $f(x) = [\![x]\!], \quad [-2, 2]$

From the graph of f, you see that the maximum value of f is 2 for $x = 2$, and the minimum value is –2 for $-2 \le x < -1$.

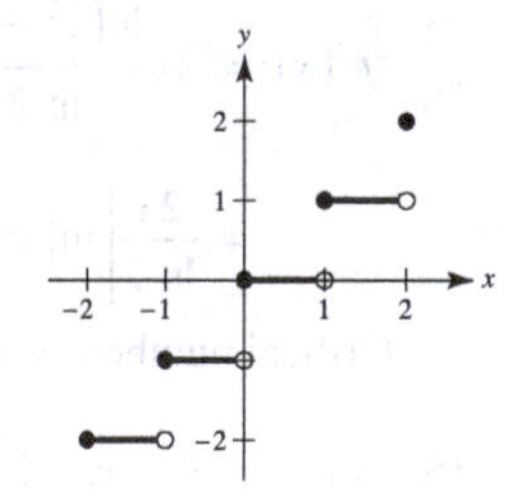

39. $y = 3 \cos x, \quad [0, 2\pi]$

$y' = -3 \sin x$

Critical number in $(0, 2\pi)$: $x = \pi$

Left endpoint: $(0, 3)$ Maximum

Critical number: $(\pi, -3)$ Minimum

Right endpoint: $(2\pi, 3)$ Maximum

41. $f(x) = \arctan x^2, [-2, 1]$

$f'(x) = \dfrac{2x}{1 + x^4}$

Critical number: $x = 0$

Left endpoint: $(-2, \arctan 4) \approx (-2, 1.326)$ Maximum

Right endpoint: $(1, \arctan 1) = \left(1, \dfrac{\pi}{4}\right) \approx (1, 0.785)$

Critical number: $(0, 0)$ Minimum

43. $h(x) = 5e^x - e^{2x}, [-1, 2]$

$h'(x) = 5e^x - 2e^{2x} = e^x\left(5 - 2e^x\right)$

$5 - 2e^x = 0 \Rightarrow e^x = \dfrac{5}{2} \Rightarrow x = \ln\left(\dfrac{5}{2}\right) \approx 0.916$

Critical number: $x = \ln\left(\dfrac{5}{2}\right)$

Left endpoint: $\left(-1, \dfrac{5}{e} - \dfrac{1}{e^2}\right) \approx (-1, 1.704)$

Right endpoint: $\left(2, 5e^2 - e^4\right) \approx (2, -17.653)$ Minimum

Critical number: $\left(\ln\left(\dfrac{5}{2}\right), \dfrac{25}{4}\right)$ Maximum

Note: $h\left(\ln\left(\dfrac{5}{2}\right)\right) = 5e^{\ln(5/2)} - e^{2\ln(5/2)}$

$= 5\left(\dfrac{5}{2}\right) - \left(\dfrac{5}{2}\right)^2 = \dfrac{25}{4}$

45. $y = e^x \sin x, [0, \pi]$

$y' = e^x \sin x + e^x \cos x = e^x(\sin x + \cos x)$

Left endpoint: $(0, 0)$ Minimum

Critical number:

$\left(\frac{3\pi}{4}, \frac{\sqrt{2}}{2}e^{3\pi/4}\right) \approx \left(\frac{3\pi}{4}, 7.46\right)$ Maximum

Right endpoint: $(\pi, 0)$ Minimum

47. $f(x) = 2x - 3$

(a) Minimum: $(0, -3)$

Maximum: $(2, 1)$

(b) Minimum: $(0, -3)$

(c) Maximum: $(2, 1)$

(d) No extrema

49. $f(x) = x^2 - 2x$

(a) Minimum: $(1, -1)$

Maximum: $(-1, 3)$

(b) Maximum: $(3, 3)$

(c) Minimum: $(1, -1)$

(d) Minimum: $(1, -1)$

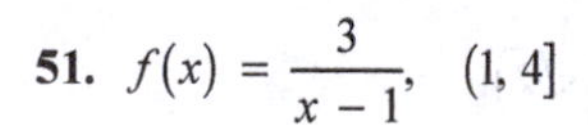

51. $f(x) = \frac{3}{x-1}, \quad (1, 4]$

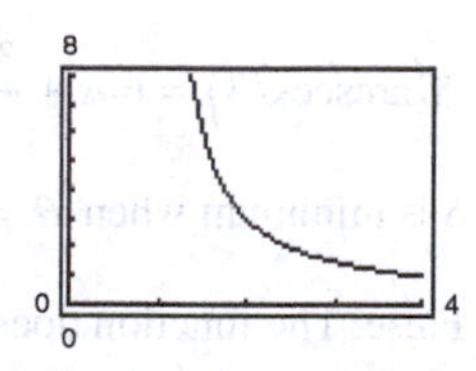

Right endpoint:

$(4, 1)$ Minimum

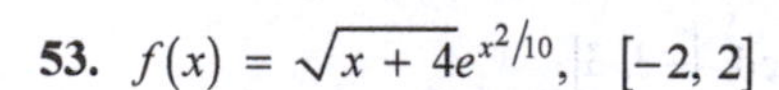

53. $f(x) = \sqrt{x + 4}e^{x^2/10}, \ [-2, 2]$

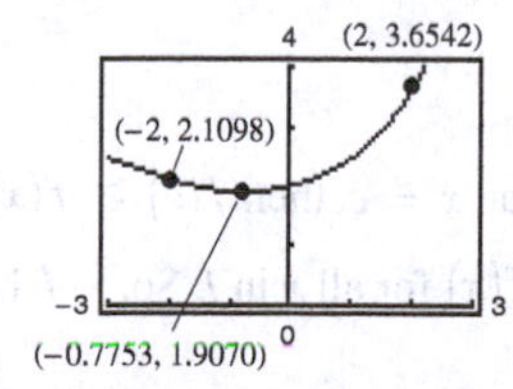

$f'(x) = \frac{(2x^2 + 8x + 5)e^{x^2/10}}{10\sqrt{x + 4}}$

Right endpoint: $(2, 3.6542)$ Maximum

Critical point: $(-0.7753, 1.9070)$ Minimum

55. (a)

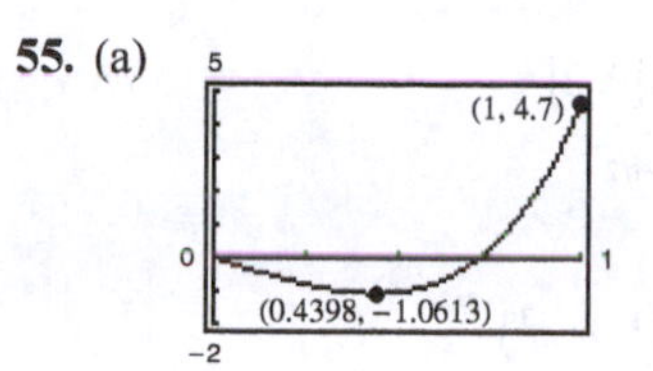

Minimum: $(0.4398, -1.0613)$

(b) $f(x) = 3.2x^5 + 5x^3 - 3.5x, \quad [0, 1]$

$f'(x) = 16x^4 + 15x^2 - 3.5$

$16x^4 + 15x^2 - 3.5 = 0$

$$x^2 = \frac{-15 \pm \sqrt{(15)^2 - 4(16)(-3.5)}}{2(16)}$$

$$= \frac{-15 \pm \sqrt{449}}{32}$$

$$x = \sqrt{\frac{-15 + \sqrt{449}}{32}} \approx 0.4398$$

Left endpoint: $(0, 0)$

Critical point: $(0.4398, -1.0613)$ Minimum

Right endpoint: $(1, 4.7)$ Maximum

57. (a)

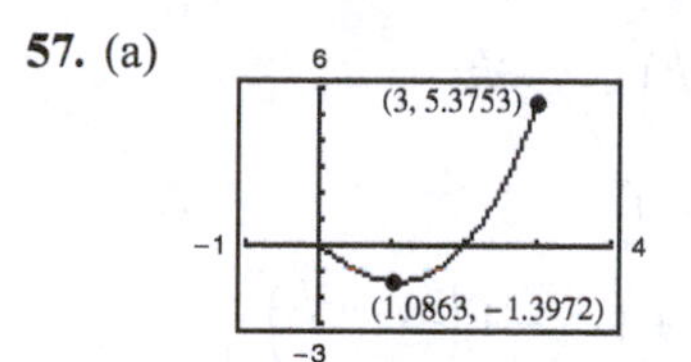

Minimum: $(1.0863, -1.3972)$

(b) $f(x) = (x^2 - 2x)\ln(x + 3), \ [0, 3]$

$$f'(x) = (x^2 - 2x) \cdot \frac{1}{x + 3} + (2x - 2)\ln(x + 3)$$

$$= \frac{x^2 - 2x + (2x^2 + 4x - 6)\ln(x + 3)}{x + 3}$$

Left endpoint: $(0, 0)$

Critical point: $(1.0863, -1.3972)$ Minimum

Right endpoint: $(3, 5.3753)$ Maximum

59. $f(x) = (1 + x^3)^{1/2}, \quad [0, 2]$

$f'(x) = \frac{3}{2}x^2(1 + x^3)^{-1/2}$

$f''(x) = \frac{3}{4}(x^4 + 4x)(1 + x^3)^{-3/2}$

$f'''(x) = -\frac{3}{8}(x^6 + 20x^3 - 8)(1 + x^3)^{-5/2}$

Setting $f''' = 0$, you have $x^6 + 20x^3 - 8 = 0$.

$x^3 = \dfrac{-20 \pm \sqrt{400 - 4(1)(-8)}}{2}$

$x = \sqrt[3]{-10 \pm \sqrt{108}} = \sqrt{3} - 1$

In the interval $[0, 2]$, choose

$x = \sqrt[3]{-10 \pm \sqrt{108}} = \sqrt{3} - 1 \approx 0.732.$

$\left|f''\left(\sqrt[3]{-10 + \sqrt{108}}\right)\right| \approx 1.47$ is the maximum value.

61. $f(x) = e^{-x^2/2}, [0, 1]$

$f'(x) = -xe^{-x^2/2}$

$f''(x) = -x\left(-xe^{-x^2/2}\right) - e^{-x^2/2}$

$= e^{-x^2/2}(x^2 - 1)$

$f'''(x) = e^{-x^2/2}(2x) + (x^2 - 1)\left(-xe^{-x^2/2}\right)$

$= xe^{-x^2/2}(3 - x^2)$

$|f''(0)| = 1$ is the maximum value.

63. $f(x) = (x + 1)^{2/3}, \quad [0, 2]$

$f'(x) = \frac{2}{3}(x + 1)^{-1/3}$

$f''(x) = -\frac{2}{9}(x + 1)^{-4/3}$

$f'''(x) = \frac{8}{27}(x + 1)^{-7/3}$

$f^{(4)}(x) = -\frac{56}{81}(x + 1)^{-10/3}$

$f^{(5)}(x) = \frac{560}{243}(x + 1)^{-13/3}$

$\left|f^{(4)}(0)\right| = \frac{56}{81}$ is the maximum value.

65. $f(x) = \tan x$

f is continuous on $[0, \pi/4]$ but not on $[0, \pi]$.

$\lim_{x \to (\pi/2)^-} \tan x = \infty.$

67. (a) Yes

(b) No

69. No. The function is not defined at $x = -2$.

71. $P = VI - RI^2 = 12I - 0.5I^2, 0 \le I \le 15$

$P = 0$ when $I = 0$.

$P = 67.5$ when $I = 15$.

$P' = 12 - I = 0$

Critical number: $I = 12$ amps

When $I = 12$ amps, $P = 72$, the maximum output.

No, a 20-amp fuse would not increase the power output. P is decreasing for $I > 12$.

73. $S = 6hs + \dfrac{3s^2}{2}\left(\dfrac{\sqrt{3} - \cos\theta}{\sin\theta}\right), \dfrac{\pi}{6} \le \theta \le \dfrac{\pi}{2}$

$\dfrac{dS}{d\theta} = \dfrac{3s^2}{2}\left(-\sqrt{3}\csc\theta\cot\theta + \csc^2\theta\right)$

$= \dfrac{3s^2}{2}\csc\theta\left(-\sqrt{3}\cot\theta + \csc\theta\right) = 0$

$\csc\theta = \sqrt{3}\cot\theta$

$\sec\theta = \sqrt{3}$

$\theta = \operatorname{arcsec}\sqrt{3} \approx 0.9553$ radians

$S\left(\dfrac{\pi}{6}\right) = 6hs + \dfrac{3s^2}{2}\left(\sqrt{3}\right)$

$S\left(\dfrac{\pi}{6}\right) = 6hs + \dfrac{3s^2}{2}\left(\sqrt{3}\right)$

$S\left(\operatorname{arcsec}\sqrt{3}\right) = 6hs + \dfrac{3s^2}{2}\left(\sqrt{2}\right)$

S is minimum when $\theta = \operatorname{arcsec}\sqrt{3} \approx 0.9553$ radian.

75. False. The function does not have a maximum on the open interval $(-3, 3)$. The maximum would be 9 if the interval were closed, $[-3, 3]$.

77. True

79. If f has a maximum value at $x = c$, then $f(c) \ge f(x)$ for all x in I. So, $-f(c) \le -f(x)$ for all x in I. So, $-f$ has a minimum value at $x = c$.

81. First do an example: Let $a = 4$ and $f(x) = 4$. Then R is the square $0 \le x \le 4, 0 \le y \le 4$.

Its area and perimeter are both $k = 16$. Claim that all real numbers $a > 2$ work. On the one hand, if $a > 2$ is given, then let $f(x) = 2a/(a - 2)$.

Then the rectangle $R = \left\{(x, y): 0 \le x \le a, 0 \le y \le \dfrac{2a}{a-2}\right\}$ has $k = \dfrac{2a^2}{a-2}$:

$$\text{Area} = a\left(\frac{2a}{a-2}\right) = \frac{2a^2}{a-2}$$

$$\text{Perimeter} = 2a + 2\left(\frac{2a}{a-2}\right) = \frac{2a(a-2) + 2(2a)}{a-2} = \frac{2a^2}{a-2}.$$

To see that a must be greater than 2, consider $R = \{(x, y): 0 \le x \le a, 0 \le y \le f(x)\}$.

f attains its maximum value on $[0, a]$ at some point $P(x_0, y_0)$, as indicated in the figure.

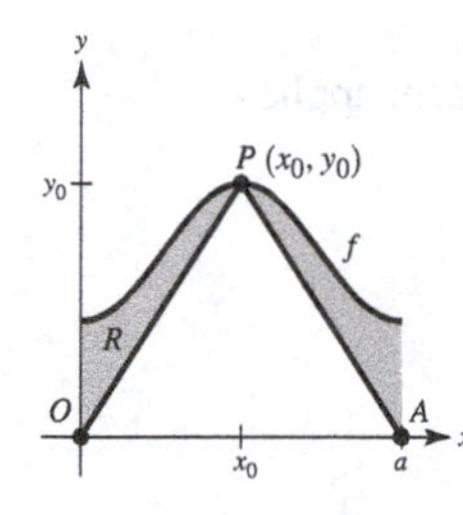

Draw segments $\overline{OP}$ and $\overline{PA}$. The region R is bounded by the rectangle $0 \le x \le a, 0 \le y \le y_0$, so area$(R) = k \le ay_0$. Furthermore, from the figure, $y_0 < \overline{OP}$ and $y_0 < \overline{PA}$. So, $k = \text{Perimeter}(R) > \overline{OP} + \overline{PA} > 2y_0$. Combining, $2y_0 < k \le ay_0 \Rightarrow a > 2$.

Section 4.2 Rolle's Theorem and the Mean Value Theorem

1. Let f be continuous on $[a, b]$ and differentiable on (a, b). Rolle's Theorem says that if $f(a) = f(b)$, then there is at least one number c in (a, b) such that $f'(c) = 0$.

3. $f(x) = \left|\dfrac{1}{x}\right|$

$f(-1) = f(1) = 1$. But, f is not continuous on $[-1, 1]$.

5. Rolle's Theorem does not apply to $f(x) = 1 - |x - 1|$ over $[0, 2]$ because f is not differentiable at $x = 1$.

7. $f(x) = x^2 - x - 2 = (x - 2)(x + 1)$

x-intercepts: $(-1, 0), (2, 0)$

$f'(x) = 2x - 1 = 0$ at $x = \frac{1}{2}$.

9. $f(x) = x\sqrt{x + 4}$

x-intercepts: $(-4, 0), (0, 0)$

$$f'(x) = x\frac{1}{2}(x + 4)^{-1/2} + (x + 4)^{1/2}$$
$$= (x + 4)^{-1/2}\left(\frac{x}{2} + (x + 4)\right)$$
$$f'(x) = \left(\frac{3}{2}x + 4\right)(x + 4)^{-1/2} = 0 \text{ at } x = -\frac{8}{3}$$

11. $f(x) = -x^2 + 3x, \quad [0, 3]$

$f(0) = -(0)^2 + 3(0)$

$f(3) = -(3)^2 + 3(3) = 0$

f is continuous on $[0, 3]$ and differentiable on $(0, 3)$. Rolle's Theorem applies.

$f'(x) = -2x + 3 = 0$

$-2x = -3 \Rightarrow x = \frac{3}{2}$

c-value: $\frac{3}{2}$

13. $f(x) = (x - 1)(x - 2)(x - 3), [1, 3]$

$f(1) = (1 - 1)(1 - 2)(1 - 3) = 0$

$f(3) = (3 - 1)(3 - 2)(3 - 3) = 0$

f is continuous on $[1, 3]$. f is differentiable on $(1, 3)$. Rolle's Theorem applies.

$f(x) = x^3 - 6x^2 + 11x - 6$

$f'(x) = 3x^2 - 12x + 11 = 0$

$$x = \frac{6 \pm \sqrt{3}}{3}$$

c-values: $\dfrac{6 - \sqrt{3}}{3}, \dfrac{6 + \sqrt{3}}{3}$

15. $f(x) = x^{2/3} - 1, [-8, 8]$

$f(-8) = (-8)^{2/3} - 1 = 3$

$f(8) = (8)^{2/3} - 1 = 3$

f is continuous on $[-8, 8]$. f is not differentiable on $(-8, 8)$ because $f'(0)$ does not exist. Rolle's Theorem does not apply.

17. $f(x) = \dfrac{x^2 - 2x - 3}{x + 2}, [-1, 3]$

$f(-1) = \dfrac{(-1)^2 - 2(-1) - 3}{-1 + 2} = 0$

$f(3) = \dfrac{3^2 - 2(3) - 3}{3 + 2} = 0$

f is continuous on $[-1, 3]$.

(**Note:** The discontinuity $x = -2$, is not in the interval.) f is differentiable on $(-1, 3)$. Rolle's Theorem applies.

$$f'(x) = \frac{(x + 2)(2x - 2) - (x^2 - 2x - 3)(1)}{(x + 2)^2} = 0$$

$$\frac{x^2 + 4x - 1}{(x + 2)^2} = 0$$

$$x = \frac{-4 \pm 2\sqrt{5}}{2} = -2 \pm \sqrt{5}$$

(**Note:** $x = -2 - \sqrt{5}$ is not in the interval.)

c-value: $-2 + \sqrt{5}$

19. $f(x) = \sin x, [0, 2\pi]$

$f(0) = \sin 0 = 0$

$f(2\pi) = \sin(2\pi) = 0$

f is continuous on $[0, 2\pi]$. f is differentiable on $(0, 2\pi)$. Rolle's Theorem applies.

$f'(x) = \cos x = 0 \Rightarrow x = \dfrac{\pi}{2}, \dfrac{3\pi}{2}$

c-values: $\dfrac{\pi}{2}, \dfrac{3\pi}{2}$

21. $f(x) = \cos \pi x, [0, 2]$

$f(0) = 1 = f(2)$

f is continuous on $[0, 2]$ and differentiable on $(0, 2)$. Rolle's Theorem applies.

$f'(x) = -\pi \sin \pi x = 0 \Rightarrow x = 1$

c-value: 1

23. $f(x) = \tan x, [0, \pi]$

$f(0) = \tan 0 = 0$

$f(\pi) = \tan \pi = 0$

f is not continuous on $[0, \pi]$ because $f(\pi/2)$ does not exist. Rolle's Theorem does not apply.

25. $f(x) = (x^2 - 2x)e^x, [0, 2]$

$f(0) = f(2) = 0$

f is continuous on $[0, 2]$ and differentiable on $(0, 2)$, so Rolle's Theorem applies.

$f'(x) = (x^2 - 2x)e^x + (2x - 2)e^x = e^x(x^2 - 2)$

$= 0 \Rightarrow x = \sqrt{2}$

c-value: $\sqrt{2} \approx 1.414$

27. $f(x) = |x| - 1, [-1, 1]$

$f(-1) = f(1) = 0$

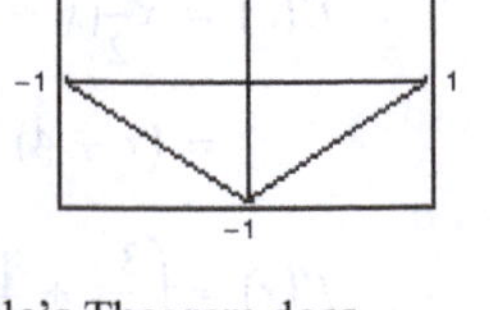

f is continuous on $[-1, 1]$. f is not differentiable on $(-1, 1)$ because $f'(0)$ does not exist. Rolle's Theorem does not apply.

29. $f(x) = x - x^{-1/3}, [0, 1]$

$f(0) = f(1) = 0$

f is continuous on $[-1, 0]$. f is differentiable on $(-1, 0)$. Rolle's Theorem applies.

$$f'(x) = \frac{1}{2} - \frac{\pi}{6}\cos\frac{\pi x}{6} = 0$$

$$\cos\frac{\pi x}{6} = \frac{3}{\pi}$$

$$x = -\frac{6}{\pi}\arccos\frac{3}{\pi}\left[\text{Value needed in } (-1, 0).\right]$$

$$\approx -0.5756 \text{ radian}$$

c-value: -0.5756

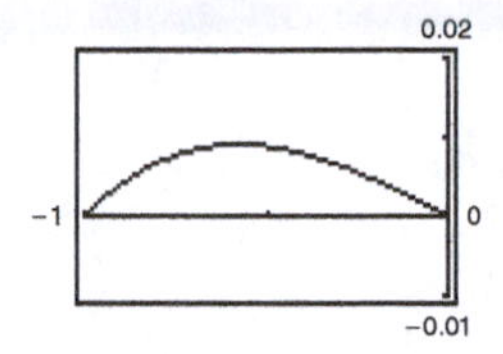

31. $f(x) = 2 + (x^2 - 4x)(2^{-x/4}), [0, 4]$

$f(0) = f(4) = 2$

f is continuous on $[0, 4]$. f is differentiable on $(0, 4)$. Rolle's Theorem applies.

$$f'(x) = (2x - 4)2^{-x/4} + (x^2 - 4x)\ln 2 \cdot 2^{-x/4}\left(-\frac{1}{4}\right)$$

$$= 2^{-x/4}\left[2x - 4 - (\ln 2)\left(\frac{x^2}{4} - x\right)\right]$$

$$= 0 \Rightarrow x \approx 1.6633$$

c-value: 1.6633

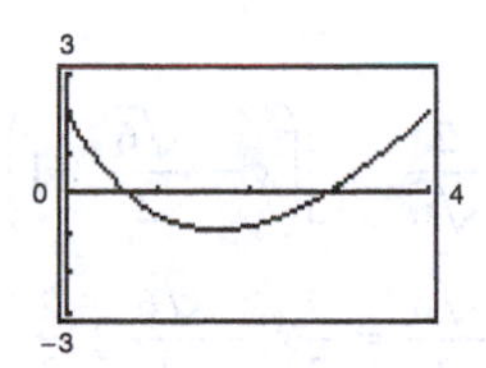

33. $f(t) = -16t^2 + 48t + 6$

(a) $f(1) = f(2) = 38$

(b) $v = f'(t)$ must be 0 at some time in $(1, 2)$.

$$f'(t) = -32t + 48 = 0$$

$$t = \tfrac{3}{2}\text{ sec}$$

35.

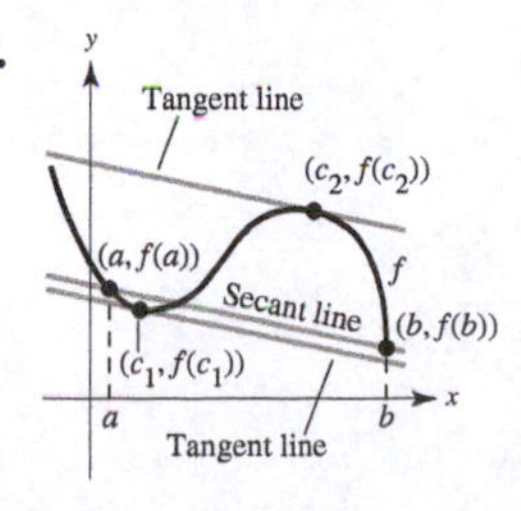

37. f is not continuous on the interval $[0, 6]$. (f is not continuous at $x = 2$.)

39. $f(x) = \dfrac{1}{x - 3}, [0, 6]$

f has a discontinuity at $x = 3$.

41. $f(x) = -x^2 + 5$

(a) Slope $= \dfrac{1 - 4}{2 + 1} = -1$

Secant line: $y - 4 = -(x + 1)$

$y = -x + 3$

$x + y - 3 = 0$

(b) $f'(x) = -2x = -1 \Rightarrow x = c = \dfrac{1}{2}$

(c) $f(c) = f\left(\dfrac{1}{2}\right) = -\dfrac{1}{4} + 5 = \dfrac{19}{4}$

Tangent line: $y - \dfrac{19}{4} = -\left(x - \dfrac{1}{2}\right)$

$4y - 19 = -4x + 2$

$4x + 4y - 21 = 0$

(d)

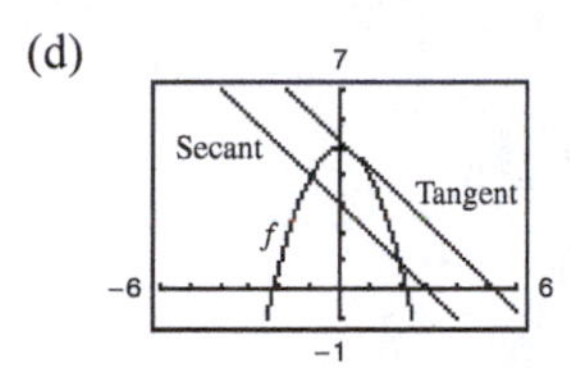

43. $f(x) = 6x^3$ is continuous on $[1, 2]$ and differentiable on $(1, 2)$.

$$\frac{f(2) - f(1)}{2 - 1} = \frac{48 - 6}{1} = 42$$

$$f'(x) = 18x^2 = 42$$

$$x^2 = \frac{7}{3} \Rightarrow x = \pm\sqrt{\frac{7}{3}}$$

On the interval $(1, 2)$, $c = \sqrt{\dfrac{7}{3}} = \dfrac{\sqrt{21}}{3}$.

$$f'\left(\frac{\sqrt{21}}{3}\right) = 42$$

45. $f(x) = x^3 + 2x$ is continuous on $[-1, 1]$ and differentiable on $(-1, 1)$.

$$\frac{f(1) - f(-1)}{1 - (-1)} = \frac{3 - (-3)}{2} = 3$$

$$f'(x) = 3x^2 + 2 = 3$$

$$3x^2 = 1$$

$$x = \pm\frac{1}{\sqrt{3}}$$

$$c = \pm\frac{\sqrt{3}}{3}$$

47. $f(x) = \dfrac{x + 2}{x - 1}$ is not continuous at $x = 1$.

The Mean Value Theorem does not apply on the interval $[-3, 3]$.

49. $f(x) = |2x + 1|$ is not differentiable at $x = -1/2$. The Mean Value Theorem does not apply on the interval $[-1, 3]$.

51. $f(x) = \sin x$ is continuous on $[0, \pi]$ and differentiable on $(0, \pi)$.

$$\frac{f(\pi) - f(0)}{\pi - 0} = \frac{0 - 0}{\pi} = 0$$

$$f'(x) = \cos x = 0$$

$$x = \frac{\pi}{2}$$

On the interval $(0, \pi)$, $c = \dfrac{\pi}{2}$ and $f'\left(\dfrac{\pi}{2}\right) = 0$.

53. $f(x) = \cos x + \tan x$ is not continuous at $x = \pi/2$. The Mean Value Theorem does not apply on the interval $[0, \pi]$.

55. $f(x) = x \log_2 x = x\dfrac{\ln x}{\ln 2}$

f is continuous on $[1, 2]$ and differentiable on $(1, 2)$.

$$\frac{f(2) - f(1)}{2 - 1} = \frac{2 - 0}{2 - 1} = 2$$

$$f'(x) = x\frac{1}{x \ln 2} + \frac{\ln x}{\ln 2} = \frac{1 + \ln x}{\ln 2} = 2$$

$$1 + \ln x = 2 \ln 2 = \ln 4$$

$$xe = 4$$

$$x = \frac{4}{e}$$

$$c = \frac{4}{e}$$

57. $f(x) = \dfrac{x}{x + 1}, \left[-\dfrac{1}{2}, 2\right]$

(a)–(c)

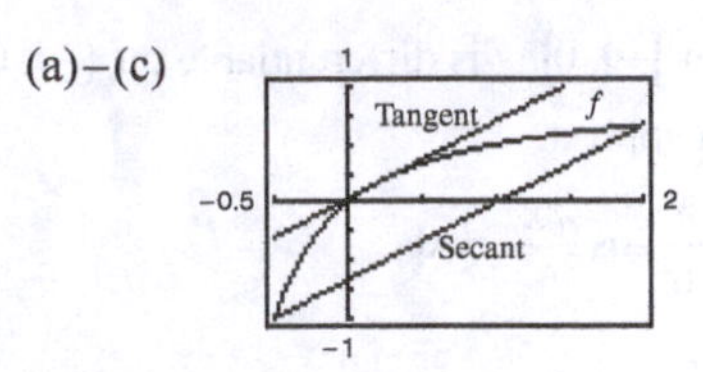

(b) Secant line:

$$\text{slope} = \frac{f(2) - f(-1/2)}{2 - (-1/2)} = \frac{2/3 - (-1)}{5/2} = \frac{2}{3}$$

$$y - \frac{2}{3} = \frac{2}{3}(x - 2)$$

$$y = \frac{2}{3}(x - 1)$$

(c) $f'(x) = \dfrac{1}{(x + 1)^2} = \dfrac{2}{3}$

$$(x + 1)^2 = \frac{3}{2}$$

$$x = -1 \pm \sqrt{\frac{3}{2}} = -1 \pm \frac{\sqrt{6}}{2}$$

In the interval $[-1/2, 2]$: $c = -1 + \left(\sqrt{6}/2\right)$

$$f(c) = \frac{-1 + \left(\sqrt{6}/2\right)}{\left[-1 + \left(\sqrt{6}/2\right)\right] + 1}$$

$$= \frac{-2 + \sqrt{6}}{\sqrt{6}}$$

$$= \frac{-2}{\sqrt{6}} + 1$$

Tangent line: $y - 1 + \dfrac{2}{\sqrt{6}} = \dfrac{2}{3}\left(x - \dfrac{\sqrt{6}}{2} + 1\right)$

$$y - 1 + \frac{\sqrt{6}}{3} = \frac{2}{3}x - \frac{\sqrt{6}}{3} + \frac{2}{3}$$

$$y = \frac{1}{3}\left(2x + 5 - 2\sqrt{6}\right)$$

59. $f(x) = \sqrt{x}, [1, 9]$

(a)–(c)

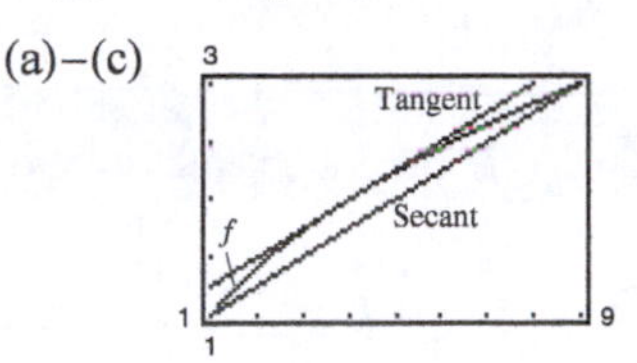

(b) Secant line:

$$\text{slope} = \frac{f(9) - f(1)}{9 - 1} = \frac{3 - 1}{8} = \frac{1}{4}$$

$$y - 1 = \frac{1}{4}(x - 1)$$

$$y = \frac{1}{4}x + \frac{3}{4}$$

(c) $f'(x) = \dfrac{1}{2\sqrt{x}} = \dfrac{1}{4}$

$$x = c = 4$$

$$f(4) = 2$$

Tangent line: $y - 2 = \dfrac{1}{4}(x - 4)$

$$y = \frac{1}{4}x + 1$$

61. $f(x) = 2e^{x/4} \cos \dfrac{\pi x}{4}, \; 0 \le x \le 2$

(a)–(c)

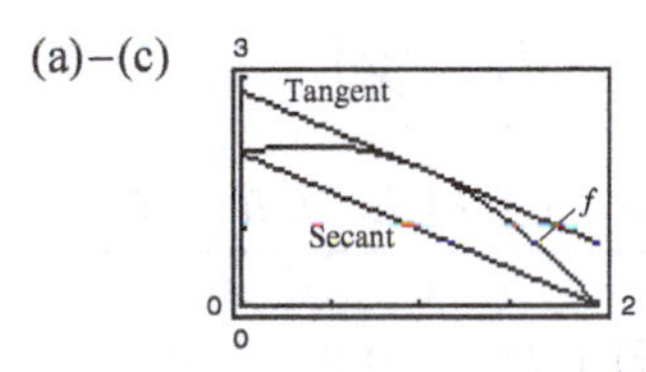

(b) Secant line:

$$\text{slope} = \frac{f(2) - f(0)}{2 - 0} = \frac{0 - 2}{2 - 0} = -1$$

$$y - 2 = -1(x - 0)$$

$$y = -x + 2$$

(c) $f'(x) = 2\left(\dfrac{1}{4}e^{x/4} \cos \dfrac{\pi x}{4}\right) + 2e^{x/4}\left(-\sin \dfrac{\pi x}{4}\right)\dfrac{\pi}{4}$

$$= e^{x/4}\left[\frac{1}{2} \cos \frac{\pi x}{4} - \frac{\pi}{2} \sin \frac{\pi x}{4}\right]$$

$$f'(c) = -1 \Rightarrow c \approx 1.0161, f(c) \approx 1.8$$

Tangent line: $y - 1.8 = -1(x - 1.0161)$

$$y = -x + 2.8161$$

63. $s(t) = -4.9t^2 + 300$

(a) $v_{\text{avg}} = \dfrac{s(3) - s(0)}{3 - 0} = \dfrac{255.9 - 300}{3} = -14.7 \text{ m/sec}$

(b) $s(t)$ is continuous on $[0, 3]$ and differentiable on $(0, 3)$. Therefore, the Mean Value Theorem applies.

$$v(t) = s'(t) = -9.8t = -14.7 \text{ m/sec}$$

$$t = \frac{-14.7}{-9.8} = 1.5 \text{ sec}$$

65. No. Let $f(x) = x^2$ on $[-1, 2]$.

$$f'(x) = 2x$$

$f'(0) = 0$ and zero is in the interval $(-1, 2)$ but $f(-1) \ne f(2)$.

67. $f(x) = \begin{cases} 0, & x = 0 \\ 1 - x, & 0 < x \le 1 \end{cases}$

No, this does not contradict Rolle's Theorem. f is not continuous on $[0, 1]$.

69. Let $S(t)$ be the position function of the plane. If $t = 0$ corresponds to 2 P.M., $S(0) = 0, S(5.5) = 2500$ and the Mean Value Theorem says that there exists a time t_0, $0 < t_0 < 5.5$, such that

$$S'(t_0) = v(t_0) = \frac{2500 - 0}{5.5 - 0} \approx 454.54.$$

Applying the Intermediate Value Theorem to the velocity function on the intervals $[0, t_0]$ and $[t_0, 5.5]$, you see that there are at least two times during the flight when the speed was 400 miles per hour. $(0 < 400 < 454.54)$

71. Let $S(t)$ be the difference in the positions of the 2 bicyclists, $S(t) = S_1(t) - S_2(t)$. Because $S(0) = S(2.25) = 0$, there must exist a time $t_0 \in (0, 2.25)$ such that $S'(t_0) = v(t_0) = 0$. At this time, $v_1(t_0) = v_2(t_0)$.

73. (a) f is continuous on $[-5, 5]$ and does not satisfy the conditions of the Mean Value Theorem. $\Rightarrow$ f is not differentiable on $(-5, 5)$. Example: $f(x) = |x|$

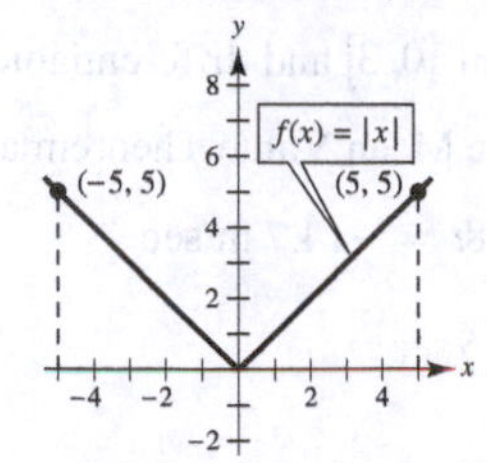

(b) f is not continuous on $[-5, 5]$.

Example:

$$f(x) = \begin{cases} 1/x, & x \neq 0 \\ 0, & x = 0 \end{cases}$$

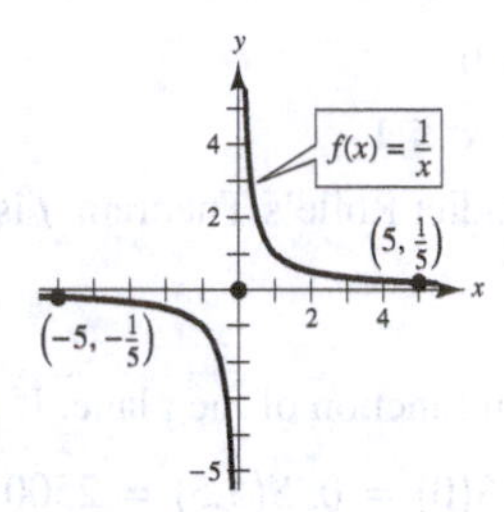

75. $f(x) = x^5 + x^3 + x + 1$

f is differentiable for all x.

$f(-1) = -2$ and $f(0) = 1$, so the Intermediate Value Theorem implies that f has at least one zero c in $[-1, 0]$, $f(c) = 0$.

Suppose f had 2 zeros, $f(c_1) = f(c_2) = 0$. Then Rolle's Theorem would guarantee the existence of a number a such that

$f'(a) = f(c_2) - f(c_1) = 0$.

But, $f'(x) = 5x^4 + 3x^2 + 1 > 0$ for all x. So, f has exactly one real solution.

77. $f(x) = 3x + 1 - \sin x$

f is differentiable for all x.

$f(-\pi) = -3\pi + 1 < 0$ and $f(0) = 1 > 0$, so the Intermediate Value Theorem implies that f has at least one zero c in $[-\pi, 0]$, $f(c) = 0$.

Suppose f had 2 zeros, $f(c_1) = f(c_2) = 0$. Then Rolle's Theorem would guarantee the existence of a number a such that

$f'(a) = f(c_2) - f(c_1) = 0$.

But $f'(x) = 3 - \cos x > 0$ for all x. So, $f(x) = 0$ has exactly one real solution.

79. $f'(x) = 0$

$f(x) = c$

$f(2) = 5$

So, $f(x) = 5$.

81. $f'(x) = 2x$

$f(x) = x^2 + c$

$f(1) = 0 \Rightarrow 0 = 1 + c \Rightarrow c = -1$

So, $f(x) = x^2 - 1$.

83. False. $f(x) = 1/x$ has a discontinuity at $x = 0$.

85. True. A polynomial is continuous and differentiable everywhere.

87. Suppose that $p(x) = x^{2n+1} + ax + b$ has two real roots x_1 and x_2. Then by Rolle's Theorem, because $p(x_1) = p(x_2) = 0$, there exists c in (x_1, x_2) such that $p'(c) = 0$. But $p'(x) = (2n + 1)x^{2n} + a \neq 0$, because $n > 0$, $a > 0$. Therefore, $p(x)$ cannot have two real roots.

89. If $p(x) = Ax^2 + Bx + C$, then

$$\begin{aligned} p'(x) = 2Ax + B &= \frac{f(b) - f(a)}{b - a} \\ &= \frac{(Ab^2 + Bb + C) - (Aa^2 + Ba + C)}{b - a} \\ &= \frac{A(b^2 - a^2) + B(b - a)}{b - a} \\ &= \frac{(b - a)[A(b + a) + B]}{b - a} \\ &= A(b + a) + B. \end{aligned}$$

So, $2Ax = A(b + a)$ and $x = (b + a)/2$ which is the midpoint of $[a, b]$.

91. Suppose $f(x)$ has two fixed points c_1 and c_2. Then, by the Mean Value Theorem, there exists c such that

$$f'(c) = \frac{f(c_2) - f(c_1)}{c_2 - c_1} = \frac{c_2 - c_1}{c_2 - c_1} = 1.$$

This contradicts the fact that $f'(x) < 1$ for all x.

93. Let $f(x) = \cos x$. f is continuous and differentiable for all real numbers. By the Mean Value Theorem, for any interval $[a, b]$, there exists c in (a, b) such that

$$\frac{f(b) - f(a)}{b - a} = f'(c)$$
$$\frac{\cos b - \cos a}{b - a} = -\sin c$$
$$\cos b - \cos a = (-\sin c)(b - a)$$
$$|\cos b - \cos a| = |-\sin c||b - a|$$
$$|\cos b - \cos a| \le |b - a| \text{ since } |-\sin c| \le 1.$$

95. Let $0 < a < b$. $f(x) = \sqrt{x}$ satisfies the hypotheses of the Mean Value Theorem on $[a, b]$. Hence, there exists c in (a, b) such that

$$f'(c) = \frac{1}{2\sqrt{c}} = \frac{f(b) - f(a)}{b - a} = \frac{\sqrt{b} - \sqrt{a}}{b - a}.$$

So, $\sqrt{b} - \sqrt{a} = (b - a)\dfrac{1}{2\sqrt{c}} < \dfrac{b - a}{2\sqrt{a}}$.

Section 4.3 Increasing and Decreasing Functions and the First Derivative Test

1. A positive derivative of a function on an open interval implies that the function is increasing on the interval. A negative derivative implies that the function is decreasing. A zero derivative implies that the function is constant.

3. (a) Increasing: $(0, 6)$ and $(8, 9)$. Largest: $(0, 6)$

(b) Decreasing: $(6, 8)$ and $(9, 10)$. Largest: $(6, 8)$

5. $y = -(x + 1)^2$

From the graph, f is increasing on $(-\infty, -1)$ and decreasing on $(-1, \infty)$.

Analytically, $y' = -2(x + 1)$.

Critical number: $x = -1$

Test intervals:	$-\infty < x < -1$	$-1 < x < \infty$
Sign of y':	$y' > 0$	$y' < 0$
Conclusion:	Increasing	Decreasing

7. $y = \dfrac{x^3}{4} - 3x$

From the graph, y is increasing on $(-\infty, -2)$ and $(2, \infty)$, and decreasing on $(-2, 2)$.

Analytically, $y' = \dfrac{3x^2}{4} - 3 = \dfrac{3}{4}(x^2 - 4) = \dfrac{3}{4}(x - 2)(x + 2)$

Critical numbers: $x = \pm 2$

Test intervals:	$-\infty < x < -2$	$-2 < x < 2$	$2 < x < \infty$
Sign of y':	$y' > 0$	$y' < 0$	$y' > 0$
Conclusion:	Increasing	Decreasing	Increasing

9. $f(x) = \dfrac{1}{(x + 1)^2}$

From the graph, f is increasing on $(-\infty, -1)$ and decreasing on $(-1, \infty)$.

Analytically, $f'(x) = \dfrac{-2}{(x + 1)^3}$.

No critical numbers. Discontinuity: $x = -1$

Test intervals:	$-\infty < x < -1$	$-1 < x < \infty$
Sign of $f'(x)$:	$f' > 0$	$f' < 0$
Conclusion:	Increasing	Decreasing

11. $g(x) = x^2 - 2x - 8$

$g'(x) = 2x - 2$

Critical number: $x = 1$

Increasing on: $(1, \infty)$

Decreasing on: $(-\infty, 1)$

Test intervals:	$-\infty < x < 1$	$1 < x < \infty$
Sign of $g'(x)$:	$g' < 0$	$g' > 0$
Conclusion:	Decreasing	Increasing

13. $y = x\sqrt{16 - x^2}$ Domain: $[-4, 4]$

$$y' = \frac{-2(x^2 - 8)}{\sqrt{16 - x^2}}$$

$$= \frac{-2}{\sqrt{16 - x^2}}(x - 2\sqrt{2})(x + 2\sqrt{2})$$

Critical numbers: $x = \pm 2\sqrt{2}$

Increasing on: $(-2\sqrt{2}, 2\sqrt{2})$

Decreasing on: $(-4, -2\sqrt{2}), (2\sqrt{2}, 4)$

Test intervals:	$-4 < x < -2\sqrt{2}$	$-2\sqrt{2} < x < 2\sqrt{2}$	$2\sqrt{2} < x < 4$
Sign of y':	$y' < 0$	$y' > 0$	$y' < 0$
Conclusion:	Decreasing	Increasing	Decreasing

15. $f(x) = \sin x - 1, \quad 0 < x < 2\pi$

$f'(x) = \cos x$

Critical numbers: $x = \frac{\pi}{2}, \frac{3\pi}{2}$

Increasing on: $\left(0, \frac{\pi}{2}\right), \left(\frac{3\pi}{2}, 2\pi\right)$

Decreasing on: $\left(\frac{\pi}{2}, \frac{3\pi}{2}\right)$

Test intervals:	$0 < x < \frac{\pi}{2}$	$\frac{\pi}{2} < x < \frac{3\pi}{2}$	$\frac{3\pi}{2} < x < 2\pi$
Sign of $f'(x)$:	$f' > 0$	$f' < 0$	$f' > 0$
Conclusion:	Increasing	Decreasing	Increasing

17. $y = x - 2\cos x, \quad 0 < x < 2\pi$

$y' = 1 + 2\sin x$

$y' = 0$: $\sin x = -\frac{1}{2}$

Critical numbers: $x = \frac{7\pi}{6}, \frac{11\pi}{6}$

Increasing on: $\left(0, \frac{7\pi}{6}\right), \left(\frac{11\pi}{6}, 2\pi\right)$

Decreasing on: $\left(\frac{7\pi}{6}, \frac{11\pi}{6}\right)$

Test intervals:	$0 < x < \frac{7\pi}{6}$	$\frac{7\pi}{6} < x < \frac{11\pi}{6}$	$\frac{11\pi}{6} < x < 2\pi$
Sign of y':	$y' > 0$	$y' < 0$	$y' > 0$
Conclusion:	Increasing	Decreasing	Increasing

19. $g(x) = e^{-x} + e^{3x}$

$g'(x) = -e^{-x} + 3e^{3x}$

Critical number: $x = -\frac{1}{4}\ln 3$

Increasing on: $\left(-\frac{1}{4}\ln 3, \infty\right)$

Decreasing on: $\left(-\infty, -\frac{1}{4}\ln 3\right)$

Test intervals:	$-\infty < x < -\frac{1}{4}\ln 3$	$-\frac{1}{4}\ln 3 < x < \infty$
Sign of $g'(x)$:	$g' < 0$	$g' > 0$
Conclusion:	Decreasing	Increasing

21. $f(x) = x^2 \ln\left(\dfrac{x}{2}\right), \quad x > 0$

$f'(x) = 2x \ln\left(\dfrac{x}{2}\right) + \dfrac{x^2}{x} = 2x \ln\left(\dfrac{x}{2}\right) + x$

Critical number: $x = \dfrac{2}{\sqrt{e}}$

Increasing on: $\left(\dfrac{2}{\sqrt{e}}, \infty\right)$

Decreasing on: $\left(0, \dfrac{2}{\sqrt{e}}\right)$

Test intervals:	$0 < x < \dfrac{2}{\sqrt{e}}$	$\dfrac{2}{\sqrt{e}} < x < \infty$
Sign of $f'(x)$:	$f' < 0$	$f' > 0$
Conclusion:	Decreasing	Increasing

23. (a) $f(x) = x^2 - 8x$

$f'(x) = 2x - 8$

Critical number: $x = 4$

(b)

Test intervals:	$-\infty < x < 4$	$4 < x < \infty$
Sign of $f'(x)$:	$f' < 0$	$f' > 0$
Conclusion:	Decreasing	Increasing

Decreasing on: $(-\infty, 4)$

Increasing on: $(4, \infty)$

(c) Relative minimum: $(4, -16)$

25. (a) $f(x) = -2x^2 + 4x + 3$

$f'(x) = -4x + 4 = 0$

Critical number: $x = 1$

(b)

Test intervals:	$-\infty < x < 1$	$1 < x < \infty$
Sign of $f'(x)$:	$f' > 0$	$f' < 0$
Conclusion:	Increasing	Decreasing

Increasing on: $(-\infty, 1)$

Decreasing on: $(1, \infty)$

(c) Relative maximum: $(1, 5)$

27. (a) $f(x) = -7x^3 + 21x + 3$

$f'(x) = -21x^2 + 21 = 0$

$x^2 = 1$

Critical numbers: $x = \pm 1$

(b)

Test intervals:	$-\infty < x < -1$	$-1 < x < 1$	$1 < x < \infty$
Sign of $f'(x)$:	$f' < 0$	$f' > 0$	$f' < 0$
Conclusion:	Decreasing	Increasing	Decreasing

Increasing on: $(-1, 1)$

Decreasing on: $(-\infty, -1), (1, \infty)$

(c) Relative minimum: $(-1, -11)$

Relative maximum: $(1, 17)$

29. (a) $f(x) = (x-1)^2(x+3) = x^3 + x^2 - 5x + 3$

$f'(x) = 3x^2 + 2x - 5 = (x-1)(3x+5)$

Critical numbers: $x = 1, -\frac{5}{3}$

(b)

Test intervals:	$-\infty < x < -\frac{5}{3}$	$-\frac{5}{3} < x < 1$	$1 < x < \infty$
Sign of f':	$f' > 0$	$f' < 0$	$f' > 0$
Conclusion:	Increasing	Decreasing	Increasing

Increasing on: $\left(-\infty, -\frac{5}{3}\right)$ and $(1, \infty)$

Decreasing on: $\left(-\frac{5}{3}, 1\right)$

(c) Relative maximum: $\left(-\frac{5}{3}, \frac{256}{27}\right)$

Relative minimum: $(1, 0)$

31. (a) $f(x) = \dfrac{x^5 - 5x}{5}$

$f'(x) = x^4 - 1$

Critical numbers: $x = -1, 1$

(b)

Test intervals:	$-\infty < x < -1$	$-1 < x < 1$	$1 < x < \infty$
Sign of $f'(x)$:	$f' > 0$	$f' < 0$	$f' > 0$
Conclusion:	Increasing	Decreasing	Increasing

Increasing on: $(-\infty, -1), (1, \infty)$

Decreasing on: $(-1, 1)$

(c) Relative maximum: $\left(-1, \dfrac{4}{5}\right)$

Relative minimum: $\left(1, -\dfrac{4}{5}\right)$

33. (a) $f(x) = x^{1/3} + 1$

$f'(x) = \dfrac{1}{3}x^{-2/3} = \dfrac{1}{3x^{2/3}}$

Critical number: $x = 0$

(b)

Test intervals:	$-\infty < x < 0$	$0 < x < \infty$
Sign of $f'(x)$:	$f' > 0$	$f' > 0$
Conclusion:	Increasing	Increasing

Increasing on: $(-\infty, \infty)$

(c) No relative extrema

35. (a) $f(x) = (x+2)^{2/3}$

$f'(x) = \dfrac{2}{3}(x+2)^{-1/3} = \dfrac{2}{3(x+2)^{1/3}}$

Critical number: $x = -2$

(b)

Test intervals:	$-\infty < x < -2$	$-2 < x < \infty$
Sign of f':	$f' < 0$	$f' > 0$
Conclusion:	Decreasing	Increasing

Decreasing on: $(-\infty, -2)$

Increasing on: $(-2, \infty)$

(c) Relative minimum: $(-2, 0)$

37. (a) $f(x) = 5 - |x - 5|$

$f'(x) = -\frac{x-5}{|x-5|} = \begin{cases} 1, & x < 5 \\ -1, & x > 5 \end{cases}$

Critical number: $x = 5$

(b)

Test intervals:	$-\infty < x < 5$	$5 < x < \infty$
Sign of $f'(x)$:	$f' > 0$	$f' < 0$
Conclusion:	Increasing	Decreasing

Increasing on: $(-\infty, 5)$

Decreasing on: $(5, \infty)$

(c) Relative maximum: $(5, 5)$

39. (a) $f(x) = 2x + \frac{1}{x}$

$f'(x) = 2 - \frac{1}{x^2} = \frac{2x^2 - 1}{x^2}$

Critical numbers: $x = \pm\frac{\sqrt{2}}{2}$

Discontinuity: $x = 0$

(b)

Test intervals:	$-\infty < x < -\frac{\sqrt{2}}{2}$	$-\frac{\sqrt{2}}{2} < x < 0$	$0 < x < \frac{\sqrt{2}}{2}$	$\frac{\sqrt{2}}{2} < x < \infty$
Sign of f':	$f' > 0$	$f' < 0$	$f' < 0$	$f' > 0$
Conclusion:	Increasing	Decreasing	Decreasing	Increasing

Increasing on: $\left(-\infty, -\frac{\sqrt{2}}{2}\right)$ and $\left(\frac{\sqrt{2}}{2}, \infty\right)$

Decreasing on: $\left(-\frac{\sqrt{2}}{2}, 0\right)$ and $\left(0, \frac{\sqrt{2}}{2}\right)$

(c) Relative maximum: $\left(-\frac{\sqrt{2}}{2}, -2\sqrt{2}\right)$

Relative minimum: $\left(\frac{\sqrt{2}}{2}, 2\sqrt{2}\right)$

41. (a) $f(x) = \frac{x^2}{x^2 - 9}$

$f'(x) = \frac{(x^2 - 9)(2x) - (x^2)(2x)}{(x^2 - 9)^2} = \frac{-18x}{(x^2 - 9)^2}$

Critical number: $x = 0$

Discontinuities: $x = -3, 3$

(b)

Test intervals:	$-\infty < x < -3$	$-3 < x < 0$	$0 < x < 3$	$3 < x < \infty$
Sign of $f'(x)$:	$f' > 0$	$f' > 0$	$f' < 0$	$f' < 0$
Conclusion:	Increasing	Increasing	Decreasing	Decreasing

Increasing on: $(-\infty, -3), (-3, 0)$

Decreasing on: $(0, 3), (3, \infty)$

(c) Relative maximum: $(0, 0)$

43. (a) $f(x) = \begin{cases} 4 - x^2, & x \le 0 \\ -2x, & x > 0 \end{cases}$

$f'(x) = \begin{cases} -2x, & x < 0 \\ -2, & x > 0 \end{cases}$

Critical number: $x = 0$

(b)

Test intervals:	$-\infty < x < 0$	$0 < x < \infty$
Sign of f':	$f' > 0$	$f' < 0$
Conclusion:	Increasing	Decreasing

Increasing on: $(-\infty, 0)$

Decreasing on: $(0, \infty)$

(c) Relative maximum: $(0, 4)$

45. $f(x) = (3 - x)e^{x-3}$

$f'(x) = (3 - x)e^{x-3} - e^{x-3}$

$= e^{x-3}(2 - x)$

Critical number: $x = 2$

Test intervals:	$-\infty < x < 2$	$2 < x < \infty$
Sign of $f'(x)$:	$f' > 0$	$f' < 0$
Conclusion:	Increasing	Decreasing

Increasing on: $(-\infty, 2)$

Decreasing on: $(2, \infty)$

Relative maximum: $\left(2, e^{-1}\right)$

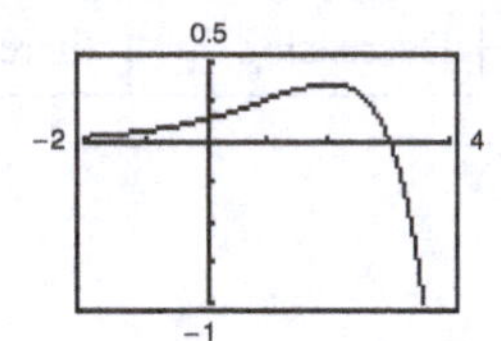

47. $f(x) = 4(x - \arcsin x), -1 \le x \le 1$

$f'(x) = 4 - \dfrac{4}{\sqrt{1 - x^2}}$

Critical number: $x = 0$

Test intervals:	$-1 \le x < 0$	$0 < x \le 1$
Sign of $f'(x)$:	$f' < 0$	$f' < 0$
Conclusion:	Decreasing	Decreasing

Decreasing on: $[-1, 1]$

No relative extrema

(Absolute maximum at $x = -1$, absolute minimum at $x = 1$)

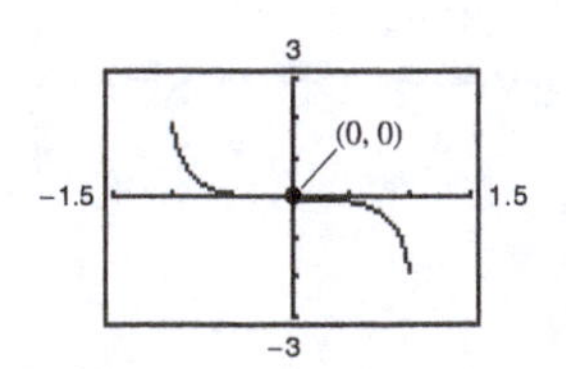

49. $g(x) = (x)3^{-x}$

$g'(x) = (1 - x \ln 3)3^{-x}$

Critical number: $x = \dfrac{1}{\ln 3} \approx 0.9102$

Test intervals:	$-\infty < x < \dfrac{1}{\ln 3}$	$\dfrac{1}{\ln 3} < x < \infty$
Sign of $f'(x)$:	$f' > 0$	$f' < 0$
Conclusion:	Increasing	Decreasing

Increasing on: $\left(-\infty, \dfrac{1}{\ln 3}\right)$

Decreasing on: $\left(\dfrac{1}{\ln 3}, \infty\right)$

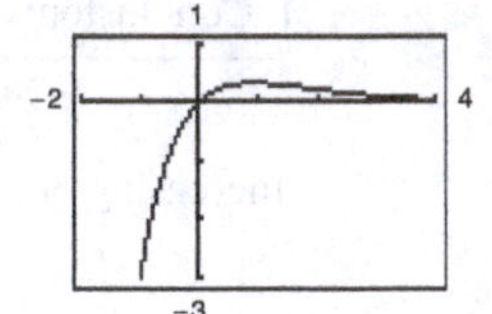

Relative maximum: $\left(\dfrac{1}{\ln 3}, \dfrac{1}{e \ln 3}\right) \approx (0.9102, 0.3349)$

51. $f(x) = x - \log_4 x = x - \dfrac{\ln x}{\ln 4}$

$f'(x) = 1 - \dfrac{1}{x \ln 4} = 0 \Rightarrow x \ln 4 = 1 \Rightarrow x = \dfrac{1}{\ln 4}$

Critical number: $x = \dfrac{1}{\ln 4}$

Test intervals:	$0 < x < \dfrac{1}{\ln 4}$	$\dfrac{1}{\ln 4} < x < \infty$
Sign of $f'(x)$:	$f' < 0$	$f' > 0$
Conclusion:	Decreasing	Increasing

Increasing on: $\left(\dfrac{1}{\ln 4}, \infty\right)$

Decreasing on: $\left(0, \dfrac{1}{\ln 4}\right)$

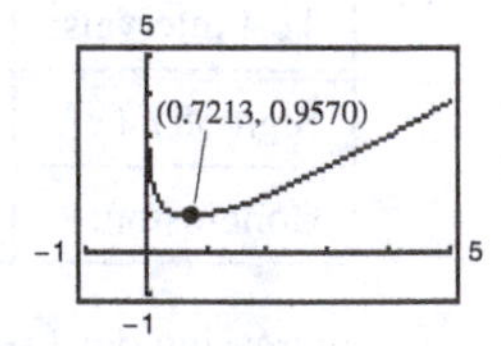

Relative minimum:

$$\left(\frac{1}{\ln 4}, \frac{1}{\ln 4} - \log_4\left(\frac{1}{\ln 4}\right)\right) = \left(\frac{1}{\ln 4}, \frac{\ln(\ln 4) + 1}{\ln 4}\right)$$

$$\approx (0.7213, 0.9570)$$

53. $g(x) = \dfrac{e^{2x}}{e^{2x} + 1}$

$$g'(x) = \frac{(e^{2x} + 1)2e^{2x} - e^{2x}(2e^{2x})}{(e^{2x} + 1)^2}$$

$$= \frac{2e^{2x}}{(e^{2x} + 1)^2}$$

No critical numbers.

Increasing on: $(-\infty, \infty)$

No relative extrema.

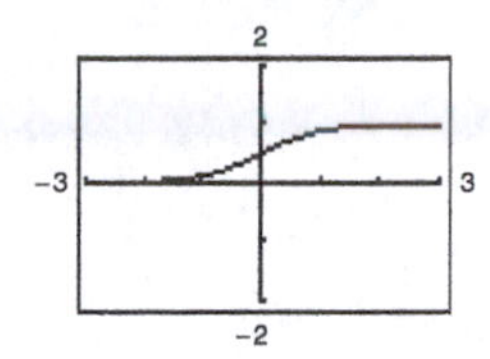

55. $f(x) = e^{-1/(x-2)} = e^{1/(2-x)}, x \neq 2$

$$f'(x) = e^{1/(2-x)}\left(\frac{1}{(2 - x)^2}\right)$$

No critical numbers.

$x = 2$ is a vertical asymptote.

Test intervals:	$-\infty < x < 2$	$2 < x < \infty$
Sign of $f'(x)$:	$f' > 0$	$f' > 0$
Conclusion:	Increasing	Increasing

Increasing on: $(-\infty, 2), (2, \infty)$

No relative extrema.

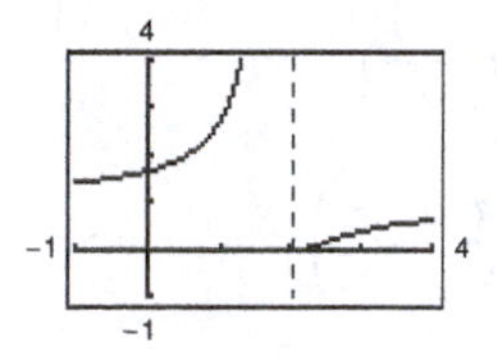

57. (a) $f(x) = x - 2\sin x, (0, 2\pi)$

$f'(x) = 1 - 2\cos x$

Critical numbers: $x = \dfrac{\pi}{3}, \dfrac{5\pi}{3}$

Test intervals:	$0 < x < \dfrac{\pi}{3}$	$\dfrac{\pi}{3} < x < \dfrac{5\pi}{3}$	$\dfrac{5\pi}{3} < x < 2\pi$
Sign of $f'(x)$:	$f' < 0$	$f' > 0$	$f' < 0$
Conclusion:	Decreasing	Increasing	Decreasing

Increasing on: $\left(\dfrac{\pi}{3}, \dfrac{5\pi}{3}\right)$

Decreasing on: $\left(0, \dfrac{\pi}{3}\right), \left(\dfrac{5\pi}{3}, 2\pi\right)$

(b) Relative minimum: $\left(\dfrac{\pi}{3}, \dfrac{\pi}{3} - \sqrt{3}\right)$

Relative maximum: $\left(\dfrac{5\pi}{3}, \dfrac{5\pi}{3} + \sqrt{3}\right)$

(c)

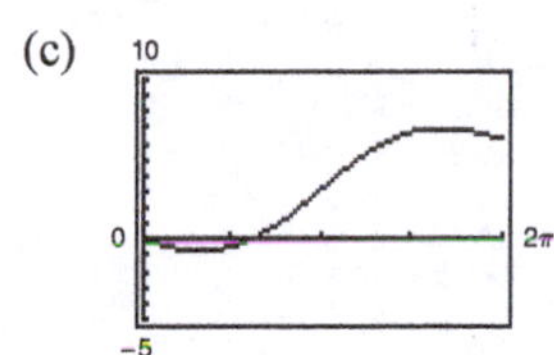

59. (a) $f(x) = \sin x + \cos x, \quad 0 < x < 2\pi$

$f'(x) = \cos x - \sin x = 0 \Rightarrow \sin x = \cos x$

Critical numbers: $x = \frac{\pi}{4}, \frac{5\pi}{4}$

Test intervals:	$0 < x < \frac{\pi}{4}$	$\frac{\pi}{4} < x < \frac{5\pi}{4}$	$\frac{5\pi}{4} < x < 2\pi$
Sign of $f'(x)$:	$f' > 0$	$f' < 0$	$f' > 0$
Conclusion:	Increasing	Decreasing	Increasing

Increasing on: $\left(0, \frac{\pi}{4}\right), \left(\frac{5\pi}{4}, 2\pi\right)$

Decreasing on: $\left(\frac{\pi}{4}, \frac{5\pi}{4}\right)$

(b) Relative maximum: $\left(\frac{\pi}{4}, \sqrt{2}\right)$

Relative minimum: $\left(\frac{5\pi}{4}, -\sqrt{2}\right)$

(c)

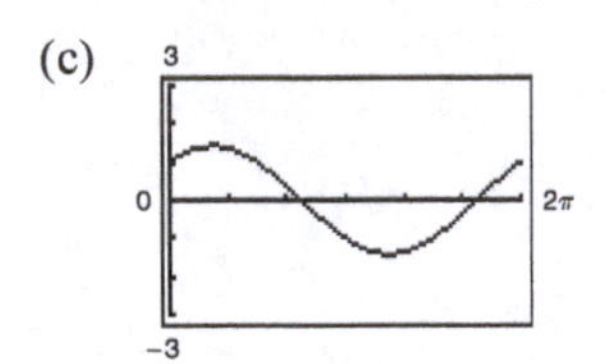

61. (a) $f(x) = \cos^2(2x), \quad 0 < x < 2\pi$

$f'(x) = -4\cos 2x \sin 2x = 0 \Rightarrow \cos 2x = 0 \text{ or } \sin 2x = 0$

Critical numbers: $x = \frac{\pi}{4}, \frac{3\pi}{4}, \frac{5\pi}{4}, \frac{7\pi}{4}, \frac{\pi}{2}, \pi, \frac{3\pi}{2}$

Test intervals:	$0 < x < \frac{\pi}{4}$	$\frac{\pi}{4} < x < \frac{\pi}{2}$	$\frac{\pi}{2} < x < \frac{3\pi}{4}$	$\frac{3\pi}{4} < x < \pi$
Sign of $f'(x)$:	$f' < 0$	$f' > 0$	$f' < 0$	$f' > 0$
Conclusion:	Decreasing	Increasing	Decreasing	Increasing

Test intervals:	$\pi < x < \frac{5\pi}{4}$	$\frac{5\pi}{4} < x < \frac{3\pi}{2}$	$\frac{3\pi}{2} < x < \frac{7\pi}{4}$	$\frac{7\pi}{4} < x < 2\pi$
Sign of $f'(x)$:	$f' < 0$	$f' > 0$	$f' < 0$	$f' > 0$
Conclusion:	Decreasing	Increasing	Decreasing	Increasing

Increasing on: $\left(\frac{\pi}{4}, \frac{\pi}{2}\right), \left(\frac{3\pi}{4}, \pi\right), \left(\frac{5\pi}{4}, \frac{3\pi}{2}\right), \left(\frac{7\pi}{4}, 2\pi\right)$

Decreasing on: $\left(0, \frac{\pi}{4}\right), \left(\frac{\pi}{2}, \frac{3\pi}{4}\right), \left(\pi, \frac{5\pi}{4}\right), \left(\frac{3\pi}{2}, \frac{7\pi}{4}\right)$

(b) Relative maxima: $\left(\frac{\pi}{2}, 1\right), (\pi, 1), \left(\frac{3\pi}{2}, 1\right)$

Relative minima: $\left(\frac{\pi}{4}, 0\right), \left(\frac{3\pi}{4}, 0\right), \left(\frac{5\pi}{4}, 0\right), \left(\frac{7\pi}{4}, 0\right)$

(c)

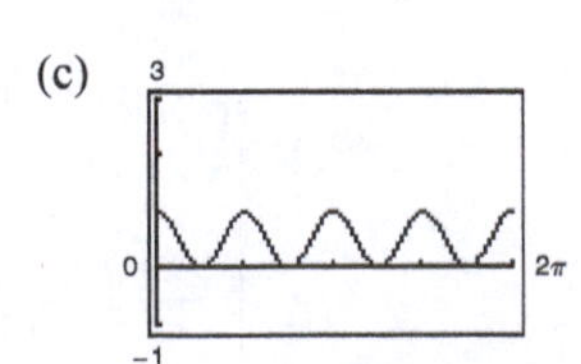

63. $f(x) = 2x\sqrt{9 - x^2}, [-3, 3]$

(a) $f'(x) = \dfrac{2(9 - 2x^2)}{\sqrt{9 - x^2}}$

(b)

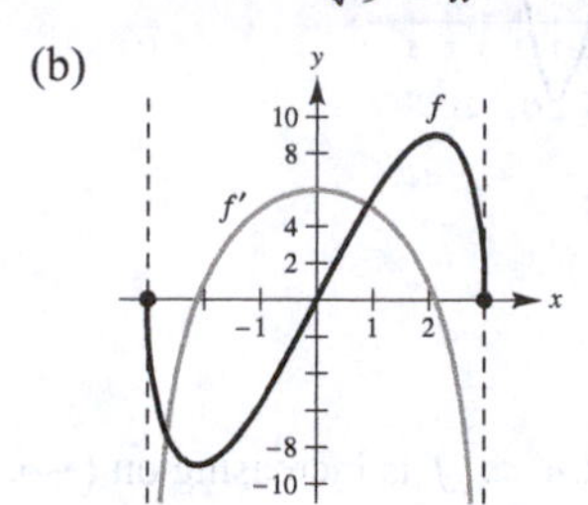

(c) $\dfrac{2(9 - 2x^2)}{\sqrt{9 - x^2}} = 0$

Critical numbers: $x = \pm\dfrac{3}{\sqrt{2}} = \pm\dfrac{3\sqrt{2}}{2}$

(d) Intervals:

$\left(-3, -\dfrac{3\sqrt{2}}{2}\right)$	$\left(-\dfrac{3\sqrt{2}}{2}, \dfrac{3\sqrt{2}}{2}\right)$	$\left(\dfrac{3\sqrt{2}}{2}, 3\right)$
$f'(x) < 0$	$f'(x) > 0$	$f'(x) < 0$
Decreasing	Increasing	Decreasing

f is increasing when f' is positive and decreasing when f' is negative.

65. $f(t) = t^2 \sin t, [0, 2\pi]$

(a) $f'(t) = t^2 \cos t + 2t \sin t = t(t \cos t + 2 \sin t)$

(b)

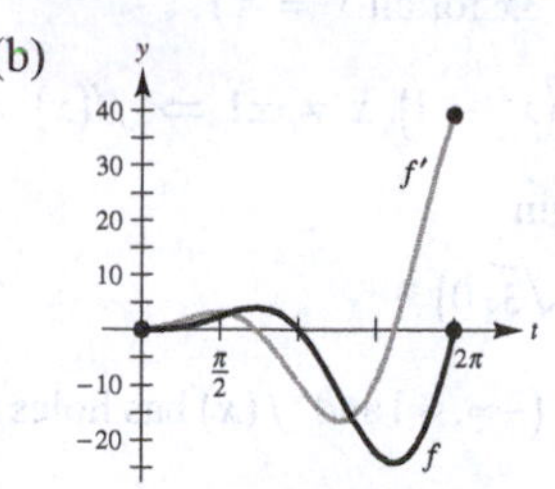

(c)

$$t(t \cos t + 2 \sin t) = 0$$
$$t = 0 \text{ or } t = -2 \tan t$$
$$t \cot t = -2$$
$$t \approx 2.2889, 5.0870 \text{ (graphing utility)}$$

Critical numbers: $t = 2.2889, 5.0870$

(d) Intervals:

$(0, 2.2889)$	$(2.2889, 5.0870)$	$(5.0870, 2\pi)$
$f'(t) > 0$	$f'(t) < 0$	$f'(t) > 0$
Increasing	Decreasing	Increasing

f is increasing when f' is positive and decreasing when f' is negative.

67. (a) $f(x) = -3 \sin \dfrac{x}{3}, [0, 6\pi]$

$f'(x) = -\cos \dfrac{x}{3}$

(b)

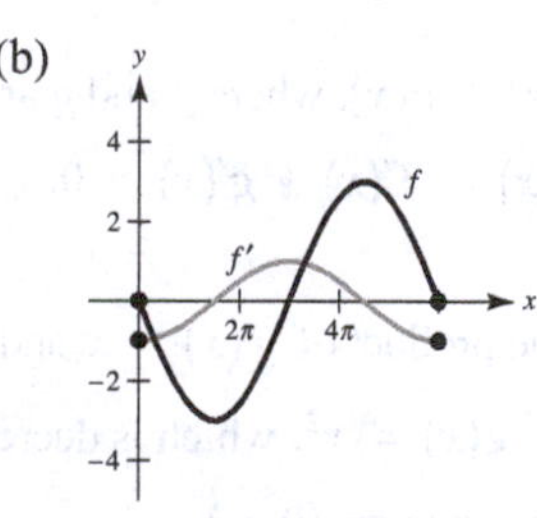

(c) Critical numbers: $x = \dfrac{3\pi}{2}, \dfrac{9\pi}{2}$

(d) Intervals:

$\left(0, \dfrac{3\pi}{2}\right)$	$\left(\dfrac{3\pi}{2}, \dfrac{9\pi}{2}\right)$	$\left(\dfrac{9\pi}{2}, 6\pi\right)$
$f' < 0$	$f' > 0$	$f' < 0$
Decreasing	Increasing	Decreasing

f is increasing when f' is positive and decreasing when f' is negative.

69. $f(x) = \dfrac{1}{2}(x^2 - \ln x), (0, 3]$

(a) $f'(x) = \dfrac{2x^2 - 1}{2x}$

(b)

(c) $\dfrac{2x^2 - 1}{2x} = 0$

Critical number: $x = \dfrac{1}{\sqrt{2}} = \dfrac{\sqrt{2}}{2}$

(d) Intervals:

$\left(0, \dfrac{\sqrt{2}}{2}\right)$	$\left(\dfrac{\sqrt{2}}{2}, 3\right)$
$f'(x) < 0$	$f'(x) > 0$
Decreasing	Increasing

(e) f is increasing when f' is positive, and decreasing when f' is negative.

71. $f(x) = \dfrac{x^5 - 4x^3 + 3x}{x^2 - 1} = \dfrac{(x^2 - 1)(x^3 - 3x)}{x^2 - 1} = x^3 - 3x, x \neq \pm 1$

$f(x) = g(x) = x^3 - 3x$ for all $x \neq \pm 1$.

$f'(x) = 3x^2 - 3 = 3(x^2 - 1), x \neq \pm 1 \Rightarrow f'(x) \neq 0$

f symmetric about origin

zeros of f: $(0, 0), (\pm\sqrt{3}, 0)$

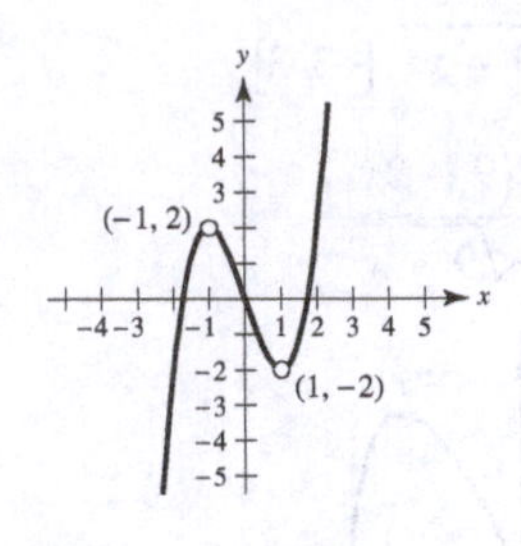

$g(x)$ is continuous on $(-\infty, \infty)$ and $f(x)$ has holes at $(-1, 2)$ and $(1, -2)$.

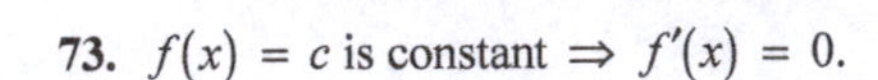

73. $f(x) = c$ is constant $\Rightarrow f'(x) = 0$.

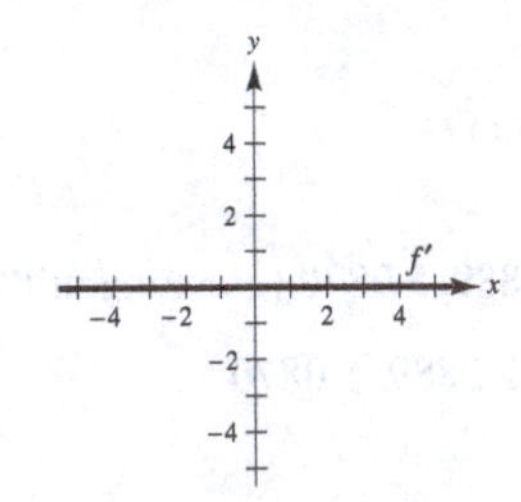

75. f is quadratic $\Rightarrow f'$ is a line.

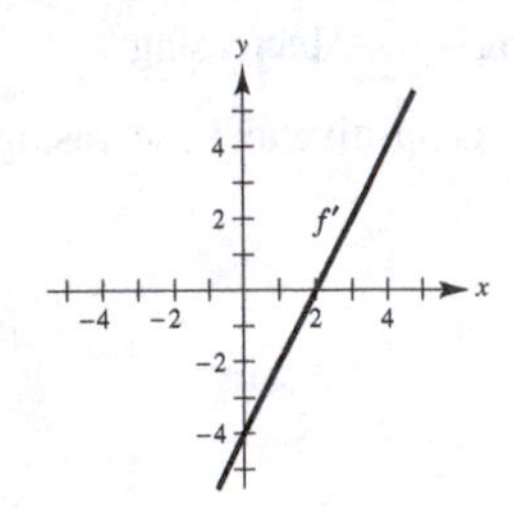

77. f has positive, but decreasing slope.

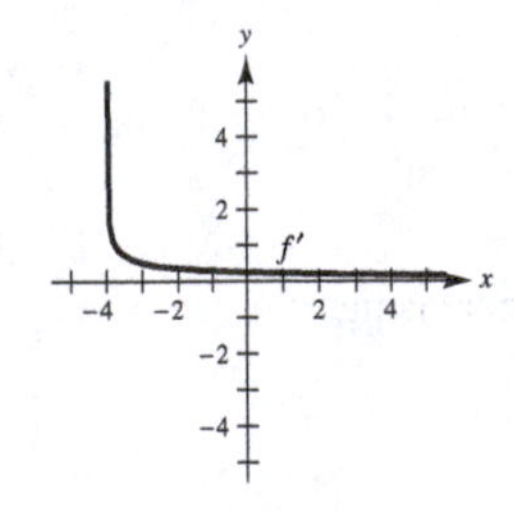

In Exercises 79–81, $f'(x) > 0$ on $(-\infty, -4)$, $f'(x) < 0$ on $(-4, 6)$ and $f'(x) > 0$ on $(6, \infty)$.

79. $g(x) = f(x) + 5$

$g'(x) = f'(x)$

$g'(0) = f'(0) < 0$

81. $g(x) = -f(x)$

$g'(x) = -f'(x)$

$g'(-6) = -f'(-6) < 0$

83. $f'(x)\begin{cases} > 0, & x < 4 \Rightarrow f \text{ is increasing on } (-\infty, 4). \\ \text{undefined}, & x = 4 \\ < 0, & x > 4 \Rightarrow f \text{ is decreasing on } (4\ \infty). \end{cases}$

Two possibilities for $f(x)$ are given below.

(a)

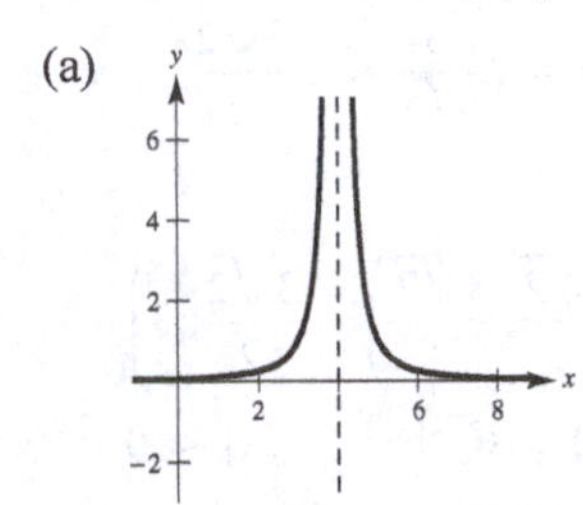

(b)

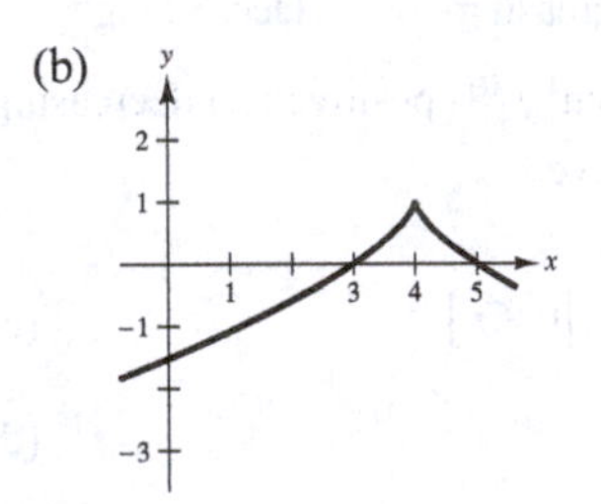

85. (a) Yes. If $h(x) = f(x) + g(x)$, where f and g are increasing, then $h'(x) = f'(x) + g'(x) > 0$. So, h is increasing.

(b) No. For example, the product of $f(x) = x$ and $g(x) = x$ is $f(x) \cdot g(x) = x^2$, which is decreasing on $(-\infty, 0)$ and increasing on $(0, \infty)$.

87. Critical number: $x = 5$

$f'(4) = -2.5 \Rightarrow f$ is decreasing at $x = 4$.

$f'(6) = 3 \Rightarrow f$ is increasing at $x = 6$.

$(5, f(5))$ is a relative minimum.

In Exercises 89, answers will vary.

Sample answers:

89. (a)

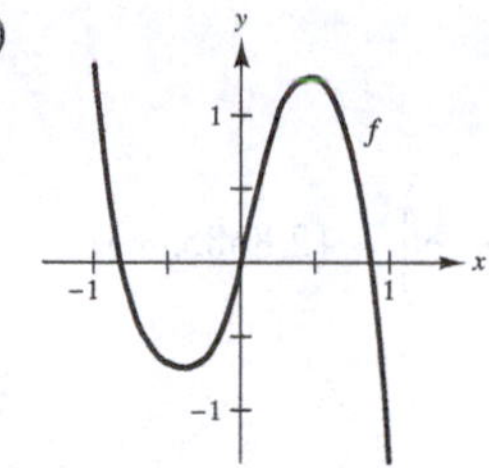

(b) The critical numbers are in intervals $(-0.50, -0.25)$ and $(0.25, 0.50)$ because the sign of f' changes in these intervals. f is decreasing on approximately $(-1, -0.40)$, $(0.48, 1)$, and increasing on $(-0.40, 0.48)$.

(c) Relative minimum when $x \approx -0.40$: $(-0.40, 0.75)$

Relative maximum when $x \approx 0.48$: $(0.48, 1.25)$

91. $s(t) = 4.9(\sin\theta)t^2$

(a) $s'(t) = 4.9(\sin\theta)(2t) = 9.8(\sin\theta)t$

speed $= |s'(t)| = |9.8(\sin\theta)t|$

(b)

θ	0	$\frac{\pi}{4}$	$\frac{\pi}{3}$	$\frac{\pi}{2}$	$\frac{2\pi}{3}$	$\frac{3\pi}{4}$	π
$\lvert s'(t)\rvert$	0	$4.9\sqrt{2}t$	$4.9\sqrt{3}t$	$9.8t$	$4.9\sqrt{3}t$	$4.9\sqrt{2}t$	0

The speed is maximum for $\theta = \frac{\pi}{2}$.

93. $C = \frac{3t}{27 + t^3}, t \geq 0$

(a)

t	0	0.5	1	1.5	2	2.5	3
$C(t)$	0	0.055	0.107	0.148	0.171	0.176	0.167

The concentration seems greatest near $t = 2.5$ hours.

(b)

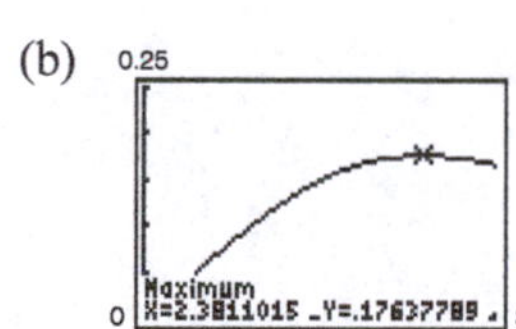

The concentration is greatest when $t \approx 2.38$ hours.

(c) $C' = \frac{(27 + t^3)(3) - (3t)(3t^2)}{(27 + t^3)^2} = \frac{3(27 - 2t^3)}{(27 + t^3)^2}$

$C' = 0$ when $t = 3/\sqrt[3]{2} \approx 2.38$ hours.

By the First Derivative Test, this is a maximum.

95. $v = k(R - r)r^2 = k(Rr^2 - r^3)$

$v' = k(2Rr - 3r^2) = kr(2R - 3r) = 0$

$r = 0$ or $\frac{2}{3}R$

Maximum when $r = \frac{2}{3}R$.

97. (a) $s(t) = 6t - t^2, t \geq 0$

$v(t) = 6 - 2t$

(b) $v(t) = 0$ when $t = 3$.

Moving in positive direction for $0 \leq t < 3$ because $v(t) > 0$ on $0 \leq t < 3$.

(c) Moving in negative direction when $t > 3$.

(d) The particle changes direction at $t = 3$.

99. (a) $s(t) = t^3 - 5t^2 + 4t, t \geq 0$

$v(t) = 3t^2 - 10t + 4$

(b) $v(t) = 0$ for $t = \dfrac{10 \pm \sqrt{100 - 48}}{6} = \dfrac{5 \pm \sqrt{13}}{3}$

Particle is moving in a positive direction on $\left[0, \dfrac{5 - \sqrt{13}}{3}\right) \approx [0, 0.4648)$ and $\left(\dfrac{5 + \sqrt{13}}{3}, \infty\right) \approx (2.8685, \infty)$

because $v > 0$ on these intervals.

(c) Particle is moving in a negative direction on $\left(\dfrac{5 - \sqrt{13}}{3}, \dfrac{5 + \sqrt{13}}{3}\right) \approx (0.4648, 2.8685)$

(d) The particle changes direction at $t = \dfrac{5 \pm \sqrt{13}}{3}$.

101. Answers will vary.

103. (a) Use a cubic polynomial

$f(x) = a_3x^3 + a_2x^2 + a_1x + a_0$

(b) $f'(x) = 3a_3x^2 + 2a_2x + a_1$.

$$\begin{aligned}
f(0) = 0:&\quad a_3(0)^3 + a_2(0)^2 + a_1(0) + a_0 = 0 \Rightarrow & a_0 &= 0\\
f'(0) = 0:&\quad 3a_3(0)^2 + 2a_2(0) + a_1 = 0 \Rightarrow & a_1 &= 0\\
f(2) = 2:&\quad a_3(2)^3 + a_2(2)^2 + a_1(2) + a_0 = 2 \Rightarrow & 8a_3 + 4a_2 &= 2\\
f'(2) = 0:&\quad 3a_3(2)^2 + 2a_2(2) + a_1 = 0 \Rightarrow & 12a_3 + 4a_2 &= 0
\end{aligned}$$

(c) The solution is $a_0 = a_1 = 0, a_2 = \frac{3}{2}, a_3 = -\frac{1}{2}$: $f(x) = -\frac{1}{2}x^3 + \frac{3}{2}x^2$.

(d)

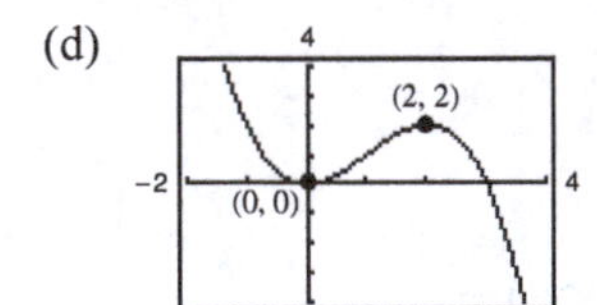

105. (a) Use a fourth degree polynomial

$f(x) = a_4x^4 + a_3x^3 + a_2x^2 + a_1x + a_0$.

(b) $f'(x) = 4a_4x^3 + 3a_3x^2 + 2a_2x + a_1$

$$\begin{aligned}
f(0) = 0:&\quad a_4(0)^4 + a_3(0)^3 + a_2(0)^2 + a_1(0) + a_0 = 0 \Rightarrow & a_0 &= 0\\
f'(0) = 0:&\quad 4a_4(0)^3 + 3a_3(0)^2 + 2a_2(0) + a_1 = 0 \Rightarrow & a_1 &= 0\\
f(4) = 0:&\quad a_4(4)^4 + a_3(4)^3 + a_2(4)^2 + a_1(4) + a_0 = 0 \Rightarrow & 256a_4 + 64a_3 + 16a_2 &= 0\\
f'(4) = 0:&\quad 4a_4(4)^3 + 3a_3(4)^2 + 2a_2(4) + a_1 = 0 \Rightarrow & 256a_4 + 48a_3 + 8a_2 &= 0\\
f(2) = 4:&\quad a_4(2)^4 + a_3(2)^3 + a_2(2)^2 + a_1(2) + a_0 = 4 \Rightarrow & 16a_4 + 8a_3 + 4a_2 &= 4\\
f'(2) = 0:&\quad 4a_4(2)^3 + 3a_3(2)^2 + 2a_2(2) + a_1 = 0 \Rightarrow & 32a_4 + 12a_3 + 4a_2 &= 0
\end{aligned}$$

(c) The solution is $a_0 = a_1 = 0, a_2 = 4, a_3 = -2, \quad a_4 = \frac{1}{4}$.

$f(x) = \frac{1}{4}x^4 - 2x^3 + 4x^2$

(d)

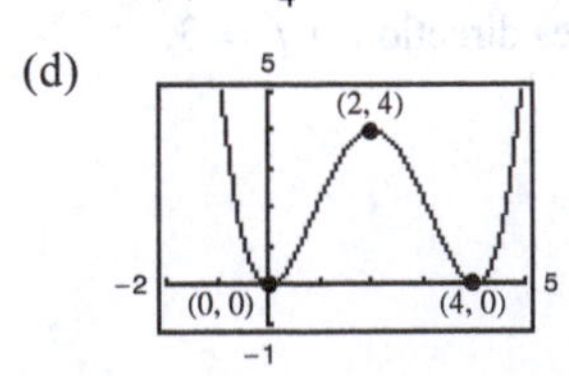

107. False. For example, $f(x) = \sin x$ has an infinite number of critical points at $x = n\pi$, where n is an integer.

109. False.

Let $f(x) = x^3$, then $f'(x) = 3x^2$ and f only has one critical number. Or, let $f(x) = x^3 + 3x + 1$, then $f'(x) = 3(x^2 + 1)$ has no critical numbers.

111. False. For example, $f(x) = x^3$ does not have a relative extrema at the critical number $x = 0$.

113. Assume that $f'(x) < 0$ for all x in the interval (a, b) and let $x_1 < x_2$ be any two points in the interval. By the Mean Value Theorem, you know there exists a number c such that $x_1 < c < x_2$, and

$$f'(c) = \frac{f(x_2) - f(x_1)}{x_2 - x_1}$$

Because $f'(c) < 0$ and $x_2 - x_1 > 0$, then $f(x_2) - f(x_1) < 0$, which implies that $f(x_2) < f(x_1)$.

So, f is decreasing on the interval.

115. Let $f(x) = (1 + x)^n - nx - 1$. Then

$$f'(x) = n(1 + x)^{n-1} - n = n\left[(1 + x)^{n-1} - 1\right] > 0$$

because $x > 0$ and $n > 1$.

So, $f(x)$ is increasing on $(0, \infty)$.

Because $f(0) = 0 \Rightarrow f(x) > 0$ on $(0, \infty)$

$(1 + x)^n - nx - 1 > 0 \Rightarrow (1 + x)^n > 1 + nx$.

117. Let x_1 and x_2 be two positive real numbers, $0 < x_1 < x_2$. Then

$$\frac{1}{x_1} > \frac{1}{x_2}$$

$$f(x_1) > f(x_2)$$

So, f is decreasing on $(0, \infty)$.

119. First observe that

$$\tan x + \cot x + \sec x + \csc x = \frac{\sin x}{\cos x} + \frac{\cos x}{\sin x} + \frac{1}{\cos x} + \frac{1}{\sin x} = \frac{\sin^2 x + \cos^2 x + \sin x + \cos x}{\sin x \cos x}$$

$$= \frac{1 + \sin x + \cos x}{\sin x \cos x}\left(\frac{\sin x + \cos x - 1}{\sin x + \cos x - 1}\right) = \frac{(\sin x + \cos x)^2 - 1}{\sin x \cos x(\sin x + \cos x - 1)}$$

$$= \frac{2 \sin x \cos x}{\sin x \cos x(\sin x + \cos x - 1)} = \frac{2}{\sin x + \cos x - 1}$$

Let $t = \sin x + \cos x - 1$. The expression inside the absolute value sign is

$$f(t) = \sin x + \cos x + \frac{2}{\sin x + \cos x - 1} = (\sin x + \cos x - 1) + 1 + \frac{2}{\sin x + \cos x - 1} = t + 1 + \frac{2}{t}$$

Because $\sin\left(x + \frac{\pi}{4}\right) = \sin x \cos\frac{\pi}{4} + \cos x \sin\frac{\pi}{4} = \frac{\sqrt{2}}{2}(\sin x + \cos x)$, $\sin x + \cos x \in \left[-\sqrt{2}, \sqrt{2}\right]$

and $t = \sin x + \cos x - 1 \in \left[-1 - \sqrt{2}, -1 + \sqrt{2}\right]$.

$$f'(t) = 1 - \frac{2}{t^2} = \frac{t^2 - 2}{t^2} = \frac{\left(t + \sqrt{2}\right)\left(t - \sqrt{2}\right)}{t^2}$$

$$f\left(-1 + \sqrt{2}\right) = -1 + \sqrt{2} + 1 + \frac{2}{-1 + \sqrt{2}} = \sqrt{2} + \frac{2}{\sqrt{2} - 1} = \frac{4 - \sqrt{2}}{\sqrt{2} - 1}\left(\frac{\sqrt{2} + 1}{\sqrt{2} + 1}\right) = \frac{4\sqrt{2} - 2 + 4 - \sqrt{2}}{1} = 2 + 3\sqrt{2}$$

For $t > 0$, f is decreasing and $f(t) > f\left(-1 + \sqrt{2}\right) = 2 + 3\sqrt{2}$

For $t < 0$, f is increasing on $\left(-\sqrt{2} - 1, -\sqrt{2}\right)$, then decreasing on $\left(-\sqrt{2}, 0\right)$. So $f(t) < f\left(-\sqrt{2}\right) = 1 - 2\sqrt{2}$.

Finally, $\left|f(t)\right| \geq 2\sqrt{2} - 1$. (You can verify this easily with a graphing utility.)

Section 4.4 Concavity and the Second Derivative Test

1. Find the second derivative of a function and form test intervals by using the values for which the second derivative is zero or does not exist and the values at which the function is not continuous. Determine the sign of the second derivative on these test intervals. If the second derivative is positive, then the graph is concave upward. If the second derivative is negative, then the graph is concave downward.

3. $f(x) = x^2 - 4x + 8$

$f'(x) = 2x - 4$

$f''(x) = 2$

$f''(x) > 0$ for all x.

Concave upward: $(-\infty, \infty)$

5. $f(x) = x^4 - 3x^3$

$f'(x) = 4x^3 - 9x^2$

$f''(x) = 12x^2 - 18x = 6x(2x - 3)$

$f''(x) = 0$ when $x = 0, \frac{3}{2}$.

Intervals:	$-\infty < x < 0$	$0 < x < \frac{3}{2}$	$\frac{3}{2} < x < \infty$
Sign of f'':	$f'' > 0$	$f'' < 0$	$f'' > 0$
Conclusion:	Concave upward	Concave downward	Concave upward

Concave upward: $(-\infty, 0), \left(\frac{3}{2}, \infty\right)$

Concave downward: $\left(0, \frac{3}{2}\right)$

7. $f(x) = \dfrac{24}{x^2 + 12}$

$f'(x) = \dfrac{-48x}{(x^2 + 12)^2}$

$f''(x) = \dfrac{-144(4 - x^2)}{(x^2 + 12)^3}$

$f''(x) = 0$ when $x = \pm 2$.

Intervals:	$-\infty < x < -2$	$-2 < x < 2$	$2 < x < \infty$
Sign of f'':	$f'' > 0$	$f'' < 0$	$f'' > 0$
Conclusion:	Concave upward	Concave downward	Concave upward

Concave upward: $(-\infty, -2), (2, \infty)$

Concave downward: $(-2, 2)$

9. $f(x) = \dfrac{x - 2}{6x + 1}$

$f'(x) = \dfrac{13}{(6x + 1)^2}$

$f''(x) = \dfrac{-156}{(6x + 1)^3}$

f is not continuous at $x = -\frac{1}{6}$.

Intervals:	$-\infty < x < -\frac{1}{6}$	$-\frac{1}{6} < x < \infty$
Sign of f'':	$f'' > 0$	$f'' < 0$
Conclusion:	Concave upward	Concave downward

Concave upward: $\left(-\infty, -\frac{1}{6}\right)$

Concave downward: $\left(-\frac{1}{6}, \infty\right)$

11. $f(x) = \dfrac{x^2 + 1}{x^2 - 1}$

$f' = \dfrac{-4x}{(x^2 - 1)^2}$

$f'' = \dfrac{4(3x^2 + 1)}{(x^2 - 1)^3}$

f is not continuous at $x = \pm 1$.

Concave upward: $(-\infty, -1), (1, \infty)$

Concave downward: $(-1, 1)$

Intervals:	$-\infty < x < -1$	$-1 < x < 1$	$1 < x < \infty$
Sign of f'':	$f'' > 0$	$f'' < 0$	$f'' > 0$
Conclusion:	Concave upward	Concave downward	Concave upward

13. $y = 2x - \tan x, \left(-\dfrac{\pi}{2}, \dfrac{\pi}{2}\right)$

$y' = 2 - \sec^2 x$

$y'' = -2\sec^2 x \tan x$

$y'' = 0$ when $x = 0$.

Concave upward: $\left(-\dfrac{\pi}{2}, 0\right)$

Concave downward: $\left(0, \dfrac{\pi}{2}\right)$

Intervals:	$-\dfrac{\pi}{2} < x < 0$	$0 < x < \dfrac{\pi}{2}$
Sign of y'':	$y'' > 0$	$y'' < 0$
Conclusion:	Concave upward	Concave downward

15. $f(x) = x^3 - 9x^2 + 24x - 18$

$f'(x) = 3x^2 - 18x + 24$

$f''(x) = 6x - 18 = 0$ when $x = 3$.

Concave upward: $(3, \infty)$

Concave downward: $(-\infty, 3)$

Point of inflection: $(3, 0)$

Intervals:	$-\infty < x < 3$	$3 < x < \infty$
Sign of f'':	$f'' < 0$	$f'' > 0$
Conclusion:	Concave downward	Concave upward

17. $f(x) = 2 - 7x^4$

$f'(x) = -28x^3$

$f''(x) = -84x^2 = 0$ when $x = 0$.

Concave downward: $(-\infty, \infty)$

No points of inflection

Intervals:	$-\infty < x < 0$	$0 < x < \infty$
Sign of f'':	$f'' < 0$	$f'' < 0$
Conclusion:	Concave downward	Concave downward

19. $f(x) = x(x - 4)^3$

$f'(x) = x\left[3(x - 4)^2\right] + (x - 4)^3 = (x - 4)^2(4x - 4)$

$f''(x) = 4(x - 1)\left[2(x - 4)\right] + 4(x - 4)^2 = 4(x - 4)\left[2(x - 1) + (x - 4)\right] = 4(x - 4)(3x - 6) = 12(x - 4)(x - 2)$

$f''(x) = 12(x - 4)(x - 2) = 0$ when $x = 2, 4$.

Intervals:	$-\infty < x < 2$	$2 < x < 4$	$4 < x < \infty$
Sign of $f''(x)$:	$f''(x) > 0$	$f''(x) < 0$	$f''(x) > 0$
Conclusion:	Concave upward	Concave downward	Concave upward

Concave upward: $(-\infty, 2), (4, \infty)$

Concave downward: $(2, 4)$

Points of inflection: $(2, -16), (4, 0)$

21. $f(x) = x\sqrt{x+3}$, Domain: $[-3, \infty)$

$$f'(x) = x\left(\frac{1}{2}\right)(x+3)^{-1/2} + \sqrt{x+3} = \frac{3(x+2)}{2\sqrt{x+3}}$$

$$f''(x) = \frac{6\sqrt{x+3} - 3(x+2)(x+3)^{-1/2}}{4(x+3)} = \frac{3(x+4)}{4(x+3)^{3/2}} = 0 \text{ when } x = -4.$$

$x = -4$ is not in the domain. f'' is not continuous at $x = -3$.

Interval:	$-3 < x < \infty$
Sign of f'':	$f'' > 0$
Conclusion:	Concave upward

Concave upward: $(-3, \infty)$

There are no points of inflection.

23. $f(x) = \dfrac{6-x}{\sqrt{x}} = 6x^{-1/2} - x^{1/2}$

$$f'(x) = \frac{-(x+6)}{2x^{3/2}}$$

$$f''(x) = \frac{x+18}{4x^{5/2}}$$

Note that the domain of f is $x > 0$.

Furthermore, $f''(x) > 0$ on $(0, \infty)$.

Concave upward: $(0, \infty)$

No points of inflection

25. $f(x) = \sin\dfrac{x}{2}, 0 \le x \le 4\pi$

$$f'(x) = \frac{1}{2}\cos\left(\frac{x}{2}\right)$$

$$f''(x) = -\frac{1}{4}\sin\left(\frac{x}{2}\right)$$

$f''(x) = 0$ when $x = 0, 2\pi, 4\pi$.

Intervals:	$0 < x < 2\pi$	$2\pi < x < 4\pi$
Sign of f'':	$f'' < 0$	$f'' > 0$
Conclusion:	Concave downward	Concave upward

Concave upward: $(2\pi, 4\pi)$

Concave downward: $(0, 2\pi)$

Point of inflection: $(2\pi, 0)$

27. $f(x) = \sec\left(x - \dfrac{\pi}{2}\right), 0 < x < 4\pi$

$$f'(x) = \sec\left(x - \frac{\pi}{2}\right)\tan\left(x - \frac{\pi}{2}\right)$$

$$f''(x) = \sec^3\left(x - \frac{\pi}{2}\right) + \sec\left(x - \frac{\pi}{2}\right)\tan^2\left(x - \frac{\pi}{2}\right) \ne 0 \text{ for any } x \text{ in the domain of } f.$$

f'' is not continuous at $x = \pi$, $x = 2\pi$, and $x = 3\pi$.

Intervals:	$0 < x < \pi$	$\pi < x < 2\pi$	$2\pi < x < 3\pi$	$3\pi < x < 4\pi$
Sign of f'':	$f'' > 0$	$f'' < 0$	$f'' > 0$	$f'' < 0$
Conclusion:	Concave upward	Concave downward	Concave upward	Concave upward

Concave upward: $(0, \pi), (2\pi, 3\pi)$

Concave downward: $(\pi, 2\pi), (3\pi, 4\pi)$

No point of inflection

29. $f(x) = 2\sin x + \sin 2x, 0 \le x \le 2\pi$

$f'(x) = 2\cos x + 2\cos 2x$

$f''(x) = -2\sin x - 4\sin 2x = -2\sin x(1 + 4\cos x)$

$f''(x) = 0$ when $x = 0, 1.823, \pi, 4.460$.

Intervals:	$0 < x < 1.823$	$1.823 < x < \pi$	$\pi < x < 4.460$	$4.460 < x < 2\pi$
Sign of f'':	$f'' < 0$	$f'' > 0$	$f'' < 0$	$f'' > 0$
Conclusion:	Concave downward	Concave upward	Concave downward	Concave upward

Concave upward: $(1.823, \pi), (4.460, 2\pi)$

Concave downward: $(0, 1.823), (\pi, 4.460)$

Points of inflection: $(1.823, 1.452), (\pi, 0), (4.46, -1.452)$

31. $y = e^{-3/x}$

$y' = \frac{3}{x^2}e^{-3/x}$

$y'' = \frac{e^{-3/x}(9 - 6x)}{x^4}$

Test intervals:	$-\infty < x < 0$	$0 < x < \frac{3}{2}$	$\frac{3}{2} < x < \infty$
Sign of y'':	$y'' > 0$	$y'' > 0$	$y'' < 0$
Conclusion:	Concave upward	Concave upward	Concave downward

$y'' = 0$ when $x = \frac{3}{2}$.

y is not defined at $x = 0$.

Point of inflection: $\left(\frac{3}{2}, e^{-2}\right)$

Concave upward: $(-\infty, 0), \left(0, \frac{3}{2}\right)$

Concave downward: $\left(\frac{3}{2}, \infty\right)$

33. $f(x) = x - \ln x$, Domain: $x > 0$

$f'(x) = 1 - \frac{1}{x}$

$f''(x) = \frac{1}{x^2}$

$f''(x) > 0$ on the entire domain of f. There are no points of inflection.

Concave upward: $(0, \infty)$

35. $f(x) = \arcsin x^{4/5}, \quad -1 \le x \le 1$

$$f'(x) = \frac{4}{5x^{1/5}\sqrt{1 - x^{8/5}}}$$

$$f''(x) = \frac{20x^{8/5} - 4}{25x^{6/5}\left(1 - x^{8/5}\right)^{3/2}}$$

$f''(x) = 0$ when $20x^{8/5} = 4 \Rightarrow x^{8/5} = \frac{1}{5} \Rightarrow x = \pm\left(\frac{1}{5}\right)^{5/8} \approx \pm 0.3657.$

f'' is undefined at $x = 0$.

Test intervals:	$-1 < x < -\left(\frac{1}{5}\right)^{5/8}$	$-\left(\frac{1}{5}\right)^{5/8} < x < 0$	$0 < x < \left(\frac{1}{5}\right)^{5/8}$	$\left(\frac{1}{5}\right)^{5/8} < x < 1$
Sign of f'':	$f'' > 0$	$f'' < 0$	$f'' < 0$	$f'' > 0$
Conclusion:	Concave upward	Concave downward	Concave downward	Concave upward

Points of inflection: $\left(\pm\left(\frac{1}{5}\right)^{5/8}, \arcsin\sqrt{\frac{1}{5}}\right) \approx (\pm 0.3657, 0.4636)$

Concave upward: $\left(-1, -\left(\frac{1}{5}\right)^{5/8}\right), \left(\left(\frac{1}{5}\right)^{5/8}, 1\right)$

Concave downward: $\left(-\left(\frac{1}{5}\right)^{5/8}, 0\right), \left(0, \left(\frac{1}{5}\right)^{5/8}\right)$

37. $f(x) = 6x - x^2$

$f'(x) = 6 - 2x$

$f''(x) = -2$

Critical number: $x = 3$

$f''(3) = -2 < 0$

Therefore, $(3, 9)$ is a relative maximum.

39. $f(x) = x^4 - 4x^3 + 2$

$f'(x) = 4x^3 - 12x^2 = 4x^2(x - 3)$

$f''(x) = 12x^2 - 24x = 12x(x - 2)$

Critical numbers: $x = 0, x = 3$

However, $f''(0) = 0$, so you must use the First Derivative Test. $f'(x) < 0$ on the intervals $(-\infty, 0)$ and $(0, 3)$; so, $(0, 2)$ is not an extremum. $f''(3) > 0$ so $(3, -25)$ is a relative minimum.

41. $f(x) = x^{2/3} - 3$

$$f'(x) = \frac{2}{3x^{1/3}}$$

$$f''(x) = -\frac{2}{9x^{4/3}}$$

Critical number: $x = 0$

However, $f''(0)$ is undefined, so you must use the First Derivative Test. Because $f'(x) < 0$ on $(-\infty, 0)$ and $f'(x) > 0$ on $(0, \infty)$, $(0, -3)$ is a relative minimum.

43. $f(x) = x + \frac{4}{x}$

$$f'(x) = 1 - \frac{4}{x^2} = \frac{x^2 - 4}{x^2}$$

$$f''(x) = \frac{8}{x^3}$$

Critical numbers: $x = \pm 2$

$f''(-2) = -1 < 0$

Therefore, $(-2, -4)$ is a relative maximum.

$f''(2) = 1 > 0$

Therefore, $(2, 4)$ is a relative minimum.

45. $f(x) = \cos x - x, 0 \le x \le 4\pi$

$f'(x) = -\sin x - 1 \le 0$

Therefore, f is non-increasing and there are no relative extrema.

47. $y = f(x) = 8x^2 - \ln x$

$f'(x) = 16x - \dfrac{1}{x}$

$f''(x) = 16 + \dfrac{1}{x^2}$

$f'(x) = 0 \Rightarrow 16x = \dfrac{1}{x} \Rightarrow 16x^2 = 1 \Rightarrow x = \pm\dfrac{1}{4}$

Critical number: $x = \dfrac{1}{4}\left(x = -\dfrac{1}{4} \text{ is not in the domain.}\right)$

$f''\left(\dfrac{1}{4}\right) > 0$

Therefore, $\left(\dfrac{1}{4}, \dfrac{1}{2} - \ln\dfrac{1}{4}\right) = \left(\dfrac{1}{4}, \dfrac{1}{2} + \ln 4\right)$ is a relative minimum.

49. $y = f(x) = \dfrac{x}{\ln x}$

Domain: $0 < x < 1, x > 1$

$f'(x) = \dfrac{(\ln x)(1) - (x)(1/x)}{(\ln x)^2} = \dfrac{\ln x - 1}{(\ln x)^2}$

$f''(x) = \dfrac{2 - \ln x}{x(\ln x)}$

Critical number: $x = e$

$f''(e) > 0$

Therefore, (e, e) is a relative minimum.

51. $f(x) = \dfrac{e^x + e^{-x}}{2}$

$f'(x) = \dfrac{e^x - e^{-x}}{2}$

$f''(x) = \dfrac{e^x + e^{-x}}{2}$

Critical number: $x = 0$

$f''(0) > 0$

Therefore, $(0, 1)$ is a relative minimum.

53. $f(x) = x^2e^{-x}$

$f'(x) = -x^2e^{-x} + 2xe^{-x} = xe^{-x}(2 - x)$

$f''(x) = -e^{-x}(2x - x^2) + e^{-x}(2 - 2x)$

$= e^{-x}(x^2 - 4x + 2)$

Critical numbers: $x = 0, 2$

$f''(0) > 0$

Therefore, $(0, 0)$ is a relative minimum.

$f''(2) < 0$

Therefore, $(2, 4e^{-2})$ is a relative maximum.

55. $f(x) = 8x(4^{-x})$

$f'(x) = -8(4^{-x})(x \ln 4 - 1)$

$f''(x) = 8(4^{-x}) \ln 4(x \ln 4 - 2)$

Critical number: $x = \dfrac{1}{\ln 4} = \dfrac{1}{2 \ln 2}$

$f''\left(\dfrac{1}{2 \ln 2}\right) < 0$

Therefore, $\left(\dfrac{1}{2 \ln 2}, \dfrac{4e^{-1}}{\ln 2}\right)$ is a relative maximum.

57. $f(x) = \text{arcsec}\, x - x$

$f'(x) = \dfrac{1}{|x|\sqrt{x^2 - 1}} - 1 = 0$ when $|x|\sqrt{x^2 - 1} = 1$.

$x^2(x^2 - 1) = 1$

$x^4 - x^2 - 1 = 0$ when $x^2 = \dfrac{1 + \sqrt{5}}{2}$

or $x = \pm\sqrt{\dfrac{1 + \sqrt{5}}{2}} = \pm 1.272$.

$f''(x) = -\dfrac{1}{x\sqrt{x^2 - 1}|x|} - \dfrac{x}{(x^2 - 1)^{3/2}|x|}$

$f''(1.272) < 0$

Therefore, $(1.272, -0.606)$ is a relative maximum.

$f''(-1.272) > 0$

Therefore, $(-1.272, 3.747)$ is a relative minimum.

59. $f(x) = 0.2x^2(x - 3)^3, [-1, 4]$

(a) $f'(x) = 0.2x(5x - 6)(x - 3)^2$

$f''(x) = (x - 3)(4x^2 - 9.6x + 3.6) = 0.4(x - 3)(10x^2 - 24x + 9)$

(b) $f''(0) < 0 \Rightarrow (0, 0)$ is a relative maximum.

$f''\left(\frac{6}{5}\right) > 0 \Rightarrow (1.2, -1.6796)$ is a relative minimum.

Points of inflection: $(3, 0), (0.4652, -0.7048), (1.9348, -0.9049)$

(c)

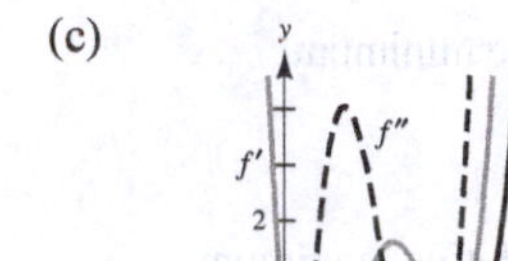

f is increasing when $f' > 0$ and decreasing when $f' < 0$. f is concave upward when $f'' > 0$ and concave downward when $f'' < 0$.

61. $f(x) = \sin x - \frac{1}{3}\sin 3x + \frac{1}{5}\sin 5x, \quad [0, \pi]$

(a) $f'(x) = \cos x - \cos 3x + \cos 5x$

$f'(x) = 0$ when $x = \frac{\pi}{6}, x = \frac{\pi}{2}, x = \frac{5\pi}{6}$.

$f''(x) = -\sin x + 3\sin 3x - 5\sin 5x$

$f''(x) = 0$ when $x = \frac{\pi}{6}, x = \frac{5\pi}{6}$,

$x \approx 1.1731, x \approx 1.9685$

(b) $f''\left(\frac{\pi}{2}\right) < 0 \Rightarrow \left(\frac{\pi}{2}, 1.53333\right)$ is a relative maximum.

Points of inflection: $\left(\frac{\pi}{6}, 0.2667\right), (1.1731, 0.9638), (1.9685, 0.9637), \left(\frac{5\pi}{6}, 0.2667\right)$

Note: $(0, 0)$ and $(\pi, 0)$ are not points of inflection because they are endpoints.

(c)

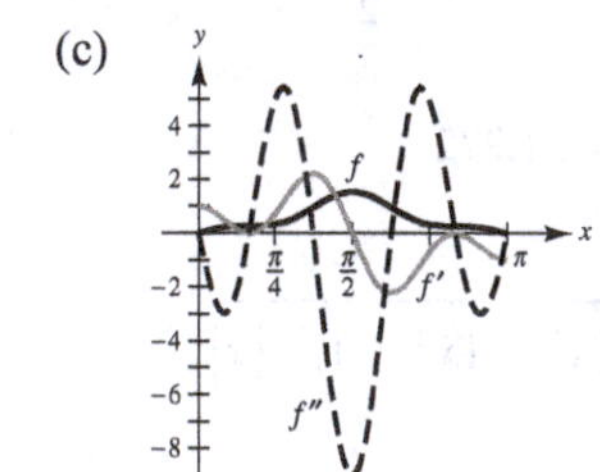

The graph of f is increasing when $f' > 0$ and decreasing when $f' < 0$. f is concave upward when $f'' > 0$ and concave downward when $f'' < 0$.

63. (a)

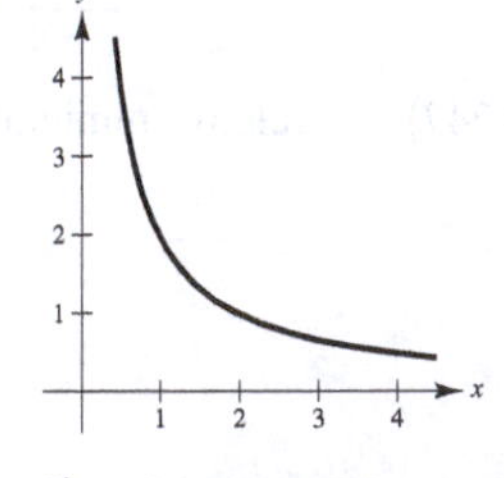

$f' < 0$ means f decreasing

(b)

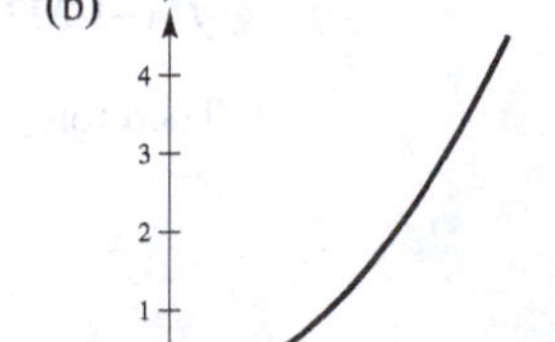

$f' > 0$ means f increasing

f' increasing means concave upward

65. (a)

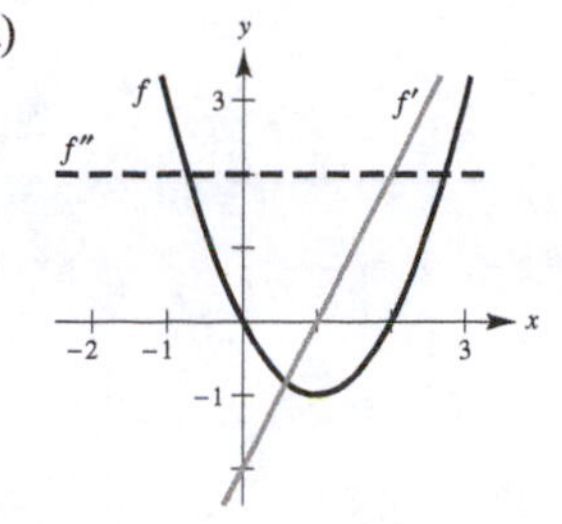

(b)

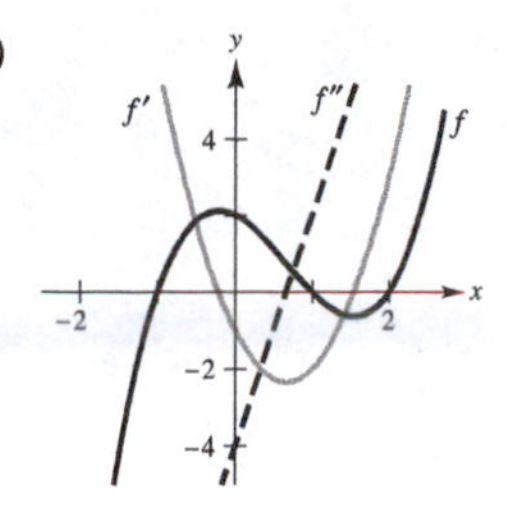

67.

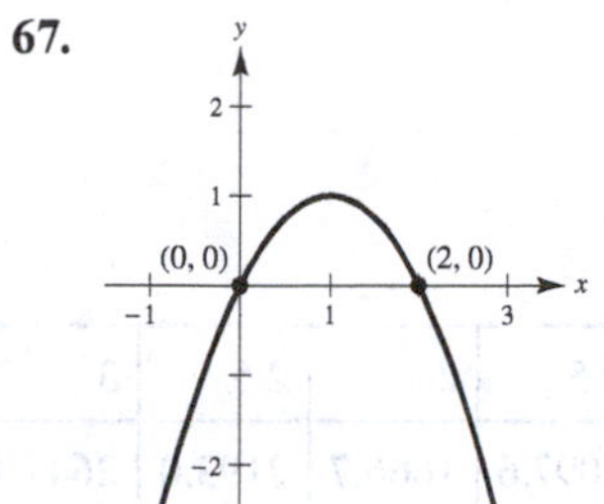

69.

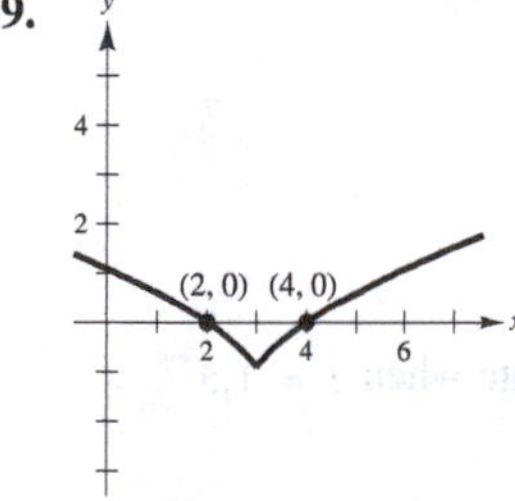

71.

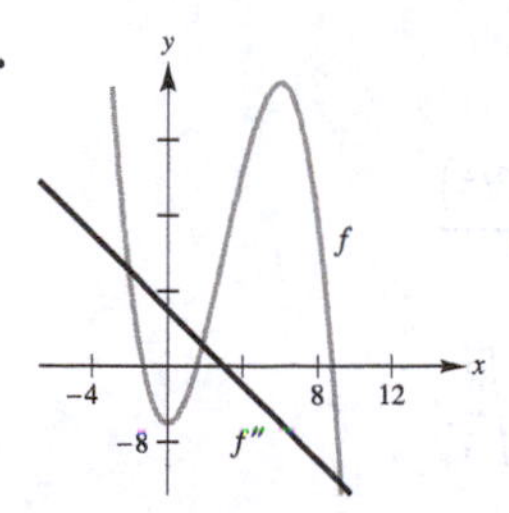

f'' is linear.

f' is quadratic.

f is cubic.

f concave upward on $(-\infty, 3)$, downward on $(3, \infty)$.

73. (a) $n = 1$:

$$f(x) = x - 2$$
$$f'(x) = 1$$
$$f''(x) = 0$$

No point of inflection

$n = 2$:

$$f(x) = (x - 2)^2$$
$$f'(x) = 2(x - 2)$$
$$f''(x) = 2$$

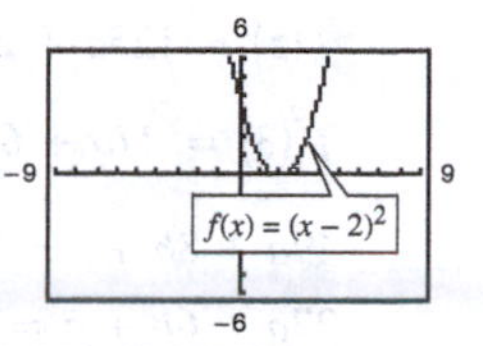

No point of inflection

Relative minimum: $(2, 0)$

$n = 3$:

$$f(x) = (x - 2)^3$$
$$f'(x) = 3(x - 2)^2$$
$$f''(x) = 6(x - 2)$$

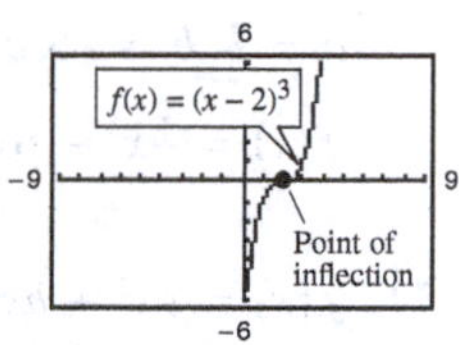

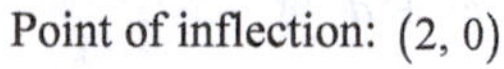

Point of inflection: $(2, 0)$

$n = 4$:

$$f(x) = (x - 2)^4$$
$$f'(x) = 4(x - 2)^3$$
$$f''(x) = 12(x - 2)^2$$

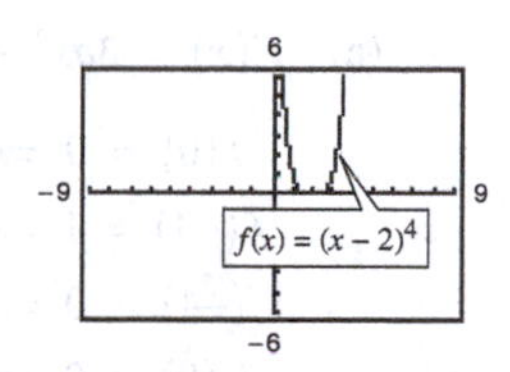

No point of inflection

Relative minimum: $(2, 0)$

Conclusion: If $n \geq 3$ and n is odd, then $(2, 0)$ is point of inflection. If $n \geq 2$ and n is even, then $(2, 0)$ is a relative minimum.

(b) Let $f(x) = (x - 2)^n$, $f'(x) = n(x - 2)^{n-1}$,

$$f''(x) = n(n - 1)(x - 2)^{n-2}.$$

For $n \geq 3$ and odd, $n - 2$ is also odd and the concavity changes at $x = 2$.

For $n \geq 4$ and even, $n - 2$ is also even and the concavity does not change at $x = 2$.

So, $x = 2$ is point of inflection if and only if $n \geq 3$ is odd.

75. $f(x) = ax^3 + bx^2 + cx + d$

Relative maximum: $(3, 3)$

Relative minimum: $(5, 1)$

Point of inflection: $(4, 2)$

$f'(x) = 3ax^2 + 2bx + c,\ f''(x) = 6ax + 2b$

$$\left.\begin{aligned} f(3) &= 27a + 9b + 3c + d = 3 \\ f(5) &= 125a + 25b + 5c + d = 1 \end{aligned}\right\} 98a + 16b + 2c = -2 \Rightarrow 49a + 8b + c = -1$$

$f'(3) = 27a + 6b + c = 0,\ f''(4) = 24a + 2b = 0$

$$\begin{array}{lcl} 49a + 8b + c = -1 & \quad & 24a + 2b = 0 \\ \underline{27a + 6b + c = \ \ 0} & & \underline{22a + 2b = -1} \\ 22a + 2b \quad\ = -1 & & 2a \qquad\ = 1 \end{array}$$

$a = \frac{1}{2}, b = -6, c = \frac{45}{2}, d = -24$

$f(x) = \frac{1}{2}x^3 - 6x^2 + \frac{45}{2}x - 24$

77. $f(x) = ax^3 + bx^2 + cx + d$

Maximum: $(-4, 1)$

Minimum: $(0, 0)$

(a) $f'(x) = 3ax^2 + 2bx + c, \qquad f''(x) = 6ax + 2b$

$$\begin{aligned} f(0) &= 0 \Rightarrow d = 0 \\ f(-4) &= 1 \Rightarrow -64a + 16b - 4c = 1 \\ f'(-4) &= 0 \Rightarrow 48a - 8b + c = 0 \\ f'(0) &= 0 \Rightarrow c = 0 \end{aligned}$$

Solving this system yields $a = \frac{1}{32}$ and $b = 6a = \frac{3}{16}$.

$f(x) = \frac{1}{32}x^3 + \frac{3}{16}x^2$

(b) The plane would be descending at the greatest rate at the point of inflection.

$f''(x) = 6ax + 2b = \frac{3}{16}x + \frac{3}{8} = 0 \Rightarrow x = -2.$

Two miles from touchdown.

79. $C = 0.5x^2 + 15x + 5000$

$\overline{C} = \frac{C}{x} = 0.5x + 15 + \frac{5000}{x}$

$\overline{C}$ = average cost per unit

$\frac{d\overline{C}}{dx} = 0.5 - \frac{5000}{x^2} = 0$ when $x = 100$

By the First Derivative Test, $\overline{C}$ is minimized when $x = 100$ units.

81. $S = \frac{5000t^2}{8 + t^2}, 0 \le t \le 3$

(a)

t	0.5	1	1.5	2	2.5	3
S	151.5	555.6	1097.6	1666.7	2193.0	2647.1

Increasing at greatest rate when $1.5 < t < 2$

(b)

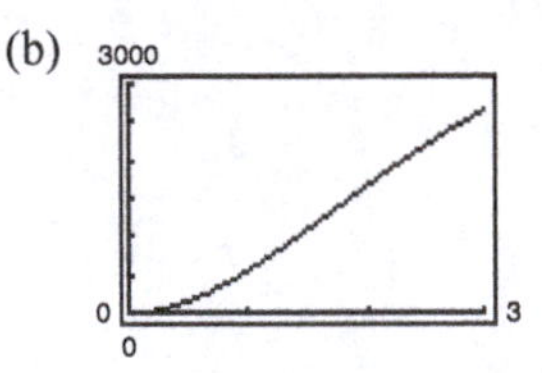

Increasing at greatest rate when $t \approx 1.5$.

(c) $S = \frac{5000t^2}{8 + t^2}$

$S'(t) = \frac{80{,}000t}{(8 + t^2)^2}$

$S''(t) = \frac{80{,}000(8 - 3t^2)}{(8 + t^2)^3}$

$S''(t) = 0$ for $t = \pm\sqrt{\frac{8}{3}}$.

So, $t = \frac{2\sqrt{6}}{3} \approx 1.633$ yrs.

83. $f(x) = 2(\sin x + \cos x), \quad f\left(\frac{\pi}{4}\right) = 2\sqrt{2}$

$f'(x) = 2(\cos x - \sin x), \quad f'\left(\frac{\pi}{4}\right) = 0$

$f''(x) = 2(-\sin x - \cos x), \quad f''\left(\frac{\pi}{4}\right) = -2\sqrt{2}$

$P_1(x) = 2\sqrt{2} + 0\left(x - \frac{\pi}{4}\right) = 2\sqrt{2}$

$P_1'(x) = 0$

$P_2(x) = 2\sqrt{2} + 0\left(x - \frac{\pi}{4}\right) + \frac{1}{2}\left(-2\sqrt{2}\right)\left(x - \frac{\pi}{4}\right)^2$

$= 2\sqrt{2} - \sqrt{2}\left(x - \frac{\pi}{4}\right)^2$

$P_2'(x) = -2\sqrt{2}\left(x - \frac{\pi}{4}\right)$

$P_2''(x) = -2\sqrt{2}$

The values of f, P_1, P_2, and their first derivatives are equal at $x = \pi/4$. The values of the second derivatives of f and P_2 are equal at $x = \pi/4$. The approximations worsen as you move away from $x = \pi/4$.

85. $f(x) = \arctan x, a = -1, \quad f(-1) = -\frac{\pi}{4}$

$f'(x) = \frac{1}{1 + x^2}, \quad f'(-1) = \frac{1}{2}$

$f''(x) = -\frac{2x}{\left(1 + x^2\right)^2}, \quad f''(-1) = \frac{1}{2}$

$P_1(x) = f(-1) + f'(-1)(x + 1) = -\frac{\pi}{4} + \frac{1}{2}(x + 1)$

$P_1'(x) = \frac{1}{2}$

$P_2(x) = f(-1) + f'(-1)(x + 1) + \frac{1}{2}f''(-1)(x + 1)^2$

$= -\frac{\pi}{4} + \frac{1}{2}(x + 1) + \frac{1}{4}(x + 1)^2$

$P_2'(x) = \frac{1}{2} + \frac{1}{2}(x + 1)$

$P_2''(x) = \frac{1}{2}$

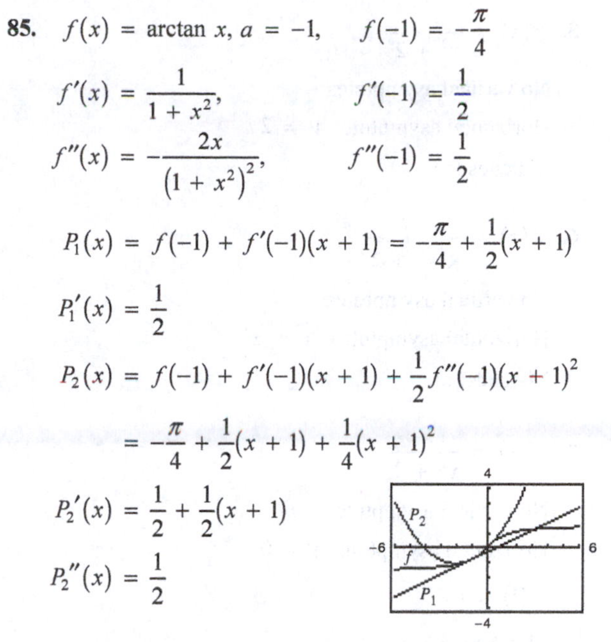

The values of f, P_1, P_2, and their first derivatives are equal when $x = -1$. The approximations worsen as you move away from $x = -1$.

87. $f(x) = x \sin\left(\frac{1}{x}\right)$

$f'(x) = x\left[-\frac{1}{x^2}\cos\left(\frac{1}{x}\right)\right] + \sin\left(\frac{1}{x}\right) = -\frac{1}{x}\cos\left(\frac{1}{x}\right) + \sin\left(\frac{1}{x}\right)$

$f''(x) = -\frac{1}{x}\left[\frac{1}{x^2}\sin\left(\frac{1}{x}\right)\right] + \frac{1}{x^2}\cos\left(\frac{1}{x}\right) - \frac{1}{x^2}\cos\left(\frac{1}{x}\right) = -\frac{1}{x^3}\sin\left(\frac{1}{x}\right) = 0$

$x = \frac{1}{\pi}$

Point of inflection: $\left(\frac{1}{\pi}, 0\right)$

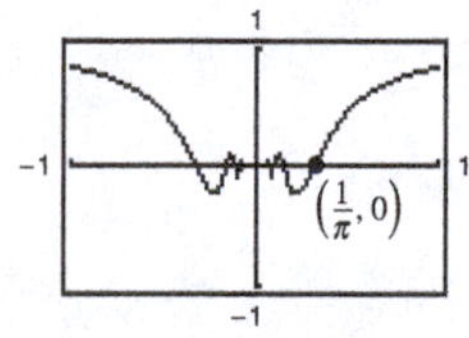

When $x > 1/\pi$, $f'' < 0$, so the graph is concave downward.

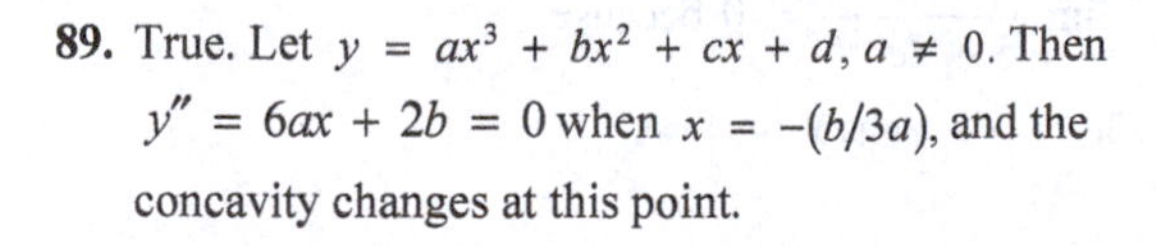

89. True. Let $y = ax^3 + bx^2 + cx + d, a \neq 0$. Then $y'' = 6ax + 2b = 0$ when $x = -(b/3a)$, and the concavity changes at this point.

91. False. Concavity is determined by f''. For example, let $f(x) = x$ and $c = 2$. $f'(c) = f'(2) > 0$, but f is not concave upward at $c = 2$.

93. f and g are concave upward on (a, b) implies that f' and g' are increasing on (a, b), and $f'' > 0$ and $g'' > 0$.

So, $(f + g)'' > 0 \Rightarrow f + g$ is concave upward on (a, b) by Theorem 3.7.

Section 4.5 Limits at Infinity

1. (a) As x increases without bound, $f(x)$ approaches -5.

(b) As x decreases without bound, $f(x)$ approaches 3.

3. The maximum number is 2, one from the left and one from the right.

5. $f(x) = \dfrac{2x^2}{x^2 + 2}$

No vertical asymptotes

Horizontal asymptote: $y = 2$

Matches (f).

6. $f(x) = \dfrac{2x}{\sqrt{x^2 + 2}}$

No vertical asymptotes

Horizontal asymptotes: $y = \pm 2$

Matches (c).

7. $f(x) = \dfrac{x}{x^2 + 2}$

No vertical asymptotes

Horizontal asymptote: $y = 0$

$f(1) < 1$

Matches (d).

8. $f(x) = 2 + \dfrac{x^2}{x^4 + 1}$

No vertical asymptotes

Horizontal asymptote: $y = 2$

Matches (a).

9. $f(x) = \dfrac{4 \sin x}{x^2 + 1}$

No vertical asymptotes

Horizontal asymptote: $y = 0$

$f(1) > 1$

Matches (b).

10. $f(x) = \dfrac{4}{1 + e^{-x}}$

No vertical asymptotes

Horizontal asymptotes: $y = 4$ and $y = 0$

Matches (e).

11. (a) $h(x) = \dfrac{f(x)}{x^2} = \dfrac{5x^3 - 3x^2 + 10x}{x^2} = 5x - 3 + \dfrac{10}{x}$

$\lim_{x\to\infty} h(x) = \infty$ (Limit does not exist)

(b) $h(x) = \dfrac{f(x)}{x^3} = \dfrac{5x^3 - 3x^2 + 10x}{x^3} = 5 - \dfrac{3}{x} + \dfrac{10}{x^2}$

$\lim_{x\to\infty} h(x) = 5$

(c) $h(x) = \dfrac{f(x)}{x^4} = \dfrac{5x^3 - 3x^2 + 10x}{x^4} = \dfrac{5}{x} - \dfrac{3}{x^2} + \dfrac{10}{x^3}$

$\lim_{x\to\infty} h(x) = 0$

13. (a) $\lim_{x\to\infty} \dfrac{x^2 + 2}{x^3 - 1} = 0$

(b) $\lim_{x\to\infty} \dfrac{x^2 + 2}{x^2 - 1} = 1$

(c) $\lim_{x\to\infty} \dfrac{x^2 + 2}{x - 1} = \infty$ (Limit does not exist)

15. (a) $\lim_{x\to\infty} \dfrac{5 - 2x^{3/2}}{3x^2 - 4} = 0$

(b) $\lim_{x\to\infty} \dfrac{5 - 2x^{3/2}}{3x^{3/2} - 4} = -\dfrac{2}{3}$

(c) $\lim_{x\to\infty} \dfrac{5 - 2x^{3/2}}{3x - 4} = -\infty$ (Limit does not exist)

17. $\lim_{x\to\infty}\left(4 + \dfrac{3}{x}\right) = 4 + 0 = 4$

19. $\lim_{x\to\infty} \dfrac{7x + 6}{9x - 4} = \lim_{x\to\infty} \dfrac{7 + 6/x}{9 - 4/x} = \dfrac{7 + 0}{9 - 0} = \dfrac{7}{9}$

21. $\lim_{x\to-\infty} \dfrac{2x^2 + x}{6x^3 + 2x^2 + x} = \lim_{x\to-\infty} \dfrac{2/x + 1/x^2}{6 + 2/x + 1/x^2}$

$= \dfrac{0 + 0}{6 + 0 + 0} = 0$

23. $\lim_{x\to-\infty} \dfrac{-4}{3 + 3e^{-2x}} = 0$ because

$3e^{-2x} \to \infty$ as $x \to -\infty$.

25. $\lim_{x\to-\infty} \dfrac{x}{\sqrt{x^2 - x}} = \lim_{x\to-\infty} \dfrac{1}{\left(\dfrac{\sqrt{x^2 - x}}{-\sqrt{x^2}}\right)} = \lim_{x\to-\infty} \dfrac{-1}{\sqrt{1 - (1/x)}} = -1$, (for $x < 0$ we have $x = -\sqrt{x^2}$)

27. $\lim_{x\to-\infty} \dfrac{2x + 1}{\sqrt{x^2 - x}} = \lim_{x\to-\infty} \dfrac{2 + \dfrac{1}{x}}{\left(\dfrac{\sqrt{x^2 - x}}{-\sqrt{x^2}}\right)} = \lim_{x\to-\infty} \dfrac{-2 - \left(\dfrac{1}{x}\right)}{\sqrt{1 - \dfrac{1}{x}}} = -2$, (for $x < 0$, $x = -\sqrt{x^2}$)

29. $\lim_{x\to\infty} \frac{\sqrt{x^2 - 1}}{2x - 1} = \lim_{x\to\infty} \frac{\sqrt{x^2 - 1}/\sqrt{x^2}}{2 - 1/x} = \lim_{x\to\infty} \frac{\sqrt{1 - 1/x^2}}{2 - 1/x} = \frac{1}{2}$

31. $\lim_{x\to\infty} \frac{x + 1}{(x^2 + 1)^{1/3}} = \lim_{x\to\infty} \frac{x + 1}{(x^2 + 1)^{1/3}}\left(\frac{1/x^{2/3}}{1/(x^2)^{1/3}}\right) = \lim_{x\to\infty} \frac{x^{1/3} + 1/x^{2/3}}{(1 + 1/x^2)^{1/3}} = \infty$

Limit does not exist.

33. $\lim_{x\to\infty} \frac{1}{2x + \sin x} = 0$

35. Because $(-1/x) \le (\sin 2x)/x \le (1/x)$ for all $x \ne 0$, you have by the Squeeze Theorem,

$$\lim_{x\to\infty} -\frac{1}{x} \le \lim_{x\to\infty} \frac{\sin 2x}{x} \le \lim_{x\to\infty} \frac{1}{x}$$

$$0 \le \lim_{x\to\infty} \frac{\sin 2x}{x} \le 0.$$

Therefore, $\lim_{x\to\infty} \frac{\sin 2x}{x} = 0.$

37. $\lim_{x\to\infty} (2 - 5e^{-x}) = 2$

39. $\lim_{x\to\infty} \log_{10}(1 + 10^{-x}) = 0$

41. $\lim_{t\to\infty} (8t^{-1} - \arctan t) = \lim_{t\to\infty}\left(\frac{8}{t}\right) - \lim_{t\to\infty} \arctan t$

$$= 0 - \frac{\pi}{2} = -\frac{\pi}{2}$$

43. $f(x) = \frac{|x|}{x + 1}$

$\lim_{x\to\infty} \frac{|x|}{x + 1} = 1$

$\lim_{x\to-\infty} \frac{|x|}{x + 1} = -1$

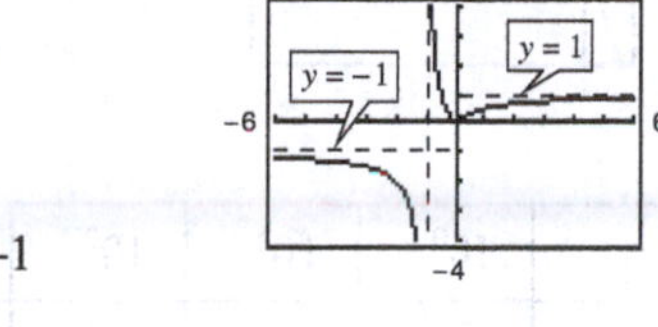

Therefore, $y = 1$ and $y = -1$ are both horizontal asymptotes.

45. $f(x) = \frac{3x}{\sqrt{x^2 + 2}}$

$\lim_{x\to\infty} f(x) = 3$

$\lim_{x\to-\infty} f(x) = -3$

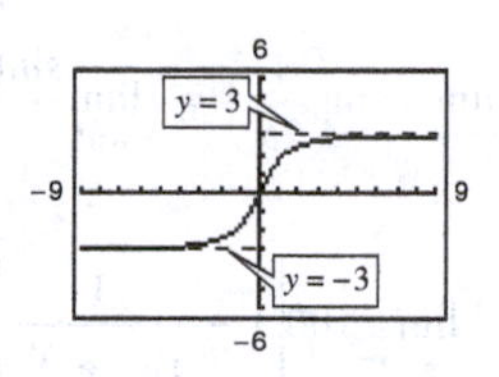

Therefore, $y = 3$ and $y = -3$ are both horizontal asymptotes.

47. $\lim_{x\to\infty} x \sin\frac{1}{x} = \lim_{t\to 0^+} \frac{\sin t}{t} = 1$

(Let $x = 1/t$.)

49. $\lim_{x\to-\infty} \left(x + \sqrt{x^2 + 3}\right) = \lim_{x\to-\infty} \left[\left(x + \sqrt{x^2 + 3}\right) \cdot \frac{x - \sqrt{x^2 + 3}}{x - \sqrt{x^2 + 3}}\right] = \lim_{x\to-\infty} \frac{-3}{x - \sqrt{x^2 + 3}} = 0$

51. $\lim_{x\to-\infty} \left(3x + \sqrt{9x^2 - x}\right) = \lim_{x\to-\infty} \left[\left(3x + \sqrt{9x^2 - x}\right) \cdot \frac{3x - \sqrt{9x^2 - x}}{3x - \sqrt{9x^2 - x}}\right]$

$$= \lim_{x\to-\infty} \frac{x}{3x - \sqrt{9x^2 - x}}$$

$$= \lim_{x\to-\infty} \frac{1}{3 - \frac{\sqrt{9x^2 - x}}{-\sqrt{x^2}}} \quad \left(\text{for } x < 0 \text{ you have } x = -\sqrt{x^2}\right)$$

$$= \lim_{x\to-\infty} \frac{1}{3 + \sqrt{9 - (1/x)}} = \frac{1}{6}$$

53.

x	10^0	10^1	10^2	10^3	10^4	10^5	10^6
$f(x)$	1	0.513	0.501	0.500	0.500	0.500	0.500

$$\lim_{x\to\infty}\left(x - \sqrt{x(x-1)}\right) = \lim_{x\to\infty}\frac{x - \sqrt{x^2 - x}}{1}\cdot\frac{x + \sqrt{x^2 - x}}{x + \sqrt{x^2 - x}} = \lim_{x\to\infty}\frac{x}{x + \sqrt{x^2 - x}} = \lim_{x\to\infty}\frac{1}{1 + \sqrt{1 - (1/x)}} = \frac{1}{2}$$

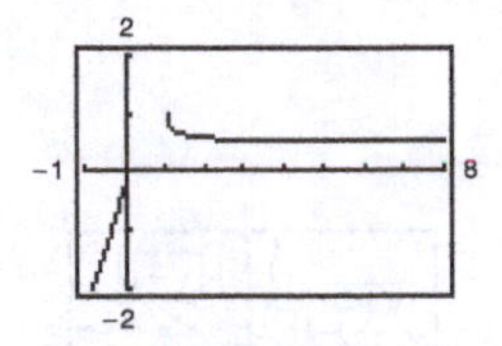

55.

x	10^0	10^1	10^2	10^3	10^4	10^5	10^6
$f(x)$	0.479	0.500	0.500	0.500	0.500	0.500	0.500

Let $x = 1/t$.

$$\lim_{x\to\infty} x\sin\left(\frac{1}{2x}\right) = \lim_{t\to0^+}\frac{\sin(t/2)}{t} = \lim_{t\to0^+}\frac{1}{2}\frac{\sin(t/2)}{t/2} = \frac{1}{2}$$

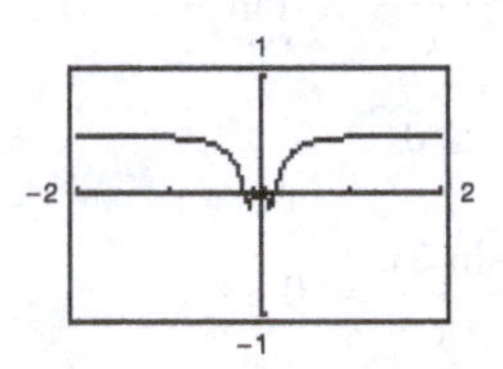

57. $\lim_{v_1/v_2\to\infty} 100\left[1 - \frac{1}{(v_1/v_2)^c}\right] = 100[1 - 0] = 100\%$

59. An infinite limit describes how a limit fails to exist. A limit at infinity deals with the end behavior of a function.

61. (a) The function is even: $\lim_{x\to-\infty} f(x) = 5$

(b) The function is odd: $\lim_{x\to-\infty} f(x) = -5$

63. (a)

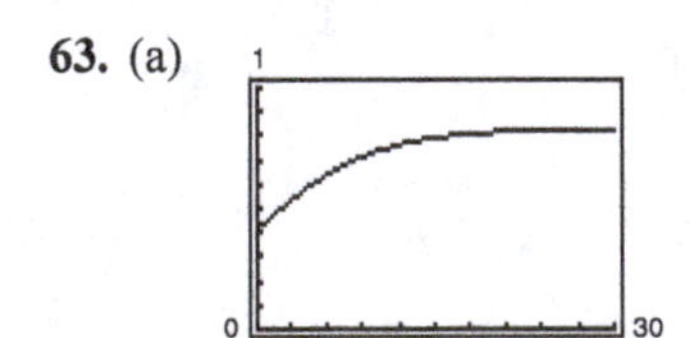

(b) $\lim_{n\to\infty}\frac{0.83}{1 + e^{-0.2n}} = 0.83 = 83\%$

The limiting proportion of correct responses is 83%.

65. $f(x) = \frac{2x^2}{x^2 + 2}$

(a) $\lim_{x\to\infty} f(x) = 2 = L$

(b)
$$f(x_1) + \varepsilon = \frac{2x_1^2}{x_1^2 + 2} + \varepsilon = 2$$
$$2x_1^2 + \varepsilon x_1^2 + 2\varepsilon = 2x_1^2 + 4$$
$$x_1^2\varepsilon = 4 - 2\varepsilon$$
$$x_1 = \sqrt{\frac{4 - 2\varepsilon}{\varepsilon}}$$
$$x_2 = -x_1 \text{ by symmetry}$$

(c) Let $M = \sqrt{\frac{4 - 2\varepsilon}{\varepsilon}} > 0$. For $x > M$:
$$x > \sqrt{\frac{4 - 2\varepsilon}{\varepsilon}}$$
$$x^2\varepsilon > 4 - 2\varepsilon$$
$$2x^2 + x^2\varepsilon + 2\varepsilon > 2x^2 + 4$$
$$\frac{2x^2}{x^2 + 2} + \varepsilon > 2$$
$$\left|\frac{2x^2}{x^2 + 2} - 2\right| > |-\varepsilon| = \varepsilon$$
$$|f(x) - L| > \varepsilon$$

(d) Similarly, $N = -\sqrt{\frac{4 - 2\varepsilon}{\varepsilon}}$.

67. $\lim_{x\to\infty} \dfrac{3x}{\sqrt{x^2+3}} = 3$

$$f(x_1) + \varepsilon = \frac{3x_1}{\sqrt{x_1^2+3}} + \varepsilon = 3$$

$$3x_1 = (3-\varepsilon)\sqrt{x_1^2+3}$$

$$9x_1^2 = (3-\varepsilon)^2(x_1^2+3)$$

$$9x_1^2 - (3-\varepsilon)^2 x_1^2 = 3(3-\varepsilon)^2$$

$$x_1^2(9 - 9 + 6\varepsilon - \varepsilon^2) = 3(3-\varepsilon)^2$$

$$x_1^2 = \frac{3(3-\varepsilon)^2}{6\varepsilon - \varepsilon^2}$$

$$x_1 = (3-\varepsilon)\sqrt{\frac{3}{6\varepsilon-\varepsilon^2}}$$

Let $M = x_1 = (3-\varepsilon)\sqrt{\dfrac{3}{6\varepsilon-\varepsilon^2}}$

(a) When $\varepsilon = 0.5$:

$$M = (3-0.5)\sqrt{\frac{3}{6(0.5)-(0.5)^2}} = \frac{5\sqrt{33}}{11}$$

(b) When $\varepsilon = 0.1$:

$$M = (3-0.1)\sqrt{\frac{3}{6(0.1)-(0.1)^2}} = \frac{29\sqrt{177}}{59}$$

69. $\lim_{x\to\infty} \dfrac{1}{x^2} = 0$. Let $\varepsilon > 0$ be given. You need $M > 0$ such that

$$|f(x) - L| = \left|\frac{1}{x^2} - 0\right| = \frac{1}{x^2} < \varepsilon \text{ whenever } x > M.$$

$$x^2 > \frac{1}{\varepsilon} \Rightarrow x > \frac{1}{\sqrt{\varepsilon}}$$

Let $M = \dfrac{1}{\sqrt{\varepsilon}}$.

For $x > M$, you have

$$x > \frac{1}{\sqrt{\varepsilon}} \Rightarrow x^2 > \frac{1}{\varepsilon} \Rightarrow \frac{1}{x^2} < \varepsilon \Rightarrow |f(x) - L| < \varepsilon.$$

71. $\lim_{x\to-\infty} \dfrac{1}{x^3} = 0$. Let $\varepsilon > 0$. You need $N < 0$ such that

$$|f(x) - L| = \left|\frac{1}{x^3} - 0\right| = \frac{-1}{x^3} < \varepsilon \text{ whenever } x < N.$$

$$\frac{-1}{x^3} < \varepsilon \Rightarrow -x^3 > \frac{1}{\varepsilon} \Rightarrow x < \frac{-1}{\varepsilon^{1/3}}$$

Let $N = \dfrac{-1}{\sqrt[3]{\varepsilon}}$.

For $x < N = \dfrac{-1}{\sqrt[3]{\varepsilon}}$,

$$\frac{1}{x} > -\sqrt[3]{\varepsilon}$$

$$-\frac{1}{x} < \sqrt[3]{\varepsilon}$$

$$-\frac{1}{x^3} < \varepsilon$$

$$\Rightarrow |f(x) - L| < \varepsilon.$$

73. line: $mx - y + 4 = 0$

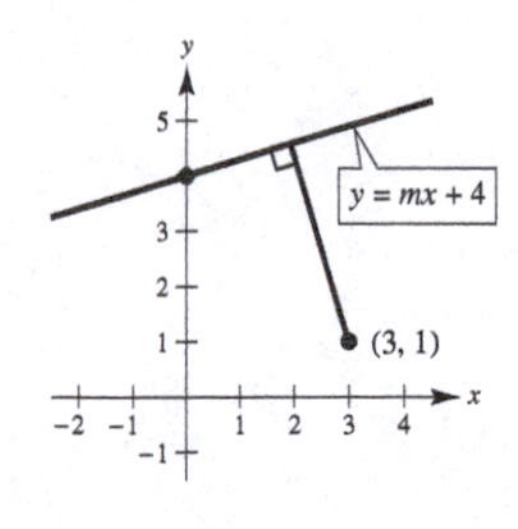

(a) $d = \dfrac{|Ax_1 + By_1 + C|}{\sqrt{A^2+B^2}} = \dfrac{|m(3) - 1(1) + 4|}{\sqrt{m^2+1}} = \dfrac{|3m+3|}{\sqrt{m^2+1}}$

(b)

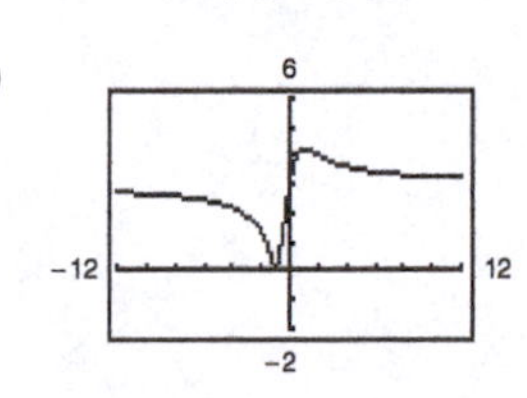

(c) $\lim_{m\to\infty} d(m) = 3 = \lim_{m\to-\infty} d(m)$

The line approaches the vertical line $x = 0$. So, the distance from $(3, 1)$ approaches 3.

75. $\lim_{x\to\infty} \dfrac{p(x)}{q(x)} = \lim_{x\to\infty} \dfrac{a_n x^n + \cdots + a_1 x + a_0}{b_m x^m + \cdots + b_1 x + b_0}$

Divide $p(x)$ and $q(x)$ by x^m.

Case 1: If $n < m$: $\lim_{x\to\infty} \dfrac{p(x)}{q(x)} = \lim_{x\to\infty} \dfrac{\dfrac{a_n}{x^{m-n}} + \cdots + \dfrac{a_1}{x^{m-1}} + \dfrac{a_0}{x^m}}{b_m + \cdots + \dfrac{b_1}{x^{m-1}} + \dfrac{b_0}{x^m}} = \dfrac{0 + \cdots + 0 + 0}{b_m + \cdots + 0 + 0} = \dfrac{0}{b_m} = 0.$

Case 2: If $m = n$: $\lim_{x\to\infty} \dfrac{p(x)}{q(x)} = \lim_{x\to\infty} \dfrac{a_n + \cdots + \dfrac{a_1}{x^{m-1}} + \dfrac{a_0}{x^m}}{b_m + \cdots + \dfrac{b_1}{x^{m-1}} + \dfrac{b_0}{x^m}} = \dfrac{a_n + \cdots + 0 + 0}{b_m + \cdots + 0 + 0} = \dfrac{a_n}{b_m}.$

Case 3: If $n > m$: $\lim_{x\to\infty} \dfrac{p(x)}{q(x)} = \lim_{x\to\infty} \dfrac{a_n x^{n-m} + \cdots + \dfrac{a_1}{x^{m-1}} + \dfrac{a_0}{x^m}}{b_m + \cdots + \dfrac{b_1}{x^{m-1}} + \dfrac{b_0}{x^m}} = \dfrac{\pm\infty + \cdots + 0}{b_m + \cdots + 0} = \pm\infty.$

Section 4.6 A Summary of Curve Sketching

1. Domain, range, asymptotes, symmetry, end behavior, differentiability, relative extrema, points of inflection, concavity, increasing and decreasing, infinite limits at infinity

3. Rational functions can have slant asymptotes. Use long division to rewrite the rational function as the sum of a first-degree polynomial and another rational function.

5. $y = \dfrac{1}{x-2} - 3$

$y' = -\dfrac{1}{(x-2)^2} \Rightarrow$ undefined when $x = 2$

$y'' = \dfrac{2}{(x-2)^3} \Rightarrow$ undefined when $x = 2$

Intercepts: $\left(\dfrac{7}{3}, 0\right), \left(0, -\dfrac{7}{2}\right)$

Vertical asymptote: $x = 2$

Horizontal asymptote: $y = -3$

	y	y'	y''	Conclusion
$-\infty < x < 2$		$-$	$-$	Decreasing, concave down
$2 < x < \infty$		$-$	$+$	Decreasing, concave up

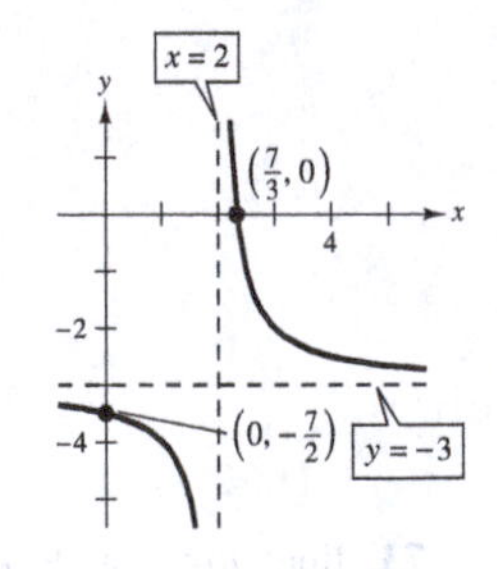

7. $y = \dfrac{x}{1-x}$

$y' = \dfrac{1}{(x-1)^2}$, undefined when $x = 1$

$y'' = -\dfrac{2}{(x-1)^3}$, undefined when $x = 1$

Horizontal asymptote: $y = -1$

Vertical asymptote: $x = 1$

	y	y'	y''	Conclusion
$-\infty < x < 1$		$+$	$+$	Increasing, concave up
$1 < x < \infty$		$+$	$-$	Increasing, concave down

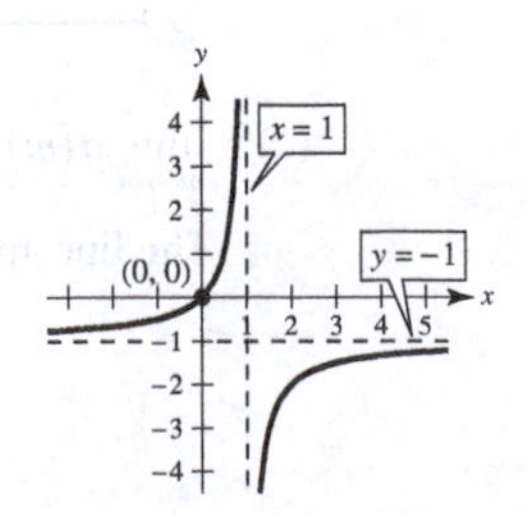

9. $y = \dfrac{x+1}{x^2-4}$

$y' = -\dfrac{x^2+2x+4}{\left(x^2-4\right)^2}$, undefined when $x = \pm 2$

$y'' = \dfrac{2\left(x^3+3x^2+12x+4\right)}{\left(x^2-4\right)^3} = 0$ when $x = \sqrt[3]{9} - \dfrac{3}{\sqrt[3]{9}} - 1 \approx -0.36$, undefined when $x = \pm 2$

Horizontal asymptote: $y = 0$

Vertical asymptote: $x = \pm 2$

	y	y'	y''	Conclusion
$-\infty < x < -2$		$-$	$-$	Decreasing, concave down
$-2 < x < \sqrt[3]{9} - \dfrac{3}{\sqrt[3]{9}} - 1$		$+$	$+$	Increasing, concave up
$x = \sqrt[3]{9} - \dfrac{3}{\sqrt[3]{9}} - 1$	-0.16	$+$	0	Point of inflection
$\sqrt[3]{9} - \dfrac{3}{\sqrt[3]{9}} - 1 < x < 2$		$-$	$-$	Decreasing, concave down
$2 < x < \infty$		$-$	$+$	Decreasing, concave up

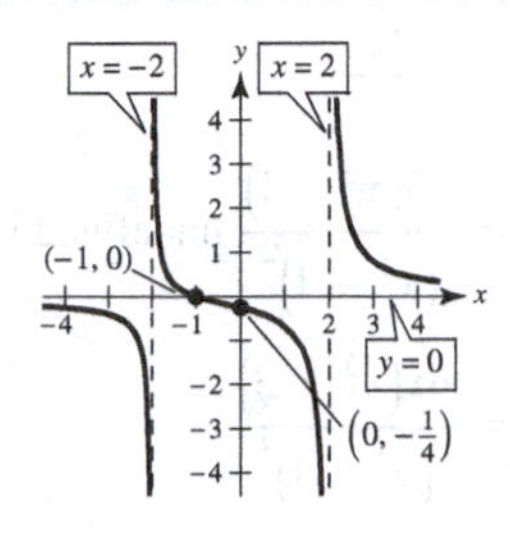

11. $y = \dfrac{x^2}{x^2+3}$

$y' = \dfrac{6x}{\left(x^2+3\right)^2} = 0$ when $x = 0$.

$y'' = \dfrac{18\left(1-x^2\right)}{\left(x^2+3\right)^3} = 0$ when $x = \pm 1$.

Horizontal asymptote: $y = 1$

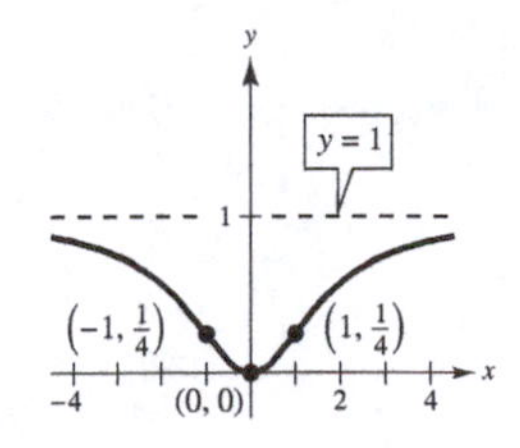

	y	y'	y''	Conclusion
$-\infty < x < -1$		$-$	$-$	Decreasing, concave down
$x = -1$	$\dfrac{1}{4}$	$-$	0	Point of inflection
$-1 < x < 0$		$-$	$+$	Decreasing, concave up
$x = 0$	0	0	$+$	Relative minimum
$0 < x < 1$		$+$	$+$	Increasing, concave up
$x = 1$	$\dfrac{1}{4}$	$+$	0	Point of inflection
$1 < x < \infty$		$+$	$-$	Increasing, concave down

13. $y = 3 + \dfrac{2}{x}$

$y' = -\dfrac{2}{x^2}$, undefined when $x = 0$

$y'' = \dfrac{4}{x^3}$, undefined when $x = 0$

Horizontal asymptote: $y = 3$

Vertical asymptote: $x = 0$

	y	y'	y''	Conclusion
$-\infty < x < 0$		$-$	$-$	Decreasing, concave down
$0 < x < \infty$		$-$	$+$	Decreasing, concave up

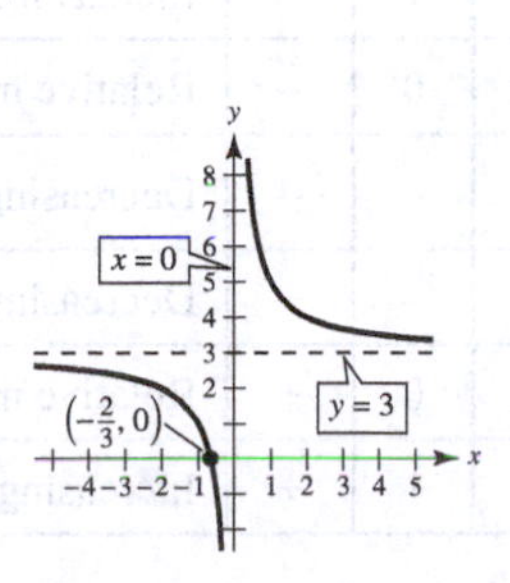

15. $f(x) = x + \dfrac{32}{x^2}$

$f'(x) = 1 - \dfrac{64}{x^3} = \dfrac{(x-4)(x^2+4x+16)}{x^3} = 0$ when $x = 4$ and undefined when $x = 0$.

$f''(x) = \dfrac{192}{x^4}$

Intercept: $\left(-2\sqrt[3]{4}, 0\right)$

Vertical asymptote: $x = 0$

Slant asymptote: $y = x$

	y	y'	y''	Conclusion
$-\infty < x < 0$		$+$	$+$	Increasing, concave up
$0 < x < 4$		$-$	$+$	Decreasing, concave up
$x = 4$	6	0	$+$	Relative minimum
$4 < x < \infty$		$+$	$+$	Increasing, concave up

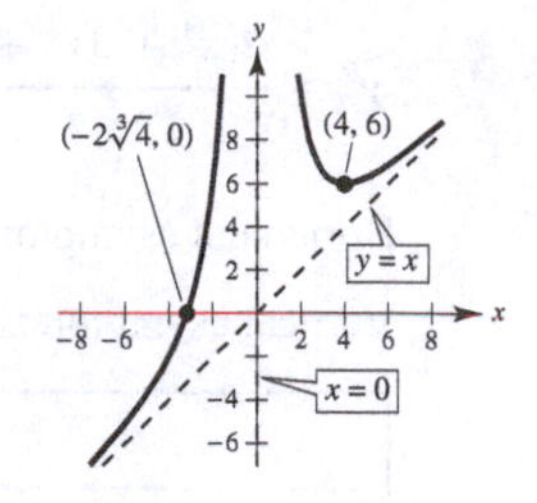

17. $y = \dfrac{3x}{x^2 - 1}$

$y' = \dfrac{-3(x^2+1)}{(x^2-1)^2}$ undefined when $x = \pm 1$

$y'' = \dfrac{6x(x^2+3)}{(x^2-1)^3}$

Intercept: $(0, 0)$

Symmetry with respect to origin

Vertical asymptotes: $x = \pm 1$

Horizontal asymptote: $y = 0$

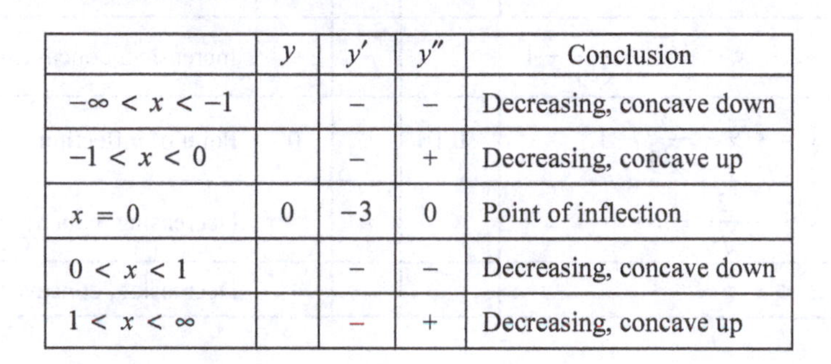

	y	y'	y''	Conclusion
$-\infty < x < -1$		$-$	$-$	Decreasing, concave down
$-1 < x < 0$		$-$	$+$	Decreasing, concave up
$x = 0$	0	-3	0	Point of inflection
$0 < x < 1$		$-$	$-$	Decreasing, concave down
$1 < x < \infty$		$-$	$+$	Decreasing, concave up

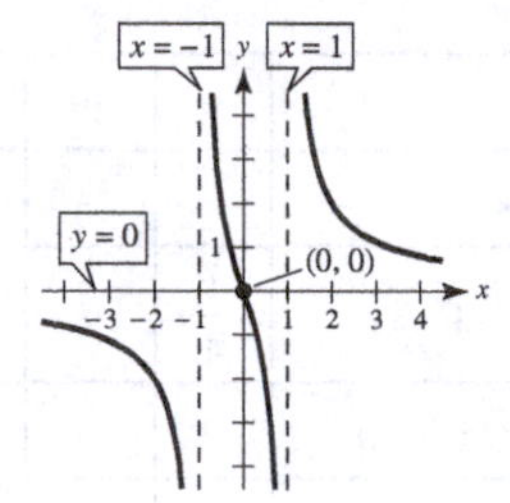

19. $y = \dfrac{x^2 - 6x + 12}{x - 4} = x - 2 + \dfrac{4}{x - 4}$

$y' = 1 - \dfrac{4}{(x-4)^2} = \dfrac{(x-2)(x-6)}{(x-4)^2} = 0$ when $x = 2, 6$ and is undefined when $x = 4$.

$y'' = \dfrac{8}{(x-4)^3}$

Vertical asymptote: $x = 4$

Slant asymptote: $y = x - 2$

	y	y'	y''	Conclusion
$-\infty < x < 2$		$+$	$-$	Increasing, concave down
$x = 2$	-2	0	$-$	Relative maximum
$2 < x < 4$		$-$	$-$	Decreasing, concave down
$4 < x < 6$		$-$	$+$	Decreasing, concave up
$x = 6$	6	0	$+$	Relative minimum
$6 < x < \infty$		$+$	$+$	Increasing, concave up

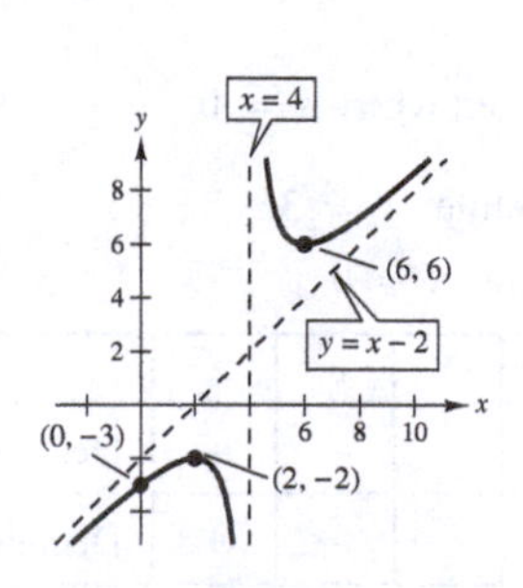

21. $y = \dfrac{x^3}{\sqrt{x^2 - 4}}$, Domain: $(-\infty, -2), (2, \infty)$

$y' = \dfrac{2x^2(x^2 - 6)}{(x^2 - 4)^{3/2}} = 0$ when $x = 0, \pm\sqrt{6}$, undefined when $x = \pm 2$

$y'' = \dfrac{2x^5 - 20x^3 + 96x}{(x^2 - 4)^{5/2}} = 0$ when $x = 0$, undefined when $x = \pm 2$

Vertical asymptotes: $x = \pm 2$

Symmetric with respect to the origin

	y	y'	y''	Conclusion
$-\infty < x < -\sqrt{6}$		+	−	Increasing, concave down
$x = -\sqrt{6}$	$-6\sqrt{3}$	0	−	Relative maximum
$-\sqrt{6} < x < -2$		−	−	Decreasing, concave down
$2 < x < \sqrt{6}$		−	+	Decreasing, concave up
$x = \sqrt{6}$	$6\sqrt{3}$	0	+	Relative minimum
$\sqrt{6} < x < \infty$		+	+	Increasing, concave up

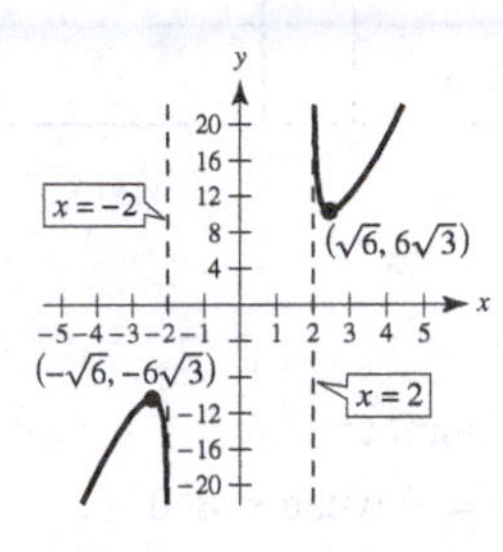

23. $y = x\sqrt{4 - x}$, Domain: $(-\infty, 4]$

$y' = \dfrac{8 - 3x}{2\sqrt{4 - x}} = 0$ when $x = \dfrac{8}{3}$ and undefined when $x = 4$.

$y'' = \dfrac{3x - 16}{4(4 - x)^{3/2}} = 0$ when $x = \dfrac{16}{3}$ and undefined when $x = 4$.

Note: $x = \dfrac{16}{3}$ is not in the domain.

	y	y'	y''	Conclusion
$-\infty < x < \frac{8}{3}$		+	−	Increasing, concave down
$x = \frac{8}{3}$	$\frac{16}{3\sqrt{3}}$	0	−	Relative maximum
$\frac{8}{3} < x < 4$		−	−	Decreasing, concave down
$x = 4$	0	Undefined	Undefined	Endpoint

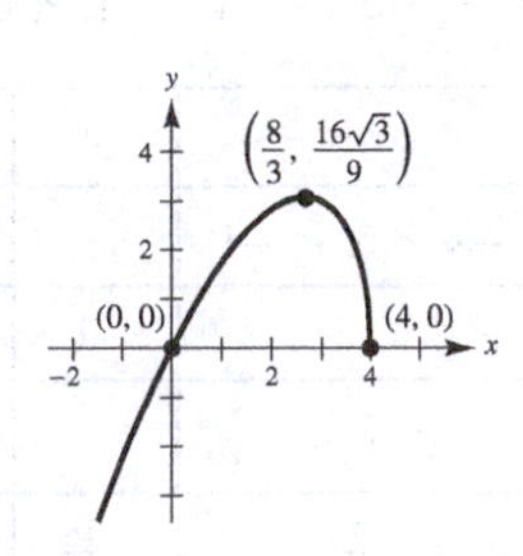

25. $y = 3x^{2/3} - 2x$

$y' = 2x^{-1/3} - 2 = \dfrac{2(1 - x^{1/3})}{x^{1/3}} = 0$ when $x = 1$ and undefined when $x = 0$.

$y'' = \dfrac{-2}{3x^{4/3}} < 0$ when $x \neq 0$.

	y	y'	y''	Conclusion
$-\infty < x < 0$		$-$	$-$	Decreasing, concave down
$x = 0$	0	Undefined	Undefined	Relative minimum
$0 < x < 1$		$+$	$-$	Increasing, concave down
$x = 1$	1	0	$-$	Relative maximum
$1 < x < \infty$		$-$	$-$	Decreasing, concave down

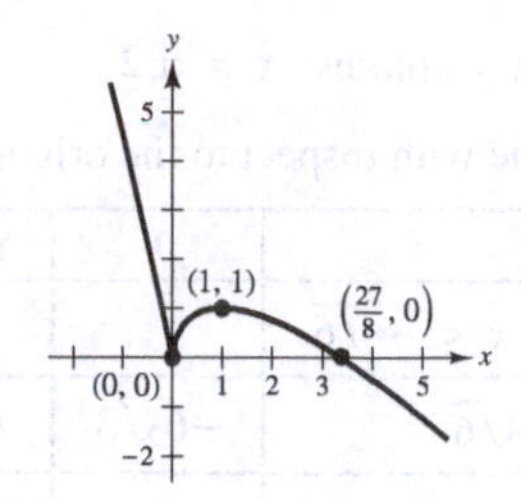

27. $y = 2 - x - x^3$

$y' = -1 - 3x^2$

No critical numbers

$y'' = -6x = 0$ when $x = 0$.

	y	y'	y''	Conclusion
$-\infty < x < 0$		$-$	$+$	Decreasing, concave up
$x = 0$	2	$-$	0	Point of inflection
$0 < x < \infty$		$-$	$-$	Decreasing, concave down

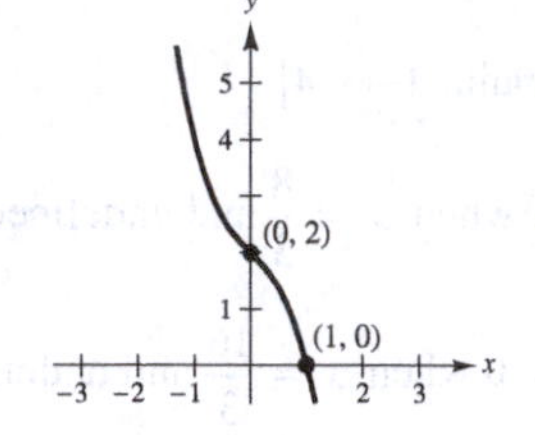

29. $y = 3x^4 + 4x^3$

$y' = 12x^3 + 12x^2 = 12x^2(x + 1) = 0$ when $x = 0, x = -1$.

$y'' = 36x^2 + 24x = 12x(3x + 2) = 0$ when $x = 0, x = -\frac{2}{3}$.

	y	y'	y''	Conclusion
$-\infty < x < -1$		$-$	$+$	Decreasing, concave up
$x = -1$	-1	0	$+$	Relative minimum
$-1 < x < -\frac{2}{3}$		$+$	$+$	Increasing, concave up
$x = -\frac{2}{3}$	$-\frac{16}{27}$	$+$	0	Point of inflection
$-\frac{2}{3} < x < 0$		$+$	$-$	Increasing, concave down
$x = 0$	0	0	0	Point of inflection
$0 < x < \infty$		$+$	$+$	Increasing, concave up

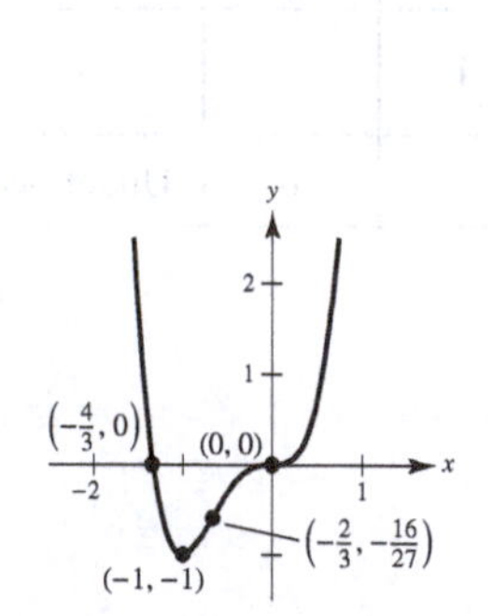

31. $xy^2 = 9$

$y^2 = \dfrac{9}{x}$

$y = \pm\dfrac{3}{\sqrt{x}}, x > 0$

Horizontal asymptote: $y = 0$

Vertical asymptote: $x = 0$

Symmetric with respect to x-axis

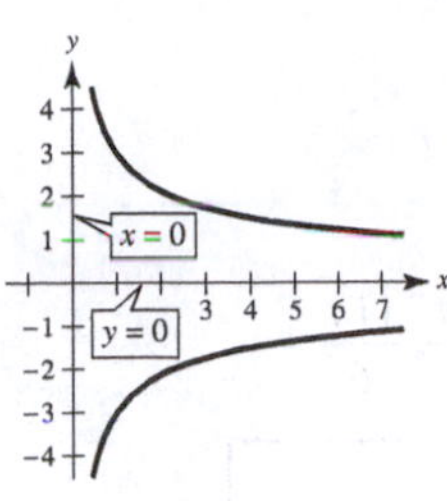

33. $y = |2x - 3|$

$y' = \dfrac{2(2x - 3)}{|2x - 3|}$ undefined at $x = \dfrac{3}{2}$.

$y'' = 0$

	y	y'	Conclusion
$-\infty < x < \frac{3}{2}$		$-$	Decreasing
$x = \frac{3}{2}$	0	Undefined	Relative minimum
$\frac{3}{2} < x < \infty$		$+$	Increasing

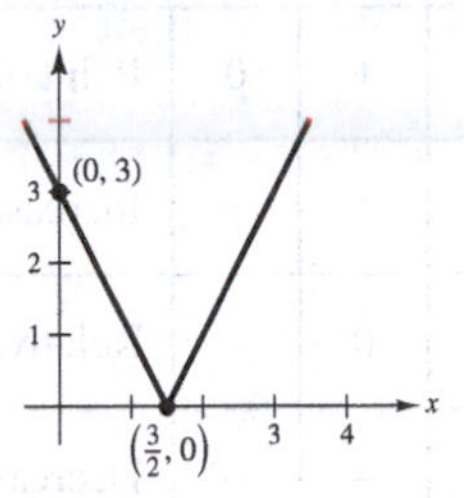

35. $f(x) = 2x - 4\sin x, 0 \le x \le 2\pi$

$f'(x) = 2 - 4\cos x = 0$ when $\cos x = \dfrac{1}{2} \Rightarrow x = \dfrac{\pi}{3}, \dfrac{5\pi}{3}$

$f''(x) = 4\sin x = 0$ when $x = 0, \pi, 2\pi$

	$f(x)$	$f'(x)$	$f''(x)$	Conclusion
$x = 0$	0	$-$	0	Endpoint, point of inflection
$0 < x < \frac{\pi}{3}$		$-$	$+$	Decreasing, concave up
$x = \frac{\pi}{3}$	$\frac{2\pi}{3} - 2\sqrt{3}$	0	$+$	Relative minimum
$\frac{\pi}{3} < x < \pi$		$+$	$+$	Increasing, concave up
$x = \pi$	2π	$+$	0	Point of inflection
$\pi < x < \frac{5\pi}{3}$		$+$	$-$	Increasing, concave down
$x = \frac{5\pi}{3}$	$\frac{10\pi}{3} + 2\sqrt{3}$	0	$-$	Relative maximum
$\frac{5\pi}{3} < x < 2\pi$		$-$	$-$	Decreasing, concave down
$x = 2\pi$	4π	$-$	0	Endpoint, point of inflection

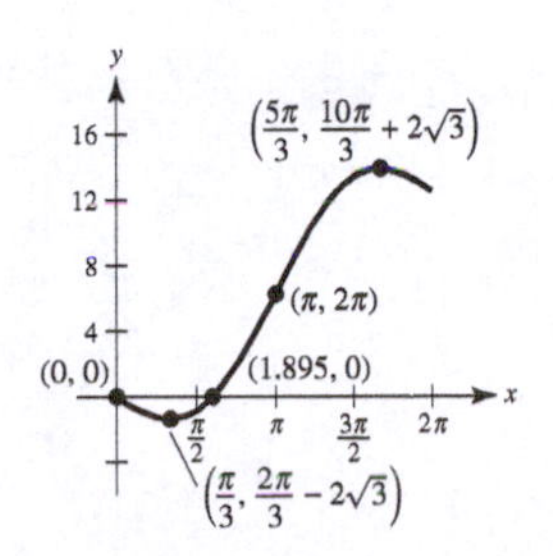

37. $y = \sin x - \frac{1}{18}\sin 3x,\ 0 \le x \le 2\pi$

$y' = \cos x\left(\frac{5}{6} + \frac{2}{3}\sin^2 x\right) = 0$ when $x = \frac{\pi}{2}, \frac{3\pi}{2}$

$y'' = \sin x\left(4\cos^2 x - 1\right) = 0$ when $x = 0, \pi, 2\pi, \frac{\pi}{6}, \frac{5\pi}{6}, \frac{7\pi}{6}, \frac{11\pi}{6}$

	y	y'	y''	Conclusion
$x = 0$	0	+	0	Endpoint, point of inflection
$0 < x < \frac{\pi}{6}$		+	+	Increasing, concave up
$x = \frac{\pi}{6}$	$\frac{4}{9}$	+	0	Point of inflection
$\frac{\pi}{6} < x < \frac{\pi}{2}$		+	−	Increasing, concave down
$x = \frac{\pi}{2}$	$\frac{19}{18}$	0	−	Relative maximum
$\frac{\pi}{2} < x < \frac{5\pi}{6}$		−	−	Decreasing, concave down
$x = \frac{5\pi}{6}$	$\frac{4}{9}$	−	0	Point of inflection
$\frac{5\pi}{6} < x < \pi$		−	+	Decreasing, concave up
$x = \pi$	0	−	0	Point of inflection
$\pi < x < \frac{7\pi}{6}$		−	−	Decreasing, concave down
$x = \frac{7\pi}{6}$	$-\frac{4}{9}$	−	0	Point of inflection
$\frac{7\pi}{6} < x < \frac{3\pi}{2}$		−	+	Decreasing, concave up
$x = \frac{3\pi}{2}$	$-\frac{19}{18}$	0	+	Relative minimum
$\frac{3\pi}{2} < x < \frac{11\pi}{6}$		+	+	Increasing, concave up
$x = \frac{11\pi}{6}$	$-\frac{4}{9}$	+	0	Point of inflection
$\frac{11\pi}{6} < x < 2\pi$		+	−	Increasing, concave down
$x = 2\pi$	0	+	0	Endpoint, point of inflection

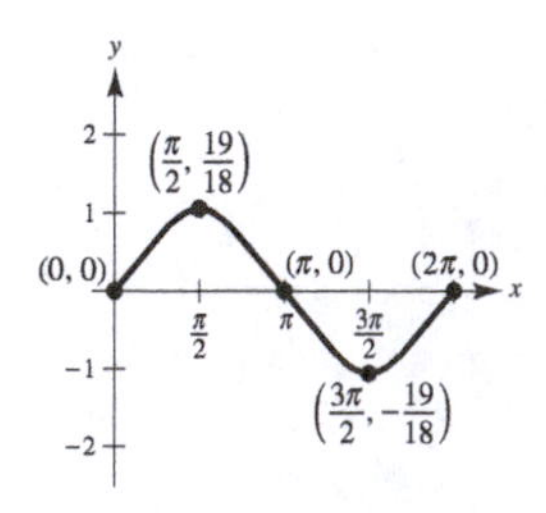

39. $y = 2(\csc x + \sec x),\ 0 < x < \dfrac{\pi}{2}$

$y' = 2(\sec x \tan x - \csc x \cot x) = 0$ when $x = \dfrac{\pi}{4}$

$y'' = 2(\sec x \tan^2 x + \sec^3 x + \csc^3 x + \cot^2 x \csc x) \neq 0$ in the given interval

	y	y'	y''	Conclusion
$x = 0$	Undef.			Vertical asymptote
$0 < x < \dfrac{\pi}{4}$		$-$	$+$	Decreasing, concave up
$x = \dfrac{\pi}{4}$	$4\sqrt{2}$	0	$+$	Relative minimum
$\dfrac{\pi}{4} < x < \dfrac{\pi}{2}$		$+$	$+$	Increasing, concave up
$x = \dfrac{\pi}{2}$	Undef.			Vertical asymptote

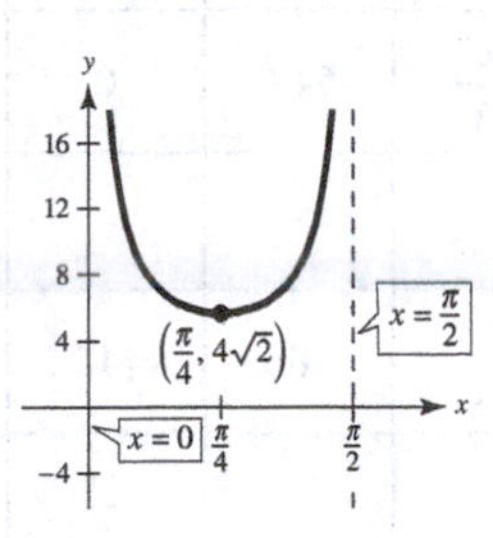

41. $g(x) = x \tan x,\ -\dfrac{3\pi}{2} < x < \dfrac{3\pi}{2}$

$g'(x) = \dfrac{x + \sin x \cos x}{\cos^2 x} = 0$ when $x = 0$

$g''(x) = \dfrac{2(\cos x + x \sin x)}{\cos^3 x} = 0$ when $x \approx \pm 2.798$

	$g(x)$	$g'(x)$	$g''(x)$	Conclusion
$x = -\dfrac{3\pi}{2}$	Undef.			Vertical asymptote
$-\dfrac{3\pi}{2} < x < -2.798$		$-$	$+$	Decreasing, concave up
$x \approx -2.798$	-1.001	$-$	0	Point of inflection
$-2.798 < x < -\dfrac{\pi}{2}$		$-$	$-$	Decreasing, concave down
$x = -\dfrac{\pi}{2}$	Undef.			Vertical asymptote
$-\dfrac{\pi}{2} < x < 0$		$-$	$+$	Decreasing, concave up
$x = 0$	0	0	$+$	Relative minimum
$0 < x < \dfrac{\pi}{2}$		$+$	$+$	Increasing, concave up
$\dfrac{\pi}{2} < x < 2.798$		$+$	$-$	Increasing, concave down
$x = 2.798$	-1.001	$+$	0	Point of inflection
$2.798 < x < \dfrac{3\pi}{2}$		$+$	$+$	Increasing, concave up
$x = \dfrac{3\pi}{2}$	Undef.			Point of inflection

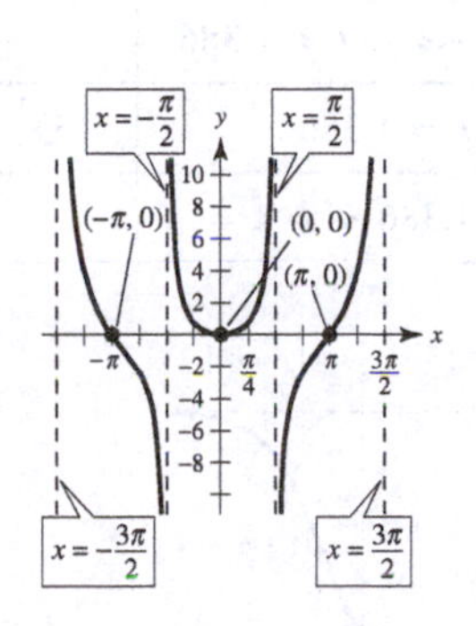

43. $f(x) = e^{3x}(2 - x)$

$f'(x) = -e^{3x} + 2(2 - x)e^{3x} = e^{3x}(5 - 3x) = 0$ when $x = \frac{5}{3}$.

$f''(x) = -3e^{3x}(-4 + 3x) = 0$ when $x = \frac{4}{3}$.

	$f(x)$	$f'(x)$	$f''(x)$	Conclusion
$-\infty < x < \frac{4}{3}$		+	+	Increasing, concave up
$x = \frac{4}{3}$	$\frac{2e^4}{3}$	54.6	0	Point of inflection
$\frac{4}{3} < x < \frac{5}{3}$		+	−	Increasing, concave down
$x = \frac{5}{3}$	$\frac{e^5}{3}$	0	−445.2	Relative maximum
$\frac{5}{3} < x < \infty$		−	−	Decreasing, concave down

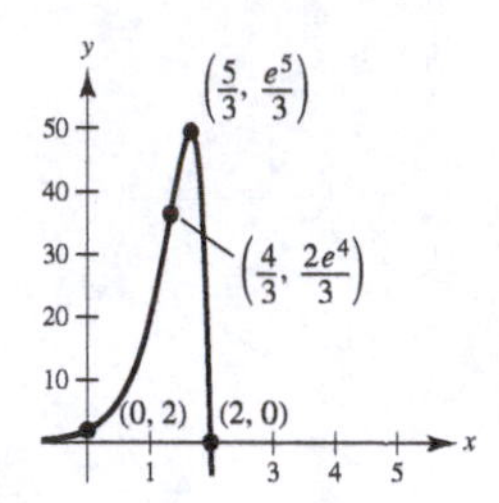

45. $g(t) = \dfrac{10}{1 + 4e^{-t}}$

$g'(t) = \dfrac{40e^{-t}}{\left(1 + 4e^{-t}\right)^2} > 0$ for all t.

$g''(t) = \dfrac{40e^{-t}\left(4e^{-t} - 1\right)}{\left(1 + 4e^{-t}\right)^3} = 0$ at $t \approx 1.386$.

$\lim_{t\to\infty} g(t) = 10 \Rightarrow t = 10$ is a horizontal asymptote.

$\lim_{t\to-\infty} g(t) = 0 \Rightarrow t = 0$ is a horizontal asymptote.

	$g(t)$	$g'(t)$	$g''(t)$	Conclusion
$-\infty < t < 1.386$		+	+	Increasing, concave up
$t = 1.386$	5	2.5	0	Point of inflection
$1.386 < t < \infty$		+	−	Increasing, concave down

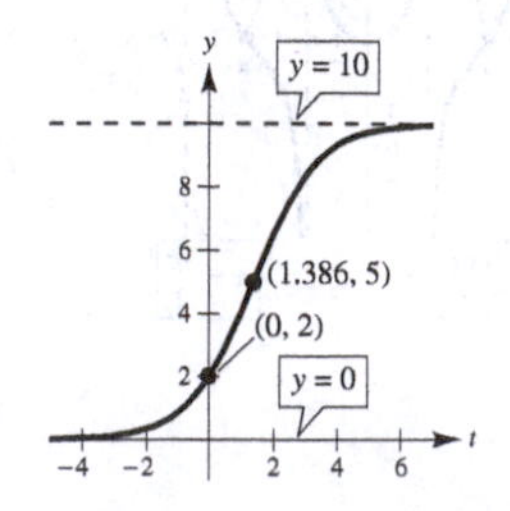

47. $y = (x - 1)\ln(x - 1)$, Domain: $x > 1$

$y' = 1 + \ln(x - 1) = 0$ when $\ln(x - 1) = -1 \Rightarrow (x - 1) = e^{-1} \Rightarrow x = 1 + e^{-1}$

$$y'' = \frac{1}{x - 1}$$

	y	y'	y''	Conclusion
$1 < x < 1 + e^{-1}$		$-$	$+$	Decreasing, concave up
$x = 1 + e^{-1}$	$-e^{-1}$	0	e	Relative minimum
$1 + e^{-1} < x < \infty$		$+$	$+$	Increasing, concave up

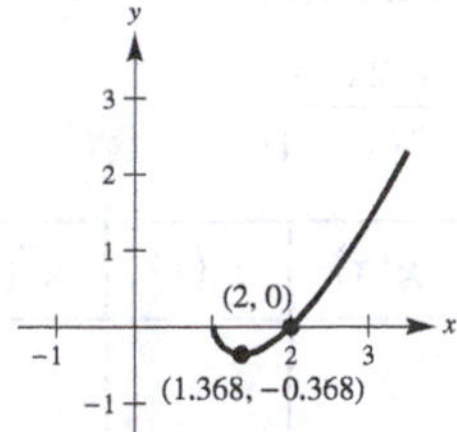

49. $g(x) = 6\arcsin\left(\frac{x - 2}{2}\right)^2$, Domain: $[0, 4]$

$$g'(x) = \frac{12(x - 2)}{\sqrt{(4x - x^2)(x^2 - 4x + 8)}} = 0 \text{ when } x = 2.$$

$$g''(x) = \frac{12(x^4 - 8x^3 + 24x^2 - 32x + 32)}{\left[(4x - x^2)(x^2 - 4x + 8)\right]^{3/2}}$$

	$g(x)$	$g'(x)$	$g''(x)$	Conclusion
$0 < x < 2$		$-$	$+$	Decreasing, concave up
$x = 2$	0	0	$+$	Relative minimum
$2 < x < 4$		$+$	$+$	Increasing, concave up

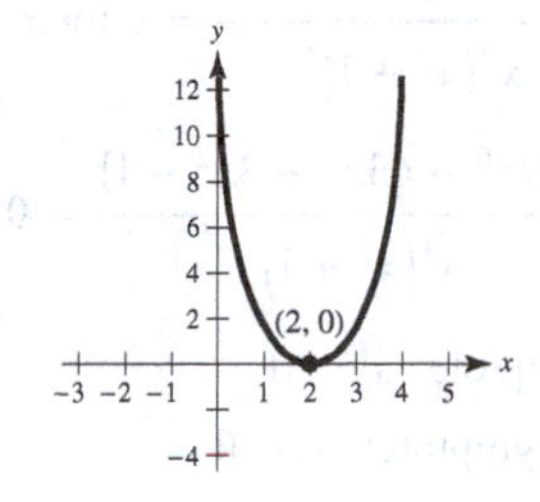

51. $f(x) = \frac{x}{3^{x-3}} = \frac{27x}{3^x}$

$$f'(x) = \frac{27(1 - x\ln 3)}{3^x} = 0 \Rightarrow x = \frac{1}{\ln 3} \approx 0.910$$

$$f''(x) = \frac{27\ln 3(x\ln 3 - 2)}{3^x} = 0 \Rightarrow x = \frac{2}{\ln 3} \approx 1.820$$

$\lim_{x\to\infty} f(x) = 0, \ \lim_{x\to-\infty} f(x) = -\infty$

Horizontal symptote: $y = 0$

Intercept: $(0, 0)$

	$f(x)$	$f'(x)$	$f''(x)$	Conclusion
$-\infty < x < 0.910$		$+$	$-$	Increasing, concave down
$x = 0.910$	9.041	0	$-$	Relative maximum
$0.910 < x < 1.820$		$-$	$-$	Decreasing, concave down
$x = 1.820$		$-$	0	Point of inflection
$1.820 < x < \infty$	6.652	$-$	$+$	Decreasing, concave up

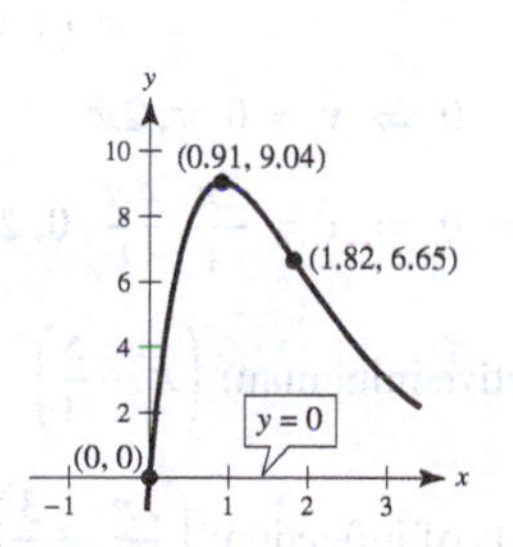

53. $g(x) = \log_4(x - x^2) = \dfrac{\ln(x - x^2)}{\ln 4}$, Domain: $0 < x < 1$

$g'(x) = \dfrac{2x - 1}{\ln 4 \cdot x(x - 1)} = 0$ when $x = \dfrac{1}{2}$.

$g''(x) = \dfrac{-2x^2 + 2x - 1}{\ln 4 \cdot x^2(x - 1)^2}$

	$g(x)$	$g'(x)$	$g''(x)$	Conclusion
$0 < x < \frac{1}{2}$		$+$	$-$	Increasing, concave down
$x = \frac{1}{2}$	-1	0	$-$	Relative maximum
$\frac{1}{2} < x < 1$		$-$	$-$	Decreasing, concave down

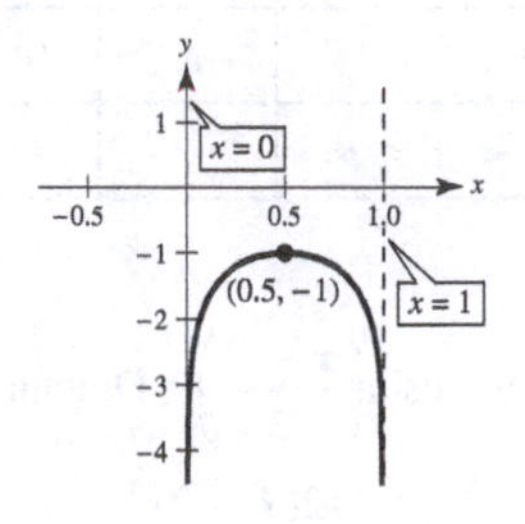

55. $f(x) = \dfrac{20x}{x^2 + 1} - \dfrac{1}{x} = \dfrac{19x^2 - 1}{x(x^2 + 1)}$

$f'(x) = \dfrac{-(19x^4 - 22x^2 - 1)}{x^2(x^2 + 1)^2} = 0$ for $x \approx \pm 1.10$

$f''(x) = \dfrac{2(19x^6 - 63x^9 - 3x^2 - 1)}{x^3(x^2 + 1)^3} = 0$ for $x \approx \pm 1.84$

Vertical asymptote: $x = 0$

Horizontal asymptote: $y = 0$

Minimum: $(-1.10, -9.05)$

Maximum: $(1.10, 9.05)$

Points of inflection: $(-1.84, -7.86)$, $(1.84, 7.86)$

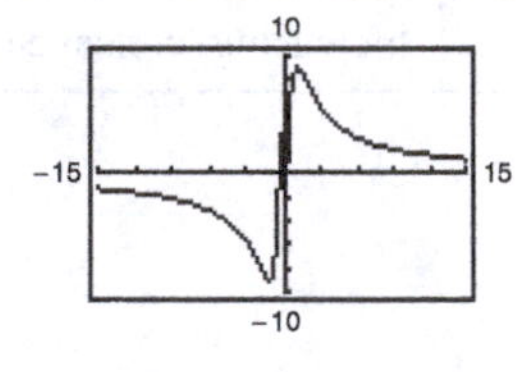

57. $y = \cos x - \dfrac{1}{4}\cos 2x,\ 0 \le x \le 2\pi$

$y = 0$ at $x \approx 1.797, 4.486$

$$y = -\sin x + \frac{1}{2}\sin 2x = -\sin x + \sin x \cos x$$
$$= \sin x(\cos x - 1)$$

$$y'' = -\cos x + \cos 2x = -\cos x + 2\cos^2 x - 1$$
$$= (2\cos x + 1)(\cos x - 1)$$

$y' = 0 \Rightarrow x = 0, \pi, 2\pi$

$y'' = 0 \Rightarrow x = \dfrac{2\pi}{3}, \dfrac{4\pi}{3}, 0, 2\pi$

Relative minimum: $\left(\pi, -\dfrac{5}{4}\right)$

Points of inflection: $\left(\dfrac{2\pi}{3}, -\dfrac{3}{8}\right), \left(\dfrac{4\pi}{3}, -\dfrac{3}{8}\right)$

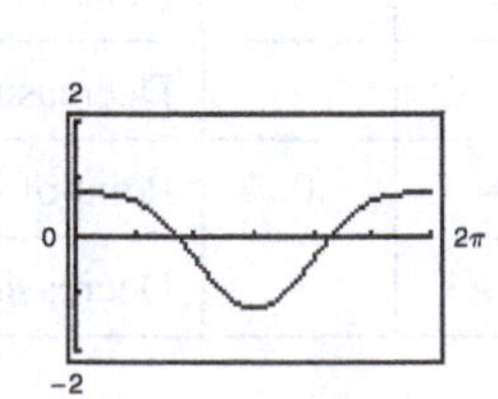

59. $y = \dfrac{x}{2} + \ln\left(\dfrac{x}{x+3}\right)$

$y' = \dfrac{1}{2} + \dfrac{3}{x(x+3)}$

$y'' = \dfrac{-3(2x+3)}{x^2(x+3)^2}$

Vertical asymptotes: $x = -3, x = 0$

Slant asymptote: $y = \dfrac{x}{2}$

61. $f(x) = 2 + (x^2 - 3)e^{-x}$

$f'(x) = -e^{-x}(x+1)(x-3)$

Critical numbers: $x = -1, x = 3$

Relative minimum: $(-1, 2 - 2e) \approx (-1, -3.4366)$

Relative maximum: $(3, 2 + 6e^{-3}) \approx (3, 2.2987)$

Horizontal asymptote: $y = 2$

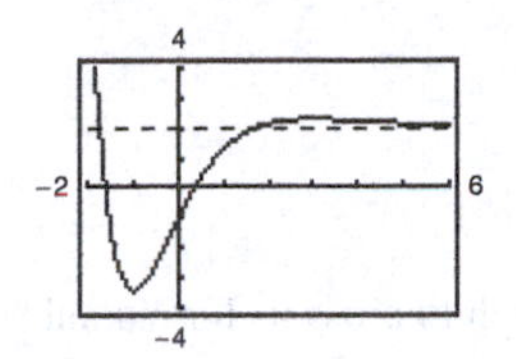

63. f is cubic.

f' is quadratic.

f'' is linear.

The zeros of f' correspond to the points where the graph of f has horizontal tangents. The zero of f'' corresponds to the point where the graph of f' has a horizontal tangent.

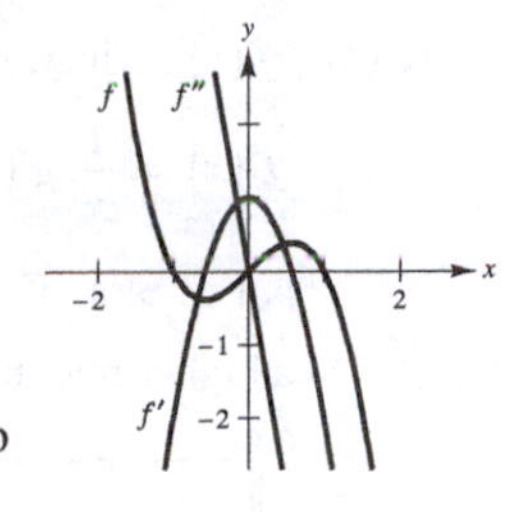

65.

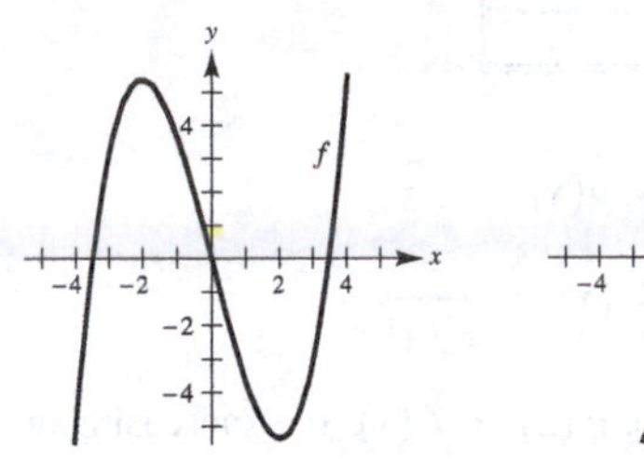

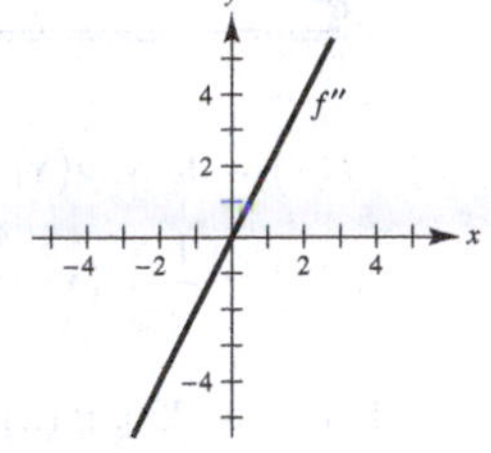

(or any vertical translation of f)

67.

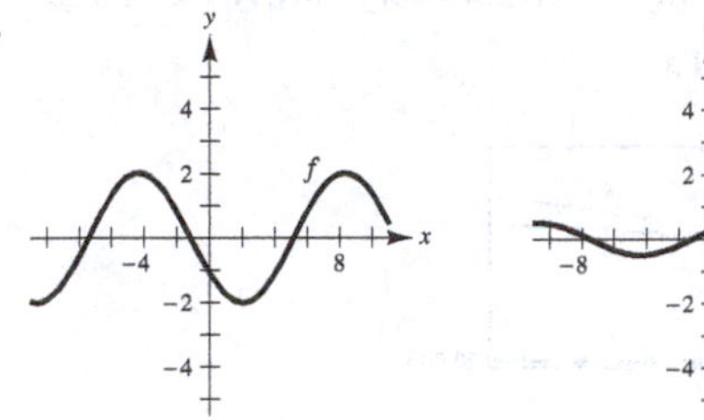

(or any vertical translation of f)

69. $f(x) = \dfrac{\cos^2 \pi x}{\sqrt{x^2+1}}, (0, 4)$

(a)

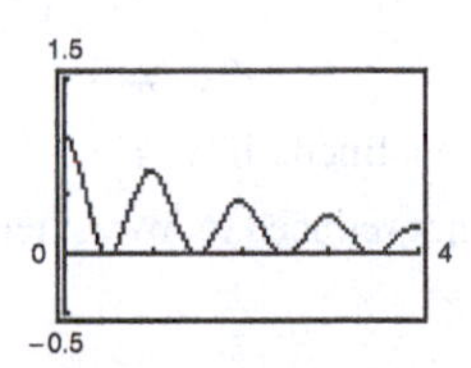

On $(0, 4)$ there seem to be 7 critical numbers: 0.5, 1.0, 1.5, 2.0, 2.5, 3.0, 3.5

(b) $f'(x) = \dfrac{-\cos \pi x(x \cos \pi x + 2\pi(x^2+1)\sin \pi x)}{(x^2+1)^{3/2}} = 0$

Critical numbers $\approx \dfrac{1}{2}, 0.97, \dfrac{3}{2}, 1.98, \dfrac{5}{2}, 2.98, \dfrac{7}{2}$.

The critical numbers where maxima occur appear to be integers in part (a), but approximating them using f' shows that they are not integers.

71. (a) $f(x) = \ln x, g(x) = \sqrt{x}$

$f'(x) = \dfrac{1}{x}, g'(x) = \dfrac{1}{2\sqrt{x}}$

For $x > 4$, $g'(x) > f'(x)$. g is increasing at a higher rate than f for "large" values of x.

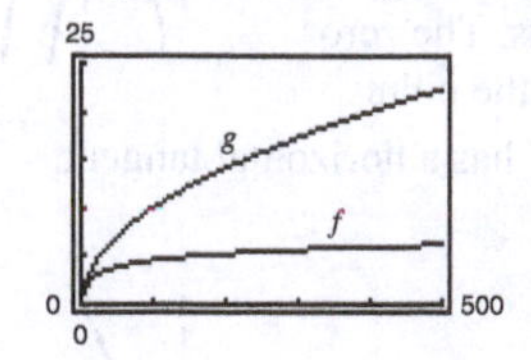

(b) $f(x) = \ln x, g(x) = \sqrt[4]{x}$

$f'(x) = \dfrac{1}{x}, g'(x) = \dfrac{1}{4\sqrt[4]{x^3}}$

For $x > 256$, $g'(x) > f'(x)$. g is increasing at a higher rate than f for "large" values of x. $f(x) = \ln x$ increases very slowly for "large" values of x.

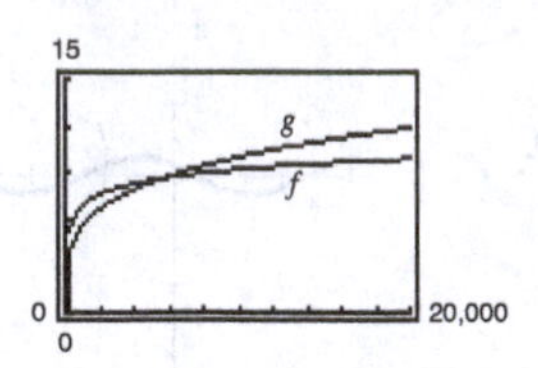

73. $x = 2$ is a critical number.

$f'(x) < 0$ for $x < 2$.

$f'(x) > 0$ for $x > 2$.

$\lim\limits_{x\to-\infty} f(x) = \lim\limits_{x\to\infty} f(x) = 6$

For example, let

$$f(x) = \frac{-6}{0.1(x-2)^2 + 1} + 6.$$

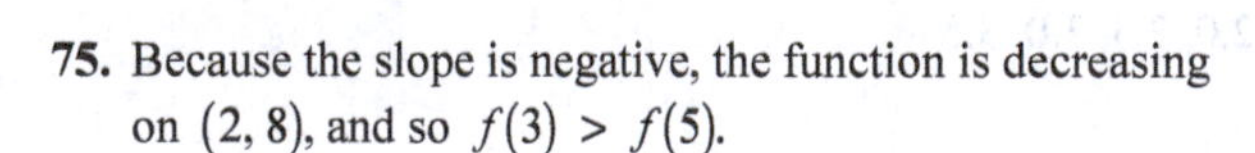

75. Because the slope is negative, the function is decreasing on $(2, 8)$, and so $f(3) > f(5)$.

77. (a)

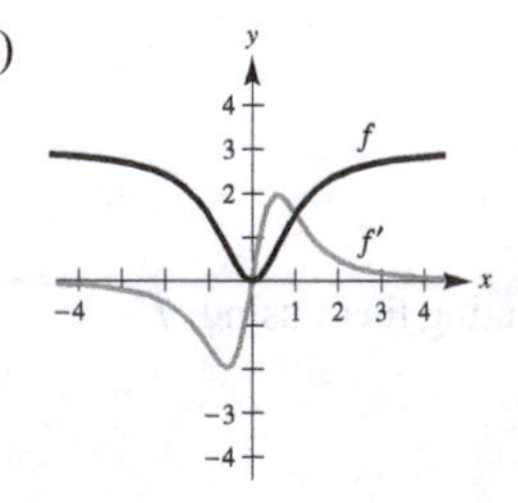

(b) $\lim\limits_{x\to\infty} f(x) = 3 \qquad \lim\limits_{x\to\infty} f'(x) = 0$

(c) Because $\lim\limits_{x\to\infty} f(x) = 3$, the graph approaches that of a horizontal line, $\lim\limits_{x\to\infty} f'(x) = 0$.

79. $f(x) = \dfrac{4(x-1)^2}{x^2 - 4x + 5}$

Vertical asymptote: none

Horizontal asymptote: $y = 4$

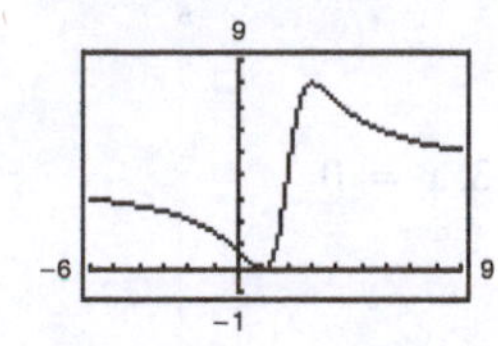

The graph crosses the horizontal asymptote $y = 4$. If a function has a vertical asymptote at $x = c$, the graph would not cross it because $f(c)$ is undefined.

81. $h(x) = \dfrac{\sin 2x}{x}$

Vertical asymptote: none

Horizontal asymptote: $y = 0$

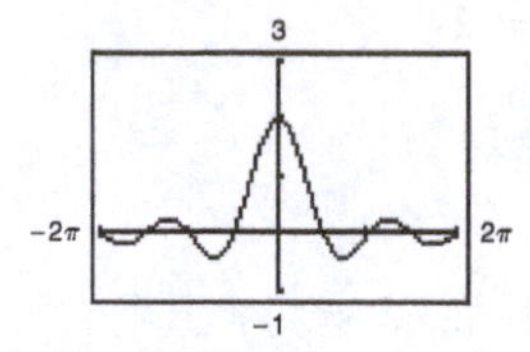

Yes, it is possible for a graph to cross its horizontal asymptote.

It is not possible to cross a vertical asymptote because the function is not continuous there.

83. $h(x) = \dfrac{6 - 2x}{3 - x}$

$$= \frac{2(3-x)}{3-x} = \begin{cases} 2, & \text{if } x \neq 3 \\ \text{Undefined}, & \text{if } x = 3 \end{cases}$$

The rational function is not reduced to lowest terms.

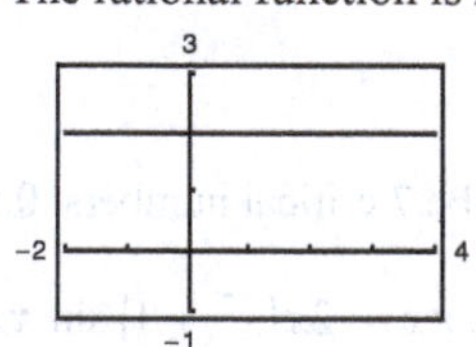

There is a hole at $(3, 2)$.

85. $x_n - \dfrac{F(x_n)}{F'(x_n)}$

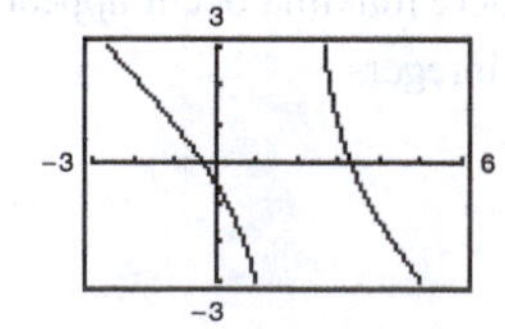

The graph appears to approach the slant asymptote $y = -x + 1$.

87. $f(x) = \dfrac{2x^3}{x^2 + 1} = 2x - \dfrac{2x}{x^2 + 1}$

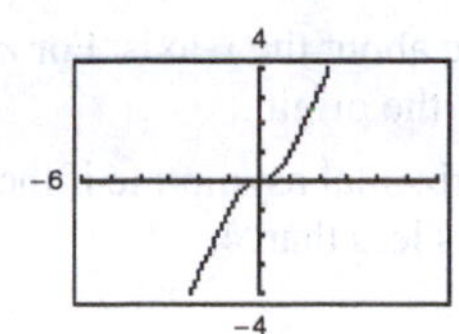

The graph appears to approach the slant asymptote $y = 2x$.

89. $f(x) = \dfrac{x^3 - 3x^2 + 2}{x(x - 3)} = x + \dfrac{2}{x(x - 3)}$

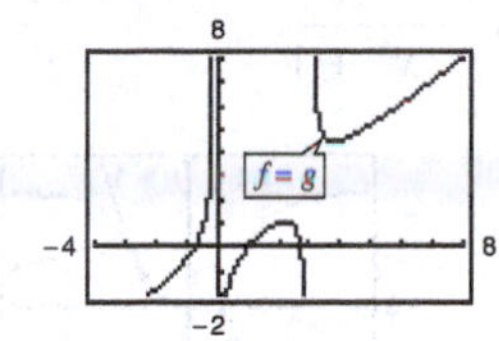

The graph appears to approach the slant asymptote $y = x$.

91. Tangent line at P: $y - y_0 = f'(x_0)(x - x_0)$

(a) Let $y = 0$: $-y_0 = f'(x_0)(x - x_0)$

$$f'(x_0)x = x_0 f'(x_0) - y_0$$

$$x = x_0 - \frac{y_0}{f'(x_0)} = x_0 - \frac{f(x_0)}{f'(x_0)}$$

x-intercept: $\left(x_0 - \dfrac{f(x_0)}{f'(x_0)}, 0\right)$

(b) Let $x = 0$: $y - y_0 = f'(x_0)(-x_0)$

$$y = y_0 - x_0 f'(x_0)$$

$$y = f(x_0) - x_0 f'(x_0)$$

y-intercept: $\left(0, f(x_0) - x_0 f'(x_0)\right)$

(c) Normal line: $y - y_0 = -\dfrac{1}{f'(x_0)}(x - x_0)$

Let $y = 0$: $-y_0 = -\dfrac{1}{f'(x_0)}(x - x_0)$

$$-y_0 f'(x_0) = -x + x_0$$

$$x = x_0 + y_0 f'(x_0) = x_0 + f(x_0)f'(x_0)$$

x-intercept: $\left(x_0 + f(x_0)f'(x_0), 0\right)$

(d) Let $x = 0$: $y - y_0 = \dfrac{-1}{f'(x_0)}(-x_0)$

$$y = y_0 + \frac{x_0}{f'(x_0)}$$

y-intercept: $\left(0, y_0 + \dfrac{x_0}{f'(x_0)}\right)$

(e) $|BC| = \left|x_0 - \dfrac{f(x_0)}{f'(x_0)} - x_0\right| = \left|\dfrac{f(x_0)}{f'(x_0)}\right|$

(f) $|PC|^2 = y_0^2 + \left(\dfrac{f(x_0)}{f'(x_0)}\right) = \dfrac{f(x_0)^2 f'(x_0)^2 + f(x_0)^2}{f'(x_0)^2}$

$$|PC| = \left|\frac{f(x_0)\sqrt{1 + \left[f'(x_0)\right]^2}}{f'(x_0)}\right|$$

(g) $|AB| = \left|x_0 - \left(x_0 + f(x_0)f'(x_0)\right)\right| = \left|f(x_0)f'(x_0)\right|$

(h) $|AP|^2 = f(x_0)^2 f'(x_0)^2 + y_0^2$

$$|AP| = \left|f(x_0)\right|\sqrt{1 + \left[f'(x_0)\right]^2}$$

93. Vertical asymptote: $x = 3$
Horizontal asymptote: $y = 0$

$$y = \frac{1}{x - 3}$$

95. Vertical asymptote: $x = 3$

Slant asymptote: $y = 3x + 2$

$$y = 3x + 2 + \frac{1}{x - 3} = \frac{3x^2 - 7x - 5}{x - 3}$$

97. False. Let $f(x) = \dfrac{2x}{\sqrt{x^2 + 2}}$, $f'(x) < 0$ for all real numbers.

99. False. The graph of a rational function (having no common factors and whose denominator is of degree 1 or greater) has a slant asymptote only when the degree of the numerator exceeds the degree of the denominator by exactly 1.

101. (a) $f'(x) < 0$ when $-3 < x < 1$.

So, f is decreasing on the interval $(-3, 1)$.

(b) The slope of the graph of f' is negative for $-7 < x < -1$.

So, the graph of f is concave downward on the interval $(-7, -1)$.

(c) $f'(x) = 0$ when $x = -3$ and $x = 1$.

Because f is decreasing on the interval $(-3, 1)$, f has a relative maximum at $x = -3$ and a relative minimum at $x = 1$.

(d) The slope of the graph of f' is 0 when $x = -1$.

So, the graph of f has a point of inflection at $x = -1$.

103. $f(x) = \dfrac{ax}{(x-b)^2}$

Answers will vary. *Sample answer*: The graph has a vertical asymptote at $x = b$. If a and b are both positive, or both negative, then the graph of f approaches ∞ as x approaches b, and the graph has a minimum at $x = -b$. If a and b have opposite signs, then the graph of f approaches $-\infty$ as x approaches b, and the graph has a maximum at $x = -b$.

105. $y = \sqrt{4 + 16x^2}$

As $x \to \infty, y \to 4x$. As $x \to -\infty, y \to -4x$.

Slant asymptotes: $y = \pm 4x$

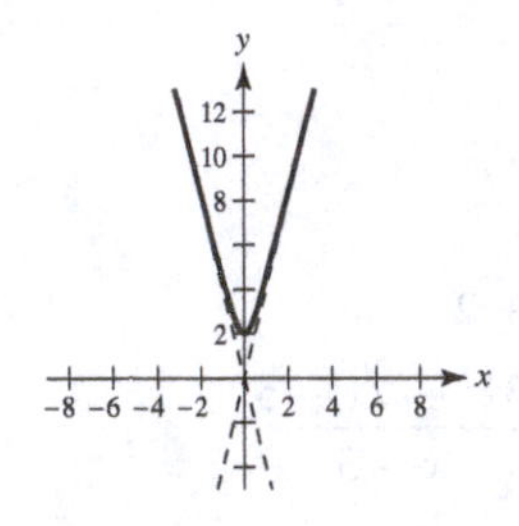

107. $f(x) = \dfrac{2x^n}{x^4 + 1}$

(a) For n even, f is symmetric about the y-axis. For n odd, f is symmetric about the origin.

(b) The x-axis will be the horizontal asymptote if the degree of the numerator is less than 4. That is, $n = 0, 1, 2, 3$.

(c) $n = 4$ gives $y = 2$ as the horizontal asymptote.

(d) There is a slant asymptote $y = 2x$ if $n = 5$: $\dfrac{2x^5}{x^4+1} = 2x - \dfrac{2x}{x^4+1}$.

(e)

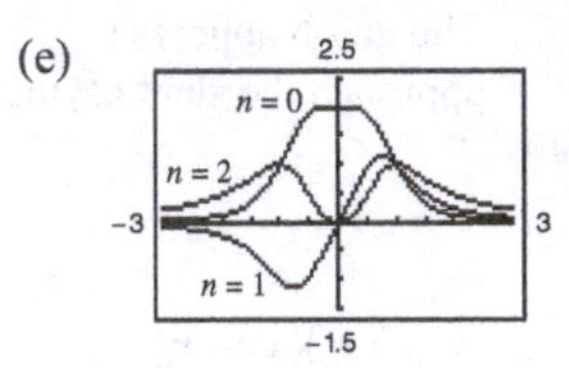

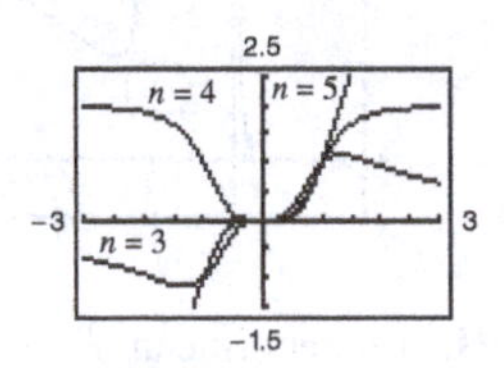

n	0	1	2	3	4	5
M	1	2	3	2	1	0
N	2	3	4	5	2	3

Section 4.7 Optimization Problems

1. A primary equation is a formula for the quantity to be optimized. A secondary equation can be solved for a variable and then substituted into the primary equation to obtain a function of a single variable. A feasible domain is the set of input values that make sense in an optimization problem.

3. (a)

First Number, x	Second Number	Product, P
10	110 – 10	10(110 – 10) = 1000
20	110 – 20	20(110 – 20) = 1800
30	110 – 30	30(110 – 30) = 2400
40	110 – 40	40(110 – 40) = 2800
50	110 – 50	50(110 – 50) = 3000
60	110 – 60	60(110 – 60) = 3000

(b) $P = x(110 - x) = 110x - x^2$

(c) $\dfrac{dP}{dx} = 110 - 2x = 0$ when $x = 55$.

$\dfrac{d^2P}{dx^2} = -2 < 0$

P is a maximum when $x = 110 - x = 55$.

(d)

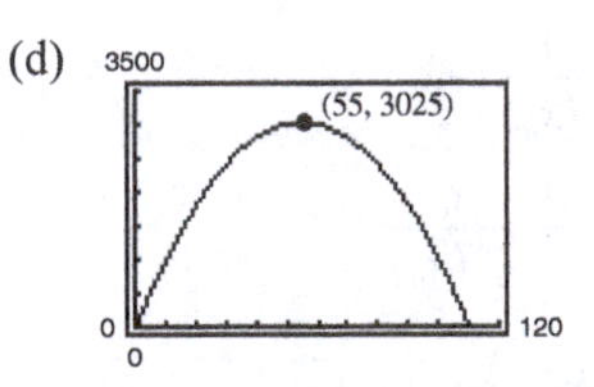

The solution appears to be $x = 55$.

5. Let x and y be two positive numbers such that $x + y = S$.

$$P = xy = x(S - x) = Sx - x^2$$

$$\frac{dP}{dx} = S - 2x = 0 \text{ when } x = \frac{S}{2}.$$

$$\frac{d^2P}{dx^2} = -2 < 0 \text{ when } x = \frac{S}{2}.$$

For $x = \frac{S}{2}, \frac{S}{2} + y = S \Rightarrow y = \frac{S}{2}.$

P is a maximum when $x = y = S/2$.

7. Let x and y be two positive numbers such that $xy = 147$.

$$S = x + 3y = \frac{147}{y} + 3y$$

$$\frac{dS}{dy} = 3 - \frac{147}{y^2} = 0 \text{ when } y = 7.$$

$$\frac{d^2S}{dy^2} = \frac{294}{y^3} > 0 \text{ when } y = 7.$$

For $y = 7$, $7x = 147 \Rightarrow x = 21$.

S is minimum when $y = 7$ and $x = 21$.

9. Let x and y be two positive numbers such that $x + 2y = 108$.

$$P = xy = y(108 - 2y) = 108y - 2y^2$$

$$\frac{dP}{dy} = 108 - 4y = 0 \text{ when } y = 27.$$

$$\frac{d^2P}{dy^2} = -4 < 0 \text{ when } y = 27.$$

For $y = 27, x + 2(27) = 108 \Rightarrow x = 54$.

P is a maximum when $x = 54$ and $y = 27$.

11. Let x be the length and y the width of the rectangle.

$$2x + 2y = 80$$

$$y = 40 - x$$

$$A = xy = x(40 - x) = 40x - x^2$$

$$\frac{dA}{dx} = 40 - 2x = 0 \text{ when } x = 20.$$

$$\frac{d^2A}{dx^2} = -2 < 0 \text{ when } x = 20.$$

A is maximum when $x = y = 20$ meters.

13. Let x be the length and y the width of the rectangle.

$$xy = 49 \Rightarrow y = \frac{49}{x}$$

$$P = 2x + 2y = 2x + 2\left(\frac{49}{x}\right) = 2x + \frac{98}{x}$$

$$\frac{dP}{dx} = 2 - 98x^{-2} = 0 \text{ when } x = 7.$$

$$\frac{d^2P}{dx^2} > 0 \text{ when } x = 7.$$

P is a minimum when $x = y = 7$ feet.

y

x

15. The distance from $(0, 3)$ to $y = x^2$ is

$$d = \sqrt{(x - 0)^2 + (y - 3)^2} = \sqrt{x^2 + (x^2 - 3)^2}.$$

Because d is smallest when d^2 is smallest, find the critical numbers of

$$f(x) = x^2 + (x^2 - 3)^2 = x^4 - 5x^2 + 9.$$

$$f'(x) = 4x^3 - 10x = 2x(2x^2 - 5) = 0 \Rightarrow x = \pm\sqrt{\tfrac{5}{2}}$$

By the First Derivative Test, the closest points are $\left(-\sqrt{\frac{5}{2}}, \frac{5}{2}\right)$ and $\left(\sqrt{\frac{5}{2}}, \frac{5}{2}\right)$.

17. $xy = 648 \Rightarrow y = \dfrac{648}{x}$

$$A = (x + 2)(y + 4)$$

$$= (x + 2)\left(\frac{648}{x} + 4\right)$$

$$= 648 + \frac{1296}{x} + 4x + 8$$

$$= 4x + 1296x^{-1} + 656$$

x + 2

x

y

y + 4

$$A'(x) = 4 - 1296x^{-2} = 4 - \frac{1296}{x^2} = 0 \Rightarrow 4x^2 = 1296 \Rightarrow x = 18$$

So, $y = \dfrac{648}{18} = 36$. By the First Derivative Test, these values yield a minimum.

The dimensions of the poster are 20 inches × 40 inches.

19. $xy = 405{,}000 \Rightarrow y = \dfrac{405{,}000}{x}$

$$S = x + 2y = x + 2\left(\frac{405{,}000}{x}\right)$$

$$S'(x) = 1 - 810{,}000x^{-2} = 0$$

$$x^2 = 810{,}000$$

$$x = 900$$

So, $y = \dfrac{405{,}000}{900} = 450$. By the First Derivative Test, these values yield a minimum.

The dimensions of the fence are 900 meters × 450 meters.

21. $16 = 2y + x + \pi\left(\dfrac{x}{2}\right)$

$$32 = 4y + 2x + \pi x$$

$$y = \frac{32 - 2x - \pi x}{4}$$

$$A = xy + \frac{\pi}{2}\left(\frac{x}{2}\right)^2 = \left(\frac{32 - 2x - \pi x}{4}\right)x + \frac{\pi x^2}{8} = 8x - \frac{1}{2}x^2 - \frac{\pi}{4}x^2 + \frac{\pi}{8}x^2$$

$$\frac{dA}{dx} = 8 - x - \frac{\pi}{2}x + \frac{\pi}{4}x = 8 - x\left(1 + \frac{\pi}{4}\right) = 0 \text{ when } x = \frac{8}{1 + (\pi/4)} = \frac{32}{4 + \pi}.$$

$$\frac{d^2A}{dx^2} = -\left(1 + \frac{\pi}{4}\right) < 0 \text{ when } x = \frac{32}{4 + \pi}.$$

$$y = \frac{32 - 2[32/(4 + \pi)] - \pi[32/(4 + \pi)]}{4} = \frac{16}{4 + \pi}$$

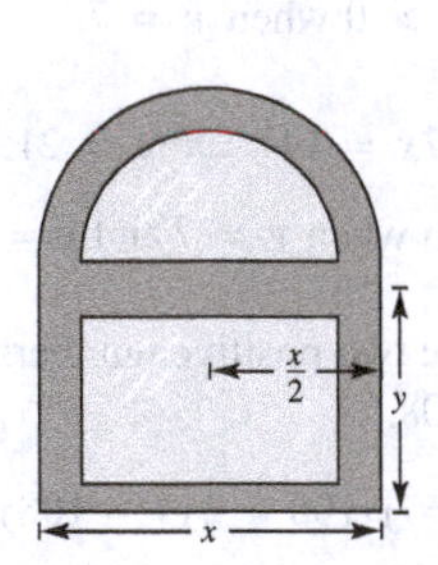

The area is maximum when $y = \dfrac{16}{4 + \pi}$ ft and $x = \dfrac{32}{4 + \pi}$ ft.

23. (a) $\dfrac{y - 2}{0 - 1} = \dfrac{0 - 2}{x - 1}$

$$y = 2 + \frac{2}{x - 1}$$

$$L = \sqrt{x^2 + y^2} = \sqrt{x^2 + \left(2 + \frac{2}{x - 1}\right)^2} = \sqrt{x^2 + 4 + \frac{8}{x - 1} + \frac{4}{(x - 1)^2}}, \quad x > 1$$

(b) L is minimum when $x \approx 2.587$ and $L \approx 4.162$.

(c) Area $= A(x) = \dfrac{1}{2}xy = \dfrac{1}{2}x\left(2 + \dfrac{2}{x - 1}\right) = x + \dfrac{x}{x - 1}$

$$A'(x) = 1 + \frac{(x - 1) - x}{(x - 1)^2} = 1 - \frac{1}{(x - 1)^2} = 0$$

$$(x - 1)^2 = 1$$

$$x - 1 = \pm 1$$

$$x = 0, 2 \text{ (select } x = 2)$$

They $y = 4$ and $A = 4$.

Vertices: $(0, 0), (2, 0), (0, 4)$

25. $A = 2xy = 2x\sqrt{25 - x^2}$ (see figure)

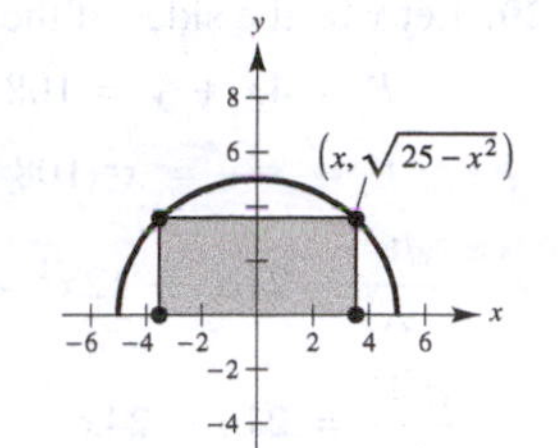

$$\frac{dA}{dx} = 2x\left(\frac{1}{2}\right)\left(\frac{-2x}{\sqrt{25 - x^2}}\right) + 2\sqrt{25 - x^2} = 2\left(\frac{25 - 2x^2}{\sqrt{25 - x^2}}\right) = 0 \text{ when } x = y = \frac{5\sqrt{2}}{2} \approx 3.54.$$

By the First Derivative Test, the inscribed rectangle of maximum area has vertices

$$\left(\pm\frac{5\sqrt{2}}{2}, 0\right), \left(\pm\frac{5\sqrt{2}}{2}, \frac{5\sqrt{2}}{2}\right).$$

Width: $\frac{5\sqrt{2}}{2}$; Length: $5\sqrt{2}$

27. (a) $P = 2x + 2\pi r = 2x + 2\pi\left(\frac{y}{2}\right) = 2x + \pi y = 200 \Rightarrow y = \frac{200 - 2x}{\pi} = \frac{2}{\pi}(100 - x)$

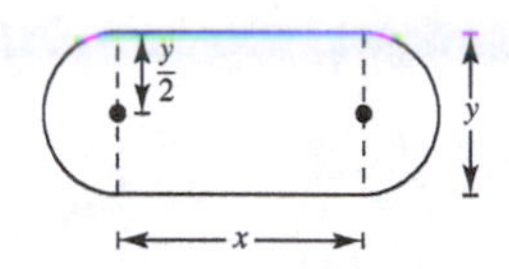

(b)

Length, x	Width, y	Area, xy
10	$\frac{2}{\pi}(100 - 10)$	$(10)\frac{2}{\pi}(100 - 10) \approx 573$
20	$\frac{2}{\pi}(100 - 20)$	$(20)\frac{2}{\pi}(100 - 20) \approx 1019$
30	$\frac{2}{\pi}(100 - 30)$	$(30)\frac{2}{\pi}(100 - 30) \approx 1337$
40	$\frac{2}{\pi}(100 - 40)$	$(40)\frac{2}{\pi}(100 - 40) \approx 1528$
50	$\frac{2}{\pi}(100 - 50)$	$(50)\frac{2}{\pi}(100 - 50) \approx 1592$
60	$\frac{2}{\pi}(100 - 60)$	$(60)\frac{2}{\pi}(100 - 60) \approx 1528$

The maximum area of the rectangle is approximately 1592 m^2.

(c) $A = xy = x\frac{2}{\pi}(100 - x) = \frac{2}{\pi}(100x - x^2)$

(d) $A' = \frac{2}{\pi}(100 - 2x)$. $A' = 0$ when $x = 50$.

Maximum value is approximately 1592 when length $= 50$ m and width $= \frac{100}{\pi}$.

(e)

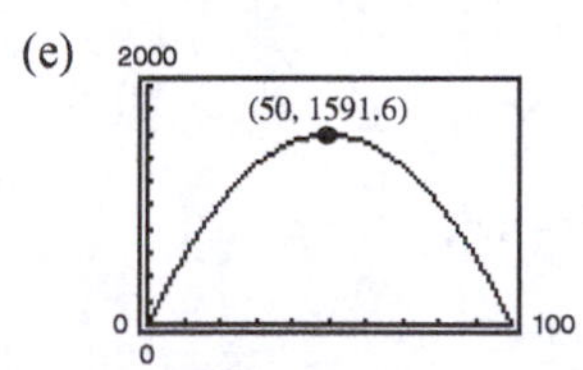

Maximum area is approximately 1591.55 m^2 ($x = 50$ m).

29. Let x be the sides of the square ends and y the length of the package.

$$P = 4x + y = 108 \Rightarrow y = 108 - 4x$$

$$V = x^2y = x^2(108 - 4x) = 108x^2 - 4x^3$$

$$\frac{dV}{dx} = 216x - 12x^2 = 12x(18 - x) = 0 \text{ when } x = 18.$$

$$\frac{d^2V}{dx^2} = 216 - 24x = -216 < 0 \text{ when } x = 18.$$

The volume is maximum when $x = 18$ in. and $y = 108 - 4(18) = 36$ in.

31. No. The volume will change because the shape of the container changes when squeezed.

33. $$V = 14 = \frac{4}{3}\pi r^3 + \pi r^2 h$$

$$h = \frac{14 - (4/3)\pi r^3}{\pi r^2} = \frac{14}{\pi r^2} - \frac{4}{3}r$$

$$S = 4\pi r^2 + 2\pi rh = 4\pi r^2 + 2\pi r\left(\frac{14}{\pi r^2} - \frac{4}{3}r\right) = 4\pi r^2 + \frac{28}{r} - \frac{8}{3}\pi r^2 = \frac{4}{3}\pi r^2 + \frac{28}{r}$$

$$\frac{dS}{dr} = \frac{8}{3}\pi r - \frac{28}{r^2} = 0 \text{ when } r = \sqrt[3]{\frac{21}{2\pi}} \approx 1.495 \text{ cm.}$$

$$\frac{d^2S}{dr^2} = \frac{8}{3}\pi + \frac{56}{r^3} > 0 \text{ when } r = \sqrt[3]{\frac{21}{2\pi}}.$$

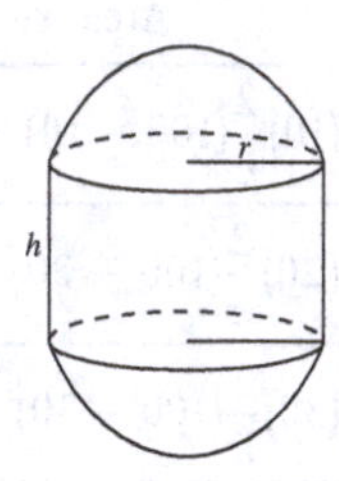

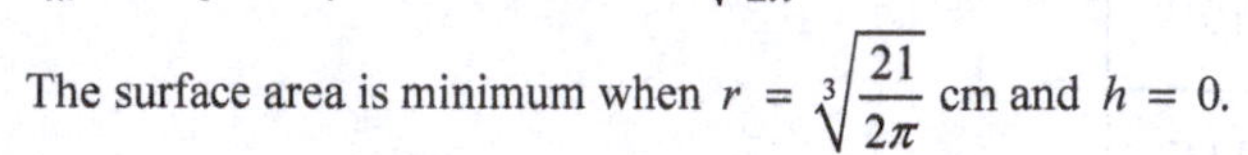

The surface area is minimum when $r = \sqrt[3]{\frac{21}{2\pi}}$ cm and $h = 0$.

The resulting solid is a sphere of radius $r \approx 1.495$ cm.

35. Let x be the length of a side of the square and y the length of a side of the triangle.

$$4x + 3y = 10$$

$$A = x^2 + \frac{1}{2}y\left(\frac{\sqrt{3}}{2}y\right) = \frac{(10 - 3y)^2}{16} + \frac{\sqrt{3}}{4}y^2$$

$$\frac{dA}{dy} = \frac{1}{8}(10 - 3y)(-3) + \frac{\sqrt{3}}{2}y = 0$$

$$-30 + 9y + 4\sqrt{3}y = 0$$

$$y = \frac{30}{9 + 4\sqrt{3}}$$

$$\frac{d^2A}{dy^2} = \frac{9 + 4\sqrt{3}}{8} > 0$$

A is minimum when $y = \dfrac{30}{9 + 4\sqrt{3}}$ and $x = \dfrac{10\sqrt{3}}{9 + 4\sqrt{3}}$.

37. Let S be the strength and k the constant of proportionality. Given $h^2 + w^2 = 20^2$, $h^2 = 20^2 - w^2$,

$$S = kwh^2$$

$$S = kw(400 - w^2) = k(400w - w^3)$$

$$\frac{dS}{dw} = k(400 - 3w^2) = 0 \text{ when } w = \frac{20\sqrt{3}}{3} \text{ in. and } h = \frac{20\sqrt{6}}{3} \text{ in.}$$

$$\frac{d^2S}{dw^2} = -6kw < 0 \text{ when } w = \frac{20\sqrt{3}}{3}.$$

These values yield a maximum.

39. $C(x) = 2k\sqrt{x^2 + 4} + k(4 - x)$

$$C'(x) = \frac{2xk}{\sqrt{x^2 + 4}} - k = 0$$

$$2x = \sqrt{x^2 + 4}$$

$$4x^2 = x^2 + 4$$

$$3x^2 = 4$$

$$x = \frac{2}{\sqrt{3}}$$

41. (a)

$$S = \sqrt{x^2 + 4},\ L = \sqrt{1 + (3 - x)^2}$$

$$\text{Time} = T = \frac{\sqrt{x^2 + 4}}{2} + \frac{\sqrt{x^2 - 6x + 10}}{4}$$

$$\frac{dT}{dx} = \frac{x}{2\sqrt{x^2 + 4}} + \frac{x - 3}{4\sqrt{x^2 - 6x + 10}} = 0$$

$$\frac{x^2}{x^2 + 4} = \frac{9 - 6x + x^2}{4(x^2 - 6x + 10)}$$

$$x^4 - 6x^3 + 9x^2 + 8x - 12 = 0$$

You need to find the roots of this equation in the interval $[0, 3]$. By using a computer or graphing utility you can determine that this equation has only one root in this interval $(x = 1)$. Testing at this value and at the endpoints, you see that $x = 1$ yields the minimum time. So, the man should row to a point 1 mile from the nearest point on the coast.

(b) $T = \dfrac{\sqrt{x^2 + 4}}{v_1} + \dfrac{\sqrt{x^2 - 6x + 10}}{v_2}$

$$\frac{dT}{dx} = \frac{x}{v_1\sqrt{x^2 + 4}} + \frac{x - 3}{v_2\sqrt{x^2 - 6x + 10}} = 0$$

Because $\dfrac{x}{\sqrt{x^2 + 4}} = \sin\theta_1$ and $\dfrac{x - 3}{\sqrt{x^2 - 6x + 10}} = -\sin\theta_2$

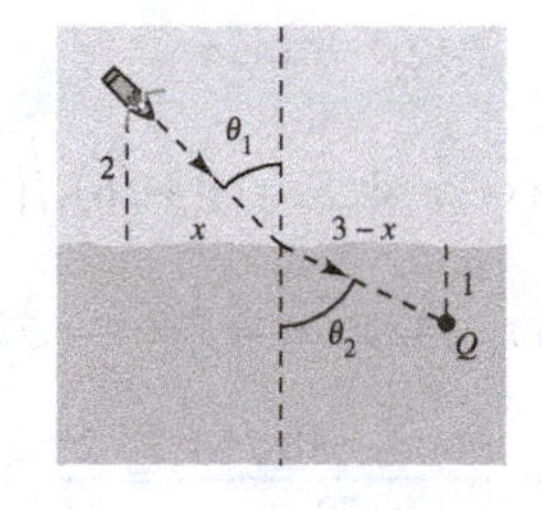

you have

$$\frac{\sin\theta_1}{v_1} - \frac{\sin\theta_2}{v_2} = 0 \Rightarrow \frac{\sin\theta_1}{v_1} = \frac{\sin\theta_2}{v_2}.$$

Because $\dfrac{d^2T}{dx^2} = \dfrac{4}{v_1(x^2 + 4)^{3/2}} + \dfrac{1}{v_2(x^2 - 6x + 10)^{3/2}} > 0$

this condition yields a minimum time.

43. $f(x) = 2 - 2\sin x$

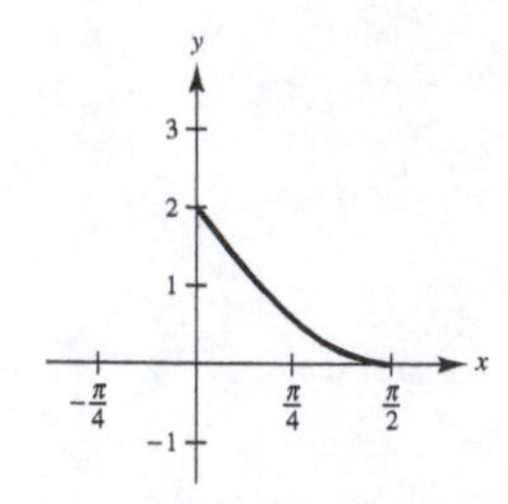

(a) Distance from origin to y-intercept is 2.

Distance from origin to x-intercept is $\pi/2 \approx 1.57$.

(b) $d = \sqrt{x^2 + y^2} = \sqrt{x^2 + (2 - 2\sin x)^2}$

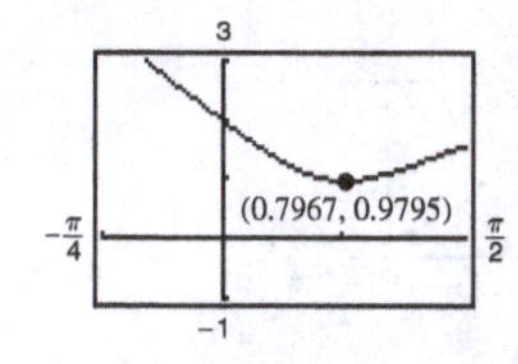

Minimum distance $= 0.9795$ at $x = 0.7967$.

(c) Let $f(x) = d^2(x) = x^2 + (2 - 2\sin x)^2$.

$f'(x) = 2x + 2(2 - 2\sin x)(-2\cos x)$

Setting $f'(x) = 0$, you obtain $x \approx 0.7967$, which corresponds to $d = 0.9795$.

45. $V = \frac{1}{3}\pi r^2 h = \frac{1}{3}\pi r^2\sqrt{144 - r^2}$

$$\frac{dV}{dr} = \frac{1}{3}\pi\left[r^2\left(\frac{1}{2}\right)(144 - r^2)^{-1/2}(-2r) + 2r\sqrt{144 - r^2}\right] = \frac{1}{3}\pi\left[\frac{288r - 3r^3}{\sqrt{144 - r^2}}\right] = \pi\left[\frac{r(96 - r^2)}{\sqrt{144 - r^2}}\right] = 0 \text{ when } r = 0, 4\sqrt{6}.$$

By the First Derivative Test, V is maximum when $r = 4\sqrt{6}$ and $h = 4\sqrt{3}$.

Area of circle: $A = \pi(12)^2 = 144\pi$

Lateral surface area of cone: $S = \pi\left(4\sqrt{6}\right)\sqrt{\left(4\sqrt{6}\right)^2 + \left(4\sqrt{3}\right)^2} = 48\sqrt{6}\pi$

Area of sector: $144\pi - 48\sqrt{6}\pi = \frac{1}{2}\theta r^2 = 72\theta$

$$\theta = \frac{144\pi - 48\sqrt{6}\pi}{72} = \frac{2\pi}{3}\left(3 - \sqrt{6}\right) \approx 1.153 \text{ radians or } 66°$$

47. Let d be the amount deposited in the bank, i be the interest rate paid by the bank, and P be the profit.

$P = (0.12)d - id$

$d = ki^2$ (because d is proportional to i^2)

$P = (0.12)(ki^2) - i(ki^2) = k(0.12i^2 - i^3)$

$\frac{dP}{di} = k(0.24i - 3i^2) = 0$ when $i = \frac{0.24}{3} = 0.08$.

$\frac{d^2P}{di^2} = k(0.24 - 6i) < 0$ when $i = 0.08$ (**Note:** $k > 0$).

The profit is a maximum when $i = 8\%$.

49. $y = \dfrac{L}{1 + ae^{-x/b}},\ a > 0,\ b > 0,\ L > 0$

$$y' = \frac{-L\left(-\frac{a}{b}e^{-x/b}\right)}{\left(1 + ae^{-x/b}\right)^2} = \frac{\frac{aL}{b}e^{-x/b}}{\left(1 + ae^{-x/b}\right)^2}$$

$$y'' = \frac{\left(1 + ae^{-x/b}\right)^2\left(\frac{-aL}{b^2}e^{-x/b}\right) - \left(\frac{aL}{b}e^{-x/b}\right)2\left(1 + ae^{-x/b}\right)\left(\frac{-a}{b}e^{-x/b}\right)}{\left(1 + ae^{-x/b}\right)^4}$$

$$= \frac{\left(1 + ae^{-x/b}\right)\left(\frac{-aL}{b^2}e^{-x/b}\right) + 2\left(\frac{aL}{b}e^{-x/b}\right)\left(\frac{a}{b}e^{-x/b}\right)}{\left(1 + ae^{-x/b}\right)^3} = \frac{Lae^{-x/b}\left(ae^{-x/b} - 1\right)}{\left(1 + ae^{-x/b}\right)^3 b^2}$$

$$y'' = 0 \text{ if } ae^{-x/b} = 1 \Rightarrow \frac{-x}{b} = \ln\left(\frac{1}{a}\right) \Rightarrow x = b \ln a$$

$$y(b \ln a) = \frac{L}{1 + ae^{-(b \ln a)/b}} = \frac{L}{1 + a(1/a)} = \frac{L}{2}$$

Therefore, the y-coordinate of the inflection point is $L/2$.

51. $S_1 = (4m - 1)^2 + (5m - 6)^2 + (10m - 3)^2$

$$\frac{dS_1}{dm} = 2(4m - 1)(4) + 2(5m - 6)(5) + 2(10m - 3)(10) = 282m - 128 = 0 \text{ when } m = \frac{64}{141}.$$

Line: $y = \dfrac{64}{141}x$

$$S = \left|4\left(\frac{64}{141}\right) - 1\right| + \left|5\left(\frac{64}{141}\right) - 6\right| + \left|10\left(\frac{64}{141}\right) - 3\right| = \left|\frac{256}{141} - 1\right| + \left|\frac{320}{141} - 6\right| + \left|\frac{640}{141} - 3\right| = \frac{858}{141} \approx 6.1 \text{ mi}$$

53. $S_3 = \dfrac{|4m - 1|}{\sqrt{m^2 + 1}} + \dfrac{|5m - 6|}{\sqrt{m^2 + 1}} + \dfrac{|10m - 3|}{\sqrt{m^2 + 1}}$

Using a graphing utility, you can see that the minimum occurs when $x \approx 0.3$.

Line: $y \approx 0.3x$

$$S_3 = \frac{|4(0.3) - 1| + |5(0.3) - 6| + |10(0.3) - 3|}{\sqrt{(0.3)^2 + 1}} \approx 4.5 \text{ mi.}$$

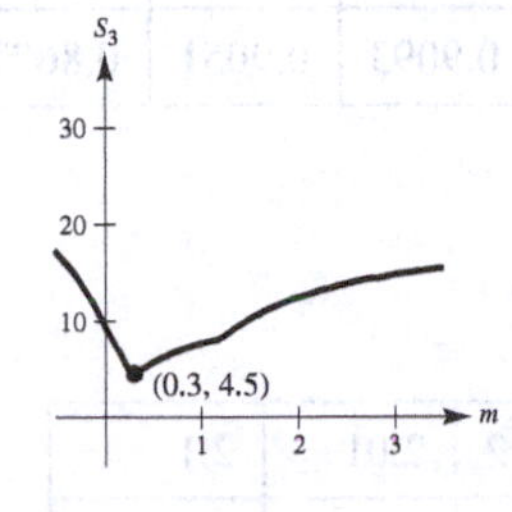

55. $16x = y^2,\ (6, 0)$

$$d = \sqrt{(x - 6)^2 + (y - 0)^2} = \sqrt{\left(\frac{y^2}{16} - 6\right)^2 + y^2}$$

Minimize d^2.

$$f(y) = \left(\frac{y^2}{16} - 6\right)^2 + y^2$$

$$f'(y) = 2\left(\frac{y^2}{16} - 6\right)\left(\frac{y}{8}\right) + 2y$$

$$= \frac{y^3}{64} + \frac{y}{2}$$

$$= \frac{y}{64}\left(y^2 + 32\right)$$

$f'(y) = 0$ when $y = 0$ and $x = 0$. By the First Derivative Test, the closest point is $(0, 0)$.

57. $f(x) = x^3 - 3x;\ x^4 + 36 \le 13x^2$

$x^4 - 13x^2 + 36 = (x^2 - 9)(x^2 - 4) = (x - 3)(x - 2)(x + 2)(x + 3) \le 0$

So, $-3 \le x \le -2$ or $2 \le x \le 3$.

$f'(x) = 3x^2 - 3 = 3(x + 1)(x - 1)$

f is increasing on $(-\infty, -1)$ and $(1, \infty)$.

So, f is increasing on $[-3, -2]$ and $[2, 3]$.

$f(-2) = -2,\ f(3) = 18$. The maximum value of f is 18.

Section 4.8 Differentials

1. The equation of the tangent line approximation to the graph of f at $(c, f(c))$ is $y = f(c) + f'(c)(x - c)$.

3. Propagated error $= f(x + \Delta x) - f(x)$,

relative error $= \left|\frac{dy}{y}\right|$, and the percent error $= \left|\frac{dy}{y}\right| \times 100$.

5. $f(x) = x^2$

$f'(x) = 2x$

Tangent line at (2, 4): $y - f(2) = f'(2)(x - 2)$

$y - 4 = 4(x - 2)$

$y = 4x - 4$

x	1.9	1.99	2	2.01	2.1
$f(x) = x^2$	3.6100	3.9601	4	4.0401	4.4100
$T(x) = 4x - 4$	3.6000	3.9600	4	4.0400	4.4000

7. $f(x) = x^5$

$f'(x) = 5x^4$

Tangent line at $(2, 32)$:

$y - f(2) = f'(2)(x - 2)$

$y - 32 = 80(x - 2)$

$y = 80x - 128$

x	1.9	1.99	2	2.01	2.1
$f(x) = x^5$	24.7610	31.2080	32	32.8080	40.8410
$T(x) = 80x - 128$	24.0000	31.2000	32	32.8000	40.0000

9. $f(x) = \sin x$

$f'(x) = \cos x$

Tangent line at $(2, \sin 2)$:

$y - f(2) = f'(2)(x - 2)$

$y - \sin 2 = (\cos 2)(x - 2)$

$y = (\cos 2)(x - 2) + \sin 2$

x	1.9	1.99	2	2.01	2.1
$f(x) = \sin x$	0.9463	0.9134	0.9093	0.9051	0.8632
$T(x) = (\cos 2)(x - 2) + \sin 2$	0.9509	0.9135	0.9093	0.9051	0.8677

11. $f(x) = 3^x$

$f'(x) = (\ln 3)3^x$

$f'(2) = 9 \ln 3$

Tangent line at $(2, 9)$:

$y - 9 = 9 \ln 3(x - 2)$

$y = 9(\ln 3)x + 9 - 18 \ln 3$

x	1.9	1.99	2	2.01	2.1
$f(x) = 3^x$	8.0636	8.9017	9	9.0994	10.0451
$T(x) = 9(\ln 3)x + 9 - 18 \ln 3$	8.0112	8.9011	9	9.0989	9.9888

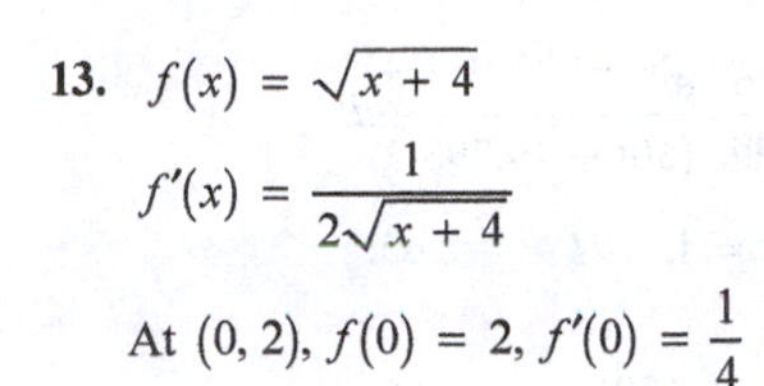

13. $f(x) = \sqrt{x + 4}$

$f'(x) = \dfrac{1}{2\sqrt{x + 4}}$

At $(0, 2)$, $f(0) = 2$, $f'(0) = \dfrac{1}{4}$

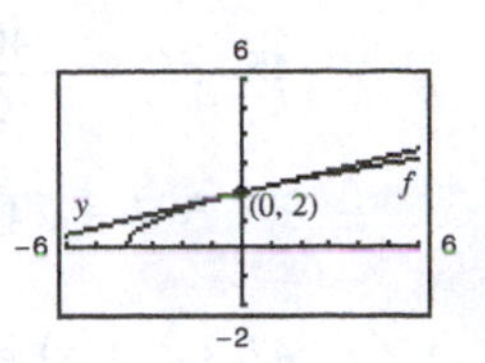

Tangent line: $y - 2 = \dfrac{1}{4}(x - 0)$

$y = \dfrac{1}{4}x + 2$

15. $y = f(x) = 0.5x^3, f'(x) = 1.5x^2, x = 1, \Delta x = dx = 0.1$

$$\begin{aligned}\Delta y &= f(x + \Delta x) - f(x) \\ &= f(1.1) - f(1) \\ &= 0.1655\end{aligned} \qquad \begin{aligned}dy &= f'(x)\,dx \\ &= 1.5x^2\,dx \\ &= 1.5(1)^2(0.1) \\ &= 0.15\end{aligned}$$

17. $y = f(x) = x^4 + 1, f'(x) = 4x^3, x = -1, \Delta x = dx = 0.01$

$$\begin{aligned}\Delta y &= f(x + \Delta x) - f(x) \\ &= f(-0.99) - f(-1) \\ &= \left[(-0.99)^4 + 1\right] - \left[(-1)^4 + 1\right] \approx -0.0394\end{aligned} \qquad \begin{aligned}dy &= f'(x)\,dx \\ &= f'(-1)(0.01) \\ &= (-4)(0.01) = -0.04\end{aligned}$$

19. $y = f(x) = x - 2x^3, f'(x) = 1 - 6x^2, x = 3, \Delta x = dx = 0.001$

$$\begin{aligned}\Delta y &= f(x + \Delta x) - f(x) \\ &= f(3.001) - f(3) \\ &= -51.05302 - (-51) \\ &= -0.05302\end{aligned} \qquad \begin{aligned}dy &= f'(x)\,dx \\ &= \left(1 - 6x^2\right)dx \\ &= \left[1 - 6(3)^2\right](0.001) \\ &= -0.053\end{aligned}$$

21. $y = 3x^2 - 4$

$dy = 6x\,dx$

23. $y = x\tan x$

$dy = \left(x\sec^2 x + \tan x\right)dx$

25. $y = \dfrac{x + 1}{2x - 1}$

$dy = \dfrac{-3}{(2x - 1)^2}\,dx$

27. $y = \sqrt{9 - x^2}$

$dy = \dfrac{1}{2}\left(9 - x^2\right)^{-1/2}(-2x)\,dx = \dfrac{-x}{\sqrt{9 - x^2}}\,dx$

29. $y = \ln\sqrt{4 - x^2} = \dfrac{1}{2}\ln\left(4 - x^2\right)$

$dy = \dfrac{1}{2}\left(\dfrac{-2x}{4 - x^2}\right)dx = \dfrac{-x}{4 - x^2}\,dx$

31. $y = x \arcsin x$

$dy = \left(\dfrac{x}{\sqrt{1 - x^2}} + \arcsin x\right)dx$

33. (a) $f(1.9) = f(2 - 0.1) \approx f(2) + f'(2)(-0.1)$
$\approx 1 + (1)(-0.1) = 0.9$

(b) $f(2.04) = f(2 + 0.04) \approx f(2) + f'(2)(0.04)$
$\approx 1 + (1)(0.04) = 1.04$

35. (a) $g(2.93) = g(3 - 0.07) \approx g(3) + g'(3)(-0.07)$
$\approx 8 + \left(-\tfrac{1}{2}\right)(-0.07) = 8.035$

(b) $g(3.1) = g(3 + 0.1) \approx g(3) + g'(3)(0.1)$
$\approx 8 + \left(-\tfrac{1}{2}\right)(0.1) = 7.95$

37. $x = 10$ in., $\Delta x = dx = \pm\frac{1}{32}$ in.

(a) $A = x^2$

$dA = 2x dx$

$\Delta A \approx dA = 2(10)\left(\pm\frac{1}{32}\right) = \pm\frac{5}{8}$ in.^{2}

(b) Percent error:

$$\frac{dA}{A} = \frac{5/8}{100} = \frac{5}{800} = \frac{1}{160} = 0.00625 = 0.625\%$$

39. $x = 15$ in., $\Delta x = dx = \pm 0.03$ in.

(a) $V = x^3$

$dV = 3x^2 dx$

$\Delta V \approx dV = 3(15)^2(\pm 0.03) = \pm 20.25$ in.^{3}

(b) $S = 6x^2$

$dS = 12x\,dx$

$\Delta S \approx dS = 12(15)(\pm 0.03) = \pm 5.4$ in.^{2}

(c) Percent error of volume:

$$\frac{dV}{V} = \frac{20.25}{15^3} = 0.006 \text{ or } 0.6\%$$

Percent error of surface area:

$$\frac{dS}{S} = \frac{5.4}{6(15)^2} = 0.004 \text{ or } 0.4\%$$

41. $T = 2.5x + 0.5x^2$, $\Delta x = dx = 26 - 25 = 1$, $x = 25$

$dT = (2.5 + x)dx = (2.5 + 25)(1) = 27.5$ mi

$$\text{Percentage change} = \frac{dT}{T} = \frac{27.5}{375} \approx 7.3\%$$

43. (a) $T = 2\pi\sqrt{L/g}$

$$dT = \frac{\pi}{g\sqrt{L/g}}dL$$

Relative error:

$$\frac{dT}{T} = \frac{(\pi\, dL)/(g\sqrt{L/g})}{2\pi\sqrt{L/g}} = \frac{dL}{2L}$$

$$= \frac{1}{2}(\text{relative error in } L)$$

$$= \frac{1}{2}(0.005) = 0.0025$$

Percentage error: $\frac{dT}{T}(100) = 0.25\% = \frac{1}{4}\%$

(b) $(0.0025)(3600)(24) = 216$ sec $= 3.6$ min

45. $$dH = -\frac{401{,}493{,}267\, e^{369{,}444/(50t+19{,}793)}}{2{,}000{,}000\,(50t + 19{,}793)^2}dt$$

At $t = 72$ and $dt = 1$, $dH \approx -2.65$.

47. Let $f(x) = \sqrt{x}$, $x = 100$, $dx = -0.6$.

$$f(x + \Delta x) \approx f(x) + f'(x)\,dx$$

$$= \sqrt{x} + \frac{1}{2\sqrt{x}}dx$$

$$f(x + \Delta x) = \sqrt{99.4}$$

$$\approx \sqrt{100} + \frac{1}{2\sqrt{100}}(-0.6) = 9.97$$

Using a calculator: $\sqrt{99.4} \approx 9.96995$

49. Let $f(x) = \sqrt[4]{x}$, $x = 625$, $dx = -1$.

$$f(x + \Delta x) \approx f(x) + f'(x)\,dx = \sqrt[4]{x} + \frac{1}{4\sqrt[4]{x^3}}dx$$

$$f(x + \Delta x) = \sqrt[4]{624} \approx \sqrt[4]{625} + \frac{1}{4\left(\sqrt[4]{625}\right)^3}(-1)$$

$$= 5 - \frac{1}{500} = 4.998$$

Using a calculator, $\sqrt[4]{624} \approx 4.9980$.

51. In general, the value of dy becomes closer to the value of Δy as Δx approaches 0. Graphs will vary.

53. Let $f(x) = \sqrt{x}$, $x = 4$, $dx = 0.02$,

$f'(x) = 1/(2\sqrt{x})$.

Then $f(4.02) \approx f(4) + f'(4)\,dx$

$$\sqrt{4.02} \approx \sqrt{4} + \frac{1}{2\sqrt{4}}(0.02) = 2 + \frac{1}{4}(0.02).$$

55. True

57. True

59. True

Review Exercises for Chapter 4

1. $f(x) = x^2 + 5x, \quad [-4, 0]$

$f'(x) = 2x + 5 = 0$ when $x = -5/2$

Critical number: $x = -5/2$

Left endpoint: $(-4, -4)$

Critical number: $(-5/2, -25/4)$ Minimum

Right endpoint: $(0, 0)$ Maximum

3. $f(x) = \sqrt{x} - 2, [0, 4]$

$f'(x) = \dfrac{1}{2\sqrt{x}}$

No critical numbers on $(0, 4)$

Left endpoint: $(0, -2)$ Minimum

Right endpoint: $(4, 0)$ Maximum

5. $f(x) = \dfrac{4x}{x^2 + 9}, [-4, 4]$

$$f'(x) = \frac{(x^2 + 9)4 - 4x(2x)}{(x^2 + 9)^2} = \frac{36 - 4x^2}{(x^2 + 9)^2}$$

$$= 0 \Rightarrow 36 - 4x^2 = 0 \Rightarrow x = \pm 3$$

Critical numbers: $x = \pm 3$

Left endpoint: $\left(-4, -\frac{16}{25}\right)$

Critical number: $\left(-3, -\frac{2}{3}\right)$ Minimum

Critical number: $\left(3, \frac{2}{3}\right)$ Maximum

Right endpoint: $\left(4, \frac{16}{25}\right)$

7. $g(x) = 2x + 5\cos x, [0, 2\pi]$

$g'(x) = 2 - 5\sin x = 0$ when $\sin x = \frac{2}{5}$.

Critical numbers: $x \approx 0.41, x \approx 2.73$

Left endpoint: $(0, 5)$

Critical number: $(0.41, 5.41)$

Critical number: $(2.73, 0.88)$ Minimum

Right endpoint: $(2\pi, 17.57)$ Maximum

9. $f(x) = x^3 - 3x - 6, [-1, 2]$

$f(-1) = (-1)^3 - 3(-1) - 6 = -4$

$f(2) = 2^3 - 3(2) - 6 = -4$

f is continuous on $[-1, 2]$. f is differentiable on $(-1, 2)$.

Rolle's Theorem applies.

$f'(x) = 3x^2 - 3 = 0 \Rightarrow x^2 = 1 \Rightarrow x = 1$

c-value: 1

11. $f(x) = \dfrac{x^2}{1 - x^2}, [-2, 2]$

$$f(-2) = \frac{(-2)^2}{1 - (-2)^2} = -\frac{4}{3}$$

$$f(2) = \frac{2^2}{1 - 2^2} = -\frac{4}{3}$$

f is not continuous on $[-2, 2]$ because $f(-1)$ does not exist.

Rolle's Theorem does not apply.

13. $f(x) = x^{2/3}, [1, 8]$

f is continuous on $[1, 8]$ and differentiable on $(1, 8)$.

$$f'(x) = \frac{2}{3}x^{-1/3}$$

$$\frac{f(b) - f(a)}{b - a} = \frac{4 - 1}{8 - 1} = \frac{3}{7}$$

$$f'(c) = \frac{2}{3}c^{-1/3} = \frac{3}{7}$$

$$c = \left(\frac{14}{9}\right)^3 = \frac{2744}{729} \approx 3.764$$

15. The Mean Value Theorem cannot be applied. f is not differentiable at $x = 5$ in $[2, 6]$.

17. $f(x) = x - \cos x, \left[-\frac{\pi}{2}, \frac{\pi}{2}\right]$

f is continuous on $\left[-\frac{\pi}{2}, \frac{\pi}{2}\right]$ and differentiable on $\left(-\frac{\pi}{2}, \frac{\pi}{2}\right)$.

$$f'(x) = 1 + \sin x$$

$$\frac{f(b) - f(a)}{b - a} = \frac{(\pi/2) - (-\pi/2)}{(\pi/2) - (-\pi/2)} = 1$$

$$f'(c) = 1 + \sin c = 1$$

$$c = 0$$

19. No; the function is discontinuous at $x = 0$ which is in the interval $[-2, 1]$.

21. $f(x) = x^2 + 3x - 12$

$f'(x) = 2x + 3$

Critical number: $x = -\frac{3}{2}$

Intervals:	$-\infty < x < -\frac{3}{2}$	$-\frac{3}{2} < x < \infty$
Sign of $f'(x)$:	$f'(x) < 0$	$f'(x) > 0$
Conclusion:	Decreasing	Increasing

23. $f(x) = (x - 1)^2(2x - 5)$

$f'(x) = 6(x - 2)(x - 1)$

Critical numbers: $x = 1, 2$

Intervals:	$-\infty < x < 1$	$1 < x < 2$	$2 < x < \infty$
Sign of $f'(x)$:	$f'(x) > 0$	$f'(x) < 0$	$f'(x) > 0$
Conclusion:	Increasing	Decreasing	Increasing

25. $h(x) = \sqrt{x}(x - 3) = x^{3/2} - 3x^{1/2}$

Domain: $(0, \infty)$

$$h'(x) = \frac{3}{2}x^{3/2} - \frac{3}{2}x^{-1/2} = \frac{3}{2}x^{-1/2}(x - 1) = \frac{3(x - 1)}{2\sqrt{x}}$$

Critical number: $x = 1$

Intervals:	$0 < x < 1$	$1 < x < \infty$
Sign of $h'(x)$:	$h'(x) < 0$	$h'(x) > 0$
Conclusion:	Decreasing	Increasing

27. $f(t) = (2 - t)2^t$

$f'(t) = (2 - t)2^t \ln 2 - 2^t = 2^t\left[(2 - t)\ln 2 - 1\right]$

$f'(t) = 0: (2 - t)\ln 2 = 1$

$$2 - t = \frac{1}{\ln 2}$$

$$t = 2 - \frac{1}{\ln 2} \approx 0.5573, \text{ Critical number}$$

Increasing on: $\left(-\infty, 2 - \dfrac{1}{\ln 2}\right)$

Decreasing on: $\left(2 - \dfrac{1}{\ln 2}, \infty\right)$

Intervals:	$-\infty < t < 2 - \dfrac{1}{\ln 2}$	$2 - \dfrac{1}{\ln 2} < t < \infty$
Sign of $f'(t)$:	$f'(t) > 0$	$f'(t) < 0$
Conclusion:	Increasing	Decreasing

29. (a) $f(x) = x^2 - 6x + 5$

$f'(x) = 2x - 6 = 0$ when $x = 3$.

(b)

Intervals:	$-\infty < x < 3$	$3 < x < \infty$
Sign of $f'(x)$:	$f'(x) < 0$	$f'(x) > 0$
Conclusion:	Decreasing	Increasing

(c) Relative minimum: $(3, -4)$

(d)

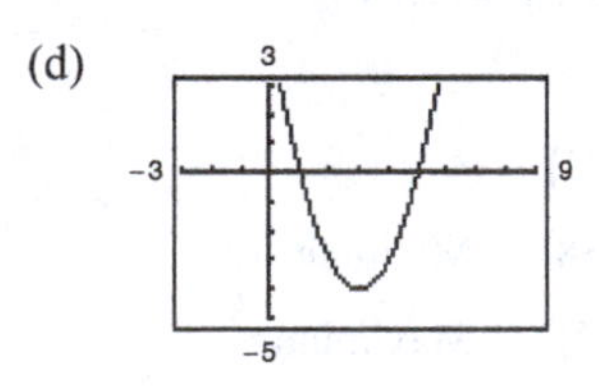

31. (a) $f(t) = \frac{1}{4}t^4 - 8t$

$f'(t) = t^3 - 8 = 0$ when $t = 2$.

(b)

Intervals:	$-\infty < t < 2$	$2 < t < \infty$
Sign of $f'(t)$:	$f'(t) < 0$	$f'(t) > 0$
Conclusion:	Decreasing	Increasing

(c) Relative minimum: $(2, -12)$

(d)

10

−2 6

−15

33. (a) $f(x) = \dfrac{x+4}{x^2}$

$f'(x) = \dfrac{x^2(1) - (x+4)(2x)}{x^4} = -\dfrac{x^2 + 8x}{x^4} = -\dfrac{x+8}{x^3}$

$f'(x) = 0$ when $x = -8$.

Discontinuity at: $x = 0$

(b)

Intervals:	$-\infty < x < -8$	$-8 < x < 0$	$0 < x < \infty$
Sign of $f'(x)$:	$f'(x) < 0$	$f'(x) > 0$	$f'(x) < 0$
Conclusion:	Decreasing	Increasing	Decreasing

(c) Relative minimum: $\left(-8, -\frac{1}{16}\right)$

(d)

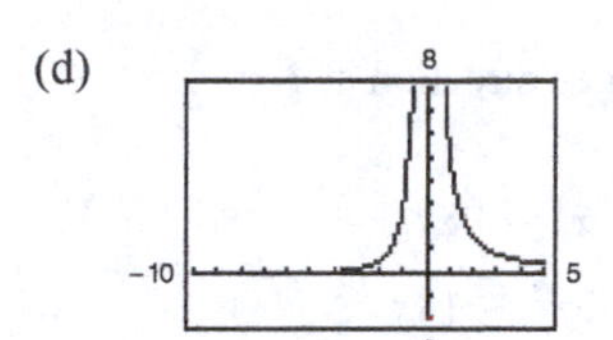

35. (a) $f(x) = \cos x - \sin x$, $(0, 2\pi)$

$f'(x) = -\sin x - \cos x = 0 \Rightarrow -\cos x = \sin x \Rightarrow \tan x = -1$

Critical numbers: $x = \dfrac{3\pi}{4}, \dfrac{7\pi}{4}$

(b)

Intervals:	$0 < x < \dfrac{3\pi}{4}$	$\dfrac{3\pi}{4} < x < \dfrac{7\pi}{4}$	$\dfrac{7\pi}{4} < x < 2\pi$
Sign of $f'(x)$:	$f'(x) < 0$	$f'(x) > 0$	$f'(x) < 0$
Conclusion:	Decreasing	Increasing	Decreasing

(c) Relative minimum: $\left(\dfrac{3\pi}{4}, -\sqrt{2}\right)$

Relative maximum: $\left(\dfrac{7\pi}{4}, \sqrt{2}\right)$

(d)

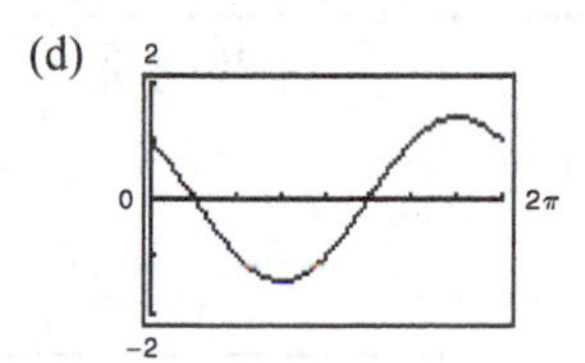

37. (a) $f(x) = \ln x - x^2, x > 0$

$f'(x) = \dfrac{1}{x} - 2x = 0 \Rightarrow \dfrac{1}{x} = 2x \Rightarrow x^2 = \dfrac{1}{2} \Rightarrow x = \dfrac{\sqrt{2}}{2}$

Critical numbers: $x = \dfrac{\sqrt{2}}{2}$

(b)

Intervals:	$0 < x < \dfrac{\sqrt{2}}{2}$	$\dfrac{\sqrt{2}}{2} < x < \infty$
Sign of $f'(x)$:	$f'(x) > 0$	$f'(x) < 0$
Conclusion:	Increasing	Decreasing

(c) Relative maximum: $\left(\dfrac{\sqrt{2}}{2}, \ln\left(\dfrac{\sqrt{2}}{2}\right) - \dfrac{1}{2}\right) \approx (0.7071, -0.8466)$

(d)

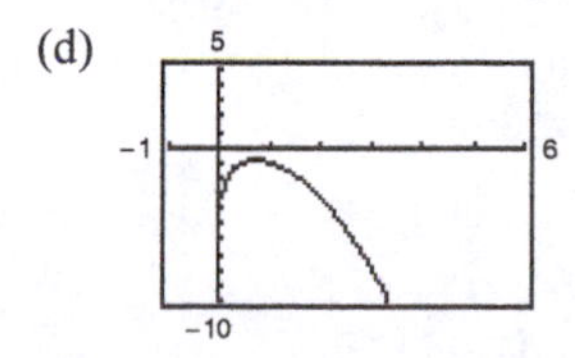

39. (a) $s(t) = 3t - 2t^2,\ t \ge 0$

$v(t) = s'(t) = 3 - 4t$

(b) $v(t) = 0$ when $t = \frac{3}{4}$

$v(t) > 0$ for $0 \le t < \frac{3}{4} \Rightarrow$ positive direction

(c) $v(t) < 0$ for $t > \frac{3}{4} \Rightarrow$ negative direction

(d) Changes direction at $t = \frac{3}{4}$

41. $f(x) = x^3 - 9x^2$

$f'(x) = 3x^2 - 18x$

$f''(x) = 6x - 18 = 0$ when $x = 3$.

Intervals:	$-\infty < x < 3$	$3 < x < \infty$
Sign of $f''(x)$:	$f''(x) < 0$	$f''(x) > 0$
Conclusion:	Concave downward	Concave upward

Point of inflection: $(3, -54)$

43. $g(x) = x\sqrt{x + 5}$, Domain: $x \ge -5$

$$g'(x) = x\left(\frac{1}{2}\right)(x + 5)^{-1/2} + (x + 5)^{1/2} = \frac{1}{2}(x + 5)^{-1/2}(x + 2(x + 5)) = \frac{3x + 10}{2\sqrt{x + 5}}$$

$$g''(x) = \frac{2\sqrt{x + 5}(3) - (3x + 10)(x + 5)^{-1/2}}{4(x + 5)} = \frac{6(x + 5) - (3x + 10)}{4(x + 5)^{3/2}} = \frac{3x + 20}{4(x + 5)^{3/2}} > 0 \text{ on } (-5, \infty).$$

Concave upward on $(-5, \infty)$

No point of inflection

45. $f(x) = x + \cos x,\ 0 \le x \le 2\pi$

$f'(x) = 1 - \sin x$

$f''(x) = -\cos x = 0$ when $x = \frac{\pi}{2}, \frac{3\pi}{2}$.

Points of inflection: $\left(\frac{\pi}{2}, \frac{\pi}{2}\right), \left(\frac{3\pi}{2}, \frac{3\pi}{2}\right)$

Intervals:	$0 < x < \frac{\pi}{2}$	$\frac{\pi}{2} < x < \frac{3\pi}{2}$	$\frac{3\pi}{2} < x < 2\pi$
Sign of $f''(x)$:	$f''(x) < 0$	$f''(x) > 0$	$f''(x) < 0$
Conclusion:	Concave downward	Concave upward	Concave downward

47. $f(x) = e^{4/x}$

$f'(x) = \dfrac{-4e^{4/x}}{x^2}$

$f''(x) = \dfrac{8(x + 2)e^{4/x}}{x^4}$

$f''(x) = 0$ when $x = -2$.

(f is undefined at $x = 0$.)

Point of inflection: $\left(-2, e^{-2}\right) \approx (-2, 0.1353)$

Intervals:	$-\infty < x < -2$	$-2 < x < 0$	$0 < x < \infty$
Sign of $f''(x)$:	$f''(x) < 0$	$f''(x) > 0$	$f''(x) > 0$
Conclusion:	Concave downward	Concave upward	Concave upward

49. $f(x) = (x + 9)^2$

$f'(x) = 2(x + 9) = 0 \Rightarrow x = -9$

$f''(x) = 2 > 0 \Rightarrow (-9, 0)$ is a relative minimum.

51. $g(x) = 2x^2(1 - x^2)$

$g'(x) = -4x(2x^2 - 1) = 0 \Rightarrow x = 0, \pm\frac{1}{\sqrt{2}}$

$g''(x) = 4 - 24x^2$

$g''(0) = 4 > 0 \quad (0, 0)$ is a relative minimum.

$g''\left(\pm\frac{1}{\sqrt{2}}\right) = -8 < 0 \quad \left(\pm\frac{1}{\sqrt{2}}, \frac{1}{2}\right)$ are relative maxima.

53. $f(x) = 2x + \frac{18}{x}$

$f'(x) = 2 - \frac{18}{x^2} = 0 \Rightarrow 2x^2 = 18 \Rightarrow x = \pm 3$

Critical numbers: $x = \pm 3$

$f''(x) = \frac{36}{x^3}$

$f''(-3) < 0 \Rightarrow (-3, -12)$ is a relative maximum.

$f''(3) > 0 \Rightarrow (3, 12)$ is a relative minimum.

55. $g(x) = x \log_5 \frac{2}{x} = x \cdot \frac{\ln\left(\frac{2}{x}\right)}{\ln 5}, x > 0$

$g'(x) = \frac{1}{\ln 5}\left[\ln\left(\frac{2}{x}\right) - 1\right]$

$g'(x) = 0 \Rightarrow \ln\left(\frac{2}{x}\right) = 1 \Rightarrow \frac{2}{x} = e \Rightarrow x = \frac{2}{e} \approx 0.7358$

Critical number: $x = \frac{2}{e}$

$g''(x) = \frac{-1}{x \ln 5}$

$g''\left(\frac{2}{e}\right) = \frac{-e}{2 \ln 5} < 0 \Rightarrow \left(\frac{2}{e}, \frac{2}{e \ln 5}\right) \approx (0.7358, 0.4572)$ is a relative maximum.

57.

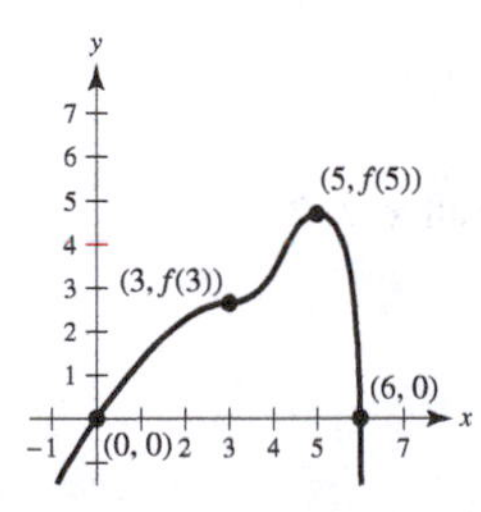

59. The first derivative is positive and the second derivative is negative. The graph is increasing and is concave down.

61. (a) $D = 0.41489t^4 - 17.1307t^3 + 249.888t^2 - 1499.45t + 3684.8$

(b)

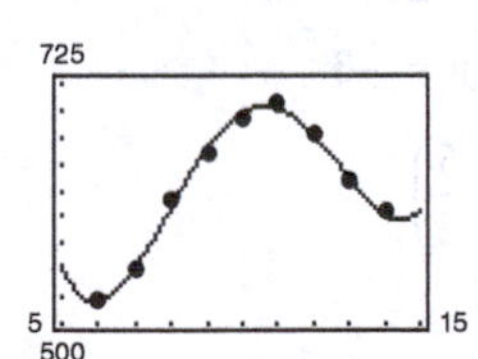

(c) The maximum occurred in 2011 $(t = 11)$.

The minimum occurred in 2006 $(t = 6)$.

(d) The outlays for national defense were increasing at the greatest rate in 2008 $(t = 8)$.

63. $\lim_{x \to \infty}\left(8 + \frac{1}{x}\right) = 8 + 0 = 8$

65. $\lim_{x \to -\infty} \frac{x^2}{1 - 8x^2} = \lim_{x \to -\infty} \frac{1}{(1/x^2) - 8} = -\frac{1}{8}$

67. $\lim_{x \to -\infty} \frac{3x^2}{x + 5} = -\infty$

69. $\lim_{x \to \infty} \frac{5 \cos x}{x} = 0$, because $|5 \cos x| \le 5$.

71. $\lim_{x\to-\infty} \dfrac{6x}{x + \cos x} = 6$

73. $\lim_{x\to\infty} \left(e^{-x} + 4 \ln \dfrac{x+1}{x}\right) = 0 + 4(0) = 0$

75. $f(x) = \dfrac{3}{x} + 4$

Discontinuity: $x = 0$

$\lim_{x\to\infty} \left(\dfrac{3}{x} + 4\right) = 4$

Vertical asymptote: $x = 0$

Horizontal asymptote: $y = 4$

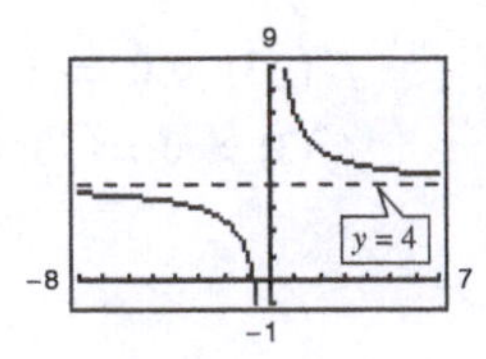

77. $f(x) = \dfrac{5}{3 + 2e^{-x}}$

$\lim_{x\to\infty} \dfrac{5}{3 + 2e^{-x}} = \dfrac{5}{3}$

$\lim_{x\to-\infty} \dfrac{5}{3 + 2e^{-x}} = 0$

Horizontal asymptotes: $y = 0, y = \dfrac{5}{3}$

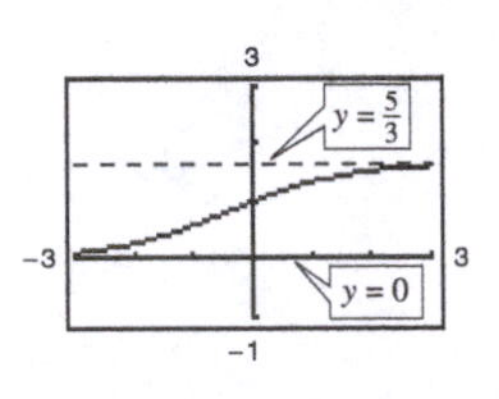

79. $S = \dfrac{100t^2}{65 + t^2}, t > 0$

(a)

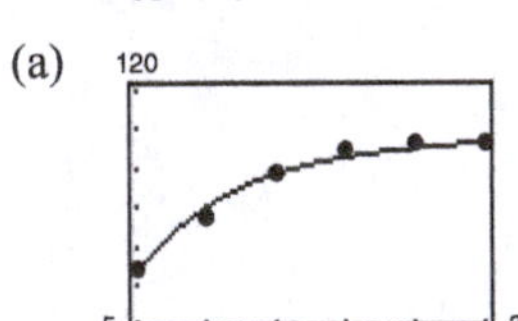

(b) Yes. $\lim_{x\to\infty} S = \dfrac{100}{1} = 100$

81. $f(x) = 4x - x^2 = x(4 - x)$

Domain: $(-\infty, \infty)$; Range: $(-\infty, 4]$

$f'(x) = 4 - 2x = 0$ when $x = 2$.

$f''(x) = -2$

Therefore, $(2, 4)$ is a relative maximum.

Intercepts: $(0, 0), (4, 0)$

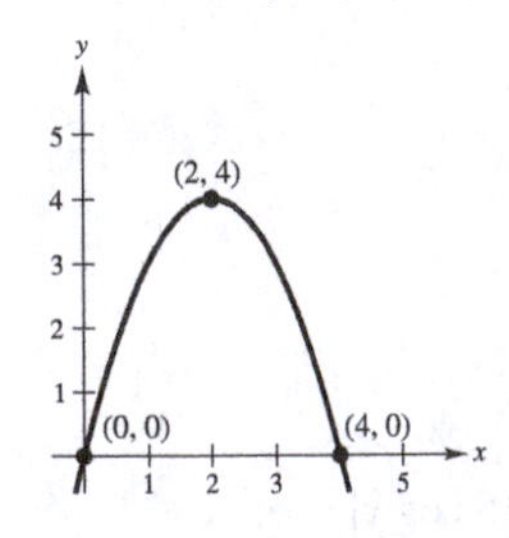

83. $f(x) = x\sqrt{16 - x^2}$

Domain: $[-4, 4]$; Range: $[-8, 8]$

$f'(x) = \dfrac{16 - 2x^2}{\sqrt{16 - x^2}} = 0$ when $x = \pm 2\sqrt{2}$ and undefined when $x = \pm 4$.

$f''(x) = \dfrac{2x(x^2 - 24)}{(16 - x^2)^{3/2}}$

$f''(-2\sqrt{2}) > 0$

Therefore, $(-2\sqrt{2}, -8)$ is a relative minimum.

$f''(2\sqrt{2}) < 0$

Therefore, $(2\sqrt{2}, 8)$ is a relative maximum.

Point of inflection: $(0, 0)$

Intercepts: $(-4, 0), (0, 0), (4, 0)$

Symmetry with respect to origin

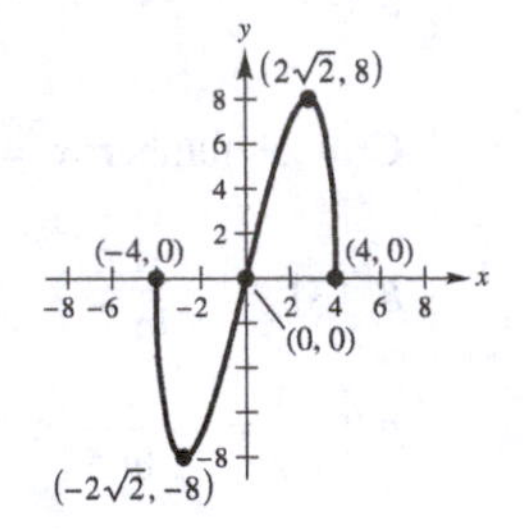

85. $f(x) = \dfrac{5 - 3x}{x - 2}$

$f'(x) = \dfrac{1}{(x - 2)^2} > 0$ for all $x \neq 2$

$f''(x) = \dfrac{-2}{(x - 2)^3}$

Concave upward on $(-\infty, 2)$

Concave downward on $(2, \infty)$

Vertical asymptote: $x = 2$

Horizontal asymptote: $y = -3$

Intercepts: $\left(\dfrac{5}{3}, 0\right), \left(0, -\dfrac{5}{2}\right)$

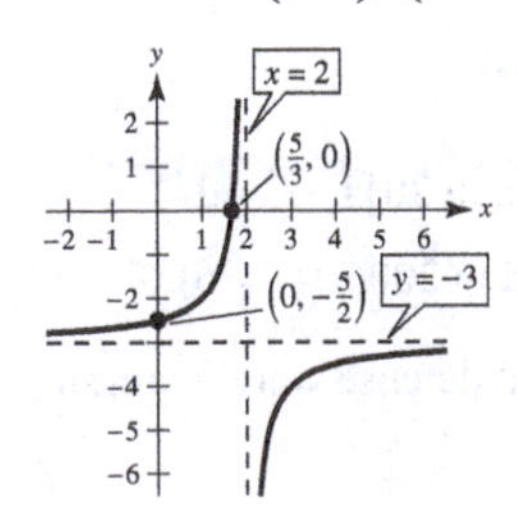

87. $f(x) = x^3 + x + \dfrac{4}{x}$

Domain: $(-\infty, 0), (0, \infty)$; Range: $(-\infty, -6], [6, \infty)$

$$f'(x) = 3x^2 + 1 - \frac{4}{x^2}$$
$$= \frac{3x^4 + x^2 - 4}{x^2} = \frac{(3x^2 + 4)(x^2 - 1)}{x^2} = 0$$

when $x = \pm 1$.

$$f''(x) = 6x + \frac{8}{x^3} = \frac{6x^4 + 8}{x^3} \neq 0$$

$f''(-1) < 0$

Therefore, $(-1, -6)$ is a relative maximum.

$f''(1) > 0$

Therefore, $(1, 6)$ is a relative minimum.

Vertical asymptote: $x = 0$

Symmetric with respect to origin

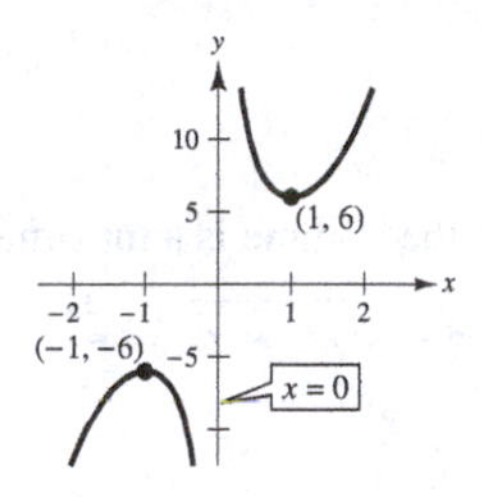

89. $f(x) = \dfrac{1}{1 + \cos x}$

Domain: $[-\pi, \pi]$; Range: $[0, \infty)$

$$f'(x) = \frac{\sin x}{(1 + \cos x)^2} = 0 \text{ when } x = 0.$$

$$f''(x) = \frac{1 + \cos x + \sin^2 x}{(1 + \cos x)^3} > 0 \text{ on } (-\pi\ \pi)$$

Therefore, $\left(0, \dfrac{1}{2}\right)$ is a relative minimum.

Concave upward on $(-\pi, \pi)$

No points of inflection

Intercept: $\left(0, \dfrac{1}{2}\right)$

Vertical asymptotes: $x = \pm\pi$

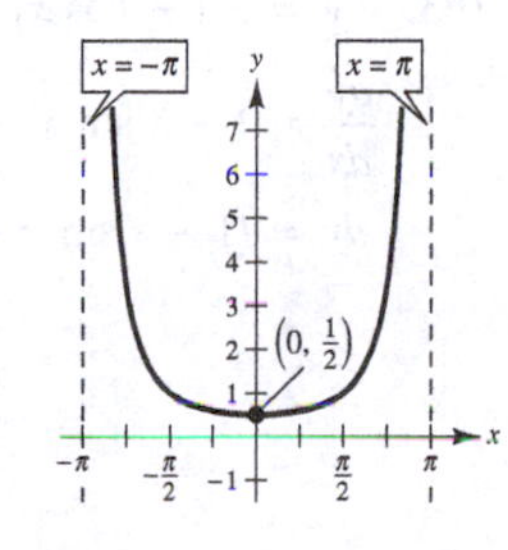

91. $f(x) = e^{1-x^3}$

Domain: $(-\infty, \infty)$; Range: $(0, \infty)$

$f'(x) = -3x^2 e^{1-x^3} = 0$ when $x = 0$.

$f''(x) = 3x(3x^3 - 2)e^{1-x^3} = 0$ when $x = 0$ and

$x = \left(\dfrac{2}{3}\right)^{1/3}$

Points of inflection: $(0, e)$, $(0.8736, 1.3956)$

$f'(x) < 0 \Rightarrow$ decreasing

y-intercept: $(0, e)$

Asymptote: $y = 0$

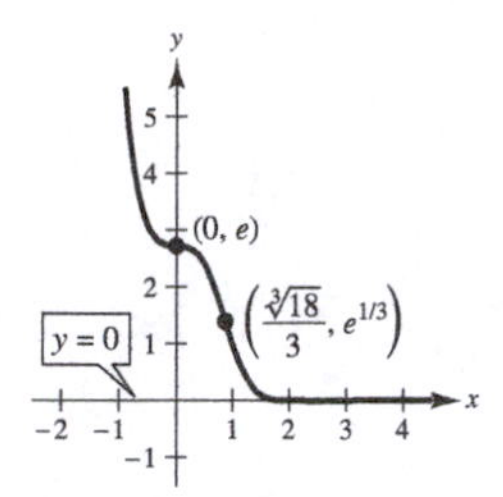

93. Let x and y be two positive numbers such that $2x + 3y = 216$.

$$P = xy = x\left(\frac{216 - 2x}{3}\right) = \frac{1}{3}x(216 - 2x) = 72x - \frac{2}{3}x^2$$

$$\frac{dP}{dx} = 72 - \frac{4}{3}x = 0 \text{ when } x = 54.$$

$$\frac{d^2P}{dx^2} = -\frac{4}{3} < 0 \text{ when } x = 54.$$

For $x = 54$, $y = \dfrac{216 - 2(54)}{3} = 36$.

P is a maximum when $x = 54$ and $y = 36$.

95. $4x + 3y = 400$ is the perimeter.

$$A = 2xy = 2x\left(\frac{400 - 4x}{3}\right) = \frac{8}{3}(100x - x^2)$$

$$\frac{dA}{dx} = \frac{8}{3}(100 - 2x) = 0 \text{ when } x = 50.$$

$$\frac{d^2A}{dx^2} = -\frac{16}{3} < 0 \text{ when } x = 50.$$

A is a maximum when $x = 50$ ft and $y = \dfrac{200}{3}$ ft.

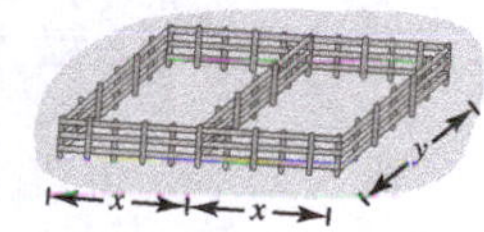

97. You have points $(0, y)$, $(x, 0)$, and $(1, 8)$. So,

$$m = \frac{y-8}{0-1} = \frac{0-8}{x-1} \text{ or } y = \frac{8x}{x-1}.$$

Let $f(x) = L^2 = x^2 + \left(\frac{8x}{x-1}\right)^2$.

$$f'(x) = 2x + 128\left(\frac{x}{x-1}\right)\left[\frac{(x-1)-x}{(x-1)^2}\right] = 0$$

$$x - \frac{64x}{(x-1)^3} = 0$$

$$x\left[(x-1)^3 - 64\right] = 0 \text{ when } x = 0, 5 \text{ (minimum)}.$$

Vertices of triangle: $(0, 0)$, $(5, 0)$, $(0, 10)$

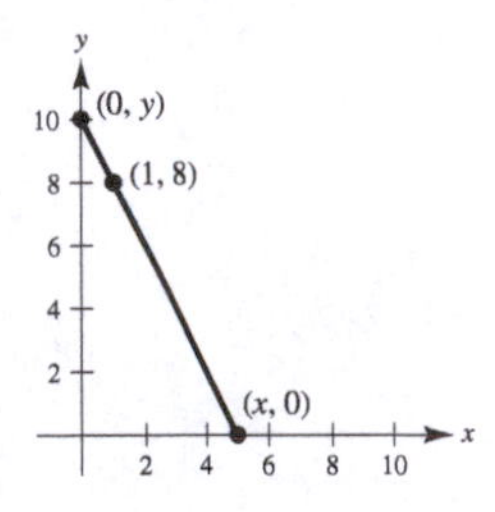

99. You can form a right triangle with vertices $(0, 0)$, $(x, 0)$ and $(0, y)$. Assume that the hypotenuse of length L passes through $(4, 6)$.

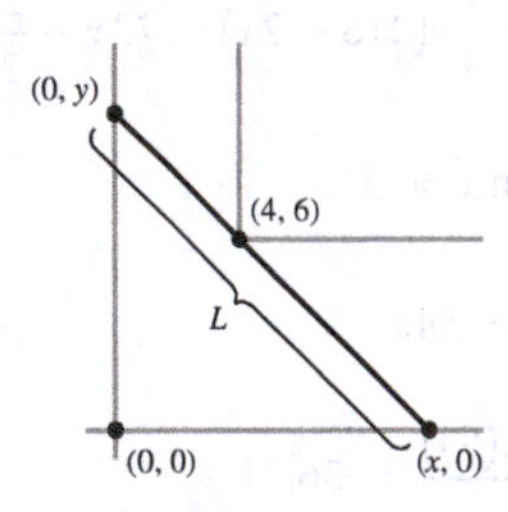

$$m = \frac{y-6}{0-4} = \frac{6-0}{4-x} \text{ or } y = \frac{6x}{x-4}$$

Let $f(x) = L^2 = x^2 + y^2 = x^2 + \left(\frac{6x}{x-4}\right)^2$.

$$f'(x) = 2x + 72\left(\frac{x}{x-4}\right)\left[\frac{-4}{(x-4)^2}\right] = 0$$

$$x\left[(x-4)^3 - 144\right] = 0 \text{ when } x = 0 \text{ or } x = 4 + \sqrt[3]{144}.$$

$$L \approx 14.05 \text{ ft}$$

101. $V = \frac{1}{3}\pi x^2 h = \frac{1}{3}\pi x^2\left(r + \sqrt{r^2 - x^2}\right)$ (see figure)

$$\frac{dV}{dx} = \frac{1}{3}\pi\left[\frac{-x^3}{\sqrt{r^2 - x^2}} + 2x\left(r + \sqrt{r^2 - x^2}\right)\right]$$

$$= \frac{\pi x}{3\sqrt{r^2 - x^2}}\left(2r^2 + 2r\sqrt{r^2 - x^2} - 3x^2\right) = 0$$

$$2r^2 + 2r\sqrt{r^2 - x^2} - 3x^2 = 0$$

$$2r\sqrt{r^2 - x^2} = 3x^2 - 2r^2$$

$$4r^2\left(r^2 - x^2\right) = 9x^4 - 12x^2r^2 + 4r^4$$

$$0 = 9x^4 - 8x^2r^2 = x^2\left(9x^2 - 8r^2\right)$$

$$x = 0, \frac{2\sqrt{2}r}{3}$$

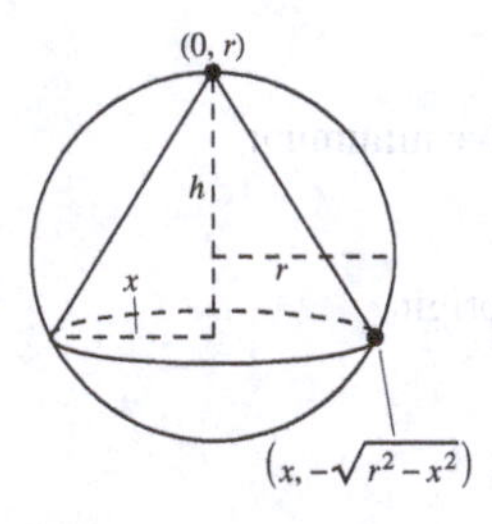

By the First Derivative Test, the volume is a maximum when $x = \frac{2\sqrt{2}r}{3}$ and $h = r + \sqrt{r^2 - x^2} = \frac{4r}{3}$.

Thus, the maximum volume is

$$V = \frac{1}{3}\pi\left(\frac{8r^2}{9}\right)\left(\frac{4r}{3}\right) = \frac{32\pi r^3}{81} \text{ cubic units.}$$

103. $y = f(x) = 4x^3, f'(x) = 12x^2, x = 2,$

$\Delta x = dx = 0.1$

$$\Delta y = f(x + \Delta x) - f(x) = f(2.1) - f(2) = 37.044 - 32 = 5.044$$

$$dy = f'(x)\,dx = \left(12x^2\right)dx = 12(2)^2(0.1) = 4.8$$

105. $y = x(1 - \cos x) = x - x\cos x$

$$\frac{dy}{dx} = 1 + x\sin x - \cos x$$

$$dy = (1 + x\sin x - \cos x)\,dx$$

107. $r = 9$ cm, $dr = \Delta r = \pm 0.025$

(a) $V = \frac{4}{3}\pi r^3$

$dV = 4\pi r^2\, dr$

$\Delta V \approx dV = 4\pi(9)^2(\pm 0.025) = \pm 8.1\pi \text{ cm}^3$

(b) $S = 4\pi r^2$

$dS = 8\pi r\, dr$

$\Delta S \approx dS = 8\pi(9)(\pm 0.025) = \pm 1.8\pi \text{ cm}^2$

(c) Percent error of volume: $\frac{dV}{V} = \frac{8.1\pi}{\frac{4}{3}\pi(9)^3} = 0.0083$, or 0.83%

Percent error of surface area: $\frac{dS}{S} = \frac{1.8\pi}{4\pi(9)^2} = 0.0056$, or 0.56%

109. $P = 100xe^{-x/400}$, x changes from 115 to 120.

$$dP = 100\left(e^{-x/400} - \frac{x}{400}e^{-x/400}\right)dx = e^{-115/400}\left(100 - \frac{115}{4}\right)(120 - 115) \approx 267.24$$

Approximate percentage change: $\frac{dP}{P}(100) = \frac{267.24}{8626.57}(100) \approx 3.1\%$

Problem Solving for Chapter 4

1. $p(x) = x^4 + ax^2 + 1$

(a) $p'(x) = 4x^3 + 2ax = 2x(2x^2 + a)$

$p''(x) = 12x^2 + 2a$

For $a \geq 0$, there is one relative minimum at $(0, 1)$.

(b) For $a < 0$, there is a relative maximum at $(0, 1)$.

(c) For $a < 0$, there are two relative minima at

$x = \pm\sqrt{-\frac{a}{2}}$.

(d) If $a < 0$, there are three critical points; if $a > 0$, there is only one critical point.

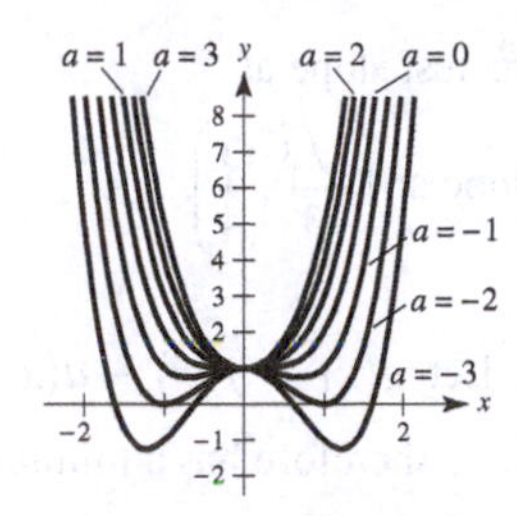

3. $f(x) = \frac{c}{x} + x^2$

$f'(x) = -\frac{c}{x^2} + 2x = 0 \Rightarrow \frac{c}{x^2} = 2x \Rightarrow x^3 = \frac{c}{2}$

$\Rightarrow x = \sqrt[3]{\frac{c}{2}}$

$f''(x) = \frac{2c}{x^3} + 2$

If $c = 0$, $f(x) = x^2$ has a relative minimum, but no relative maximum.

If $c > 0$, $x = \sqrt[3]{\frac{c}{2}}$ is a relative minimum, because

$f''\left(\sqrt[3]{\frac{c}{2}}\right) > 0.$

If $c < 0$, $x = \sqrt[3]{\frac{c}{2}}$ is a relative minimum, too.

Answer: All c.

5. Set $\dfrac{f(b) - f(a) - f'(a)(b - a)}{(b - a)^2} = k.$

Define

$F(x) = f(x) - f(a) - f'(a)(x - a) - k(x - a)^2.$

$F(a) = 0,$

$F(b) = f(b) - f(a) - f'(a)(b - a) - k(b - a)^2 = 0$

F is continuous on $[a, b]$ and differentiable on (a, b).

There exists c_1, $a < c_1 < b$, satisfying $F'(c_1) = 0$.

$F'(x) = f'(x) - f'(a) - 2k(x - a)$ satisfies the hypothesis of Rolle's Theorem on $[a, c_1]$:

$F'(a) = 0, F'(c_1) = 0.$

There exists c_2, $a < c_2 < c_1$ satisfying $F''(c_2) = 0$.

Finally, $F''(x) = f''(x) - 2k$ and $F''(c_2) = 0$ implies that

$k = \dfrac{f''(c_2)}{2}.$

$$\text{So, } k = \frac{f(b) - f(a) - f'(a)(b - a)}{(b - a)^2}$$

$$= \frac{f''(c_2)}{2} \Rightarrow f(b)$$

$$= f(a) + f'(a)(b - a) + \frac{1}{2}f''(c_2)(b - a)^2.$$

7. Distance $= \sqrt{4^2 + x^2} + \sqrt{(4 - x)^2 + 4^2} = f(x)$

$$f'(x) = \frac{x}{\sqrt{4^2 + x^2}} - \frac{4 - x}{\sqrt{(4 - x)^2 + 4^2}} = 0$$

$$x\sqrt{(4 - x)^2 + 4^2} = -(x - 4)\sqrt{4^2 + x^2}$$

$$x^2\left[16 - 8x + x^2 + 16\right] = \left(x^2 - 8x + 16\right)\left(16 + x^2\right)$$

$$32x^2 - 8x^3 + x^4 = x^4 - 8x^3 + 32x^2 - 128x + 256$$

$$128x = 256$$

$$x = 2$$

The bug should head towards the midpoint of the opposite side.

Without Calculus: Imagine opening up the cube:

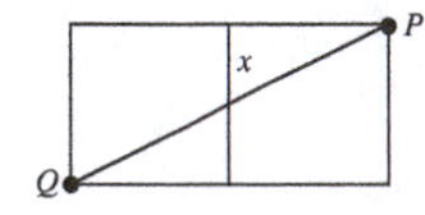

The shortest distance is the line PQ, passing through the midpoint.

9. f continuous at $x = 0$: $1 = b$

f continuous at $x = 1$: $a + 1 = 5 + c$

f differentiable at $x = 1$: $a = 2 + 4 = 6$. So, $c = 2$.

$$f(x) = \begin{cases} 1, & x = 0 \\ 6x + 1, & 0 < x \le 1 \\ x^2 + 4x + 2, & 1 < x \le 3 \end{cases}$$

$$= \begin{cases} 6x + 1, & 0 \le x \le 1 \\ x^2 + 4x + 2, & 1 < x \le 3 \end{cases}$$

11. Let $h(x) = g(x) - f(x)$, which is continuous on $[a, b]$ and differentiable on (a, b). $h(a) = 0$ and $h(b) = g(b) - f(b)$.

By the Mean Value Theorem, there exists c in (a, b) such that

$$h'(c) = \frac{h(b) - h(a)}{b - a} = \frac{g(b) - f(b)}{b - a}.$$

Because $h'(c) = g'(c) - f'(c) > 0$ and $b - a > 0$, $g(b) - f(b) > 0 \Rightarrow g(b) > f(b)$.

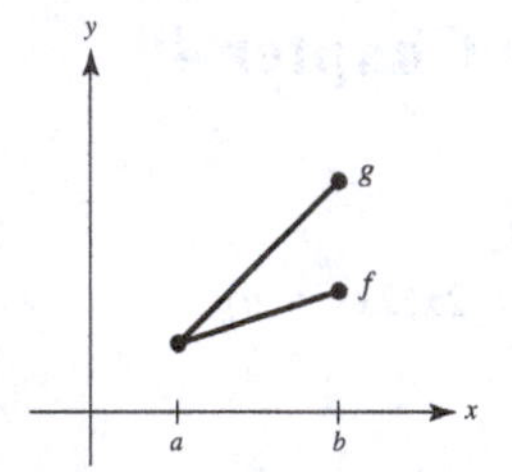

13. $y = \left(1 + x^2\right)^{-1}$

$$y' = \frac{-2x}{\left(1 + x^2\right)^2}$$

$$y'' = \frac{2\left(3x^2 - 1\right)}{\left(x^2 + 1\right)^3} = 0 \Rightarrow x = \pm\frac{1}{\sqrt{3}} = \pm\frac{\sqrt{3}}{3}$$

y'': $+++$ on $\left(-\infty, -\frac{\sqrt{3}}{3}\right)$, $----$ on $\left(-\frac{\sqrt{3}}{3}, 0\right)$, $----$ on $\left(0, \frac{\sqrt{3}}{3}\right)$, $+++$ on $\left(\frac{\sqrt{3}}{3}, \infty\right)$

The tangent line has greatest slope at $\left(-\dfrac{\sqrt{3}}{3}, \dfrac{3}{4}\right)$ and least slope at $\left(\dfrac{\sqrt{3}}{3}, \dfrac{3}{4}\right)$.

15. Assume $y_1 < d < y_2$. Let $g(x) = f(x) - d(x - a)$. g is continuous on $[a, b]$ and therefore has a minimum $(c, g(c))$ on $[a, b]$. The point c cannot be an endpoint of $[a, b]$ because

$g'(a) = f'(a) - d = y_1 - d < 0$

$g'(b) = f'(b) - d = y_2 - d > 0.$

So, $a < c < b$ and $g'(c) = 0 \Rightarrow f'(c) = d$.

17.
$$\begin{aligned} p(x) &= ax^3 + bx^2 + cx + d \\ p'(x) &= 3ax^2 + 2bx + c \\ p''(x) &= 6ax + 2b \\ 6ax + 2b &= 0 \\ x &= -\frac{b}{3a} \end{aligned}$$

The sign of $p''(x)$ changes at $x = -b/3a$. Therefore, $(-b/3a, p(-b/3a))$ is a point of inflection.

$$p\left(-\frac{b}{3a}\right) = a\left(-\frac{b^3}{27a^3}\right) + b\left(\frac{b^2}{9a^2}\right) + c\left(-\frac{b}{3a}\right) + d = \frac{2b^3}{27a^2} - \frac{bc}{3a} + d$$

When $p(x) = x^3 - 3x^2 + 2$, $a = 1$, $b = -3$, $c = 0$, and $d = 2$.

$$x_0 = \frac{-(-3)}{3(1)} = 1$$

$$y_0 = \frac{2(-3)^3}{27(1)^2} - \frac{(-3)(0)}{3(1)} + 2 = -2 - 0 + 2 = 0$$

The point of inflection of $p(x) = x^3 - 3x^2 + 2$ is $(x_0, y_0) = (1, 0)$.

19. $f(x) = \sin(\ln x)$

(a) Domain: $x > 0$ or $(0, \infty)$

(b) $f(x) = 1 = \sin(\ln x) \Rightarrow \ln x = \frac{\pi}{2} + 2k\pi$.

Two values are $x = e^{\pi/2}, e^{(\pi/2)+2\pi}$.

(c) $f(x) = -1 = \sin(\ln x) \Rightarrow \ln x = \frac{3\pi}{2} + 2k\pi$.

Two values are $x = e^{-\pi/2}, e^{3\pi/2}$.

(d) Because the range of the sine function is $[-1, 1]$, parts (b) and (c) show that the range of f is $[-1, 1]$.

(e) $f'(x) = \frac{1}{x}\cos(\ln x)$

$f'(x) = 0 \Rightarrow \cos(\ln x) = 0 \Rightarrow \ln x = \frac{\pi}{2} + k\pi \Rightarrow x = e^{\pi/2}$ on $[1, 10]$.

$$\left.\begin{aligned} f(e^{\pi/2}) &= 1 \\ f(1) &= 0 \\ f(10) &\approx 0.7440 \end{aligned}\right\} \text{Maximum is 1 at } x = e^{\pi/2} \approx 4.8105.$$

(f)

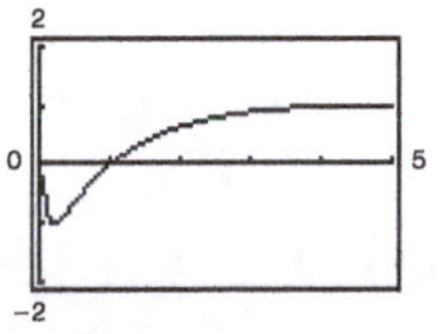

$\lim_{x \to 0^+} f(x)$ seems to be $-\frac{1}{2}$. (This is incorrect.)

(g) For the points $x = e^{\pi/2}$, $x = e^{-3\pi/2}$, $x = e^{-7\pi/2}, \ldots$ you have $f(x) = 1$.

For the points $x = e^{-\pi/2}$, $x = e^{-5\pi/2}$, $x = e^{-9\pi/2}, \ldots$ you have $f(x) = -1$.

That is, as $x \to 0^+$, there is an infinite number of points where $f(x) = 1$, and an infinite number where $f(x) = -1$.

So, $\lim_{x \to 0^+} \sin(\ln x)$ does not exist.

You can verify this by graphing $f(x)$ on small intervals close to the origin.

CHAPTER 5
Integration

CHAPTER 5
Integration

Section 5.1 Antiderivatives and Indefinite Integration

1. A function F is an antiderivative of f on an interval I if $F'(x) = f(x)$ for all x in I.

3. The particular solution results from knowing the value of $y = F(x)$ for one value of x. Using the initial condition in the general solution, you can solve for C to obtain the particular solution.

5. $\frac{d}{dx}\left(\frac{2}{x^3} + C\right) = \frac{d}{dx}\left(2x^{-3} + C\right) = -6x^{-4} = \frac{-6}{x^4}$

7. $\frac{dy}{dt} = 9t^2$

$y = 3t^3 + C$

Check: $\frac{d}{dt}\left[3t^3 + C\right] = 9t^2$

9. $\frac{dy}{dx} = x^{3/2}$

$y = \frac{2}{5}x^{5/2} + C$

Check: $\frac{d}{dx}\left[\frac{2}{5}x^{5/2} + C\right] = x^{3/2}$

	Given	*Rewrite*	*Integrate*	*Simplify*
11.	$\int \sqrt[3]{x}\,dx$	$\int x^{1/3}\,dx$	$\frac{x^{4/3}}{4/3} + C$	$\frac{3}{4}x^{4/3} + C$
13.	$\int \frac{1}{x\sqrt{x}}\,dx$	$\int x^{-3/2}\,dx$	$\frac{x^{-1/2}}{-1/2} + C$	$-\frac{2}{\sqrt{x}} + C$

15. $\int (x + 7)\,dx = \frac{x^2}{2} + 7x + C$

Check: $\frac{d}{dx}\left[\frac{x^2}{2} + 7x + C\right] = x + 7$

17. $\int \left(x^{3/2} + 2x + 1\right)dx = \frac{2}{5}x^{5/2} + x^2 + x + C$

Check: $\frac{d}{dx}\left(\frac{2}{5}x^{5/2} + x^2 + x + C\right) = x^{3/2} + 2x + 1$

19. $\int (x + 1)(3x - 2)\,dx = \int \left(3x^2 + x - 2\right)dx$

$= x^3 + \frac{1}{2}x^2 - 2x + C$

Check: $\frac{d}{dx}\left(x^3 + \frac{1}{2}x^2 - 2x + C\right) = 3x^2 + x - 2$

$= (x + 1)(3x - 2)$

21. $\int \frac{1}{x^5}\,dx = \int x^{-5}\,dx = \frac{x^{-4}}{-4} + C = \frac{-1}{4x^4} + C$

Check: $\frac{d}{dx}\left(\frac{-1}{4x^4} + C\right) = \frac{d}{dx}\left(-\frac{1}{4}x^{-4} + C\right)$

$= -\frac{1}{4}\left(-4x^{-5}\right) = \frac{1}{x^5}$

23. $\int \frac{x + 6}{\sqrt{x}}\,dx = \int \left(x^{1/2} + 6x^{-1/2}\right)dx$

$= \frac{x^{3/2}}{3/2} + 6\frac{x^{1/2}}{1/2} + C$

$= \frac{2}{3}x^{3/2} + 12x^{1/2} + C$

$= \frac{2}{3}x^{1/2}(x + 18) + C$

Check: $\frac{d}{dx}\left(\frac{2}{3}x^{3/2} + 12x^{1/2} + C\right)$

$= \frac{2}{3}\left(\frac{3}{2}x^{1/2}\right) + 12\left(\frac{1}{2}x^{-1/2}\right)$

$= x^{1/2} + 6x^{-1/2} = \frac{x + 6}{\sqrt{x}}$

25. $\int (5\cos x + 4\sin x)\,dx = 5\sin x - 4\cos x + C$

Check:

$\frac{d}{dx}(5\sin x - 4\cos x + C) = 5\cos x + 4\sin x$

27. $\int (\sec^2\theta - \sin\theta)\,d\theta = \tan\theta + \cos\theta + C$

Check: $\frac{d}{d\theta}(\tan\theta + \cos\theta + C) = \sec^2\theta - \sin\theta$

29. $\int (\tan^2 y + 1)\,dy = \int \sec^2 y\,dy = \tan y + C$

Check: $\frac{d}{dy}(\tan y + C) = \sec^2 y = \tan^2 y + 1$

31. $\int (2\sin x - 5e^x)\,dx = -2\cos x - 5e^x + C$

Check: $\frac{d}{dx}(-2\cos x - 5e^x + C) = 2\sin x - 5e^x$

33. $\int (2x - 4^x)\,dx = x^2 - \frac{4^x}{\ln 4} + C$

Check: $\frac{d}{dx}\left(x^2 - \frac{4^x}{\ln 4} + C\right) = 2x - 4^x$

35. $\int \left(x - \frac{5}{x}\right)dx = \frac{x^2}{2} - 5\ln|x| + C$

Check: $\frac{d}{dx}\left(\frac{x^2}{2} - 5\ln|x| + C\right) = x - \frac{5}{x}$

37. $f'(x) = 6x,\ f(0) = 8$

$f(x) = \int 6x\,dx = 3x^2 + C$

$f(0) = 8 = 3(0)^2 + C \Rightarrow C = 8$

$f(x) = 3x^2 + 8$

39. $f''(x) = 2$

$f'(2) = 5$

$f(2) = 10$

$f'(x) = \int 2\,dx = 2x + C_1$

$f'(2) = 4 + C_1 = 5 \Rightarrow C_1 = 1$

$f'(x) = 2x + 1$

$f(x) = \int (2x + 1)\,dx = x^2 + x + C_2$

$f(2) = 6 + C_2 = 10 \Rightarrow C_2 = 4$

$f(x) = x^2 + x + 4$

41. $f''(x) = x^{-3/2}$

$f'(4) = 2$

$f(0) = 0$

$f'(x) = \int x^{-3/2}\,dx = -2x^{-1/2} + C_1 = -\frac{2}{\sqrt{x}} + C_1$

$f'(4) = -\frac{2}{2} + C_1 = 2 \Rightarrow C_1 = 3$

$f'(x) = -\frac{2}{\sqrt{x}} + 3$

$f(x) = \int (-2x^{-1/2} + 3)\,dx = -4x^{1/2} + 3x + C_2$

$f(0) = 0 + 0 + C_2 = 0 \Rightarrow C_2 = 0$

$f(x) = -4x^{1/2} + 3x = -4\sqrt{x} + 3x$

43. $f''(x) = e^x$

$f'(0) = 2$

$f(0) = 5$

$f'(x) = \int e^x\,dx = e^x + C_1$

$f'(0) = 2 = e^0 + C_1 \Rightarrow C_1 = 1$

$f'(x) = e^x + 1$

$f(x) = \int (e^x + 1)\,dx = e^x + x + C_2$

$f(0) = 5 = e^0 + 0 + C_2 \Rightarrow C_2 = 4$

$f(x) = e^x + x + 4$

45. (a) Answers will vary. *Sample answer.*

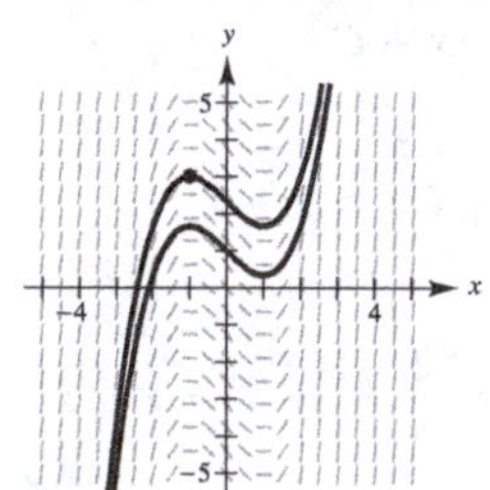

(b) $\frac{dy}{dx} = x^2 - 1,\ (-1, 3)$

$y = \frac{x^3}{3} - x + C$

$3 = \frac{(-1)^3}{3} - (-1) + C$

$C = \frac{7}{3}$

$y = \frac{x^3}{3} - x + \frac{7}{3}$

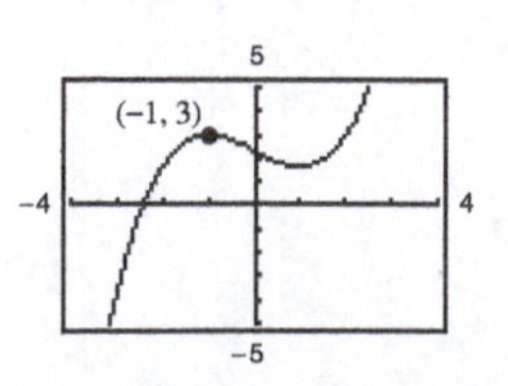

47. (a)

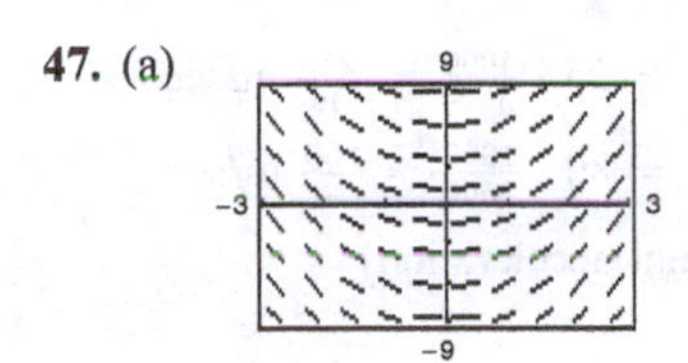

(b) $\dfrac{dy}{dx} = 2x, (-2, -2)$

$y = \int 2x\,dx = x^2 + C$

$-2 = (-2)^2 + C = 4 + C \Rightarrow C = -6$

$y = x^2 - 6$

(c)

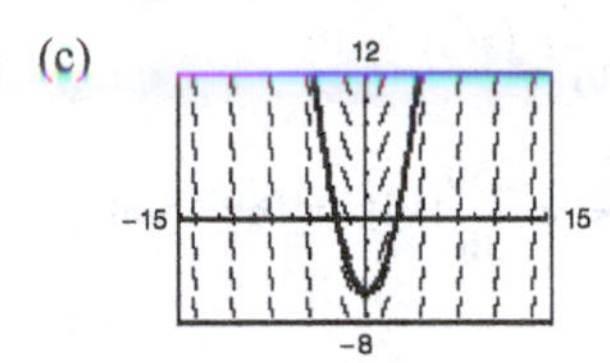

49. $f'(x) = 4$

$f(x) = 4x + C$

Answers will vary.

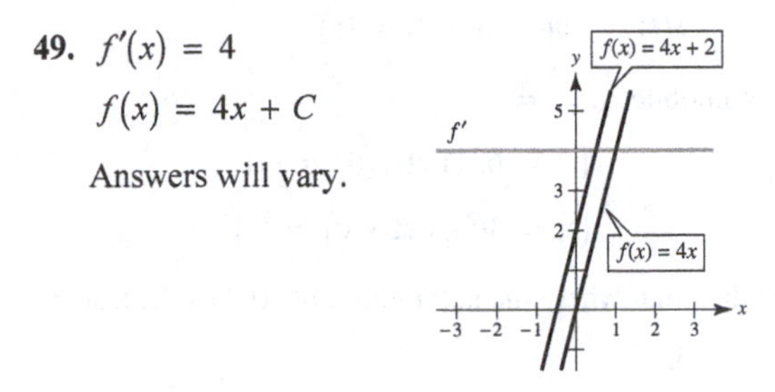

51. $f(x) = \tan^2 x \Rightarrow f'(x) = 2\tan x \cdot \sec^2 x$

$g(x) = \sec^2 x \Rightarrow g'(x) = 2\sec x \cdot \sec x \tan x = f'(x)$

The derivatives are the same, so f and g differ by a constant. In fact, $\tan^2 x + 1 = \sec^2 x$.

53. $f''(x) = 2x$

$f'(x) = x^2 + C$

$f'(2) = 0 \Rightarrow 4 + C = 0 \Rightarrow C = -4$

$f(x) = \dfrac{x^3}{3} - 4x + C_1$

$f(2) = 0 \Rightarrow \dfrac{8}{3} - 8 + C_1 = 0 \Rightarrow C_1 = \dfrac{16}{3}$

$f(x) = \dfrac{x^3}{3} - 4x + \dfrac{16}{3}$

55. (a) $h(t) = \int (1.5t + 5)\,dt = 0.75t^2 + 5t + C$

$h(0) = 0 + 0 + C = 12 \Rightarrow C = 12$

$h(t) = 0.75t^2 + 5t + 12$

(b) $h(6) = 0.75(6)^2 + 5(6) + 12 = 69$ cm

57. $a(t) = -32$ ft/sec^2

$v(t) = \int -32\,dt = -32t + C_1$

$v(0) = 60 = C_1$

$s(t) = \int (-32t + 60)\,dt = -16t^2 + 60t + C_2$

$s(0) = 6 = C_2$

$s(t) = -16t^2 + 60t + 6$, Position function

The ball reaches its maximum height when

$v(t) = -32t + 60 = 0$

$32t = 60$

$t = \frac{15}{8}$ seconds.

$s\left(\frac{15}{8}\right) = -16\left(\frac{15}{8}\right)^2 + 60\left(\frac{15}{8}\right) + 6 = 62.25$ feet

59. $v_0 = 16$ ft/sec

$s_0 = 64$ ft

(a) $s(t) = -16t^2 + 16t + 64 = 0$

$-16\left(t^2 - t - 4\right) = 0$

$t = \dfrac{1 \pm \sqrt{17}}{2}$

Choosing the positive value,

$t = \dfrac{1 + \sqrt{17}}{2} \approx 2.562$ seconds.

(b) $v(t) = s'(t) = -32t + 16$

$v\left(\dfrac{1 + \sqrt{17}}{2}\right) = -32\left(\dfrac{1 + \sqrt{17}}{2}\right) + 16$

$= -16\sqrt{17} \approx -65.970$ ft/sec

61. From Exercise 60, $f(t) = -4.9t^2 + v_0 t + 2$.

If $f(t) = 200 = -4.9t^2 + v_0 t + 2$,

then

$v(t) = -9.8t + v_0 = 0$

for this t value. So, $t = v_0/9.8$ and you solve

$-4.9\left(\dfrac{v_0}{9.8}\right)^2 + v_0\left(\dfrac{v_0}{9.8}\right) + 2 = 200$

$\dfrac{-4.9v_0^2}{(9.8)^2} + \left(\dfrac{v_0^2}{9.8}\right) = 198$

$-4.9v_0^2 + 9.8v_0^2 = (9.8)^2 198$

$4.9v_0^2 = (9.8)^2 198$

$v_0^2 = 3880.8$

$v_0 \approx 62.3$ m/sec.

63. $a = -1.6$

$v(t) = \int -1.6\,dt = -1.6t + v_0 = -1.6t$, because the stone was dropped, $v_0 = 0$.

$s(t) = \int (-1.6t)\,dt = -0.8t^2 + s_0$

$s(20) = 0 \Rightarrow -0.8(20)^2 + s_0 = 0$

$s_0 = 320$

So, the height of the cliff is 320 meters.

$v(t) = -1.6t$

$v(20) = -32$ m/sec

65. $x(t) = t^3 - 6t^2 + 9t - 2, \quad 0 \le t \le 5$

(a) $v(t) = x'(t) = 3t^2 - 12t + 9$

$= 3(t^2 - 4t + 3) = 3(t - 1)(t - 3)$

$a(t) = v'(t) = 6t - 12 = 6(t - 2)$

(b) $v(t) > 0$ when $0 < t < 1$ or $3 < t < 5$.

(c) $a(t) = 6(t - 2) = 0$ when $t = 2$.

$v(2) = 3(1)(-1) = -3$

67. $v(t) = \dfrac{1}{\sqrt{t}} = t^{-1/2} \quad t > 0$

$x(t) = \int v(t)dt = 2t^{1/2} + C$

$x(1) = 4 = 2(1) + C \Rightarrow C = 2$

Position function: $x(t) = 2t^{1/2} + 2$

Acceleration function: $a(t) = v'(t) = -\dfrac{1}{2}t^{-3/2} = \dfrac{-1}{2t^{3/2}}$

69. (a) $v(0) = 25 \text{ km/h} = 25 \cdot \frac{1000}{3600} = \frac{250}{36}$ m/sec

$v(13) = 80 \text{ km/h} = 80 \cdot \frac{1000}{3600} = \frac{800}{36}$ m/sec

$a(t) = a$ (constant acceleration)

$v(t) = at + C$

$v(0) = \frac{250}{36} \Rightarrow v(t) = at + \frac{250}{36}$

$v(13) = \frac{800}{36} = 13a + \frac{250}{36}$

$\frac{550}{36} = 13a$

$a = \frac{550}{468} = \frac{275}{234} \approx 1.175 \text{ m/sec}^2$

(b) $s(t) = a\dfrac{t^2}{2} + \dfrac{250}{36}t \quad (s(0) = 0)$

$s(13) = \dfrac{275}{234}\dfrac{(13)^2}{2} + \dfrac{250}{36}(13) \approx 189.58$ m

71. Truck: $v(t) = 30$

$s(t) = 30t$ (Let $s(0) = 0$.)

Automobile: $a(t) = 6$

$v(t) = 6t$ (Let $v(0) = 0$.)

$s(t) = 3t^2$ (Let $s(0) = 0$.)

At the point where the automobile overtakes the truck:

$30t = 3t^2$

$0 = 3t^2 - 30t$

$0 = 3t(t - 10)$ when $t = 10$ sec.

(a) $s(10) = 3(10)^2 = 300$ ft

(b) $v(10) = 6(10) = 60 \text{ ft/sec} \approx 41 \text{ mi/h} \quad kt = 160$

73. False. f has an infinite number of antiderivatives, each differing by a constant.

75. $\dfrac{d}{dx}\left[[s(x)]^2 + [c(x)]^2\right] = 2s(x)s'(x) + 2c(x)c'(x)$

$= 2s(x)c(x) - 2c(x)s(x) = 0$

So, $[s(x)]^2 + [c(x)]^2 = k$ for some constant k. Because, $s(0) = 0$ and $c(0) = 1$, $k = 1$.

So, $[s(x)]^2 + [c(x)]^2 = k$ for some constant k. Because, $s(0) = 0$ and $c(0) = 1$, $k = 1$.

Therefore, $[s(x)]^2 + [c(x)]^2 = 1$. [Note that $s(x) = \sin x$ and $c(x) = \cos x$ satisfy these properties.]

77. $\dfrac{d}{dx}(\ln|x| + C) = \dfrac{1}{x} + 0 = \dfrac{1}{x}$

79. $f(x + y) = f(x)f(y) - g(x)g(y)$

$g(x + y) = f(x)g(y) + g(x)f(y)$

$f'(0) = 0$

[Note: $f(x) = \cos x$ and $g(x) = \sin x$ satisfy these conditions]

$f'(x + y) = f(x)f'(y) - g(x)g'(y)$ (Differentiate with respect to y)

$g'(x + y) = f(x)g'(y) + g(x)f'(y)$ (Differentiate with respect to y)

Letting $y = 0$, $f'(x) = f(x)f'(0) - g(x)g'(0) = -g(x)g'(0)$

$g'(x) = f(x)g'(0) + g(x)f'(0) = f(x)g'(0)$

So, $2f(x)f'(x) = -2f(x)g(x)g'(0)$

$2g(x)g'(x) = 2g(x)f(x)g'(0)$.

Adding, $2f(x)f'(x) + 2g(x)g'(x) = 0$.

Integrating, $f(x)^2 + g(x)^2 = C$.

Clearly $C \neq 0$, for if $C = 0$, then $f(x)^2 = -g(x)^2 \Rightarrow f(x) = g(x) = 0$, which contradicts that f, g are nonconstant.

$$\begin{aligned} \text{Now, } C = f(x + y)^2 + g(x + y)^2 &= \left(f(x)f(y) - g(x)g(y)\right)^2 + \left(f(x)g(y) + g(x)f(y)\right)^2 \\ &= f(x)^2 f(y)^2 + g(x)^2 g(y)^2 + f(x)^2 g(y)^2 + g(x)^2 f(y)^2 \\ &= \left[f(x)^2 + g(x)^2\right]\left[f(y)^2 + g(y)^2\right] = C^2 \end{aligned}$$

So, $C = 1$ and you have $f(x)^2 + g(x)^2 = 1$.

Section 5.2 Area

1. For $\sum_{i=3}^{8} (i - 4)$, the index of summation is i, the upper bound of summation is 8, and the lower bound of summation is 3.

3. You can use the line $y = x$ bounded by $x = a$ and $x = b$. The sum of the areas of these inscribed rectangles is the lower sum.

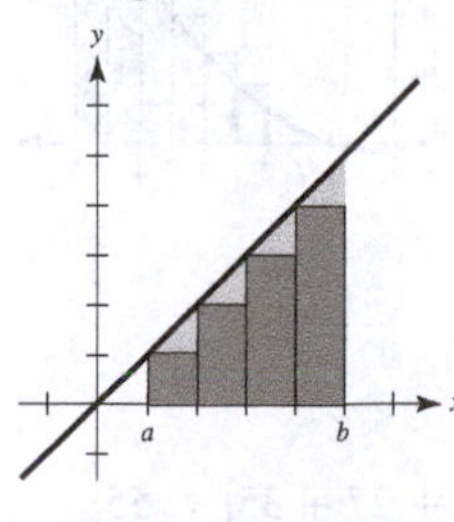

The sum of the areas of these circumscribed rectangles is the upper sum.

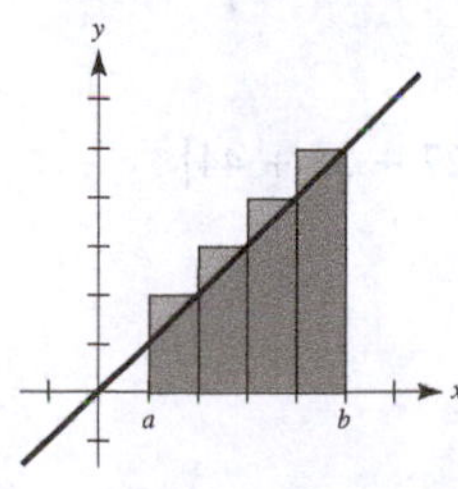

You can see that the rectangles do not contain all of the area in the first graph and the rectangles in the second graph cover more than the area of the region. The exact value of the area lies between these two sums.

5. $$\begin{aligned} \sum_{i=1}^{6}(3i + 2) &= 3\sum_{i=1}^{6} i + \sum_{i=1}^{6} 2 \\ &= 3(1 + 2 + 3 + 4 + 5 + 6) + 12 \\ &= 75 \end{aligned}$$

7. $\displaystyle\sum_{k=0}^{4}\frac{1}{k^2 + 1} = 1 + \frac{1}{2} + \frac{1}{5} + \frac{1}{10} + \frac{1}{17} = \frac{158}{85}$

9. $\displaystyle\sum_{k=0}^{7} c = c + c + c + c + c + c + c + c = 8c$

11. $\displaystyle\sum_{i=1}^{11}\frac{1}{5i}$

13. $\displaystyle\sum_{j=1}^{6}\left[7\left(\frac{j}{6}\right) + 5\right]$

15. $\displaystyle\frac{2}{n}\sum_{i=1}^{n}\left[\left(\frac{2i}{n}\right)^3 - \left(\frac{2i}{n}\right)\right]$

17. $\displaystyle\sum_{i=1}^{12} 7 = 7(12) = 84$

19. $\sum_{i=1}^{24} 4i = 4\sum_{i=1}^{24} i = 4\left[\frac{24(25)}{2}\right] = 1200$

21. $\sum_{i=1}^{20} (i-1)^2 = \sum_{i=1}^{19} i^2 = \left[\frac{19(20)(39)}{6}\right] = 2470$

23.
$$\begin{aligned}\sum_{i=1}^{7} i(i+3)^2 &= \sum_{i=1}^{7} i(i^2+6i+9)\\ &= \sum_{i=1}^{7}\left(i^3+6i^2+9i\right)\\ &= \frac{7^2(7+1)^2}{4} + 6\cdot\frac{7(7+1)(14+1)}{6} + 9\cdot\frac{7(7+1)}{2}\\ &= 784 + 840 + 252\\ &= 1876\end{aligned}$$

25. $\sum_{i=1}^{n}\frac{2i+1}{n^2} = \frac{1}{n^2}\sum_{i=1}^{n}(2i+1) = \frac{1}{n^2}\left[2\frac{n(n+1)}{2} + n\right] = \frac{n+2}{n} = 1 + \frac{2}{n} = S(n)$

$$\begin{aligned}S(10) &= \frac{12}{10} = 1.2\\ S(100) &= 1.02\\ S(1000) &= 1.002\\ S(10{,}000) &= 1.0002\end{aligned}$$

27.
$$\begin{aligned}\sum_{k=1}^{n}\frac{6k(k-1)}{n^3} &= \frac{6}{n^3}\sum_{k=1}^{n}\left(k^2-k\right) = \frac{6}{n^3}\left[\frac{n(n+1)(2n+1)}{6} - \frac{n(n+1)}{2}\right]\\ &= \frac{6}{n^2}\left[\frac{2n^2+3n+1-3n-3}{6}\right] = \frac{1}{n^2}\left[2n^2-2\right] = 2 - \frac{2}{n^2} = S(n)\end{aligned}$$

$$\begin{aligned}S(10) &= 1.98\\ S(100) &= 1.9998\\ S(1000) &= 1.999998\\ S(10{,}000) &= 1.99999998\end{aligned}$$

29.

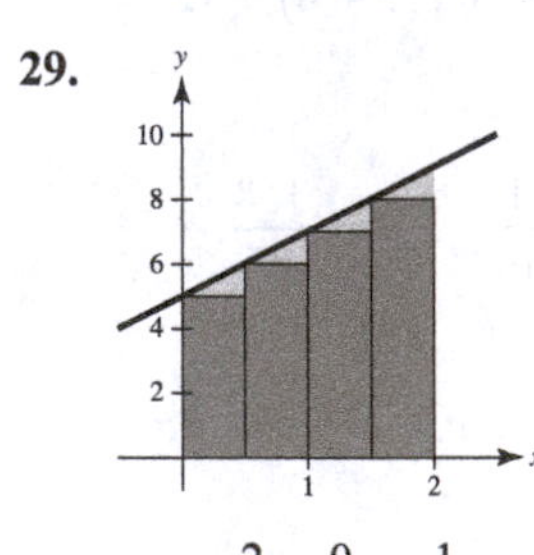

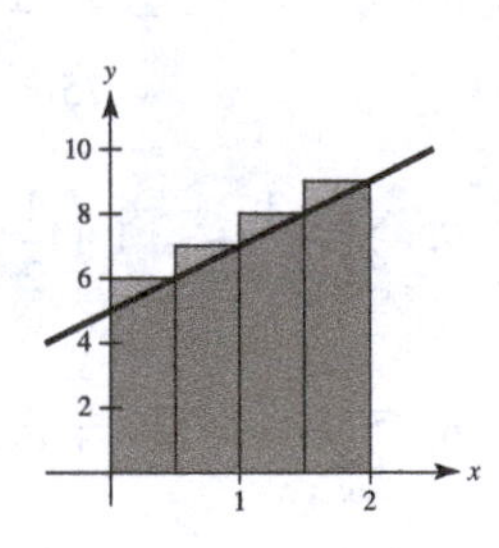

$\Delta x = \frac{2-0}{4} = \frac{1}{2}$

Left endpoints: Area $\approx \frac{1}{2}[5+6+7+8] = \frac{26}{2} = 13$

Right endpoints: Area $\approx \frac{1}{2}[6+7+8+9] = \frac{30}{2} = 15$

$13 < \text{Area} < 15$

31.

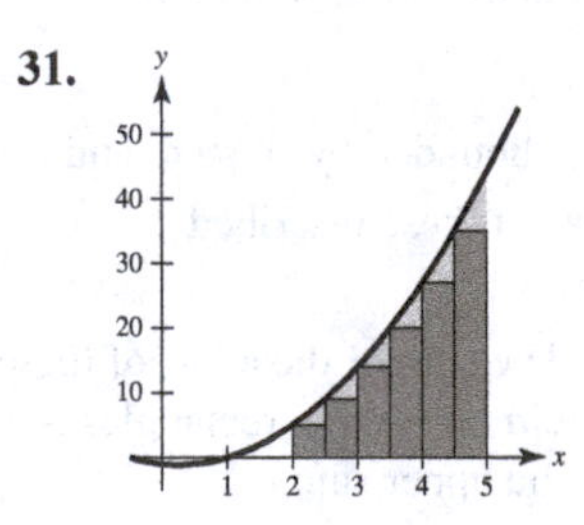

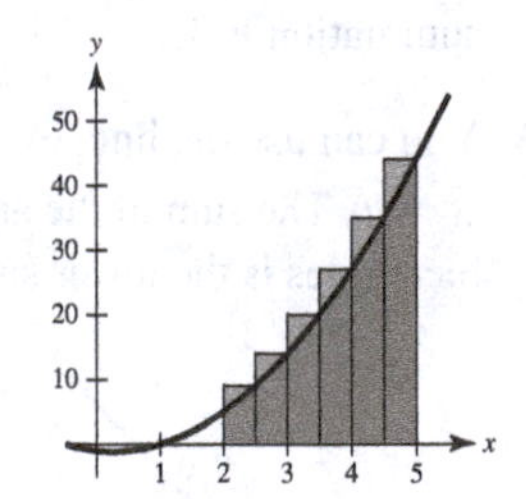

$\Delta x = \frac{5-2}{6} = \frac{1}{2}$

Left endpoints:

Area $\approx \frac{1}{2}[5+9+14+20+27+35] = 55$

Right endpoints:

$$\begin{aligned}\text{Area} &\approx \tfrac{1}{2}[9+14+20+27+35+44]\\ &= \tfrac{149}{2}\\ &= 74.5\end{aligned}$$

$55 < \text{Area} < 74.5$

33.

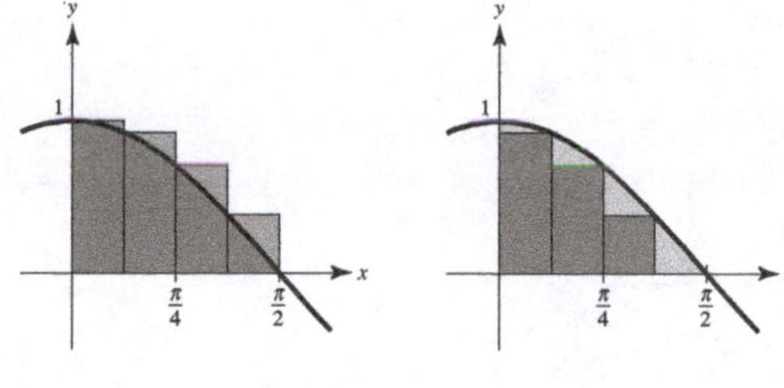

$$\Delta x = \frac{\frac{\pi}{2} - 0}{4} = \frac{\pi}{8}$$

Left endpoints:

$$\text{Area} \approx \frac{\pi}{8}\left[\cos(0) + \cos\left(\frac{\pi}{8}\right) + \cos\left(\frac{\pi}{4}\right) + \cos\left(\frac{3\pi}{8}\right)\right] \approx 1.1835$$

Right endpoints:

$$\text{Area} \approx \frac{\pi}{8}\left[\cos\left(\frac{\pi}{8}\right) + \cos\left(\frac{\pi}{4}\right) + \cos\left(\frac{3\pi}{8}\right) + \cos\left(\frac{\pi}{2}\right)\right] \approx 0.7908$$

$$0.7908 < \text{Area} < 1.1835$$

35. $S = \left[3 + 4 + \frac{9}{2} + 5\right](1) = \frac{33}{2} = 16.5$

$s = \left[1 + 3 + 4 + \frac{9}{2}\right](1) = \frac{25}{2} = 12.5$

37.

$$S(4) = \sqrt{\frac{1}{4}}\left(\frac{1}{4}\right) + \sqrt{\frac{1}{2}}\left(\frac{1}{4}\right) + \sqrt{\frac{3}{4}}\left(\frac{1}{4}\right) + \sqrt{1}\left(\frac{1}{4}\right) = \frac{1 + \sqrt{2} + \sqrt{3} + 2}{8} \approx 0.768$$

$$s(4) = 0\left(\frac{1}{4}\right) + \sqrt{\frac{1}{4}}\left(\frac{1}{4}\right) + \sqrt{\frac{1}{2}}\left(\frac{1}{4}\right) + \sqrt{\frac{3}{4}}\left(\frac{1}{4}\right) = \frac{1 + \sqrt{2} + \sqrt{3}}{8} \approx 0.518$$

39.

$$S(5) = 1\left(\frac{1}{5}\right) + \frac{1}{6/5}\left(\frac{1}{5}\right) + \frac{1}{7/5}\left(\frac{1}{5}\right) + \frac{1}{8/5}\left(\frac{1}{5}\right) + \frac{1}{9/5}\left(\frac{1}{5}\right) = \frac{1}{5} + \frac{1}{6} + \frac{1}{7} + \frac{1}{8} + \frac{1}{9} \approx 0.746$$

$$s(5) = \frac{1}{6/5}\left(\frac{1}{5}\right) + \frac{1}{7/5}\left(\frac{1}{5}\right) + \frac{1}{8/5}\left(\frac{1}{5}\right) + \frac{1}{9/5}\left(\frac{1}{5}\right) + \frac{1}{2}\left(\frac{1}{5}\right) = \frac{1}{6} + \frac{1}{7} + \frac{1}{8} + \frac{1}{9} + \frac{1}{10} \approx 0.646$$

41. $f(x) = 3x, [0, 4]$

$$\Delta x = \frac{b - a}{n} = \frac{4 - 0}{n} = \frac{4}{n}$$

Left endpoints for lower sum:

$$m_i = 0 + (i - 1)\Delta x = (i - 1)\frac{4}{n}$$

Right endpoints for upper sum: $M_i = i\Delta x = i\left(\frac{4}{n}\right)$

$$s(n) = \sum_{i=1}^{n} f(m_i)\Delta x = \sum_{i=1}^{n} 3(i - 1)\frac{4}{n}\left(\frac{4}{n}\right)$$

$$= \frac{48}{n^2}\sum_{i=1}^{n}(i - 1) = \frac{48}{n^2}\left[\frac{n(n + 1)}{2} - n\right]$$

$$= \frac{48}{n^2}\left(\frac{n^2}{2} - \frac{n}{2}\right) = 24 - \frac{24}{n}$$

$$S(n) = \sum_{i=1}^{n} f(M_i)\Delta x = \sum_{i=1}^{n} 3i\left(\frac{4}{n}\right)\left(\frac{4}{n}\right)$$

$$= \frac{48}{n^2}\sum_{i=1}^{n} i = \frac{48}{n^2}\cdot\frac{n(n + 1)}{2} = 24 + \frac{24}{n}$$

43. $f(x) = 5x^2, [0, 1]$

$$\Delta x = \frac{b - a}{n} = \frac{1 - 0}{n} = \frac{1}{n}$$

Right endpoints for upper sum: $M_i = i\Delta x = i\left(\frac{1}{n}\right)$

Left endpoints for lower sum:

$$m_i = (i - 1)\Delta x = (i - 1)\left(\frac{1}{n}\right)$$

$$s(n) = \sum_{i=1}^{n} f(m_i)\Delta x = \sum_{i=1}^{n} 5\left[(i - 1)\frac{1}{n}\right]^2\left(\frac{1}{n}\right)$$

$$= \frac{5}{n^3}\sum_{i=1}^{n}(i - 1)^2 = \frac{5}{n^3}\sum_{i=1}^{n}\left(i^2 - 2i + 1\right)$$

$$= \frac{5}{n^3}\left[\frac{n(n + 1)(2n + 1)}{6} - 2\frac{n(n + 1)}{2} + n\right]$$

$$= \frac{5}{n^3}\left[\frac{2n^3 + 3n^2 + n}{6} - \left(n^2 + n\right) + n\right]$$

$$= \frac{5}{6}\left[2 - \frac{3}{n} + \frac{1}{n^2}\right]$$

$$S(n) = \sum_{i=1}^{n} f(M_i)\Delta x = \sum_{i=1}^{n} 5\left(i\frac{1}{n}\right)^2\left(\frac{1}{n}\right)$$

$$= \frac{5}{n^3}\sum_{i=1}^{n} i^2 = \frac{5}{n^3}\frac{n(n + 1)(2n + 1)}{6}$$

$$= \frac{5}{6}\left(\frac{2n^3 + 3n^2 + n}{n^3}\right) = \frac{5}{6}\left(2 + \frac{3}{n} + \frac{1}{n^2}\right)$$

45. (a)

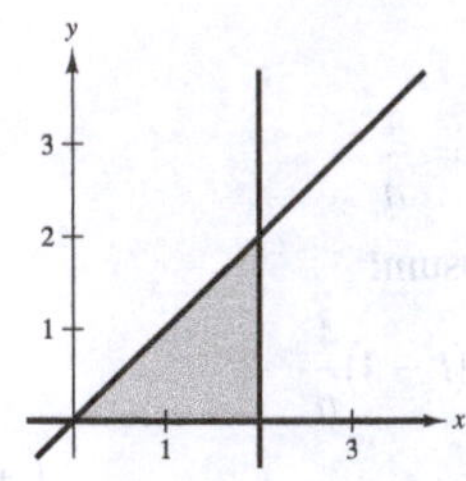

(b) $\Delta x = \frac{2-0}{n} = \frac{2}{n}$

Endpoints: $0 < 1\left(\frac{2}{n}\right) < 2\left(\frac{2}{n}\right) < \ldots < (n-1)\left(\frac{2}{n}\right) < n\left(\frac{2}{n}\right) = 2$

(c) Because $y = x$ is increasing, $f(m_i) = f(x_{i-1})$ on $[x_{i-1}, x_i]$.

$$s(n) = \sum_{i=1}^{n} f(x_{i-1})\Delta x = \sum_{i=1}^{n} f\left(\frac{2i-2}{n}\right)\left(\frac{2}{n}\right) = \sum_{i=1}^{n}\left[(i-1)\left(\frac{2}{n}\right)\right]\left(\frac{2}{n}\right)$$

(d) $f(M_i) = f(x_i)$ on $[x_{i-1}, x_i]$

$$S(n) = \sum_{i=1}^{n} f(x_i)\,\Delta x = \sum_{i=1}^{n} f\left(\frac{2i}{n}\right)\frac{2}{n} = \sum_{i=1}^{n}\left[i\left(\frac{2}{n}\right)\right]\left(\frac{2}{n}\right)$$

(e)

x	5	10	50	100
$s(n)$	1.6	1.8	1.96	1.98
$S(n)$	2.4	2.2	2.04	2.02

(f) $$\lim_{n\to\infty}\sum_{i=1}^{n}\left[(i-1)\left(\frac{2}{n}\right)\right]\left(\frac{2}{n}\right) = \lim_{n\to\infty}\frac{4}{n^2}\sum_{i=1}^{n}(i-1) = \lim_{n\to\infty}\frac{4}{n^2}\left[\frac{n(n+1)}{2} - n\right] = \lim_{n\to\infty}\left[\frac{2(n+1)}{n} - \frac{4}{n}\right] = 2$$

$$\lim_{n\to\infty}\sum_{i=1}^{n}\left[i\left(\frac{2}{n}\right)\right]\left(\frac{2}{n}\right) = \lim_{n\to\infty}\frac{4}{n^2}\sum_{i=1}^{n} i = \lim_{n\to\infty}\left(\frac{4}{n^2}\right)\frac{n(n+1)}{2} = \lim_{n\to\infty}\frac{2(n+1)}{n} = 2$$

47. $y = -4x + 5$ on $[0, 1]$. $\left(\textbf{Note: } \Delta x = \frac{1}{n}\right)$

$$s(n) = \sum_{i=1}^{n} f\left(\frac{i}{n}\right)\left(\frac{1}{n}\right) = \sum_{i=1}^{n}\left[-4\left(\frac{i}{n}\right) + 5\right]\left(\frac{1}{n}\right)$$
$$= -\frac{4}{n^2}\sum_{i=1}^{n} i + 5$$
$$= -\frac{4}{n^2}\frac{n(n+1)}{2} + 5$$
$$= -2\left(1 + \frac{1}{n}\right) + 5$$

Area $= \lim_{n\to\infty} s(n) = 3$

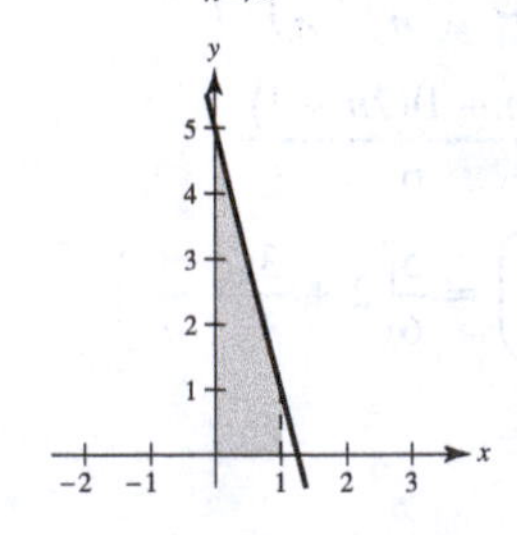

49. $y = x^2 + 2$ on $[0, 1]$. $\left(\textbf{Note: } \Delta x = \frac{1}{n}\right)$

$$S(n) = \sum_{i=1}^{n} f\left(\frac{i}{n}\right)\left(\frac{1}{n}\right)$$
$$= \sum_{i=1}^{n}\left[\left(\frac{i}{n}\right)^2 + 2\right]\left(\frac{1}{n}\right)$$
$$= \left[\frac{1}{n^3}\sum_{i=1}^{n} i^2\right] + 2$$
$$= \frac{n(n+1)(2n+1)}{6n^3} + 2 = \frac{1}{6}\left(2 + \frac{3}{n} + \frac{1}{n^2}\right) + 2$$

Area $= \lim_{n\to\infty} S(n) = \frac{7}{3}$

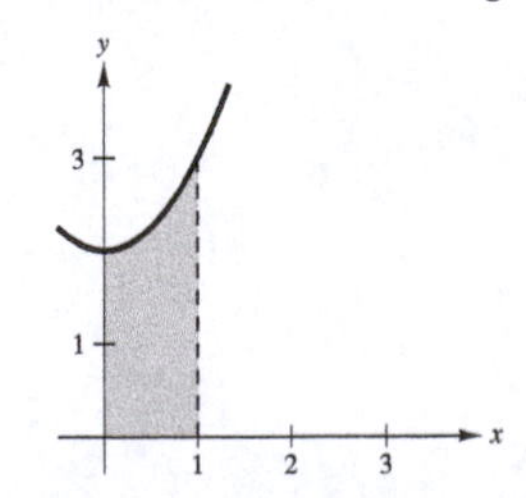

51. $y = 25 - x^2$ on $[1, 4]$. $\left(\textbf{Note: } \Delta x = \dfrac{3}{n}\right)$

$$s(n) = \sum_{i=1}^{n} f\left(1 + \frac{3i}{n}\right)\left(\frac{3}{n}\right) = \sum_{i=1}^{n}\left[25 - \left(1 + \frac{3i}{n}\right)^2\right]\left(\frac{3}{n}\right)$$
$$= \frac{3}{n}\sum_{i=1}^{n}\left[24 - \frac{9i^2}{n^2} - \frac{6i}{n}\right]$$
$$= \frac{3}{n}\left[24n - \frac{9}{n^2}\frac{n(n+1)(2n+1)}{6} - \frac{6}{n}\frac{n(n+1)}{2}\right]$$
$$= 72 - \frac{9}{2n^2}(n+1)(2n+1) - \frac{9}{n}(n+1)$$

$$\text{Area} = \lim_{n\to\infty} s(n) = 72 - 9 - 9 = 54$$

53. $y = 27 - x^3$ on $[1, 3]$. $\left(\textbf{Note: } \Delta x = \dfrac{3-1}{n} = \dfrac{2}{n}\right)$

$$s(n) = \sum_{i=1}^{n} f\left(1 + \frac{2i}{n}\right)\left(\frac{2}{n}\right) = \sum_{i=1}^{n}\left[27 - \left(1 + \frac{2i}{n}\right)^3\right]\left(\frac{2}{n}\right)$$
$$= \frac{2}{n}\sum_{i=1}^{n}\left[26 - \frac{8i^3}{n^3} - \frac{12i^2}{n^2} - \frac{6i}{n}\right]$$
$$= \frac{2}{n}\left[26n - \frac{8}{n^3}\frac{n^2(n+1)^2}{4} - \frac{12}{n^2}\frac{n(n+1)(2n+1)}{6} - \frac{6}{n}\frac{n(n+1)}{2}\right]$$
$$= 52 - \frac{4}{n^2}(n+1)^2 - \frac{4}{n^2}(n+1)(2n+1) - \frac{6n+1}{n}$$

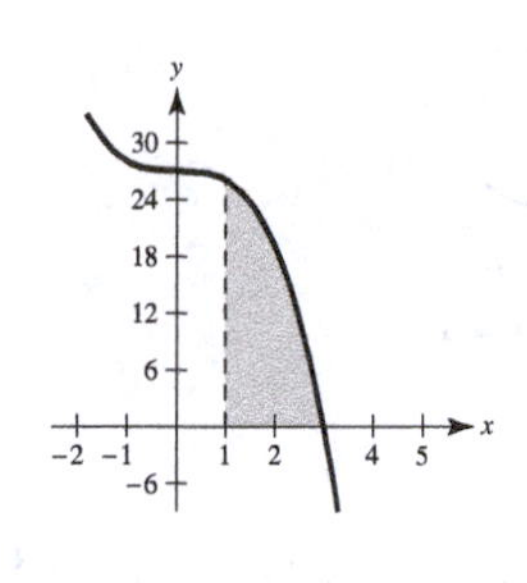

$$\text{Area} = \lim_{n\to\infty} s(n) = 52 - 4 - 8 - 6 = 34$$

55. $y = x^2 - x^3$ on $[-1, 1]$. $\left(\textbf{Note: } \Delta x = \dfrac{1-(-1)}{n} = \dfrac{2}{n}\right)$

Because y both increases and decreases on $[-1, 1]$, $T(n)$ is neither an upper nor a lower sum.

$$T(n) = \sum_{i=1}^{n} f\left(-1 + \frac{2i}{n}\right)\left(\frac{2}{n}\right) = \sum_{i=1}^{n}\left[\left(-1 + \frac{2i}{n}\right)^2 - \left(-1 + \frac{2i}{n}\right)^3\right]\left(\frac{2}{n}\right)$$
$$= \sum_{i=1}^{n}\left[\left(1 - \frac{4i}{n} + \frac{4i^2}{n^2}\right) - \left(-1 + \frac{6i}{n} - \frac{12i^2}{n^2} + \frac{8i^3}{n^3}\right)\right]\left(\frac{2}{n}\right)$$
$$= \sum_{i=1}^{n}\left[2 - \frac{10i}{n} + \frac{16i^2}{n^2} - \frac{8i^3}{n^3}\right]\left(\frac{2}{n}\right) = \frac{4}{n}\sum_{i=1}^{n} 1 - \frac{20}{n^2}\sum_{i=1}^{n} i + \frac{32}{n^3}\sum_{i=1}^{n} i^2 - \frac{16}{n^4}\sum_{i=1}^{n} i^3$$
$$= \frac{4}{n}(n) - \frac{20}{n^2}\cdot\frac{n(n+1)}{2} + \frac{32}{n^3}\cdot\frac{n(n+1)(2n+1)}{6} - \frac{16}{n^4}\cdot\frac{n^2(n+1)^2}{4}$$
$$= 4 - 10\left(1 + \frac{1}{n}\right) + \frac{16}{3}\left(2 + \frac{3}{n} + \frac{1}{n^2}\right) - 4\left(1 + \frac{2}{n} + \frac{1}{n^2}\right)$$

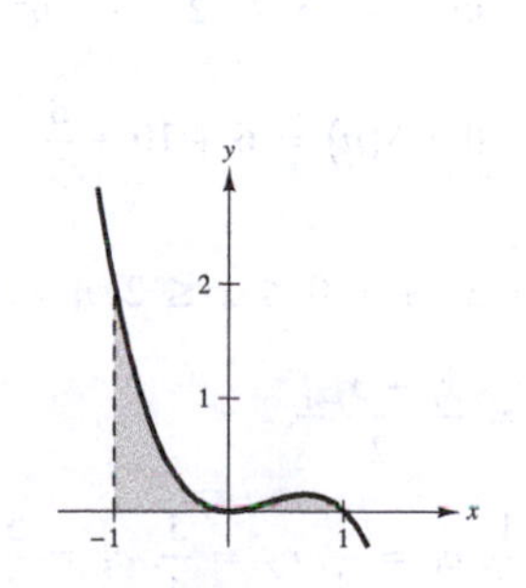

$$\text{Area} = \lim_{n\to\infty} T(n) = 4 - 10 + \frac{32}{3} - 4 = \frac{2}{3}$$

57. $f(y) = 4y,\ 0 \le y \le 2$ $\left(\textbf{Note: } \Delta y = \dfrac{2-0}{n} = \dfrac{2}{n}\right)$

$$S(n) = \sum_{i=1}^{n} f(m_i)\Delta y$$
$$= \sum_{i=1}^{n} f\left(\frac{2i}{n}\right)\left(\frac{2}{n}\right)$$
$$= \sum_{i=1}^{n} 4\left(\frac{2i}{n}\right)\left(\frac{2}{n}\right)$$
$$= \frac{16}{n^2}\sum_{i=1}^{n} i$$
$$= \left(\frac{16}{n^2}\right)\cdot\frac{n(n+1)}{2} = \frac{8(n+1)}{n} = 8 + \frac{8}{n}$$

$$\text{Area} = \lim_{n\to\infty} S(n) = \lim_{n\to\infty}\left(8 + \frac{8}{n}\right) = 8$$

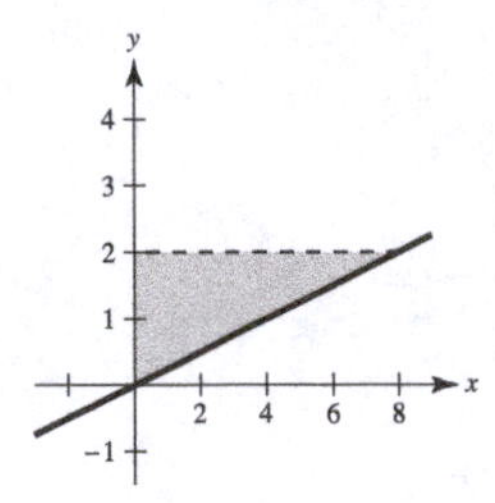

59. $f(y) = y^2,\ 0 \le y \le 5$ $\left(\textbf{Note: } \Delta y = \dfrac{5-0}{n} = \dfrac{5}{n}\right)$

$$S(n) = \sum_{i=1}^{n} f\left(\frac{5i}{n}\right)\left(\frac{5}{n}\right)$$
$$= \sum_{i=1}^{n}\left(\frac{5i}{n}\right)^2\left(\frac{5}{n}\right)$$
$$= \frac{125}{n^3}\sum_{i=1}^{n} i^2$$
$$= \frac{125}{n^3}\cdot\frac{n(n+1)(2n+1)}{6}$$
$$= \frac{125}{n^2}\left(\frac{2n^2+3n+1}{6}\right) = \frac{125}{3} + \frac{125}{2n} + \frac{125}{6n^2}$$

$$\text{Area } \lim_{n\to\infty} S(n) = \lim_{n\to\infty}\left(\frac{125}{3} + \frac{125}{2n} + \frac{125}{6n^2}\right) = \frac{125}{3}$$

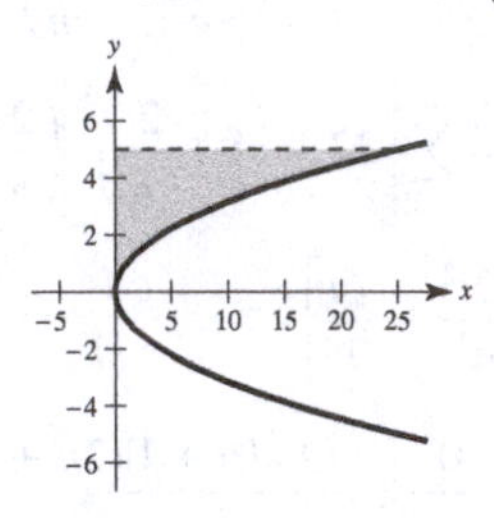

61. $g(y) = 4y^2 - y^3,\ 1 \le y \le 3.$ $\left(\text{Note: } \Delta y = \dfrac{3-1}{n} = \dfrac{2}{n}\right)$

$$S(n) = \sum_{i=1}^{n} g\left(1 + \frac{2i}{n}\right)\left(\frac{2}{n}\right)$$
$$= \sum_{i=1}^{n}\left[4\left(1+\frac{2i}{n}\right)^2 - \left(1+\frac{2i}{n}\right)^3\right]\frac{2}{n}$$
$$= \frac{2}{n}\sum_{i=1}^{n} 4\left[1 + \frac{4i}{n} + \frac{4i^2}{n^2}\right] - \left[1 + \frac{6i}{n} + \frac{12i^2}{n^2} + \frac{8i^3}{n^3}\right]$$
$$= \frac{2}{n}\sum_{i=1}^{n}\left[3 + \frac{10i}{n} + \frac{4i^2}{n^2} - \frac{8i^3}{n^3}\right]$$
$$= \frac{2}{n}\left[3n + \frac{10}{n}\frac{n(n+1)}{2} + \frac{4}{n^2}\frac{n(n+1)(2n+1)}{6} - \frac{8}{n^2}\frac{n^2(n+1)^2}{4}\right]$$

$$\text{Area} = \lim_{n\to\infty} S(n) = 6 + 10 + \frac{8}{3} - 4 = \frac{44}{3}$$

63. $f(x) = x^2 + 3,\ 0 \le x \le 2,\ n = 4$

Let $c_i = \dfrac{x_i + x_{i-1}}{2}$.

$$\Delta x = \frac{1}{2},\ c_1 = \frac{1}{4},\ c_2 = \frac{3}{4},\ c_3 = \frac{5}{4},\ c_4 = \frac{7}{4}$$

$$\text{Area} \approx \sum_{i=1}^{n} f(c_i)\Delta x = \sum_{i=1}^{4}\left[c_i^2 + 3\right]\left(\frac{1}{2}\right) = \frac{1}{2}\left[\left(\frac{1}{16}+3\right) + \left(\frac{9}{16}+3\right) + \left(\frac{25}{16}+3\right) + \left(\frac{49}{16}+3\right)\right] = \frac{69}{8}$$

65. $f(x) = \tan x,\ 0 \le x \le \dfrac{\pi}{4},\ n = 4$

Let $c_i = \dfrac{x_i + x_{i-1}}{2}.$

$\Delta x = \dfrac{\pi}{16},\ c_1 = \dfrac{\pi}{32},\ c_2 = \dfrac{3\pi}{32},\ c_3 = \dfrac{5\pi}{32},\ c_4 = \dfrac{7\pi}{32}$

$$\text{Area} \approx \sum_{i=1}^{n} f(c_i)\Delta x = \sum_{i=1}^{4}(\tan c_i)\left(\frac{\pi}{16}\right) = \frac{\pi}{16}\left(\tan\frac{\pi}{32} + \tan\frac{3\pi}{32} + \tan\frac{5\pi}{32} + \tan\frac{7\pi}{32}\right) \approx 0.345$$

67. $f(x) = \ln x,\ 1 \le x \le 5,\ n = 4$

Let $c_i = \dfrac{x_i + x_{i-1}}{2},\ \Delta x = 1$

$c_1 = \dfrac{3}{2},\ c_2 = \dfrac{5}{2},\ c_3 = \dfrac{7}{2},\ c_4 = \dfrac{9}{2}$

$$\text{Area} \approx \sum_{i=1}^{n} f(c_i)\Delta x = \sum_{i=1}^{4}\left[\ln(c_i)\right](1) \approx 0.40547 + 0.91629 + 1.25276 + 1.50408 \approx 4.0786$$

69.

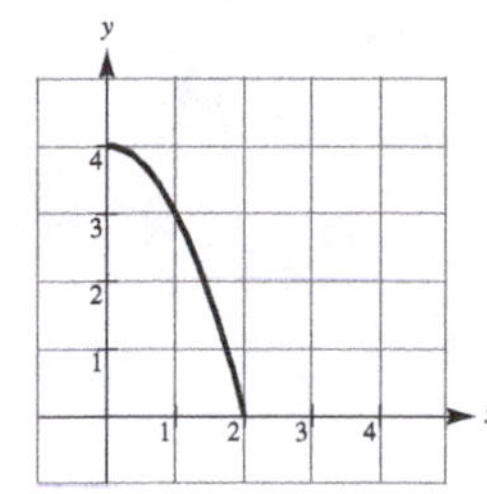

(b) $A \approx 6$ square units

71. In general, an overestimate on one side of the midpoint compensates for an underestimate on the other side.

73. 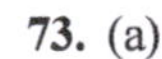

(a)

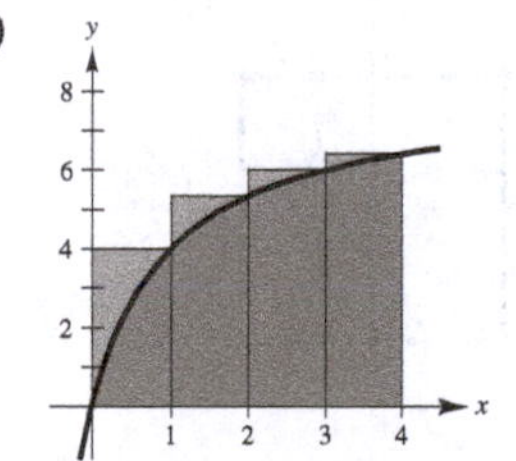

Lower sum:

$s(4) = 0 + 4 + 5\frac{1}{3} + 6 = 15\frac{1}{3} = \frac{46}{3} \approx 15.333$

(b)

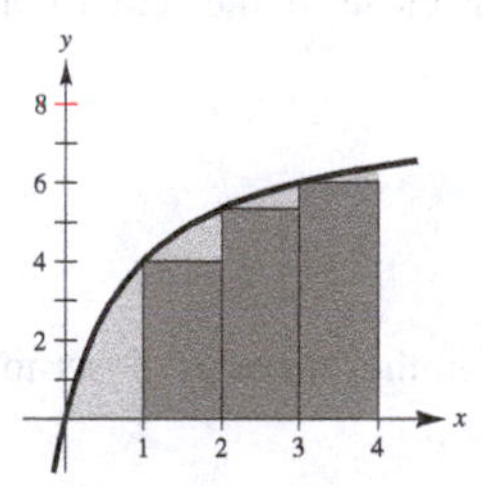

Upper sum:

$S(4) = 4 + 5\frac{1}{3} + 6 + 6\frac{2}{5} = 21\frac{11}{15} = \frac{326}{15} \approx 21.733$

(c)

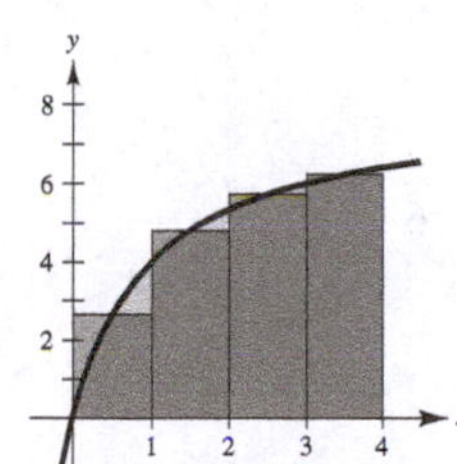

Midpoint Rule: $M(4) = 2\frac{2}{3} + 4\frac{4}{5} + 5\frac{5}{7} + 6\frac{2}{9} = \frac{6112}{315} \approx 19.403$

(d) In each case, $\Delta x = 4/n$. The lower sum uses left end-points, $(i - 1)(4/n)$.

The upper sum uses right endpoints, $(i)(4/n)$. The Midpoint Rule uses midpoints, $\left(i - \frac{1}{2}\right)(4/n)$.

(e)

N	4	8	20	100	200
$s(n)$	15.333	17.368	18.459	18.995	19.06
$S(n)$	21.733	20.568	19.739	19.251	19.188
$M(n)$	19.403	19.201	19.137	19.125	19.125

(f) $s(n)$ increases because the lower sum approaches the exact value as n increases. $S(n)$decreases because the upper sum approaches the exact value as n increases. Because of the shape of the graph, the lower sum is always smaller than the exact value, whereas the upper sum is always larger.

75. True. (Theorem 5.2 (2))

77. Suppose there are n rows and $n + 1$ columns in the figure. The stars on the left total $1 + 2 + \cdots + n$, as do the stars on the right. There are $n(n + 1)$ stars in total, so

$$2[1 + 2 + \cdots + n] = n(n + 1)$$
$$1 + 2 + \cdots + n = \tfrac{1}{2}(n)(n + 1).$$

79. If n is even, the number of seats is $\sum_{i=1}^{n/2} 2i = 2\frac{(n/2)(n/2 + 1)}{2} = \frac{n^2 + 2n}{4}$.

For example, if $n = 8$, then the number of seats is $8 + 6 + 4 + 2 = 20 = \frac{8^2 + 2(8)}{4}$.

If n is odd, the number of seats can be calculated from the case above:

$$\frac{(n + 1)^2 + 2(n + 1)}{4} - \frac{n + 1}{2} = \frac{(n + 1)^2}{4}.$$

For example, if $n = 9$, then the number of seats is $(10 + 8 + 6 + 4 + 2) - (5) = 30 - 5 = 25$.

81. Assume that the dartboard has corners at $(\pm 1, \pm 1)$.

A point (x, y) in the square is closer to the center than the top edge if

$$\sqrt{x^2 + y^2} \le 1 - y$$
$$x^2 + y^2 \le 1 - 2y + y^2$$
$$y \le \tfrac{1}{2}(1 - x^2).$$

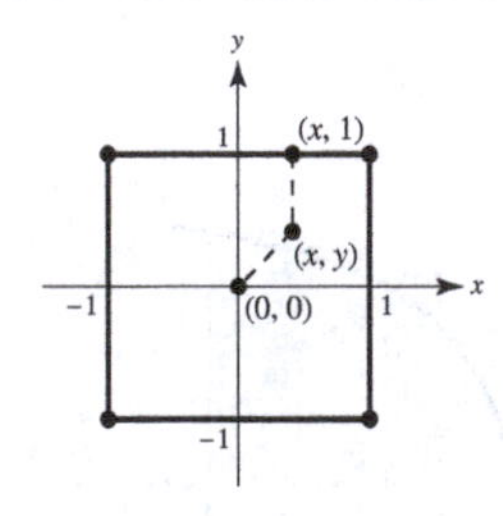

By symmetry, a point (x, y) in the square is closer to the center than the right edge if $x \le \tfrac{1}{2}(1 - y^2)$.

In the first quadrant, the parabolas $y = \tfrac{1}{2}(1 - x^2)$ and $x = \tfrac{1}{2}(1 - y^2)$ intersect at $(\sqrt{2} - 1, \sqrt{2} - 1)$.

There are 8 equal regions that make up the total region, as indicated in the figure.

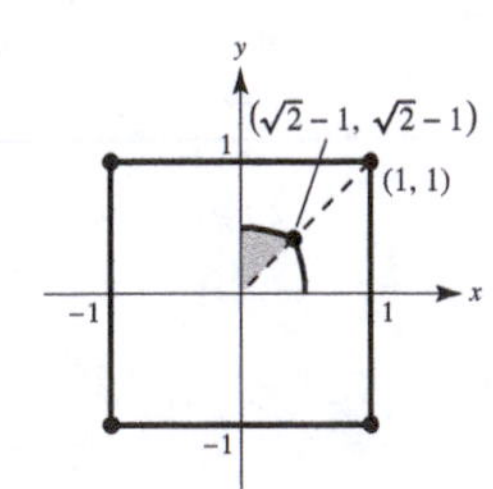

Area of shaded region $S = \int_0^{\sqrt{2}-1} \left[\frac{1}{2}(1 - x^2) - x\right] dx = \frac{2\sqrt{2}}{3} - \frac{5}{6}$

$$\text{Probability} = \frac{8S}{\text{Area square}} = 2\left[\frac{2\sqrt{2}}{3} - \frac{5}{6}\right] = \frac{4\sqrt{2}}{3} - \frac{5}{3}$$

Section 5.3 Riemann Sums and Definite Integrals

1. A Riemann sum represents the sum of all of the sub regions for a function f defined on the interval $[a, b]$.

3. $f(x) = \sqrt{x},\ y = 0,\ x = 0,\ x = 3,\ c_i = \dfrac{3i^2}{n^2}$

$$\Delta x_i = \frac{3i^2}{n^2} - \frac{3(i-1)^2}{n^2} = \frac{3}{n^2}(2i-1)$$

$$\begin{aligned}
\lim_{n\to\infty}\sum_{i=1}^{n} f(c_i)\Delta x_i &= \lim_{n\to\infty}\sum_{i=1}^{n}\sqrt{\frac{3i^2}{n^2}}\frac{3}{n^2}(2i-1)\\
&= \lim_{n\to\infty}\frac{3\sqrt{3}}{n^3}\sum_{i=1}^{n}\left(2i^2 - i\right)\\
&= \lim_{n\to\infty}\frac{3\sqrt{3}}{n^3}\left[2\frac{n(n+1)(2n+1)}{6} - \frac{n(n+1)}{2}\right]\\
&= \lim_{n\to\infty}3\sqrt{3}\left[\frac{(n+1)(2n+1)}{3n^2} - \frac{n+1}{2n^2}\right]\\
&= 3\sqrt{3}\left[\frac{2}{3} - 0\right] = 2\sqrt{3} \approx 3.464
\end{aligned}$$

5. $y = 8$ on $[2, 6]$. $\left(\textbf{Note: } \Delta x = \dfrac{6-2}{n} = \dfrac{4}{n},\ \|\Delta\| \to 0 \text{ as } n \to \infty\right)$

$$\sum_{i=1}^{n} f(c_i)\Delta x_i = \sum_{i=1}^{n} f\left(2 + \frac{4i}{n}\right)\left(\frac{4}{n}\right) = \sum_{i=1}^{n} 8\left(\frac{4}{n}\right) = \sum_{i=1}^{n}\frac{32}{n} = \frac{1}{n}\sum_{i=1}^{n} 32 = \frac{1}{n}(32n) = 32$$

$$\int_2^6 8\,dx = \lim_{n\to\infty} 32 = 32$$

7. $y = x^3$ on $[-1, 1]$. $\left(\textbf{Note: } \Delta x = \dfrac{1-(-1)}{n} = \dfrac{2}{n},\ \|\Delta\| \to 0 \text{ as } n \to \infty\right)$

$$\begin{aligned}
\sum_{i=1}^{n} f(c_i)\Delta x_i &= \sum_{i=1}^{n} f\left(-1 + \frac{2i}{n}\right)\left(\frac{2}{n}\right)\\
&= \sum_{i=1}^{n}\left(-1 + \frac{2i}{n}\right)^3\left(\frac{2}{n}\right)\\
&= \sum_{i=1}^{n}\left[-1 + \frac{6i}{n} - \frac{12i^2}{n^2} + \frac{8i^3}{n^3}\right]\left(\frac{2}{n}\right)\\
&= -2 + \frac{12}{n^2}\sum_{i=1}^{n} i - \frac{24}{n^3}\sum_{i=1}^{n} i^2 + \frac{16}{n^4}\sum_{i=1}^{n} i^3\\
&= -2 + 6\left(1 + \frac{1}{n}\right) - 4\left(2 + \frac{3}{n} + \frac{1}{n^2}\right) + 4\left(1 + \frac{2}{n} + \frac{1}{n^2}\right) = \frac{2}{n}
\end{aligned}$$

$$\int_{-1}^{1} x^3\,dx = \lim_{n\to\infty}\frac{2}{n} = 0$$

9. $y = x^2 + 1$ on $[1, 2]$. $\left(\textbf{Note: } \Delta x = \dfrac{2-1}{n} = \dfrac{1}{n}, \|\Delta\| \to 0 \text{ as } n \to \infty\right)$

$$\begin{aligned}
\sum_{i=1}^{n} f(c_i)\,\Delta x_i &= \sum_{i=1}^{n} f\left(1 + \frac{i}{n}\right)\left(\frac{1}{n}\right)\\
&= \sum_{i=1}^{n}\left[\left(1 + \frac{i}{n}\right)^2 + 1\right]\left(\frac{1}{n}\right)\\
&= \sum_{i=1}^{n}\left[1 + \frac{2i}{n} + \frac{i^2}{n^2} + 1\right]\left(\frac{1}{n}\right)\\
&= 2 + \frac{2}{n^2}\sum_{i=1}^{n} i + \frac{1}{n^3}\sum_{i=1}^{n} i^2 = 2 + \left(1 + \frac{1}{n}\right) + \frac{1}{6}\left(2 + \frac{3}{n} + \frac{1}{n^2}\right) = \frac{10}{3} + \frac{3}{2n} + \frac{1}{6n^2}
\end{aligned}$$

$$\int_1^2 \left(x^2 + 1\right) dx = \lim_{n\to\infty}\left(\frac{10}{3} + \frac{3}{2n} + \frac{1}{6n^2}\right) = \frac{10}{3}$$

11. $\displaystyle \lim_{\|\Delta\|\to 0}\sum_{i=1}^{n}(3c_i + 10)\,\Delta x_i = \int_{-1}^{5}(3x + 10)\,dx$

on the interval $[-1, 5]$.

13. $\displaystyle \lim_{\|\Delta\|\to 0}\sum_{i=1}^{n}\left(1 + \frac{3}{c_i}\right)\Delta x_i = \int_1^5 \left(1 + \frac{3}{x}\right) dx$

on the interval $[1, 5]$.

15. $\displaystyle \int_0^4 5\,dx$

17. $\displaystyle \int_{-4}^{4} \left(4 - |x|\right) dx$

19. $\displaystyle \int_{-5}^{5} \left(25 - x^2\right) dx$

21. $\displaystyle \int_0^{\pi/2} \cos x\,dx$

23. $\displaystyle \int_0^2 y^3\,dy$

25. $\displaystyle \int_1^4 \frac{2}{x}\,dx$

27. Rectangle

$A = bh = 3(4)$

$A = \displaystyle \int_0^3 4\,dx = 12$

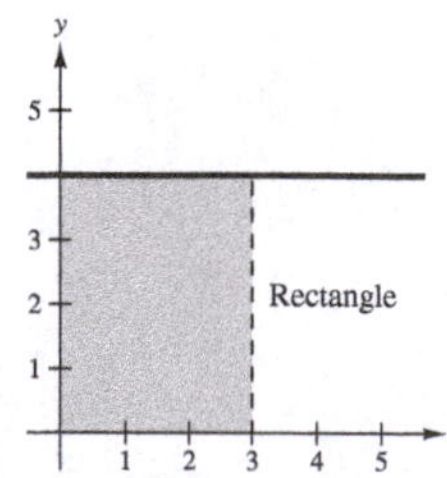

29. Triangle

$A = \frac{1}{2}bh = \frac{1}{2}(4)(4) = 8$

$A = \displaystyle \int_0^4 x\,dx = 8$

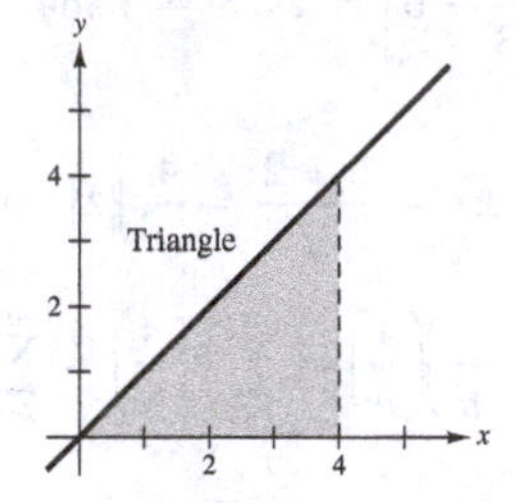

31. Trapezoid

$A = \dfrac{b_1 + b_2}{2}h = \left(\dfrac{4 + 10}{2}\right)2$

$= 14$

$A = \displaystyle \int_0^2 (3x + 4)\,dx = 14$

33. Triangle

$A = \frac{1}{2}bh = \frac{1}{2}(2)(1) = 1$

$A = \displaystyle \int_{-1}^{1} \left(1 - |x|\right) dx = 1$

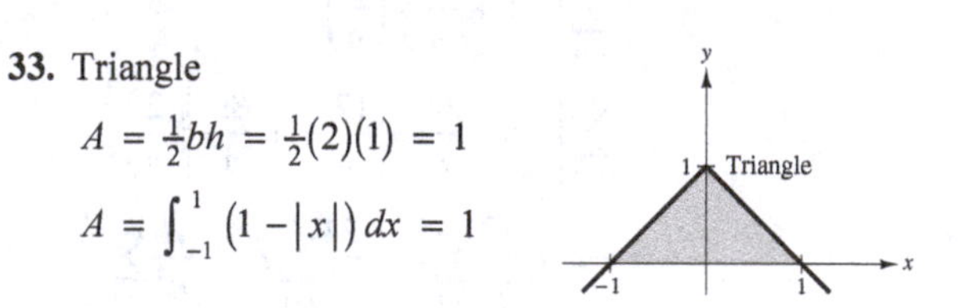

35. Semicircle

$A = \dfrac{1}{2}\pi r^2 = \dfrac{1}{2}\pi(7)^2$

$= \dfrac{49\pi}{2}$

$A = \displaystyle \int_{-7}^{7} \sqrt{49 - x^2}\,dx$

$= \dfrac{49\pi}{2}$

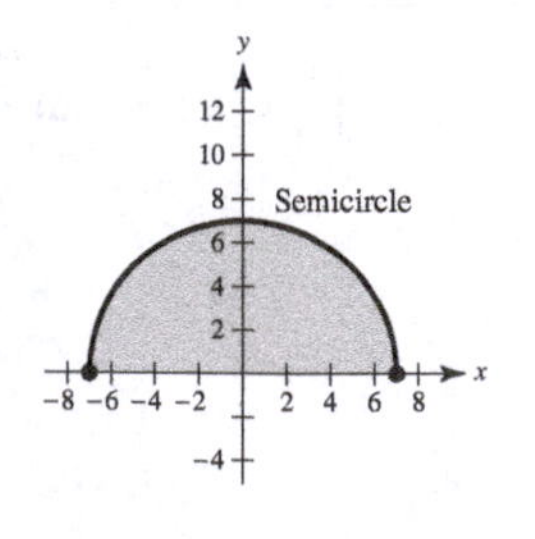

In Exercises 37–43, $\int_2^6 x^3\,dx = 320, \int_2^6 x\,dx = 16, \int_2^6 dx = 4.$

37. $\int_6^2 x^3\,dx = -\int_2^6 x^3\,dx = -320$

39. $\int_2^6 \frac{1}{4}x^3\,dx = \frac{1}{4}\int_2^6 x^3\,dx = \frac{1}{4}(320) = 80$

41. $\int_2^6 (x - 14)\,dx = \int_2^6 x\,dx - 14\int_2^6 dx = 16 - 14(4) = -40$

43. $\int_2^6 \left(2x^3 - 10x + 7\right)dx = 2\int_2^6 x^3\,dx - 10\int_2^6 x\,dx + 7\int_2^6 dx = 2(320) - 10(16) + 7(4) = 508$

45. (a) $\int_0^7 f(x)\,dx = \int_0^5 f(x)\,dx + \int_5^7 f(x)\,dx = 10 + 3 = 13$

(b) $\int_5^0 f(x)\,dx = -\int_0^5 f(x)\,dx = -10$

(c) $\int_5^5 f(x)\,dx = 0$

(d) $\int_0^5 3f(x)\,dx = 3\int_0^5 f(x)\,dx = 3(10) = 30$

47. (a) $\int_2^6 \left[f(x) + g(x)\right]dx = \int_2^6 f(x)\,dx + \int_2^6 g(x)\,dx$
$= 10 + (-2) = 8$

(b) $\int_2^6 \left[g(x) - f(x)\right]dx = \int_2^6 g(x)\,dx - \int_2^6 f(x)\,dx$
$= -2 - 10 = -12$

(c) $\int_2^6 2g(x)\,dx = 2\int_2^6 g(x)\,dx = 2(-2) = -4$

(d) $\int_2^6 3f(x)\,dx = 3\int_2^6 f(x)\,dx = 3(10) = 30$

49. Lower estimate: $[24 + 12 - 4 - 20 - 36](2) = -48$

Upper estimate: $[32 + 24 + 12 - 4 - 20](2) = 88$

51. (a) Quarter circle below x-axis:
$-\frac{1}{4}\pi r^2 = -\frac{1}{4}\pi(2)^2 = -\pi$

(b) Triangle: $\frac{1}{2}bh = \frac{1}{2}(4)(2) = 4$

(c) Triangle + Semicircle below x-axis:
$-\frac{1}{2}(2)(1) - \frac{1}{2}\pi(2)^2 = -(1 + 2\pi)$

(d) Sum of parts (b) and (c): $4 - (1 + 2\pi) = 3 - 2\pi$

(e) Sum of absolute values of (b) and (c):
$4 + (1 + 2\pi) = 5 + 2\pi$

(f) Answers to (d) plus
$2(10) = 20$: $(3 - 2\pi) + 20 = 23 - 2\pi$

53. (a) $\int_0^5 \left[f(x) + 2\right]dx = \int_0^5 f(x)\,dx + \int_0^5 2\,dx$
$= 4 + 10 = 14$

(b) $\int_{-2}^3 f(x + 2)\,dx = \int_0^5 f(x)\,dx = 4$ (Let $u = x + 2$.)

(c) $\int_{-5}^5 f(x)\,dx = 2\int_0^5 f(x)\,dx = 2(4) = 8$ (f even)

(d) $\int_{-5}^5 f(x)\,dx = 0$ (f odd)

55. $f(x) = \begin{cases} 4, & x < 4 \\ x, & x \ge 4 \end{cases}$

$\int_0^8 f(x)\,dx = 4(4) + 4(4) + \frac{1}{2}(4)(4) = 40$

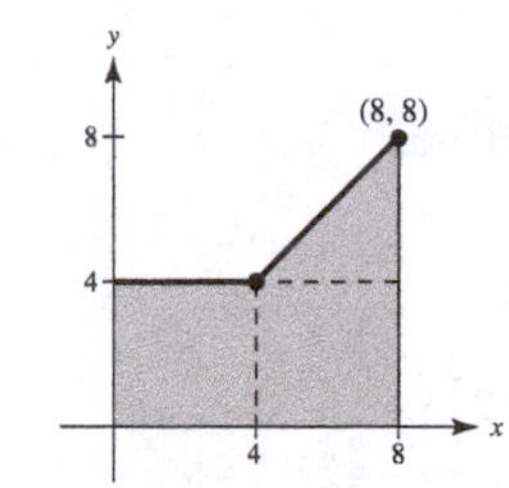

57.

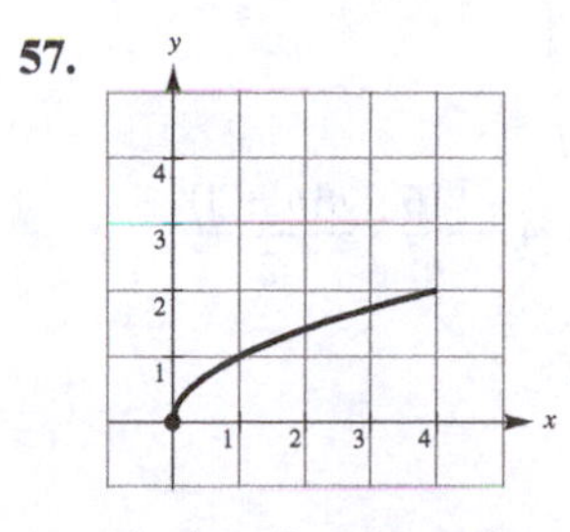

(a) $A \approx 5$ square units

59.

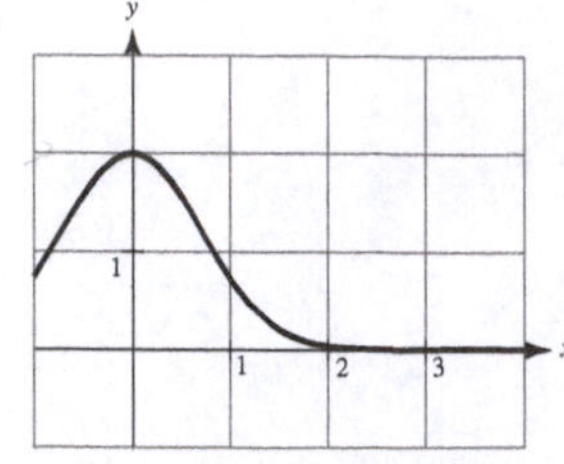

(c) $A \approx 2$ square units

61. Answers will vary. *Sample answer:*

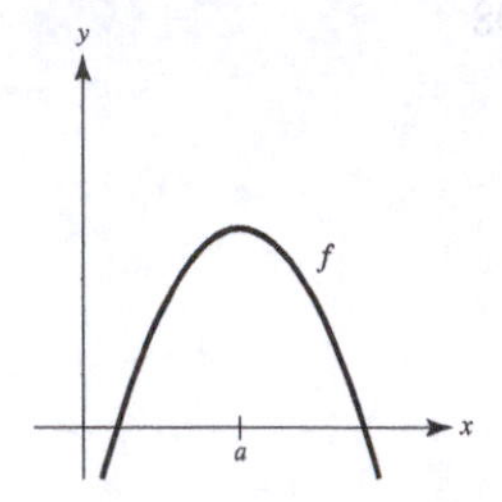

$\int_a^a f(x)\,dx = 0$ because there is no region.

63.

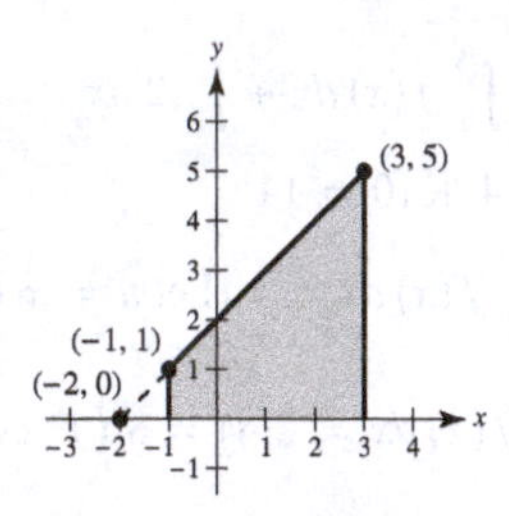

Method 1, difference of two triangles:

$$\int_{-1}^{3} (x + 2)\,dx = \int_{-2}^{3} (x + 2)\,dx - \int_{-2}^{1} (x + 2)\,dx$$
$$= \frac{1}{2}(5)(5) - \frac{1}{2}(1)(1)$$
$$= \frac{25}{2} - \frac{1}{2}$$
$$= 12$$

Method 2, limit definition of area:

$$\Delta x = \frac{4}{n}, x_i = -1 + i\Delta x$$

$$S(n) = \sum_{i=1}^{n} \left[(-1 + i\Delta x) + 2\right]\Delta x$$
$$= \sum_{i=1}^{n} \left[1 + i\left(\frac{4}{n}\right)\right]\left(\frac{4}{n}\right)$$
$$= \frac{4}{n}\sum_{i=1}^{n} 1 + \frac{16}{n^2}\sum_{i=1}^{n} i = 4 + \frac{16}{n^2} \cdot \frac{n(n+1)}{2}$$

$$\lim_{n\to\infty} S(n) = 4 + 8 = 12$$

65. $\int_{-2}^{1} f(x)\,dx + \int_{1}^{5} f(x)\,dx = \int_{-2}^{5} f(x)\,dx$

$a = -2,\ b = 5$

67. Answers will vary.

Sample answer:

$a = \pi,\ b = 2\pi$

$\int_{\pi}^{2\pi} \sin x\,dx < 0$

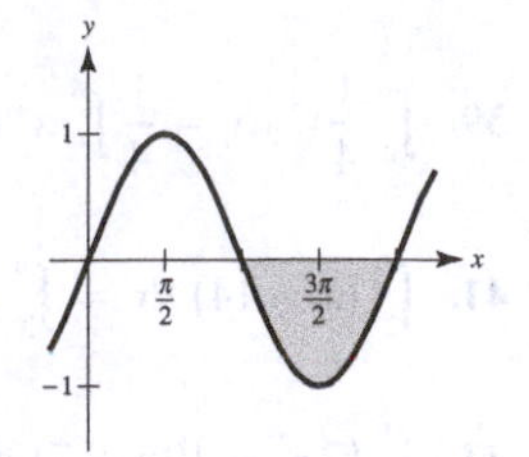

69. True

71. True

73. False

$\int_0^2 (-x)\,dx = -2$

75. $f(x) = x^2 + 3x,\ [0, 8]$

$x_0 = 0,\ x_1 = 1,\ x_2 = 3,\ x_3 = 7,\ x_4 = 8$

$\Delta x_1 = 1,\ \Delta x_2 = 2,\ \Delta x_3 = 4,\ \Delta x_4 = 1$

$c_1 = 1, c_2 = 2,\ c_3 = 5,\ c_4 = 8$

$$\sum_{i=1}^{4} f(c_i)\Delta x = f(1)\Delta x_1 + f(2)\Delta x_2 + f(5)\Delta x_3 + f(8)\Delta x_4$$
$$= (4)(1) + (10)(2) + (40)(4) + (88)(1)$$
$$= 272$$

77. $\Delta x = \frac{b-a}{n},\ c_i = a + i(\Delta x) = a + i\left(\frac{b-a}{n}\right)$

$$\int_0^b x\,dx = \lim_{\|\Delta\|\to 0} \sum_{i=1}^{n} f(c_i)\Delta x$$
$$= \lim_{n\to\infty} \sum_{i=1}^{n} \left[a + i\left(\frac{b-a}{n}\right)\right]\left(\frac{b-a}{n}\right)$$
$$= \lim_{n\to\infty} \left[\left(\frac{b-a}{n}\right)\sum_{i=1}^{n} a + \left(\frac{b-a}{n}\right)^2 \sum_{i=1}^{n} i\right]$$
$$= \lim_{n\to\infty} \left[\frac{b-a}{n}(an) + \left(\frac{b-a}{n}\right)^2 \frac{n(n+1)}{2}\right]$$
$$= \lim_{n\to\infty} \left[a(b-a) + \frac{(b-a)^2}{n}\frac{n+1}{2}\right]$$
$$= a(b-a) + \frac{(b-a)^2}{2} = (b-a)\left[a + \frac{b-a}{2}\right]$$
$$= \frac{(b-a)(a+b)}{2} = \frac{b^2 - a^2}{2}$$

79. $f(x) = \begin{cases} 1, & x \text{ is rational} \\ 0, & x \text{ is irrational} \end{cases}$

is not integrable on the interval $[0, 1]$. As $\|\Delta\| \to 0$, $f(c_i) = 1$ or $f(c_i) = 0$ in each subinterval because there are an infinite number of both rational and irrational numbers in any interval, no matter how small.

81. The function f is nonnegative between $x = -1$ and $x = 1$.

So, $\int_a^b (1 - x^2)\, dx$

is a maximum for $a = -1$ and $b = 1$.

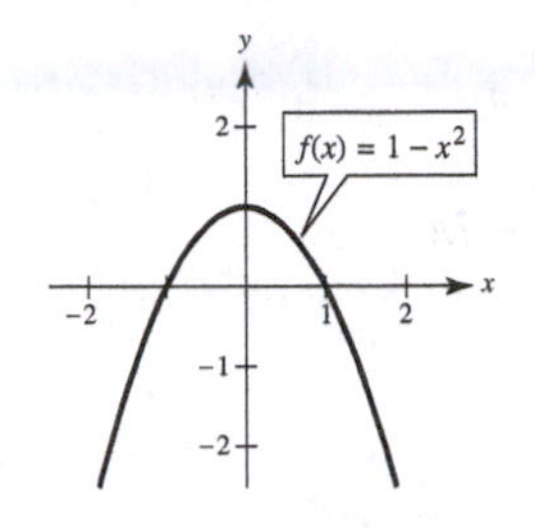

83. Answers will vary. *Sample answer:*

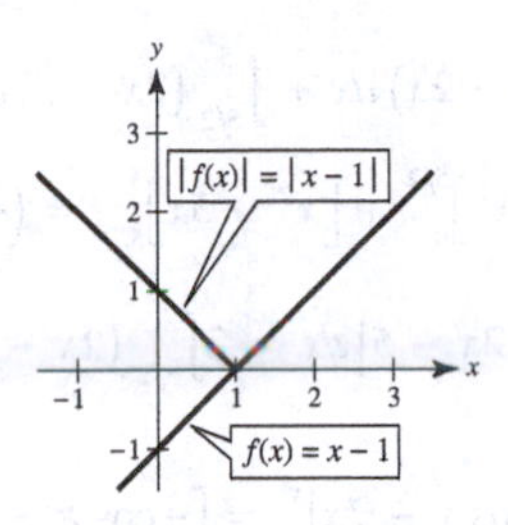

The integrals are equal when f is always greater than or equal to 0 on $[a, b]$.

85. Let $f(x) = x^2$, $0 \le x \le 1$, and $\Delta x_i = 1/n$. The appropriate Riemann Sum is

$$\sum_{i=1}^{n} f(c_i)\Delta x_i = \sum_{i=1}^{n}\left(\frac{i}{n}\right)^2 \frac{1}{n} = \frac{1}{n^3}\sum_{i=1}^{n} i^2.$$

$$\lim_{n\to\infty} \frac{1}{n^3}\left[1^2 + 2^2 + 3^2 + \cdots + n^2\right] = \lim_{n\to\infty} \frac{1}{n^3} \cdot \frac{n(2n+1)(n+1)}{6} = \lim_{n\to\infty} \frac{2n^2 + 3n + 1}{6n^2} = \lim_{n\to\infty}\left(\frac{1}{3} + \frac{1}{2n} + \frac{1}{6n^2}\right) = \frac{1}{3}$$

Section 5.4 The Fundamental Theorem of Calculus

1. Find an antiderivative F of the function and evaluate the difference of the antiderivative at the upper limit of integration and the lower limit of integration, $F(b) - F(a)$.

3. The average value of a function on an interval is the integral of the function on $[a, b]$ times $\dfrac{1}{b - a}$.

5. $f(x) = \dfrac{4}{x^2 + 1}$

$\displaystyle\int_0^{\pi} \frac{4}{x^2 + 1}\, dx$ is positive.

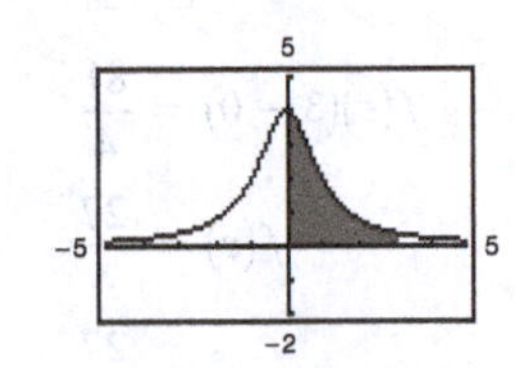

7. $f(x) = x\sqrt{x^2 + 1}$

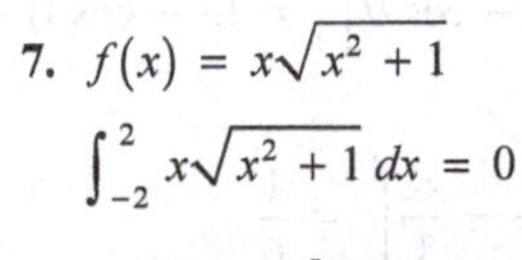

$\displaystyle\int_{-2}^{2} x\sqrt{x^2 + 1}\, dx = 0$

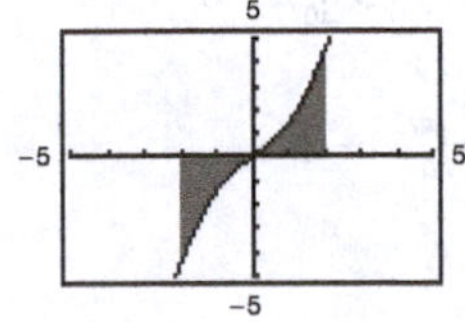

9. $\displaystyle\int_{-1}^{0} (2x - 1)\, dx = \left[x^2 - x\right]_{-1}^{0}$

$= 0 - \left((-1)^2 - (-1)\right) = -(1 + 1) = -2$

11. $\displaystyle\int_0^1 (2t - 1)^2\, dt = \int_0^1 \left(4t^2 - 4t + 1\right) dt = \left[\tfrac{4}{3}t^3 - 2t^2 + t\right]_0^1 = \tfrac{4}{3} - 2 + 1 = \tfrac{1}{3}$

13. $\displaystyle\int_1^2 \left(\frac{3}{x^2} - 1\right) dx = \left[-\frac{3}{x} - x\right]_1^2 = \left(-\frac{3}{2} - 2\right) - (-3 - 1) = \frac{1}{2}$

15. $\int_1^4 \frac{u-2}{\sqrt{u}}\,du = \int_1^4 \left(u^{1/2} - 2u^{-1/2}\right) du = \left[\frac{2}{3}u^{3/2} - 4u^{1/2}\right]_1^4 = \left[\frac{2}{3}\left(\sqrt{4}\right)^3 - 4\sqrt{4}\right] - \left[\frac{2}{3} - 4\right] = \frac{2}{3}$

17. $\int_{-1}^1 \left(\sqrt[3]{t} - 2\right) dt = \left[\frac{3}{4}t^{4/3} - 2t\right]_{-1}^1 = \left(\frac{3}{4} - 2\right) - \left(\frac{3}{4} + 2\right) = -4$

19. $\int_{-1}^0 \left(t^{1/3} - t^{2/3}\right) dt = \left[\frac{3}{4}t^{4/3} - \frac{3}{5}t^{5/3}\right]_{-1}^0 = 0 - \left(\frac{3}{4} + \frac{3}{5}\right) = -\frac{27}{20}$

21. $\int_0^5 |2x - 5|\,dx = \int_0^{5/2} (5 - 2x)\,dx + \int_{5/2}^5 (2x - 5)\,dx$ (split up the integral at the zero $x = \frac{5}{2}$)

$= \left[5x - x^2\right]_0^{5/2} + \left[x^2 - 5x\right]_{5/2}^5 = \left(\frac{25}{2} - \frac{25}{4}\right) - 0 + (25 - 25) - \left(\frac{25}{4} - \frac{25}{2}\right) = 2\left(\frac{25}{2} - \frac{25}{4}\right) = \frac{25}{2}$

Note: By Symmetry, $\int_0^5 |2x - 5|\,dx = 2\int_{5/2}^5 (2x - 5)\,dx$.

23. $\int_0^\pi (\sin x - 7)\,dx = \left[-\cos x - 7x\right]_0^\pi = \left[-\cos\pi - 7(\pi)\right] - (-\cos 0 - 0) = 1 - 7\pi + 1 = 2 - 7\pi$

25. $\int_0^{\pi/4} \frac{1 - \sin^2\theta}{\cos^2\theta}\,d\theta = \int_0^{\pi/4} d\theta = \left[\theta\right]_0^{\pi/4} = \frac{\pi}{4}$

27. $\int_{-\pi/6}^{\pi/6} \sec^2 x\,dx = \left[\tan x\right]_{-\pi/6}^{\pi/6} = \frac{\sqrt{3}}{3} - \left(-\frac{\sqrt{3}}{3}\right) = \frac{2\sqrt{3}}{3}$

29. $\int_{-\pi/3}^{\pi/3} 4\sec\theta\tan\theta\,d\theta = \left[4\sec\theta\right]_{-\pi/3}^{\pi/3} = 4(2) - 4(2) = 0$

31. $\int_0^2 \left(2^x + 6\right) dx = \left[\frac{2^x}{\ln 2} + 6x\right]_0^2 = \left(\frac{4}{\ln 2} + 12\right) - \left(\frac{1}{\ln 2} + 0\right) = \frac{3}{\ln 2} + 12$

33. $\int_{-1}^1 \left(e^\theta + \sin\theta\right) d\theta = \left[e^\theta - \cos\theta\right]_{-1}^1 = (e - \cos 1) - \left[e^{-1} - \cos(-1)\right] = e - \frac{1}{e}$

35. $A = \int_0^1 \left(x - x^2\right) dx = \left[\frac{x^2}{2} - \frac{x^3}{3}\right]_0^1 = \frac{1}{6}$

37. $A = \int_0^{\pi/2} \cos x\,dx = \left[\sin x\right]_0^{\pi/2} = 1$

39. Because $y > 0$ on $[0, 2]$,

Area $= \int_0^2 \left(5x^2 + 2\right) dx = \left[\frac{5}{3}x^3 + 2x\right]_0^2$

$= \frac{40}{3} + 4 = \frac{52}{3}.$

41. Because $y > 0$ on $[0, 8]$,

Area $= \int_0^8 \left(1 + x^{1/3}\right) dx = \left[x + \frac{3}{4}x^{4/3}\right]_0^8$

$= 8 + \frac{3}{4}(16) = 20.$

43. Because $y > 0$ on $[1, e]$,

Area $= \int_1^e \frac{4}{x}\,dx = \left[4\ln x\right]_1^e = 4\ln e - 4\ln 1 = 4.$

45. $\int_0^3 x^3\,dx = \left[\frac{x^4}{4}\right]_0^3 = \frac{81}{4}$

$f(c)(3 - 0) = \frac{81}{4}$

$f(c) = \frac{27}{4}$

$c^3 = \frac{27}{4}$

$c = \frac{3}{\sqrt[3]{4}} = \frac{3}{2}\sqrt[3]{2} \approx 1.8899$

47. $\int_1^4 \left(5 - \frac{1}{x}\right) dx = \left[5x - \ln x\right]_1^4$

$= (20 - \ln 4) - (5 - 0) = 15 - \ln 4$

$f(c)(4 - 1) = 15 - \ln 4$

$\left(5 - \frac{1}{c}\right)(3) = 15 - \ln 4$

$15 - \frac{3}{c} = 15 - \ln 4$

$\frac{3}{c} = \ln 4$

$c = \frac{3}{\ln 4} \approx 2.1640$

49. $\int_{-\pi/4}^{\pi/4} 2\sec^2 x\, dx = \left[2 \tan x\right]_{-\pi/4}^{\pi/4} = 2(1) - 2(-1) = 4$

$f(c)\left[\frac{\pi}{4} - \left(-\frac{\pi}{4}\right)\right] = 4$

$2\sec^2 c = \frac{8}{\pi}$

$\sec^2 c = \frac{4}{\pi}$

$\sec c = \pm\frac{2}{\sqrt{\pi}}$

$c = \pm\text{arcsec}\left(\frac{2}{\sqrt{\pi}}\right)$

$= \pm\arccos\frac{\sqrt{\pi}}{2} \approx \pm 0.4817$

51. $f(x) = 4 - x^2, [-2, 2]$

$\frac{1}{2 - (-2)}\int_{-2}^{2} (4 - x^2)\, dx = \frac{1}{4}\left[4x - \frac{1}{3}x^3\right]_{-2}^{2}$

$= \frac{1}{4}\left[\left(8 - \frac{8}{3}\right) - \left(-8 + \frac{8}{3}\right)\right]$

$= \frac{8}{3}$

Average value $= \frac{8}{3}$

$4 - x^2 = \frac{8}{3}$ when $x^2 = 4 - \frac{8}{3}$ or

$x = \pm\sqrt{\frac{4}{3}} = \pm\frac{2\sqrt{3}}{3} \approx \pm 1.1547$

53. $\frac{1}{1 - (-1)}\int_{-1}^{1} 2e^x\, dx = \int_{-1}^{1} e^x\, dx$

$= \left[e^x\right]_{-1}^{1} = e - e^{-1} \approx 2.3504$

Average value $= e - e^{-1} \approx 2.3504$

$2e^x = e - e^{-1}$

$e^x = \frac{1}{2}\left(e - e^{-1}\right)$

$x = \ln\left(\frac{e - e^{-1}}{2}\right) \approx 0.1614$

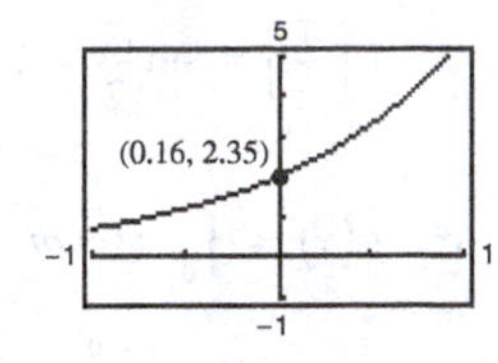

55. $f(x) = \sin x, [0, \pi]$

$\frac{1}{\pi - 0}\int_0^{\pi} \sin x\, dx = \left[-\frac{1}{\pi}\cos x\right]_0^{\pi} = \frac{2}{\pi}$

Average value $= \frac{2}{\pi}$

$\sin x = \frac{2}{\pi}$

$x \approx 0.690, 2.451$

57. (a) $F(x) = k\sec^2 x$

$F(0) = k = 500$

$F(x) = 500\sec^2 x$

(b) $\frac{1}{\pi/3 - 0}\int_0^{\pi/3} 500\sec^2 x\, dx = \frac{1500}{\pi}\left[\tan x\right]_0^{\pi/3}$

$= \frac{1500}{\pi}\left(\sqrt{3} - 0\right)$

≈ 826.99 newtons

≈ 827 newtons

59. $P = \frac{2}{\pi}\int_0^{\pi/2} \sin\theta\, d\theta$

$= \left[-\frac{2}{\pi}\cos\theta\right]_0^{\pi/2}$

$= -\frac{2}{\pi}(0 - 1)$

$= \frac{2}{\pi} \approx 63.7\%$

61. $F(x) = \int_1^x \frac{20}{v^2}\, dv = \int_1^x 20v^{-2}\, dv = -\frac{20}{v}\Big]_1^x$

$= -\frac{20}{x} + 20 = 20\left(1 - \frac{1}{x}\right)$

$F(2) = 20\left(\frac{1}{2}\right) = 10$

$F(5) = 20\left(\frac{4}{5}\right) = 16$

$F(8) = 20\left(\frac{7}{8}\right) = \frac{35}{2}$

63. $F(x) = \int_0^x \cos\theta\, d\theta = [\sin\theta]_0^x = \sin x - \sin 0 = \sin x$

$F(0) = 0$

$F\left(\frac{\pi}{4}\right) = \sin\frac{\pi}{4} = \frac{\sqrt{2}}{2}$

$F\left(\frac{\pi}{2}\right) = \sin\frac{\pi}{2} = 1$

65. $g(x) = \int_0^x f(t)\, dt$

(a) $g(0) = \int_0^0 f(t)\, dt = 0$

$g(2) = \int_0^2 f(t)\, dt \approx 4 + 2 + 1 = 7$

$g(4) = \int_0^4 f(t)\, dt \approx 7 + 2 = 9$

$g(6) = \int_0^6 f(t)\, dt \approx 9 + (-1) = 8$

$g(8) = \int_0^8 f(t)\, dt \approx 8 - 3 = 5$

(b) g increasing on (0, 4) and decreasing on (4, 8)

(c) g is a maximum of 9 at $x = 4$.

(d)

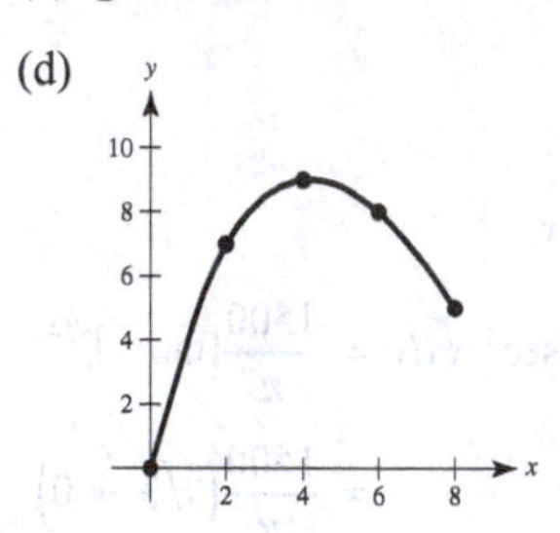

67. (a) $F(x) = \int_0^x (t + 2)\, dt = \left[\frac{t^2}{2} + 2t\right]_0^x = \frac{1}{2}x^2 + 2x$

(b) $\frac{d}{dx}\left[\frac{1}{2}x^2 + 2x\right] = x + 2$

69. (a) $F(x) = \int_8^x \sqrt[3]{t}\, dt = \left[\frac{3}{4}t^{4/3}\right]_8^x = \frac{3}{4}\left(x^{4/3} - 16\right)$

$= \frac{3}{4}x^{4/3} - 12$

(b) $\frac{d}{dx}\left[\frac{3}{4}x^{4/3} - 12\right] = x^{1/3} = \sqrt[3]{x}$

71. (a) $F(x) = \int_{\pi/4}^x \sec^2 t\, dt = [\tan t]_{\pi/4}^x = \tan x - 1$

(b) $\frac{d}{dx}[\tan x - 1] = \sec^2 x$

73. (a) $F(x) = \int_{-1}^x e^t\, dt = \left[e^t\right]_{-1}^x = e^x - e^{-1}$

(b) $\frac{d}{dx}\left(e^x - e^{-1}\right) = e^x$

75. $F(x) = \int_{-2}^x \left(t^2 - 2t\right) dt$

$F'(x) = x^2 - 2x$

77. $F(x) = \int_{-1}^x \sqrt{t^4 + 1}\, dt$

$F'(x) = \sqrt{x^4 + 1}$

79. $F(x) = \int_1^x \sqrt{t} \csc t\, dt$

$F'(x) = \sqrt{x} \csc x$

81. $F(x) = \int_x^{x+2} (4t + 1)\, dt$

$= \left[2t^2 + t\right]_x^{x+2}$

$= \left[2(x + 2)^2 + (x + 2)\right] - \left[2x^2 + x\right]$

$= 8x + 10$

$F'(x) = 8$

Alternate solution:

$F(x) = \int_x^{x+2} (4t + 1)\, dt$

$= \int_x^0 (4t + 1)\, dt + \int_0^{x+2} (4t + 1)\, dt$

$= -\int_0^x (4t + 1)\, dt + \int_0^{x+2} (4t + 1)\, dt$

$F'(x) = -(4x + 1) + 4(x + 2) + 1 = 8$

83. $F(x) = \int_0^{\sin x} \sqrt{t}\, dt = \left[\frac{2}{3}t^{3/2}\right]_0^{\sin x} = \frac{2}{3}(\sin x)^{3/2}$

$F'(x) = (\sin x)^{1/2} \cos x = \cos x\sqrt{\sin x}$

Alternate solution:

$F(x) = \int_0^{\sin x} \sqrt{t}\, dt$

$F'(x) = \sqrt{\sin x}\frac{d}{dx}(\sin x) = \sqrt{\sin x}(\cos x)$

85. $F(x) = \int_0^{x^3} \sin t^2\, dt$

$F'(x) = \sin\left(x^3\right)^2 \cdot 3x^2 = 3x^2 \sin x^6$

87. $g(x) = \int_0^x f(t)\,dt$

$g(0) = 0, g(1) \approx \frac{1}{2}, g(2) \approx 1, g(3) \approx \frac{1}{2}, g(4) = 0$

g has a relative maximum at $x = 2$.

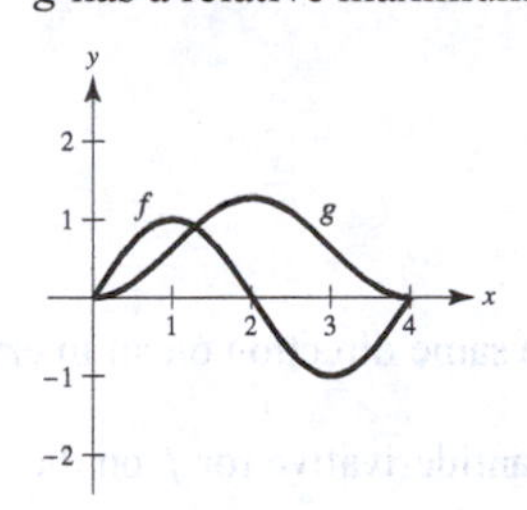

89. Let $c(t)$ be the amount of water that is flowing out of the tank. Then $c'(t) = 500 - 5t$ liters per minute is the rate of flow.

$$\int_0^{18} c'(t)dt = \int_0^{18} (500 - 5t)\,dt$$

$$= \left[500t - \frac{5t^2}{2}\right]_0^{18}$$

$$= 9000 - 810 = 8190 \text{ L}$$

91. The distance traveled is $\int_0^8 v(t)\,dt$. The area under the curve from $0 \le t \le 8$ is approximately $(18 \text{ squares})(30) \approx 540$ feet.

93. (a) $v(t) = 5t - 7,\ 0 \le t \le 3$

$$\text{Displacement} = \int_0^3 (5t - 7)\,dt = \left[\frac{5t^2}{2} - 7t\right]_0^3 = \frac{45}{2} - 21 = \frac{3}{2} \text{ ft to the right}$$

(b) Total distance traveled $= \int_0^3 |5t - 7|\,dt$

$$= \int_0^{7/5} (7 - 5t)\,dt + \int_{7/5}^3 (5t - 7)\,dt$$

$$= \left[7t - \frac{5t^2}{2}\right]_0^{7/5} + \left[\frac{5t^2}{2} - 7t\right]_{7/5}^3$$

$$= 7\left(\frac{7}{5}\right) - \frac{5}{2}\left(\frac{7}{5}\right)^2 + \left(\frac{5}{2}(9) - 21\right) - \left(\frac{5}{2}\left(\frac{7}{5}\right)^2 - 7\left(\frac{7}{5}\right)\right)$$

$$= \frac{49}{5} - \frac{49}{10} + \frac{45}{2} - 21 - \frac{49}{10} + \frac{49}{5} = \frac{113}{10} \text{ ft}$$

95. (a) $v(t) = t^3 - 10t^2 + 27t - 18 = (t - 1)(t - 3)(t - 6),\ 1 \le t \le 7$

$$\text{Displacement} = \int_1^7 \left(t^3 - 10t^2 + 27t - 18\right)dt$$

$$= \left[\frac{t^4}{4} - \frac{10t^3}{3} + \frac{27t^2}{2} - 18t\right]_1^7$$

$$= \left[\frac{7^4}{4} - \frac{10(7^3)}{3} + \frac{27(7^2)}{2} - 18(7)\right] - \left[\frac{1}{4} - \frac{10}{3} + \frac{27}{2} - 18\right]$$

$$= -\frac{91}{12} - \left(-\frac{91}{12}\right) = 0$$

(b) Total distance traveled $= \int_1^7 |v(t)|\,dt$

$$= \int_1^3 \left(t^3 - 10t^2 + 27t - 18\right)dt - \int_3^6 \left(t^3 - 10t^2 + 27t - 18\right)dt + \int_6^7 \left(t^3 - 10t^2 + 27t - 18\right)dt$$

Evaluating each of these integrals, you obtain

Total distance $= \frac{16}{3} - \left(-\frac{63}{4}\right) + \frac{125}{12} = \frac{63}{2}$ ft

97. (a) $v(t) = \frac{1}{\sqrt{t}}, 1 \le t \le 4$

Because $v(t) > 0$,

Displacement = Total Distance

$$\text{Displacement} = \int_1^4 t^{-1/2}\,dt = \left[2t^{1/2}\right]_1^4 = 4 - 2 = 2 \text{ ft to the right}$$

(b) Total distance = 2 ft

99. The displacement and total distance traveled are equal when the particle is always moving in the same direction on an interval.

101. The Fundamental Theorem of Calculus requires that f be continuous on $[a, b]$ and that F be an antiderivative for f on the entire interval. On an interval containing c, the function $f(x) = \frac{1}{x - c}$ is not continuous at c.

103. $x(t) = t^3 - 6t^2 + 9t - 2$

$x'(t) = 3t^2 - 12t + 9 = 3(t^2 - 4t + 3) = 3(t - 3)(t - 1)$

$$\text{Total distance} = \int_0^5 |x'(t)|\,dt = \int_0^5 3|(t - 3)(t - 1)|\,dt$$

$$= 3\int_0^1 (t^2 - 4t + 3)\,dt - 3\int_1^3 (t^2 - 4t + 3)\,dt + 3\int_3^5 (t^2 - 4t + 3)\,dt = 4 + 4 + 20 = 28 \text{ units}$$

105. The function $f(x) = x^{-2}$ is not continuous on $[-1, 1]$.

$$\int_{-1}^1 x^{-2}\,dx = \int_{-1}^0 x^{-2}\,dx + \int_0^1 x^{-2}\,dx$$

Each of these integrals is infinite. $f(x) = x^{-2}$ has a nonremovable discontinuity at $x = 0$.

107. The function $f(x) = \sec^2 x$ is not continuous on $\left[\frac{\pi}{4}, \frac{3\pi}{4}\right]$.

$$\int_{\pi/4}^{3\pi/4} \sec^2 x\,dx = \int_{\pi/4}^{\pi/2} \sec^2 x\,dx + \int_{\pi/2}^{3\pi/4} \sec^2 x\,dx$$

Each of these integrals is infinite. $f(x) = \sec^2 x$ has a nonremovable discontinuity at $x = \frac{\pi}{2}$.

109. True

111. $f(x) = \int_0^{1/x} \frac{1}{t^2 + 1}\,dt + \int_0^x \frac{1}{t^2 + 1}\,dt$

By the Second Fundamental Theorem of Calculus, you have

$$f'(x) = \frac{1}{(1/x)^2 + 1}\left(-\frac{1}{x^2}\right) + \frac{1}{x^2 + 1}$$

$$= -\frac{1}{1 + x^2} + \frac{1}{x^2 + 1} = 0.$$

Because $f'(x) = 0$, $f(x)$ must be constant.

113. $G(x) = \int_0^x \left[s\int_0^s f(t)\,dt\right] ds$

(a) $G(0) = \int_0^0 \left[s\int_0^s f(t)\,dt\right] ds = 0$

(b) Let $F(s) = s\int_0^s f(t)\,dt$.

$G(x) = \int_0^x F(s)\,ds$

$G'(x) = F(x) = x\int_0^x f(t)\,dt$

$G'(0) = 0\int_0^0 f(t)\,dt = 0$

(c) $G''(x) = x \cdot f(x) + \int_0^x f(t)\,dt$

(d) $G''(0) = 0 \cdot f(0) + \int_0^0 f(t)\,dt = 0$

115. $I(f) - J(f) = \int_0^1 x^2 f(x)\,dx - \int_0^1 xf(x)^2\,dx.$

Observe that

$$\frac{x^3}{4} - x\left(f(x) - \frac{x}{2}\right)^2 = \frac{x^3}{4} - x\left(f(x)^2 - xf(x) + \frac{x^2}{4}\right) = \frac{x^3}{4} - xf(x)^2 + x^2f(x) - \frac{x^3}{4} = x^2f(x) - xf(x)^2$$

So, $I(f) - J(f) = \int_0^1 \left[x^2f(x) - xf(x)^2\right]dx = \int_0^1 \left[\frac{x^3}{4} - x\left(f(x) - \frac{x}{2}\right)^2\right]dx \le \int_0^1 \frac{x^3}{4}\,dx = \frac{1}{16}$

Furthermore, $6 + f(x) = \frac{x}{2}$. Then $I(f) = \int_0^1 x^2\left(\frac{x}{2}\right)dx = \frac{1}{8}$ and $J(f) = \int_0^1 x\left(\frac{x^2}{4}\right) = \frac{1}{16}$

So $I(f) - J(f) = \frac{1}{8} - \frac{1}{16} = \frac{1}{16}$

The maximum value is $\frac{1}{16}$.

Section 5.5 Integration by Substitution

1. Multiply and divide by a constant, if necessary.

3. The integral of $\left[g(x)\right]^n g'(x)$ is $\frac{\left[g(x)\right]^{n+1}}{n+1} + C,\ n \ne -1.$

Recall the power rule for polynomials.

$\int f(g(x))g'(x)\,dx$	$u = g(x)$	$du = g'(x)\,dx$
5. $\int (5x^2 + 1)^2(10x)\,dx$	$5x^2 + 1$	$10x\ dx$
7. $\int \tan^2 x \sec^2 x\,dx$	$\tan x$	$\sec^2 x\,dx$

9. $\int (1 + 6x)^4(6)\,dx = \frac{(1 + 6x)^5}{5} + C$

Check: $\frac{d}{dx}\left[\frac{(1 + 6x)^5}{5} + C\right] = 6(1 + 6x)^4$

11. $\int \sqrt{25 - x^2}(-2x)\,dx = \frac{(25 - x^2)^{3/2}}{3/2} + C$

$$= \frac{2}{3}(25 - x^2)^{3/2} + C$$

Check:

$$\frac{d}{dx}\left[\frac{2}{3}(25 - x^2)^{3/2} + C\right] = \frac{2}{3}\left(\frac{3}{2}\right)(25 - x^2)^{1/2}(-2x) = \sqrt{25 - x^2}(-2x)$$

13. $\int x^3(x^4 + 3)^2\,dx = \frac{1}{4}\int (x^4 + 3)^2(4x^3)\,dx = \frac{1}{4}\frac{(x^4 + 3)^3}{3} + C = \frac{(x^4 + 3)^3}{12} + C$

Check: $\frac{d}{dx}\left[\frac{(x^4 + 3)^3}{12} + C\right] = \frac{3(x^4 + 3)^2}{12}(4x^3) = (x^4 + 3)^2(x^3)$

15. $\int x^2(2x^3 - 1)^4\,dx = \frac{1}{6}\int (2x^3 - 1)^4(6x^2)\,dx = \frac{1}{6}\left[\frac{1}{5}(2x^3 - 1)^5\right] + C = \frac{(2x^3 - 1)^5}{30} + C$

Check: $\frac{d}{dx}\left[\frac{(2x^3 - 1)^5}{30} + C\right] = 5\frac{(2x^3 - 1)^4(6x^2)}{30} = x^2(2x^3 - 1)^4$

17. $\int t\sqrt{t^2+2}\,dt = \frac{1}{2}\int (t^2+2)^{1/2}(2t)\,dt = \frac{1}{2}\frac{(t^2+2)^{3/2}}{3/2} + C = \frac{(t^2+2)^{3/2}}{3} + C$

Check: $\frac{d}{dt}\left[\frac{(t^2+2)^{3/2}}{3} + C\right] = \frac{3/2(t^2+2)^{1/2}(2t)}{3} = (t^2+2)^{1/2}t$

19. $\int \frac{7x}{(1-x^2)^3}\,dx = \frac{7}{-2}\int (1-x^2)^{-3}(-2x)\,dx = \frac{7}{-2}\frac{(1-x^2)^{-2}}{(-2)} + C = \frac{7}{4(1-x^2)^2} + C$

Check: $\frac{d}{dx}\left[\frac{7}{4(1-x^2)^2} + C\right] = \frac{d}{dx}\left[\frac{7}{4}(1-x^2)^{-2} + C\right] = \frac{7}{4}(-2)(1-x^2)^{-3}(-2x) = 7x(1-x^2)^{-3} = \frac{7x}{(1-x^2)^3}$

21. $\int \frac{x^2}{(1+x^3)^2}\,dx = \frac{1}{3}\int (1+x^3)^{-2}(3x^2)\,dx = \frac{1}{3}\left[\frac{(1+x^3)^{-1}}{-1}\right] + C = -\frac{1}{3(1+x^3)} + C$

Check: $\frac{d}{dx}\left[-\frac{1}{3(1+x^3)} + C\right] = -\frac{1}{3}(-1)(1+x^3)^{-2}(3x^2) = \frac{x^2}{(1+x^3)^2}$

23. $\int \frac{x}{\sqrt{1-x^2}}\,dx = -\frac{1}{2}\int (1-x^2)^{-1/2}(-2x)\,dx = -\frac{1}{2}\frac{(1-x^2)^{1/2}}{1/2} + C = -\sqrt{1-x^2} + C$

Check: $\frac{d}{dx}\left[-(1-x^2)^{1/2} + C\right] = -\frac{1}{2}(1-x^2)^{-1/2}(-2x) = \frac{x}{\sqrt{1-x^2}}$

25. $\int \left(1+\frac{1}{t}\right)^3\left(\frac{1}{t^2}\right)dt = -\int \left(1+\frac{1}{t}\right)^3\left(-\frac{1}{t^2}\right)dt = -\frac{\left[1+\left(\frac{1}{t}\right)\right]^4}{4} + C$

Check: $\frac{d}{dt}\left[-\frac{[1+(1/t)]^4}{4} + C\right] = -\frac{1}{4}(4)\left(1+\frac{1}{t}\right)^3\left(-\frac{1}{t^2}\right) = \frac{1}{t^2}\left(1+\frac{1}{t}\right)^3$

27. $\int \frac{1}{\sqrt{2x}}\,dx = \frac{1}{2}\int (2x)^{-1/2}2\,dx = \frac{1}{2}\left[\frac{(2x)^{1/2}}{1/2}\right] + C = \sqrt{2x} + C$

Alternate Solution: $\int \frac{1}{\sqrt{2x}}\,dx = \frac{1}{\sqrt{2}}\int x^{-1/2}\,dx = \frac{1}{\sqrt{2}}\frac{x^{1/2}}{(1/2)} + C = \sqrt{2x} + C$

Check: $\frac{d}{dx}\left[\sqrt{2x} + C\right] = \frac{1}{2}(2x)^{-1/2}(2) = \frac{1}{\sqrt{2x}}$

29. $y = \int \left[4x + \frac{4x}{\sqrt{16-x^2}}\right]dx = 4\int x\,dx - 2\int (16-x^2)^{-1/2}(-2x)\,dx = 4\left(\frac{x^2}{2}\right) - 2\left[\frac{(16-x^2)^{1/2}}{1/2}\right] + C = 2x^2 - 4\sqrt{16-x^2} + C$

31. $y = \int \frac{x+1}{(x^2+2x-3)^2}\,dx = \frac{1}{2}\int (x^2+2x-3)^{-2}(2x+2)\,dx = \frac{1}{2}\left[\frac{(x^2+2x-3)^{-1}}{-1}\right] + C = -\frac{1}{2(x^2+2x-3)} + C$

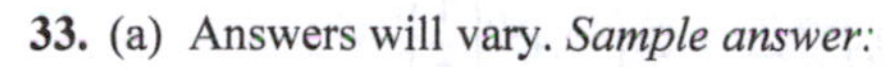

33. (a) Answers will vary. *Sample answer:*

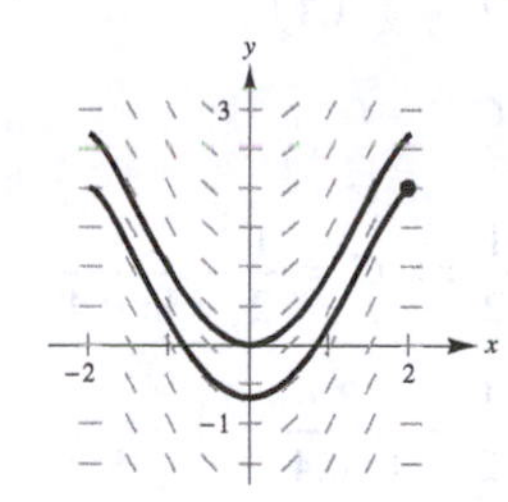

(b) $\dfrac{dy}{dx} = x\sqrt{4 - x^2},\ (2, 2)$

$$y = \int x\sqrt{4 - x^2}\,dx = -\frac{1}{2}\int \left(4 - x^2\right)^{1/2}(-2x\,dx)$$

$$= -\frac{1}{2}\cdot\frac{2}{3}\left(4 - x^2\right)^{3/2} + C$$

$$= -\frac{1}{3}\left(4 - x^2\right)^{3/2} + C$$

$$(2, 2):\ 2 = -\frac{1}{3}\left(4 - 2^2\right)^{3/2} + C \Rightarrow C = 2$$

$$y = -\frac{1}{3}\left(4 - x^2\right)^{3/2} + 2$$

35. $\int \pi \sin \pi x\,dx = -\cos \pi x + C$

37. $\int \cos 6x\,dx = \frac{1}{6}\int \cos 6x(6)\,dx = \frac{1}{6}\sin 6x + C$

39. $\int \frac{1}{\theta^2}\cos\frac{1}{\theta}\,d\theta = -\int \cos\frac{1}{\theta}\left(-\frac{1}{\theta^2}\right)d\theta = -\sin\frac{1}{\theta} + C$

41. $\int \sin 2x \cos 2x\,dx = \frac{1}{2}\int (\sin 2x)(2\cos 2x)\,dx = \frac{1}{2}\frac{(\sin 2x)^2}{2} + C = \frac{1}{4}\sin^2 2x + C$ OR

$\int \sin 2x \cos 2x\,dx = -\frac{1}{2}\int (\cos 2x)(-2\sin 2x)\,dx = -\frac{1}{2}\frac{(\cos 2x)^2}{2} + C_1 = -\frac{1}{4}\cos^2 2x + C_1$ OR

$\int \sin 2x \cos 2x\,dx = \frac{1}{2}\int 2\sin 2x \cos 2x\,dx = \frac{1}{2}\int \sin 4x\,dx = -\frac{1}{8}\cos 4x + C_2$

43. $\int \frac{\csc^2 x}{\cot^3 x}\,dx = -\int (\cot x)^{-3}\left(-\csc^2 x\right)dx$

$$= -\frac{(\cot x)^{-2}}{-2} + C = \frac{1}{2\cot^2 x} + C = \frac{1}{2}\tan^2 x + C = \frac{1}{2}\left(\sec^2 x - 1\right) + C = \frac{1}{2}\sec^2 x + C_1$$

45. $\int e^{-x^3}\left(-3x^2\right)dx = e^{-x^3} + C$

47. $\int e^x\left(e^x + 1\right)^2 dx = \frac{\left(e^x + 1\right)^3}{3} + C$

49. $\int \frac{5 - e^x}{e^{2x}}\,dx = \int 5e^{-2x}\,dx - \int e^{-x}\,dx$

$$= -\frac{5}{2}e^{-2x} + e^{-x} + C$$

51. $\int e^{\sin \pi x}\cos \pi x\,dx = \frac{1}{\pi}\int e^{\sin \pi x}(\pi \cos \pi x)\,dx$

$$= \frac{1}{\pi}e^{\sin \pi x} + C$$

53. $\int e^{-x}\sec^2\left(e^{-x}\right)dx = -\int \sec^2\left(e^{-x}\right)\left(-e^{-x}\right)dx$

$$= -\tan\left(e^{-x}\right) + C$$

55. $\int 3^{x/2}\,dx = 2\int 3^{x/2}\left(\frac{1}{2}\right)dx = 2\frac{3^{x/2}}{\ln 3} + C = \frac{2}{\ln 3}3^{x/2} + C$

57. $f'(x) = 2x\left(4x^2 - 10\right)^2,\ (2, 10)$

$$f(x) = \frac{\left(4x^2 - 10\right)^3}{12} + C = \frac{2\left(2x^2 - 5\right)^3}{3} + C$$

$$f(2) = \frac{2(8 - 5)^3}{3} + C = 18 + C = 10 \Rightarrow C = -8$$

$$f(x) = \frac{2}{3}\left(2x^2 - 5\right)^3 - 8$$

59. $f(x) = \int -\sin\frac{x}{2}\,dx = 2\cos\frac{x}{2} + C$

Because $f(0) = 6 = 2\cos\left(\frac{0}{2}\right) + C$, $C = 4$. So,

$f(x) = 2\cos\frac{x}{2} + 4.$

61. $f(x) = \int 2e^{-x/4}\,dx = -8\int e^{-x/4}\left(-\frac{1}{4}\right)dx$

$= -8e^{-x/4} + C$

$f(0) = 1 = -8 + C \Rightarrow C = 9$

$f(x) = -8e^{-x/4} + 9$

63. $f(x) = \int 0.4^{x/3}\,dx = 3\int 0.4^{x/3}\left(\frac{1}{3}\right)dx$

$= \frac{3}{\ln 0.4}0.4^{x/3} + C$

$f(0) = \frac{3}{\ln 0.4} + C = \frac{1}{2} \Rightarrow C = \frac{1}{2} - \frac{3}{\ln 0.4}$

$f(x) = \frac{3}{\ln 0.4}0.4^{x/3} + \frac{1}{2} - \frac{3}{\ln 0.4}$

65. $u = x + 6$, $x = u - 6$, $dx = du$

$$\begin{aligned}\int x\sqrt{x+6}\,dx &= \int (u-6)\sqrt{u}\,du\\ &= \int \left(u^{3/2} - 6u^{1/2}\right)du\\ &= \frac{2}{5}u^{5/2} - 4u^{3/2} + C\\ &= \frac{2u^{3/2}}{5}(u - 10) + C\\ &= \frac{2}{5}(x+6)^{3/2}\left[(x+6) - 10\right] + C\\ &= \frac{2}{5}(x+6)^{3/2}(x-4) + C\end{aligned}$$

67. $u = 1 - x$, $x = 1 - u$, $dx = -du$

$$\begin{aligned}\int x^2\sqrt{1-x}\,dx &= -\int (1-u)^2\sqrt{u}\,du = -\int \left(u^{1/2} - 2u^{3/2} + u^{5/2}\right)du\\ &= -\left(\frac{2}{3}u^{3/2} - \frac{4}{5}u^{5/2} + \frac{2}{7}u^{7/2}\right) + C = -\frac{2u^{3/2}}{105}\left(35 - 42u + 15u^2\right) + C\\ &= -\frac{2}{105}(1-x)^{3/2}\left[35 - 42(1-x) + 15(1-x)^2\right] + C = -\frac{2}{105}(1-x)^{3/2}\left(15x^2 + 12x + 8\right) + C\end{aligned}$$

69. $u = 2x - 1$, $x = \frac{1}{2}(u+1)$, $dx = \frac{1}{2}\,du$

$$\begin{aligned}\int \frac{x^2 - 1}{\sqrt{2x-1}}\,dx &= \int \frac{\left[(1/2)(u+1)\right]^2 - 1}{\sqrt{u}}\frac{1}{2}\,du = \frac{1}{8}\int u^{-1/2}\left[\left(u^2 + 2u + 1\right) - 4\right]du\\ &= \frac{1}{8}\int \left(u^{3/2} + 2u^{1/2} - 3u^{-1/2}\right)du = \frac{1}{8}\left(\frac{2}{5}u^{5/2} + \frac{4}{3}u^{3/2} - 6u^{1/2}\right) + C\\ &= \frac{u^{1/2}}{60}\left(3u^2 + 10u - 45\right) + C = \frac{\sqrt{2x-1}}{60}\left[3(2x-1)^2 + 10(2x-1) - 45\right] + C\\ &= \frac{1}{60}\sqrt{2x-1}\left(12x^2 + 8x - 52\right) + C = \frac{1}{15}\sqrt{2x-1}\left(3x^2 + 2x - 13\right) + C\end{aligned}$$

71. $u = 2x$, $du = 2\,dx \Rightarrow dx = \frac{du}{2}$

$$\int \cos^3 2x \sin 2x\,dx = \frac{1}{2}\int \cos^3 u \sin u\,du = \frac{1}{2}\left(-\frac{\cos^4 u}{4}\right) + C = -\frac{1}{8}\cos^4 2x + C$$

73. Let $u = x^2 + 1$, $du = 2x\,dx$.

$$\int_{-1}^{1} x\left(x^2+1\right)^3 dx = \frac{1}{2}\int_{-1}^{1}\left(x^2+1\right)^3(2x)\,dx = \left[\frac{1}{8}\left(x^2+1\right)^4\right]_{-1}^{1} = 0$$

75. Let $u = x^3 + 1, du = 3x^2\,dx$.

$$\int_1^2 2x^2\sqrt{x^3+1}\,dx = 2\cdot\frac{1}{3}\int_1^2 (x^3+1)^{1/2}(3x^2)\,dx = \left[\frac{(x^3+1)^{3/2}}{3/2}\right]_1^2 = \frac{4}{9}\left[(x^3+1)^{3/2}\right]_1^2 = \frac{4}{9}\left[27 - 2\sqrt{2}\right] = 12 - \frac{8}{9}\sqrt{2}$$

77. Let $u = 2x + 1, du = 2\,dx$.

$$\int_0^4 \frac{1}{\sqrt{2x+1}}\,dx = \frac{1}{2}\int_0^4 (2x+1)^{-1/2}(2)\,dx = \left[\sqrt{2x+1}\right]_0^4 = \sqrt{9} - \sqrt{1} = 2$$

79. Let $u = 1 + \sqrt{x}, du = \frac{1}{2\sqrt{x}}\,dx$.

$$\int_1^9 \frac{1}{\sqrt{x}\left(1+\sqrt{x}\right)^2}\,dx = 2\int_1^9 \left(1+\sqrt{x}\right)^{-2}\left(\frac{1}{2\sqrt{x}}\right)dx = \left[-\frac{2}{1+\sqrt{x}}\right]_1^9 = -\frac{1}{2} + 1 = \frac{1}{2}$$

81. Let $u = x^2, du = 2xdx$,

$$\int_3^4 4xe^{x^2}\,dx = 2\int_9^{16} e^u\,du = \left[2e^u\right]_9^{16} = 2e^{16} - 2e^9$$

83. $$\int_1^3 \frac{e^{3/x}}{x^2}\,dx = -\frac{1}{3}\int_1^3 e^{3/x}\left(-\frac{3}{x^2}\right)dx = \left[-\frac{1}{3}e^{3/x}\right]_1^3 = -\frac{1}{3}\left(e - e^3\right) = \frac{e}{3}\left(e^2 - 1\right)$$

85. $u = x + 1, x = u - 1, dx = du$

When $x = 0, u = 1$. When $x = 7, u = 8$.

$$\text{Area} = \int_0^7 x\sqrt[3]{x+1}\,dx = \int_1^8 (u-1)\sqrt[3]{u}\,du$$

$$= \int_1^8 \left(u^{4/3} - u^{1/3}\right)du = \left[\frac{3}{7}u^{7/3} - \frac{3}{4}u^{4/3}\right]_1^8$$

$$= \left(\frac{384}{7} - 12\right) - \left(\frac{3}{7} - \frac{3}{4}\right) = \frac{1209}{28}$$

87. $$\text{Area} = \int_{\pi/2}^{2\pi/3} \sec^2\left(\frac{x}{2}\right)dx = 2\int_{\pi/2}^{2\pi/3} \sec^2\left(\frac{x}{2}\right)\left(\frac{1}{2}\right)dx$$

$$= \left[2\tan\left(\frac{x}{2}\right)\right]_{\pi/2}^{2\pi/3} = 2\left(\sqrt{3} - 1\right)$$

89. $$\int_0^5 e^x\,dx = \left[e^x\right]_0^5$$

$$= e^5 - 1$$

$$\approx 147.413$$

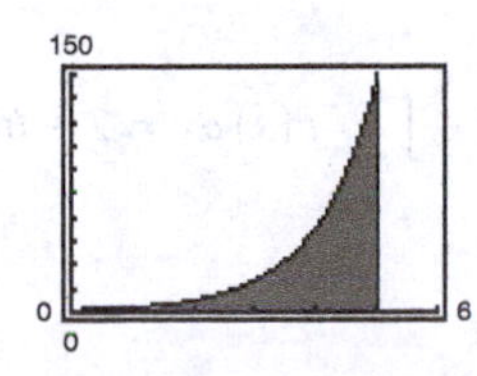

91. $$\int_0^{\sqrt{6}} xe^{-x^2/4}\,dx = \left[-2e^{-x^2/4}\right]_0^{\sqrt{6}} = -2e^{-3/2} + 2 \approx 1.554$$

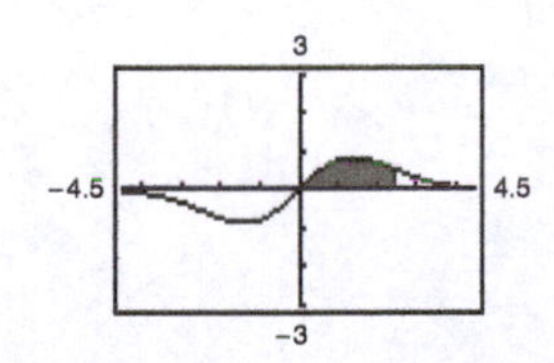

93. $f(x) = x^2\left(x^2 + 1\right)$ is even.

$$\int_{-2}^2 x^2\left(x^2+1\right)dx = 2\int_0^2 \left(x^4 + x^2\right)dx = 2\left[\frac{x^5}{5} + \frac{x^3}{3}\right]_0^2$$

$$= 2\left[\frac{32}{5} + \frac{8}{3}\right] = \frac{272}{15}$$

95. $f(x) = \sin x\cos x$ is odd.

$$\int_{-\pi/2}^{\pi/2} \sin x\cos x\,dx = 0$$

97. $\int_0^6 x^2\,dx = 72$ and $y = x^2$ is even.

(a) $\int_{-6}^6 x^2\,dx = 2\int_0^6 x^2\,dx = 2(72) = 144$

(b) $\int_{-6}^0 x^2\,dx = \int_0^6 x^2\,dx = 72$

(c) $\int_0^6 -2x^2\,dx = -2\int_0^6 x^2\,dx = -2(72) = -144$

(d) $\int_{-6}^6 3x^2\,dx = 3\int_{-6}^6 x^2\,dx = 3(144) = 432$

99. $\int_{-3}^{3}\left(x^3 + 4x^2 - 3x - 6\right) dx = \int_{-3}^{3}\left(x^3 - 3x\right) dx + \int_{-3}^{3}\left(4x^2 - 6\right) dx = 0 + 2\int_{0}^{3}\left(4x^2 - 6\right) dx = 2\left[\frac{4}{3}x^3 - 6x\right]_0^3 = 36$

101. (a) $\int x^2\sqrt{x^3 + 1}\, dx$ is easier to compute. Use substitution with $u = x^3 + 1$.

(b) $\int \cot^3 2x \csc^2 2x\, dx$ is easier to compute. Use substitution with $u = \cot 2x$.

(c) $\int e^{4x-3} dx$ is easier to compute. Use substitution with $u = 4x - 3$.

103. $\frac{dV}{dt} = \frac{k}{(t + 1)^2}$

$V(t) = \int \frac{k}{(t + 1)^2}\, dt = -\frac{k}{t + 1} + C$

$V(0) = -k + C = 500{,}000$

$V(1) = -\frac{1}{2}k + C = 400{,}000$

Solving this system yields $k = -200{,}000$ and $C = 300{,}000$. So, $V(t) = \frac{200{,}000}{t + 1} + 300{,}000$.

When $t = 4$, $V(4) = \$340{,}000$.

105. $\frac{1}{b - a}\int_a^b \left[74.50 + 43.75 \sin \frac{\pi t}{6}\right] dt = \frac{1}{b - a}\left[74.50t - \frac{262.5}{\pi} \cos \frac{\pi t}{6}\right]_a^b$

(a) $\frac{1}{3}\left[74.50t - \frac{262.5}{\pi} \cos \frac{\pi t}{6}\right]_0^3 = \frac{1}{3}\left(223.5 + \frac{262.5}{\pi}\right) \approx 102.352$ thousand units

(b) $\frac{1}{3}\left[74.50t - \frac{262.5}{\pi} \cos \frac{\pi t}{6}\right]_3^6 = \frac{1}{3}\left(447 + \frac{262.5}{\pi} - 223.5\right) \approx 102.352$ thousand units

(c) $\frac{1}{12}\left[74.50t - \frac{262.5}{\pi} \cos \frac{\pi t}{6}\right]_0^{12} = \frac{1}{12}\left(894 - \frac{262.5}{\pi} + \frac{262.5}{\pi}\right) = 74.5$ thousand units

107. (a)

(b) g is nonnegative because the graph of f is positive at the beginning, and generally has more positive sections than negative ones.

(c) The points on g that correspond to the extrema of f are points of inflection of g.

(d) No, some zeros of f, like $x = \pi/2$, do not correspond to an extrema of g. The graph of g continues to increase after $x = \pi/2$ because f remains above the x-axis.

(e) The graph of h is that of g shifted 2 units downward.

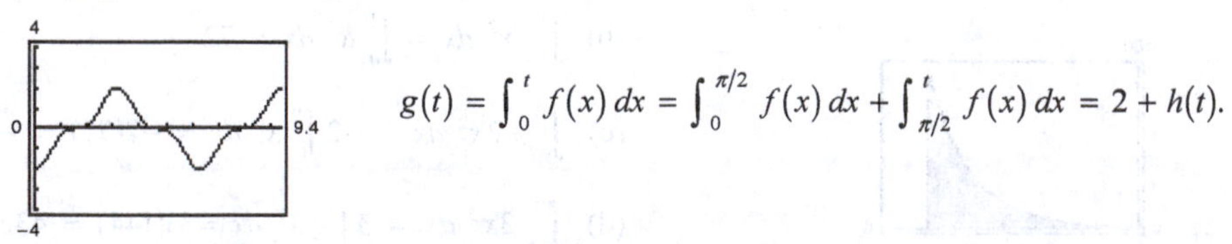

$g(t) = \int_0^t f(x)\, dx = \int_0^{\pi/2} f(x)\, dx + \int_{\pi/2}^{t} f(x)\, dx = 2 + h(t).$

109. (a) Let $u = 1 - x$ and $du = -dx$.

$\int_0^1 x^3(1 - x)^8\, dx = \int_1^0 (1 - u)^3(u)^8(-du) = \int_0^1 (1 - x)^3 x^8\, dx$

(b) Let $u = 1 - x$ and $du = -dx$.

$\int_0^1 x^a(1 - x)^b\, dx = \int_1^0 (1 - u)^a u^b(-du) = \int_0^1 (1 - x)^a x^b\, dx$

111. $u = 1 - x,\ x = 1 - u,\ dx = -du$

When $x = a,\ u = 1 - a$. When $x = b,\ u = 1 - b$.

$$P_{a,b} = \int_a^b \frac{15}{4}x\sqrt{1-x}\,dx = \frac{15}{4}\int_{1-a}^{1-b} -(1-u)\sqrt{u}\,du$$

$$= \frac{15}{4}\int_{1-a}^{1-b}\left(u^{3/2} - u^{1/2}\right)du = \frac{15}{4}\left[\frac{2}{5}u^{5/2} - \frac{2}{3}u^{3/2}\right]_{1-a}^{1-b} = \frac{15}{4}\left[\frac{2u^{3/2}}{15}(3u-5)\right]_{1-a}^{1-b} = \left[-\frac{(1-x)^{3/2}}{2}(3x+2)\right]_a^b$$

(a) $$P_{0.50,\,0.75} = \left[-\frac{(1-x)^{3/2}}{2}(3x+2)\right]_{0.50}^{0.75} = 0.353 = 35.3\%$$

(b) $$P_{0,b} = \left[-\frac{(1-x)^{3/2}}{2}(3x+2)\right]_0^b = -\frac{(1-b)^{3/2}}{2}(3b+2) + 1 = 0.5$$

$$(1-b)^{3/2}(3b+2) = 1$$

$$b \approx 0.586 = 58.6\%$$

113. True. Let $u = x^3 + 5$.

115. True

$$\int_{-10}^{10}\left(ax^3 + bx^2 + cx + d\right)dx = \int_{-10}^{10}\left(ax^3 + cx\right)dx + \int_{-10}^{10}\left(bx^2 + d\right)dx = 0 + 2\int_0^{10}\left(bx^2 + d\right)dx$$

Odd Even

117. True

$$4\int \sin x\cos x\,dx = 2\int \sin 2x\,dx = -\cos 2x + C$$

119. Let $u = cx,\ du = c\,dx$:

$$c\int_a^b f(cx)\,dx = c\int_{ca}^{cb} f(u)\frac{du}{c} = \int_{ca}^{cb} f(u)\,du = \int_{ca}^{cb} f(x)\,dx$$

121. Because f is odd, $f(-x) = -f(x)$. Then

$$\int_{-a}^{a} f(x)\,dx = \int_{-a}^{0} f(x)\,dx + \int_0^a f(x)\,dx$$

$$= -\int_0^{-a} f(x)\,dx + \int_0^a f(x)\,dx.$$

Let $x = -u,\ dx = -du$ in the first integral.

When $x = 0,\ u = 0$. When $x = -a,\ u = a$.

$$\int_{-a}^{1} f(x)\,dx = -\int_0^a f(-u)(-du) + \int_0^a f(x)\,dx$$

$$= -\int_0^a f(u)\,du + \int_0^a f(x)\,dx = 0$$

123. Let $f(x) = a_0 + a_1x + a_2x^2 + \cdots + a_nx^n$.

$$\int_0^1 f(x)\,dx = \left[a_0x + a_1\frac{x^2}{2} + a_2\frac{x^3}{3} + \cdots + a_n\frac{x^{n+1}}{n+1}\right]_0^1$$

$$= a_0 + \frac{a_1}{2} + \frac{a_2}{3} + \cdots + \frac{a_n}{n+1} = 0 \text{ (Given)}$$

By the Mean Value Theorem for Integrals, there exists c in $[0, 1]$ such that

$$\int_0^1 f(x)\,dx = f(c)(1-0)$$

$$0 = f(c).$$

So the equation has at least one real zero.

Section 5.6 Indeterminate Forms and L'Hôpital's Rule

1. L'Hôpital's Rule allows you to address limits of the form $0/0$ and ∞/∞.

3. $\lim_{x\to 0} \dfrac{\sin 4x}{\sin 3x} \approx 1.3333 \left(\text{exact: } \dfrac{4}{3}\right)$

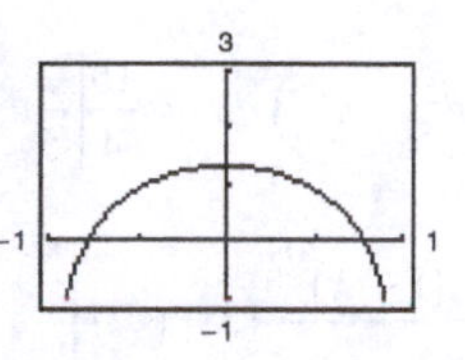

x	−0.1	−0.01	−0.001	0.001	0.01	0.1
$f(x)$	1.3177	1.3332	1.3333	1.3333	1.3332	1.3177

5. $\lim_{x\to\infty} x^5 e^{-x/100} \approx 0$

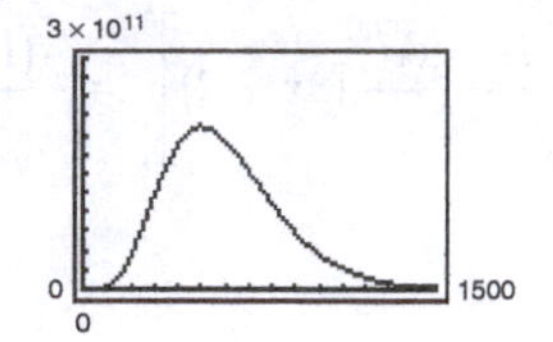

x	1	10	10^2	10^3	10^4	10^5
$f(x)$	0.9900	90,484	3.7×10^9	4.5×10^{10}	0	0

7. (a) $\lim_{x\to 4} \dfrac{3(x-4)}{x^2-16} = \lim_{x\to 4} \dfrac{3(x-4)}{(x-4)(x+4)} = \lim_{x\to 4} \dfrac{3}{x+4} = \dfrac{3}{8}$

(b) $\lim_{x\to 4} \dfrac{3(x-4)}{x^2-16} = \lim_{x\to 4} \dfrac{d/dx[3(x-4)]}{d/dx[x^2-16]} = \lim_{x\to 4} \dfrac{3}{2x} = \dfrac{3}{8}$

9. (a) $\lim_{x\to 6} \dfrac{\sqrt{x+10}-4}{x-6} = \lim_{x\to 6} \dfrac{\sqrt{x+10}-4}{x-6} \cdot \dfrac{\sqrt{x+10}+4}{\sqrt{x+10}+4} = \lim_{x\to 6} \dfrac{(x+10)-16}{(x-6)\left(\sqrt{x+10}+4\right)} = \lim_{x\to 6} \dfrac{1}{\sqrt{x+10}+4} = \dfrac{1}{8}$

(b) $\lim_{x\to 6} \dfrac{\sqrt{x+10}-4}{x-6} = \lim_{x\to 6} \dfrac{d/dx\left[\sqrt{x+10}-4\right]}{d/dx[x-6]} = \lim_{x\to 6} \dfrac{\frac{1}{2}(x+10)^{-1/2}}{1} = \dfrac{1}{8}$

11. (a) $\lim_{x\to 0} \left(\dfrac{2-2\cos x}{6x}\right) = \lim_{x\to 0} \dfrac{(1-\cos x)}{3x} \cdot \dfrac{(1+\cos x)}{(1+\cos x)} = \lim_{x\to 0} \dfrac{1-\cos^2 x}{3x(1+\cos x)} = \lim_{x\to 0} \dfrac{\sin^2 x}{3x(1+\cos x)}$

$= \lim_{x\to 0} \dfrac{\sin x}{x} \cdot \dfrac{\sin x}{3(1+\cos x)} = (1)(0) = 0$

(b) $\lim_{x\to 0} \dfrac{2-2\cos x}{6x} = \lim_{x\to 0} \dfrac{2\sin x}{6} = 0$

13. (a) $\lim_{x\to\infty} \dfrac{5x^2-3x+1}{3x^2-5} = \lim_{x\to\infty} \dfrac{5-(3/x)+(1/x^2)}{3-(5/x^2)} = \dfrac{5}{3}$

(b) $\lim_{x\to\infty} \dfrac{5x^2-3x+1}{3x^2-5} = \lim_{x\to\infty} \dfrac{(d/dx)[5x^2-3x+1]}{(d/dx)[3x^2-5]} = \lim_{x\to\infty} \dfrac{10x-3}{6x} = \lim_{x\to\infty} \dfrac{(d/dx)[10x-3]}{(d/dx)[6x]} = \lim_{x\to\infty} \dfrac{10}{6} = \dfrac{5}{3}$

15. $\lim_{x\to 3} \dfrac{x^2-2x-3}{x-3} = \lim_{x\to 3} \dfrac{2x-2}{1} = 4$

17. $\lim_{x\to 0} \dfrac{\sqrt{25-x^2}-5}{x} = \lim_{x\to 0} \dfrac{\frac{1}{2}(25-x^2)^{-1/2}(-2x)}{1}$

$= \lim_{x\to 0} \dfrac{-x}{\sqrt{25-x^2}} = 0$

19. $\lim_{x\to 0^+} \dfrac{e^x-(1+x)}{x^3} = \lim_{x\to 0^+} \dfrac{e^x-1}{3x^2} = \lim_{x\to 0^+} \dfrac{e^x}{6x} = \infty$

21. $\lim_{x\to 1} \dfrac{x^{11}-1}{x^4-1} = \lim_{x\to 1} \dfrac{11x^{10}}{4x^3} = \dfrac{11}{4}$

23. $\lim_{x\to 0} \dfrac{\sin 3x}{\sin 5x} = \lim_{x\to 0} \dfrac{3\cos 3x}{5\cos 5x} = \dfrac{3}{5}$

25. $\lim_{x\to\infty} \frac{7x^3 - 2x + 1}{6x^3 + 1} = \lim_{x\to\infty} \frac{21x^2 - 2}{18x^2}$

$= \lim_{x\to\infty} \frac{42x}{36x} = \frac{42}{36} = \frac{7}{6}$

27. $\lim_{x\to\infty} \frac{x^2 + 4x + 7}{x - 6} = \lim_{x\to\infty} \frac{2x + 4}{1} = \infty$

29. $\lim_{x\to\infty} \frac{x^3}{e^{x/2}} = \lim_{x\to\infty} \frac{3x^2}{(1/2)e^{x/2}}$

$= \lim_{x\to\infty} \frac{6x}{(1/4)e^{x/2}} = \lim_{x\to\infty} \frac{6}{(1/8)e^{x/2}} = 0$

31. $\lim_{x\to\infty} \frac{x}{\sqrt{x^2 + 1}} = \lim_{x\to\infty} \frac{1}{\sqrt{1 + (1/x^2)}} = 1$

Note: L'Hôpital's Rule does not work on this limit. See Exercise 83.

33. $\lim_{x\to\infty} \frac{\cos x}{x} = 0$ by Squeeze Theorem

$\left(\frac{\cos x}{x} \le \frac{1}{x}, \text{ for } x > 0\right)$

35. $\lim_{x\to\infty} \frac{\ln x}{x^2} = \lim_{x\to\infty} \frac{1/x}{2x} = \lim_{x\to\infty} \frac{1}{2x^2} = 0$

37. $\lim_{x\to\infty} \frac{e^x}{x^4} = \lim_{x\to\infty} \frac{e^x}{4x^3}$

$= \lim_{x\to\infty} \frac{e^x}{12x^2}$

$= \lim_{x\to\infty} \frac{e^x}{24x}$

$= \lim_{x\to\infty} \frac{e^x}{24} = \infty$

39. $\lim_{x\to 0} \frac{\sin 5x}{\tan 9x} = \lim_{x\to 0} \frac{5\cos 5x}{9\sec^2 9x} = \frac{5}{9}$

41. $\int_x^{-3} \sin\theta \, d\theta = [-\cos\theta]_x^{-3} = -\cos(-3) + \cos x$

$\lim_{x\to -3^-} \frac{\int_x^{-3} \sin\theta \, d\theta}{x + 3} = \lim_{x\to -3^-} \frac{\cos x - \cos(-3)}{x + 3}$

$= \lim_{x\to -3^-} \frac{-\sin x}{1} = -\sin(-3)$

≈ 0.1411

43. (a) $\lim_{x\to\infty} x \ln x$, not indeterminate

(b) $\lim_{x\to\infty} x \ln x = (\infty)(\infty) = \infty$

(c)

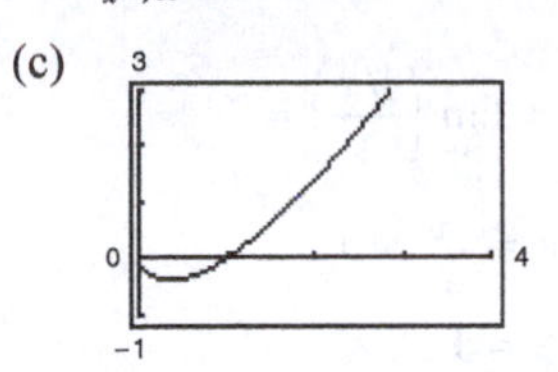

45. (a) $\lim_{x\to\infty} \left(x \sin \frac{1}{x}\right) = (\infty)(0)$

(b) $\lim_{x\to\infty} x \sin \frac{1}{x} = \lim_{x\to\infty} \frac{\sin(1/x)}{1/x}$

$= \lim_{x\to\infty} \frac{(-1/x^2)\cos(1/x)}{-1/x^2}$

$= \lim_{x\to\infty} \cos\left(\frac{1}{x}\right) = 1$

(c)

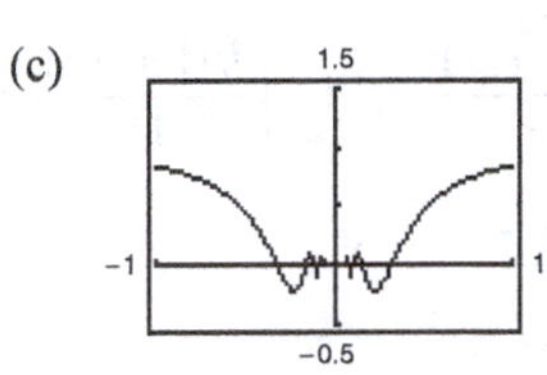

47. (a) $\lim_{x\to 0^+} (e^x + x)^{2/x} = 1^\infty$

(b) Let $y = \lim_{x\to 0^+} (e^x + x)^{2/x}$.

$\ln y = \lim_{x\to 0^+} \frac{2\ln(e^x + x)}{x}$

$= \lim_{x\to 0^+} \frac{2(e^x + 1)/(e^x + x)}{1} = 4$

So, $\ln y = 4 \Rightarrow y = e^4 \approx 54.598$.

Therefore, $\lim_{x\to 0^+} (e^x + x)^{2/x} = e^4$.

(c)

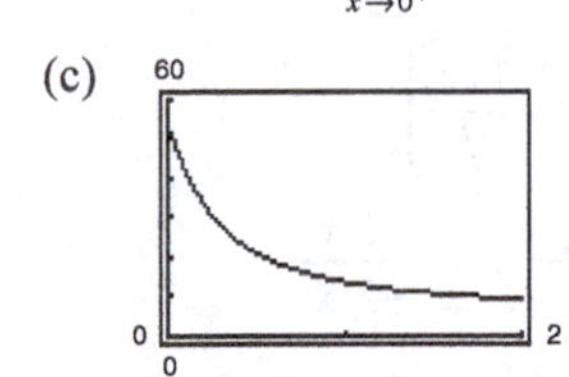

49. (a) $\lim_{x\to\infty} x^{1/x} = \infty^0$

(b) Let $y = \lim_{x\to\infty} x^{1/x}$.

$$\ln y = \lim_{x\to\infty} \frac{\ln x}{x} = \lim_{x\to\infty}\left(\frac{1/x}{1}\right) = 0$$

So, $\ln y = 0 \Rightarrow y = e^0 = 1$.

Therefore, $\lim_{x\to\infty} x^{1/x} = 1$.

(c)

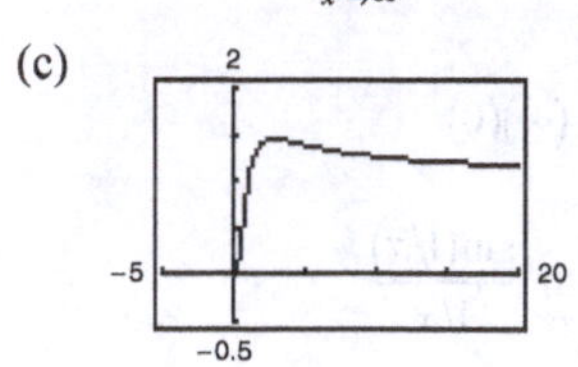

51. (a) $\lim_{x\to 0^+} (1 + x)^{1/x} = 1^\infty$

(b) Let $y = \lim_{x\to 0^+} (1 + x)^{1/x}$.

$$\ln y = \lim_{x\to 0^+} \frac{\ln(1 + x)}{x} = \lim_{x\to 0^+}\left(\frac{1/(1 + x)}{1}\right) = 1$$

So, $\ln y = 1 \Rightarrow y = e^1 = e$.

Therefore, $\lim_{x\to 0^+} (1 + x)^{1/x} = e$.

(c)

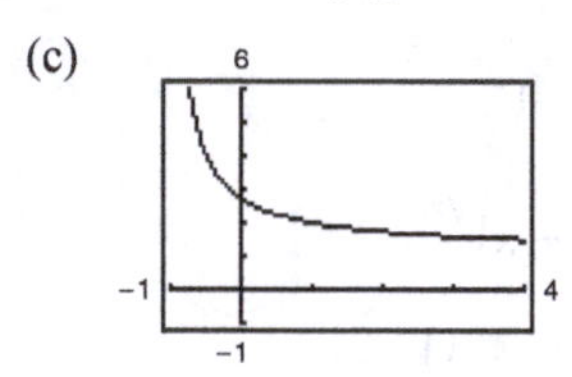

53. (a) $\lim_{x\to 0^+} \left[3(x)^{x/2}\right] = 0^0$

(b) Let $y = \lim_{x\to 0^+} 3(x)^{x/2}$.

$$\begin{aligned}\ln y &= \lim_{x\to 0^+}\left[\ln 3 + \frac{x}{2}\ln x\right]\\ &= \lim_{x\to 0^+}\left[\ln 3 + \frac{\ln x}{2/x}\right]\\ &= \lim_{x\to 0^+} \ln 3 + \lim_{x\to 0^+} \frac{1/x}{-2/x^2}\\ &= \lim_{x\to 0^+} \ln 3 - \lim_{x\to 0^+} \frac{x}{2}\\ &= \ln 3\end{aligned}$$

So, $\lim_{x\to 0^+} 3(x)^{x/2} = 3$.

(c)

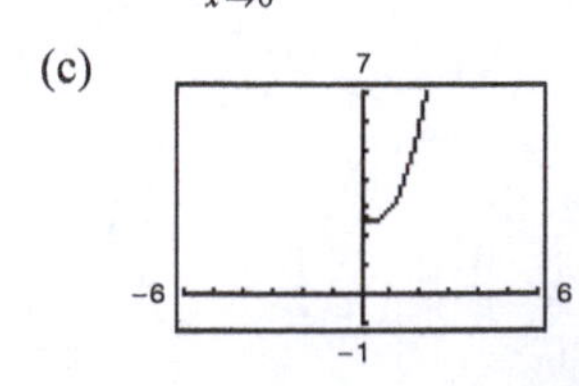

55. (a) $\lim_{x\to 1^+} (\ln x)^{x-1} = 0^0$

(b) Let $y = (\ln x)^{x-1}$.

$$\ln y = \ln\left[(\ln x)^{x-1}\right] = (x - 1)\ln(\ln x) = \frac{\ln(\ln x)}{(x - 1)^{-1}}$$

$$\begin{aligned}\lim_{x\to 1^+} \ln y &= \lim_{x\to 1^+} \frac{\ln(\ln x)}{(x - 1)^{-1}}\\ &= \lim_{x\to 1^+} \frac{1/(x\ln x)}{-(x - 1)^{-2}}\\ &= \lim_{x\to 1^+} \frac{-(x - 1)^2}{x\ln x}\\ &= \lim_{x\to 1^+} \frac{-2(x - 1)}{1 + \ln x} = 0\end{aligned}$$

Because $\lim_{x\to 1^+} \ln y = 0$, $\lim_{x\to 1^+} y = 1$.

(c)

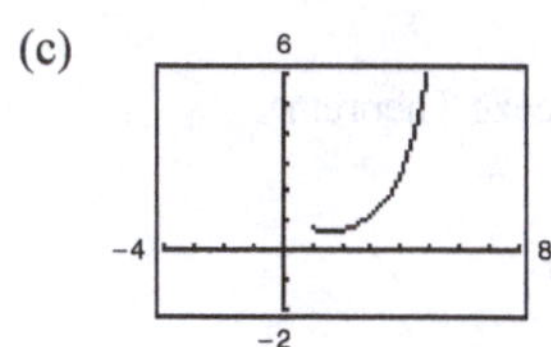

57. (a) $\lim_{x\to 2^+}\left(\frac{8}{x^2 - 4} - \frac{x}{x - 2}\right) = \infty - \infty$

(b)
$$\begin{aligned}\lim_{x\to 2^+}\left(\frac{8}{x^2 - 4} - \frac{x}{x - 2}\right) &= \lim_{x\to 2^+} \frac{8 - x(x + 2)}{x^2 - 4}\\ &= \lim_{x\to 2^+} \frac{(2 - x)(4 + x)}{(x + 2)(x - 2)}\\ &= \lim_{x\to 2^+} \frac{-(x + 4)}{x + 2} = \frac{-3}{2}\end{aligned}$$

(c)

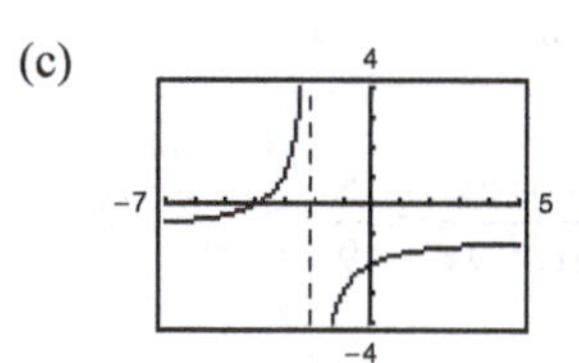

59. (a) $\lim_{x\to 1^+}\left(\frac{3}{\ln x} - \frac{2}{x - 1}\right) = \infty - \infty$

(b)
$$\begin{aligned}\lim_{x\to 1^+}\left(\frac{3}{\ln x} - \frac{2}{x - 1}\right) &= \lim_{x\to 1^+} \frac{3x - 3 - 2\ln x}{(x - 1)\ln x}\\ &= \lim_{x\to 1^+} \frac{3 - (2/x)}{\left[(x - 1)/x\right] + \ln x} = \infty\end{aligned}$$

(c)

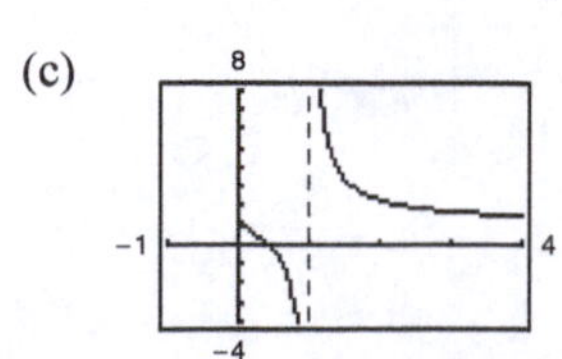

61. (a) $\lim_{x\to\infty}(e^x - x) = \infty - \infty$

(b) $\lim_{x\to\infty}(e^x - x) = \lim_{x\to\infty} x\left(\frac{e^x}{x} - 1\right)$

Now, $\lim_{x\to\infty} x = \infty$ and

$$\lim_{x\to\infty}\left(\frac{e^x}{x} - 1\right) = \lim_{x\to\infty}\frac{e^x}{1} = \infty.$$

So, $\lim_{x\to\infty} x\left(\frac{e^x}{x} - 1\right) = \infty$ and $\lim_{x\to\infty}(e^x - x) = \infty$.

(c)

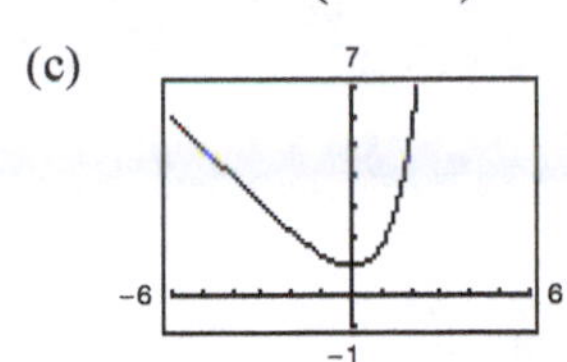

63. (a) Let $f(x) = x^2 - 25$ and $g(x) = x - 5$.

(b) Let $f(x) = (x - 5)^2$ and $g(x) = x^2 - 25$.

(c) Let $f(x) = x^2 - 25$ and $g(x) = (x - 5)^3$.

(Answers will vary.)

65. (a) Yes: $\frac{0}{0}$ (b) No: $\frac{0}{-1}$ (c) Yes: $\frac{\infty}{\infty}$

(d) Yes: $\frac{0}{0}$ (e) No: $\frac{-1}{0}$ (f) Yes: $\frac{0}{0}$

67.

x	10	10^2	10^4	10^6	10^8	10^{10}
$\frac{(\ln x)^4}{x}$	2.811	4.498	0.720	0.036	0.001	0.000

69. $\lim_{x\to\infty}\frac{x^2}{e^{5x}} = \lim_{x\to\infty}\frac{2x}{5e^{5x}} = \lim_{x\to\infty}\frac{2}{25e^{5x}} = 0$

71.
$$\begin{aligned}\lim_{x\to\infty}\frac{(\ln x)^3}{x} &= \lim_{x\to\infty}\frac{3(\ln x)^2(1/x)}{1}\\ &= \lim_{x\to\infty}\frac{3(\ln x)^2}{x}\\ &= \lim_{x\to\infty}\frac{6(\ln x)(1/x)}{1}\\ &= \lim_{x\to\infty}\frac{6(\ln x)}{x} = \lim_{x\to\infty}\frac{6}{x} = 0\end{aligned}$$

73.
$$\begin{aligned}\lim_{x\to\infty}\frac{(\ln x)^n}{x^m} &= \lim_{x\to\infty}\frac{n(\ln x)^{n-1}/x}{mx^{m-1}}\\ &= \lim_{x\to\infty}\frac{n(\ln x)^{n-1}}{mx^m}\\ &= \lim_{x\to\infty}\frac{n(n-1)(\ln x)^{n-2}}{m^2x^m}\\ &= \cdots = \lim_{x\to\infty}\frac{n!}{m^n x^m} = 0\end{aligned}$$

75. $y = x^{1/x}, x > 0$

Horizontal asymptote: $y = 1$ (See Exercise 49.)

$$\ln y = \frac{1}{x}\ln x$$

$$\left(\frac{1}{y}\right)\frac{dy}{dx} = \frac{1}{x}\left(\frac{1}{x}\right) + (\ln x)\left(-\frac{1}{x^2}\right)$$

$$\frac{dy}{dx} = x^{1/x}\left(\frac{1}{x^2}\right)(1 - \ln x) = x^{(1/x)-2}(1 - \ln x) = 0$$

Critical number: $x = e$

Intervals:	$(0, e)$	(e, ∞)
Sign of dy/dx:	+	−
$y = f(x)$:	Increasing	Decreasing

Relative maximum: $(e, e^{1/e})$

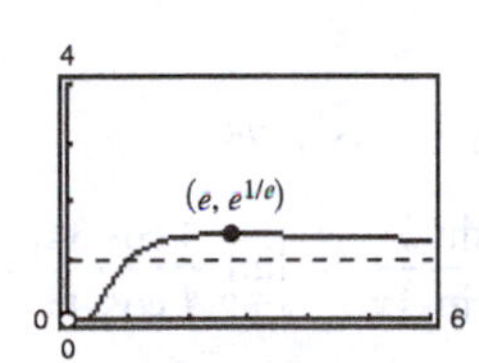

77. $y = 2xe^{-x}$

$$\lim_{x\to\infty}\frac{2x}{e^x} = \lim_{x\to\infty}\frac{2}{e^x} = 0$$

Horizontal asymptote: $y = 0$

$$\begin{aligned}\frac{dy}{dx} &= 2x(-e^{-x}) + 2e^{-x}\\ &= 2e^{-x}(1 - x) = 0\end{aligned}$$

Critical number: $x = 1$

Intervals:	$(-\infty, 1)$	$(1, \infty)$
Sign of dy/dx:	+	−
$y = f(x)$:	Increasing	Decreasing

Relative maximum: $\left(1, \frac{2}{e}\right)$

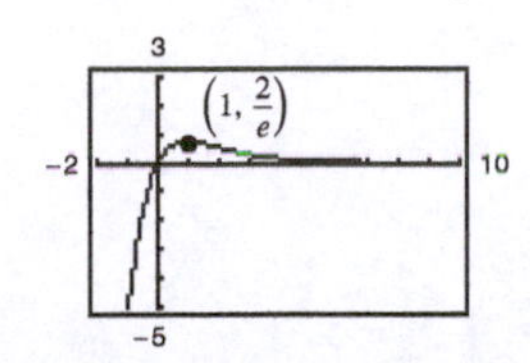

79. $\lim_{x\to 2} \dfrac{3x^2 + 4x + 1}{x^2 - x - 2} = \dfrac{21}{0}$

Limit is not of the form $\dfrac{0}{0}$ or $\dfrac{\infty}{\infty}$.

L'Hôpital's Rule does not apply.

81. $\lim_{x\to\infty} \dfrac{e^{-x}}{1 + e^{-x}} = \dfrac{0}{1 + 0} = 0$

Limit is not of the form $\dfrac{0}{0}$ or $\dfrac{\infty}{\infty}$.

L'Hôpital's Rule does not apply.

83. (a) Applying L'Hôpital's Rule twice results in the original limit, so L'Hôpital's Rule fails:

$$\lim_{x\to\infty} \frac{x}{\sqrt{x^2+1}} = \lim_{x\to\infty} \frac{1}{x/\sqrt{x^2+1}} = \lim_{x\to\infty} \frac{\sqrt{x^2+1}}{x} = \lim_{x\to\infty} \frac{x/\sqrt{x^2+1}}{1} = \lim_{x\to\infty} \frac{x}{\sqrt{x^2+1}}$$

(b) $\lim_{x\to\infty} \dfrac{x}{\sqrt{x^2+1}} = \lim_{x\to\infty} \dfrac{x/x}{\sqrt{x^2+1}/x} = \lim_{x\to\infty} \dfrac{1}{\sqrt{1+1/x^2}} = \dfrac{1}{\sqrt{1+0}} = 1$

(c)

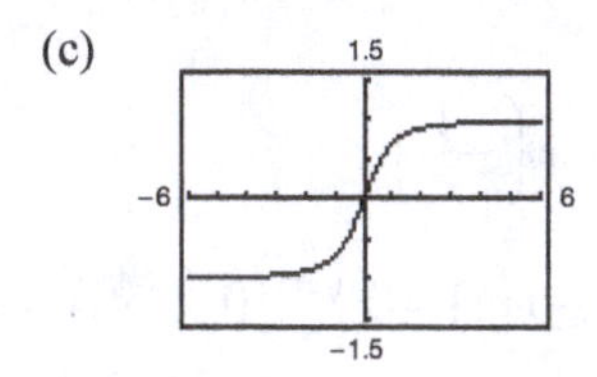

85. $f(x) = \sin(3x), g(x) = \sin(4x)$

$f'(x) = 3\cos(3x), g'(x) = 4\cos(4x)$

$y_1 = \dfrac{f(x)}{g(x)} = \dfrac{\sin 3x}{\sin 4x},$

$y_2 = \dfrac{f'(x)}{g'(x)} = \dfrac{3\cos 3x}{4\cos 4x}$

As $x \to 0$, $y_1 \to 0.75$ and $y_2 \to 0.75$

By L'Hôpital's Rule, $\lim_{x\to 0} \dfrac{\sin 3x}{\sin 4x} = \lim_{x\to 0} \dfrac{3\cos 3x}{4\cos 4x} = \dfrac{3}{4}$

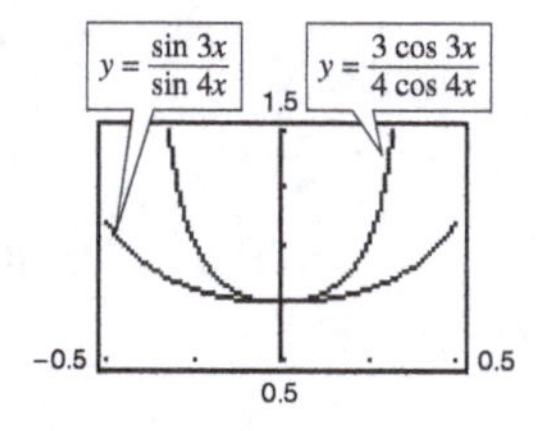

87. $I = \dfrac{V\left(1 - e^{-Rt/L}\right)}{R}$

$\lim_{R\to 0^+} I = \dfrac{0}{0}$

$\lim_{R\to\infty} V\left(\dfrac{1 - e^{-Rt/L}}{R}\right) = \lim_{R\to\infty} V\left[\dfrac{(t/L)e^{-Rt/L}}{1}\right] = V\left(\dfrac{t}{L}\right) = \dfrac{Vt}{L}$

89. Let N be a fixed value for n. Then

$$\lim_{x\to\infty} \frac{x^{N-1}}{e^x} = \lim_{x\to\infty} \frac{(N-1)x^{N-2}}{e^x} = \lim_{x\to\infty} \frac{(N-1)(N-2)x^{N-3}}{e^x} = \cdots = \lim_{x\to\infty}\left[\frac{(N-1)!}{e^x}\right] = 0. \quad \text{(See Exercise 74.)}$$

91. $f(x) = x^3, g(x) = x^2 + 1, [0, 1]$

$$\frac{f(b) - f(a)}{g(b) - g(a)} = \frac{f'(c)}{g'(c)}$$

$$\frac{f(1) - f(0)}{g(1) - g(0)} = \frac{3c^2}{2c}$$

$$\frac{1}{1} = \frac{3c}{2}$$

$$c = \frac{2}{3}$$

93. $f(x) = \sin x, g(x) = \cos x, \left[0, \frac{\pi}{2}\right]$

$$\frac{f(\pi/2) - f(0)}{g(\pi/2) - g(0)} = \frac{f'(c)}{g'(c)}$$

$$\frac{1}{-1} = \frac{\cos c}{-\sin c}$$

$$-1 = -\cot c$$

$$c = \frac{\pi}{4}$$

95. False. A limit of the form $\frac{\infty}{0}$ is equal to ∞ (or $-\infty$).

97. True

99. True

101. Area of triangle: $\frac{1}{2}(2x)(1 - \cos x) = x - x\cos x$

Shaded area: Area of rectangle − Area under curve

$$2x(1 - \cos x) - 2\int_0^x (1 - \cos t)\,dt = 2x(1 - \cos x) - 2\big[t - \sin t\big]_0^x$$
$$= 2x(1 - \cos x) - 2(x - \sin x)$$
$$= 2\sin x - 2x\cos x$$

Ratio: $$\lim_{x\to 0} \frac{x - x\cos x}{2\sin x - 2x\cos x} = \lim_{x\to 0} \frac{1 + x\sin x - \cos x}{2\cos x + 2x\sin x - 2\cos x}$$
$$= \lim_{x\to 0} \frac{1 + x\sin x - \cos x}{2x\sin x}$$
$$= \lim_{x\to 0} \frac{x\cos x + \sin x + \sin x}{2x\cos x + 2\sin x}$$
$$= \lim_{x\to 0} \frac{x\cos x + 2\sin x}{2x\cos x + 2\sin x} \cdot \frac{1/\cos x}{1/\cos x} = \lim_{x\to 0} \frac{x + 2\tan x}{2x + 2\tan x} = \lim_{x\to 0} \frac{1 + 2\sec^2 x}{2 + 2\sec^2 x} = \frac{3}{4}$$

103. $$\lim_{x\to 0} \frac{4x - 2\sin 2x}{2x^3} = \lim_{x\to 0} \frac{4 - 4\cos 2x}{6x^2} = \lim_{x\to 0} \frac{8\sin 2x}{12x} = \lim_{x\to 0} \frac{16\cos 2x}{12} = \frac{16}{12} = \frac{4}{3}$$

Let $c = \frac{4}{3}$.

105. $$\lim_{x\to 0} \frac{a - \cos bx}{x^2} = 2$$

Near $x = 0$, $\cos bx \approx 1$ and $x^2 \approx 0 \Rightarrow a = 1$.

Using L'Hôpital's Rule, $\lim_{x\to 0} \frac{1 - \cos bx}{x^2} = \lim_{x\to 0} \frac{b\sin bx}{2x} = \lim_{x\to 0} \frac{b^2\cos bx}{2} = 2.$

So, $b^2 = 4$ and $b = \pm 2$.

Answer: $a = 1, b = \pm 2$

107. (a) $\lim_{h\to 0} \dfrac{f(x+h)-f(x-h)}{2h} = \lim_{h\to 0} \dfrac{f'(x+h)(1) - f'(x-h)(-1)}{2}$

$$= \lim_{h\to 0}\left[\frac{f'(x+h)+f'(x-h)}{2}\right]$$

$$= \frac{f'(x)+f'(x)}{2}$$

$$= f'(x)$$

(b)

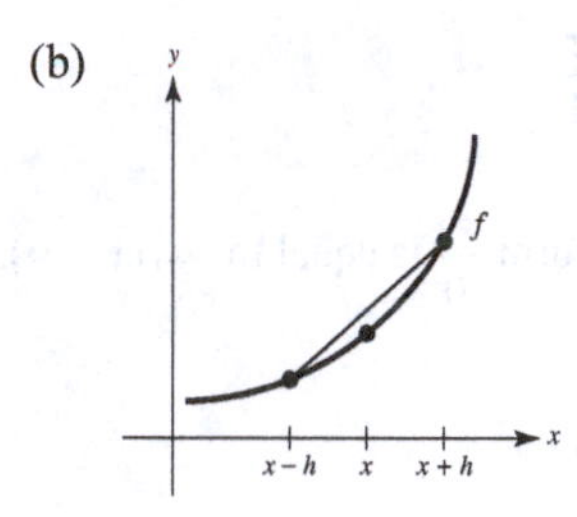

Graphically, the slope of the line joining $(x-h, f(x-h))$ and $(x+h, f(x+h))$ is approximately $f'(x)$.

So, $\lim_{h\to 0} \dfrac{f(x+h)-f(x-h)}{2h} = f'(x)$.

109. (a) $\lim_{x\to 0^+} (-x \ln x)$ is the form $0 \cdot \infty$.

(b) $\lim_{x\to 0^+} \dfrac{-\ln x}{1/x} = \lim_{x\to 0^+} \dfrac{-1/x}{-1/x^2} = \lim_{x\to 0^+} (x) = 0$

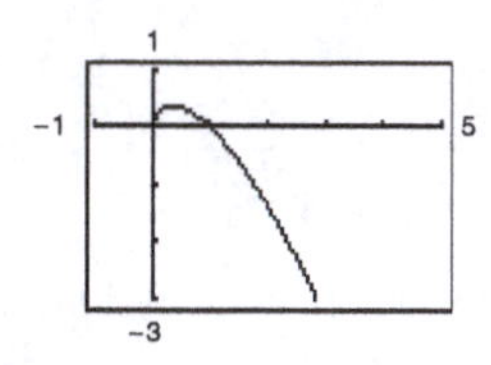

111. $\lim_{x\to a} f(x)^{g(x)}$

$$y = f(x)^{g(x)}$$

$$\ln y = g(x)\ln f(x)$$

$$\lim_{x\to a} g(x)\ln f(x) = (-\infty)(-\infty) = \infty$$

As $x \to a$, $\ln y \Rightarrow \infty$, and therefore $y = \infty$. So,

$$\lim_{x\to a} f(x)^{g(x)} = \infty.$$

113. (a) $\lim_{x\to 0^+} x^{(\ln 2)/(1+\ln x)}$ is of form 0^0.

Let $y = x^{(\ln 2)/(1+\ln x)}$

$$\ln y = \frac{\ln 2}{1+\ln x}\ln x$$

$$\lim_{x\to 0^+} \ln y = \frac{\ln 2(1/x)}{1/x} = \ln 2.$$

So, $\lim_{x\to 0^+} x^{(\ln 2)/(1+\ln x)} = 2$.

(b) $\lim_{x\to\infty} x^{(\ln 2)/(1+\ln x)}$ is of form ∞^0.

Let $y = x^{(\ln 2)/(1+\ln x)}$

$$\ln y = \frac{\ln 2}{1+\ln x}\ln x$$

$$\lim_{x\to\infty} \ln y = \frac{\ln 2(1/x)}{1/x} = \ln 2.$$

So, $\lim_{x\to\infty} x^{(\ln 2)/(1+\ln x)} = 2$.

(c) $\lim_{x\to 0}(x+1)^{(\ln 2)/(x)}$ is of form 1^∞.

Let $y = (x+1)^{(\ln 2)/(x)}$

$$\ln y = \frac{\ln 2}{x}\ln(x+1)$$

$$\lim_{x\to 0} \ln y = \lim_{x\to 0} \frac{(\ln 2)1/(x+1)}{1} = \ln 2.$$

So, $\lim_{x\to 0} (x+1)^{(\ln 2)/(x)} = 2$.

115. (a) $h(x) = \dfrac{x + \sin x}{x}$

$\lim_{x \to \infty} h(x) = 1$

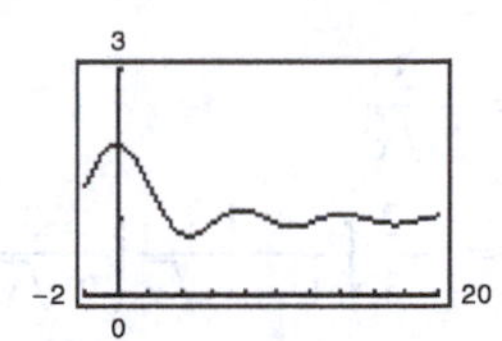

(b) $h(x) = \dfrac{x + \sin x}{x} = \dfrac{x}{x} + \dfrac{\sin x}{x} = 1 + \dfrac{\sin x}{x}, x > 0$

So, $\lim_{x \to \infty} h(x) = \lim_{x \to \infty}\left[1 + \dfrac{\sin x}{x}\right] = 1 + 0 = 1.$

(c) No. $h(x)$ is not an indeterminate form.

117. Let $f(x) = \left[\dfrac{1}{x} \cdot \dfrac{a^x - 1}{a - 1}\right]^{1/x}.$

For $a > 1$ and $x > 0$, $\ln f(x) = \dfrac{1}{x}\left[\ln \dfrac{1}{x} + \ln(a^x - 1) - \ln(a - 1)\right] = -\dfrac{\ln x}{x} + \dfrac{\ln(a^x - 1)}{x} - \dfrac{\ln(a - 1)}{x}.$

As $x \to \infty$, $\dfrac{\ln x}{x} \to 0$, $\dfrac{\ln(a - 1)}{x} \to 0$, and $\dfrac{\ln(a^x - 1)}{x} = \dfrac{\ln\left[(1 - a^{-x})a^x\right]}{x} = \dfrac{\ln(1 - a^{-x})}{x} + \ln a \to \ln a.$

So, $\ln f(x) \to \ln a.$

For $0 < a < 1$ and $x > 0$, $\ln f(x) = \dfrac{-\ln x}{x} + \dfrac{\ln(1 - a^x)}{x} - \dfrac{\ln(1 - a)}{x} \to 0$ as $x \to \infty$.

Combining these results, $\lim_{x \to \infty} f(x) = \begin{cases} a & \text{if} \quad a > 1 \\ 1 & \text{if} \quad 0 < a < 1 \end{cases}.$

Section 5.7 The Natural Logarithmic Function: Integration

1. No. To use the Log Rule, look for quotients in which the numerator is the derivative of the denominator, with rewriting in mind. This integral requires the General Power Rule:

$$\int \frac{x}{(x^2 - 4)^3}\,dx = \frac{1}{2}\int (x^2 - 4)^{-3} 2x\,dx.$$

3. Some ways to alter an integrand include rewriting using a trigonometric identity, multiplying and dividing by the same quantity, adding and subtracting the same quantity, and using long division.

5. $\int \dfrac{5}{x}\,dx = 5\int \dfrac{1}{x}\,dx = 5 \ln|x| + C$

7. $u = 2x + 5, du = 2\,dx$

$$\int \frac{1}{2x + 5}\,dx = \frac{1}{2}\int \frac{1}{2x + 5}(2)\,dx = \frac{1}{2}\ln|2x + 5| + C$$

9. $u = x^2 - 3, du = 2x\,dx$

$$\int \frac{x}{x^2 - 3}\,dx = \frac{1}{2}\int \frac{1}{x^2 - 3}(2x)\,dx = \frac{1}{2}\ln|x^2 - 3| + C$$

11. $u = x^4 + 3x, du = (4x^3 + 3)\,dx$

$$\int \frac{4x^3 + 3}{x^4 + 3x}\,dx = \int \frac{1}{x^4 + 3x}(4x^3 + 3)\,dx = \ln|x^4 + 3x| + C$$

13. $\int \dfrac{x^2 - 7}{7x}\,dx = \dfrac{1}{7}\int x\,dx - \int \dfrac{1}{x}\,dx = \dfrac{1}{14}x^2 - \ln|x| + C$

15. $u = x^3 + 3x^2 + 9x, du = 3(x^2 + 2x + 3)\,dx$

$$\int \frac{x^2 + 2x + 3}{x^3 + 3x^2 + 9x}\,dx = \frac{1}{3}\int \frac{3(x^2 + 2x + 3)}{x^3 + 3x^2 + 9x}\,dx = \frac{1}{3}\ln\left|x^3 + 3x^2 + 9x\right| + C$$

17. $$\int \frac{x^2 - 3x + 2}{x + 1}\,dx = \int\left(x - 4 + \frac{6}{x + 1}\right)dx = \frac{x^2}{2} - 4x + 6\ln|x + 1| + C$$

19. $$\int \frac{x^3 - 3x^2 + 5}{x - 3}\,dx = \int\left(x^2 + \frac{5}{x - 3}\right)dx = \frac{x^3}{3} + 5\ln|x - 3| + C$$

21. $$\int \frac{x^4 + x - 4}{x^2 + 2}\,dx = \int\left(x^2 - 2 + \frac{x}{x^2 + 2}\right)dx = \frac{x^3}{3} - 2x + \frac{1}{2}\ln(x^2 + 2) + C = \frac{x^3}{3} - 2x + \ln\sqrt{x^2 + 2} + C$$

23. $u = \ln x, du = \frac{1}{x}dx$

$$\int \frac{(\ln x)^2}{x}\,dx = \frac{1}{3}(\ln x)^3 + C$$

25. $u = 1 - 3\sqrt{x}, du = \frac{-3}{2\sqrt{x}}$

$$\int \frac{1}{\sqrt{x}(1 - 3\sqrt{x})}\,dx = -\frac{2}{3}\int \frac{1}{1 - 3\sqrt{x}}\left(\frac{-3}{2\sqrt{x}}\right)dx = -\frac{2}{3}\ln\left|1 - 3\sqrt{x}\right| + C$$

27. $$\begin{aligned}\int \frac{6x}{(x - 5)^2}\,dx &= 6\int \frac{x}{(x - 5)^2}\,dx \\ &= 6\int \frac{x - 5}{(x - 5)^2}\,dx + 6\int \frac{5}{(x - 5)^2}\,dx \\ &= 6\int \frac{1}{x - 5}\,dx + 30\int (x - 5)^{-2}\,dx \\ &= 6\ln|x - 5| - 30(x - 5)^{-1} + C \\ &= 6\ln|x - 5| - \frac{30}{x - 5} + C\end{aligned}$$

29. $u = 1 + \sqrt{2x}, du = \frac{1}{\sqrt{2x}}\,dx \Rightarrow (u - 1)\,du = dx$

$$\begin{aligned}\int \frac{1}{1 + \sqrt{2x}}\,dx &= \int \frac{(u - 1)}{u}\,du = \int\left(1 - \frac{1}{u}\right)du \\ &= u - \ln|u| + C_1 \\ &= \left(1 + \sqrt{2x}\right) - \ln\left|1 + \sqrt{2x}\right| + C_1 \\ &= \sqrt{2x} - \ln\left(1 + \sqrt{2x}\right) + C\end{aligned}$$

where $C = C_1 + 1$.

31. $u = \sqrt{x} - 3, du = \frac{1}{2\sqrt{x}}\,dx \Rightarrow 2(u + 3)du = dx$

$$\begin{aligned}\int \frac{\sqrt{x}}{\sqrt{x} - 3}\,dx &= 2\int \frac{(u + 3)^2}{u}\,du \\ &= 2\int \frac{u^2 + 6u + 9}{u}\,du = 2\int\left(u + 6 + \frac{9}{u}\right)du \\ &= 2\left[\frac{u^2}{2} + 6u + 9\ln|u|\right] + C_1 \\ &= u^2 + 12u + 18\ln|u| + C_1 \\ &= \left(\sqrt{x} - 3\right)^2 + 12\left(\sqrt{x} - 3\right) + 18\ln\left|\sqrt{x} - 3\right| + C_1 \\ &= x + 6\sqrt{x} + 18\ln\left|\sqrt{x} - 3\right| + C\end{aligned}$$

where $C = C_1 - 27$.

33. $\int \cot\left(\frac{\theta}{3}\right) d\theta = 3\int \cot\left(\frac{\theta}{3}\right)\left(\frac{1}{3}\right) d\theta = 3\ln\left|\sin\frac{\theta}{3}\right| + C$

35. $\int \csc 2x\, dx = \frac{1}{2}\int (\csc 2x)(2)\, dx$

$= -\frac{1}{2}\ln\left|\csc 2x + \cot 2x\right| + C$

37. $u = 1 + \sin t,\ du = \cos t\, dt$

$\int \frac{\cos t}{1 + \sin t}\, dt = \ln\left|1 + \sin t\right| + C$

39. $u = \sec x - 1,\ du = \sec x \tan x\, dx$

$\int \frac{\sec x \tan x}{\sec x - 1}\, dx = \ln\left|\sec x - 1\right| + C$

41. $\int e^{-x}\tan\left(e^{-x}\right) dx = -\int \tan\left(e^{-x}\right)\left(-e^{-x}\right) dx$

$= -\left(-\ln\left|\cos\left(e^{-x}\right)\right|\right) + C$

$= \ln\left|\cos\left(e^{-x}\right)\right| + C$

43. $y = \int \frac{3}{2 - x}\, dx = -3\int \frac{1}{x - 2}\, dx = -3\ln\left|x - 2\right| + C$

$(1, 0)$: $0 = -3\ln\left|1 - 2\right| + C \Rightarrow C = 0$

$y = -3\ln\left|x - 2\right|$

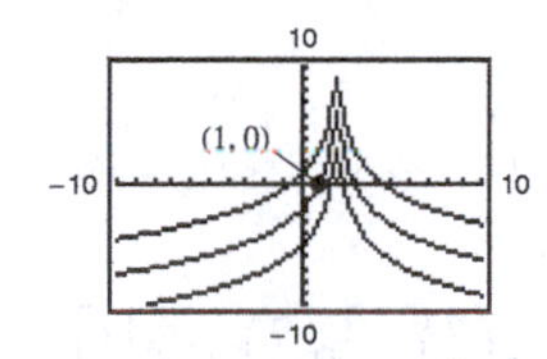

45. $y = \int \frac{2x}{x^2 - 9}\, dx = \ln\left|x^2 - 9\right| + C$

$(0, 4)$: $4 = \ln\left|0 - 9\right| + C \Rightarrow C = 4 - \ln 9$

$y = \ln\left|x^2 - 9\right| + 4 - \ln 9$

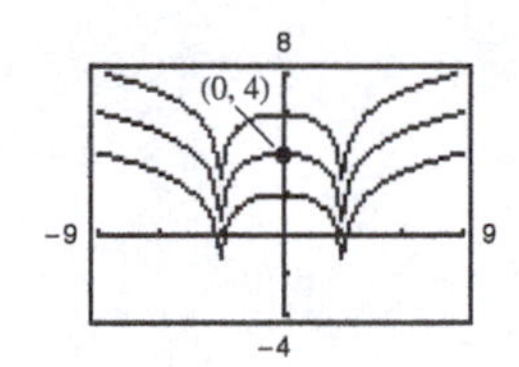

47. $f''(x) = \frac{2}{x^2} = 2x^{-2},\quad x > 0$

$f'(x) = \frac{-2}{x} + C$

$f'(1) = 1 = -2 + C \Rightarrow C = 3$

$f'(x) = \frac{-2}{x} + 3$

$f(x) = -2\ln x + 3x + C_1$

$f(1) = 1 = -2(0) + 3 + C_1 \Rightarrow C_1 = -2$

$f(x) = -2\ln x + 3x - 2$

49. $\frac{dy}{dx} = \frac{1}{x + 2},\ (0, 1)$

(a)

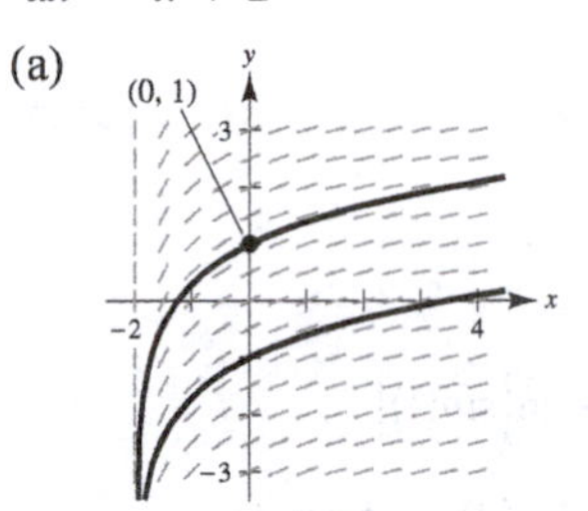

(b) $y = \int \frac{1}{x + 2}\, dx = \ln\left|x + 2\right| + C$

$y(0) = 1 \Rightarrow 1 = \ln 2 + C \Rightarrow C = 1 - \ln 2$

So, $y = \ln\left|x + 2\right| + 1 - \ln 2 = \ln\left(\frac{x + 2}{2}\right) + 1.$

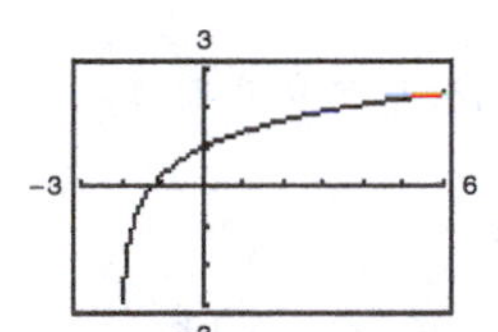

51. $\int_0^4 \frac{5}{3x + 1}\, dx = \left[\frac{5}{3}\ln\left|3x + 1\right|\right]_0^4 = \frac{5}{3}\ln 13 \approx 4.275$

53. $u = 1 + \ln x,\ du = \frac{1}{x}\, dx$

$\int_1^e \frac{(1 + \ln x)^2}{x}\, dx = \left[\frac{1}{3}(1 + \ln x)^3\right]_1^e = \frac{7}{3}$

55. $\int_0^2 \frac{x^2 - 2}{x + 1}\, dx = \int_0^2 \left(x - 1 - \frac{1}{x + 1}\right) dx$

$= \left[\frac{1}{2}x^2 - x - \ln\left|x + 1\right|\right]_0^2 = -\ln 3$

≈ -1.099

57. $\int_1^2 \frac{1 - \cos\theta}{\theta - \sin\theta}\,d\theta = \left[\ln|\theta - \sin\theta|\right]_1^2$

$= \ln\left|\frac{2 - \sin 2}{1 - \sin 1}\right| \approx 1.929$

59. $\int \frac{1 - \sqrt{x}}{1 + \sqrt{x}}\,dx = 4\sqrt{x} - x - 4\ln(1 + \sqrt{x}) + C$

Note: In Exercises 61–63, you can use the Second Fundamental Theorem of Calculus or integrate the function.

61. $F(x) = \int_1^x \frac{1}{t}\,dt$

$F'(x) = \frac{1}{x}$

63. $F(x) = \int_1^{4x} \cot t\,dt$

$= \left[\ln|\sin t|\right]_1^{4x}$

$= \ln|\sin 4x| - \ln|\sin(1)|$

$F'(x) = \frac{1}{\sin 4x}(\cos 4x)(4) = 4\cot 4x$

Alternate Solution:

Using the Second Fundamental Theorem of Calculus:

$F'(x) = \cot(4x)\frac{d}{dx}(4x) = 4\cot 4x$

65. $A = \int_1^3 \frac{6}{x}\,dx = \left[6\ln|x|\right]_1^3 = 6\ln 3$

67. $A = \int_0^1 \csc(x + 1)\,dx$

$= \left[-\ln|\csc(x + 1) + \cot(x + 1)|\right]_0^1$

$= -\ln|\csc 2 + \cot 2| + \ln|\csc 1 + \cot 1|$

≈ 1.048

69. $A = \int_1^4 \frac{x^2 + 4}{x}\,dx = \int_1^4 \left(x + \frac{4}{x}\right)dx$

$= \left[\frac{x^2}{2} + 4\ln x\right]_1^4$

$= (8 + 4\ln 4) - \frac{1}{2}$

$= \frac{15}{2} + 8\ln 2$

≈ 13.045

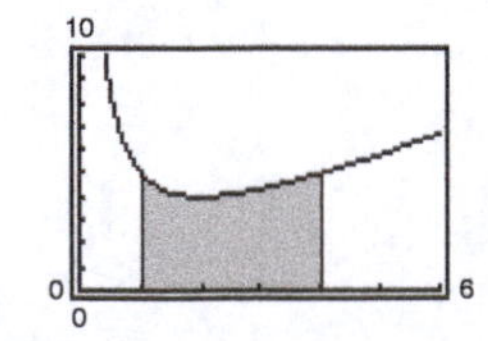

71. $\int_0^2 2\sec\frac{\pi x}{6}\,dx = \frac{12}{\pi}\int_0^2 \sec\left(\frac{\pi x}{6}\right)\frac{\pi}{6}\,dx$

$= \frac{12}{\pi}\left[\ln\left|\sec\frac{\pi x}{6} + \tan\frac{\pi x}{6}\right|\right]_0^2$

$= \frac{12}{\pi}\left(\ln\left|\sec\frac{\pi}{3} + \tan\frac{\pi}{3}\right| - \ln|1 + 0|\right)$

$= \frac{12}{\pi}\ln(2 + \sqrt{3})$

≈ 5.03041

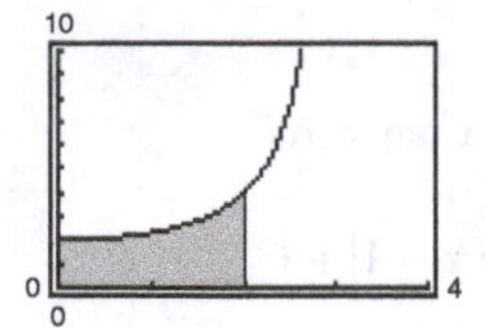

73. Average value $= \frac{1}{4 - 2}\int_2^4 \frac{8}{x^2}\,dx = 4\int_2^4 x^{-2}\,dx$

$= \left[-4\frac{1}{x}\right]_2^4 = -4\left(\frac{1}{4} - \frac{1}{2}\right) = 1$

75. Average value $= \frac{1}{e - 1}\int_1^e \frac{2\ln x}{x}\,dx = \frac{2}{e - 1}\left[\frac{(\ln x)^2}{2}\right]_1^e$

$= \frac{1}{e - 1}(1 - 0) = \frac{1}{e - 1} \approx 0.582$

77. $n = 4, \Delta x = \frac{3 - 1}{4} = \frac{1}{2}$

Midpoint approximation:

$\int_1^3 \frac{12}{x}\,dx \approx \frac{1}{2}\left[f\left(\frac{5}{4}\right) + f\left(\frac{7}{4}\right) + f\left(\frac{9}{4}\right) + f\frac{11}{4}\right]$

$\approx \frac{1}{2}[9.6 + 6.8571 + 5.3333 + 4.3636]$

≈ 13.077

79.

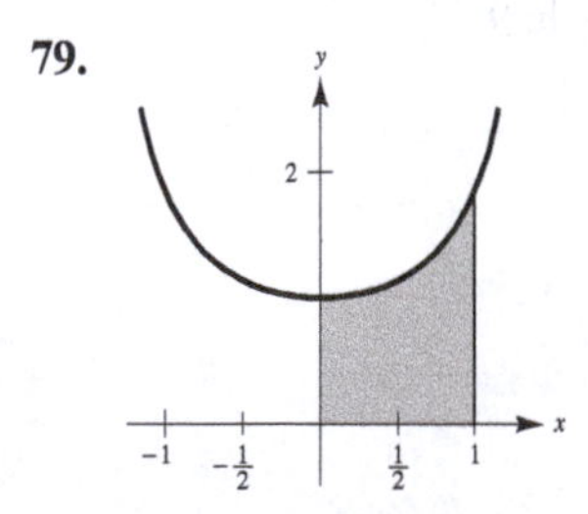

$A \approx 1.25$; Matches (d)

80.

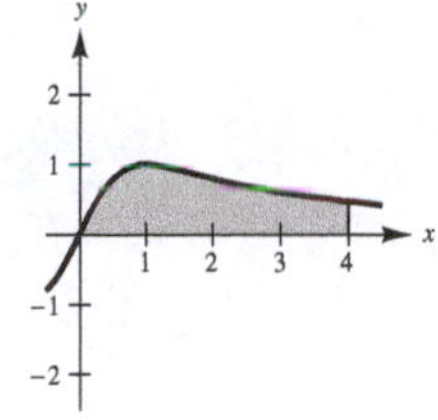

$A \approx 3$; Matches (a)

81. Let $f(t) = \ln t$ on $[x, y]$, $0 < x < y$.

By the Mean Value Theorem,

$$\frac{f(y) - f(x)}{y - x} = f'(c), \quad x < c < y,$$

$$\frac{\ln y - \ln x}{y - x} = \frac{1}{c}.$$

Because $0 < x < c < y$, $\frac{1}{x} > \frac{1}{c} > \frac{1}{y}$.

So, $\frac{1}{y} < \frac{\ln y - \ln x}{y - x} < \frac{1}{x}$.

83.
$$\int_1^x \frac{3}{t}\,dt = \int_{1/4}^x \frac{1}{t}\,dt$$
$$\left[3 \ln|t|\right]_1^x = \left[\ln|t|\right]_{1/4}^x$$
$$3 \ln x = \ln x - \ln\left(\frac{1}{4}\right)$$
$$2 \ln x = -\ln\left(\frac{1}{4}\right) = \ln 4$$
$$\ln x = \frac{1}{2} \ln 4 = \ln 2$$
$$x = 2$$

85. $\int \cot u\,du = \int \frac{\cos u}{\sin u}\,du = \ln|\sin u| + C$

Alternate solution:

$$\frac{d}{du}\left[\ln|\sin u| + C\right] = \frac{1}{\sin u} \cos u + C = \cot u + C$$

87. $-\ln|\cos x| + C = \ln\left|\frac{1}{\cos x}\right| + C = \ln|\sec x| + C$

89.
$$\ln|\sec x + \tan x| + C = \ln\left|\frac{(\sec x + \tan x)(\sec x - \tan x)}{(\sec x - \tan x)}\right| + C$$
$$= \ln\left|\frac{\sec^2 x - \tan^2 x}{\sec x - \tan x}\right| + C$$
$$= \ln\left|\frac{1}{\sec x - \tan x}\right| + C$$
$$= -\ln|\sec x - \tan x| + C$$

91.
$$P(t) = \int \frac{3000}{1 + 0.25t}\,dt = (3000)(4)\int \frac{0.25}{1 + 0.25t}\,dt = 12{,}000 \ln|1 + 0.25t| + C$$
$$P(0) = 12{,}000 \ln|1 + 0.25(0)| + C = 1000$$
$$C = 1000$$
$$P(t) = 12{,}000 \ln|1 + 0.25t| + 1000 = 1000\left[12 \ln|1 + 0.25t| + 1\right]$$
$$P(3) = 1000\left[12(\ln 1.75) + 1\right] \approx 7715$$

93.
$$t = \frac{10}{\ln 2}\int_{250}^{300} \frac{1}{T - 100}\,dT = \frac{10}{\ln 2}\left[\ln(T - 100)\right]_{250}^{300} = \frac{10}{\ln 2}[\ln 200 - \ln 150] = \frac{10}{\ln 2}\left[\ln\left(\frac{4}{3}\right)\right] \approx 4.1504 \text{ min}$$

95. $f(x) = \dfrac{x}{1 + x^2}$

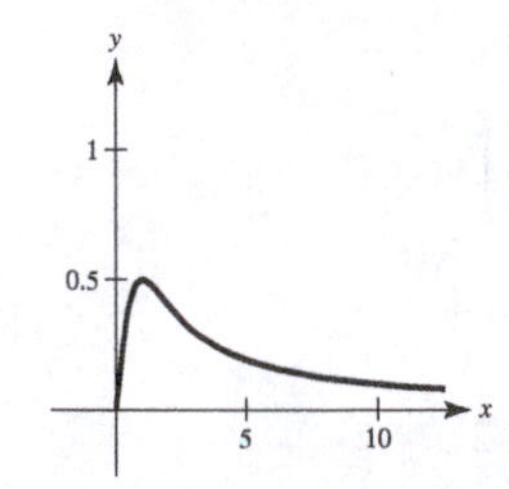

(a) $y = \dfrac{1}{2}x$ intersects $f(x) = \dfrac{x}{1 + x^2}$:

$$\frac{1}{2}x = \frac{x}{1 + x^2}$$
$$1 + x^2 = 2$$
$$x = 1$$

$$A = \int_0^1 \left(\left[\frac{x}{1 + x^2}\right] - \frac{1}{2}x\right) dx = \left[\frac{1}{2}\ln(x^2 + 1) - \frac{x^2}{4}\right]_0^1 = \frac{1}{2}\ln 2 - \frac{1}{4}$$

(b) $f'(x) = \dfrac{(1 + x^2) - x(2x)}{(1 + x^2)^2} = \dfrac{1 - x^2}{(1 + x^2)^2}$

$f'(0) = 1$

So, for $0 < m < 1$, the graphs of f and $y = mx$ enclose a finite region.

(c)

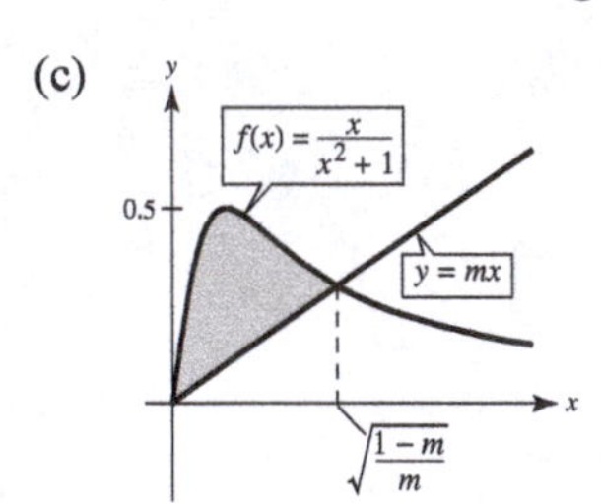

$f(x) = \dfrac{x}{x^2 + 1}$ intersects $y = mx$:

$$\frac{x}{1 + x^2} = mx$$
$$1 = m + mx^2$$
$$x^2 = \frac{1 - m}{m}$$
$$x = \sqrt{\frac{1 - m}{m}}$$

$$A = \int_0^{\sqrt{(1-m)/m}} \left(\frac{x}{1 + x^2} - mx\right) dx, \quad 0 < m < 1 = \left[\frac{1}{2}\ln(1 + x^2) - \frac{mx^2}{2}\right]_0^{\sqrt{(1-m)/m}} = \frac{1}{2}\ln\left(1 + \frac{1 - m}{m}\right) - \frac{1}{2}m\left(\frac{1 - m}{m}\right)$$
$$= \frac{1}{2}\ln\left(\frac{1}{m}\right) - \frac{1}{2}(1 - m) = \frac{1}{2}\left[m - \ln(m) - 1\right]$$

97. True

99. True

$$\int \frac{1}{x}\,dx = \ln|x| + C_1 = \ln|x| + \ln|C| = \ln|Cx|, C \neq 0$$

101. $\dfrac{d}{dx}\ln|x| = \dfrac{1}{x}$ implies that

$$\int \frac{1}{x}\,dx = \ln|x| + C.$$

The second formula follows by the Chain Rule.

Section 5.8 Inverse Trigonometric Functions: Integration

1. (a) No

(b) Yes. Use the formula involving the arcsecant function.

3. $\int \frac{dx}{\sqrt{9 - x^2}} = \arcsin\left(\frac{x}{3}\right) + C$

5. $\int \frac{1}{x\sqrt{4x^2 - 1}}\,dx = \int \frac{2}{2x\sqrt{(2x)^2 - 1}}\,dx$

$= \operatorname{arcsec}|2x| + C$

7. $\int \frac{1}{\sqrt{1 - (x+1)^2}}\,dx = \arcsin(x + 1) + C$

9. Let $u = t^2$, $du = 2t\,dt$.

$$\int \frac{t}{\sqrt{1 - t^4}}\,dt = \frac{1}{2}\int \frac{1}{\sqrt{1 - (t^2)^2}}(2t)\,dt = \frac{1}{2}\arcsin t^2 + C$$

11. $\int \frac{t}{t^4 + 25}\,dt = \frac{1}{2}\int \frac{1}{(t^2)^2 + 5^2}(2)\,dt = \frac{1}{2}\frac{1}{5}\arctan\left(\frac{t^2}{5}\right) + C = \frac{1}{10}\arctan\left(\frac{t^2}{5}\right) + C$

13. Let $u = e^{2x}$, $du = 2e^{2x}\,dx$.

$$\int \frac{e^{2x}}{4 + e^{4x}}\,dx = \frac{1}{2}\int \frac{2e^{2x}}{4 + (e^{2x})^2}\,dx = \frac{1}{4}\arctan\frac{e^{2x}}{2} + C$$

15. Let $u = \csc x$, $du = -\csc x \cot x\,dx$, $a = 5$.

$$\int \frac{-\csc x \cot x}{\sqrt{25 - \csc^2 x}}\,dx = \arcsin\left(\frac{\csc x}{5}\right) + C$$

17. $\int \frac{1}{\sqrt{x}\sqrt{1 - x}}\,dx$, $u = \sqrt{x}$, $x = u^2$, $dx = 2u\,du$

$$\int \frac{1}{u\sqrt{1 - u^2}}(2u\,du) = 2\int \frac{du}{\sqrt{1 - u^2}} = 2\arcsin u + C = 2\arcsin\sqrt{x} + C$$

19. $\int \frac{x - 3}{x^2 + 1}\,dx = \frac{1}{2}\int \frac{2x}{x^2 + 1}\,dx - 3\int \frac{1}{x^2 + 1}\,dx = \frac{1}{2}\ln(x^2 + 1) - 3\arctan x + C$

21. $\int \frac{x + 5}{\sqrt{9 - (x-3)^2}}\,dx = \int \frac{x - 3 + 8}{\sqrt{9 - (x-3)^2}}\,dx = \int \frac{x - 3}{\sqrt{9 - (x-3)^2}}\,dx + \int \frac{8}{\sqrt{9 - (x-3)^2}}\,dx$

For the first integral on the right, use the substitution $u = 9 - (x - 3)^2$, which gives $du = -2(x - 3)\,dx$. Then you have

$$\begin{aligned}
\int \frac{x + 5}{\sqrt{9 - (x-3)^2}}\,dx &= \int \frac{x - 3}{\sqrt{9 - (x-3)^2}}\,dx + \int \frac{8}{\sqrt{9 - (x-3)^2}}\,dx \\
&= -\frac{1}{2}\int \frac{-2(x - 3)}{\left(9 - (x-3)^2\right)^{1/2}}\,dx + 8\int \frac{1}{\sqrt{3^2 - (x-3)^2}}\,dx + C \\
&= -\frac{1}{2}\frac{\left(9 - (x-3)^2\right)^{1/2}}{(1/2)} + 8\arcsin\frac{x - 3}{3} + C \\
&= -\sqrt{9 - (x-3)^2} + 8\arcsin\frac{x - 3}{3} + C \\
&= -\sqrt{6x - x^2} + 8\arcsin\left(\frac{x}{3} - 1\right) + C.
\end{aligned}$$

23. Let $u = 3x$, $du = 3\,dx$.

$$\int_0^{1/6} \frac{3}{\sqrt{1 - 9x^2}}\,dx = \int_0^{1/6} \frac{1}{\sqrt{1 - (3x)^2}}(3)\,dx = \Big[\arcsin(3x)\Big]_0^{1/6} = \frac{\pi}{6}$$

25. Let $u = 2x$, $du = 2\,dx$.

$$\int_0^{\sqrt{3}/2} \frac{1}{1 + 4x^2}\,dx = \frac{1}{2}\int_0^{\sqrt{3}/2} \frac{2}{1 + (2x)^2}\,dx$$

$$= \left[\frac{1}{2}\arctan(2x)\right]_0^{\sqrt{3}/2} = \frac{\pi}{6}$$

27. Let $u = x + 2$, $du = dx$, $a = 3$.

$$\int_1^7 \frac{1}{9 + (x + 2)^2}\,dx = \left[\frac{1}{3}\arctan\left(\frac{x + 2}{3}\right)\right]_1^7$$

$$= \frac{1}{3}\arctan 3 - \frac{1}{3}\arctan 1$$

$$= \frac{1}{3}\arctan 3 - \frac{\pi}{12} \approx 0.155$$

29. Let $u = e^x$, $du = e^x\,dx$.

$$\int_0^{\ln 5} \frac{e^x}{1 + e^{2x}}\,dx = \Big[\arctan(e^x)\Big]_0^{\ln 5} = \arctan 5 - \frac{\pi}{4} \approx 0.588$$

31. Let $u = \cos x$, $du = -\sin x\,dx$.

$$\int_{\pi/2}^{\pi} \frac{\sin x}{1 + \cos^2 x}\,dx = -\int_{\pi/2}^{\pi} \frac{-\sin x}{1 + \cos^2 x}\,dx = \Big[-\arctan(\cos x)\Big]_{\pi/2}^{\pi} = \frac{\pi}{4}$$

33. Let $u = \arcsin x$, $du = \dfrac{1}{\sqrt{1 - x^2}}\,dx$.

$$\int_0^{1/\sqrt{2}} \frac{\arcsin x}{\sqrt{1 - x^2}}\,dx = \left[\frac{1}{2}\arcsin^2 x\right]_0^{1/\sqrt{2}} = \frac{\pi^2}{32} \approx 0.308$$

35. $$\int_0^2 \frac{dx}{x^2 - 2x + 2} = \int_0^2 \frac{1}{1 + (x - 1)^2}\,dx = \Big[\arctan(x - 1)\Big]_0^2 = \frac{\pi}{2}$$

37. $$\int \frac{dx}{\sqrt{-2x^2 + 8x + 4}} = \int \frac{dx}{\sqrt{12 - 2(x^2 - 4x + 4)}} = \frac{1}{\sqrt{2}}\int \frac{dx}{\sqrt{6 - (x - 2)^2}}$$

$$= \frac{1}{\sqrt{2}}\arcsin\left(\frac{x - 2}{\sqrt{6}}\right) + C = \frac{\sqrt{2}}{2}\arcsin\left[\frac{\sqrt{6}}{6}(x - 2)\right] + C$$

39. $$\int \frac{1}{\sqrt{-x^2 - 4x}}\,dx = \int \frac{1}{\sqrt{4 - (x + 2)^2}}\,dx = \arcsin\left(\frac{x + 2}{2}\right) + C$$

41. $$\int_2^3 \frac{2x - 3}{\sqrt{4x - x^2}}\,dx = \int_2^3 \frac{2x - 4}{\sqrt{4x - x^2}}\,dx + \int_2^3 \frac{1}{\sqrt{4x - x^2}}\,dx$$

$$= -\int_2^3 (4x - x^2)^{-1/2}(4 - 2x)\,dx + \int_2^3 \frac{1}{\sqrt{4 - (x - 2)^2}}\,dx$$

$$= \left[-2\sqrt{4x - x^2} + \arcsin\left(\frac{x - 2}{2}\right)\right]_2^3 = 4 - 2\sqrt{3} + \frac{\pi}{6} \approx 1.059$$

43. Let $u = \sqrt{e^t - 3}$. Then $u^2 + 3 = e^t$, $2u\,du = e^t\,dt$, and $\dfrac{2u\,du}{u^2 + 3} = dt$.

$$\int \sqrt{e^t - 3}\,dt = \int \frac{2u^2}{u^2 + 3}\,du = \int 2\,du - \int 6\frac{1}{u^2 + 3}\,du$$

$$= 2u - 2\sqrt{3}\arctan\frac{u}{\sqrt{3}} + C = 2\sqrt{e^t - 3} - 2\sqrt{3}\arctan\sqrt{\frac{e^t - 3}{3}} + C$$

45. $\displaystyle\int_1^3 \frac{dx}{\sqrt{x}(1 + x)}$

Let $u = \sqrt{x}$, $u^2 = x$, $2u\,du = dx$, $1 + x = 1 + u^2$.

$$\int_1^{\sqrt{3}} \frac{2u\,du}{u(1 + u^2)} = \int_1^{\sqrt{3}} \frac{2}{1 + u^2}\,du$$

$$= \Big[2\arctan(u)\Big]_1^{\sqrt{3}}$$

$$= 2\left(\frac{\pi}{3} - \frac{\pi}{4}\right) = \frac{\pi}{6}$$

47. (a) $\displaystyle\int \frac{1}{\sqrt{1 - x^2}}\,dx = \arcsin x + C, \quad u = x$

(b) $\displaystyle\int \frac{x}{\sqrt{1 - x^2}}\,dx = -\sqrt{1 - x^2} + C, \quad u = 1 - x^2$

(c) $\displaystyle\int \frac{1}{x\sqrt{1 - x^2}}\,dx$ cannot be evaluated using the basic integration rules.

49. (a) $\displaystyle\int \sqrt{x - 1}\,dx = \frac{2}{3}(x - 1)^{3/2} + C, \quad u = x - 1$

(b) Let $u = \sqrt{x - 1}$. Then $x = u^2 + 1$ and $dx = 2u\,du$.

$$\int x\sqrt{x - 1}\,dx = \int (u^2 + 1)(u)(2u)\,du$$

$$= 2\int (u^4 + u^2)\,du$$

$$= 2\left(\frac{u^5}{5} + \frac{u^3}{3}\right) + C$$

$$= \frac{2}{15}u^3(3u^2 + 5) + C$$

$$= \frac{2}{15}(x - 1)^{3/2}\left[3(x - 1) + 5\right] + C$$

$$= \frac{2}{15}(x - 1)^{3/2}(3x + 2) + C$$

(c) Let $u = \sqrt{x - 1}$. Then $x = u^2 + 1$ and $dx = 2u\,du$.

$$\int \frac{x}{\sqrt{x - 1}}\,dx = \int \frac{u^2 + 1}{u}(2u)\,du$$

$$= 2\int (u^2 + 1)\,du$$

$$= 2\left(\frac{u^3}{3} + u\right) + C$$

$$= \frac{2}{3}u(u^2 + 3) + C$$

$$= \frac{2}{3}\sqrt{x - 1}(x + 2) + C$$

Note: In (b) and (c), substitution was necessary *before* the basic integration rules could be used.

51.
$$\frac{d}{dx}\left[\arcsin x^3 + C\right] = \frac{3x^2}{\sqrt{1 - x^6}}$$

$$\frac{d}{dx}\left[\arccos\sqrt{1 - x^6} + C\right] = \frac{-\frac{1}{2}(1 - x^6)^{-1/2}(-6x^5)}{\sqrt{1 - (1 - x^6)}}$$

$$= \frac{3x^5}{\sqrt{1 - x^6}\cdot x^3}$$

$$= \frac{3x^2}{\sqrt{1 - x^6}}$$

53. No. Graphing $f(x) = \arcsin x$ and $g(x) = -\arccos x$, you can see that the graph of f is the graph of g shifted vertically.

55. (a)

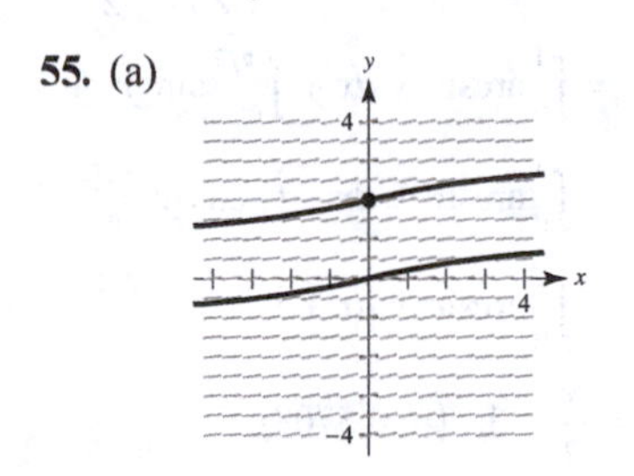

(b) $y' = \dfrac{2}{9 + x^2}, \quad (0, 2)$

$$y = \int \frac{2}{9 + x^2}\,dx = \frac{2}{3}\arctan\left(\frac{x}{3}\right) + C$$

$$2 = C$$

$$y = \frac{2}{3}\arctan\left(\frac{x}{3}\right) + 2$$

57. $\dfrac{dy}{dx} = \dfrac{10}{x\sqrt{x^2 - 1}}, \quad (3, 0)$

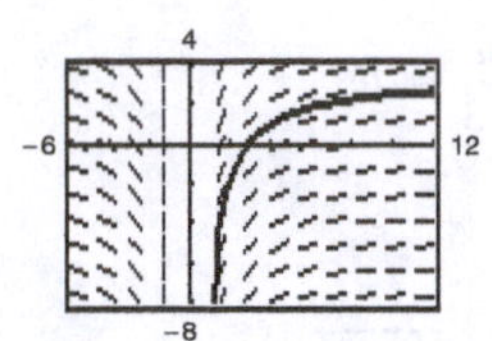

59. $\dfrac{dy}{dx} = \dfrac{2y}{\sqrt{16 - x^2}}, \quad (0, 2)$

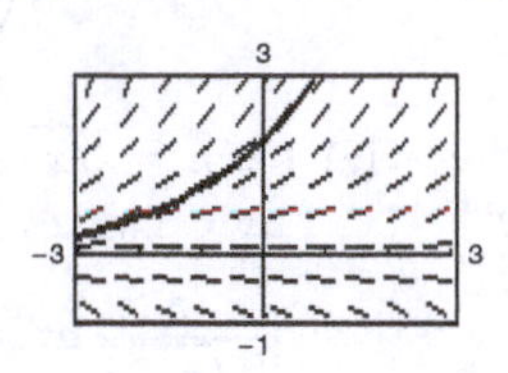

61. $y' = \dfrac{1}{\sqrt{4 - x^2}}, \quad (0, \pi)$

$$y = \int \frac{1}{\sqrt{4 - x^2}}\,dx = \arcsin\left(\frac{x}{2}\right) + C$$

When $x = 0$, $y = \pi \Rightarrow C = \pi$

$$y = \arcsin\left(\frac{x}{2}\right) + \pi$$

63. Area $= \displaystyle\int_0^1 \frac{2}{\sqrt{4 - x^2}}\,dx = \left[2\arcsin\left(\frac{x}{2}\right)\right]_0^1$

$$= 2\arcsin\left(\frac{1}{2}\right) - 2\arcsin(0) = 2\left(\frac{\pi}{6}\right) = \frac{\pi}{3}$$

65. Area $= \displaystyle\int_{-\pi/2}^{\pi/2} \frac{3\cos x}{1 + \sin^2 x}\,dx = 3\int_{-\pi/2}^{\pi/2} \frac{1}{1 + \sin^2 x}(\cos x\,dx)$

$$= \left[3\arctan(\sin x)\right]_{-\pi/2}^{\pi/2}$$

$$= 3\arctan(1) - 3\arctan(-1)$$

$$= \frac{3\pi}{4} + \frac{3\pi}{4} = \frac{3\pi}{2}$$

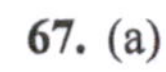

67. (a)

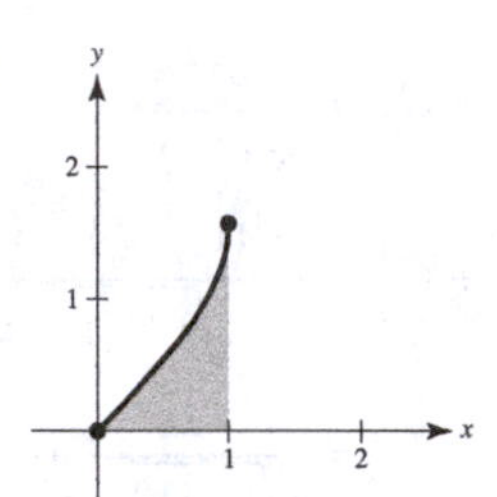

Shaded area is given by $\int_0^1 \arcsin x\,dx$.

(b) $\int_0^1 \arcsin x\,dx \approx 0.5708$

(c) Divide the rectangle into two regions.

Area rectangle $= (\text{base})(\text{height}) = 1\left(\dfrac{\pi}{2}\right) = \dfrac{\pi}{2}$

Area rectangle $= \displaystyle\int_0^1 \arcsin x\,dx + \int_0^{\pi/2} \sin y\,dy$

$$\frac{\pi}{2} = \int_0^1 \arcsin x\,dx + (-\cos y)\Big]_0^{\pi/2}$$

$$= \int_0^1 \arcsin x\,dx + 1$$

So, $\displaystyle\int_0^1 \arcsin x\,dx = \frac{\pi}{2} - 1$, (≈ 0.5708).

69. $F(x) = \dfrac{1}{2}\displaystyle\int_x^{x+2} \frac{2}{t^2 + 1}\,dt$

(a) $F(x)$ represents the average value of $f(x)$ over the interval $[x, x + 2]$. Maximum at $x = -1$, because the graph is greatest on $[-1, 1]$.

(b) $F(x) = [\arctan t]_x^{x+2}$

$$= \arctan(x + 2) - \arctan x$$

$$F'(x) = \frac{1}{1 + (x + 2)^2} - \frac{1}{1 + x^2}$$

$$= \frac{(1 + x^2) - (x^2 + 4x + 5)}{(x^2 + 1)(x^2 + 4x + 5)}$$

$$= \frac{-4(x + 1)}{(x^2 + 1)(x^2 + 4x + 5)}$$

$= 0$ when $x = -1$.

71. False, $\displaystyle\int \frac{dx}{3x\sqrt{9x^2 - 16}} = \frac{1}{12}\,\text{arcsec}\frac{|3x|}{4} + C$

73. $\dfrac{d}{dx}\left[\arcsin\left(\dfrac{u}{a}\right) + C\right] = \dfrac{1}{\sqrt{1 - (u^2/a^2)}}\left(\dfrac{u'}{a}\right)$

$$= \frac{u'}{\sqrt{a^2 - u^2}}$$

So, $\displaystyle\int \frac{du}{\sqrt{a^2 - 4^2}} = \arcsin\left(\frac{u}{d}\right) + C.$

75. Assume $u > 0$.

$$\frac{d}{dx}\left[\frac{1}{a}\operatorname{arcsec}\frac{u}{a} + C\right] = \frac{1}{a}\left[\frac{u'/a}{(u/a)\sqrt{(u/a)^2 - 1}}\right] = \frac{1}{a}\left[\frac{u'}{u\sqrt{(u^2 - a^2)/a^2}}\right] = \frac{u'}{u\sqrt{u^2 - a^2}}.$$

The case $u < 0$ is handled in a similar manner.

So, $\displaystyle\int \frac{du}{u\sqrt{u^2 - a^2}} = \int \frac{u'}{u\sqrt{u^2 - a^2}}\,dx = \frac{1}{a}\operatorname{arcsec}\frac{|u|}{a} + C.$

77. (a) Area $= \displaystyle\int_0^1 \frac{1}{1 + x^2}\,dx$

(b) Trapezoidal Rule: $n = 8, b - a = 1 - 0 = 1$

Area ≈ 0.7847

(c) Because $\displaystyle\int_0^1 \frac{1}{1 + x^2}\,dx = \left[\arctan x\right]_0^1 = \frac{\pi}{4}$, you can use the Trapezoidal Rule to approximate $\pi/4$, and therefore, π.

For example, using $n = 200$, you obtain $\pi \approx 4(0.785397) = 3.141588$.

Section 5.9 Hyperbolic Functions

1. The name *hyperbolic function* came from comparing the area of a semicircular region with the area of a region bounded by a hyperbola.

3. $\sinh^2 x = \dfrac{-1 + \cosh 2x}{2}$

5. (a) $\sinh 3 = \dfrac{e^3 - e^{-3}}{2} \approx 10.018$

(b) $\tanh(-2) = \dfrac{\sinh(-2)}{\cosh(-2)} = \dfrac{e^{-2} - e^2}{e^{-2} + e^2} \approx -0.964$

7. (a) $\operatorname{csch}(\ln 2) = \dfrac{2}{e^{\ln 2} - e^{-\ln 2}} = \dfrac{2}{2 - (1/2)} = \dfrac{4}{3}$

(b) $\coth(\ln 5) = \dfrac{\cosh(\ln 5)}{\sinh(\ln 5)} = \dfrac{e^{\ln 5} + e^{-\ln 5}}{e^{\ln 5} - e^{-\ln 5}}$

$= \dfrac{5 + (1/5)}{5 - (1/5)} = \dfrac{13}{12}$

9. (a) $\cosh^{-1} 2 = \ln\left(2 + \sqrt{3}\right) \approx 1.317$

(b) $\operatorname{sech}^{-1}\dfrac{2}{3} = \ln\left(\dfrac{1 + \sqrt{1 - (4/9)}}{2/3}\right) \approx 0.962$

11. $\sinh x + \cosh x = \dfrac{e^x - e^{-x}}{2} + \dfrac{e^x + e^{-x}}{2} = e^x$

13. $\tanh^2 x + \operatorname{sech}^2 x = \left(\dfrac{e^x - e^{-x}}{e^x + e^{-x}}\right)^2 + \left(\dfrac{2}{e^x + e^{-x}}\right)^2 = \dfrac{e^{2x} - 2 + e^{-2x} + 4}{\left(e^x + e^{-x}\right)^2} = \dfrac{e^{2x} + 2 + e^{-2x}}{e^{2x} + 2 + e^{-2x}} = 1$

15. $\dfrac{1 + \cosh 2x}{2} = \dfrac{1 + \left(e^{2x} + e^{-2x}\right)/2}{2} = \dfrac{e^{2x} + 2 + e^{-2x}}{4} = \left(\dfrac{e^x + e^{-x}}{2}\right)^2 = \cosh^2 x$

17. $2\sinh x\cosh x = 2\left(\dfrac{e^x - e^{-x}}{2}\right)\left(\dfrac{e^x + e^{-x}}{2}\right) = \dfrac{e^{2x} - e^{-2x}}{2} = \sinh 2x$

19. $\sinh x = \dfrac{3}{2}$

$\cosh^2 x - \left(\dfrac{3}{2}\right)^2 = 1 \Rightarrow \cosh^2 x = \dfrac{13}{4} \Rightarrow \cosh x = \dfrac{\sqrt{13}}{2}$

$\tanh x = \dfrac{3/2}{\sqrt{13}/2} = \dfrac{3\sqrt{13}}{13}$

$\operatorname{csch} x = \dfrac{1}{3/2} = \dfrac{2}{3}$

$\operatorname{sech} x = \dfrac{1}{\sqrt{13}/2} = \dfrac{2\sqrt{13}}{13}$

$\coth x = \dfrac{1}{3/\sqrt{13}} = \dfrac{\sqrt{13}}{3}$

21. $\lim_{x\to\infty} \sinh x = \infty$

23. $\lim_{x\to 0} \dfrac{\sinh x}{x} = \lim_{x\to 0} \dfrac{e^x - e^{-x}}{2x} = 1$

25. $f(x) = \sinh(9x)$

$f'(x) = 9\cosh(9x)$

27. $y = \operatorname{sech}(5x^2)$

$y' = -\operatorname{sech}(5x^2)\tanh(5x^2)(10x)$

$= -10x\operatorname{sech}(5x^2)\tanh(5x^2)$

29. $f(x) = \ln(\sinh x)$

$f'(x) = \dfrac{1}{\sinh x}(\cosh x) = \coth x$

31. $h(t) = \dfrac{t}{6}\sinh(-3t)$

$h'(t) = \dfrac{t}{6}\cosh(-3t)(-3) + \dfrac{1}{6}\sinh(-3t)$

$= -\dfrac{t}{2}\cosh(-3t) + \dfrac{1}{6}\sinh(-3t)$

33. $f(t) = \arctan(\sinh t)$

$f'(t) = \dfrac{1}{1+\sinh^2 t}(\cosh t) = \dfrac{\cosh t}{\cosh^2 t} = \operatorname{sech} t$

35. $y = \sinh(1 - x^2),\ (1, 0)$

$y' = \cosh(1 - x^2)(-2x)$

$y'(1) = -2$

Tangent line: $y - 0 = -2(x - 1)$

$y = -2x + 2$

37. $y = (\cosh x - \sinh x)^2,\ (0, 1)$

$y' = 2(\cosh x - \sinh x)(\sinh x - \cosh x)$

At $(0, 1)$, $y' = 2(1)(-1) = -2$.

Tangent line: $y - 1 = -2(x - 0)$

$y = -2x + 1$

39. $g(x) = x\operatorname{sech} x$

$g'(x) = \operatorname{sech} x - x\operatorname{sech} x\tanh x$

$= \operatorname{sech} x(1 - x\tanh x) = 0$

$x\tanh x = 1$

Using a graphing utility, $x \approx \pm 1.1997$.

By the First Derivative Test, $(1.1997, 0.6627)$ is a relative maximum and $(-1.1997, -0.6627)$ is a relative minimum.

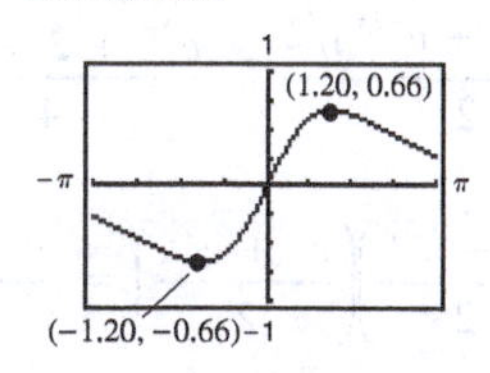

41. $f(x) = \sin x \sinh x - \cos x \cosh x, \quad -4 \le x \le 4$

$f'(x) = \sin x \cosh x + \cos x \sinh x - \cos x \sinh x + \sin x \cosh x$

$= 2 \sin x \cosh x = 0$ when $x = 0, \pm\pi$.

Relative maxima: $(\pm\pi, \cosh \pi)$

Relative minimum: $(0, -1)$

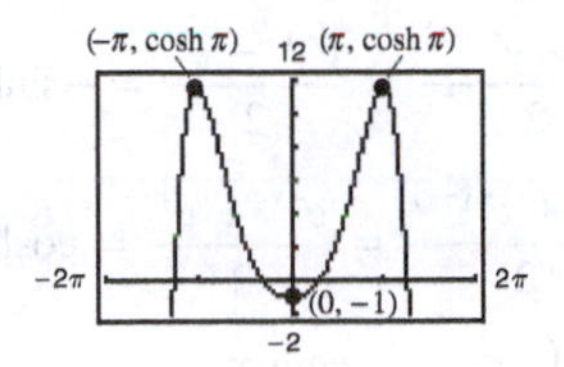

43. (a) $y = 10 + 15 \cosh \dfrac{x}{15}, \quad -15 \le x \le 15$

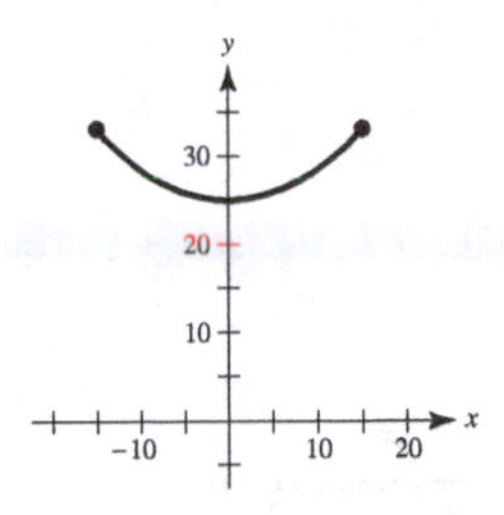

(b) At $x = \pm 15$, $y = 10 + 15 \cosh(1) \approx 33.146$.

At $x = 0$, $y = 10 + 15 \cosh(0) = 25$.

(c) $y' = \sinh \dfrac{x}{15}$. At $x = 15$, $y' = \sinh(1) \approx 1.175$.

45. $\displaystyle\int \cosh 4x\, dx = \frac{1}{4}\int \cosh 4x(4\, dx)$

$\displaystyle= \frac{1}{4} \sinh 4x + C$

47. Let $u = 1 - 2x$, $du = -2\, dx$.

$\displaystyle\int \sinh(1 - 2x)\, dx = -\tfrac{1}{2}\int \sinh(1 - 2x)(-2)\, dx$

$= -\tfrac{1}{2}\cosh(1 - 2x) + C$

49. Let $u = \cosh(x - 1)$, $du = \sinh(x - 1)\, dx$.

$\displaystyle\int \cosh^2(x - 1) \sinh(x - 1)\, dx = \tfrac{1}{3}\cosh^3(x - 1) + C$

51. Let $u = \sinh x$, $du = \cosh x\, dx$.

$\displaystyle\int \frac{\cosh x}{\sinh x}\, dx = \ln|\sinh x| + C$

53. Let $u = \dfrac{x^2}{2}$, $du = x\, dx$.

$\displaystyle\int x \operatorname{csch}^2 \frac{x^2}{2}\, dx = \int \left(\operatorname{csch}^2 \frac{x^2}{2}\right) x\, dx = -\coth \frac{x^2}{2} + C$

55. $\displaystyle\int_0^{\ln 2} \tanh x\, dx = \int_0^{\ln 2} \frac{\sinh x}{\cosh x}\, dx, \quad (u = \cosh x)$

$= \left[\ln(\cosh x)\right]_0^{\ln 2}$

$= \ln(\cosh(\ln 2)) - \ln(\cosh(0))$

$\displaystyle= \ln\left(\frac{5}{4}\right) - 0 = \ln\left(\frac{5}{4}\right)$

Note: $\displaystyle\cosh(\ln 2) = \frac{e^{\ln 2} + e^{-\ln 2}}{2} = \frac{2 + (1/2)}{2} = \frac{5}{4}$

57. $\displaystyle\int_3^4 \operatorname{csch}^2(x - 2)\, dx = \left[-\coth(x - 2)\right]_3^4$

$= -\coth 2 + \coth 1$

59. $\displaystyle\int_{5/3}^2 \operatorname{csch}(3x - 4) \coth(3x - 4)\, dx = \frac{1}{3}\int_{5/3}^2 \operatorname{csch}(3x - 4) \coth(3x - 4)(3\, dx)$

$\displaystyle= -\frac{1}{3}\left[\operatorname{csch}(3x - 4)\right]_{5/3}^2$

$\displaystyle= -\frac{1}{3}\operatorname{csch} 2 + \frac{1}{3}\operatorname{csch} 1$

61. The graph of $y_1 = \cosh x$ lies above the graph of $y_2 = \sinh x$. They do not intersect.

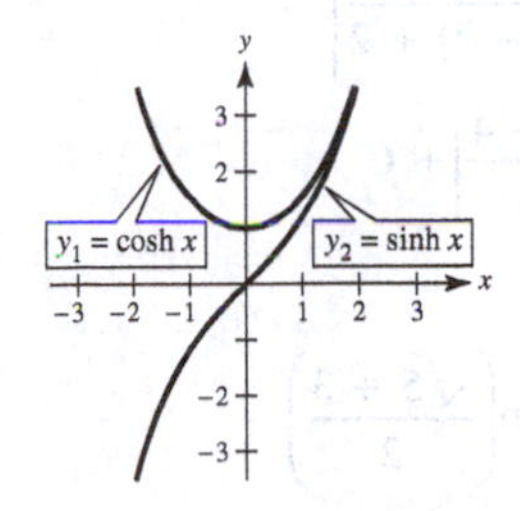

63. $\sinh(-x) = \dfrac{e^{-x} - e^{-(-x)}}{2} = \dfrac{e^{-x} - e^{x}}{2} = -\sinh x \Rightarrow \text{odd}$

$\cosh(-x) = \dfrac{e^{-x} + e^{-(-x)}}{2} = \dfrac{e^{-x} + e^{x}}{2} = \cosh x \Rightarrow \text{even}$

$\tanh(-x) = \dfrac{\sinh(-x)}{\cosh(-x)} = -\dfrac{\sinh x}{\cosh x} = -\tanh x \Rightarrow \text{odd}$

$\coth(-x) = \dfrac{\cosh(-x)}{\sinh(-x)} = \dfrac{\cosh x}{-\sinh x} = -\coth x \Rightarrow \text{odd}$

$\operatorname{csch}(-x) = \dfrac{1}{\sinh(-x)} = \dfrac{1}{-\sinh x} = -\operatorname{csch} x \Rightarrow \text{odd}$

$\operatorname{sech}(-x) = \dfrac{1}{\cosh(-x)} = \dfrac{1}{\cosh x} = \operatorname{sech} x \Rightarrow \text{even}$

65. $y = \cosh^{-1}(3x)$

$y' = \dfrac{3}{\sqrt{9x^2 - 1}}$

67. $y = \tanh^{-1}\sqrt{x}$

$y' = \dfrac{1}{1 - \left(\sqrt{x}\right)^2}\left(\dfrac{1}{2}x^{-1/2}\right) = \dfrac{1}{2\sqrt{x}(1 - x)}$

69. $y = \sinh^{-1}(\tan x)$

$y' = \dfrac{1}{\sqrt{\tan^2 x + 1}}(\sec^2 x) = |\sec x|$

71. $y = \operatorname{sech}^{-1}(\sin x), 0 < x < \pi/2$

$y' = \dfrac{-\cos x}{\sin x\sqrt{1 - \sin^2 x}} = \dfrac{-\cos x}{\sin x(\cos x)} = -\csc x$

73. $y = 2x\sinh^{-1}(2x) - \sqrt{1 + 4x^2}$

$y' = 2x\left(\dfrac{2}{\sqrt{1 + 4x^2}}\right) + 2\sinh^{-1}(2x) - \dfrac{4x}{\sqrt{1 + 4x^2}}$

$= 2\sinh^{-1}(2x)$

75. $\displaystyle\int \frac{1}{3 - 9x^2}\,dx = \frac{1}{3}\int \frac{1}{3 - (3x)^2}(3)\,dx$

$= \dfrac{1}{3}\dfrac{1}{2\sqrt{3}}\ln\left|\dfrac{\sqrt{3} + 3x}{\sqrt{3} - 3x}\right| + C$

$= \dfrac{\sqrt{3}}{18}\ln\left|\dfrac{1 + \sqrt{3}x}{1 - \sqrt{3}x}\right| + C$

77. $\displaystyle\int \frac{1}{\sqrt{1 + e^{2x}}}\,dx = \int \frac{e^x}{e^x\sqrt{1 + \left(e^x\right)^2}}\,dx$

$= -\operatorname{csch}^{-1}\left(e^x\right) + C$

$= -\ln\left(\dfrac{1 + \sqrt{1 + e^{2x}}}{e^x}\right) + C$

$= \ln\left(\dfrac{e^x}{1 + \sqrt{1 + e^{2x}}}\right) + C$

$= \ln\left(\dfrac{-e^x + e^x\sqrt{1 + e^{2x}}}{e^{2x}}\right) + C$

$= \ln\left(\sqrt{1 + e^{2x}} - 1\right) - x + C$

79. Let $u = \sqrt{x}$, $du = \dfrac{1}{2\sqrt{x}}\,dx$.

$\displaystyle\int \frac{1}{\sqrt{x}\sqrt{1 + x}}\,dx = 2\int \frac{1}{\sqrt{1 + \left(\sqrt{x}\right)^2}}\left(\frac{1}{2\sqrt{x}}\right)dx$

$= 2\sinh^{-1}\sqrt{x} + C$

$= 2\ln\left(\sqrt{x} + \sqrt{1 + x}\right) + C$

81. $\displaystyle\int \frac{-1}{4x - x^2}\,dx = \int \frac{1}{(x - 2)^2 - 4}\,dx$

$= \dfrac{1}{4}\ln\left|\dfrac{(x - 2) - 2}{(x - 2) + 2}\right|$

$= \dfrac{1}{4}\ln\left|\dfrac{x - 4}{x}\right| + C$

83. $\displaystyle\int_3^7 \frac{1}{\sqrt{x^2 - 4}}\,dx = \left[\ln\left(x + \sqrt{x^2 - 4}\right)\right]_3^7 = \ln\left(7 + \sqrt{45}\right) - \ln\left(3 + \sqrt{5}\right) = \ln\left(\frac{7 + \sqrt{45}}{3 + \sqrt{5}}\right) = \ln\left(\frac{\sqrt{5} + 3}{2}\right)$

85. $\int_{-1}^{1} \frac{1}{16 - 9x^2}\,dx = \frac{1}{3}\int_{-1}^{1} \frac{1}{4^2 - (3x)^2}(3)\,dx = \left[\frac{1}{3}\frac{1}{4}\frac{1}{2}\ln\left|\frac{4 + 3x}{4 - 3x}\right|\right]_{-1}^{1} = \frac{1}{24}\left[\ln(7) - \ln\left(\frac{1}{7}\right)\right]$

$$= \frac{1}{24}[\ln 7 - \ln 1 + \ln 7] = \frac{1}{12}\ln 7$$

87. $y = \int \frac{x^3 - 21x}{5 + 4x - x^2}\,dx = \int\left(-x - 4 + \frac{20}{5 + 4x - x^2}\right)dx$

$$= \int(-x - 4)\,dx + 20\int \frac{1}{3^2 - (x - 2)^2}\,dx$$

$$= -\frac{x^2}{2} - 4x + \frac{20}{6}\ln\left|\frac{3 + (x - 2)}{3 - (x - 2)}\right| + C$$

$$= -\frac{x^2}{2} - 4x + \frac{10}{3}\ln\left|\frac{1 + x}{5 - x}\right| + C$$

$$= \frac{-x^2}{2} - 4x - \frac{10}{3}\ln\left|\frac{5 - x}{x + 1}\right| + C$$

89. $A = 2\int_0^4 \operatorname{sech}\frac{x}{2}\,dx$

$$= 2\int_0^4 \frac{2}{e^{x/2} + e^{-x/2}}\,dx$$

$$= 4\int_0^4 \frac{e^{x/2}}{\left(e^{x/2}\right)^2 + 1}\,dx$$

$$= \left[8\arctan\left(e^{x/2}\right)\right]_0^4$$

$$= 8\arctan\left(e^2\right) - 2\pi \approx 5.207$$

91. $A = \int_0^2 \frac{5x}{\sqrt{x^4 + 1}}\,dx$

$$= \frac{5}{2}\int_0^2 \frac{2x}{\sqrt{\left(x^2\right)^2 + 1}}\,dx$$

$$= \left[\frac{5}{2}\ln\left(x^2 + \sqrt{x^4 + 1}\right)\right]_0^2$$

$$= \frac{5}{2}\ln\left(4 + \sqrt{17}\right) \approx 5.237$$

93. (a) $y = a\operatorname{sech}^{-1}\frac{x}{a} - \sqrt{a^2 - x^2}, \quad a > 0$

$$\frac{dy}{dx} = \frac{-1}{(x/a)\sqrt{1 - \left(x^2/a^2\right)}} + \frac{x}{\sqrt{a^2 - x^2}} = \frac{-a^2}{x\sqrt{a^2 - x^2}} + \frac{x}{\sqrt{a^2 - x^2}} = \frac{x^2 - a^2}{x\sqrt{a^2 - x^2}} = \frac{-\sqrt{a^2 - x^2}}{x}$$

(b) Equation of tangent line through $P = (x_0, y_0)$: $y - a\operatorname{sech}^{-1}\frac{x_0}{a} + \sqrt{a^2 - x_0^2} = -\frac{\sqrt{a^2 - x_0^2}}{x_0}(x - x_0)$

When $x = 0$, $y = a\operatorname{sech}^{-1}\frac{x_0}{a} - \sqrt{a^2 - x_0^2} + \sqrt{a^2 - x_0^2} = a\operatorname{sech}^{-1}\frac{x_0}{a}$.

So, Q is the point $\left[0, a\operatorname{sech}^{-1}(x_0/a)\right]$.

Distance from P to Q: $d = \sqrt{(x_0 - 0)^2 + \left(y_0 - a\operatorname{sech}^{-1}(x_0/a)\right)} = \sqrt{x_0^2 + \left(-\sqrt{a^2 - x_0^2}\right)^2} = \sqrt{a^2} = a$

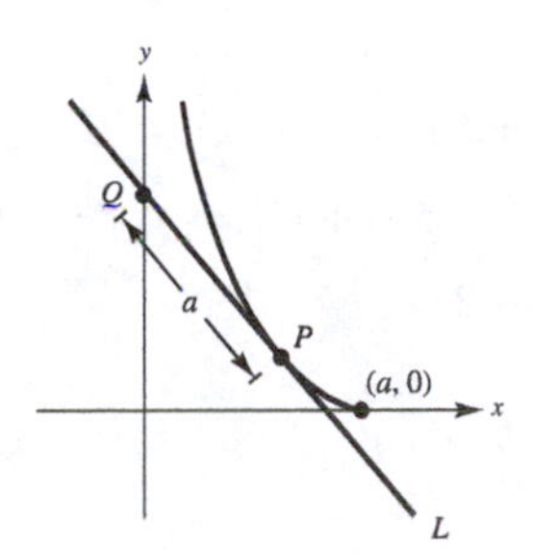

95. Let $u = \tanh^{-1}x, \; -1 < x < 1$

$$\tanh u = x.$$

$$\frac{\sinh u}{\cosh u} = \frac{e^u - e^{-u}}{e^u + e^{-u}} = x$$

$$e^u - e^{-u} = xe^u + xe^{-u}$$

$$e^{2u} - 1 = xe^{2u} + x$$

$$e^{2u}(1 - x) = 1 + x$$

$$e^{2u} = \frac{1 + x}{1 - x}$$

$$2u = \ln\left(\frac{1 + x}{1 - x}\right)$$

$$u = \frac{1}{2}\ln\left(\frac{1 + x}{1 - x}\right), \; -1 < x < 1$$

97. Let $y = \arcsin(\tanh x)$. Then,

$$\sin y = \tanh x = \frac{e^x - e^{-x}}{e^x + e^{-x}}$$

and $\tan y = \dfrac{e^x - e^{-x}}{2} = \sinh x.$

So, $y = \arctan(\sinh x)$. Therefore,

$\arctan(\sinh x) = \arcsin(\tanh x).$

99. $y = \cosh x = \dfrac{e^x + e^{-x}}{2}$

$$y' = \frac{e^x - e^{-x}}{2} = \sinh x$$

101. $y = \operatorname{sech} x = \dfrac{2}{e^x + e^{-x}}$

$$y' = -2\left(e^x + e^{-x}\right)^{-2}\left(e^x - e^{-x}\right)$$

$$= \left(\frac{-2}{e^x + e^{-x}}\right)\left(\frac{e^x - e^{-x}}{e^x + e^{-x}}\right) = -\operatorname{sech} x \tanh x$$

103. $y = \sinh^{-1} x$

$$\sinh y = x$$

$$(\cosh y)y' = 1$$

$$y' = \frac{1}{\cosh y} = \frac{1}{\sqrt{\sinh^2 y + 1}} = \frac{1}{\sqrt{x^2 + 1}}$$

105. $y = c \cosh \dfrac{x}{c}$

Let $P(x_1, y_1)$ be a point on the catenary.

$$y' = \sinh \frac{x}{c}$$

The slope at P is $\sinh(x_1/c)$. The equation of line L is

$$y - c = \frac{-1}{\sinh(x_1/c)}(x - 0).$$

When $y = 0$, $c = \dfrac{x}{\sinh(x_1/c)} \Rightarrow x = c\sinh\left(\dfrac{x_1}{c}\right).$

The length of L is

$$\sqrt{c^2 \sinh^2\left(\frac{x_1}{c}\right) + c^2} = c \cdot \cosh\frac{x_1}{c} = y_1,$$

the ordinate y_1 of the point P.

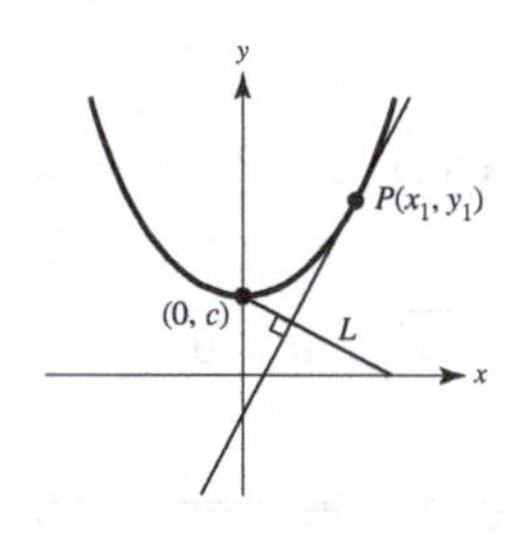

Review Exercises for Chapter 5

1. $\int \left(4x^2 + x + 3\right) dx = \frac{4}{3}x^3 + \frac{1}{2}x^2 + 3x + C$

3. $\int \dfrac{x^4 + 8}{x^3}\, dx = \int \left(x + 8x^{-3}\right) dx = \dfrac{1}{2}x^2 - \dfrac{4}{x^2} + C$

5. $\int \left(5 - e^x\right) dx = 5x - e^x + C$

7. $f'(x) = -6x, \; f(1) = -2$

$$f(x) = -3x^2 + C$$

$$f(1) = -2 = -3(1)^2 + C \Rightarrow C = 1$$

$$f(x) = -3x^2 + 1$$

9. $f''(x) = 24x, \; f'(-1) = 7, \; f(1) = -4$

$$f'(x) = 12x^2 + C_1$$

$$f'(-1) = 7 = 12(-1)^2 + C_1 \Rightarrow C_1 = -5$$

$$f'(x) = 12x^2 - 5$$

$$f(x) = 4x^3 - 5x + C_2$$

$$f(1) = -4 = 4(1)^3 - 5(1) + C_2 \Rightarrow C_2 = -3$$

$$f(x) = 4x^3 - 5x - 3$$

11. $a(t) = -32$

$v(t) = -32t + 96$

$s(t) = -16t^2 + 96t$

(a) $v(t) = -32t + 96 = 0$ when $t = 3$ sec.

$s(3) = -144 + 288 = 144$ ft

(b) $v(t) = -32t + 96 = \frac{96}{2}$ when $t = \frac{3}{2}$ sec.

(c) $s\left(\frac{3}{2}\right) = -16\left(\frac{9}{4}\right) + 96\left(\frac{3}{2}\right) = 108$ ft

13. $\sum_{i=1}^{5}(5i - 3) = 2 + 7 + 12 + 17 + 22 = 60$

15. $\frac{1}{5(3)} + \frac{2}{5(4)} + \frac{3}{5(5)} + \dots + \frac{10}{5(12)} = \sum_{i=1}^{10} \frac{i}{5(i + 2)}$

17. $\sum_{i=1}^{20} 2i = 2\left(\frac{20(21)}{2}\right) = 420$

19. $\sum_{i=1}^{20} (i + 1)^2 = \sum_{i=1}^{20}\left(i^2 + 2i + 1\right)$

$= \frac{20(21)(41)}{6} = 2\frac{20(21)}{2} + 20$

$= 2870 + 420 + 20 = 3310$

21. $f(x) = 4x + 1, \quad [2, 3]$

$\Delta x = \frac{3 - 2}{n} = \frac{1}{n}$

Right endpoints for upper sum: $M_i = 2 + i\Delta x = 2 + i\left(\frac{1}{n}\right)$

Left endpoints for lower sum: $m_i = 2 + (i - 1)\Delta x = 2 + (i - 1)\left(\frac{1}{n}\right)$

$$S(n) = \sum_{i=1}^{n} f(Mi)\Delta x = \sum_{i=1}^{n}\left[4\left(2 + i\left(\frac{1}{n}\right)\right) + 1\right]\left(\frac{1}{n}\right)$$

$$= \frac{1}{n}\sum_{i=1}^{n} 9 + \frac{4}{n^2}\sum_{i=1}^{n} i = \frac{1}{n}(9n) + \frac{4}{n^2}\left[\frac{n(n + 1)}{2}\right] = 11 + \frac{2}{n}$$

$$s(n) = \sum_{i=1}^{n} f(m_i)\Delta x = \sum_{i=1}^{n}\left[4\left(2 + (i - 1)\left(\frac{1}{n}\right)\right) + 1\right]\left(\frac{1}{n}\right)$$

$$= \frac{1}{n}\sum_{i=1}^{n} 9 + \frac{4}{n^2}\sum_{i=1}^{4}(i - 1) = 9 + \frac{4}{n^2}\left[\frac{n(n - 1)}{2}\right] = 11 - \frac{2}{n}$$

23. $y = 8 - 2x$, $\Delta x = \frac{3}{n}$, right endpoints

$$\text{Area} = \lim_{n\to\infty}\sum_{i=1}^{n} f(c_i)\Delta x$$

$$= \lim_{n\to\infty}\sum_{i=1}^{n}\left(8 - 2\left(\frac{3i}{n}\right)\right)\frac{3}{n}$$

$$= \lim_{n\to\infty}\frac{3}{n}\sum_{i=1}^{n}\left(8 - \frac{6i}{n}\right)$$

$$= \lim_{n\to\infty}\frac{3}{n}\left[8n - \frac{6}{n}\frac{n(n + 1)}{2}\right]$$

$$= \lim_{n\to\infty}\left[24 - 9\frac{n + 1}{n}\right] = 24 - 9 = 15$$

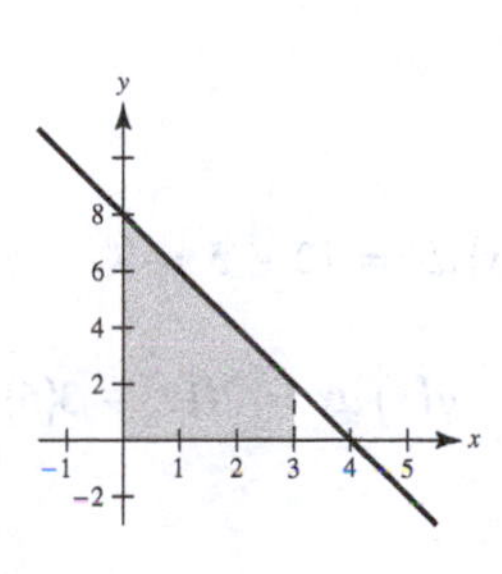

25. $y = 5 - x^2,\ \Delta x = \dfrac{3}{n}$

$$\text{Area} = \lim_{n\to\infty}\sum_{i=1}^{n} f(c_i)\Delta x$$
$$= \lim_{n\to\infty}\sum_{i=1}^{n}\left[5 - \left(-2 + \frac{3i}{n}\right)^2\right]\left(\frac{3}{n}\right)$$
$$= \lim_{n\to\infty}\frac{3}{n}\sum_{i=1}^{n}\left[1 + \frac{12i}{n} - \frac{9i^2}{n^2}\right]$$
$$= \lim_{n\to\infty}\frac{3}{n}\left[n + \frac{12}{n}\frac{n(n+1)}{2} - \frac{9}{n^2}\frac{n(n+1)(2n+1)}{6}\right]$$
$$= \lim_{n\to\infty}\left[3 + 18\frac{n+1}{n} - \frac{9}{2}\frac{(n+1)(2n+1)}{n^2}\right] = 3 + 18 - 9 = 12$$

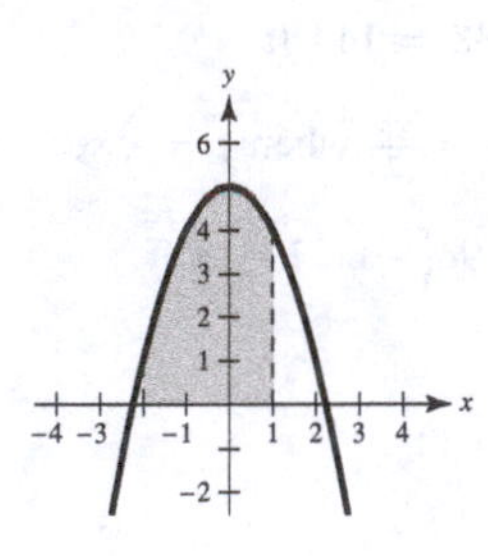

27. $f(x) = 16 - x^2,\ [0, 4],\ n = 4$

Let $c_i = \dfrac{x_i + x_{i-1}}{2}$.

$\Delta x = 1,\ c_1 = \dfrac{1}{2},\ c_2 = \dfrac{3}{2},\ c_3 = \dfrac{5}{2},\ c_4 = \dfrac{7}{2}$

$$\text{Area} \approx \sum_{i=1}^{n} f(c_i)\Delta x = \sum_{i=1}^{4}\left[16 - c_i^2\right](1) = \left[\left(16 - \frac{1}{4}\right) + \left(16 - \frac{9}{4}\right) + \left(16 - \frac{25}{4}\right) + \left(16 - \frac{49}{4}\right)\right] = 43$$

29. $y = 6x$ on $[-3, 5]$. $\left(\textbf{Note: } \Delta x = \dfrac{5 - (-3)}{n} = \dfrac{8}{n}\right)$

$$S(n) = \sum_{i=1}^{n} f\left(-3 + \frac{8i}{n}\right)\left(\frac{8}{n}\right)$$
$$= \frac{8}{n}\sum_{i=1}^{n} 6\left(-3 + \frac{8i}{n}\right)$$
$$= \frac{8}{n}\sum_{i=1}^{n} -18 + \left(\frac{8}{n}\right)^2\sum_{i=1}^{n} 6i$$
$$= \frac{8}{n}(-18n) + \frac{384}{n^2}\cdot\left(\frac{n(n+1)}{2}\right)$$

$$\lim_{n\to\infty} S(n) = -144 + 192 = 48$$

31.

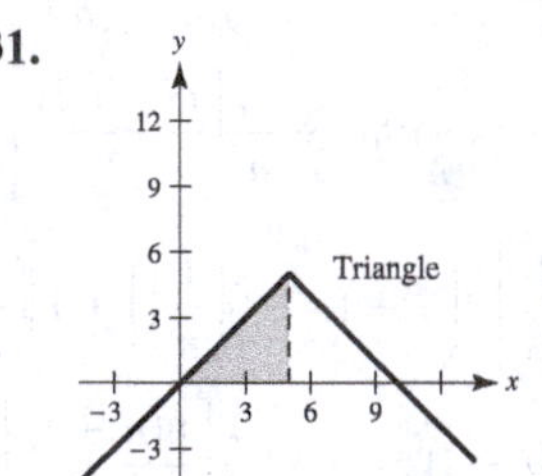

$$\int_0^5 (5 - |x - 5|)\,dx = \tfrac{1}{2}(5)(5) = \tfrac{25}{2}$$

(triangle)

33. $\int_4^8 f(x)dx = 12,\ \int_4^8 g(x)dx = 5$

(a) $\int_4^8 [f(x) - g(x)]\,dx = \int_4^8 f(x)\,dx - \int_4^8 g(x)\,dx = 12 - 5 = 7$

(b) $\int_4^8 [2f(x) - 3g(x)]\,dx = 2\int_4^8 f(x)\,dx - 3\int_4^8 g(x)\,dx = 2(12) - 3(5) = 9$

35. $\int_0^8 (3 + x)\,dx = \left[3x + \dfrac{x^2}{2}\right]_0^8 = 24 + \dfrac{64}{2} = 56$

37. $\int_4^9 x\sqrt{x}\,dx = \int_4^9 x^{3/2}\,dx = \left[\dfrac{2}{5}x^{5/2}\right]_4^9 = \dfrac{2}{5}\left[\left(\sqrt{9}\right)^5 - \left(\sqrt{4}\right)^5\right] = \dfrac{2}{5}(243 - 32) = \dfrac{422}{5}$

39. $\int_0^2 (x + e^x)\,dx = \left[\frac{x^2}{2} + e^x\right]_0^2 = 2 + e^2 - 1 = 1 + e^2$

41. $A = \int_0^6 (8 - x)\,dx = \left[8x - \frac{x^2}{2}\right]_0^6 = (48 - 18) - 0 = 30$

43. $A = \int_1^3 \frac{2}{x}\,dx = [2 \ln x]_1^3 = 2 \ln 3 - 2 \ln 1 = \ln 9$

45. $\frac{1}{3 - 1}\int_1^3 3x^2\,dx = \frac{1}{2}\left[x^3\right]_1^3 = \frac{1}{2}(27 - 1) = 13$

$3x^2 = 13$

$x^2 = \frac{13}{3}$

$x = \sqrt{\frac{13}{3}}$

47. Average value: $\frac{1}{9 - 4}\int_4^9 \frac{1}{\sqrt{x}}\,dx = \left[\frac{1}{5}2\sqrt{x}\right]_4^9$

$= \frac{2}{5}(3 - 2) = \frac{2}{5}$

$\frac{2}{5} = \frac{1}{\sqrt{x}}$

$\sqrt{x} = \frac{5}{2}$

$x = \frac{25}{4}$

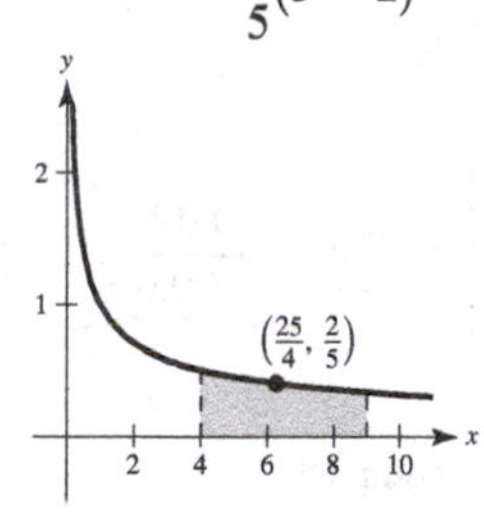

49. $F'(x) = x^2\sqrt{1 + x^3}$

51. $F'(x) = x^2 + 3x + 2$

53. $u = x^3 + 3,\ du = 3x^2\,dx$

$\int \frac{x^2}{\sqrt{x^3 + 3}}\,dx = \int (x^3 + 3)^{-1/2} x^2\,dx$

$= \frac{1}{3}\int (x^3 + 3)^{-1/2} 3x^2\,dx$

$= \frac{2}{3}(x^3 + 3)^{1/2} + C$

55. $u = 1 - 3x^2,\ du = -6x\,dx$

$\int x(1 - 3x^2)^4\,dx = -\frac{1}{6}\int (1 - 3x^2)^4(-6x\,dx)$

$= -\frac{1}{30}(1 - 3x^2)^5 + C$

$= \frac{1}{30}(3x^2 - 1)^5 + C$

57. $\int \frac{\cos\theta}{\sqrt{1 - \sin\theta}}\,d\theta = -\int (1 - \sin\theta)^{-1/2}(-\cos\theta)\,d\theta$

$= -2(1 - \sin\theta)^{1/2} + C$

$= -2\sqrt{1 - \sin\theta} + C$

59. $\int xe^{-3x^2}\,dx = -\frac{1}{6}\int e^{-3x^2}(-6x)\,dx = -\frac{1}{6}e^{-3x^2} + C$

61. $\int (x + 1)5^{(x+1)^2}\,dx = \frac{1}{2}\int 5^{(x+1)^2}\,2(x + 1)\,dx$

$= \frac{1}{2 \ln 5}5^{(x+1)^2} + C$

63. (a) Answers will vary. *Sample answer:*

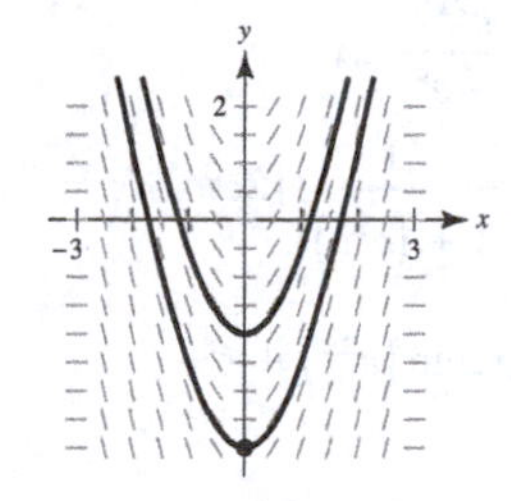

(b) $\frac{dy}{dx} = x\sqrt{9 - x^2},\ (0, -4)$

$y = \int (9 - x^2)^{1/2}\,x\,dx = \frac{-1}{2}\frac{(9 - x^2)^{3/2}}{3/2} + C = -\frac{1}{3}(9 - x^2)^{3/2} + C$

$-4 = -\frac{1}{3}(9 - 0)^{3/2} + C = -\frac{1}{3}(27) + C \Rightarrow C = 5$

$y = -\frac{1}{3}(9 - x^2)^{3/2} + 5$

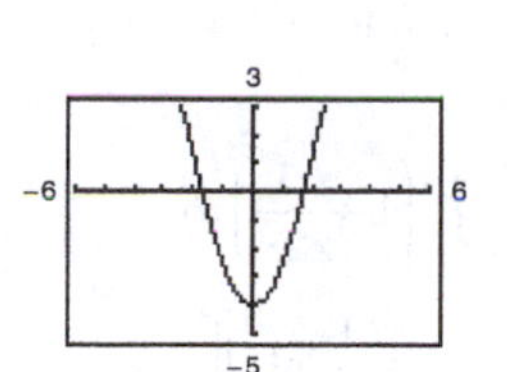

65. $\int_0^3 \frac{1}{\sqrt{1+x}}\,dx = \int_0^3 (1+x)^{-1/2}\,dx = \left[2(1+x)^{1/2}\right]_0^3 = 4 - 2 = 2$

67. $u = 1 - y,\ y = 1 - u,\ dy = -du$

When $y = 0, u = 1$. When $y = 1, u = 0$.

$$2\pi\int_0^1 (y+1)\sqrt{1-y}\,dy = 2\pi\int_1^0 -[(1-u)+1]\sqrt{u}\,du = 2\pi\int_1^0 \left(u^{3/2} - 2u^{1/2}\right)du = 2\pi\left[\frac{2}{5}u^{5/2} - \frac{4}{3}u^{3/2}\right]_1^0 = \frac{28\pi}{15}$$

69. $\int_0^\pi \cos\left(\frac{x}{2}\right)dx = 2\int_0^\pi \cos\left(\frac{x}{2}\right)\frac{1}{2}\,dx = \left[2\sin\left(\frac{x}{2}\right)\right]_0^\pi = 2$

71. $\lim_{x\to 1}\left[\frac{(\ln x)^2}{x-1}\right] = \lim_{x\to 1}\left[\frac{2(1/x)\ln x}{1}\right] = 0$

73. $\lim_{x\to\infty}\frac{e^{2x}}{x^2} = \lim_{x\to\infty}\frac{2e^{2x}}{2x} = \lim_{x\to\infty}\frac{4e^{2x}}{2} = \infty$

75. $y = \lim_{x\to\infty}(\ln x)^{2/x}$

$\ln y = \lim_{x\to\infty}\frac{2\ln(\ln x)}{x} = \lim_{x\to\infty}\left[\frac{2/(x\ln x)}{1}\right] = 0$

Because $\ln y = 0$, $y = 1$.

77. $\lim_{n\to\infty} 1000\left(1 + \frac{0.09}{n}\right)^n = 1000\lim_{n\to\infty}\left(1 + \frac{0.09}{n}\right)^n$

Let $y = \lim_{n\to\infty}\left(1 + \frac{0.09}{n}\right)^n$.

$$\ln y = \lim_{n\to\infty} n\ln\left(1 + \frac{0.09}{n}\right) = \lim_{n\to\infty}\frac{\ln\left(1 + \frac{0.09}{n}\right)}{\frac{1}{n}} = \lim_{n\to\infty}\left(\frac{\frac{-0.09/n^2}{1 + (0.09/n)}}{-\frac{1}{n^2}}\right) = \lim_{n\to\infty}\frac{0.09}{1 + \left(\frac{0.09}{n}\right)} = 0.09$$

So, $\ln y = 0.09 \Rightarrow y = e^{0.09}$ and $\lim_{n\to\infty} 1000\left(1 + \frac{0.09}{n}\right)^n = 1000e^{0.09} \approx 1094.17$.

79. $u = 7x - 2,\ du = 7\,dx$

$$\int\frac{1}{7x-2}\,dx = \frac{1}{7}\int\frac{1}{7x-2}(7)\,dx = \frac{1}{7}\ln|7x-2| + C$$

81. $\int\frac{\sin x}{1+\cos x}\,dx = -\int\frac{-\sin x}{1+\cos x}\,dx$

$= -\ln|1+\cos x| + C$

83. Let $u = e^{2x} + e^{-2x},\ du = \left(2e^{2x} - e^{-2x}\right)dx$.

$$\int\frac{e^{2x} - e^{-2x}}{e^{2x} + e^{-2x}}\,dx = \frac{1}{2}\int\frac{2e^{2x} - 2e^{-2x}}{e^{2x} + e^{-2x}}\,dx$$

$= \frac{1}{2}\ln\left(e^{2x} + e^{-2x}\right) + C$

85. $\int_1^4 \frac{2x+1}{2x}\,dx = \int_1^4\left(1 + \frac{1}{2x}\right)dx$

$= \left[x + \frac{1}{2}\ln|x|\right]_1^4$

$= 4 + \frac{1}{2}\ln 4 - 1 = 3 + \ln 2$

87. $\int_0^{\pi/3}\sec\theta\,d\theta = \left[\ln|\sec\theta + \tan\theta|\right]_0^{\pi/3} = \ln\left(2 + \sqrt{3}\right)$

89. Let $u = e^{2x},\ du = 2e^{2x}\,dx$.

$$\int\frac{1}{e^{2x} + e^{-2x}}\,dx = \int\frac{e^{2x}}{1 + e^{4x}}\,dx$$

$= \frac{1}{2}\int\frac{1}{1 + \left(e^{2x}\right)^2}\left(2e^{2x}\right)dx$

$= \frac{1}{2}\arctan\left(e^{2x}\right) + C$

91. Let $u = x^2,\ du = 2x\,dx$.

$$\int\frac{x}{\sqrt{1-x^4}}\,dx = \frac{1}{2}\int\frac{1}{\sqrt{1-\left(x^2\right)^2}}(2x)\,dx$$

$= \frac{1}{2}\arcsin x^2 + C$

93. Let $u = \arctan\left(\frac{x}{2}\right)$, $du = \frac{2}{4 + x^2}\,dx$.

$$\int \frac{\arctan(x/2)}{4 + x^2}\,dx = \frac{1}{2}\int\left(\arctan\frac{x}{2}\right)\left(\frac{2}{4 + x^2}\right)dx = \frac{1}{4}\left(\arctan\frac{x}{2}\right)^2 + C$$

95. $y = \operatorname{sech}(4x - 1)$

$$y' = -\operatorname{sech}(4x - 1)\tanh(4x - 1)(4) = -4\operatorname{sech}(4x - 1)\tanh(4x - 1)$$

97. $y = \sinh^{-1}(4x)$

$$y' = \frac{4}{\sqrt{(4x)^2 + 1}} = \frac{4}{\sqrt{16x^2 + 1}}$$

99. Let $u = x^3$, $du = 3x^2\,dx$.

$$\int x^2(\operatorname{sech} x^3)^2\,dx = \frac{1}{3}\int(\operatorname{sech} x^3)^2(3x^2)\,dx = \frac{1}{3}\tanh x^3 + C$$

101. Let $u = \frac{2}{3}x$, $du = \frac{2}{3}\,dx$.

$$\int \frac{1}{9 - 4x^2}\,dx = \int \frac{1/9}{1 - \left(\frac{4}{9}x^2\right)}\,dx = \frac{1}{6}\tanh^{-1}\left(\frac{2}{3}x\right) + C$$

Alternate solution:

$$\int \frac{1}{3^2 - (2x)^2}\,dx = \frac{1}{12}\ln\left|\frac{3 + 2x}{3 - 2x}\right| + C$$

Problem Solving for Chapter 5

1. (a) $L(1) = \int_1^1 \frac{1}{t}\,dt = 0$

(b) $L'(x) = \frac{1}{x}$ by the Second Fundamental Theorem of Calculus.

$L'(1) = 1$

(c) $L(x) = 1 = \int_1^x \frac{1}{t}\,dt$ for $x \approx 2.718$

$$\int_1^{2.718} \frac{1}{t}\,dt = 0.999896$$

(**Note:** The exact value of x is e, the base of the natural logarithm function.)

(d) First show that $\int_1^{x_1} \frac{1}{t}\,dt = \int_{1/x_1}^1 \frac{1}{t}\,dt$. To see this, let $u = \frac{t}{x_1}$ and $du = \frac{1}{x_1}\,dt$.

Then $\int_1^{x_1} \frac{1}{t}\,dt = \int_{1/x_1}^1 \frac{1}{ux_1}(x_1\,du) = \int_{1/x_1}^1 \frac{1}{u}\,du = \int_{1/x_1}^1 \frac{1}{t}\,dt$.

Now,

$$L(x_1 x_2) = \int_1^{x_1x_2} \frac{1}{t}\,dt = \int_{1/x_1}^{x_2} \frac{1}{u}\,du \left(\text{using } u = \frac{t}{x_1}\right) = \int_{1/x_1}^1 \frac{1}{u}\,du + \int_1^{x_2} \frac{1}{u}\,du = \int_1^{x_1} \frac{1}{u}\,du + \int_1^{x_2} \frac{1}{u}\,du = L(x_1) + L(x_2).$$

3. (a) Let $A = \int_0^b \frac{f(x)}{f(x) + f(b - x)}\,dx$. Let $u = b - x$, $du = -dx$.

$$A = \int_b^0 \frac{f(b - u)}{f(b - u) + f(u)}(-du) = \int_0^b \frac{f(b - u)}{f(b - u) + f(u)}\,du = \int_0^b \frac{f(b - x)}{f(b - x) + f(x)}\,dx$$

Then, $2A = \int_0^b \frac{f(x)}{f(x) + f(b - x)}\,dx + \int_0^b \frac{f(b - x)}{f(b - x) + f(x)}\,dx = \int_0^b 1\,dx = b$. So, $A = \frac{b}{2}$.

(b) $b = 1 \Rightarrow \int_0^1 \frac{\sin x}{\sin(1 - x) + \sin x}\,dx = \frac{1}{2}$

(c) $b = 3$, $f(x) = \sqrt{x}$

$$\int_0^3 \frac{\sqrt{x}}{\sqrt{x} + \sqrt{3 - x}}\,dx = \frac{3}{2}$$

5. (a) $\int_{-1}^{1} \cos x\,dx \approx \cos\left(-\frac{1}{\sqrt{3}}\right) + \cos\left(\frac{1}{\sqrt{3}}\right) = 2\cos\left(\frac{1}{\sqrt{3}}\right) \approx 1.6758$

$\int_{-1}^{1} \cos x\,dx = \sin x\Big]_{-1}^{1} = 2\sin(1) \approx 1.6829$

Error: $|1.6829 - 1.6758| = 0.0071$

(b) $\int_{-1}^{1} \frac{1}{1+x^2}\,dx \approx \frac{1}{1+(1/3)} + \frac{1}{1+(1/3)} = \frac{3}{2}$ (**Note:** exact answer is $\pi/2 \approx 1.5708$)

(c) Let $p(x) = ax^3 + bx^2 + cx + d$.

$\int_{-1}^{1} p(x)\,dx = \left[\frac{ax^4}{4} + \frac{bx^3}{3} + \frac{cx^2}{2} + dx\right]_{-1}^{1} = \frac{2b}{3} + 2d$

$p\left(-\frac{1}{\sqrt{3}}\right) + p\left(\frac{1}{\sqrt{3}}\right) = \left(\frac{b}{3} + d\right) + \left(\frac{b}{3} + d\right) = \frac{2b}{3} + 2d$

7. Let d be the distance traversed and a be the uniform acceleration. You can assume that $v(0) = 0$ and $s(0) = 0$.

Then

$a(t) = a$

$v(t) = at$

$s(t) = \frac{1}{2}at^2.$

$s(t) = d$ when $t = \sqrt{\frac{2d}{a}}.$

The highest speed is $v = a\sqrt{\frac{2d}{a}} = \sqrt{2ad}.$

The lowest speed is $v = 0$.

The mean speed is $\frac{1}{2}\left(\sqrt{2ad} + 0\right) = \sqrt{\frac{ad}{2}}.$

The time necessary to traverse the distance d at the mean speed is $t = \frac{d}{\sqrt{ad/2}} = \sqrt{\frac{2d}{a}}$

which is the same as the time calculated above.

9. Consider $F(x) = [f(x)]^2 \Rightarrow F'(x) = 2f(x)f'(x)$.

So,

$$\begin{aligned}\int_a^b f(x)f'(x)\,dx &= \int_a^b \tfrac{1}{2}F'(x)\,dx \\ &= \left[\tfrac{1}{2}F(x)\right]_a^b \\ &= \tfrac{1}{2}\left[F(b) - F(a)\right] \\ &= \tfrac{1}{2}\left[f(b)^2 - f(a)^2\right].\end{aligned}$$

11. Consider $\int_0^1 x^5\,dx = \frac{x^6}{6}\Big]_0^1 = \frac{1}{6}.$

The corresponding Riemann Sum using right endpoints is

$S(n) = \frac{1}{n}\left[\left(\frac{1}{n}\right)^5 + \left(\frac{2}{n}\right)^5 + \cdots + \left(\frac{n}{n}\right)^5\right]$ So,

$= \frac{1}{n^6}\left[1^5 + 2^5 + \cdots + n^5\right].$

$\lim_{n\to\infty} S(n) = \lim_{n\to\infty} \frac{1^5 + 2^5 + \cdots + n^5}{n^6} = \frac{1}{6}.$

13. (a)

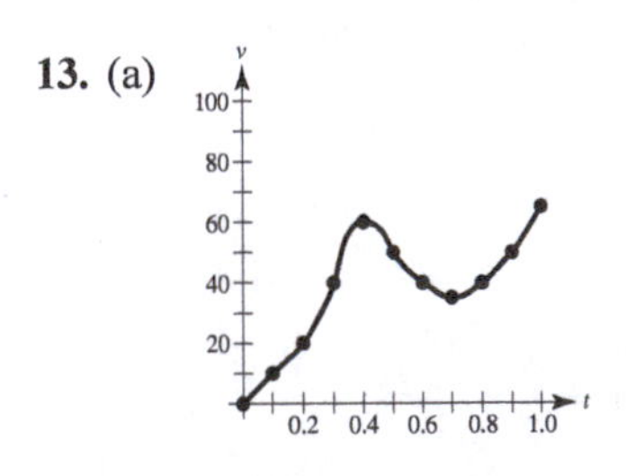

(b) v is increasing (positive acceleration) on $(0, 0.4)$ and $(0.7, 1.0)$.

(c) Average acceleration $= \frac{v(0.4) - v(0)}{0.4 - 0} = \frac{60 - 0}{0.4} = 150$ mi/h^2

(d) This integral is the total distance traveled in miles.

$\int_0^1 v(t)\,dt \approx \frac{1}{5}[10 + 40 + 50 + 35 + 50] = 37$ miles

(e) One approximation is

$a(0.8) \approx \frac{v(0.9) - v(0.8)}{0.9 - 0.8} = \frac{50 - 40}{0.1} = 100$ mi/h^2

(other answers possible)

15. (a) $(1+i)^3 = 1 + 3i + 3i^2 + i^3 \Rightarrow (1+i)^3 - i^3 = 3i^2 + 3i + 1$

(b) $3i^2 + 3i + 1 = (i+1)^3 - i^3$

$$\sum_{i=1}^{n}\left(3i^2 + 3i + 1\right) = \sum_{i=1}^{n}\left[(i+1)^3 - i^3\right] = \left(2^3 - 1^3\right) + \left(3^3 - 2^3\right) + \cdots + \left[\left((n+1)^3 - n^3\right)\right] = (n+1)^3 - 1$$

So, $(n+1)^3 = \sum_{i=1}^{n}\left(3i^2 + 3i + 1\right) + 1$.

(c) $(n+1)^3 - 1 = \sum_{i=1}^{n}\left(3i^2 + 3i + 1\right) = \sum_{i=1}^{n} 3i^2 + \frac{3(n)(n+1)}{2} + n$

$$\Rightarrow \sum_{i=1}^{n} 3i^2 = n^3 + 3n^2 + 3n - \frac{3n(n+1)}{2} - n$$

$$= \frac{2n^3 + 6n^2 + 6n - 3n^2 - 3n - 2n}{2}$$

$$= \frac{2n^3 + 3n^2 + n}{2}$$

$$= \frac{n(n+1)(2n+1)}{2}$$

$$\Rightarrow \sum_{i=1}^{n} i^2 = \frac{n(n+1)(2n+1)}{6}$$

17. Let $u = 1 + \sqrt{x}$, $\sqrt{x} = u - 1$, $x = u^2 - 2u + 1$, $dx = (2u - 2)\,du$.

$$\text{Area} = \int_1^4 \frac{1}{\sqrt{x} + x}\,dx = \int_2^3 \frac{2u - 2}{(u-1) + \left(u^2 - 2u + 1\right)}\,du = \int_2^3 \frac{2(u-1)}{u^2 - u}\,du$$

$$= \int_2^3 \frac{2}{u}\,du = \left[2\ln|u|\right]_2^3 = 2\ln 3 - 2\ln 2$$

$$= 2\ln\left(\frac{3}{2}\right) \approx 0.8109$$

19. (a) (i) $y = e^x$

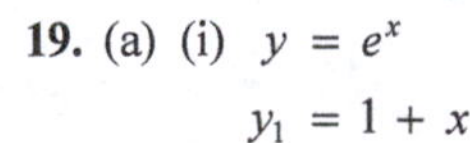

$y_1 = 1 + x$

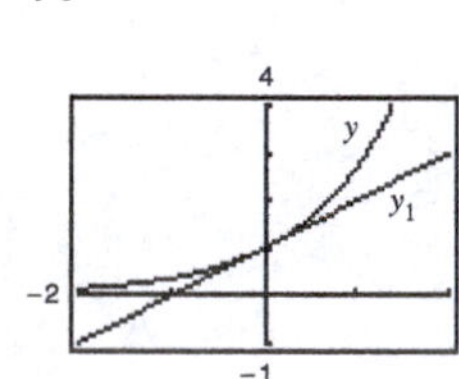

(ii) $y = e^x$

$y_2 = 1 + x + \left(\frac{x^2}{2}\right)$

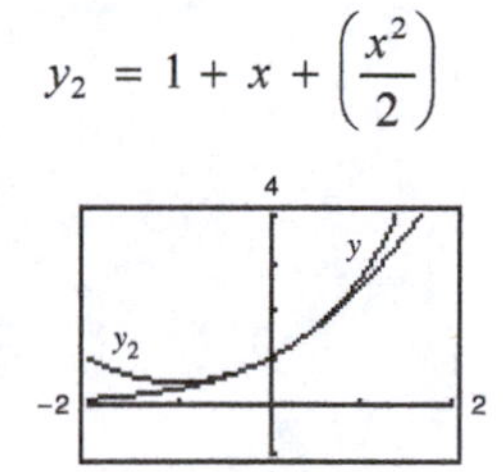

(iii) $y = e^x$

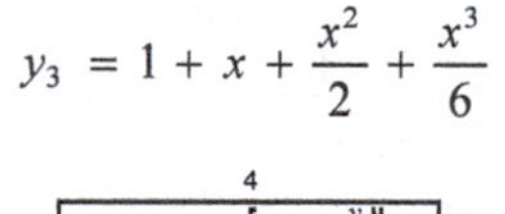

$y_3 = 1 + x + \frac{x^2}{2} + \frac{x^3}{6}$

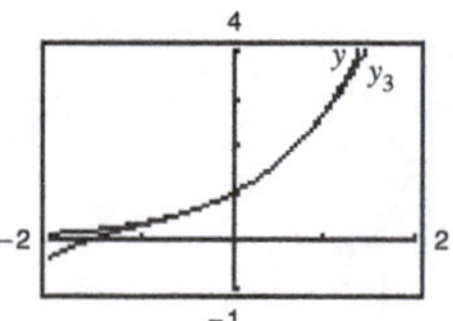

(b) n^{th} term is $x^n/n!$ in polynomial: $y_4 = 1 + x + \frac{x^2}{2!} + \frac{x^3}{3!} + \frac{x^4}{4!}$

(c) Conjecture: $e^x = 1 + x + \frac{x^2}{2!} + \frac{x^3}{3!} + \cdots$

CHAPTER 6
Differential Equations

CHAPTER 6
Differential Equations

Section 6.1 Slope Fields and Euler's Method

1. A function $f(x)$ is a solution of a differential equation if the equation is satisfied when y and its derivatives are replaced by $f(x)$ and its derivatives.

3. The line segments show the general shape of all the solutions of a differential equation and give a visual perspective of the directions of the solutions of the differential equation.

5. Differential equation: $y' = 5y$

Solution: $y = Ce^{5x}$

Check: $y' = 5Ce^{5x} = 5y$

7. Differential equation: $y'' + y = 0$

Solution: $y = C_1 \sin x - C_2 \cos x$

$y' = C_1 \cos x + C_2 \sin x$

$y'' = -C_1 \sin x + C_2 \cos x$

Check: $y'' + y = (-C_1 \sin x + C_2 \cos x) + (C_1 \sin x - C_2 \cos x) = 0$

9. Differential equation: $y'' + y = \tan x$

Solution: $y = -\cos x \ln|\sec x + \tan x|$

$$y' = (-\cos x)\frac{1}{\sec x + \tan x}(\sec x \cdot \tan x + \sec^2 x) + \sin x \ln|\sec x + \tan x|$$

$$= \frac{(-\cos x)}{\sec x + \tan x}(\sec x)(\tan x + \sec x) + \sin x \ln|\sec x + \tan x|$$

$$= -1 + \sin x \ln|\sec x + \tan x|$$

$$y'' = (\sin x)\frac{1}{\sec x + \tan x}(\sec x \cdot \tan x + \sec^2 x) + \cos x \ln|\sec x + \tan x|$$

$$= (\sin x)(\sec x) + \cos x \ln|\sec x + \tan x|$$

Check: $y'' + y = (\sin x)(\sec x) + \cos x \ln|\sec x + \tan x| - \cos x \ln|\sec x + \tan x| = \tan x.$

11. $y = \sin x \cos x - \cos^2 x$

$y' = -\sin^2 x + \cos^2 x + 2 \cos x \sin x = -1 + 2\cos^2 x + \sin 2x$

Differential equation:

$2y + y' = 2(\sin x \cos x - \cos^2 x) + (-1 + 2\cos^2 x + \sin 2x) = 2 \sin x \cos x - 1 + \sin 2x = 2 \sin 2x - 1$

Initial condition $\left(\frac{\pi}{4}, 0\right)$: $\sin \frac{\pi}{4} \cos \frac{\pi}{4} - \cos^2 \frac{\pi}{4} = \frac{\sqrt{2}}{2} \cdot \frac{\sqrt{2}}{2} - \left(\frac{\sqrt{2}}{2}\right)^2 = 0$

13. $y = 4e^{-6x^2}$

$y' = 4e^{-6x^2}(-12x) = -48xe^{-6x^2}$

Differential equation: $y' = -12xy = -12x\left(4e^{-6x^2}\right) = -48xe^{-6x^2}$

Initial condition $(0, 4)$: $4e^0 = 4$

In Exercises 15–21, the differential equation is $y^{(4)} - 16y = 0$.

15. $y = 3\cos 2x$

$y^{(4)} = 48\cos 2x$

$y^{(4)} - 16y = 48\cos 2x - 48\cos 2x = 0,$

Yes

17. $y = 3\cos x$

$y^{(4)} = 3\cos x$

$y^{(4)} - 16y = -45\cos x \neq 0,$

No

19. $y = e^{-2x}$

$y^{(4)} = 16e^{-2x}$

$y^{(4)} - 16y = 16e^{-2x} - 16e^{-2x} = 0,$

Yes

21. $y = \ln x + e^{2x} + Cx^4$

$y^{(4)} = 16e^{2x} - \dfrac{6}{x^4} + 24C$

$y^{(4)} - 16y = 16e^{2x} - \dfrac{6}{x^4} + 24C - \ln x - e^{2x} - Cx^4 \neq 0,$

No

In Exercises 23–29, the differential equation is $xy' - 2y = x^3e^x$.

23. $y = x^2 + e^x, y' = 2x + e^x$

$xy' - 2y = x(2x + e^x) - 2(x^2 + e^x)$

$= xe^x - 2e^x \neq x^3e^x$

No

25. $y = x^2e^x, y' = x^2e^x + 2xe^x = e^x(x^2 + 2x)$

$xy' - 2y = x(e^x(x^2 + 2x)) - 2(x^2e^x) = x^3e^x,$

Yes

27. $y = e^x - \sin x, y' = e^x - \cos x$

$xy' - 2y = x(e^x - \cos x) - 2(e^x - \sin x)$

$= xe^x - x\cos x - 2e^x + 2\sin x \neq x^3e^x$

No

29. $y = 2e^x \ln x, y' = 2e^x \ln x + \dfrac{2}{x}e^x$

$xy' - 2y = x\left(2e^x \ln x + \dfrac{2}{x}e^x\right) - 2(2e^x \ln x) \neq x^3e^x$

No

31. $y = Ce^{-x/2}$ passes through $(0, 3)$.

$3 = Ce^0 = C \Rightarrow C = 3$

Particular solution: $y = 3e^{-x/2}$

33. $y^2 = Cx^3$ passes through $(4, 4)$.

$16 = C(64) \Rightarrow C = \frac{1}{4}$

Particular solution: $y^2 = \frac{1}{4}x^3$ or $4y^2 = x^3$

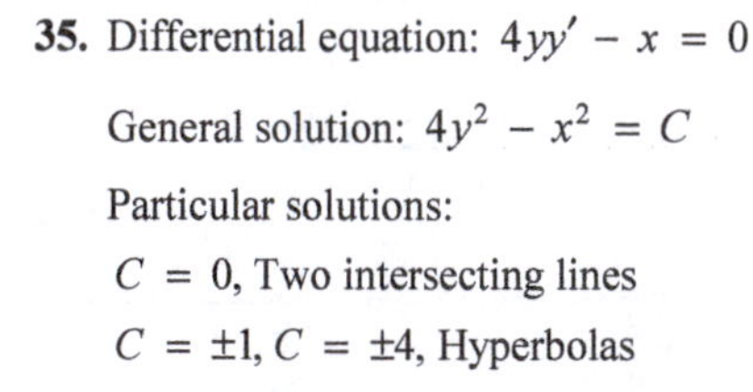

35. Differential equation: $4yy' - x = 0$

General solution: $4y^2 - x^2 = C$

Particular solutions:

$C = 0$, Two intersecting lines

$C = \pm 1, C = \pm 4$, Hyperbolas

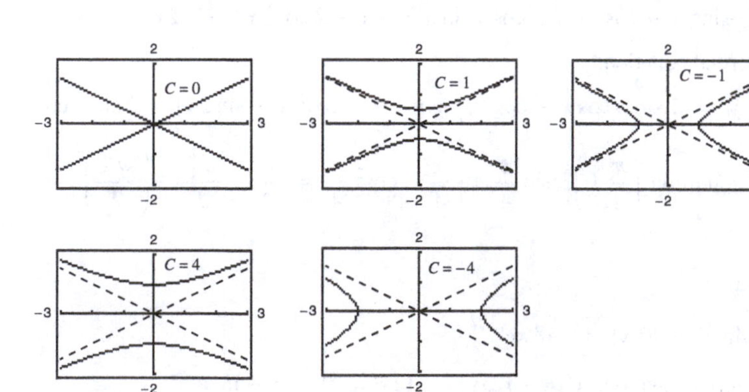

37. Differential equation: $y' + 6y = 0$

General Solution: $y = Ce^{-6x}$

$y' + 6y = \left(-6Ce^{-6x}\right) + 6\left(Ce^{-6x}\right) = 0$

Initial condition $(0, 3)$: $3 = Ce^0 = C$

Particular solution: $y = 3e^{-6x}$

39. Differential equation: $y'' + 9y = 0$

General solution: $y = C_1 \sin 3x + C_2 \cos 3x$

$$y' = 3C_1 \cos 3x - 3C_2 \sin 3x,$$
$$y'' = -9C_1 \sin 3x - 9C_2 \cos 3x$$
$$y'' + 9y = (-9C_1 \sin 3x - 9C_2 \cos 3x) + 9(C_1 \sin 3x + C_2 \cos 3x) = 0$$

Initial conditions $\left(\frac{\pi}{6}, 2\right)$ and $y' = 1$ when $x = \frac{\pi}{6}$:

$$2 = C_1 \sin\left(\frac{\pi}{2}\right) + C_2 \cos\left(\frac{\pi}{2}\right) \Rightarrow C_1 = 2$$
$$y' = 3C_1 \cos 3x - 3C_2 \sin 3x$$
$$1 = 3C_1 \cos\left(\frac{\pi}{2}\right) - 3C_2 \sin\left(\frac{\pi}{2}\right) = -3C_2 \Rightarrow C_2 = -\frac{1}{3}$$

Particular solution: $y = 2 \sin 3x - \frac{1}{3} \cos 3x$

41. Differential equation: $x^2y'' - 3xy' + 3y = 0$

General solution: $y = C_1x + C_2x^3$

$$y' = C_1 + 3C_2x^2,\ y'' = 6C_2x$$
$$x^2y'' - 3xy' + 3y = x^2(6C_2x) - 3x\left(C_1 + 3C_2x^2\right) + 3\left(C_1x + C_2x^3\right) = 0$$

Initial conditions $(2, 0)$ and $y' = 4$ when $x = 2$:

$$0 = 2C_1 + 8C_2$$
$$y' = C_1 + 3C_2x^2$$
$$4 = C_1 + 12C_2$$
$$\left.\begin{aligned} C_1 + 4C_2 &= 0 \\ C_1 + 12C_2 &= 4 \end{aligned}\right\} C_2 = \tfrac{1}{2},\ C_1 = -2$$

Particular solution: $y = -2x + \frac{1}{2}x^3$

43. $\frac{dy}{dx} = 12x^2$

$$y = \int 12x^2\,dx = 4x^3 + C$$

45. $\frac{dy}{dx} = \frac{x}{1 + x^2}$

$$y = \int \frac{x}{1 + x^2}\,dx = \frac{1}{2}\ln\left(1 + x^2\right) + C$$

$\left(u = 1 + x^2,\ du = 2x\,dx\right)$

47. $\frac{dy}{dx} = \sin 2x$

$$y = \int \sin 2x\,dx = -\frac{1}{2}\cos 2x + C$$

$(u = 2x,\ du = 2\,dx)$

49. $\dfrac{dy}{dx} = x\sqrt{x-6}$

Let $u = \sqrt{x-6}$, then $x = u^2 + 6$ and $dx = 2u\,du$.

$$\begin{aligned} y = \int x\sqrt{x-6}\,dx &= \int (u^2+6)(u)(2u)\,du \\ &= 2\int (u^4 + 6u^2)\,du \\ &= 2\left(\frac{u^5}{5} + 2u^3\right) + C \\ &= \frac{2}{5}(x-6)^{5/2} + 4(x-6)^{3/2} + C \\ &= \frac{2}{5}(x-6)^{3/2}(x - 6 + 10) + C \\ &= \frac{2}{5}(x-6)^{3/2}(x+4) + C \end{aligned}$$

51. $\dfrac{dy}{dx} = xe^{x^2}$

$$y = \int xe^{x^2}\,dx = \frac{1}{2}e^{x^2} + C$$

$(u = x^2,\ du = 2x\,dx)$

53.

x	-4	-2	0	2	4	8
y	2	0	4	4	6	8
dy/dx	-4	Undef.	0	1	$\frac{4}{3}$	2

55.

x	-4	-2	0	2	4	8
y	2	0	4	4	6	8
dy/dx	$-2\sqrt{2}$	-2	0	0	$-2\sqrt{2}$	-8

57. $\dfrac{dy}{dx} = \sin 2x$

For $x = 0$, $\dfrac{dy}{dx} = 0$. Matches (b).

58. $\dfrac{dy}{dx} = \dfrac{1}{2}\cos x$

For $x = 0$, $\dfrac{dy}{dx} = \dfrac{1}{2}$. Matches (c).

59. $\dfrac{dy}{dx} = e^{-2x}$

As $x \to \infty$, $\dfrac{dy}{dx} \to 0$. Matches (d).

60. $\dfrac{dy}{dx} = \dfrac{x}{x^2+1}$

For $x = 0$, $\dfrac{dy}{dx} = 0$ and for $x = 3$, $\dfrac{dy}{dx} = \dfrac{3}{10} > 0$.

Matches (a).

61. (a), (b)

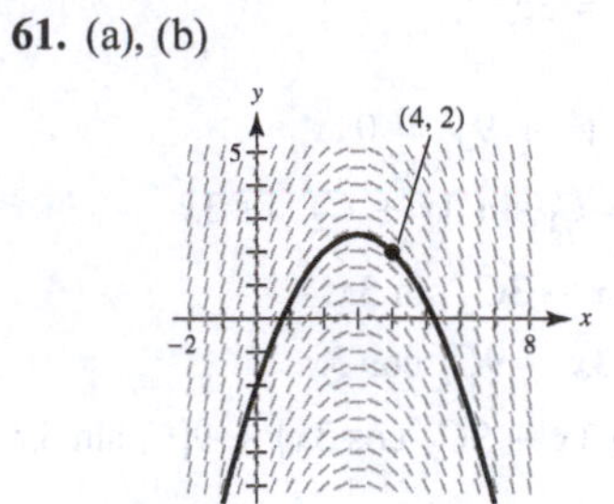

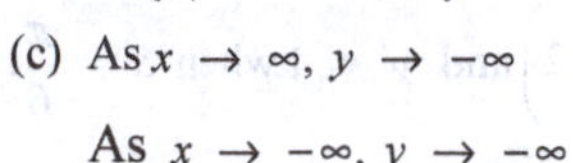

(c) As $x \to \infty$, $y \to -\infty$

As $x \to -\infty$, $y \to -\infty$

63. (a), (b)

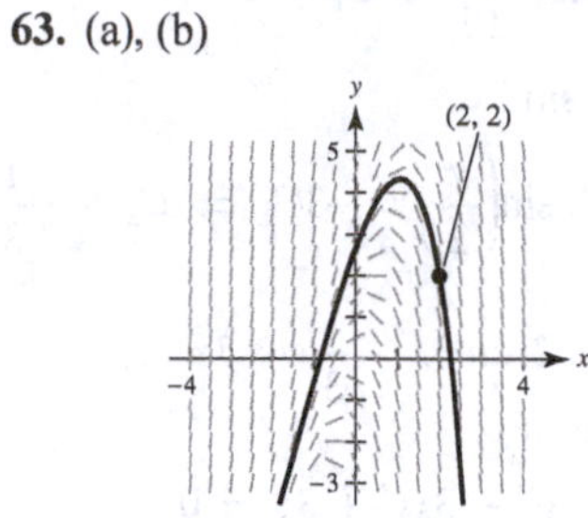

(c) As $x \to \infty$, $y \to -\infty$

As $x \to -\infty$, $y \to -\infty$

65. (a) $y' = \dfrac{1}{x}$, $(1, 0)$

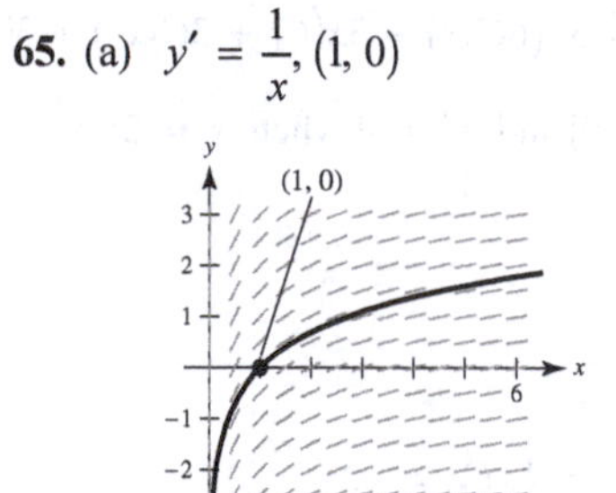

As $x \to \infty$, $y \to \infty$

[Note: The solution is $y = \ln x$.]

(b) $y' = \dfrac{1}{x}$, $(2, -1)$

As $x \to \infty$, $y \to \infty$

67. $\frac{dy}{dx} = 0.25y,\ y(0) = 4$

(a), (b)

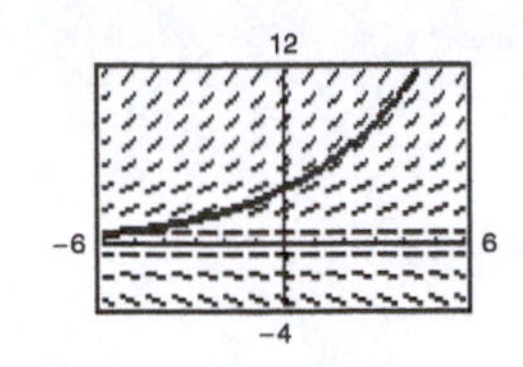

69. $\frac{dy}{dx} = 0.02y(10 - y),\ y(0) = 2$

(a), (b)

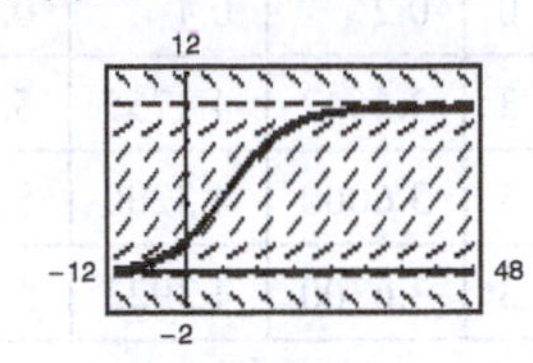

71. $\frac{dy}{dx} = 0.4y(3 - x),\ y(0) = 1$

(a), (b)

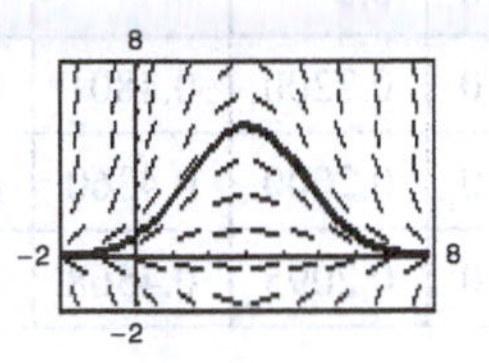

73. $y' = x + y,\quad y(0) = 2,\quad n = 10,\quad h = 0.1$

$y_1 = y_0 + hF(x_0, y_0) = 2 + (0.1)(0 + 2) = 2.2$

$y_2 = y_1 + hF(x_1, y_1) = 2.2 + (0.1)(0.1 + 2.2) = 2.43$, etc.

n	0	1	2	3	4	5	6	7	8	9	10
x_n	0	0.1	0.2	0.3	0.4	0.5	0.6	0.7	0.8	0.9	1.0
y_n	2	2.2	2.43	2.693	2.992	3.332	3.715	4.146	4.631	5.174	5.781

75. $y' = 3x - 2y,\quad y(0) = 3,\quad n = 10,\quad h = 0.05$

$y_1 = y_0 + hF(x_0, y_0) = 3 + (0.05)(3(0) - 2(3)) = 2.7$

$y_2 = y_1 + hF(x_1, y_1) = 2.7 + (0.05)(3(0.05) + 2(2.7)) = 2.4375$, etc.

n	0	1	2	3	4	5	6	7	8	9	10
x_n	0	0.05	0.1	0.15	0.2	0.25	0.3	0.35	0.4	0.45	0.5
y_n	3	2.7	2.438	2.209	2.010	1.839	1.693	1.569	1.464	1.378	1.308

77. $y' = e^{xy},\quad y(0) = 1,\quad n = 10,\quad h = 0.1$

$y_1 = y_0 + hF(x_0, y_0) = 1 + (0.1)e^{0(1)} = 1.1$

$y_2 = y_1 + hF(x_1, y_1) = 1.1 + (0.1)e^{(0.1)(1.1)} \approx 1.2116$, etc.

n	0	1	2	3	4	5	6	7	8	9	10
x_n	0	0.1	0.2	0.3	0.4	0.5	0.6	0.7	0.8	0.9	1.0
y_n	1	1.1	1.212	1.339	1.488	1.670	1.900	2.213	2.684	3.540	5.958

79. $\frac{dy}{dx} = y, y = 3e^x, (0, 3)$

x	0	0.2	0.4	0.6	0.8	1
$y(x)$ (exact)	3	3.6642	4.4755	5.4664	6.6766	8.1548
$y(x)(h = 0.2)$	3	3.6000	4.3200	5.1840	6.2208	7.4650
$y(x)(h = 0.1)$	3	3.6300	4.3923	5.3147	6.4308	7.7812

81. $\frac{dy}{dx} = y + \cos x, y = \frac{1}{2}(\sin x - \cos x + e^x), \quad (0, 0)$

x	0	0.2	0.4	0.6	0.8	1
$y(x)$ (exact)	0	0.2200	0.4801	0.7807	1.1231	1.5097
$y(x)(h = 0.2)$	0	0.2000	0.4360	0.7074	1.0140	1.3561
$y(x)(h = 0.1)$	0	0.2095	0.4568	0.7418	1.0649	1.4273

83. $\frac{dy}{dt} = -\frac{1}{2}(y - 72), \quad (0, 140), h = 0.1$

(a)

t	0	1	2	3
Euler	140	112.7	96.4	86.6

(b) $y = 72 + 68e^{-t/2}$ exact

t	0	1	2	3
Exact	140	113.24	97.016	87.173

(c) $\frac{dy}{dt} = -\frac{1}{2}(y - 72), \quad (0, 140), h = 0.05$

t	0	1	2	3
Euler	140	112.98	96.7	86.9

The approximations are better using $h = 0.05$.

85. Euler's Method produces an exact solution to an initial value problem when the exact solution is a line.

87. False. Consider Example 2. $y = x^3$ is a solution to $xy' - 3y = 0$, but $y = x^3 + 1$ is not a solution.

89. $\frac{dy}{dx} = -2y, y(0) = 4, y = 4e^{-2x}$

(a)

x	0	0.2	0.4	0.6	0.8	1
y	4	2.6813	1.7973	1.2048	0.8076	0.5413
y_1	4	2.5600	1.6384	1.0486	0.6711	0.4295
y_2	4	2.4000	1.4400	0.8640	0.5184	0.3110
e_1	0	0.1213	0.1589	0.1562	0.1365	0.1118
e_2	0	0.2813	0.3573	0.3408	0.2892	0.2303
r		0.4312	0.4447	0.4583	0.4720	0.4855

(b) If h is halved, then the error is approximately halved $(r \approx 0.5)$.

(c) When $h = 0.05$, the errors will again be approximately halved.

91. (a) $L\frac{dI}{dt} + RI = E(t)$

$4\frac{dI}{dt} + 12I = 24$

$\frac{dI}{dt} = \frac{1}{4}(24 - 12I)$

$= 6 - 3I$

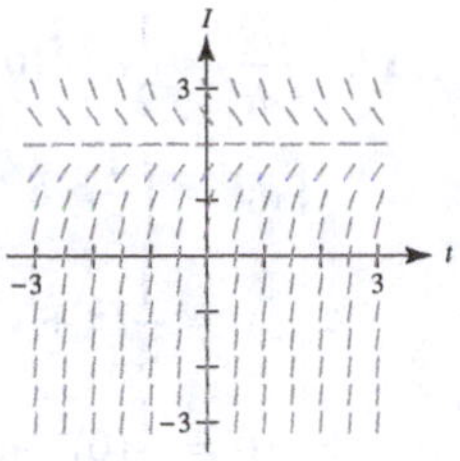

(b) As $t \to \infty$, $I \to 2$. That is, $\lim_{t\to\infty} I(t) = 2$.

In fact, $I = 2$ is a solution to the differential equation.

93. $y = A\sin\omega t$

$y' = A\omega\cos\omega t$

$y'' = -A\omega^2\sin\omega t$

$y'' + 16y = 0$

$-A\omega^2\sin\omega t + 16A\sin\omega t = 0$

$A\sin\omega t\left[16 - \omega^2\right] = 0$

If $A \neq 0$, then $\omega = \pm 4$

95. $f(x) + f''(x) = -xg(x)f'(x), \quad g(x) \geq 0$

$2f(x)f'(x) + 2f'(x)f''(x) = -2xg(x)\left[f'(x)\right]^2$

$\frac{d}{dx}\left[f(x)^2 + f'(x)^2\right] = -2x\,g(x)\left[f'(x)\right]^2$

For $x < 0$, $-2x\,g(x)\left[f'(x)\right]^2 \geq 0$

For $x > 0$, $-2x\,g(x)\left[f'(x)\right]^2 \leq 0$

So, $f(x)^2 + f'(x)^2$ is increasing for $x < 0$ and decreasing for $x > 0$.

$f(x)^2 + f'(x)^2$ has a maximum at $x = 0$. So, it is bounded by its value at $x = 0$, $f(0)^2 + f'(0)^2$. So, f (and f') is bounded.

Section 6.2 Growth and Decay

1. In the model $y = Ce^{kt}$, C represents the initial value of y (when $t = 0$). k is the proportionality constant.

3. $\frac{dy}{dx} = x + 3$

$y = \int(x + 3)\,dx = \frac{x^2}{2} + 3x + C$

5. $\frac{dy}{dx} = y + 3$

$\frac{dy}{y + 3} = dx$

$\int\frac{1}{y + 3}\,dy = \int dx$

$\ln|y + 3| = x + C_1$

$y + 3 = e^{x + C_1} = Ce^x$

$y = Ce^x - 3$

7. $y' = \frac{5x}{y}$

$yy' = 5x$

$\int yy'\,dx = \int 5x\,dx$

$\int y\,dy = \int 5x\,dx$

$\frac{1}{2}y^2 = \frac{5}{2}x^2 + C_1$

$y^2 - 5x^2 = C$

9. $y' = \sqrt{x}y$

$\frac{y'}{y} = \sqrt{x}$

$\int\frac{y'}{y}\,dx = \int\sqrt{x}\,dx$

$\int\frac{dy}{y} = \int\sqrt{x}\,dx$

$\ln|y| = \frac{2}{3}x^{3/2} + C_1$

$y = e^{(2/3)x^{3/2} + C_1} = e^{C_1}e^{(2/3)x^{3/2}} = Ce^{\left(2x^{3/2}\right)/3}$

11. $(1 + x^2)y' - 2xy = 0$

$$y' = \frac{2xy}{1 + x^2}$$

$$\frac{y'}{y} = \frac{2x}{1 + x^2}$$

$$\int \frac{y'}{y}\,dx = \int \frac{2x}{1 + x^2}\,dx$$

$$\int \frac{dy}{y} = \int \frac{2x}{1 + x^2}\,dx$$

$$\ln|y| = \ln(1 + x^2) + C_1$$

$$\ln|y| = \ln(1 + x^2) + \ln C$$

$$\ln|y| = \ln[C(1 + x^2)]$$

$$y = C(1 + x^2)$$

13. $\dfrac{dQ}{dt} = \dfrac{k}{t^2}$

$$\int \frac{dQ}{dt}\,dt = \int \frac{k}{t^2}\,dt$$

$$\int dQ = -\frac{k}{t} + C$$

$$Q = -\frac{k}{t} + C$$

15. (a)

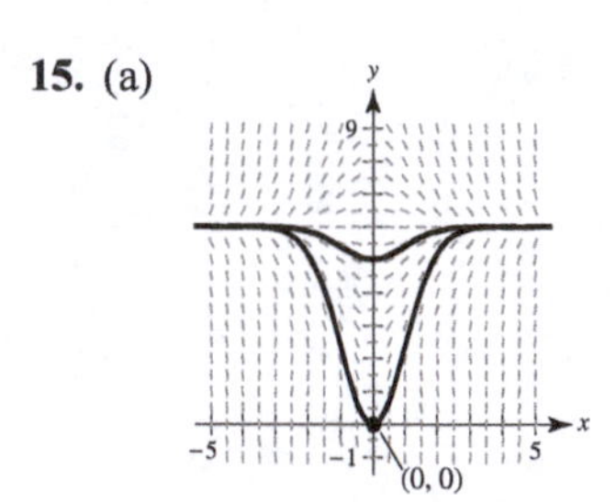

(b) $\dfrac{dy}{dx} = x(6 - y), \quad (0, 0)$

$$\frac{dy}{y - 6} = -x\,dx$$

$$\ln|y - 6| = \frac{-x^2}{2} + C$$

$$y - 6 = e^{-x^2/2 + C} = C_1 e^{-x^2/2}$$

$$y = 6 + C_1 e^{-x^2/2}$$

$$(0, 0):\ 0 = 6 + C_1 \Rightarrow C_1 = -6$$

$$y = 6 - 6e^{-x^2/2}$$

17. $\dfrac{dy}{dt} = \dfrac{1}{2}t, \quad (0, 10)$

$$\int dy = \int \frac{1}{2}t\,dt$$

$$y = \frac{1}{4}t^2 + C$$

$$10 = \frac{1}{4}(0)^2 + C \Rightarrow C = 10$$

$$y = \frac{1}{4}t^2 + 10$$

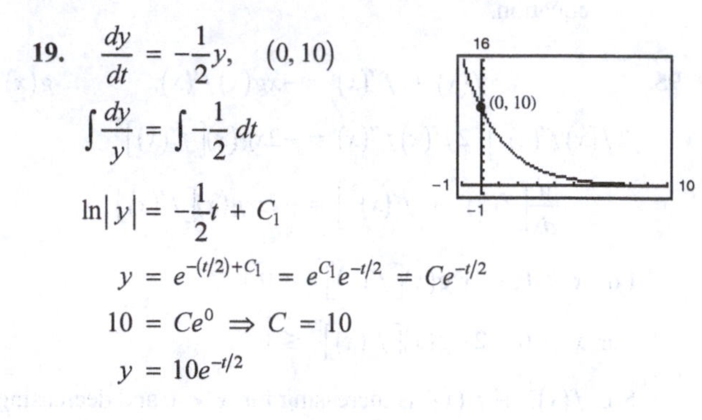

19. $\dfrac{dy}{dt} = -\dfrac{1}{2}y, \quad (0, 10)$

$$\int \frac{dy}{y} = \int -\frac{1}{2}\,dt$$

$$\ln|y| = -\frac{1}{2}t + C_1$$

$$y = e^{-(t/2) + C_1} = e^{C_1}e^{-t/2} = Ce^{-t/2}$$

$$10 = Ce^0 \Rightarrow C = 10$$

$$y = 10e^{-t/2}$$

21. $\dfrac{dN}{dt} = kN$

$N = Ce^{kt}$ (Theorem 6.1)

$(0, 250)$: $C = 250$

$(1, 400)$: $400 = 250e^k \Rightarrow k = \ln\dfrac{400}{250} = \ln\dfrac{8}{5}$

$N = 250e^{\ln(8/5)t} \approx 250e^{0.4700t}$

When $t = 4$, $N = 250e^{4\ln(8/5)} = 250e^{\ln(8/5)^4}$

$$= 250\left(\frac{8}{5}\right)^4 = \frac{8192}{5}.$$

23. $y = Ce^{kt}$, $(0, 2)$, $(4, 3)$

$$C = 2$$

$$y = 2e^{kt}$$

$$3 = 2e^{4k}$$

$$k = \frac{\ln(3/2)}{4}$$

$$y = 2e^{[(1/4)\ln(3/2)]t} \approx 2e^{0.1014t}$$

25. $y = Ce^{kt}, \quad (1, 5), (5, 2)$

$$5 = Ce^{k} \Rightarrow 10 = 2Ce^{k}$$
$$2 = Ce^{5k} \Rightarrow 10 = 5Ce^{k}$$
$$2Ce^{k} = 5Ce^{5k}$$
$$2e^{k} = 5e^{5k}$$
$$\frac{2}{5} = e^{4k}$$
$$k = \frac{1}{4}\ln\left(\frac{2}{5}\right) = \ln\left(\frac{2}{5}\right)^{1/4}$$
$$C = 5e^{-k} = 5e^{-1/4\ln(2/5)} = 5\left(\frac{2}{5}\right)^{-1/4} = 5\left(\frac{5}{2}\right)^{1/4}$$
$$y = 5\left(\frac{5}{2}\right)^{1/4} e^{[1/4\ln(2/5)]t} \approx 6.2872\, e^{-0.2291t}$$

27. $\dfrac{dy}{dx} = \dfrac{1}{2}xy$

$\dfrac{dy}{dx} > 0$ when $xy > 0$. Quadrants I and III.

29. Because the initial quantity is 20 grams,

$y = 20e^{kt}$.

Because the half-life is 1599 years,

$$10 = 20e^{k(1599)}$$
$$k = \frac{1}{1599}\ln\left(\frac{1}{2}\right).$$

So, $y = 20e^{[\ln(1/2)/1599]t}$.

When $t = 1000$, $y = 20e^{[\ln(1/2)/1599](1000)} \approx 12.96$ g.

When $t = 10{,}000$, $y \approx 0.26$ g.

31. Because the half-life is 1599 years,

$$\frac{1}{2} = 1e^{k(1599)}$$
$$k = \frac{1}{1599}\ln\left(\frac{1}{2}\right).$$

Because there are 0.1 gram after 10,000 years,

$$0.1 = Ce^{[\ln(1/2)/1599](10{,}000)}$$
$$C \approx 7.63.$$

So, the initial quantity is approximately 7.63 g.

When $t = 1000$, $y = 7.63e^{[\ln(1/2)/1599](1000)} \approx 4.95$ g.

33. Because the initial quantity is 5 grams, $C = 5$.

Because the half-life is 5715 years,

$$2.5 = 5e^{k(5715)}$$
$$k = \frac{1}{5715}\ln\left(\frac{1}{2}\right).$$

When $t = 1000$ years, $y = 5e^{[\ln(1/2)/5715](1000)} \approx 4.43$ g.

When $t = 10{,}000$ years, $y = 5e^{[\ln(1/2)/5715](10{,}000)}$

≈ 1.49 g.

35. Because the half-life is 24,100 years,

$$\frac{1}{2} = 1e^{k(24{,}100)}$$
$$k = \frac{1}{24{,}100}\ln\left(\frac{1}{2}\right).$$

Because there are 2.1 grams after 1000 years,

$$2.1 = Ce^{[\ln(1/2)/24{,}100](1000)}$$
$$C \approx 2.161.$$

So, the initial quantity is approximately 2.161 g.

When $t = 10{,}000$, $y = 2.161e^{[\ln(1/2)/24{,}100](10{,}000)}$

≈ 1.62 g.

37. $y = Ce^{kt}$

$$\frac{1}{2}C = Ce^{k(1599)}$$
$$k = \frac{1}{1599}\ln\left(\frac{1}{2}\right)$$

When $t = 100$, $y = Ce^{[\ln(1/2)/1599](100)} \approx 0.9576C$

Therefore, 95.76% remains after 100 years.

39. Because $A = 1000e^{0.12t}$, the time to double is given by

$$2000 = 1000e^{0.12t}$$
$$2 = e^{0.12t}$$
$$t = \frac{\ln 2}{0.12} \approx 5.78 \text{ years}$$

The amount after 10 years is

$A = 1000e^{0.12(10)} \approx \$3320.12$.

41. Because $A = 150e^{rt}$ and $A = 300$ when $t = 15$, you have

$$300 = 150e^{r(15)}$$
$$2 = e^{15r}$$
$$r = \frac{\ln 2}{15} \approx 0.0462 \text{ or } 4.62\%.$$

The amount after 10 years is

$A = 150e^{0.0462(10)} \approx \$238.09$.

43. Because $A = 900e^{rt}$ and $A = 1845.25$ when $t = 10$, you have

$$1845.25 = 900e^{r(10)}$$
$$2.0503 \approx e^{10r}$$
$$\ln(2.0503) = 10r$$
$$r \approx 0.0718 \text{ or } 7.18\%.$$

The time to double is given by

$$1800 = 900e^{0.0718t}$$
$$2 = e^{0.0718t}$$
$$t = \frac{\ln 2}{0.0718} \approx 9.65 \text{ years.}$$

45. $1{,}000{,}000 = P\left(1 + \frac{0.075}{12}\right)^{(12)(20)}$

$$P = 1{,}000{,}000\left(1 + \frac{0.075}{12}\right)^{-240} \approx \$224{,}174.18$$

47. $1{,}000{,}000 = P\left(1 + \frac{0.08}{12}\right)^{(12)(35)}$

$$P = 1{,}000{,}000\left(1 + \frac{0.08}{12}\right)^{-420} = \$61{,}377.75$$

49. (a) $2000 = 1000(1 + 0.07)^t$

$$2 = 1.07^t$$
$$\ln 2 = t \ln 1.07$$
$$t = \frac{\ln 2}{\ln 1.07} \approx 10.24 \text{ yr}$$

(b) $2000 = 1000\left(1 + \frac{0.07}{12}\right)^{12t}$

$$2 = \left(1 + \frac{0.007}{12}\right)^{12t}$$
$$\ln 2 = 12t \ln\left(1 + \frac{0.07}{12}\right)$$
$$t = \frac{\ln 2}{12 \ln(1 + (0.07/12))} \approx 9.93 \text{ yr}$$

(c) $2000 = 1000\left(1 + \frac{0.07}{365}\right)^{365t}$

$$2 = \left(1 + \frac{0.07}{365}\right)^{365t}$$
$$\ln 2 = 365t \ln\left(1 + \frac{0.07}{365}\right)$$
$$t = \frac{\ln 2}{365 \ln(1 + (0.07/365))} \approx 9.90 \text{ yr}$$

(d) $2000 = 1000e^{(0.07)t}$

$$2 = e^{0.07t}$$
$$\ln 2 = 0.07t$$
$$t = \frac{\ln 2}{0.07} \approx 9.90 \text{ yr}$$

51. (a) $P = Ce^{kt} = Ce^{-0.011t}$

$P(5) = 2.0 = Ce^{-0.011(5)} \Rightarrow C = 2.0e^{0.011(5)} \approx 2.113$

$P = 2.113e^{-0.011t}$

(b) For 2030, $t = 20$ and $P = 2.113e^{-0.011(20)} \approx 1.70$ million

(c) Because $k < 0$, the population is decreasing.

53. (a) $P = Ce^{kt} = Ce^{0.012t}$

$P(5) = 6.8 = Ce^{0.012(5)} \Rightarrow C = 6.8e^{-0.012(5)} \approx 6.404$

$P = 6.404e^{0.012t}$

(b) For 2030, $t = 20$ and $P = 6.404e^{0.012(20)} \approx 8.14$ million

(c) Because $k > 0$, the population is increasing.

55. (a) $N = 100.1596(1.2455)^t$

(b) $N = 400$ when $t = 6.3$ hr (graphing utility)

Analytically, $400 = 100.1596(1.2455)^t$

$$1.2455^t = \frac{400}{100.1596} = 3.9936$$

$$t \ln 1.2455 = \ln 3.9936$$

$$t = \frac{\ln 3.9936}{\ln 1.2455} \approx 6.3 \text{ hr}$$

57. (a) $19 = 30(1 - e^{20k})$

$$30e^{20k} = 11$$

$$k = \frac{\ln(11/30)}{20} \approx -0.0502$$

$$N \approx 30(1 - e^{-0.0502t})$$

(b) $25 = 30(1 - e^{-0.0502t})$

$$e^{-0.0502t} = \frac{1}{6}$$

$$t = \frac{-\ln 6}{-0.0502} \approx 36 \text{ days}$$

59. (a) Because the population increases by a constant each month, the rate of change from month to month will always be the same. So, the slope is constant, and the model is linear.

(b) Although the percentage increase is constant each month, the rate of growth is not constant. The rate of change of y is given by

$$\frac{dy}{dt} = ry$$

which is an exponential model.

61. (a) Using a graphing utility, $M_1 = 2335.3e^{0.0407t}$.

(b) Using a graphing utility, $M_2 = 206.9t + 1685$.

(c) One way to determine which model fits the data better is to use a graphing utility to graph the data with both models.

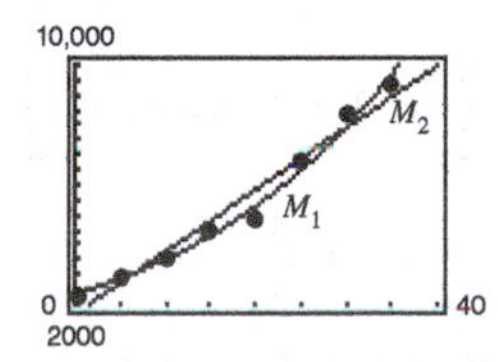

The exponential model fits the data better because the graph is closer to the data values than is the graph of the linear model.

(d) $15{,}000 = 2335.3e^{0.0407t}$

$$6.423 = e^{0.0407t}$$

$$t = \frac{\ln 6.423}{0.0407} \approx 46 \text{ years, or year 2026}$$

Yes. The exponential model indicates a reasonably slow growth rate.

63. $\beta(I) = 10 \log_{10}\frac{I}{I_0}$, $I_0 = 10^{-16}$

(a) $\beta(10^{-14}) = 10 \log_{10}\frac{10^{-14}}{10^{-16}} = 20$ decibels

(b) $\beta(10^{-9}) = 10 \log_{10}\frac{10^{-9}}{10^{-16}} = 70$ decibels

(c) $\beta(10^{-4}) = 10 \log_{10}\frac{10^{-4}}{10^{-16}} = 120$ decibels

65. (a)

$$\frac{dy}{dt} = k(y - 80)$$

$$\int \frac{1}{y - 80}\,dy = \int k\,dt$$

$$\ln(y - 80) = kt + C_1$$

$$y = 80 + e^{kt + C_1} = 80 + Ce^{kt}$$

When $t = 0$, $y = 1500$, so $C = 1420$.

$$y = 80 + 1420e^{kt}$$

When $t = 1$, $y = 1120 = 80 + 1420e^{k(1)}$

$$\Rightarrow e^k = \frac{1120 - 80}{1420} = \frac{52}{71}$$

$$\Rightarrow k = \ln\left(\frac{52}{71}\right) \approx -0.3114$$

So, $y = 80 + 1420e^{-0.3114t}$

(b) At $t = 6$, $y = 80 + 1420e^{-0.3114(6)} \approx 299.2°$ F.

67. False. The half-life of radium is 1599 years.

Section 6.3 Separation of Variables

1. (a) $y = 2x^5y' - y' = (2x^5 - 1)\dfrac{dy}{dx}$

$$\frac{dx}{(2x^5 - 1)} = \frac{dy}{y}$$

Separable

(b) Not separable

3. $$\frac{dy}{dx} = \frac{x}{y}$$

$$\int y\,dy = \int x\,dx$$

$$\frac{y^2}{2} = \frac{x^2}{2} + C_1$$

$$y^2 - x^2 = C$$

5. $$\frac{dy}{dx} = \frac{x-1}{y^3}$$

$$\int y^3\,dy = \int (x-1)\,dx$$

$$\frac{1}{4}y^4 = \frac{1}{2}x^2 - x + C_1$$

$$y^4 - 2x^2 + 4x = C$$

7. $$\frac{dr}{ds} = \frac{4}{9}r$$

$$\int \frac{dr}{r} = \int \frac{4}{9}\,ds$$

$$\ln|r| = \frac{4}{9}s + C_1$$

$$r = e^{4/9s + C_1}$$

$$r = Ce^{4/9s}$$

9. $(2 + x)y' = 3y$

$$\int \frac{dy}{y} = \int \frac{3}{2+x}\,dx$$

$$\ln|y| = 3\ln|2 + x| + \ln C = \ln\left|C(2+x)^3\right|$$

$$y = C(x+2)^3$$

11. $$y^2\frac{dy}{dx} = \sin 9x$$

$$\int y^2\,dy = \int \sin 9x\,dx$$

$$\frac{y^3}{3} = -\frac{1}{9}\cos 9x + C_1$$

$$y^3 = C - \frac{1}{3}\cos 9x$$

13. $\sqrt{1 - 4x^2}\,y' = x$

$$dy = \frac{x}{\sqrt{1-4x^2}}\,dx$$

$$\int dy = \int \frac{x}{\sqrt{1-4x^2}}\,dx$$

$$= -\frac{1}{8}\int (1 - 4x^2)^{-1/2}(-8x\,dx)$$

$$y = -\frac{1}{4}\sqrt{1-4x^2} + C$$

15. $y \ln x - xy' = 0$

$$\int \frac{dy}{y} = \int \frac{\ln x}{x}\,dx \quad \left(u = \ln x,\ du = \frac{dx}{x}\right)$$

$$\ln|y| = \frac{1}{2}(\ln x)^2 + C_1$$

$$y = e^{(1/2)(\ln x)^2 + C_1} = Ce^{(\ln x)^2/2}$$

17. $yy' - 2e^x = 0$

$$y\frac{dy}{dx} = 2e^x$$

$$\int y\,dy = \int 2e^x\,dx$$

$$\frac{y^2}{2} = 2e^x + C$$

Initial condition $(0, 3)$: $\dfrac{9}{2} = 2 + C \Rightarrow C = \dfrac{5}{2}$

Particular solution: $\dfrac{y^2}{2} = 2e^x + \dfrac{5}{2}$

$$y^2 = 4e^x + 5$$

19. $y(x + 1) + y' = 0$

$$\int \frac{dy}{y} = -\int (x+1)\,dx$$

$$\ln|y| = -\frac{(x+1)^2}{2} + C_1$$

$$y = Ce^{-(x+1)^2/2}$$

Initial condition $(-2, 1)$: $1 = Ce^{-1/2}$, $C = e^{1/2}$

Particular solution: $y = e^{\left[1-(x+1)^2\right]/2} = e^{-(x^2+2x)/2}$

21. $y(1 + x^2)y' = x(1 + y^2)$

$$\frac{y}{1 + y^2}\,dy = \frac{x}{1 + x^2}\,dx$$

$$\frac{1}{2}\ln(1 + y^2) = \frac{1}{2}\ln(1 + x^2) + C_1$$

$$\ln(1 + y^2) = \ln(1 + x^2) + \ln C = \ln[C(1 + x^2)]$$

$$1 + y^2 = C(1 + x^2)$$

Initial condition $(0, \sqrt{3})$: $1 + 3 = C \Rightarrow C = 4$

Particular solution: $1 + y^2 = 4(1 + x^2)$

$$y^2 = 3 + 4x^2$$

23. $\dfrac{du}{dv} = uv\sin v^2$

$$\int \frac{du}{u} = \int (\sin v^2)v\,dv$$

$$\ln|u| = -\frac{1}{2}\cos v^2 + C$$

Initial condition $(e^2, 0)$: $\ln e^2 = -\frac{1}{2}\cos 0 + C$

$$2 = -\frac{1}{2} + C$$

$$C = \frac{5}{2}$$

Particular solution: $\ln u = -\frac{1}{2}\cos v^2 + \frac{5}{2}$

$$u = e^{(5 - \cos v^2)/2}$$

25. $dP - kP\,dt = 0$

$$\int \frac{dP}{P} = k\int dt$$

$$\ln|P| = kt + C_1$$

$$P = Ce^{kt}$$

Initial condition: $P(0) = P_0$, $P_0 = Ce^0 = C$

Particular solution: $P = P_0e^{kt}$

27. $y' = \dfrac{dy}{dx} = \dfrac{x}{4y}$

$$\int 4y\,dy = \int x\,dx$$

$$2y^2 = \frac{x^2}{2} + C$$

Initial condition $(0, 2)$: $2(2^2) = 0 + C \Rightarrow C = 8$

Particular solution: $2y^2 = \dfrac{x^2}{2} + 8$

$$4y^2 - x^2 = 16$$

29. $y' = \dfrac{dy}{dx} = \dfrac{y}{2x}$

$$\int \frac{2}{y}\,dy = \int \frac{1}{x}\,dx$$

$$2\ln|y| = \ln|x| + C_1 = \ln|x| + \ln C$$

$$y^2 = Cx$$

Initial condition $(9, 1)$: $1 = 9C \Rightarrow C = \frac{1}{9}$

Particular solution: $y^2 = \frac{1}{9}x$

$$9y^2 - x = 0$$

$$y = \frac{1}{3}\sqrt{x}$$

31. $m = \dfrac{dy}{dx} = \dfrac{0 - y}{(x + 2) - x} = -\dfrac{y}{2}$

$$\int \frac{dy}{y} = \int -\frac{1}{2}\,dx$$

$$\ln|y| = -\frac{1}{2}x + C_1$$

$$y = Ce^{-x/2}$$

33. $\dfrac{dy}{dx} = x$

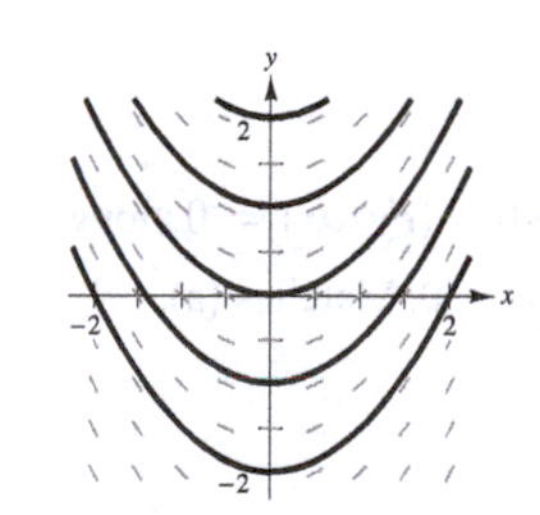

$$y = \int x\,dx = \frac{1}{2}x^2 + C$$

35. $\dfrac{dy}{dx} = 4 - y$

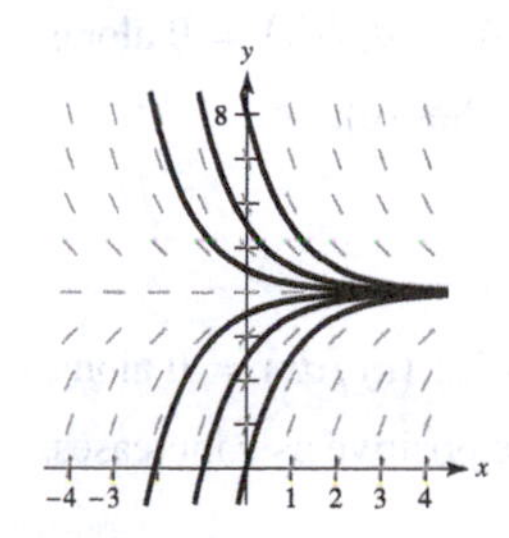

$$\int \frac{dy}{4 - y} = \int dx$$

$$\ln|4 - y| = -x + C_1$$

$$4 - y = e^{-x + C_1}$$

$$y = 4 + Ce^{-x}$$

37. (a) Euler's Method gives $y \approx 0.1602$ when $x = 1$.

(b) $$\frac{dy}{dx} = -6xy$$

$$\int \frac{dy}{y} = \int -6x$$

$$\ln|y| = -3x^2 + C_1$$

$$y = Ce^{-3x^2}$$

$$y(0) = 5 \Rightarrow C = 5$$

$$y = 5e^{-3x^2}$$

(c) At $x = 1$, $y = 5e^{-3(1)} \approx 0.2489$.

Error: $0.2489 - 0.1602 \approx 0.0887$

39. $\dfrac{dy}{dt} = ky, \quad y = Ce^{kt}$

Initial amount: $y(0) = y_0 = C$

Half-life: $\dfrac{y_0}{2} = y_0 e^{k(1599)}$

$$k = \frac{1}{1599}\ln\left(\frac{1}{2}\right)$$

$$y = Ce^{[\ln(1/2)/1599]t}$$

When $t = 50$, $y = 0.9786C$ or 97.86%.

41. (a) $\dfrac{dy}{dx} = k(y - 4)$

(b) The direction field satisfies $(dy/dx) = 0$ along $y = 4$; but not along $y = 0$. Matches (a).

42. (a) $\dfrac{dy}{dx} = k(x - 4)$

(b) The direction field satisfies $(dy/dx) = 0$ along $x = 4$. Matches (b).

43. (a) $\dfrac{dy}{dx} = ky(y - 4)$

(b) The direction field satisfies $(dy/dx) = 0$ along $y = 0$ and $y = 4$. Matches (c).

44. (a) $\dfrac{dy}{dx} = ky^2$

(b) The direction field satisfies $(dy/dx) = 0$ along $y = 0$, and grows more positive as y increases. Matches (d).

45. (a) $$\frac{dw}{dt} = k(1200 - w)$$

$$\int \frac{dw}{1200 - w} = \int k\,dt$$

$$\ln|1200 - w| = -kt + C_1$$

$$1200 - w = e^{-kt+C_1} = Ce^{-kt}$$

$$w = 1200 - Ce^{-kt}$$

$$w(0) = 60 = 1200 - C \Rightarrow C = 1200 - 60 = 1140$$

$$w = 1200 - 1140e^{-kt}$$

(b)

1400

0 0 10

$k = 0.8$

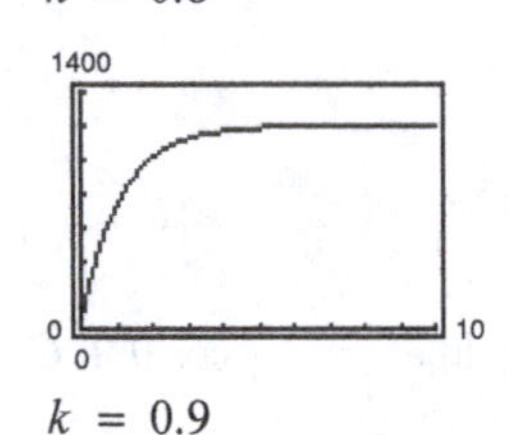

$k = 0.9$

1400

0 0 10

$k = 1$

(c) $k = 0.8$: $t = 1.31$ years

$k = 0.9$: $t = 1.16$ years

$k = 1.0$: $t = 1.05$ years

(d) Maximum weight: 1200 pounds

$$\lim_{x \to \infty} w = 1200$$

47. Given family (hyperbolas): $3x^2 - y^2 = C$

$$6x - 2yy' = 0$$

$$y' = \frac{3x}{y}$$

Orthogonal trajectory: $y' = \dfrac{y}{3x}$

$$\int \frac{1}{y}\,dy = \int \frac{1}{3x}\,dx$$

$$3\ln|y| = \ln|x| + \ln K = \ln Kx$$

$$y^3 = Kx \Rightarrow y = \frac{K}{\sqrt[3]{x}}$$

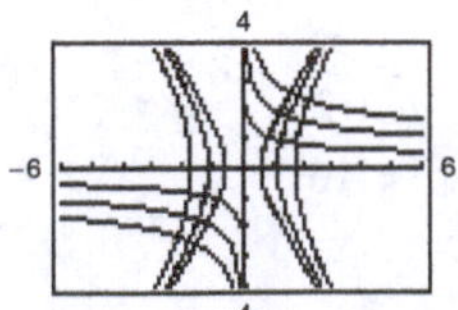

49. Given family (parabolas): $x^2 = Cy$

$$2x = Cy'$$

$$y' = \frac{2x}{C} = \frac{2x}{x^2/y} = \frac{2y}{x}$$

Orthogonal trajectory (ellipses): $y' = -\frac{x}{2y}$

$$2\int y\,dy = -\int x\,dx$$

$$y^2 = -\frac{x^2}{2} + K_1$$

$$x^2 + 2y^2 = K$$

51. Given family: $y^2 = Cx^3$

$$2yy' = 3Cx^2$$

$$y' = \frac{3Cx^2}{2y} = \frac{3x^2}{2y}\left(\frac{y^2}{x^3}\right) = \frac{3y}{2x}$$

Orthogonal trajectory (ellipses): $y' = -\frac{2x}{3y}$

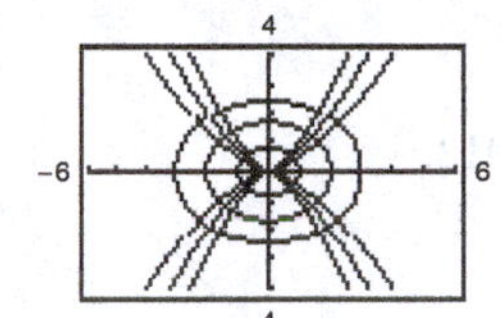

$$3\int y\,dy = -2\int x\,dx$$

$$\frac{3y^2}{2} = -x^2 + K_1$$

$$3y^2 + 2x^2 = K$$

53.

$$\frac{dN}{dt} = kN(500 - N)$$

$$\int \frac{dN}{N(500 - N)} = \int k\,dt$$

$$\frac{1}{500}\int\left[\frac{1}{N} + \frac{1}{500 - N}\right]dN = \int k\,dt$$

$$\ln|N| - \ln|500 - N| = 500(kt + C_1)$$

$$\frac{N}{500 - N} = e^{500kt + C_2} = Ce^{500kt}$$

$$N = \frac{500Ce^{500kt}}{1 + Ce^{500kt}}$$

When $t = 0$, $N = 100$.

So, $100 = \frac{500C}{1 + C} \Rightarrow C = 0.25$.

Therefore, $N = \frac{125e^{500kt}}{1 + 0.25e^{500kt}}$.

When $t = 4$, $N = 200$.

So, $200 = \frac{125e^{2000k}}{1 + 0.25e^{2000k}} \Rightarrow k = \frac{\ln(8/3)}{2000} \approx 0.00049$.

Therefore, $N = \frac{125e^{0.2452t}}{1 + 0.25e^{0.2452t}} = \frac{500}{1 + 4e^{-0.2452t}}$.

55. The general solution is $y = 1 - Ce^{-kt}$.

Because $y = 0$ when $t = 0$, it follows that $C = 1$.

Because $y = 0.75$ when $t = 1$, you have

$$0.75 = 1 - e^{-k(1)}$$

$$-0.25 = -e^{-k}$$

$$0.25 = e^{-k}$$

$$\ln 0.25 = -k$$

$$k = \ln 0.25 = \ln 4 \approx 1.386.$$

So, $y \approx 1 - e^{-1.386t}$.

Note: This can be written as $y = 1 - 4^{-x}$.

57. The general solution is $y = -\frac{1}{kt + C}$.

Because $y = 45$ when $t = 0$, it follows that

$45 = -\frac{1}{C}$ and $C = -\frac{1}{45}$.

Therefore, $y = -\frac{1}{kt - (1/45)} = \frac{45}{1 - 45kt}$.

Because $y = 4$ when $t = 2$, you have

$$4 = \frac{45}{1 - 45k(2)} \Rightarrow k = -\frac{41}{360}.$$

So, $y = \frac{45}{1 + (41/8)t} = \frac{360}{8 + 41t}$.

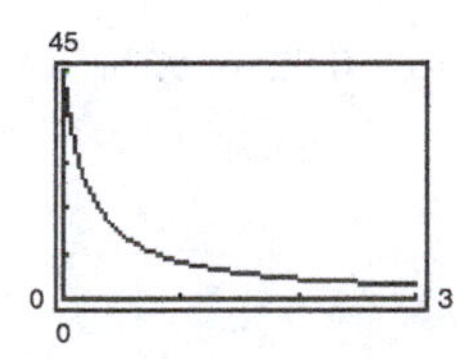

59. Because $y = 100$ when $t = 0$, it follows that

$100 = 500e^{-C}$, which implies that $C = \ln 5$.

So, you have $y = 500e^{(-\ln 5)e^{-kt}}$.

Because $y = 150$ when $t = 2$, it follows that

$$150 = 500e^{(-\ln 5)e^{-2k}}$$

$$e^{-2k} = \frac{\ln 0.3}{\ln 0.2}$$

$$k = -\frac{1}{2}\ln\frac{\ln 0.3}{\ln 0.2}$$

$$\approx 0.1452.$$

So, y is given by $y = 500e^{-1.6904e^{-0.1451t}}$.

61. From Example 8, the general solution is $y = 60e^{-Ce^{-kt}}$.

Because $y = 8$ when $t = 0$, $8 = 60e^{-C} \Rightarrow C = \ln\frac{15}{2} \approx 2.0149$.

Because $y = 15$ when $t = 3$,

$$15 = 60e^{-2.0149e^{-3k}}$$

$$\frac{1}{4} = e^{-2.0149e^{-3k}}$$

$$\ln\frac{1}{4} = -2.0149e^{-3k}$$

$$k = -\frac{1}{3}\ln\left(\frac{\ln(1/4)}{-2.0149}\right) \approx 0.1246.$$

So, $y = 60e^{-2.0149e^{-0.1246t}}$. When $t = 10$, $y \approx 34$ beavers.

63. Following Example 9, the differential equation is $\frac{dy}{dt} = ky(1 - y)(2 - y)$ and its general solution is $\frac{y(2 - y)}{(1 - y)^2} = Ce^{2kt}$.

$$y = \frac{1}{2} \text{ when } t = 0 \Rightarrow \frac{(1/2)(3/2)}{(1/2)^2} = C \Rightarrow C = 3$$

$$y = 0.75 = \frac{3}{4} \text{ when } t = 4 \Rightarrow \frac{(3/4)(5/4)}{(1/4)^2} = 15 = 3e^{2k(4)} \Rightarrow 5 = e^{8k} \Rightarrow k = \frac{1}{8}\ln 5 \approx 0.2012.$$

So, the particular solution is $\frac{y(2 - y)}{(1 - y)^2} = 3e^{0.4024t}$.

Using a symbolic algebra utility or graphing utility, you find that when $t = 10$,

$\frac{y(2 - y)}{(1 - y)^2} = 3e^{0.4024(10)}$ and $y \approx 0.92$, or 92%.

65. (a)
$$\frac{dQ}{dt} = -\frac{Q}{20}$$

$$\int\frac{dQ}{Q} = \int -\frac{1}{20}dt$$

$$\ln|Q| = -\frac{1}{20}t + C_1$$

$$Q = e^{-(1/20)t + C_1} = Ce^{-(1/20)t}$$

Because $Q = 25$ when $t = 0$, you have $25 = C$.

So, the particular solution is $Q = 25e^{-(1/20)t}$.

(b) When $Q = 15$, you have $15 = 25e^{-(1/20)t}$.

$$\frac{3}{5} = e^{-(1/20)t}$$

$$\ln\left(\frac{3}{5}\right) = -\frac{1}{20}t$$

$$-20\ln\left(\frac{3}{5}\right) = t$$

$t \approx 10.217$ minutes

67. (a)
$$\frac{dy}{dt} = ky$$

$$\int\frac{dy}{y} = \int k\,dt$$

$$\ln y = kt + C_1$$

$$y = e^{kt + C_1} = Ce^{kt}$$

(b) $y(0) = 20 \Rightarrow C = 20$

$$y(1) = 16 = 20e^k \Rightarrow k = \ln\frac{16}{20} = \ln\left(\frac{4}{5}\right)$$

$$y = 20e^{t\ln(4/5)}$$

When 75% has changed:

$$5 = 20e^{t\ln(4/5)}$$

$$\frac{1}{4} = e^{t\ln(4/5)}$$

$$t = \frac{\ln(1/4)}{\ln(4/5)} \approx 6.2 \text{ hours}$$

69. The general solution is $y = Ce^{kt}$.

Because $y = 0.60C$ when $t = 1$, you have

$0.60C = Ce^{k} \Rightarrow k = \ln 0.60 \approx -0.5108.$

So, $y = Ce^{-0.5108t}$.

When $y = 0.20C$, you have

$$\begin{aligned} 0.20C &= Ce^{-0.5108t} \\ \ln 0.20 &= -0.5108t \\ t &\approx 3.15 \text{ hours.} \end{aligned}$$

71. $$\begin{aligned} \int \frac{1}{kP + N}\,dP &= \int dt \\ \frac{1}{k}\ln|kP + N| &= t + C_1 \\ kP + N &= C_2 e^{kt} \\ P &= Ce^{kt} - \frac{N}{k} \end{aligned}$$

73. $$\begin{aligned} \frac{dA}{dt} &= rA + P \\ \frac{dA}{rA + P} &= dt \\ \int \frac{dA}{rA + P} &= \int dt \\ \frac{1}{r}\ln(rA + P) &= t + C_1 \\ \ln(rA + P) &= rt + C_2 \\ rA + P &= e^{rt + C_2} \\ A &= \frac{C_3 e^{rt} - P}{r} \\ A &= Ce^{rt} - \frac{P}{r} \end{aligned}$$

When $t = 0$: $A = 0$

$$\begin{aligned} 0 &= C - \frac{P}{r} \Rightarrow C = \frac{P}{r} \\ A &= \frac{P}{r}\left(e^{rt} - 1\right) \end{aligned}$$

75. From Exercise 73,

$$A = \frac{P}{r}\left(e^{rt} - 1\right).$$

Because $A = 260{,}000{,}000$ when $t = 8$ and $r = 0.0725$, you have

$$\begin{aligned} P &= \frac{Ar}{e^{rt} - 1} = \frac{(260{,}000{,}000)(0.0725)}{e^{(0.0725)(8)} - 1} \\ &\approx \$23{,}981{,}015.77. \end{aligned}$$

77. $\dfrac{dy}{dt} = 0.02y \ln\left(\dfrac{5000}{y}\right)$

(a)

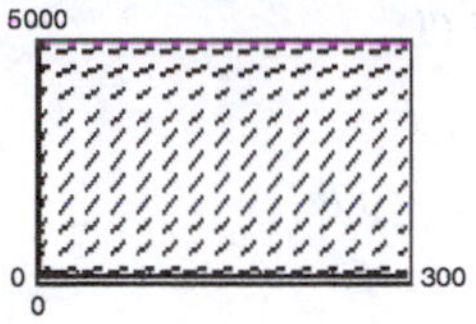

(b) As $t \to \infty$, $y \to L = 5000$.

(c) Using a computer algebra system or separation of variables, the general solution is

$$y = 5000e^{-Ce^{-kt}} = 5000e^{-Ce^{-0.02t}}.$$

Using the initial condition $y(0) = 500$, you obtain

$$500 = 5000e^{-C} \Rightarrow C = \ln 10 \approx 2.3026.$$

So, $y = 5000e^{-2.3026e^{-0.02t}}$.

(d)

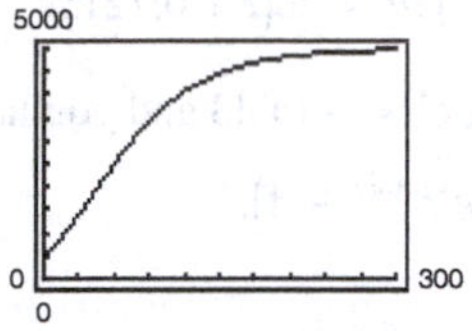

The graph is concave upward on $(0, 41.7)$ and concave downward on $(41.7, \infty)$.

79. Yes. Rewrite the equation as

$$\begin{aligned} \frac{dy}{dx} &= f(x)\big(g(y) - h(y)\big) \\ \frac{dy}{g(y) - h(y)} &= f(x)\,dx. \end{aligned}$$

81. $$\begin{aligned} y(1 + x)\,dx + x\,dy &= 0 \\ x\,dy &= -y(1 + x)\,dx \\ \frac{1}{y}\,dy &= -\frac{(1 + x)}{x}\,dx \end{aligned}$$

Separable

83. $\dfrac{dy}{dx} + xy = 5$

Not separable

85. (a)

$$\frac{dv}{dt} = k(W - v)$$

$$\int \frac{dv}{W - v} = \int k\, dt$$

$$-\ln|W - v| = kt + C_1$$

$$v = W - Ce^{-kt}$$

Initial conditions:

$W = 20$, $v = 0$ when $t = 0$ and $v = 10$ when $t = 0.5$ so, $C = 20$, $k = \ln 4$.

Particular solution:

$$v = 20\left(1 - e^{-(\ln 4)t}\right) = 20\left(1 - \left(\frac{1}{4}\right)^t\right)$$

or

$$v = 20\left(1 - e^{-1.386t}\right)$$

(b) $s = \int 20\left(1 - e^{-1.386t}\right) dt \approx 20\left(t + 0.7215e^{-1.386t}\right) + C$

Because $s(0) = 0$, $C \approx -14.43$ and you have

$$s \approx 20t + 14.43\left(e^{-1.386t} - 1\right).$$

87. $f(x, y) = x^3 - 4xy^2 + y^3$

$$f(tx, ty) = t^3x^3 - 4txt^2y^2 + t^3y^3 = t^3\left(x^3 - 4xy^2 + y^3\right)$$

Homogeneous of degree 3

89. $f(x, y) = e^{x/y}$

$$f(tx, ty) = e^{tx/ty} = e^{x/y}$$

Homogenous of degree 0

91. $f(x, y) = 2 \ln xy$

$$f(tx, ty) = 2\ln[txty] = 2\ln\left[t^2xy\right] = 2\left(\ln t^2 + \ln xy\right)$$

Not homogeneous

93. $f(x, y) = 2 \ln \frac{x}{y}$

$$f(tx, ty) = 2 \ln \frac{tx}{ty} = 2 \ln \frac{x}{y}$$

Homogeneous of degree 0

95. $(x + y)dx - 2x\, dy = 0$, $y = ux$, $dy = x\, du + u\, dx$

$$(x + ux)dx - 2x(x\, du + u\, dx) = 0$$

$$(1 + u)dx - 2x\, du - 2u\, dx = 0$$

$$(1 - u)dx = 2x\, du$$

$$\frac{1}{x}dx = \frac{2}{1 - u}du$$

$$\int \frac{1}{x}dx = 2\int \frac{1}{1 - u}du$$

$$\ln|x| + \ln C = -2\ln|1 - u|$$

$$\ln|Cx| = \ln|1 - u|^{-2}$$

$$|Cx| = \frac{1}{(1 - u)^2} = \frac{1}{\left[1 - (y/x)\right]^2}$$

$$|Cx| = \frac{x^2}{(x - y)^2}$$

$$|x| = C(x - y)^2$$

97. $(x - y)dx - (x + y)dy = 0$, $y = ux$, $dy = x\, du + u\, dx$

$$(x - ux)dx - (x + ux)(x\, du + u\, dx) = 0$$

$$(1 - u)dx - (1 + u)(x\, du + u\, dx) = 0$$

$$\left(1 - 2u - u^2\right)dx = x(1 + u)du$$

$$-\frac{dx}{x} = \frac{1 + u}{u^2 + 2u - 1}du$$

$$-\int \frac{dx}{x} = \int \frac{u + 1}{u^2 + 2u - 1}du$$

$$-\ln|x| + \ln C = \frac{1}{2}\ln\left|u^2 + 2u - 1\right|$$

$$\ln\left|\frac{C}{x}\right| = \ln\left|u^2 + 2u - 1\right|^{1/2}$$

$$\frac{C^2}{x^2} = \left|u^2 + 2u - 1\right|$$

$$\frac{C}{x^2} = \left|\left(\frac{y}{x}\right)^2 + 2\left(\frac{y}{x}\right) - 1\right|$$

$$C = \left|y^2 + 2yx - x^2\right|$$

99. $xy\,dx + (y^2 - x^2)dy = 0,\ y = ux,\ dy = x\,du + u\,dx$

$$x(ux)\,dx + \left[(ux)^2 - x^2\right](x\,du + u\,dx) = 0$$
$$u\,dx + (u^2 - 1)(x\,du + u\,dx) = 0$$
$$u^3\,dx = -(u^2 - 1)x\,du$$
$$\frac{dx}{x} = \frac{1 - u^2}{u^3}\,du$$
$$\int \frac{dx}{x} = \int \left(u^{-3} - \frac{1}{u}\right) du$$
$$\ln|x| + \ln|C_1| = -\frac{1}{2u^2} - \ln|u|$$
$$\ln|C_1 xu| = -\frac{1}{2u^2}$$
$$\ln|C_1 y| = -\frac{1}{2(y/x)^2}$$
$$= -\frac{x^2}{2y^2}$$
$$y = Ce^{-x^2/(2y^2)}$$

101. False. $\frac{dy}{dx} = \frac{x}{y}$ is separable, but $y = 0$ is not a solution.

103. True

$$x^2 + y^2 = 2Cy \qquad\qquad x^2 + y^2 = 2Kx$$
$$\frac{dy}{dx} = \frac{x}{C - y} \qquad\qquad \frac{dy}{dx} = \frac{K - x}{y}$$
$$\frac{x}{C - y} \cdot \frac{K - x}{y} = \frac{Kx - x^2}{Cy - y^2}$$
$$= \frac{2Kx - 2x^2}{2Cy - 2y^2}$$
$$= \frac{x^2 + y^2 - 2x^2}{x^2 + y^2 - 2y^2}$$
$$= \frac{y^2 - x^2}{x^2 - y^2}$$
$$= -1$$

Section 6.4 The Logistic Equation

1. The carrying capacity is the maximum population that can be sustained over time.

3. $y = \frac{12}{1 + e^{-x}}$

Because $y(0) = 6$, it matches (c) or (d).

Because (d) approaches its horizontal asymptote slower than (c), it matches (d).

4. $y = \frac{12}{1 + 3e^{-x}}$

Because $y(0) = \frac{12}{4} = 3$, it matches (a).

5. $y = \frac{12}{1 + \frac{1}{2}e^{-x}}$

Because $y(0) = \frac{12}{\left(\frac{3}{2}\right)} = 8$, it matches (b).

6. $y = \frac{12}{1 + e^{-2x}}$

Because $y(0) = 6$, it matches (c) or (d).

Because y approaches $L = 12$ faster for (c), it matches (c).

7. $y = \frac{8}{1 + e^{-2t}} = 8(1 + e^{-2t})^{-1};\ L = 8,\ k = 2,\ b = 1$

$$\frac{dy}{dt} = -8(1 + e^{-2t})^{-2}(-2e^{-2t})$$
$$= \frac{8}{(1 + e^{-2t})} \cdot \frac{2e^{-2t}}{(1 + e^{-2t})}$$
$$= 2y\left(\frac{e^{-2t}}{1 + e^{-2t}}\right)$$
$$= 2y\left(1 - \frac{8}{8(1 + e^{-2t})}\right)$$
$$= 2y\left(1 - \frac{y}{8}\right)$$

$$y(0) = \frac{8}{1 + e^0} = 4$$

9. $y = 12(1 + 6e^{-t})^{-1}$; $L = 12$, $k = 1$, $b = 6$

$$y' = -12(1 + 6e^{-t})^{-2}(-6e^{-t})$$
$$= \left(\frac{12}{1 + 6e^{-t}}\right)\left(\frac{6e^{-t}}{1 + 6e^{-t}}\right)$$
$$= y\left(1 - \frac{1}{1 + 6e^{-t}}\right)$$
$$= y\left(1 - \frac{12}{12(1 + 6e^{-t})}\right)$$
$$= y\left(1 - \frac{y}{12}\right)$$

$$y(0) = \frac{12}{1 + 6} = \frac{12}{7}$$

11. $P(t) = \dfrac{2100}{1 + 29e^{-0.75t}}$

(a) $k = 0.75$

(b) $L = 2100$

(c) $P(0) = \dfrac{2100}{1 + 29} = 70$

(d)
$$1050 = \frac{2100}{1 + 29e^{-0.75t}}$$
$$1 + 29e^{-0.75t} = 2$$
$$e^{-0.75t} = \frac{1}{29}$$
$$-0.75t = \ln\left(\frac{1}{29}\right) = -\ln 29$$
$$t = \frac{\ln 29}{0.75} \approx 4.4897 \text{ years}$$

(e) $\dfrac{dP}{dt} = 0.75P\left(1 - \dfrac{P}{2100}\right)$

13. $P(t) = \dfrac{6000}{1 + 4999e^{-0.8t}}$

(a) $k = 0.8$

(b) $L = 6000$

(c) $P(0) = \dfrac{6000}{1 + 4999} = \dfrac{6}{5}$

(d)
$$3000 = \frac{6000}{1 + 4999e^{-0.8t}}$$
$$1 + 4999e^{-0.8t} = 2$$
$$e^{-0.8t} = \frac{1}{4999}$$
$$-0.8t = \ln\left(\frac{1}{4999}\right) = -\ln 4999$$
$$t = \frac{\ln 4999}{0.8} \approx 10.65 \text{ years}$$

(e) $\dfrac{dP}{dt} = 0.8P\left(1 - \dfrac{P}{6000}\right)$

15. $\dfrac{dP}{dt} = 3P\left(1 - \dfrac{P}{100}\right)$

(a) $k = 3$

(b) $L = 100$

(c)

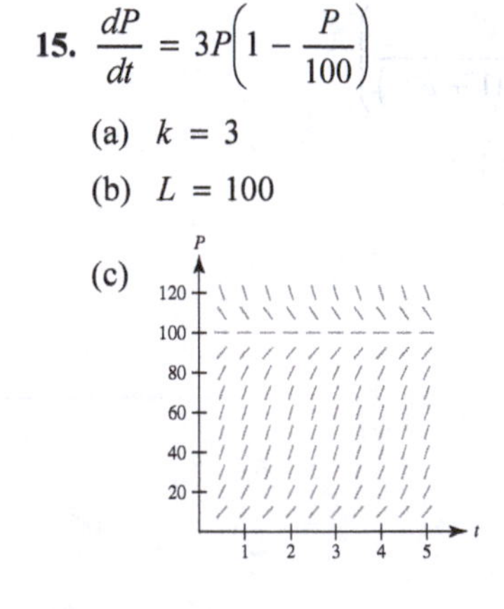

(d)

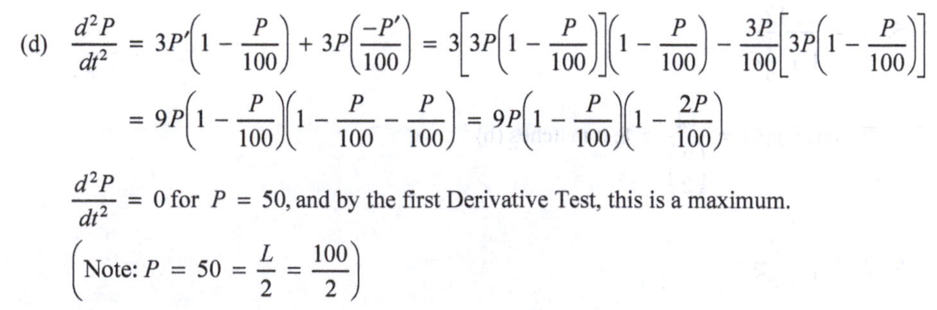

$$\frac{d^2P}{dt^2} = 3P'\left(1 - \frac{P}{100}\right) + 3P\left(\frac{-P'}{100}\right) = 3\left[3P\left(1 - \frac{P}{100}\right)\right]\left(1 - \frac{P}{100}\right) - \frac{3P}{100}\left[3P\left(1 - \frac{P}{100}\right)\right]$$
$$= 9P\left(1 - \frac{P}{100}\right)\left(1 - \frac{P}{100} - \frac{P}{100}\right) = 9P\left(1 - \frac{P}{100}\right)\left(1 - \frac{2P}{100}\right)$$

$\dfrac{d^2P}{dt^2} = 0$ for $P = 50$, and by the first Derivative Test, this is a maximum.

$\left(\text{Note: } P = 50 = \dfrac{L}{2} = \dfrac{100}{2}\right)$

17. $\dfrac{dP}{dt} = 0.1P - 0.0004P^2 = 0.1P(1 - 0.004P)$

$= 0.1P\left(1 - \dfrac{P}{250}\right)$

(a) $k = 0.1 = \dfrac{1}{10}$

(b) $L = 250$

(c)

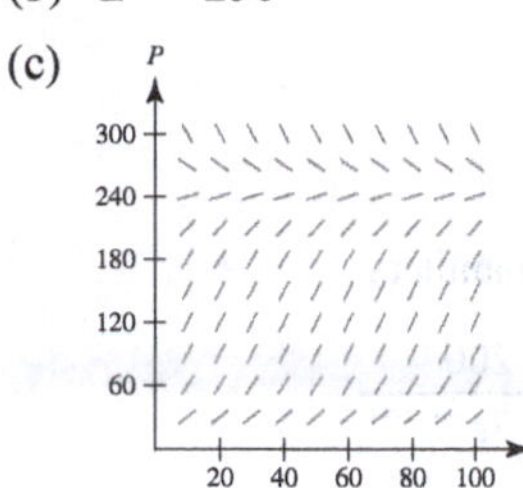

(d) $P = \dfrac{250}{2} = 125$

(Same argument as in Exercise 13)

19. $\dfrac{dy}{dt} = y\left(1 - \dfrac{y}{36}\right), \quad y(0) = 4$

$k = 1, L = 36$

$y = \dfrac{L}{1 + be^{-kt}} = \dfrac{36}{1 + be^{-t}}$

$(0, 4)$: $4 = \dfrac{36}{1 + b} \Rightarrow b = 8$

Solution: $y = \dfrac{36}{1 + 8e^{-t}}$

$y(5) = \dfrac{36}{1 + 8e^{-5}} \approx 34.16$

$y(100) = \dfrac{36}{1 + 8e^{-100}} \approx 36.00$

21. $\dfrac{dy}{dt} = \dfrac{4y}{5} - \dfrac{y^2}{150} = \dfrac{4}{5}y\left(1 - \dfrac{y}{120}\right), \quad y(0) = 8$

$k = \dfrac{4}{5} = 0.8, L = 120$

$y = \dfrac{L}{1 + be^{-kt}} = \dfrac{120}{1 + be^{-0.8t}}$

$y(0) = 8$: $8 = \dfrac{120}{1 + b} \Rightarrow b = 14$

Solution: $y = \dfrac{120}{1 + 14e^{-0.8t}}$

$y(5) = \dfrac{120}{1 + 14e^{-0.8(5)}} \approx 95.51$

$y(100) = \dfrac{120}{1 + 14e^{-0.8(100)}} \approx 120.0$

23. $L = 250$ and $y(0) = 350$

Matches (c).

24. $L = 100$ and $y(0) = 100$

Matches (d).

25. $L = 250$ and $y(0) = 50$

Matches (b).

26. $L = 100$ and $y(0) = 50$

Matches (a).

27. $\dfrac{dy}{dt} = 0.2y\left(1 - \dfrac{y}{1000}\right)$

(a)

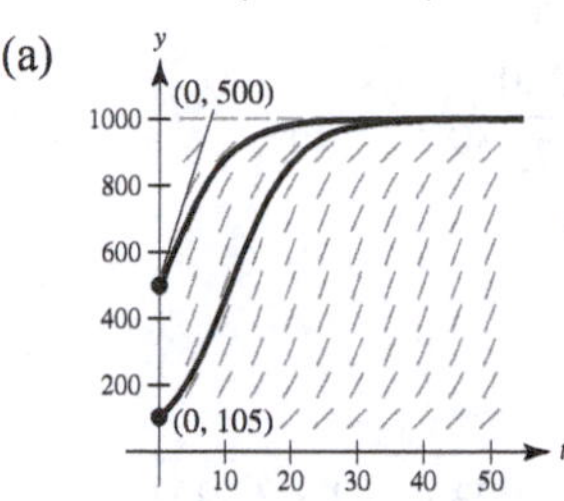

(b) $k = 0.2, L = 1000$

$y = \dfrac{1000}{1 + be^{-0.2t}}$

$y(0) = 105 = \dfrac{1000}{1 + b}$

$1 + b = \dfrac{1000}{105} = \dfrac{200}{21}$

$b = \dfrac{179}{21} \approx 8.524$

$y = \dfrac{1000}{1 + (179/21)e^{-0.2t}}$

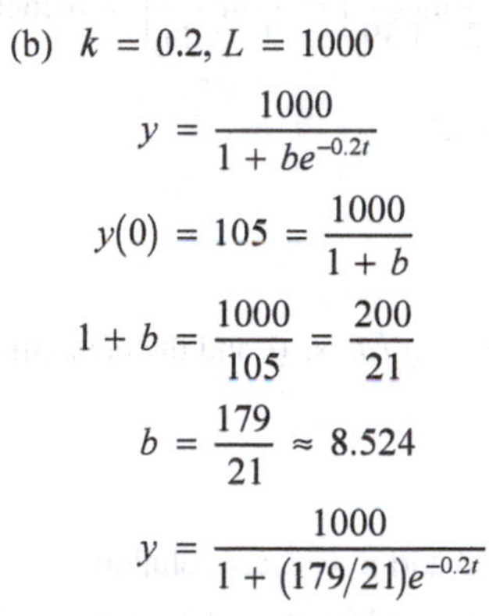

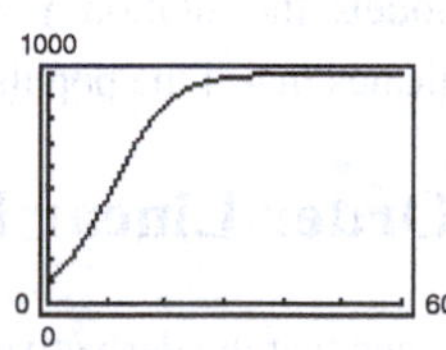

29. No, it is not possible to determine b. However, $L = 2500$ and $k = 0.75$. You need an initial condition to determine b.

31. $\frac{dy}{dt} = ky\left(1 - \frac{y}{L}\right), \quad y(0) < L$

$$\frac{d^2y}{dt^2} = ky'\left(1 - \frac{y}{L}\right) + ky\left(-\frac{y'}{L}\right) = k^2y\left(1 - \frac{y}{L}\right)^2 + ky\left[\frac{-ky\left(1 - \frac{y}{L}\right)}{L}\right] = k^2\left(1 - \frac{y}{L}\right)y\left[\left(1 - \frac{y}{L}\right) - \frac{y}{L}\right] = k^2\left(1 - \frac{y}{L}\right)y\left(1 - \frac{2y}{L}\right)$$

So, $\frac{d^2y}{dt^2} = 0$ when $1 - \frac{2y}{L} = 0 \Rightarrow y = \frac{L}{2}$.

By the First Derivative Test, this is a maximum.

33. (a) $P = \frac{L}{1 + be^{-kt}}, \; L = 200, \; P(0) = 25$

$$25 = \frac{200}{1 + b} \Rightarrow b = 7$$

$$39 = \frac{200}{1 + 7e^{-k(2)}}$$

$$1 + 7e^{-2k} = \frac{200}{39}$$

$$e^{-2k} = \frac{23}{39}$$

$$k = -\frac{1}{2}\ln\left(\frac{23}{39}\right) = \frac{1}{2}\ln\left(\frac{39}{23}\right) \approx 0.2640$$

$$P = \frac{200}{1 + 7e^{-0.2640t}}$$

(b) For $t = 5$, $P \approx 70$ panthers.

(c)
$$100 = \frac{200}{1 + 7e^{-0.264t}}$$

$$1 + 7e^{-0.264t} = 2$$

$$-0.264t = \ln\left(\frac{1}{7}\right)$$

$$t \approx 7.37 \text{ years}$$

(d)
$$\frac{dP}{dt} = kP\left(1 - \frac{P}{L}\right)$$

$$= 0.264P\left(1 - \frac{P}{200}\right), \; P(0) = 25$$

Using Euler's Method, $P \approx 65.6$ when $t = 5$.

(e) P is increasing most rapidly where $P = 200/2 = 100$, corresponds to $t \approx 7.37$ years.

35. False. If $y > L$, then $dy/dt < 0$ and the population decreases.

37. $\frac{dy}{dt} = ky\left(1 - \frac{y}{L}\right)$

The functions $y = 0$ and $y = L$ are solutions.

In terms of population models, the solution $y = 0$ indicates that if the population is 0, it will stay 0.

The solution $y = L$ indicates that if the population is at carrying capacity, it will remain at carrying capacity.

Section 6.5 First-Order Linear Differential Equations

1. The term "first-order" means that the derivative in the equation is first order.

3. $x^3y' + xy = e^x + 1$

$$y' + \frac{1}{x^2}y = \frac{1}{x^3}(e^x + 1)$$

Linear

5. $y' - y\sin x = xy^2$

Not linear, because of the xy^2-term.

7. $\frac{dy}{dx} + \left(\frac{1}{x}\right)y = 6x + 2$

Integrating factor: $e^{\int(1/x)\,dx} = e^{\ln x} = x$

$$xy = \int x(6x + 2)\,dx = 2x^3 + x^2 + C$$

$$y = 2x^2 + x + \frac{C}{x}$$

9. $y' + 2xy = 10x$

Integrating factor: $e^{\int 2x\,dx} = e^{x^2}$

$$ye^{x^2} = \int 10xe^{x^2}\,dx = 5e^{x^2} + C$$

$$y = 5 + Ce^{-x^2}$$

11. $(y + 1)\cos x\,dx = dy$

$y' = (y + 1)\cos x = y\cos x + \cos x$

$y' - (\cos x)y = \cos x$

Integrating factor: $e^{\int -\cos x\,dx} = e^{-\sin x}$

$$y'e^{-\sin x} - (\cos x)e^{-\sin x}y = (\cos x)e^{-\sin x}$$

$$ye^{-\sin x} = \int (\cos x)e^{-\sin x}\,dx$$

$$= -e^{-\sin x} + C$$

$$y = -1 + Ce^{\sin x}$$

13. $y' + 3y = e^{3x}$

Integrating factor: $e^{\int 3\,dx} = e^{3x}$

$$ye^{3x} = \int e^{3x}e^{3x}dx = \int e^{6x}dx = \frac{1}{6}e^{6x} + C$$

$$y = \frac{1}{6}e^{3x} + Ce^{-3x}$$

15. (a) Answers will vary.

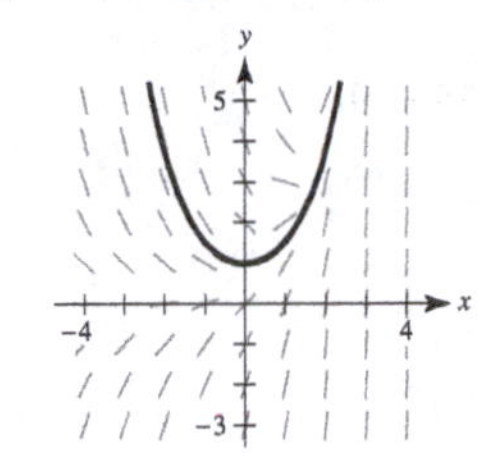

(b) $\dfrac{dy}{dx} = e^x - y$

$\dfrac{dy}{dx} + y = e^x$ Integrating factor: $e^{\int dx} = e^x$

$$e^x y' + e^x y = e^{2x}$$

$$(ye^x) = \int e^{2x}dx$$

$$ye^x = \frac{1}{2}e^{2x} + C$$

$$y(0) = 1 \Rightarrow 1 = \frac{1}{2} + C \Rightarrow C = \frac{1}{2}$$

$$ye^x = \frac{1}{2}e^{2x} + \frac{1}{2}$$

$$y = \frac{1}{2}e^x + \frac{1}{2}e^{-x} = \frac{1}{2}(e^x + e^{-x})$$

(c)

17. $y' + y = 6e^x$

Integrating factor: $e^{\int dx} = e^x$

$$ye^x = \int 6e^{2x}\,dx$$

$$y = \frac{1}{e}x(3e^{2x} + C) = 3e^x + Ce^{-x}$$

Initial condition: $y(0) = 3,\ 3 = 3e^0 + Ce^0,\ C = 0$

Particular solution: $y = 3e^x$

19. $y' + y\tan x = \sec x + \cos x$

Integrating factor: $e^{\int \tan x\,dx} = e^{\ln|\sec x|} = \sec x$

$$y\sec x = \int \sec x(\sec x + \cos x)\,dx = \tan x + x + C$$

$$y = \sin x + x\cos x + C\cos x$$

Initial condition: $y(0) = 1,\ 1 = C$

Particular solution: $y = \sin x + (x + 1)\cos x$

21. $y' + \left(\dfrac{1}{x}\right)y = 0$

Integrating factor: $e^{\int (1/x)\,dx} = e^{\ln|x|} = x$

Separation of variables:

$$\frac{dy}{dx} = -\frac{y}{x}$$

$$\int \frac{1}{y}\,dy = \int -\frac{1}{x}\,dx$$

$$\ln y = -\ln x + \ln C$$

$$\ln xy = \ln C$$

$$xy = C$$

Initial condition: $y(2) = 2,\ C = 4$

Particular solution: $xy = 4$

23. $x\,dy = (x + y + 2)\,dx$

$$\frac{dy}{dx} = \frac{x + y + 2}{x} = \frac{y}{x} + 1 + \frac{2}{x}$$

$\dfrac{dy}{dx} - \dfrac{1}{x}y = 1 + \dfrac{2}{x}$ Linear

$$u(x) = e^{\int -(1/x)\,dx} = \frac{1}{x}$$

$$y = x\int \left(1 + \frac{2}{x}\right)\frac{1}{x}\,dx = x\int\left(\frac{1}{x} + \frac{2}{x^2}\right)dx$$

$$= x\left[\ln|x| + \frac{-2}{x} + C\right]$$

$$= -2 + x\ln|x| + Cx$$

$$y(1) = 10 = -2 + C \Rightarrow C = 12$$

$$y = -2 + x\ln|x| + 12x$$

25. $$\frac{dP}{dt} = kP + N, N \text{ constant}$$

$$\frac{dP}{kP + N} = dt$$

$$\int \frac{1}{kP + N}\,dP = \int dt$$

$$\frac{1}{k}\ln(kP + N) = t + C_1$$

$$\ln(kP + N) = kt + C_2$$

$$kP + N = e^{kt + C_2}$$

$$P = \frac{C_3e^{kt} - N}{k}$$

$$P = Ce^{kt} - \frac{N}{k}$$

When $t = 0$: $P = P_0$

$$P_0 = C - \frac{N}{k} \Rightarrow C = P_0 + \frac{N}{k}$$

$$P = \left(P_0 + \frac{N}{k}\right)e^{kt} - \frac{N}{k}$$

27. (a) $A = \frac{P}{r}\left(e^{rt} - 1\right)$

$$A = \frac{275{,}000}{0.06}\left(e^{0.08(10)} - 1\right) \approx \$4{,}212{,}796.94$$

(b) $A = \frac{550{,}000}{0.05}\left(e^{0.059(25)} - 1\right) \approx \$31{,}424{,}909.75$

29. (a) $\frac{dN}{dt} = k(75 - N)$

(b) $N' + kN = 75k$

Integrating factor: $e^{\int k\,dt} = e^{kt}$

$$N'e^{kt} + kNe^{kt} = 75\,ke^{kt}$$

$$\left(Ne^{kt}\right)' = 75\,ke^{kt}$$

$$Ne^{kt} = \int 75\,ke^{kt} = 75\,e^{kt} + C$$

$$N = 75 + Ce^{-kt}$$

(c) For $t = 1, N = 20$:

$$20 = 75 + Ce^{-k} \Rightarrow -55 = Ce^{-k}$$

For $t = 20, N = 35$:

$$35 = 75 + Ce^{-20k} \Rightarrow -40 = Ce^{-20k}$$

$$\frac{55}{40} = \frac{Ce^{-k}}{Ce^{-20k}} \Rightarrow e^{19k} = \frac{11}{8} \Rightarrow k = \frac{1}{19}\ln\left(\frac{11}{8}\right) \approx 0.0168$$

$$Ce^{-k} = -55$$

$$C = -55e^{k} \approx -55.9296$$

$$N = 75 - 55.9296\,e^{-0.0168t}$$

31. From Example 3, $v = \frac{-mg}{k}\left(1 - e^{-kt/m}\right)$, solution

$g = 9.8 \text{ m/sec}^2$, $mg = (9.8)(4) = 39.2$, $v(5) = -31 = \frac{-39.2}{k}\left(1 - e^{-k(5)/4}\right)$

Using a graphing utility, $k \approx 0.7984$, and $v = \frac{-39.2}{0.7984}\left(1 - e^{-0.7984t/4}\right) = -49.0982\left(1 - e^{-0.1996t}\right)$

The limiting value is $\lim_{t\to\infty} v = -49.0982$ meters per second.

33. $L\frac{dI}{dt} + RI = E_0,\ I' + \frac{R}{L}I = \frac{E_0}{L}$

Integrating factor: $e^{\int (R/L)\,dt} = e^{Rt/L}$

$$I\,e^{Rt/L} = \int \frac{E_0}{L}e^{Rt/L}\,dt = \frac{E_0}{R}e^{Rt/L} + C$$

$$I = \frac{E_0}{R} + Ce^{-Rt/L}$$

35. Let Q be the number of pounds of concentrate in the solution at any time t. Because the number of gallons of solution in the tank at any time t is $v_0 + (r_1 - r_2)t$ and because the tank loses r_2 gallons of solution per minute, it must lose concentrate at the rate

$$\left[\frac{Q}{v_0 + (r_1 - r_2)t}\right]r_2.$$

The solution gains concentrate at the rate r_1q_1. Therefore, the net rate of change is

$$\frac{dQ}{dt} = q_1r_1 - \left[\frac{Q}{v_0 + (r_1 - r_2)t}\right]r_2$$

or $$\frac{dQ}{dt} + \frac{r_2Q}{v_0 + (r_1 - r_2)t} = q_1r_1.$$

37. (a) From Exercise 35,

$$\frac{dQ}{dt} + \frac{r_2 Q}{u_0 + (r_1 - r_2)t} = q_1 r_1$$

You have $Q(0) = q_0 = 25$, $q_1 = 0$, $v_0 = 200$, and $r_1 = r_2 = 10$.

Hence, the linear differential equation is $\frac{dQ}{dt} + \frac{1}{20}Q = 0$.

By separating variables,

$$\int \frac{dQ}{Q} = -\int \frac{1}{20}\,dt$$

$$\ln Q = -\frac{1}{20}t + \ln C_1$$

$$Q = Ce^{-\frac{1}{20}t}.$$

The initial condition $Q(0) = 25$ implies that $C = 25$. Hence, $Q = 25e^{-\frac{1}{20}t}$.

(b) $15 = 25e^{-\frac{1}{20}t} \Rightarrow \frac{3}{5} = e^{-\frac{1}{20}t} \Rightarrow \ln\left(\frac{3}{5}\right) = -\frac{1}{20}t \Rightarrow t = -20\ln\left(\frac{3}{5}\right) \approx 10.2$ min

(c) $\lim_{t\to\infty} Q = \lim_{t\to\infty} 25e^{-\frac{1}{20}t} = 0$

39. $y' + P(x)y = Q(x)$

Integrating factor: $u = e^{\int P(x)\,dx}$

$$y'u + P(x)yu = Q(x)u$$

$$(uy)' = Q(x)u$$

so $u'(x) = P(x)u$

Answer (a)

41. Separation of variables:

$$\frac{dy}{dx} = x - 3xy = x(1 - 3y)$$

$$\frac{dy}{1 - 3y} = x\,dx$$

Integrating factor. $P(x) = 3x$, $Q(x) = x$, $u(x) = e^{\int 3x\,dx}$

43. $y' - 2x = 0$

$$\int dy = \int 2x\,dx$$

$$y = x^2 + C$$

Matches (c).

44. $y' - 2y = 0$

$$\int \frac{dy}{y} = \int 2\,dx$$

$$\ln y = 2x + C_1$$

$$y = Ce^{2x}$$

Matches (d).

45. $y' - 2xy = 0$

$$\int \frac{dy}{y} = \int 2x\,dx$$

$$\ln y = x^2 + C_1$$

$$y = Ce^{x^2}$$

Matches (a).

46. $y' - 2xy = x$

$$\int \frac{dy}{2y + 1} = \int x\,dx$$

$$\frac{1}{2}\ln(2y + 1) = \frac{1}{2}x^2 + C_1$$

$$2y + 1 = C_2 e^{x^2}$$

$$y = -\frac{1}{2} + Ce^{x^2}$$

Matches (b).

47. (a) 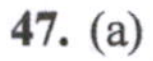

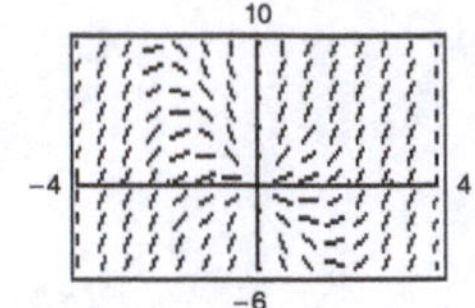

(b) $\frac{dy}{dx} - \frac{1}{x}y = x^2$

Integrating factor: $e^{-\int 1/x\,dx} = e^{-\ln x} = \frac{1}{x}$

$$\frac{1}{x}y' - \frac{1}{x^2}y = x$$

$$\left(\frac{1}{x}y\right) = \int x\,dx = \frac{x^2}{2} + C$$

$$y = \frac{x^3}{2} + Cx$$

$$(-2, 4):\ 4 = \frac{-8}{2} - 2C \Rightarrow C = -4 \Rightarrow y = \frac{x^3}{2} - 4x = \frac{1}{2}x(x^2 - 8)$$

$$(2, 8):\ 8 = \frac{8}{2} + 2C \Rightarrow C = 2 \Rightarrow y = \frac{x^3}{2} + 2x = \frac{1}{2}x(x^2 + 4)$$

(c)

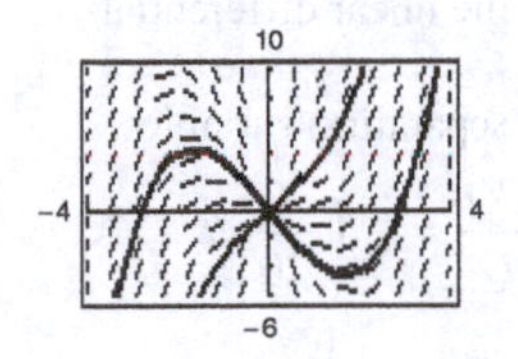

49. $e^{2x+y}\,dx - e^{x-y}\,dy = 0$

Separation of variables:

$$e^{2x}e^{y}\,dx = e^{x}e^{-y}\,dy$$

$$\int e^{x}\,dx = \int e^{-2y}\,dy$$

$$e^{x} = -\tfrac{1}{2}e^{-2y} + C_1$$

$$2e^{x} + e^{-2y} = C$$

51. $(y\cos x - \cos x)\,dx + dy = 0$

Separation of variables:

$$\int \cos x\,dx = \int \frac{-1}{y - 1}\,dy$$

$$\sin x = -\ln(y - 1) + \ln C$$

$$\ln(y - 1) = -\sin x + \ln C$$

$$y = Ce^{-\sin x} + 1$$

53. $(2y - e^{x})\,dx + x\,dy = 0$

Linear: $y' + \left(\frac{2}{x}\right)y = \frac{1}{x}e^{x}$

Integrating factor: $e^{\int (2/x)\,dx} = e^{\ln x^2} = x^2$

$$yx^2 = \int x^2\frac{1}{x}e^{x}\,dx = e^{x}(x - 1) + C$$

$$y = \frac{e^{x}}{x^2}(x - 1) + \frac{C}{x^2}$$

55. $3(y - 4x^2)\,dx = -x\,dy$

$$x\frac{dy}{dx} = -3y + 12x^2$$

$$y' + \frac{3}{x}y = 12x$$

Integrating factor: $e^{\int (3/x)\,dx} = e^{3\ln x} = x^3$

$$y'x^3 + \frac{3}{x}x^3y = 12x(x^3) = 12x^4$$

$$yx^3 = \int 12x^4\,dx = \frac{12}{5}x^5 + C$$

$$y = \frac{12}{5}x^2 + \frac{C}{x^3}$$

57. $y' + 3x^2y = x^2y^3$

$n = 3,\ Q = x^2,\ P = 3x^2$

$$y^{-2}e^{\int(-2)3x^2\,dx} = \int(-2)x^2e^{\int(-2)3x^2\,dx}\,dx$$

$$y^{-2}e^{-2x^3} = -\int 2x^2e^{-2x^3}\,dx$$

$$y^{-2}e^{-2x^3} = \frac{1}{3}e^{-2x^3} + C$$

$$y^{-2} = \frac{1}{3} + Ce^{2x^3}$$

$$\frac{1}{y^2} + Ce^{2x^3} + \frac{1}{3}$$

59. $y' + \left(\dfrac{1}{x}\right)y = xy^2$

$n = 2, Q = x, P = x^{-1}$

$e^{\int -(1/x)\,dx} = e^{-\ln|x|} = x^{-1}$

$$y^{-1}x^{-1} = \int -x(x^{-1})\,dx = -x + C$$

$$\frac{1}{y} = -x^2 + Cx$$

$$y = \frac{1}{Cx - x^2}$$

61. $xy' + y = xy^3$

$y' + \dfrac{1}{x}y = y^3$

$n = 3, Q = 1, P = \dfrac{1}{x}, \qquad e^{\int \frac{-2}{x}\,dx} = e^{-2\ln x} = x^{-2}$

$$y^{-2}x^{-2} = \int -2x^{-2}\,dx + C = 2x^{-1} + C$$

$$y^{-2} = 2x + Cx^2$$

$$y^2 = \frac{1}{2x + Cx^2} \quad \text{or} \quad \frac{1}{y^2} = 2x + Cx^2$$

63. $y' - y = e^x\sqrt[3]{y}, n = \frac{1}{3}, Q = e^x, P = -1$

$e^{\int -(2/3)\,dx} = e^{-(2/3)x}$

$$y^{2/3}e^{-(2/3)x} = \int \tfrac{2}{3}e^x e^{-(2/3)x}\,dx = \int \tfrac{2}{3}e^{(1/3)x}\,dx$$

$$y^{2/3}e^{-(2/3)x} = 2e^{(1/3)x} + C$$

$$y^{2/3} = 2e^x + Ce^{2x/3}$$

65. False. The equation contains $\sqrt{y}$.

Section 6.6 Predator-Prey Differential Equations

1. An autonomous differential equation does not depend explicitly on the time t.

3. $\dfrac{dx}{dt} = ax - bxy = 0.9x - 0.05xy$

$\dfrac{dy}{dt} = -my + nxy = -0.6y + 0.008xy$

$\dfrac{dx}{dt} = \dfrac{dy}{dt} = 0 \Rightarrow 0.9x - 0.05xy = x(0.9 - 0.05y) = 0$

$-0.6y + 0.008xy = y(-0.6 + 0.008x) = 0$

If $x = 0$, then $y = 0$.

If $y = \dfrac{0.9}{0.05} = \dfrac{90}{5} = 18$, then $x = \dfrac{0.6}{0.008} = \dfrac{600}{8} = 75$.

Solutions: $(0, 0)$ and $(75, 18)$

5. (a)

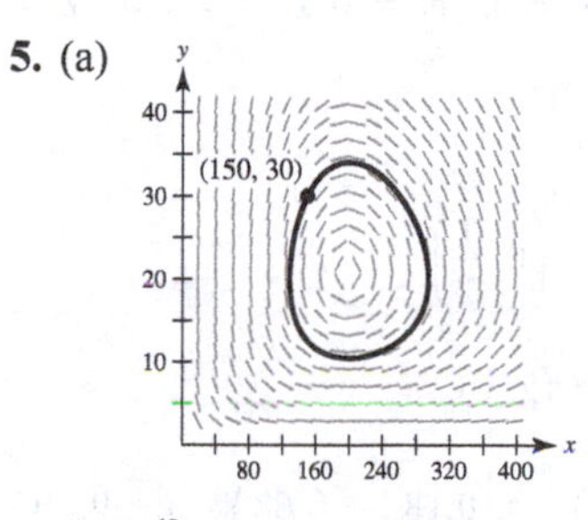

(b)

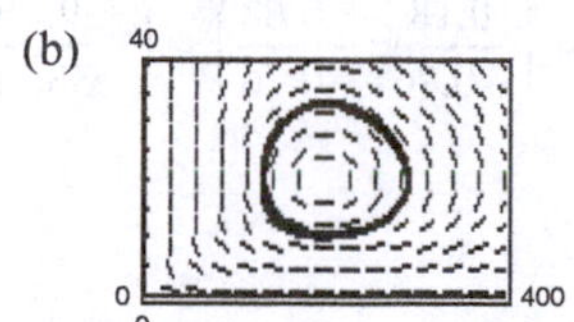

7. (a) The initial conditions are $x(0) = 40$ and $y(0) = 20$.

(b)

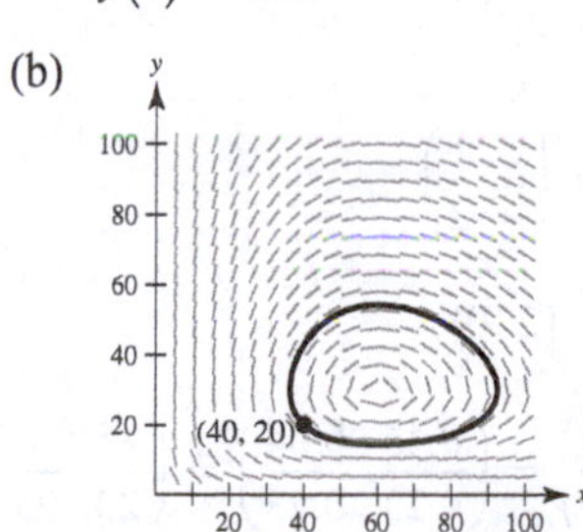

9. Critical points are $(x, y) = (0,0)$ and

$$(x, y) = \left(\frac{m}{n}, \frac{a}{b}\right) = \left(\frac{0.3}{0.006}, \frac{0.8}{0.04}\right) = (50, 20).$$

11.

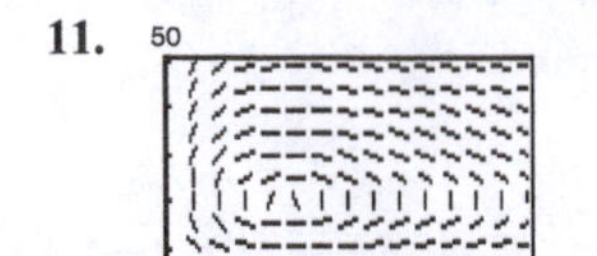

13. Critical points are $(x, y) = (0,0)$ and

$$(x, y) = \left(\frac{m}{n}, \frac{a}{b}\right) = \left(\frac{0.4}{0.00004}, \frac{0.1}{0.00008}\right)$$
$$= (10{,}000, 1250).$$

15.

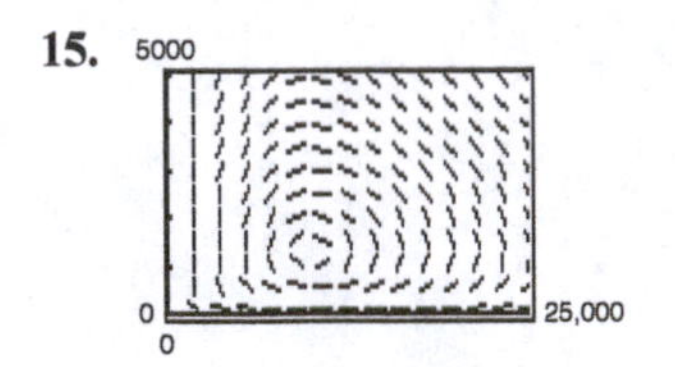

17. Using $x(0) = 50$ and $y(0) = 20$, you obtain the constant solutions $x = 50$ and $y = 20$.

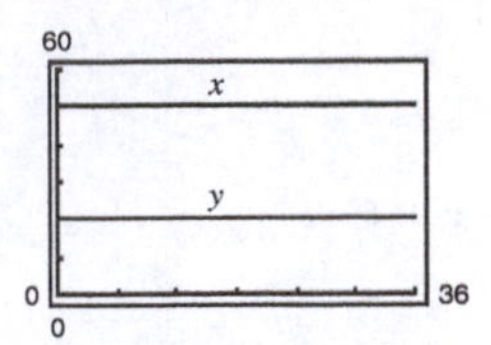

The slope field is the same, but the solution curve reduces to a single point at $(50, 20)$.

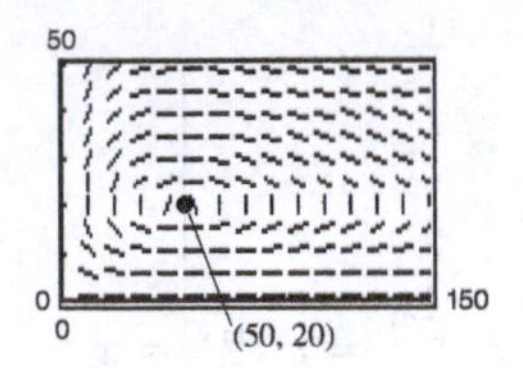

19. $\dfrac{dx}{dt} = ax - bx^2 - cxy = 2x - 3x^2 - 2xy$

$\dfrac{dy}{dt} = my - ny^2 - pxy = 2y - 3y^2 - 2xy$

From Example 5, you have $(0, 0)$, $\left(0, \frac{m}{n}\right) = \left(0, \frac{2}{3}\right)$, $\left(\frac{a}{b}, 0\right) = \left(\frac{2}{3}, 0\right)$ and

$$\left(\frac{an - mc}{bn - cp}, \frac{bm - ap}{bn - cp}\right) = \left(\frac{6 - 4}{9 - 4}, \frac{6 - 4}{9 - 4}\right) = \left(\frac{2}{5}, \frac{2}{5}\right).$$

21. $\dfrac{dx}{dt} = ax - bx^2 - cxy = 0.15x - 0.6x^2 - 0.75xy$

$\dfrac{dy}{dt} = my - ny^2 - pxy = 0.15y - 12y^2 - 0.45xy$

From Example 5, you have $(0, 0)$, $\left(0, \frac{m}{n}\right) = \left(0, \frac{1}{8}\right)$, $\left(\frac{a}{b}, 0\right) = \left(\frac{1}{4}, 0\right)$ and

$$\left(\frac{an - mc}{bn - cp}, \frac{bm - ap}{bn - cp}\right) = \left(\frac{0.18 - 0.1125}{0.72 - 0.3375}, \frac{0.09 - 0.0675}{0.72 - 0.3375}\right) = \left(\frac{3}{17}, \frac{1}{17}\right) \approx (0.1765, 0.0588).$$

23. $a = 0.8$, $b = 0.4$, $c = 0.1$, $m = 0.3$, $n = 0.6$, $p = 0.1$

Four critical points:

$(0, 0)$

$\left(0, \frac{m}{n}\right) = \left(0, \frac{0.3}{0.6}\right) = \left(0, \frac{1}{2}\right)$

$\left(\frac{a}{b}, 0\right) = \left(\frac{0.8}{0.4}, 0\right) = (2, 0)$

$\left(\frac{an - mc}{bn - cp}, \frac{bm - ap}{bn - cp}\right) = \left(\frac{0.45}{0.23}, \frac{0.04}{0.23}\right) = \left(\frac{45}{23}, \frac{4}{23}\right)$

25. $a = 0.8$, $b = 0.4$, $c = 1$, $m = 0.3$, $n = 0.6$, $p = 1$

Four critical points:

$(0, 0)$

$\left(0, \frac{m}{n}\right) = \left(0, \frac{0.3}{0.6}\right) = \left(0, \frac{1}{2}\right)$

$\left(\frac{a}{b}, 0\right) = \left(\frac{0.8}{0.4}, 0\right) = (2, 0)$

$\left(\frac{an - mc}{bn - cp}, \frac{bm - ap}{bn - cp}\right) = \left(\frac{0.18}{-0.76}, \frac{-0.68}{-0.76}\right) = \left(-\frac{9}{38}, \frac{17}{19}\right)$

27. Assuming the initial conditions are the critical points

$(x(0), y(0)) = \left(\frac{45}{23}, \frac{4}{23}\right)$

you obtain constant solutions.

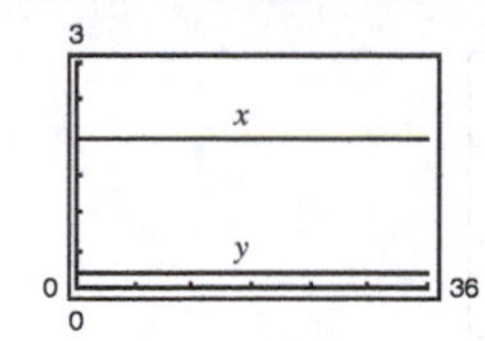

29. Solve the equations

$$\frac{dx}{dt} = ax - bxy = 0$$

$$\frac{dy}{dt} = -my + nxy = 0$$

to obtain critical points

$(0, 0)$ and $\left(\frac{m}{n}, \frac{a}{b}\right)$.

The solutions will be constant for these initial conditions.

31. (a) If $y = 0$, then $\frac{dx}{dt} = ax\left(1 - \frac{x}{L}\right)$, which is a logistic equation.

(b) $\frac{dx}{dt} = 0.4x\left(1 - \frac{x}{100}\right) - 0.01xy$

$\frac{dy}{dt} = -0.3y + 0.005xy$

$(0, 0)$ is a critical point. If $y = 0$, then $x = 100$ and $(100, 0)$ is a critical point. If $x,\ y \neq 0$, then

$0.4\left(1 - \frac{x}{100}\right) = 0.01y$

$-0.3 + 0.05x = 0.$

So, $x = \frac{0.3}{0.005} = 60$ and $0.4\left(1 - \frac{60}{100}\right) = 0.01y \Rightarrow y = 16.$

The third critical point is $(60, 16)$.

(c)

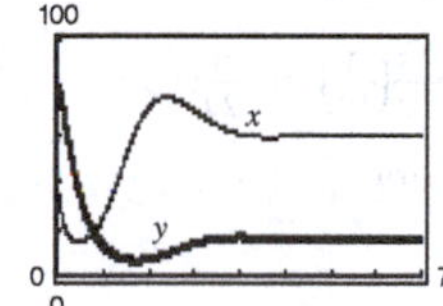

(d)

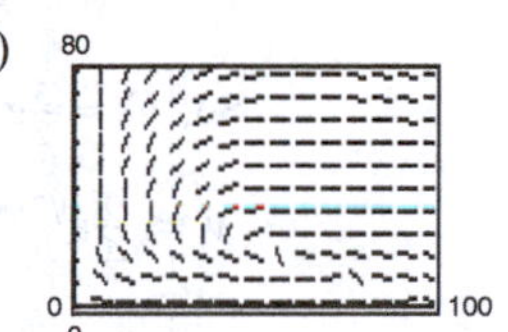

(e)

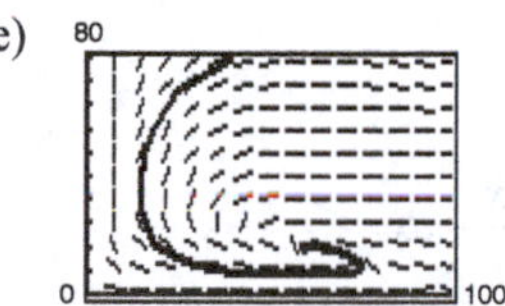

Review Exercises for Chapter 6

1. $y = x^3, y' = 3x^2$

$2xy' + 4y = 2x(3x^2) + 4(x^3) = 10x^3.$

Yes, it is a solution.

3. $\frac{dy}{dx} = 4x^2 + 7$

$$y = \int (4x^2 + 7)\,dx = \frac{4x^3}{3} + 7x + C$$

5. $\frac{dy}{dx} = \cos 2x$

$$y = \int \cos 2x\,dx = \frac{1}{2}\sin 2x + C$$

7. $\frac{dy}{dx} = e^{2-x}$

$$y = \int e^{2-x}\,dx = -e^{2-x} + C$$

9. $\frac{dy}{dx} = 2x - y$

x	-4	-2	0	2	4	8
y	2	0	4	4	6	8
dy/dx	-10	-4	-4	0	2	8

11. $y' = 2x^2 - x, \quad (0, 2)$

(a) and (b)

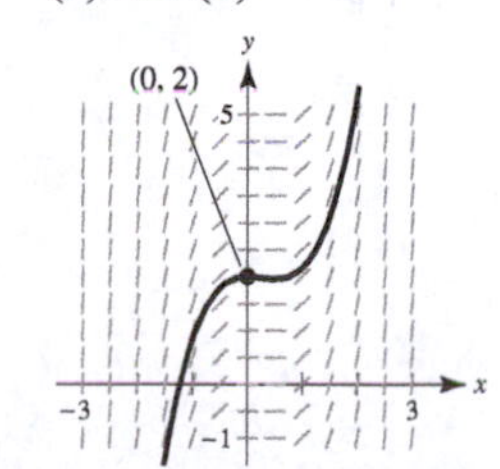

13. $y' = x - y, y(0) = 4, n = 10, h = 0.05$

$y_1 = y_0 + hF(x_0, y_0) = 4 + (0.05)(0 - 4) = 3.8$

$y_2 = y_1 + hF(x_1, y_1) = 3.8 + (0.05)(0.05 - 3.8) = 3.6125$, etc.

n	0	1	2	3	4	5	6	7	8	9	10
x_n	0	0.05	0.1	0.15	0.2	0.25	0.3	0.35	0.4	0.45	0.5
y_n	4	3.8	3.6125	3.437	3.273	3.119	2.975	2.842	2.717	2.601	2.494

15. $\frac{dy}{dx} = 6x - x^3$

$$y = \int (6x - x^3)\, dx = 3x^2 - \frac{x^4}{4} + C$$

17. $\frac{dy}{dx} = (y - 1)^2$

$$\int (y - 1)^{-2}\, dy = \int dx$$

$$-(y - 1)^{-1} = x + C$$

$$-(y - 1) = \frac{1}{x + C}$$

$$y = \frac{-1}{x + C} + 1$$

19. $(2 + x)y' - xy = 0$

$$(2 + x)\frac{dy}{dx} = xy$$

$$\frac{1}{y}\, dy = \frac{x}{2 + x} dx$$

$$\frac{1}{y}\, dy = \left(1 - \frac{2}{2 + x}\right) dx$$

$$\ln|y| = x - 2\ln|2 + x| + C_1$$

$$y = Ce^x(2 + x)^{-2} = \frac{Ce^x}{(2 + x)^2}$$

21. $\sqrt{x + 1}\, y' - y = 0$

$$\sqrt{x + 1}\frac{dy}{dx} = y$$

$$\frac{1}{y}\, dy = (x + 1)^{-1/2}\, dx$$

$$\ln y = 2\sqrt{x + 1} + C_1$$

$$y = e^{2\sqrt{x+1}+C_1} = Ce^{2\sqrt{x+1}}$$

23. $\frac{dy}{dt} = \frac{k}{t^3}$

$$\int dy = \int kt^{-3}\, dt$$

$$y = -\frac{k}{2t^2} + C$$

25. $y = Ce^{kt}$

$$\left(0, \tfrac{3}{4}\right): \tfrac{3}{4} = C$$

$$(5, 5): 5 = \tfrac{3}{4}e^{k(5)}$$

$$\tfrac{20}{3} = e^{5k}$$

$$k = \tfrac{1}{5}\ln\left(\tfrac{20}{3}\right)$$

$$y = \tfrac{3}{4}e^{[\ln(20/3)/5]t} \approx \tfrac{3}{4}e^{0.379t}$$

27. $y = Ce^{kt}$

$$\left(2, \tfrac{3}{2}\right): \tfrac{3}{2} = Ce^{2k} \Rightarrow C = \tfrac{3}{2}e^{-2k}$$

$$(4, 5): 5 = Ce^{4k} = \left(\tfrac{3}{2}e^{-2k}\right)e^{4k} = \tfrac{3}{2}e^{2k}$$

$$\tfrac{10}{3} = e^{2k} \Rightarrow k = \tfrac{1}{2}\ln\left(\tfrac{10}{3}\right)$$

So, $C = \tfrac{3}{2}e^{-2(1/2)\ln(10/3)} = \tfrac{3}{2}\left(\tfrac{3}{10}\right) = \tfrac{9}{20}$.

$$y = \tfrac{9}{20}e^{1/2\ln(10/3)t} \approx \tfrac{9}{20}e^{0.602t}$$

29. $\frac{dP}{dh} = kP, \quad P(0) = 30$

$$P(h) = 30e^{kh}$$

$$P(18{,}000) = 30e^{18{,}000k} = 15$$

$$k = \frac{\ln(1/2)}{18{,}000} = \frac{-\ln 2}{18{,}000}$$

$$P(h) = 30e^{-(h\ln 2)/18{,}000}$$

$$P(35{,}000) = 30e^{-(35{,}000\ln 2)/18{,}000} \approx 7.79 \text{ inches}$$

31. $P = Ce^{0.0185t}$

$$2C = Ce^{0.0185t}$$

$$2 = e^{0.0185t}$$

$$\ln 2 = 0.0185t$$

$$t = \frac{\ln 2}{0.0185} \approx 37.5 \text{ years}$$

33. $S = Ce^{k/t}$

(a) $S = 5$ when $t = 1$.

$$5 = Ce^{k}$$
$$\lim_{t\to\infty} Ce^{k/t} = C = 30$$
$$5 = 30e^{k}$$
$$k = \ln\tfrac{1}{6} \approx -1.7918$$
$$S = 30e^{-1.7918/t}$$

(b) When $t = 5$, $S \approx 20.9646$ which is 20,965 units.

(c)

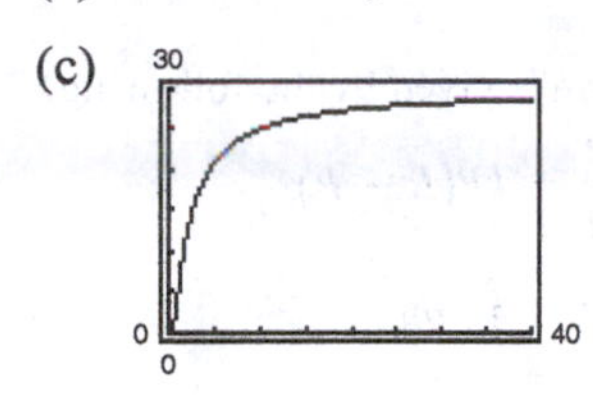

35. $\dfrac{dy}{dx} = \dfrac{5x}{y}$

$$\int y\,dy = \int 5x\,dx$$
$$\frac{y^2}{2} = \frac{5x^2}{2} + C_1$$
$$y^2 = 5x^2 + C$$

37. $y'e^{y-3x} = e^{x+2y}$

$$y'e^{y}e^{-3x} = e^{x}e^{2y}$$
$$\int e^{-y}\,dy = \int e^{4x}\,dx$$
$$-e^{-y} = \frac{1}{4}e^{4x} + C$$
$$e^{-y} = C - \frac{1}{4}e^{4x}$$
$$-y = \ln\left(C - \frac{1}{4}e^{4x}\right)$$
$$y = -\ln\left(C - \frac{1}{4}e^{4x}\right)$$

39. $y^3y' - 3x = 0,\ y(2) = 2$

$$y^3\frac{dy}{dx} = 3x$$
$$\int y^3\,dy = \int 3x\,dx$$
$$\frac{y^4}{4} = \frac{3x^2}{2} + C_1$$
$$y^4 = 6x^2 + C$$

Initial condition: $y(2) = 2: 16 = 24 + C$

$$C = -8$$

Particular solution: $y^4 = 6x^2 - 8$

41. $y^3(x^4 + 1)y' - x^3(y^4 + 1) = 0,\ y(0) = 1$

$$y^3(x^4 + 1)\frac{dy}{dx} = x^3(y^4 + 1)$$
$$\int \frac{y^3}{y^4 + 1}\,dy = \int \frac{x^3}{x^4 + 1}\,dx$$
$$\frac{1}{4}\ln(y^4 + 1) = \frac{1}{4}\ln(x^4 + 1) + \frac{1}{4}\ln C_1$$
$$\ln(y^4 + 1) = \ln\left[C(x^4 + 1)\right]$$
$$y^4 + 1 = C(x^4 + 1)$$

Initial condition: $y(0) = 1: 1 + 1 = C(0 + 1)$

$$C = 2$$

Particular solution: $y^4 + 1 = 2(x^4 + 1)$

$$y^4 = 2x^4 + 1$$

43. $\dfrac{dy}{dx} = \dfrac{-4x}{y}$

$$\int y\,dy = \int -4x\,dx$$
$$\frac{y^2}{2} = -2x^2 + C_1$$
$$4x^2 + y^2 = C \quad \text{ellipses}$$

45. $y' = \dfrac{2x}{y}$

$$y\,dy = 2x\,dx$$
$$\int y\,dy = \int 2x\,dx$$
$$\frac{y^2}{2} = x^2 + C$$
$$(1, 3):\ \frac{9}{2} = 1 + C \Rightarrow C = \frac{7}{2}$$
$$\frac{y^2}{2} = x^2 + \frac{7}{2}$$
$$y^2 = 2x^2 + 7$$

47. Given family (hyperbolas): $5x^2 - 4y^2 = C$

$$10x - 8yy' = 0$$
$$y' = \frac{5x}{4y}$$

Orthogonal trajectory:

$$y' = \frac{-4y}{5x}$$
$$\int \frac{5}{y}\,dy = \int \frac{-4}{x}\,dx$$
$$5\ln|y| = -4\ln|x| + K_1$$
$$= -4\ln|x| + \ln K$$
$$y^5 = Kx^{-4}$$
$$y = Kx^{-4/5}$$

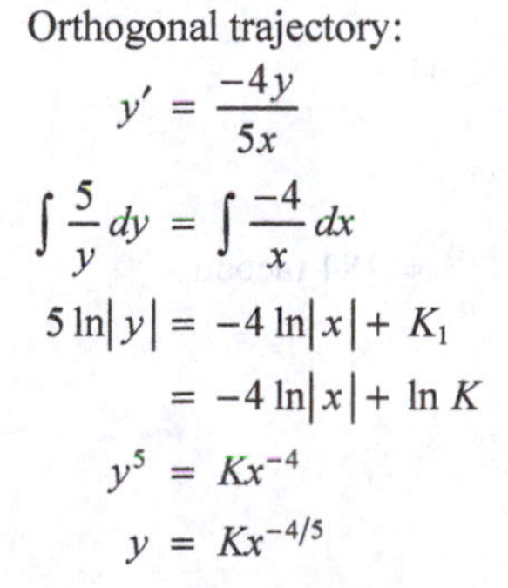

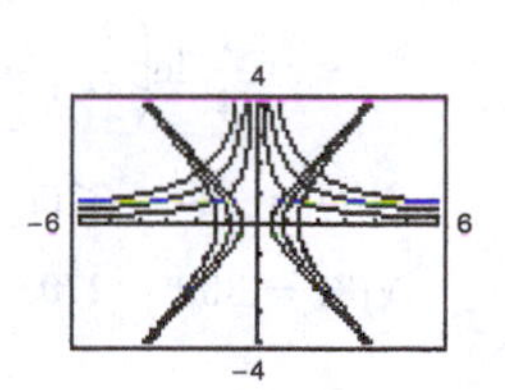

49. $P(t) = \dfrac{5250}{1 + 34e^{-0.55t}}$

(a) $k = 0.55$

(b) $L = 5250$

(c) $P(0) = \dfrac{5250}{1 + 34} = 150$

(d) $2625 = \dfrac{5250}{1 + 34e^{-0.55t}}$

$1 + 34e^{-0.55t} = 2$

$e^{-0.55t} = \dfrac{1}{34}$

$t = \dfrac{-1}{0.55}\ln\left(\dfrac{1}{34}\right) \approx 6.41$ yr

(e) $\dfrac{dP}{dt} = 0.55P\left(1 - \dfrac{P}{5250}\right)$

51. $\dfrac{dy}{dt} = y\left(1 - \dfrac{y}{80}\right), \quad (0, 8)$

$k = 1, L = 80$

$y = \dfrac{L}{1 + be^{-kt}} = \dfrac{80}{1 + be^{-t}}$

$y(0) = 8: \quad 8 = \dfrac{80}{1 + b} \Rightarrow b = 9$

Solution: $y = \dfrac{80}{1 + 9e^{-t}}$

53. $\dfrac{dN}{dt} = k(380 - N(t))$

$\int \dfrac{dN}{380 - N} = \int k\,dt$

$\ln|380 - N| = kt + C_1$

$380 - N = Ce^{kt}$

$N = 380 - Ce^{kt}$

$N(0) = 110 = 380 - C \Rightarrow C = 270$

$N = 380 - 270e^{kt}$

$N(4) = 150 = 380 - 270e^{k(4)}$

$e^{k(4)} = \dfrac{150 - 380}{-270} = \dfrac{23}{27}$

$k = \dfrac{1}{4}\ln\left(\dfrac{23}{27}\right) \approx -0.04$

$N(t) = 380 - 270e^{-0.04t}$

$N(8) = 380 - 270e^{-0.04(8)} \approx 184$ racoons

55. $\dfrac{dS}{dt} = k(L - S)$

$\int \dfrac{dS}{L - S} = \int k\,dt$

$-\ln|L - S| = kt + C_1$

$L - S = e^{-kt - C_1}$

$S = L + Ce^{-kt}$

Because $S = 0$ when $t = 0$, you have

$0 = L + C \Rightarrow C = -L$. So, $S = L(1 - e^{-kt})$.

57. The differential equation is given by the following.

$\dfrac{dP}{dn} = kP(L - P)$

$\int \dfrac{1}{P(L - P)}\,dP = \int k\,dn$

$\dfrac{1}{L}\left[\ln|P| - \ln|L - P|\right] = kn + C_1$

$\dfrac{P}{L - P} = Ce^{Lkn}$

$P = \dfrac{CLe^{Lkn}}{1 + Ce^{Lkn}} = \dfrac{CL}{e^{-Lkn} + C}$

59. $y' - y = 10$

$P(x) = -1, Q(x) = 10$

$u(x) = e^{\int -dx} = e^{-x}$

$y = \dfrac{1}{e^{-x}}\int 10e^{-x}\,dx = e^{x}\left(-10e^{-x} + C\right) = -10 + Ce^{x}$

61. $4y' = e^{x/y} + y$

$y' - \dfrac{1}{4}y = \dfrac{1}{4}e^{x/4}$

$P(x) = -\dfrac{1}{4}, Q(x) = \dfrac{1}{4}e^{x/4}$

$u(x) = e^{\int -(1/4)\,dx} = e^{-(1/4)x}$

$y = \dfrac{1}{e^{-(1/4)x}}\int \dfrac{1}{4}e^{x/4}e^{-(1/4)x}\,dx = e^{(1/4)x}\left(\dfrac{1}{4}x + C\right)$

$= \dfrac{1}{4}xe^{x/4} + Ce^{x/4}$

63. $(x - 2)y' + y = 1, x > 2$

$\dfrac{dy}{dx} + \dfrac{1}{x - 2}y = \dfrac{1}{x - 2}$

$P(x) = \dfrac{1}{x - 2}, Q(x) = \dfrac{1}{x - 2}$

$u(x) = e^{\int (1/x-2)\,dx} = e^{\ln|x-2|} = x - 2$

$y = \dfrac{1}{x - 2}\int\left(\dfrac{1}{x - 2}\right)(x - 2)\,dx = \dfrac{1}{x - 2}(x + C)$

65. $y' + 5y = e^{5x}$

Integrating factor: $e^{\int 5\,dx} = e^{5x}$

$$ye^{5x} = \int e^{10x}\,dx = \tfrac{1}{10}e^{10x} + C$$

$$y = \tfrac{1}{10}e^{5x} + Ce^{-5x}$$

67. (a)

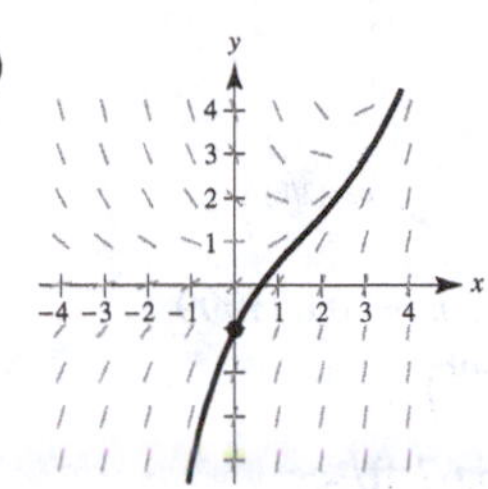

(b)

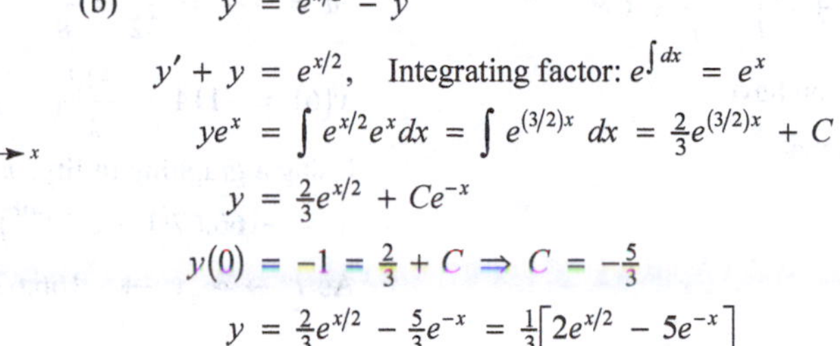

$$y' = e^{x/2} - y$$

$$y' + y = e^{x/2}, \quad \text{Integrating factor: } e^{\int dx} = e^x$$

$$ye^x = \int e^{x/2}e^x dx = \int e^{(3/2)x}\,dx = \tfrac{2}{3}e^{(3/2)x} + C$$

$$y = \tfrac{2}{3}e^{x/2} + Ce^{-x}$$

$$y(0) = -1 = \tfrac{2}{3} + C \Rightarrow C = -\tfrac{5}{3}$$

$$y = \tfrac{2}{3}e^{x/2} - \tfrac{5}{3}e^{-x} = \tfrac{1}{3}\left[2e^{x/2} - 5e^{-x}\right]$$

(c)

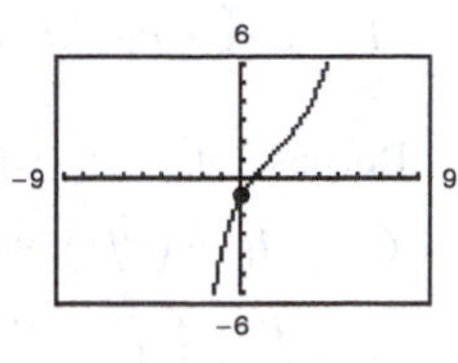

69. (a)

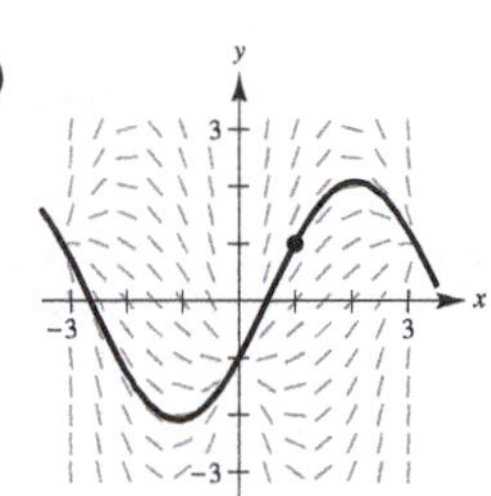

(b)

$$\frac{dy}{dx} = \csc x + y \cot x$$

$$\frac{dy}{dx} - (\cot x)y = \csc x$$

Integrating factor: $e^{\int -\cot x\,dx} = e^{-\ln|\sin x|} = \csc x$

$$\csc x \cdot y' - \csc x \cot x \cdot y = \csc^2 x$$

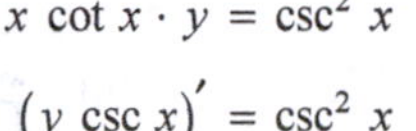

$$(y \csc x)' = \csc^2 x$$

$$y \csc x = \int \csc^2 x\,dx = -\cot x + C$$

$$y = -\cos x + C \sin x$$

$$y(1) = 1 \Rightarrow 1 = -\cos 1 + C \sin 1 \Rightarrow C = \frac{1 + \cos 1}{\sin 1} \approx 1.8305$$

$$y = -\cos x + 1.8305 \sin x$$

(c)

71. $y' + 5y = e^{5x}, y(0) = 3$

$P(x) = 5, Q(x) = e^{5x}$

$u(x) = e^{\int 5\,dx} = e^{5x}$

$$y = \frac{1}{e^{5x}}\int \left(e^{5x}\right)\left(e^{5x}\right) dx$$

$$= \frac{1}{e^{5x}}\int e^{10x}\,dx$$

$$= \frac{1}{e^{5x}}\int \left(\frac{1}{10}e^{10x} + C\right)$$

$$= \frac{1}{10}e^{5x} + Ce^{-5x}$$

Initial condition:

$$y(0) = 3 : 3 = \frac{1}{10}e^0 + Ce^0 \Rightarrow C = \frac{29}{10}$$

Particular solution: $y = \dfrac{1}{10}e^{5x} + \dfrac{29}{10}e^{-5x}$

73. $(3y + 5)\cos x\,dx = dy, \quad y(\pi) = 0$

$$\frac{dy}{dx} - 3\cos x\,y = 5\cos x$$

$P(x) = -3\cos x, Q(x) = 5\cos x$

$u(x) = e^{\int -3\cos x\,dx} = e^{-3\sin x}$

$$\frac{dy}{dx}e^{-3\sin x} - 3\cos x\left(e^{-3\sin x}\right)y = 5\cos x\left(e^{-3\sin x}\right)$$

$$ye^{-3\sin x} = \int 5e^{-3\sin x}\cos x\,dx$$

$$= \frac{-5}{3}e^{-3\sin x} + C$$

$$y = -\frac{5}{3} + Ce^{3\sin x}$$

Initial condition: $y(\pi) = 0$: $0 = -\dfrac{5}{3} + C \Rightarrow C = \dfrac{5}{3}$

Particular solution: $y = -\dfrac{5}{3} + \dfrac{5}{3}e^{3\sin x}$

75. $\dfrac{dA}{dt} - rA = -P$

For this linear differential equation, you have $P(t) = -r$ and $Q(t) = -P$. Therefore, the integrating factor is $u(x) = e^{\int -r\,dt} = e^{-rt}$ and the solution is

$$A = e^{rt}\int -Pe^{-rt}dt = e^{rt}\left(\frac{P}{r}e^{-rt} + C\right) = \frac{P}{r} + Ce^{rt}.$$

Because $A = A_0$ when $t = 0$, you have $C = A_0 - (P/r)$ which implies that

$$A = \frac{P}{r} + \left(A_0 - \frac{P}{r}\right)e^{rt}.$$

77. From Example 3, Section 6.5,

$$\frac{dv}{dt} + \frac{kv}{m} = -g$$

$$v = \frac{-mg}{k}\left(1 - e^{-kt/m}\right).$$

$g = -32,\ mg = 12,\ v(6) = -114,$

$$m = \frac{-12}{g} = \frac{12}{32} = \frac{3}{8}$$

$$v(6) = -114 = \frac{-12}{k}\left(1 - e^{-6k/(3/8)}\right)$$

Using a graphing utility, $k \approx 0.071999$.

$v = -166.67\left(1 - e^{-0.192t}\right)$

As $t \to \infty$, $v \to -166.67$ ft/sec.

79. (a) $\dfrac{dx}{dt} = ax - bxy = 0.3x - 0.02xy$

$\dfrac{dy}{dt} = -my + nxy = -0.4y + 0.01xy$

(b) $x' = y' = 0$ when $(x, y) = (0, 0)$ and

$$(x, y) = \left(\frac{m}{n}, \frac{a}{b}\right) = \left(\frac{0.4}{0.01}, \frac{0.3}{0.02}\right) = (40, 14).$$

(c)

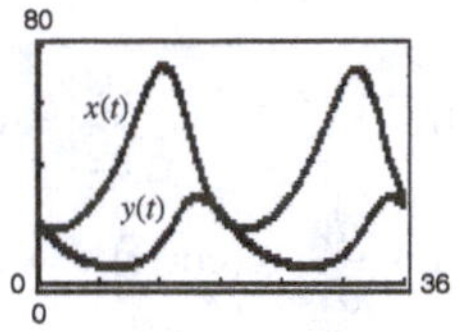

81. (a) $\dfrac{dx}{dt} = ax - bx^2 - cxy = 3x - x^2 - xy$

$\dfrac{dy}{dt} = my - ny^2 - pxy = 2y - y^2 - 0.5xy$

(b) $x' = y' = 0$ when $(x, y) = (0, 0)$,

$$(x, y) = \left(0, \frac{m}{n}\right) = (0, 2),\ (x, y) = \left(\frac{a}{b}, 0\right) = (3, 0),$$

$$(x, y) = \left(\frac{an - mc}{bn - cp}, \frac{bm - ap}{bn - cp}\right) = \left(\frac{1}{1/2}, \frac{1/2}{1/2}\right) = (2, 1).$$

(c)

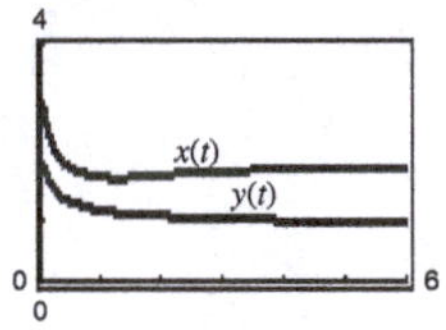

Problem Solving for Chapter 6

1. (a)

$$\frac{dy}{dt} = y^{1.01}$$

$$\int y^{-1.01}\,dy = \int dt$$

$$\frac{y^{-0.01}}{-0.01} = t + C_1$$

$$\frac{1}{y^{0.01}} = -0.01t + C$$

$$y^{0.01} = \frac{1}{C - 0.01t}$$

$$y = \frac{1}{(C - 0.01t)^{100}}$$

$$y(0) = 1: 1 = \frac{1}{C^{100}} \Rightarrow C = 1$$

So, $y = \dfrac{1}{(1 - 0.01t)^{100}}$.

For $T = 100$, $\displaystyle\lim_{t \to T^-} y = \infty$.

(b) $\displaystyle\int y^{-(1+\varepsilon)}\,dy = \int k\,dt$

$$\frac{y^{-\varepsilon}}{-\varepsilon} = kt + C_1$$

$$y^{-\varepsilon} = -\varepsilon kt + C$$

$$y = \frac{1}{(C - \varepsilon kt)^{1/\varepsilon}}$$

$$y(0) = y_0 = \frac{1}{C^{1/\varepsilon}} \Rightarrow C^{1/\varepsilon} = \frac{1}{y_0} \Rightarrow C = \left(\frac{1}{y_0}\right)^{\varepsilon}$$

So, $y = \dfrac{1}{\left(\dfrac{1}{y_0^{\varepsilon}} - \varepsilon kt\right)^{1/\varepsilon}}$.

For $t \to \dfrac{1}{y_0^{\varepsilon}\varepsilon k}$, $y \to \infty$.

3. (a) $y' = x - y, y(0) = 1, h = 0.1$

Using the modified Euler Method, you obtain:

x	y
0	1.0
0.1	0.91
0.2	0.83805
$\vdots$	$\vdots$
1.0	0.73708

(b) 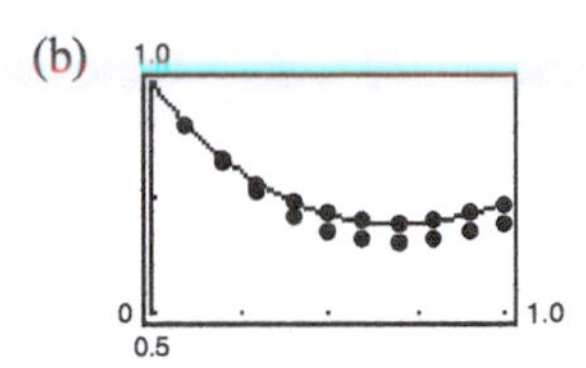

The modified Euler Method is more accurate.

```
[                      ]
[.1      .9            ]
[                      ]
[.2      .82           ]
[                      ]
[.3      .758          ]
[                      ]
[.4      .7122         ]
[                      ]
[.5      .68098        ]
[                      ]
[.6      .662882       ]
[                      ]
[.7      .6565938      ]
[                      ]
[.8      .66093442     ]
[                      ]
[.9      .674840978    ]
[                      ]
[1.0     .6973568802]
```

```
[x              y      ]
[                      ]
[0              1      ]
[                      ]
[.1      .9100000000]
[                      ]
[.2      .8380500000]
[                      ]
[.3      .7824352500]
[                      ]
[.4      .7416039013]
[                      ]
[.5      .7141515307]
[                      ]
[.6      .6988071353]
[                      ]
[.7      .6944204575]
[                      ]
[.8      .6999505140]
[                      ]
[.9      .7144552152]
[                      ]
[1.0     .7370819698]
```

5. $k = \left(\frac{1}{12}\right)^2 \pi$

$g = 32$

$x^2 + (y - 6)^2 = 36$ Equation of tank

$$x^2 = 36 - (y - 6)^2 = 12y - y^2$$

Area of cross section: $A(h) = (12h - h^2)\pi$

$$A(h)\frac{dh}{dt} = -k\sqrt{2gh}$$

$$(12h - h^2)\pi\frac{dh}{dt} = -\frac{1}{144}\pi\sqrt{64h}$$

$$(12h - h^2)\frac{dh}{dt} = -\frac{1}{18}h^{1/2}$$

$$\int\left(18h^{3/2} - 216h^{1/2}\right)dh = \int dt$$

$$\frac{36}{5}h^{5/2} - 144h^{3/2} = t + C$$

$$\frac{h^{3/2}}{5}(36h - 720) = t + C$$

When $h = 6, t = 0$ and $C = \frac{6^{3/2}}{5}(-504) \approx -1481.45$.

The tank is completely drained when

$h = 0 \Rightarrow t = 1481.45 \text{ sec} \approx 24 \text{ min}, 41 \text{ sec}$

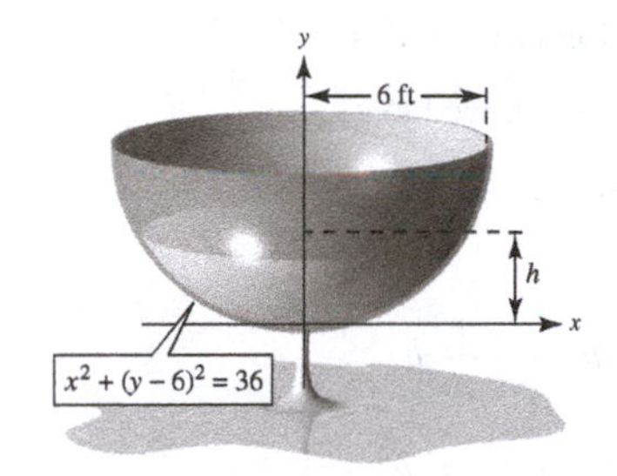

7. $A(h)\frac{dh}{dt} = -k\sqrt{2gh}$

$$\pi 64\frac{dh}{dt} = \frac{-\pi}{36}8\sqrt{h}$$

$$\int h^{-1/2}\,dh = \int\frac{-1}{288}\,dt$$

$$2\sqrt{h} = \frac{-t}{288} + C$$

$$h = 20: 2\sqrt{20} = C = 4\sqrt{5}$$

$$2\sqrt{h} = \frac{-t}{288} + 4\sqrt{5}$$

$$h = 0 \Rightarrow t = 4\sqrt{5}(288)$$

$$\approx 2575.95 \text{ sec} \approx 42 \text{ min}, 56 \text{ sec}$$

9. $\dfrac{ds}{dt} = 3.5 - 0.019s$

(a) $\displaystyle\int \frac{-ds}{3.5 - 0.019s} = -\int dt$

$$\frac{1}{0.019}\ln|3.5 - 0.019s| = -t + C_1$$

$$\ln|3.5 - 0.019s| = -0.019t + C_2$$

$$3.5 - 0.019s = C_3e^{-0.019t}$$

$$0.019s = 3.5 - C_3e^{-0.019t}$$

$$s = 184.21 - Ce^{-0.019t}$$

(b)

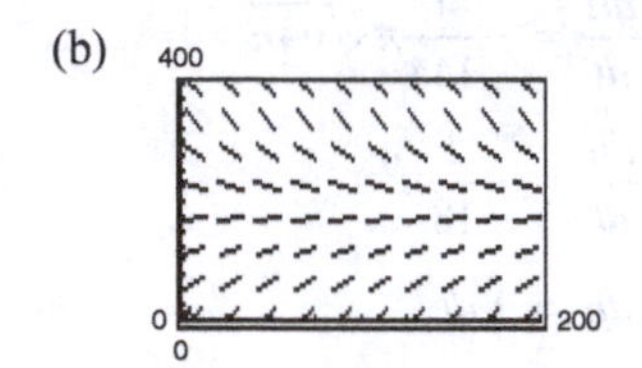

(c) As $t \to \infty$, $Ce^{-0.019t} \to 0$, and $s \to 184.21$.

11. (a) $\displaystyle\int \frac{dC}{C} = \int -\frac{R}{V}\,dt$

$$\ln|C| = -\frac{R}{V}t + K_1$$

$$C = Ke^{-Rt/V}$$

Since $C = C_0$ when $t = 0$, it follows that $K = C_0$ and the function is $C = C_0e^{-Rt/V}$.

(b) Finally, as $t \to \infty$, we have

$$\lim_{t\to\infty} C = \lim_{t\to\infty} C_0e^{-Rt/V} = 0.$$

13. (a) $\displaystyle\int \frac{1}{Q - RC}\,dC = \int \frac{1}{V}\,dt$

$$-\frac{1}{R}\ln|Q - RC| = \frac{t}{V} + K_1$$

$$Q - RC = e^{-R[(t/V)+K_1]}$$

$$C = \frac{1}{R}\left(Q - e^{-R[(t/V)+K_1]}\right)$$

$$= \frac{1}{R}\left(Q - Ke^{-Rt/V}\right)$$

Because $C = 0$ when $t = 0$, it follows that $K = Q$ and you have $C = \dfrac{Q}{R}\left(1 - e^{-Rt/V}\right)$.

(b) As $t \to \infty$, the limit of C is Q/R.

CHAPTER 7
Applications of Integration

CHAPTER 7
Applications of Integration

Section 7.1 Area of a Region Between Two Curves

1. In variable x, the area of the region between two graphs is the area under the graph of the top function minus the area under the graph of the bottom function.

3. The points of intersection are used to determine the vertical lines that bound the region.

5. $A = \int_0^6 \left[0 - (x^2 - 6x)\right] dx = -\int_0^6 (x^2 - 6x)\, dx$

7.
$$\begin{aligned} A &= \int_0^3 \left[(-x^2 + 2x + 3) - (x^2 - 4x + 3)\right] dx \\ &= \int_0^3 (-2x^2 + 6x)\, dx \end{aligned}$$

9. $A = 2\int_{-1}^{0} 3(x^3 - x)\, dx = 6\int_{-1}^{0} (x^3 - x)\, dx$

or $-6\int_0^1 (x^3 - x)\, dx$

11. $\int_0^4 \left[(x + 1) - \dfrac{x}{2}\right] dx$

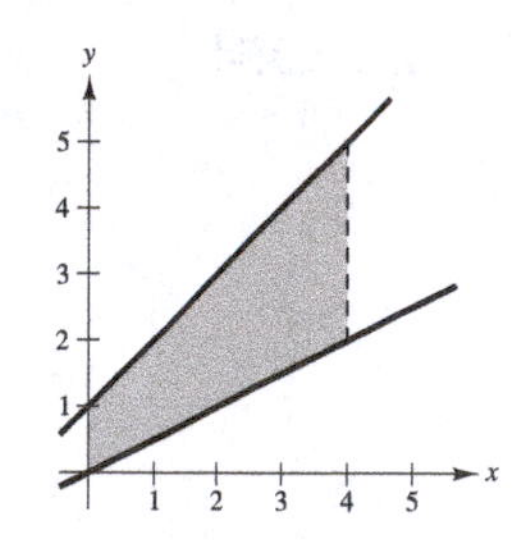

13. $\int_{-2}^{1} \left[(2 - y) - y^2\right] dy$

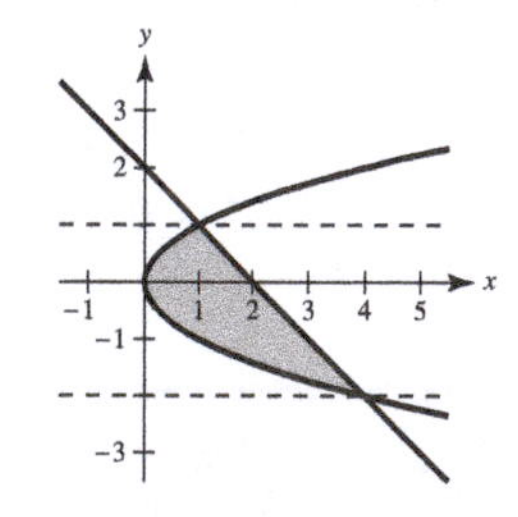

15.
$$\begin{aligned} A &= \int_0^1 \left[(-x + 2) - (x^2 - 1)\right] dx \\ &= \int_0^1 (-x^2 - x + 3)\, dx \\ &= \left[\frac{-x^3}{3} - \frac{x^2}{2} + 3x\right]_0^1 \\ &= \left(-\frac{1}{3} - \frac{1}{2} + 3\right) - 0 = \frac{13}{6} \end{aligned}$$

17. The points of intersection are given by:

$$\begin{aligned} x^2 + 2x &= x + 2 \\ x^2 + x - 2 &= 0 \\ (x + 2)(x - 1) &= 0 \quad \text{when } x = -2, 1 \end{aligned}$$

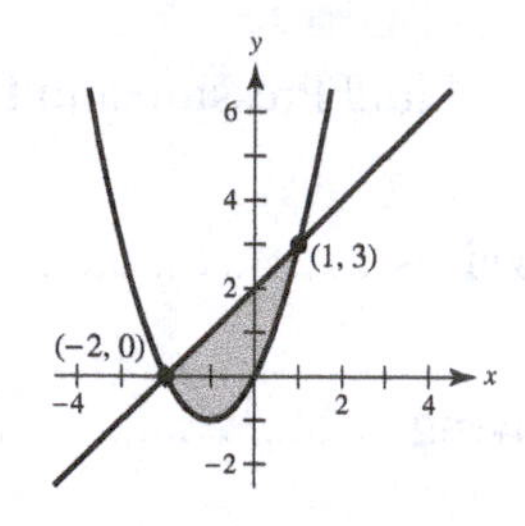

$$\begin{aligned} A &= \int_{-2}^{1} \left[g(x) - f(x)\right] dx \\ &= \int_{-2}^{1} \left[(x + 2) - (x^2 + 2x)\right] dx \\ &= \left[\frac{-x^3}{3} - \frac{x^2}{2} + 2x\right]_{-2}^{1} \\ &= \left(-\frac{1}{3} - \frac{1}{2} + 2\right) - \left(\frac{8}{3} - 2 - 4\right) = \frac{9}{2} \end{aligned}$$

19.

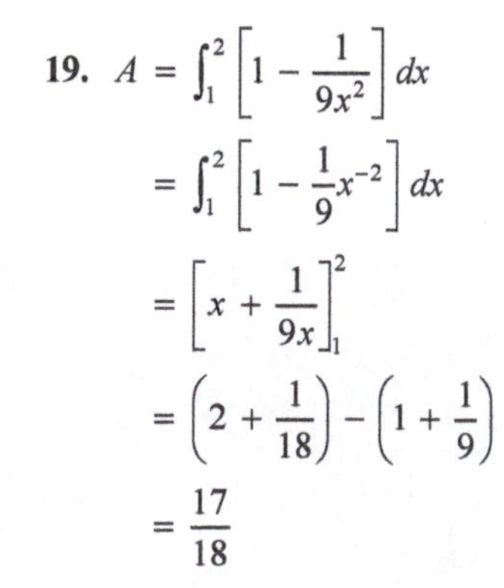

$$\begin{aligned} A &= \int_1^2 \left[1 - \frac{1}{9x^2}\right] dx \\ &= \int_1^2 \left[1 - \frac{1}{9}x^{-2}\right] dx \\ &= \left[x + \frac{1}{9x}\right]_1^2 \\ &= \left(2 + \frac{1}{18}\right) - \left(1 + \frac{1}{9}\right) \\ &= \frac{17}{18} \end{aligned}$$

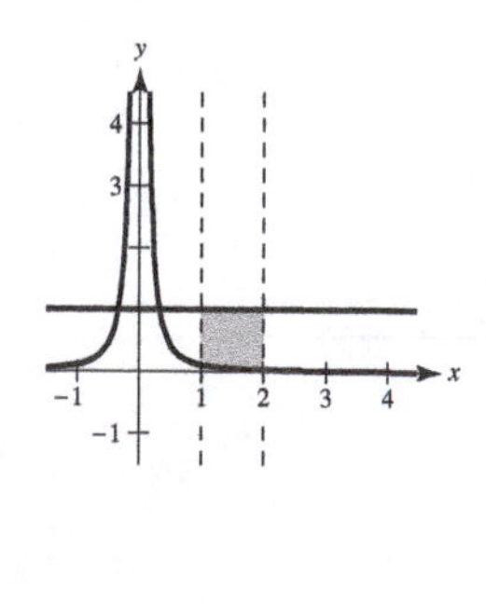

21. The points of intersection are given by:

$$x^5 + 2 = x + 2$$
$$x^5 = x \text{ when } x = -1, 0, 1$$

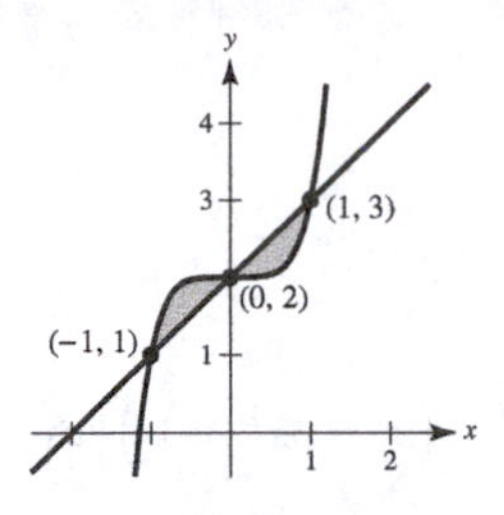

$$A = \int_{-1}^{0}\left[\left(x^5 + 2\right) - (x + 2)\right]dx + \int_{0}^{1}\left[(x + 2) - \left(x^5 + 2\right)\right]dx$$
$$= \int_{-1}^{0}\left(x^5 - x\right)dx + \int_{0}^{1}\left(x - x^5\right)dx$$
$$= \left[\frac{x^6}{6} - \frac{x^2}{2}\right]_{-1}^{0} + \left[\frac{x^2}{2} - \frac{x^6}{6}\right]_{0}^{1}$$
$$= \left(-\frac{1}{6} + \frac{1}{2}\right) + \left(\frac{1}{2} - \frac{1}{6}\right) = \frac{2}{3}$$

23. The points of intersection are given by:

$$y^2 = y + 2$$
$$(y - 2)(y + 1) = 0 \quad \text{when } y = -1, 2$$

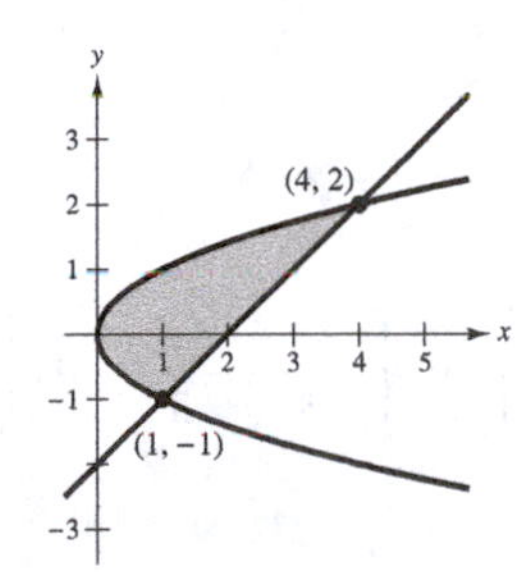

$$A = \int_{-1}^{2}\left[g(y) - f(y)\right]dy$$
$$= \int_{-1}^{2}\left[(y + 2) - y^2\right]dy$$
$$= \left[2y + \frac{y^2}{2} - \frac{y^3}{3}\right]_{-1}^{2} = \frac{9}{2}$$

25.

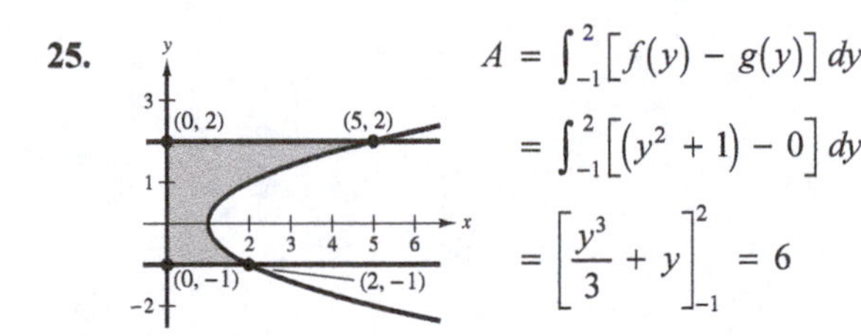

$$A = \int_{-1}^{2}\left[f(y) - g(y)\right]dy$$
$$= \int_{-1}^{2}\left[\left(y^2 + 1\right) - 0\right]dy$$
$$= \left[\frac{y^3}{3} + y\right]_{-1}^{2} = 6$$

27. $y = \dfrac{10}{x} \Rightarrow x = \dfrac{10}{y}$

$$A = \int_{2}^{10}\frac{10}{y}\,dy$$
$$= \left[10 \ln y\right]_{2}^{10}$$
$$= 10(\ln 10 - \ln 2)$$
$$= 10 \ln 5 \approx 16.0944$$

29. (a)

$$x = 4 - y^2$$
$$x = y - 2$$
$$4 - y^2 = y - 2$$
$$y^2 + y - 6 = 0$$
$$(y + 3)(y - 2) = 0$$

Intersection points: $(0, 2)$ and $(-5, -3)$

$$A = \int_{-5}^{0}\left[(x + 2) + \sqrt{4 - x}\right]dx + \int_{0}^{4}2\sqrt{4 - x}\,dx$$
$$= \frac{61}{6} + \frac{32}{3} = \frac{125}{6}$$

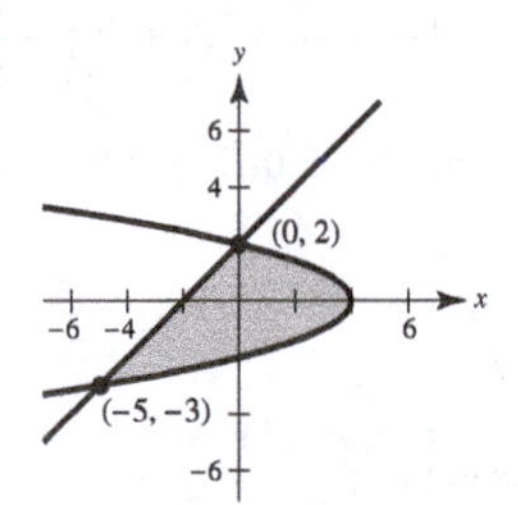

(b) $A = \int_{-3}^{2}\left[\left(4 - y^2\right) - (y - 2)\right]dy = \frac{125}{6}$

(c) The second method is simpler. Explanations will vary.

31. (a)

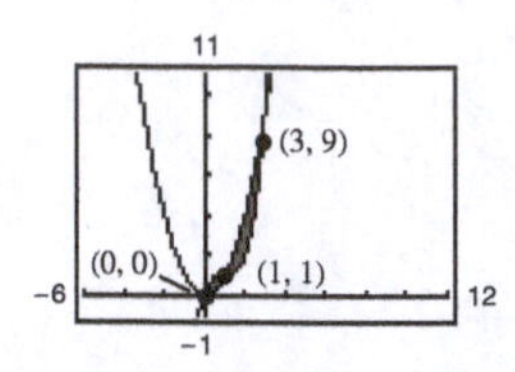

(b) The points of intersection are given by:

$$x^3 - 3x^2 + 3x = x^2$$

$$x(x-1)(x-3) = 0 \quad \text{when } x = 0, 1, 3$$

$$A = \int_0^1 [f(x) - g(x)]\,dx + \int_1^3 [g(x) - f(x)]\,dx = \int_0^1 [(x^3 - 3x^2 + 3x) - x^2]\,dx + \int_1^3 [x^2 - (x^3 - 3x^2 + 3x)]\,dx$$

$$= \int_0^1 (x^3 - 4x^2 + 3x)\,dx + \int_1^3 (-x^3 + 4x^2 - 3x)\,dx = \left[\frac{x^4}{4} - \frac{4}{3}x^3 + \frac{3}{2}x^2\right]_0^1 + \left[\frac{-x^4}{4} + \frac{4}{3}x^3 - \frac{3}{2}x^2\right]_1^3 = \frac{5}{12} + \frac{8}{3} = \frac{37}{12}$$

(c) Numerical approximation: $0.417 + 2.667 \approx 3.083$

33. (a) $f(x) = x^4 - 4x^2, \quad g(x) = x^2 - 4$

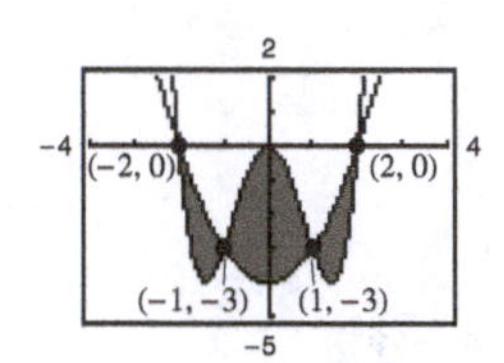

(b) The points of intersection are given by:

$$x^4 - 4x^2 = x^2 - 4$$

$$x^4 - 5x^2 + 4 = 0$$

$$(x^2 - 4)(x^2 - 1) = 0 \quad \text{when } x = \pm 2, \pm 1$$

By symmetry: $A = 2\int_0^1 [(x^4 - 4x^2) - (x^2 - 4)]\,dx + 2\int_1^2 [(x^2 - 4) - (x^4 - 4x^2)]\,dx$

$$= 2\int_0^1 (x^4 - 5x^2 + 4)\,dx + 2\int_1^2 (-x^4 + 5x^2 - 4)\,dx = 2\left[\frac{x^5}{5} - \frac{5x^3}{3} + 4x\right]_0^1 + 2\left[-\frac{x^5}{5} + \frac{5x^3}{3} - 4x\right]_1^2$$

$$= 2\left[\frac{1}{5} - \frac{5}{3} + 4\right] + 2\left[\left(-\frac{32}{5} + \frac{40}{3} - 8\right) - \left(-\frac{1}{5} + \frac{5}{3} - 4\right)\right] = 8$$

(c) Numerical approximation: $5.067 + 2.933 = 8.0$

35. (a)

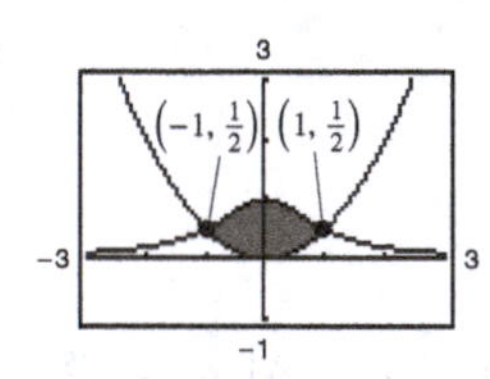

(b) The points of intersection are given by:

$$\frac{1}{1 + x^2} = \frac{x^2}{2}$$

$$x^4 + x^2 - 2 = 0$$

$$(x^2 + 2)(x^2 - 1) = 0 \quad \text{when } x = \pm 1$$

$$A = 2\int_0^1 [f(x) - g(x)]\,dx = 2\int_0^1 \left[\frac{1}{1 + x^2} - \frac{x^2}{2}\right] dx = 2\left[\arctan x - \frac{x^3}{6}\right]_0^1 = 2\left(\frac{\pi}{4} - \frac{1}{6}\right) = \frac{\pi}{2} - \frac{1}{3} \approx 1.237$$

(c) Numerical approximation: 1.237

37. $A = \int_0^{2\pi} \left[(2 - \cos x) - \cos x\right] dx$

$= 2\int_0^{2\pi} (1 - \cos x)\, dx$

$= 2\left[x - \sin x\right]_0^{2\pi} = 4\pi \approx 12.566$

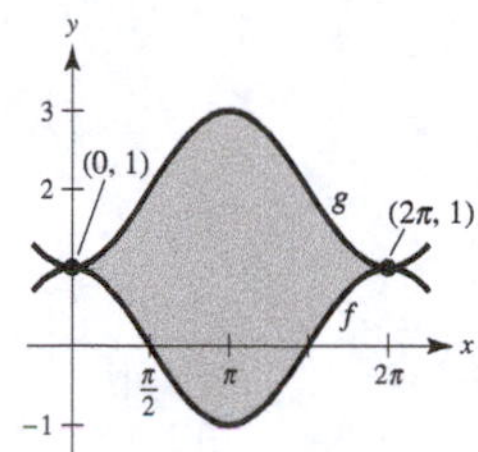

39. $A = 2\int_0^{\pi/3} \left[f(x) - g(x)\right] dx$

$= 2\int_0^{\pi/3} (2 \sin x - \tan x)\, dx$

$= 2\left[-2 \cos x + \ln\left|\cos x\right|\right]_0^{\pi/3} = 2(1 - \ln 2) \approx 0.614$

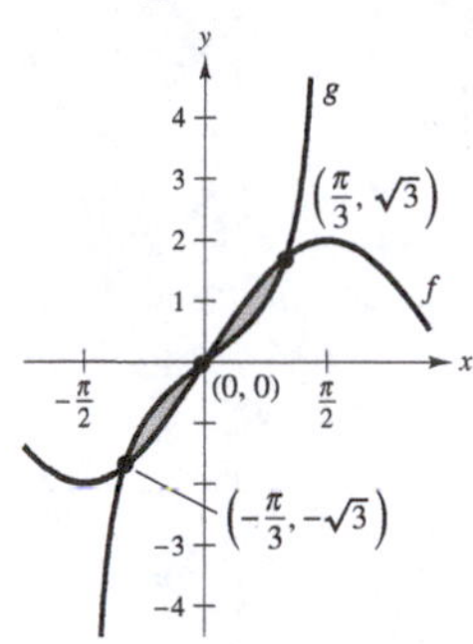

41. $A = \int_0^1 \left[xe^{-x^2} - 0\right] dx$

$= \left[-\frac{1}{2}e^{-x^2}\right]_0^1 = \frac{1}{2}\left(1 - \frac{1}{e}\right) \approx 0.316$

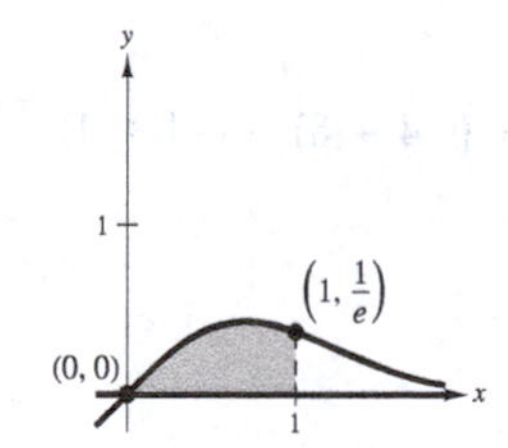

43. (a)

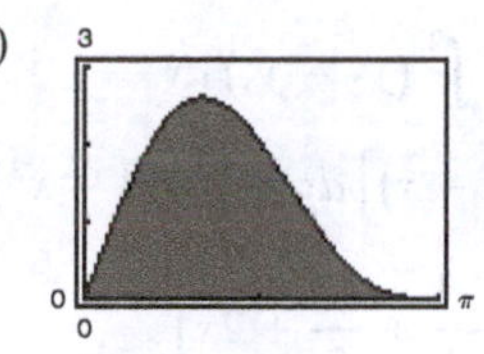

(b) $A = \int_0^{\pi} (2 \sin x + \sin 2x)\, dx$

$= \left[-2 \cos x - \frac{1}{2} \cos 2x\right]_0^{\pi}$

$= \left(2 - \frac{1}{2}\right) - \left(-2 - \frac{1}{2}\right) = 4$

(c) Numerical approximation: 4.0

45. (a)

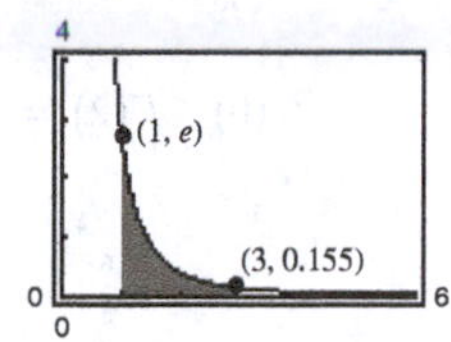

(b) $A = \int_1^3 \frac{1}{x^2} e^{1/x}\, dx$

$= \left[-e^{-1/x}\right]_1^3$

$= e - e^{1/3}$

(c) Numerical approximation: 1.323

47. (a)

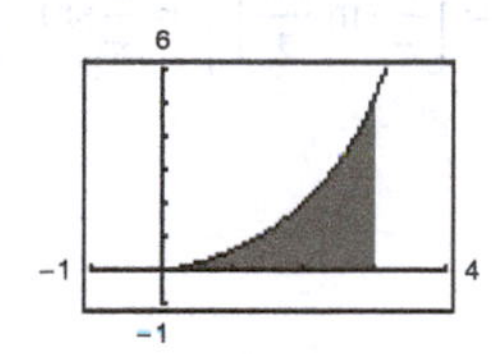

(b) The integral

$$A = \int_0^3 \sqrt{\frac{x^3}{4 - x}}\, dx$$

does not have an elementary antiderivative.

(c) $A \approx 4.7721$

49. (a)

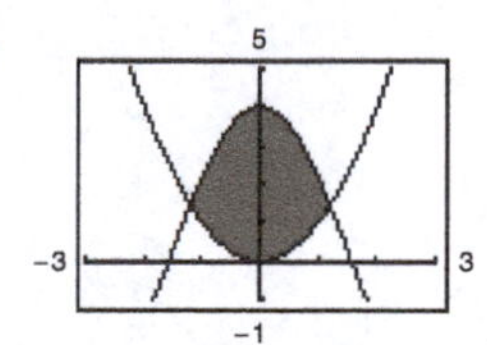

(b) The intersection points are difficult to determine by hand.

(c) Area $= \int_{-c}^{c} \left[4 \cos x - x^2\right] dx \approx 6.3043$ where $c \approx 1.201538$.

51. $A = \int_0^1 (y_1 - y_3)\,dx + \int_1^2 (y_2 - y_3)\,dx$

$= \int_0^1 \left[(x^2 + 2) - (2 - x)\right] dx + \int_1^2 \left[(4 - x^2) - (2 - x)\right] dx$

$= \left[\frac{x^3}{3} + \frac{x^2}{2}\right]_0^1 + \left[-\frac{x^3}{3} + \frac{x^2}{2} + 2x\right]_1^2$

$= \left(\frac{1}{3} + \frac{1}{2}\right) + \left(-\frac{8}{3} + 2 + 4\right) - \left(-\frac{1}{3} + \frac{1}{2} + 2\right)$

$= 2$

53. $F(x) = \int_0^x \left(\frac{1}{2}t + 1\right) dt = \left[\frac{t^2}{4} + t\right]_0^x = \frac{x^2}{4} + x$

(a) $F(0) = 0$

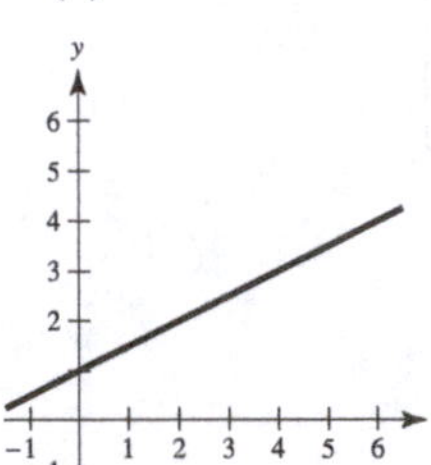

(b) $F(2) = \frac{2^2}{4} + 2 = 3$

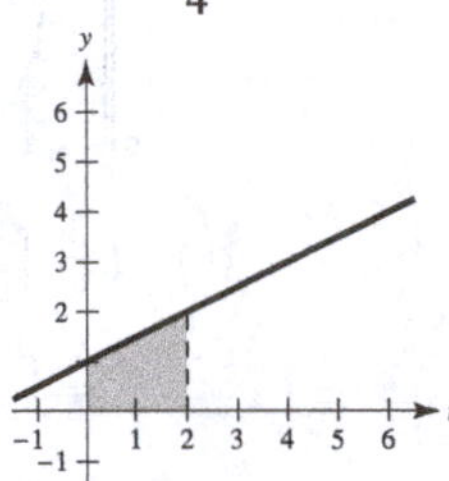

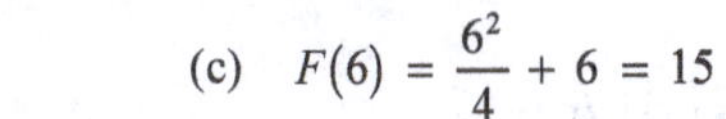

(c) $F(6) = \frac{6^2}{4} + 6 = 15$

55. $F(\alpha) = \int_{-1}^{\alpha} \cos\frac{\pi\theta}{2}\,d\theta = \left[\frac{2}{\pi}\sin\frac{\pi\theta}{2}\right]_{-1}^{\alpha} = \frac{2}{\pi}\sin\frac{\pi\alpha}{2} + \frac{2}{\pi}$

(a) $F(-1) = 0$

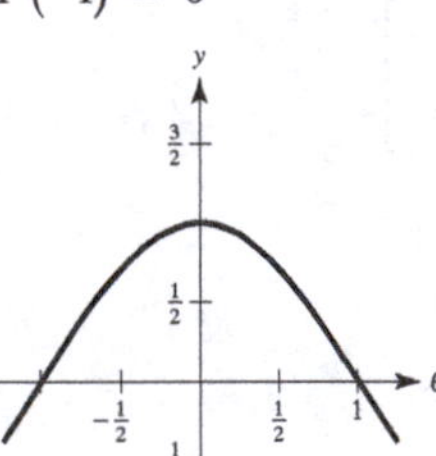

(b) $F(0) = \frac{2}{\pi} \approx 0.6366$

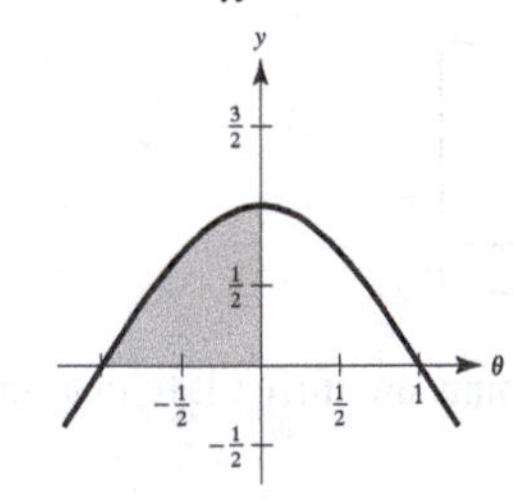

(c) $F\left(\frac{1}{2}\right) = \frac{2 + \sqrt{2}}{\pi} \approx 1.0868$

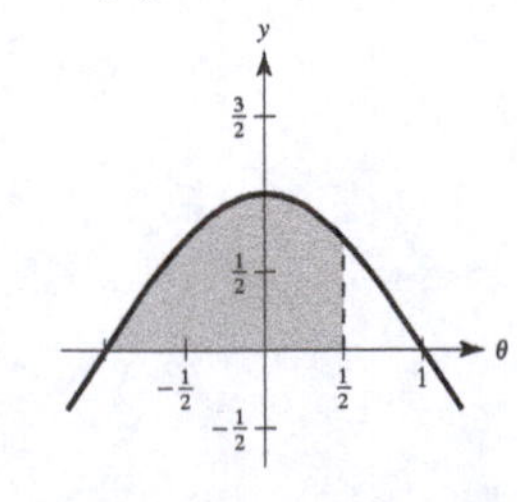

57.

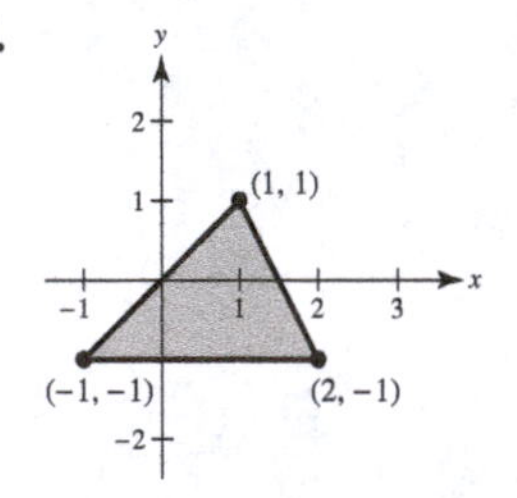

$A = \int_{-1}^{1} (x + 1)\,dx + \int_1^2 (-2x + 3 + 1)\,dx$

$= \left[\frac{x^2}{2} + x\right]_{-1}^{1} + \left[-x^2 + 4x\right]_1^2$

$= \left(\frac{1}{2} + 1\right) - \left(\frac{1}{2} - 1\right) + (-4 + 8) - (-1 + 4)$

$= 2 + 1 = 3$

59.

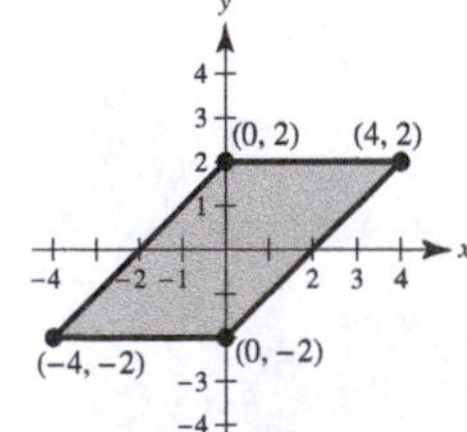

Left boundary line: $y = x + 2 \Leftrightarrow x = y - 2$

Right boundary line: $y = x - 2 \Leftrightarrow x = y + 2$

$A = \int_{-2}^{2} \left[(y + 2) - (y - 2)\right] dy$

$= \int_{-2}^{2} 4\,dy = \left[4y\right]_{-2}^{2} = 8 - (-8) = 16$

61. $f(x) = 2x^3 - 1$

$f'(x) = 6x^2$

At $(1, 1)$, $f'(1) = 6$.

Tangent line: $y - 1 = 6(x - 1)$ or $y = 6x - 5$

The tangent line intersects $f(x) = 2x^3 - 1$ at $(-2, -17)$.

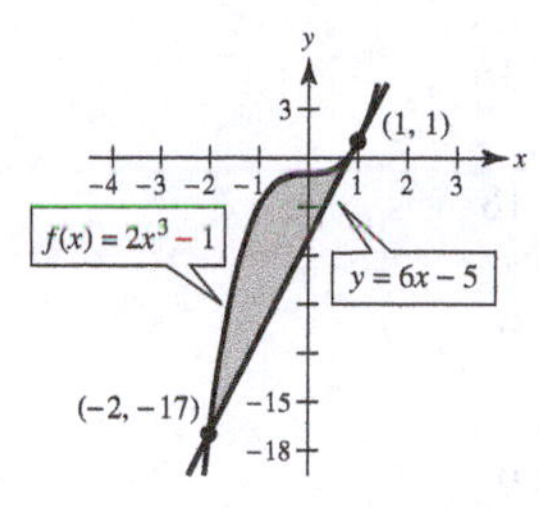

$$A = \int_{-2}^{1} \left[(2x^3 - 1) - (6x - 5)\right] dx = \left[\frac{x^4}{2} - 3x^2 + 4x\right]_{-2}^{1} = \left(\frac{1}{2} - 3 + 4\right) - (8 - 12 - 8) = \frac{27}{2}$$

63. $f(x) = \dfrac{1}{x^2 + 1}$

$f'(x) = -\dfrac{2x}{(x^2 + 1)^2}$

At $\left(1, \dfrac{1}{2}\right)$, $f'(1) = -\dfrac{1}{2}$.

Tangent line: $y - \dfrac{1}{2} = -\dfrac{1}{2}(x - 1)$ or $y = -\dfrac{1}{2}x + 1$

The tangent line intersects $f(x) = \dfrac{1}{x^2 + 1}$ at $(0, 1)$.

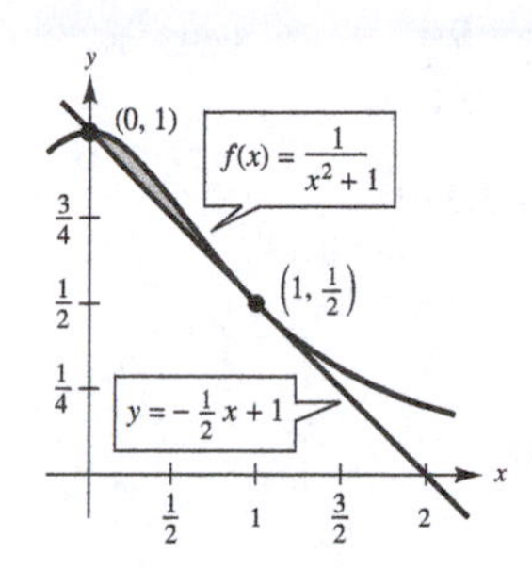

$$A = \int_0^1 \left[\frac{1}{x^2 + 1} - \left(-\frac{1}{2}x + 1\right)\right] dx = \left[\arctan x + \frac{x^2}{4} - x\right]_0^1 = \frac{\pi - 3}{4} \approx 0.0354$$

65. $x^4 - 2x^2 + 1 \le 1 - x^2$ on $[-1, 1]$

$$A = \int_{-1}^{1} \left[(1 - x^2) - (x^4 - 2x^2 + 1)\right] dx$$

$$= \int_{-1}^{1} (x^2 - x^4)\, dx$$

$$= \left[\frac{x^3}{3} - \frac{x^5}{5}\right]_{-1}^{1} = \frac{4}{15}$$

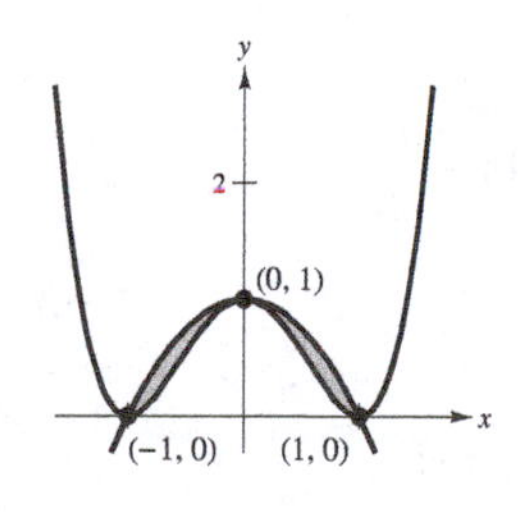

You can use a single integral because $x^4 - 2x^2 + 1 \le 1 - x^2$ on $[-1, 1]$.

67. (a) $\int_0^5 \left[v_1(t) - v_2(t)\right] dt = 10$ means that Car 1 traveled 10 more meters than Car 2 on the interval $0 \le t \le 5$.

$\int_0^{10} \left[v_1(t) - v_2(t)\right] dt = 30$ means that Car 1 traveled 30 more meters than Car 2 on the interval $0 \le t \le 10$.

$\int_{20}^{30} \left[v_1(t) - v_2(t)\right] dt = -5$ means that Car 2 traveled 5 more meters than Car 1 on the interval $20 \le t \le 30$.

(b) No, it is not possible because you do not know the initial distance between the cars.

(c) At $t = 10$, Car 1 is ahead by 30 meters.

(d) At $t = 20$, Car 1 is ahead of Car 2 by 13 meters. From part (a), at $t = 30$, Car 1 is ahead by $13 - 5 = 8$ meters.

69. $A = \int_{-3}^{3} (9 - x^2)\, dx = 36$

$$\int_{-\sqrt{9-b}}^{\sqrt{9-b}} \left[(9 - x^2) - b\right] dx = 18$$

$$\int_{0}^{\sqrt{9-b}} \left[(9 - b) - x^2\right] dx = 9$$

$$\left[(9 - b)x - \frac{x^3}{3}\right]_0^{\sqrt{9-b}} = 9$$

$$\frac{2}{3}(9 - b)^{3/2} = 9$$

$$(9 - b)^{3/2} = \frac{27}{2}$$

$$9 - b = \frac{9}{\sqrt[3]{4}}$$

$$b = 9 - \frac{9}{\sqrt[3]{4}} \approx 3.330$$

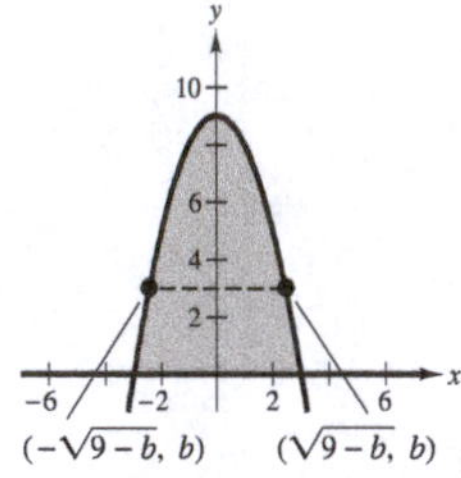

71. Area of triangle OAB is $\frac{1}{2}(4)(4) = 8$.

$$4 = \int_0^a (4 - x)\, dx = \left[4x - \frac{x^2}{2}\right]_0^a = 4a - \frac{a^2}{2}$$

$$a^2 - 8a + 8 = 0$$

$$a = 4 \pm 2\sqrt{2}$$

Because $0 < a < 4$, select $a = 4 - 2\sqrt{2} \approx 1.172$.

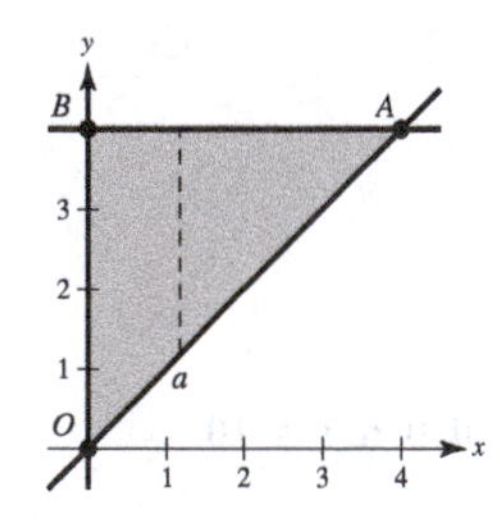

73. $\lim_{\|\Delta\| \to 0} \sum_{i=1}^{n} (x_i - x_i^2)\, \Delta x$

where $x_i = \frac{i}{n}$ and $\Delta x = \frac{1}{n}$ is the same as

$$\int_0^1 (x - x^2)\, dx = \left[\frac{x^2}{2} - \frac{x^3}{3}\right]_0^1 = \frac{1}{6}.$$

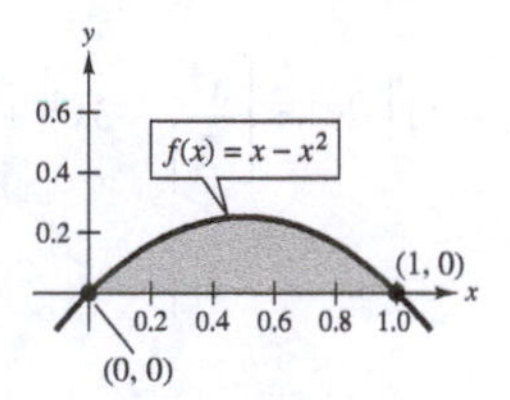

75. R_1 projects the greater revenue because the area under the curve is greater.

$$\int_0^5 \left[(7.21 + 0.58t) - (7.21 + 0.45t)\right] dt$$

$$= \int_0^5 0.13t\, dt = \left[\frac{0.13t^2}{2}\right]_0^5 = \$1.625 \text{ million}$$

77. (a) $y_1 = 0.0124x^2 - 0.385x + 7.85$

(b)

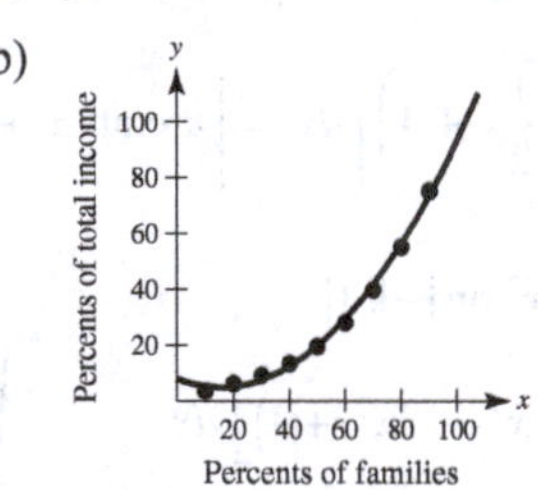

(c)

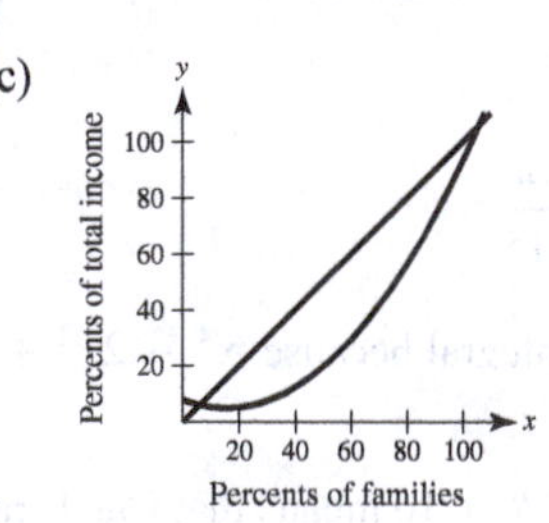

(d) Income inequality $= \int_0^{100} [x - y_1]\, dx \approx 2006.7$

79. (a) $A = 2\left[\int_0^5 \left(1 - \frac{1}{3}\sqrt{5 - x}\right) dx + \int_5^{5.5} (1 - 0)\, dx\right]$

$$= 2\left(\left[x + \frac{2}{9}(5 - x)^{3/2}\right]_0^5 + \left[x\right]_5^{5.5}\right)$$

$$= 2\left(5 - \frac{10\sqrt{5}}{9} + 5.5 - 5\right) \approx 6.031 \text{ m}^2$$

(b) $V = 2A \approx 2(6.031) \approx 12.062 \text{ m}^3$

(c) $5000\,V \approx 5000(12.062) = 60{,}310$ pounds

81. Line: $y = \dfrac{-3}{7\pi}x$

$$A = \int_0^{7\pi/6}\left[\sin x + \frac{3x}{7\pi}\right]dx$$

$$= \left[-\cos x + \frac{3x^2}{14\pi}\right]_0^{7\pi/6}$$

$$= \frac{\sqrt{3}}{2} + \frac{7\pi}{24} + 1$$

$$\approx 2.7823$$

83. True. The region has been shifted C units upward (if $C > 0$), or C units downward (if $C < 0$).

85. False. Let $f(x) = x$ and $g(x) = 2x - x^2$, f and g intersect at $(1, 1)$, the midpoint of $[0, 2]$, but

$$\int_a^b \left[f(x) - g(x)\right]dx = \int_0^2\left[x - \left(2x - x^2\right)\right]dx = \tfrac{2}{3} \neq 0.$$

87. You want to find c such that:

$$\int_0^b \left[\left(2x - 3x^3\right) - c\right]dx = 0$$

$$\left[x^2 - \tfrac{3}{4}x^4 - cx\right]_0^b = 0$$

$$b^2 - \tfrac{3}{4}b^4 - cb = 0$$

But, $c = 2b - 3b^3$ because (b, c) is on the graph.

$$b^2 - \tfrac{3}{4}b^4 - \left(2b - 3b^3\right)b = 0$$

$$4 - 3b^2 - 8 + 12b^2 = 0$$

$$9b^2 = 4$$

$$b = \tfrac{2}{3}$$

$$c = \tfrac{4}{9}$$

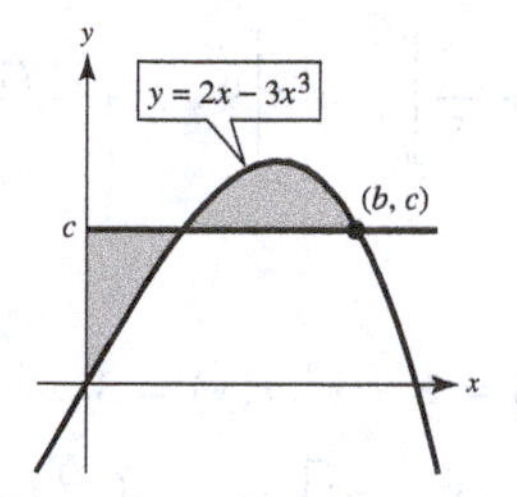

Section 7.2 Volume: The Disk Method

1. Integrate the square of the radius of the disk over the interval, and then multiply by π.

3. You need more than one integral when the solid of revolution is formed by two or more distinct solids.

5. $V = \pi\int_1^4 \left(\sqrt{x}\right)^2 dx = \pi\int_1^4 x\,dx = \pi\left[\dfrac{x^2}{2}\right]_1^4 = \dfrac{15\pi}{2}$

7. $V = \pi\int_0^1\left[\left(x^2\right)^2 - \left(x^5\right)^2\right]dx = \pi\int_0^1\left(x^4 - x^{10}\right)dx$

$$= \pi\left[\frac{x^5}{5} - \frac{x^{11}}{11}\right]_0^1 = \pi\left(\frac{1}{5} - \frac{1}{11}\right) = \frac{6\pi}{55}$$

9. $y = x^2 \Rightarrow x = \sqrt{y}$

$$V = \pi\int_0^4\left(\sqrt{y}\right)^2 dy = \pi\int_0^4 y\,dy = \pi\left[\frac{y^2}{2}\right]_0^4 = 8\pi$$

11. $y = x^{2/3} \Rightarrow x = y^{3/2}$

$$V = \pi\int_0^1\left(y^{3/2}\right)^2 dy = \pi\int_0^1 y^3\,dy = \pi\left[\frac{y^4}{4}\right]_0^1 = \frac{\pi}{4}$$

13. $y = \sqrt{x}, y = 0, x = 3$

(a) $R(x) = \sqrt{x}, r(x) = 0$

$$V = \pi\int_0^3\left(\sqrt{x}\right)^2 dx = \pi\int_0^3 x\,dx = \pi\left[\frac{x^2}{2}\right]_0^3 = \frac{9\pi}{2}$$

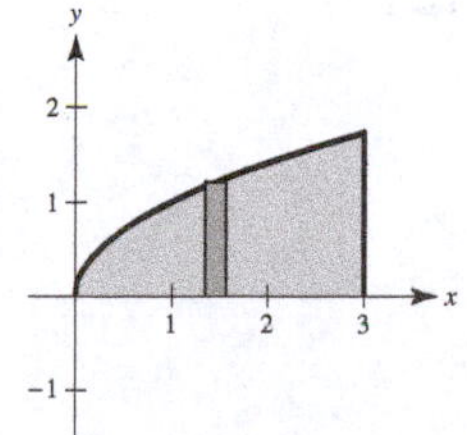

(b) $R(y) = 3, r(y) = y^2$

$$V = \pi\int_0^{\sqrt{3}}\left[3^2 - \left(y^2\right)^2\right] dy = \pi\int_0^{\sqrt{3}}\left(9 - y^4\right) dy = \pi\left[9y - \frac{y^5}{5}\right]_0^{\sqrt{3}} = \pi\left[9\sqrt{3} - \frac{9}{5}\sqrt{3}\right] = \frac{36\sqrt{3}\pi}{5}$$

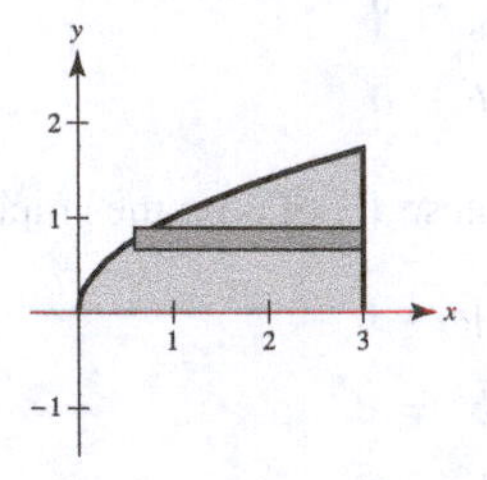

(c) $R(y) = 3 - y^2, r(y) = 0$

$$V = \pi\int_0^{\sqrt{3}}\left(3 - y^2\right)^2 dy = \pi\int_0^{\sqrt{3}}\left(9 - 6y^2 + y^4\right) dy$$

$$= \pi\left[9y - 2y^3 + \frac{y^5}{5}\right]_0^{\sqrt{3}} = \pi\left[9\sqrt{3} - 6\sqrt{3} + \frac{9\sqrt{3}}{5}\right]$$

$$= \frac{24\sqrt{3}\pi}{5}$$

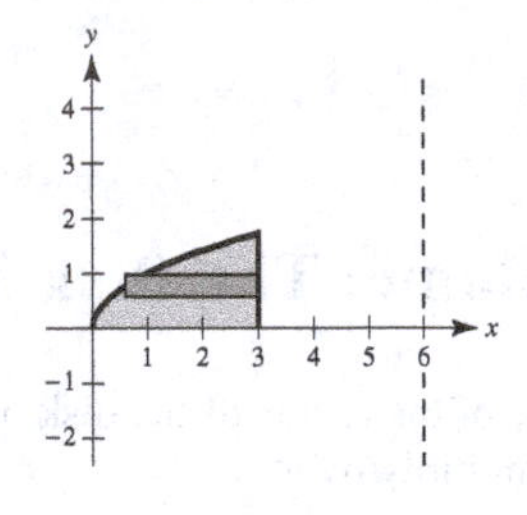

(d) $R(y) = 3 + \left(3 - y^2\right) = 6 - y^2, r(y) = 3$

$$V = \pi\int_0^{\sqrt{3}}\left[\left(6 - y^2\right)^2 - 3^2\right] dy = \pi\int_0^{\sqrt{3}}\left(y^4 - 12y^2 + 27\right) dy$$

$$= \pi\left[\frac{y^5}{5} - 4y^3 + 27y\right]_0^{\sqrt{3}} = \pi\left[\frac{9\sqrt{3}}{5} - 12\sqrt{3} + 27\sqrt{3}\right]$$

$$= \frac{84\sqrt{3}\pi}{5}$$

15. $y = x^2, y = 4x - x^2$ intersect at $(0, 0)$ and $(2, 4)$.

(a) $R(x) = 4x - x^2, r(x) = x^2$

$$V = \pi\int_0^2\left[\left(4x - x^2\right)^2 - x^4\right] dx$$

$$= \pi\int_0^2\left(16x^2 - 8x^3\right) dx$$

$$= \pi\left[\frac{16}{3}x^3 - 2x^4\right]_0^2 = \frac{32\pi}{3}$$

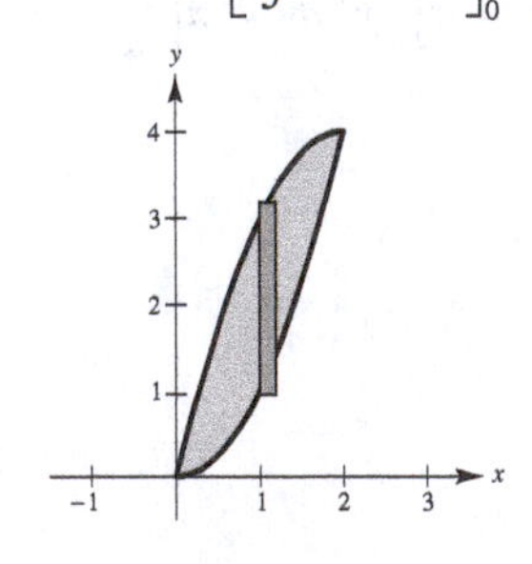

(b) $R(x) = 6 - x^2, r(x) = 6 - \left(4x - x^2\right)$

$$V = \pi\int_0^2\left[\left(6 - x^2\right)^2 - \left(6 - 4x + x^2\right)^2\right] dx$$

$$= 8\pi\int_0^2\left(x^3 - 5x^2 + 6x\right) dx$$

$$= 8\pi\left[\frac{x^4}{4} - \frac{5}{3}x^3 + 3x^2\right]_0^2 = \frac{64\pi}{3}$$

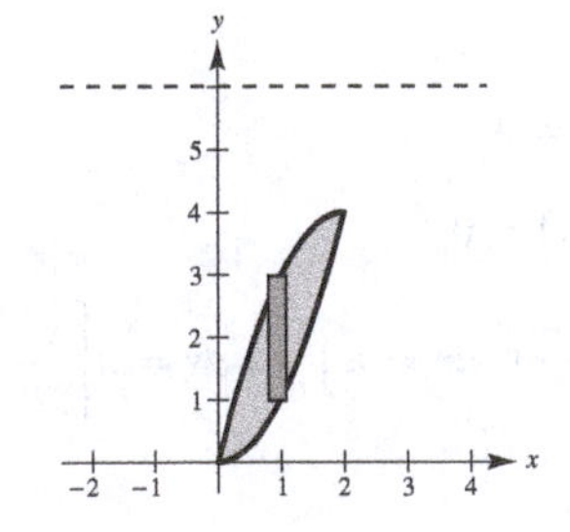

17. $R(x) = 4 - x,\ r(x) = 1$

$$V = \pi\int_0^3\left[(4 - x)^2 - (1)^2\right] dx$$
$$= \pi\int_0^3\left(x^2 - 8x + 15\right) dx$$
$$= \pi\left[\frac{x^3}{3} - 4x^2 + 15x\right]_0^3$$
$$= 18\pi$$

19. $R(x) = 4,\ r(x) = 4 - \dfrac{2}{1 + x}$

$$V = \pi\int_0^4\left[4^2 - \left(4 - \frac{2}{1 + x}\right)^2\right] dx$$
$$= \pi\int_0^4\left[\frac{16}{1 + x} - \frac{4}{(1 + x)^2}\right] dx$$
$$= \pi\left[16\ln(1 + x) + \frac{4}{1 + x}\right]_0^4$$
$$= \pi\left[\left(16\ln 5 + \frac{4}{5}\right) - 4\right]$$
$$= \pi\left(16\ln 5 - \frac{16}{5}\right)$$

21. $y = x$

$R(y) = 5 - y,\ r(y) = 0$

$$V = \pi\int_0^4(5 - y)^2\, dy$$
$$= \pi\int_0^4\left(25 - 10y + y^2\right) dy$$
$$= \pi\left[25y - 5y^2 + \frac{y^3}{3}\right]_0^4$$
$$= \pi\left[100 - 80 + \frac{64}{3}\right]$$
$$= \frac{124\pi}{3}$$

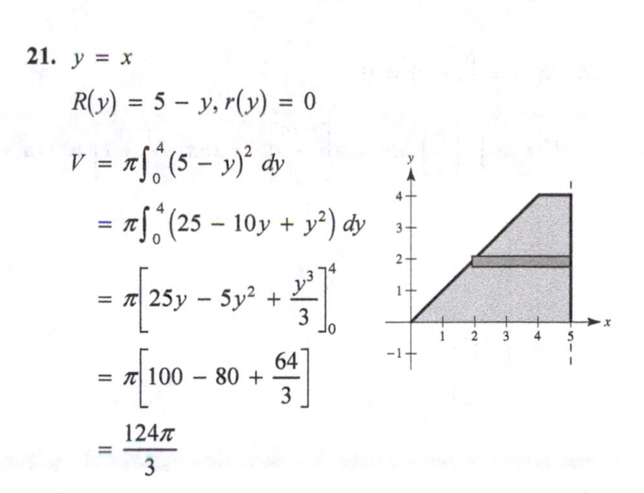

23. $x = y^2$

$R(y) = 5 - y^2,\ r(y) = 1$

$$V = \pi\int_{-2}^2\left[\left(5 - y^2\right)^2 - 1\right] dy$$
$$= 2\pi\int_0^2\left[y^4 - 10y^2 + 24\right] dy$$
$$= 2\pi\left[\frac{y^5}{5} - \frac{10y^3}{3} + 24y\right]_0^2$$
$$= 2\pi\left[\frac{32}{5} - \frac{80}{3} + 48\right] = \frac{832\pi}{15}$$

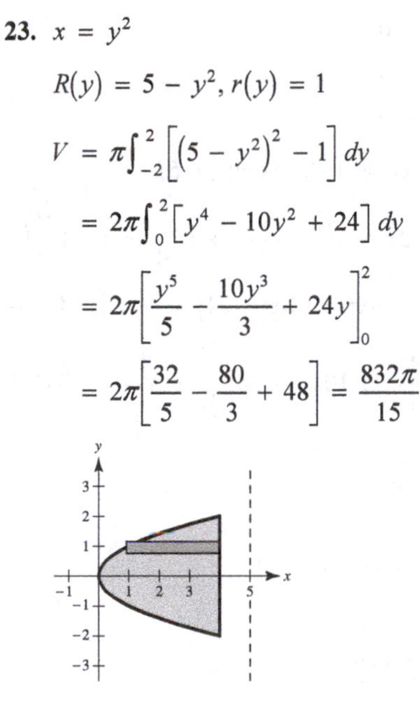

25. $R(x) = \dfrac{1}{\sqrt{3x + 5}},\ r(x) = 0$

$$V = \pi\int_0^2\left(\frac{1}{\sqrt{3x + 5}}\right)^2 dx$$
$$= \pi\int_0^2\frac{1}{3x + 5}\, dx$$
$$= \frac{\pi}{3}\left[\ln(3x + 5)\right]_0^2$$
$$= \frac{\pi}{3}(\ln 11 - \ln 5) = \frac{\pi}{3}\ln\frac{11}{5}$$

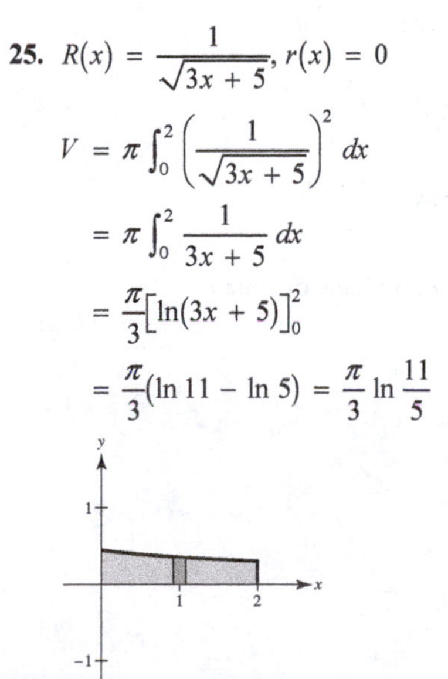

27. $R(x) = \frac{6}{x},\ r(x) = 0$

$$V = \pi\int_1^3 \left(\frac{6}{x}\right)^2 dx = \pi\left[-\frac{36}{x}\right]_1^3 = 36\pi\left[-\frac{1}{3} + 1\right] = 24\pi$$

29. $R(x) = e^{-3x},\ r(x) = 0$

$$V = \pi\int_0^2 \left(e^{-3x}\right)^2 dx$$
$$= \pi\int_0^2 e^{-6x}\,dx$$
$$= \pi\left[\left(-\frac{1}{6}\right)e^{-6x}\right]_0^2$$
$$= -\frac{1}{6}\pi\left(e^{-12} - 1\right)$$
$$= \frac{\pi\left(1 - e^{-12}\right)}{6}$$

31.

$$x^2 + 1 = -x^2 + 2x + 5$$
$$2x^2 - 2x - 4 = 0$$
$$x^2 - x - 2 = 0$$
$$(x - 2)(x + 1) = 0$$

The curves intersect at $(-1, 2)$ and $(2, 5)$.

$$V = \pi\int_0^2\left[\left(5 + 2x - x^2\right)^2 - \left(x^2 + 1\right)^2\right]dx + \pi\int_2^3\left[\left(x^2 + 1\right)^2 - \left(5 + 2x - x^2\right)^2\right]dx$$
$$= \pi\int_0^2\left(-4x^3 - 8x^2 + 20x + 24\right)dx + \pi\int_2^3\left(4x^3 + 8x^2 - 20x - 24\right)dx$$
$$= \pi\left[-x^4 - \frac{8}{3}x^3 + 10x^2 + 24x\right]_0^2 + \pi\left[x^4 + \frac{8}{3}x^3 - 10x^2 - 24x\right]_2^3$$
$$= \pi\frac{152}{3} + \pi\frac{125}{3} = \frac{277\pi}{3}$$

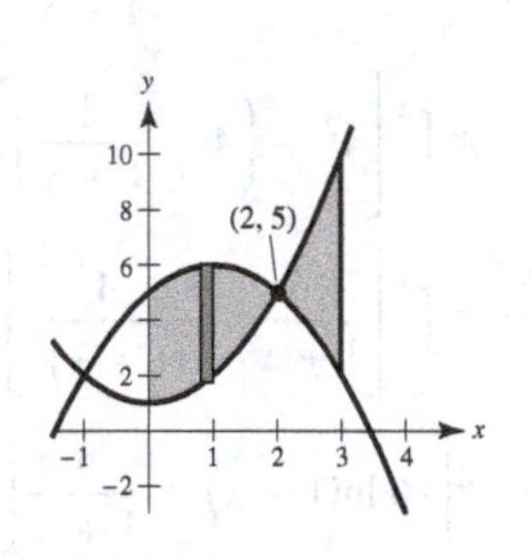

33. $y = 6 - 3x \Rightarrow x = \frac{1}{3}(6 - y)$

$$V = \pi\int_0^6\left[\frac{1}{3}(6 - y)\right]^2 dy$$
$$= \frac{\pi}{9}\int_0^6\left[36 - 12y + y^2\right]dy$$
$$= \frac{\pi}{9}\left[36y - 6y^2 + \frac{y^3}{3}\right]_0^6$$
$$= \frac{\pi}{9}\left[216 - 216 + \frac{216}{3}\right]$$
$$= 8\pi = \frac{1}{3}\pi r^2 h, \text{ Volume of cone}$$

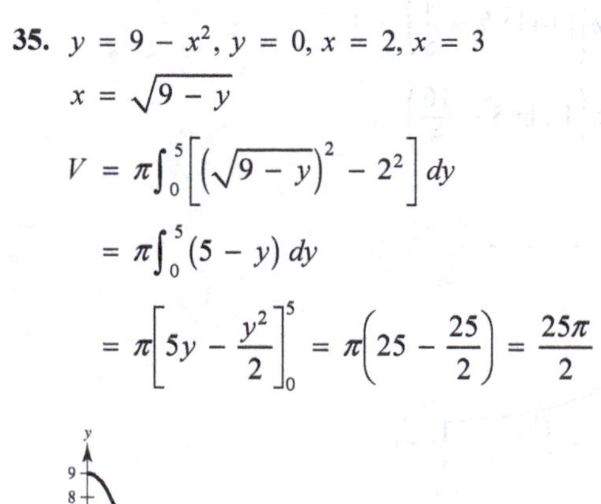

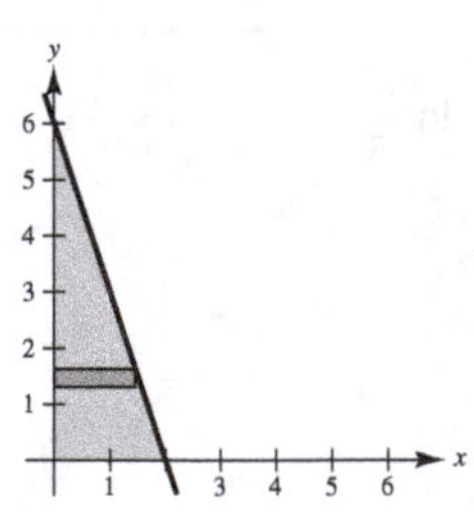

35. $y = 9 - x^2,\ y = 0,\ x = 2,\ x = 3$

$$x = \sqrt{9 - y}$$
$$V = \pi\int_0^5\left[\left(\sqrt{9 - y}\right)^2 - 2^2\right]dy$$
$$= \pi\int_0^5 (5 - y)\,dy$$
$$= \pi\left[5y - \frac{y^2}{2}\right]_0^5 = \pi\left(25 - \frac{25}{2}\right) = \frac{25\pi}{2}$$

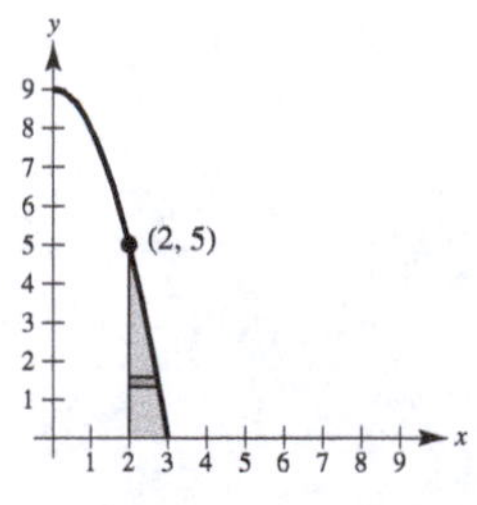

37. $V = \pi\int_0^{\pi} (\sin x)^2\, dx$

$= \pi\int_0^{\pi} \dfrac{1 - \cos 2x}{2}\, dx$

$= \dfrac{\pi}{2}\left[x - \dfrac{1}{2}\sin 2x\right]_0^{\pi} = \dfrac{\pi}{2}[\pi] = \dfrac{\pi^2}{2}$

Numerical approximation: 4.9348

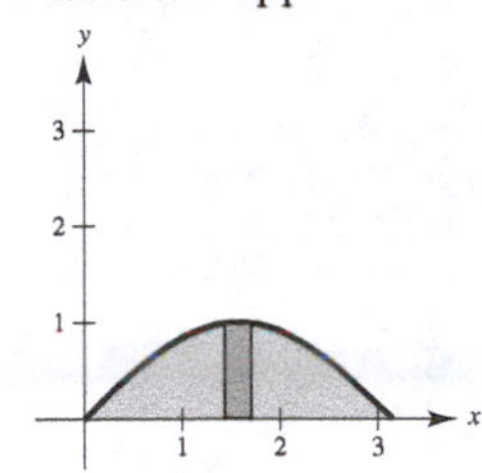

39. 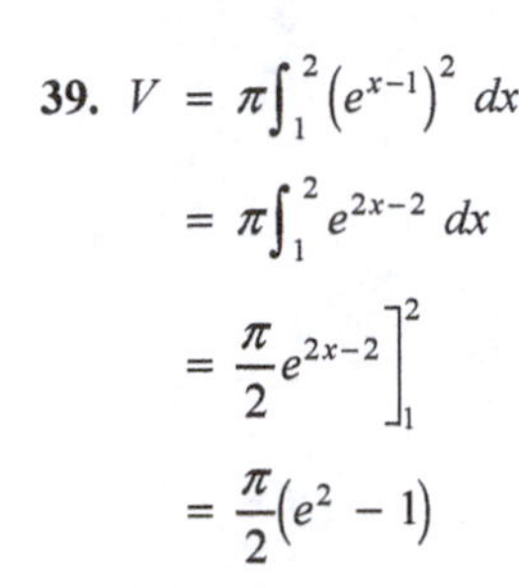

$V = \pi\int_1^2 \left(e^{x-1}\right)^2 dx$

$= \pi\int_1^2 e^{2x-2}\, dx$

$= \dfrac{\pi}{2}e^{2x-2}\Big]_1^2$

$= \dfrac{\pi}{2}\left(e^2 - 1\right)$

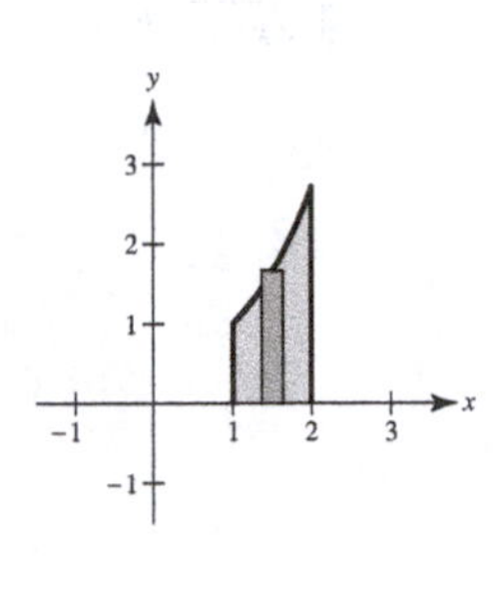

Numerical approximation: 10.0359

41. $V = \pi\int_0^1 x^2\, dx = \pi\left[\dfrac{x^3}{3}\right]_0^1 = \dfrac{\pi}{3}$

43. $V = \pi\int_0^1 y^2\, dy = \pi\dfrac{y^3}{3}\Big]_0^1 = \dfrac{\pi}{3}$

45. $V = \pi\int_0^1 \left(x^2 - x^4\right) dx$

$= \pi\left[\dfrac{x^3}{3} - \dfrac{x^5}{5}\right]_0^1$

$= \pi\left(\dfrac{1}{3} - \dfrac{1}{5}\right)$

$= \dfrac{2\pi}{15}$

47. $V = \pi\int_0^1 (1 - y)\, dy$

$= \pi\left[y - \dfrac{y^2}{2}\right]_0^1 = \pi\left(1 - \dfrac{1}{2}\right) = \dfrac{\pi}{2}$

49. $V = \pi\int_0^2 \left[e^{-x^2}\right]^2 dx \approx 1.9686$

51. $V = \pi\int_0^5 \left[2\arctan(0.2x)\right]^2 dx$

≈ 15.4115

53. (a) $\pi\int_0^{\pi/2} \sin^2 x\, dx$ represents the volume of the solid generated by revolving the region bounded by $y = \sin x$, $y = 0$, $x = 0$, $x = \pi/2$ about the x-axis.

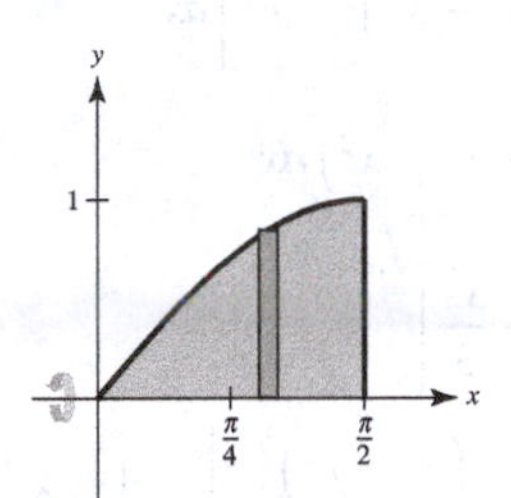

(b) $\pi\int_2^4 y^4\, dy$ represents the volume of the solid generated by revolving the region bounded by $x = y^2$, $x = 0$, $y = 2$, $y = 4$ about the y-axis.

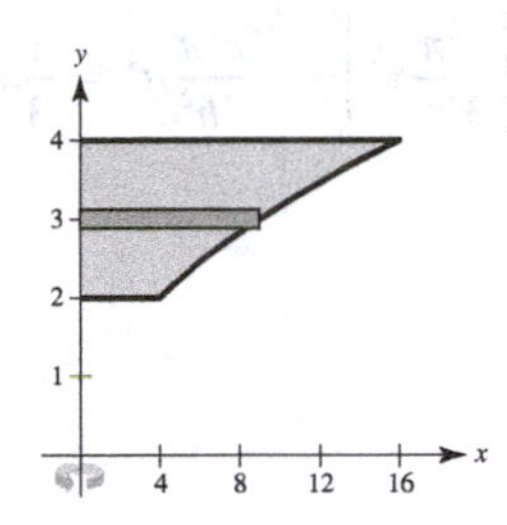

55.

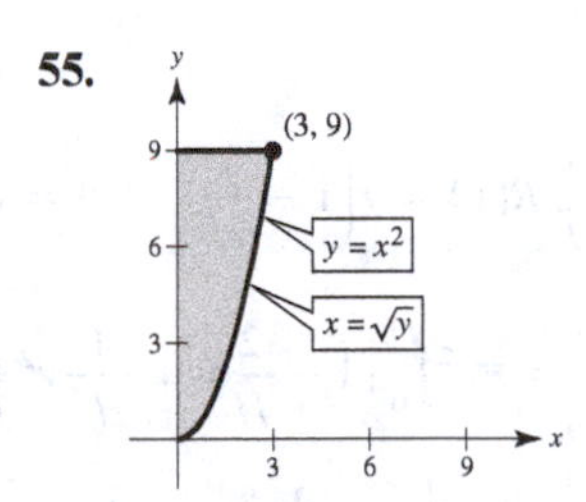

(a) Around x-axis:

$V = \pi\int_0^3 \left[9^2 - \left(x^2\right)^2\right] dx = \frac{972}{5}\pi = 194.4\pi$

(b) Around y-axis:

$V = \pi\int_0^9 \left(\sqrt{y}\right)^2 dy = \frac{81}{2}\pi = 40.5\pi$

(c) Around $x = 3$:

$V = \pi(3^2)9 - \int_0^9 \pi\left(\sqrt{y} - 3\right)^2 dy = 81\pi - \dfrac{27}{2}\pi$

$= \dfrac{135\pi}{2} \approx 67.5\pi$

So, $b < c < a$.

57. $V = \pi \int_1^3 \left(\sqrt{x}\right)^2 \, dx = \pi \dfrac{x^2}{2}\Big]_1^3 = 4\pi$

Let $1 < c < 3$.

$$\pi \int_1^c x \, dx = \pi \frac{x^2}{2}\Big]_1^c = \frac{\pi c^2}{2} - \frac{\pi}{2} = 2\pi \Rightarrow c^2 - 1 = 4 \Rightarrow c = \sqrt{5}$$

59. $V = \pi \int_{-\sqrt{R^2-r^2}}^{\sqrt{R^2-r^2}} \left[\left(\sqrt{R^2 - x^2}\right)^2 - r^2\right] dx$

$$= 2\pi \int_0^{\sqrt{R^2-r^2}} \left(R^2 - r^2 - x^2\right) dx$$

$$= 2\pi \left[\left(R^2 - r^2\right)x - \frac{x^3}{3}\right]_0^{\sqrt{R^2-r^2}}$$

$$= 2\pi \left[\left(R^2 - r^2\right)^{3/2} - \frac{\left(R^2 - r^2\right)^{3/2}}{3}\right] = \frac{4}{3}\pi\left(R^2 - r^2\right)^{3/2}$$

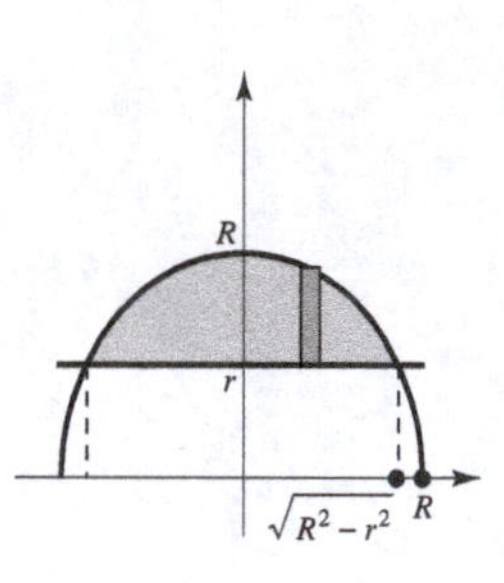

61. $R(x) = \dfrac{r}{h}x, \; r(x) = 0$

$$V = \pi \int_0^h \frac{r^2}{h^2} x^2 \, dx = \left[\frac{r^2 \pi}{3h^2} x^3\right]_0^h = \frac{r^2 \pi}{3h^2} h^3 = \frac{1}{3}\pi r^2 h$$

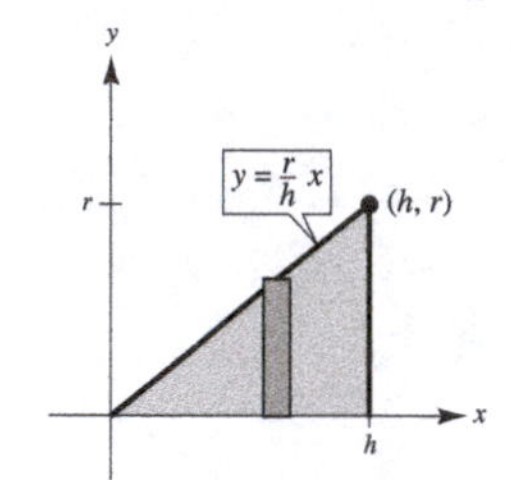

63. $x = r - \dfrac{r}{H}y = r\left(1 - \dfrac{y}{H}\right), \; R(y) = r\left(1 - \dfrac{y}{H}\right), \; r(y) = 0$

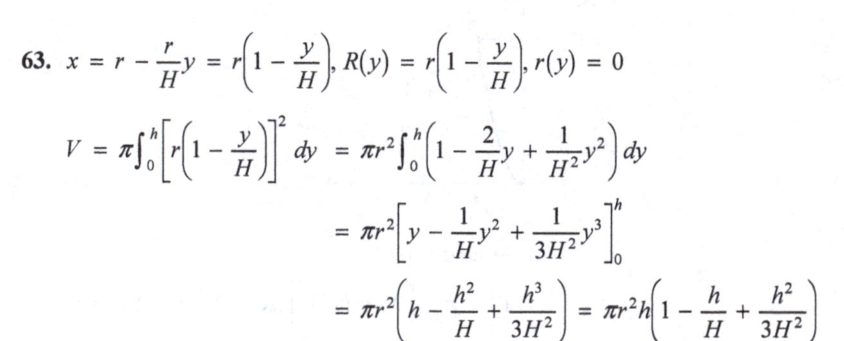

$$V = \pi \int_0^h \left[r\left(1 - \frac{y}{H}\right)\right]^2 dy = \pi r^2 \int_0^h \left(1 - \frac{2}{H}y + \frac{1}{H^2}y^2\right) dy$$

$$= \pi r^2 \left[y - \frac{1}{H}y^2 + \frac{1}{3H^2}y^3\right]_0^h$$

$$= \pi r^2 \left(h - \frac{h^2}{H} + \frac{h^3}{3H^2}\right) = \pi r^2 h\left(1 - \frac{h}{H} + \frac{h^2}{3H^2}\right)$$

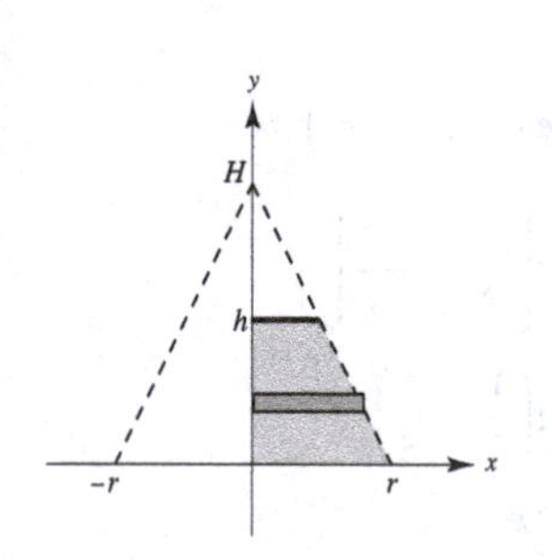

65.

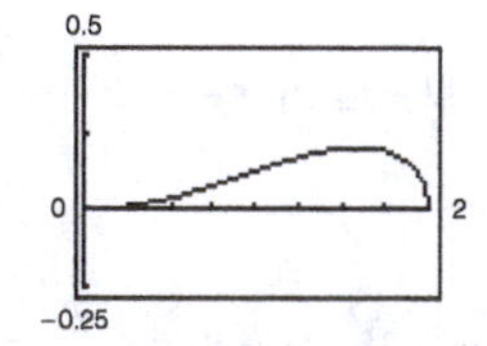

$$V = \pi \int_0^2 \left(\frac{1}{8} x^2 \sqrt{2 - x}\right)^2 dx = \frac{\pi}{64} \int_0^2 x^4 (2 - x) \, dx = \frac{\pi}{64}\left[\frac{2x^5}{5} - \frac{x^6}{6}\right]_0^2 = \frac{\pi}{30} \text{ m}^3$$

67. (a) $R(x) = \frac{3}{5}\sqrt{25 - x^2},\ r(x) = 0$

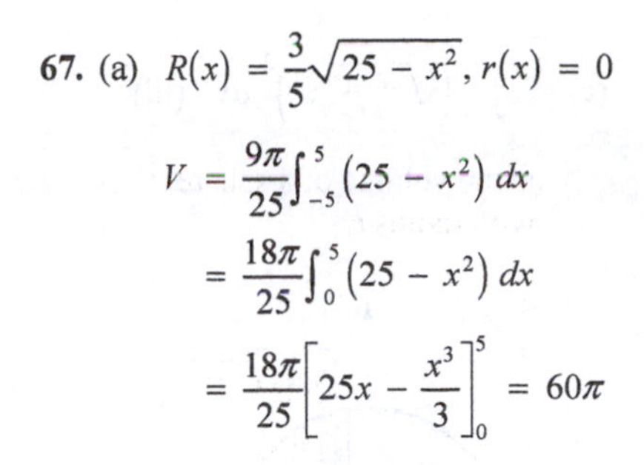

$$V = \frac{9\pi}{25}\int_{-5}^{5}\left(25 - x^2\right)dx$$
$$= \frac{18\pi}{25}\int_{0}^{5}\left(25 - x^2\right)dx$$
$$= \frac{18\pi}{25}\left[25x - \frac{x^3}{3}\right]_0^5 = 60\pi$$

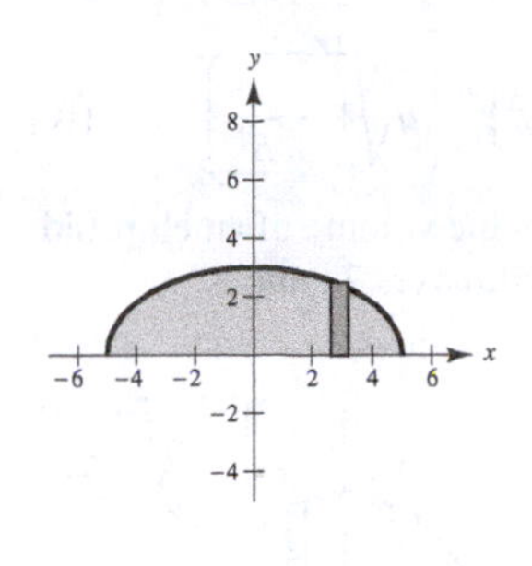

(b) $R(y) = \frac{5}{3}\sqrt{9 - y^2},\ r(y) = 0,\ x \geq 0$

$$V = \frac{25\pi}{9}\int_{0}^{3}\left(9 - y^2\right)dy$$
$$= \frac{25\pi}{9}\left[9y - \frac{y^3}{3}\right]_0^3$$
$$= 50\pi$$

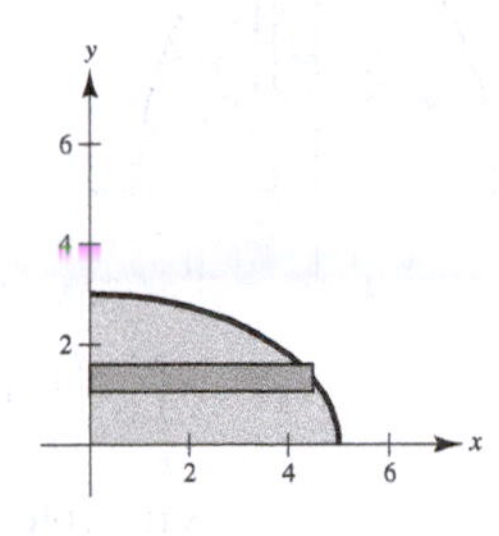

69. (a) First find where $y = b$ intersects the parabola:

$$b = 4 - \frac{x^2}{4}$$
$$x^2 = 16 - 4b = 4(4 - b)$$
$$x = 2\sqrt{4 - b}$$

$$V = \int_0^{2\sqrt{4-b}} \pi\left[4 - \frac{x^2}{4} - b\right]^2 dx + \int_{2\sqrt{4-b}}^{4} \pi\left[b - 4 + \frac{x^2}{4}\right]^2 dx$$
$$= \int_0^4 \pi\left[4 - \frac{x^2}{4} - b\right]^2 dx$$
$$= \pi\int_0^4\left[\frac{x^4}{16} - 2x^2 + \frac{bx^2}{2} + b^2 - 8b + 16\right]dx$$
$$= \pi\left[\frac{x^5}{80} - \frac{2x^3}{3} + \frac{bx^3}{6} + b^2x - 8bx + 16x\right]_0^4$$
$$= \pi\left(\frac{64}{5} - \frac{128}{3} + \frac{32}{3}b + 4b^2 - 32b + 64\right) = \pi\left(4b^2 - \frac{64}{3}b + \frac{512}{15}\right)$$

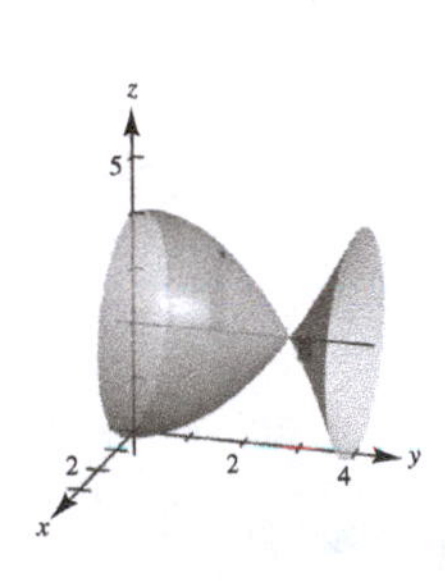

(b) Graph of $V(b) = \pi\left(4b^2 - \frac{64}{3}b + \frac{512}{15}\right)$

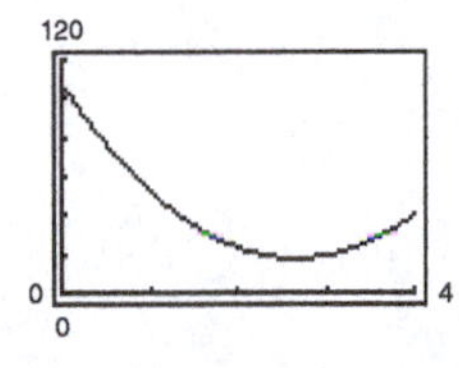

Minimum volume is 17.87 for $b = 2.67$.

(c) $V'(b) = \pi\left(8b - \frac{64}{3}\right) = 0 \Rightarrow b = \frac{64/3}{8} = \frac{8}{3} = 2\frac{2}{3}$

$V''(b) = 8\pi > 0 \Rightarrow b = \frac{8}{3}$ is a relative minimum.

71. (a) $\pi\int_0^h r^2\,dx$ (ii)

is the volume of a right circular cylinder with radius r and height h.

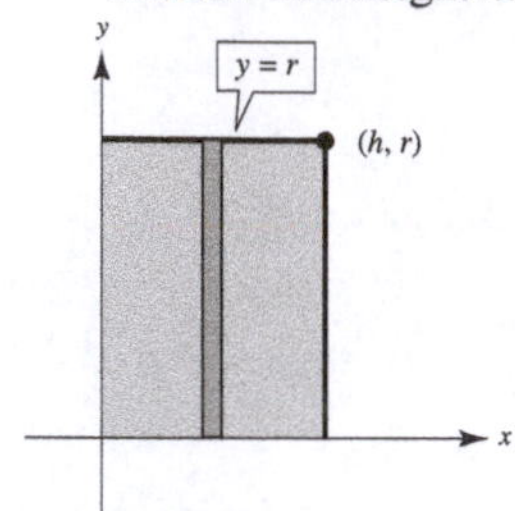

(b) $\pi\int_{-b}^{b}\left(a\sqrt{1-\dfrac{x^2}{b^2}}\right)^2 dx$ (iv)

is the volume of an ellipsoid with axes $2a$ and $2b$.

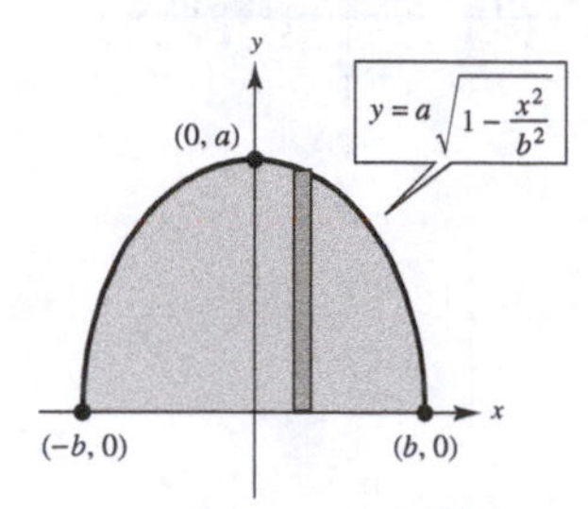

(c) $\pi\int_{-r}^{r}\left(\sqrt{r^2-x^2}\right)^2 dx$ (iii)

is the volume of a sphere with radius r.

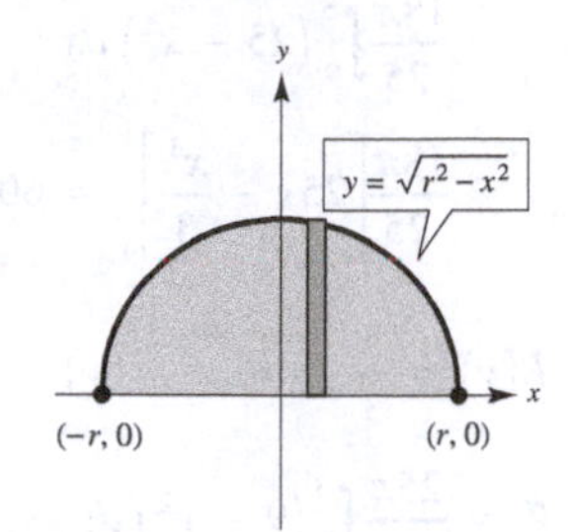

(d) $\pi\int_0^h\left(\dfrac{rx}{h}\right)^2 dx$ (i)

is the volume of a right circular cone with the radius of the base as r and height h.

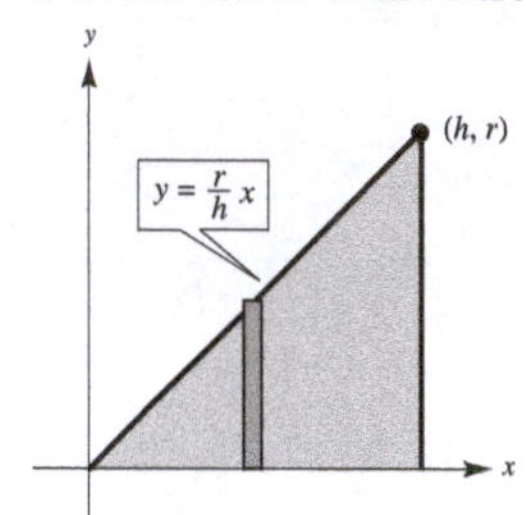

(e) $\pi\int_{-r}^{r}\left[\left(R+\sqrt{r^2-x^2}\right)^2-\left(R-\sqrt{r^2-x^2}\right)^2\right]dx$ (v)

is the volume of a torus with the radius of its circular cross section as r and the distance from the axis of the torus to the center of its cross section as R.

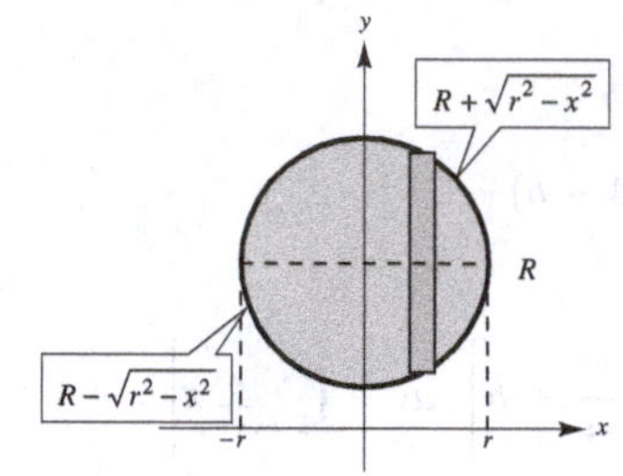

73.

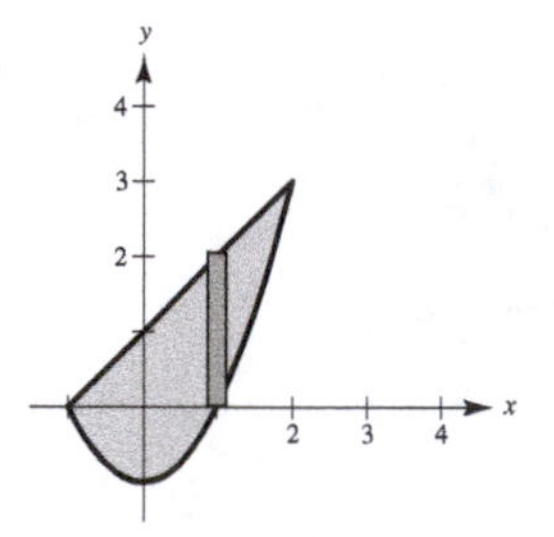

Base of cross section $= (x+1)-(x^2-1) = 2+x-x^2$

(a) $A(x) = b^2 = \left(2+x-x^2\right)^2 = 4+4x-3x^2-2x^3+x^4$

$$V = \int_{-1}^{2}\left(4+4x-3x^2-2x^3+x^4\right)dx = \left[4x+2x^2-x^3-\frac{1}{2}x^4+\frac{1}{5}x^5\right]_{-1}^{2} = \frac{81}{10}$$

$2+x-x^2$

$2+x-x^2$

(b) $A(x) = bh = \left(2+x-x^2\right)1$

$$V = \int_{-1}^{2}\left(2+x-x^2\right)dx = \left[2x+\frac{x^2}{2}-\frac{x^3}{3}\right]_{-1}^{2} = \frac{9}{2}$$

1

$2+x-x^2$

75. The cross sections are squares. By symmetry, you can set up an integral for an eighth of the volume and multiply by 8.

$$A(y) = b^2 = \left(\sqrt{r^2 - y^2}\right)^2$$

$$V = 8\int_0^r (r^2 - y^2)\,dy$$

$$= 8\left[r^2y - \frac{1}{3}y^3\right]_0^r$$

$$= \frac{16}{3}r^3$$

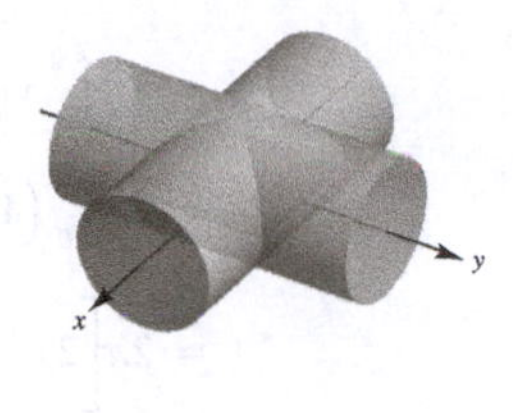

77. (a) Because the cross sections are isosceles right triangles:

$$A(x) = \frac{1}{2}bh = \frac{1}{2}\left(\sqrt{r^2 - y^2}\right)\left(\sqrt{r^2 - y^2}\right) = \frac{1}{2}(r^2 - y^2)$$

$$V = \frac{1}{2}\int_{-r}^{r}(r^2 - y^2)\,dy = \int_0^r (r^2 - y^2)\,dy = \left[r^2y - \frac{y^3}{3}\right]_0^r = \frac{2}{3}r^3$$

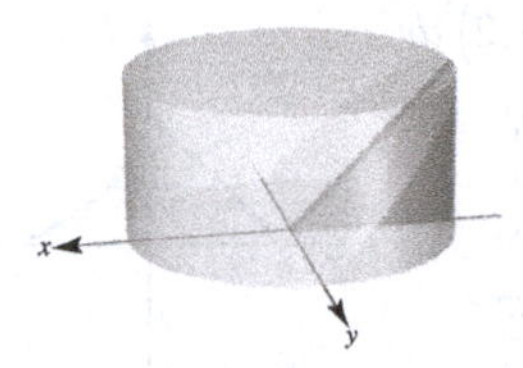

(b) $A(x) = \frac{1}{2}bh = \frac{1}{2}\sqrt{r^2 - y^2}\left(\sqrt{r^2 - y^2}\tan\theta\right) = \frac{\tan\theta}{2}(r^2 - y^2)$

$$V = \frac{\tan\theta}{2}\int_{-r}^{r}(r^2 - y^2)\,dy = \tan\theta\int_0^r (r^2 - y^2)\,dy = \tan\theta\left[r^2y - \frac{y^3}{3}\right]_0^r = \frac{2}{3}r^3\tan\theta$$

As $\theta \to 90°$, $V \to \infty$.

Section 7.3 Volume: The Shell Method

1. Determine the distance from the center of a representative rectangle to the axis of revolution, and find the height of the rectangle. Then use the formula

$$V = 2\pi\int_c^d p(y)h(y)\,dy$$

for a horizontal axis of revolution.

Use

$$V = 2\pi\int_a^b p(x)h(x)\,dx$$

for a vertical axis of revolution.

3. $p(x) = x$, $h(x) = x$

$$V = 2\pi\int_0^2 x(x)\,dx = \left[\frac{2\pi x^3}{3}\right]_0^2 = \frac{16\pi}{3}$$

5. $p(x) = x$, $h(x) = \sqrt{x}$

$$V = 2\pi\int_0^4 x\sqrt{x}\,dx = 2\pi\int_0^4 x^{3/2}\,dx = \left[\frac{4\pi}{5}x^{5/2}\right]_0^4 = \frac{128\pi}{5}$$

7. $p(x) = x$, $h(x) = \frac{1}{4}x^2$

$$V = 2\pi\int_0^4 x\left(\frac{1}{4}x^2\right)dx$$

$$= \frac{\pi}{2}\left[\frac{x^4}{4}\right]_0^4$$

$$= 32\pi$$

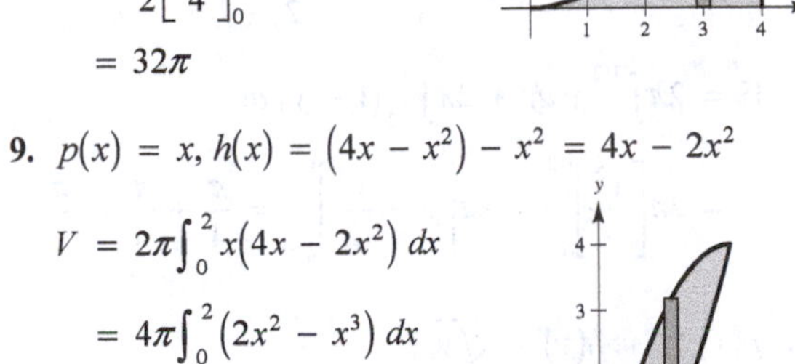

9. $p(x) = x$, $h(x) = (4x - x^2) - x^2 = 4x - 2x^2$

$$V = 2\pi\int_0^2 x(4x - 2x^2)\,dx$$

$$= 4\pi\int_0^2 (2x^2 - x^3)\,dx$$

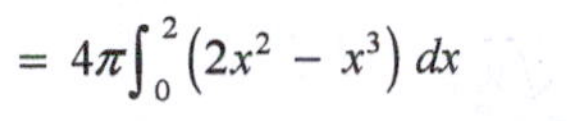

$$= 4\pi\left[\frac{2}{3}x^3 - \frac{1}{4}x^4\right]_0^2 = \frac{16\pi}{3}$$

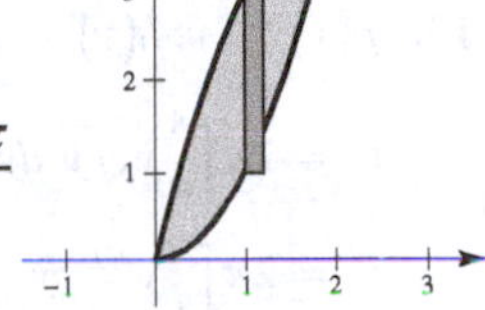

11. $p(x) = x, h(x) = \sqrt{2x - 5}$

$$V = 2\pi \int_{5/2}^{4} x\sqrt{2x - 5}\, dx$$

Let $u = 2x - 5, x = \frac{1}{2}(u + 5), du = 2\, dx.$

When $x = \frac{5}{2}, u = 0.$

When $x = 4, u = 3.$

$$V = 2\pi \int_0^3 \frac{1}{2}(u + 5)u^{1/2} \frac{du}{2}$$

$$= \frac{\pi}{2} \int_0^3 \left(u^{3/2} + 5u^{1/2}\right) du$$

$$= \frac{\pi}{2}\left[\frac{2}{5}u^{5/2} + \frac{10}{3}u^{3/2}\right]_0^3$$

$$= \frac{\pi}{2}\left[\frac{2}{5}\left(3^{5/2}\right) + \frac{10}{3}\left(3^{3/2}\right)\right]$$

$$= \frac{\pi}{2}\left[\frac{18}{5}\sqrt{3} + 10\sqrt{3}\right]$$

$$= 34\pi\frac{\sqrt{3}}{5}$$

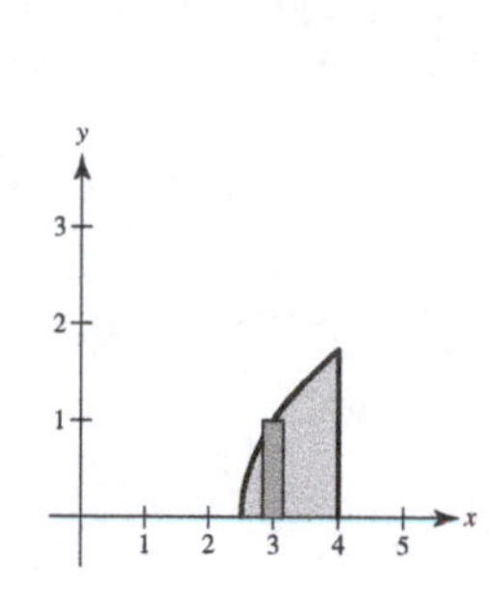

13. $p(y) = y, h(y) = 2 - y$

$$V = 2\pi\int_0^2 y(2 - y)\, dy$$

$$= 2\pi\int_0^2 \left(2y - y^2\right) dy$$

$$= 2\pi\left[y^2 - \frac{y^3}{3}\right]_0^2 = \frac{8\pi}{3}$$

15. $p(y) = y$ and $h(y) = 1$ if $0 \le y < \frac{1}{2}.$

$p(y) = y$ and $h(y) = \frac{1}{y} - 1$ if $\frac{1}{2} \le y \le 1.$

$$V = 2\pi\int_0^{1/2} y\, dy + 2\pi\int_{1/2}^{1} (1 - y)\, dy$$

$$= 2\pi\left[\frac{y^2}{2}\right]_0^{1/2} + 2\pi\left[y - \frac{y^2}{2}\right]_{1/2}^{1} = \frac{\pi}{4} + \frac{\pi}{4} = \frac{\pi}{2}$$

17. $p(y) = y, h(y) = \sqrt[3]{y}$

$$V = 2\pi\int_0^8 y\sqrt[3]{y}\, dy$$

$$= 2\pi\int_0^8 y^{4/3}\, dy$$

$$= \left[2\pi\left(\frac{3}{7}\right)y^{7/3}\right]_0^8$$

$$= \frac{6\pi}{7}\left(2^7\right) = \frac{768\pi}{7}$$

19. $p(y) = y, h(y) = (4 - y) - (y) = 4 - 2y$

$$V = 2\pi\int_0^2 y(4 - 2y)\, dy$$

$$= 2\pi\int_0^2 \left(4y - 2y^2\right) dy$$

$$= 2\pi\left[2y^2 - \frac{2}{3}y^3\right]_0^2$$

$$= 2\pi\left(8 - \frac{16}{3}\right) = \frac{16\pi}{3}$$

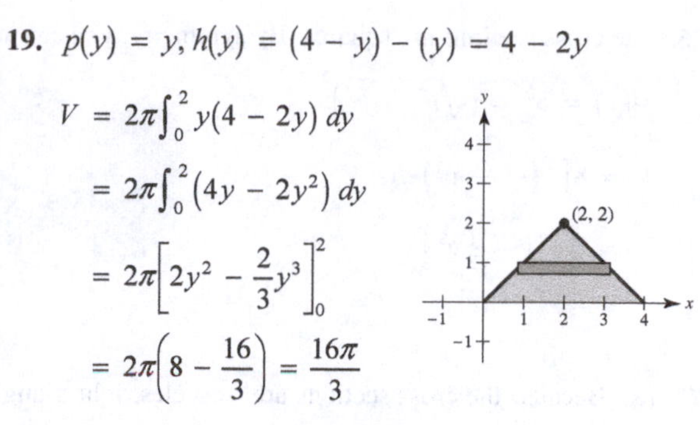

21. $y = 1 - \sqrt{x} \Rightarrow \sqrt{x} = 1 - y \Rightarrow x = (1 - y)^2$

$p(y) = y, h(y) = (1 - y)^2 - (y - 1) = y^2 - 3y + 2$

$$V = 2\pi\int_0^1 y\left(y^2 - 3y + 2\right) dy$$

$$= 2\pi\left[\frac{y^4}{4} - y^3 + y^2\right]_0^1$$

$$= 2\pi\left[\frac{1}{4} - 1 + 1\right] = \frac{\pi}{2}$$

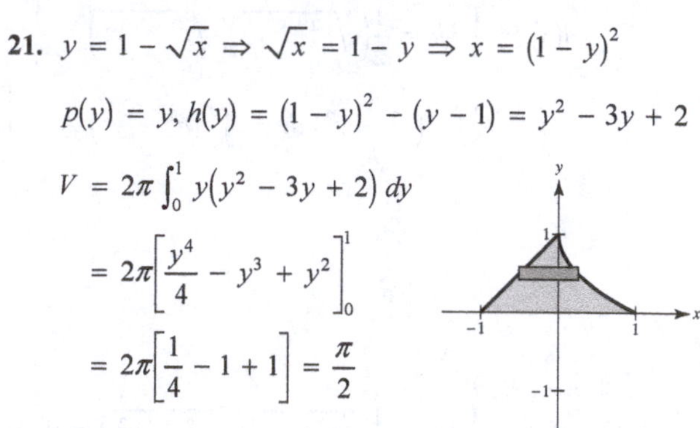

23. $p(x) = 4 - x, h(x) = 2x - x^2$

$$V = 2\pi\int_0^2 (4 - x)\left(2x - x^2\right) dx$$

$$= 2\pi\int_0^2 \left(8x - 6x^2 + x^3\right) dx$$

$$= 2\pi\left[4x^2 - 2x^3 + \frac{x^4}{4}\right]_0^2$$

$$= 2\pi[16 - 16 + 4] = 8\pi$$

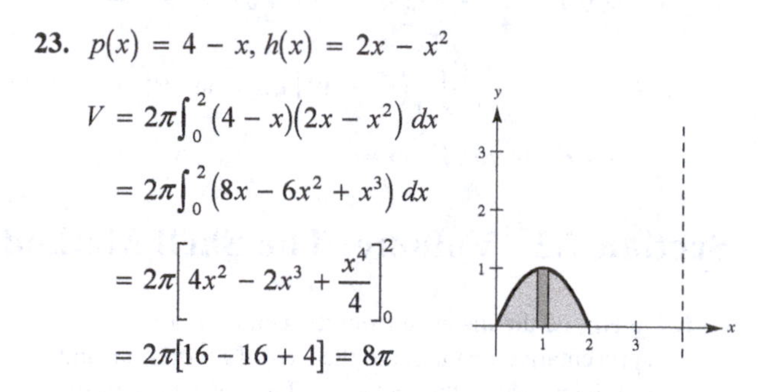

25. $3x - x^2 = x^2 \Rightarrow 3x = 2x^2 \Rightarrow x = 0, \frac{3}{2}$

$p(x) = 2 - x, h(x) = \left(3x - x^2\right) - x^2 = 3x - 2x^2$

$$V = 2\pi\int_0^{3/2} (2 - x)\left(3x - 2x^2\right) dx$$

$$= 2\pi\int_0^{3/2} \left(2x^3 - 7x^2 + 6x\right) dx$$

$$= 2\pi\left[\frac{x^4}{2} - \frac{7x^3}{3} + 3x^2\right]_0^{3/2} = \frac{45\pi}{16}$$

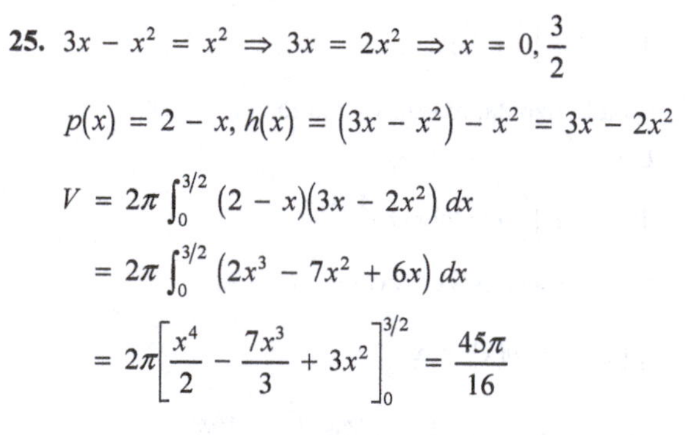

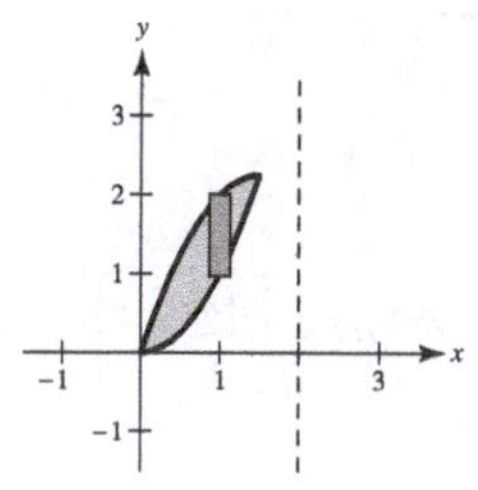

27. The shell method would be easier:

$$V = 2\pi\int_0^4 \left[4 - (y-2)^2\right]y\,dy$$

Using the disk method:

$$V = \pi\int_0^4 \left[\left(2 + \sqrt{4-x}\right)^2 - \left(2 - \sqrt{4-x}\right)^2\right]dx$$

$$\left[\textbf{Note: } V = \frac{128\pi}{3}\right]$$

29. (a) **Disk**

$R(x) = x^3, r(x) = 0$

$$V = \pi\int_0^2 x^6\,dx = \pi\left[\frac{x^7}{7}\right]_0^2 = \frac{128\pi}{7}$$

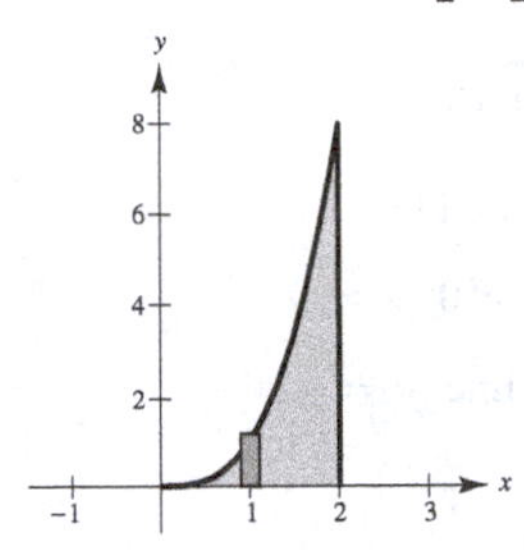

(b) **Shell**

$p(x) = x, h(x) = x^3$

$$V = 2\pi\int_0^2 x^4\,dx = 2\pi\left[\frac{x^5}{5}\right]_0^2 = \frac{64\pi}{5}$$

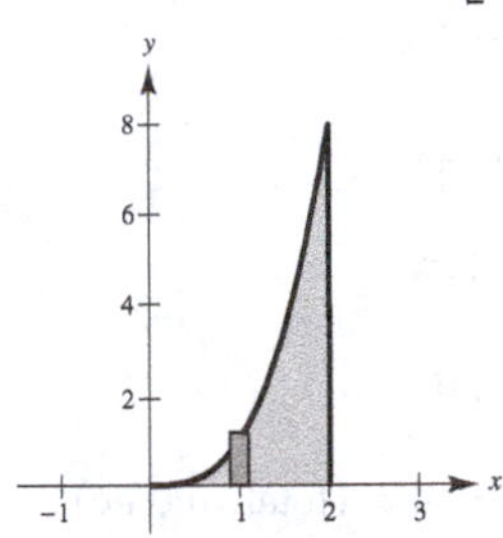

(c) **Shell**

$p(x) = 4 - x, h(x) = x^3$

$$V = 2\pi\int_0^2 (4-x)x^3\,dx$$

$$= 2\pi\int_0^2 \left(4x^3 - x^4\right)dx$$

$$= 2\pi\left[x^4 - \frac{1}{5}x^5\right]_0^2 = \frac{96\pi}{5}$$

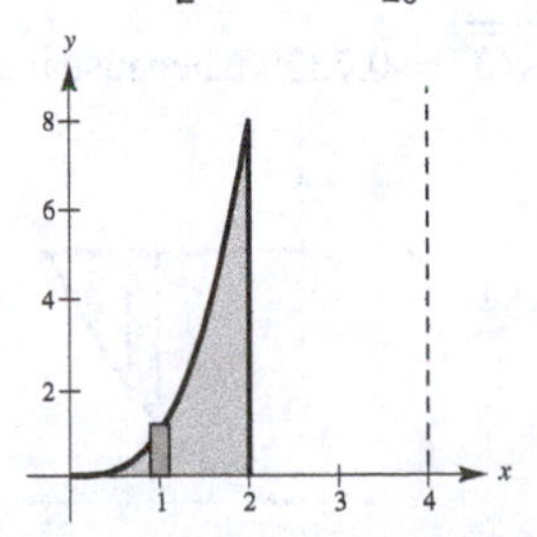

31. (a) **Shell**

$p(y) = y, h(y) = \left(a^{1/2} - y^{1/2}\right)^2$

$$V = 2\pi\int_0^a y\left(a - 2a^{1/2}y^{1/2} + y\right)dy$$

$$= 2\pi\int_0^a \left(ay - 2a^{1/2}y^{3/2} + y^2\right)dy$$

$$= 2\pi\left[\frac{a}{2}y^2 - \frac{4a^{1/2}}{5}y^{5/2} + \frac{y^3}{3}\right]_0^a$$

$$= 2\pi\left(\frac{a^3}{2} - \frac{4a^3}{5} + \frac{a^3}{3}\right) = \frac{\pi a^3}{15}$$

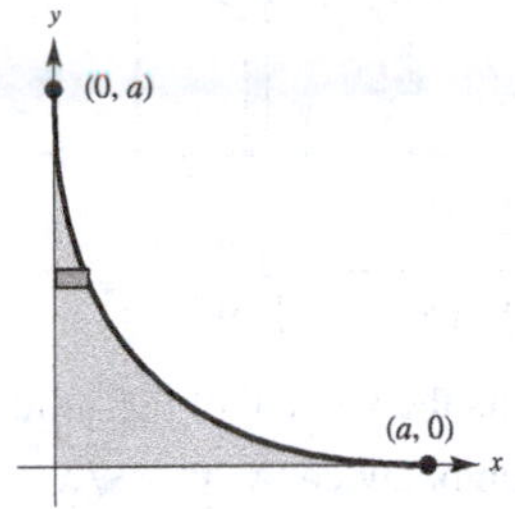

(b) Same as part (a) by symmetry

(c) **Shell**

$p(x) = a - x, h(x) = \left(a^{1/2} - x^{1/2}\right)^2$

$$V = 2\pi\int_0^a (a-x)\left(a^{1/2} - x^{1/2}\right)^2 dx$$

$$= 2\pi\int_0^a \left(a^2 - 2a^{3/2}x^{1/2} + 2a^{1/2}x^{3/2} - x^2\right)dx$$

$$= 2\pi\left[a^2x - \frac{4}{3}a^{3/2}x^{3/2} + \frac{4}{5}a^{1/2}x^{5/2} - \frac{1}{3}x^3\right]_0^a$$

$$= \frac{4\pi a^3}{15}$$

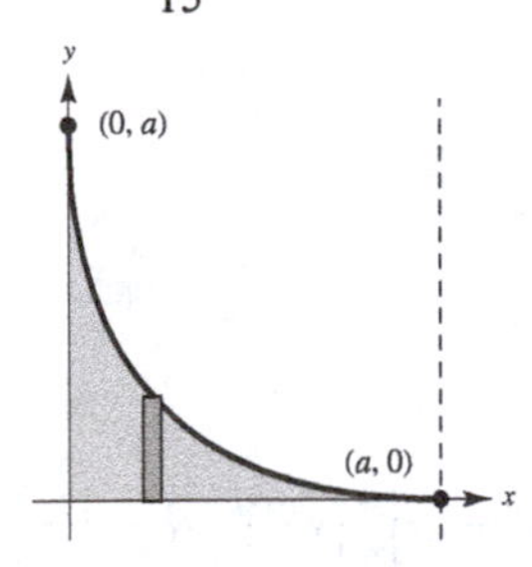

33. (a)

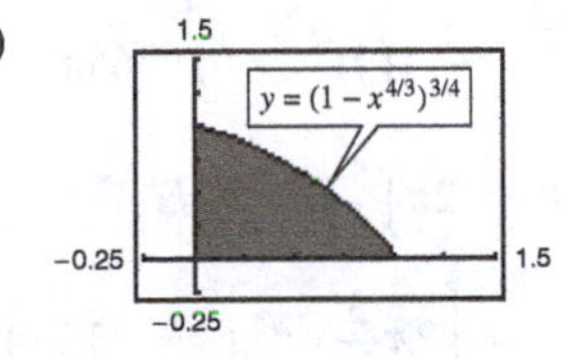

(b) $x^{4/3} + y^{4/3} = 1, x = 0, y = 0$

$$y = \left(1 - x^{4/3}\right)^{3/4}$$

$$V = 2\pi\int_0^1 x\left(1 - x^{4/3}\right)^{3/4} dx \approx 1.5056$$

35. (a)

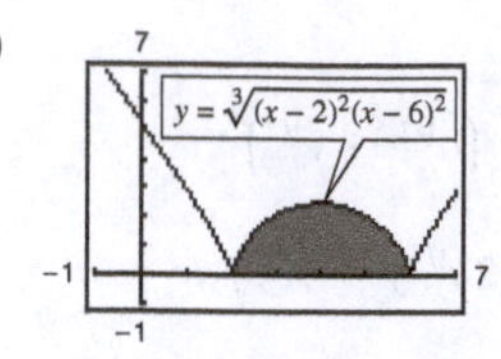

(b) $V = 2\pi\int_2^6 x\sqrt[3]{(x-2)^2(x-6)^2}\,dx \approx 187.249$

37. (a) radius $= k$

height $= b$

(b) radius $= b$

height $= k$

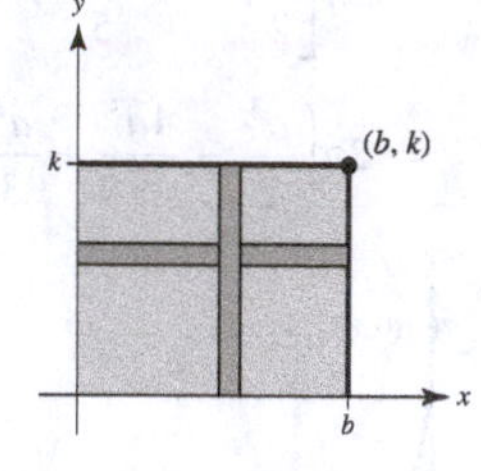

39. $\pi\int_1^5 (x-1)\,dx = \pi\int_1^5 \left(\sqrt{x-1}\right)^2 dx$

This integral represents the volume of the solid generated by revolving the region bounded by $y = \sqrt{x-1}$, $y = 0$, and $x = 5$ about the x-axis by using the disk method.

$2\pi\int_0^2 y\left[5 - (y^2+1)\right]dy$

represents this same volume by using the shell method.

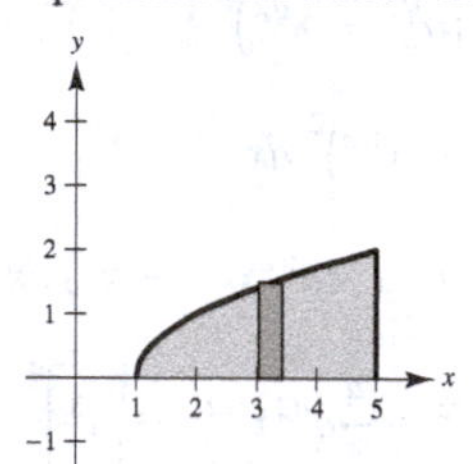

Disk method

41.

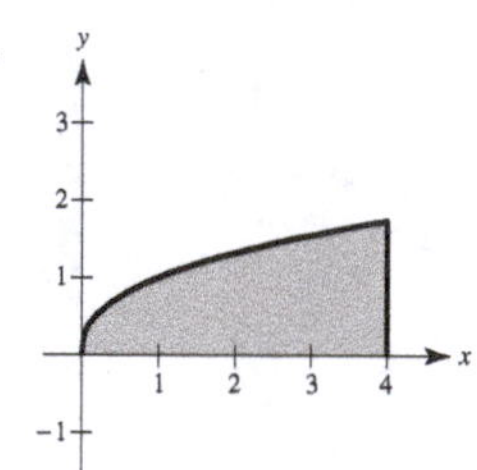

(a) Around x-axis: $V = \pi\int_0^4 \left(x^{2/5}\right)^2 dx = \left[\pi\frac{5}{9}x^{9/5}\right]_0^4$

$= \frac{5}{9}\pi(4)^{9/5} \approx 6.7365\pi$

(b) Around y-axis: $V = 2\pi\int_0^4 x\left(x^{2/5}\right)dx$

$= \left[2\pi\frac{5}{12}x^{12/5}\right]_0^4 \approx 23.2147\pi$

(c) Around $x = 4$:

$V = 2\pi\int_0^4 (4-x)x^{2/5}\,dx \approx 16.5819\pi$

So, $(a) < (c) < (b)$.

43. $2\pi\int_0^2 x^3\,dx = 2\pi\int_0^2 x\left(x^2\right)dx$

(a) Plane region bounded by

$y = x^2, y = 0, x = 0, x = 2$

(b) Revolved about the y-axis

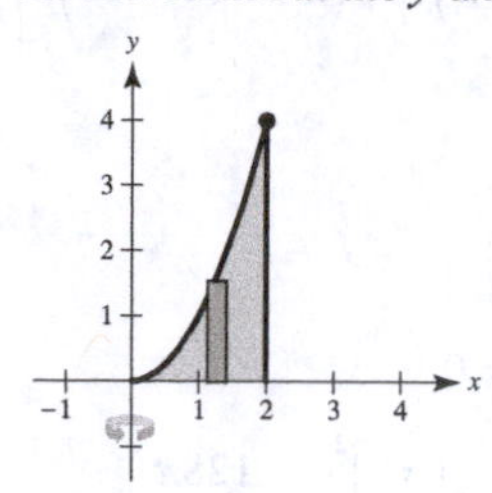

Other answers possible.

45. $2\pi\int_0^6 (y+2)\sqrt{6-y}\,dy$

(a) Plane region bounded by

$x = \sqrt{6-y}, x = 0, y = 0$

(b) Revolved around line $y = -2$

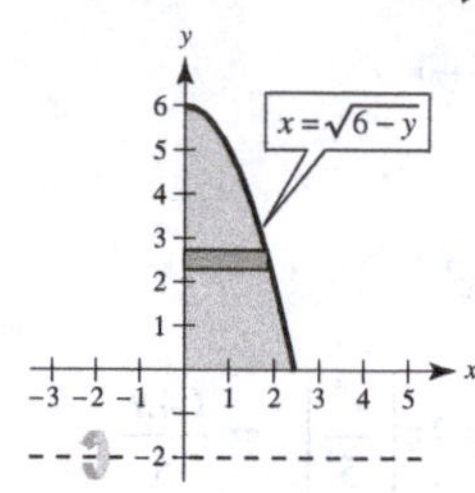

Other answers possible.

47. $p(x) = x, h(x) = 2 - \frac{1}{2}x^2$

$V = 2\pi\int_0^2 x\left(2 - \frac{1}{2}x^2\right)dx$

$= 2\pi\int_0^2 \left(2x - \frac{1}{2}x^3\right)dx$

$= 2\pi\left[x^2 - \frac{1}{8}x^4\right]_0^2 = 4\pi$ (total volume)

Now find x_0 such that:

$\pi = 2\pi\int_0^{x_0} \left(2x - \frac{1}{2}x^3\right)dx$

$1 = 2\left[x^2 - \frac{1}{8}x^4\right]_0^{x_0}$

$1 = 2x_0^2 - \frac{1}{4}x_0^4$

$x_0^4 - 8x_0^2 + 4 = 0$

$x_0^2 = 4 \pm 2\sqrt{3}$ (Quadratic Formula)

Take $x_0 = \sqrt{4 - 2\sqrt{3}} \approx 0.73205$, because the other root is too large.

Diameter:

$2\sqrt{4 - 2\sqrt{3}} \approx 1.464$

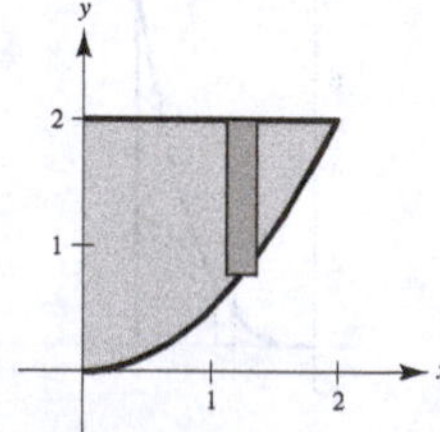

49. $V = 4\pi\int_{-1}^{1}(2 - x)\sqrt{1 - x^2}\,dx$

$= 8\pi\int_{-1}^{1}\sqrt{1 - x^2}\,dx - 4\pi\int_{-1}^{1}x\sqrt{1 - x^2}\,dx$

$= 8\pi\left(\frac{\pi}{2}\right) + 2\pi\int_{-1}^{1}x(1 - x^2)^{1/2}(-2)\,dx$

$= 4\pi^2 + \left[2\pi\left(\frac{2}{3}\right)(1 - x^2)^{3/2}\right]_{-1}^{1} = 4\pi^2$

51. (a) $\frac{d}{dx}[\sin x - x\cos x + C] = \cos x + x\sin x - \cos x$

$= x\sin x$

So, $\int x\sin x\,dx = \sin x - x\cos x + C.$

(b) (i) $p(x) = x,\ h(x) = \sin x$

$V = 2\pi\int_0^{\pi/2} x\sin x\,dx$

$= 2\pi[\sin x - x\cos x]_0^{\pi/2}$

$= 2\pi[(1 - 0) - 0] = 2\pi$

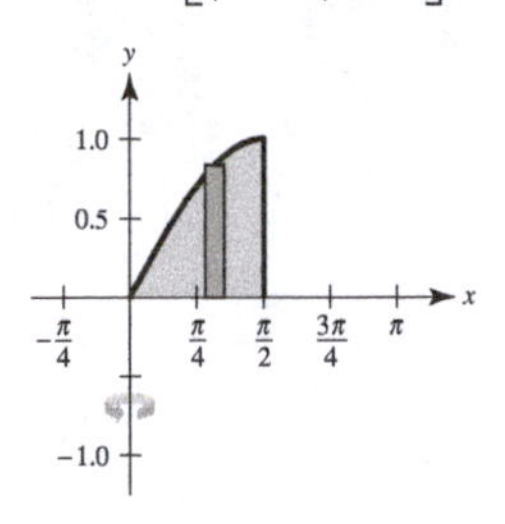

(ii) $p(x) = x,\ h(x) = 2\sin x - (-\sin x) = 3\sin x$

$V = 2\pi\int_0^{\pi} x(3\sin x)\,dx$

$= 6\pi\int_0^{\pi} x\sin x\,dx$

$= 6\pi[\sin x - x\cos x]_0^{\pi}$

$= 6\pi(\pi) = 6\pi^2$

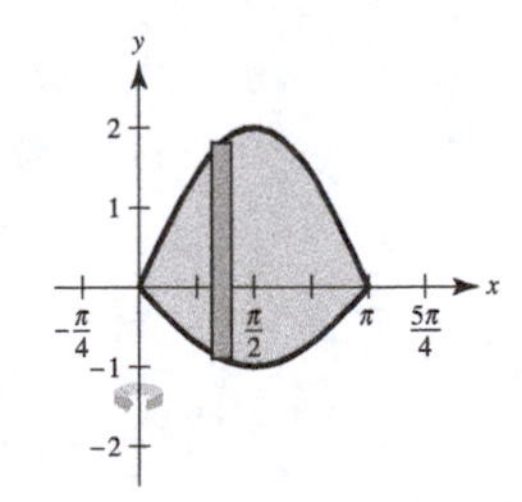

53. Disk Method

$R(y) = \sqrt{r^2 - y^2}$

$r(y) = 0$

$V = \pi\int_{r-h}^{r}(r^2 - y^2)\,dy$

$= \pi\left[r^2y - \frac{y^3}{3}\right]_{r-h}^{r}$

$= \frac{1}{3}\pi h^2(3r - h)$

55. (a) Area of region $= \int_0^b\left[ab^n - ax^n\right]dx$

$= \left[ab^n x - a\frac{x^{n+1}}{n+1}\right]_0^b$

$= ab^{n+1} - a\frac{b^{n+1}}{n+1}$

$= ab^{n+1}\left(1 - \frac{1}{n+1}\right)$

$= ab^{n+1}\left(\frac{n}{n+1}\right)$

$R_1(n) = \frac{ab^{n+1}[n/(n+1)]}{(ab^n)b} = \frac{n}{n+1}$

(b) $\lim_{n\to\infty} R_1(n) = \lim_{n\to\infty}\frac{n}{n+1} = 1$

$\lim_{n\to\infty}(ab^n)b = \infty$

(c) **Disk Method:**

$V = 2\pi\int_0^b x(ab^n - ax^n)\,dx$

$= 2\pi a\int_0^b(xb^n - x^{n+1})\,dx$

$= 2\pi a\left[\frac{b^n}{2}x^2 - \frac{x^{n+2}}{n+2}\right]_0^b$

$= 2\pi a\left[\frac{b^{n+2}}{2} - \frac{b^{n+2}}{n+2}\right] = \pi ab^{n+2}\left(\frac{n}{n+2}\right)$

$R_2(n) = \frac{\pi ab^{n+2}[n/(n+2)]}{(\pi b^2)(ab^n)} = \left(\frac{n}{n+2}\right)$

(d) $\lim_{n\to\infty} R_2(n) = \lim_{n\to\infty}\left(\frac{n}{n+2}\right) = 1$

$\lim_{n\to\infty}(\pi b^2)(ab^n) = \infty$

(e) As $n \to \infty$, the graph approaches the line $x = b$.

57. Top line: $y - 50 = \dfrac{40-50}{20-0}(x-0) = -\dfrac{1}{2}x \Rightarrow y = -\dfrac{1}{2}x + 50$

Bottom line: $y - 40 = \dfrac{0-40}{40-20}(x-20) = -2(x-20) \Rightarrow y = -2x + 80$

$$V = 2\pi\int_0^{20} x\left(-\frac{1}{2}x + 50\right)dx + 2\pi\int_{20}^{40} x(-2x+80)\,dx$$

$$= 2\pi\int_0^{20}\left(-\frac{1}{2}x^2 + 50x\right)dx + 2\pi\int_{20}^{40}\left(-2x^2 + 80x\right)dx$$

$$= 2\pi\left[-\frac{x^3}{6} + 25x^2\right]_0^{20} + 2\pi\left[-\frac{2x^3}{3} + 40x^2\right]_{20}^{40} = 2\pi\left(\frac{26{,}000}{3}\right) + 2\pi\left(\frac{32{,}000}{3}\right) \approx 121{,}475 \text{ ft}^3$$

59. $V_1 = \pi\displaystyle\int_{1/4}^{c}\frac{1}{x^2}\,dx = \pi\left[-\frac{1}{x}\right]_{1/4}^{c} = \pi\left[-\frac{1}{c} + 4\right] = \frac{4c-1}{c}\pi$

$V_2 = 2\pi\displaystyle\int_{1/4}^{c} x\left(\frac{1}{x}\right)dx = \left[2\pi x\right]_{1/4}^{c} = 2\pi\left(c - \frac{1}{4}\right)$

$$V_1 = V_2 \Rightarrow \frac{4c-1}{c}\pi = 2\pi\left(c - \frac{1}{4}\right)$$

$$4c - 1 = 2c\left(c - \frac{1}{4}\right)$$

$$4c^2 - 9c + 2 = 0$$

$$(4c-1)(c-2) = 0$$

$$c = 2 \quad \left(c = \frac{1}{4} \text{ yields no volume.}\right)$$

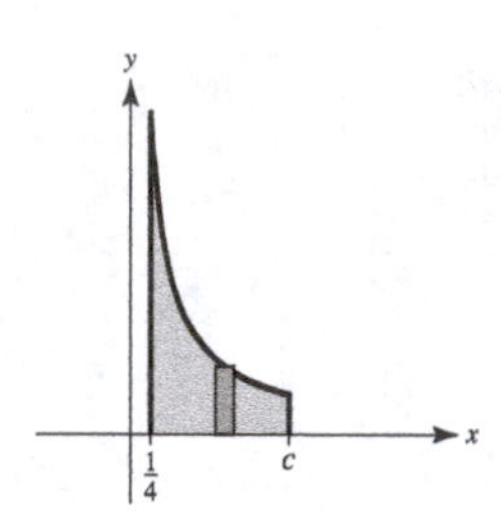

61. $y^2 = x(4-x)^2, \quad 0 \le x \le 4$

$y_1 = \sqrt{x(4-x)^2} = (4-x)\sqrt{x}$

$y_2 = -\sqrt{x(4-x)^2} = -(4-x)\sqrt{x}$

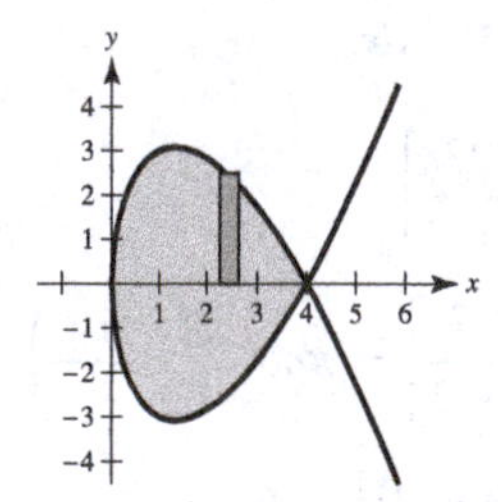

(a) $V = \pi\displaystyle\int_0^4 x(4-x)^2\,dx = \pi\int_0^4\left(x^3 - 8x^2 + 16x\right)dx = \pi\left[\frac{x^4}{4} - \frac{8x^3}{3} + 8x^2\right]_0^4 = \frac{64\pi}{3}$

(b) $V = 4\pi\displaystyle\int_0^4 x(4-x)\sqrt{x}\,dx = 4\pi\int_0^4\left(4x^{3/2} - x^{5/2}\right)dx = 4\pi\left[\frac{8}{5}x^{5/2} - \frac{2}{7}x^{7/2}\right]_0^4 = \frac{2048\pi}{35}$

(c) $V = 4\pi\displaystyle\int_0^4 (4-x)(4-x)\sqrt{x}\,dx = 4\pi\int_0^4\left(16\sqrt{x} - 8x^{3/2} + x^{5/2}\right)dx = 4\pi\left[\frac{32}{3}x^{3/2} - \frac{16}{5}x^{5/2} + \frac{2}{7}x^{7/2}\right]_0^4 = \frac{8192\pi}{105}$

Section 7.4 Arc Length and Surfaces of Revolution

1. The graph of a function f is rectifiable between $(a, f(a))$ and $(b, f(b))$ if f' is continuous on $[a, b]$.

3. The function is $f(x) = 2x^2$ (or $f(x) = 2x^2 + C$), since $f'(x) = 4x$.

5. (a) $d = \sqrt{(5-2)^2 + (3-1)^2} = \sqrt{9+4} = \sqrt{13}$

(b) $m = \dfrac{3-1}{5-2} = \dfrac{2}{3}$

Line $y - 1 = \frac{2}{3}(x - 2)$ or $y = \frac{2}{3}x - \frac{1}{3}$

$$y' = \frac{2}{3}$$

$$s = \int_2^5 \sqrt{1 + \left(\frac{2}{3}\right)^2}\,dx = \left[\frac{\sqrt{13}}{3}x\right]_2^5 = \sqrt{13}$$

7. $y = \frac{2}{3}(x^2+1)^{3/2}$

$$y' = (x^2+1)^{1/2}(2x), \quad 0 \le x \le 1$$

$$1 + (y')^2 = 1 + 4x^2(x^2+1)$$

$$= 4x^4 + 4x^2 + 1 = (2x^2+1)^2$$

$$s = \int_0^1 \sqrt{1 + (y')^2}\,dx$$

$$= \int_0^1 (2x^2+1)\,dx = \left[\frac{2x^3}{3} + x\right]_0^1 = \frac{5}{3}$$

9. $y = \frac{2}{3}x^{3/2} + 1$

$$y' = x^{1/2}, \quad 0 \le x \le 1$$

$$s = \int_0^1 \sqrt{1+x}\,dx$$

$$= \left[\tfrac{2}{3}(1+x)^{3/2}\right]_0^1 = \tfrac{2}{3}\left(\sqrt{8} - 1\right) \approx 1.219$$

11. $y = \frac{3}{2}x^{2/3}$

$$y' = \frac{1}{x^{1/3}}, \quad 1 \le x \le 8$$

$$s = \int_1^8 \sqrt{1 + \left(\frac{1}{x^{1/3}}\right)^2}\,dx = \int_1^8 \sqrt{\frac{x^{2/3}+1}{x^{2/3}}}\,dx = \frac{3}{2}\int_1^8 \sqrt{x^{2/3}+1}\left(\frac{2}{3x^{1/3}}\right)dx$$

$$= \frac{3}{2}\left[\frac{2}{3}(x^{2/3}+1)^{3/2}\right]_1^8 = 5\sqrt{5} - 2\sqrt{2} \approx 8.352$$

13. $y = \dfrac{x^5}{10} + \dfrac{1}{6x^3}, \quad 2 \le x \le 5$

$$y' = \frac{x^4}{2} - \frac{1}{2x^4} = \frac{1}{2}\left(x^4 - \frac{1}{x^4}\right)$$

$$1 + (y')^2 = 1 + \frac{1}{4}\left(x^4 - \frac{1}{x^4}\right)^2 = 1 + \frac{1}{4}\left(x^8 - 2 + \frac{1}{x^8}\right)$$

$$= \frac{1}{4}\left(x^8 + 2 + \frac{1}{x^8}\right) = \frac{1}{4}\left(x^4 + \frac{1}{x^4}\right)^2$$

$$s = \int_2^5 \sqrt{1+(y')^2}\,dx = \int_2^5 \frac{1}{2}\left(x^4 + \frac{1}{x^4}\right)dx$$

$$= \frac{1}{2}\left[\frac{x^5}{5} - \frac{1}{3x^3}\right]_2^5 = \frac{1}{2}\left[\left(625 - \frac{1}{375}\right) - \left(\frac{32}{5} - \frac{1}{24}\right)\right]$$

$$= \frac{618{,}639}{2000} \approx 309.320$$

15. $y = \ln(\sin x), \quad \left[\frac{\pi}{4}, \frac{3\pi}{4}\right]$

$$y' = \frac{1}{\sin x}\cos x = \cot x$$

$$1 + (y')^2 = 1 + \cot^2 x = \csc^2 x$$

$$s = \int_{\pi/4}^{3\pi/4} \csc x\, dx$$

$$= \Big[\ln|\csc x - \cot x|\Big]_{\pi/4}^{3\pi/4}$$

$$= \ln\left(\sqrt{2} + 1\right) - \ln\left(\sqrt{2} - 1\right) \approx 1.763$$

17. $y = \frac{1}{2}\left(e^x + e^{-x}\right)$

$$y' = \frac{1}{2}\left(e^x - e^{-x}\right), \quad [0, 2]$$

$$1 + (y')^2 = \left[\frac{1}{2}\left(e^x + e^{-x}\right)\right]^2, \quad [0, 2]$$

$$s = \int_0^2 \sqrt{\left[\frac{1}{2}\left(e^x + e^{-x}\right)\right]^2}\, dx$$

$$= \frac{1}{2}\int_0^2 \left(e^x + e^{-x}\right) dx$$

$$= \frac{1}{2}\Big[e^x - e^{-x}\Big]_0^2 = \frac{1}{2}\left(e^2 - \frac{1}{e^2}\right) \approx 3.627$$

19. $x = \frac{1}{3}\left(y^2 + 2\right)^{3/2}, \quad 0 \le y \le 4$

$$\frac{dx}{dy} = y\left(y^2 + 2\right)^{1/2}$$

$$s = \int_0^4 \sqrt{1 + y^2\left(y^2 + 2\right)}\, dy = \int_0^4 \sqrt{y^4 + 2y^2 + 1}\, dy$$

$$= \int_0^4 \left(y^2 + 1\right) dy = \left[\frac{y^3}{3} + y\right]_0^4 = \frac{64}{3} + 4 = \frac{76}{3}$$

21. (a) $y = 4 - x^2, \quad [0, 2]$

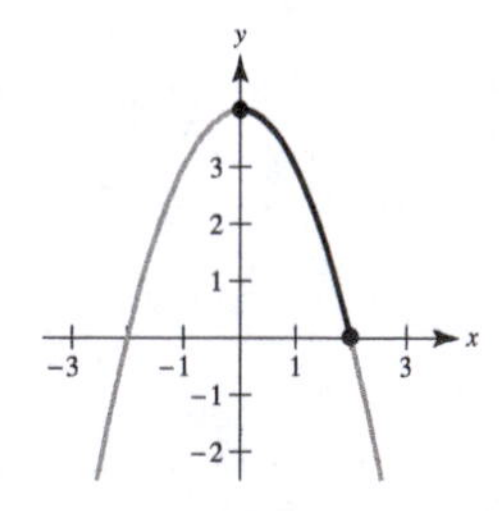

(b) $y' = -2x$

$$1 + (y')^2 = 1 + 4x^2$$

$$L = \int_0^2 \sqrt{1 + 4x^2}\, dx$$

(c) $L \approx 4.647$

23. (a) $y = \frac{1}{x}, \quad [1, 3]$

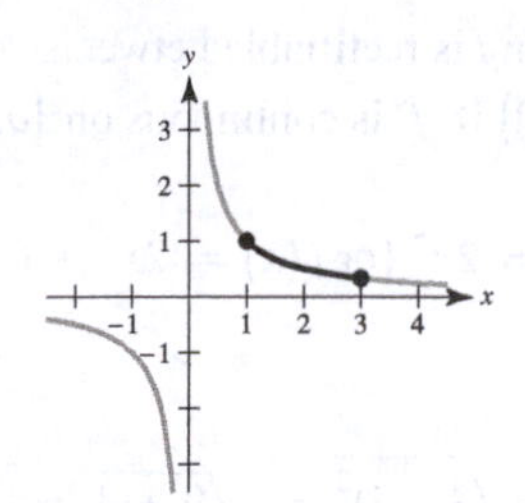

(b) $y' = -\frac{1}{x^2}$

$$1 + (y')^2 = 1 + \frac{1}{x^4}$$

$$L = \int_1^3 \sqrt{1 + \frac{1}{x^4}}\, dx$$

(c) $L \approx 2.147$

25. (a) $y = \sin x, \quad [0, \pi]$

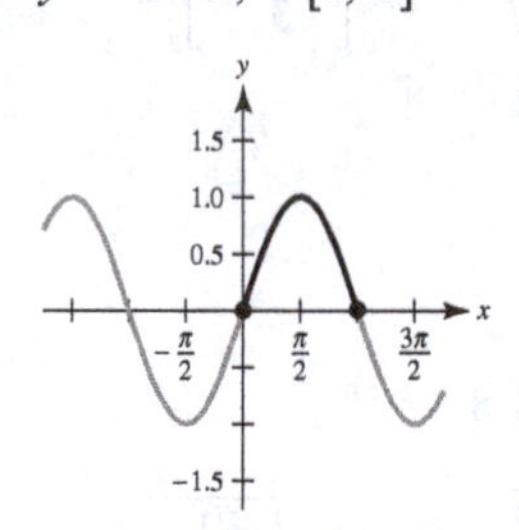

(b) $y' = \cos x$

$$1 + (y')^2 = 1 + \cos^2 x$$

$$L = \int_0^\pi \sqrt{1 + \cos^2 x}\, dx$$

(c) $L \approx 3.820$

27. (a) $y = 2\arctan x, \quad [0, 1]$

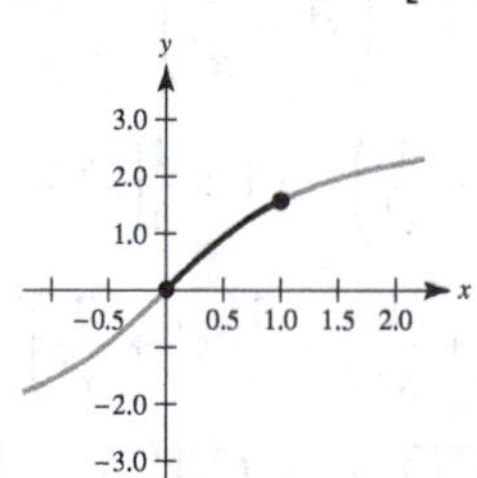

(b) $y' = \frac{2}{1 + x^2}$

$$L = \int_0^1 \sqrt{1 + \frac{4}{\left(1 + x^2\right)^2}}\, dx$$

(c) $L \approx 1.871$

29. (a) $x = e^{-y}, \quad 0 \le y \le 2$

$y = -\ln x$

$1 \ge x \ge e^{-2} \approx 0.135$

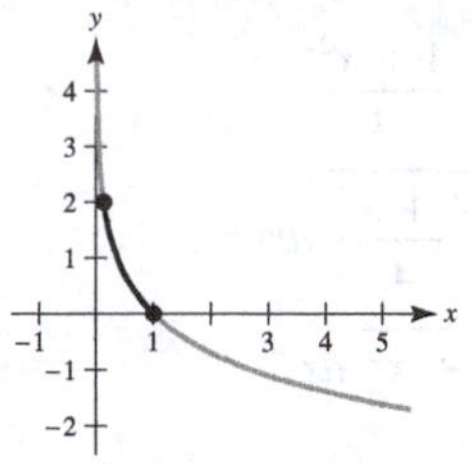

(b) $y' = -\dfrac{1}{x}$

$1 + (y')^2 = 1 + \dfrac{1}{x^2}$

$L = \displaystyle\int_{e^{-2}}^{1} \sqrt{1 + \frac{1}{x^2}}\, dx$

(c) $L \approx 2.221$

Alternatively, you can do all the computations with respect to y.

(a) $x = e^{-y}, \quad 0 \le y \le 2$

(b) $\dfrac{dx}{dy} = -e^{-y}$

$1 + \left(\dfrac{dx}{dy}\right)^2 = 1 + e^{-2y}$

$L = \displaystyle\int_0^2 \sqrt{1 + e^{-2y}}\, dy$

(c) $L \approx 2.221$

31. $y = x^3, \quad [0, 4]$

(a) $d = \sqrt{(4 - 0)^2 + (64 - 0)^2} \approx 64.125$

(b) $d = \sqrt{(1 - 0)^2 + (1 - 0)^2} + \sqrt{(2 - 1)^2 + (8 - 1)^2} + \sqrt{(3 - 2)^2 + (27 - 8)^2} + \sqrt{(4 - 3)^2 + (64 - 27)^2} \approx 64.525$

(c) 64.672

33. $y = 20 \cosh \dfrac{x}{20}, \quad -20 \le x \le 20$

$y' = \sinh \dfrac{x}{20}$

$1 + (y')^2 = 1 + \sinh^2 \dfrac{x}{20} = \cosh^2 \dfrac{x}{20}$

$$L = \int_{-20}^{20} \cosh \frac{x}{20}\, dx = 2\int_0^{20} \cosh \frac{x}{20}\, dx = \left[2(20) \sinh \frac{x}{20}\right]_0^{20} = 40 \sinh(1) \approx 47.008 \text{ m}$$

35. $y = 693.8597 - 68.7672 \cosh 0.0100333x$

$y' = -0.6899619478 \sinh 0.0100333x$

$$s = \int_{-299.2239}^{299.2239} \sqrt{1 + (-0.6899619478 \sinh 0.0100333x)^2}\, dx \approx 1480$$

(Use a graphing utility.)

37. $y = \sqrt{9 - x^2}$

$y' = \dfrac{-x}{\sqrt{9 - x^2}}$

$1 + (y')^2 = \dfrac{9}{9 - x^2}$

$$s = \int_0^2 \sqrt{\frac{9}{9 - x^2}}\, dx = \int_0^2 \frac{3}{\sqrt{9 - x^2}}\, dx = \left[3 \arcsin \frac{x}{3}\right]_0^2 = 3\left(\arcsin \frac{2}{3} - \arcsin 0\right) = 3 \arcsin \frac{2}{3} \approx 2.1892$$

39. $y = \frac{x^3}{3}$

$y' = x^2, \quad [0, 3]$

$$S = 2\pi\int_0^3 \frac{x^3}{3}\sqrt{1 + x^4}\,dx$$
$$= \frac{\pi}{6}\int_0^3 \left(1 + x^4\right)^{1/2}\left(4x^3\right) dx$$
$$= \left[\frac{\pi}{9}\left(1 + x^4\right)^{3/2}\right]_0^3$$
$$= \frac{\pi}{9}\left(82\sqrt{82} - 1\right) \approx 258.85$$

41. $y = \frac{x^3}{6} + \frac{1}{2x}$

$y' = \frac{x^2}{2} - \frac{1}{2x^2}$

$1 + (y')^2 = \left(\frac{x^2}{2} + \frac{1}{2x^2}\right)^2, \quad [1, 2]$

$$S = 2\pi\int_1^2 \left(\frac{x^3}{6} + \frac{1}{2x}\right)\left(\frac{x^2}{2} + \frac{1}{2x^2}\right) dx$$
$$= 2\pi\int_1^2 \left(\frac{x^5}{12} + \frac{x}{3} + \frac{1}{4x^3}\right) dx$$
$$= 2\pi\left[\frac{x^6}{72} + \frac{x^2}{6} - \frac{1}{8x^2}\right]_1^2 = \frac{47\pi}{16}$$

43. $y = \sqrt{4 - x^2}$

$y' = \frac{1}{2}\left(4 - x^2\right)^{-1/2}(-2x) = \frac{-x}{\sqrt{4 - x^2}}, \quad -1 \le x \le 1$

$1 + (y')^2 = 1 + \frac{x^2}{4 - x^2} = \frac{4}{4 - x^2}$

$$S = 2\pi\int_{-1}^1 \sqrt{4 - x^2}\cdot\sqrt{\frac{4}{4 - x^2}}\,dx$$
$$= 4\pi\int_{-1}^1 dx = 4\pi\left[x\right]_{-1}^1 = 8\pi$$

45. $y = \sqrt[3]{x} + 2$

$y' = \frac{1}{3x^{2/3}}, \quad [1, 8]$

$$S = 2\pi\int_1^8 x\sqrt{1 + \frac{1}{9x^{4/3}}}\,dx$$
$$= \frac{2\pi}{3}\int_1^8 x^{1/3}\sqrt{9x^{4/3} + 1}\,dx$$
$$= \frac{\pi}{18}\int_1^8 \left(9x^{4/3} + 1\right)^{1/2}\left(12x^{1/3}\right) dx$$
$$= \left[\frac{\pi}{27}\left(9x^{4/3} + 1\right)^{3/2}\right]_1^8$$
$$= \frac{\pi}{27}\left(145\sqrt{145} - 10\sqrt{10}\right) \approx 199.48$$

47. $y = 1 - \frac{x^2}{4}$

$y' = -\frac{x}{2}, \quad 0 \le x \le 2$

$1 + (y')^2 = 1 + \frac{x^2}{4} = \frac{4 + x^2}{4}$

$$S = 2\pi\int_0^2 x\sqrt{\frac{4 + x^2}{4}}\,dx$$
$$= \pi\int_0^2 x\sqrt{4 + x^2}\,dx$$
$$= \frac{1}{2}\pi\int_0^2 \left(4 + x^2\right)^{1/2}(2x)\,dx$$
$$= \frac{1}{2}\pi\left[\frac{2}{3}\left(4 + x^2\right)^{3/2}\right]_0^2$$
$$= \frac{\pi}{3}\left(8^{3/2} - 4^{3/2}\right)$$
$$= \frac{\pi}{3}\left(16\sqrt{2} - 8\right) \approx 15.318$$

49. $y = \sin x$

$y' = \cos x, \quad [0, \pi]$

$S = 2\pi\int_0^\pi \sin x\sqrt{1 + \cos^2 x}\,dx \approx 14.4236$

51. $\int_0^2 \sqrt{1 + \left[\frac{d}{dx}\left(\frac{5}{x^2 + 1}\right)\right]^2}\,dx$

$s \approx 5$

Matches (b)

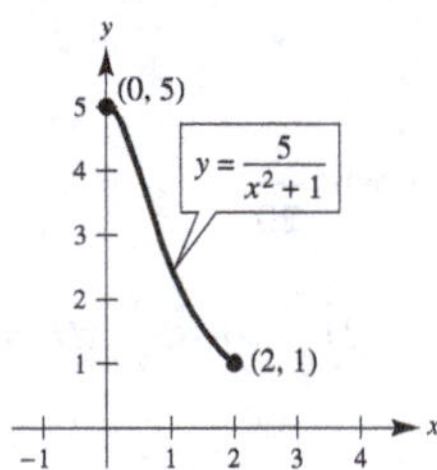

52. $\int_0^{\pi/4} \sqrt{1 + \left[\frac{d}{dx}(\tan x)\right]^2}\,dx$

$s \approx 1$

Matches (e)

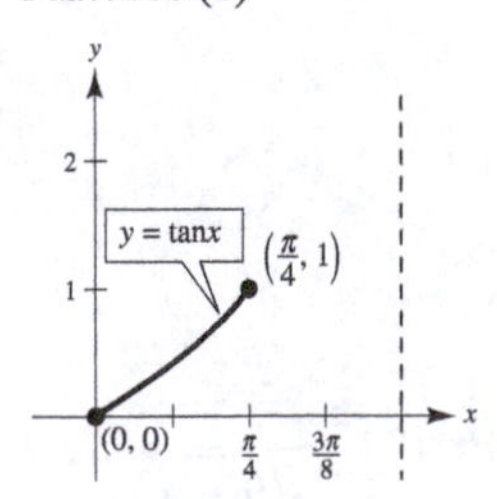

53. $f(x) = \frac{1}{4}e^x + e^{-x}, [a, b]$

Integral: $\int_a^b \left(\frac{1}{4}e^x + e^{-x}\right) dx = \left[\frac{e^x}{4} - e^{-x}\right]_a^b = \frac{e^b}{4} - e^{-b} + \frac{e^a}{4} - e^{-a}$

Arc length:
$$f'(x) = \frac{1}{4}e^x - e^{-x}$$
$$1 + f'(x)^2 = 1 + \frac{1}{16}e^{2x} - \frac{1}{2} + e^{-2x}$$
$$= \frac{1}{16}e^{2x} + \frac{1}{2} + e^{-2x}$$
$$= \left(\frac{1}{4}e^x + e^{-x}\right)^2$$
$$s = \int_a^b \left(\frac{1}{4}e^x + e^{-x}\right) dx = \frac{e^b}{4} - e^{-b} + \frac{e^a}{4} - e^{-a}$$

They have the same value.

55. (a)

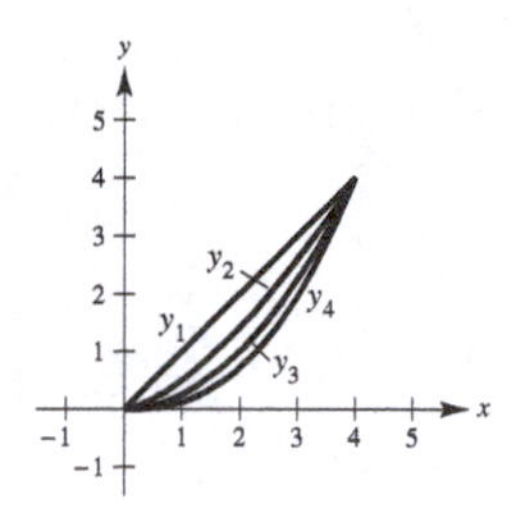

(b) y_1, y_2, y_3, y_4

(c) $y_1' = 1, \quad s_1 = \int_0^4 \sqrt{2}\, dx \approx 5.657$

$y_2' = \frac{3}{4}x^{1/2}, \quad s_2 = \int_0^4 \sqrt{1 + \frac{9x}{16}}\, dx \approx 5.759$

$y_3' = \frac{1}{2}x, \quad s_3 = \int_0^4 \sqrt{1 + \frac{x^2}{4}}\, dx \approx 5.916$

$y_4' = \frac{5}{16}x^{3/2}, \quad s_4 = \int_0^4 \sqrt{1 + \frac{25}{256}x^3}\, dx \approx 6.063$

57.
$$y = \frac{3x}{4}, \quad y' = \frac{3}{4}$$
$$1 + (y')^2 = 1 + \frac{9}{16} = 25/16$$
$$S = 2\pi\int_0^4 x\sqrt{\frac{25}{16}}\, dx = \frac{5\pi}{2}\left[\frac{x^2}{2}\right]_0^4 = 20\pi$$

59.
$$y = \sqrt{9 - x^2}$$
$$y' = \frac{-x}{\sqrt{9 - x^2}}$$
$$\sqrt{1 + (y')^2} = \frac{3}{\sqrt{9 - x^2}}$$
$$S = 2\pi\int_0^2 \frac{3x}{\sqrt{9 - x^2}}\, dx$$
$$= -3\pi\int_0^2 \frac{-2x}{\sqrt{9 - x^2}}\, dx$$
$$= \left[-6\pi\sqrt{9 - x^2}\right]_0^2$$
$$= 6\pi\left(3 - \sqrt{5}\right) \approx 14.40$$

See figure in Exercise 60.

61. (a) Approximate the volume by summing six disks of thickness 3 and circumference C_i equal to the average of the given circumferences:

$$V \approx \sum_{i=1}^{6} \pi r_i^2(3) = \sum_{i=1}^{6} \pi\left(\frac{C_i}{2\pi}\right)^2(3) = \frac{3}{4\pi}\sum_{i=1}^{6} C_i^2$$
$$= \frac{3}{4\pi}\left[\left(\frac{50 + 65.5}{2}\right)^2 + \left(\frac{65.5 + 70}{2}\right)^2 + \left(\frac{70 + 66}{2}\right)^2 + \left(\frac{66 + 58}{2}\right)^2 + \left(\frac{58 + 51}{2}\right)^2 + \left(\frac{51 + 48}{2}\right)^2\right]$$
$$= \frac{3}{4\pi}\left[57.75^2 + 67.75^2 + 68^2 + 62^2 + 54.5^2 + 49.5^2\right] = \frac{3}{4\pi}(21813.625) = 5207.62 \text{ in.}^3$$

(b) The lateral surface area of a frustum of a right circular cone is $\pi s(R + r)$. For the first frustum:

$$S_1 \approx \pi\left[3^2 + \left(\frac{65.5 - 50}{2\pi}\right)^2\right]^{1/2}\left[\frac{50}{2\pi} + \frac{65.5}{2\pi}\right]$$

$$= \left(\frac{50 + 65.5}{2}\right)\left[9 + \left(\frac{65.5 - 50}{2\pi}\right)^2\right]^{1/2}.$$

Adding the six frustums together:

$$S \approx \left(\frac{50 + 65.5}{2}\right)\left[9 + \left(\frac{15.5}{2\pi}\right)^2\right]^{1/2} + \left(\frac{65.5 + 70}{2}\right)\left[9 + \left(\frac{4.5}{2\pi}\right)^2\right]^{1/2}$$

$$+ \left(\frac{70 + 66}{2}\right)\left[9 + \left(\frac{4}{2\pi}\right)^2\right]^{1/2} + \left(\frac{66 + 58}{2}\right)\left[9 + \left(\frac{8}{2\pi}\right)^2\right]^{1/2}$$

$$+ \left(\frac{58 + 51}{2}\right)\left[9 + \left(\frac{7}{2\pi}\right)^2\right]^{1/2} + \left(\frac{51 + 48}{2}\right)\left[9 + \left(\frac{3}{2\pi}\right)^2\right]^{1/2}$$

$$\approx 224.30 + 208.96 + 208.54 + 202.06 + 174.41 + 150.37 = 1168.64$$

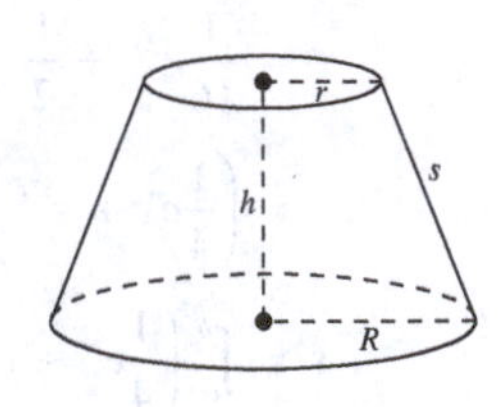

(c) $r = 0.00401y^3 - 0.1416y^2 + 1.232y + 7.943$

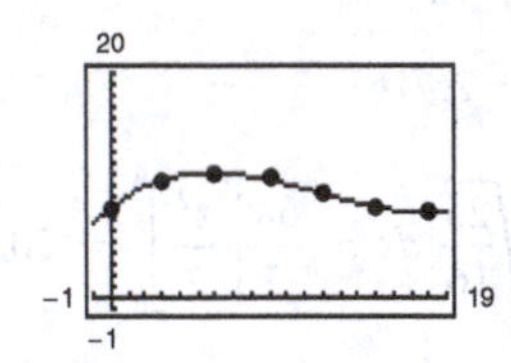

(d) $V = \int_0^{18} \pi r^2 dy \approx 5275.9 \text{ in.}^3$

$S = \int_0^{18} 2\pi r(y)\sqrt{1 + r'(y)^2}\, dy \approx 1179.5 \text{ in.}^2$

63. (a) $V = \pi\int_1^b \frac{1}{x^2}\, dx = \left[-\frac{\pi}{x}\right]_1^b = \pi\left(1 - \frac{1}{b}\right)$

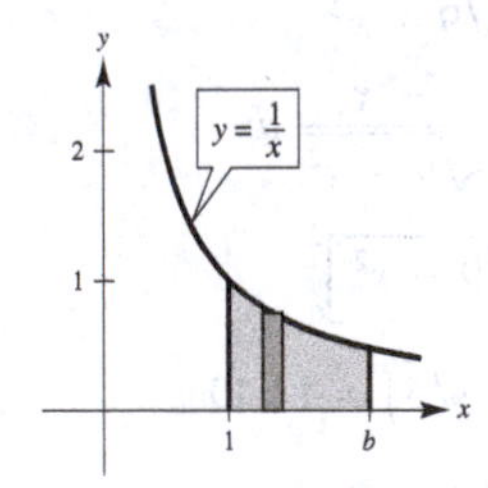

(b) $S = 2\pi\int_1^b \frac{1}{x}\sqrt{1 + \left(-\frac{1}{x^2}\right)^2}\, dx = 2\pi\int_1^b \frac{1}{x}\sqrt{1 + \frac{1}{x^4}}\, dx = 2\pi\int_1^b \frac{\sqrt{x^4 + 1}}{x^3}\, dx$

(c) $\lim_{b\to\infty} V = \lim_{b\to\infty} \pi\left(1 - \frac{1}{b}\right) = \pi$

(d) Because $\frac{\sqrt{x^4 + 1}}{x^3} > \frac{\sqrt{x^4}}{x^3} = \frac{1}{x} > 0$ on $[1, b]$, you have $\int_1^b \frac{\sqrt{x^4 + 1}}{x^3}\, dx > \int_1^b \frac{1}{x}\, dx = [\ln x]_1^b = \ln b$

and $\lim_{b\to\infty} \ln b \to \infty$. So, $\lim_{b\to\infty} 2\pi\int_1^b \frac{\sqrt{x^4 + 1}}{x^3}\, dx = \infty$.

65. $y = \frac{1}{3}\left(x^{3/2} - 3x^{1/2} + 2\right)$

When $x = 0$, $y = \frac{2}{3}$. So, the fleeing object has traveled $\frac{2}{3}$ unit when it is caught.

$$y' = \frac{1}{3}\left(\frac{3}{2}x^{1/2} - \frac{3}{2}x^{-1/2}\right) = \left(\frac{1}{2}\right)\frac{x-1}{x^{1/2}}$$

$$1 + (y')^2 = 1 + \frac{(x-1)^2}{4x} = \frac{(x+1)^2}{4x}$$

$$s = \int_0^1 \frac{x+1}{2x^{1/2}}\,dx = \frac{1}{2}\int_0^1 \left(x^{1/2} + x^{-1/2}\right)dx$$

$$= \frac{1}{2}\left[\frac{2}{3}x^{3/2} + 2x^{1/2}\right]_0^1 = \frac{4}{3} = 2\left(\frac{2}{3}\right)$$

The pursuer has traveled twice the distance that the fleeing object has traveled when it is caught.

67. $x^{2/3} + y^{2/3} = 4$

$$y^{2/3} = 4 - x^{2/3}$$

$$y = \left(4 - x^{2/3}\right)^{3/2}, \quad 0 \le x \le 8$$

$$y' = \frac{3}{2}\left(4 - x^{2/3}\right)^{1/2}\left(-\frac{2}{3}x^{-1/3}\right) = \frac{-\left(4 - x^{2/3}\right)^{1/2}}{x^{1/3}}$$

$$1 + (y')^2 = 1 + \frac{4 - x^{2/3}}{x^{2/3}} = \frac{4}{x^{2/3}}$$

$$S = 2\pi\int_0^8 \left(4 - x^{2/3}\right)^{3/2}\sqrt{\frac{4}{x^{2/3}}}\,dx = 4\pi\int_0^8 \frac{\left(4 - x^{2/3}\right)^{3/2}}{x^{1/3}}\,dx = \left[-\frac{12\pi}{5}\left(4 - x^{2/3}\right)^{5/2}\right]_0^8 = \frac{384\pi}{5}$$

[Surface area of portion above the x-axis]

69. $y = kx^2$, $y' = 2kx$

$$1 + (y')^2 = 1 + 4k^2x^2$$

$$h = kw^2 \Rightarrow k = \frac{h}{w^2} \Rightarrow 1 + (y') = 1 + \frac{4h^2}{w^4}x^2$$

By symmetry, $C = 2\int_0^w \sqrt{1 + \frac{4h^2}{w^4}x^2}\,dx.$

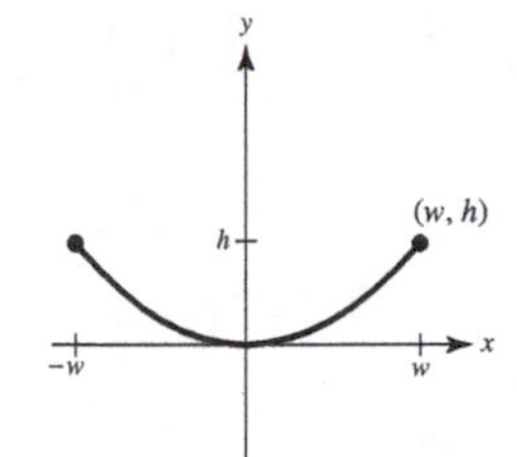

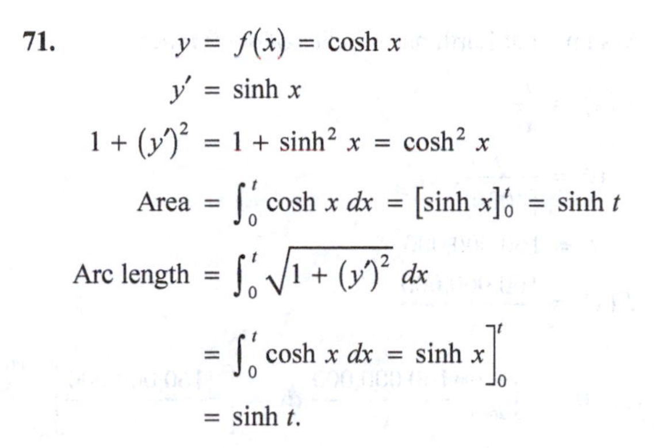

71.

$$y = f(x) = \cosh x$$

$$y' = \sinh x$$

$$1 + (y')^2 = 1 + \sinh^2 x = \cosh^2 x$$

$$\text{Area} = \int_0^t \cosh x\,dx = [\sinh x]_0^t = \sinh t$$

$$\text{Arc length} = \int_0^t \sqrt{1 + (y')^2}\,dx$$

$$= \int_0^t \cosh x\,dx = \sinh x\Big]_0^t$$

$$= \sinh t.$$

Another curve with this property is $g(x) = 1$.

$$\text{Area} = \int_0^t dx = t$$

$$\text{Arc length} = t$$

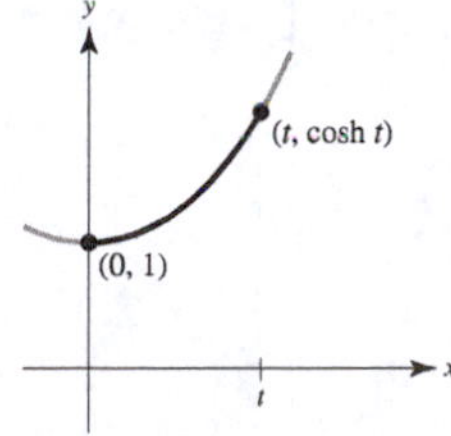

Section 7.5 Work

1. Work is done by a force when it moves an object.

3. Hooke's Law says that the force needed to extend or compress a spring by some distance d is proportional to that distance, $F = kd$.

5. $W = Fd = 1200(40) = 48{,}000$ ft-lb

7. $W = Fd = 112(8) = 896$ joules (Newton-meters)

9. $F(x) = kx$

$5 = k(3)$

$k = \frac{5}{3}$

$F(x) = \frac{5}{3}x$

$$W = \int_0^7 F(x)\,dx = \int_0^7 \frac{5}{3}x\,dx = \left[\frac{5}{6}x^2\right]_0^7 = \frac{245}{6} \text{ in.-lb}$$

≈ 40.833 in.-lb ≈ 3.403 ft-lb

11. $F(x) = kx$

$20 = k(9)$

$k = \frac{20}{9}$

$$W = \int_0^{12} \frac{20}{9}x\,dx = \left[\frac{10}{9}x^2\right]_0^{12} = 160 \text{ in.-lb} = \frac{40}{3} \text{ ft-lb}$$

13. $W = 18 = \int_0^{1/3} kx\,dx = \left[\frac{kx^2}{2}\right]_0^{1/3} = \frac{k}{18} \Rightarrow k = 324$

$$W = \int_{1/3}^{7/12} 324x\,dx = \left[162x^2\right]_{1/3}^{7/12} = 37.125 \text{ ft-lb}$$

[**Note:** 4 inches $= \frac{1}{3}$ foot]

15. Assume that Earth has a radius of 4000 miles.

$F(x) = \frac{k}{x^2}$

$5 = \frac{k}{(4000)^2}$

$k = 80{,}000{,}000$

$F(x) = \frac{80{,}000{,}000}{x^2}$

(a) $$W = \int_{4000}^{4100} \frac{80{,}000{,}000}{x^2}\,dx = \left[\frac{-80{,}000{,}000}{x}\right]_{4000}^{4100}$$

≈ 487.8 mi-tons $\approx 5.15 \times 10^9$ ft-lb

(b) $$W = \int_{4000}^{4300} \frac{80{,}000{,}000}{x^2}\,dx$$

≈ 1395.3 mi-ton $\approx 1.47 \times 10^{10}$ ft-ton

17. Assume that Earth has a radius of 4000 miles.

$F(x) = \frac{k}{x^2}$

$10 = \frac{k}{(4000)^2}$

$k = 160{,}000{,}000$

$F(x) = \frac{160{,}000{,}000}{x^2}$

(a) $$W = \int_{4000}^{15{,}000} \frac{160{,}000{,}000}{x^2}\,dx = \left[-\frac{160{,}000{,}000}{x}\right]_{4000}^{15{,}000} \approx -10{,}666.667 + 40{,}000$$

$= 29{,}333.333$ mi-ton

$\approx 2.93 \times 10^4$ mi-ton

$\approx 3.10 \times 10^{11}$ ft-lb

(b) $$W = \int_{4000}^{26{,}000} \frac{160{,}000{,}000}{x^2}\,dx = \left[-\frac{160{,}000{,}000}{x}\right]_{4000}^{26{,}000} \approx -6{,}153.846 + 40{,}000$$

$= 33{,}846.154$ mi-ton

$\approx 3.38 \times 10^4$ mi-ton

$\approx 3.57 \times 10^{11}$ ft-lb

19. Weight of each layer: $62.4(20)\,\Delta y$

Distance: $4 - y$

(a) $W = \int_2^4 62.4(20)(4 - y)\,dy = \left[4992y - 624y^2\right]_2^4 = 2496$ ft-lb

(b) $W = \int_0^4 62.4(20)(4 - y)\,dy = \left[4992y - 624y^2\right]_0^4 = 9984$ ft-lb

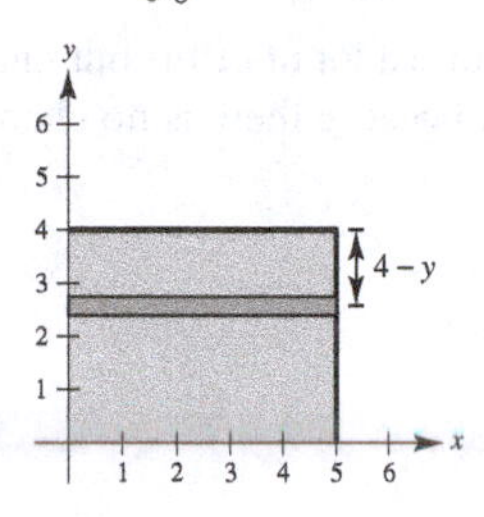

21. Volume of disk: $\pi(2)^2\,\Delta y = 4\pi\,\Delta y$

Weight of disk of water: $9800(4\pi)\,\Delta y$

Distance the disk of water is moved: $5 - y$

$$W = \int_0^4 (5 - y)(9800)4\pi\,dy = 39{,}200\pi\int_0^4 (5 - y)\,dy$$

$$= 39{,}200\pi\left[5y - \frac{y^2}{2}\right]_0^4$$

$$= 39{,}200\pi(12) = 470{,}400\pi \text{ newton–meters}$$

23. Volume of disk: $\pi\left(\frac{2}{3}y\right)^2\,\Delta y$

Weight of disk: $62.4\pi\left(\frac{2}{3}y\right)^2\,\Delta y$

Distance: $6 - y$

$$W = \frac{4(62.4)\pi}{9}\int_0^6 (6 - y)y^2\,dy$$

$$= \frac{4}{9}(62.4)\pi\left[2y^3 - \frac{1}{4}y^4\right]_0^6$$

$$= 2995.2\pi \text{ ft-lb}$$

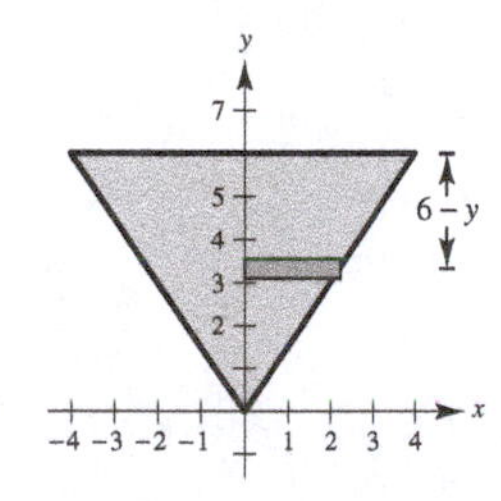

25. Volume of disk: $\pi\left(\sqrt{36 - y^2}\right)^2\,\Delta y$

Weight of disk: $62.4\pi\left(36 - y^2\right)\Delta y$

Distance: y

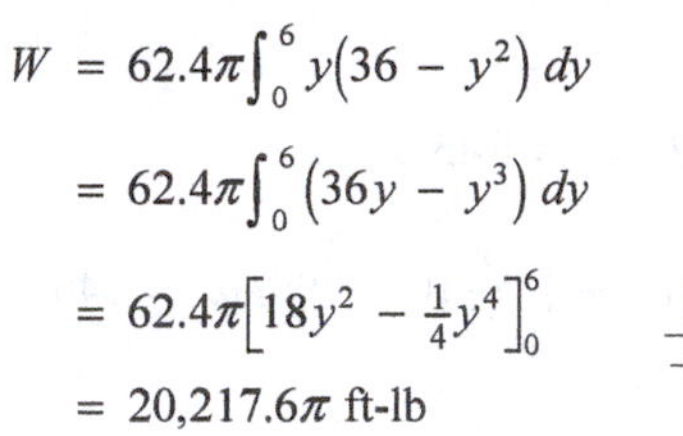

$$W = 62.4\pi\int_0^6 y\left(36 - y^2\right)dy$$

$$= 62.4\pi\int_0^6 \left(36y - y^3\right)dy$$

$$= 62.4\pi\left[18y^2 - \tfrac{1}{4}y^4\right]_0^6$$

$$= 20{,}217.6\pi \text{ ft-lb}$$

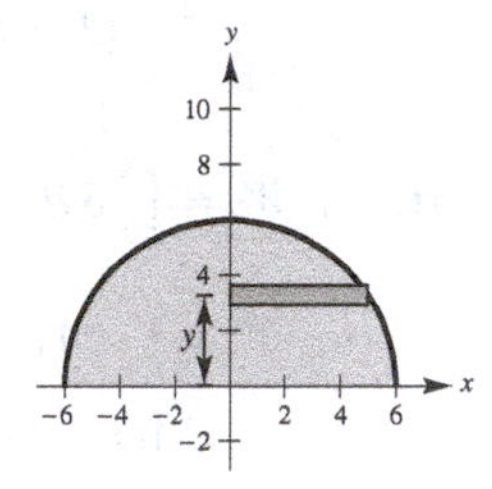

27. Volume of layer: $V = lwh = 4(2)\sqrt{(9/4) - y^2}\,\Delta y$

Weight of layer: $W = 42(8)\sqrt{(9/4) - y^2}\,\Delta y$

Distance: $\frac{13}{2} - y$

$$W = \int_{-1.5}^{1.5} 42(8)\sqrt{\tfrac{9}{4} - y^2}\left(\tfrac{13}{2} - y\right)dy$$

$$= 336\left[\frac{13}{2}\int_{-1.5}^{1.5}\sqrt{\tfrac{9}{4} - y^2}\,dy - \int_{-1.5}^{1.5}\sqrt{\tfrac{9}{4} - y^2}\,y\,dy\right]$$

The second integral is zero because the integrand is odd and the limits of integration are symmetric to the origin. The first integral represents the area of a semicircle of radius $\frac{3}{2}$. So, the work is

$$W = 336\left(\tfrac{13}{2}\right)\pi\left(\tfrac{3}{2}\right)^2\left(\tfrac{1}{2}\right)$$

$$= 2457\pi \text{ ft-lb.}$$

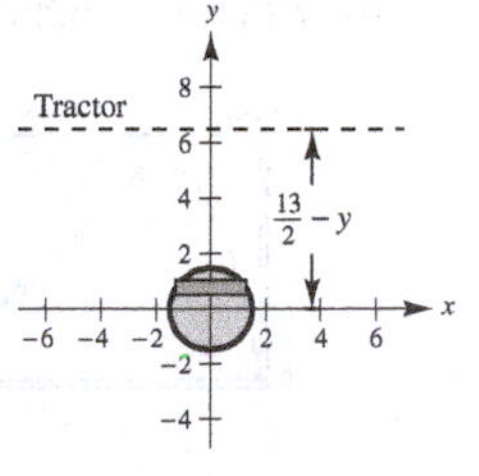

29. Weight of section of chain: $3\,\Delta y$

Distance: $20 - y$. $\Delta W = (\text{force increment})(\text{distance}) = (3\,\Delta y)(20 - y)$

$$W = \int_0^{20} (20 - y)3\,dy = 3\left[20y - \frac{y^2}{2}\right]_0^{20} = 3\left[400 - \frac{400}{2}\right] = 600 \text{ ft-lb}$$

31. The lower 10 feet of fence are raised 10 feet with a constant force.

$W_1 = 3(10)(10) = 300$ ft-lb

The top 10 feet are raised with a variable force.

Weight of section: $3\,\Delta y$

Distance: $10 - y$

$$W_2 = \int_0^{10} 3(10 - y)\,dy = 3\left[10y - \frac{y^2}{2}\right]_0^{10} = 150 \text{ ft-lb}$$

$$W = W_1 + W_2 = 300 + 150 = 450 \text{ ft-lb}$$

33. Weight of section of chain: $3\,\Delta y$

Distance: $15 - 2y$

$$W = 3\int_0^{7.5} (15 - 2y)\,dy = \left[-\tfrac{3}{4}(15 - 2y)^2\right]_0^{7.5} = \tfrac{3}{4}(15)^2 = 168.75 \text{ ft-lb}$$

35. No. Something can require a lot of effort but take no work. There is no work because there is no change in distance.

37. $$W = \int_a^b G\frac{m_1 m_2}{x^2}\,dx = G\,m_1 m_2 \int_a^b x^{-2}\,dx = G\,m_1 m_2\left[-\frac{1}{x}\right]_a^b = G m_1 m_2\left(\frac{1}{a} - \frac{1}{b}\right)$$

39. $$F(x) = \frac{k}{(2 - x)^2}$$

$$W = \int_{-2}^{1} \frac{k}{(2 - x)^2}\,dx = \left[\frac{k}{2 - x}\right]_{-2}^{1} = k\left(1 - \frac{1}{4}\right) = \frac{3k}{4} \text{ (units of work)}$$

41. (a) $W = \int_0^9 6\,dx = 54$ ft-lb

(b) $W = \int_0^7 20\,dx + \int_7^9 (-10x + 90)\,dx = 140 + 20 = 160$ ft-lb

(c) $W = \int_0^9 \frac{1}{27}x^2\,dx = \left[\frac{x^3}{81}\right]_0^9 = 9$ ft-lb

(d) $W = \int_0^9 \sqrt{x}\,dx = \left[\frac{2}{3}x^{3/2}\right]_0^9 = \frac{2}{3}(27) = 18$ ft-lb

43. $$p = \frac{k}{V}$$

$$1000 = \frac{k}{2}$$

$$k = 2000$$

$$W = \int_2^3 \frac{2000}{V}\,dV = \left[2000 \ln|V|\right]_2^3 = 2000 \ln\left(\frac{3}{2}\right) \approx 810.93 \text{ ft-lb}$$

45. $W = \int_0^5 1000\left[1.8 - \ln(x + 1)\right]dx \approx 3249.44$ ft-lb

47. $W = \int_0^5 100x\sqrt{125 - x^3}\,dx \approx 10{,}330.3$ ft-lb

49. (a) $W \approx \frac{2 - 0}{3(6)}\left[0 + 4(20{,}000) + 2(22{,}000) + 4(15{,}000) + 2(10{,}000) + 4(5000) + 0\right] \approx 24888.889$ ft-lb

(b) $F(x) = -16{,}261.36x^4 + 85{,}295.45x^3 - 157{,}738.64x^2 + 104{,}386.36x - 32.4675$

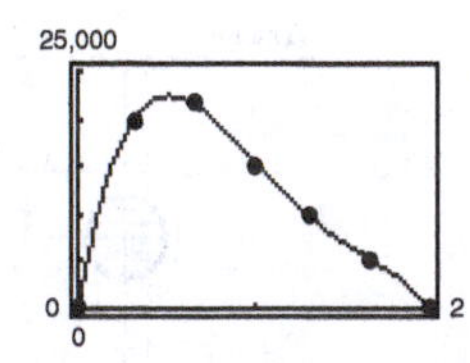

(c) $F(x)$ is a maximum when $x \approx 0.524$ feet.

(d) $W = \int_0^2 F(x)\,dx \approx 25{,}180.5$ ft-lb

Section 7.6 Moments, Centers of Mass, and Centroids

1. Weight is a force that is dependent on gravity. Mass is a measure of a body's resistance to change in motion and is independent of the gravitational system in which the body is located. The weight (or force) of an object is its mass times the acceleration due to gravity, $F = mg$.

3. A planar lamina is a flat plate of material of constant density. The center of mass of a planar lamina is its balancing point.

5. $\bar{x} = \dfrac{7(-5) + 3(0) + 5(3)}{7 + 3 + 5} = \dfrac{-20}{15} = -\dfrac{4}{3}$

7. $\bar{x} = \dfrac{1(6) + 3(10) + 2(3) + 9(2) + 5(4)}{1 + 3 + 2 + 9 + 5} = \dfrac{80}{20} = 4$

9.
$$48x = 72(L - x) = 72(10 - x)$$
$$48x = 720 - 72x$$
$$120x = 720$$
$$x = 6 \text{ ft}$$

11. $\bar{x} = \dfrac{5(2) + 1(-3) + 3(1)}{5 + 1 + 3} = \dfrac{10}{9}$

$\bar{y} = \dfrac{5(2) + 1(1) + 3(-4)}{5 + 1 + 3} = -\dfrac{1}{9}$

$(\bar{x}, \bar{y}) = \left(\dfrac{10}{9}, -\dfrac{1}{9}\right)$

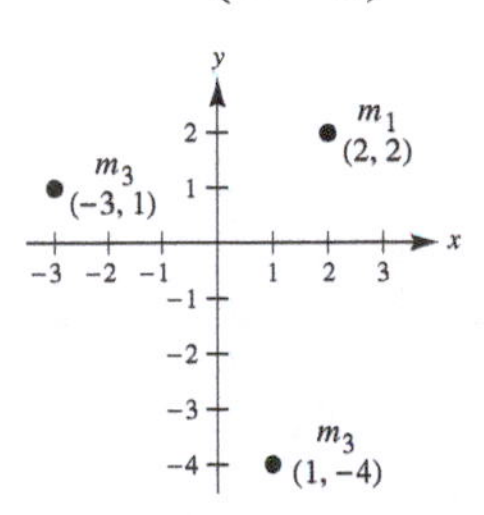

13. $\bar{x} = \dfrac{12(2) + 6(-1) + (9/2)(6) + 15(2)}{12 + 6 + (9/2) + 15} = \dfrac{75}{37.5} = 2$

$\bar{y} = \dfrac{12(3) + 6(5) + (9/2)(8) + 15(-2)}{12 + 6 + (9/2) + 15} = \dfrac{72}{37.5} = \dfrac{48}{25}$

$(\bar{x}, \bar{y}) = \left(2, \dfrac{48}{25}\right)$

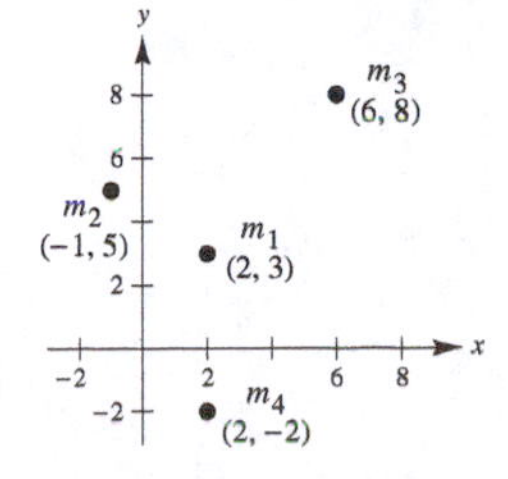

15. $m = \rho\displaystyle\int_0^2 \frac{x}{2}\,dx = \left[\rho\frac{x^2}{4}\right]_0^2 = \rho$

$M_x = \rho\displaystyle\int_0^2 \frac{1}{2}\left(\frac{x}{2}\right)^2 dx$

$= \dfrac{\rho}{8}\left[\dfrac{x^3}{3}\right]_0^2 = \dfrac{\rho}{3}$

$\bar{y} = \dfrac{M_x}{m} = \dfrac{\rho/3}{\rho} = \dfrac{1}{3}$

$M_y = \rho\displaystyle\int_0^2 x\left(\frac{x}{2}\right) dx = \frac{\rho}{2}\left[\frac{x^3}{3}\right]_0^2 = \frac{4}{3}\rho$

$\bar{x} = \dfrac{M_y}{m} = \dfrac{4/3\rho}{\rho} = \dfrac{4}{3}$

$(\bar{x}, \bar{y}) = \left(\dfrac{4}{3}, \dfrac{1}{3}\right)$

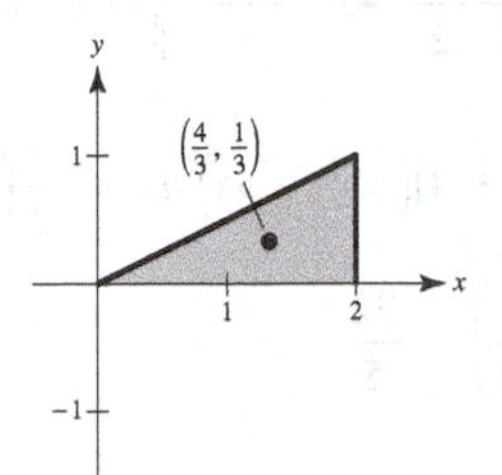

17. $m = \rho\displaystyle\int_0^4 \sqrt{x}\,dx = \left[\frac{2\rho}{3}x^{3/2}\right]_0^4 = \frac{16\rho}{3}$

$M_x = \rho\displaystyle\int_0^4 \frac{\sqrt{x}}{2}\left(\sqrt{x}\right) dx = \left[\rho\frac{x^2}{4}\right]_0^4 = 4\rho$

$\bar{y} = \dfrac{M_x}{m} = 4\rho\left(\dfrac{3}{16\rho}\right) = \dfrac{3}{4}$

$M_y = \rho\displaystyle\int_0^4 x\sqrt{x}\,dx = \left[\rho\frac{2}{5}x^{5/2}\right]_0^4 = \frac{64\rho}{5}$

$\bar{x} = \dfrac{M_y}{m} = \dfrac{64\rho}{5}\left(\dfrac{3}{16\rho}\right) = \dfrac{12}{5}$

$(\bar{x}, \bar{y}) = \left(\dfrac{12}{5}, \dfrac{3}{4}\right)$

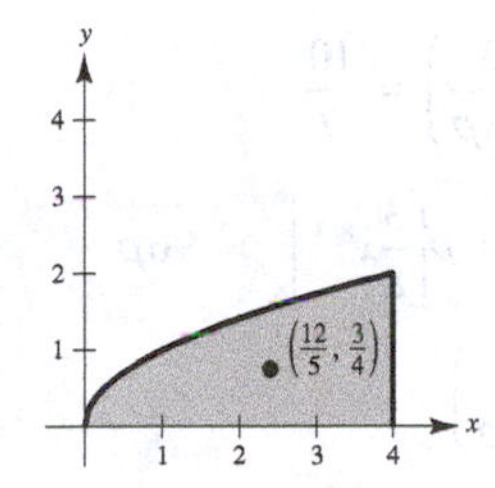

19. $m = \rho\int_0^1 \left(x^2 - x^3\right) dx = \rho\left[\frac{x^3}{3} - \frac{x^4}{4}\right]_0^1 = \frac{\rho}{12}$

$M_x = \rho\int_0^1 \frac{\left(x^2 + x^3\right)}{2}\left(x^2 - x^3\right) dx = \frac{\rho}{2}\int_0^1 \left(x^4 - x^6\right) dx = \frac{\rho}{2}\left[\frac{x^5}{5} - \frac{x^7}{7}\right]_0^1 = \frac{\rho}{35}$

$\overline{y} = \frac{M_x}{m} = \frac{\rho}{35}\left(\frac{12}{\rho}\right) = \frac{12}{35}$

$M_y = \rho\int_0^1 x\left(x^2 - x^3\right) dx = \rho\int_0^1 \left(x^3 - x^4\right) dx = \rho\left[\frac{x^4}{4} - \frac{x^5}{5}\right]_0^1 = \frac{\rho}{20}$

$\overline{x} = \frac{M_y}{m} = \frac{\rho}{20}\left(\frac{12}{\rho}\right) = \frac{3}{5}$

$\left(\overline{x}, \overline{y}\right) = \left(\frac{3}{5}, \frac{12}{35}\right)$

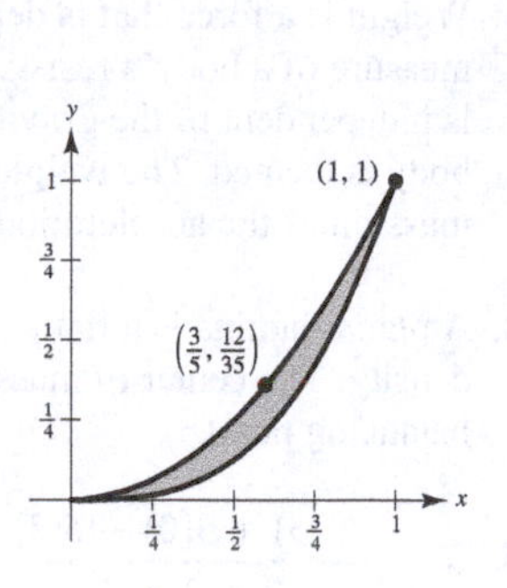

21. $m = \rho\int_0^3 \left[\left(-x^2 + 4x + 2\right) - (x + 2)\right] dx = -\rho\left[\frac{x^3}{3} + \frac{3x^2}{2}\right]_0^3 = \frac{9\rho}{2}$

$M_x = \rho\int_0^3 \left[\frac{\left(-x^2 + 4x + 2\right) + (x + 2)}{2}\right]\left[\left(-x^2 + 4x + 2\right) - (x + 2)\right] dx$

$= \frac{\rho}{2}\int_0^3 \left(-x^2 + 5x + 4\right)\left(-x^2 + 3x\right) dx = \frac{\rho}{2}\int_0^3 \left(x^4 - 8x^3 + 11x^2 + 12x\right) dx = \frac{\rho}{2}\left[\frac{x^5}{5} - 2x^4 + \frac{11x^3}{3} + 6x^2\right]_0^3 = \frac{99\rho}{5}$

$\overline{y} = \frac{M_x}{m} = \frac{99\rho}{5}\left(\frac{2}{9\rho}\right) = \frac{22}{5}$

$M_y = \rho\int_0^3 x\left[\left(-x^2 + 4x - 2\right) - (x + 2)\right] dx = \rho\int_0^3 \left(-x^3 + 3x^2\right) dx = \rho\left[-\frac{x^4}{4} + x^3\right]_0^3 = \frac{27\rho}{4}$

$\overline{x} = \frac{M_y}{m} = \frac{27\rho}{4}\left(\frac{2}{9\rho}\right) = \frac{3}{2}$

$\left(\overline{x}, \overline{y}\right) = \left(\frac{3}{2}, \frac{22}{5}\right)$

23. $m = \rho\int_0^8 x^{2/3}\, dx = \rho\left[\frac{3}{5}x^{5/3}\right]_0^8 = \frac{96\rho}{5}$

$M_x = \rho\int_0^8 \frac{x^{2/3}}{2}\left(x^{2/3}\right) dx = \frac{\rho}{2}\left[\frac{3}{7}x^{7/3}\right]_0^8 = \frac{192\rho}{7}$

$\overline{y} = \frac{M_x}{m} = \frac{192\rho}{7}\left(\frac{5}{96\rho}\right) = \frac{10}{7}$

$M_y = \rho\int_0^8 x\left(x^{2/3}\right) dx = \rho\left[\frac{3}{8}x^{8/3}\right]_0^8 = 96\rho$

$\overline{x} = \frac{M_y}{m} = 96\rho\left(\frac{5}{96\rho}\right) = 5$

$\left(\overline{x}, \overline{y}\right) = \left(5, \frac{10}{7}\right)$

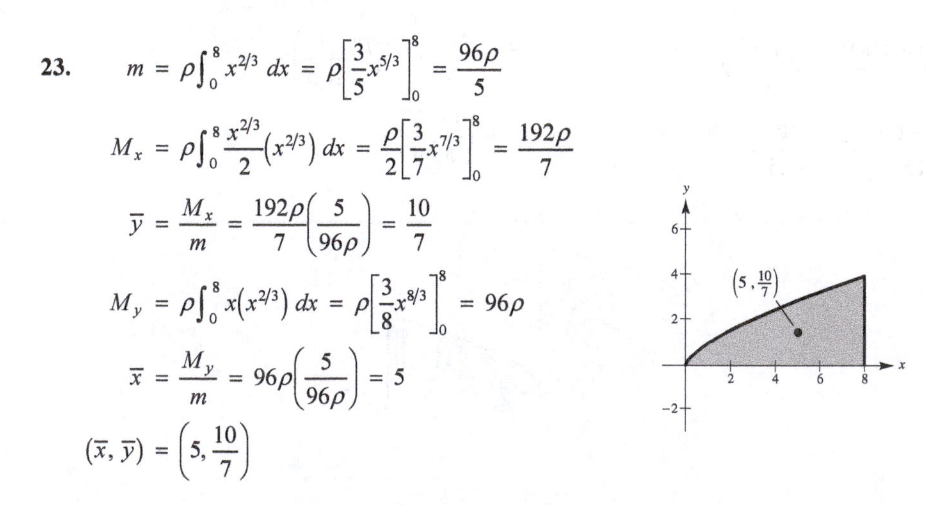

25. $m = 2\rho\int_0^2 (4 - y^2)\,dy = 2\rho\left[4y - \frac{y^3}{3}\right]_0^2 = \frac{32\rho}{3}$

$M_y = 2\rho\int_0^2\left(\frac{4-y^2}{2}\right)(4 - y^2)\,dy = \rho\left[16y - \frac{8}{3}y^3 + \frac{y^5}{5}\right]_0^2 = \frac{256\rho}{15}$

$\bar{x} = \frac{M_y}{m} = \frac{256\rho}{15}\left(\frac{3}{32\rho}\right) = \frac{8}{5}$

By symmetry, M_x and $\bar{y} = 0$.

$(\bar{x}, \bar{y}) = \left(\frac{8}{5}, 0\right)$

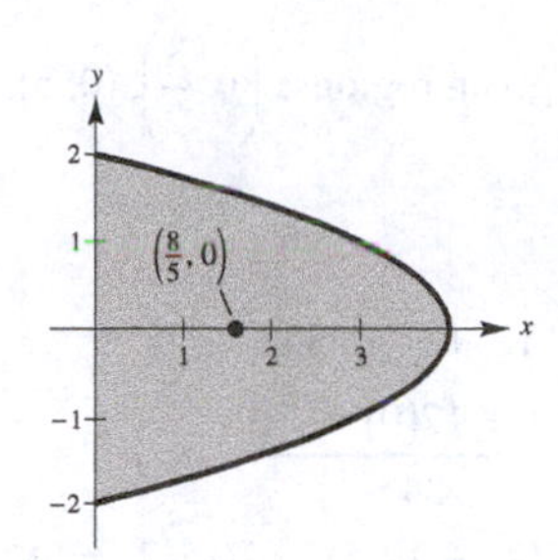

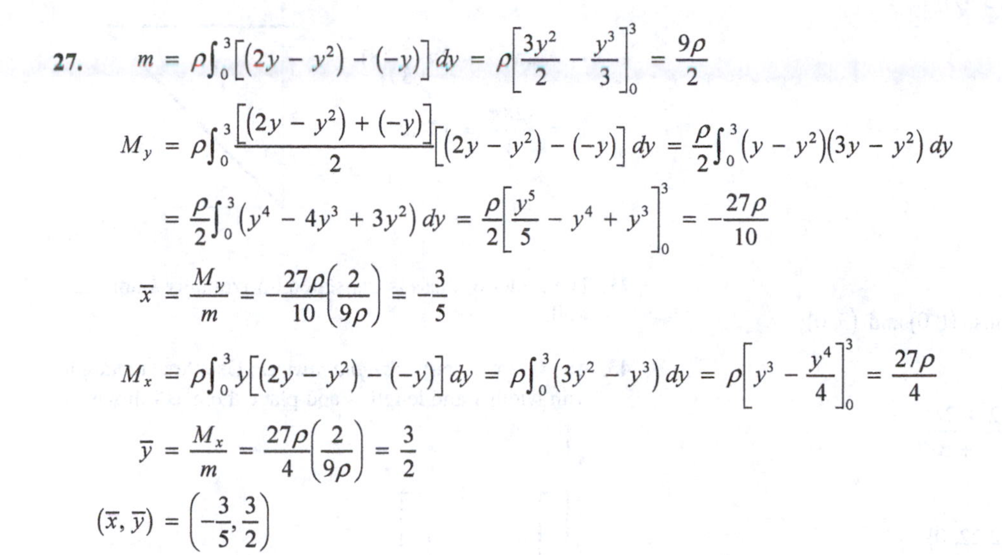

27. $m = \rho\int_0^3\left[(2y - y^2) - (-y)\right]dy = \rho\left[\frac{3y^2}{2} - \frac{y^3}{3}\right]_0^3 = \frac{9\rho}{2}$

$M_y = \rho\int_0^3 \frac{\left[(2y - y^2) + (-y)\right]}{2}\left[(2y - y^2) - (-y)\right]dy = \frac{\rho}{2}\int_0^3 (y - y^2)(3y - y^2)\,dy$

$= \frac{\rho}{2}\int_0^3 (y^4 - 4y^3 + 3y^2)\,dy = \frac{\rho}{2}\left[\frac{y^5}{5} - y^4 + y^3\right]_0^3 = -\frac{27\rho}{10}$

$\bar{x} = \frac{M_y}{m} = -\frac{27\rho}{10}\left(\frac{2}{9\rho}\right) = -\frac{3}{5}$

$M_x = \rho\int_0^3 y\left[(2y - y^2) - (-y)\right]dy = \rho\int_0^3 (3y^2 - y^3)\,dy = \rho\left[y^3 - \frac{y^4}{4}\right]_0^3 = \frac{27\rho}{4}$

$\bar{y} = \frac{M_x}{m} = \frac{27\rho}{4}\left(\frac{2}{9\rho}\right) = \frac{3}{2}$

$(\bar{x}, \bar{y}) = \left(-\frac{3}{5}, \frac{3}{2}\right)$

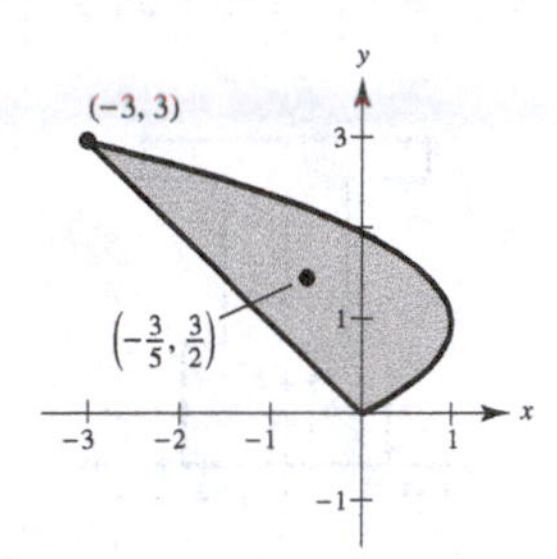

29. $m = \rho\int_{-20}^{20} 5\sqrt[3]{400 - x^2}\,dx \approx 1239.76\rho$

$M_x = \rho\int_{-20}^{20} \frac{5\sqrt[3]{400 - x^2}}{2}\left(5\sqrt[3]{400 - x^2}\right)dx$

$= \frac{25\rho}{2}\int_{-20}^{20} (400 - x^2)^{2/3}\,dx \approx 20064.27$

$\bar{y} = \frac{M_x}{m} \approx 16.18$

$\bar{x} = 0$ by symmetry. Therefore, the centroid is $(0, 16.2)$.

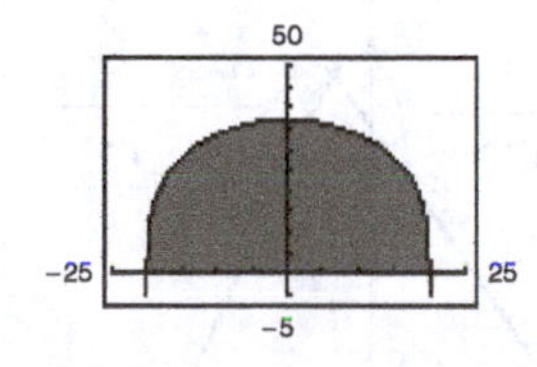

31. Centroids of the given regions: $(1, 0)$ and $(3, 0)$

Area: $A = 4 + \pi$

$\bar{x} = \frac{4(1) + \pi(3)}{4 + \pi} = \frac{4 + 3\pi}{4 + \pi}$

$\bar{y} = \frac{4(0) + \pi(0)}{4 + \pi} = 0$

$(\bar{x}, \bar{y}) = \left(\frac{4 + 3\pi}{4 + \pi}, 0\right) \approx (1.88, 0)$

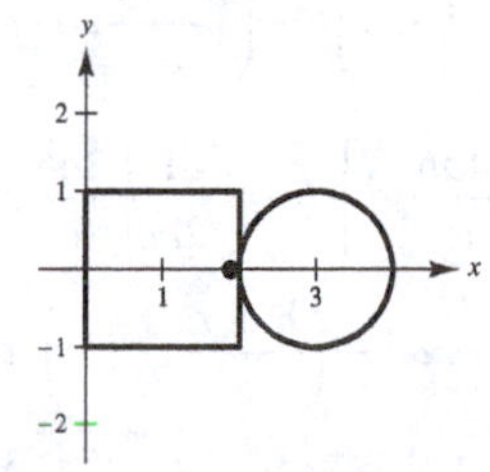

33. Centroids of the given regions: $\left(0, \frac{3}{2}\right)$, $(0, 5)$, and $\left(0, \frac{15}{2}\right)$

Area: $A = 15 + 12 + 7 = 34$

$$\bar{x} = \frac{15(0) + 12(0) + 7(0)}{34} = 0$$

$$\bar{y} = \frac{15(3/2) + 12(5) + 7(15/2)}{34} = \frac{135}{34}$$

$$(\bar{x}, \bar{y}) = \left(0, \frac{135}{34}\right) \approx (0, 3.97)$$

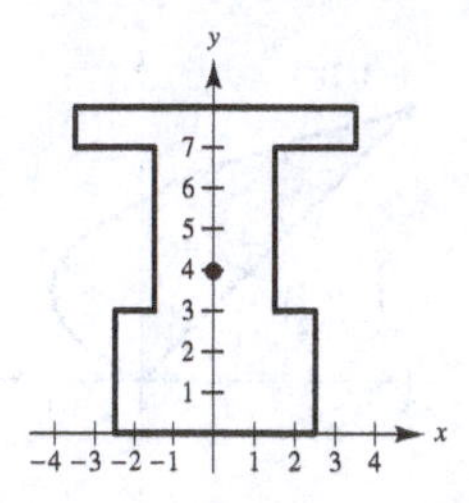

35. Centroids of the given regions: $(1, 0)$ and $(3, 0)$

Mass: $4 + 2\pi$

$$\bar{x} = \frac{4(1) + 2\pi(3)}{4 + 2\pi} = \frac{2 + 3\pi}{2 + \pi}$$

$$\bar{y} = 0$$

$$(\bar{x}, \bar{y}) = \left(\frac{2 + 3\pi}{2 + \pi}, 0\right) \approx (2.22, 0)$$

37. $r = 5$ is distance between center of circle and y-axis.

$A \approx \pi(4)^2 = 16\pi$ is the area of circle. So,

$V = 2\pi rA = 2\pi(5)(16\pi) = 160\pi^2 \approx 1579.14.$

39. $A = \frac{1}{2}(4)(4) = 8$

$$\bar{y} = \left(\frac{1}{8}\right)\frac{1}{2}\int_0^4 (4 + x)(4 - x)\,dx = \frac{1}{16}\left[16x - \frac{x^3}{3}\right]_0^4 = \frac{8}{3}$$

$$r = \bar{y} = \frac{8}{3}$$

$$V = 2\pi rA = 2\pi\left(\frac{8}{3}\right)(8) = \frac{128\pi}{3} \approx 134.04$$

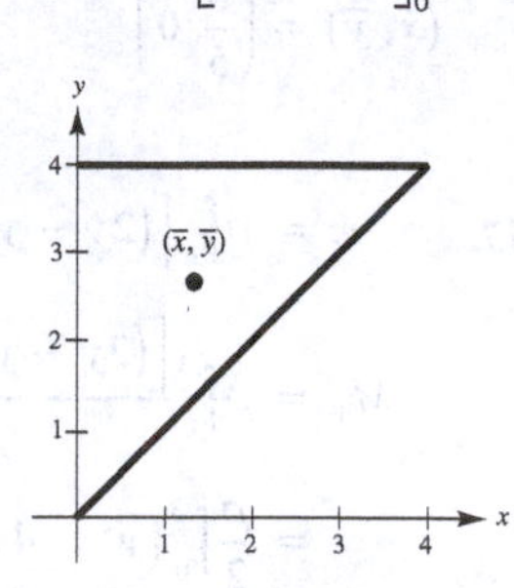

41. The center of mass is translated horizontally k units as well.

43. Answers will vary. *Sample answer*: Use three rectangles with width 1 and length 4 and place them as follows.

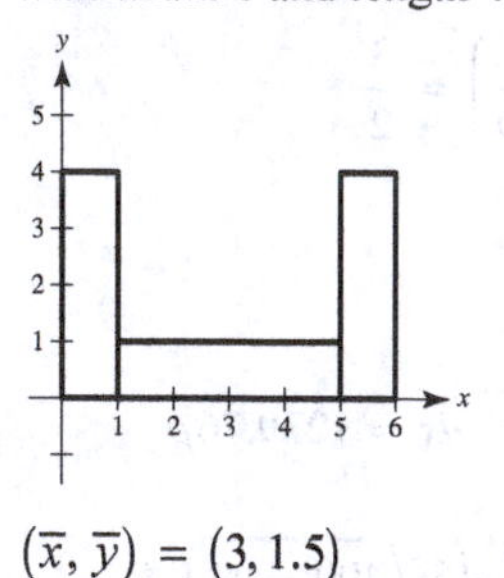

$(\bar{x}, \bar{y}) = (3, 1.5)$

45.
$$A = \frac{1}{2}(2a)c = ac$$

$$\frac{1}{A} = \frac{1}{ac}$$

$$\bar{x} = \left(\frac{1}{ac}\right)\frac{1}{2}\int_0^c \left[\left(\frac{b - a}{c}y + a\right)^2 - \left(\frac{b + a}{c}y - a\right)^2\right] dy$$

$$= \frac{1}{2ac}\int_0^c \left[\frac{4ab}{c}y - \frac{4ab}{c^2}y^2\right] dy = \frac{1}{2ac}\left[\frac{2ab}{c}y^2 - \frac{4ab}{3c^2}y^3\right]_0^c = \frac{1}{2ac}\left(\frac{2}{3}abc\right) = \frac{b}{3}$$

$$\bar{y} = \frac{1}{ac}\int_0^c y\left[\left(\frac{b - a}{c}y + a\right) - \left(\frac{b + a}{c}y - a\right)\right] dy$$

$$= \frac{1}{ac}\int_0^c y\left(-\frac{2a}{c}y + 2a\right) dy = \frac{2}{c}\int_0^c \left(y - \frac{y^2}{c}\right) dy = \frac{2}{c}\left[\frac{y^2}{2} - \frac{y^3}{3c}\right]_0^c = \frac{c}{3}$$

$$(\bar{x}, \bar{y}) = \left(\frac{b}{3}, \frac{c}{3}\right)$$

From elementary geometry, $(b/3, c/3)$ is the point of intersection of the medians.

47. $A = \frac{c}{2}(a + b)$

$$\frac{1}{A} = \frac{2}{c(a + b)}$$

$$\bar{x} = \frac{2}{c(a + b)}\int_0^c x\left(\frac{b - a}{c}x + a\right) dx = \frac{2}{c(a + b)}\int_0^c \left(\frac{b - a}{c}x^2 + ax\right) dx = \frac{2}{c(a + b)}\left[\frac{b - a}{c}\frac{x^3}{3} + \frac{ax^2}{2}\right]_0^c$$

$$= \frac{2}{c(a + b)}\left[\frac{(b - a)c^2}{3} + \frac{ac^2}{2}\right] = \frac{2}{c(a + b)}\left[\frac{2bc^2 - 2ac^2 + 3ac^2}{6}\right] = \frac{c(2b + a)}{3(a + b)} = \frac{(a + 2b)c}{3(a + b)}$$

$$\bar{y} = \frac{2}{c(a + b)}\frac{1}{2}\int_0^c \left(\frac{b - a}{c}x + a\right)^2 dx = \frac{1}{c(a + b)}\int_0^c \left[\left(\frac{b - a}{c}\right)^2 x^2 + \frac{2a(b - a)}{c}x + a^2\right] dx$$

$$= \frac{1}{c(a + b)}\left[\left(\frac{b - a}{c}\right)^2 \frac{x^3}{3} + \frac{2a(b - a)}{c}\frac{x^2}{2} + a^2x\right]_0^c = \frac{1}{c(a + b)}\left[\frac{(b - a)^2 c}{3} + ac(b - a) + a^2c\right]$$

$$= \frac{1}{3c(a + b)}\left[(b^2 - 2ab + a^2)c + 3ac(b - a) + 3a^2c\right]$$

$$= \frac{1}{3(a + b)}\left[b^2 - 2ab + a^2 + 3ab - 3a^2 + 3a^2\right] = \frac{a^2 + ab + b^2}{3(a + b)}$$

So, $(\bar{x}, \bar{y}) = \left(\frac{(a + 2b)c}{3(a + b)}, \frac{a^2 + ab + b^2}{3(a + b)}\right)$.

The one line passes through $\left(0, \frac{a}{2}\right)$ and $\left(c, \frac{b}{2}\right)$. Its equation is $y = \frac{b - a}{2c}x + \frac{a}{2}$. The other line passes through $(0, -b)$ and $(c, a + b)$. Its equation is $y = \frac{a + 2b}{c}x - b$. $(\bar{x}, \bar{y})$ is the point of intersection of these two lines.

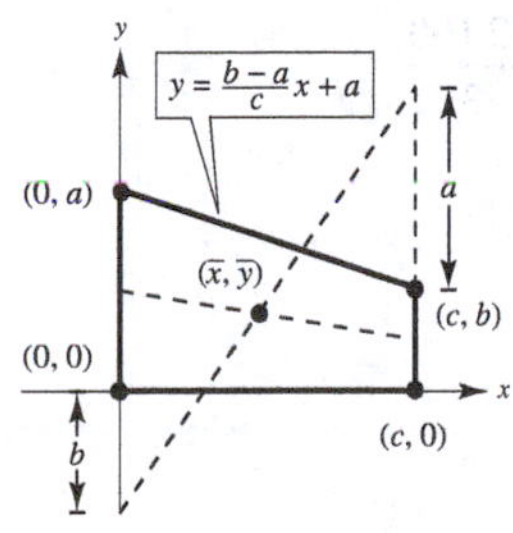

49. $\bar{x} = 0$ by symmetry.

$$A = \frac{1}{2}\pi ab$$

$$\frac{1}{A} = \frac{2}{\pi ab}$$

$$\bar{y} = \frac{2}{\pi ab}\frac{1}{2}\int_{-a}^{a}\left(\frac{b}{a}\sqrt{a^2 - x^2}\right)^2 dx$$

$$= \frac{1}{\pi ab}\left(\frac{b^2}{a^2}\right)\left[a^2x - \frac{x^3}{3}\right]_{-a}^{a} = \frac{b}{\pi a^3}\left(\frac{4a^3}{3}\right) = \frac{4b}{3\pi}$$

$$(\bar{x}, \bar{y}) = \left(0, \frac{4b}{3\pi}\right)$$

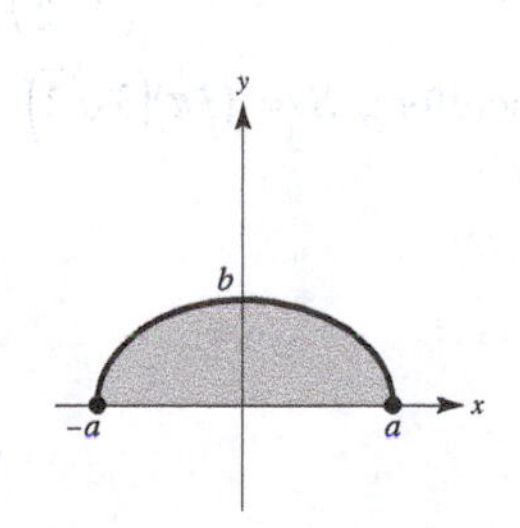

51. (a)

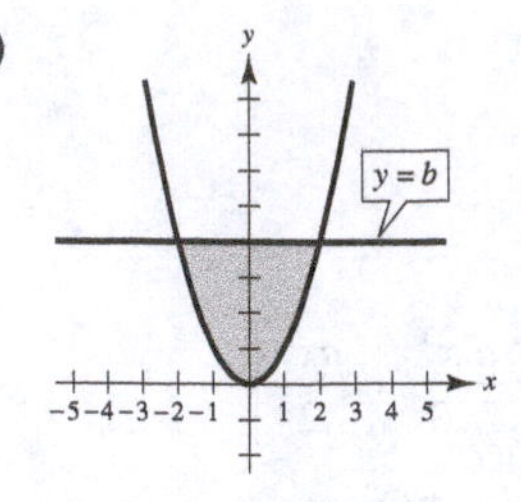

(b) $\bar{x} = 0$ by symmetry.

(c) $M_y = \int_{-\sqrt{b}}^{\sqrt{b}} x(b - x^2)\,dx = 0$ because $bx - x^3$ is odd.

(d) $\bar{y} > \frac{b}{2}$ because there is more area above $y = \frac{b}{2}$ than below.

(e) $M_x = \int_{-\sqrt{b}}^{\sqrt{b}} \frac{(b + x^2)(b - x^2)}{2}\,dx = \int_{-\sqrt{b}}^{\sqrt{b}} \frac{b^2 - x^4}{2}\,dx = \frac{1}{2}\left[b^2x - \frac{x^5}{5}\right]_{-\sqrt{b}}^{\sqrt{b}} = b^2\sqrt{b} - \frac{b^2\sqrt{b}}{5} = \frac{4b^2\sqrt{b}}{5}$

$A = \int_{-\sqrt{b}}^{\sqrt{b}} (b - x^2)\,dx = \left[bx - \frac{x^3}{3}\right]_{-\sqrt{b}}^{\sqrt{b}} = \left(b\sqrt{b} - \frac{b\sqrt{b}}{3}\right)2 = 4\frac{b\sqrt{b}}{3}$

$\bar{y} = \frac{M_x}{A} = \frac{4b^2\sqrt{b}/5}{4b\sqrt{b}/3} = \frac{3}{5}b$

53. (a) $\bar{x} = 0$ by symmetry.

$A = 2\int_0^{40} f(x)\,dx = \frac{2(40)}{3(4)}\left[30 + 4(29) + 2(26) + 4(20) + 0\right] = \frac{20}{3}(278) = \frac{5560}{3}$

$M_x = \int_{-40}^{40} \frac{f(x)^2}{2}\,dx = \frac{40}{3(4)}\left[30^2 + 4(29)^2 + 2(26)^2 + 4(20)^2 + 0\right] = \frac{10}{3}(7216) = \frac{72{,}160}{3}$

$\bar{y} = \frac{M_x}{A} = \frac{72{,}160/3}{5560/3} = \frac{72{,}160}{5560} \approx 12.98$

$(\bar{x}, \bar{y}) = (0, 12.98)$

(b) $y = (-1.02 \times 10^{-5})x^4 - 0.0019x^2 + 29.28$ (Use nine data points.)

(c) $\bar{y} = \frac{M_x}{A} \approx \frac{23{,}697.68}{1843.54} \approx 12.85$

$(\bar{x}, \bar{y}) = (0, 12.85)$

55. The centroid of the line joining $(0, 3)$ and $(3, 0)$ is $\left(\frac{3}{2}, \frac{3}{2}\right)$. The distance traveled by the centroid is $2\pi\left(\frac{3}{2}\right) = 3\pi$.

The arc length of C is $3\sqrt{2}$. Therefore, $S = (3\pi)(3\sqrt{2}) = 9\sqrt{2}\,\pi$.

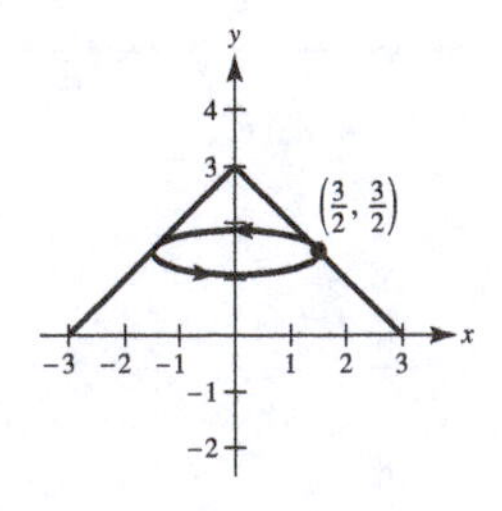

57. $A = \int_0^1 x^n\,dx = \left[\frac{x^{n+1}}{n+1}\right]_0^1 = \frac{1}{n+1}$

$m = \rho A = \frac{\rho}{n+1}$

$M_x = \frac{\rho}{2}\int_0^1 (x^n)^2\,dx = \left[\frac{\rho}{2}\cdot\frac{x^{2n+1}}{2n+1}\right]_0^1 = \frac{\rho}{2(2n+1)}$

$M_y = \rho\int_0^1 x(x^n)\,dx = \left[\rho\cdot\frac{x^{n+2}}{n+2}\right]_0^1 = \frac{\rho}{n+2}$

$\bar{x} = \frac{M_y}{m} = \frac{n+1}{n+2}$

$\bar{y} = \frac{M_x}{m} = \frac{n+1}{2(2n+1)} = \frac{n+1}{4n+2}$

Centroid: $\left(\frac{n+1}{n+2}, \frac{n+1}{4n+2}\right)$

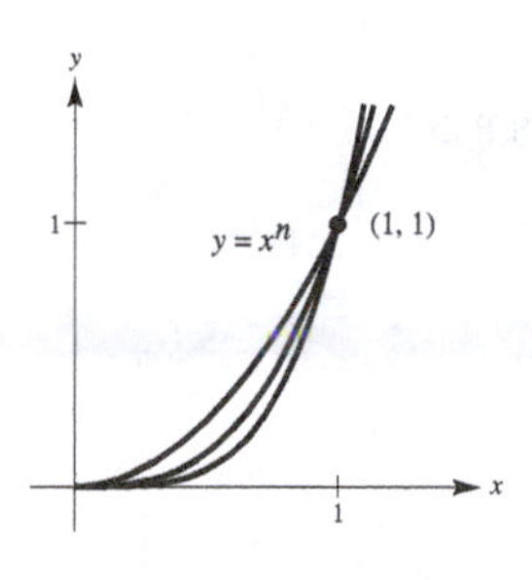

As $n \to \infty$, $(\bar{x}, \bar{y}) \to \left(1, \frac{1}{4}\right)$. The graph approaches the x-axis and the line $x = 1$ as $n \to \infty$.

59. Let T be the shaded triangle with vertices $(-1, 4)$, $(1, 4)$, and $(0, 3)$. Let U be the large triangle with vertices $(-4, 4)$, $(4, 4)$, and $(0, 0)$. V consists of the region U minus the region T.

Centroid of T: $\left(0, \frac{11}{3}\right)$; Area $= 1$

Centroid of U: $\left(0, \frac{8}{3}\right)$; Area $= 16$

Area: $V = 16 - 1 = 15$

$\bar{x} = 0$ by symmetry.

$15\bar{y} + 1\left(\frac{11}{3}\right) = 16\left(\frac{8}{3}\right)$

$15\bar{y} = \frac{117}{3}$

$\bar{y} = \frac{13}{5}$

$(\bar{x}, \bar{y}) = \left(0, \frac{13}{5}\right)$

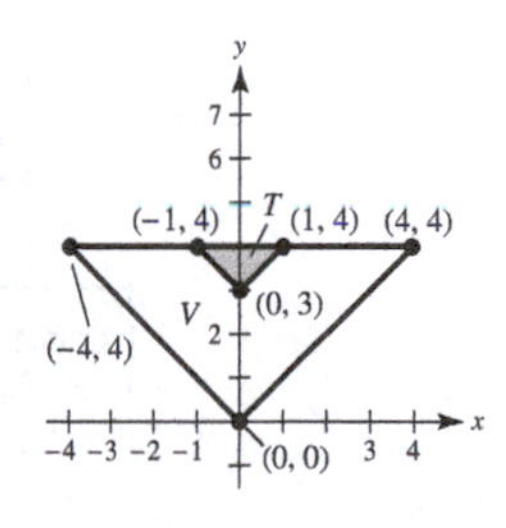

Section 7.7 Fluid Pressure and Fluid Force

1. Fluid pressure is the force per unit area over the surface of a body submerged in a fluid.

3. $F = PA = [62.4(8)]3 = 1497.6$ lb

5. $F = PA = [62.4(8)]10 = 4992$ lb

7. The weight-density of ethyl alcohol is 49.4 lb/ft^3.

$P = wh = (49.4)(5) = 247$ lb/ft^2

Fluid force $= F = PA = (247)(9) = 2223$ lb

9. $h(y) = 3 - y$

$L(y) = 4$

$F = 62.4\int_0^3 (3 - y)(4)\,dy$

$= 249.6\int_0^3 (3 - y)\,dy$

$= 249.6\left[3y - \frac{y^2}{2}\right]_0^3 = 1123.2$ lb

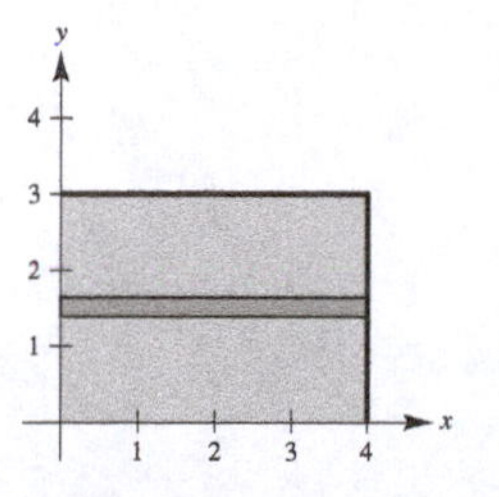

11. $h(y) = 3 - y$

$L(y) = 2\left(\dfrac{y}{3} + 1\right)$

$$F = 2(62.4)\int_0^3 (3 - y)\left(\frac{y}{3} + 1\right) dy$$

$$= 124.8\int_0^3 \left(3 - \frac{y^2}{3}\right) dy$$

$$= 124.8\left[3y - \frac{y^3}{9}\right]_0^3 = 748.8 \text{ lb}$$

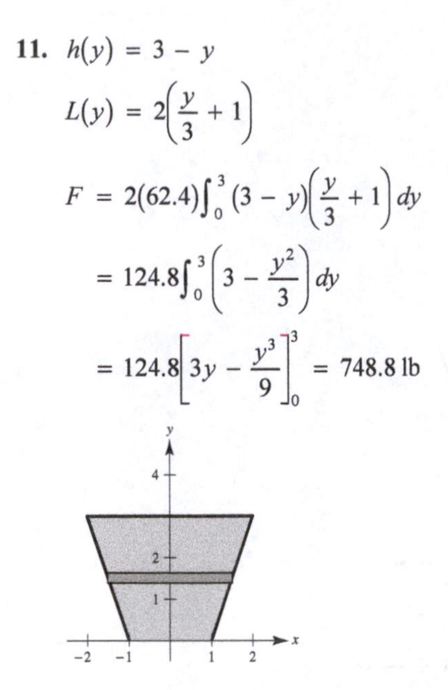

13. $h(y) = 4 - y$

$L(y) = 2\sqrt{y}$

$$F = 2(62.4)\int_0^4 (4 - y)\sqrt{y}\, dy$$

$$= 124.8\int_0^4 \left(4y^{1/2} - y^{3/2}\right) dy$$

$$= 124.8\left[\frac{8y^{3/2}}{3} - \frac{2y^{5/2}}{5}\right]_0^4 = 1064.96 \text{ lb}$$

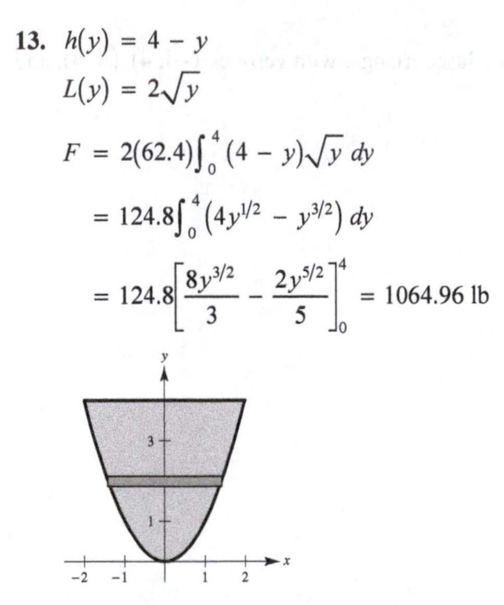

15. $h(y) = 4 - y$

$L(y) = 2$

$$F = 9800\int_0^2 2(4 - y)\, dy$$

$$= 9800\left[8y - y^2\right]_0^2 = 117{,}600 \text{ newtons}$$

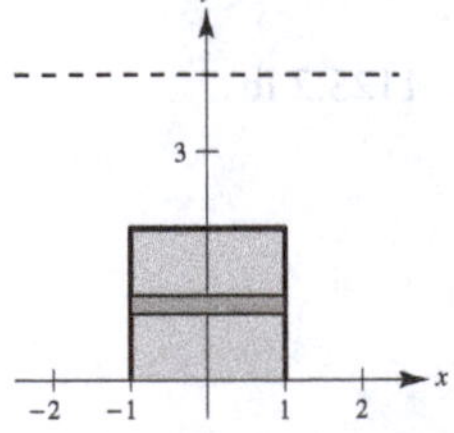

17. $h(y) = 12 - y$

$L(y) = 6 - \dfrac{2y}{3}$

$$F = 9800\int_0^9 (12 - y)\left(6 - \frac{2y}{3}\right) dy$$

$$= 9800\left[72y - 7y^2 + \frac{2y^3}{9}\right]_0^9$$

$$= 2{,}381{,}400 \text{ newtons}$$

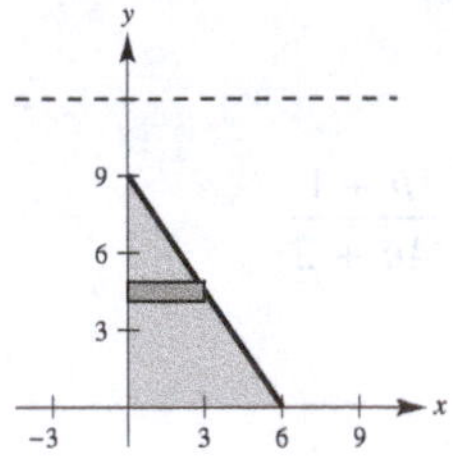

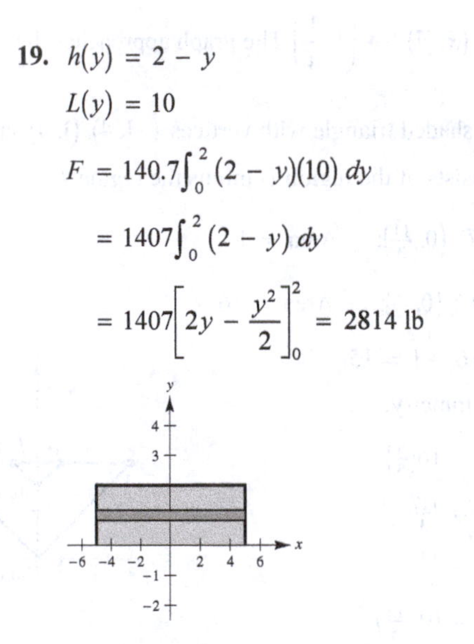

19. $h(y) = 2 - y$

$L(y) = 10$

$$F = 140.7\int_0^2 (2 - y)(10)\, dy$$

$$= 1407\int_0^2 (2 - y)\, dy$$

$$= 1407\left[2y - \frac{y^2}{2}\right]_0^2 = 2814 \text{ lb}$$

21. $h(y) = 4 - y$

$L(y) = 6$

$$F = 140.7\int_0^4 (4 - y)(6)\, dy$$

$$= 844.2\int_0^4 (4 - y)\, dy$$

$$= 844.2\left[4y - \frac{y^2}{2}\right]_0^4 = 6753.6 \text{ lb}$$

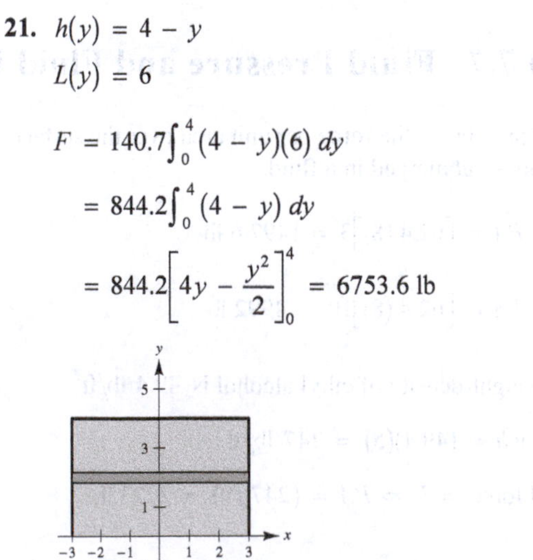

23. $h(y) = -y$

$L(y) = 2\left(\frac{1}{2}\right)\sqrt{9 - 4y^2}$

$$F = 42\int_{-3/2}^{0}(-y)\sqrt{9 - 4y^2}\,dy$$

$$= \frac{42}{8}\int_{-3/2}^{0}\left(9 - 4y^2\right)^{1/2}(-8y)\,dy$$

$$= \left[\left(\frac{21}{4}\right)\left(\frac{2}{3}\right)\left(9 - 4y^2\right)^{3/2}\right]_{-3/2}^{0}$$

$$= 94.5 \text{ lb}$$

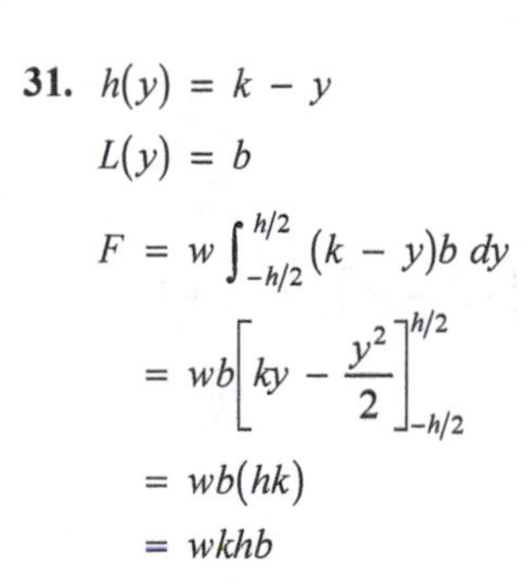

25. You use horizontal representative rectangles because you are measuring total force against a region between two depths.

27. If the fluid force is one-half of 1123.2 lb, and the height of the water is b, then

$h(y) = b - y$

$L(y) = 4$

$$F = 62.4\int_0^b (b - y)(4)\,dy = \frac{1}{2}(1123.2)$$

$$\int_0^b (b - y)\,dy = 2.25$$

$$\left[by - \frac{y^2}{2}\right]_0^b = 2.25$$

$$b^2 - \frac{b^2}{2} = 2.25$$

$$b^2 = 4.5 \Rightarrow b \approx 2.12 \text{ ft.}$$

The pressure increases with increasing depth.

29. $h(y) = k - y$

$L(y) = 2\sqrt{r^2 - y^2}$

$$F = w\int_{-r}^{r}(k - y)\sqrt{r^2 - y^2}(2)\,dy$$

$$= w\left[2k\int_{-r}^{r}\sqrt{r^2 - y^2}\,dy + \int_{-r}^{r}\sqrt{r^2 - y^2}(-2y)\,dy\right]$$

The second integral is zero because its integrand is odd and the limits of integration are symmetric to the origin. The first integral is the area of a semicircle with radius r.

$$F = w\left[(2k)\frac{\pi r^2}{2} + 0\right]$$

$$= wk\pi r^2$$

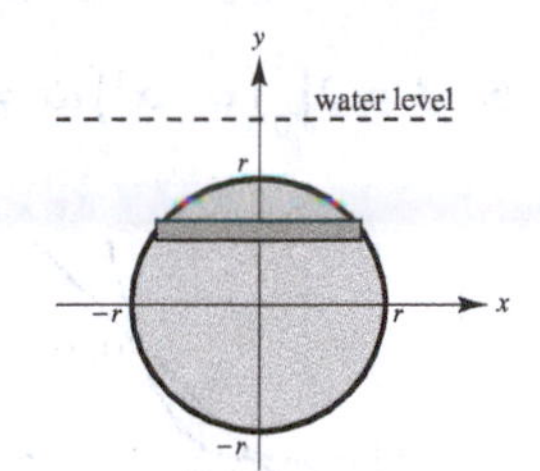

31. $h(y) = k - y$

$L(y) = b$

$$F = w\int_{-h/2}^{h/2}(k - y)b\,dy$$

$$= wb\left[ky - \frac{y^2}{2}\right]_{-h/2}^{h/2}$$

$$= wb(hk)$$

$$= wkhb$$

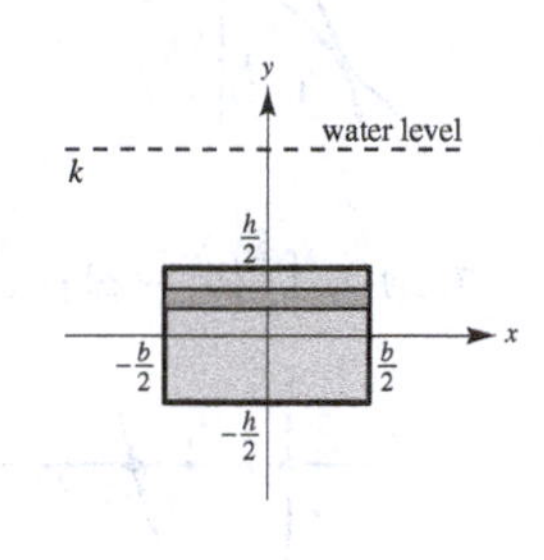

33. From Exercise 31:

$$F = 64(15)(1)(1) = 960 \text{ lb}$$

35. $h(y) = 4 - y$

$$F = 64.0\int_0^4 (4 - y)L(y)\,dy$$

Using the Midpoint Rule with $n = 4$, you have $\Delta x = \frac{4}{4} = 1$ and

$$F \approx 64.0\left[3.5\left(\frac{0 + 5}{2}\right) + 2.5\left(\frac{5 + 9}{2}\right) + 1.5\left(\frac{9 + 10.25}{2}\right) + 0.5\left(\frac{10.25 + 10.5}{2}\right)\right] = 64.0(45.875) = 2936 \text{ lb}$$

Review Exercises for Chapter 7

1. $$A = \int_{-2}^{2}\left[\left(6 - \frac{x^2}{2}\right) - \frac{3}{4}x\right]dx$$

$$= \left[6x - \frac{x^3}{6} - \frac{3x^2}{8}\right]_{-2}^{2}$$

$$= \left(12 - \frac{4}{3} - \frac{3}{2}\right) - \left(-12 + \frac{4}{3} - \frac{3}{2}\right) = \frac{64}{3}$$

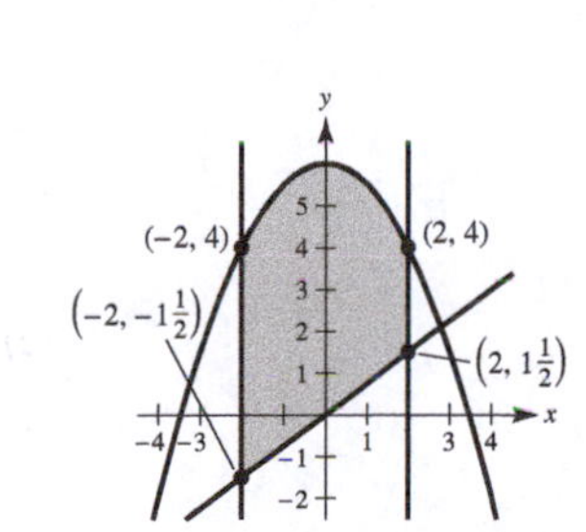

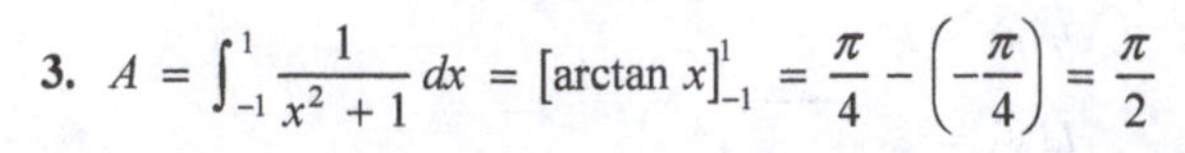

3. $A = \int_{-1}^{1} \frac{1}{x^2+1}\, dx = \left[\arctan x\right]_{-1}^{1} = \frac{\pi}{4} - \left(-\frac{\pi}{4}\right) = \frac{\pi}{2}$

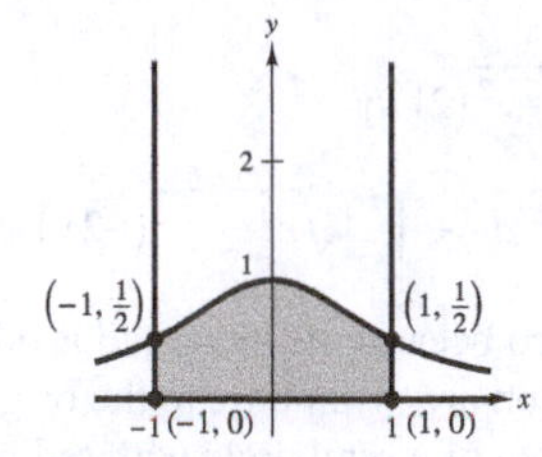

5. $A = 2\int_{0}^{1} \left(x - x^3\right) dx = 2\left[\frac{1}{2}x^2 - \frac{1}{4}x^4\right]_{0}^{1} = \frac{1}{2}$

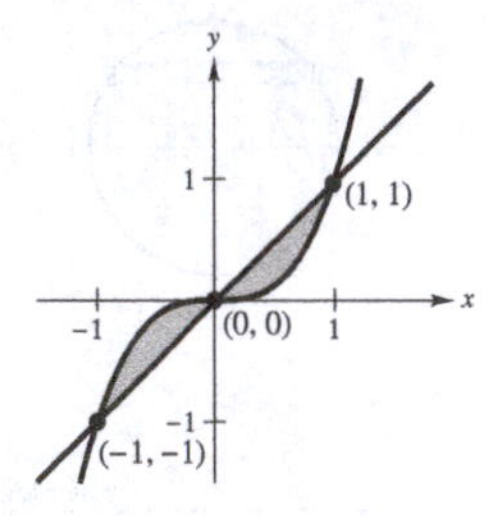

7. $A = \int_{0}^{2} \left(e^2 - e^x\right) dx = \left[xe^2 - e^x\right]_{0}^{2} = e^2 + 1$

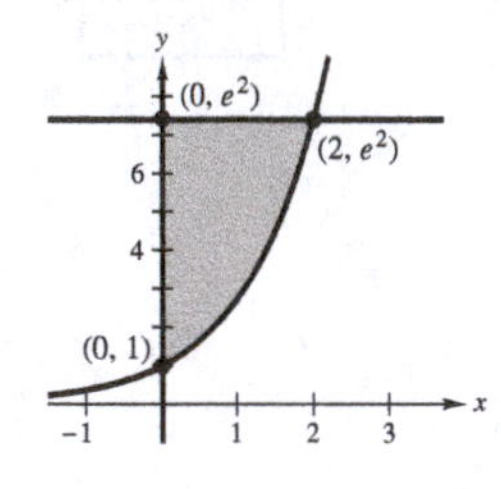

9. $A = \int_{\pi/4}^{5\pi/4} (\sin x - \cos x)\, dx$

$= \left[-\cos x - \sin x\right]_{\pi/4}^{5\pi/4}$

$= \left(\frac{1}{\sqrt{2}} + \frac{1}{\sqrt{2}}\right) - \left(-\frac{1}{\sqrt{2}} - \frac{1}{\sqrt{2}}\right)$

$= \frac{4}{\sqrt{2}} = 2\sqrt{2}$

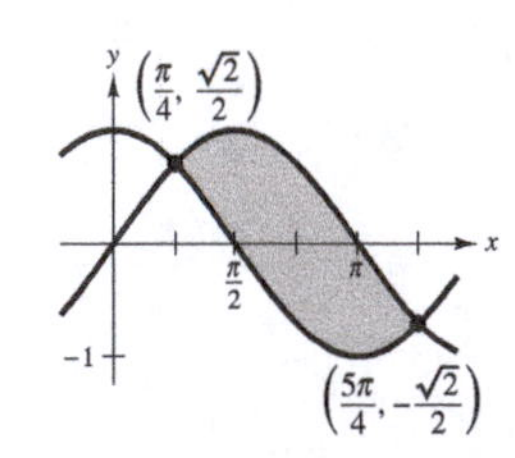

11. (a)

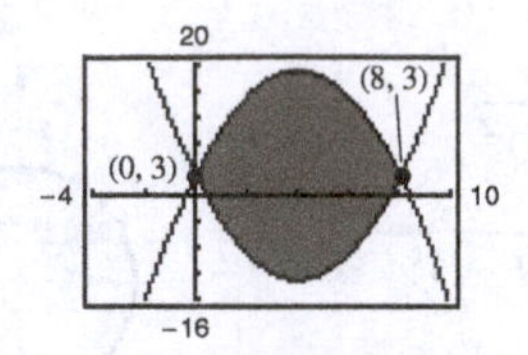

(b) Points of intersection:

$x^2 - 8x + 3 = 3 + 8x - x^2$

$2x^2 - 16x = 0$ when $x = 0, 8$

$A = \int_{0}^{8} \left[\left(3 + 8x - x^2\right) - \left(x^2 - 8x + 3\right)\right] dx$

≈ 170.6667

13. (a)

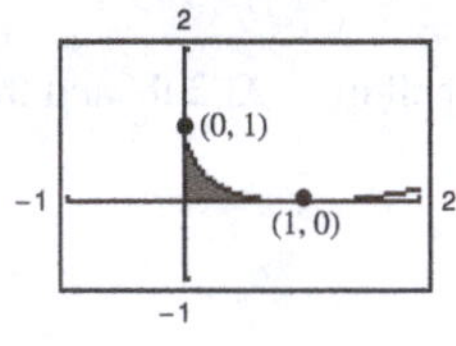

(b) $y = \left(1 - \sqrt{x}\right)^2$

$A = \int_{0}^{1} \left(1 - \sqrt{x}\right)^2 dx \approx 0.1667$

15. $F(x) = \int_{0}^{x} (3t + 1)\, dt = \left[\frac{3t^2}{2} + t\right]_{0}^{x} = \frac{3x^2}{2} + x$

(a) $F(0) = 0$

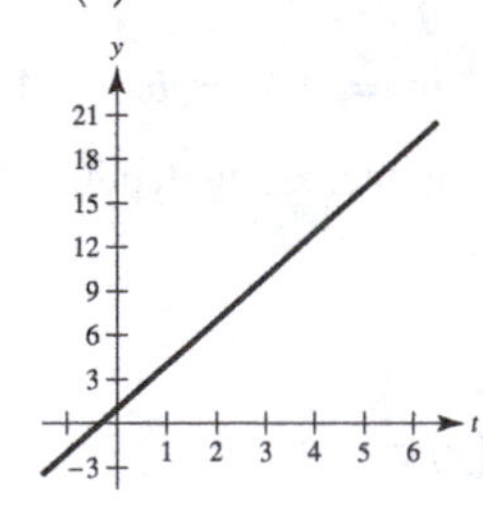

(b) $F(2) = \frac{3(2)^2}{2} + 2 = 8$

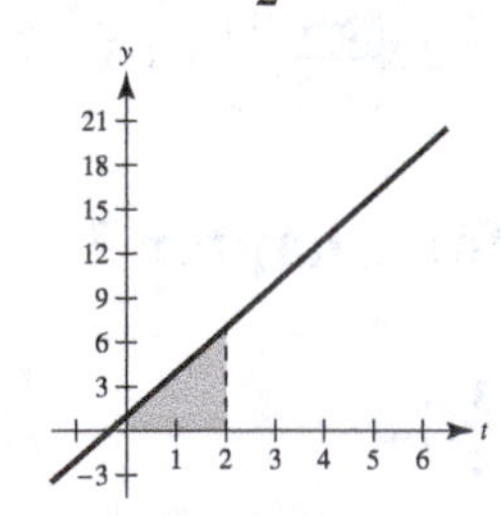

(c) $F(6) = \frac{3(6^2)}{2} + 6 = 60$

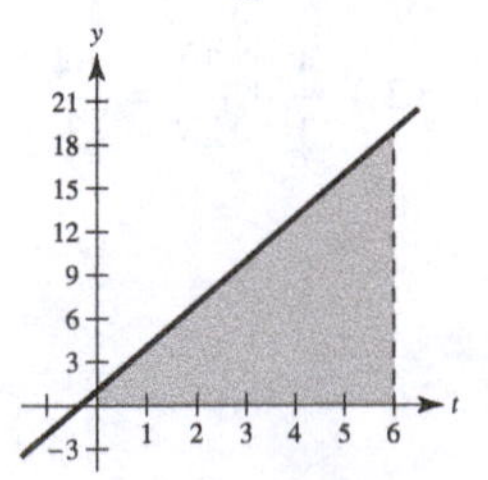

17. R_1 projects more revenue.

$$\int_0^5 \left[(2.98 + 0.65t) - (2.98 + 0.56t)\right] dt = \int_0^5 0.09t\, dt$$
$$= \left[\frac{0.09t^2}{2}\right]_0^5 = \$1.125 \text{ million}$$

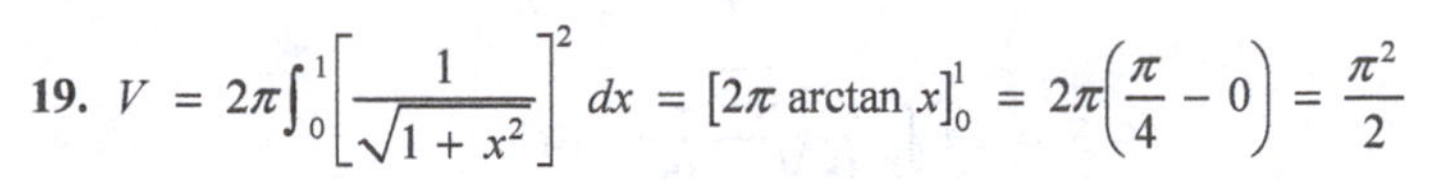

19. $V = 2\pi\int_0^1\left[\frac{1}{\sqrt{1 + x^2}}\right]^2 dx = \left[2\pi \arctan x\right]_0^1 = 2\pi\left(\frac{\pi}{4} - 0\right) = \frac{\pi^2}{2}$

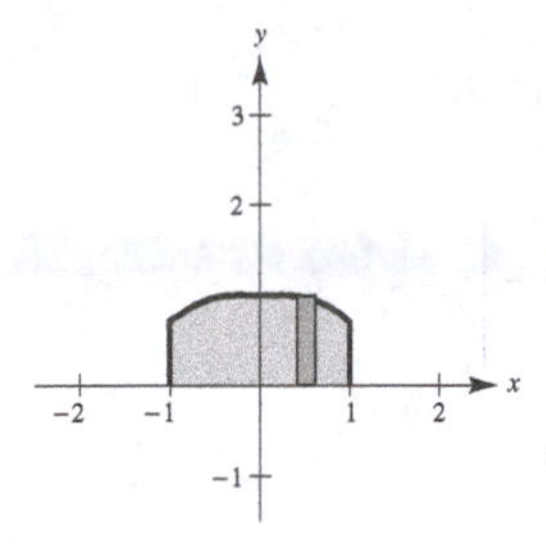

21. $V = 2\pi\int_0^1 \frac{x}{x^4 + 1}\, dx = \pi\int_0^1 \frac{(2x)}{\left(x^2\right)^2 + 1}\, dx = \left[\pi \arctan\left(x^2\right)\right]_0^1 = \pi\left(\frac{\pi}{4} - 0\right) = \frac{\pi^2}{4}$

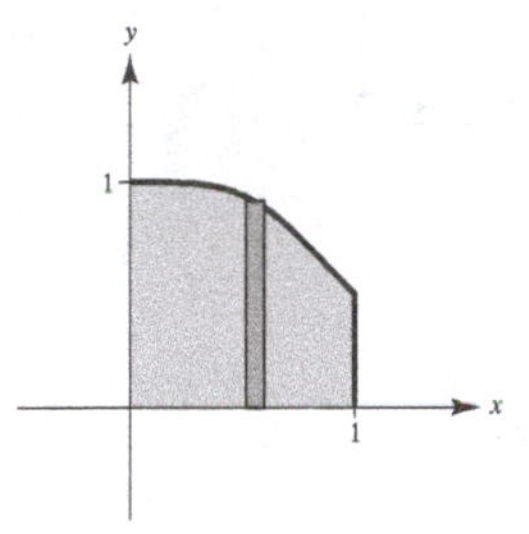

23. (a) **Disk**

$V = \pi\int_0^3 x^2\, dx = \left[\frac{\pi x^3}{3}\right]_0^3 = 9\pi$

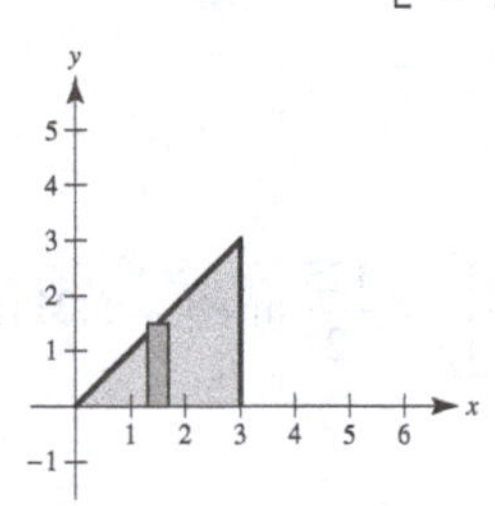

(b) **Shell**

$V = 2\pi\int_0^3 x(x)\, dx = 2\pi\left[\frac{x^3}{3}\right]_0^3 = 18\pi$

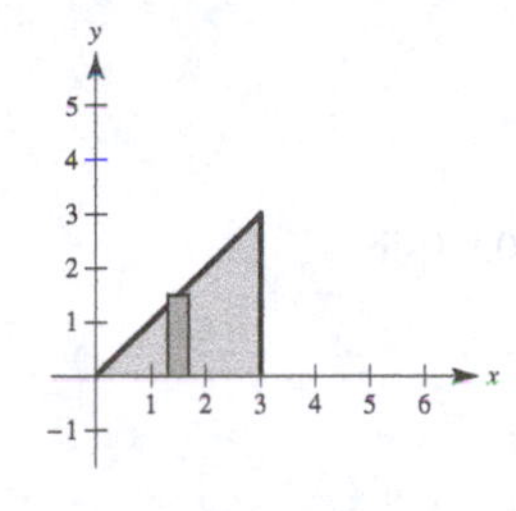

(c) **Shell**

$V = 2\pi\int_0^3 (3 - x)x\, dx = 2\pi\left[\frac{3x^2}{2} - \frac{x^3}{3}\right]_0^3 = 9\pi$

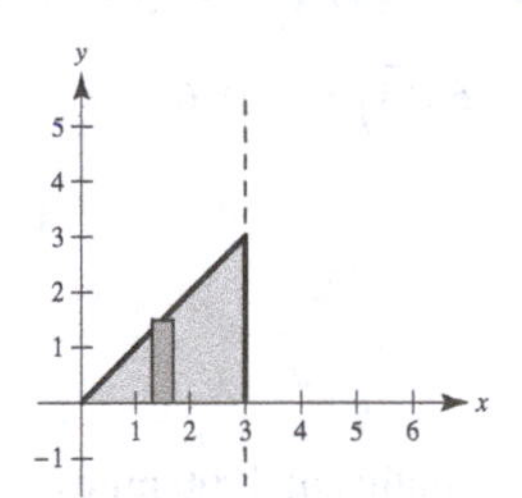

(d) **Shell**

$V = 2\pi\int_0^3 (6 - x)x\, dx = 2\pi\left[3x^2 - \frac{x^3}{3}\right]_0^3 = 36\pi$

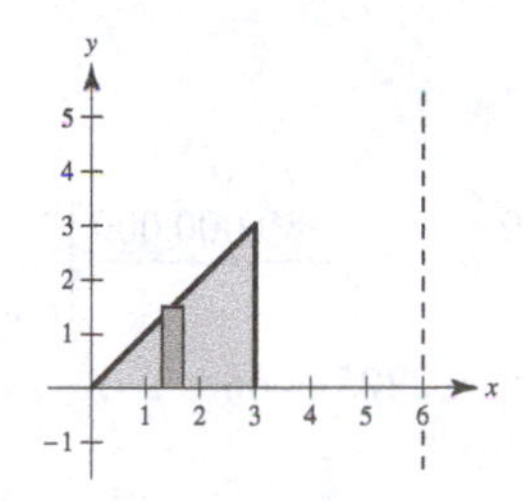

25. The volume of the spheroid is given by:

$$V = 4\pi\int_0^4 x\left(\tfrac{3}{4}\right)\sqrt{16 - x^2}\,dx$$

$$= \left[3\pi\left(-\tfrac{1}{2}\right)\left(\tfrac{2}{3}\right)\left(16 - x^2\right)^{3/2}\right]_0^4$$

$$= 64\pi$$

$$\tfrac{1}{4}V = 16\pi$$

Disk:
$$\pi\int_{-3}^{y_0} \tfrac{16}{9}\left(9 - y^2\right) dy = 16\pi$$

$$\tfrac{1}{9}\int_{-3}^{y_0} \left(9 - y^2\right) dy = 1$$

$$\left[9y - \tfrac{1}{3}y^3\right]_{-3}^{y_0} = 9$$

$$\left(9y_0 - \tfrac{1}{3}y_0^3\right) - (-27 + 9) = 9$$

$$y_0^3 - 27y_0 - 27 = 0$$

By Newton's Method, $y_0 \approx -1.042$ and the depth of the gasoline is $3 - 1.042 = 1.958$ feet.

27.
$$f(x) = \tfrac{4}{5}x^{5/4},\ [0, 4]$$

$$f'(x) = x^{1/4}$$

$$1 + \left[f'(x)\right]^2 = 1 + \sqrt{x}$$

$$u = 1 + \sqrt{x}$$

$$x = (u - 1)^2$$

$$dx = 2(u - 1)\,du$$

$$s = \int_0^4 \sqrt{1 + \sqrt{x}}\,dx = 2\int_1^3 \sqrt{u}(u - 1)\,du$$

$$= 2\int_1^3 \left(u^{3/2} - u^{1/2}\right) du$$

$$= 2\left[\tfrac{2}{5}u^{5/2} - \tfrac{2}{3}u^{3/2}\right]_1^3 = \tfrac{4}{15}\left[u^{3/2}(3u - 5)\right]_1^3$$

$$= \tfrac{8}{15}\left(1 + 6\sqrt{3}\right) = 6.076$$

29. $y = \dfrac{x^3}{18}, 3 \le x \le 6$

$$y' = \frac{x^2}{6}$$

$$1 + (y')^2 = 1 + \frac{x^4}{36} = \frac{\left(36 + x^4\right)}{36}$$

$$S = 2\pi\int_3^6 \frac{x^3}{18}\sqrt{\frac{36 + x^4}{36}}\,dx$$

$$= \frac{\pi}{54}\int_3^6 \sqrt{36 + x^4}\,x^3\,dx$$

$$= \frac{1}{4}\cdot\frac{\pi}{54}\left[\frac{\left(36 + x^4\right)^{3/2}}{(3/2)}\right]_3^6$$

$$= \frac{\pi}{324}\left[\left(36 + x^4\right)^{3/2}\right]_3^6$$

$$= \frac{\pi}{324}\left(1332^{3/2} - 117^{3/2}\right)$$

$$\approx 459.098$$

31.
$$y = \frac{x^2}{2} + 4, 0 \le x \le 2$$

$$y' = x$$

$$1 + (y')^2 = 1 + x^2$$

$$S = 2\pi\int_0^2 x\sqrt{1 + x^2}\,dx$$

$$= \frac{2\pi}{3}\left[\left(x^2 + 1\right)^{3/2}\right]_0^2$$

$$= \frac{2\pi}{3}\left(5^{3/2} - 1\right) \approx 21.322$$

33.
$$F = kx$$

$$5 = k(1)$$

$$F = 5x$$

$$W = \int_0^5 5x\,dx = \left[\frac{5x^2}{2}\right]_0^5 = \frac{125}{2}\text{ in.-lb} \approx 5.21\text{ ft-lb}$$

35. Assume that Earth has a radius of 4000 miles.

$$F(x) = \frac{k}{x^2}$$

$$5 = \frac{k}{4000^2} \Rightarrow k = 80{,}000{,}000$$

$$F(x) = \frac{80{,}000{,}000}{x^2}$$

$$W = \int_{4000}^{4200} \frac{80{,}000{,}000}{x^2}\,dx = \left[\frac{-80{,}000{,}000}{x}\right]_{4000}^{4200} = \frac{20{,}000}{21} \approx 952.38\text{ mi-tons} = 1.109 \times 10^{10}\text{ ft-lb}$$

(**Note:** One metric ton = 2205 pounds)

37. Weight of section of chain: $4\,\Delta x$

Distance moved: $10 - x$

$$W = 4\int_0^{10}(10 - x)\,dx = 4\left[10x - \frac{x^2}{2}\right]_0^{10} = 200 \text{ ft-lb}$$

39. $\rho = \dfrac{k}{V}$

$$500 = \frac{k}{1} \Rightarrow k = 500$$

$$W = \int_1^4 \frac{500}{V}\,dV = \left[500 \ln V\right]_1^4 = 500 \ln 4$$

$$= 1000 \ln 2 \approx 693.15 \text{ ft-lb}$$

41. $\bar{x} = \dfrac{8(-1) + 12(2) + 6(5) + 14(7)}{8 + 12 + 6 + 14} = \dfrac{144}{40} = \dfrac{18}{5} = 3.6$

43. $A = \int_{-1}^{3}\left[(2x + 3) - x^2\right]dx = \left[x^2 + 3x - \frac{1}{3}x^3\right]_{-1}^{3} = \dfrac{32}{3}$

$$\frac{1}{A} = \frac{3}{32}$$

$$\bar{x} = \frac{3}{32}\int_{-1}^{3} x(2x + 3 - x^2)\,dx = \frac{3}{32}\int_{-1}^{3}(3x + 2x^2 - x^3)\,dx = \frac{3}{32}\left[\frac{3}{2}x^2 + \frac{2}{3}x^3 - \frac{1}{4}x^4\right]_{-1}^{3} = 1$$

$$\bar{y} = \left(\frac{3}{32}\right)\frac{1}{2}\int_{-1}^{3}\left[(2x + 3)^2 - x^4\right]dx = \frac{3}{64}\int_{-1}^{3}\left(9 + 12x + 4x^2 - x^4\right)dx$$

$$= \frac{3}{64}\left[9x + 6x^2 + \frac{4}{3}x^3 - \frac{1}{5}x^5\right]_{-1}^{3} = \frac{17}{5}$$

$$(\bar{x}, \bar{y}) = \left(1, \frac{17}{5}\right)$$

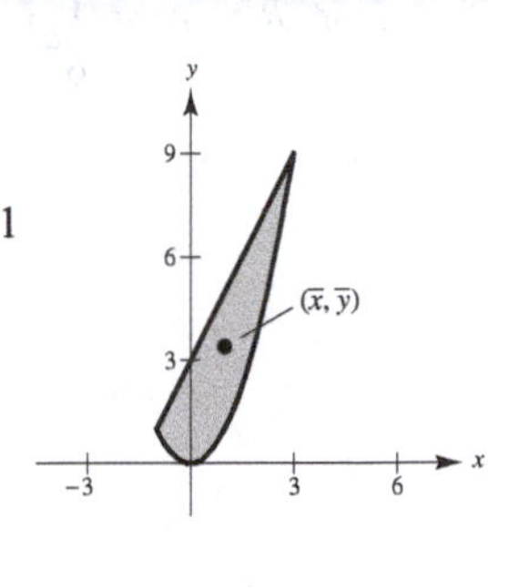

45. Answers will vary. *Sample answer*:

The centroid of the triangle is $\left(\frac{1}{3}, 0\right)$.

The centroid of the circle is $(2, 0)$.

$$A = 1 + \pi$$

$$\bar{x} = \frac{1(1/3) + \pi(2)}{1 + \pi} = \frac{1 + 6\pi}{3(1 + \pi)} \approx 1.596$$

$$(\bar{x}, \bar{y}) \approx (1.596, 0)$$

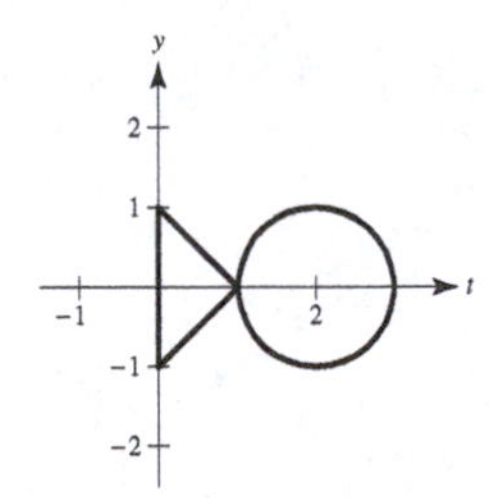

47. The weight density of water is 62.4 lb/ft^3.

$$P = wh = (62.4)(3) = 187.2 \text{ lb/ft}^2$$

Fluid force $= F = PA = (187.2)(2) = 374.4$ lb

49. $h(y) = 9 - y$

$$L(y) = 4 - \frac{4}{3}y$$

$$F = 64\int_0^3 (9 - y)\left(4 - \frac{4}{3}y\right)dy$$

$$= 64\int_0^3\left(36 - 16y + \frac{4}{3}y^2\right)dy$$

$$= 64\left[36y - 8y^2 + \frac{4}{9}y^3\right]_0^3$$

$$= 64\left[36(3) - 8(9) + 4(3)\right] = 64(48)$$

$$= 3072 \text{ lb}$$

51. From Exercise 29 in Section 7.7:

$$F = wk\left(\pi r^2\right) = (64.0)(1600)\left(\pi(1.5)^2\right) \approx 723{,}822.95 \text{ lb}$$

Problem Solving for Chapter 7

1. $T = \frac{1}{2}c(c^2) = \frac{1}{2}c^3$

$R = \int_0^c (cx - x^2)\,dx$

$= \left[\frac{cx^2}{2} - \frac{x^3}{3}\right]_0^c$

$= \frac{c^3}{2} - \frac{c^3}{3} = \frac{c^3}{6}$

$$\lim_{c\to 0^+} \frac{T}{R} = \lim_{c\to 0^+} \frac{\frac{1}{2}c^3}{\frac{1}{6}c^3} = 3$$

3. $R = \int_0^1 x(1 - x)\,dx = \left[\frac{x^2}{2} - \frac{x^3}{3}\right]_0^1 = \frac{1}{2} - \frac{1}{3} = \frac{1}{6}$

Let (c, mc) be the intersection of the line and the parabola.

Then, $mc = c(1 - c) \Rightarrow m = 1 - c$ or $c = 1 - m$.

$$\frac{1}{2}\left(\frac{1}{6}\right) = \int_0^{1-m} (x - x^2 - mx)\,dx$$

$$\frac{1}{12} = \left[\frac{x^2}{2} - \frac{x^3}{3} - m\frac{x^2}{2}\right]_0^{1-m}$$

$$= \frac{(1 - m)^2}{2} - \frac{(1 - m)^3}{3} - m\frac{(1 - m)^2}{2}$$

$$1 = 6(1 - m)^2 - 4(1 - m)^3 - 6m(1 - m)^2$$

$$= (1 - m)^2(6 - 4(1 - m) - 6m)$$

$$= (1 - m)^2(2 - 2m)$$

$$\frac{1}{2} = (1 - m)^3$$

$$\left(\frac{1}{2}\right)^{1/3} = 1 - m$$

$$m = 1 - \left(\frac{1}{2}\right)^{1/3} \approx 0.2063$$

So, $y = 0.2063x$.

5. $\bar{y} = 0$ by symmetry.

For the trapezoid:

$$m = [(4)(6) - (1)(6)]\rho = 18\rho$$

$$M_y = \rho\int_0^6 x\left[\left(\frac{1}{6}x + 1\right) - \left(-\frac{1}{6}x - 1\right)\right]dx = \rho\int_0^6 \left(\frac{1}{3}x^2 + 2x\right)dx = \rho\left[\frac{x^3}{9} + x^2\right]_0^6 = 60\rho$$

For the semicircle:

$$m = \left(\frac{1}{2}\right)(\pi)(2)^2\rho = 2\pi\rho$$

$$M_y = \rho\int_6^8 x\left[\sqrt{4 - (x - 6)^2} - \left(-\sqrt{4 - (x - 6)^2}\right)\right]dx = 2\rho\int_6^8 x\sqrt{4 - (x - 6)^2}\,dx$$

Let $u = x - 6$, then $x = u + 6$ and $dx = du$. When $x = 6$, $u = 0$. When $x = 8$, $u = 2$.

$$M_y = 2\rho\int_0^2 (u + 6)\sqrt{4 - u^2}\,du = 2\rho\int_0^2 u\sqrt{4 - u^2}\,du + 12\rho\int_0^2 \sqrt{4 - u^2}\,du$$

$$= 2\rho\left[\left(-\frac{1}{2}\right)\left(\frac{2}{3}\right)(4 - u^2)^{3/2}\right]_0^2 + 12\rho\left[\frac{\pi(2)^2}{4}\right] = \frac{16\rho}{3} + 12\pi\rho = \frac{4\rho(4 + 9\pi)}{3}$$

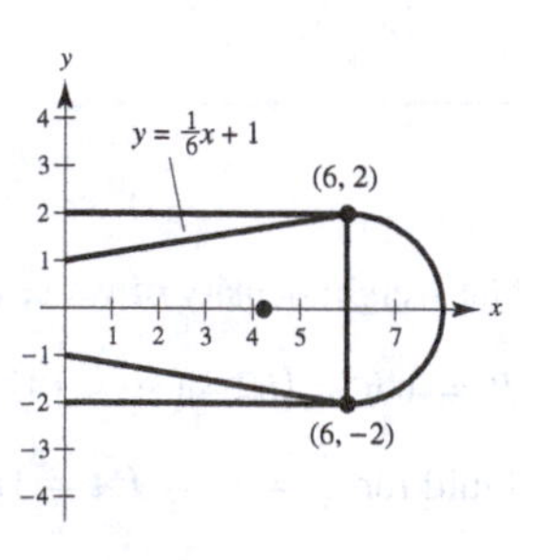

So, you have: $\bar{x}(18\rho + 2\pi\rho) = 60\rho + \frac{4\rho(4 + 9\pi)}{3}$

$$\bar{x} = \frac{180\rho + 4\rho(4 + 9\pi)}{3} \cdot \frac{1}{2\rho(9 + \pi)} = \frac{2(9\pi + 49)}{3(\pi + 9)}$$

The centroid of the blade is $\left(\frac{2(9\pi + 49)}{3(\pi + 9)}, 0\right)$.

7. By the Theorem of Pappus,

$$V = 2\pi rA = 2\pi\left[d + \tfrac{1}{2}\sqrt{w^2 + l^2}\right]lw.$$

9. $f'(x)^2 = e^x$

$f'(x) = e^{x/2}$

$f(x) = 2e^{x/2} + C$

$f(0) = 0 \Rightarrow C = -2$

$f(x) = 2e^{x/2} - 2$

11. Let ρ_f be the density of the fluid and ρ_0 the density of the iceberg. The buoyant force is

$$F = \rho_f g\int_{-h}^{0} A(y)\,dy$$

where $A(y)$ is a typical cross section and g is the acceleration due to gravity. The weight of the object is

$$W = \rho_0 g\int_{-h}^{L-h} A(y)\,dy.$$

$F = W$

$$\rho_f g\int_{-h}^{0} A(y)\,dy = \rho_0 g\int_{-h}^{L-h} A(y)\,dy$$

$$\frac{\rho_0}{\rho_f} = \frac{\text{submerged volume}}{\text{total volume}} = \frac{0.92 \times 10^3}{1.03 \times 10^3}$$

$= 0.893$ or 89.3%

13. (a) $\bar{y} = 0$ by symmetry

$$M_y = 2\int_1^6 x\frac{1}{x^4}\,dx = 2\int_1^6 \frac{1}{x^3}\,dx = \frac{35}{36}$$

$$m = 2\int_1^6 \frac{1}{x^4}\,dx = \frac{215}{324}$$

$$\bar{x} = \frac{35/36}{215/324} = \frac{63}{43} \qquad (\bar{x}, \bar{y}) = \left(\frac{63}{43}, 0\right)$$

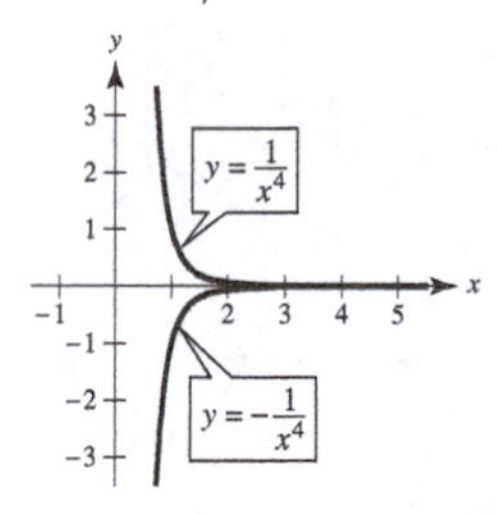

(b) $M_y = 2\int_1^b \frac{1}{x^3}\,dx = \frac{b^2 - 1}{b^2}$

$$m = 2\int_1^b \frac{1}{x^4}\,dx = \frac{2(b^3 - 1)}{3b^3}$$

$$\bar{x} = \frac{(b^2 - 1)/b^2}{2(b^3 - 1)/3b^3} = \frac{3b(b + 1)}{2(b^2 + b + 1)}$$

$$(\bar{x}, \bar{y}) = \left(\frac{3b(b + 1)}{2(b^2 + b + 1)}, 0\right)$$

(c) $\lim_{b\to\infty} \bar{x} = \frac{3b(b + 1)}{2(b^2 + b + 1)} = \frac{3}{2}$

$$(\bar{x}, \bar{y}) = \left(\frac{3}{2}, 0\right)$$

15. Point of equilibrium: $50 - 0.5x = 0.125x$

$x = 80,\ p = 10$

$(P_0, x_0) = (10, 80)$

$$\text{Consumer surplus} = \int_0^{80} \left[(50 - 0.5x) - 10\right]dx = 1600$$

$$\text{Producer surplus} = \int_0^{80} (10 - 0.125x)\,dx = 400$$

17. Use Exercise 25, Section 7.7, which gives $F = wkhb$ for a rectangle plate.

Wall at shallow end

From Exercise 25: $F = 62.4(2)(4)(20) = 9984$ lb

Wall at deep end

From Exercise 25: $F = 62.4(4)(8)(20) = 39{,}936$ lb

Side wall

From Exercise 25: $F_1 = 62.4(2)(4)(40) = 19{,}968$ lb

$$F_2 = 62.4\int_0^4 (8 - y)(10y)\,dy$$

$$= 624\int_0^4 (8y - y^2)\,dy = 624\left[4y^2 - \frac{y^3}{3}\right]_0^4$$

$= 26{,}624$ lb

Total force: $F_1 + F_2 = 46{,}592$ lb

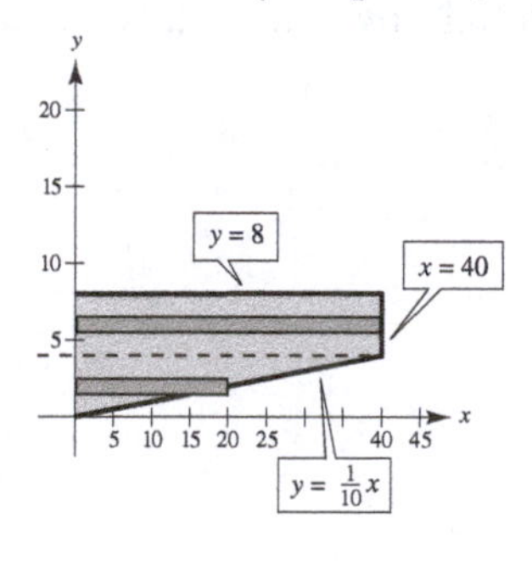

CHAPTER 8
Integration Techniques and Improper Integrals

CHAPTER 8
Integration Techniques and Improper Integrals

Section 8.1 Basic Integration Rules

1. Use long division to rewrite the function as the sum of a polynomial and a proper rational function.

3. (a) $\frac{d}{dx}\left[2\sqrt{x^2+1}+C\right] = 2\left(\frac{1}{2}\right)\left(x^2+1\right)^{-1/2}(2x) = \frac{2x}{\sqrt{x^2+1}}$

(b) $\frac{d}{dx}\left[\sqrt{x^2+1}+C\right] = \frac{1}{2}\left(x^2+1\right)^{-1/2}(2x) = \frac{x}{\sqrt{x^2+1}}$

(c) $\frac{d}{dx}\left[\frac{1}{2}\sqrt{x^2+1}+C\right] = \frac{1}{2}\left(\frac{1}{2}\right)\left(x^2+1\right)^{-1/2}(2x) = \frac{x}{2\sqrt{x^2+1}}$

(d) $\frac{d}{dx}\left[\ln\left(x^2+1\right)+C\right] = \frac{2x}{x^2+1}$

$\int \frac{x}{\sqrt{x^2+1}}\,dx$ matches (b).

4. (a) $\frac{d}{dx}\left[\ln\sqrt{x^2+1}+C\right] = \frac{1}{2}\left(\frac{2x}{x^2+1}\right) = \frac{x}{x^2+1}$

(b) $\frac{d}{dx}\left[\frac{2x}{\left(x^2+1\right)^2}+C\right] = \frac{\left(x^2+1\right)^2(2)-(2x)(2)\left(x^2+1\right)(2x)}{\left(x^2+1\right)^4} = \frac{2\left(1-3x^2\right)}{\left(x^2+1\right)^3}$

(c) $\frac{d}{dx}\left[\arctan x + C\right] = \frac{1}{1+x^2}$

(d) $\frac{d}{dx}\left[\ln\left(x^2+1\right)+C\right] = \frac{2x}{x^2+1}$

$\int \frac{1}{x^2+1}\,dx$ matches (c).

5. $\int (5x-3)^4\,dx$

$u = 5x-3,\ du = 5\,dx,\ n = 4$

Use $\int u^n\,du$.

7. $\int \frac{1}{\sqrt{x}\left(1-2\sqrt{x}\right)}\,dx$

$u = 1-2\sqrt{x},\ du = -\frac{1}{\sqrt{x}}\,dx$

Use $\int \frac{du}{u}$.

9. $\int \frac{3}{\sqrt{1-t^2}}\,dt$

$u = t,\ du = dt,\ a = 1$

Use $\int \frac{du}{\sqrt{a^2-u^2}}$.

11. $\int t\sin t^2\,dt$

$u = t^2,\ du = 2t\,dt$

Use $\int \sin u\,du$.

13. $\int (\cos x)e^{\sin x}\,dx$

$u = \sin x,\ du = \cos x\,dx$

Use $\int e^u\,du$.

15. Let $u = x - 5$, $du = dx$.

$$\int 14(x-5)^6\,dx = 14\int (x-5)^6\,dx = 2(x-5)^7 + C$$

17. Let $u = z - 10$, $du = dz$.

$$\int \frac{7}{(z-10)^7}\,dz = 7\int (z-10)^{-7}\,dz = -\frac{7}{6(z-10)^6} + C$$

19. $$\int \left[z^2 + \frac{1}{(1-z)^6}\right] dz = \int \left[z^2 + (1-z)^{-6}\right] dz$$
$$= \frac{z^3}{3} + \frac{(1-z)^{-5}}{5} + C$$
$$= \frac{z^3}{3} + \frac{1}{5(1-z)^5} + C$$
$$= \frac{z^3}{3} - \frac{1}{5(z-1)^5} + C$$

21. Let $u = -t^3 + 9t + 1$,

$du = \left(-3t^2 + 9\right) dt = -3\left(t^2 - 3\right) dt$.

$$\int \frac{t^2 - 3}{-t^3 + 9t + 1}\,dt = -\frac{1}{3}\int \frac{-3\left(t^2 - 3\right)}{-t^3 + 9t + 1}\,dt$$
$$= -\frac{1}{3}\ln\left|-t^3 + 9t + 1\right| + C$$

23. $$\int \frac{x^2}{x-1}\,dx = \int (x+1)\,dx + \int \frac{1}{x-1}\,dx$$
$$= \frac{1}{2}x^2 + x + \ln|x-1| + C$$

25. $$\int \frac{x+2}{x+1}\,dx = \int \frac{x+1+1}{x+1}\,dx$$
$$= \int \left(1 + \frac{1}{x+1}\right) dx$$
$$= x + \ln|x+1| + C$$

27. $$\int \left(5 + 4x^2\right)^2 dx = \int \left(25 + 40x^2 + 16x^4\right) dx$$
$$= 25x + \frac{40}{3}x^3 + \frac{16}{5}x^5 + C$$
$$= \frac{x}{15}\left(48x^4 + 200x^2 + 375\right) + C$$

29. Let $u = 2\pi x^2$, $du = 4\pi x\,dx$.

$$\int x\left(\cos 2\pi x^2\right) dx = \frac{1}{4\pi}\int \left(\cos 2\pi x^2\right)(4\pi x)\,dx$$
$$= \frac{1}{4\pi}\sin 2\pi x^2 + C$$

31. Let $u = \cos x$, $du = -\sin x\,dx$.

$$\int \frac{\sin x}{\sqrt{\cos x}}\,dx = -\int (\cos x)^{-1/2}(-\sin x)\,dx$$
$$= -2\sqrt{\cos x} + C$$

33. Let $u = 1 + e^x$, $du = e^x\,dx$.

$$\int \frac{2}{e^{-x} + 1}\,dx = 2\int \left(\frac{2}{e^{-x} + 1}\right)\left(\frac{e^x}{e^x}\right) dx$$
$$= 2\int \frac{e^x}{1 + e^x}\,dx = 2\ln\left(1 + e^x\right) + C$$

35. $$\int \frac{\ln x^2}{x}\,dx = 2\int (\ln x)\frac{1}{x}\,dx$$
$$= 2\frac{(\ln x)^2}{2} + C = (\ln x)^2 + C$$

37. $$\int \frac{1 + \cos\alpha}{\sin\alpha}\,d\alpha = \int \csc\alpha\,d\alpha + \int \cot\alpha\,d\alpha$$
$$= -\ln|\csc\alpha + \cot\alpha| + \ln|\sin\alpha| + C$$

39. Let $u = 4t + 1$, $du = 4\,dt$.

$$\int \frac{-1}{\sqrt{1 - (4t+1)^2}}\,dt = -\frac{1}{4}\int \frac{4}{\sqrt{1 - (4t+1)^2}}\,dt$$
$$= -\frac{1}{4}\arcsin(4t + 1) + C$$

41. Let $u = \cos\left(\dfrac{2}{t}\right)$, $du = \dfrac{2\sin(2/t)}{t^2}\,dt$.

$$\int \frac{\tan(2/t)}{t^2}\,dt = \frac{1}{2}\int \frac{1}{\cos(2/t)}\left[\frac{2\sin(2/t)}{t^2}\right] dt$$
$$= \frac{1}{2}\ln\left|\cos\left(\frac{2}{t}\right)\right| + C$$

43. $$\int \frac{6}{z\sqrt{9z^2 - 25}}\,dz = \int \frac{6}{3z\sqrt{(3z)^2 - 5^2}}(3dz)$$
$$= \frac{6}{5}\operatorname{arcsec}\left(\frac{|3z|}{5}\right) + C$$

45. $$\int \frac{4}{4x^2 + 4x + 65}\,dx = \int \frac{1}{\left[x + (1/2)\right]^2 + 16}\,dx$$
$$= \frac{1}{4}\arctan\left[\frac{x + (1/2)}{4}\right] + C$$
$$= \frac{1}{4}\arctan\left(\frac{2x + 1}{8}\right) + C$$

47. $\dfrac{ds}{dt} = \dfrac{t}{\sqrt{1 - t^4}}, \left(0, -\dfrac{1}{2}\right)$

(a)

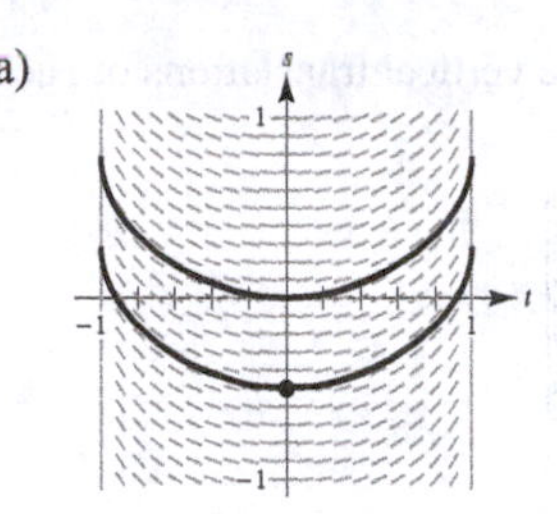

(b) $u = t^2, du = 2t\,dt$

$$\int \frac{t}{\sqrt{1 - t^4}}\,dt = \frac{1}{2}\int \frac{2t}{\sqrt{1 - (t^2)^2}}\,dt$$

$$= \frac{1}{2}\arcsin t^2 + C$$

$$\left(0, -\frac{1}{2}\right): -\frac{1}{2} = \frac{1}{2}\arcsin 0 + C \Rightarrow C = -\frac{1}{2}$$

$$s = \frac{1}{2}\arcsin t^2 - \frac{1}{2}$$

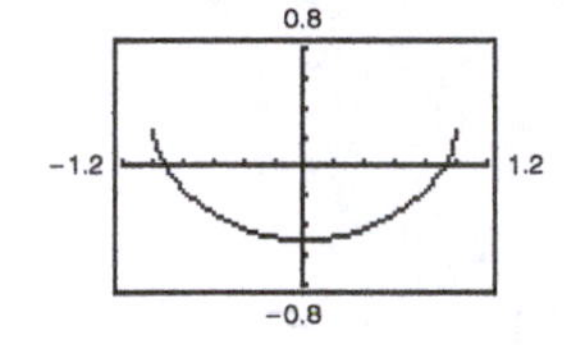

49.

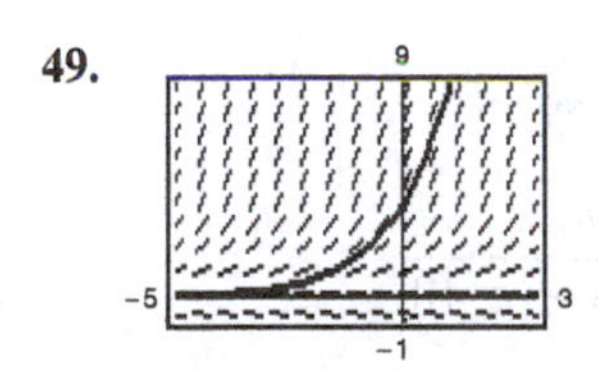

$y = 4e^{0.8x}$

51. $\dfrac{dy}{dx} = \left(e^x + 5\right)^2 = e^{2x} + 10e^x + 25$

$$y = \int \left(e^{2x} + 10e^x + 25\right) dx$$

$$= \frac{1}{2}e^{2x} + 10e^x + 25x + C$$

53. $\dfrac{dr}{dt} = \dfrac{10e^t}{\sqrt{1 - e^{2t}}}$

$$r = \int \frac{10e^t}{\sqrt{1 - \left(e^t\right)^2}}\,dt$$

$$= 10\arcsin\left(e^t\right) + C$$

55. $\dfrac{dy}{dx} = \dfrac{\sec^2 x}{4 + \tan^2 x}$

Let $u = \tan x, du = \sec^2 x\,dx$.

$$y = \int \frac{\sec^2 x}{4 + \tan^2 x}\,dx = \frac{1}{2}\arctan\left(\frac{\tan x}{2}\right) + C$$

57. $\displaystyle\int_{2/3}^{1} (2 - 3t)^4\,dt = \int_{2/3}^{1} (3t - 2)^4\,dt$

$$= \left[\frac{1}{3}\frac{(3t - 2)^5}{5}\right]_{2/3}^{1}$$

$$= \frac{1}{15}(1 - 0) = \frac{1}{15}$$

59. Let $u = 2x, du = 2\,dx$.

$$\int_0^{\pi/4} \cos 2x\,dx = \tfrac{1}{2}\int_0^{\pi/4} \cos 2x(2)\,dx$$

$$= \left[\tfrac{1}{2}\sin 2x\right]_0^{\pi/4} = \tfrac{1}{2}$$

61. Let $u = -x^2, du = -2x\,dx$.

$$\int_0^1 xe^{-x^2}\,dx = -\tfrac{1}{2}\int_0^1 e^{-x^2}(-2x)\,dx = \left[-\tfrac{1}{2}e^{-x^2}\right]_0^1$$

$$= \tfrac{1}{2}\left(1 - e^{-1}\right) \approx 0.316$$

63. $\displaystyle\int_2^3 \frac{\ln(x + 1)^3}{x + 1}\,dx = 3\int_2^3 \ln(x + 1)\frac{1}{x + 1}\,dx$

$$= \left[3\frac{\left[\ln(x + 1)\right]^2}{2}\right]_2^3$$

$$= \frac{3}{2}\left[\left[\ln(x + 1)\right]^2\right]_2^3$$

$$= \frac{3}{2}\left[(\ln 4)^2 - (\ln 3)^2\right] \approx 1.072$$

65. Let $u = x^2 + 36, du = 2x\,dx$.

$$\int_0^8 \frac{2x}{\sqrt{x^2 + 36}}\,dx = \int_0^8 \left(x^2 + 36\right)^{-1/2}(2x)\,dx$$

$$= 2\left[\left(x^2 + 36\right)^{1/2}\right]_0^8 = 8$$

67. $\displaystyle\int_3^5 \frac{2t}{t^2 - 4t + 4}\,dt = \int_3^5 \frac{2t - 4}{t^2 - 4t + 4}\,dt + \int_3^5 \frac{4}{(t - 2)^2}\,dt$

$$= \left[\ln\left(t^2 - 4t + 4\right) - \frac{4}{t - 2}\right]_3^5$$

$$= \left(\ln 9 - \frac{4}{3}\right) - (\ln 1 - 4)$$

$$= \ln 9 + \frac{8}{3} \approx 4.864$$

69. Let $u = 3x$, $du = 3\,dx$.

$$\int_0^{2/\sqrt{3}} \frac{1}{4 + 9x^2}\,dx = \frac{1}{3}\int_0^{2/\sqrt{3}} \frac{3}{4 + (3x)^2}\,dx$$
$$= \left[\frac{1}{6}\arctan\left(\frac{3x}{2}\right)\right]_0^{2/\sqrt{3}}$$
$$= \frac{\pi}{18} \approx 0.175$$

71. $$\int_{-4}^{0} 3^{1-x}\,dx = \left[\frac{-1}{\ln 3}3^{1-x}\right]_{-4}^{0}$$
$$= -\frac{1}{\ln 3}\left(3 - 3^5\right)$$
$$= \frac{240}{\ln 3} \approx 218.457$$

73. $$A = \int_0^{3/2} (-4x + 6)^{3/2}\,dx$$
$$= -\frac{1}{4}\int_0^{3/2} (6 - 4x)^{3/2}(-4)\,dx$$
$$= -\frac{1}{4}\left[\frac{2}{5}(6 - 4x)^{5/2}\right]_0^{3/2}$$
$$= -\frac{1}{10}\left(0 - 6^{5/2}\right)$$
$$= \frac{18}{5}\sqrt{6} \approx 8.8182$$

75. $$y^2 = x^2\left(1 - x^2\right)$$
$$y = \pm\sqrt{x^2\left(1 - x^2\right)}$$
$$A = 4\int_0^1 x\sqrt{1 - x^2}\,dx$$
$$= -2\int_0^1 \left(1 - x^2\right)^{1/2}(-2x)\,dx$$
$$= -\frac{4}{3}\left[\left(1 - x^2\right)^{3/2}\right]_0^1$$
$$= -\frac{4}{3}(0 - 1) = \frac{4}{3}$$

77. $$\int \frac{1}{x^2 + 4x + 13}\,dx = \frac{1}{3}\arctan\left(\frac{x + 2}{3}\right) + C$$

The antiderivatives are vertical translations of each other.

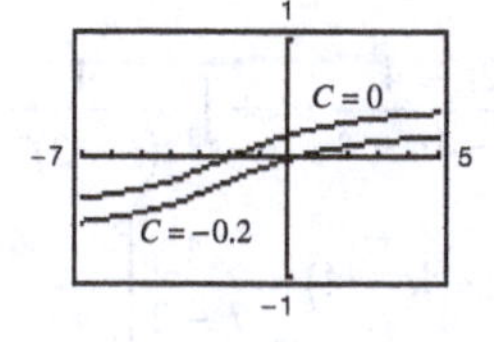

79. $$\int \frac{1}{1 + \sin\theta}\,d\theta = \tan\theta - \sec\theta + C \quad \left(\text{or } \frac{-2}{1 + \tan(\theta/2)}\right)$$

The antiderivatives are vertical translations of each other.

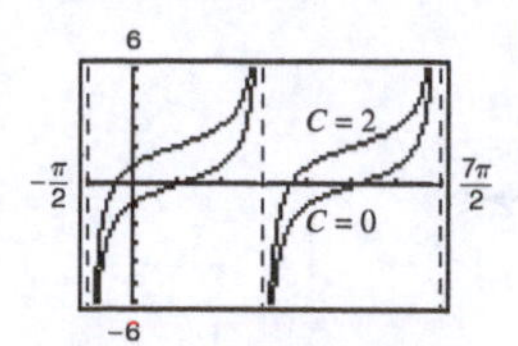

81. No. When $u = x^2$, it does not follow that $x = \sqrt{u}$ because x is negative on $[-1, 0)$.

83. $$\sin x + \cos x = a\sin(x + b)$$
$$\sin x + \cos x = a\sin x\cos b + a\cos x\sin b$$
$$\sin x + \cos x = (a\cos b)\sin x + (a\sin b)\cos x$$

Equate coefficients of like terms to obtain the following.

$1 = a\cos b$ and $1 = a\sin b$

So, $a = 1/\cos b$. Now, substitute for a in $1 = a\sin b$.

$$1 = \left(\frac{1}{\cos b}\right)\sin b$$
$$1 = \tan b \Rightarrow b = \frac{\pi}{4}$$

Because $b = \frac{\pi}{4}$, $a = \frac{1}{\cos(\pi/4)} = \sqrt{2}$. So,

$$\sin x + \cos x = \sqrt{2}\sin\left(x + \frac{\pi}{4}\right).$$

$$\int \frac{dx}{\sin x + \cos x} = \int \frac{dx}{\sqrt{2}\sin(x + (\pi/4))}$$
$$= \frac{1}{\sqrt{2}}\int \csc\left(x + \frac{\pi}{4}\right)dx$$
$$= -\frac{1}{\sqrt{2}}\ln\left|\csc\left(x + \frac{\pi}{4}\right) + \cot\left(x + \frac{\pi}{4}\right)\right| + C$$

85. (a) They are equivalent because

$e^{x+C_1} = e^x \cdot e^{C_1} = Ce^x, C = e^{C_1}$.

(b) They differ by a constant.

$\sec^2 x + C_1 = \left(\tan^2 x + 1\right) + C_1 = \tan^2 x + C$

87. $\int_0^2 \frac{4x}{x^2+1}\,dx \approx 3$

Matches (a).

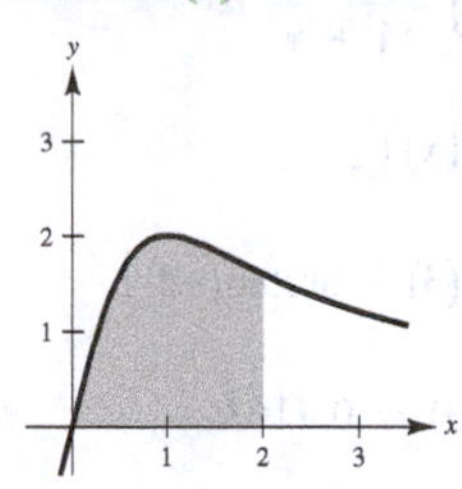

88. $\int_0^2 \frac{4}{x^2+1}\,dx \approx 4$

Matches (d).

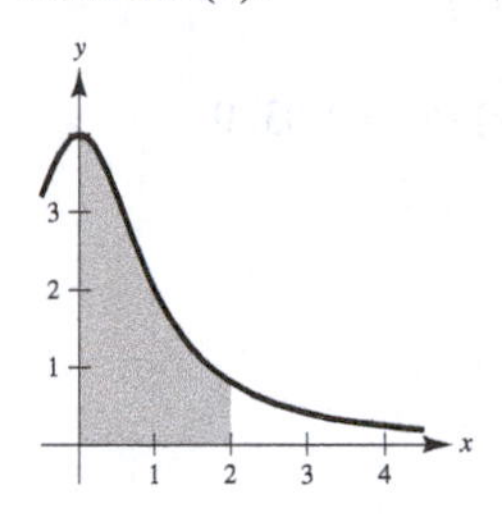

89. (a) $y = 2\pi x^2, \quad 0 \le x \le 2$

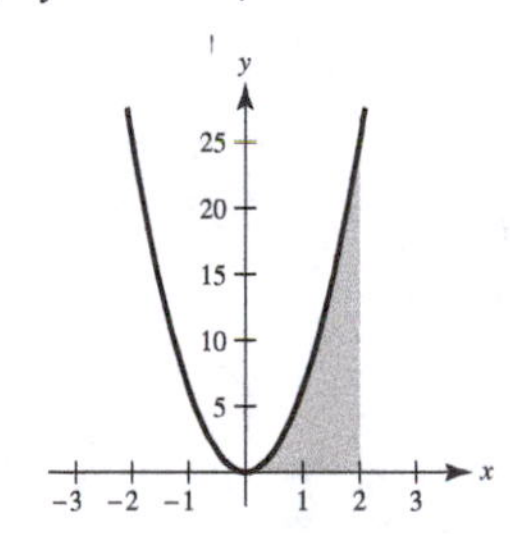

(b) $y = \sqrt{2}x, \quad 0 \le x \le 2$

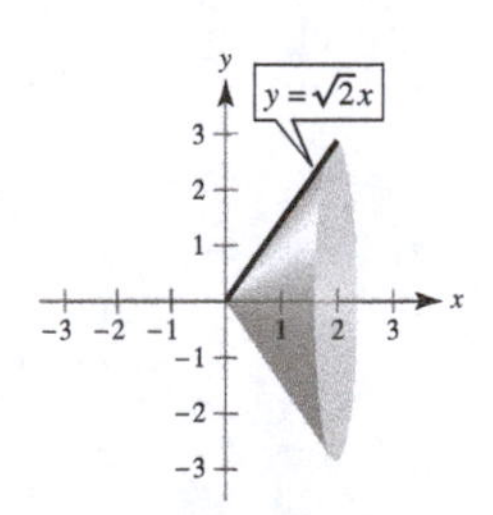

(c) $y = x, \quad 0 \le x \le 2$

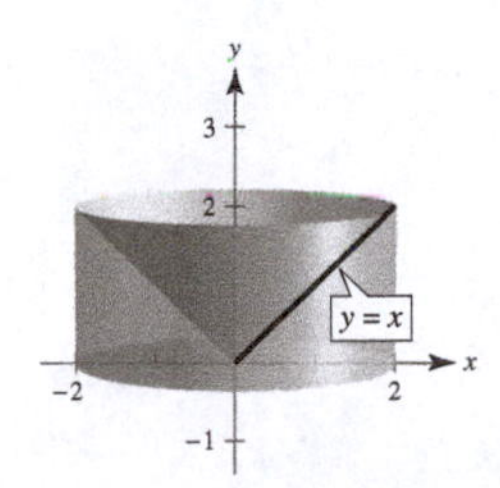

91. (a) **Shell Method:**

Let $u = -x^2$, $du = -2x\,dx$.

$$V = 2\pi\int_0^1 xe^{-x^2}\,dx$$
$$= -\pi\int_0^1 e^{-x^2}(-2x)\,dx$$
$$= \left[-\pi e^{-x^2}\right]_0^1$$
$$= \pi\left(1 - e^{-1}\right) \approx 1.986$$

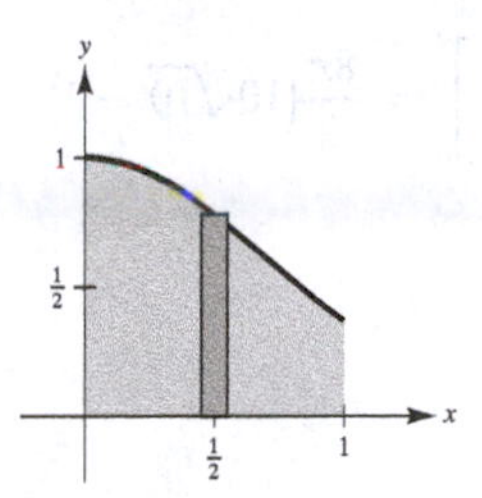

(b) **Shell Method:**

$$V = 2\pi\int_0^b xe^{-x^2}\,dx$$
$$= \left[-\pi e^{-x^2}\right]_0^b$$
$$= \pi\left(1 - e^{-b^2}\right) = \frac{4}{3}$$
$$e^{-b^2} = \frac{3\pi - 4}{3\pi}$$
$$b = \sqrt{\ln\left(\frac{3\pi}{3\pi - 4}\right)} \approx 0.743$$

93. $y = f(x) = \ln(\sin x)$

$$f'(x) = \frac{\cos x}{\sin x}$$
$$s = \int_{\pi/4}^{\pi/2}\sqrt{1 + \frac{\cos^2 x}{\sin^2 x}}\,dx$$
$$= \int_{\pi/4}^{\pi/2}\sqrt{\frac{\sin^2 x + \cos^2 x}{\sin^2 x}}\,dx$$
$$= \int_{\pi/4}^{\pi/2}\frac{1}{\sin x}\,dx = \int_{\pi/4}^{\pi/2}\csc x\,dx$$
$$= \left[-\ln\left|\csc x + \cot x\right|\right]_{\pi/4}^{\pi/2}$$
$$= -\ln(1) + \ln\left(\sqrt{2} + 1\right)$$
$$= \ln\left(\sqrt{2} + 1\right) \approx 0.8814$$

95. $y = 2\sqrt{x}$

$y' = \frac{1}{\sqrt{x}}$

$1 + (y')^2 = 1 + \frac{1}{x} = \frac{x+1}{x}$

$S = 2\pi\int_0^9 2\sqrt{x}\sqrt{\frac{x+1}{x}}\,dx$

$= 2\pi\int_0^9 2\sqrt{x+1}\,dx$

$= \left[4\pi\left(\frac{2}{3}\right)(x+1)^{3/2}\right]_0^9 = \frac{8\pi}{3}\left(10\sqrt{10} - 1\right)$

≈ 256.545

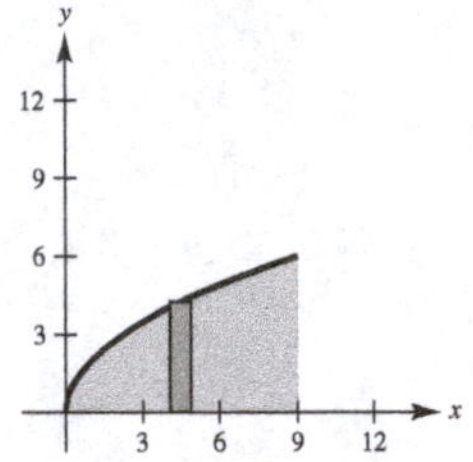

97. Average value $= \frac{1}{b-a}\int_a^b f(x)\,dx$

$= \frac{1}{3-(-3)}\int_{-3}^3 \frac{1}{1+x^2}\,dx$

$= \frac{1}{6}\left[\arctan(x)\right]_{-3}^3$

$= \frac{1}{6}\left[\arctan(3) - \arctan(-3)\right]$

$= \frac{1}{3}\arctan(3) \approx 0.4163$

99. $y = \tan(\pi x)$

$y' = \pi\sec^2(\pi x)$

$1 + (y')^2 = 1 + \pi^2\sec^4(\pi x)$

$s = \int_0^{1/4}\sqrt{1 + \pi^2\sec^4(\pi x)}\,dx \approx 1.0320$

101. (a) $\int \cos^3 x\,dx = \int\left(1 - \sin^2 x\right)\cos x\,dx = \sin x - \frac{\sin^3 x}{3} + C = \frac{1}{3}\sin x\left(\cos^2 x + 2\right) + C$

(b) $\int \cos^5 x\,dx = \int\left(1 - \sin^2 x\right)^2\cos x\,dx = \int\left(1 - 2\sin^2 x + \sin^4 x\right)\cos x\,dx$

$= \sin x - \frac{2}{3}\sin^3 x + \frac{\sin^5 x}{5} + C = \frac{1}{15}\sin x\left(3\cos^4 x + 4\cos^2 x + 8\right) + C$

(c) $\int \cos^7 x\,dx = \int\left(1 - \sin^2 x\right)^3\cos x\,dx$

$= \int\left(1 - 3\sin^2 x + 3\sin^4 x - \sin^6 x\right)\cos x\,dx$

$= \sin x - \sin^3 x + \frac{3}{5}\sin^5 x - \frac{1}{7}\sin^7 x + C$

$= \frac{1}{35}\sin x\left(5\cos^6 x + 6\cos^4 x + 8\cos^2 x + 16\right) + C$

(d) $\int \cos^{15} x\,dx = \int\left(1 - \sin^2 x\right)^7\cos x\,dx$

You would expand $\left(1 - \sin^2 x\right)^7$.

103. Let $f(x) = \frac{1}{2}\left(x\sqrt{x^2+1} + \ln\left|x + \sqrt{x^2+1}\right|\right) + C.$

$$f'(x) = \frac{1}{2}\left(x\frac{1}{2}(x^2+1)^{-1/2}(2x) + \sqrt{x^2+1} + \frac{1}{x+\sqrt{x^2+1}}\left(1 + \frac{1}{2}(x^2+1)^{-1/2}(2x)\right)\right)$$

$$= \frac{1}{2}\left(\frac{x^2}{\sqrt{x^2+1}} + \sqrt{x^2+1} + \frac{1}{x+\sqrt{x^2+1}}\left(1 + \frac{x}{\sqrt{x^2+1}}\right)\right)$$

$$= \frac{1}{2}\left(\frac{x^2 + (x^2+1)}{\sqrt{x^2+1}} + \frac{1}{x+\sqrt{x^2+1}}\left(\frac{\sqrt{x^2+1}+x}{\sqrt{x^2+1}}\right)\right)$$

$$= \frac{1}{2}\left(\frac{2x^2+1}{\sqrt{x^2+1}} + \frac{1}{\sqrt{x^2+1}}\right) = \frac{1}{2}\left(\frac{2(x^2+1)}{\sqrt{x^2+1}}\right) = \sqrt{x^2+1}$$

So, $\int \sqrt{x^2+1}\,dx = \frac{1}{2}\left(x\sqrt{x^2+1} + \ln\left|x+\sqrt{x^2+1}\right|\right) + C.$

Let $g(x) = \frac{1}{2}\left(x\sqrt{x^2+1} + \operatorname{arcsinh}(x)\right).$

$$g'(x) = \frac{1}{2}\left(x\frac{1}{2}(x^2+1)^{-1/2}(2x) + \sqrt{x^2+1} + \frac{1}{\sqrt{x^2+1}}\right)$$

$$= \frac{1}{2}\left(\frac{x^2}{\sqrt{x^2+1}} + \sqrt{x^2+1} + \frac{1}{\sqrt{x^2+1}}\right)$$

$$= \frac{1}{2}\left(\frac{x^2 + (x^2+1) + 1}{\sqrt{x^2+1}}\right)$$

$$= \frac{1}{2}\left(\frac{2(x^2+1)}{\sqrt{x^2+1}}\right) = \sqrt{x^2+1}$$

So, $\int \sqrt{x^2+1}\,dx = \frac{1}{2}\left(x\sqrt{x^2+1} + \operatorname{arcsinh}(x)\right) + C.$

Section 8.2 Integration by Parts

1. Integration by parts is based on the formula for the derivative of a product.

3. Let u be that single term and $dv = dx$.

5. $\int xe^{9x}\,dx$

$u = x,\ dv = e^{9x}\,dx$

7. $\int (\ln x)^2\,dx$

$u = (\ln x)^2,\ dv = dx$

9. $\int x\sec^2 x\,dx$

$u = x,\ dv = \sec^2 x\,dx$

11. $dv = x^3\,dx \Rightarrow v = \int x^3\,dx = \frac{x^4}{4}$

$u = \ln x \Rightarrow du = \frac{1}{x}\,dx$

$$\int x^3 \ln x\,dx = uv - \int v\,du$$

$$= (\ln x)\frac{x^4}{4} - \int\left(\frac{x^4}{4}\right)\frac{1}{x}\,dx$$

$$= \frac{x^4}{4}\ln x - \frac{1}{4}\int x^3\,dx$$

$$= \frac{x^4}{4}\ln x - \frac{1}{16}x^4 + C$$

$$= \frac{1}{16}x^4(4\ln x - 1) + C$$

13. $dv = \sin 4x\, dx \Rightarrow v = \int \sin 4x\, dx = -\frac{1}{4}\cos 4x$

$u = 2x + 1 \Rightarrow du = 2\, dx$

$$\int (2x+1)\sin 4x\, dx = uv - \int v\, du$$
$$= (2x+1)\left(-\frac{1}{4}\cos 4x\right) - \int\left(-\frac{1}{4}\cos 4x\right)(2\, dx)$$
$$= -\frac{1}{4}(2x+1)\cos 4x + \frac{1}{8}\sin 4x + C$$

15. $dv = e^{4x}\, dx \Rightarrow v = \int e^{4x}\, dx = \frac{1}{4}e^{4x}\, dx$

$u = x \Rightarrow du = dx$

$$\int xe^{4x}\, dx = x\left(\frac{1}{4}e^{4x}\right) - \int\left(\frac{1}{4}e^{4x}\right)dx = \frac{x}{4}e^{4x} - \frac{1}{16}e^{4x} + C = \frac{e^{4x}}{16}(4x-1) + C$$

17. Use integration by parts three times.

(1) $dv = e^x\, dx \Rightarrow v = \int e^x\, dx = e^x$

$u = x^3 \Rightarrow du = 3x^2\, dx$

(2) $dv = e^x\, dx \Rightarrow v = \int e^x\, dx = e^x$

$u = x^2 \Rightarrow du = 2x\, dx$

(3) $dv = e^x\, dx \Rightarrow v = \int e^x\, dx = e^x$

$u = x \Rightarrow du = dx$

$$\int x^3e^x\, dx = x^3e^x - 3\int x^2e^x\, dx = x^3e^x - 3x^2e^x + 6\int xe^x\, dx$$
$$= x^3e^x - 3x^2e^x + 6xe^x - 6e^x + C = e^x\left(x^3 - 3x^2 + 6x - 6\right) + C$$

19. $dv = t\, dt \Rightarrow v = \int t\, dt = \frac{t^2}{2}$

$u = \ln(t+1) \Rightarrow du = \frac{1}{t+1}\, dt$

$$\int t\ln(t+1)\, dt = \frac{t^2}{2}\ln(t+1) - \frac{1}{2}\int\frac{t^2}{t+1}\, dt$$
$$= \frac{t^2}{2}\ln(t+1) - \frac{1}{2}\int\left(t - 1 + \frac{1}{t+1}\right)dt$$
$$= \frac{t^2}{2}\ln(t+1) - \frac{1}{2}\left[\frac{t^2}{2} - t + \ln(t+1)\right] + C$$
$$= \frac{1}{4}\left[2\left(t^2-1\right)\ln\left|t+1\right| - t^2 + 2t\right] + C$$

21. Let $u = \ln x$, $du = \frac{1}{x}\, dx$.

$$\int\frac{(\ln x)^2}{x}\, dx = \int(\ln x)^2\left(\frac{1}{x}\right)dx = \frac{(\ln x)^3}{3} + C$$

23. $dv = \frac{1}{(2x+1)^2}\, dx \Rightarrow v = \int(2x+1)^{-2}\, dx$

$= -\frac{1}{2(2x+1)}$

$u = xe^{2x} \Rightarrow du = \left(2xe^{2x} + e^{2x}\right)dx$

$= e^{2x}(2x+1)\, dx$

$$\int\frac{xe^{2x}}{(2x+1)^2}\, dx = -\frac{xe^{2x}}{2(2x+1)} + \int\frac{e^{2x}}{2}\, dx$$
$$= \frac{-xe^{2x}}{2(2x+1)} + \frac{e^{2x}}{4} + C$$
$$= \frac{e^{2x}}{4(2x+1)} + C$$

25. $dv = \sqrt{x-5}\,dx \Rightarrow v = \int (x-5)^{1/2}\,dx = \frac{2}{3}(x-5)^{3/2}$

$u = x \Rightarrow du = dx$

$$\begin{aligned}\int x\sqrt{x-5}\,dx &= x\tfrac{2}{3}(x-5)^{3/2} - \int \tfrac{2}{3}(x-5)^{3/2}\,dx \\ &= \tfrac{2}{3}x(x-5)^{3/2} - \tfrac{4}{15}(x-5)^{5/2} + C \\ &= \tfrac{2}{15}(x-5)^{3/2}(5x - 2(x-5)) + C \\ &= \tfrac{2}{15}(x-5)^{3/2}(3x+10) + C\end{aligned}$$

27. $dv = \csc^2 x\,dx \Rightarrow v = \int \csc^2 x\,dx = -\cot x$

$u = x \Rightarrow du = dx$

$$\int x\csc^2 x\,dx = uv - \int v\,du = x(-\cot x) - \int(-\cot x)\,dx = -x\cot x + \ln|\sin x| + C$$

29. Use integration by parts three times.

(1) $u = x^3,\ du = 3x^2\,dx,\ dv = \sin x\,dx,\ v = -\cos x$

$$\int x^3 \sin dx = -x^3\cos x + 3\int x^2\cos x\,dx$$

(2) $u = x^2,\ du = 2x\,dx,\ dv = \cos x\,dx,\ v = \sin x$

$$\int x^3\sin x\,dx = -x^3\cos x + 3\left(x^2\sin x - 2\int x\sin x\,dx\right) = -x^3\cos x + 3x^2\sin x - 6\int x\sin x\,dx$$

(3) $u = x,\ du = dx,\ dv = \sin x\,dx,\ v = -\cos x$

$$\begin{aligned}\int x^3\sin x\,dx &= -x^3\cos x + 3x^2\sin x - 6\left(-x\cos x + \int\cos x\,dx\right) \\ &= -x^3\cos x + 3x^2\sin x + 6x\cos x - 6\sin x + C \\ &= \left(6x - x^3\right)\cos x + \left(3x^2 - 6\right)\sin x + C\end{aligned}$$

31. $dv = dx \Rightarrow v = \int dx = x$

$u = \arctan x \Rightarrow du = \dfrac{1}{1+x^2}\,dx$

$$\int \arctan x\,dx = x\arctan x - \int\frac{x}{1+x^2}\,dx = x\arctan x - \frac{1}{2}\ln\left(1+x^2\right) + C$$

33. Use integration by parts twice.

(1) $dv = e^{-3x}\,dx \Rightarrow v = \int e^{-3x}\,dx = -\frac{1}{3}e^{-3x}$

$u = \sin 5x \Rightarrow du = 5\cos 5x\,dx$

$$\int e^{-3x}\sin 5x\,dx = \sin 5x\left(-\tfrac{1}{3}e^{-3x}\right) - \int\left(-\tfrac{1}{3}e^{-3x}\right)5\cos x\,dx = -\tfrac{1}{3}e^{-3x}\sin 5x + \tfrac{5}{3}\int e^{-3x}\cos 5x\,dx$$

(2) $dv = e^{-3x}\,dx \Rightarrow v = \int e^{-3x}\,dx = -\frac{1}{3}e^{-3x}$

$u = \cos 5x \Rightarrow du = -5\sin 5x\,dx$

$$\begin{aligned}\int e^{-3x}\sin 5x\,dx &= -\tfrac{1}{3}e^{-3x}\sin 5x + \tfrac{5}{3}\left[\left(-\tfrac{1}{3}e^{-3x}\cos 5x\right) - \int\left(-\tfrac{1}{3}e^{-3x}\right)(-5\sin 5x)\,dx\right] \\ &= -\tfrac{1}{3}e^{-3x}\sin 5x - \tfrac{5}{9}e^{-3x}\cos 5x - \tfrac{25}{9}\int e^{-3x}\sin 5x\,dx\end{aligned}$$

$$\left(1 + \tfrac{25}{9}\right)\int e^{-3x}\sin 5x\,dx = -\tfrac{1}{3}e^{-3x}\sin 5x - \tfrac{5}{9}e^{-3x}\cos 5x$$

$$\int e^{-3x}\sin 5x\,dx = \tfrac{9}{34}\left(-\tfrac{1}{3}e^{-3x}\sin 5x - \tfrac{5}{9}e^{-3x}\cos 5x\right) + C = -\tfrac{3}{34}e^{-3x}\sin 5x - \tfrac{5}{34}e^{-3x}\cos 5x + C$$

35. $dv = dx \quad \Rightarrow \quad v = x$

$u = \ln x \quad \Rightarrow \quad du = \frac{1}{x}\,dx$

$y' = \ln x$

$y = \int \ln x\,dx = x \ln x - \int x\left(\frac{1}{x}\right) dx = x \ln x - x + C = x(-1 + \ln x) + C$

37. Use integration by parts twice.

(1) $dv = \frac{1}{\sqrt{3 + 5t}}\,dt \quad \Rightarrow \quad v = \int (3 + 5t)^{-1/2}\,dt = \frac{2}{5}(3 + 5t)^{1/2}$

$u = t^2 \quad \Rightarrow \quad du = 2t\,dt$

$$\int \frac{t^2}{\sqrt{3 + 5t}}\,dt = \frac{2}{5}t^2(3 + 5t)^{1/2} - \int \frac{2}{5}(3 + 5t)^{1/2} 2t\,dt$$

$$= \frac{2}{5}t^2(3 + 5t)^{1/2} - \frac{4}{5}\int t(3 + 5t)^{1/2}\,dt$$

(2) $dv = (3 + 5t)^{1/2}\,dt \quad \Rightarrow \quad v = \int (3 + 5t)^{1/2}\,dt = \frac{2}{15}(3 + 5t)^{3/2}$

$u = t \quad \Rightarrow \quad du = dt$

$$\int \frac{t^2}{\sqrt{3 + 5t}}\,dt = \frac{2}{5}t^2(3 + 5t)^{1/2} - \frac{4}{5}\left[\frac{2}{15}t(3 + 5t)^{3/2} - \int \frac{2}{15}(3 + 5t)^{3/2}\,dt\right]$$

$$= \frac{2}{5}t^2(3 + 5t)^{1/2} - \frac{8}{75}t(3 + 5t)^{3/2} + \frac{8}{75}\int (3 + 5t)^{3/2}\,dt$$

$$= \frac{2}{5}t^2(3 + 5t)^{1/2} - \frac{8}{75}t(3 + 5t)^{3/2} + \frac{16}{1875}(3 + 5t)^{5/2} + C$$

$$= \frac{2}{1875}\sqrt{3 + 5t}\left(375t^2 - 100t(3 + 5t) + 8(3 + 5t)^2\right) + C$$

$$= \frac{2}{625}\sqrt{3 + 5t}\left(25t^2 - 20t + 24\right) + C$$

39. (a)

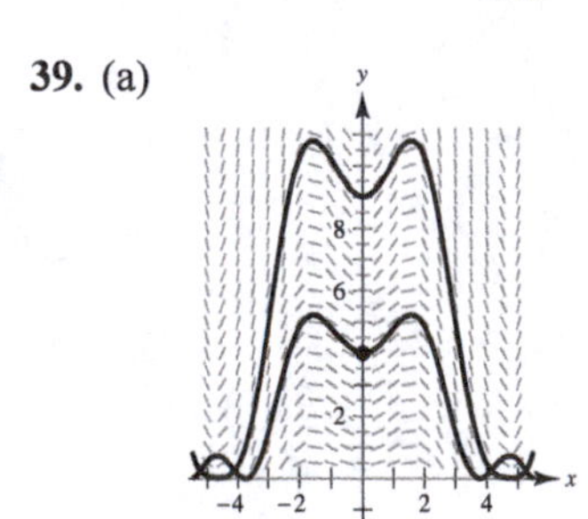

(b) $\frac{dy}{dx} = x\sqrt{y}\cos x, \quad (0, 4)$

$\int \frac{dy}{\sqrt{y}} = \int x \cos x\,dx$

$\int y^{-1/2}\,dy = \int x \cos x\,dx \qquad (u = x,\ du = dx,\ dv = \cos x\,dx,\ v = \sin x)$

$2y^{1/2} = x \sin x - \int \sin x\,dx = x \sin x + \cos x + C$

$(0, 4)$: $2(4)^{1/2} = 0 + 1 + C \Rightarrow C = 3$

$2\sqrt{y} = x \sin x + \cos x + 3$

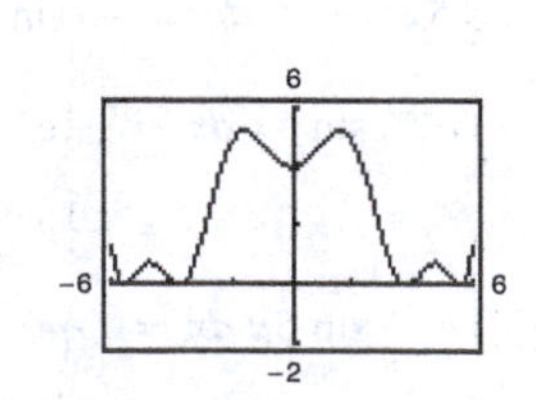

41. $\dfrac{dy}{dx} = \dfrac{x}{y}e^{x/8},\ y(0) = 2$

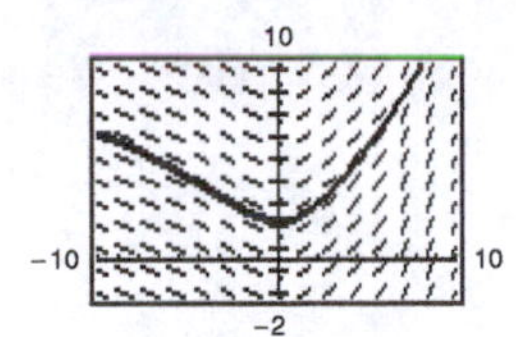

43. $u = x,\ du = dx,\ dv = e^{x/2}\,dx,\ v = 2e^{x/2}$

$$\int xe^{x/2}\,dx = 2xe^{x/2} - \int 2e^{x/2}\,dx$$
$$= 2xe^{x/2} - 4e^{x/2} + C$$

So,

$$\int_0^3 xe^{x/2}\,dx = \left[2xe^{x/2} - 4e^{x/2}\right]_0^3$$
$$= \left(6e^{3/2} - 4e^{3/2}\right) - (-4)$$
$$= 4 + 2e^{3/2} \approx 12.963$$

45. $u = x,\ du = dx,\ dv = \cos 2x\,dx,\ v = \dfrac{1}{2}\sin 2x$

$$\int x\cos 2x\,dx = \frac{1}{2}x\sin 2x - \int\frac{1}{2}\sin 2x\,dx$$
$$= \frac{1}{2}x\sin 2x + \frac{1}{4}\cos 2x + C$$

So,

$$\int_0^{\pi/4} x\cos 2x\,dx = \left[\frac{1}{2}x\sin 2x + \frac{1}{4}\cos 2x\right]_0^{\pi/4}$$
$$= \left(\frac{\pi}{8}(1) + 0\right) - \left(0 + \frac{1}{4}\right)$$
$$= \frac{\pi}{8} - \frac{1}{4} \approx 0.143$$

47. $u = \arccos x,\ du = -\dfrac{1}{\sqrt{1 - x^2}}\,dx,\ dv = dx,\ v = x$

$$\int \arccos x\,dx = x\arccos x + \int\frac{x}{\sqrt{1 - x^2}}\,dx$$
$$= x\arccos x - \sqrt{1 - x^2} + C$$

So,

$$\int_0^{1/2} \arccos x = \left[x\arccos x - \sqrt{1 - x^2}\right]_0^{1/2}$$
$$= \frac{1}{2}\arccos\left(\frac{1}{2}\right) - \sqrt{\frac{3}{4}} + 1$$
$$= \frac{\pi}{6} - \frac{\sqrt{3}}{2} + 1 \approx 0.658.$$

49. Use integration by parts twice.

(1) $dv = e^x\,dx \Rightarrow v = \int e^x\,dx = e^x$ (2) $dv = e^x\,dx \Rightarrow v = \int e^x\,dx = e^x$

$u = \sin x \Rightarrow du = \cos x\,dx$ $u = \cos x \Rightarrow du = -\sin x\,dx$

$$\int e^x\sin x\,dx = e^x\sin x - \int e^x\cos x\,dx = e^x\sin x - e^x\cos x - \int e^x\sin x\,dx$$
$$2\int e^x\sin x\,dx = e^x(\sin x - \cos x)$$
$$\int e^x\sin x\,dx = \frac{e^x}{2}(\sin x - \cos x) + C$$

So, $\displaystyle\int_0^1 e^x\sin x\,dx = \left[\frac{e^x}{2}(\sin x - \cos x)\right]_0^1 = \frac{e}{2}(\sin 1 - \cos 1) + \frac{1}{2} = \frac{e(\sin 1 - \cos 1) + 1}{2} \approx 0.909.$

51. $dv = x\,dx,\ v = \dfrac{x^2}{2},\ u = \operatorname{arcsec} x,\ du = \dfrac{1}{x\sqrt{x^2 - 1}}\,dx$

$$\int x\operatorname{arcsec} x\,dx = \frac{x^2}{2}\operatorname{arcsec} x - \int\frac{x^2/2}{x\sqrt{x^2 - 1}}\,dx = \frac{x^2}{2}\operatorname{arcsec} x - \frac{1}{4}\int\frac{2x}{\sqrt{x^2 - 1}}\,dx = \frac{x^2}{2}\operatorname{arcsec} x - \frac{1}{2}\sqrt{x^2 - 1} + C$$

So,

$$\int_2^4 x\operatorname{arcsec} x\,dx = \left[\frac{x^2}{2}\operatorname{arcsec} x - \frac{1}{2}\sqrt{x^2 - 1}\right]_2^4 = \left(8\operatorname{arcsec} 4 - \frac{\sqrt{15}}{2}\right) - \left(\frac{2\pi}{3} - \frac{\sqrt{3}}{2}\right) = 8\operatorname{arcsec} 4 - \frac{\sqrt{15}}{2} + \frac{\sqrt{3}}{2} - \frac{2\pi}{3} \approx 7.380.$$

53. $\int x^2 e^{2x}\,dx = x^2\left(\frac{1}{2}e^{2x}\right) - (2x)\left(\frac{1}{4}e^{2x}\right) + 2\left(\frac{1}{8}e^{2x}\right) + C$

$= \frac{1}{2}x^2e^{2x} - \frac{1}{2}xe^{2x} + \frac{1}{4}e^{2x} + C$

$= \frac{1}{4}e^{2x}\left(2x^2 - 2x + 1\right) + C$

Alternate signs	u and its derivatives	v' and its antiderivatives
$+$	x^2	e^{2x}
$-$	$2x$	$\frac{1}{2}e^{2x}$
$+$	2	$\frac{1}{4}e^{2x}$
$-$	0	$\frac{1}{8}e^{2x}$

55. $\int (x + 2)^2 \sin x\,dx = (x + 2)^2(-\cos x) - 2(x + 2)(-\sin x) + 2(\cos x) + C$

$= -\cos x\,(x + 2)^2 + 2\sin x\,(x + 2) + 2\cos x + C$

Alternate signs	u and its derivatives	v' and its antiderivatives
$+$	$(x + 2)^2$	$\sin x$
$-$	$2(x + 2)$	$-\cos x$
$+$	2	$-\sin x$
$-$	0	$\cos x$

57. $\int (6 + x)\sqrt{4x + 9}\,dx = (6 + x)\left(\frac{1}{6}\right)(4x + 9)^{3/2} - \frac{1}{60}(4x + 9)^{5/2} + C$

$= \frac{1}{60}(4x + 9)^{3/2}\left[60 + 10x - 4x - 9\right] + C$

$= \frac{1}{60}(4x + 9)^{3/2}(51 + 6x) + C$

$= \frac{1}{20}(4x + 9)^{3/2}(17 + 2x) + C$

Alternate signs	u and its derivatives	v' and its antiderivatives
$+$	$6 + x$	$(4x + 9)^{1/2}$
$-$	1	$\frac{1}{6}(4x + 9)^{3/2}$
$+$	0	$\frac{1}{60}(4x + 9)^{5/2}$

59. Answers will vary. *Sample answer*: $\int x^3 \sin x\,dx$.

It takes three applications of integration by parts for the term x^3 to become a constant.

Other possible answers: $\int x^3 \cos x\,dx$, $\int x^3 e^x\,dx$

61. (a) No

Substitution

(b) Yes

$u = \ln x,\ dv = x\,dx$

(c) Yes

$u = x^2,\ dv = e^{-3x}\,dx$

(d) No

Substitution

(e) Yes. Let $u = x$ and

$$dv = \frac{1}{\sqrt{x+1}}\,dx.$$

(Substitution also works. Let $u = \sqrt{x+1}$.)

(f) No

Substitution

63. $u = \sqrt{x} \Rightarrow u^2 = x \Rightarrow 2u\,du = dx$

$$\int \sin\sqrt{x}\,dx = \int \sin u(2u\,du) = 2\int u\sin u\,du$$

Integration by parts:

$w = u,\ dw = du,\ dv = \sin u\,du,\ v = -\cos u$

$$\begin{aligned} 2\int u\sin u\,du &= 2\left(-u\cos u + \int \cos u\,du\right) \\ &= 2(-u\cos u + \sin u) + C \\ &= 2\left(-\sqrt{x}\cos\sqrt{x} + \sin\sqrt{x}\right) + C \end{aligned}$$

65. $u = x^2,\ du = 2x\,dx$

$$\int x^5 e^{x^2}\,dx = \frac{1}{2}\int e^{x^2}x^4\,2x\,dx = \frac{1}{2}\int e^u u^2\,du$$

Integration by parts twice.

(1) $w = u^2,\ dw = 2u\,du,\ dv = e^u\,du,\ v = e^u$

$$\begin{aligned} \frac{1}{2}\int e^u u^2\,du &= \frac{1}{2}\left[u^2e^u - \int 2ue^u\,du\right] \\ &= \frac{1}{2}u^2e^u - \int ue^u\,du \end{aligned}$$

(2) $w = u,\ dw = du,\ dv = e^u\,du,\ v = e^u$

$$\begin{aligned} \frac{1}{2}\int e^u u^2\,du &= \frac{1}{2}u^2e^u - \left(ue^u - \int e^u\,du\right) \\ &= \frac{1}{2}u^2e^u - ue^u + e^u + C \\ &= \frac{1}{2}x^4e^{x^2} - x^2e^{x^2} + e^{x^2} + C \\ &= \frac{e^{x^2}}{2}\left(x^4 - 2x^2 + 2\right) + C \end{aligned}$$

67. (a) $dv = \dfrac{x}{\sqrt{4+x^2}}\,dx \quad\Rightarrow\quad v = \int\left(4+x^2\right)^{-1/2}x\,dx = \sqrt{4+x^2}$

$u = x^2 \quad\Rightarrow\quad du = 2x\,dx$

$$\begin{aligned} \int \frac{x^3}{\sqrt{4+x^2}}\,dx &= x^2\sqrt{4+x^2} - 2\int x\sqrt{4+x^2}\,dx \\ &= x^2\sqrt{4+x^2} - \frac{2}{3}\left(4+x^2\right)^{3/2} + C = \frac{1}{3}\sqrt{4+x^2}\left(x^2-8\right) + C \end{aligned}$$

(b) $u = 4 + x^2 \Rightarrow x^2 = u - 4$ and $2x\,dx = du \Rightarrow x\,dx = \frac{1}{2}du$

$$\begin{aligned} \int \frac{x^3}{\sqrt{4+x^2}}\,dx &= \int \frac{x^2}{\sqrt{4+x^2}}x\,dx = \int\left(\frac{u-4}{\sqrt{u}}\right)\frac{1}{2}\,du \\ &= \frac{1}{2}\int\left(u^{1/2} - 4u^{-1/2}\right)du = \frac{1}{2}\left(\frac{2}{3}u^{3/2} - 8u^{1/2}\right) + C \\ &= \frac{1}{3}u^{1/2}(u-12) + C \\ &= \frac{1}{3}\sqrt{4+x^2}\left[\left(4+x^2\right) - 12\right] + C = \frac{1}{3}\sqrt{4+x^2}\left(x^2-8\right) + C \end{aligned}$$

69. $n = 0$: $\int \ln x\, dx = x(\ln x - 1) + C$

$n = 1$: $\int x \ln x\, dx = \frac{x^2}{4}(2 \ln x - 1) + C$

$n = 2$: $\int x^2 \ln x\, dx = \frac{x^3}{9}(3 \ln x - 1) + C$

$n = 3$: $\int x^3 \ln x\, dx = \frac{x^4}{16}(4 \ln x - 1) + C$

$n = 4$: $\int x^4 \ln x\, dx = \frac{x^5}{25}(5 \ln x - 1) + C$

In general, $\int x^n \ln x\, dx = \frac{x^{n+1}}{(n+1)^2}\left[(n+1)\ln x - 1\right] + C.$

71. $dv = \sin x\, dx \Rightarrow v = -\cos x$

$u = x^n \Rightarrow du = nx^{n-1}\, dx$

$\int x^n \sin x\, dx = -x^n \cos x + n\int x^{n-1} \cos x\, dx$

73. $dv = x^n\, dx \Rightarrow v = \frac{x^{n+1}}{n+1}$

$u = \ln x \Rightarrow du = \frac{1}{x}\, dx$

$$\int x^n \ln x\, dx = \frac{x^{n+1}}{n+1}\ln x - \int \frac{x^n}{n+1}\, dx$$
$$= \frac{x^{n+1}}{n+1}\ln x - \frac{x^{n+1}}{(n+1)^2} + C$$
$$= \frac{x^{n+1}}{(n+1)^2}\left[(n+1)\ln x - 1\right] + C$$

75. Use integration by parts twice.

(1) $dv = e^{ax}\, dx \Rightarrow v = \frac{1}{a}e^{ax}$

$u = \sin bx \Rightarrow du = b \cos bx\, dx$

(2) $dv = e^{ax}\, dx \Rightarrow v = \frac{1}{a}e^{ax}$

$u = \cos bx \Rightarrow du = -b \sin bx\, dx$

$$\int e^{ax} \sin bx\, dx = \frac{e^{ax} \sin bx}{a} - \frac{b}{a}\int e^{ax} \cos bx\, dx$$
$$= \frac{e^{ax} \sin bx}{a} - \frac{b}{a}\left(\frac{e^{ax} \cos bx}{a} + \frac{b}{a}\int e^{ax} \sin bx\, dx\right) = \frac{e^{ax} \sin bx}{a} - \frac{b}{a^2}e^{ax} \cos bx - \frac{b^2}{a^2}\int e^{ax} \sin bx\, dx$$

Therefore, $\left(1 + \frac{b^2}{a^2}\right)\int e^{ax} \sin bx\, dx = \frac{e^{ax}(a \sin bx - b \cos bx)}{a^2}$

$$\int e^{ax} \sin bx\, dx = \frac{e^{ax}(a \sin bx - b \cos bx)}{a^2 + b^2} + C.$$

77. $n = 2$ (Use formula in Exercise 67.)

$$\int x^2 \sin x\, dx = -x^2 \cos x + 2\int x \cos x\, dx$$
$$= -x^2 \cos x + 2\left[x \sin x - \int \sin x\, dx\right] \text{ (Use formula in Exercise 68; } (n = 1).)$$
$$= -x^2 \cos x + 2x \sin x + 2 \cos x + C$$

79. $n = 5$ (Use formula in Exercise 69.)

$$\int x^5 \ln x\, dx = \frac{x^6}{6^2}(-1 + 6 \ln x) + C = \frac{x^6}{36}(-1 + 6 \ln x) + C$$

81. $a = -3, b = 4$ (Use formula in Exercise 71.)

$$\int e^{-3x}\sin 4x\,dx = \frac{e^{-3x}(-3\sin 4x - 4\cos 4x)}{(-3)^2 + (4)^2} + C = \frac{-e^{-3x}(3\sin 4x + 4\cos 4x)}{25} + C$$

83.

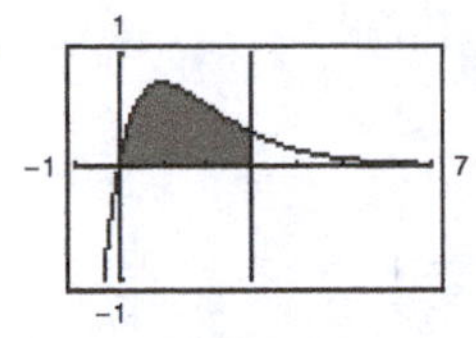

$$dv = e^{-x}\,dx \implies v = \int e^{-x}\,dx = -e^{-x}$$

$$u = 2x \implies du = 2\,dx$$

$$\int 2xe^{-x}dx = 2x(-e^{-x}) - \int -2e^{-x}\,dx$$

$$= -2xe^{-x} - 2e^{-x} + C$$

$$A = \int_0^3 2xe^{-x}\,dx = \left[-2xe^{-x} - 2e^{-x}\right]_0^3$$

$$= \left(-6e^{-3} - 2e^{-3}\right) - (-2)$$

$$= 2 - 8e^{-3} \approx 1.602$$

85.

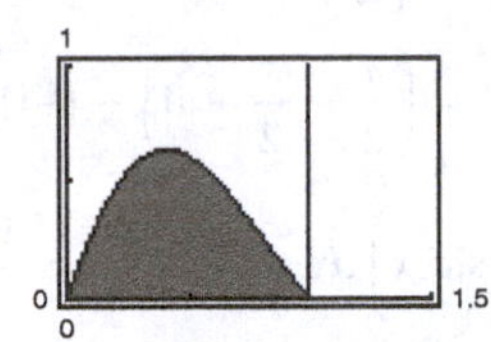

$$A = \int_0^1 e^{-x}\sin(\pi x)\,dx$$

$$= \left[\frac{e^{-x}(-\sin \pi x - \pi\cos \pi x)}{1 + \pi^2}\right]_0^1$$

$$= \frac{1}{1 + \pi^2}\left(\frac{\pi}{e} + \pi\right)$$

$$= \frac{\pi}{1 + \pi^2}\left(\frac{1}{e} + 1\right)$$

≈ 0.395 (See Exercise 71.)

87. (a) $dv = dx \implies v = x$

$$u = \ln x \implies du = \frac{1}{x}\,dx$$

$$A = \int_1^e \ln x\,dx = \left[x\ln x - x\right]_1^e = 1$$ (Use integration by parts once.)

(b) $R(x) = \ln x, r(x) = 0$

$$V = \pi\int_1^e (\ln x)^2\,dx$$

$$= \pi\left[x(\ln x)^2 - 2x\ln x + 2x\right]_1^e$$ (Use integration by parts twice, see Exercise 3.)

$$= \pi(e - 2) \approx 2.257$$

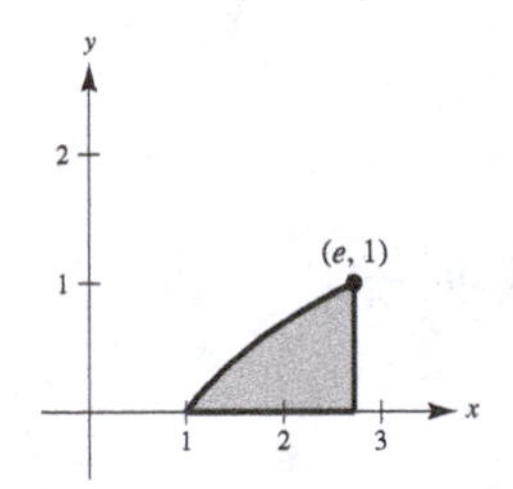

(c) $p(x) = x, h(x) = \ln x$

$$V = 2\pi\int_1^e x\ln x\,dx = 2\pi\left[\frac{x^2}{4}(-1 + 2\ln x)\right]_1^e = \frac{(e^2 + 1)\pi}{2} \approx 13.177$$ (See Exercise 73.)

(d) $$\bar{x} = \frac{\int_1^e x\ln x\,dx}{1} = \frac{e^2 + 1}{4} \approx 2.097$$

$$\bar{y} = \frac{\frac{1}{2}\int_1^e (\ln x)^2\,dx}{1} = \frac{e - 2}{2} \approx 0.359$$

$$(\bar{x}, \bar{y}) = \left(\frac{e^2 + 1}{4}, \frac{e - 2}{2}\right) \approx (2.097, 0.359)$$

89. In Example 6, you showed that the centroid of an equivalent region was $(1, \pi/8)$. By symmetry, the centroid of this region is $(\pi/8, 1)$. You can also solve this problem directly.

$$A = \int_0^1 \left(\frac{\pi}{2} - \arcsin x\right) dx = \left[\frac{\pi}{2}x - x \arcsin x - \sqrt{1 - x^2}\right]_0^1 \qquad \text{(Example 3)}$$

$$= \left(\frac{\pi}{2} - \frac{\pi}{2} - 0\right) - (-1) = 1$$

$$\bar{x} = \frac{M_y}{A} = \int_0^1 x\left(\frac{\pi}{2} - \arcsin x\right) dx = \frac{\pi}{8}, \quad \bar{y} = \frac{M_x}{A} = \int_0^1 \frac{(\pi/2) + \arcsin x}{2}\left(\frac{\pi}{2} - \arcsin x\right) dx = 1$$

91. Average value $= \dfrac{1}{\pi}\displaystyle\int_0^\pi e^{-4t}(\cos 2t + 5 \sin 2t)\, dt$

$$= \frac{1}{\pi}\left[e^{-4t}\left(\frac{-4 \cos 2t + 2 \sin 2t}{20}\right) + 5e^{-4t}\left(\frac{-4 \sin 2t - 2 \cos 2t}{20}\right)\right]_0^\pi \qquad \text{(From Exercises 71 and 72)}$$

$$= \frac{7}{10\pi}\left(1 - e^{-4\pi}\right) \approx 0.223$$

93. $c(t) = 100{,}000 + 4000t,\ r = 5\% = 0.05,\ t_1 = 10$

$$P = \int_0^{10} (100{,}000 + 4000t)e^{-0.05t}\, dt = 4000\int_0^{10} (25 + t)e^{-0.05t}\, dt$$

Let $u = 25 + t,\ dv = e^{-0.05t}dt,\ du = dt,\ v = -\dfrac{100}{5}e^{-0.05t}$.

$$P = 4000\left\{\left[(25 + t)\left(-\frac{100}{5}e^{-0.05t}\right)\right]_0^{10} + \frac{100}{5}\int_0^{10} e^{-0.05t}dt\right\} = 4000\left\{\left[(25 + t)\left(-\frac{100}{5}e^{-0.05t}\right)\right]_0^{10} - \left[\frac{10{,}000}{25}e^{-0.05t}\right]_0^{10}\right\} \approx \$931{,}265$$

95. $$\int_{-\pi}^{\pi} x \sin nx\, dx = \left[-\frac{x}{n}\cos nx + \frac{1}{n^2}\sin nx\right]_{-\pi}^{\pi} = -\frac{\pi}{n}\cos \pi n - \frac{\pi}{n}\cos(-\pi n) = -\frac{2\pi}{n}\cos \pi n = \begin{cases} -(2\pi/n), & \text{if } n \text{ is even} \\ (2\pi/n), & \text{if } n \text{ is odd} \end{cases}$$

97. Let $u = x,\ dv = \sin\left(\dfrac{n\pi}{2}x\right) dx,\ du = dx,\ v = -\dfrac{2}{n\pi}\cos\left(\dfrac{n\pi}{2}x\right)$.

$$I_1 = \int_0^1 x \sin\left(\frac{n\pi}{2}x\right) dx = \left[\frac{-2x}{n\pi}\cos\left(\frac{n\pi}{2}x\right)\right]_0^1 + \frac{2}{n\pi}\int_0^1 \cos\left(\frac{n\pi}{2}x\right) dx$$

$$= -\frac{2}{n\pi}\cos\left(\frac{n\pi}{2}\right) + \left[\left(\frac{2}{n\pi}\right)^2 \sin\left(\frac{n\pi}{2}x\right)\right]_0^1 = -\frac{2}{n\pi}\cos\left(\frac{n\pi}{2}\right) + \left(\frac{2}{n\pi}\right)^2 \sin\left(\frac{n\pi}{2}\right)$$

Let $u = (-x + 2),\ dv = \sin\left(\dfrac{n\pi}{2}x\right) dx,\ du = -dx,\ v = -\dfrac{2}{n\pi}\cos\left(\dfrac{n\pi}{2}x\right)$.

$$I_2 = \int_1^2 (-x + 2) \sin\left(\frac{n\pi}{2}x\right) dx = \left[\frac{-2(-x + 2)}{n\pi}\cos\left(\frac{n\pi}{2}x\right)\right]_1^2 - \frac{2}{n\pi}\int_1^2 \cos\left(\frac{n\pi}{2}x\right) dx$$

$$= \frac{2}{n\pi}\cos\left(\frac{n\pi}{2}\right) - \left[\left(\frac{2}{n\pi}\right)^2 \sin\left(\frac{n\pi}{2}x\right)\right]_1^2 = \frac{2}{n\pi}\cos\left(\frac{n\pi}{2}\right) + \left(\frac{2}{n\pi}\right)^2 \sin\left(\frac{n\pi}{2}\right)$$

$$h(I_1 + I_2) = b_n = h\left[\left(\frac{2}{n\pi}\right)^2 \sin\left(\frac{n\pi}{2}\right) + \left(\frac{2}{n\pi}\right)^2 \sin\left(\frac{n\pi}{2}\right)\right] = \frac{8h}{(n\pi)^2}\sin\left(\frac{n\pi}{2}\right)$$

99. For any integrable function, $\int f(x)dx = C + \int f(x)dx$, but this cannot be used to imply that $C = 0$.

Section 8.3 Trigonometric Integrals

1. The integral $\int \sin^8 x\, dx$ requires more steps. The other integral can be found by u-substitution ($u = \sin x$, $du = \cos x\, dx$).

3. Let $u = \cos x$, $du = -\sin x\, dx$.

$$\int \cos^5 x \sin x\, dx = -\int \cos^5 x(-\sin x)\, dx = -\frac{\cos^6 x}{6} + C$$

5.
$$\begin{aligned}\int \cos^3 x \sin^4 x\, dx &= \int \cos x\left(1 - \sin^2 x\right)\sin^4 x\, dx \\ &= \int\left(\sin^4 x - \sin^6 x\right)\cos x\, dx \\ &= \frac{\sin^5 x}{5} - \frac{\sin^7 x}{7} + C\end{aligned}$$

7.
$$\begin{aligned}\int \sin^3 x \cos^2 x\, dx &= \int\left(1 - \cos^2 x\right)\cos^2 x \sin x\, dx \\ &= \int\left(\cos^2 x - \cos^4 x\right)\sin x\, dx \\ &= -\int\left(\cos^2 x - \cos^4 x\right)(-\sin x)\, dx \\ &= -\frac{\cos^3 x}{3} + \frac{\cos^5 x}{5} + C\end{aligned}$$

9.
$$\begin{aligned}\int \sin^3 2\theta\sqrt{\cos 2\theta}\, d\theta &= \int\left(1 - \cos^2 2\theta\right)\sqrt{\cos 2\theta}\sin 2\theta\, d\theta \\ &= \int\left[(\cos 2\theta)^{1/2} - (\cos 2\theta)^{5/2}\right]\sin 2\theta\, d\theta \\ &= -\tfrac{1}{2}\int\left[(\cos 2\theta)^{1/2} - (\cos 2\theta)^{5/2}\right](-2\sin 2\theta)\, d\theta \\ &= -\tfrac{1}{2}\left[\tfrac{2}{3}(\cos 2\theta)^{3/2} - \tfrac{2}{7}(\cos 2\theta)^{7/2}\right] + C \\ &= -\tfrac{1}{3}(\cos 2\theta)^{3/2} + \tfrac{1}{7}(\cos 2\theta)^{7/2} + C\end{aligned}$$

11.
$$\int \cos^2 3x\, dx = \int \frac{1 + \cos 6x}{2}\, dx = \frac{1}{2}\left(x + \frac{1}{6}\sin 6x\right) + C = \frac{1}{12}(6x + \sin 6x) + C$$

13. $dv = \cos^2 x\, dx = \dfrac{1 + \cos 2x}{2}\, dx \Rightarrow v = \dfrac{1}{2}x + \dfrac{\sin 2x}{4}$

$u = 8x \Rightarrow du = 8\, dx$

$$\begin{aligned}\int 8x\cos^2 x\, dx &= uv - \int v\, du \\ &= 8x\left(\frac{1}{2}x + \frac{\sin 2x}{4}\right) - \int\left(\frac{1}{2}x + \frac{\sin 2x}{4}\right)(8\, dx) \\ &= 4x^2 + 2x\sin 2x - 2x^2 + \cos 2x + C \\ &= 2x^2 + 2x\sin 2x + \cos 2x + C\end{aligned}$$

15. $\displaystyle\int_0^{\pi/2} \cos^3 x\, dx = \frac{2}{3}$ $\quad(n = 3, \text{odd})$

17. $\displaystyle\int_0^{\pi/2} \sin^2 x\, dx = \frac{1}{2}\left(\frac{\pi}{2}\right) = \frac{\pi}{4}$ $\quad(n = 2, \text{even})$

19. $\displaystyle\int_0^{\pi/2} \sin^{10} x\, dx = \left(\frac{1}{2}\right)\left(\frac{3}{4}\right)\left(\frac{5}{6}\right)\left(\frac{7}{8}\right)\left(\frac{9}{10}\right)\left(\frac{\pi}{2}\right) = \frac{63}{512}\pi$ $\quad(n = 10, \text{even})$

21. $\displaystyle\int \sec 4x\, dx = \tfrac{1}{4}\int \sec 4x(4\, dx) = \tfrac{1}{4}\ln\left|\sec 4x + \tan 4x\right| + C$

23. $dv = \sec^2 \pi x\, dx \quad \Rightarrow \quad v = \frac{1}{\pi} \tan \pi x$

$u = \sec \pi x \quad \Rightarrow \quad du = \pi \sec \pi x \tan \pi x\, dx$

$$\int \sec^3 \pi x\, dx = \frac{1}{\pi} \sec \pi x \tan \pi x - \int \sec \pi x \tan^2 \pi x\, dx = \frac{1}{\pi} \sec \pi x \tan \pi x - \int \sec \pi x (\sec^2 \pi x - 1)\, dx$$

$$2\int \sec^3 \pi x\, dx = \frac{1}{\pi}(\sec \pi x \tan \pi x + \ln|\sec \pi x + \tan \pi x|) + C_1$$

$$\int \sec^3 \pi x\, dx = \frac{1}{2\pi}(\sec \pi x \tan \pi x + \ln|\sec \pi x + \tan \pi x|) + C$$

25.
$$\int \tan^5 \frac{x}{2}\, dx = \int \left(\sec^2 \frac{x}{2} - 1\right) \tan^3 \frac{x}{2}\, dx$$
$$= \int \tan^3 \frac{x}{2} \sec^2 \frac{x}{2}\, dx - \int \tan^3 \frac{x}{2}\, dx$$
$$= \frac{\tan^4 \frac{x}{2}}{2} - \int \left(\sec^2 \frac{x}{2} - 1\right) \tan \frac{x}{2}\, dx$$
$$= \frac{1}{2} \tan^4 \frac{x}{2} - \tan^2 \frac{x}{2} - 2 \ln\left|\cos \frac{x}{2}\right| + C$$

27. Let $u = \sec 2t$, $du = 2 \sec 2t \tan 2t\, dt$.

$$\int \tan^3 2t \cdot \sec^3 2t\, dt = \int (\sec^2 2t - 1) \sec^3 2t \cdot \tan 2t\, dt$$
$$= \int (\sec^4 2t - \sec^2 2t)(\sec 2t \tan 2t)\, dt = \frac{\sec^5 2t}{10} - \frac{\sec^3 2t}{6} + C$$

29.
$$\int \sec^6 4x \tan 4x\, dx = \frac{1}{4}\int \sec^5 4x (4 \sec 4x \tan 4x)\, dx$$
$$= \frac{\sec^6 4x}{24} + C$$

31.
$$\int \sec^5 x \tan^3 x\, dx = \int \sec^4 x \tan^2 x (\sec x \tan x)\, dx$$
$$= \int \sec^4 x(\sec^2 x - 1)(\sec x \tan x)\, dx$$
$$= \int (\sec^6 x - \sec^4 x)(\sec x \tan x)\, dx$$
$$= \frac{\sec^7 x}{7} - \frac{\sec^5 x}{5} + C$$

33.
$$\int \frac{\tan^2 x}{\sec x}\, dx = \int \frac{(\sec^2 x - 1)}{\sec x}\, dx$$
$$= \int (\sec x - \cos x)\, dx$$
$$= \ln|\sec x + \tan x| - \sin x + C$$

35.
$$r = \int \sin^4(\pi\theta)\, d\theta = \frac{1}{4}\int \left[1 - \cos(2\pi\theta)\right]^2 d\theta$$
$$= \frac{1}{4}\int \left[1 - 2\cos(2\pi\theta) + \cos^2(2\pi\theta)\right] d\theta$$
$$= \frac{1}{4}\int \left[1 - 2\cos(2\pi\theta) + \frac{1 + \cos(4\pi\theta)}{2}\right] d\theta$$
$$= \frac{1}{4}\left[\theta - \frac{1}{\pi}\sin(2\pi\theta) + \frac{\theta}{2} + \frac{1}{8\pi}\sin(4\pi\theta)\right] + C$$
$$= \frac{1}{32\pi}\left[12\pi\theta - 8\sin(2\pi\theta) + \sin(4\pi\theta)\right] + C$$

37.
$$y = \int \tan^3 3x \sec 3x\, dx$$
$$= \int (\sec^2 3x - 1) \sec 3x \tan 3x\, dx$$
$$= \tfrac{1}{3}\int \sec^2 3x (3 \sec 3x \tan 3x)\, dx - \tfrac{1}{3}\int 3 \sec 3x \tan 3x\, dx$$
$$= \tfrac{1}{9}\sec^3 3x - \tfrac{1}{3}\sec 3x + C$$

39. (a)

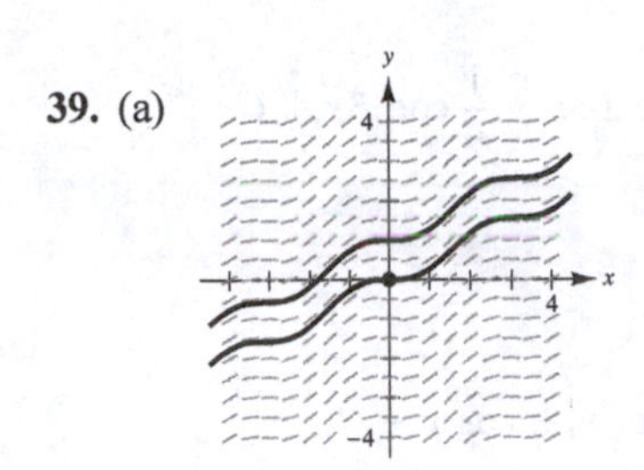

(b)

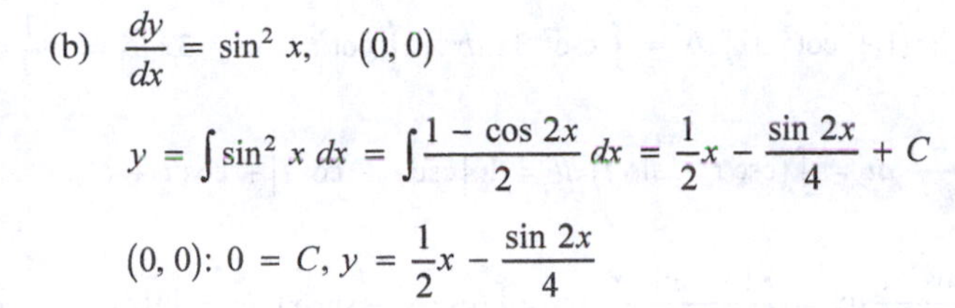

$$\frac{dy}{dx} = \sin^2 x, \quad (0, 0)$$

$$y = \int \sin^2 x\, dx = \int \frac{1 - \cos 2x}{2}\, dx = \frac{1}{2}x - \frac{\sin 2x}{4} + C$$

$$(0, 0)\text{: } 0 = C,\ y = \frac{1}{2}x - \frac{\sin 2x}{4}$$

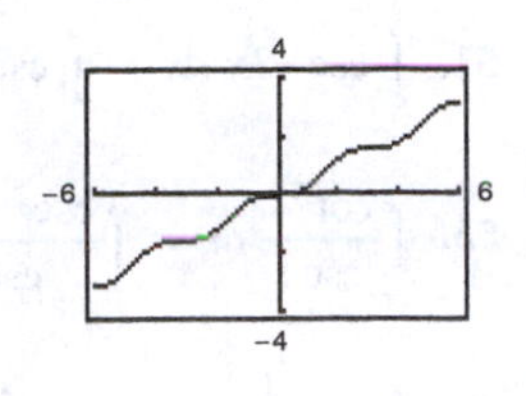

41. $\dfrac{dy}{dx} = \dfrac{3 \sin x}{y},\ y(0) = 2$

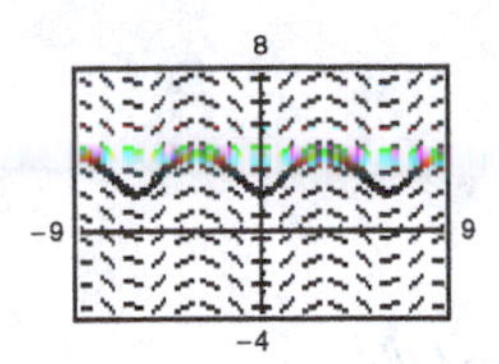

43.
$$\begin{aligned}\int \cos 2x \cos 6x\, dx &= \frac{1}{2}\int \left[\cos((2 - 6)x) + \cos((2 + 6)x)\right] dx\\
&= \frac{1}{2}\int \left[\cos(-4x) + \cos 8x\right] dx\\
&= \frac{1}{2}\int (\cos 4x + \cos 8x)\, dx\\
&= \frac{1}{2}\left[\frac{\sin 4x}{4} + \frac{\sin 8x}{8}\right] + C\\
&= \frac{\sin 4x}{8} + \frac{\sin 8x}{16} + C\\
&= \frac{1}{16}(2 \sin 4x + \sin 8x) + C\end{aligned}$$

45.
$$\begin{aligned}\int \sin 2t \cos 9t\, dt &= \frac{1}{2}\int \left[\sin[(2 - 9)t] + \sin[(2 + 9)t]\right] dt\\
&= \frac{1}{2}\int \left[\sin(-7t) + \sin(11t)\right] dt\\
&= -\frac{1}{2}\int (\sin 7t + \sin 11t)\, dt\\
&= \frac{1}{14}\cos 7t - \frac{1}{22}\cos 11t + C\end{aligned}$$

47.
$$\begin{aligned}\int \sin \theta \sin 3\theta\, d\theta &= \tfrac{1}{2}\int (\cos 2\theta - \cos 4\theta)\, d\theta\\
&= \tfrac{1}{2}\left(\tfrac{1}{2}\sin 2\theta - \tfrac{1}{4}\sin 4\theta\right) + C\\
&= \tfrac{1}{8}(2 \sin 2\theta - \sin 4\theta) + C\end{aligned}$$

49.
$$\begin{aligned}\int \cot^3 2x\, dx &= \int \left(\csc^2 2x - 1\right) \cot 2x\, dx\\
&= -\frac{1}{2}\int \cot 2x\left(-2 \csc^2 2x\right) dx - \frac{1}{2}\int \frac{2 \cos 2x}{\sin 2x}\, dx\\
&= -\frac{1}{4}\cot^2 2x - \frac{1}{2}\ln\left|\sin 2x\right| + C\\
&= \frac{1}{4}\left(\ln\left|\csc^2 2x\right| - \cot^2 2x\right) + C\end{aligned}$$

51. $\int \csc^4 3x\,dx = \int \csc^2 3x(1 + \cot^2 3x)\,dx = \int \csc^2 3x\,dx + \int \cot^2 3x \csc^2 3x\,dx = -\frac{1}{3}\cot 3x - \frac{1}{9}\cot^3 3x + C$

53. $\int \frac{\cot^2 t}{\csc t}\,dt = \int \frac{\csc^2 t - 1}{\csc t}\,dt = \int (\csc t - \sin t)\,dt = \ln|\csc t - \cot t| + \cos t + C$

55. $\int \frac{1}{\sec x \tan x}\,dx = \int \frac{\cos^2 x}{\sin x}\,dx = \int \frac{1 - \sin^2 x}{\sin x}\,dx = \int (\csc x - \sin x)\,dx = \ln|\csc x - \cot x| + \cos x + C$

57. $\int (\tan^4 t - \sec^4 t)\,dt = \int (\tan^2 t + \sec^2 t)(\tan^2 t - \sec^2 t)\,dt, \qquad (\tan^2 t - \sec^2 t = -1)$

$= -\int (\tan^2 t + \sec^2 t)\,dt = -\int (2\sec^2 t - 1)\,dt = -2\tan t + t + C$

59. $\int_{-\pi}^{\pi} \sin^2 x\,dx = 2\int_0^{\pi} \frac{1 - \cos 2x}{2}\,dx = \left[x - \frac{1}{2}\sin 2x\right]_0^{\pi} = \pi$

61. $\int_0^{\pi/4} 6\tan^3 x\,dx = 6\int_0^{\pi/4} (\sec^2 x - 1)\tan x\,dx$

$= 6\int_0^{\pi/4} \left[\tan x \sec^2 x - \tan x\right]dx$

$= 6\left[\frac{\tan^2 x}{2} + \ln|\cos x|\right]_0^{\pi/4}$

$= 6\left[\frac{1}{2} + \ln\left(\frac{\sqrt{2}}{2}\right)\right] = 6\left(\frac{1}{2} - \ln\sqrt{2}\right)$

$= 3(1 - \ln 2)$

63. Let $u = 1 + \sin t$, $du = \cos t\,dt$.

$\int_0^{\pi/2} \frac{\cos t}{1 + \sin t}\,dt = \left[\ln|1 + \sin t|\right]_0^{\pi/2} = \ln 2$

65. $\int_{-\pi/2}^{\pi/2} 3\cos^3 x\,dx = 3\int_{-\pi/2}^{\pi/2} (1 - \sin^2 x)\cos x\,dx$

$= 3\left[\sin x - \frac{\sin^3 x}{3}\right]_{-\pi/2}^{\pi/2}$

$= 3\left[\left(1 - \frac{1}{3}\right) - \left(-1 + \frac{1}{3}\right)\right] = 4$

67. (a) Let $u = \tan 3x$, $du = 3\sec^2 3x\,dx$.

$\int \sec^4 3x \tan^3 3x\,dx = \int \sec^2 3x \tan^3 3x \sec^2 3x\,dx = \frac{1}{3}\int (\tan^2 3x + 1)\tan^3 3x(3\sec^2 3x)\,dx$

$= \frac{1}{3}\int (\tan^5 3x + \tan^3 3x)(3\sec^2 3x)\,dx = \frac{\tan^6 3x}{18} + \frac{\tan^4 3x}{12} + C_1$

Or let $u = \sec 3x$, $du = 3\sec 3x \tan 3x\,dx$.

$\int \sec^4 3x \tan^3 3x\,dx = \int \sec^3 3x \tan^2 3x \sec 3x \tan 3x\,dx$

$= \frac{1}{3}\int \sec^3 3x(\sec^2 3x - 1)(3\sec 3x \tan 3x)\,dx = \frac{\sec^6 3x}{18} - \frac{\sec^4 3x}{12} + C$

(b)

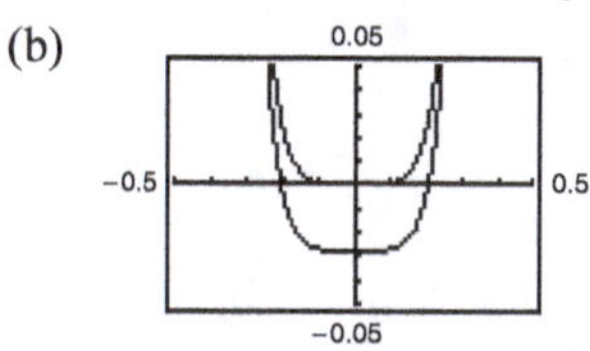

(c) $\frac{\sec^6 3x}{18} - \frac{\sec^4 3x}{12} + C = \frac{(1 + \tan^2 3x)^3}{18} - \frac{(1 + \tan^2 3x)^2}{12} + C$

$= \frac{1}{18}\tan^6 3x + \frac{1}{6}\tan^4 3x + \frac{1}{6}\tan^2 3x + \frac{1}{18} - \frac{1}{12}\tan^4 3x - \frac{1}{6}\tan^2 3x - \frac{1}{12} + C$

$= \frac{\tan^6 3x}{18} + \frac{\tan^4 3x}{12} + \left(\frac{1}{18} - \frac{1}{12}\right) + C$

$= \frac{\tan^6 3x}{18} + \frac{\tan^4 3x}{12} + C_2$

69. (a) $\int \sin x \cos x\,dx = \dfrac{\sin^2 x}{2} + C$

(b) $-\int \cos x(-\sin x)\,dx = -\dfrac{\cos^2 x}{2} + C$

(c) $dv = \cos x\,dx \Rightarrow v = \sin x$

$u = \sin x \Rightarrow du = \cos x\,dx$

$\int \sin x \cos x\,dx = \sin^2 x - \int \sin x \cos x\,dx$

$2\int \sin x \cos x\,dx = \sin^2 x$

$\int \sin x \cos x\,dx = \dfrac{\sin^2 x}{2} + C$

(Answers will vary.)

(d) $\int \sin x \cos x\,dx = \int \dfrac{1}{2}\sin 2x\,dx = -\dfrac{1}{4}\cos 2x + C$

The answers all differ by a constant.

71. $A = \int_0^{\pi/2}\left(\sin x - \sin^3 x\right)dx$

$= \int_0^{\pi/2}\sin x\,dx - \int_0^{\pi/2}\sin^3 x\,dx$

$= \left[-\cos x\right]_0^{\pi/2} - \frac{2}{3}$ (Wallis's Formula)

$= 1 - \frac{2}{3} = \frac{1}{3}$

73. $A = \int_{-\pi/4}^{\pi/4}\left[\cos^2 x - \sin^2 x\right]dx$

$= \int_{-\pi/4}^{\pi/4}\cos 2x\,dx$

$= \left[\dfrac{\sin 2x}{2}\right]_{-\pi/4}^{\pi/4} = \dfrac{1}{2} + \dfrac{1}{2} = 1$

75. Disks

$R(x) = \tan x,\ r(x) = 0$

$V = 2\pi\int_0^{\pi/4}\tan^2 x\,dx$

$= 2\pi\int_0^{\pi/4}\left(\sec^2 x - 1\right)dx$

$= 2\pi\left[\tan x - x\right]_0^{\pi/4}$

$= 2\pi\left(1 - \dfrac{\pi}{4}\right) \approx 1.348$

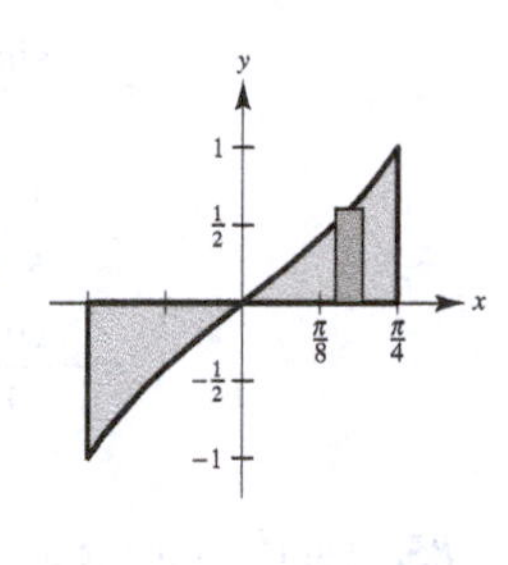

77. (a) $V = \pi\int_0^{\pi}\sin^2 x\,dx = \dfrac{\pi}{2}\int_0^{\pi}(1 - \cos 2x)\,dx = \dfrac{\pi}{2}\left[x - \dfrac{1}{2}\sin 2x\right]_0^{\pi} = \dfrac{\pi^2}{2}$

(b) $A = \int_0^{\pi}\sin x\,dx = \left[-\cos x\right]_0^{\pi} = 1 + 1 = 2$

Let $u = x$, $dv = \sin x\,dx$, $du = dx$, $v = -\cos x$.

$\bar{x} = \dfrac{1}{A}\int_0^{\pi}x\sin x\,dx = \dfrac{1}{2}\left[\left[-x\cos x\right]_0^{\pi} + \int_0^{\pi}\cos x\,dx\right] = \dfrac{1}{2}\left[-x\cos x + \sin x\right]_0^{\pi} = \dfrac{\pi}{2}$

$\bar{y} = \dfrac{1}{2A}\int_0^{\pi}\sin^2 x\,dx = \dfrac{1}{8}\int_0^{\pi}(1 - \cos 2x)\,dx = \dfrac{1}{8}\left[x - \dfrac{1}{2}\sin 2x\right]_0^{\pi} = \dfrac{\pi}{8}$

$(\bar{x}, \bar{y}) = \left(\dfrac{\pi}{2}, \dfrac{\pi}{8}\right)$

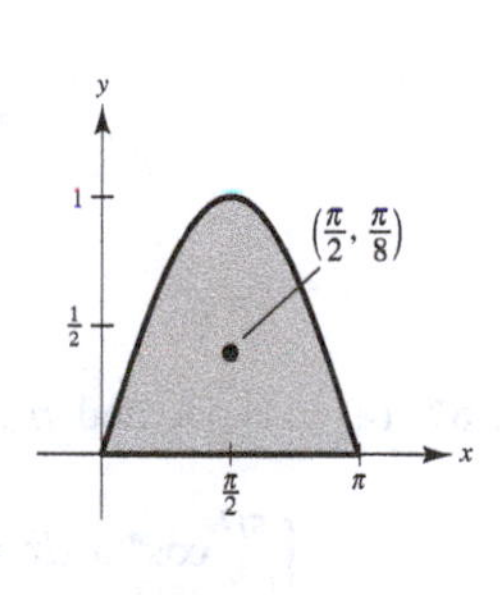

79. $dv = \sin x\,dx \Rightarrow v = -\cos x$

$u = \sin^{n-1}x \Rightarrow du = (n-1)\sin^{n-2}x\cos x\,dx$

$\int \sin^n x\,dx = -\sin^{n-1}x\cos x + (n-1)\int \sin^{n-2}x\cos^2 x\,dx = -\sin^{n-1}x\cos x + (n-1)\int \sin^{n-2}x\left(1 - \sin^2 x\right)dx$

$= -\sin^{n-1}x\cos x + (n-1)\int \sin^{n-2}x\,dx - (n-1)\int \sin^n x\,dx$

Therefore, $n\int \sin^n x\,dx = -\sin^{n-1}x\cos x + (n-1)\int \sin^{n-2}x\,dx$

$$\int \sin^n x\,dx = \frac{-\sin^{n-1}x\cos x}{n} + \frac{n-1}{n}\int \sin^{n-2}x\,dx.$$

81. Let $u = \sin^{n-1} x$, $du = (n-1)\sin^{n-2} x \cos x\, dx$, $dv = \cos^m x \sin x\, dx$, $v = \dfrac{-\cos^{m+1} x}{m+1}$.

$$\begin{aligned}
\int \cos^m x \sin^n x\, dx &= \frac{-\sin^{n-1} x \cos^{m+1} x}{m+1} + \frac{n-1}{m+1}\int \sin^{n-2} x \cos^{m+2} x\, dx \\
&= \frac{-\sin^{n-1} x \cos^{m+1} x}{m+1} + \frac{n-1}{m+1}\int \sin^{n-2} x \cos^m x\left(1 - \sin^2 x\right) dx \\
&= \frac{-\sin^{n-1} x \cos^{m+1} x}{m+1} + \frac{n-1}{m+1}\int \sin^{n-2} x \cos^m x\, dx - \frac{n-1}{m+1}\int \sin^n x \cos^m x\, dx \\
\frac{m+n}{m+1}\int \cos^m x \sin^n x\, dx &= \frac{-\sin^{n-1} x \cos^{m+1} x}{m+1} + \frac{n-1}{m+1}\int \sin^{n-2} x \cos^m x\, dx \\
\int \cos^m x \sin^n x\, dx &= \frac{-\cos^{m+1} x \sin^{n-1} x}{m+n} + \frac{n-1}{m+n}\int \cos^m x \sin^{n-2} x\, dx
\end{aligned}$$

83.
$$\begin{aligned}
\int \sin^5 x\, dx &= -\frac{\sin^4 x \cos x}{5} + \frac{4}{5}\int \sin^3 x\, dx \\
&= -\frac{\sin^4 x \cos x}{5} + \frac{4}{5}\left(-\frac{\sin^2 x \cos x}{3} + \frac{2}{3}\int \sin x\, dx\right) \\
&= -\frac{1}{5}\sin^4 x \cos x - \frac{4}{15}\sin^2 x \cos x - \frac{8}{15}\cos x + C \\
&= -\frac{\cos x}{15}\left(3\sin^4 x + 4\sin^2 x + 8\right) + C
\end{aligned}$$

85.
$$\begin{aligned}
\int \sin^4 x \cos^2 x\, dx &= -\frac{\cos^3 x \sin^3 x}{6} + \frac{1}{2}\int \cos^2 x \sin^2 x\, dx \\
&= -\frac{\cos^3 x \sin^3 x}{6} + \frac{1}{2}\left(-\frac{\cos^3 x \sin x}{4} + \frac{1}{4}\int \cos^2 x\, dx\right) \\
&= -\frac{1}{6}\cos^3 x \sin^3 x - \frac{1}{8}\cos^3 x \sin x + \frac{1}{8}\left(\frac{\cos x \sin x}{2} + \frac{x}{2}\right) + C \\
&= -\frac{1}{48}\left(8\cos^3 x \sin^3 x + 6\cos^3 x \sin x - 3\cos x \sin x - 3x\right) + C
\end{aligned}$$

87. (a) n is odd and $n \geq 3$.

$$\begin{aligned}
\int_0^{\pi/2} \cos^n x\, dx &= \left[\frac{\cos^{n-1} x \sin x}{n}\right]_0^{\pi/2} + \frac{n-1}{n}\int_0^{\pi/2} \cos^{n-2} x\, dx \\
&= \frac{n-1}{n}\left(\left[\frac{\cos^{n-3} x \sin x}{n-2}\right]_0^{\pi/2} + \frac{n-3}{n-2}\int_0^{\pi/2} \cos^{n-4} x\, dx\right) \\
&= \frac{n-1}{n}\cdot\frac{n-3}{n-2}\left(\left[\frac{\cos^{n-5} x \sin x}{n-4}\right]_0^{\pi/2} + \frac{n-5}{n-4}\int_0^{\pi/2} \cos^{n-6} x\, dx\right) \\
&= \frac{n-1}{n}\cdot\frac{n-3}{n-2}\cdot\frac{n-5}{n-4}\int_0^{\pi/2} \cos^{n-6} x\, dx \\
&= \frac{n-1}{n}\cdot\frac{n-3}{n-2}\cdot\frac{n-5}{n-4}\cdots\int_0^{\pi/2} \cos x\, dx \\
&= \left[\frac{n-1}{n}\cdot\frac{n-3}{n-2}\cdot\frac{n-5}{n-4}\cdots(\sin x)\right]_0^{\pi/2} \\
&= \frac{n-1}{n}\cdot\frac{n-3}{n-2}\cdot\frac{n-5}{n-4}\cdots 1 \quad \text{(Reverse the order.)} \\
&= (1)\left(\frac{2}{3}\right)\left(\frac{4}{5}\right)\left(\frac{6}{7}\right)\cdots\left(\frac{n-1}{n}\right) = \left(\frac{2}{3}\right)\left(\frac{4}{5}\right)\left(\frac{6}{7}\right)\cdots\left(\frac{n-1}{n}\right)
\end{aligned}$$

(b) n is even and $n \geq 2$.

$$\int_0^{\pi/2} \cos^n x\,dx = \frac{n-1}{n}\cdot\frac{n-3}{n-2}\cdot\frac{n-5}{n-4}\cdots\int_0^{\pi/2}\cos^2 x\,dx \quad \text{(From part (a))}$$

$$= \left[\frac{n-1}{n}\cdot\frac{n-3}{n-2}\cdot\frac{n-5}{n-4}\cdots\left(\frac{x}{2}+\frac{1}{4}\sin 2x\right)\right]_0^{\pi/2}$$

$$= \frac{n-1}{n}\cdot\frac{n-3}{n-2}\cdot\frac{n-5}{n-4}\cdots\frac{\pi}{4} \quad \text{(Reverse the order.)}$$

$$= \left(\frac{\pi}{2}\cdot\frac{1}{2}\right)\left(\frac{3}{4}\right)\left(\frac{5}{6}\right)\cdots\left(\frac{n-1}{n}\right) = \left(\frac{1}{2}\right)\left(\frac{3}{4}\right)\left(\frac{5}{6}\right)\cdots\left(\frac{n-1}{n}\right)\left(\frac{\pi}{2}\right)$$

89. $f(x) = \sum_{i=1}^{N} a_i \sin(ix)$

(a) $$f(x)\sin(nx) = \left[\sum_{i=1}^{N} a_i \sin(ix)\right]\sin(nx)$$

$$\int_{-\pi}^{\pi} f(x)\sin(nx)\,dx = \int_{-\pi}^{\pi}\left[\sum_{i=1}^{N} a_i \sin(ix)\right]\sin(nx)\,dx$$

$$= \int_{-\pi}^{\pi} a_n \sin^2(nx)\,dx \quad \text{(by Exercise 89)}$$

$$= \int_{-\pi}^{\pi} a_n \frac{1-\cos(2nx)}{2}\,dx = \left[\frac{a_n}{2}\left(x - \frac{\sin(2nx)}{2n}\right)\right]_{-\pi}^{\pi} = \frac{a_n}{2}(\pi+\pi) = a_n\pi$$

So, $a_n = \frac{1}{\pi}\int_{-\pi}^{\pi} f(x)\sin(nx)\,dx.$

(b) $f(x) = x$

$$a_1 = \frac{1}{\pi}\int_{-\pi}^{\pi} x\sin x\,dx = 2$$

$$a_2 = \frac{1}{\pi}\int_{-\pi}^{\pi} x\sin 2x\,dx = -1$$

$$a_3 = \frac{1}{\pi}\int_{-\pi}^{\pi} x\sin 3x\,dx = \frac{2}{3}$$

Section 8.4 Trigonometric Substitution

1. (a) Use $x = 3\tan\theta$.

(b) Use $x = 2\sin\theta$.

(c) Use $x = 5\sin\theta$.

(d) Use $x = 5\sec\theta$.

3. Let $x = 4\sin\theta$, $dx = 4\cos\theta\,d\theta$, $\sqrt{16-x^2} = 4\cos\theta$.

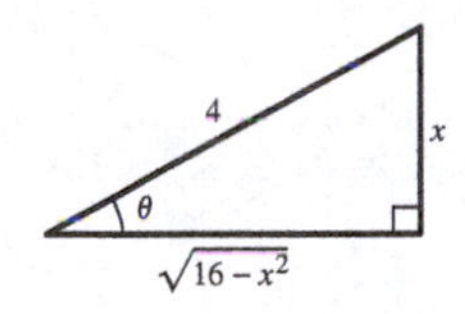

$$\int \frac{1}{(16-x^2)^{3/2}}\,dx = \int \frac{4\cos\theta}{(4\cos\theta)^3}\,d\theta = \frac{1}{16}\int \sec^2\theta\,d\theta = \frac{1}{16}\tan\theta + C = \frac{1}{16}\left(\frac{x}{\sqrt{16-x^2}}\right) + C$$

5. Same substitution as in Exercise 5

$$\int \frac{\sqrt{16 - x^2}}{x}\,dx = \int \frac{4\cos\theta}{4\sin\theta} 4\cos\theta\,d\theta$$
$$= 4\int \frac{\cos^2\theta}{\sin\theta}\,d\theta$$
$$= 4\int \frac{1 - \sin^2\theta}{\sin\theta}\,d\theta$$
$$= 4\int (\csc\theta - \sin\theta)\,d\theta$$
$$= -4\ln\left|\csc\theta + \cot\theta\right| + 4\cos\theta + C$$
$$= -4\ln\left|\frac{4}{x} + \frac{\sqrt{16 - x^2}}{x}\right| + 4\frac{\sqrt{16 - x^2}}{4} + C$$
$$= -4\ln\left|\frac{4 + \sqrt{16 - x^2}}{x}\right| + \sqrt{16 - x^2} + C$$
$$= 4\ln\left|\frac{4 - \sqrt{16 - x^2}}{x}\right| + \sqrt{16 - x^2} + C$$

7. Let $x = 5\sec\theta$, $dx = 5\sec\theta\tan\theta\,d\theta$,

$\sqrt{x^2 - 25} = 5\tan\theta$.

$$\int \frac{1}{\sqrt{x^2 - 25}}\,dx = \int \frac{5\sec\theta\tan\theta}{5\tan\theta}\,d\theta$$
$$= \int \sec\theta\,d\theta$$
$$= \ln\left|\sec\theta + \tan\theta\right| + C$$
$$= \ln\left|\frac{x}{5} + \frac{\sqrt{x^2 - 25}}{5}\right| + C$$
$$= \ln\left|x + \sqrt{x^2 - 25}\right| + C$$

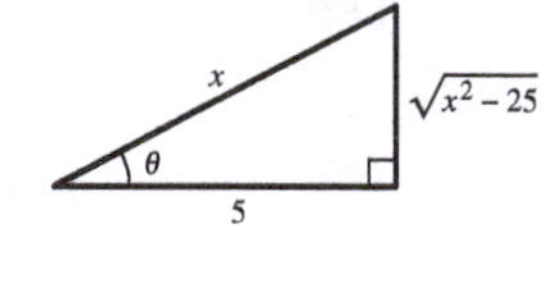

9. Same substitution as in Exercise 9

$$\int x^3\sqrt{x^2 - 25}\,dx = \int (5\sec\theta)^3(5\tan\theta)(5\sec\theta\tan\theta)\,d\theta$$
$$= 3125\int \sec^4\theta\tan^2\theta\,d\theta$$
$$= 3125\int \left(1 + \tan^2\theta\right)\tan^2\theta\sec^2\theta\,d\theta$$
$$= 3125\int \left(\tan^2\theta + \tan^4\theta\right)\sec^2\theta\,d\theta$$
$$= 3125\left[\frac{\tan^3\theta}{3} + \frac{\tan^5\theta}{5}\right] + C$$
$$= 3125\left[\frac{\left(x^2 - 25\right)^{3/2}}{125(3)} + \frac{\left(x^2 - 25\right)^{5/2}}{5^5(5)}\right] + C$$
$$= \frac{1}{15}\left(x^2 - 25\right)^{3/2}\left[125 + 3\left(x^2 - 25\right)\right] + C$$
$$= \frac{1}{15}\left(x^2 - 25\right)^{3/2}\left(50 + 3x^2\right) + C$$

11. Let $x = 2\tan\theta$, $dx = 2\sec^2\theta\,d\theta$, $\sqrt{4+x^2} = 2\sec\theta$.

$$\begin{aligned}\int \frac{x}{2}\sqrt{4+x^2}\,dx &= \int \tan\theta(2\sec\theta)\left(2\sec^2\theta\right)d\theta\\ &= 4\int \sec^2\theta\,(\sec\theta\tan\theta)\,d\theta\\ &= \frac{4}{3}\sec^3\theta + C\\ &= \frac{4}{3}\cdot\frac{\left(4+x^2\right)^{3/2}}{8} + C\\ &= \frac{1}{6}\left(4+x^2\right)^{3/2} + C\end{aligned}$$

13. Same substitution as in Exercise 11

$$\begin{aligned}\int \frac{4}{\left(4+x^2\right)^2}\,dx &= \int \frac{4}{\left(2\sec\theta\right)^4}\left(2\sec^2\theta\right)d\theta\\ &= \frac{1}{2}\int \cos^2\theta\,d\theta\\ &= \frac{1}{2}\int \frac{1+\cos 2\theta}{2}\,d\theta\\ &= \frac{1}{4}\theta + \frac{1}{8}\sin 2\theta + C = \frac{1}{4}\theta + \frac{1}{4}\sin\theta\cos\theta + C\\ &= \frac{1}{4}\arctan\frac{x}{2} + \frac{1}{4}\left(\frac{x}{\sqrt{x^2+4}}\right)\left(\frac{2}{\sqrt{x^2+4}}\right) + C\\ &= \frac{1}{4}\left(\arctan\frac{x}{2} + \frac{2x}{x^2+4}\right) + C\end{aligned}$$

15. Let $u = 4x$, $a = 7$, $du = 4\,dx$.

$$\begin{aligned}\int \sqrt{49-16x^2}\,dx &= \frac{1}{4}\int \sqrt{7^2-(4x)^2}\,4\,dx\\ &= \frac{1}{4}\left(\frac{1}{2}\right)\left(4x\sqrt{49-16x^2} + 49\arcsin\frac{4x}{4}\right) + C\\ &= \frac{1}{2}x\sqrt{49-16x^2} + \frac{49}{8}\arcsin\frac{4x}{7} + C\end{aligned}$$

17. Let $u = \sqrt{5}\,x$, $du = \sqrt{5}\,dx$, $a = 6$.

$$\begin{aligned}\int \sqrt{36-5x^2}\,dx &= \frac{1}{\sqrt{5}}\int \sqrt{36-5x^2}\left(\sqrt{5}\,dx\right)\\ &= \frac{1}{2\sqrt{5}}\left(\sqrt{5}x\sqrt{36-5x^2} + 36\arcsin\frac{\sqrt{5}x}{6}\right) + C\end{aligned}$$

19. Let $x = 2\sin\theta$, $dx = 2\cos\theta\,d\theta$,

$\sqrt{4 - x^2} = 2\cos\theta$.

$$\begin{aligned}\int\sqrt{16 - 4x^2}\,dx &= 2\int\sqrt{4 - x^2}\,dx\\ &= 2\int 2\cos\theta(2\cos\theta\,d\theta)\\ &= 8\int\cos^2\theta\,d\theta\\ &= 4\int(1 + \cos 2\theta)\,d\theta\\ &= 4\left(\theta + \frac{1}{2}\sin 2\theta\right) + C\\ &= 4\theta + 4\sin\theta\cos\theta + C\\ &= 4\arcsin\left(\frac{x}{2}\right) + x\sqrt{4 - x^2} + C\end{aligned}$$

21. Let $x = \sin\theta$, $dx = \cos\theta\,d\theta$, $\sqrt{1 - x^2} = \cos\theta$.

$$\begin{aligned}\int\frac{\sqrt{1 - x^2}}{x^4}\,dx &= \int\frac{\cos\theta(\cos\theta\,d\theta)}{\sin^4\theta}\\ &= \int\cot^2\theta\csc^2\theta\,d\theta\\ &= -\frac{1}{3}\cot^3\theta + C\\ &= -\frac{\left(1 - x^2\right)^{3/2}}{3x^3} + C\end{aligned}$$

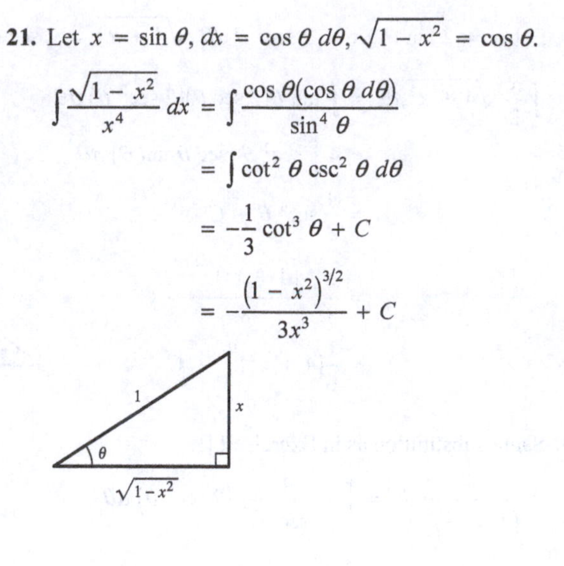

23. Let $2x = 3\tan\theta \Rightarrow x = \frac{3}{2}\tan\theta$, $dx = \frac{3}{2}\sec^2\theta\,d\theta$, $\sqrt{4x^2 + 9} = 3\sec\theta$.

$$\begin{aligned}\int\frac{1}{x\sqrt{4x^2 + 9}}\,dx &= \int\frac{(3/2)\sec^2\theta\,d\theta}{(3/2)\tan\theta\,3\sec\theta}\\ &= \frac{1}{3}\int\csc\theta\,d\theta\\ &= -\frac{1}{3}\ln\left|\csc\theta + \cot\theta\right| + C\\ &= -\frac{1}{3}\ln\left|\frac{\sqrt{4x^2 + 9} + 3}{2x}\right| + C\end{aligned}$$

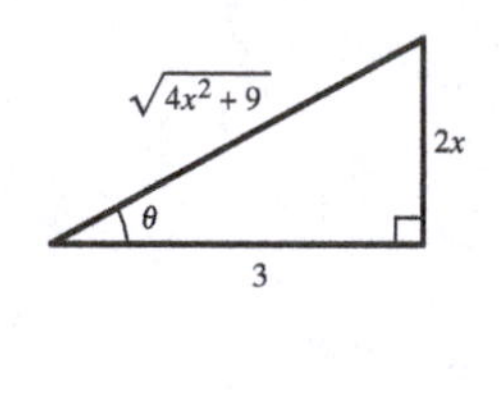

25. Let $x = \sqrt{3}\tan\theta$, $dx = \sqrt{3}\sec^2\theta\,d\theta$,

$x^2 + 3 = 3\tan^2\theta + 3 = 3\sec^2\theta$.

$$\begin{aligned}\int\frac{-3}{\left(x^2 + 3\right)^{3/2}}\,dx &= \int\frac{-3}{3\sqrt{3}\sec^3\theta}\left(\sqrt{3}\sec^2\theta\right)d\theta\\ &= -\int\cos\theta\,d\theta\\ &= -\sin\theta + C\\ &= -\frac{x}{\sqrt{x^2 + 3}} + C\end{aligned}$$

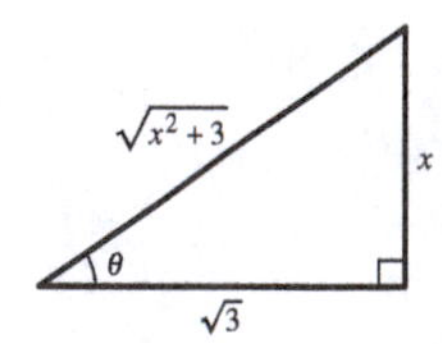

27. Let $e^x = \sin\theta$, $e^x\,dx = \cos\theta\,d\theta$, $\sqrt{1 - e^{2x}} = \cos\theta$.

$$\begin{aligned}\int e^x\sqrt{1 - e^{2x}}\,dx &= \int\cos^2\theta\,d\theta\\ &= \frac{1}{2}\int(1 + \cos 2\theta)\,d\theta\\ &= \frac{1}{2}\left(\theta + \frac{\sin 2\theta}{2}\right)\\ &= \frac{1}{2}(\theta + \sin\theta\cos\theta) + C\\ &= \frac{1}{2}\left(\arcsin e^x + e^x\sqrt{1 - e^{2x}}\right) + C\end{aligned}$$

29. Let $x = \sqrt{2}\tan\theta$, $dx = \sqrt{2}\sec^2\theta\,d\theta$, $x^2 + 2 = 2\sec^2\theta$.

$$\begin{aligned}\int \frac{1}{4 + 4x^2 + x^4}\,dx &= \int \frac{1}{\left(x^2 + 2\right)^2}\,dx = \int \frac{\sqrt{2}\sec^2\theta\,d\theta}{4\sec^4\theta}\\ &= \frac{\sqrt{2}}{4}\int \cos^2\theta\,d\theta = \frac{\sqrt{2}}{4}\left(\frac{1}{2}\right)\int (1 + \cos 2\theta)\,d\theta\\ &= \frac{\sqrt{2}}{8}\left(\theta + \frac{1}{2}\sin 2\theta\right) + C = \frac{\sqrt{2}}{8}(\theta + \sin\theta\cos\theta) + C\\ &= \frac{\sqrt{2}}{8}\left(\arctan\frac{x}{\sqrt{2}} + \frac{x}{\sqrt{x^2 + 2}}\cdot\frac{\sqrt{2}}{\sqrt{x^2 + 2}}\right) = \frac{1}{4}\left(\frac{x}{x^2 + 2} + \frac{1}{\sqrt{2}}\arctan\frac{x}{\sqrt{2}}\right) + C\end{aligned}$$

31. Use integration by parts. Because $x > \frac{1}{2}$,

$$u = \operatorname{arcsec} 2x \Rightarrow du = \frac{1}{x\sqrt{4x^2 - 1}}\,dx,\ dv = dx \Rightarrow v = x$$

$$\int \operatorname{arcsec} 2x\,dx = x\operatorname{arcsec} 2x - \int \frac{1}{\sqrt{4x^2 - 1}}\,dx$$

$$2x = \sec\theta,\ dx = \frac{1}{2}\sec\theta\tan\theta\,d\theta,\ \sqrt{4x^2 - 1} = \tan\theta$$

$$\begin{aligned}\int \operatorname{arcsec} 2x\,dx &= x\operatorname{arcsec} 2x - \int \frac{(1/2)\sec\theta\tan\theta\,d\theta}{\tan\theta} = x\operatorname{arcsec} 2x - \frac{1}{2}\int \sec\theta\,d\theta\\ &= x\operatorname{arcsec} 2x - \frac{1}{2}\ln\left|\sec\theta + \tan\theta\right| + C = x\operatorname{arcsec} 2x - \frac{1}{2}\ln\left|2x + \sqrt{4x^2 - 1}\right| + C.\end{aligned}$$

33. $\displaystyle\int \frac{x}{\sqrt{4x - x^2}}\,dx = \int \frac{x}{\sqrt{4 - (x - 2)^2}}\,dx$

Let $x - 2 = 2\sin\theta$, $dx = 2\cos\theta\,d\theta$,

$$\sqrt{4 - (x - 2)^2} = \sqrt{4 - 4\sin^2\theta} = \sqrt{4\cos^2\theta} = 2\cos\theta.$$

$$\begin{aligned}\int \frac{x}{\sqrt{4 - (x - 2)^2}}\,dx &= \int \frac{2 + 2\sin\theta}{2\cos\theta}2\cos\theta\,d\theta\\ &= \int (2 + 2\sin\theta)\,d\theta\\ &= 2\theta - 2\cos\theta + C\\ &= 2\arcsin\frac{x - 2}{2} - 2\frac{\sqrt{4 - (x - 2)^2}}{2} + C\\ &= 2\arcsin\frac{x - 2}{2} - \sqrt{4x - x^2} + C\end{aligned}$$

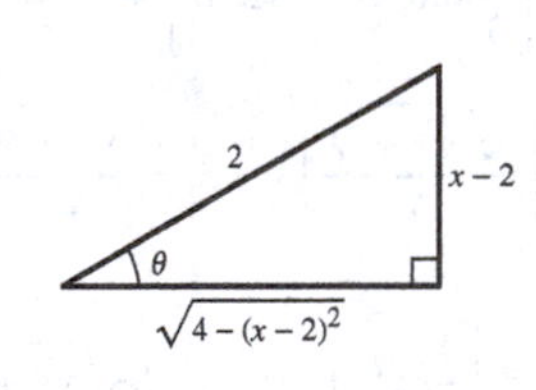

35. $x^2 + 6x + 12 = x^2 + 6x + 9 + 3 = (x+3)^2 + \left(\sqrt{3}\right)^2$

Let $x + 3 = \sqrt{3}\tan\theta$, $dx = \sqrt{3}\sec^2\theta\, d\theta$.

$$\sqrt{x^2 + 6x + 12} = \sqrt{(x+3)^2 + \left(\sqrt{3}\right)^2} = \sqrt{3}\sec\theta$$

$$\int \frac{x}{\sqrt{x^2 + 6x + 12}}\,dx = \int \frac{\sqrt{3}\tan\theta - 3}{\sqrt{3}\sec\theta}\sqrt{3}\sec^2\theta\,d\theta$$

$$= \int \sqrt{3}\sec\theta\tan\theta\,d\theta - 3\int \sec\theta\,d\theta$$

$$= \sqrt{3}\sec\theta - 3\ln\left|\sec\theta + \tan\theta\right| + C$$

$$= \sqrt{3}\left(\frac{\sqrt{x^2 + 6x + 12}}{\sqrt{3}}\right) - 3\ln\left|\frac{\sqrt{x^2 + 6x + 12}}{\sqrt{3}} + \frac{x+3}{\sqrt{3}}\right| + C$$

$$= \sqrt{x^2 + 6x + 12} - 3\ln\left|\sqrt{x^2 + 6x + 12} + (x+3)\right| + C$$

37. Let $t = \sin\theta$, $dt = \cos\theta\,d\theta$, $1 - t^2 = \cos^2\theta$.

(a) $\displaystyle\int \frac{t^2}{\left(1 - t^2\right)^{3/2}}\,dt = \int \frac{\sin^2\theta\cos\theta\,d\theta}{\cos^3\theta} = \int \tan^2\theta\,d\theta = \int\left(\sec^2\theta - 1\right)d\theta = \tan\theta - \theta + C = \frac{t}{\sqrt{1 - t^2}} - \arcsin t + C$

So, $\displaystyle\int_0^{\sqrt{3}/2} \frac{t^2}{\left(1 - t^2\right)^{3/2}}\,dt = \left[\frac{t}{\sqrt{1 - t^2}} - \arcsin t\right]_0^{\sqrt{3}/2} = \frac{\sqrt{3}/2}{\sqrt{1/4}} - \arcsin\frac{\sqrt{3}}{2} = \sqrt{3} - \frac{\pi}{3} \approx 0.685.$

(b) When $t = 0$, $\theta = 0$. When $t = \sqrt{3}/2$, $\theta = \pi/3$. So,

$$\int_0^{\sqrt{3}/2} \frac{t^2}{\left(1 - t^2\right)^{3/2}}\,dt = \left[\tan\theta - \theta\right]_0^{\pi/3} = \sqrt{3} - \frac{\pi}{3} \approx 0.685.$$

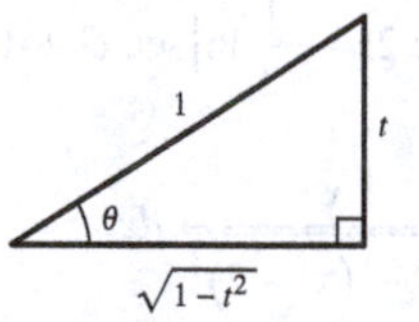

39. (a) Let $x = 3\tan\theta$, $dx = 3\sec^2\theta\,d\theta$, $\sqrt{x^2 + 9} = 3\sec\theta$.

$$\int \frac{x^3}{\sqrt{x^2 + 9}}\,dx = \int \frac{\left(27\tan^3\theta\right)\left(3\sec^2\theta\,d\theta\right)}{3\sec\theta}$$

$$= 27\int\left(\sec^2\theta - 1\right)\sec\theta\tan\theta\,d\theta$$

$$= 27\left[\frac{1}{3}\sec^3\theta - \sec\theta\right] + C = 9\left[\sec^3\theta - 3\sec\theta\right] + C$$

$$= 9\left[\left(\frac{\sqrt{x^2 + 9}}{3}\right)^3 - 3\left(\frac{\sqrt{x^2 + 9}}{3}\right)\right] + C = \frac{1}{3}\left(x^2 + 9\right)^{3/2} - 9\sqrt{x^2 + 9} + C$$

So, $\displaystyle\int_0^3 \frac{x^3}{\sqrt{x^2 + 9}}\,dx = \left[\frac{1}{3}\left(x^2 + 9\right)^{3/2} - 9\sqrt{x^2 + 9}\right]_0^3$

$$= \left(\frac{1}{3}\left(54\sqrt{2}\right) - 27\sqrt{2}\right) - (9 - 27) = 18 - 9\sqrt{2} = 9\left(2 - \sqrt{2}\right) \approx 5.272.$$

(b) When $x = 0$, $\theta = 0$. When $x = 3$, $\theta = \pi/4$. So,

$$\int_0^3 \frac{x^3}{\sqrt{x^2 + 9}}\,dx = 9\left[\sec^3\theta - 3\sec\theta\right]_0^{\pi/4} = 9\left(2\sqrt{2} - 3\sqrt{2}\right) - 9(1 - 3) = 9\left(2 - \sqrt{2}\right) \approx 5.272.$$

41. (a) Let $x = 3\sec\theta$, $dx = 3\sec\theta\tan\theta\,d\theta$, $\sqrt{x^2 - 9} = 3\tan\theta$.

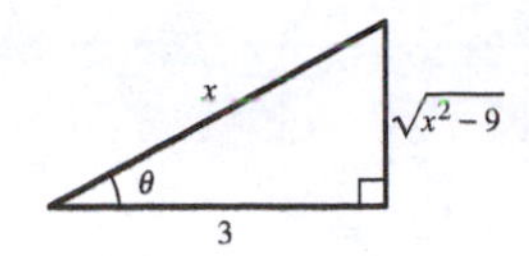

$$\begin{aligned}
\int \frac{x^2}{\sqrt{x^2-9}}\,dx &= \int \frac{9\sec^2\theta}{3\tan\theta}3\sec\theta\tan\theta\,d\theta \\
&= 9\int \sec^3\theta\,d\theta \\
&= 9\left(\frac{1}{2}\sec\theta\tan\theta + \frac{1}{2}\int \sec\theta\,d\theta\right) \quad \text{(8.3 Exercise 102 or Example 5, Section 8.2)} \\
&= \frac{9}{2}\left(\sec\theta\tan\theta + \ln\left|\sec\theta + \tan\theta\right|\right) \\
&= \frac{9}{2}\left(\frac{x}{3}\cdot\frac{\sqrt{x^2-9}}{3} + \ln\left|\frac{x}{3} + \frac{\sqrt{x^2-9}}{3}\right|\right)
\end{aligned}$$

So,

$$\begin{aligned}
\int_4^6 \frac{x^2}{\sqrt{x^2-9}}\,dx &= \frac{9}{2}\left[\frac{x\sqrt{x^2-9}}{9} + \ln\left|\frac{x}{3} + \frac{\sqrt{x^2-9}}{3}\right|\right]_4^6 \\
&= \frac{9}{2}\left[\left(\frac{6\sqrt{27}}{9} + \ln\left|2 + \frac{\sqrt{27}}{3}\right|\right) - \left(\frac{4\sqrt{7}}{9} + \ln\left|\frac{4}{3} + \frac{\sqrt{7}}{3}\right|\right)\right] \\
&= 9\sqrt{3} - 2\sqrt{7} + \frac{9}{2}\left[\ln\left(\frac{6+\sqrt{27}}{3}\right) - \ln\left(\frac{4+\sqrt{7}}{3}\right)\right] \\
&= 9\sqrt{3} - 2\sqrt{7} + \frac{9}{2}\ln\left(\frac{6+3\sqrt{3}}{4+\sqrt{7}}\right) \approx 12.644.
\end{aligned}$$

(b) When $x = 4$, $\theta = \operatorname{arcsec}\left(\frac{4}{3}\right)$. When $x = 6$, $\theta = \operatorname{arcsec}(2) = \frac{\pi}{3}$.

$$\begin{aligned}
\int_4^6 \frac{x^2}{\sqrt{x^2-9}}\,dx &= \frac{9}{2}\left[\sec\theta\tan\theta + \ln\left|\sec\theta + \tan\theta\right|\right]_{\operatorname{arcsec}(4/3)}^{\pi/3} \\
&= \frac{9}{2}\left(2\cdot\sqrt{3} + \ln\left|2 + \sqrt{3}\right|\right) - \frac{9}{2}\left(\frac{4}{3}\left(\frac{\sqrt{7}}{3}\right) + \ln\left|\frac{4}{3} + \frac{\sqrt{7}}{3}\right|\right) \\
&= 9\sqrt{3} - 2\sqrt{7} + \frac{9}{2}\ln\left(\frac{6+3\sqrt{3}}{4+\sqrt{7}}\right) \approx 12.644
\end{aligned}$$

43. Substitution: $u = x^2 + 1$, $du = 2x\,dx$

45. (a) u-substitution: Let $u = 1 - x^2$, $du = -2x\,dx$.

$$\int \frac{x}{\sqrt{1-x^2}}\,dx = -\frac{1}{2}\int (1-x^2)^{-1/2}(-2x)\,dx$$
$$= -\frac{1}{2}(1-x^2)^{1/2}(2) + C = -\sqrt{1-x^2} + C$$

Trigonometric substitution

Let $x = \sin\theta$, $dx = \cos\theta\,d\theta$, $a = 1$, $\sqrt{1-x^2} = \cos\theta$.

$$\int \frac{x}{\sqrt{1-x^2}}\,dx = \int \frac{\sin\theta}{\cos\theta}\cos\theta\,d\theta = \int \sin\theta\,d\theta = -\cos\theta + C = -\sqrt{1-x^2} + C$$

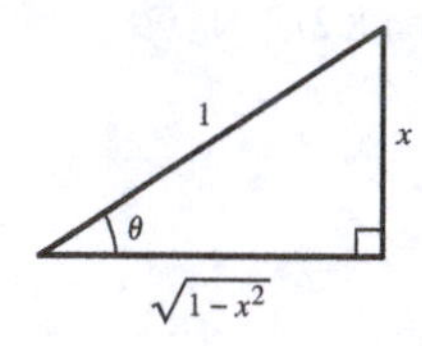

The answers are equivalent.

(b) $$\int \frac{x^2}{x^2+9}\,dx = \int \frac{x^2+9-9}{x^2+9}\,dx = \int\left(1 - \frac{9}{x^2+9}\right)dx = x - 3\arctan\left(\frac{x}{3}\right) + C$$

Let $x = 3\tan\theta$, $x^2 + 9 = 9\sec^2\theta$, $dx = 3\sec^2\theta\,d\theta$.

$$\int \frac{x^2}{x^2+9}\,dx = \int \frac{9\tan^2\theta}{9\sec^2\theta}3\sec^2\theta\,d\theta$$
$$= 3\int \tan^2\theta\,d\theta = 3\int\left(\sec^2\theta - 1\right)d\theta$$
$$= 3\tan\theta - 3\theta + C_1$$
$$= x - 3\arctan\left(\frac{x}{3}\right) + C_1$$

The answers are equivalent.

47. True

$$\int \frac{dx}{\sqrt{1-x^2}} = \int \frac{\cos\theta\,d\theta}{\cos\theta} = \int d\theta$$

49. False

$$\int_0^{\sqrt{3}} \frac{dx}{\left(\sqrt{1+x^2}\right)^3} = \int_0^{\pi/3} \frac{\sec^2\theta\,d\theta}{\sec^3\theta} = \int_0^{\pi/3} \cos\theta\,d\theta$$

51. $$A = 4\int_0^a \frac{b}{a}\sqrt{a^2-x^2}\,dx$$
$$= \frac{4b}{a}\int_0^a \sqrt{a^2-x^2}\,dx$$
$$= \left[\frac{4b}{a}\left(\frac{1}{2}\right)\left(a^2\arcsin\frac{x}{a} + x\sqrt{a^2-x^2}\right)\right]_0^a$$
$$= \frac{2b}{a}\left(a^2\left(\frac{\pi}{2}\right)\right) = \pi ab$$

Note: See Theorem 8.2 for $\int\sqrt{a^2-x^2}\,dx$.

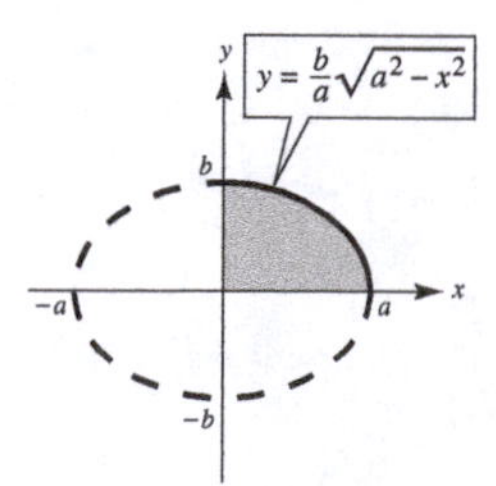

53. $y = \ln x, y' = \dfrac{1}{x}, 1 + (y')^2 = 1 + \dfrac{1}{x^2} = \dfrac{x^2 + 1}{x^2}$

Let $x = \tan\theta, dx = \sec^2\theta\, d\theta, \sqrt{x^2 + 1} = \sec\theta.$

$$s = \int_1^5 \sqrt{\frac{x^2+1}{x^2}}\, dx = \int_1^5 \frac{\sqrt{x^2+1}}{x}\, dx$$
$$= \int_a^b \frac{\sec\theta}{\tan\theta}\sec^2\theta\, d\theta = \int_a^b \frac{\sec\theta}{\tan\theta}\left(1 + \tan^2\theta\right) d\theta$$
$$= \int_a^b (\csc\theta + \sec\theta\tan\theta)\, d\theta = \Big[-\ln\left|\csc\theta + \cot\theta\right| + \sec\theta\Big]_a^b$$
$$= \left[-\ln\left|\frac{\sqrt{x^2+1}}{x} + \frac{1}{x}\right| + \sqrt{x^2+1}\right]_1^5$$
$$= \left[-\ln\left(\frac{\sqrt{26}+1}{5}\right) + \sqrt{26}\right] - \left[-\ln\left(\sqrt{2}+1\right) + \sqrt{2}\right]$$
$$= \ln\left[\frac{5\left(\sqrt{2}+1\right)}{\sqrt{26}+1}\right] + \sqrt{26} - \sqrt{2} \approx 4.367 \text{ or } \ln\left[\frac{\sqrt{26}-1}{5\left(\sqrt{2}-1\right)}\right] + \sqrt{26} - \sqrt{2}$$

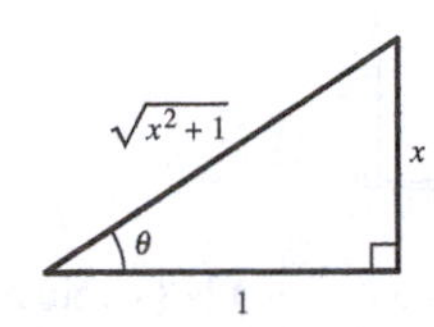

55. Let $x - 3 = \sin\theta, dx = \cos\theta\, d\theta, \sqrt{1 - (x-3)^2} = \cos\theta.$

Shell Method:

$$V = 4\pi\int_2^4 x\sqrt{1-(x-3)^2}\, dx$$
$$= 4\pi\int_{-\pi/2}^{\pi/2} (3 + \sin\theta)\cos^2\theta\, d\theta$$
$$= 4\pi\left[\frac{3}{2}\int_{-\pi/2}^{\pi/2}(1 + \cos 2\theta)\, d\theta + \int_{-\pi/2}^{\pi/2}\cos^2\theta\sin\theta\, d\theta\right]$$
$$= 4\pi\left[\frac{3}{2}\left(\theta + \frac{1}{2}\sin 2\theta\right) - \frac{1}{3}\cos^3\theta\right]_{-\pi/2}^{\pi/2} = 6\pi^2$$

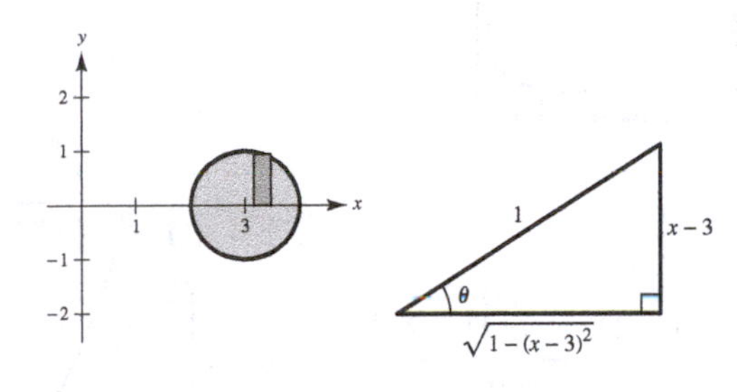

57. Let $x = 3\tan\theta, dx = 3\sec^2\theta\, d\theta, \sqrt{x^2 + 9} = 3\sec\theta.$

$$A = 2\int_0^4 \frac{3}{\sqrt{x^2+9}}\, dx = 6\int_0^4 \frac{dx}{\sqrt{x^2+9}} = 6\int_a^b \frac{3\sec^2\theta\, d\theta}{3\sec\theta}$$
$$= 6\int_a^b \sec\theta\, d\theta = \Big[6\ln\left|\sec\theta + \tan\theta\right|\Big]_a^b = \left[6\ln\left|\frac{\sqrt{x^2+9}+x}{3}\right|\right]_0^4 = 6\ln 3$$

$\bar{x} = 0$ (by symmetry)

$$\bar{y} = \frac{1}{2}\left(\frac{1}{A}\right)\int_{-4}^4 \left(\frac{3}{\sqrt{x^2+9}}\right)^2 dx = \frac{9}{12\ln 3}\int_{-4}^4 \frac{1}{x^2+9}\, dx = \frac{3}{4\ln 3}\left[\frac{1}{3}\arctan\frac{x}{3}\right]_{-4}^4 = \frac{2}{4\ln 3}\arctan\frac{4}{3} \approx 0.422$$

$$(\bar{x}, \bar{y}) = \left(0, \frac{1}{2\ln 3}\arctan\frac{4}{3}\right) \approx (0, 0.422)$$

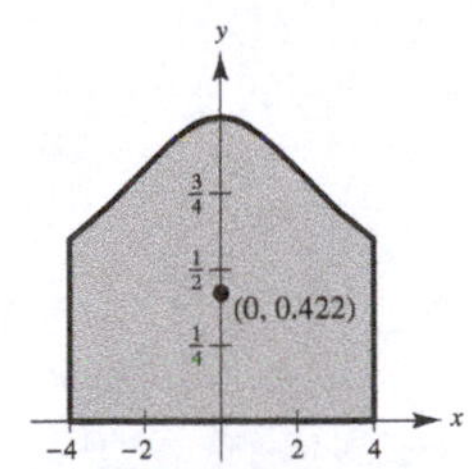

59. (a) Place the center of the circle at $(0, 1)$; $x^2 + (y-1)^2 = 1$. The depth d satisfies $0 \le d \le 2$. The volume is

$$V = 3 \cdot 2\int_0^d \sqrt{1-(y-1)^2}\,dy = 6 \cdot \frac{1}{2}\left[\arcsin(y-1) + (y-1)\sqrt{1-(y-1)^2}\right]_0^d \quad \text{(Theorem 8.2 (1))}$$

$$= 3\left[\arcsin(d-1) + (d-1)\sqrt{1-(d-1)^2} - \arcsin(-1)\right] = \frac{3\pi}{2} + 3\arcsin(d-1) + 3(d-1)\sqrt{2d-d^2}.$$

(b)

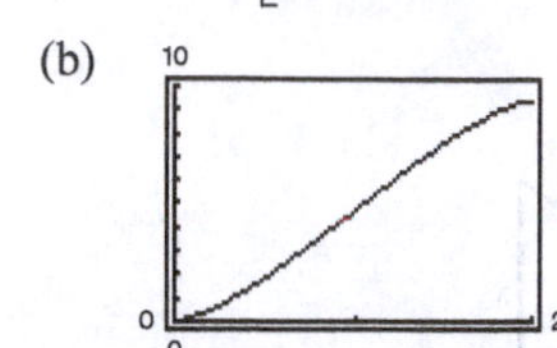

(c) The full tank holds $3\pi \approx 9.4248$ cubic meters. The horizontal lines

$$y = \frac{3\pi}{4},\ y = \frac{3\pi}{2},\ y = \frac{9\pi}{4}$$

intersect the curve at $d = 0.596, 1.0, 1.404$. The dipstick would have these markings on it.

(d) $V = 6\int_0^d \sqrt{1-(y-1)^2}\,dy$

$$\frac{dV}{dt} = \frac{dV}{dd} \cdot \frac{dd}{dt} = 6\sqrt{1-(d-1)^2} \cdot d'(t) = \frac{1}{4} \Rightarrow d'(t) = \frac{1}{24\sqrt{1-(d-1)^2}}$$

(e) The minimum occurs at $d = 1$, which is the widest part of the tank.

61. (a)

$$m = \frac{dy}{dx} = \frac{y - \left(y + \sqrt{144-x^2}\right)}{x - 0} = -\frac{\sqrt{144-x^2}}{x}$$

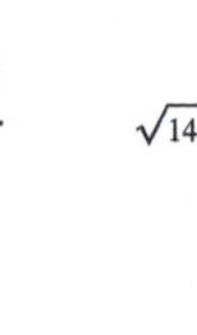
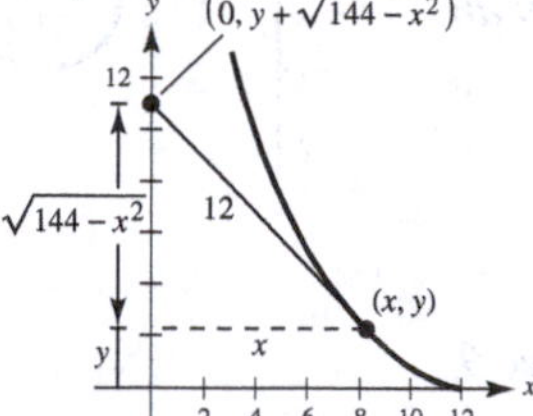

(b) $y = -\int \frac{\sqrt{144-x^2}}{x}\,dx$

Let $x = 12\sin\theta$, $dx = 12\cos\theta\,d\theta$, $\sqrt{144-x^2} = 12\cos\theta$.

$$y = -\int \frac{12\cos\theta}{12\sin\theta} 12\cos\theta\,d\theta = -12\int \frac{1-\sin^2\theta}{\sin\theta}\,d\theta$$

$$= -12\int(\csc\theta - \sin\theta)\,d\theta = -12\ln|\csc\theta - \cot\theta| - 12\cos\theta + C$$

$$= -12\ln\left|\frac{12}{x} - \frac{\sqrt{144-x^2}}{x}\right| - 12\left(\frac{\sqrt{144-x^2}}{12}\right) + C = -12\ln\left|\frac{12-\sqrt{144-x^2}}{x}\right| - \sqrt{144-x^2} + C$$

When $x = 12$, $y = 0 \Rightarrow C = 0$. So, $y = -12\ln\left(\frac{12-\sqrt{144-x^2}}{x}\right) - \sqrt{144-x^2}$.

Note: $\frac{12-\sqrt{144-x^2}}{x} > 0$ for $0 < x \le 12$

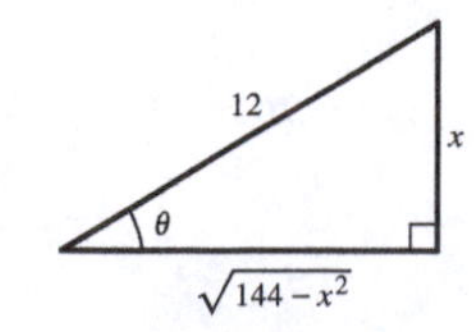

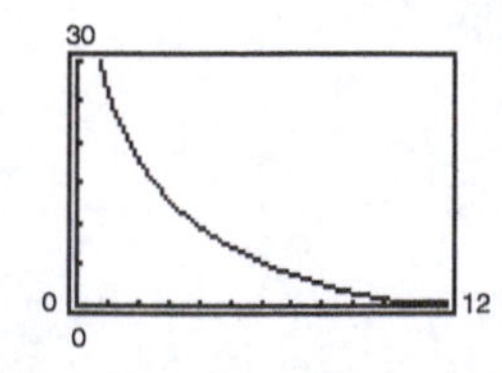

(c) Vertical asymptote: $x = 0$

(d) $y + \sqrt{144 - x^2} = 12 \Rightarrow y = 12 - \sqrt{144 - x^2}$

So, $$12 - \sqrt{144 - x^2} = -12 \ln\left(\frac{12 - \sqrt{144 - x^2}}{x}\right) - \sqrt{144 - x^2}$$

$$-1 = \ln\left(\frac{12 - \sqrt{144 - x^2}}{x}\right)$$

$$xe^{-1} = 12 - \sqrt{144 - x^2}$$

$$\left(xe^{-1} - 12\right)^2 = \left(-\sqrt{144 - x^2}\right)^2$$

$$x^2e^{-2} - 24xe^{-1} + 144 = 144 - x^2$$

$$x^2\left(e^{-2} + 1\right) - 24xe^{-1} = 0$$

$$x\left[x\left(e^{-2} + 1\right) - 24e^{-1}\right] = 0$$

$$x = 0 \text{ or } x = \frac{24e^{-1}}{e^{-2} + 1} \approx 7.77665.$$

Therefore, $$s = \int_{7.77665}^{12} \sqrt{1 + \left(-\frac{\sqrt{144 - x^2}}{x}\right)^2}\,dx = \int_{7.77665}^{12} \sqrt{\frac{x^2 + \left(144 - x^2\right)}{x^2}}\,dx$$

$$= \int_{7.77665}^{12} \frac{12}{x}\,dx = \Big[12 \ln|x|\Big]_{7.77665}^{12} = 12(\ln 12 - \ln 7.77665) \approx 5.2 \text{ meters.}$$

63. (a) Area of representative rectangle: $2\sqrt{1 - y^2}\,\Delta y$

Force: $2(62.4)(3 - y)\sqrt{1 - y^2}\,\Delta y$

$$F = 124.8\int_{-1}^{1} (3 - y)\sqrt{1 - y^2}\,dy$$

$$= 124.8\left[3\int_{-1}^{1} \sqrt{1 - y^2}\,dy - \int_{-1}^{1} y\sqrt{1 - y^2}\,dy\right]$$

$$= 124.8\left[\frac{3}{2}\left(\arcsin y + y\sqrt{1 - y^2}\right) + \frac{1}{2}\left(\frac{2}{3}\right)\left(1 - y^2\right)^{3/2}\right]_{-1}^{1} = (62.4)3\left[\arcsin 1 - \arcsin(-1)\right] = 187.2\pi \text{ lb}$$

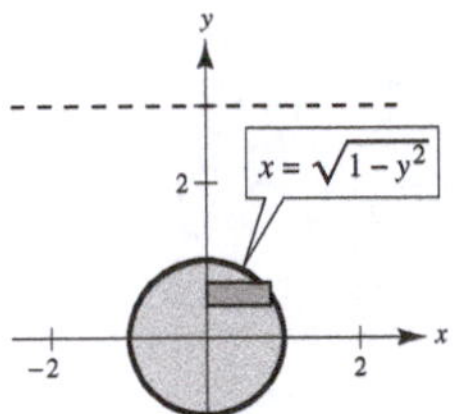

(b) $$F = 124.8\int_{-1}^{1} (d - y)\sqrt{1 - y^2}\,dy = 124.8d\int_{-1}^{1} \sqrt{1 - y^2}\,dy - 124.8\int_{-1}^{1} y\sqrt{1 - y^2}\,dy$$

$$= 124.8\left(\frac{d}{2}\right)\left[\arcsin y + y\sqrt{1 - y^2}\right]_{-1}^{1} - 124.8(0) = 62.4\pi d \text{ lb}$$

65. Let $u = a\sin\theta$, $du = a\cos\theta\,d\theta$, $\sqrt{a^2 - u^2} = a\cos\theta$.

$$\int\sqrt{a^2 - u^2}\,du = \int a^2\cos^2\theta\,d\theta = a^2\int\frac{1 + \cos 2\theta}{2}\,d\theta$$

$$= \frac{a^2}{2}\left(\theta + \frac{1}{2}\sin 2\theta\right) + C = \frac{a^2}{2}(\theta + \sin\theta\cos\theta) + C$$

$$= \frac{a^2}{2}\left[\arcsin\frac{u}{a} + \left(\frac{u}{a}\right)\left(\frac{\sqrt{a^2 + u^2}}{a}\right)\right] + C = \frac{1}{2}\left(a^2\arcsin\frac{u}{a} + u\sqrt{a^2 - u^2}\right) + C$$

Let $u = a\sec\theta$, $du = a\sec\theta\tan\theta\,d\theta$, $\sqrt{u^2 - a^2} = a\tan\theta$.

$$\int\sqrt{u^2 - a^2}\,du = \int a\tan\theta(a\sec\theta\tan\theta)\,d\theta = a^2\int\tan^2\theta\sec\theta\,d\theta$$

$$= a^2\int(\sec^2\theta - 1)\sec\theta\,d\theta = a^2\int(\sec^3\theta - \sec\theta)\,d\theta$$

$$= a^2\left[\frac{1}{2}\sec\theta\tan\theta + \frac{1}{2}\int\sec\theta\,d\theta\right] - a^2\int\sec\theta\,d\theta = a^2\left[\frac{1}{2}\sec\theta\tan\theta - \frac{1}{2}\ln\left|\sec\theta + \tan\theta\right|\right]$$

$$= \frac{a^2}{2}\left[\frac{u}{a}\cdot\frac{\sqrt{u^2 - a^2}}{a} - \ln\left|\frac{u}{a} + \frac{\sqrt{u^2 - a^2}}{a}\right|\right] + C_1 = \frac{1}{2}\left[u\sqrt{u^2 - a^2} - a^2\ln\left|u + \sqrt{u^2 - a^2}\right|\right] + C$$

Let $u = a\tan\theta$, $du = a\sec^2\theta\,d\theta$, $\sqrt{u^2 + a^2} = a\sec\theta$.

$$\int\sqrt{u^2 + a^2}\,du = \int(a\sec\theta)(a\sec^2\theta)\,d\theta$$

$$= a^2\int\sec^3\theta\,d\theta = a^2\left[\frac{1}{2}\sec\theta\tan\theta + \frac{1}{2}\ln\left|\sec\theta + \tan\theta\right|\right] + C_1$$

$$= \frac{a^2}{2}\left[\frac{\sqrt{u^2 + a^2}}{a}\cdot\frac{u}{a} + \ln\left|\frac{\sqrt{u^2 + a^2}}{a} + \frac{u}{a}\right|\right] + C_1 = \frac{1}{2}\left[u\sqrt{u^2 + a^2} + a^2\ln\left|u + \sqrt{u^2 + a^2}\right|\right] + C$$

67. Large circle: $x^2 + y^2 = 25$

$y = \sqrt{25 - x^2}$, upper half

From the right triangle, the center of the small circle is $(0, 4)$.

$x^2 + (y - 4)^2 = 9$

$y = 4 + \sqrt{9 - x^2}$, upper half

$$A = 2\int_0^3\left[\left(4 + \sqrt{9 - x^2}\right) - \sqrt{25 - x^2}\right]dx$$

$$= 2\left[4x + \frac{1}{2}\left[9\arcsin\left(\frac{x}{3}\right) + x\sqrt{9 - x^2}\right] - \frac{1}{2}\left[25\arcsin\left(\frac{x}{5}\right) + x\sqrt{25 - x^2}\right]\right]_0^3$$

$$= 2\left[12 + \frac{9}{2}\arcsin(1) - \frac{25}{2}\arcsin\frac{3}{5} - 6\right]$$

$$= 12 + \frac{9\pi}{2} - 25\arcsin\frac{3}{5} \approx 10.050$$

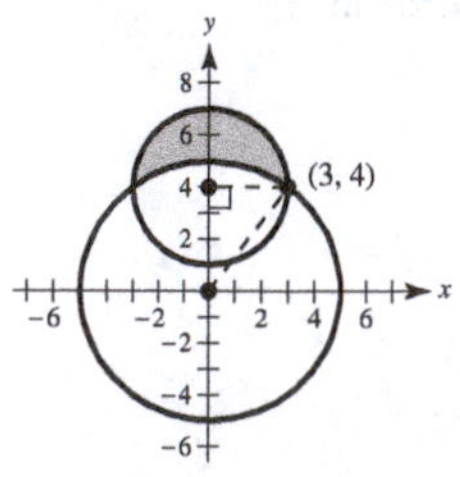

69. Let $I = \int_0^1 \frac{\ln(x+1)}{x^2+1}\,dx$

Let $x = \frac{1-u}{1+u}$, $dx = \frac{-2}{(1+u)^2}\,du$

$x + 1 = \frac{2}{1+u}$, $x^2 + 1 = \frac{2+2u^2}{(1+u)^2}$

$$I = \int_1^0 \frac{\ln\left(\frac{2}{1+u}\right)}{\left(\frac{2+2u^2}{(1+u)^2}\right)}\left(\frac{-2}{(1+u)^2}\right)du$$

$$= \int_1^0 \frac{-\ln\left(\frac{2}{1+u}\right)}{1+u^2}\,du = \int_0^1 \frac{\ln\left(\frac{2}{1+u}\right)}{1+u^2}\,du = \int_0^1 \frac{\ln 2}{1+u^2} - \int_0^1 \frac{\ln(1+u)}{1+u^2}\,du = (\ln 2)\left[\arctan u\right]_0^1 - I$$

$$\Rightarrow 2I = \ln 2\left(\frac{\pi}{4}\right)$$

$$I = \frac{\pi}{8}\ln 2 \approx 0.272198$$

Section 8.5 Partial Fractions

1. (a) $\frac{4}{x^2 - 8x} = \frac{4}{x(x-8)} = \frac{A}{x} + \frac{B}{x-8}$

(b) $\frac{2x^2+1}{(x-3)^3} = \frac{A}{x-3} + \frac{B}{(x-3)^2} + \frac{C}{(x-3)^3}$

(c) $\frac{2x-3}{x^3+10x} = \frac{2x-3}{x(x^2+10)} = \frac{A}{x} + \frac{Bx+C}{x^2+10}$

(d) $\frac{2x-1}{x(x^2+1)^2} = \frac{A}{x} + \frac{Bx+C}{x^2+1} + \frac{Dx+E}{(x^2+1)^2}$

3. $\frac{1}{x^2-9} = \frac{1}{(x-3)(x+3)} = \frac{A}{x+3} + \frac{B}{x-3}$

$$1 = A(x-3) + B(x+3)$$

When $x = 3$, $1 = 6B \Rightarrow B = \frac{1}{6}$.

When $x = -3$, $1 = -6A \Rightarrow A = -\frac{1}{6}$.

$$\int \frac{1}{x^2-9}\,dx = -\frac{1}{6}\int \frac{1}{x+3}\,dx + \frac{1}{6}\int \frac{1}{x-3}\,dx$$

$$= -\frac{1}{6}\ln|x+3| + \frac{1}{6}\ln|x-3| + C$$

$$= \frac{1}{6}\ln\left|\frac{x-3}{x+3}\right| + C$$

5. $\frac{5}{x^2+3x-4} = \frac{5}{(x+4)(x-1)} = \frac{A}{x+4} + \frac{B}{x-1}$

$$5 = A(x-1) + B(x+4)$$

When $x = 1$, $5 = 5B \Rightarrow B = 1$.

When $x = -4$, $5 = -5A \Rightarrow A = -1$.

$$\int \frac{5}{x^2+3x-4}\,dx = \int \frac{-1}{x+4}\,dx + \int \frac{1}{x-1}\,dx = -\ln|x+4| + \ln|x-1| + C = \ln\left|\frac{x-1}{x+4}\right| + C$$

7. $\dfrac{x^2 + 12x + 12}{x(x+2)(x-2)} = \dfrac{A}{x} + \dfrac{B}{x+2} + \dfrac{C}{x-2}$

$x^2 + 12x + 12 = A(x+2)(x-2) + Bx(x-2) + Cx(x+2)$

When $x = 0, 12 = -4A \Rightarrow A = -3$.

When $x = -2, -8 = 8B \Rightarrow B = -1$.

When $x = 2, 40 = 8C \Rightarrow C = 5$.

$$\int \frac{x^2 + 12x + 12}{x^3 - 4x}\,dx = 5\int \frac{1}{x-2}\,dx - \int \frac{1}{x+2}\,dx - 3\int \frac{1}{x}\,dx = 5\ln|x-2| - \ln|x+2| - 3\ln|x| + C$$

9. $\dfrac{2x^3 - 4x^2 - 15x + 5}{x^2 - 2x - 8} = 2x + \dfrac{x+5}{(x-4)(x+2)} = 2x + \dfrac{A}{x-4} + \dfrac{B}{x+2}$

$x + 5 = A(x+2) + B(x-4)$

When $x = 4, 9 = 6A \Rightarrow A = \dfrac{3}{2}$.

When $x = -2, 3 = -6B \Rightarrow B = -\dfrac{1}{2}$.

$$\int \frac{2x^3 - 4x^2 - 15x + 5}{x^2 - 2x - 8}\,dx = \int\left(2x + \frac{3/2}{x-4} - \frac{1/2}{x+2}\right)dx = x^2 + \frac{3}{2}\ln|x-4| - \frac{1}{2}\ln|x+2| + C$$

11. $\dfrac{4x^2 + 2x - 1}{x^2(x+1)} = \dfrac{A}{x} + \dfrac{B}{x^2} + \dfrac{C}{x+1}$

$4x^2 + 2x - 1 = Ax(x+1) + B(x+1) + Cx^2$

When $x = 0, B = -1$.

When $x = -1, C = 1$.

When $x = 1, A = 3$.

$$\int \frac{4x^2 + 2x - 1}{x^3 + x^2}\,dx = \int\left(\frac{3}{x} - \frac{1}{x^2} + \frac{1}{x+1}\right)dx = 3\ln|x| + \frac{1}{x} + \ln|x+1| + C = \frac{1}{x} + \ln|x^4 + x^3| + C$$

13. $\dfrac{x^2 - 6x + 2}{x^3 + 2x^2 + x} = \dfrac{x^2 - 6x + 2}{x(x^2 + 2x + 1)} = \dfrac{x^2 - 6x + 2}{x(x+1)^2} = \dfrac{A}{x} + \dfrac{B}{x+1} + \dfrac{C}{(x+1)^2}$

$x^2 - 6x + 2 = A(x+1)^2 + Bx(x+1) + Cx$

When $x = 0, A = 2$.

When $x = -1, C = -9$.

When $x = 1, -3 = 2(4) + 2B - 9 \Rightarrow B = -1$.

$$\int \frac{x^2 - 6x + 2}{x^3 + 2x^2 + x}\,dx = \int\left(\frac{2}{x} - \frac{1}{x+1} - \frac{9}{(x+1)^2}\right)dx = 2\ln|x| - \ln|x+1| + \frac{9}{x+1} + C = \ln\left|\frac{x^2}{x+1}\right| + \frac{9}{x+1} + C$$

15. $\dfrac{9 - x^2}{7x^3 + x} = \dfrac{9 - x^2}{x(7x^2 + 1)} = \dfrac{A}{x} + \dfrac{Bx + C}{7x^2 + 1}$

$9 - x^2 = A(7x^2 + 1) + Bx^2 + Cx$

When $x = 0$, $A = 9$.

When $x = 1$, $8 = 9(8) + B + C \Rightarrow B + C = -64$.

When $x = -1$, $8 = 9(8) + B - C \Rightarrow B - C = -64$

$2B = -128$, $B = -64$ and $C = 0$.

$$\int \frac{9 - x^2}{7x^3 + x}\,dx = \int\left(\frac{9}{x} - \frac{64x}{7x^2 + 1}\right)dx = 9\ln|x| - \frac{32}{7}\ln(7x^2 + 1) + C$$

17. $\dfrac{x^2}{x^4 - 2x^2 - 8} = \dfrac{A}{x - 2} + \dfrac{B}{x + 2} + \dfrac{Cx + D}{x^2 + 2}$

$$x^2 = A(x + 2)(x^2 + 2) + B(x - 2)(x^2 + 2) + (Cx + D)(x + 2)(x - 2)$$

When $x = 2$, $4 = 24A$.

When $x = -2$, $4 = -24B$.

When $x = 0$, $0 = 4A - 4B - 4D$.

When $x = 1$, $1 = 9A - 3B - 3C - 3D$.

Solving these equations you have $A = \dfrac{1}{6}$, $B = -\dfrac{1}{6}$, $C = 0$, $D = \dfrac{1}{3}$.

$$\int \frac{x^2}{x^4 - 2x^2 - 8}\,dx = \frac{1}{6}\left(\int \frac{1}{x - 2}\,dx - \int \frac{1}{x + 2}\,dx + 2\int \frac{1}{x^2 + 2}\,dx\right) = \frac{1}{6}\left(\ln\left|\frac{x - 2}{x + 2}\right| + \sqrt{2}\arctan\frac{x}{\sqrt{2}}\right) + C$$

19. $\dfrac{x^2 + 5}{(x + 1)(x^2 - 2x + 3)} = \dfrac{A}{x + 1} + \dfrac{Bx + C}{x^2 - 2x + 3}$

$$\begin{aligned} x^2 + 5 &= A(x^2 - 2x + 3) + (Bx + C)(x + 1) \\ &= (A + B)x^2 + (-2A + B + C)x + (3A + C) \end{aligned}$$

When $x = -1$, $A = 1$.

By equating coefficients of like terms, you have $A + B = 1$, $-2A + B + C = 0$, $3A + C = 5$.

Solving these equations you have $A = 1$, $B = 0$, $C = 2$.

$$\int \frac{x^2 + 5}{x^3 - x^2 + x + 3}\,dx = \int \frac{1}{x + 1}\,dx + 2\int \frac{1}{(x - 1)^2 + 2}\,dx = \ln|x + 1| + \sqrt{2}\arctan\left(\frac{x - 1}{\sqrt{2}}\right) + C$$

21. $\dfrac{3}{4x^2 + 5x + 1} = \dfrac{3}{(4x + 1)(x + 1)} = \dfrac{A}{4x + 1} + \dfrac{B}{x + 1}$

$$3 = A(x + 1) + B(4x + 1)$$

When $x = -1$, $3 = -3B \Rightarrow B = -1$.

When $-\dfrac{1}{4}$, $3 = \dfrac{3}{4}A \Rightarrow A = 4$.

$$\int_0^2 \frac{3}{4x^2 + 5x + 1}\,dx = \int_0^2 \frac{4}{4x + 1}\,dx + \int_0^2 \frac{-1}{x + 1}\,dx = \Big[\ln|4x + 1| - \ln|x + 1|\Big]_0^2 = \ln 9 - \ln 3 = 2\ln 3 - \ln 3 = \ln 3$$

23. $\dfrac{x+1}{x(x^2+1)} = \dfrac{A}{x} + \dfrac{Bx+C}{x^2+1}$

$$x + 1 = A(x^2+1) + (Bx+C)x$$

When $x = 0$, $A = 1$.

When $x = 1$, $2 = 2A + B + C$.

When $x = -1$, $0 = 2A + B - C$.

Solving these equations we have $A = 1$, $B = -1$, $C = 1$.

$$\int_1^2 \frac{x+1}{x(x^2+1)}\,dx = \int_1^2 \frac{1}{x}\,dx - \int_1^2 \frac{x}{x^2+1}\,dx + \int_1^2 \frac{1}{x^2+1}\,dx$$

$$= \left[\ln|x| - \frac{1}{2}\ln(x^2+1) + \arctan x\right]_1^2$$

$$= \frac{1}{2}\ln\frac{8}{5} - \frac{\pi}{4} + \arctan 2 \approx 0.557$$

25. Let $u = \cos x$, $du = -\sin x\,dx$.

$$\frac{1}{u(u+1)} = \frac{A}{u} + \frac{B}{u+1}$$

$$1 = A(u+1) + Bu$$

When $u = 0$, $A = 1$.

When $u = -1$, $B = -1$.

$$\int \frac{\sin x}{\cos x + \cos^2 x}\,dx = -\int \frac{1}{u(u+1)}\,du$$

$$= \int \frac{1}{u+1}\,du - \int \frac{1}{u}\,du$$

$$= \ln|u+1| - \ln|u| + C$$

$$= \ln\left|\frac{u+1}{u}\right| + C$$

$$= \ln\left|\frac{\cos x + 1}{\cos x}\right| + C$$

$$= \ln|1 + \sec x| + C$$

27. Let $u = \tan x$, $du = \sec^2 x\,dx$.

$$\frac{1}{u^2 + 5u + 6} = \frac{1}{(u+3)(u+2)} = \frac{A}{u+3} + \frac{B}{u+2}$$

$$1 = A(u+2) + B(u+3)$$

When $u = -2$, $1 = B$.

When $u = -3$, $1 = -A \Rightarrow A = -1$.

$$\int \frac{\sec^2 x}{\tan^2 x + 5\tan x + 6}\,dx = \int \frac{1}{u^2 + 5u + 6}\,du$$

$$= \int \frac{-1}{u+3}\,du + \int \frac{1}{u+2}\,du$$

$$= -\ln|u+3| + \ln|u+2| + C$$

$$= \ln\left|\frac{\tan x + 2}{\tan x + 3}\right| + C$$

29. Let $u = e^x$, $du = e^x\,dx$.

$$\frac{1}{(u-1)(u+4)} = \frac{A}{u-1} + \frac{B}{u+4}$$

$$1 = A(u+4) + B(u-1)$$

When $u = 1$, $A = \frac{1}{5}$.

When $u = -4$, $B = -\frac{1}{5}$.

$$\int \frac{e^x}{(e^x-1)(e^x+4)}\,dx = \int \frac{1}{(u-1)(u+4)}\,du$$

$$= \frac{1}{5}\left(\int \frac{1}{u-1}\,du - \int \frac{1}{u+4}\,du\right)$$

$$= \frac{1}{5}\ln\left|\frac{u-1}{u+4}\right| + C$$

$$= \frac{1}{5}\ln\left|\frac{e^x-1}{e^x+4}\right| + C$$

31. Let $u = \sqrt{x}$, $u^2 = x$, $2u\,du = dx$.

$$\int \frac{\sqrt{x}}{x-4}\,dx = \int \frac{u(2u)\,du}{u^2-4} = \int\left(\frac{2u^2-8}{u^2-4} + \frac{8}{u^2-4}\right)du = \int\left(2 + \frac{8}{u^2-4}\right)du$$

$$\frac{8}{u^2-4} = \frac{8}{(u-2)(u+2)} = \frac{A}{u-2} + \frac{B}{u+2}$$

$$8 = A(u+2) + B(u-2)$$

When $u = -2$, $8 = -4B \Rightarrow B = -2$.

When $u = 2$, $8 = 4A \Rightarrow A = 2$.

$$\int\left(2 + \frac{8}{u^2-4}\right)du = 2u + \int\left(\frac{2}{u-2} - \frac{2}{u+2}\right)du = 2u + 2\ln|u-2| - 2\ln|u+2| + C = 2\sqrt{x} + 2\ln\left|\frac{\sqrt{x}-2}{\sqrt{x}+2}\right| + C$$

33. $\dfrac{1}{x(a + bx)} = \dfrac{A}{x} + \dfrac{B}{a + bx}$

$$1 = A(a + bx) + Bx$$

When $x = 0, 1 = aA \Rightarrow A = 1/a.$

When $x = -a/b, 1 = -(a/b)B \Rightarrow B = -b/a.$

$$\int \frac{1}{x(a + bx)}\,dx = \frac{1}{a}\int\left(\frac{1}{x} - \frac{b}{a + bx}\right)dx$$

$$= \frac{1}{a}\left(\ln|x| - \ln|a + bx|\right) + C$$

$$= \frac{1}{a}\ln\left|\frac{x}{a + bx}\right| + C$$

35. $\dfrac{x}{(a + bx)^2} = \dfrac{A}{a + bx} + \dfrac{B}{(a + bx)^2}$

$$x = A(a + bx) + B$$

When $x = -a/b, B = -a/b.$

When $x = 0, 0 = aA + B \Rightarrow A = 1/b.$

$$\int \frac{x}{(a + bx)^2}\,dx = \int\left(\frac{1/b}{a + bx} + \frac{-a/b}{(a + bx)^2}\right)dx$$

$$= \frac{1}{b}\int \frac{1}{a + bx}\,dx - \frac{a}{b}\int \frac{1}{(a + bx)^2}\,dx$$

$$= \frac{1}{b^2}\ln|a + bx| + \frac{a}{b^2}\left(\frac{1}{a + bx}\right) + C$$

$$= \frac{1}{b^2}\left(\frac{a}{a + bx} + \ln|a + bx|\right) + C$$

37. Substitution: $u = x^2 + 2x - 8$

39. Trigonometric substitution (tan) or inverse tangent rule

41. $\dfrac{12}{x^2 + 5x + 6} = \dfrac{12}{(x + 2)(x + 3)} = \dfrac{A}{x + 2} + \dfrac{B}{x + 3}$

$$12 = A(x + 3) + B(x + 2)$$

When $x = -3, B = -12.$

When $x = -2, A = 12.$

$$A = \int_0^4 \frac{12}{x^2 + 5x + 6}\,dx$$

$$= \int_0^4\left(\frac{12}{x + 2} - \frac{12}{x + 3}\right)dx$$

$$= 12\Big[\ln|x + 2| - \ln|x + 3|\Big]_0^4$$

$$= 12\big[(\ln 6 - \ln 7) - (\ln 2 - \ln 3)\big]$$

$$= 12\ln\left(\frac{6(3)}{7(2)}\right) = 12\ln\frac{9}{7} \approx 3.016$$

43. $\dfrac{15}{9 - x^2} = \dfrac{-15}{(x - 3)(x + 3)} = \dfrac{A}{x - 3} + \dfrac{B}{x + 3}$

$$-15 = A(x + 3) + B(x - 3)$$

When $x = -3, -15 = -6B \Rightarrow B = \dfrac{5}{2}.$

When $x = 3, -15 = 6A \Rightarrow A = -\dfrac{5}{2}.$

$$A = \int_0^2 \frac{15}{9 - x^2}\,dx$$

$$= \int_0^2\left(\frac{-5/2}{x - 3} + \frac{5/2}{x + 3}\right)dx$$

$$= \frac{5}{2}\Big[\ln|x + 3| - \ln|x - 3|\Big]_0^2$$

$$= \frac{5}{2}\big[(\ln 5 - \ln 1) - (\ln 3 - \ln 3)\big] = \frac{5}{2}\ln 5 \approx 4.024$$

45. Average cost $= \dfrac{1}{80 - 75}\displaystyle\int_{75}^{80} \frac{124p}{(10 + p)(100 - p)}\,dp$

$$= \frac{1}{5}\int_{75}^{80}\left(\frac{-124}{(10 + p)11} + \frac{1240}{(100 - p)11}\right)dp$$

$$= \frac{1}{5}\left[\frac{-124}{11}\ln(10 + p) - \frac{1240}{11}\ln(100 - p)\right]_{75}^{80}$$

$$\approx \frac{1}{5}(24.51) = 4.9$$

Approximately \$490,000

47. (a) $V = \pi\int_0^3 \left(\frac{2x}{x^2+1}\right)^2 dx = 4\pi\int_0^3 \frac{x^2}{(x^2+1)^2}\,dx$

$$= 4\pi\int_0^3 \left(\frac{1}{x^2+1} - \frac{1}{(x^2+1)^2}\right)dx \qquad \text{(partial fractions)}$$

$$= 4\pi\left[\arctan x - \frac{1}{2}\left(\arctan x + \frac{x}{x^2+1}\right)\right]_0^3 \qquad \text{(trigonometric substitution)}$$

$$= 2\pi\left[\arctan x - \frac{x}{x^2+1}\right]_0^3 = 2\pi\left(\arctan 3 - \frac{3}{10}\right) \approx 5.963$$

(b) $A = \int_0^3 \frac{2x}{x^2+1}\,dx = \left[\ln(x^2+1)\right]_0^3 = \ln 10$

$$\bar{x} = \frac{1}{A}\int_0^3 \frac{2x^2}{x^2+1}\,dx = \frac{1}{\ln 10}\int_0^3 \left(2 - \frac{2}{x^2+1}\right)dx = \frac{1}{\ln 10}\left[2x - 2\arctan x\right]_0^3 = \frac{2}{\ln 10}(3 - \arctan 3) \approx 1.521$$

$$\bar{y} = \frac{1}{A}\left(\frac{1}{2}\right)\int_0^3 \left(\frac{2x}{x^2+1}\right)^2 dx = \frac{2}{\ln 10}\int_0^3 \frac{x^2}{(x^2+1)^2}\,dx$$

$$= \frac{2}{\ln 10}\int_0^3 \left(\frac{1}{x^2+1} - \frac{1}{(x^2+1)^2}\right)dx \qquad \text{(partial fractions)}$$

$$= \frac{2}{\ln 10}\left[\arctan x - \frac{1}{2}\left(\arctan x + \frac{x}{x^2+1}\right)\right]_0^3 \qquad \text{(trigonometric substitution)}$$

$$= \frac{2}{\ln 10}\left[\frac{1}{2}\arctan x - \frac{x}{2(x^2+1)}\right]_0^3 = \frac{1}{\ln 10}\left[\arctan x - \frac{x}{x^2+1}\right]_0^3 = \frac{1}{\ln 10}\left(\arctan 3 - \frac{3}{10}\right) \approx 0.412$$

$(\bar{x}, \bar{y}) \approx (1.521, 0.412)$

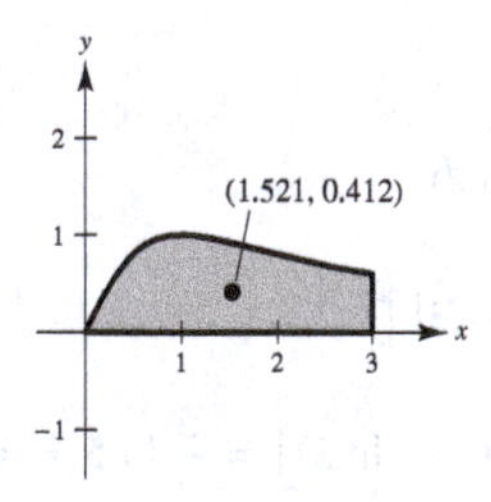

49. $\dfrac{1}{(x+1)(n-x)} = \dfrac{A}{x+1} + \dfrac{B}{n-x}, A = B = \dfrac{1}{n+1}$

$$\frac{1}{n+1}\int\left(\frac{1}{x+1} + \frac{1}{n-x}\right)dx = kt + C$$

$$\frac{1}{n+1}\ln\left|\frac{x+1}{n-x}\right| = kt + C$$

When $t = 0, x = 0, C = \dfrac{1}{n+1}\ln\dfrac{1}{n}$.

$$\frac{1}{n+1}\ln\left|\frac{x+1}{n-x}\right| = kt + \frac{1}{n+1}\ln\frac{1}{n}$$

$$\frac{1}{n+1}\left[\ln\left|\frac{x+1}{n-x}\right| - \ln\frac{1}{n}\right] = kt$$

$$\ln\frac{nx+n}{n-x} = (n+1)kt$$

$$\frac{nx+n}{n-x} = e^{(n+1)kt}$$

$$x = \frac{n\left[e^{(n+1)kt} - 1\right]}{n + e^{(n+1)kt}} \qquad \textbf{Note: } \lim_{t\to\infty} x = n$$

51. $\dfrac{x}{1+x^4} = \dfrac{Ax+B}{x^2+\sqrt{2}x+1} + \dfrac{Cx+D}{x^2-\sqrt{2}x+1}$

$$x = (Ax+B)\left(x^2 - \sqrt{2}x + 1\right) + (Cx+D)\left(x^2 + \sqrt{2}x + 1\right)$$

$$= (A+C)x^3 + \left(B + D - \sqrt{2}A + \sqrt{2}C\right)x^2 + \left(A + C - \sqrt{2}B + \sqrt{2}D\right)x + (B+D)$$

$0 = A + C \Rightarrow C = -A$

$0 = B + D - \sqrt{2}A + \sqrt{2}C \qquad -2\sqrt{2}A = 0 \Rightarrow A = 0$ and $C = 0$

$1 = A + C - \sqrt{2}B + \sqrt{2}D \qquad -2\sqrt{2}B = 1 \Rightarrow B = -\dfrac{\sqrt{2}}{4}$ and $D = \dfrac{\sqrt{2}}{4}$

$0 = B + D \Rightarrow D = -B$

So, $\displaystyle\int_0^1 \frac{x}{1+x^4}\,dx = \int_0^1\left(\frac{-\sqrt{2}/4}{x^2+\sqrt{2}x+1} + \frac{\sqrt{2}/4}{x^2-\sqrt{2}x+1}\right)dx$

$$= \frac{\sqrt{2}}{4}\int_0^1\left[\frac{-1}{\left[x+\left(\sqrt{2}/2\right)\right]^2 + (1/2)} + \frac{1}{\left[x-\left(\sqrt{2}/2\right)\right]^2 + (1/2)}\right]dx$$

$$= \frac{\sqrt{2}}{4}\cdot\frac{1}{1/\sqrt{2}}\left[-\arctan\left(\frac{x+\left(\sqrt{2}/2\right)}{1/\sqrt{2}}\right) + \arctan\left(\frac{x-\left(\sqrt{2}/2\right)}{1/\sqrt{2}}\right)\right]_0^1$$

$$= \frac{1}{2}\left[-\arctan\left(\sqrt{2}x+1\right) + \arctan\left(\sqrt{2}x-1\right)\right]_0^1$$

$$= \frac{1}{2}\left[\left(-\arctan\left(\sqrt{2}+1\right) + \arctan\left(\sqrt{2}-1\right)\right) - \left(-\arctan 1 + \arctan(-1)\right)\right]$$

$$= \frac{1}{2}\left[\arctan\left(\sqrt{2}-1\right) - \arctan\left(\sqrt{2}+1\right) + \frac{\pi}{4} + \frac{\pi}{4}\right].$$

Because $\arctan x - \arctan y = \arctan\left[(x-y)/(1+xy)\right]$, you have:

$$\int_0^1 \frac{x}{1+x^4}\,dx = \frac{1}{2}\left[\arctan\left(\frac{\left(\sqrt{2}-1\right)-\left(\sqrt{2}+1\right)}{1+\left(\sqrt{2}-1\right)\left(\sqrt{2}+1\right)}\right) + \frac{\pi}{2}\right] = \frac{1}{2}\left[\arctan\left(\frac{-2}{2}\right) + \frac{\pi}{2}\right] = \frac{1}{2}\left(-\frac{\pi}{4} + \frac{\pi}{2}\right) = \frac{\pi}{8}$$

53. The answer is 3984. Use the division algorithm to write $p(x) = \left(x^3 - x\right)q(x) + r(x)$, where the degree of $r(x)$ is less than 3, and the degree of $q(x)$ is less than 1989. Hence,

$$\frac{d^{1992}}{dx^{1992}}\left[\frac{p(x)}{x^3 - 3}\right] = \frac{d^{1992}}{dx^{1992}}\left[\frac{\left(x^3 - x\right)q(x) + r(x)}{x^3 - x}\right] = \frac{d^{1992}}{dx^{1992}}\left[\frac{r(x)}{x^3 - x}\right].$$

Using partial fractions, $\dfrac{r(x)}{x^3 - x} = \dfrac{r(x)}{(x-1)x(x+1)} = \dfrac{A}{x-1} + \dfrac{B}{x} + \dfrac{C}{x+1}$

Since $p(x)$ has no common factors with $x^3 - x$, then $r(x)$ does not either. Hence, A, B, and C are all non zero.

$$\frac{d^{1992}}{dx^{1992}}\left[\frac{r(x)}{x^3 - x}\right] = \frac{d^{1992}}{dx^{1992}}\left[\frac{A}{x-1} + \frac{B}{x} + \frac{C}{x+1}\right]$$

$$= 1992!\left[\frac{A}{(x-1)^{1993}} + \frac{B}{x^{1993}} + \frac{C}{(x+1)^{1993}}\right]$$

$$= 1992!\left[\frac{Ax^{1993}(x+1)^{1993} + B(x-1)^{1993}(x+1)^{1993} + Cx^{1993}(x-1)^{1993}}{\left(x^3 - x\right)^{1993}}\right]$$

Now expand the numerator to obtain an expression of the form

$(A + B + C)x^{3986} + 1993(A - C)x^{3985} + 1993(996A - B + 996C)x^{3984} + \ldots$

From $A = C = 1$ and $B = -2$, you see that the degree could be 3984. A lower degree would imply that $A + B + C = 0$, $A - C = 0$, and $996A - B + 996C = 0$, which means $A = B = C$, a contradiction.

Section 8.6 Numerical Integration

1. No. The integral can easily be evaluated using basic integration rules:

$$\int_0^2 \left(e^x + 5x\right)dx = \left[e^x + \frac{5}{2}x^2\right]_0^2 = e^2 + 10 - 1 = e^2 + 9.$$

3. Exact: $\int_0^2 x^2\,dx = \left[\frac{1}{3}x^3\right]_0^2 = \frac{8}{3} \approx 2.6667$

Trapezoidal: $\int_0^2 x^2\,dx \approx \frac{1}{4}\left[0 + 2\left(\frac{1}{2}\right)^2 + 2(1)^2 + 2\left(\frac{3}{2}\right)^2 + (2)^2\right] = \frac{11}{4} = 2.7500$

Simpson's: $\int_0^2 x^2\,dx \approx \frac{1}{6}\left[0 + 4\left(\frac{1}{2}\right)^2 + 2(1)^2 + 4\left(\frac{3}{2}\right)^2 + (2)^2\right] = \frac{8}{3} \approx 2.6667$

5. Exact: $\int_3^4 \frac{1}{x-2}\,dx = \ln|x - 2|\Big]_3^4 = \ln 2 - \ln 1 = \ln 2 - 0 = \ln 2 \approx 0.6931$

Trapezoidal Rule: $\int_3^4 \frac{1}{x-2}\,dx \approx \frac{1}{8}\left[1 + 2\left(\frac{4}{5}\right) + 2\left(\frac{2}{3}\right) + 2\left(\frac{4}{7}\right) + \frac{1}{2}\right] \approx 0.6970$

Simpson's Rule: $\int_3^4 \frac{1}{x-2}\,dx \approx \frac{1}{12}\left[1 + 4\left(\frac{4}{5}\right) + 2\left(\frac{2}{3}\right) + 4\left(\frac{4}{7}\right) + \frac{1}{2}\right] \approx 0.6933$

7. Exact: $\int_1^3 x^3\,dx = \left[\frac{x^4}{4}\right]_1^3 = \frac{81}{4} - \frac{1}{4} = 20$

Trapezoidal: $\int_1^3 x^3\,dx \approx \frac{1}{6}\left[1 + 2\left(\frac{4}{3}\right)^3 + 2\left(\frac{5}{3}\right)^3 + 2(2)^3 + 2\left(\frac{7}{3}\right)^3 + 2\left(\frac{8}{3}\right)^3 + 27\right] \approx 20.2222$

Simpson's: $\int_1^3 x^3\,dx \approx \frac{1}{9}\left[1 + 4\left(\frac{4}{3}\right)^3 + 2\left(\frac{5}{3}\right)^3 + 4(2)^3 + 2\left(\frac{7}{3}\right)^3 + 4\left(\frac{8}{3}\right)^3 + 27\right] = 20.0000$

9. Exact: $\int_4^9 \sqrt{x}\,dx = \left[\frac{2}{3}x^{3/2}\right]_4^9 = 18 - \frac{16}{3} = \frac{38}{3} \approx 12.6667$

Trapezoidal: $\int_4^9 \sqrt{x}\,dx \approx \frac{5}{16}\left[2 + 2\sqrt{\frac{37}{8}} + 2\sqrt{\frac{21}{4}} + 2\sqrt{\frac{47}{8}} + 2\sqrt{\frac{26}{4}} + 2\sqrt{\frac{57}{8}} + 2\sqrt{\frac{31}{4}} + 2\sqrt{\frac{67}{8}} + 3\right] \approx 12.6640$

Simpson's: $\int_4^9 \sqrt{x}\,dx \approx \frac{5}{24}\left[2 + 4\sqrt{\frac{37}{8}} + \sqrt{21} + 4\sqrt{\frac{47}{8}} + \sqrt{26} + 4\sqrt{\frac{57}{8}} + \sqrt{31} + 4\sqrt{\frac{67}{8}} + 3\right] \approx 12.6667$

11. Exact: $\int_0^1 \frac{2}{(x+2)^2}\,dx = \left[\frac{-2}{(x+2)}\right]_0^1 = \frac{-2}{3} + \frac{2}{2} = \frac{1}{3}$

Trapezoidal: $\int_0^1 \frac{2}{(x+2)^2}\,dx \approx \frac{1}{8}\left[\frac{1}{2} + 2\left(\frac{2}{((1/4)+2)^2}\right) + 2\left(\frac{2}{((1/2)+2)^2}\right) + 2\left(\frac{2}{((3/4)+2)^2}\right) + \frac{2}{9}\right]$

$$= \frac{1}{8}\left[\frac{1}{2} + 2\left(\frac{32}{81}\right) + 2\left(\frac{8}{25}\right) + 2\left(\frac{32}{121}\right) + \frac{2}{9}\right] \approx 0.3352$$

Simpson's: $\int_0^1 \frac{2}{(x+2)^2}\,dx \approx \frac{1}{12}\left[\frac{1}{2} + 4\left(\frac{2}{((1/4)+2)^2}\right) + 2\left(\frac{2}{((1/2)+2)^2}\right) + 4\left(\frac{2}{((3/4)+2)^2}\right) + \frac{2}{9}\right]$

$$= \frac{1}{12}\left[\frac{1}{2} + 4\left(\frac{32}{81}\right) + 2\left(\frac{8}{25}\right) + 4\left(\frac{32}{121}\right) + \frac{2}{9}\right] \approx 0.3334$$

13. Exact: $\int_0^2 xe^{-x}\,dx = -\left[e^{-x}(x+1)\right]_0^2 = -3e^{-2} + 1 \approx 0.5940$

Trapezoidal: $\int_0^2 xe^{-x}\,dx \approx \frac{1}{4}\left[0 + e^{-1/2} + 2e^{-1} + 3e^{-3/2} + 2e^{-2}\right] \approx \frac{2.2824}{4} \approx 0.5706$

Simpson's: $\int_0^2 2xe^{-x}\,dx \approx \frac{1}{6}\left[0 + 2e^{-1/2} + 2e^{-1} + 6e^{-3/2} + 2e^{-2}\right] \approx \frac{3.5583}{6} \approx 0.5930$

15. Trapezoidal: $\int_0^2 \sqrt{1+x^3}\,dx \approx \frac{1}{4}\left[1 + 2\sqrt{1 + \left(\frac{1}{8}\right)} + 2\sqrt{2} + 2\sqrt{1 + \left(\frac{27}{8}\right)} + 3\right] \approx 3.2833$

Simpson's: $\int_0^2 \sqrt{1+x^3}\,dx \approx \frac{1}{6}\left[1 + 4\sqrt{1 + \left(\frac{1}{8}\right)} + 2\sqrt{2} + 4\sqrt{1 + \left(\frac{27}{8}\right)} + 3\right] \approx 3.2396$

Graphing utility: 3.2413

17. Trapezoidal: $\int_0^1 \frac{1}{1+x^2}\,dx \approx \frac{1}{8}\left[\frac{1}{1+0} + \frac{2}{1+\left(\frac{1}{4}\right)^2} + \frac{2}{1+\left(\frac{1}{2}\right)^2} + \frac{2}{1+\left(\frac{3}{4}\right)^2} + \frac{1}{1+1^2}\right] \approx 0.7828$

Simpson's: $\int_0^1 \frac{1}{1+x^2}\,dx \approx \frac{1}{12}\left[\frac{1}{1+0} + \frac{4}{1+\left(\frac{1}{4}\right)^2} + \frac{2}{1+\left(\frac{1}{2}\right)^2} + \frac{4}{1+\left(\frac{3}{4}\right)^2} + \frac{1}{1+1^2}\right] \approx 0.7854$

Graphing utility: 0.7854

19. Trapezoidal: $\int_0^4 \sqrt{x}e^x\,dx = \frac{1}{2}\left[0 + 2e + 2\sqrt{2}e^2 + 2\sqrt{3}e^3 + 2e^4\right] \approx 102.5553$

Simpson's: $\int_0^4 \sqrt{x}e^x\,dx = \frac{1}{3}\left[0 + 4e + 2\sqrt{2}e^2 + 4\sqrt{3}e^3 + 2e^4\right] \approx 93.3752$

Graphing utility: 92.7437

21. Trapezoidal:

$$\int_0^{\sqrt{\pi/2}} \sin(x^2)\,dx \approx \frac{\sqrt{\pi/2}}{8}\left[\sin 0 + 2\sin\left(\frac{\sqrt{\pi/2}}{4}\right)^2 + 2\sin\left(\frac{\sqrt{\pi/2}}{2}\right)^2 + 2\sin\left(\frac{3\sqrt{\pi/2}}{4}\right)^2 + \sin\left(\sqrt{\frac{\pi}{2}}\right)^2\right] \approx 0.5495$$

Simpson's: $\int_0^{\sqrt{\pi/2}} \sin(x^2)\,dx \approx \frac{\sqrt{\pi/2}}{12}\left[\sin 0 + 4\sin\left(\frac{\sqrt{\pi/2}}{4}\right)^2 + 2\sin\left(\frac{\sqrt{\pi/2}}{2}\right)^2 + 4\sin\left(\frac{3\sqrt{\pi/2}}{4}\right)^2 + \sin\left(\sqrt{\frac{\pi}{2}}\right)^2\right] \approx 0.5483$

Graphing utility: 0.5493

23. Trapezoidal: $\int_0^{\pi/4} x\tan x\,dx \approx \frac{\pi}{32}\left[0 + 2\left(\frac{\pi}{16}\right)\tan\left(\frac{\pi}{16}\right) + 2\left(\frac{2\pi}{16}\right)\tan\left(\frac{2\pi}{16}\right) + 2\left(\frac{3\pi}{16}\right)\tan\left(\frac{3\pi}{16}\right) + \frac{\pi}{4}\right] \approx 0.1940$

Simpson's: $\int_0^{\pi/4} x\tan x\,dx \approx \frac{\pi}{48}\left[0 + 4\left(\frac{\pi}{16}\right)\tan\left(\frac{\pi}{16}\right) + 2\left(\frac{2\pi}{16}\right)\tan\left(\frac{2\pi}{16}\right) + 4\left(\frac{3\pi}{16}\right)\tan\left(\frac{3\pi}{16}\right) + \frac{\pi}{4}\right] \approx 0.1860$

Graphing utility: 0.1858

25. $f(x) = x^2 + 2x$

$f'(x) = 2x + 2$

$f''(x) = 2$

$f'''(x) = 0$

$f^{(4)}(x) = 0$

(a) Trapezoidal Rule: Because $|f''(x)|$ is maximum for all x in $[0, 2]$ and $|f''(x)| = 2$, you have

$$|\text{Error}| \le \frac{(2-0)^3}{12(4)^2}(2) = \frac{1}{12} \approx 0.0833.$$

(b) Simpson's Rule: Because $|f^{(4)}(x)|$ is maximum for all x in $[0, 2]$ and $f^{(4)}(x) = 0$, you have $|\text{Error}| \le \frac{(2-0)^5}{180(4)^4}(0) = 0.$

27. $f(x) = (x-1)^{-2}$

$f'(x) = -2(x-1)^{-3}$

$f''(x) = 6(x-1)^{-4}$

$f'''(x) = -24(x-1)^{-5}$

$f^{(4)}(x) = 120(x-1)^{-6}$

(a) Trapezoidal: Error $\le \dfrac{(4-2)^3}{12(4^2)}(6) = \dfrac{1}{4}$ because $|f''(x)|$ is a maximum of 6 at $x = 2$.

(b) Simpson's: Error $\le \dfrac{(4-2)^5}{180(4^4)}(120) = \dfrac{1}{12}$ because $|f^{(4)}(x)|$ is a maximum of 120 at $x = 2$.

29. $f(x) = x^{-1}, \quad 1 \le x \le 3$

$f'(x) = -x^{-2}$

$f''(x) = 2x^{-3}$

$f'''(x) = -6x^{-4}$

$f^{(4)}(x) = 24x^{-5}$

(a) Maximum of $|f''(x)| = |2x^{-3}|$ is 2.

Trapezoidal:

Error $\le \dfrac{2^3}{12n^2}(2) \le 0.00001$, $n^2 \ge 133{,}333.33$,

$n \ge 365.15$ Let $n = 366$.

(b) Maximum of $|f^{(4)}(x)| = |24x^{-5}|$ is 24.

Simpson's: Error $\le \dfrac{2^5}{180n^4}(24) \le 0.00001$,

$n^4 \ge 426{,}666.67$, $n \ge 25.56$ Let $n = 26$.

31. $f(x) = (x+2)^{1/2}, \quad 0 \le x \le 2$

$f'(x) = \dfrac{1}{2}(x+2)^{-1/2}$

$f''(x) = -\dfrac{1}{4}(x+2)^{-3/2}$

$f'''(x) = \dfrac{3}{8}(x+2)^{-5/2}$

$f^{(4)}(x) = \dfrac{-15}{16}(x+2)^{-7/2}$

(a) Maximum of $|f''(x)| = \left|\dfrac{-1}{4(x+2)^{3/2}}\right|$ is $\dfrac{\sqrt{2}}{16} \approx 0.0884$.

Trapezoidal:

Error $\le \dfrac{(2-0)^3}{12n^2}\left(\dfrac{\sqrt{2}}{16}\right) \le 0.00001$

$n^2 \ge \dfrac{8\sqrt{2}}{12(16)}10^5 = \dfrac{\sqrt{2}}{24}10^5$

$n \ge 76.8$. Let $n = 77$.

(b) Maximum of $|f^{(4)}(x)| = \left|\dfrac{-15}{16(x+2)^{7/2}}\right|$ is $\dfrac{15\sqrt{2}}{256} \approx 0.0829$.

Simpson's:

Error $\le \dfrac{2^5}{180n^4}\left(\dfrac{15\sqrt{2}}{256}\right) \le 0.00001$

$n^4 \ge \dfrac{32(15)\sqrt{2}}{180(256)}10^5$

$= \dfrac{\sqrt{2}}{96}10^5$

$n \ge 6.2$. Let $n = 8$ (even).

33. $f(x) = \tan(x^2)$

(a) $f''(x) = 2\sec^2(x^2)\left[1 + 4x^2\tan(x^2)\right]$ in $[0, 1]$.

$|f''(x)|$ is maximum when $x = 1$ and $|f''(1)| \approx 49.5305$.

Trapezoidal: Error $\le \dfrac{(1-0)^3}{12n^2}(49.5305) \le 0.00001$, $n^2 \ge 412{,}754.17$, $n \ge 642.46$; let $n = 643$.

(b) $f^{(4)}(x) = 8\sec^2(x^2)\left[12x^2 + (3 + 32x^4)\tan(x^2) + 36x^2\tan^2(x^2) + 48x^4\tan^3(x^2)\right]$ in $[0, 1]$

$|f^{(4)}(x)|$ is maximum when $x = 1$ and $|f^{(4)}(1)| \approx 9184.4734$.

Simpson's: Error $\le \dfrac{(1-0)^5}{180n^4}(9184.4734) \le 0.00001$, $n^4 \ge 5{,}102{,}485.22$, $n \ge 47.53$; let $n = 48$.

35. $n = 4,\ b - a = 4 - 0 = 4$

Trapezoidal: $\int_0^4 f(x)\,dx \approx \frac{4}{8}\left[3 + 2(7) + 2(9) + 2(7) + 0\right] = \frac{1}{2}(49) = \frac{49}{2} = 24.5$

Simpson's: $\int_0^4 f(x)\,dx \approx \frac{4}{12}\left[3 + 4(7) + 2(9) + 4(7) + 0\right] = \frac{77}{3} \approx 25.67$

37. $A = \int_0^{\pi/2} \sqrt{x}\cos x\,dx$

Simpson's Rule: $n = 14$

$$\int_0^{\pi/2} \sqrt{x}\cos x\,dx \approx \frac{\pi}{84}\left[\sqrt{0}\cos 0 + 4\sqrt{\frac{\pi}{28}}\cos\frac{\pi}{28} + 2\sqrt{\frac{\pi}{14}}\cos\frac{\pi}{14} + 4\sqrt{\frac{3\pi}{28}}\cos\frac{3\pi}{28} + \cdots + \sqrt{\frac{\pi}{2}}\cos\frac{\pi}{2}\right] \approx 0.701$$

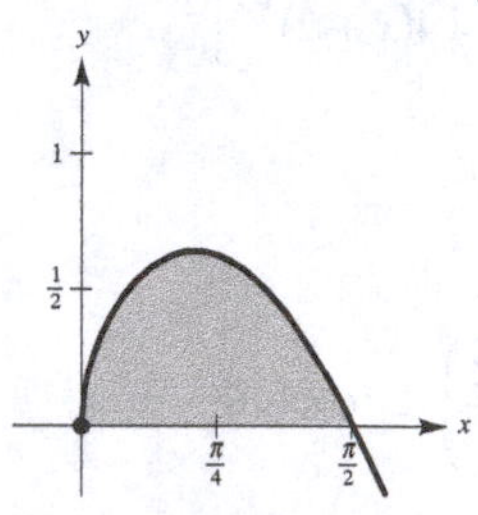

39. The Trapezoidal Rule is the average of the left-hand Riemann Sum and the right-hand Rieman Sum, $T_n = \frac{1}{2}(L_n + R_n)$.

41. Area $\approx \frac{1000}{2(10)}\left[125 + 2(125) + 2(120) + 2(112) + 2(90) + 2(90) + 2(95) + 2(88) + 2(75) + 2(35)\right] = 89{,}250\text{ m}^2$

43. $W = \int_0^5 100x\sqrt{125 - x^3}\,dx$

Simpson's Rule: $n = 12$

$$\int_0^5 100x\sqrt{125 - x^3}\,dx \approx \frac{5}{3(12)}\left[0 + 400\left(\frac{5}{12}\right)\sqrt{125 - \left(\frac{5}{12}\right)^3} + 200\left(\frac{10}{12}\right)\sqrt{125 - \left(\frac{10}{12}\right)^3} + 400\left(\frac{15}{12}\right)\sqrt{125 - \left(\frac{15}{12}\right)^3} + \cdots + 0\right] \approx 10{,}233.58\text{ ft-lb}$$

45. $\int_0^t \sin\sqrt{x}\,dx = 2,\ n = 10$

By trial and error, you obtain $t \approx 2.477$.

47. The quadratic polynomial

$$p(x) = \frac{(x - x_2)(x - x_3)}{(x_1 - x_2)(x_1 - x_3)}y_1 + \frac{(x - x_1)(x - x_3)}{(x_2 - x_1)(x_2 - x_3)}y_2 + \frac{(x - x_1)(x - x_2)}{(x_3 - x_1)(x_3 - x_2)}y_3$$

passes through the three points.

Section 8.7 Integration by Tables and Other Integration Techniques

1. Use Formula 40.

3. By Formula 6: $(a = 5, b = 1)$

$$\int \frac{x^2}{5 + x}\,dx = \left[-\frac{x}{2}(10 - x) + 25\ln|5 + x|\right] + C$$

5. By Formula 44: $\int \frac{1}{x^2\sqrt{1 - x^2}}\,dx = -\frac{\sqrt{1 - x^2}}{x} + C$

7. By Formulas 51 and 49:

$$\int \cos^4 3x\,dx = \frac{1}{3}\int \cos^4 3x\,(3)\,dx$$
$$= \frac{1}{3}\left[\frac{\cos^3 3x \sin 3x}{4} + \frac{3}{4}\int \cos^2 3x\,dx\right]$$
$$= \frac{1}{12}\cos^3 3x \sin 3x + \frac{1}{4}\cdot\frac{1}{3}\int \cos^2 3x\,(3)\,dx$$
$$= \frac{1}{12}\cos^3 3x \sin 3x + \frac{1}{12}\cdot\frac{1}{2}(3x + \sin 3x \cos 3x) + C$$
$$= \frac{1}{24}\left(2\cos^3 3x \sin 3x + 3x + \sin 3x \cos 3x\right) + C$$

9. By Formula 57: $\displaystyle\int \frac{1}{\sqrt{x}\left(1 - \cos\sqrt{x}\right)}\,dx = 2\int \frac{1}{1 - \cos\sqrt{x}}\left(\frac{1}{2\sqrt{x}}\right)dx = -2\left(\cot\sqrt{x} + \csc\sqrt{x}\right) + C$

$$u = \sqrt{x},\ du = \frac{1}{2\sqrt{x}}\,dx$$

11. By Formula 84:

$$\int \frac{1}{1 + e^{2x}}\,dx = 2x - \frac{1}{2}\ln\left(1 + e^{2x}\right) + C$$

13. By Formula 89: $(n = 6)$

$$\int x^6 \ln x\,dx = \frac{x^7}{49}(-1 + 7\ln x) + C$$

15. (a) Let $u = \frac{1}{3}x \Rightarrow du = \frac{1}{3}\,dx$

$$\int \ln\frac{x}{3}\,dx = 3\int \ln u\,du$$

By Formula 87:

$$\int \ln\frac{x}{3}\,dx = 3\left[\frac{1}{3}x\left(-1 + \ln\frac{x}{3}\right)\right] + C$$
$$= x\left(\ln\frac{x}{3} - 1\right) + C$$

(b) Integration by parts:

$$dv = dx \quad\Rightarrow\quad v = \int dx = x$$
$$u = \ln\frac{x}{3} \quad\Rightarrow\quad du = \frac{1}{x/3}\cdot\frac{1}{3} = \frac{1}{x}\,dx$$

$$\int \ln\frac{x}{3}\,dx = x\ln\frac{x}{3} - \int x\cdot\frac{1}{x}\,dx$$
$$= x\ln\frac{x}{3} - \int dx$$
$$= x\ln\frac{x}{3} - x + C$$
$$= x\left(\ln\frac{x}{3} - 1\right) + C$$

17. (a) By Formula 12: $(a = -1, b = 1)$

$$\int \frac{1}{x^2(x-1)}\,dx = \frac{-1}{(-1)}\left(\frac{1}{x} + \frac{1}{(-1)}\ln\left|\frac{x}{-1+x}\right|\right) + C$$
$$= \frac{1}{x} - \ln\left|\frac{x}{x-1}\right| + C$$
$$= \frac{1}{x} + \ln\left|\frac{x-1}{x}\right| + C$$

(b) Partial fractions:

$$\frac{1}{x^2(x-1)} = \frac{A}{x} + \frac{B}{x^2} + \frac{C}{x-1}$$
$$1 = Ax(x-1) + B(x-1) + Cx^2$$

When $x = 1, C = 1.$

When $x = 0, B = -1.$

When $x = -1, 1 = 2A + 2 + 1 \Rightarrow A = -1.$

$$\int \frac{1}{x^2(x-1)}\,dx = \int\left(\frac{-1}{x} + \frac{-1}{x^2} + \frac{1}{x-1}\right)dx$$
$$= -\ln|x| + \frac{1}{x} + \ln|x-1| + C$$
$$= \frac{1}{x} + \ln\left|\frac{x-1}{x}\right| + C$$

19. By Formula 80:

$$\int x \operatorname{arccsc}(x^2+1)\,dx = \frac{1}{2}\int \operatorname{arccsc}(x^2+1)(2x)\,dx$$
$$= \frac{1}{2}\left[(x^2+1)\operatorname{arccsc}(x^2+1) + \ln\left|x^2+1+\sqrt{(x^2+1)^2-1}\right|\right] + C$$
$$= \frac{1}{2}(x^2+1)\operatorname{arccsc}(x^2+1) + \frac{1}{2}\ln\left(x^2+1+\sqrt{x^4+2x^2}\right) + C$$

21. Let $u = x^2$, $du = 2x\,dx$.

$$\int \frac{2}{x^3\sqrt{x^4-1}}\,dx = \int \frac{2x}{x^4\sqrt{x^4-1}}\,dx = \int \frac{du}{u^2\sqrt{u^2-1}}$$

By Formula 35: $\displaystyle\int \frac{du}{u^2\sqrt{u^2-1}} = \frac{\sqrt{u^2-1}}{u} = \frac{\sqrt{x^4-1}}{x^2} + C$

23. By Formula 4: $a = 7, b = -6$

$$\int \frac{x}{(7-6x)^2}\,dx = \frac{1}{(-6)^2}\left(\frac{7}{7-6x} + \ln|7-6x|\right) + C = \frac{1}{36}\left(\frac{7}{7-6x} + \ln|7-6x|\right) + C$$

25. By Formula 76: $u = e^x$, $du = e^x\,dx$

$$\int e^x \arccos e^x\,dx = e^x \arccos e^x - \sqrt{1-e^{2x}} + C$$

27. By Formula 73: $\displaystyle\int \frac{x}{1-\sec x^2}\,dx = \frac{1}{2}\int \frac{2x}{1-\sec x^2}\,dx = \frac{1}{2}\left(x^2 + \cot x^2 + \csc x^2\right) + C$

29. By Formula 14: $u = \sin\theta$, $du = \cos\theta\,d\theta$

$$\int \frac{\cos\theta}{3+2\sin\theta+\sin^2\theta}\,d\theta = \frac{\sqrt{2}}{2}\arctan\left(\frac{1+\sin\theta}{\sqrt{2}}\right) + C \quad \left(b^2 = 4 < 12 = 4ac\right)$$

31. By Formula 35: $\displaystyle\int \frac{1}{x^2\sqrt{2+9x^2}}\,dx = 3\int \frac{3}{(3x)^2\sqrt{\left(\sqrt{2}\right)^2+(3x)^2}}\,dx = -\frac{3\sqrt{2+9x^2}}{6x} + C = -\frac{\sqrt{2+9x^2}}{2x} + C$

33. By Formula 3: $u = \ln x$, $du = \dfrac{1}{x}\,dx$

$$\int \frac{\ln x}{x(3+2\ln x)}\,dx = \frac{1}{4}\left(2\ln|x| - 3\ln\left|3+2\ln|x|\right|\right) + C$$

35. By Formulas 1, 23, and 35: $\displaystyle\int \frac{x}{\left(x^2-6x+10\right)^2}\,dx = \frac{1}{2}\int \frac{2x-6+6}{\left(x^2-6x+10\right)^2}\,dx$

$$= \frac{1}{2}\int\left(x^2-6x+10\right)^{-2}(2x-6)\,dx + 3\int \frac{1}{\left[(x-3)^2+1\right]^2}\,dx$$
$$= -\frac{1}{2\left(x^2-6x+10\right)} + \frac{3}{2}\left[\frac{x-3}{x^2-6x+10} + \arctan(x-3)\right] + C$$
$$= \frac{3x-10}{2\left(x^2-6x+10\right)} + \frac{3}{2}\arctan(x-3) + C$$

37. By Formula 31: $u = x^2 - 3$, $du = 2x\,dx$

$$\int \frac{x}{\sqrt{x^4 - 6x^2 + 5}}\,dx = \frac{1}{2}\int \frac{2x}{\sqrt{\left(x^2 - 3\right)^2 - 4}}\,dx = \frac{1}{2}\ln\left|x^2 - 3 + \sqrt{x^4 - 6x^2 + 5}\right| + C$$

39. By Formula 8: $u = e^x$, $du = e^x\,dx$

$$\int \frac{e^{3x}}{\left(1 + e^x\right)^3}\,dx = \int \frac{\left(e^x\right)^2}{\left(1 + e^x\right)^3}\left(e^x\right)dx = \frac{2}{1 + e^x} - \frac{1}{2\left(1 + e^x\right)^2} + \ln\left|1 + e^x\right| + C$$

41. By Formula 21: $a = 1, b = 1, u = x, du = dx$

$$\int_0^1 \frac{x}{\sqrt{1 + x}}\,dx = \left[\frac{-2(2 - x)}{3(1)^2}\sqrt{1 + x}\right]_0^1 = \left[\frac{2}{3}(x - 2)\sqrt{1 + x}\right]_0^1 = \frac{2}{3}(-1)\sqrt{2} - \frac{2}{3}(-2)\sqrt{1} = -\frac{2}{3}\left(\sqrt{2} - 2\right) \approx 0.3905$$

43. By Formula 89: $n = 4$

$$\int_1^2 x^4 \ln x\,dx = \left[\frac{x^5}{25}(-1 + 5\ln x)\right]_1^2 = \frac{32}{25}[-1 + 5\ln 2] - \frac{1}{25}[-1 + 0] = -\frac{31}{25} + \frac{32}{5}\ln 2 \approx 3.1961$$

45. By Formula 23: $u = \sin x$, $du = \cos x$

$$\int_{-\pi/2}^{\pi/2} \frac{\cos x}{1 + \sin^2 x}\,dx = \left[\arctan(\sin x)\right]_{-\pi/2}^{\pi/2} = \arctan(1) - \arctan(-1) = \frac{\pi}{2}$$

47. By Formulas 54 and 55:

$$\begin{aligned}\int t^3 \cos t\,dt &= t^3 \sin t - 3\int t^2 \sin t\,dt\\ &= t^3 \sin t - 3\left(-t^2 \cos t + 2\int t\cos t\,dt\right)\\ &= t^3 \sin t + 3t^2 \cos t - 6\left(t\sin t - \int \sin t\,dt\right)\\ &= t^3 \sin t + 3t^2 \cos t - 6t\sin t - 6\cos t + C\end{aligned}$$

So,

$$\begin{aligned}\int_0^{\pi/2} t^3 \cos t\,dt &= \left[t^3 \sin t + 3t^2 \cos t - 6t\sin t - 6\cos t\right]_0^{\pi/2}\\ &= \left(\frac{\pi^3}{8} - 3\pi\right) + 6 = \frac{\pi^3}{8} + 6 - 3\pi \approx 0.4510.\end{aligned}$$

49. $$\frac{u^2}{(a + bu)^2} = \frac{1}{b^2} - \frac{(2a/b)u + \left(a^2/b^2\right)}{(a + bu)^2} = \frac{1}{b^2} + \frac{A}{a + bu} + \frac{B}{(a + bu)^2}$$

$$-\frac{2a}{b}u - \frac{a^2}{b^2} = A(a + bu) + B = (aA + B) + bAu$$

Equating the coefficients of like terms you have $aA + B = -a^2/b^2$ and $bA = -2a/b$. Solving these equations you have $A = -2a/b^2$ and $B = a^2/b^2$.

$$\begin{aligned}\int \frac{u^2}{(a + bu)^2}\,du &= \frac{1}{b^2}\int du - \frac{2a}{b^2}\left(\frac{1}{b}\right)\int \frac{1}{a + bu}b\,du + \frac{a^2}{b^2}\left(\frac{1}{b}\right)\int \frac{1}{(a + bu)^2}b\,du = \frac{1}{b^2}u - \frac{2a}{b^3}\ln|a + bu| - \frac{a^2}{b^3}\left(\frac{1}{a + bu}\right) + C\\ &= \frac{1}{b^3}\left(bu - \frac{a^2}{a + bu} - 2a\ln|a + bu|\right) + C\end{aligned}$$

51. When you have $u^2 + a^2$:

$$u = a\tan\theta$$
$$du = a\sec^2\theta\,d\theta$$
$$u^2 + a^2 = a^2\sec^2\theta$$

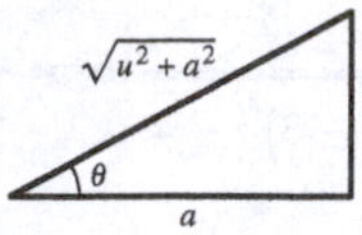

$$\int \frac{1}{(u^2+a^2)^{3/2}}\,du = \int \frac{a\sec^2\theta\,d\theta}{a^3\sec^3\theta} = \frac{1}{a^2}\int \cos\theta\,d\theta = \frac{1}{a^2}\sin\theta + C = \frac{u}{a^2\sqrt{u^2+a^2}} + C$$

When you have $u^2 - a^2$:

$$u = a\sec\theta$$
$$du = a\sec\theta\tan\theta\,d\theta$$
$$u^2 - a^2 = a^2\tan^2\theta$$

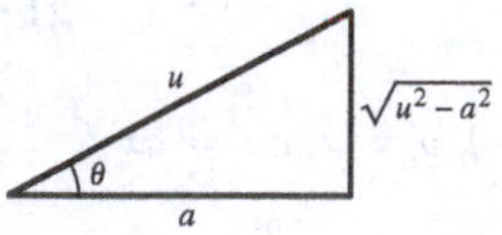

$$\int \frac{1}{(u^2-a^2)^{3/2}}\,du = \int \frac{a\sec\theta\tan\theta\,d\theta}{a^3\tan^3\theta} = \frac{1}{a^2}\int \frac{\cos\theta}{\sin^2\theta}\,d\theta = \frac{1}{a^2}\int \csc\theta\cot\theta\,d\theta = -\frac{1}{a^2}\csc\theta + C = \frac{-u}{a^2\sqrt{u^2-a^2}} + C$$

53.
$$\int(\arctan u)\,du = u\arctan u - \frac{1}{2}\int\frac{2u}{1+u^2}\,du$$
$$= u\arctan u - \frac{1}{2}\ln(1+u^2) + C$$
$$= u\arctan u - \ln\sqrt{1+u^2} + C$$

$w = \arctan u,\ dv = du,\ dw = \dfrac{du}{1+u^2},\ v = u$

55.
$$\int\frac{1}{2-3\sin\theta}\,d\theta = \int\left[\frac{\frac{2\,du}{1+u^2}}{2-3\left(\frac{2u}{1+u^2}\right)}\right],\ u = \tan\frac{\theta}{2}$$
$$= \int\frac{2}{2(1+u^2)-6u}\,du$$
$$= \int\frac{1}{u^2-3u+1}\,du$$
$$= \int\frac{1}{\left(u-\frac{3}{2}\right)^2-\frac{5}{4}}\,du$$
$$= \frac{1}{\sqrt{5}}\ln\left|\frac{\left(u-\frac{3}{2}\right)-\frac{\sqrt{5}}{2}}{\left(u-\frac{3}{2}\right)+\frac{\sqrt{5}}{2}}\right| + C$$
$$= \frac{1}{\sqrt{5}}\ln\left|\frac{2u-3-\sqrt{5}}{2u-3+\sqrt{5}}\right| + C$$
$$= \frac{1}{\sqrt{5}}\ln\left|\frac{2\tan\left(\frac{\theta}{2}\right)-3-\sqrt{5}}{2\tan\left(\frac{\theta}{2}\right)-3+\sqrt{5}}\right| + C$$

57.
$$\int_0^{\pi/2}\frac{1}{1+\sin\theta+\cos\theta}\,d\theta = \int_0^1\left[\frac{\frac{2\,du}{1+u^2}}{1+\frac{2u}{1+u^2}+\frac{1-u^2}{1+u^2}}\right]$$
$$= \int_0^1\frac{1}{1+u}\,du$$
$$= \Big[\ln|1+u|\Big]_0^1$$
$$= \ln 2$$

$u = \tan\dfrac{\theta}{2}$

59.
$$\int\frac{\sin\theta}{3-2\cos\theta}\,d\theta = \frac{1}{2}\int\frac{2\sin\theta}{3-2\cos\theta}\,d\theta$$
$$= \frac{1}{2}\ln|u| + C$$
$$= \frac{1}{2}\ln(3-2\cos\theta) + C$$

$u = 3 - 2\cos\theta,\ du = 2\sin\theta\,d\theta$

61.
$$\int\frac{\sin\sqrt{\theta}}{\sqrt{\theta}}\,d\theta = 2\int\sin\sqrt{\theta}\left(\frac{1}{2\sqrt{\theta}}\right)d\theta$$
$$= -2\cos\sqrt{\theta} + C$$

$u = \sqrt{\theta},\ du = \dfrac{1}{2\sqrt{\theta}}\,d\theta$

63. By Formula 21: $a = 3, b = 1$

$$A = \int_0^6 \frac{x}{\sqrt{x+3}}\,dx = \left[\frac{-2(6-x)}{3}\sqrt{x+3}\right]_0^6 = 4\sqrt{3} \approx 6.928 \text{ square units}$$

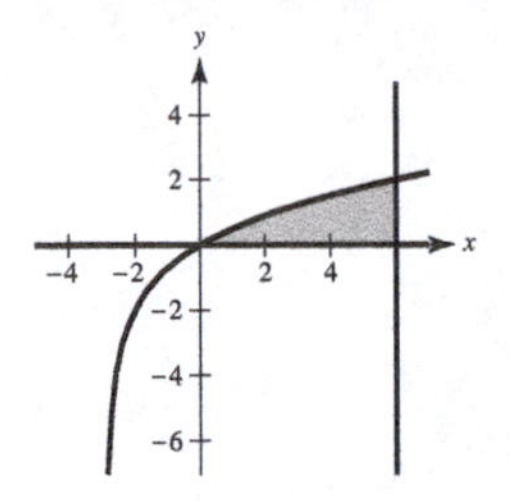

65. (a) $n = 1$: $u = \ln x, du = \frac{1}{x}\,dx, dv = x\,dx, v = \frac{x^2}{2}$

$$\int x \ln x\,dx = \frac{x^2}{2}\ln x - \int\left(\frac{x^2}{2}\right)\frac{1}{x}\,dx = \frac{x^2}{2}\ln x - \frac{x^2}{4} + C$$

$n = 2$: $u = \ln x, du = \frac{1}{x}\,dx, dv = x^2\,dx, v = \frac{x^3}{3}$

$$\int x^2 \ln x\,dx = \frac{x^3}{3}\ln x - \int\left(\frac{x^3}{3}\right)\frac{1}{x}\,dx = \frac{x^3}{3}\ln x - \frac{x^3}{9} + C$$

$n = 3$: $u = \ln x, du = \frac{1}{x}\,dx, dv = x^3\,dx, v = \frac{x^4}{4}$

$$\int x^3 \ln x\,dx = \frac{x^4}{4}\ln x - \int\left(\frac{x^4}{4}\right)\frac{1}{x}\,dx = \frac{x^4}{4}\ln x - \frac{x^4}{16} + C$$

(b) $\int x^n \ln x\,dx = \frac{x^{n+1}}{n+1}\ln x - \frac{x^{n+1}}{(n+1)^2} + C$

(c) Let $u = \ln x, du = \frac{1}{x}\,dx, dv = x^n\,dx, v = \frac{x^{n+1}}{n+1}$

$$\int x^n \ln x\,dx = (\ln x)\frac{x^{n+1}}{n+1} - \int \frac{x^{n+1}}{n+1}\left(\frac{1}{x}\,dx\right) = \frac{x^{n+1}}{n+1}\ln x - \frac{x^{n+1}}{(n+1)^2} + C$$

67. $W = \int_0^5 2000xe^{-x}\,dx$

$= -2000\int_0^5 -xe^{-x}\,dx$

$= 2000\int_0^5 (-x)e^{-x}(-1)\,dx$

$= 2000\left[(-x)e^{-x} - e^{-x}\right]_0^5$

$= 2000\left(-\frac{6}{e^5} + 1\right)$

≈ 1919.145 ft-lb

69. $\frac{1}{2-0}\int_0^2 \frac{5000}{1+e^{4.8-1.9t}}\,dt = \frac{2500}{-1.9}\int_0^2 \frac{-1.9\,dt}{1+e^{4.8-1.9t}}$

$$= -\frac{2500}{1.9}\left[(4.8-1.9t) - \ln\left(1+e^{4.8-1.9t}\right)\right]_0^2$$

$$= -\frac{2500}{1.9}\left[\left(1-\ln(1+e)\right) - \left(4.8-\ln\left(1+e^{4.8}\right)\right)\right]$$

$$= \frac{2500}{1.9}\left[3.8 + \ln\left(\frac{1+e}{1+e^{4.8}}\right)\right] \approx 401.4$$

71.

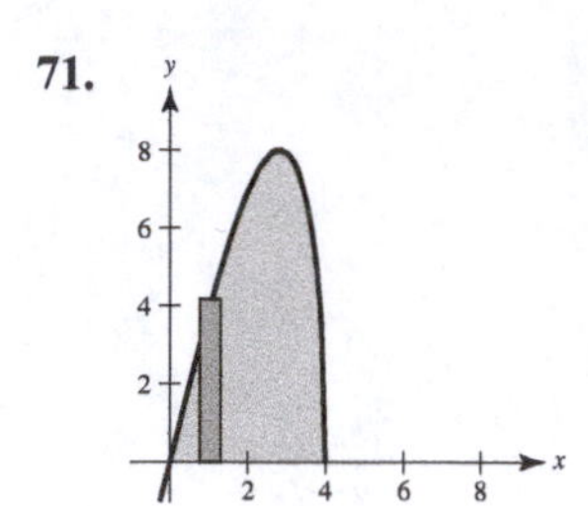

$V = 2\pi\int_0^4 x\left(x\sqrt{16-x^2}\right)dx = 2\pi\int_0^4 x^2\sqrt{16-x^2}\,dx$

By Formula 38: $(a = 4)$

$V = 2\pi\left[\frac{1}{8}\left(x\left(2x^2-16\right)\sqrt{16-x^2} + 256\arcsin\left(\frac{x}{4}\right)\right)\right]_0^4 = 2\pi\left[32\left(\frac{\pi}{2}\right)\right] = 32\pi^2$

73. Let $I = \int_0^{\pi/2} \frac{dx}{1+(\tan x)^{\sqrt{2}}}$.

For $x = \frac{\pi}{2} - u$, $dx = -du$, and

$I = \int_{\pi/2}^0 \frac{-du}{1+\left(\tan(\pi/2-u)\right)^{\sqrt{2}}} = \int_0^{\pi/2}\frac{du}{1+(\cot u)^{\sqrt{2}}} = \int_0^{\pi/2}\frac{(\tan u)^{\sqrt{2}}}{(\tan u)^{\sqrt{2}}+1}\,du.$

$2I = \int_0^{\pi/2}\frac{dx}{1+(\tan x)^{\sqrt{2}}} + \int_0^{\pi/2}\frac{(\tan x)^{\sqrt{2}}}{(\tan x)^{\sqrt{2}}+1}\,dx = \int_0^{\pi/2}dx = \frac{\pi}{2}$

So, $I = \frac{\pi}{4}$.

Section 8.8 Improper Integrals

1. An integral is improper if one or both of the limits of integration are infinite, or the function has a finite number of infinite discontinuities on the interval of integration.

3. To evaluate the improper integral $\int_a^\infty f(x)\,dx$, find the limit as $b \to \infty$ when f is continuous on $\left[a^a, \infty\right)$ or find the limit as $a \to -\infty$ when f is continuous on $(-\infty, b]$.

5. $\int_0^1 \frac{dx}{5x-3}$ is improper because $5x - 3 = 0$ when $x = \frac{3}{5}$, and $0 \le \frac{3}{5} \le 1$.

7. $\int_0^1 \frac{2x-5}{x^2-5x+6}\,dx = \int_0^1 \frac{2x-5}{(x-2)(x-3)}\,dx$ is not improper because $\frac{2x-5}{(x-2)(x-3)}$ is continuous on $[0, 1]$.

9. $\int_0^2 e^{-x}\,dx$ is not improper because $f(x) = e^{-x}$ is continuous on $[0, 2]$.

11. $\int_{-\infty}^{\infty} \frac{\sin x}{4+x^2}\,dx$ is improper because the limits of integration are $-\infty$ and ∞.

13. Infinite discontinuity at $x = 0$.

$$\int_0^4 \frac{1}{\sqrt{x}}\,dx = \lim_{b\to 0^+} \int_b^4 \frac{1}{\sqrt{x}}\,dx$$
$$= \lim_{b\to 0^+} \left[2\sqrt{x}\right]_b^4$$
$$= \lim_{b\to 0^+} \left(4 - 2\sqrt{b}\right) = 4$$

Converges

15. Infinite discontinuity at $x = 1$.

$$\int_0^2 \frac{1}{(x-1)^2}\,dx = \int_0^1 \frac{1}{(x-1)^2}\,dx + \int_1^2 \frac{1}{(x-1)^2}\,dx$$
$$= \lim_{b\to 1^-} \int_0^b \frac{1}{(x-1)^2}\,dx + \lim_{c\to 1^+} \int_c^2 \frac{1}{(x-1)^2}\,dx$$
$$= \lim_{b\to 1^-} \left[-\frac{1}{x-1}\right]_0^b + \lim_{c\to 1^+} \left[-\frac{1}{x-1}\right]_c^2$$
$$= (\infty - 1) + (-1 + \infty)$$

Diverges

17. $$\int_2^\infty \frac{1}{x^3}\,dx = \int_2^\infty x^{-3}\,dx$$
$$= \lim_{b\to\infty} \int_2^b x^{-3}\,dx$$
$$= \lim_{b\to\infty} \left[\frac{x^{-2}}{-2}\right]_2^b$$
$$= \lim_{b\to\infty} \left[-\frac{1}{2x^2}\right]_2^b$$
$$= \lim_{b\to\infty} \left[-\frac{1}{2b^2} + \frac{1}{8}\right] = \frac{1}{8}$$

19. $$\int_1^\infty \frac{3}{\sqrt[3]{x}}\,dx = \lim_{b\to\infty} \int_1^b 3x^{-1/3}\,dx$$
$$= \lim_{b\to\infty} \left[\frac{9}{2}x^{2/3}\right]_1^b = \infty$$

Diverges

21. $$\int_0^\infty e^{x/3}\,dx = \lim_{b\to\infty} \int_0^b e^{x/3}\,dx$$
$$= \lim_{b\to\infty} \left[3e^{x/3}\right]_0^b$$
$$= \lim_{b\to\infty} \left(3e^{b/3} - 3\right)$$
$$= \infty$$

Diverges

23. $$\int_0^\infty x^2 e^{-x}\,dx = \lim_{b\to\infty} \int_0^b x^2 e^{-x}\,dx$$
$$= \lim_{b\to\infty} \left[-e^{-x}\left(x^2 + 2x + 2\right)\right]_0^b$$
$$= \lim_{b\to\infty} \left(-\frac{b^2 + 2b + 2}{e^b} + 2\right) = 2$$

Because $\lim_{b\to\infty} \left(-\frac{b^2 + 2b + 2}{e^b}\right) = 0$ by L'Hôpital's Rule.

25. $$\int_4^\infty \frac{1}{x(\ln x)^3}\,dx = \lim_{b\to\infty} \int_4^b (\ln x)^{-3} \frac{1}{x}\,dx$$
$$= \lim_{b\to\infty} \left[-\frac{1}{2}(\ln x)^{-2}\right]_4^b$$
$$= \lim_{b\to\infty} \left[-\frac{1}{2}(\ln b)^{-2} + \frac{1}{2}(\ln 4)^{-2}\right]$$
$$= \frac{1}{2}\frac{1}{(2\ln 2)^2} = \frac{1}{2(\ln 4)^2}$$

27. $$\int_{-\infty}^\infty \frac{4}{16 + x^2}\,dx = \int_{-\infty}^0 \frac{4}{16 + x^2}\,dx + \int_0^\infty \frac{4}{16 + x^2}\,dx$$
$$= \lim_{b\to -\infty} \int_b^0 \frac{4}{16 + x^2}\,dx + \lim_{c\to\infty} \int_0^c \frac{4}{16 + x^2}\,dx$$
$$= \lim_{b\to -\infty} \left[\arctan\left(\frac{x}{4}\right)\right]_b^0 + \lim_{c\to\infty} \left[\arctan\left(\frac{x}{4}\right)\right]_0^c$$
$$= \lim_{b\to -\infty} \left[0 - \arctan\left(\frac{b}{4}\right)\right] + \lim_{c\to\infty} \left[\arctan\left(\frac{c}{4}\right) - 0\right]$$
$$= -\left(-\frac{\pi}{2}\right) + \frac{\pi}{2} = \pi$$

29. $$\int_0^\infty \frac{1}{e^x + e^{-x}}\,dx = \lim_{b\to\infty} \int_0^b \frac{e^x}{1 + e^{2x}}\,dx$$
$$= \lim_{b\to\infty} \left[\arctan\left(e^x\right)\right]_0^b$$
$$= \frac{\pi}{2} - \frac{\pi}{4} = \frac{\pi}{4}$$

31. $$\int_0^\infty \cos \pi x\,dx = \lim_{b\to\infty} \left[\frac{1}{\pi}\sin \pi x\right]_0^b$$

Diverges because $\sin \pi b$ does not approach a limit as $b \to \infty$.

33. $\int_0^1 \frac{1}{x^2}\,dx = \lim_{b\to 0^+}\left[\frac{-1}{x}\right]_b^1 = \lim_{b\to 0^+}\left(-1 + \frac{1}{b}\right) = -1 + \infty$

Diverges

35. $\int_0^2 \frac{1}{\sqrt[3]{x-1}}\,dx = \int_0^1 \frac{1}{\sqrt[3]{x-1}}\,dx + \int_1^2 \frac{1}{\sqrt[3]{x-1}}\,dx = \lim_{b\to 1^-}\left[\frac{3}{2}(x-1)^{2/3}\right]_0^b + \lim_{c\to 1^+}\left[\frac{3}{2}(x-1)^{2/3}\right]_c^2 = \frac{-3}{2} + \frac{3}{2} = 0$

37. $\int_0^1 x\ln x\,dx = \lim_{b\to 0^+}\left[\frac{x^2}{2}\ln|x| - \frac{x^2}{4}\right]_b^1$

$= \lim_{b\to 0^+}\left(\frac{-1}{4} - \frac{b^2\ln b}{2} + \frac{b^2}{4}\right) = \frac{-1}{4}$

because $\lim_{b\to 0^+}\left(b^2\ln b\right) = 0$ by L'Hôpital's Rule.

39. $\int_0^{\pi/2} \tan\theta\,d\theta = \lim_{b\to(\pi/2)^-}\left[\ln|\sec\theta|\right]_0^b = \infty$

Diverges

41. $\int_2^4 \frac{2}{x\sqrt{x^2-4}}\,dx = \lim_{b\to 2^+}\int_b^4 \frac{2}{x\sqrt{x^2-4}}\,dx$

$= \lim_{b\to 2^+}\left[\operatorname{arcsec}\left|\frac{x}{2}\right|\right]_b^4$

$= \lim_{b\to 2^+}\left(\operatorname{arcsec} 2 - \operatorname{arcsec}\left(\frac{b}{2}\right)\right)$

$= \frac{\pi}{3} - 0 = \frac{\pi}{3}$

43. $\int_3^5 \frac{1}{\sqrt{x^2-9}}\,dx = \lim_{b\to 3^+}\left[\ln\left|x + \sqrt{x^2-9}\right|\right]_b^5$

$= \lim_{b\to 3^+}\left[\ln 9 - \ln\left(b + \sqrt{b^2-9}\right)\right]$

$= \ln 9 - \ln 3$

$= \ln\frac{9}{3} = \ln 3$

45. $\int_3^\infty \frac{1}{x\sqrt{x^2-9}}\,dx = \lim_{b\to 3^+}\int_b^5 \frac{1}{x\sqrt{x^2-9}}\,dx + \lim_{c\to\infty}\int_5^c \frac{1}{x\sqrt{x^2-9}}\,dx$

$= \lim_{b\to 3^+}\left[\frac{1}{3}\operatorname{arcsec}\frac{x}{3}\right]_b^5 + \lim_{c\to\infty}\left[\frac{1}{3}\operatorname{arcsec}\left(\frac{x}{3}\right)\right]_5^c$

$= \lim_{b\to 3^+}\left[\frac{1}{3}\operatorname{arcsec}\left(\frac{5}{3}\right) - \frac{1}{3}\operatorname{arcsec}\left(\frac{b}{3}\right)\right] + \lim_{c\to\infty}\left[\frac{1}{3}\operatorname{arcsec}\left(\frac{c}{3}\right) - \frac{1}{3}\operatorname{arcsec}\left(\frac{5}{3}\right)\right] = -0 + \frac{1}{3}\left(\frac{\pi}{2}\right) = \frac{\pi}{6}$

47. $\int_0^\infty \frac{4}{\sqrt{x}(x+6)}\,dx = \int_0^1 \frac{4}{\sqrt{x}(x+6)}\,dx + \int_1^\infty \frac{4}{\sqrt{x}(x+6)}\,dx$

Let $u = \sqrt{x}$, $u^2 = x$, $2u\,du = dx$.

$\int \frac{4}{\sqrt{x}(x+6)}\,dx = \int \frac{4(2u\,du)}{u(u^2+6)} = 8\int \frac{du}{u^2+6} = \frac{8}{\sqrt{6}}\arctan\left(\frac{u}{\sqrt{6}}\right) + C = \frac{8}{\sqrt{6}}\arctan\left(\frac{\sqrt{x}}{\sqrt{6}}\right) + C$

So, $\int_0^\infty \frac{4}{\sqrt{x}(x+6)}\,dx = \lim_{b\to 0^+}\left[\frac{8}{\sqrt{6}}\arctan\left(\frac{\sqrt{x}}{\sqrt{6}}\right)\right]_b^1 + \lim_{c\to\infty}\left[\frac{8}{\sqrt{6}}\arctan\left(\frac{\sqrt{x}}{\sqrt{6}}\right)\right]_1^c$

$= \left[\frac{8}{\sqrt{6}}\arctan\left(\frac{1}{\sqrt{6}}\right) - \frac{8}{\sqrt{6}}(0)\right] + \left[\frac{8}{\sqrt{6}}\left(\frac{\pi}{2}\right) - \frac{8}{\sqrt{6}}\arctan\left(\frac{1}{\sqrt{6}}\right)\right] = \frac{8\pi}{2\sqrt{6}} = \frac{2\pi\sqrt{6}}{3}.$

49. If $p = 1$, $\int_1^\infty \frac{1}{x}\,dx = \lim_{b\to\infty}\int_1^b \frac{1}{x}\,dx = \lim_{b\to\infty}\left[\ln x\right]_1^b = \lim_{b\to\infty}(\ln b) = \infty$.

Diverges. For $p \neq 1$,

$$\int_1^\infty \frac{1}{x^p}\,dx = \lim_{b\to\infty}\left[\frac{x^{1-p}}{1-p}\right]_1^b = \lim_{b\to\infty}\left(\frac{b^{1-p}}{1-p} - \frac{1}{1-p}\right).$$

This converges to $\frac{1}{p-1}$ if $1 - p < 0$ or $p > 1$.

51. For $n = 1$:

$$\int_0^\infty xe^{-x}\,dx = \lim_{b\to\infty}\int_0^b xe^{-x}\,dx$$
$$= \lim_{b\to\infty}\left[-e^{-x}x - e^{-x}\right]_0^b \qquad \text{(Parts: } u = x,\ dv = e^{-x}\,dx\text{)}$$
$$= \lim_{b\to\infty}\left(-e^{-b}b - e^{-b} + 1\right)$$
$$= \lim_{b\to\infty}\left(\frac{-b}{e^b} - \frac{1}{e^b} + 1\right) = 1 \qquad \text{(L'Hôpital's Rule)}$$

Assume that $\int_0^\infty x^n e^{-x}\,dx$ converges. Then for $n + 1$ you have

$$\int x^{n+1}e^{-x}\,dx = -x^{n+1}e^{-x} + (n+1)\int x^n e^{-x}\,dx$$

by parts $\left(u = x^{n+1},\ du = (n+1)x^n\,dx,\ dv = e^{-x}\,dx,\ v = -e^{-x}\right)$.

So,

$$\int_0^\infty x^{n+1}e^{-x}\,dx = \lim_{b\to\infty}\left[-x^{n+1}e^{-x}\right]_0^b + (n+1)\int_0^\infty x^n e^{-x}\,dx = 0 + (n+1)\int_0^\infty x^n e^{-x}\,dx, \text{ which converges.}$$

53. $\int_0^1 \frac{1}{\sqrt[6]{x}}\,dx = \int_0^1 \frac{1}{x^{1/6}}\,dx$ converges by Exercise 50. $\left(p = \frac{1}{6}\right)$

55. $\int_1^\infty \frac{1}{x^5}\,dx$ converges by Exercise 49. $(p = 5)$

57. Because $\frac{1}{x^2+5} \leq \frac{1}{x^2}$ on $[1, \infty)$ and $\int_1^\infty \frac{1}{x^2}\,dx$ converges by Exercise 49, $\int_1^\infty \frac{1}{x^2+5}\,dx$ converges.

59. $\int_1^\infty \frac{2}{x^2}\,dx$ converges, and $\frac{1 - \sin x}{x^2} \leq \frac{2}{x^2}$ on $[1, \infty)$, so $\int_1^\infty \frac{1 - \sin x}{x^2}\,dx$ converges.

61. $\int_{-1}^1 \frac{1}{x^3}\,dx = \int_{-1}^0 \frac{1}{x^3}\,dx + \int_0^1 \frac{1}{x^3}\,dx$

These two integrals diverge by Exercise 50.

63.
$$A = \int_{-\infty}^{-1} -\frac{7}{(x-1)^3}\,dx$$
$$= \lim_{b\to-\infty}\int_b^{-1} -7(x-1)^{-3}\,dx$$
$$= \lim_{b\to-\infty}\left[\frac{7}{2}(x-1)^{-2}\right]_b^{-1}$$
$$= \lim_{b\to-\infty}\left[\frac{7}{2(-2)^2} - \frac{7}{2(b-1)^2}\right]$$
$$= \frac{7}{8} - 0 = \frac{7}{8}$$

65.
$$A = \int_{-\infty}^\infty \frac{1}{x^2+1}\,dx$$
$$= \lim_{b\to-\infty}\int_b^0 \frac{1}{x^2+1}\,dx + \lim_{b\to\infty}\int_0^b \frac{1}{x^2+1}\,dx$$
$$= \lim_{b\to-\infty}\left[\arctan(x)\right]_b^0 + \lim_{b\to\infty}\left[\arctan(x)\right]_0^b$$
$$= \lim_{b\to-\infty}\left[0 - \arctan(b)\right] + \lim_{b\to\infty}\left[\arctan(b) - 0\right]$$
$$= -\left(-\frac{\pi}{2}\right) + \frac{\pi}{2} = \pi$$

67. (a) $A = \int_0^\infty e^{-x}\,dx = \lim_{b\to\infty}\left[-e^{-x}\right]_0^b = 0 - (-1) = 1$

(b) **Disk:** $V = \pi\int_0^\infty \left(e^{-x}\right)^2 dx = \lim_{b\to\infty} \pi\left[-\frac{1}{2}e^{-2x}\right]_0^b = \frac{\pi}{2}$

(c) **Shell:** $V = 2\pi\int_0^\infty xe^{-x}\,dx = \lim_{b\to\infty} 2\pi\left[-e^{-x}(x+1)\right]_0^b = 2\pi$

69. $y = \sqrt{16 - x^2}, \quad 0 \le x \le 4$

$y' = \dfrac{-x}{\sqrt{16 - x^2}}$

$s = \int_0^4 \sqrt{1 + \frac{x^2}{16 - x^2}}\,dx = \int_0^4 \frac{4}{\sqrt{16 - x^2}}\,dx = \lim_{t\to 4^-}\int_0^t \frac{4}{\sqrt{16 - x^2}}\,dx = \lim_{t\to 4^-}\left[4\arcsin\left(\frac{x}{4}\right)\right]_0^t = \lim_{t\to 4^-} 4\arcsin\left(\frac{t}{4}\right) = 2\pi$

71. (a) $F(x) = \dfrac{K}{x^2},\ 5 = \dfrac{K}{(4000)^2},\ K = 80{,}000{,}000$

$W = \int_{4000}^\infty \frac{80{,}000{,}000}{x^2}\,dx = \lim_{b\to\infty}\left[\frac{-80{,}000{,}000}{x}\right]_{4000}^b = 20{,}000 \text{ mi-ton}$

(b) $\dfrac{W}{2} = 10{,}000 = \left[\dfrac{-80{,}000{,}000}{x}\right]_{4000}^b = \dfrac{-80{,}000{,}000}{b} + 20{,}000$

$\dfrac{80{,}000{,}000}{b} = 10{,}000$

$b = 8000$

Therefore, the rocket has traveled 4000 miles above Earth's surface.

73. (a) $\int_{-\infty}^\infty f(t)\,dt = \int_0^\infty \frac{1}{9}e^{-t/9}\,dt$

$= \lim_{b\to\infty}\left[-e^{-t/9}\right]_0^b = 1$

(b) $P(0 \le x \le 6) = \int_0^6 \frac{1}{9}e^{-t/9}\,dt$

$= \left[-e^{-t/9}\right]_0^6$

$= -e^{-2/3} + 1$

$\approx 0.4866 = 48.66\%$

75. (a)

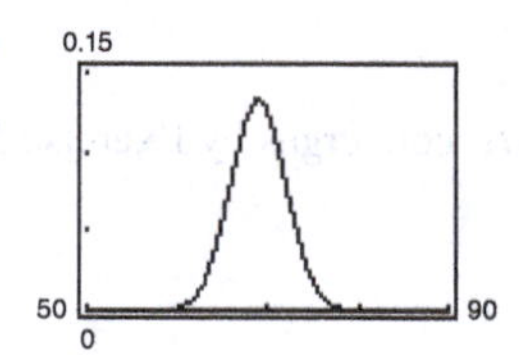

Using a graphing utility, the area under the curve is approximately 1.

(b) $P(72 \le x \le \infty) = \int_{72}^\infty f(x)\,dx \approx 0.1587$

(c) $0.5 - P(69 \le x \le 72) = 0.5 - \int_{69}^{72} f(x)\,dx$

$= 0.5 - 0.3413 = 0.1587$

The answers are the same by symmetry:

$0.5 = \int_{69}^\infty f(x)\,dx = \int_{69}^{72} f(x)\,dx + \int_{72}^\infty f(x)\,dx.$

77. (a) $C = 700{,}000 + \int_0^5 25{,}000e^{-0.06t}\,dt = 700{,}000 - \left[\frac{25{,}000}{0.06}e^{-0.06t}\right]_0^5 \approx \$807{,}992.41$

(b) $C = 700{,}000 + \int_0^{10} 25{,}000e^{-0.06t}\,dt = 700{,}000 - \left[\frac{25{,}000}{0.06}e^{-0.06t}\right]_0^{10} \approx \$887{,}995.15$

(c) $C = 700{,}000 + \int_0^\infty 25{,}000e^{-0.06t}\,dt = 700{,}000 - \lim_{b\to\infty}\left[\frac{25{,}000}{0.06}e^{-0.06t}\right]_0^b \approx \$1{,}116{,}666.67$

79. Let $K = \dfrac{2\pi NI\,r}{k}$. Then

$$P = K\int_c^{\infty} \frac{1}{(r^2 + x^2)^{3/2}}\,dx.$$

Let $x = r\tan\theta$, $dx = r\sec^2\theta\,d\theta$, $\sqrt{r^2 + x^2} = r\sec\theta$.

$$\int \frac{1}{(r^2 + x^2)^{3/2}}\,dx = \int \frac{r\sec^2\theta\,d\theta}{r^3\sec^3\theta} = \frac{1}{r^2}\int \cos\theta\,d\theta = \frac{1}{r^2}\sin\theta + C = \frac{1}{r^2}\frac{x}{\sqrt{r^2 + x^2}} + C$$

So,

$$P = K\frac{1}{r^2}\lim_{b\to\infty}\left[\frac{x}{\sqrt{r^2 + x^2}}\right]_c^b = \frac{K}{r^2}\left[1 - \frac{c}{\sqrt{r^2 + c^2}}\right]$$

$$= \frac{K\left(\sqrt{r^2 + c^2} - c\right)}{r^2\sqrt{r^2 + c^2}} = \frac{2\pi NI\left(\sqrt{r^2 + c^2} - c\right)}{kr\sqrt{r^2 + c^2}}.$$

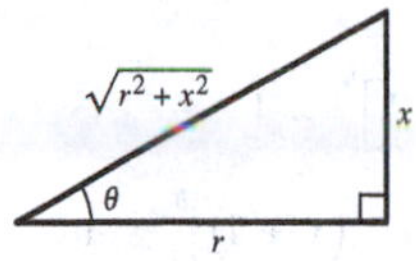

81. False. $f(x) = 1/(x + 1)$ is continuous on $[0, \infty)$, $\lim_{x\to\infty} 1/(x + 1) = 0$, but

$$\int_0^{\infty} \frac{1}{x + 1}\,dx = \lim_{b\to\infty}\Big[\ln|x + 1|\Big]_0^b = \infty.$$

Diverges

83. True

85. True

87. (a) $\displaystyle\int_{-\infty}^{\infty} \sin x\,dx = \int_{-\infty}^{0} \sin x\,dx + \int_0^{\infty} \sin x\,dx$

$$= \lim_{b\to-\infty}\int_b^0 \sin x\,dx + \lim_{c\to\infty}\int_0^c \sin x\,dx$$

$$= \lim_{b\to-\infty}\left[-\cos x\right]_b^0 + \lim_{c\to\infty}\left[-\cos x\right]_0^c$$

Because $\lim_{b\to-\infty}[-\cos b]$ diverges, as does $\lim_{c\to\infty}[-\cos c]$,

$\displaystyle\int_{-\infty}^{\infty} \sin x\,dx$ diverges.

(b) $\displaystyle\lim_{a\to\infty}\int_{-a}^{a} \sin x\,dx = \lim_{a\to\infty}\left[-\cos x\right]_{-a}^{a}$

$$= \lim_{a\to\infty}\left[-\cos(a) + \cos(-a)\right] = 0$$

(c) The definition of $\displaystyle\int_{-\infty}^{\infty} f(x)\,dx$ is not

$\displaystyle\lim_{a\to\infty}\int_{-a}^{a} f(x)\,dx.$

89. You know that $\displaystyle\int_a^b f(x)\,dx \le \int_a^b |f(x)|\,dx$

Therefore,

$$\int_a^{\infty} f(x)\,dx = \lim_{b\to\infty}\int_a^b f(x)\,dx$$

$$\le \lim_{b\to\infty}\int_a^b |f(x)|\,dx, \text{ which coverges.}$$

So, $\displaystyle\int_a^{\infty} f(x)\,dx$ converges.

91. $f(t) = 1$

$$F(s) = \int_0^{\infty} e^{-st}\,dt = \lim_{b\to\infty}\left[-\frac{1}{s}e^{-st}\right]_0^b = \frac{1}{s},\ s > 0$$

93. $f(t) = t^2$

$$F(s) = \int_0^{\infty} t^2 e^{-st}\,dt = \lim_{b\to\infty}\left[\frac{1}{s^3}\left(-s^2t^2 - 2st - 2\right)e^{-st}\right]_0^b$$

$$= \frac{2}{s^3},\ s > 0$$

95. $f(t) = \cos at$

$$F(s) = \int_0^{\infty} e^{-st}\cos at\,dt$$

$$= \lim_{b\to\infty}\left[\frac{e^{-st}}{s^2 + a^2}(-s\cos at + a\sin at)\right]_0^b$$

$$= 0 + \frac{s}{s^2 + a^2} = \frac{s}{s^2 + a^2},\ s > 0$$

97. $f(t) = \cosh at$

$$F(s) = \int_0^\infty e^{-st} \cosh at\, dt = \int_0^\infty e^{-st}\left(\frac{e^{at} + e^{-at}}{2}\right) dt = \frac{1}{2}\int_0^\infty \left[e^{t(-s+a)} + e^{t(-s-a)}\right] dt$$

$$= \lim_{b\to\infty} \frac{1}{2}\left[\frac{1}{(-s+a)}e^{t(-s+a)} + \frac{1}{(-s-a)}e^{t(-s-a)}\right]_0^b = 0 - \frac{1}{2}\left[\frac{1}{(-s+a)} + \frac{1}{(-s-a)}\right]$$

$$= \frac{-1}{2}\left[\frac{1}{(-s+a)} + \frac{1}{(-s-a)}\right] = \frac{s}{s^2 - a^2},\ s > |a|$$

99. $\Gamma(n) = \int_0^\infty x^{n-1}e^{-x}\, dx$

(a) $\Gamma(1) = \int_0^\infty e^{-x}\, dx = \lim_{b\to\infty}\left[-e^{-x}\right]_0^b = 1$

$\Gamma(2) = \int_0^\infty xe^{-x}\, dx = \lim_{b\to\infty}\left[-e^{-x}(x+1)\right]_0^b = 1$

$\Gamma(3) = \int_0^\infty x^2e^{-x}\, dx = \lim_{b\to\infty}\left[-x^2e^{-x} - 2xe^{-x} - 2e^{-x}\right]_0^b = 2$

(b) $\Gamma(n+1) = \int_0^\infty x^ne^{-x}\, dx = \lim_{b\to\infty}\left[-x^ne^{-x}\right]_0^b + \lim_{b\to\infty} n\int_0^b x^{n-1}e^{-x}\, dx = 0 + n\Gamma(n) \qquad \left(u = x^n,\ dv = e^{-x}\, dx\right)$

(c) $\Gamma(n) = (n-1)!$

101. $$\int_0^\infty\left(\frac{1}{\sqrt{x^2+1}} - \frac{c}{x+1}\right) dx = \lim_{b\to\infty}\int_0^b\left(\frac{1}{\sqrt{x^2+1}} - \frac{c}{x+1}\right) dx$$

$$= \lim_{b\to\infty}\left[\ln\left|x + \sqrt{x^2+1}\right| - c\ln|x+1|\right]_0^b$$

$$= \lim_{b\to\infty}\left[\ln\left(b + \sqrt{b^2+1}\right) - \ln(b+1)^c\right] = \lim_{b\to\infty}\ln\left[\frac{b + \sqrt{b^2+1}}{(b+1)^c}\right]$$

This limit exists for $c = 1$, and you have

$$\lim_{b\to\infty}\ln\left[\frac{b + \sqrt{b^2+1}}{(b+1)}\right] = \ln 2.$$

103. $f(x) = \begin{cases} x\ln x, & 0 < x \le 2 \\ 0, & x = 0\end{cases}$

$V = \pi\int_0^2 (x\ln x)^2\, dx$

Let $u = \ln x$, $e^u = x$, $e^u\, du = dx$.

$$V = \pi\int_{-\infty}^{\ln 2} e^{2u}u^2\left(e^u\, du\right)$$

$$= \pi\int_{-\infty}^{\ln 2} e^{3u}u^2\, du$$

$$= \lim_{b\to-\infty}\left[\pi\left[\frac{u^2}{3} - \frac{2u}{9} + \frac{2}{27}\right]e^{3u}\right]_b^{\ln 2}$$

$$= 8\pi\left[\frac{(\ln 2)^2}{3} - \frac{2\ln 2}{9} + \frac{2}{27}\right] \approx 2.0155$$

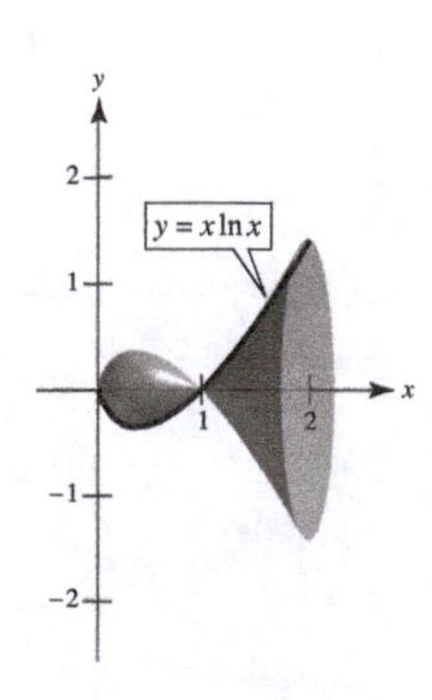

105. $u = \sqrt{x}, u^2 = x, 2u\,du = dx$

$$\int_0^1 \frac{\sin x}{\sqrt{x}}\,dx = \int_0^1 \frac{\sin(u^2)}{u}(2u\,du) = \int_0^1 2\sin(u^2)\,du$$

Trapezoidal Rule $(n = 5)$: 0.6278

107. Assume $a < b$. The proof is similar if $a > b$.

$$\int_{-\infty}^a f(x)\,dx + \int_a^\infty f(x)\,dx = \lim_{c\to-\infty}\int_c^a f(x)\,dx + \lim_{d\to\infty}\int_a^d f(x)\,dx$$

$$= \lim_{c\to-\infty}\int_c^a f(x)\,dx + \lim_{d\to\infty}\left[\int_a^b f(x)\,dx + \int_b^d f(x)\,dx\right]$$

$$= \lim_{c\to-\infty}\int_c^a f(x)\,dx + \int_a^b f(x)\,dx + \lim_{d\to\infty}\int_b^d f(x)\,dx$$

$$= \lim_{c\to-\infty}\left[\int_c^a f(x)\,dx + \int_a^b f(x)\,dx\right] + \lim_{d\to\infty}\int_b^d f(x)\,dx$$

$$= \lim_{c\to-\infty}\int_c^b f(x)\,dx + \lim_{d\to\infty}\int_b^d f(x)\,dx$$

$$= \int_{-\infty}^b f(x)\,dx + \int_b^\infty f(x)\,dx$$

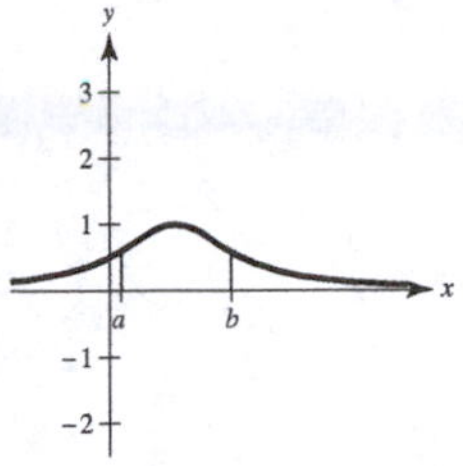

Review Exercises for Chapter 8

1. $$\int x^2\sqrt{x^3 - 27}\,dx = \frac{1}{3}\int (x^3 - 27)^{1/2}\,3x^2\,dx$$
$$= \frac{1}{3}\frac{(x^3 - 27)^{3/2}}{(3/2)} + C$$
$$= \frac{2}{9}(x^3 - 27)^{3/2} + C$$

3. $$\int \csc^2\left(\frac{x+8}{4}\right)dx = 4\int \csc^2\left(\frac{x+8}{4}\right)\frac{1}{4}\,dx$$
$$= -4\cot\left(\frac{x+8}{4}\right) + C$$

5. Let $u = \ln(2x)$, $du = \frac{1}{x}\,dx$.

$$\int_1^e \frac{\ln(2x)}{x}\,dx = \int_{\ln 2}^{1+\ln 2} u\,du$$
$$= \left[\frac{u^2}{2}\right]_{\ln 2}^{1+\ln 2}$$
$$= \frac{1}{2}\left[1 + 2\ln 2 + (\ln 2)^2 - (\ln 2)^2\right]$$
$$= \frac{1}{2} + \ln 2 \approx 1.1931$$

7. $$\int \frac{100}{\sqrt{100 - x^2}}\,dx = 100\arcsin\left(\frac{x}{10}\right) + C$$

9. Let $u = x$, $du = dx$, $dv = e^{1-x}\,dx$, $v = -e^{1-x}$.

$$\int xe^{1-x}\,dx = -xe^{1-x} + \int e^{1-x}\,dx = -xe^{1-x} - e^{1-x} + C$$

11. $$\int e^{2x}\sin 3x\,dx = -\frac{1}{3}e^{2x}\cos 3x + \frac{2}{3}\int e^{2x}\cos 3x\,dx$$
$$= -\frac{1}{3}e^{2x}\cos 3x + \frac{2}{3}\left(\frac{1}{3}e^{2x}\sin 3x - \frac{2}{3}\int e^{2x}\sin 3x\,dx\right)$$
$$\frac{13}{9}\int e^{2x}\sin 3x\,dx = -\frac{1}{3}e^{2x}\cos 3x + \frac{2}{9}e^{2x}\sin 3x$$
$$\int e^{2x}\sin 3x\,dx = \frac{e^{2x}}{13}(2\sin 3x - 3\cos 3x) + C$$

(1) $dv = \sin 3x\,dx \Rightarrow v = -\frac{1}{3}\cos 3x$
$u = e^{2x} \Rightarrow du = 2e^{2x}\,dx$

(2) $dv = \cos 3x\,dx \Rightarrow v = \frac{1}{3}\sin 3x$
$u = e^{2x} \Rightarrow du = 2e^{2x}\,dx$

13. Let $u = x$, $du = dx$, $dv = \sec^2 x\,dx$, $v = \tan x$.

$$\int x \sec^2 x\,dx = x \tan x - \int \tan x\,dx = x \tan x + \ln|\cos x| + C$$

15.
$$\int x \arcsin 2x\,dx = \frac{x^2}{2}\arcsin 2x - \int \frac{x^2}{\sqrt{1-4x^2}}\,dx$$
$$= \frac{x^2}{2}\arcsin 2x - \frac{1}{4}\int \frac{(2x)^2}{\sqrt{1-(2x)^2}}\,dx$$
$$= \frac{x^2}{2}\arcsin 2x - \frac{1}{4}\left(\frac{1}{2}\right)\left[-(2x)\sqrt{1-4x^2} + \arcsin 2x\right] + C \text{ (by Formula 43 of Integration Tables)}$$
$$= \frac{1}{8}\left[\left(4x^2 - 1\right)\arcsin 2x + 2x\sqrt{1-4x^2}\right] + C$$

$$dv = x\,dx \quad \Rightarrow \quad v = \frac{x^2}{2}$$

$$u = \arcsin 2x \quad \Rightarrow \quad du = \frac{2}{\sqrt{1-4x^2}}\,dx$$

17.
$$\int \sin x \cos^4 x\,dx = -\frac{\cos^5 x}{5} + C \qquad (u = \cos x,\ du = -\sin x\,dx)$$

19.
$$\int \cos^3(\pi x - 1)\,dx = \int\left[1 - \sin^2(\pi x - 1)\right]\cos(\pi x - 1)\,dx$$
$$= \frac{1}{\pi}\left[\sin(\pi x - 1) - \frac{1}{3}\sin^3(\pi x - 1)\right] + C$$
$$= \frac{1}{3\pi}\sin(\pi x - 1)\left[3 - \sin^2(\pi x - 1)\right] + C$$
$$= \frac{1}{3\pi}\sin(\pi x - 1)\left[3 - \left(1 - \cos^2(\pi x - 1)\right)\right] + C$$
$$= \frac{1}{3\pi}\sin(\pi x - 1)\left[2 + \cos^2(\pi x - 1)\right] + C$$

21.
$$\int \sec^4\left(\frac{x}{2}\right)dx = \int\left[\tan^2\left(\frac{x}{2}\right) + 1\right]\sec^2\left(\frac{x}{2}\right)dx$$
$$= \int \tan^2\left(\frac{x}{2}\right)\sec^2\left(\frac{x}{2}\right)dx + \int \sec^2\left(\frac{x}{2}\right)dx$$
$$= \frac{2}{3}\tan^3\left(\frac{x}{2}\right) + 2\tan\left(\frac{x}{2}\right) + C = \frac{2}{3}\left[\tan^3\left(\frac{x}{2}\right) + 3\tan\left(\frac{x}{2}\right)\right] + C$$

23. Let $u = x^2$, $du = 2x\,dx$.

$$\int x \tan^4 x^2\,dx = \frac{1}{2}\int \tan^4 u\,du$$
$$= \frac{1}{2}\int \tan^2 u\left(\sec^2 u - 1\right)du$$
$$= \frac{1}{2}\int \tan^2 u \sec^2 u\,du - \frac{1}{2}\int \tan^2 u\,du$$
$$= \frac{1}{2}\int \tan^2 u \sec^2 u\,du - \frac{1}{2}\int\left(\sec^2 u - 1\right)du$$
$$= \frac{1}{2}\cdot\frac{\tan^3 u}{3} - \frac{1}{2}\tan u + \frac{1}{2}u + C$$
$$= \frac{1}{6}\tan^3 x^2 - \frac{1}{2}\tan x^2 + \frac{x^2}{2} + C$$

25. $\displaystyle\int \frac{1}{1-\sin\theta}\,d\theta = \int \frac{1}{1-\sin\theta}\cdot\frac{1+\sin\theta}{1+\sin\theta}\,d\theta = \int \frac{1+\sin\theta}{\cos^2\theta}\,d\theta = \int\left(\sec^2\theta + \sec\theta\tan\theta\right)d\theta = \tan\theta + \sec\theta + C$

27. $$A = \int_{\pi/4}^{3\pi/4} \sin^4 x\,dx = \int\left(\frac{1-\cos 2x}{2}\right)^2 dx = \int\left(\frac{1}{4}\cos^2 2x - \frac{1}{2}\cos 2x + \frac{1}{4}\right)dx$$
$$= \left[\frac{1}{32}\sin 4x - \frac{1}{4}\sin 2x + \frac{3x}{8}\right]_{\pi/4}^{3\pi/4}$$
$$= \left(0 + \frac{1}{4} + \frac{9\pi}{32}\right) - \left(0 - \frac{1}{4} + \frac{3\pi}{32}\right)$$
$$\approx 1.0890$$

29. $x = 2\sin\theta,\ dx = 2\cos\theta\,d\theta,\ \sqrt{4-x^2} = 2\cos\theta$

$$\int \frac{-12}{x^2\sqrt{4-x^2}}\,dx = \int \frac{-24\cos\theta\,d\theta}{(4\sin^2\theta)(2\cos\theta)}$$
$$= -3\int \csc^2\theta\,d\theta$$
$$= 3\cot\theta + C$$
$$= \frac{3\sqrt{4-x^2}}{x} + C$$

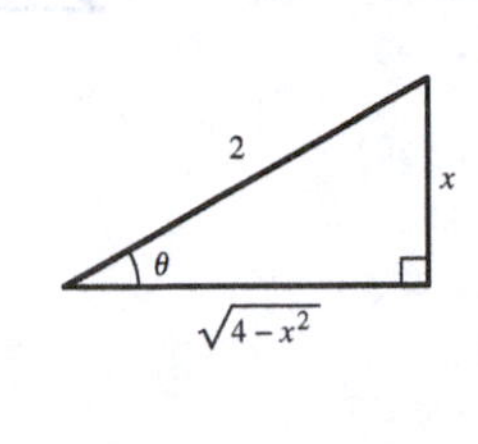

31. $x = 2\tan\theta,\ dx = 2\sec^2\theta\,d\theta,\ 4 + x^2 = 4\sec^2\theta$

$$\int \frac{x^3}{\sqrt{4+x^2}}\,dx = \int \frac{8\tan^3\theta}{2\sec\theta}\,2\sec^2\theta\,d\theta$$
$$= 8\int \tan^3\theta\sec\theta\,d\theta$$
$$= 8\int\left(\sec^2\theta - 1\right)\tan\theta\sec\theta\,d\theta$$
$$= 8\left[\frac{\sec^3\theta}{3} - \sec\theta\right] + C$$
$$= 8\left[\frac{(x^2+4)^{3/2}}{24} - \frac{\sqrt{x^2+4}}{2}\right] + C$$
$$= \sqrt{x^2+4}\left[\frac{1}{3}(x^2+4) - 4\right] + C$$
$$= \frac{1}{3}x^2\sqrt{x^2+4} - \frac{8}{3}\sqrt{x^2+4} + C$$
$$= \frac{1}{3}(x^2+4)^{1/2}(x^2-8) + C$$

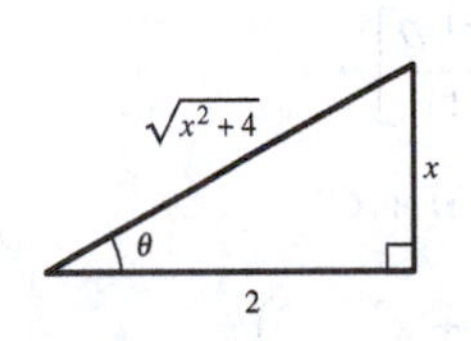

33. $x = 4\tan\theta,\ dx = 4\sec^2\theta\,d\theta,\ \sqrt{16 + x^2} = 4\sec\theta$

$$\begin{aligned}\int \frac{6x^3}{\sqrt{16+x^2}}\,dx &= \int \frac{6(4\tan\theta)^3}{4\sec\theta}\,4\sec^2\theta\,d\theta\\ &= 384\int \tan^3\theta\sec\theta\,d\theta\\ &= 384\int(\sec^2\theta - 1)\sec\theta\tan\theta\,d\theta\\ &= 384\left[\frac{\sec^3\theta}{3} - \sec\theta\right] + C\\ &= \frac{384}{3}\cdot\frac{(16+x^2)^{3/2}}{64} - \frac{384\sqrt{16+x^2}}{4} + C\\ &= 2\sqrt{x^2+16}(16+x^2-48) + C\\ &= 2\sqrt{x^2+16}(x^2-32) + C\end{aligned}$$

$$\begin{aligned}\int_0^1 \frac{6x^3}{\sqrt{16+x^2}}\,dx &= \left[2\sqrt{x^2+16}\,(x^2-32)\right]_0^1\\ &= 2\sqrt{17}(-31) - 2(4)(-32)\\ &= 256 - 62\sqrt{17}\end{aligned}$$

35. (a) Let $x = 2\tan\theta,\ dx = 2\sec^2\theta\,d\theta$.

$$\begin{aligned}\int \frac{x^3}{\sqrt{4+x^2}}\,dx &= \int \frac{8\tan^3\theta}{2\sec\theta}\,2\sec^2\theta\,d\theta\\ &= 8\int \tan^3\theta\sec\theta\,d\theta\\ &= 8\int \frac{\sin^3\theta}{\cos^4\theta}\,d\theta\\ &= 8\int(1-\cos^2\theta)\cos^{-4}\theta\sin\theta\,d\theta\\ &= 8\int(\cos^{-4}\theta - \cos^{-2}\theta)\sin\theta\,d\theta\\ &= 8\left[\frac{\cos^{-3}\theta}{3} - \frac{\cos^{-1}\theta}{-1}\right] + C\\ &= \frac{8}{3}\sec\theta(\sec^2\theta - 3) + C\\ &= \frac{8}{3}\left(\frac{\sqrt{4+x^2}}{2}\right)\left(\frac{4+x^2}{4} - 3\right) + C\\ &= \frac{1}{3}\sqrt{4+x^2}(x^2-8) + C\end{aligned}$$

(b)
$$\begin{aligned}\int \frac{x^3}{\sqrt{4+x^2}}\,dx &= \int \frac{x^2}{\sqrt{4+x^2}}\,x\,dx\\ &= \int \frac{(u^2-4)u\,du}{u}\\ &= \int (u^2-4)\,du\\ &= \frac{1}{3}u^3 - 4u + C\\ &= \frac{u}{3}(u^2-12) + C\\ &= \frac{\sqrt{4+x^2}}{3}(x^2-8) + C\end{aligned}$$

$u^2 = 4 + x^2,\ 2u\,du = 2x\,dx$

(c)
$$\begin{aligned}\int \frac{x^3}{\sqrt{4+x^2}}\,dx &= x^2\sqrt{4+x^2} - \int 2x\sqrt{4+x^2}\,dx\\ &= x^2\sqrt{4+x^2} - \frac{2}{3}(4+x^2)^{3/2} + C = \frac{\sqrt{4+x^2}}{3}(x^2-8) + C\end{aligned}$$

$dv = \dfrac{x}{\sqrt{4+x^2}}\,dx \Rightarrow v = \sqrt{4+x^2}$

$u = x^2 \Rightarrow du = 2x\,dx$

37. $\dfrac{x-8}{x^2-x-6} = \dfrac{x-8}{(x-3)(x+2)} = \dfrac{A}{x-3} + \dfrac{B}{x+2}$

$x - 8 = A(x+2) + B(x-3)$

When $x = -2, -10 = -5B \Rightarrow B = 2$.

When $x = 3, -5 = 5A \Rightarrow A = -1$.

$$\int \frac{x-8}{x^2-x-6}\,dx = \int\left(\frac{-1}{x-3} + \frac{2}{x+2}\right)dx = -\ln|x-3| + 2\ln|x+2| + C = \ln\left|\frac{(x+2)^2}{x-3}\right| + C$$

39. $\dfrac{x^2+2x}{(x-1)(x^2+1)} = \dfrac{A}{x-1} + \dfrac{Bx+C}{x^2+1}$

$x^2 + 2x = A(x^2+1) + (Bx+C)(x-1)$

When $x = 1,\quad 3 = 2A \Rightarrow A = \dfrac{3}{2}$.

When $x = 0,\quad 0 = A - C \Rightarrow C = \dfrac{3}{2}$.

When $x = 2,\quad 8 = 5A + 2B + C \Rightarrow B = -\dfrac{1}{2}$.

$$\begin{aligned}\int \frac{x^2+2x}{x^3-x^2+x-1}\,dx &= \frac{3}{2}\int\frac{1}{x-1}\,dx - \frac{1}{2}\int\frac{x-3}{x^2+1}\,dx\\ &= \frac{3}{2}\int\frac{1}{x-1}\,dx - \frac{1}{4}\int\frac{2x}{x^2+1}\,dx + \frac{3}{2}\int\frac{1}{x^2+1}\,dx\\ &= \frac{3}{2}\ln|x-1| - \frac{1}{4}\ln|x^2+1| + \frac{3}{2}\arctan x + C\\ &= \frac{1}{4}\left[6\ln|x-1| - \ln(x^2+1) + 6\arctan x\right] + C\end{aligned}$$

41. $\dfrac{x^2}{x^2-2x+1} = 1 + \dfrac{2x-1}{x^2-2x+1} = 1 + \dfrac{A}{x-1} + \dfrac{B}{(x-1)^2}$

$2x - 1 = A(x-1) + B$

When $x = 1, B = 1$.

When $x = 0, -1 = -A + B = -A + 1 \Rightarrow A = 2$.

$$\begin{aligned}\int\frac{x^2}{x^2-2x+1}\,dx &= \int\left(1 + \frac{2}{x-1} + \frac{1}{(x-1)^2}\right)dx\\ &= x + 2\ln|x-1| + \frac{1}{1-x} + C\end{aligned}$$

43. Let $u = e^x$, $du = e^x\,dx$.

$$\frac{4}{(e^{2x}-1)(e^x+3)} = \frac{4}{(u^2-1)(u+3)} = \frac{A}{u-1} + \frac{B}{u+1} + \frac{C}{u+3}$$

$$4 = A(u+1)(u+3) + B(u-1)(u+3) + C(u-1)(u+1)$$

When $u = 1$, $4 = 8A \Rightarrow A = \frac{1}{2}$.

When $u = -1$, $4 = -4B \Rightarrow B = -1$.

When $u = -3$, $4 = 8C \Rightarrow C = \frac{1}{2}$.

$$\begin{aligned}\int \frac{4e^x}{(e^{2x}-1)(e^x+3)}\,dx &= \int \frac{4}{(u^2-1)(u+3)}\,du \\ &= \int \left(\frac{1/2}{u-1} - \frac{1}{u+1} + \frac{1/2}{u+3}\right)du \\ &= \frac{1}{2}\ln|u-1| - \ln|u+1| + \frac{1}{2}\ln|u+3| + C \\ &= \frac{1}{2}\ln|e^x-1| - \ln|e^x+1| + \frac{1}{2}\ln|e^x+3| + C\end{aligned}$$

45. Trapezoidal Rule $(n = 4)$: $\int_2^3 \frac{2}{1+x^2}\,dx$

$$\approx \frac{1}{8}\left[\frac{2}{1+2^2} + 2\left(\frac{2}{1+(9/4)^2}\right) + 2\left(\frac{2}{1+(5/2)^2}\right) + 2\left(\frac{2}{1+(11/4)^2}\right) + \frac{2}{1+3^2}\right] \approx 0.2848$$

Simpson's Rule $(n = 4)$: $\int_2^3 \frac{2}{1+x^2}\,dx$

$$\approx \frac{1}{12}\left[\frac{2}{1+2^2} + 4\left(\frac{2}{1+(9/4)^2}\right) + 2\left(\frac{2}{1+(5/2)^2}\right) + 4\left(\frac{2}{1+(11/4)^2}\right) + \frac{2}{1+3^2}\right] \approx 0.2838$$

Graphing utility: 0.2838

47. Trapezoidal Rule $(n = 4)$: $\int_0^{\pi/2} \sqrt{x}\cos x\,dx \approx \frac{\pi}{16}\left[0 + \frac{\sqrt{2\pi}}{2}\sqrt{\frac{\sqrt{2}}{4} + \frac{1}{2}} + \frac{\sqrt{2\pi}}{2} + \frac{\sqrt{6\pi}}{2}\sqrt{-\frac{\sqrt{2}}{4} + \frac{1}{2}} + 0\right] \approx 0.6366$

Simpson's Rule $(n = 4)$: $\int_0^{\pi/2} \sqrt{x}\cos x\,dx \approx \frac{\pi}{24}\left[0 + \sqrt{2\pi}\sqrt{\frac{\sqrt{2}}{4} + \frac{1}{2}} + \frac{\sqrt{2\pi}}{2} + \sqrt{6\pi}\sqrt{-\frac{\sqrt{2}}{4} + \frac{1}{2}} + 0\right] \approx 0.6845$

Graphing utility: 0.7041

49. Using Formula 4: $(a = 4, b = 5)$

$$\int \frac{x}{(4+5x)^2}\,dx = \frac{1}{25}\left(\frac{4}{4+5x} + \ln|4+5x|\right) + C$$

51. Let $u = x^2$, $du = 2x\,dx$.

$$\begin{aligned}\int_0^{\sqrt{\pi}/2} \frac{x}{1+\sin x^2}\,dx &= \frac{1}{2}\int_0^{\pi/4} \frac{1}{1+\sin u}\,du \\ &= \frac{1}{2}\Big[\tan u - \sec u\Big]_0^{\pi/4} \\ &= \frac{1}{2}\Big[\left(1-\sqrt{2}\right) - (0-1)\Big] \\ &= 1 - \frac{\sqrt{2}}{2}\end{aligned}$$

53. $\int \frac{x}{x^2 + 4x + 8}\,dx = \frac{1}{2}\left[\ln\left|x^2 + 4x + 8\right| - 4\int \frac{1}{x^2 + 4x + 8}\,dx\right]$ (Formula 15)

$= \frac{1}{2}\left[\ln\left|x^2 + 4x + 8\right|\right] - 2\left[\frac{2}{\sqrt{32 - 16}}\arctan\left(\frac{2x + 4}{\sqrt{32 - 16}}\right)\right] + C$ (Formula 14)

$= \frac{1}{2}\ln\left|x^2 + 4x + 8\right| - \arctan\left(1 + \frac{x}{2}\right) + C$

55. $\int \frac{1}{\sin \pi x \cos \pi x}\,dx = \frac{1}{\pi}\int \frac{1}{\sin \pi x \cos \pi x}(\pi)\,dx \quad (u = \pi x)$

$= \frac{1}{\pi}\ln\left|\tan \pi x\right| + C$ (Formula 58)

57. $\int \theta \sin\theta \cos\theta\,d\theta = \frac{1}{2}\int \theta \sin 2\theta\,d\theta$

$= -\frac{1}{4}\theta\cos 2\theta + \frac{1}{4}\int \cos 2\theta\,d\theta = -\frac{1}{4}\theta\cos 2\theta + \frac{1}{8}\sin 2\theta + C = \frac{1}{8}(\sin 2\theta - 2\theta\cos 2\theta) + C$

$dv = \sin 2\theta\,d\theta \Rightarrow v = -\frac{1}{2}\cos 2\theta$

$u = \theta \Rightarrow du = d\theta$

59. $\int \frac{x^{1/4}}{1 + x^{1/2}}\,dx = 4\int \frac{u(u^3)}{1 + u^2}\,du$

$= 4\int\left(u^2 - 1 + \frac{1}{u^2 + 1}\right)du$

$= 4\left(\frac{1}{3}u^3 - u + \arctan u\right) + C$

$= \frac{4}{3}\left[x^{3/4} - 3x^{1/4} + 3\arctan(x^{1/4})\right] + C$

$u = \sqrt[4]{x},\ x = u^4,\ dx = 4u^3\,du$

61. $\int \sqrt{1 + \cos x}\,dx = \int \frac{\sqrt{1 + \cos x}}{1}\cdot\frac{\sqrt{1 - \cos x}}{\sqrt{1 - \cos x}}\,dx$

$= \int \frac{\sin x}{\sqrt{1 - \cos x}}\,dx$

$= \int (1 - \cos x)^{-1/2}(\sin x)\,dx$

$= 2\sqrt{1 - \cos x} + C$

$u = 1 - \cos x,\ du = \sin x\,dx$

63. $\int \cos x \ln(\sin x)\,dx = \sin x \ln(\sin x) - \int \cos x\,dx = \sin x \ln(\sin x) - \sin x + C$

$dv = \cos x\,dx \Rightarrow v = \sin x$

$u = \ln(\sin x) \Rightarrow du = \frac{\cos x}{\sin x}\,dx$

65. $y = \int \frac{25}{x^2 - 25}\,dx = 25\left(\frac{1}{10}\right)\ln\left|\frac{x - 5}{x + 5}\right| + C = \frac{5}{2}\ln\left|\frac{x - 5}{x + 5}\right| + C$

(Formula 24)

67. $y = \int \ln(x^2 + x)\,dx = x\ln\left|x^2 + x\right| - \int \frac{2x^2 + x}{x^2 + x}\,dx$

$= x\ln\left|x^2 + x\right| - \int \frac{2x + 1}{x + 1}\,dx$

$= x\ln\left|x^2 + x\right| - \int 2\,dx + \int \frac{1}{x + 1}\,dx$

$= x\ln\left|x^2 + x\right| - 2x + \ln\left|x + 1\right| + C$

$dv = dx \Rightarrow v = x$

$u = \ln(x^2 + x) \Rightarrow du = \frac{2x + 1}{x^2 + x}\,dx$

69. $\int_2^{\sqrt{5}} x(x^2 - 4)^{3/2}\,dx = \left[\frac{1}{5}(x^2 - 4)^{5/2}\right]_2^{\sqrt{5}} = \frac{1}{5}$

71. $\int_1^4 \frac{\ln x}{x}\,dx = \left[\frac{1}{2}(\ln x)^2\right]_1^4 = \frac{1}{2}(\ln 4)^2 \approx 0.961$

73. $\int (x^2 - 4)\sin x\,dx = (x^2 - 4)(-\cos x) - 2x(-\sin x) + 2\cos x + C$

$$\begin{aligned}\int_2^{\pi} (x^2 - 4)\sin x\,dx &= \left[(4 - x^2)\cos x + 2x\sin x + 2\cos x\right]_2^{\pi}\\ &= \left[(4 - \pi^2)(-1) - 2\right] - \left[4\sin 2 + 2\cos 2\right]\\ &= \pi^2 - 4\sin 2 - 2\cos 2 - 6 \approx 1.0647\end{aligned}$$

Alternate signs	u and its derivatives	v' and its antiderivatives
+	$x^2 - 4$	$\sin x$
−	$2x$	$-\cos x$
+	2	$-\sin x$
−	0	$\cos x$

75. $A = \int_0^{3/2} x\sqrt{3 - 2x}\,dx$

Let $3 - 2x = u$, $-2\,dx = du$, $x = \frac{3 - u}{2}$

$$\begin{aligned}A &= \int_3^0 \left(\frac{3 - u}{2}\right)u^{1/2}\left(-\frac{1}{2}\,du\right)\\ &= \frac{1}{4}\int_0^3 \left(3u^{1/2} - u^{3/2}\right)du\\ &= \frac{1}{4}\left[2u^{3/2} - \frac{2}{5}u^{5/2}\right]_0^3\\ &= \frac{1}{4}\left[2\left(3^{3/2}\right) - \frac{2}{5}\left(3^{5/2}\right)\right]\\ &= \frac{1}{4}\left[6\sqrt{3} - \frac{2}{5}9\sqrt{3}\right]\\ &= 3\frac{\sqrt{3}}{5} \approx 1.0392\end{aligned}$$

77. By symmetry, $\bar{x} = 0$, $A = \frac{1}{2}\pi$.

$$\bar{y} = \frac{2}{\pi}\left(\frac{1}{2}\right)\int_{-1}^{1}\left(\sqrt{1 - x^2}\right)^2 dx = \frac{1}{\pi}\left[x - \frac{1}{3}x^3\right]_{-1}^{1} = \frac{4}{3\pi}$$

$(\bar{x}, \bar{y}) = \left(0, \frac{4}{3\pi}\right)$

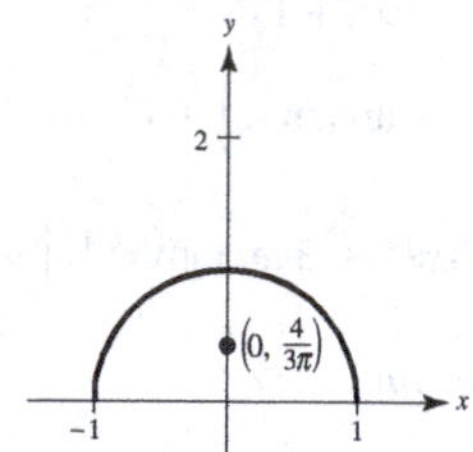

79. $\int_0^{16} \frac{1}{\sqrt[4]{x}}\,dx = \lim_{b\to 0^+}\left[\frac{4}{3}x^{3/4}\right]_b^{16} = \frac{32}{3}$

81. $\int_1^{\infty} x^2 \ln x\,dx = \lim_{b\to\infty}\left[\frac{x^3}{9}(-1 + 3\ln x)\right]_1^b = \infty$

Diverges

83. Let $u = \ln x$, $du = \frac{1}{x}\,dx$, $dv = x^{-2}\,dx$, $v = -x^{-1}$.

$$\int \frac{\ln x}{x^2}\,dx = \frac{-\ln x}{x} + \int \frac{1}{x^2}\,dx = \frac{-\ln x}{x} - \frac{1}{x} + C$$

$$\int_1^{\infty} \frac{\ln x}{x^2}\,dx = \lim_{b\to\infty}\left[\frac{-\ln x}{x} - \frac{1}{x}\right]_1^b = \lim_{b\to\infty}\left(\frac{-\ln b}{b} - \frac{1}{b}\right) - (-1) = 0 + 1 = 1$$

85. $\int_2^\infty \frac{1}{x\sqrt{x^2-4}}\,dx = \int_2^3 \frac{1}{x\sqrt{x^2-4}}\,dx + \int_3^\infty \frac{1}{x\sqrt{x^2-4}}\,dx$

$$= \lim_{b\to 2^+}\left[\frac{1}{2}\operatorname{arcsec}\left(\frac{x}{2}\right)\right]_b^3 + \lim_{c\to\infty}\left[\frac{1}{2}\operatorname{arcsec}\left(\frac{x}{2}\right)\right]_3^c$$

$$= \frac{1}{2}\operatorname{arcsec}\left(\frac{3}{2}\right) - \frac{1}{2}(0) + \frac{1}{2}\left(\frac{\pi}{2}\right) - \frac{1}{2}\operatorname{arcsec}\left(\frac{3}{2}\right)$$

$$= \frac{\pi}{4}$$

87. $\int_0^{t_0} 500{,}000e^{-0.05t}\,dt = \left[\frac{500{,}000}{-0.05}e^{-0.05t}\right]_0^{t_0} = \frac{-500{,}000}{0.05}\left(e^{-0.05t_0} - 1\right) = 10{,}000{,}000\left(1 - e^{-0.05t_0}\right)$

(a) $t_0 = 20$: $6,321,205.59

(b) $t_0 \to \infty$: $10,000,000

89. (a) $P(13 \le x < \infty) = \frac{1}{0.95\sqrt{2\pi}}\int_{13}^\infty e^{-(x-12.9)^2/2(0.95)^2}\,dx \approx 0.4581$

(b) $P(15 \le x < \infty) = \frac{1}{0.95\sqrt{2\pi}}\int_{15}^\infty e^{-(x-12.9)^2/2(0.95)^2}\,dx \approx 0.0135$

Problem Solving for Chapter 8

1. (a) $\int_{-1}^1 (1-x^2)\,dx = \left[x - \frac{x^3}{3}\right]_{-1}^1 = 2\left(1 - \frac{1}{3}\right) = \frac{4}{3}$

$\int_{-1}^1 (1-x^2)^2\,dx = \int_{-1}^1 (1 - 2x^2 + x^4)\,dx = \left[x - \frac{2x^3}{3} + \frac{x^5}{5}\right]_{-1}^1 = 2\left(1 - \frac{2}{3} + \frac{1}{5}\right) = \frac{16}{15}$

(b) Let $x = \sin u$, $dx = \cos u\,du$, $1 - x^2 = 1 - \sin^2 u = \cos^2 u$.

$$\int_{-1}^1 (1-x^2)^n\,dx = \int_{-\pi/2}^{\pi/2} (\cos^2 u)^n \cos u\,du$$

$$= \int_{-\pi/2}^{\pi/2} \cos^{2n+1} u\,du$$

$$= 2\left[\frac{2}{3}\cdot\frac{4}{5}\cdot\frac{6}{7}\cdots\frac{(2n)}{(2n+1)}\right] \qquad \text{(Wallis's Formula)}$$

$$= 2\left[\frac{2^2\cdot 4^2\cdot 6^2\cdots(2n)^2}{2\cdot 3\cdot 4\cdot 5\cdots(2n)(2n+1)}\right]$$

$$= \frac{2(2^{2n})(n!)^2}{(2n+1)!} = \frac{2^{2n+1}(n!)^2}{(2n+1)!}$$

3. (a) $R < I < T < L$

(b) $S(4) = \frac{4-0}{3(4)}\left[f(0) + 4f(1) + 2f(2) + 4f(3) + f(4)\right] \approx \frac{1}{3}\left[4 + 4(2) + 2(1) + 4\left(\frac{1}{2}\right) + \frac{1}{4}\right] \approx 5.417$

5. $u = \tan\frac{x}{2}$, $\cos x = \frac{1-u^2}{1+u^2}$,

$2 + \cos x = 2 + \frac{1-u^2}{1+u^2} = \frac{3+u^2}{1+u^2}$

$dx = \frac{2\,du}{1+u^2}$

$$\int_0^{\pi/2} \frac{1}{2+\cos x}\,dx = \int_0^1 \left(\frac{1+u^2}{3+u^2}\right)\left(\frac{2}{1+u^2}\right)du$$
$$= \int_0^1 \frac{2}{3+u^2}\,du$$
$$= \left[2\frac{1}{\sqrt{3}}\arctan\left(\frac{u}{\sqrt{3}}\right)\right]_0^1$$
$$= \frac{2}{\sqrt{3}}\arctan\left(\frac{1}{\sqrt{3}}\right)$$
$$= \frac{2}{\sqrt{3}}\left(\frac{\pi}{6}\right) = \frac{\pi\sqrt{3}}{9} \approx 0.6046$$

7. Let $u = cx$, $du = c\,dx$.

$$\int_0^b e^{-c^2x^2}\,dx = \int_0^{cb} e^{-u^2}\frac{du}{c} = \frac{1}{c}\int_0^{cb} e^{-u^2}\,du$$

As $b \to \infty$, $cb \to \infty$. So, $\int_0^\infty e^{-c^2x^2}dx = \frac{1}{c}\int_0^\infty e^{-x^2}\,dx$.

$\bar{x} = 0$ by symmetry.

$$\bar{y} = \frac{M_x}{m} = \frac{2\int_0^\infty \frac{\left(e^{-c^2x^2}\right)}{2}\,dx}{2\int_0^\infty e^{-c^2x^2}\,dx}$$
$$= \left(\frac{1}{2}\right)\frac{\int_0^\infty e^{-2c^2x^2}\,dx}{\int_0^\infty e^{-c^2x^2}\,dx}$$
$$= \left(\frac{1}{2}\right)\frac{\frac{1}{\sqrt{2}c}\int_0^\infty e^{-x^2}\,dx}{\frac{1}{c}\int_0^\infty e^{-x^2}\,dx}$$
$$= \frac{1}{2\sqrt{2}} = \frac{\sqrt{2}}{4}$$

So, $(\bar{x}, \bar{y}) = \left(0, \frac{\sqrt{2}}{4}\right)$.

9. (a) Let $y = f^{-1}(x)$, $f(y) = x$, $dx = f'(y)\,dy$.

$$\int f^{-1}(x)\,dx = \int yf'(y)\,dy$$
$$= yf(y) - \int f(y)\,dy \qquad \left[\begin{array}{l} u = y,\ du = dy \\ dv = f'(y)\,dy,\ v = f(y) \end{array}\right]$$
$$= xf^{-1}(x) - \int f(y)\,dy$$

(b) $f^{-1}(x) = \arcsin x = y$, $f(x) = \sin x$

$$\int \arcsin x\,dx = x\arcsin x - \int \sin y\,dy = x\arcsin x + \cos y + C = x\arcsin x + \sqrt{1-x^2} + C$$

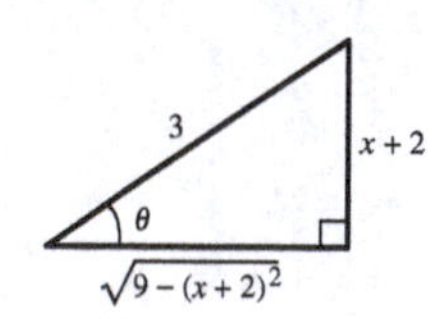

(c) $f(x) = e^x$, $f^{-1}(x) = \ln x = y \quad x = 1 \Leftrightarrow y = 0;\ x = e \Leftrightarrow y = 1$

$$\int_1^e \ln x\,dx = [x\ln x]_1^e - \int_0^1 e^y dy = e - [e^y]_0^1 = e - (e-1) = 1$$

11. $\dfrac{N(x)}{D(x)} = \dfrac{P_1}{x - c_1} + \dfrac{P_2}{x - c_2} + \cdots + \dfrac{P_n}{x - c_n}$

$N(x) = P_1(x - c_2)(x - c_3)\ldots(x - c_n) + P_2(x - c_1)(x - c_3)\ldots(x - c_n) + \cdots + P_n(x - c_1)(x - c_2)\ldots(x - c_{n-1})$

Let $x = c_1$: $N(c_1) = P_1(c_1 - c_2)(c_1 - c_3)\ldots(c_1 - c_n)$

$$P_1 = \frac{N(c_1)}{(c_1 - c_2)(c_1 - c_3)\ldots(c_1 - c_n)}$$

Let $x = c_2$: $N(c_2) = P_2(c_2 - c_1)(c_2 - c_3)\ldots(c_2 - c_n)$

$$P_2 = \frac{N(c_2)}{(c_2 - c_1)(c_2 - c_3)\ldots(c_2 - c_n)}$$

$\vdots$

Let $x = c_n$: $N(c_n) = P_n(c_n - c_1)(c_n - c_2)\ldots(c_n - c_{n-1})$

$$P_n = \frac{N(c_n)}{(c_n - c_1)(c_n - c_2)\ldots(c_n - c_{n-1})}$$

If $D(x) = (x - c_1)(x - c_2)(x - c_3)\ldots(x - c_n)$, then by the Product Rule

$D'(x) = (x - c_2)(x - c_3)\ldots(x - c_n) + (x - c_1)(x - c_3)\ldots(x - c_n) + \cdots + (x - c_1)(x - c_2)(x - c_3)\ldots(x - c_{n-1})$

and

$D'(c_1) = (c_1 - c_2)(c_1 - c_3)\ldots(c_1 - c_n)$

$D'(c_2) = (c_2 - c_1)(c_2 - c_3)\ldots(c_2 - c_n)$

$\vdots$

$D'(c_n) = (c_n - c_1)(c_n - c_2)\ldots(c_n - c_{n-1})$.

So, $P_k = N(c_k)/D'(c_k)$ for $k = 1, 2, \ldots, n$.

13. (a) Let $x = \dfrac{\pi}{2} - u$, $dx = du$.

$$I = \int_0^{\pi/2} \frac{\sin x}{\cos x + \sin x}\,dx = \int_{\pi/2}^{0} \frac{\sin\left(\frac{\pi}{2} - u\right)}{\cos\left(\frac{\pi}{2} - u\right) + \sin\left(\frac{\pi}{2} - u\right)}(-du) = \int_0^{\pi/2} \frac{\cos u}{\sin u + \cos u}\,du$$

So,

$$2I = \int_0^{\pi/2} \frac{\sin x}{\cos x + \sin x}\,dx + \int_0^{\pi/2} \frac{\cos x}{\sin x + \cos x}\,dx = \int_0^{\pi/2} 1\,dx = \frac{\pi}{2} \Rightarrow I = \frac{\pi}{4}.$$

(b) $$I = \int_{\pi/2}^{0} \frac{\sin^n\left(\frac{\pi}{2} - u\right)}{\cos^n\left(\frac{\pi}{2} - u\right) + \sin^n\left(\frac{\pi}{2} - u\right)}(-du) = \int_0^{\pi/2} \frac{\cos^n u}{\sin^n u + \cos^n u}\,du$$

So, $2I = \int_0^{\pi/2} 1\,dx = \dfrac{\pi}{2} \Rightarrow I = \dfrac{\pi}{4}$.

15. $s(t) = \int\left[-32t + 12{,}000 \ln \frac{50{,}000}{50{,}000 - 400t}\right] dt = -16t^2 + 12{,}000\int\left[\ln 50{,}000 - \ln(50{,}000 - 400t)\right] dt$

$$= 16t^2 + 12{,}000t \ln 50{,}000 - 12{,}000\left[t \ln(50{,}000 - 400t) - \int \frac{-400t}{50{,}000 - 400t}\,dt\right]$$

$$= -16t^2 + 12{,}000t \ln \frac{50{,}000}{50{,}000 - 400t} + 12{,}000t\int\left[1 - \frac{50{,}000}{50{,}000 - 400t}\right] dt$$

$$= -16t^2 + 12{,}000t \ln \frac{50{,}000}{50{,}000 - 400t} + 12{,}000t + 1{,}500{,}000 \ln(50{,}000 - 400t) + C$$

$s(0) = 1{,}500{,}000 \ln 50{,}000 + C = 0$

$C = -1{,}500{,}000 \ln 50{,}000$

$$s(t) = -16t^2 + 12{,}000t\left[1 + \ln \frac{50{,}000}{50{,}000 - 400t}\right] + 1{,}500{,}000 \ln \frac{50{,}000 - 400t}{50{,}000}$$

When $t = 100$, $s(100) \approx 557{,}168.626$ feet.

17. Let $u = (x - a)(x - b)$, $du = [(x - a) + (x - b)]\,dx$, $dv = f''(x)\,dx$, $v = f'(x)$.

$$\int_a^b (x - a)(x - b)\,dx = \left[(x - a)(x - b)f'(x)\right]_a^b - \int_a^b [(x - a) + (x - b)]f'(x)\,dx$$

$$= -\int_a^b (2x - a - b)f'(x)\,dx \qquad \begin{pmatrix} u = 2x - a - b \\ dv = f'(x)\,dx \end{pmatrix}$$

$$= \left[-(2x - a - b)f(x)\right]_a^b + \int_a^b 2f(x)\,dx = 2\int_a^b f(x)\,dx$$

19. $\frac{1}{2}V = \int_0^{\arcsin(c)} \pi(c - \sin x)^2\,dx + \int_{\arcsin(c)}^{\pi/2} \pi(\sin x - c)^2\,dx = \frac{2c^2\pi - 8c + \pi}{4}\pi = f(c)$

$f'(c) = \frac{4c\pi - 8}{4}\pi = 0 \Rightarrow c = \frac{2}{\pi}$

For $c = 0$, $\frac{1}{2}V = \frac{\pi^2}{4} \approx 2.4674$.

For $c = 1$, $\frac{1}{2}V = \frac{\pi}{4}(3\pi - 8) \approx 1.1190$.

For $c = \frac{2}{\pi}$, $\frac{1}{2}V = \frac{\pi^2 - 8}{4} \approx 0.4674$.

(a) Maximum: $c = 0$

(b) Minimum: $c = \frac{2}{\pi}$

CHAPTER 9
Infinite Series

CHAPTER 9
Infinite Series

Section 9.1 Sequences

1. A sequence is defined recursively if you are given the first few terms, and the other terms are defined using previous terms.

3. g grows faster. In general, factorials grow faster than exponential functions.

5. $a_n = 3^n$

$a_1 = 3^1 = 3$

$a_2 = 3^2 = 9$

$a_3 = 3^3 = 27$

$a_4 = 3^4 = 81$

$a_5 = 3^5 = 243$

7. $a_n = \sin \dfrac{n\pi}{2}$

$a_1 = \sin \dfrac{\pi}{2} = 1$

$a_2 = \sin \pi = 0$

$a_3 = \sin \dfrac{3\pi}{2} = -1$

$a_4 = \sin 2\pi = 0$

$a_5 = \sin \dfrac{5\pi}{2} = 1$

9. $a_n = (-1)^{n+1}\left(\dfrac{2}{n}\right)$

$a_1 = \dfrac{2}{1} = 2$

$a_2 = -\dfrac{2}{2} = -1$

$a_3 = \dfrac{2}{3}$

$a_4 = -\dfrac{2}{4} = -\dfrac{1}{2}$

$a_5 = \dfrac{2}{5}$

11. $a_1 = 3,\ a_{k+1} = 2(a_k - 1)$

$a_2 = 2(a_1 - 1)$
$= 2(3 - 1) = 4$

$a_3 = 2(a_2 - 1)$
$= 2(4 - 1) = 6$

$a_4 = 2(a_3 - 1)$
$= 2(6 - 1) = 10$

$a_5 = 2(a_4 - 1)$
$= 2(10 - 1) = 18$

13. $a_n = \dfrac{10}{n+1},\ a_1 = \dfrac{10}{1+1} = 5,\ a_2 = \dfrac{10}{3}$

Matches (c).

14. $a_n = \dfrac{10n}{n+1},\ a_1 = \dfrac{10}{2} = 5,\ a_2 = \dfrac{20}{3}$

Matches (a).

15. $a_n = (-1)^n,\ a_1 = -1,\ a_2 = 1,\ a_3 = -1, \ldots$

Matches (d).

16. $a_n = \dfrac{(-1)^n}{n},\ a_1 = \dfrac{-1}{1} = -1,\ a_2 = \dfrac{1}{2}.$

Matches (b).

17. $\dfrac{(n+1)!}{(n-1)!} = \dfrac{(n+1)n(n-1)!}{(n-1)!} = (n+1)n = n^2 + n$

19. $\dfrac{n!}{(n-3)!} = \dfrac{n(n-1)(n-2)(n-3)!}{(n-3)!} = n(n-1)(n-2)$

21. $\lim\limits_{n\to\infty} \dfrac{n+1}{n} = 1$

23. $\lim\limits_{n\to\infty} \dfrac{2n}{\sqrt{n^2+1}} = \lim\limits_{n\to\infty} \dfrac{2}{\sqrt{1 + (1/n^2)}} = \dfrac{2}{1} = 2$

25.

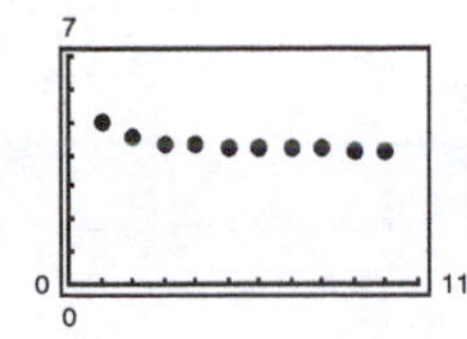

The graph seems to indicate that the sequence converges to 4. Analytically,

$$\lim_{n\to\infty} a_n = \lim_{n\to\infty} \frac{4n+1}{n} = \lim_{x\to\infty} \frac{4x+1}{x} = 4.$$

27.

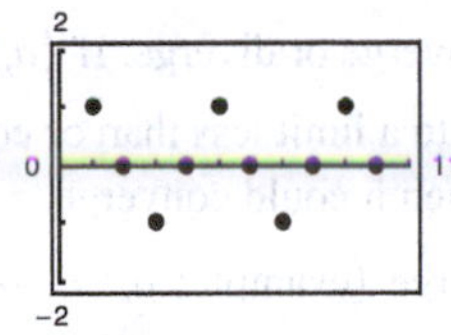

The graph seems to indicate that the sequence diverges. Analytically, the sequence is

$\{a_n\} = \{1, 0, -1, 0, 1, \ldots\}$.

So, $\lim_{n\to\infty} a_n$ does not exist.

29. $\lim_{n\to\infty} \dfrac{5}{n+2} = 0$, converges

31. $\lim_{n\to\infty} (-1)^n\left(\dfrac{n}{n+1}\right)$

does not exist (oscillates between −1 and 1), diverges.

33. $a_n = \dfrac{3n+\sqrt{n}}{4n} = \dfrac{3}{4} + \dfrac{1}{4\sqrt{n}}$

$\lim_{n\to\infty} a_n = \dfrac{3}{4} + 0 = \dfrac{3}{4}$, converges

35. $\lim_{n\to\infty} \dfrac{\ln(n^3)}{2n} = \lim_{n\to\infty} \dfrac{3}{2}\dfrac{\ln(n)}{n}$

$= \lim_{n\to\infty} \dfrac{3}{2}\left(\dfrac{1}{n}\right) = 0$, converges

(L'Hôpital's Rule)

37. $\lim_{n\to\infty} \dfrac{(n+1)!}{n!} = \lim_{n\to\infty} (n+1) = \infty$, diverges

39. $\lim_{n\to\infty} \dfrac{n^p}{e^n} = 0$, converges

$(p > 0, n \geq 2)$

41. $\lim_{n\to\infty} 2^{1/n} = 2^0 = 1$, converges

43. $\lim_{n\to\infty} \dfrac{\sin n}{n} = \lim_{n\to\infty} (\sin n)\dfrac{1}{n} = 0$,

converges (because $\sin n$ is bounded)

45. $a_n = -4 + 6n$

47. $a_n = n^2 - 3$

49. $a_n = \dfrac{n+1}{n+2}$

51. $a_n = 1 + \dfrac{1}{n} = \dfrac{n+1}{n}$

53. $a_n = 4 - \dfrac{1}{n} < 4 - \dfrac{1}{n+1} = a_{n+1}$

Monotonic; $|a_n| < 4$, bounded

55. $a_n = ne^{-n/2}$

$a_1 = 0.6065$

$a_2 = 0.7358$

$a_3 = 0.6694$

Not monotonic; $|a_n| \leq 0.7358$, bounded

57. $a_n = \left(\frac{2}{3}\right)^n > \left(\frac{2}{3}\right)^{n+1} = a_{n+1}$

Monotonic; $|a_n| \leq \frac{2}{3}$, bounded

59. $a_n = \sin\left(\dfrac{n\pi}{6}\right)$

$a_1 = 0.500$

$a_2 = 0.8660$

$a_3 = 1.000$

$a_4 = 0.8660$

Not monotonic; $|a_n| \leq 1$, bounded

61. (a) $a_n = 7 + \dfrac{1}{n}$

$\left|7 + \dfrac{1}{n}\right| \leq 8 \Rightarrow \{a_n\}$, bounded

$a_n = 7 + \dfrac{1}{n} > 7 + \dfrac{1}{n+1} = a_{n+1} \Rightarrow \{a_n\}$, monotonic

Therefore, $\{a_n\}$ converges.

(b)

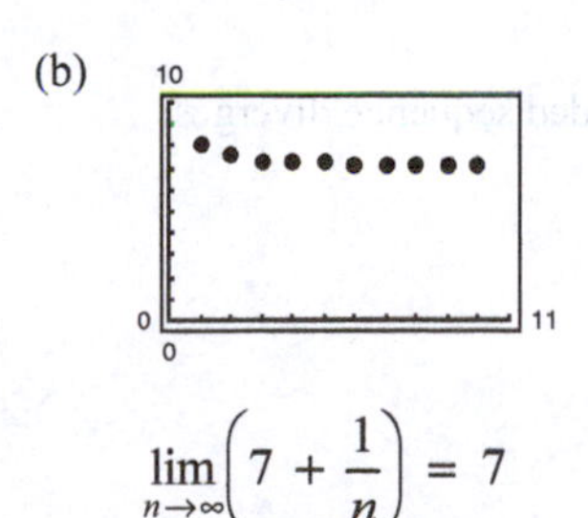

$$\lim_{n\to\infty}\left(7 + \frac{1}{n}\right) = 7$$

63. (a) $a_n = \frac{1}{3}\left(1 - \frac{1}{3^n}\right)$

$\left|\frac{1}{3}\left(1 - \frac{1}{3^n}\right)\right| < \frac{1}{3} \Rightarrow \{a_n\}$, bounded

$a_n = \frac{1}{3}\left(1 - \frac{1}{3^n}\right) < \frac{1}{3}\left(1 - \frac{1}{3^{n+1}}\right)$

$= a_{n+1} \Rightarrow \{a_n\}$, monotonic

Therefore, $\{a_n\}$ converges.

(b)

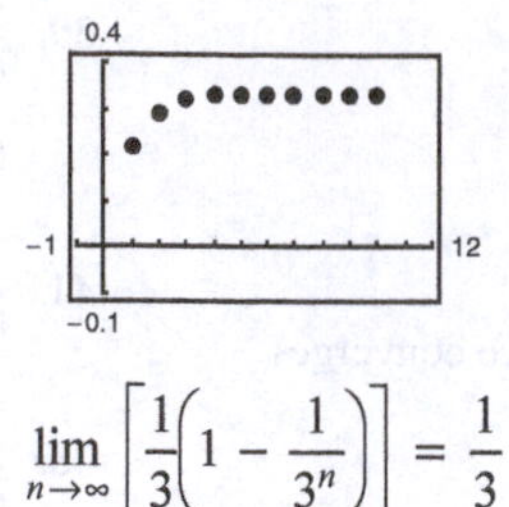

$$\lim_{n\to\infty}\left[\frac{1}{3}\left(1 - \frac{1}{3^n}\right)\right] = \frac{1}{3}$$

65. $A_n = P\left(1 + \frac{r}{12}\right)^n$

(a) Because $P > 0$ and $\left(1 + \frac{r}{12}\right) > 1$, the sequence diverges. $\lim_{n\to\infty} A_n = \infty$

(b) $P = 10{,}000$, $r = 0.055$, $A_n = 10{,}000\left(1 + \frac{0.055}{12}\right)^n$

$A_0 = 10{,}000$
$A_1 = 10{,}045.83$
$A_2 = 10{,}091.88$
$A_3 = 10{,}138.13$
$A_4 = 10{,}184.60$
$A_5 = 10{,}231.28$
$A_6 = 10{,}278.17$
$A_7 = 10{,}325.28$
$A_8 = 10{,}372.60$
$A_9 = 10{,}420.14$
$A_{10} = 10{,}467.90$

67. $P_n = 25{,}000(1.045)^n$

$P_1 = \$26{,}125.00$
$P_2 = \$27{,}300.63$
$P_3 = \$28{,}529.15$
$P_4 = \$29{,}812.97$
$P_5 = \$31{,}154.55$

69. (a) $a_n = 10 - \frac{1}{n}$

(b) Impossible. The sequence converges by Theorem 9.5.

(c) $a_n = \frac{3n}{4n + 1}$

(d) Impossible. An unbounded sequence diverges.

71. The sequence $\{a_n\}$ could converge or diverge. If $\{a_n\}$ is increasing, then it converges to a limit less than or equal to 1. If $\{a_n\}$ is decreasing, then it could converge (example: $a_n = 1/n$) or diverge (example: $a_n = -n$).

73. $a_n = \sqrt[n]{n} = n^{1/n}$

$a_1 = 1^{1/1} = 1$
$a_2 = \sqrt{2} \approx 1.4142$
$a_3 = \sqrt[3]{3} \approx 1.4422$
$a_4 = \sqrt[4]{4} \approx 1.4142$
$a_5 = \sqrt[5]{5} \approx 1.3797$
$a_6 = \sqrt[6]{6} \approx 1.3480$

Let $y = \lim_{n\to\infty} n^{1/n}$.

$$\ln y = \lim_{n\to\infty}\left(\frac{1}{n}\ln n\right) = \lim_{n\to\infty}\frac{\ln n}{n} = \lim_{n\to\infty}\frac{1/n}{1} = 0$$

Because $\ln y = 0$, you have $y = e^0 = 1$. Therefore, $\lim_{n\to\infty}\sqrt[n]{n} = 1$.

75. Because $\lim_{n\to\infty} s_n = L > 0$, there exists for each $\varepsilon > 0$, an integer N such that $|s_n - L| < \varepsilon$ for every $n > N$.

Let $\varepsilon = L > 0$ and you have, $|s_n - L| < L, -L < s_n - L < L$, or $0 < s_n < 2L$ for each $n > N$.

77. False. For example, $a_n = (-1)^n$ diverges, but does not approach ∞ or $-\infty$ (it alternates between two values).

79. True

81. $a_{n+2} = a_n + a_{n+1}$

(a) $a_1 = 1$ | $a_7 = 8 + 5 = 13$
$a_2 = 1$ | $a_8 = 13 + 8 = 21$
$a_3 = 1 + 1 = 2$ | $a_9 = 21 + 13 = 34$
$a_4 = 2 + 1 = 3$ | $a_{10} = 34 + 21 = 55$
$a_5 = 3 + 2 = 5$ | $a_{11} = 55 + 34 = 89$
$a_6 = 5 + 3 = 8$ | $a_{12} = 89 + 55 = 144$

(b) $b_n = \dfrac{a_{n+1}}{a_n}, n \geq 1$

$b_1 = \dfrac{1}{1} = 1$ | $b_6 = \dfrac{13}{8} = 1.625$
$b_2 = \dfrac{2}{1} = 2$ | $b_7 = \dfrac{21}{13} \approx 1.6154$
$b_3 = \dfrac{3}{2} = 1.5$ | $b_8 = \dfrac{34}{21} \approx 1.6190$
$b_4 = \dfrac{5}{3} \approx 1.6667$ | $b_9 = \dfrac{55}{34} \approx 1.6176$
$b_5 = \dfrac{8}{5} = 1.6$ | $b_{10} = \dfrac{89}{55} \approx 1.6182$

(c) $1 + \dfrac{1}{b_{n-1}} = 1 + \dfrac{1}{a_n/a_{n-1}}$

$$= 1 + \frac{a_{n-1}}{a_n} = \frac{a_n + a_{n-1}}{a_n} = \frac{a_{n+1}}{a_n} = b_n$$

(d) If $\lim_{n\to\infty} b_n = \rho$, then $\lim_{n\to\infty}\left(1 + \dfrac{1}{b_{n-1}}\right) = \rho$.

Because $\lim_{n\to\infty} b_n = \lim_{n\to\infty} b_{n-1}$, you have

$$1 + (1/\rho) = \rho.$$

$$\rho + 1 = \rho^2$$
$$0 = \rho^2 - \rho - 1$$
$$\rho = \frac{1 \pm \sqrt{1 + 4}}{2} = \frac{1 \pm \sqrt{5}}{2}$$

Because a_n, and therefore b_n, is positive,

$$\rho = \frac{1 + \sqrt{5}}{2} \approx 1.6180.$$

83. (a) $a_1 = \sqrt{2} \approx 1.4142$

$$a_2 = \sqrt{2 + \sqrt{2}} \approx 1.8478$$
$$a_3 = \sqrt{2 + \sqrt{2 + \sqrt{2}}} \approx 1.9616$$
$$a_4 = \sqrt{2 + \sqrt{2 + \sqrt{2 + \sqrt{2}}}} \approx 1.9904$$
$$a_5 = \sqrt{2 + \sqrt{2 + \sqrt{2 + \sqrt{2 + \sqrt{2}}}}} \approx 1.9976$$

(b) $a_n = \sqrt{2 + a_{n-1}}, \quad n \geq 2, a_1 = \sqrt{2}$

(c) First use mathematical induction to show that $a_n \leq 2$; clearly $a_1 \leq 2$. So assume $a_k \leq 2$. Then

$$a_k + 2 \leq 4$$
$$\sqrt{a_k + 2} \leq 2$$
$$a_{k+1} \leq 2.$$

Now show that $\{a_n\}$ is an increasing sequence. Because $a_n \geq 0$ and $a_n \leq 2$,

$$(a_n - 2)(a_n + 1) \leq 0$$
$$a_n^2 - a_n - 2 \leq 0$$
$$a_n^2 \leq a_n + 2$$
$$a_n \leq \sqrt{a_n + 2}$$
$$a_n \leq a_{n+1}.$$

Because $\{a_n\}$ is a bounding increasing sequence, it converges to some number L, by Theorem 9.5.

$$\lim_{n\to\infty} a_n = L \Rightarrow \sqrt{2 + L} = L \Rightarrow 2 + L = L^2 \Rightarrow L^2 - L - 2 = 0$$
$$\Rightarrow (L - 2)(L + 1) = 0 \Rightarrow L = 2 \quad (L \neq -1)$$

85. (a)

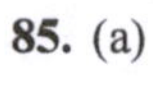

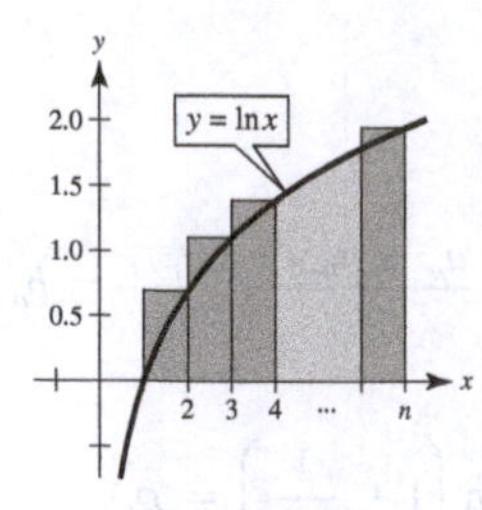

$$\int_1^n \ln x\,dx < \ln 2 + \ln 3 + \cdots + \ln n$$
$$= \ln(1 \cdot 2 \cdot 3 \cdots n) = \ln(n!)$$

(b)

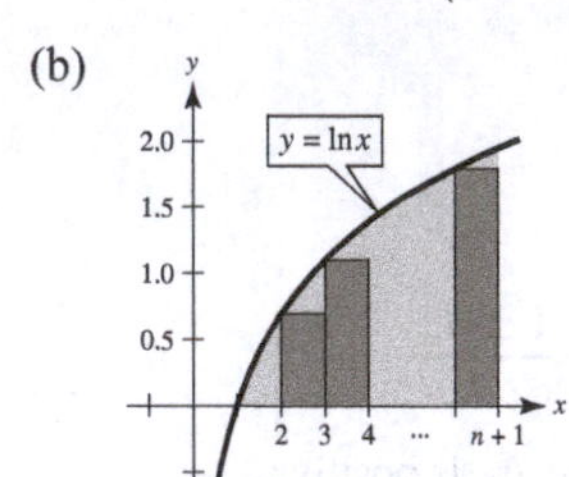

$$\int_1^{n+1} \ln x\,dx > \ln 2 + \ln 3 + \cdots + \ln n = \ln(n!)$$

(c) $\int \ln x\,dx = x \ln x - x + C$

$$\int_1^n \ln x\,dx = n \ln n - n + 1 = \ln n^n - n + 1$$

From part (a): $\ln n^n - n + 1 < \ln(n!)$

$$e^{\ln n^n - n + 1} < n!$$
$$\frac{n^n}{e^{n-1}} < n!$$

$$\int_1^{n+1} \ln x\,dx = (n+1)\ln(n+1) - (n+1) + 1$$
$$= \ln(n+1)^{n+1} - n$$

From part (b): $\ln(n+1)^{n+1} - n > \ln(n!)$

$$e^{\ln(n+1)^{n+1} - n} > n!$$
$$\frac{(n+1)^{n+1}}{e^n} > n!$$

(d) $$\frac{n^n}{e^{n-1}} < n! < \frac{(n+1)^{n+1}}{e^n}$$

$$\frac{n}{e^{1-(1/n)}} < \sqrt[n]{n!} < \frac{(n+1)^{(n+1)/n}}{e}$$

$$\frac{1}{e^{1-(1/n)}} < \frac{\sqrt[n]{n!}}{n} < \frac{(n+1)^{1+(1/n)}}{ne}$$

$$\lim_{n\to\infty} \frac{1}{e^{1-(1/n)}} = \frac{1}{e}$$

$$\lim_{n\to\infty} \frac{(n+1)^{1+(1/n)}}{ne} = \lim_{n\to\infty} \frac{(n+1)}{n}\frac{(n+1)^{1/n}}{e} = (1)\frac{1}{e} = \frac{1}{e}$$

By the Squeeze Theorem, $\lim_{n\to\infty} \frac{\sqrt[n]{n!}}{n} = \frac{1}{e}$.

(e) $n = 20$: $\frac{\sqrt[20]{20!}}{20} \approx 0.4152$

$n = 50$: $\frac{\sqrt[50]{50!}}{50} \approx 0.3897$

$n = 100$: $\frac{\sqrt[100]{100!}}{100} \approx 0.3799$

$\frac{1}{e} \approx 0.3679$

87. For a given $\varepsilon > 0$, you must find $M > 0$ such that $|a_n - L| = |r^n| \varepsilon$ whenever $n > M$. That is, $n \ln|r| < \ln(\varepsilon)$ or

$$n > \frac{\ln(\varepsilon)}{\ln|r|} \text{ (because } \ln|r| < 0 \text{ for } |r| < 1).$$

So, let $\varepsilon > 0$ be given. Let M be an integer satisfying

$$M > \frac{\ln(\varepsilon)}{\ln|r|}.$$

For $n > M$, you have

$$n > \frac{\ln(\varepsilon)}{\ln|r|}$$
$$n \ln|r| < \ln(\varepsilon)$$
$$\ln|r|^n < \ln(\varepsilon)$$
$$|r|^n < \varepsilon$$
$$|r^n - 0| < \varepsilon.$$

So, $\lim_{n\to\infty} r^n = 0$ for $-1 < r < 1$.

89. If $\{a_n\}$ is bounded, monotonic and nonincreasing, then $a_1 \ge a_2 \ge a_3 \ge \cdots \ge a_n \ge \cdots$. Then $-a_1 \le -a_2 \le -a_3 \le \cdots \le -a_n \le \cdots$ is a bounded, monotonic, nondecreasing sequence which converges by the first half of the theorem. Because $\{-a_n\}$ converges, then so does $\{a_n\}$.

91. $T_n = n! + 2^n$

Use mathematical induction to verify the formula.

$T_0 = 1 + 1 = 2$

$T_1 = 1 + 2 = 3$

$T_2 = 2 + 4 = 6$

Assume $T_k = k! + 2^k$. Then

$$\begin{aligned} T_{k+1} &= (k + 1 + 4)T_k - 4(k + 1)T_{k-1} + \big(4(k + 1) - 8\big)T_{k-2} \\ &= (k + 5)\left[k! + 2^k\right] - 4(k + 1)\big((k - 1)! + 2^{k-1}\big) + (4k - 4)\big((k - 2)! + 2^{k-2}\big) \\ &= \left[(k + 5)(k)(k - 1) - 4(k + 1)(k - 1) + 4(k - 1)\right](k - 2)! + \left[(k + 5)4 - 8(k + 1) + 4(k - 1)\right]2^{k-2} \\ &= \left[k^2 + 5k - 4k - 4 + 4\right](k - 1)! + 8 \cdot 2^{k-2} \\ &= (k + 1)! + 2^{k+1}. \end{aligned}$$

By mathematical induction, the formula is valid for all n.

Section 9.2 Series and Convergence

1. $\lim_{n\to\infty} a_n = 5$ means that the limit of the sequence $\{a_n\}$ is 5.

$\sum_{n=1}^{\infty} a_n = a_1 + a_2 + \cdots = 5$ means that the limit of the partial sums is 5.

3. You cannot make a conclusion. The series may either converge or diverge.

5. $S_1 = 1$

$S_2 = 1 + \frac{1}{4} = 1.2500$

$S_3 = 1 + \frac{1}{4} + \frac{1}{9} \approx 1.3611$

$S_4 = 1 + \frac{1}{4} + \frac{1}{9} + \frac{1}{16} \approx 1.4236$

$S_5 = 1 + \frac{1}{4} + \frac{1}{9} + \frac{1}{16} + \frac{1}{25} \approx 1.4636$

7. $S_1 = 3$

$S_2 = 3 - \frac{9}{2} = -1.5$

$S_3 = 3 - \frac{9}{2} + \frac{27}{4} = 5.25$

$S_4 = 3 - \frac{9}{2} + \frac{27}{4} - \frac{81}{8} = -4.875$

$S_5 = 3 - \frac{9}{2} + \frac{27}{4} - \frac{81}{8} + \frac{243}{16} = 10.3125$

9. $S_1 = 3$

$S_2 = 3 + \frac{3}{2} = 4.5$

$S_3 = 3 + \frac{3}{2} + \frac{3}{4} = 5.250$

$S_4 = 3 + \frac{3}{2} + \frac{3}{4} + \frac{3}{8} = 5.625$

$S_5 = 3 + \frac{3}{2} + \frac{3}{4} + \frac{3}{8} + \frac{3}{16} = 5.8125$

11. $\sum_{n=0}^{\infty} 5\left(\frac{5}{2}\right)^n$

Geometric series

$r = \frac{5}{2} > 1$

Diverges by Theorem 9.6

13. $\sum_{n=1}^{\infty} \frac{n}{n + 1}$

$\lim_{n\to\infty} \frac{n}{n + 1} = 1 \neq 0$

Diverges by Theorem 9.9

15. $\sum_{n=1}^{\infty} \frac{n^3 + 1}{n^3 + n^2}$

$\lim_{n\to\infty} \frac{n^3 + 1}{n^3 + n^2} = 1 \neq 0$

Diverges by Theorem 9.9

17. $\sum_{n=1}^{\infty} \frac{4^n + 3}{4^{n+1}}$

$\lim_{n\to\infty} \frac{4^n + 3}{4^{n+1}} = \frac{1}{4} \neq 0$

Diverges by Theorem 9.9

19. $\sum_{n=0}^{\infty} \left(\frac{5}{6}\right)^n$

Geometric series with $r = \frac{5}{6} < 1$

Converges by Theorem 9.6

21. $\sum_{n=0}^{\infty} (0.9)^n$

Geometric series with $r = 0.9 < 1$

Converges by Theorem 9.6

23. $\sum_{n=1}^{\infty} \frac{1}{n(n+1)} = \sum_{n=1}^{\infty} \left(\frac{1}{n} - \frac{1}{n+1}\right) = \left(1 - \frac{1}{2}\right) + \left(\frac{1}{2} - \frac{1}{3}\right) + \left(\frac{1}{3} - \frac{1}{4}\right) + \left(\frac{1}{4} - \frac{1}{5}\right) + \cdots, \quad S_n = 1 - \frac{1}{n+1}$

$$\sum_{n=1}^{\infty} \frac{1}{n(n+1)} = \lim_{n\to\infty} S_n = \lim_{n\to\infty}\left(1 - \frac{1}{n+1}\right) = 1$$

25. (a) $\sum_{n=1}^{\infty} \frac{6}{n(n+3)} = 2\sum_{n=1}^{\infty} \left(\frac{1}{n} - \frac{1}{n+3}\right) = 2\left[\left(1 - \frac{1}{4}\right) + \left(\frac{1}{2} - \frac{1}{5}\right) + \left(\frac{1}{3} - \frac{1}{6}\right) + \left(\frac{1}{4} - \frac{1}{7}\right) + \cdots\right]$

$$\left(S_n = 2\left[1 + \frac{1}{2} + \frac{1}{3} - \left(\frac{1}{n+1} + \frac{1}{n+2} + \frac{1}{n+3}\right)\right]\right) = 2\left(1 + \frac{1}{2} + \frac{1}{3}\right) = \frac{11}{3} \approx 3.667$$

(b)

n	5	10	20	50	100
S_n	2.7976	3.1643	3.3936	3.5513	3.6078

(c) 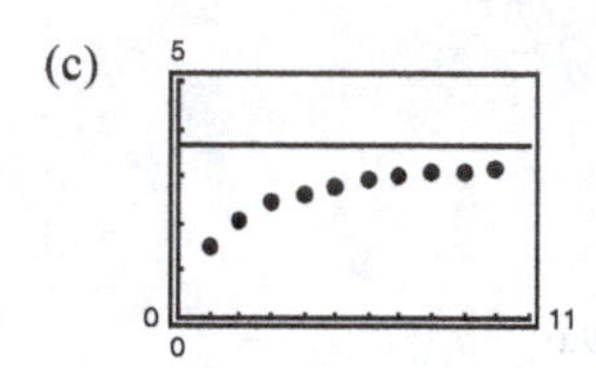

(d) The terms of the series decrease in magnitude slowly. So, the sequence of partial sums approaches the sum slowly.

27. (a) $\sum_{n=1}^{\infty} 2(0.9)^{n-1} = \sum_{n=0}^{\infty} 2(0.9)^n = \frac{2}{1 - 0.9} = 20$

(b)

n	5	10	20	50	100
S_n	8.1902	13.0264	17.5685	19.8969	19.9995

(c) 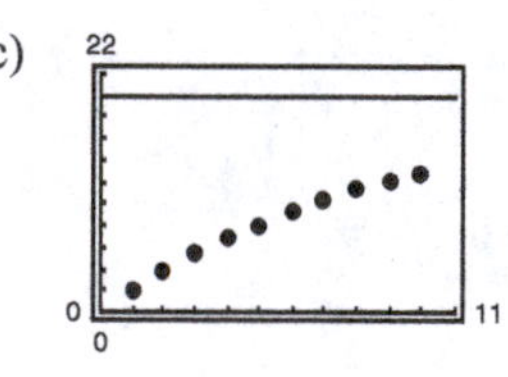

(d) The terms of the series decrease in magnitude slowly. So, the sequence of partial sums approaches the sum slowly.

29. $\sum_{n=0}^{\infty} 5\left(\frac{2}{3}\right)^n = \frac{5}{1 - (2/3)} = 15$

31. $\sum_{n=1}^{\infty} \frac{4}{n(n+2)} = 2\sum_{n=1}^{\infty} \left(\frac{1}{n} - \frac{1}{n+2}\right)$

$$S_n = 2\left[\left(1 - \frac{1}{3}\right) + \left(\frac{1}{2} - \frac{1}{4}\right) + \left(\frac{1}{3} - \frac{1}{5}\right) + \cdots + \left(\frac{1}{n-1} - \frac{1}{n+1}\right) + \left(\frac{1}{n} - \frac{1}{n+2}\right)\right] = 2\left(1 + \frac{1}{2} - \frac{1}{n+1} - \frac{1}{n+2}\right)$$

$$\sum_{n=1}^{\infty} \frac{4}{n(n+2)} = \lim_{n\to\infty} S_n = \lim_{n\to\infty} 2\left(1 + \frac{1}{2} - \frac{1}{n+1} - \frac{1}{n+2}\right) = 3$$

33. $\sum_{n=0}^{\infty} 8\left(\frac{3}{4}\right)^n = \frac{8}{1-(3/4)} = 32$

35. $\sum_{n=0}^{\infty}\left(\frac{1}{2^n} - \frac{1}{3^n}\right) = \sum_{n=0}^{\infty}\left(\frac{1}{2}\right)^n - \sum_{n=0}^{\infty}\left(\frac{1}{3}\right)^n$

$= \frac{1}{1-(1/2)} - \frac{1}{1-(1/3)} = 2 - \frac{3}{2} = \frac{1}{2}$

37. Note that $\sin(1) \approx 0.8415 < 1$. The series $\sum_{n=1}^{\infty}\left[\sin(1)\right]^n$ is geometric with $r = \sin(1) < 1$. So,

$$\sum_{n=1}^{\infty}\left[\sin(1)\right]^n = \sin(1)\sum_{n=0}^{\infty}\left[\sin(1)\right]^n = \frac{\sin(1)}{1-\sin(1)} \approx 5.3080.$$

39. (a) $0.\overline{4} = \sum_{n=0}^{\infty}\frac{4}{10}\left(\frac{1}{10}\right)^n$

(b) Geometric series with $a = \frac{4}{10}$ and $r = \frac{1}{10}$

$S = \frac{a}{1-r} = \frac{4/10}{1-(1/10)} = \frac{4}{9}$

41. (a) $0.\overline{12} = \sum_{n=0}^{\infty}\frac{12}{100}\left(\frac{1}{100}\right)^n$

(b) Geometric series with $a = \frac{12}{100}$ and $r = \frac{1}{100}$

$S = \frac{a}{1-r} = \frac{12/100}{1-1/100} = \frac{12}{99} = \frac{4}{33}$

43. (a) $0.0\overline{75} = \sum_{n=0}^{\infty}\frac{3}{40}\left(\frac{1}{100}\right)^n$

(b) Geometric series with $a = \frac{3}{40}$ and $r = \frac{1}{100}$

$S = \frac{a}{1-r} = \frac{3/40}{99/100} = \frac{5}{66}$

45. $\sum_{n=0}^{\infty}(1.075)^n$

Geometric series with $r = 1.075$

Diverges by Theorem 9.6

47. $\sum_{n=1}^{\infty}\frac{n+1}{2n-1}$

$\lim_{n\to\infty}\frac{n+1}{2n-1} = \frac{1}{2} \neq 0$

Diverges by Theorem 9.9

49. $\sum_{n=1}^{\infty}\left(\frac{1}{n} - \frac{1}{n+2}\right)$

$$S_n = \left(1-\frac{1}{3}\right) + \left(\frac{1}{2}-\frac{1}{4}\right) + \left(\frac{1}{3}-\frac{1}{5}\right) + \cdots + \left(\frac{1}{n-1}-\frac{1}{n+1}\right) + \left(\frac{1}{n}-\frac{1}{n+2}\right) = 1 + \frac{1}{2} - \frac{1}{n+1} - \frac{1}{n+2}$$

$$\sum_{n=1}^{\infty}\left(\frac{1}{n}-\frac{1}{n+2}\right) = \lim_{n\to\infty} S_n = \lim_{n\to\infty}\left(1+\frac{1}{2}-\frac{1}{n+1}-\frac{1}{n+2}\right) = \frac{3}{2}, \text{ converges}$$

51. $\sum_{n=1}^{\infty}\frac{3^n}{n^3}$

$\lim_{n\to\infty}\frac{3^n}{n^3} = \lim_{n\to\infty}\frac{(\ln 3)3^n}{3n^2}$

$= \lim_{n\to\infty}\frac{(\ln 3)^2 3^n}{6n} = \lim_{n\to\infty}\frac{(\ln 3)^3 3^n}{6} = \infty$

(by L'Hôpital's Rule); diverges by Theorem 9.9

53. Because $n > \ln(n)$, the terms $a_n = \frac{n}{\ln(n)}$ do not approach 0 as $n \to \infty$. So, the series $\sum_{n=2}^{\infty}\frac{n}{\ln(n)}$ diverges.

55. For $k \neq 0$,

$\lim_{n\to\infty}\left(1+\frac{k}{n}\right)^n = \lim_{n\to\infty}\left[\left(1+\frac{k}{n}\right)^{n/k}\right]^k = e^k \neq 0.$

For $k = 0$, $\lim_{n\to\infty}(1+0)^n = 1 \neq 0$.

So, $\sum_{n=1}^{\infty}\left[1+\frac{k}{n}\right]^n$ diverges.

57. $\lim_{n\to\infty}\arctan n = \frac{\pi}{2} \neq 0$

So, $\sum_{n=1}^{\infty}\arctan n$ diverges.

59. Yes, the new series will still diverge.

61. $\sum_{n=1}^{\infty}(3x)^n = (3x)\sum_{n=0}^{\infty}(3x)^n$

Geometric series: converges for $|3x| < 1 \Rightarrow |x| < \frac{1}{3}$

$f(x) = (3x)\sum_{n=0}^{\infty}(3x)^n = (3x)\frac{1}{1-3x} = \frac{3x}{1-3x},\ |x| < \frac{1}{3}$

63. $\sum_{n=1}^{\infty}(x-1)^n = (x-1)\sum_{n=0}^{\infty}(x-1)^n$

Geometric series: converges for $|x-1| < 1 \Rightarrow 0 < x < 2$

$f(x) = (x-1)\sum_{n=0}^{\infty}(x-1)^n = (x-1)\frac{1}{1-(x-1)} = \frac{x-1}{2-x}, \quad 0 < x < 2$

65. $\sum_{n=0}^{\infty}(-1)^n x^n = \sum_{n=0}^{\infty}(-x)^n$

Geometric series: converges for

$|-x| < 1 \Rightarrow |x| < 1 \Rightarrow -1 < x < 1$

$f(x) = \sum_{n=0}^{\infty}(-x)^n = \frac{1}{1+x}, \quad -1 < x < 1$

67. (a) x is the common ratio.

(b) $1 + x + x^2 + \cdots = \sum_{n=0}^{\infty} x^n = \frac{1}{1-x}, \quad |x| < 1$

(c) $y_1 = \frac{1}{1-x}$

$y_2 = S_3 = 1 + x + x^2$

$y_3 = S_5 = 1 + x + x^2 + x^3 + x^4$

Answers will vary.

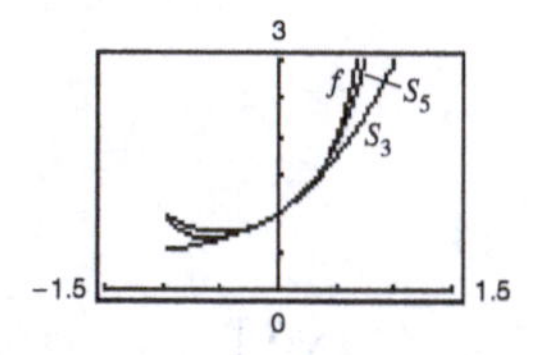

69. $\frac{1}{n(n+1)} < 0.0001$

$10{,}000 < n^2 + n$

$0 < n^2 + n - 10{,}000$

$n = \frac{-1 \pm \sqrt{1^2 - 4(1)(-10{,}000)}}{2}$

Choosing the positive value for n you have $n \approx 99.5012$. The first *term* that is less than 0.0001 is $n = 100$.

$\left(\frac{1}{8}\right)^n < 0.0001$

$10{,}000 < 8^n$

This inequality is true when $n = 5$. This series converges at a faster rate.

71. $\sum_{i=0}^{n-1} 8000(0.95)^i = \frac{8000\left[1 - 0.95^n\right]}{1 - 0.95}$

$= 160{,}000\left[1 - 0.95^n\right], \quad n > 0$

73. $\sum_{i=0}^{\infty} 200(0.75)^i = 800$ million dollars

75. $D_1 = 16$

$D_2 = \underbrace{0.81(16)}_{\text{up}} + \underbrace{0.81(16)}_{\text{down}} = 32(0.81)$

$D_3 = 16(0.81)^2 + 16(0.81)^2 = 32(0.81)^2$

$\vdots$

$D = 16 + 32(0.81) + 32(0.81)^2 + \cdots$

$= -16 + \sum_{n=0}^{\infty} 32(0.81)^n = -16 + \frac{32}{1 - 0.81}$

≈ 152.42 feet

77. $P(n) = \frac{1}{2}\left(\frac{1}{2}\right)^n$

$P(2) = \frac{1}{2}\left(\frac{1}{2}\right)^2 = \frac{1}{8}$

$\sum_{n=0}^{\infty} \frac{1}{2}\left(\frac{1}{2}\right)^n = \frac{1/2}{1 - (1/2)} = 1$

79. (a) $\sum_{n=1}^{\infty}\left(\frac{1}{2}\right)^n = \sum_{n=0}^{\infty}\frac{1}{2}\left(\frac{1}{2}\right)^n = \frac{1}{2}\frac{1}{(1-(1/2))} = 1$

(b) No, the series is not geometric.

(c) $\sum_{n=1}^{\infty} n\left(\frac{1}{2}\right)^n = 2$

81. (a) $64 + 32 + 16 + 8 + 4 + 2 = 126 \text{ in.}^2$

(b) $\displaystyle\sum_{n=0}^{\infty} 64\left(\frac{1}{2}\right)^n = \frac{64}{1-(1/2)} = 128 \text{ in.}^2$

Note: This is one-half of the area of the original square

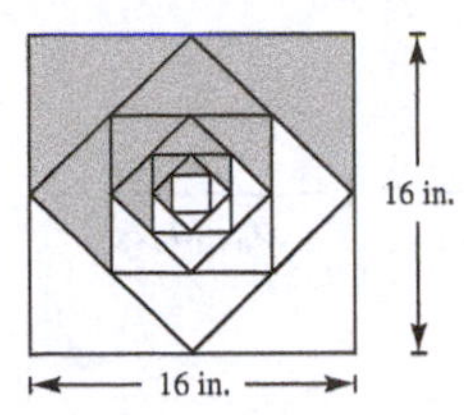

83. $\displaystyle\sum_{n=1}^{20} 100{,}000\left(\frac{1}{1.06}\right)^n = \frac{100{,}000}{1.06}\sum_{i=0}^{19}\left(\frac{1}{1.06}\right)^i = \frac{100{,}000}{1.06}\left[\frac{1-1.06^{-20}}{1-1.06^{-1}}\right] \left(n = 20, r = 1.06^{-1}\right) \approx \$1{,}146{,}992.12$

The \$2,000,000 sweepstakes has a present value of \$1,146,992.12. After accruing interest over the 20-year period, it attains its full value.

85. $\displaystyle w = \sum_{i=0}^{n-1} 0.01(2)^i = \frac{0.01(1-2^n)}{1-2} = 0.01(2^n - 1)$

(a) When $n = 29$: $w = \$5{,}368{,}709.11$

(b) When $n = 30$: $w = \$10{,}737{,}418.23$

(c) When $n = 31$: $w = \$21{,}474{,}836.47$

87. $P = 50, r = 2\% = 0.02, t = 20$

(a) $\displaystyle A = 50\left(\frac{12}{0.02}\right)\left[\left(1 + \frac{0.02}{12}\right)^{12(20)} - 1\right] \approx \$14{,}739.84$

(b) $\displaystyle A = \frac{50\left(e^{0.02(20)} - 1\right)}{e^{0.02/12} - 1} \approx \$14{,}742.45$

89. $P = 1050, r = 0.9\% = 0.009, t = 35$

(a) $\displaystyle A = 1050\left(\frac{12}{0.009}\right)\left[\left(1 + \frac{0.009}{12}\right)^{12(35)} - 1\right]$

$\approx \$518{,}136.56$

(b) $\displaystyle A = \frac{1050\left[e^{0.009(35)} - 1\right]}{e^{0.009/12} - 1} \approx \$518{,}168.67$

91. False. $\displaystyle\lim_{n\to\infty} \frac{1}{n} = 0$, but $\displaystyle\sum_{n=1}^{\infty} \frac{1}{n}$ diverges.

93. False; $\displaystyle\sum_{n=1}^{\infty} ar^n = \left(\frac{a}{1-r}\right) - a$

The formula requires that the geometric series begins with $n = 0$.

95. True

$$\begin{aligned}
0.74999\ldots &= 0.74 + \frac{9}{10^3} + \frac{9}{10^4} + \cdots \\
&= 0.74 + \frac{9}{10^3}\sum_{n=0}^{\infty}\left(\frac{1}{10}\right)^n \\
&= 0.74 + \frac{9}{10^3} \cdot \frac{1}{1-(1/10)} \\
&= 0.74 + \frac{9}{10^3} \cdot \frac{10}{9} \\
&= 0.74 + \frac{1}{100} = 0.75
\end{aligned}$$

97. Let $\displaystyle\sum a_n = \sum_{n=0}^{\infty} 1$ and $\displaystyle\sum b_n = \sum_{n=0}^{\infty} (-1)$.

Both are divergent series.

$$\sum(a_n + b_n) = \sum_{n=0}^{\infty}[1 + (-1)] = \sum_{n=0}^{\infty}[1 - 1] = 0$$

99. (a) $\dfrac{1}{a_{n+1}a_{n+2}} - \dfrac{1}{a_{n+2}a_{n+3}} = \dfrac{a_{n+3} - a_{n+1}}{a_{n+1}a_{n+2}a_{n+3}} = \dfrac{a_{n+2}}{a_{n+1}a_{n+2}a_{n+3}} = \dfrac{1}{a_{n+1}a_{n+3}}$

(b) $$S_n = \sum_{k=0}^{n} \frac{1}{a_{k+1}a_{k+3}}$$
$$= \sum_{k=0}^{n} \left[\frac{1}{a_{k+1}a_{k+2}} - \frac{1}{a_{k+2}a_{k+3}}\right]$$
$$= \left[\frac{1}{a_1a_2} - \frac{1}{a_2a_3}\right] + \left[\frac{1}{a_2a_3} - \frac{1}{a_3a_4}\right] + \cdots + \left[\frac{1}{a_{n+1}a_{n+2}} - \frac{1}{a_{n+2}a_{n+3}}\right] = \frac{1}{a_1a_2} - \frac{1}{a_{n+2}a_{n+3}} = 1 - \frac{1}{a_{n+2}a_{n+3}}$$
$$\sum_{n=0}^{\infty} \frac{1}{a_{n+1}a_{n+3}} = \lim_{n\to\infty} S_n = \lim_{n\to\infty}\left[1 - \frac{1}{a_{n+2}a_{n+3}}\right] = 1$$

101. $$\frac{1}{r} + \frac{1}{r^2} + \frac{1}{r^3} + \cdots = \sum_{n=0}^{\infty} \frac{1}{r}\left(\frac{1}{r}\right)^n = \frac{1/r}{1 - (1/r)} = \frac{1}{r-1} \quad \left(\text{since}\left|\frac{1}{r}\right| < 1\right)$$

This is a geometric series which converges if

$\left|\dfrac{1}{r}\right| < 1 \Leftrightarrow |r| > 1.$

103. The series is telescoping:

$$S_n = \sum_{k=1}^{n} \frac{6^k}{\left(3^{k+1} - 2^{k+1}\right)\left(3^k - 2^k\right)} = \sum_{k=1}^{n}\left[\frac{3^k}{3^k - 2^k} - \frac{3^{k+1}}{3^{k+1} - 2^{k+1}}\right] = 3 - \frac{3^{n+1}}{3^{n+1} - 2^{n+1}}$$

$$\lim_{n\to\infty} S_n = 3 - 1 = 2$$

Section 9.3 The Integral Test and *p*-Series

1. f must be positive, continuous, and decreasing for $x \ge 1$, and $a_n = f(n)$.

3. $\displaystyle\sum_{n=1}^{\infty} \frac{1}{n+3}$

Let

$f(x) = \dfrac{1}{x+3},\ f'(x) = -\dfrac{1}{(x+3)^2} < 0$ for $x \ge 1$.

f is positive, continuous, and decreasing for $x \ge 1$.

$$\int_1^{\infty} \frac{1}{x+3}\,dx = \left[\ln(x+3)\right]_1^{\infty} = \infty$$

So, the series diverges by Theorem 9.10.

5. $\displaystyle\sum_{n=1}^{\infty} \frac{1}{2^n}$

Let $f(x) = \dfrac{1}{2^x},\ f'(x) = -(\ln 2)2^{-x} < 0$ for $x \ge 1$.

f is positive, continuous, and decreasing for $x \ge 1$.

$$\int_1^{\infty} \frac{1}{2^x}\,dx = \left[\frac{-1}{(\ln 2)\,2^x}\right]_1^{\infty} = \frac{1}{2\ln 2}$$

So, the series converges by Theorem 9.10.

7. $\displaystyle\sum_{n=1}^{\infty} e^{-n}$

Let $f(x) = e^{-x},\ f'(x) = -e^{-x} < 0$ for $x \ge 1$.

f is positive, continuous, and decreasing for $x \ge 1$.

$$\int_1^{\infty} e^{-x}\,dx = \left[-e^{-x}\right]_1^{\infty} = \frac{1}{e}$$

So, the series converges by Theorem 9.10.

9. $\displaystyle\sum_{n=1}^{\infty} \frac{\ln(n+1)}{n+1}$

Let

$f(x) = \dfrac{\ln(x+1)}{x+1},\ f'(x) = \dfrac{1 - \ln(x+1)}{(x+1)^2} < 0$ for $x \ge 2$.

f is positive, continuous, and decreasing for $x \ge 2$.

$$\int_1^{\infty} \frac{\ln(x+1)}{x+1}\,dx = \left[\frac{\left[\ln(x+1)\right]^2}{2}\right]_1^{\infty} = \infty$$

So, the series diverges by Theorem 9.10.

11. $\sum_{n=1}^{\infty} \frac{1}{2n+1}$

Let

$$f(x) = \frac{1}{2x+1},\ f'(x) = -\frac{2}{(2x+1)^2} < 0 \text{ for } x \geq 1.$$

f is positive, continuous, and decreasing for $x \geq 1$.

$$\int_1^{\infty} \frac{1}{2x+1}\,dx = \left[\ln\sqrt{2x+1}\right]_1^{\infty} = \infty$$

So, the series diverges by Theorem 9.10.

13. $\sum_{n=1}^{\infty} \frac{\arctan n}{n^2+1}$

Let $f(x) = \frac{\arctan x}{x^2+1}$,

$$f'(x) = \frac{1 - 2x\arctan x}{\left(x^2+1\right)^2} < 0 \text{ for } x \geq 1.$$

f is positive, continuous, and decreasing for $x \geq 1$.

$$\int_1^{\infty} \frac{\arctan x}{x^2+1}\,dx = \left[\frac{(\arctan x)^2}{2}\right]_1^{\infty} = \frac{3\pi^2}{32}$$

So, the series converges by Theorem 9.10.

15. $\sum_{n=1}^{\infty} \frac{\ln n}{n^2}$

Let $f(x) = \frac{\ln x}{x^2},\ f'(x) = \frac{1 - 2\ln x}{x^3}$.

f is positive, continuous, and decreasing for $x > e^{1/2} \approx 1.6$.

$$\int_1^{\infty} \frac{\ln x}{x^2}\,dx = \left[\frac{-(\ln x + 1)}{x}\right]_1^{\infty} = 1$$

So, the series converges by Theorem 9.10.

17. $\sum_{n=1}^{\infty} \frac{1}{(2n+3)^3}$

Let $f(x) = (2x+3)^{-3},\ f'(x) = \frac{-6}{(2x+3)^4} < 0$

f is positive, continuous, and decreasing for $x \geq 1$.

$$\int_1^{\infty} (2x+3)^{-3}\,dx = \left[\frac{-1}{4(2x+3)^2}\right]_1^{\infty} = \frac{1}{100}$$

So, the series converges by Theorem 9.10.

19. $\sum_{n=1}^{\infty} \frac{4n}{2n^2+1}$

Let $f(x) = \frac{4x}{2x^2+1},\ f'(x) = \frac{-4\left(2x^2-1\right)}{\left(2x^2+1\right)^2} < 0$

for $x \geq 1$.

f is positive, continuous, and decreasing for $x \geq 1$.

$$\int_1^{\infty} \frac{4x}{2x^2+1}\,dx = \left[\ln\left(2x^2+1\right)\right]_1^{\infty} = \infty$$

So, the series diverges by Theorem 9.10.

21. $\sum_{n=1}^{\infty} \frac{n}{n^4+1}$

Let $f(x) = \frac{x}{x^4+1},\ f'(x) = \frac{1 - 3x^4}{\left(x^4+1\right)^2} < 0 \text{ for } x > 1.$

f is positive, continuous, and decreasing for $x > 1$.

$$\int_1^{\infty} \frac{x}{x^4+1}\,dx = \left[\frac{1}{2}\arctan\left(x^2\right)\right]_1^{\infty} = \frac{\pi}{8}$$

So, the series converges by Theorem 9.10.

23. $\sum_{n=1}^{\infty} \frac{n^{k-1}}{n^k + c}$

Let

$$f(x) = \frac{x^{k-1}}{x^k + c},\quad f'(x) = \frac{x^{k-2}\left[c(k-1) - x^k\right]}{\left(x^k + c\right)^2} < 0$$

for $x > \sqrt[k]{c(k-1)}$.

f is positive, continuous, and decreasing for $x > \sqrt[k]{c(k-1)}$.

$$\int_1^{\infty} \frac{x^{k-1}}{x^k + c}\,dx = \left[\frac{1}{k}\ln\left(x^k + c\right)\right]_1^{\infty} = \infty$$

So, the series diverges by Theorem 9.10.

25. Let $f(x) = \frac{(-1)^x}{x},\ f(n) = a_n$.

The function f is not positive for $x \geq 1$.

27. Let $f(x) = \frac{2 + \sin x}{x},\ f(n) = a_n$.

The function f is not decreasing for $x \geq 1$.

29. $\sum_{n=1}^{\infty} \frac{1}{n^7}$

Let $f(x) = \frac{1}{x^7}$.

f is positive, continuous and decreasing for $x \geq 1$.

$\int_1^{\infty} \frac{1}{x^7}\,dx = \left[\frac{-1}{6x^6}\right]_1^{\infty} = \frac{1}{6}$

Converges by Theorem 9.10

31. $\sum_{n=1}^{\infty} \frac{1}{n^{0.9}}$

Let $f(x) = \frac{1}{x^{0.9}}$.

f is positive, continuous, and decreasing for $x \geq 1$.

$\int_1^{\infty} \frac{1}{x^{0.9}}\,dx = \left[10x^{0.1}\right]_1^{\infty} = \infty$

Diverges by Theorem 9.10

33. $\sum_{n=1}^{\infty} \frac{1}{\sqrt[5]{n}} = \sum_{n=1}^{\infty} \frac{1}{n^{1/5}}$

Divergent p-series with $p = \frac{1}{5} < 1$

35. $\sum_{n=1}^{\infty} \frac{1}{n^{3/2}}$

Convergent p-series with $p = \frac{3}{2} > 1$

37. $\sum_{n=1}^{\infty} \frac{1}{n^{1.03}}$

Convergent p-series with $p = 1.03 > 1$

39. (a)

n	5	10	20	50	100
S_n	3.7488	3.75	3.75	3.75	3.75

The partial sums approach the sum 3.75 very rapidly.

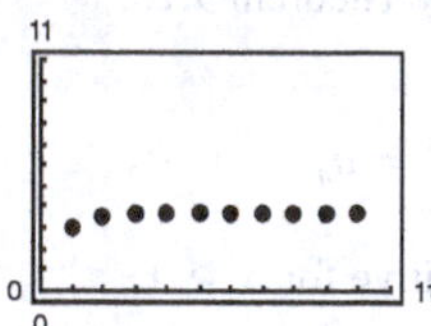

(b)

n	5	10	20	50	100
S_n	1.4636	1.5498	1.5962	1.6251	1.635

The partial sums approach the sum $\pi^2/6 \approx 1.6449$ slower than the series in part (a).

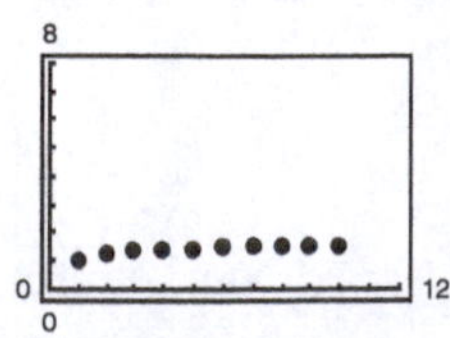

41. Your friend is not correct. The series

$$\sum_{n=10,000}^{\infty} \frac{1}{n} = \frac{1}{10,000} + \frac{1}{10,001} + \cdots$$

is the harmonic series, starting with the 10,000th term, and therefore diverges.

43. (a)

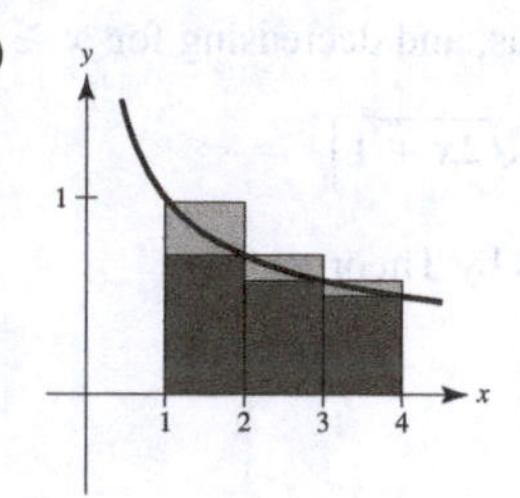

$$\sum_{n=1}^{\infty} \frac{1}{\sqrt{n}} > \int_1^{\infty} \frac{1}{\sqrt{x}}\,dx$$

The area under the rectangle is greater than the area under the curve.

Because $\int_1^{\infty} \frac{1}{\sqrt{x}}\,dx = \left[2\sqrt{x}\right]_1^{\infty} = \infty$, diverges,

$\sum_{n=1}^{\infty} \frac{1}{\sqrt{n}}$ diverges.

(b)

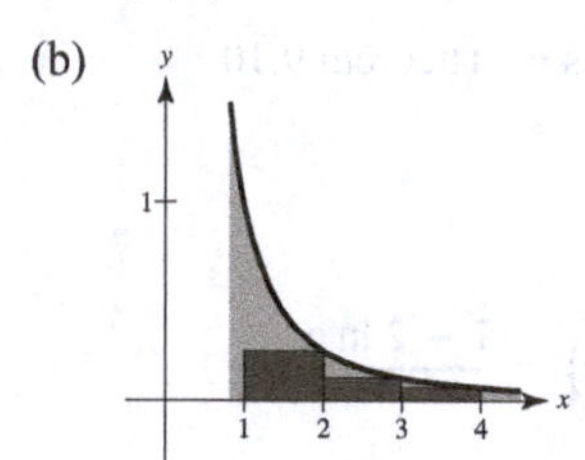

$$\sum_{n=2}^{\infty} \frac{1}{n^2} < \int_1^{\infty} \frac{1}{x^2}\,dx$$

The area under the rectangles is less than the area under the curve.

Because $\int_1^{\infty} \frac{1}{x^2}\,dx = \left[-\frac{1}{x}\right]_1^{\infty} = 1$, converges,

$\sum_{n=2}^{\infty} \frac{1}{n^2}$ converges $\left(\text{and so does } \sum_{n=1}^{\infty} \frac{1}{n^2}\right)$.

45. $\sum_{n=2}^{\infty} \frac{1}{n(\ln n)^p}$

If $p = 1$, then the series diverges by the Integral Test. If $p \neq 1$,

$$\int_2^{\infty} \frac{1}{x(\ln x)^p}\,dx = \int_2^{\infty} (\ln x)^{-p}\frac{1}{x}\,dx = \left[\frac{(\ln x)^{-p+1}}{-p+1}\right]_2^{\infty}.$$

Converges for $-p + 1 < 0$ or $p > 1$

47. $\sum_{n=1}^{\infty} \frac{n}{\left(1 + n^2\right)^p}$

If $p = 1$, $\sum_{n=1}^{\infty} \frac{n}{1 + n^2}$ diverges (see Example 1). Let

$$f(x) = \frac{x}{\left(1 + x^2\right)^p}, \; p \neq 1$$

$$f'(x) = \frac{1 - (2p - 1)x^2}{\left(1 + x^2\right)^{p+1}}.$$

For a fixed $p > 0$, $p \neq 1$, $f'(x)$ is eventually negative. f is positive, continuous, and eventually decreasing.

$$\int_1^{\infty} \frac{x}{\left(1 + x^2\right)^p}\,dx = \left[\frac{1}{\left(x^2 + 1\right)^{p-1}(2 - 2p)}\right]_1^{\infty}$$

For $p > 1$, this integral converges. For $0 < p < 1$, it diverges.

49. $\sum_{n=1}^{\infty} \left(\frac{3}{p}\right)^n$, Geometric series.

Converges for $\left|\frac{3}{p}\right| < 1 \Rightarrow |p| > 3 \Rightarrow p > 3$

51.

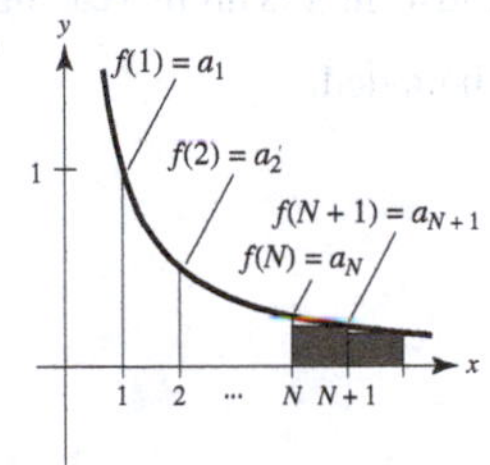

$$S_N = \sum_{n=1}^{N} a_n = a_1 + a_2 + \cdots + a_N$$

$$R_N = S - S_N = \sum_{n=N+1}^{\infty} a_n > 0$$

$$R_N = S - S_N = \sum_{n=N+1}^{\infty} a_n = a_{N+1} + a_{N+2} + \cdots$$

$$\leq \int_N^{\infty} f(x)\,dx$$

So, $0 \leq R_n \leq \int_N^{\infty} f(x)\,dx$

53. $S_3 = 1 + \frac{1}{2^4} + \frac{1}{3^4} \approx 1.0748$

$$0 \leq R_3 \leq \int_3^{\infty} \frac{1}{x^4}\,dx = \left[-\frac{1}{3x^3}\right]_3^{\infty} = \frac{1}{81} \approx 0.0123$$

$$1.0748 \leq \sum_{n=1}^{\infty} \frac{1}{n^4} \leq 1.0748 + 0.0123 = 1.0871$$

55. $S_8 = \frac{1}{2} + \frac{1}{5} + \cdots + \frac{1}{65} \approx 0.9597$

$$0 \leq R_8 \leq \int_8^{\infty} \frac{1}{x^2 + 1}\,dx = \left[\arctan x\right]_8^{\infty} \approx 0.1244$$

$$0.9597 \leq \sum_{n=1}^{\infty} \frac{1}{n^2 + 1} \leq 0.9597 + 0.1244 = 1.0841$$

57. $S_4 = \frac{1}{e} + \frac{2}{e^4} + \frac{3}{e^9} + \frac{4}{e^{16}} \approx 0.4049$

$$0 \leq R_4 \leq \int_4^{\infty} xe^{-x^2}\,dx = \left[-\frac{1}{2}e^{-x^2}\right]_4^{\infty} = \frac{e^{-16}}{2} \approx 5.6 \times 10^{-8}$$

$$0.4049 \leq \sum_{n=1}^{\infty} ne^{-n^2} \leq 0.4049 + 5.6 \times 10^{-8}$$

59. $0 \leq R_N \leq \int_N^{\infty} \frac{1}{x^4}\,dx = \left[-\frac{1}{3x^3}\right]_N^{\infty} = \frac{1}{3N^3} < 0.001$

$$\frac{1}{N^3} < 0.003$$

$$N^3 > 333.33$$

$$N > 6.93$$

$$N \geq 7$$

61. $R_N \leq \int_N^{\infty} e^{-x/2}\,dx = \left[-2e^{-x/2}\right]_N^{\infty} = \frac{2}{e^{N/2}} < 0.001$

$$\frac{2}{e^{N/2}} < 0.001$$

$$e^{N/2} > 2000$$

$$\frac{N}{2} > \ln 2000$$

$$N > 2 \ln 2000 \approx 15.2$$

$$N \geq 16$$

63. (a) $\sum_{n=2}^{\infty} \frac{1}{n^{1.1}}$. This is a convergent p-series with $p = 1.1 > 1$. $\sum_{n=2}^{\infty} \frac{1}{n \ln n}$ is a divergent series. Use the Integral Test.

$f(x) = \frac{1}{x \ln x}$ is positive, continuous, and decreasing for $x \geq 2$.

$$\int_2^{\infty} \frac{1}{x \ln x}\,dx = \left[\ln|\ln x|\right]_2^{\infty} = \infty$$

(b) $\sum_{n=2}^{6} \frac{1}{n^{1.1}} = \frac{1}{2^{1.1}} + \frac{1}{3^{1.1}} + \frac{1}{4^{1.1}} + \frac{1}{5^{1.1}} + \frac{1}{6^{1.1}} \approx 0.4665 + 0.2987 + 0.2176 + 0.1703 + 0.1393$

$\sum_{n=2}^{6} \frac{1}{n \ln n} = \frac{1}{2 \ln 2} + \frac{1}{3 \ln 3} + \frac{1}{4 \ln 4} + \frac{1}{5 \ln 5} + \frac{1}{6 \ln 6} \approx 0.7213 + 0.3034 + 0.1803 + 0.1243 + 0.0930$

For $n \geq 4$, the terms of the convergent series **seem** to be larger than those of the divergent series.

(c) $\frac{1}{n^{1.1}} < \frac{1}{n \ln n}$

$n \ln n < n^{1.1}$

$\ln n < n^{0.1}$

This inequality holds when $n \geq 3.5 \times 10^{15}$. Or, $n > e^{40}$. Then $\ln e^{40} = 40 < \left(e^{40}\right)^{0.1} = e^4 \approx 55$.

65. (a) Let $f(x) = 1/x$. f is positive, continuous, and decreasing on $[1, \infty)$.

$S_n - 1 \leq \int_1^n \frac{1}{x}\,dx$

$S_n - 1 \leq \ln n$

So, $S_n \leq 1 + \ln n$. Similarly,

$S_n \geq \int_1^{n+1} \frac{1}{x}\,dx = \ln(n + 1)$.

So, $\ln(n + 1) \leq S_n \leq 1 + \ln n$.

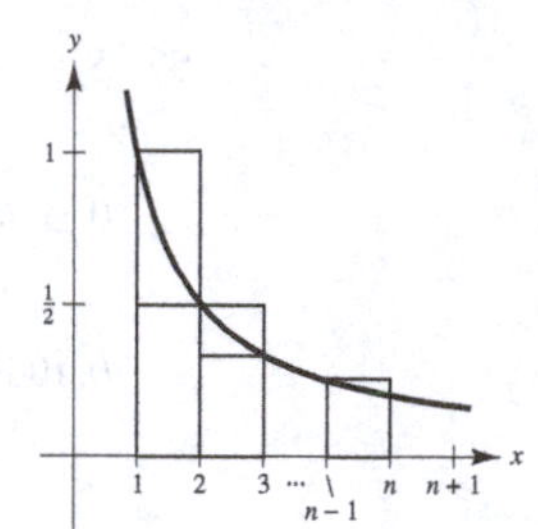

(b) Because $\ln(n + 1) \leq S_n \leq 1 + \ln n$, you have $\ln(n + 1) - \ln n \leq S_n - \ln n \leq 1$. Also, because $\ln x$ is an increasing function, $\ln(n + 1) - \ln n > 0$ for $n \geq 1$. So, $0 \leq S_n - \ln n \leq 1$ and the sequence $\{a_n\}$ is bounded.

(c) $a_n - a_{n+1} = [S_n - \ln n] - [S_{n+1} - \ln(n + 1)] = \int_n^{n+1} \frac{1}{x}\,dx - \frac{1}{n + 1} \geq 0$

So, $a_n \geq a_{n+1}$ and the sequence is decreasing.

(d) Because the sequence is bounded and monotonic, it converges to a limit, γ.

(e) $a_{100} = S_{100} - \ln 100 \approx 0.5822$ (Actually $\gamma \approx 0.577216$.)

67. $\sum_{n=2}^{\infty} x^{\ln n}$

(a) $x = 1$: $\sum_{n=2}^{\infty} 1^{\ln n} = \sum_{n=2}^{\infty} 1$, diverges

(b) $x = \frac{1}{e}$: $\sum_{n=2}^{\infty} \left(\frac{1}{e}\right)^{\ln n} = \sum_{n=2}^{\infty} e^{-\ln n} = \sum_{n=2}^{\infty} \frac{1}{n}$, diverges

(c) Let x be given, $x > 0$. Put

$x = e^{-p} \Leftrightarrow \ln x = -p$.

$\sum_{n=2}^{\infty} x^{\ln n} = \sum_{n=2}^{\infty} e^{-p \ln n} = \sum_{n=2}^{\infty} n^{-p} = \sum_{n=2}^{\infty} \frac{1}{n^p}$

This series converges for $p > 1 \Rightarrow x < \frac{1}{e}$.

69. Let $f(x) = \frac{1}{3x - 2}$, $f'(x) = \frac{-3}{(3x - 2)^2} < 0$ for $x \geq 1$

f is positive, continuous, and decreasing for $x \geq 1$.

$\int_1^{\infty} \frac{1}{3x - 2}\,dx = \left[\frac{1}{3} \ln|3x - 2|\right]_1^{\infty} = \infty$

So, the series $\sum_{n=1}^{\infty} \frac{1}{3n - 2}$

diverges by Theorem 9.10.

71. $\sum_{n=1}^{\infty} \frac{1}{n\sqrt[4]{n}} = \sum_{n=1}^{\infty} \frac{1}{n^{5/4}}$

p-series with $p = \frac{5}{4}$

Converges by Theorem 9.11

73. $\sum_{n=0}^{\infty} \left(\frac{2}{3}\right)^n$

Geometric series with $r = \frac{2}{3}$

Converges by Theorem 9.6

75. $\sum_{n=1}^{\infty} \frac{n}{\sqrt{3n^2 + 3}}$

$$\lim_{n \to \infty} \frac{n}{\sqrt{3n^2 + 3}} = \frac{1}{\sqrt{3}} \neq 0$$

Diverges by Theorem 9.9

77. $\sum_{n=1}^{\infty} \left(1 + \frac{1}{n}\right)^n$

$$\lim_{n \to \infty} \left(1 + \frac{1}{n}\right)^n = e \neq 0$$

Fails nth-Term Test

Diverges by Theorem 9.9

79. $\sum_{n=2}^{\infty} \frac{1}{n(\ln n)^3}$

Let $f(x) = \frac{1}{x(\ln x)^3}$.

f is positive, continuous, and decreasing for $x \geq 2$.

$$\int_2^{\infty} \frac{1}{x(\ln x)^3}\,dx = \int_2^{\infty} (\ln x)^{-3}\frac{1}{x}\,dx$$

$$= \left[\frac{(\ln x)^{-2}}{-2}\right]_2^{\infty}$$

$$= \left[-\frac{1}{2(\ln x)^2}\right]_2^{\infty}$$

$$= \frac{1}{2(\ln 2)^2}$$

Converges by Theorem 9.10 (See Exercise 45.)

Section 9.4 Comparisons of Series

1. Yes. The Direct Comparison Test requires that $0 \leq a_n \leq b_n$ for all n greater than some integer N. In this case, $N = 6$. The beginning terms do not affect the convergence or divergence of a series.

3. (a) $\sum_{n=1}^{\infty} \frac{6}{n^{3/2}} = \frac{6}{1} + \frac{6}{2^{3/2}} + \cdots;\ S_1 = 6$

$\sum_{n=1}^{\infty} \frac{6}{n^{3/2} + 3} = \frac{6}{4} + \frac{6}{2^{3/2} + 3} + \cdots;\ S_1 = \frac{3}{2}$

$\sum_{n=1}^{\infty} \frac{6}{n\sqrt{n^2 + 0.5}} = \frac{6}{1\sqrt{1.5}} + \frac{6}{2\sqrt{4.5}} + \cdots;\ S_1 = \frac{6}{\sqrt{1.5}} \approx 4.9$

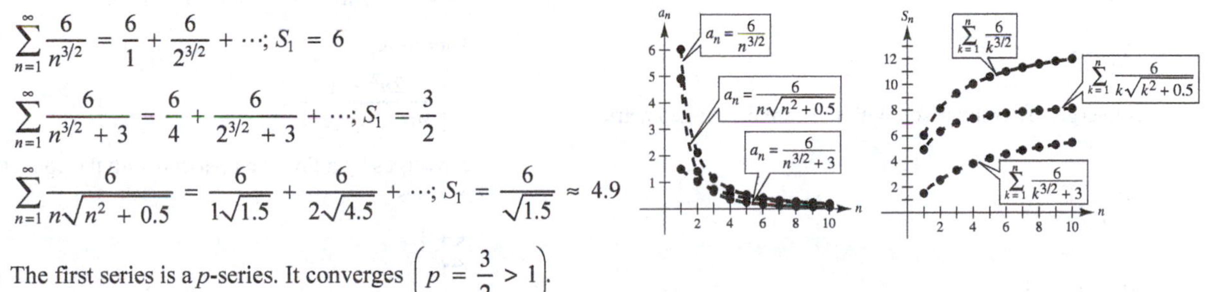

(b) The first series is a p-series. It converges $\left(p = \frac{3}{2} > 1\right)$.

(c) The magnitudes of the terms of the other two series are less than the corresponding terms at the convergent p-series. So, the other two series converge.

(d) The smaller the magnitude of the terms, the smaller the magnitude of the terms of the sequence of partial sums.

5. $\frac{1}{2n - 1} > \frac{1}{2n} > 0$ for $n \geq 1$

Therefore,

$$\sum_{n=1}^{\infty} \frac{1}{2n - 1}$$

diverges by comparison with the divergent p-series

$$\frac{1}{2}\sum_{n=1}^{\infty} \frac{1}{n}.$$

7. $\frac{1}{\sqrt{n} - 1} > \frac{1}{\sqrt{n}}$ for $n \geq 2$

Therefore,

$$\sum_{n=2}^{\infty} \frac{1}{\sqrt{n} - 1}$$

diverges by comparison with the divergent p-series

$$\sum_{n=2}^{\infty} \frac{1}{\sqrt{n}}.$$

9. For $n \geq 3$, $\dfrac{\ln n}{n+1} > \dfrac{1}{n+1} > 0$.

Therefore,

$$\sum_{n=1}^{\infty} \frac{\ln n}{n+1}$$

diverges by comparison with the divergent series

$$\sum_{n=1}^{\infty} \frac{1}{n+1}.$$

Note: $\displaystyle\sum_{n=1}^{\infty} \frac{1}{n+1}$ diverges by the Integral Test.

11. For $n > 3$, $\dfrac{1}{n^2} > \dfrac{1}{n!} > 0$.

Therefore,

$$\sum_{n=0}^{\infty} \frac{1}{n!}$$

converges by comparison with the convergent p-series

$$\sum_{n=1}^{\infty} \frac{1}{n^2}.$$

13. $0 < \dfrac{1}{e^{n^2}} \leq \dfrac{1}{e^n}$

Therefore,

$$\sum_{n=0}^{\infty} \frac{1}{e^{n^2}}$$

converges by comparison with the convergent geometric series

$$\sum_{n=0}^{\infty} \left(\frac{1}{e}\right)^n.$$

15. $\dfrac{\sin^2 n}{n^3} \leq \dfrac{1}{n^3}$ for $n \geq 1$.

Therefore,

$$\sum_{n=1}^{\infty} \frac{\sin^2 n}{n^3}$$

converges by comparison with the convergent p-series

$$\sum_{n=1}^{\infty} \frac{1}{n^3}.$$

17. $\displaystyle\lim_{n\to\infty} \frac{n/(n^2+1)}{1/n} = \lim_{n\to\infty} \frac{n^2}{n^2+1} = 1$

Therefore,

$$\sum_{n=1}^{\infty} \frac{n}{n^2+1}$$

diverges by a limit comparison with the divergent p-series

$$\sum_{n=1}^{\infty} \frac{1}{n}.$$

19. $\displaystyle\lim_{n\to\infty} \frac{1/\sqrt{n^2+1}}{1/n} = \lim_{n\to\infty} \frac{n}{\sqrt{n^2+1}} = 1$

Therefore,

$$\sum_{n=0}^{\infty} \frac{1}{\sqrt{n^2+1}}$$

diverges by a limit comparison with the divergent p-series

$$\sum_{n=1}^{\infty} \frac{1}{n}.$$

21. $\displaystyle\lim_{n\to\infty} \frac{\frac{2n^2-1}{3n^5+2n+1}}{1/n^3} = \lim_{n\to\infty} \frac{2n^5-n^3}{3n^5+2n+1} = \frac{2}{3}$

Therefore,

$$\sum_{n=1}^{\infty} \frac{2n^2-1}{3n^5+2n+1}$$

converges by a limit comparison with the convergent p-series

$$\sum_{n=1}^{\infty} \frac{1}{n^3}.$$

23. $\displaystyle\lim_{n\to\infty} \frac{1/\left(n\sqrt{n^2+1}\right)}{1/n^2} = \lim_{n\to\infty} \frac{n^2}{n\sqrt{n^2+1}} = 1$

Therefore,

$$\sum_{n=1}^{\infty} \frac{1}{n\sqrt{n^2+1}}$$

converges by a limit comparison with the convergent p-series

$$\sum_{n=1}^{\infty} \frac{1}{n^2}.$$

25. $\lim_{n\to\infty} \frac{(n^{k-1})/(n^k+1)}{1/n} = \lim_{n\to\infty} \frac{n^k}{n^k+1} = 1$

Therefore,

$$\sum_{n=1}^{\infty} \frac{n^{k-1}}{n^k+1}$$

diverges by a limit comparison with the divergent p-series

$$\sum_{n=1}^{\infty} \frac{1}{n}.$$

27. $\sum_{n=1}^{\infty} \frac{\sqrt[3]{n}}{n} = \sum_{n=1}^{\infty} \frac{1}{n^{2/3}}$

Diverges;

p-series with $p = \frac{2}{3}$

29. $\sum_{n=1}^{\infty} \frac{1}{5^n+1}$

Converges;

Direct comparison with convergent geometric series

$$\sum_{n=1}^{\infty} \left(\frac{1}{5}\right)^n$$

31. $\sum_{n=1}^{\infty} \frac{2n}{3n-2}$

Diverges; n^{th}-Term Test

$$\lim_{n\to\infty} \frac{2n}{3n-2} = \frac{2}{3} \neq 0$$

33. $\sum_{n=1}^{\infty} \frac{n}{(n^2+1)^2}$

Converges; Integral Test

35. $\lim_{n\to\infty} \frac{a_n}{1/n} = \lim_{n\to\infty} na_n$. By given conditions $\lim_{n\to\infty} na_n$ is finite and nonzero. Therefore,

$$\sum_{n=1}^{\infty} a_n$$

diverges by a limit comparison with the p-series

$$\sum_{n=1}^{\infty} \frac{1}{n}.$$

37. $\frac{1}{2} + \frac{2}{5} + \frac{3}{10} + \frac{4}{17} + \frac{5}{26} + \cdots = \sum_{n=1}^{\infty} \frac{n}{n^2+1},$

which diverges because the degree of the numerator is only one less than the degree of the denominator.

39. $\sum_{n=1}^{\infty} \frac{1}{n^3+1}$

converges because the degree of the numerator is three less than the degree of the denominator.

41. $\lim_{n\to\infty} n\left(\frac{n^3}{5n^4+3}\right) = \lim_{n\to\infty} \frac{n^4}{5n^4+3} = \frac{1}{5} \neq 0$

Therefore, $\sum_{n=1}^{\infty} \frac{n^3}{5n^4+3}$ diverges.

43. $\frac{1}{200} + \frac{1}{400} + \frac{1}{600} + \cdots = \sum_{n=1}^{\infty} \frac{1}{200n}$

diverges, (harmonic)

45. $\frac{1}{201} + \frac{1}{204} + \frac{1}{209} + \frac{1}{216} = \sum_{n=1}^{\infty} \frac{1}{200+n^2}$

converges

47. Some series diverge or converge very slowly. You cannot decide convergence or divergence of a series by comparing the first few terms.

49. (a) $\sum_{n=1}^{\infty} \frac{1}{(2n-1)^2} = \sum_{n=1}^{\infty} \frac{1}{4n^2-4n+1}$

converges because the degree of the numerator is two less than the degree of the denominator. (See Exercise 36.)

(b)

n	5	10	20	50	100
S_n	1.1839	1.2087	1.2212	1.2287	1.2312

(c) $\sum_{n=3}^{\infty} \frac{1}{(2n-1)^2} = \frac{\pi^2}{8} - S_2 \approx 0.1226$

(d) $\sum_{n=10}^{\infty} \frac{1}{(2n-1)^2} = \frac{\pi^2}{8} - S_9 \approx 0.0277$

51. $\frac{x_1}{10} + \frac{x_2}{10} + \cdots \le \frac{9}{10} + \frac{9}{10^2} + \cdots = \sum_{n=0}^{\infty} \frac{9}{10}\left(\frac{1}{10}\right)^n,$

which is a convergent geometric series $\left(r = \frac{1}{10}\right)$.

53. False. Let $a_n = \frac{1}{n^3}$ and $b_n = \frac{1}{n^2}$. $0 < a_n \le b_n$ and both

$\sum_{n=1}^{\infty} \frac{1}{n^3}$ and $\sum_{n=1}^{\infty} \frac{1}{n^2}$ converge.

55. True

57. True

59. Because $\sum_{n=1}^{\infty} b_n$ converges, $\lim_{n\to\infty} b_n = 0$. There exists N such that $b_n < 1$ for $n > N$. So, $a_n b_n < a_n$ for $n > N$ and $\sum_{n=1}^{\infty} a_n b_n$ converges by comparison to the convergent series $\sum_{i=1}^{\infty} a_n$.

61. $\sum \frac{1}{n^2}$ and $\sum \frac{1}{n^3}$ both converge, and therefore, so does

$\sum\left(\frac{1}{n^2}\right)\left(\frac{1}{n^3}\right) = \sum \frac{1}{n^5}.$

63. Suppose $\lim_{n\to\infty} \frac{a_n}{b_n} = 0$ and Σb_n converges.

From the definition of limit of a sequence, there exists $M > 0$ such that

$$\left|\frac{a_n}{b_n} - 0\right| < 1$$

whenever $n > M$. So, $a_n < b_n$ for $n > M$. From the Comparison Test, Σa_n converges.

65. (a) Let $\sum a_n = \sum \frac{1}{(n+1)^3}$, and $\sum b_n = \sum \frac{1}{n^2}$ converges.

$$\lim_{n\to\infty} \frac{a_n}{b_n} = \lim_{n\to\infty} \frac{1/\left[(n+1)^3\right]}{1/\left(n^2\right)} = \lim_{n\to\infty} \frac{n^2}{(n+1)^3} = 0$$

By Exercise 63, $\sum_{n=1}^{\infty} \frac{1}{(n+1)^3}$ converges.

(b) Let $\sum a_n = \sum \frac{1}{\sqrt{n}\pi^n}$, and $\sum b_n = \sum \frac{1}{\pi^n}$ converges.

$$\lim_{n\to\infty} \frac{a_n}{b_n} = \lim_{n\to\infty} \frac{1/\left(\sqrt{n}\pi^n\right)}{1/\left(\pi^n\right)} = \lim_{n\to\infty} \frac{1}{\sqrt{n}} = 0$$

By Exercise 63, $\sum_{n=1}^{\infty} \frac{1}{\sqrt{n}\pi^n}$ converges.

(c) Let $\sum a_n = \frac{\ln n}{n^3}$, and $\sum b_n = \frac{1}{n^2}$ converges.

$$\lim_{n\to\infty} \frac{a_n}{b_n} = \lim_{n\to\infty} \frac{\ln n/n^3}{1/n^2} = \lim_{n\to\infty} \frac{\ln n}{n} = 0$$

By Exercise 63, $\sum_{n=2}^{\infty} \frac{\ln n}{n}$ converges.

(d) Let $\sum a_n = \frac{n^2+1}{e^n}$, and $\sum b_n = \frac{n^3}{e^n}$ converges.

$$\lim_{n\to\infty} \frac{a_n}{b_n} = \frac{\left(n^2+1\right)/e^n}{n^3/e^n} = \lim_{n\to\infty} \frac{n^2+1}{n^3} = 0$$

By Exercise 63, $\sum_{n=1}^{\infty} \frac{n^2+1}{e^n}$ converges.

67. Because $\lim_{n\to\infty} a_n = 0$, the terms of $\Sigma \sin(a_n)$ are positive for sufficiently large n. Because

$\lim_{n\to\infty} \frac{\sin(a_n)}{a_n} = 1$ and $\sum a_n$

converges, so does $\Sigma \sin(a_n)$.

69. First note that $f(x) = \ln x - x^{1/4} = 0$ when $x \approx 5503.66$. That is,

$\ln n < n^{1/4}$ for $n > 5504$

which implies that

$\frac{\ln n}{n^{3/2}} < \frac{1}{n^{5/4}}$ for $n > 5504$.

Because $\sum_{n=1}^{\infty} \frac{1}{n^{5/4}}$ is a convergent p-series,

$\sum_{n=1}^{\infty} \frac{\ln n}{n^{3/2}}$

converges by direct comparison.

71. The series diverges. For $n > 1$,

$$n < 2^n$$
$$n^{1/n} < 2$$
$$\frac{1}{n^{1/n}} > \frac{1}{2}$$
$$\frac{1}{n^{(n+1)/n}} > \frac{1}{2n}$$

Because $\sum \frac{1}{2n}$ diverges, so does $\sum \frac{1}{n^{(n+1)/n}}$.

Section 9.5 Alternating Series

1. The series must diverge by the nth-Term Test for Divergence.

3. $\sum a_n$ is absolutely convergent if $\sum |a_n|$ converges. $\sum a_n$ is conditionally convergent if $\sum |a_n|$ diverges, but $\sum a_n$ converges.

5. $\displaystyle\sum_{n=1}^{\infty} \frac{(-1)^{n-1}}{2n-1} = \frac{\pi}{4} \approx 0.7854$

(a)

n	1	2	3	4	5	6	7	8	9	10
S_n	1	0.6667	0.8667	0.7238	0.8349	0.7440	0.8209	0.7543	0.8131	0.7605

(b)

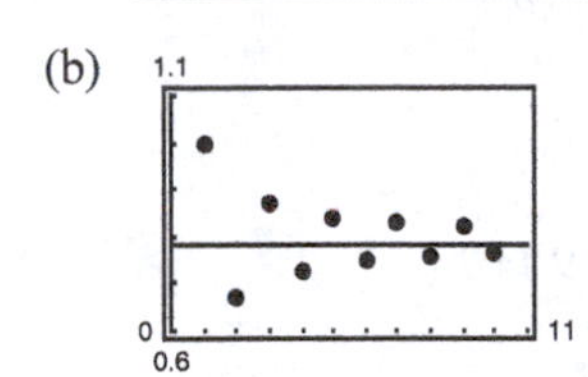

(c) The points alternate sides of the horizontal line $y = \frac{\pi}{4}$ that represents the sum of the series.

The distance between successive points and the line decreases.

(d) The distance in part (c) is always less than the magnitude of the next term of the series.

7. $\displaystyle\sum_{n=1}^{\infty} \frac{(-1)^{n-1}}{n^2} = \frac{\pi^2}{12} \approx 0.8225$

(a)

n	1	2	3	4	5	6	7	8	9	10
S_n	1	0.75	0.8611	0.7986	0.8386	0.8108	0.8312	0.8156	0.8280	0.8180

(b)

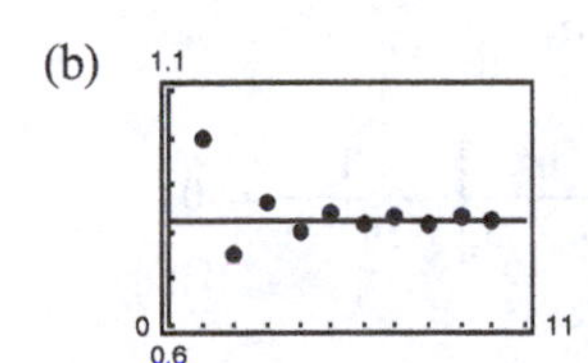

(c) The points alternate sides of the horizontal line $y = \frac{\pi^2}{12}$ that represents the sum of the series.

The distance between successive points and the line decreases.

(d) The distance in part (c) is always less than the magnitude of the next term in the series.

9. $\sum_{n=1}^{\infty} \frac{(-1)^{n+1}}{n+1}$

$$a_{n+1} = \frac{1}{n+2} < \frac{1}{n+1} = a_n$$

$$\lim_{n\to\infty} a_n = \lim_{n\to\infty} \frac{1}{n+1} = 0$$

Converges by Theorem 9.14

11. $\sum_{n=1}^{\infty} \frac{(-1)^n}{3^n}$

$$a_{n+1} = \frac{1}{3^{n+1}} < \frac{1}{3^n} = a_n$$

$$\lim_{n\to 0} \frac{1}{3^n} = 0$$

Converges by Theorem 9.14

(**Note:** $\sum_{n=1}^{\infty} \left(\frac{-1}{3}\right)^n$ is a convergent geometric series)

13. $\sum_{n=1}^{\infty} \frac{(-1)^n(5n-1)}{4n+1}$

$$\lim_{n\to\infty} \frac{5n-1}{4n+1} = \frac{5}{4}$$

Diverges by nth-Term test

15. $\sum_{n=1}^{\infty} \frac{(-1)^n n}{\ln(n+1)}$

$$\lim_{n\to\infty} \frac{n}{\ln(n+1)} = \infty$$

Diverges by nth-Term test

17. $\sum_{n=1}^{\infty} \frac{(-1)^n}{\sqrt{n}}$

$$a_{n+1} = \frac{1}{\sqrt{n+1}} < \frac{1}{\sqrt{n}} = a_n$$

$$\lim_{n\to\infty} \frac{1}{\sqrt{n}} = 0$$

Converges by Theorem 9.14

19. $\sum_{n=1}^{\infty} \frac{(-1)^{n+1}(n+1)}{\ln(n+1)}$

$$\lim_{n\to\infty} \frac{n+1}{\ln(n+1)} = \lim_{n\to\infty} \frac{1}{1/(n+1)} = \lim_{n\to\infty} (n+1) = \infty$$

Diverges by the nth-Term Test

21. $\sum_{n=1}^{\infty} \sin\left[\frac{(2n-1)\pi}{2}\right] = \sum_{n=1}^{\infty} (-1)^{n+1}$

Diverges by the nth-Term Test

23. $\sum_{n=0}^{\infty} \frac{(-1)^n}{n!}$

$$a_{n+1} = \frac{1}{(n+1)!} < \frac{1}{n!} = a_n$$

$$\lim_{n\to\infty} \frac{1}{n!} = 0$$

Converges by Theorem 9.14

25. $\sum_{n=1}^{\infty} \frac{(-1)^{n+1}\sqrt{n}}{n+2}$

$$a_{n+1} = \frac{\sqrt{n+1}}{(n+1)+2} < \frac{\sqrt{n}}{n+2} \text{ for } n \ge 2$$

$$\lim_{n\to\infty} \frac{\sqrt{n}}{n+2} = 0$$

Converges by Theorem 9.14

27. $\sum_{n=1}^{\infty} \frac{(-1)^{n+1} n!}{1 \cdot 3 \cdot 5 \cdots (2n-1)}$

$$a_{n+1} = \frac{(n+1)!}{1 \cdot 3 \cdot 5 \cdots (2n-1)(2n+1)} = \frac{n!}{1 \cdot 3 \cdot 5 \cdots (2n-1)} \cdot \frac{n+1}{2n+1} = a_n\left(\frac{n+1}{2n+1}\right) < a_n$$

$$\lim_{n\to\infty} a_n = \lim_{n\to\infty} \frac{n!}{1 \cdot 3 \cdot 5 \cdots (2n-1)} = \lim_{n\to\infty} \frac{1 \cdot 2 \cdot 3 \cdots n}{1 \cdot 3 \cdot 5 \cdots (2n-1)} = \lim_{n\to\infty} 2\left[\frac{3}{3} \cdot \frac{4}{5} \cdot \frac{5}{7} \cdots \frac{n}{2n-3}\right] \cdot \frac{1}{2n-1} = 0$$

Converges by Theorem 9.14

29. $\sum_{n=1}^{\infty} \frac{(-1)^{n+1}(2)}{e^n - e^{-n}} = \sum_{n=1}^{\infty} \frac{(-1)^{n+1}(2e^n)}{e^{2n} - 1}$

Let $f(x) = \frac{2e^x}{e^{2x} - 1}$. Then

$$f'(x) = \frac{-2e^x(e^{2x} + 1)}{(e^{2x} - 1)^2} < 0.$$

So, $f(x)$ is decreasing. Therefore, $a_{n+1} < a_n$, and

$$\lim_{n\to\infty} \frac{2e^n}{e^{2n} - 1} = \lim_{n\to\infty} \frac{2e^n}{2e^{2n}} = \lim_{n\to\infty} \frac{1}{e^n} = 0.$$

The series converges by Theorem 9.14.

31. $S_5 = \sum_{n=0}^{5} \frac{(-1)^n 5}{n!} = \frac{11}{6}$

$|R_5| = |S - S_5| \le a_6 = \frac{5}{720} = \frac{1}{144}$

$\frac{11}{6} - \frac{1}{144} \le S \le \frac{11}{6} + \frac{1}{144}$

$1.8264 \le S \le 1.8403$

33. $S_6 = \sum_{n=1}^{6} \frac{(-1)^{n+1} 2}{n^3} \approx 1.7996$

$|R_6| = |S - S_6| \le a_7 = \frac{2}{7^3} \approx 0.0058$

$1.7796 - 0.0058 \le S \le 1.7796 + 0.0058$

$1.7938 \le S \le 1.8054$

35. $\sum_{n=1}^{\infty} \frac{(-1)^{n+1}}{n^3}$

By Theorem 9.15,

$$|R_N| \le a_{N+1} = \frac{1}{(N + 1)^3} < 0.001$$

$$\Rightarrow (N + 1)^3 > 1000 \Rightarrow N + 1 > 10.$$

Use 10 terms.

37. $\sum_{n=1}^{\infty} \frac{(-1)^{n+1}}{2n^3 - 1}$

By Theorem 9.15,

$$|R_N| \le a_{N+1} = \frac{1}{2(N + 1)^3 - 1} < 0.001$$

$$\Rightarrow 2(N + 1)^3 - 1 > 1000.$$

By trial and error, this inequality is valid when $N = 7\left[2(8^3) - 1 = 1024\right]$.

Use 7 terms.

39. $\sum_{n=0}^{\infty} \frac{(-1)^n}{n!}$

By Theorem 9.15,

$$|R_N| \le a_{N+1} = \frac{1}{(N + 1)!} < 0.001$$

$$\Rightarrow (N + 1)! > 1000.$$

By trial and error, this inequality is valid when $N = 6(7! = 5040)$. Use 7 terms since the sum begins with $n = 0$.

41. $\sum_{n=1}^{\infty} \frac{(-1)^n}{2^n}$

$\sum_{n=1}^{\infty} \frac{1}{2^n}$ is a convergent geometric series.

Therefore, $\sum_{n=1}^{\infty} \frac{(-1)^n}{2^n}$ converges absolutely.

43. $\sum_{n=1}^{\infty} \frac{(-1)^n}{n!}$

$\frac{1}{n!} < \frac{1}{n^2}$ for $n \ge 4$

and $\sum_{n=1}^{\infty} \frac{1}{n^2}$ is a convergent p-series.

So, $\sum_{n=1}^{\infty} \frac{1}{n!}$ converges, and $\sum_{n=1}^{\infty} \frac{(-1)^n}{n!}$ converges absolutely.

45. $\sum_{n=1}^{\infty} \frac{(-1)^{n+1}}{\sqrt{n}}$

The given series converges by the Alternating Series Test, but does not converge absolutely because

$$\sum_{n=1}^{\infty} \frac{1}{\sqrt{n}}$$

is a divergent p-series. Therefore, the series converges conditionally.

47. $\sum_{n=1}^{\infty} \frac{(-1)^{n+1} n^2}{(n + 1)^2}$

$$\lim_{n\to\infty} \frac{n^2}{(n + 1)^2} = 1$$

Therefore, the series diverges by the nth-Term Test.

49. $\sum_{n=2}^{\infty} \frac{(-1)^n}{n \ln n}$

The series converges by the Alternating Series Test.

Let $f(x) = \frac{1}{x \ln x}$.

$\int_2^{\infty} \frac{1}{x \ln x}\,dx = \left[\ln(\ln x)\right]_2^{\infty} = \infty$

By the Integral Test, $\sum_{n=2}^{\infty} \frac{1}{n \ln n}$ diverges.

So, the series $\sum_{n=2}^{\infty} \frac{(-1)^n}{n \ln n}$ converges conditionally.

51. $\sum_{n=2}^{\infty} \frac{(-1)^n n}{n^3 - 5}$

$\sum_{n=2}^{\infty} \frac{n}{n^3 - 5}$ converges by a limit comparison to the p-series $\sum_{n=2}^{\infty} \frac{1}{n^2}$. Therefore, the given series converges absolutely.

53. $\sum_{n=0}^{\infty} \frac{(-1)^n}{(2n+1)!}$

$\sum_{n=0}^{\infty} \frac{1}{(2n+1)!}$

is convergent by comparison to the convergent geometric series $\sum_{n=0}^{\infty} \left(\frac{1}{2}\right)^n$

Because $\frac{1}{(2n+1)!} < \frac{1}{2^n}$ for $n > 0$.

Therefore, the given series converges absolutely.

55. $\sum_{n=0}^{\infty} \frac{\cos n\pi}{n+1} = \sum_{n=0}^{\infty} \frac{(-1)^n}{n+1}$

The given series converges by the Alternating Series Test, but $\sum_{n=0}^{\infty} \frac{|\cos n\pi|}{n+1} = \sum_{n=0}^{\infty} \frac{1}{n+1}$ diverges by a limit comparison to the divergent harmonic series,

$\sum_{n=1}^{\infty} \frac{1}{n}$.

$\lim_{n\to\infty} \frac{|\cos n\pi|/(n+1)}{1/n} = 1$, therefore, the given series converges conditionally.

57. $\sum_{n=1}^{\infty} \frac{\cos(n\pi/3)}{n^2}$

$\sum_{n=1}^{\infty} \frac{\cos(n\pi/3)}{n^2}$ converges by a limit comparison to the p-series $\sum_{n=1}^{\infty} \frac{1}{n^2}$ because $\left|\frac{\cos(n\pi/3)}{n^2}\right| \le \frac{1}{n^2}$.

Therefore, the given series converges absolutely.

59. $\sum_{n=1}^{\infty} \frac{(-1)^n}{n} = -1 + \frac{1}{2} - \frac{1}{3} + \cdots + \frac{1}{50} - \frac{1}{51} + \cdots$

$S_{50} = -1 + \frac{1}{2} - \frac{1}{3} + \cdots + \frac{1}{50} > \sum_{n=1}^{\infty} \frac{(-1)^n}{n}$

S_{50} is an overestimate of the sum because the next term $\left(-\frac{1}{51}\right)$ is negative.

61. (a) False. For example, let $a_n = \frac{(-1)^n}{n}$.

Then $\sum a_n = \sum \frac{(-1)^n}{n}$ converges

and $\sum(-a_n) = \sum \frac{(-1)^{n+1}}{n}$ converges.

But, $\sum |a_n| = \sum \frac{1}{n}$ diverges.

(b) True. For if $\sum |a_n|$ converged, then so would $\sum a_n$ by Theorem 9.16.

63. $\sum_{n=1}^{\infty} (-1)^n \frac{1}{n^p}$

If $p = 0$, then $\sum_{n=1}^{\infty} (-1)^n$ diverges.

If $p < 0$, then $\sum_{n=1}^{\infty} (-1)^n n^{-p}$ diverges.

If $p > 0$, then $\lim_{n\to\infty} \frac{1}{n^p} = 0$ and

$a_{n+1} = \frac{1}{(n+1)^p} < \frac{1}{n^p} = a_n$.

Therefore, the series converges for $p > 0$.

65. Because $\sum_{n=1}^{\infty}|a_n|$ converges you have $\lim_{n\to\infty}|a_n| = 0$.

So, there must exist an $N > 0$ such that $|a_N| < 1$ for all $n > N$ and it follows that $a_n^2 \le |a_n|$ for all $n > N$.

So, by the Comparison Test, $\sum_{n=1}^{\infty} a_n^2$ converges.

Let $a_n = 1/n$ to see that the converse is false.

67. $\sum_{n=1}^{\infty}\frac{1}{n^2}$ converges, and so does $\sum_{n=1}^{\infty}\frac{1}{n^4}$.

69. (a) No, the series does not satisfy $a_{n+1} \le a_n$ for all n.

For example, $\frac{1}{9} < \frac{1}{8}$.

(b) Yes, the series converges.

$$S_{2n} = \frac{1}{2} - \frac{1}{3} + \cdots + \frac{1}{2^n} - \frac{1}{3^n}$$

$$= \left(\frac{1}{2} + \cdots + \frac{1}{2^n}\right) - \left(\frac{1}{3} + \cdots + \frac{1}{3^n}\right)$$

$$= \left(1 + \frac{1}{2} + \cdots + \frac{1}{2^n}\right) - \left(1 + \frac{1}{3} + \cdots + \frac{1}{3^n}\right)$$

As $n \to \infty$,

$$S_{2n} \to \frac{1}{1-(1/2)} - \frac{1}{1-(1/3)} = 2 - \frac{3}{2} = \frac{1}{2}.$$

71. $\sum_{n=1}^{\infty}\frac{8}{\sqrt[3]{n}} = 8\sum_{n=1}^{\infty}\frac{1}{n^{1/3}}$

Divergent p-series

73. $\sum_{n=1}^{\infty}\frac{3^n}{n^2}$

$$\lim_{n\to\infty}\frac{3^n}{n^2} = \infty$$

Diverges by nth-Term Test

75. $\sum_{n=1}^{\infty}\left(\frac{9}{8}\right)^n$

Geometric series with $r = \frac{9}{8} > 1$

Diverges

77. $\sum_{n=1}^{\infty} 100e^{-n/2}$

Geometric series with $r = \frac{1}{\sqrt{e}} < 1$

Converges

79. $\sum_{n=1}^{\infty}\frac{(-1)^{n+1}4}{3n^2 - 1}$

$$a_{n+1} = \frac{1}{3(n+1)^2 - 1} < \frac{1}{3n^2 - 1} = a_n$$

$$\lim_{n\to\infty}\frac{1}{3n^2 - 1} = 0$$

Converges (absolutely) by the Alternating Series Test

81. You cannot arbitrarily change 0 to $1 - 1$.

Section 9.6 The Ratio and Root Tests

1. Converges (Ratio Test)

3. Diverges (Ratio Test)

5. Inconclusive (See Root Test)

7. $$\frac{9^{n+1}(n-1)!}{9^n(n-2)!} = \frac{9\cdot 9^n(n-1)(n-2)!}{9^n(n-2)!} = 9(n-1)$$

9. $\sum_{n=1}^{\infty} n\left(\frac{3}{4}\right)^n = 1\left(\frac{3}{4}\right) + 2\left(\frac{9}{16}\right) + \cdots$

$S_1 = \frac{3}{4}$, $S_2 \approx 1.875$

Matches (d).

10. $\sum_{n=1}^{\infty}\left(\frac{3}{4}\right)^n\left(\frac{1}{n!}\right) = \frac{3}{4} + \frac{9}{16}\left(\frac{1}{2}\right) + \cdots$

$S_1 = \frac{3}{4}$, $S_2 = 1.03$

Matches (c).

11. $\sum_{n=1}^{\infty}\frac{(-3)^{n+1}}{n!} = 9 - \frac{3^3}{2} + \cdots$

$S_1 = 9$

Matches (f).

12. $\sum_{n=1}^{\infty}\frac{(-1)^{n-1}4}{(2n)!} = \frac{4}{2} - \frac{4}{24} + \cdots$

$S_1 = 2$

Matches (b).

13. $\sum_{n=1}^{\infty}\left(\frac{4n}{5n-3}\right)^n = \frac{4}{2} + \left(\frac{8}{7}\right)^2 + \cdots$

$S_1 = 2, S_2 = 3.31$

Matches (a).

14. $\sum_{n=0}^{\infty} 4e^{-n} = 4 + \frac{4}{e} + \cdots$

$S_1 = 4$

Matches (e).

15. (a) Ratio Test: $\lim_{n\to\infty}\left|\frac{a_{n+1}}{a_n}\right| = \lim_{n\to\infty}\frac{(n+1)^3(1/2)^{n+1}}{n^3(1/2)^n} = \lim_{n\to\infty}\left(\frac{n+1}{n}\right)^3\frac{1}{2} = \frac{1}{2} < 1$, converges

(b)

n	5	10	15	20	25
S_n	13.7813	24.2363	25.8468	25.9897	25.9994

(c)

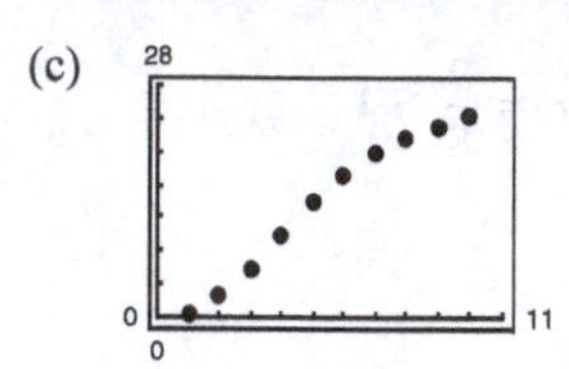

(d) The sum is approximately 26.

(e) The more rapidly the terms of the series approach 0, the more rapidly the sequence of partial sums approaches the sum of the series.

17. $\sum_{n=1}^{\infty}\frac{1}{8^n}$

$\lim_{n\to\infty}\left|\frac{a_n+1}{a_n}\right| = \lim_{n\to\infty}\left|\frac{1/8^{n+1}}{1/8^n}\right| = \lim_{n\to\infty}\frac{8^n}{8^{n+1}} = \frac{1}{8} < 1$

Therefore, the series converges by the Ratio Test.

19. $\sum_{n=1}^{\infty}\frac{(n-1)!}{4^n}$

$\lim_{n\to\infty}\left|\frac{a_{n+1}}{a_n}\right| = \lim_{n\to\infty}\left|\frac{n!/4^{n+1}}{(n-1)!/4^n}\right|$

$= \lim_{n\to\infty}\frac{n(n-1)!}{4(n-1)!}$

$= \lim_{n\to\infty}\frac{n}{4} = \infty > 1$

Therefore, the series diverges by the Ratio Test.

21. $\sum_{n=0}^{\infty}(n+2)\left(\frac{9}{7}\right)^{n+1}$

$\lim_{n\to\infty}\left|\frac{a_{n+1}}{a_n}\right| = \lim_{n\to\infty}\left|\frac{(n+3)(9/7)^{n+2}}{(n+2)(9/7)^{n+1}}\right|$

$= \lim_{n\to\infty}\left(\frac{n+3}{n+2}\right)\left(\frac{9}{7}\right) = \frac{9}{7} > 1$

Therefore, the series diverges by the Ratio Test.

23. $\sum_{n=1}^{\infty}\frac{9^n}{n^5}$

$\lim_{n\to\infty}\left|\frac{a_n+1}{a_n}\right| = \lim_{n\to\infty}\left|\frac{9^{(n+1)}/(n+1)^5}{9^n/n^5}\right|$

$= \lim_{n\to\infty} 9\left(\frac{n+1}{n}\right)^5 = 9 > 1$

Therefore, the series diverges by the Ratio Test.

25. $\sum_{n=1}^{\infty}\frac{n^3}{3^n}$

$\lim_{n\to\infty}\left|\frac{a_{n+1}}{a_n}\right| = \lim_{n\to\infty}\left|\frac{(n+1)^3/3^{(n+1)}}{n^3/3^n}\right|$

$= \lim_{n\to\infty}\left(\frac{n+1}{n}\right)^3\frac{1}{3} = \frac{1}{3} < 1$

Therefore, the series converges by the Ratio Test.

27. $\sum_{n=0}^{\infty}\frac{(-1)^n 2^n}{n!}$

$\lim_{n\to\infty}\left|\frac{a_{n+1}}{a_n}\right| = \lim_{n\to\infty}\left|\frac{2^{n+1}}{(n+1)!}\cdot\frac{n!}{2^n}\right|$

$= \lim_{n\to\infty}\frac{2}{n+1} = 0$

Therefore, by the Ratio Test, the series converges.

29. $\sum_{n=1}^{\infty} \frac{n^2}{(n+1)(n^2+2)}$

$$\lim_{n\to\infty}\left|\frac{a_{n+1}}{a_n}\right| = \lim_{n\to\infty}\left[\frac{(n+1)^2}{(n+2)\left((n+1)^2+2\right)}\right] \Big/ \frac{n^2}{(n+1)(n^2+2)} = \lim_{n\to\infty}\frac{(n+1)^2(n+1)(n^2+1)}{n^2(n+2)(n^2+2n+3)} = 1$$

The Ratio Test is inconclusive.

Using the Limit Comparison Test with $\sum_{n=1}^{\infty}\frac{1}{n}$, the given series diverges.

31. $\sum_{n=0}^{\infty}\frac{e^n}{n!}$

$$\lim_{n\to\infty}\left|\frac{a_{n+1}}{a_n}\right| = \lim_{n\to\infty}\frac{e^{n+1}/(n+1)!}{e^n/n!} = \lim_{n\to\infty} e\left(\frac{n!}{(n+1)!}\right) = \lim_{n\to\infty}\frac{e}{n+1} = 0$$

Therefore, the series converges by the Ratio Test.

33. $\sum_{n=0}^{\infty}\frac{6^n}{(n+1)^n}$

$$\lim_{n\to\infty}\left|\frac{a_{n+1}}{a_n}\right| = \lim_{n\to\infty}\frac{6^{n+1}/(n+2)^{n+1}}{6^n/(n+1)^n} = \lim_{n\to\infty}\frac{6}{n+2}\left(\frac{n+1}{n+2}\right)^n = 0\left(\frac{1}{e}\right) = 0.$$

To find $\lim_{n\to\infty}\left(\frac{n+1}{n+2}\right)^n$: Let $y = \left(\frac{n+1}{n+2}\right)^n$

$$\ln y = n\ln\left(\frac{n+1}{n+2}\right) = \frac{\ln(n+1) - \ln(n+2)}{1/n}$$

$$\lim_{n\to\infty}[\ln y] = \lim_{n\to\infty}\left[\frac{1/(n+1) - 1/(n+2)}{-1/n^2}\right] = \lim_{n\to\infty}\left[\frac{-n^2\left[(n+2)-(n+1)\right]}{(n+1)(n+2)}\right] = -1$$

by L'Hôpital's Rule. So, $y \to \frac{1}{e}$.

Therefore, the series converges by the Ratio Test.

35. $\sum_{n=0}^{\infty}\frac{5^n}{2^n+1}$

$$\lim_{n\to\infty}\left|\frac{a_{n+1}}{a_n}\right| = \lim_{n\to\infty}\frac{5^{n+1}/(2^{n+1}+1)}{5^n/(2^n+1)} = \lim_{n\to\infty}\frac{5(2^n+1)}{(2^{n+1}+1)} = \lim_{n\to\infty}\frac{5(1+1/2^n)}{2+1/2^n} = \frac{5}{2} > 1$$

Therefore, the series diverges by the Ratio Test.

37. $\sum_{n=0}^{\infty}\frac{(-1)^{n+1}n!}{1\cdot 3\cdot 5\cdots(2n+1)}$

$$\lim_{n\to\infty}\left|\frac{a_{n+1}}{a_n}\right| = \lim_{n\to\infty}\left|\frac{(n+1)!}{1\cdot 3\cdot 5\cdots(2n+1)(2n+3)}\cdot\frac{1\cdot 3\cdot 5\cdots(2n+1)}{n!}\right| = \lim_{n\to\infty}\frac{n+1}{2n+3} = \frac{1}{2}$$

Therefore, by the Ratio Test, the series converges.

Note: The first few terms of this series are $-1 + \frac{1}{1\cdot 3} - \frac{2!}{1\cdot 3\cdot 5} + \frac{3!}{1\cdot 3\cdot 5\cdot 7} - \cdots$.

39. $\sum_{n=1}^{\infty}\left(\frac{n}{2n+1}\right)^n$

$$\lim_{n\to\infty}\sqrt[n]{|a_n|} = \lim_{n\to\infty}\sqrt[n]{\left(\frac{n}{2n+1}\right)^n} = \lim_{n\to\infty}\frac{n}{2n+1} = \frac{1}{2}$$

Therefore, by the Root Test, the series converges.

41. $\sum_{n=1}^{\infty}\left(\frac{3n+2}{n+3}\right)^n$

$$\lim_{n\to\infty}\sqrt[n]{|a_n|} = \lim_{n\to\infty}\sqrt[n]{\left(\frac{3n+2}{n+3}\right)^n}$$
$$= \lim_{n\to\infty}\frac{3n+2}{n+3} = 3 > 1$$

Therefore, the series diverges by the Root Test.

43. $\sum_{n=2}^{\infty}\frac{(-1)^n}{(\ln n)^n}$

$$\lim_{n\to\infty}\sqrt[n]{|a_n|} = \lim_{n\to\infty}\sqrt[n]{\left|\frac{(-1)^n}{(\ln n)^n}\right|} = \lim_{n\to\infty}\frac{1}{|\ln n|} = 0$$

Therefore, by the Root Test, the series converges.

45. $\sum_{n=1}^{\infty}\left(2\sqrt[n]{n}+1\right)^n$

$$\lim_{n\to\infty}\sqrt[n]{|a_n|} = \lim_{n\to\infty}\sqrt[n]{\left(2\sqrt[n]{n}+1\right)^n} = \lim_{n\to\infty}\left(2\sqrt[n]{n}+1\right)$$

To find $\lim_{n\to\infty}\sqrt[n]{n}$, let $y = \lim_{n\to\infty}\sqrt[n]{n}$. Then

$$\ln y = \lim_{n\to\infty}\left(\ln\sqrt[n]{n}\right)$$
$$= \lim_{n\to\infty}\frac{1}{n}\ln n = \lim_{n\to\infty}\frac{\ln n}{n} = \lim_{n\to\infty}\frac{1/n}{1} = 0.$$

So, $\ln y = 0$, so $y = e^0 = 1$ and

$$\lim_{n\to\infty}\left(2\sqrt[n]{n}+1\right) = 2(1)+1 = 3.$$

Therefore, by the Root Test, the series diverges.

47. $\sum_{n=1}^{\infty}\frac{n}{3^n}$

$$\lim_{n\to\infty}\sqrt[n]{|a_n|} = \lim_{n\to\infty}\left(\frac{n}{3^n}\right)^{1/n} = \lim_{n\to\infty}\frac{n^{1/n}}{3} = \frac{1}{3}$$

Therefore, the series converges by the Root Test.

Note: You can use L'Hôpital's Rule to show $\lim_{n\to\infty} n^{1/n} = 1$:

Let $y = n^{1/n}$, $\ln y = \frac{1}{n}\ln n = \frac{\ln n}{n}$

$$\lim_{n\to\infty}\frac{\ln n}{n} = \lim_{n\to\infty}\frac{1/n}{1} = 0 \Rightarrow y \to 1$$

49. $\sum_{n=1}^{\infty}\left(\frac{1}{n}-\frac{1}{n^2}\right)^n$

$$\lim_{n\to\infty}\sqrt[n]{|a_n|} = \lim_{n\to\infty}\sqrt[n]{\left(\frac{1}{n}-\frac{1}{n^2}\right)^n}$$
$$= \lim_{n\to\infty}\left(\frac{1}{n}-\frac{1}{n^2}\right) = 0 - 0 = 0 < 1$$

Therefore, by the Root Test, the series converges.

51. $\sum_{n=2}^{\infty}\frac{n}{(\ln n)^n}$

$$\lim_{n\to\infty}\sqrt[n]{|a_n|} = \lim_{n\to\infty}\sqrt[n]{\frac{n}{(\ln n)^n}} = \lim_{n\to\infty}\frac{n^{1/n}}{\ln n} = 0$$

Therefore, by the Root Test, the series converges.

53. $\sum_{n=1}^{\infty}\frac{(-1)^{n+1}5}{n}$

$$a_{n+1} = \frac{5}{n+1} < \frac{5}{n} = a_n$$
$$\lim_{n\to\infty}\frac{5}{n} = 0$$

Therefore, by the Alternating Series Test, the series converges (conditional convergence).

55. $\sum_{n=1}^{\infty}\frac{3}{n\sqrt{n}} = 3\sum_{n=1}^{\infty}\frac{1}{n^{3/2}}$

This is a convergent p-series.

57. $\sum_{n=1}^{\infty}\frac{5n}{2n-1}$

$$\lim_{n\to\infty}\frac{5n}{2n-1} = \frac{5}{2}$$

Therefore, the series diverges by the nth-Term Test

59. $\sum_{n=1}^{\infty}\frac{(-1)^n 3^{n-2}}{2^n} = \sum_{n=1}^{\infty}\frac{(-1)^n 3^n 3^{-2}}{2^n} = \sum_{n=1}^{\infty}\frac{1}{9}\left(-\frac{3}{2}\right)^n$

Because $|r| = \frac{3}{2} > 1$, this is a divergent geometric series.

61. $\sum_{n=1}^{\infty}\frac{10n+3}{n2^n}$

$$\lim_{n\to\infty}\frac{(10n+3)/n2^n}{1/2^n} = \lim_{n\to\infty}\frac{10n+3}{n} = 10$$

Therefore, the series converges by a Limit Comparison Test with the geometric series

$$\sum_{n=0}^{\infty}\left(\frac{1}{2}\right)^n.$$

63. $\left|\dfrac{\cos n}{3^n}\right| \le \dfrac{1}{3^n}$

Therefore the series $\displaystyle\sum_{n=1}^{\infty}\left|\frac{\cos n}{3^n}\right|$ converges by Direct comparison with the convergent geometric series $\displaystyle\sum_{n=1}^{\infty}\frac{1}{3^n}$.

So, $\displaystyle\sum\frac{\cos n}{3^n}$ converges.

65. $\displaystyle\sum_{n=1}^{\infty}\frac{n!}{n7^n}$

$$\lim_{n\to\infty}\left|\frac{a_{n+1}}{a_n}\right| = \lim_{n\to\infty}\left|\frac{(n+1)!/(n+1)7^{n+1}}{n!/n7^n}\right| = \lim_{n\to\infty}\frac{(n+1)!\,n}{(n+1)\,n!}7 = \lim_{n\to\infty}7n = \infty$$

Therefore, the series diverges by the Ratio Test.

67. $\displaystyle\sum_{n=1}^{\infty}\frac{(-1)^n 3^{n-1}}{n!}$

$$\lim_{n\to\infty}\left|\frac{a_{n+1}}{a_n}\right| = \lim_{n\to\infty}\left|\frac{3^n}{(n+1)!}\cdot\frac{n!}{3^{n-1}}\right| = \lim_{n\to\infty}\frac{3}{n+1} = 0$$

Therefore, by the Ratio Test, the series converges. (Absolutely)

69. $\displaystyle\sum_{n=1}^{\infty}\frac{(-3)^n}{3\cdot5\cdot7\cdots(2n+1)}$

$$\lim_{n\to\infty}\left|\frac{a_{n+1}}{a_n}\right| = \lim_{n\to\infty}\left|\frac{(-3)^{n+1}}{3\cdot5\cdot7\cdots(2n+1)(2n+3)}\cdot\frac{3\cdot5\cdot7\cdots(2n+1)}{(-3)^n}\right| = \lim_{n\to\infty}\frac{3}{2n+3} = 0$$

Therefore, by the Ratio Test, the series converges.

71. (a) and (c) are the same.

$$\sum_{n=1}^{\infty}\frac{n5^n}{n!} = \sum_{n=0}^{\infty}\frac{(n+1)5^{n+1}}{(n+1)!}$$
$$= 5 + \frac{(2)(5)^2}{2!} + \frac{(3)(5)^3}{3!} + \frac{(4)(5)^4}{4!} + \cdots$$

73. (a) and (b) are the same.

$$\sum_{n=0}^{\infty}\frac{(-1)^n}{(2n+1)!} = \sum_{n=1}^{\infty}\frac{(-1)^{n-1}}{(2n-1)!}$$
$$= 1 - \frac{1}{3!} + \frac{1}{5!} - \cdots$$

75. Replace n with $n+1$.

$$\sum_{n=1}^{\infty}\frac{n}{7^n} = \sum_{n=0}^{\infty}\frac{n+1}{7^{n+1}}$$

77. $$\lim_{n\to\infty}\left|\frac{a_{n+1}}{a_n}\right| = \lim_{n\to\infty}\left|\frac{(4n-1)/(3n+2)a_n}{a_n}\right|$$
$$= \lim_{n\to\infty}\frac{4n-1}{3n+2} = \frac{4}{3} > 1$$

The series diverges by the Ratio Test.

79. $$\lim_{n\to\infty}\left|\frac{a_{n+1}}{a_n}\right| = \lim_{n\to\infty}\left|\frac{(\sin n+1)/(\sqrt{n})a_n}{a_n}\right|$$
$$= \lim_{n\to\infty}\frac{\sin n+1}{\sqrt{n}} = 0 < 1$$

The series converges by the Ratio Test.

81. $$\lim_{n\to\infty}\left|\frac{a_{n+1}}{a_n}\right| = \lim_{n\to\infty}\left|\frac{(1+(1)/(n))a_n}{a_n}\right| = \lim_{n\to\infty}\left(1+\frac{1}{n}\right) = 1$$

The Ratio Test is inconclusive.

But, $\lim\limits_{n\to\infty} a_n \ne 0$, so the series diverges.

83. $\displaystyle\lim_{n\to\infty}\left|\frac{a_{n+1}}{a_n}\right| = \lim_{n\to\infty}\left|\frac{\dfrac{1\cdot 2\cdots n(n+1)}{1\cdot 3\cdots(2n-1)(2n+1)}}{\dfrac{1\cdot 2\cdots n}{1\cdot 3\cdots(2n-1)}}\right|$

$\displaystyle = \lim_{n\to\infty}\frac{n+1}{2n+1} = \frac{1}{2} < 1$

The series converges by the Ratio Test.

85. $\displaystyle\sum_{n=3}^{\infty}\frac{1}{(\ln n)^n}$

$\displaystyle\lim_{n\to\infty}\sqrt[n]{|a_n|} = \lim_{n\to\infty}\sqrt[n]{\frac{1}{(\ln n)^n}} = \lim_{n\to\infty}\frac{1}{\ln n} = 0$

Therefore, by the Root Test, the series converges.

87. $\displaystyle\sum_{n=0}^{\infty}2\left(\frac{x}{3}\right)^n$

$\displaystyle\lim_{n\to\infty}\left|\frac{a_{n+1}}{a_n}\right| = \lim_{n\to\infty}\left|\frac{2(x/3)^{n+1}}{2(x/3)^n}\right| = \lim_{n\to\infty}\left|\frac{x}{3}\right| = \left|\frac{x}{3}\right|$

For the series to converge, $\left|\dfrac{x}{3}\right| < 1 \Rightarrow -3 < x < 3.$

For $x = 3$, $\displaystyle\sum_{n=0}^{\infty}2(1)^n$ diverges.

For $x = -3$, $\displaystyle\sum_{n=0}^{\infty}2(-1)^n$ diverges.

89. $\displaystyle\sum_{n=1}^{\infty}\frac{(-1)^n(x+1)^n}{n}$

$\displaystyle\lim_{n\to\infty}\left|\frac{a_{n+1}}{a_n}\right| = \lim_{n\to\infty}\left|\frac{(x+1)^{n+1}/(n+1)}{x^n/n}\right|$

$\displaystyle = \lim_{n\to\infty}\left|\frac{n}{n+1}(x+1)\right| = |x+1|$

For the series to converge,

$|x+1| < 1 \Rightarrow -1 < x+1 < 1$

$\Rightarrow -2 < x < 0.$

For $x = 0$, $\displaystyle\sum_{n=1}^{\infty}\frac{(-1)^n}{n}$ converges.

For $x = -2$, $\displaystyle\sum_{n=1}^{\infty}\frac{(-1)^n(-1)^n}{n} = \sum_{n=1}^{\infty}\frac{1}{n}$ diverges.

91. $\displaystyle\sum_{n=0}^{\infty}n!\left(\frac{x}{2}\right)^n$

$\displaystyle\lim_{n\to\infty}\left|\frac{a_{n+1}}{a_n}\right| = \lim_{n\to\infty}\frac{(n+1)!\left|\dfrac{x}{2}\right|^{n+1}}{n!\left|\dfrac{x}{2}\right|^n}$

$\displaystyle = \lim_{n\to\infty}(n+1)\left|\frac{x}{2}\right| = \infty$

The series converges only at $x = 0$.

93. The Ratio Test will be inconclusive when a_n is a rational function of n. In this case,

$\displaystyle\lim_{n\to\infty}\frac{a_{n+1}}{a_n} = 1.$

95. No. Let $a_n = \dfrac{1}{n+10{,}000}$.

The series $\displaystyle\sum_{n=1}^{\infty}\frac{1}{n+10{,}000}$ diverges.

97. Assume that

$\displaystyle\lim_{n\to\infty}|a_{n+1}/a_n| = L > 1$ or that $\displaystyle\lim_{n\to\infty}|a_{n+1}/a_n| = \infty.$

Then there exists $N > 0$ such that $|a_{n+1}/a_n| > 1$ for all $n > N$. Therefore,

$|a_{n+1}| > |a_n|,\ n > N \Rightarrow \displaystyle\lim_{n\to\infty}a_n \neq 0 \Rightarrow \sum a_n$ diverges.

99. $\displaystyle\sum_{n=1}^{\infty}\frac{1}{n^{3/2}}$

$\displaystyle\lim_{n\to\infty}\left|\frac{a_{n+1}}{a_n}\right| = \lim_{n\to\infty}\left|\frac{1}{(n+1)^{3/2}}\cdot\frac{n^{3/2}}{1}\right| = \lim_{n\to\infty}\left(\frac{n}{n+1}\right)^{3/2} = 1$

101. $\displaystyle\sum_{n=1}^{\infty}\frac{1}{n^4}$

$\displaystyle\lim_{n\to\infty}\left|\frac{a_{n+1}}{a_n}\right| = \lim_{n\to\infty}\left|\frac{1}{(n+1)^4}\cdot\frac{n^4}{1}\right| = \lim_{n\to\infty}\left(\frac{n}{n+1}\right)^4 = 1$

103. $\displaystyle\sum_{n=1}^{\infty}\frac{1}{n^p}$, p-series

$\displaystyle\lim_{n\to\infty}\sqrt[n]{|a_n|} = \lim_{n\to\infty}\sqrt[n]{\frac{1}{n^p}} = \lim_{n\to\infty}\frac{1}{n^{p/n}} = 1$

So, the Root Test is inconclusive.

Note: $\displaystyle\lim_{n\to\infty}n^{p/n} = 1$ because if $y = n^{p/n}$, then

$\ln y = \dfrac{p}{n}\ln n$ and $\dfrac{p}{n}\ln n \to 0$ as $n \to \infty$.

So $y \to 1$ as $n \to \infty$.

105. $\sum_{n=1}^{\infty} \frac{(n!)^2}{(xn)!}$, x positive integer

(a) $x = 1$: $\sum \frac{(n!)^2}{n!} = \sum n!$, diverges

(b) $x = 2$: $\sum \frac{(n!)^2}{(2n)!}$ converges by the Ratio Test:

$$\lim_{n\to\infty} \frac{[(n+1)!]^2}{(2n+2)!} \bigg/ \frac{(n!)^2}{(2n)!} = \lim_{n\to\infty} \frac{(n+1)^2}{(2n+2)(2n+1)} = \frac{1}{4} < 1$$

(c) $x = 3$: $\sum \frac{(n!)^2}{(3n)!}$ converges by the Ratio Test:

$$\lim_{n\to\infty} \frac{[(n+1)!]^2}{(3n+3)!} \bigg/ \frac{(n!)^2}{(3n)!} = \lim_{n\to\infty} \frac{(n+1)^2}{(3n+3)(3n+2)(3n+1)} = 0 < 1$$

(d) Use the Ratio Test:

$$\lim_{n\to\infty} \frac{[(n+1)!]^2}{[x(n+1)]!} \bigg/ \frac{(n!)^2}{(xn)!} = \lim_{n\to\infty} (n+1)^2 \frac{(xn)!}{(xn+x)!}$$

The cases $x = 1, 2, 3$ were solved above. For $x > 3$, the limit is 0. So, the series converges for all integers $x \geq 2$.

107. First prove Abel's Summation Theorem:

If the partial sums of $\sum a_n$ are bounded and if $\{b_n\}$ decreases to zero, then $\sum a_n b_n$ converges.

Let $S_k = \sum_{i=1}^{k} a_i$. Let M be a bound for $\{|S_k|\}$.

$$\begin{aligned} a_1b_1 + a_2b_2 + \cdots + a_nb_n &= S_1b_1 + (S_2 - S_1)b_2 + \cdots + (S_n - S_{n-1})b_n \\ &= S_1(b_1 - b_2) + S_2(b_2 - b_3) + \cdots + S_{n-1}(b_{n-1} - b_n) + S_nb_n \\ &= \sum_{i=1}^{n-1} S_i(b_i - b_{i+1}) + S_nb_n \end{aligned}$$

The series $\sum_{i=1}^{\infty} S_i(b_i - b_{i+1})$ is absolutely convergent because $|S_i(b_i - b_{i+1})| \leq M(b_i - b_{i+1})$ and $\sum_{i=1}^{\infty}(b_i - b_{i+1})$ converges to b_1.

Also, $\lim_{n\to\infty} S_nb_n = 0$ because $\{S_n\}$ bounded and $b_n \to 0$. Thus, $\sum_{n=1}^{\infty} a_nb_n = \lim_{n\to\infty}\sum_{i=1}^{n} a_ib_i$ converges.

Now let $b_n = \frac{1}{n}$ to finish the problem.

Section 9.7 Taylor Polynomials and Approximations

1. The graph of the approximating polynomial P and the elementary function f both pass through the point $(c, f(c))$ and the slopes of P and f agree at $(c, f(c))$. Depending on the degree of P, the nth derivatives of P and f agree at $(c, f(c))$.

3. The accuracy is represented by the remainder of the Taylor polynomial. The remainder is

$$R_n(x) = \frac{f^{(n+1)}(z)(x-c)^{n+1}}{(n+1)!}.$$

5. $y = -\frac{1}{2}x^2 + 1$

Parabola

Matches (d)

6. $y = \frac{1}{8}x^4 - \frac{1}{2}x^2 + 1$

y-axis symmetry

Three relative extrema

Matches (c)

7. $y = e^{-1/2}\left[(x+1)+1\right]$

Linear

Matches (a)

8. $y = e^{-1/2}\left[\frac{1}{3}(x-1)^3 - (x-1) + 1\right]$

Cubic

Matches (b)

9. $f(x) = \dfrac{\sqrt{x}}{4}, C = 4, f(4) = \dfrac{1}{2}$

$f'(x) = \dfrac{1}{8\sqrt{x}}, f'(4) = \dfrac{1}{16}$

$P_1(x) = f(4) + f'(4)(x-4)$

$= \dfrac{1}{2} + \dfrac{1}{16}(x-4)$

$= \dfrac{1}{16}x + \dfrac{1}{4}$

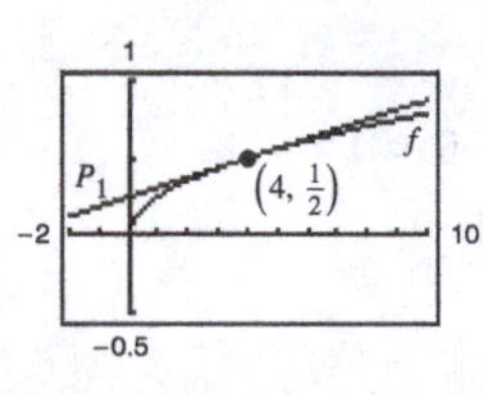

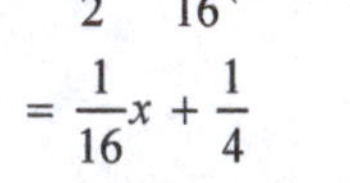

P_1 is the first-degree Taylor polynomial for f at 4.

11. $f(x) = \sec x \qquad f\left(\dfrac{\pi}{6}\right) = \dfrac{2\sqrt{3}}{3}$

$f'(x) = \sec x \tan x \qquad f'\left(\dfrac{\pi}{6}\right) = \dfrac{2}{3}$

$P_1(x) = f\left(\dfrac{\pi}{6}\right) + f'\left(\dfrac{\pi}{6}\right)\left(x - \dfrac{\pi}{6}\right)$

$P_1(x) = \dfrac{2\sqrt{3}}{3} + \dfrac{2}{3}\left(x - \dfrac{\pi}{6}\right)$

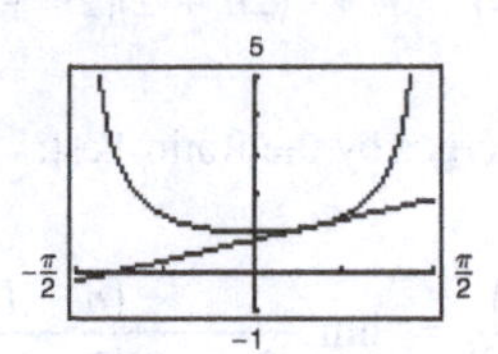

P_1 is the first degree Taylor polynomial for f at $\dfrac{\pi}{6}$.

13. $f(x) = \dfrac{4}{\sqrt{x}} = 4x^{-1/2} \qquad f(1) = 4$

$f'(x) = -2x^{-3/2} \qquad f'(1) = -2$

$f''(x) = 3x^{-5/2} \qquad f''(1) = 3$

$P_2 = f(1) + f'(1)(x-1) + \dfrac{f''(1)}{2}(x-1)^2$

$= 4 - 2(x-1) + \dfrac{3}{2}(x-1)^2$

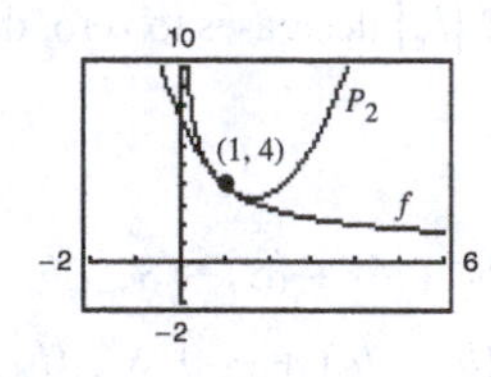

x	0	0.8	0.9	1.0	1.1	1.2	2
$f(x)$	Error	4.4721	4.2164	4.0	3.8139	3.6515	2.8284
$P_2(x)$	7.5	4.46	4.215	4.0	3.815	3.66	3.5

15. $f(x) = \cos x$

$P_2(x) = 1 - \frac{1}{2}x^2$

$P_4(x) = 1 - \frac{1}{2}x^2 + \frac{1}{24}x^4$

$P_6(x) = 1 - \frac{1}{2}x^2 + \frac{1}{24}x^4 - \frac{1}{720}x^6$

(a)

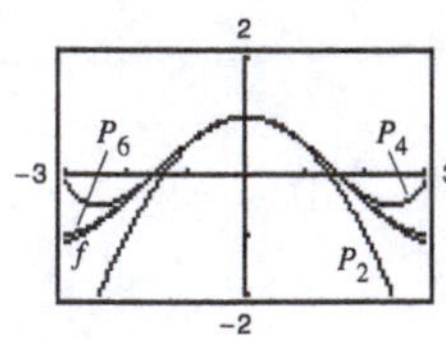

(b) $f'(x) = -\sin x \qquad P_2'(x) = -x$

$f''(x) = -\cos x \qquad P_2''(x) = -1$

$f''(0) = P_2''(0) = -1$

$f'''(x) = \sin x \qquad P_4'''(x) = x$

$f^{(4)}(x) = \cos x \qquad P_4^{(4)}(x) = 1$

$f^{(4)}(0) = 1 = P_4^{(4)}(0)$

$f^{(5)}(x) = -\sin x \qquad P_6^{(5)}(x) = -x$

$f^{(6)}(x) = -\cos x \qquad P^{(6)}(x) = -1$

$f^{(6)}(0) = -1 = P_6^{(6)}(0)$

(c) In general, $f^{(n)}(0) = P_n^{(n)}(0)$ for all n.

17. $f(x) = e^{4x}$ $\quad f(0) = 1$

$f'(x) = 4e^{4x}$ $\quad f'(0) = 4$

$f''(x) = 16e^{4x}$ $\quad f''(0) = 16$

$f'''(x) = 64e^{4x}$ $\quad f'''(0) = 64$

$f^{(4)}(x) = 256e^{4x}$ $\quad f^{(4)}(0) = 256$

$$P_4(x) = f(0) + f'(0)x + \frac{f''(0)}{2!}x^2 + \frac{f'''(0)}{3!}x^3 + \frac{f^{(4)}(0)}{4!}x^4 = 1 + 4x + 8x^2 + \frac{32}{3}x^3 + \frac{32}{3}x^4$$

19. $f(x) = \sin x$ $\quad f(0) = 0$

$f'(x) = \cos x$ $\quad f'(0) = 1$

$f''(x) = -\sin x$ $\quad f''(0) = 0$

$f'''(x) = -\cos x$ $\quad f'''(0) = -1$

$f^{(4)}(x) = \sin x$ $\quad f^{(4)}(0) = 0$

$f^{(5)}(x) = \cos x$ $\quad f^{(5)}(0) = 1$

$$P_5(x) = 0 + (1)x + \frac{0}{2!}x^2 + \frac{-1}{3!}x^3 + \frac{0}{4!}x^4 + \frac{1}{5!}x^5 = x - \frac{1}{6}x^3 + \frac{1}{120}x^5$$

21. $f(x) = xe^x$ $\quad f(0) = 0$

$f'(x) = xe^x + e^x$ $\quad f'(0) = 1$

$f''(x) = xe^x + 2e^x$ $\quad f''(0) = 2$

$f'''(x) = xe^x + 3e^x$ $\quad f'''(0) = 3$

$f^{(4)}(x) = xe^x + 4e^x$ $\quad f^{(4)}(0) = 4$

$$P_4(x) = 0 + x + \frac{2}{2!}x^2 + \frac{3}{3!}x^3 + \frac{4}{4!}x^4$$

$$= x + x^2 + \frac{1}{2}x^3 + \frac{1}{6}x^4$$

23. $f(x) = \dfrac{1}{1-x}$ $\quad f(0) = 1$

$f'(x) = \dfrac{1}{(1-x)^2}$ $\quad f'(0) = 1$

$f''(x) = \dfrac{2}{(1-x)^3}$ $\quad f''(0) = 2$

$f'''(x) = \dfrac{6}{(1-x)^4}$ $\quad f'''(0) = 6$

$f^{(4)}(x) = \dfrac{24}{(1-x)^5}$ $\quad f^{(4)}(0) = 24$

$f^{(5)}(x) = \dfrac{120}{(1-x)^6}$ $\quad f^{(5)}(0) = 120$

$$P_5(x) = 1 + x + \frac{2x^2}{2!} + \frac{6x^3}{3!} + \frac{24x^4}{4!} + \frac{120x^5}{5!}$$

$$= 1 + x + x^2 + x^3 + x^4 + x^5$$

25. $f(x) = \sec x$ $\quad f(0) = 1$

$f'(x) = \sec x \tan x$ $\quad f'(0) = 0$

$f''(x) = \sec^3 x + \sec x \tan^2 x$ $\quad f''(0) = 1$

$$P_2(x) = 1 + 0x + \frac{1}{2!}x^2 = 1 + \frac{1}{2}x^2$$

27. $f(x) = \dfrac{2}{x} = 2x^{-1}$ $\quad f(1) = 2$

$f'(x) = -2x^{-2}$ $\quad f'(1) = -2$

$f''(x) = 4x^{-3}$ $\quad f''(1) = 4$

$f'''(x) = -12x^{-4}$ $\quad f'''(1) = -12$

$$P_3(x) = 2 - 2(x-1) + \frac{4}{2!}(x-1)^2 - \frac{12}{3!}(x-1)^3$$

$$= 2 - 2(x-1) + 2(x-1)^2 - 2(x-1)^3$$

29. $f(x) = \sqrt{x}$ $\quad f(4) = 2$

$f'(x) = \dfrac{1}{2\sqrt{x}}$ $\quad f'(4) = \dfrac{1}{4}$

$f''(x) = -\dfrac{1}{4x^{3/2}}$ $\quad f''(4) = -\dfrac{1}{32}$

$$P_2(x) = 2 + \frac{1}{4}(x-4) - \frac{1/32}{2}(x-4)^2$$

$$= 2 + \frac{1}{4}(x-4) - \frac{1}{64}(x-4)^2$$

31. $f(x) = \ln x \qquad f(2) = \ln 2$

$f'(x) = \dfrac{1}{x} = x^{-1} \qquad f'(2) = 1/2$

$f''(x) = -x^{-2} \qquad f''(2) = -1/4$

$f'''(x) = 2x^{-3} \qquad f'''(2) = 1/4$

$f^{(4)}(x) = -6x^{-4} \qquad f^{(4)}(2) = -3/8$

$$P_4(x) = \ln 2 + \frac{1}{2}(x-2) - \frac{1/4}{2!}(x-2)^2 + \frac{1/4}{3!}(x-2)^3 - \frac{3/8}{4!}(x-2)^4$$

$$= \ln 2 + \frac{1}{2}(x-2) - \frac{1}{8}(x-2)^2 + \frac{1}{24}(x-2)^3 - \frac{1}{64}(x-2)^4$$

33. (a) $P_3(x) = \pi x + \dfrac{\pi^3}{3}x^3$

(b) $Q_3(x) = 1 + 2\pi\left(x - \dfrac{1}{4}\right) + 2\pi^2\left(x - \dfrac{1}{4}\right)^2 + \dfrac{8}{3}\pi^3\left(x - \dfrac{1}{4}\right)^3$

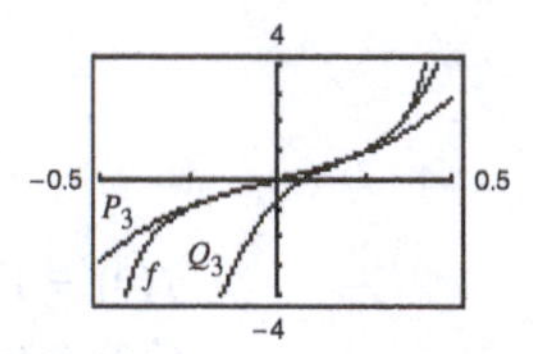

35. $f(x) = \sin x$

$P_1(x) = x$

$P_3(x) = x - \frac{1}{6}x^3$

$P_5(x) = x - \frac{1}{6}x^3 + \frac{1}{120}x^5$

(a)

x	0.00	0.25	0.50	0.75	1.00
$\sin x$	0.0000	0.2474	0.4794	0.6816	0.8415
$P_1(x)$	0.0000	0.2500	0.5000	0.7500	1.0000
$P_3(x)$	0.0000	0.2474	0.4792	0.6797	0.8333
$P_5(x)$	0.0000	0.2474	0.4794	0.6817	0.8417

(b)

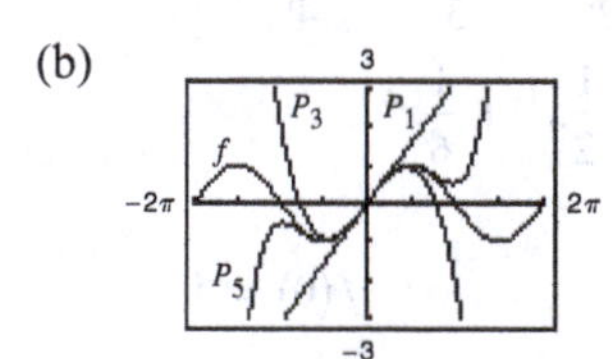

(c) As the distance increases, the accuracy decreases.

37. $f(x) = \arcsin x$

(a) $P_3(x) = x + \dfrac{x^3}{6}$

(b)

x	−0.75	−0.50	−0.25	0	0.25	0.50	0.75
$f(x)$	−0.848	−0.524	−0.253	0	0.253	0.524	0.848
$P_3(x)$	−0.820	−0.521	−0.253	0	0.253	0.521	0.820

(c)

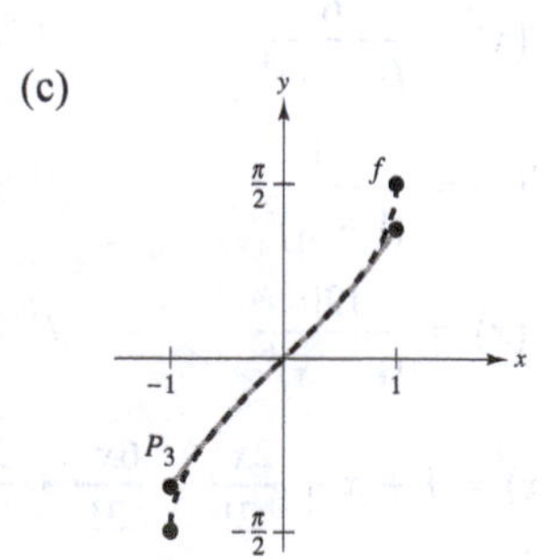

39. $f(x) = e^{4x} \approx 1 + 4x + 8x^2 + \frac{32}{3}x^3 + \frac{32}{3}x^4$

$f\left(\frac{1}{4}\right) \approx 2.7083$

41. $f(x) = \frac{1}{x^2} \approx \frac{1}{4} + \frac{1}{4}(x+2) + \frac{3}{16}(x+2)^2 + \frac{1}{8}(x+2)^3 + \frac{5}{64}(x+2)^4$

$f(-2.1) \approx \frac{1}{4} + \frac{1}{4}(-0.1) + \frac{3}{16}(-0.1)^2 + \frac{1}{8}(-0.1)^3 + \frac{5}{64}(-0.1)^4 \approx 0.227$

43. $f(x) = \ln x \approx \ln(2) + \frac{1}{2}(x-2) - \frac{1}{8}(x-2)^2 + \frac{1}{24}(x-2)^3 - \frac{1}{64}(x-2)^4$

$f(2.1) \approx 0.7419$

45. $f(x) = \cos x$; $f^{(5)}(x) = -\sin x \Rightarrow$ Max on $[0, 0.3]$ is 1.

$R_4(x) \le \frac{1}{5!}(0.3)^5 = 2.025 \times 10^{-5}$

Note: you could use $R_5(x)$: $f^{(6)}(x) = -\cos x$, max on $[0, 0.3]$ is 1.

$R_5(x) \le \frac{1}{6!}(0.3)^6 = 1.0125 \times 10^{-6}$

Exact error: $\left|\cos(0.3) - \left(1 - \frac{0.3^2}{2!} + \frac{0.3^4}{4!}\right)\right| \approx 0.000001 = 1.0 \times 10^{-6}$

47. $f(x) = \sinh x$; $f^{(6)}(x) = \sinh x \Rightarrow$ Max on $[0, 0.2]$ is $\sinh(0.2) \approx 0.2013$.

$R_5 \le \frac{(0.2)^6}{6!}(0.2013) \approx 1.8 \times 10^{-8}$

Exact error: $\left|\sinh(0.2) - \left(0.2 + \frac{(0.2)^3}{3!} + \frac{(0.2)^5}{5!}\right)\right| \approx 2.5 \times 10^{-9}$

49. $f(x) = \arcsin x$; $f^{(4)}(x) = \frac{x(6x^2+9)}{(1-x^2)^{7/2}} \Rightarrow$ Max on $[0, 0.4]$ is $f^{(4)}(0.4) \approx 7.3340$.

$R_3(x) \le \frac{7.3340}{4!}(0.4)^4 \approx 0.00782 = 7.82 \times 10^{-3}$

Exact error: $\left|\arcsin(0.4) - \left(0.4 + \frac{(0.4)^3}{2.3}\right)\right| \approx 0.00085 = 8.5 \times 10^{-4}$ [Note: You could use R_4.]

51. $g(x) = \sin x$

$\left|g^{(n+1)}(x)\right| \le 1$ for all x and all n.

$R_n(x) = \left|\frac{f^{(n+1)}(z)\,x^{n+1}}{(n+1)!}\right| \le \frac{1}{(n+1)!}(0.3)^{n+1} < 0.001$

By trial and error, $n = 3$.

53. $f(x) = e^x$

$f^{(n+1)}(x) = e^x$

Max on $[0, 0.6]$ is $e^{0.6} \approx 1.8221$.

$R_n = \left|\frac{f^{(n+1)}(z)\,x^{n+1}}{(n+1)!}\right| \le \frac{1.8221}{(n+1)!}(0.6)^{n+1} < 0.001$

By trial and error, $n = 5$.

55. $f(x) = \frac{1}{x-2}$

$\left|f^{(n+1)}(x)\right| = \left|\frac{\pm(n+1)!}{(x-2)^{n+2}}\right| \le \frac{(n+1)!}{2^{n+2}}$ on $[0, 0.15]$

$\left|R_n(x)\right| = \left|\frac{f^{(n+1)}(z)x^{n+1}}{(n+1)!}\right| \le \frac{(n+1)!(0.15)^{n+1}}{2^{n+2}(n+1)!}$

$= \frac{(0.15)^{n+1}}{2^{n+2}} < 0.001$

By trial and error, $n = 2$.

57. $f(x) = \ln(x + 1)$

$$f^{(n+1)}(x) = \frac{(-1)^n n!}{(x + 1)^{n+1}} \Rightarrow \text{Max on } [0, 0.5] \text{ is } n!.$$

$$R_n \le \frac{n!}{(n + 1)!}(0.5)^{n+1} = \frac{(0.5)^{n+1}}{n + 1} < 0.0001$$

By trial and error, $n = 9$. (See Example 9.) Using 9 terms, $\ln(1.5) \approx 0.4055$.

59. $$f(x) = e^x \approx 1 + x + \frac{x^2}{2} + \frac{x^3}{6}, x < 0$$

$$R_3(x) = \frac{e^z}{4!}x^4 < 0.001$$

$$e^z x^4 < 0.024$$

$$\left|xe^{z/4}\right| < 0.3936$$

$$|x| < \frac{0.3936}{e^{z/4}} < 0.3936, z < 0$$

$$-0.3936 < x < 0$$

61. $f(x) = \cos x \approx 1 - \frac{x^2}{2!} + \frac{x^4}{4!}$, fifth degree polynomial

$\left|f^{(n+1)}(x)\right| \le 1$ for all x and all n.

$$\left|R_5(x)\right| \le \frac{1}{6!}|x|^6 < 0.001$$

$$|x|^6 < 0.72$$

$$|x| < 0.9467$$

$$-0.9467 < x < 0.9467$$

Note: Use a graphing utility to graph $y = \cos x - \left(1 - x^2/2 + x^4/24\right)$ in the viewing window $[-0.9467, 0.9467] \times [-0.001, 0.001]$ to verify the answer.

63. The tangent line is the same as the first Taylor polynomial for the function centered at the point.

65. $f(x) = e^x$

From Example 2,

$$P_4(x) = 1 + x + \frac{1}{2}x^2 + \frac{1}{6}x^3 + \frac{1}{24}x^4.$$

To obtain the fourth Maclaurin polynomial of $g(x) = e^{2x}$, replace x with $2x$:

$$Q_4(x) = 1 + (2x) + \frac{1}{2}(2x)^2 + \frac{1}{6}(2x)^3 + \frac{1}{24}(2x)^4$$

$$= 1 + 2x + 2x^2 + \frac{4}{3}x^3 + \frac{2}{3}x^4$$

67. (a) $f(x) = e^x$

$$P_4(x) = 1 + x + \frac{1}{2}x^2 + \frac{1}{6}x^3 + \frac{1}{24}x^4$$

$$g(x) = xe^x$$

$$Q_5(x) = x + x^2 + \frac{1}{2}x^3 + \frac{1}{6}x^4 + \frac{1}{24}x^5$$

$$Q_5(x) = x\,P_4(x)$$

(b) $f(x) = \sin x$

$$P_5(x) = x - \frac{x^3}{3!} + \frac{x^5}{5!}$$

$$g(x) = x \sin x$$

$$Q_6(x) = x\,P_5(x) = x^2 - \frac{x^4}{3!} + \frac{x^6}{5!}$$

(c) $g(x) = \frac{\sin x}{x} = \frac{1}{x}P_5(x) = 1 - \frac{x^2}{3!} + \frac{x^4}{5!}$

69. (a) $Q_2(x) = -1 + \frac{\pi^2(x + 2)^2}{32}$

(b) $R_2(x) = -1 + \frac{\pi^2(x - 6)^2}{32}$

(c) No. The polynomial will be linear. Horizontal translations of the result in part (a) are possible only at $x = -2 + 8n$ (where n is an integer) because the period of f is 8.

71. Let f be an even function and P_n be the nth Maclaurin polynomial for f. Because f is even, f' is odd, f'' is even, f''' is odd, etc. All of the odd derivatives of f are odd and so, all of the odd powers of x will have coefficients of zero. P_n will only have terms with even powers of x.

73. Because f has continuous first and second derivative and f has a relative maximum at $x = c$, you know that $f'(c) = 0$ and $f''(c) < 0$. The second Taylor polynomial for f centered at $x = c$ is

$$P_2(x) = f(c) + f'(c)(x - c) + \frac{f''(c)}{2!}(x - c)^2$$

$$= f(c) + \frac{f''(c)}{2}(x - c)^2.$$

Now, $P_2'(x) = f''(c)(x - c)$ and $P_2''(x) = f''(c)$. You have $P_2'(c) = 0$ and $P_2''(c) = f''(c) < 0$, so $P_2(x)$ has a relative maximum at c.

Section 9.8 Power Series

1. A Maclaurin polynomial approximates a function, whereas a power series exactly represents a function. The Maclaurin polynomial has a finite number of terms and a power series has an infinite number of terms.

3. $\left|\dfrac{x-2}{5}\right| < 1 \Rightarrow |x-2| < 5 \Rightarrow -5 < x-2 < 5$

The radius is 5.

5. Centered at 0

7. Centered at 2

9. $\displaystyle\sum_{n=0}^{\infty}(-1)^n\frac{x^n}{n+1}$

$$L = \lim_{n\to\infty}\left|\frac{u_{n+1}}{u_n}\right| = \lim_{n\to\infty}\left|\frac{(-1)^{n+1}x^{n+1}}{n+2}\cdot\frac{n+1}{(-1)^n x^n}\right|$$

$$= \lim_{n\to\infty}\left|\frac{n+1}{n+2}\right||x| = |x|$$

$|x| < 1 \Rightarrow R = 1$

11. $\displaystyle\sum_{n=1}^{\infty}\frac{(4x)^n}{n^2}$

$$L = \lim_{n\to\infty}\left|\frac{u_{n+1}}{u_n}\right|$$

$$= \lim_{n\to\infty}\left|\frac{(4x)^{n+1}/(n+1)^2}{(4x)^n/n^2}\right| = \lim_{n\to\infty}\left|\frac{n^2}{(n+1)^2}(4x)\right| = 4|x|$$

$4|x| < 1 \Rightarrow R = \dfrac{1}{4}$

13. $\displaystyle\sum_{n=0}^{\infty}\frac{x^{2n}}{(2n)!}$

$$L = \lim_{n\to\infty}\left|\frac{u_{n+1}}{u_n}\right|$$

$$= \lim_{n\to\infty}\left|\frac{x^{(2n+2)}/(2n+2)!}{x^{2n}/(2n)!}\right|$$

$$= \lim_{n\to\infty}\left|\frac{x^2}{(2n+2)(2n+1)}\right| = 0$$

So, the series converges for all $x \Rightarrow R = \infty$.

15. $\displaystyle\sum_{n=0}^{\infty}\left(\frac{x}{4}\right)^n$

Because the series is geometric, it converges only if

$\left|\dfrac{x}{4}\right| < 1$, or $-4 < x < 4$.

17. $\displaystyle\sum_{n=1}^{\infty}\frac{(-1)^n x^n}{n}$

$$\lim_{n\to\infty}\left|\frac{u_{n+1}}{u_n}\right| = \lim_{n\to\infty}\left|\frac{(-1)^{n+1}x^{n+1}}{n+1}\cdot\frac{n}{(-1)^n x^n}\right|$$

$$= \lim_{n\to\infty}\left|\frac{nx}{n+1}\right| = |x|$$

Interval: $-1 < x < 1$

When $x = 1$, the alternating series $\displaystyle\sum_{n=1}^{\infty}\frac{(-1)^n}{n}$ converges.

When $x = -1$, the p-series $\displaystyle\sum_{n=1}^{\infty}\frac{1}{n}$ diverges.

Therefore, the interval of convergence is $(-1, 1]$.

19. $\displaystyle\sum_{n=0}^{\infty}\frac{x^{5n}}{n!}$

$$\lim_{n\to\infty}\left|\frac{u_{n+1}}{u_n}\right| = \lim_{n\to\infty}\left|\frac{x^{5(n+1)}/(n+1)!}{5^n/n!}\right| = \lim_{n\to\infty}\left|\frac{x^5}{n+1}\right| = 0$$

The series converges for all x. The interval of convergence is $(-\infty, \infty)$.

21. $\displaystyle\sum_{n=0}^{\infty}(2n)!\left(\frac{x}{3}\right)^n$

$$\lim_{n\to\infty}\left|\frac{u_{n+1}}{u_n}\right| = \lim_{n\to\infty}\left|\frac{(2n+2)!(x/3)^{n+1}}{(2n)!(x/3)^n}\right|$$

$$= \left|\frac{(2n+2)(2n+1)x}{3}\right| = \infty$$

The series converges only for $x = 0$.

23. $\displaystyle\sum_{n=1}^{\infty}\frac{(-1)^{n+1}x^n}{6^n}$

Because the series is geometric, it converges only if

$\left|\dfrac{x}{6}\right| < 1 \Rightarrow |x| < 6$ or $-6 < x < 6$.

25. $\sum_{n=1}^{\infty} \frac{(-1)^{n+1}(x-4)^n}{n9^n}$

$$\lim_{n\to\infty}\left|\frac{u_{n+1}}{u_n}\right| = \lim_{n\to\infty}\left|\frac{(-1)^{n+2}(x-4)^{n+1}/((n+1)9^{n+1})}{(-1)^n(x-4)^n/(n9^n)}\right|$$

$$= \lim_{n\to\infty}\left|\frac{n}{n+1}\frac{(x-4)}{9}\right| = \frac{1}{9}|x-4|$$

$R = 9$

Interval: $-5 < x < 13$

When $x = 13$, $\sum_{n=1}^{\infty}\frac{(-1)^{n+1}9^n}{n9^n} = \sum_{n=1}^{\infty}\frac{(-1)^{n+1}}{n}$ converges.

When $x = -5$, $\sum_{n=1}^{\infty}\frac{(-1)^{n+1}(-9)^n}{n9^n} = \sum_{n=1}^{\infty}\frac{-1}{n}$ diverges.

Therefore, the interval of convergence is $(-5, 13]$.

27. $\sum_{n=0}^{\infty}\frac{(-1)^{n+1}(x-1)^{n+1}}{n+1}$

$$\lim_{n\to\infty}\left|\frac{u_{n+1}}{u_n}\right| = \lim_{n\to\infty}\left|\frac{(-1)^{n+2}(x-1)^{n+2}}{n+2}\cdot\frac{n+1}{(-1)^{n+1}(x-1)^{n+1}}\right|$$

$$= \lim_{n\to\infty}\left|\frac{(n+1)(x-1)}{n+2}\right| = |x-1|$$

$R = 1$

Center: $x = 1$

Interval: $-1 < x - 1 < 1$ or $0 < x < 2$

When $x = 0$, the series $\sum_{n=0}^{\infty}\frac{1}{n+1}$ diverges by the integral test.

When $x = 2$, the alternating series $\sum_{n=0}^{\infty}\frac{(-1)^{n+1}}{n+1}$ converges.

Therefore, the interval of convergence is $(0, 2]$.

29. $\sum_{n=1}^{\infty}\left(\frac{x-3}{3}\right)^{n-1}$ is geometric. It converges if

$$\left|\frac{x-3}{3}\right| < 1 \Rightarrow |x-3| < 3 \Rightarrow 0 < x < 6.$$

Therefore, the interval of convergence is $(0, 6)$.

31. $\sum_{n=1}^{\infty}\frac{n}{n+1}(-2x)^{n-1}$

$$\lim_{n\to\infty}\left|\frac{u_{n+1}}{u_n}\right| = \lim_{n\to\infty}\left|\frac{(n+1)(-2x)^n}{n+2}\cdot\frac{n+1}{n(-2x)^{n-1}}\right|$$

$$= \lim_{n\to\infty}\left|\frac{(-2x)(n+1)^2}{n(n+2)}\right| = 2|x|$$

$R = \frac{1}{2}$

Interval: $-\frac{1}{2} < x < \frac{1}{2}$

When $x = -\frac{1}{2}$, the series $\sum_{n=1}^{\infty}\frac{n}{n+1}$ diverges by the nth Term Test.

When $x = \frac{1}{2}$, the alternating series $\sum_{n=1}^{\infty}\frac{(-1)^{n-1}n}{n+1}$ diverges.

Therefore, the interval of convergence is $\left(-\frac{1}{2}, \frac{1}{2}\right)$.

33. $\sum_{n=0}^{\infty}\frac{x^{3n+1}}{(3n+1)!}$

$$\lim_{n\to\infty}\left|\frac{u_{n+1}}{u_n}\right| = \lim_{n\to\infty}\left|\frac{x^{3n+4}/(3n+4)!}{x^{3n+1}/(3n+1)!}\right|$$

$$= \lim_{n\to\infty}\left|\frac{x^3}{(3n+4)(3n+3)(3n+2)}\right| = 0$$

Therefore, the interval of convergence is $(-\infty, \infty)$.

35. $\sum_{n=1}^{\infty}\frac{2\cdot 3\cdot 4\cdots(n+1)x^n}{n!} = \sum_{n=1}^{\infty}(n+1)x^n$

$$\lim_{n\to\infty}\left|\frac{a_{n+1}}{a_n}\right| = \lim_{n\to\infty}\left|\frac{(n+2)x^{n+1}}{(n+1)x^n}\right| = \lim_{n\to\infty}\left|\frac{n+2}{n+1}x\right| = |x|$$

Converges if $|x| < 1 \Rightarrow -1 < x < 1$.

At $x = \pm 1$, diverges.

Therefore the interval of convergence is $(-1, 1)$.

37. $\sum_{n=1}^{\infty} \frac{(-1)^{n+1} 3 \cdot 7 \cdot 11 \cdots (4n-1)(x-3)^n}{4^n}$

$$\lim_{n\to\infty}\left|\frac{u_{n+1}}{u_n}\right| = \lim_{n\to\infty}\left|\frac{(-1)^{n+2} \cdot 3 \cdot 7 \cdot 11 \cdots (4n-1)(4n+3)(x-3)^{n+1}}{4^{n+1}} \cdot \frac{4^n}{(-1)^{n+1} \cdot 3 \cdot 7 \cdot 11 \cdots (4n-1)(x-3)^n}\right|$$

$$= \lim_{n\to\infty}\left|\frac{(4n+3)(x-3)}{4}\right| = \infty$$

$R = 0$

Center: $x = 3$

Therefore, the series converges only for $x = 3$.

39. $\sum_{n=1}^{\infty} \frac{(x-c)^{n-1}}{c^{n-1}}$

$$\lim_{n\to\infty}\left|\frac{u_{n+1}}{u_n}\right| = \lim_{n\to\infty}\left|\frac{(x-c)^n}{c^n} \cdot \frac{c^{n-1}}{(x-c)^{n-1}}\right| = \frac{1}{c}|x-c|$$

Converges if $\frac{1}{c}|x-c| < 1 \Rightarrow R = c$.

41. $\sum_{n=0}^{\infty} \left(\frac{x}{k}\right)^n$

Because the series is geometric, it converges only if $|x/k| < 1$ or $-k < x < k$.

Therefore, the interval of convergence is $(-k, k)$.

43. $\sum_{n=1}^{\infty} \frac{k(k+1)\cdots(k+n-1)x^n}{n!}$

$$\lim_{n\to\infty}\left|\frac{u_{n+1}}{u_n}\right| = \lim_{n\to\infty}\left|\frac{k(k+1)\cdots(k+n-1)(k+n)x^{n+1}}{(n+1)!} \cdot \frac{n!}{k(k+1)\cdots(k+n-1)x^n}\right| = \lim_{n\to\infty}\left|\frac{(k+n)x}{n+1}\right| = |x|$$

$R = 1$

When $x = \pm 1$, the series diverges and the interval of convergence is $(-1, 1)$.

$$\left[\frac{k(k+1)\cdots(k+n-1)}{1 \cdot 2 \cdots n} \geq 1\right]$$

45. $\sum_{n=0}^{\infty} \frac{x^n}{n!} = 1 + \frac{x}{1} + \frac{x^2}{2} + \cdots = \sum_{n=1}^{\infty} \frac{x^{n-1}}{(n-1)!}$

Replace n with $n - 1$.

47. $\sum_{n=2}^{\infty} \frac{x^{n-1}}{(7n-1)!} = \sum_{n=1}^{\infty} \frac{x^n}{[7(n+1)-1]!} = \sum_{n=1}^{\infty} \frac{x^n}{(7n+6)!}$

Replace n with $n + 1$.

49. (a) $f(x) = \sum_{n=0}^{\infty} \left(\frac{x}{3}\right)^n, (-3, 3)$ (Geometric)

(b) $f'(x) = \sum_{n=1}^{\infty} \frac{n}{3}\left(\frac{x}{3}\right)^{n-1}, (-3, 3)$

(c) $f''(x) = \sum_{n=2}^{\infty} \frac{n(n-1)}{9}\left(\frac{x}{3}\right)^{n-2}, (-3, 3)$

(d) $\int f(x)\,dx = \sum_{n=0}^{\infty} \frac{3}{n+1}\left(\frac{x}{3}\right)^{n+1}, [-3, 3)$

$$\left[\sum \frac{3}{n+1}\left(\frac{-3}{3}\right)^{n+1} = \sum \frac{(-1)^{n+1}3}{n+1}, \text{ converges}\right]$$

51. (a) $f(x) = \sum_{n=0}^{\infty} \frac{(-1)^{n+1}(x-1)^{n+1}}{n+1}, (0, 2]$

(b) $f'(x) = \sum_{n=0}^{\infty} (-1)^{n+1}(x-1)^n, (0, 2)$

(c) $f''(x) = \sum_{n=1}^{\infty} (-1)^{n+1} n(x-1)^{n-1}, (0, 2)$

(d) $\int f(x)\,dx = \sum_{n=1}^{\infty} \frac{(-1)^{n+1}(x-1)^{n+2}}{(n+1)(n+2)}, [0, 2]$

53. Answers will vary. *Sample answer*:

$$\sum_{n=1}^{\infty} \left(\frac{x}{3}\right)^n$$

Note that the series diverges for $x = \pm 3$.

55. Answers will vary.

$\sum_{n=1}^{\infty} \frac{x^n}{n}$ converges for $-1 \le x < 1$. At $x = -1$, the convergence is conditional because $\sum \frac{1}{n}$ diverges.

$\sum_{n=1}^{\infty} \frac{x^n}{n^2}$ converges for $-1 \le x \le 1$. At $x = \pm 1$, the convergence is absolute.

57. (a) $f(x) = \sum_{n=0}^{\infty} \frac{x^{2n+1}}{(2n+1)!}$

$$\lim_{n\to\infty}\left|\frac{u_{n+1}}{u_n}\right| = \lim_{n\to\infty}\left|\frac{x^{2n+3}}{(2n+3)!} \cdot \frac{(2n+1)!}{x^{2n+1}}\right| = \lim_{n\to\infty}\left|\frac{x^2}{(2n+2)(2n+3)}\right| = 0$$

Therefore, the interval of convergence is $(-\infty\ \infty)$.

$g(x) = \sum_{n=0}^{\infty} \frac{(-1)^n x^{2n}}{(2n)!}$

$$\lim_{n\to\infty}\left|\frac{u_{n+1}}{u_n}\right| = \lim_{n\to\infty}\left|\frac{(-1)^{n+1}x^{2n+2}}{(2n+2)!} \cdot \frac{(2n)!}{(-1)^n x^{2n}}\right| = \lim_{n\to\infty} \frac{1}{2n+2} = 0$$

Therefore, the interval of convergence is $(-\infty\ \infty)$.

(b) $f'(x) = \sum_{n=0}^{\infty} \frac{(-1)^n x^{2n}}{(2n)!} = g(x)$

$$g'(x) = \sum_{n=1}^{\infty} \frac{(-1)^n x^{2n-1}}{(2n-1)!} = \sum_{n=0}^{\infty} \frac{(-1)^{n+1} x^{2n+1}}{(2n+1)!} = -\sum_{n=0}^{\infty} \frac{(-1)^n x^{2n+1}}{(2n+1)!} = -f(x)$$

(c) $f(x) = \sin x$ and $g(x) = \cos x$

59. $y = \sum_{n=0}^{\infty} \frac{(-1)^n x^{2n+1}}{(2n+1)!} = \sum_{n=1}^{\infty} \frac{(-1)^{n-1} x^{2n-1}}{(2n-1)!}$

$$y' = \sum_{n=0}^{\infty} \frac{(-1)^n (2n+1) x^{2n}}{(2n+1)!} = \sum_{n=0}^{\infty} \frac{(-1)^n x^{2n}}{(2n)!}$$

$$y'' = \sum_{n=1}^{\infty} \frac{(-1)^n (2n) x^{2n-1}}{(2n)!} = \sum_{n=1}^{\infty} \frac{(-1) x^{2n-1}}{(2n-1)!}$$

$$y'' + y = \sum_{n=1}^{\infty} \frac{(-1)^n x^{2n-1}}{(2n-1)!} + \sum_{n=1}^{\infty} \frac{(-1)^{n-1} x^{2n-1}}{(2n-1)!} = 0$$

61. $y = \sum_{n=0}^{\infty} \frac{x^{2n+1}}{(2n+1)!} = \sum_{n=1}^{\infty} \frac{x^{2n-1}}{(2n-1)!}$

$$y' = \sum_{n=0}^{\infty} \frac{(2n+1)x^{2n}}{(2n+1)!} = \sum_{n=0}^{\infty} \frac{x^{2n}}{(2n)!}$$

$$y'' = \sum_{n=1}^{\infty} \frac{(2n)x^{2n-1}}{(2n)!} = \sum_{n=1}^{\infty} \frac{x^{2n-1}}{(2n-1)!} = y$$

$$y'' - y = 0$$

63. $y = \sum_{n=0}^{\infty} \frac{x^{2n}}{2^n n!} \qquad y' = \sum_{n=1}^{\infty} \frac{2nx^{2n-1}}{2^n n!} \qquad y'' = \sum_{n=1}^{\infty} \frac{2n(2n-1)x^{2n-2}}{2^n n!}$

$$y'' - xy' - y = \sum_{n=1}^{\infty} \frac{2n(2n-1)x^{2n-2}}{2^n n!} - \sum_{n=1}^{\infty} \frac{2nx^{2n}}{2^n n!} - \sum_{n=0}^{\infty} \frac{x^{2n}}{2^n n!}$$

$$= \sum_{n=1}^{\infty} \frac{2n(2n-1)x^{2n-2}}{2^n n!} - \sum_{n=0}^{\infty} \frac{(2n+1)x^{2n}}{2^n n!}$$

$$= \sum_{n=0}^{\infty}\left[\frac{(2n+2)(2n+1)x^{2n}}{2^{n+1}(n+1)!} - \frac{(2n+1)x^{2n}}{2^n n!} \cdot \frac{2(n+1)}{2(n+1)}\right]$$

$$= \sum_{n=0}^{\infty} \frac{2(n+1)x^{2n}\left[(2n+1) - (2n+1)\right]}{2^{n+1}(n+1)!} = 0$$

65. $J_0(x) = \sum_{k=0}^{\infty} \frac{(-1)^k x^{2k}}{2^{2k}(k!)^2}$

(a) $\lim_{k\to\infty}\left|\frac{u_{k+1}}{u_k}\right| = \lim_{k\to\infty}\left|\frac{(-1)^{k+1}x^{2k+2}}{2^{2k+2}[(k+1)!]^2}\cdot\frac{2^{2k}(k!)^2}{(-1)^k x^{2k}}\right| = \lim_{k\to\infty}\left|\frac{(-1)x^2}{2^2(k+1)^2}\right| = 0$

Therefore, the interval of convergence is $-\infty < x < \infty$.

(b)

$$J_0 = \sum_{k=0}^{\infty}(-1)^k \frac{x^{2k}}{4^k(k!)^2}$$

$$J_0' = \sum_{k=1}^{\infty}(-1)^k \frac{2kx^{2k-1}}{4^k(k!)^2} = \sum_{k=0}^{\infty}(-1)^{k+1}\frac{(2k+2)x^{2k+1}}{4^{k+1}[(k+1)!]^2}$$

$$J_0'' = \sum_{k=1}^{\infty}(-1)^k \frac{2k(2k-1)x^{2k-2}}{4^k(k!)^2} = \sum_{k=0}^{\infty}(-1)^{k+1}\frac{(2k+2)(2k+1)x^{2k}}{4^{k+1}[(k+1)!]^2}$$

$$x^2J_0'' + xJ_0' + x^2J_0 = \sum_{k=0}^{\infty}(-1)^{k+1}\frac{2(2k+1)x^{2k+2}}{4^{k+1}(k+1)!k!} + \sum_{k=0}^{\infty}(-1)^{k+1}\frac{2x^{2k+2}}{4^{k+1}(k+1)!k!} + \sum_{k=0}^{\infty}(-1)^k\frac{x^{2k+2}}{4^k(k!)^2}$$

$$= \sum_{k=0}^{\infty}\frac{(-1)^k x^{2k+2}}{4^k(k!)^2}\left[(-1)\frac{2(2k+1)}{4(k+1)} + (-1)\frac{2}{4(k+1)} + 1\right]$$

$$= \sum_{k=0}^{\infty}\frac{(-1)^k x^{2k+2}}{4^k(k!)^2}\left[\frac{-4k-2}{4k+4} - \frac{2}{4k+4} + \frac{4k+4}{4k+4}\right] = 0$$

(c) $P_6(x) = 1 - \frac{x^2}{4} + \frac{x^4}{64} - \frac{x^6}{2304}$

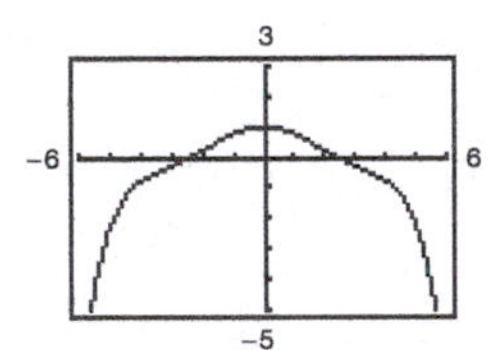

(d) $\int_0^1 J_0\,dx = \int_0^1 \sum_{k=0}^{\infty}\frac{(-1)^k x^{2k}}{4^k(k!)^2}\,dx = \left[\sum_{k=0}^{\infty}\frac{(-1)^k x^{2k+1}}{4^k(k!)^2(2k+1)}\right]_0^1 = \sum_{k=0}^{\infty}\frac{(-1)^k}{4^k(k!)^2(2k+1)} = 1 - \frac{1}{12} + \frac{1}{320} \approx 0.92$

(integral is approximately 0.9197304101)

67. $\sum_{n=0}^{\infty}\left(\frac{x}{4}\right)^n, \quad (-4, 4)$

(a) $\sum_{n=0}^{\infty}\left(\frac{(5/2)}{4}\right)^n = \sum_{n=0}^{\infty}\left(\frac{5}{8}\right)^n = \frac{1}{1-5/8} = \frac{8}{3}$

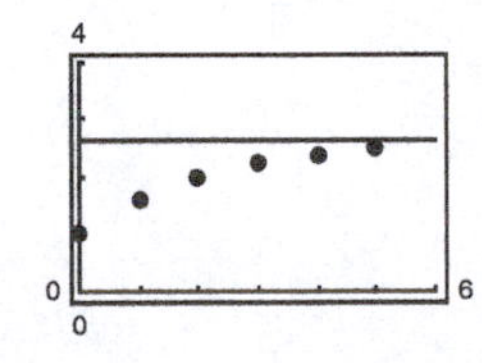

(b) $\sum_{n=0}^{\infty}\left(\frac{(-5/2)}{4}\right)^n = \sum_{n=0}^{\infty}\left(-\frac{5}{8}\right)^n = \frac{1}{1+5/8} = \frac{8}{13}$

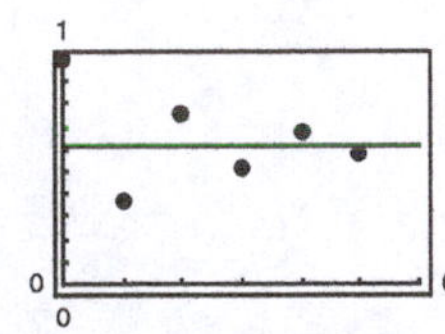

(c) The alternating series converges more rapidly. The partial sums of the series of positive terms approaches the sum from below. The partial sums of the alternating series alternate sides of the horizontal line representing the sum.

(d)

M	10	100	1000	10,000
N	5	14	24	35

69. $f(x) = \sum_{n=0}^{\infty} (-1)^n \frac{\pi^{2n}x^{2n}}{(2n)!}$

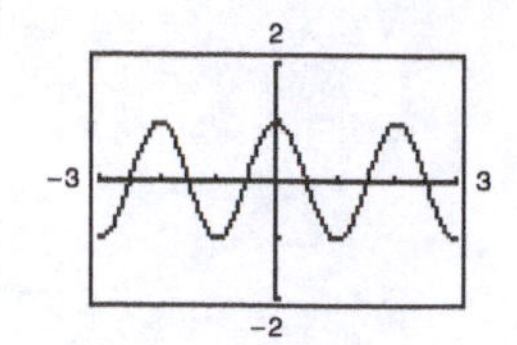

The function is $f(x) = \cos(\pi x)$.

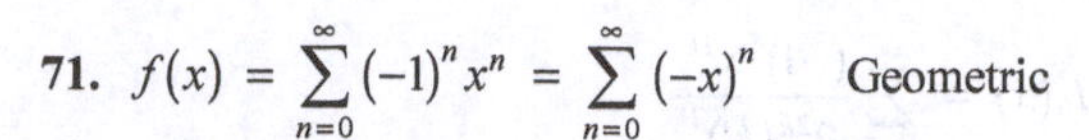

71. $f(x) = \sum_{n=0}^{\infty} (-1)^n x^n = \sum_{n=0}^{\infty} (-x)^n$ Geometric

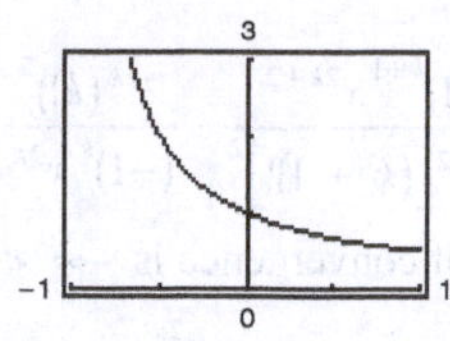

The function is

$$f(x) = \frac{1}{1-(-x)} = \frac{1}{1+x} \text{ for } -1 < x < 1.$$

73. False;

$\sum_{n=1}^{\infty} \frac{(-1)^n x^n}{n2^n}$ converges for $x = 2$ but diverges for $x = -2$.

75. True; the radius of convergence is $R = 1$ for both series.

77. $$\lim_{n\to\infty}\left|\frac{a_{n+1}}{a_n}\right| = \lim_{n\to\infty}\left|\frac{(n+1+p)!}{(n+1)!(n+1+q)!}x^{n+1} \Big/ \frac{(n+p)!}{n!(n+q)!}x^n\right| = \lim_{n\to\infty}\left|\frac{(n+1+p)x}{(n+1)(n+1+q)}\right| = 0$$

So, the series converges for all x: $R = \infty$.

79. (a) $f(x) = \sum_{n=0}^{\infty} c_n x^n,\ c_{n+3} = c_n$

$$= c_0 + c_1x + c_2x^2 + c_0x^3 + c_1x^4 + c_2x^5 + c_0x^6 + \cdots$$

$$S_{3n} = c_0\left(1 + x^3 + \cdots + x^{3n}\right) + c_1x\left(1 + x^3 + \cdots + x^{3n}\right) + c_2x^2\left(1 + x^3 + \cdots + x^{3n}\right)$$

$$\lim_{n\to\infty} S_{3n} = c_0\sum_{n=0}^{\infty} x^{3n} + c_1x\sum_{n=0}^{\infty} x^{3n} + c_2x^2\sum_{n=0}^{\infty} x^{3n}$$

Each series is geometric, $R = 1$, and the interval of convergence is $(-1, 1)$.

(b) For $|x| < 1$, $f(x) = c_0\frac{1}{1-x^3} + c_1x\frac{1}{1-x^3} + c_2x^2\frac{1}{1-x^3} = \frac{c_0 + c_1x + c_2x^2}{1-x^3}$.

81. At $x = x_0 + R$, $\sum_{n=0}^{\infty} c_n(x - x_0)^n = \sum_{n=0}^{\infty} c_nR^n$, diverges.

At $x = x_0 - R$, $\sum_{n=0}^{\infty} c_n(x - x_0)^n = \sum_{n=0}^{\infty} c_n(-R)^n$, converges.

Furthermore, at $x = x_0 - R$,

$\sum_{n=0}^{\infty} \left|c_n(x - x_0)^n\right| = \sum_{n=0}^{\infty} c_nR^n$, diverges.

So, the series converges conditionally at $x_0 - R$.

Section 9.9 Representation of Functions by Power Series

1. You need to algebraically manipulate $\dfrac{b}{c-x}$ so that it resembles the form $\dfrac{a}{1-r}$.

3. (a) $$\frac{1}{4-x} = \frac{1/4}{1-(x/4)}$$
$$= \frac{a}{1-r} = \sum_{n=0}^{\infty}\left(\frac{1}{4}\right)\left(\frac{x}{4}\right)^n = \sum_{n=0}^{\infty}\frac{x^n}{4^{n+1}}$$

This series converges on $(-4, 4)$.

(b)
$$\begin{array}{r l}
 & \frac{1}{4} + \frac{x}{16} + \frac{x^2}{64} + \cdots \\
4 - x\,\big) & 1 \\
 & 1 - \frac{x}{4} \\
 & \quad \frac{x}{4} \\
 & \quad \frac{x}{4} - \frac{x^2}{16} \\
 & \qquad \frac{x^2}{16} \\
 & \qquad \frac{x^2}{16} - \frac{x^3}{64} \\
 & \qquad \vdots
\end{array}$$

5. (a) $$\frac{4}{3+x} = \frac{4/3}{1-(-x/3)} = \frac{a}{1-r}$$
$$= \sum_{n=0}^{\infty}\frac{4}{3}\left(\frac{-x}{3}\right)^n = \sum_{n=0}^{\infty}\frac{4(-1)^n x^n}{3^{n+1}}$$

The series converges on $(-3, 3)$.

(b)
$$\begin{array}{r l}
 & \frac{4}{3} - \frac{4}{9}x + \frac{4x^2}{27} - \cdots \\
3 + x\,\big) & 4 \\
 & 4 + \frac{4}{3}x \\
 & \quad -\frac{4}{3}x \\
 & \quad -\frac{4}{3}x - \frac{4x^2}{9} \\
 & \qquad \frac{4x^2}{9} \\
 & \qquad \frac{4x^2}{9} + \frac{4x^3}{27} \\
 & \qquad\quad -\frac{4x^3}{27} \\
 & \qquad\quad \vdots
\end{array}$$

7. $$\frac{1}{6-x} = \frac{1}{5-(x-1)} = \frac{1/5}{1-\left(\frac{x-1}{5}\right)} = \frac{a}{1-r}$$
$$= \sum_{n=0}^{\infty}\frac{1}{5}\left(\frac{x-1}{5}\right)^n = \sum_{n=0}^{\infty}\frac{(x-n)^n}{5^{n+1}}$$

Interval of convergence:

$$\left|\frac{x-1}{5}\right| < 1 \Rightarrow |x-1| < 5 \Rightarrow (-4, 6)$$

9. $$\frac{1}{1-3x} = \frac{a}{1-r} = \sum_{n=0}^{\infty}(3x)^n$$

Interval of convergence: $|3x| < 1 \Rightarrow \left(-\frac{1}{3}, \frac{1}{3}\right)$

11. $$\frac{5}{2x-3} = \frac{5}{-9+2(x+3)} = \frac{-5/9}{1-\frac{2}{9}(x+3)} = \frac{a}{1-r}$$
$$= -\frac{5}{9}\sum_{n=0}^{\infty}\left(\frac{2}{9}(x+3)\right)^n, \quad \left|\frac{2}{9}(x+3)\right| < 1$$
$$= -5\sum_{n=0}^{\infty}\frac{2^n}{9^{n+1}}(x+3)^n$$

Interval of convergence: $\left|\frac{2}{9}(x+3)\right| < 1 \Rightarrow \left(-\frac{15}{2}, \frac{3}{2}\right)$

13. $$\frac{2}{5x+4} = \frac{2}{-1+5(x+1)} = \frac{-2}{1-5(x+1)} = \frac{a}{1-r}$$
$$= \sum_{n=0}^{\infty}-2\left[5(x-1)\right]^n$$

Interval of convergence:

$$|5(x+1)| < 1 \Rightarrow |x+1| < \frac{1}{5} \Rightarrow \left(-\frac{6}{5}, -\frac{4}{5}\right)$$

15. $$\frac{4x}{x^2+2x-3} = \frac{3}{x+3} + \frac{1}{x-1}$$
$$= \frac{1}{1-(-x/3)} + \frac{-1}{1-x}$$
$$= \sum_{n=0}^{\infty}\left(-\frac{x}{3}\right)^n - \sum_{n=0}^{\infty}x^n = \sum_{n=0}^{\infty}\left[\frac{1}{(-3)^n} - 1\right]x^n$$

Interval of convergence: $\left|\frac{x}{3}\right| < 1$ and $|x| < 1 \Rightarrow (-1, 1)$

17. $\dfrac{2}{1-x^2} = \dfrac{1}{1-x} + \dfrac{1}{1+x} = \sum_{n=0}^{\infty}\left(1+(-1)^n\right)x^n = 2\sum_{n=0}^{\infty} x^{2n}$

Interval of convergence: $\left|x^2\right| < 1$ or $(-1, 1)$ because $\lim_{n\to\infty}\left|\dfrac{u_{n+1}}{u_n}\right| = \lim_{n\to\infty}\left|\dfrac{2x^{2n+2}}{2x^{2n}}\right| = \left|x^2\right|$

19. $\dfrac{1}{1+x} = \sum_{n=0}^{\infty}(-1)^n x^n$

$\dfrac{1}{1-x} = \sum_{n=0}^{\infty}(-1)^n(-x)^n = \sum_{n=0}^{\infty}(-1)^{2n}x^n = \sum_{n=0}^{\infty}x^n$

$h(x) = \dfrac{-2}{x^2-1} = \dfrac{1}{1+x} + \dfrac{1}{1-x} = \sum_{n=0}^{\infty}(-1)^n x^n + \sum_{n=0}^{\infty}x^n = \sum_{n=0}^{\infty}\left[(-1)^n + 1\right]x^n$

$= 2 + 0x + 2x^2 + 0x^3 + 2x^4 + 0x^5 + 2x^6 + \cdots = 2\sum_{n=0}^{\infty}x^{2n}, (-1, 1)$ (See Exercise 17.)

21. By taking the first derivative, you have $\dfrac{d}{dx}\left[\dfrac{1}{x+1}\right] = \dfrac{-1}{(x+1)^2}$. Therefore,

$\dfrac{-1}{(x+1)^2} = \dfrac{d}{dx}\left[\sum_{n=0}^{\infty}(-1)^n x^n\right] = \sum_{n=1}^{\infty}(-1)^n nx^{n-1} = \sum_{n=0}^{\infty}(-1)^{n+1}(n+1)x^n, (-1, 1).$

23. By integrating, you have $\int\dfrac{1}{x+1}\,dx = \ln(x+1)$. Therefore,

$\ln(x+1) = \int\left[\sum_{n=0}^{\infty}(-1)^n x^n\right]dx = C + \sum_{n=0}^{\infty}\dfrac{(-1)^n x^{n+1}}{n+1}, -1 < x \le 1.$

To solve for C, let $x = 0$ and conclude that $C = 0$. Therefore,

$\ln(x+1) = \sum_{n=0}^{\infty}\dfrac{(-1)^n x^{n+1}}{n+1}, (-1, 1].$

25. $\dfrac{1}{x^2+1} = \sum_{n=0}^{\infty}(-1)^n\left(x^2\right)^n = \sum_{n=0}^{\infty}(-1)^n x^{2n}, (-1, 1)$

27. Because, $\dfrac{1}{x+1} = \sum_{n=0}^{\infty}(-1)^n x^n$, you have $\dfrac{1}{4x^2+1} = \sum_{n=0}^{\infty}(-1)^n\left(4x^2\right)^n = \sum_{n=0}^{\infty}(-1)^n 4^n x^{2n} = \sum_{n=0}^{\infty}(-1)^n(2x)^{2n}, \left(-\dfrac{1}{2}, \dfrac{1}{2}\right).$

29. $x - \dfrac{x^2}{2} \le \ln(x+1) \le x - \dfrac{x^2}{2} + \dfrac{x^3}{3}$

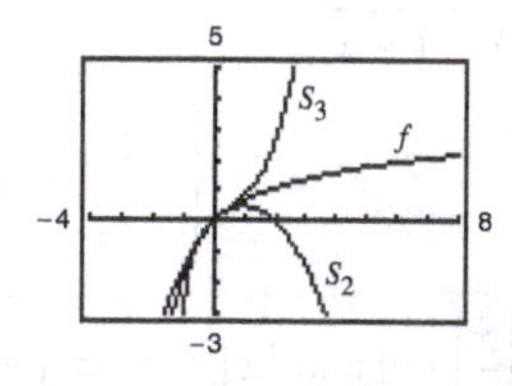

x	0.0	0.2	0.4	0.6	0.8	1.0
$S_2 = x - \dfrac{x^2}{2}$	0.000	0.180	0.320	0.420	0.480	0.500
$\ln(x+1)$	0.000	0.182	0.336	0.470	0.588	0.693
$S_3 = x - \dfrac{x^2}{2} + \dfrac{x^3}{3}$	0.000	0.183	0.341	0.492	0.651	0.833

31. $\sum_{n=1}^{\infty} \frac{(-1)^{n+1}(x-1)^n}{n} = \frac{(x-1)}{1} - \frac{(x-1)^2}{2} + \frac{(x-1)^3}{3} - \cdots$

(a)

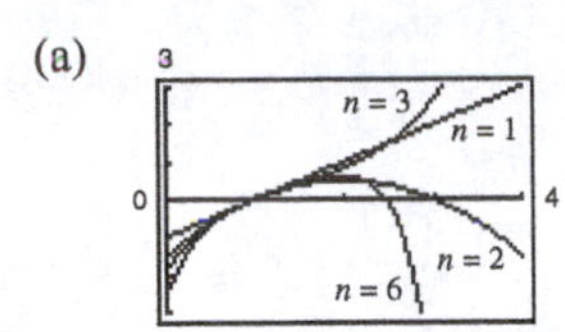

(b) From Example 4, $\sum_{n=1}^{\infty} \frac{(-1)^{n+1}(x-1)^n}{n} = \sum_{n=0}^{\infty} \frac{(-1)^n(x-1)^{n+1}}{n+1} = \ln x,\ 0 < x \le 2,\ R = 1.$

(c) $x = 0.5$: $\sum_{n=1}^{\infty} \frac{(-1)^{n+1}(-1/2)^n}{n} = \sum_{n=1}^{\infty} \frac{-(1/2)^n}{n} \approx -0.693147$

(d) This is an approximation of $\ln\left(\frac{1}{2}\right)$. The error is approximately 0. [The error is less than the first omitted term, $1/\left(51 \cdot 2^{51}\right) \approx 8.7 \times 10^{-18}$.]

In Exercises 33 and 35, arctan $x = \sum_{n=0}^{\infty} (-1)^n \frac{x^{2n+1}}{2n+1}$.

33. $\arctan \frac{1}{4} = \sum_{n=0}^{\infty} (-1)^n \frac{(1/4)^{2n+1}}{2n+1} = \sum_{n=0}^{\infty} \frac{(-1)^n}{(2n+1)4^{2n+1}} = \frac{1}{4} - \frac{1}{192} + \frac{1}{5120} + \cdots$

Because $\frac{1}{5120} < 0.001$, you can approximate the series by its first two terms: $\arctan \frac{1}{4} \approx \frac{1}{4} - \frac{1}{192} \approx 0.245$.

35. $\frac{\arctan x^2}{x} = \frac{1}{x}\sum_{n=0}^{\infty} (-1)^n \frac{\left(x^2\right)^{2n+1}}{2n+1} = \sum_{n=0}^{\infty} (-1)^n \frac{x^{4n+1}}{2n+1}$

$\int \frac{\arctan x^2}{x}\,dx = \sum_{n=0}^{\infty} (-1)^n \frac{x^{4n+2}}{(4n+2)(2n+1)} + C$ (Note: $C = 0$)

$\int_0^{1/2} \frac{\arctan x^2}{x}\,dx = \sum_{n=0}^{\infty} (-1)^n \frac{1}{(4n+2)(2n+1)2^{4n+2}} = \frac{1}{8} - \frac{1}{1152} + \cdots$

Because $\frac{1}{1152} < 0.001$, you can approximate the series by its first term: $\int_0^{1/2} \frac{\arctan x^2}{x}\,dx \approx 0.125$.

In Exercises 37 and 39, use $\frac{1}{1-x} = \sum_{n=0}^{\infty} x^n, |x| < 1$.

37. $\frac{1}{(1-x)^2} = \frac{d}{dx}\left[\frac{1}{1-x}\right] = \frac{d}{dx}\left[\sum_{n=0}^{\infty} x^n\right] = \sum_{n=1}^{\infty} nx^{n-1}, |x| < 1$

39. $\frac{1+x}{(1-x)^2} = \frac{1}{(1-x)^2} + \frac{x}{(1-x)^2}$

$= \sum_{n=1}^{\infty} n\left(x^{n-1} + x^n\right), \quad |x| < 1$

$= \sum_{n=0}^{\infty} (2n+1)x^n, \quad |x| < 1$

41. $P(n) = \left(\frac{1}{2}\right)^n$

$E(n) = \sum_{n=1}^{\infty} nP(n) = \sum_{n=1}^{\infty} n\left(\frac{1}{2}\right)^n = \frac{1}{2}\sum_{n=1}^{\infty} n\left(\frac{1}{2}\right)^{n-1}$

$= \frac{1}{2}\frac{1}{\left[1-(1/2)\right]^2} = 2$

Because the probability of obtaining a head on a single toss is $\frac{1}{2}$, it is expected that, on average, a head will be obtained in two tosses.

43. Let $\arctan x + \arctan y = \theta$. Then,

$$\tan(\arctan x + \arctan y) = \tan\theta$$

$$\frac{\tan(\arctan x) + \tan(\arctan y)}{1 - \tan(\arctan x)\tan(\arctan y)} = \tan\theta$$

$$\frac{x + y}{1 - xy} = \tan\theta$$

$$\arctan\left(\frac{x + y}{1 - xy}\right) = \theta.$$

Therefore, $\arctan x + \arctan y = \arctan\left(\dfrac{x + y}{1 - xy}\right)$ for $xy \neq 1$.

45. (a) $2\arctan\dfrac{1}{2} = \arctan\dfrac{1}{2} + \arctan\dfrac{1}{2} = \arctan\left[\dfrac{\frac{1}{2} + \frac{1}{2}}{1 - (1/2)^2}\right] = \arctan\dfrac{4}{3}$

$$2\arctan\frac{1}{2} - \arctan\frac{1}{7} = \arctan\frac{4}{3} + \arctan\left(-\frac{1}{7}\right) = \arctan\left[\frac{(4/3) - (1/7)}{1 + (4/3)(1/7)}\right] = \arctan\frac{25}{25} = \arctan 1 = \frac{\pi}{4}$$

(b) $\pi = 8\arctan\dfrac{1}{2} - 4\arctan\dfrac{1}{7} \approx 8\left[\dfrac{1}{2} - \dfrac{(0.5)^3}{3} + \dfrac{(0.5)^5}{5} - \dfrac{(0.5)^7}{7}\right] - 4\left[\dfrac{1}{7} - \dfrac{(1/7)^3}{3} + \dfrac{(1/7)^5}{5} - \dfrac{(1/7)^7}{7}\right] \approx 3.14$

47. From Exercise 23, you have $\ln(x + 1) = \displaystyle\sum_{n=0}^{\infty}\frac{(-1)^n x^{n+1}}{n + 1} = \sum_{n=1}^{\infty}\frac{(-1)^{n-1}x^n}{n} = \sum_{n=1}^{\infty}\frac{(-1)^{n+1}x^n}{n}$.

So, $\displaystyle\sum_{n=1}^{\infty}(-1)^{n+1}\frac{1}{2^n n} = \sum_{n=1}^{\infty}\frac{(-1)^{n+1}(1/2)^n}{n} = \ln\left(\frac{1}{2} + 1\right) = \ln\frac{3}{2} \approx 0.4055.$

49. From Exercise 47, you have $\displaystyle\sum_{n=1}^{\infty}(-1)^{n+1}\frac{2^n}{5^n n} = \sum_{n=1}^{\infty}\frac{(-1)^{n+1}(2/5)^n}{n} = \ln\left(\frac{2}{5} + 1\right) = \ln\frac{7}{5} \approx 0.3365.$

51. From Exercise 50, you have $\displaystyle\sum_{n=0}^{\infty}(-1)^n\frac{1}{2^{2n+1}(2n + 1)} = \sum_{n=0}^{\infty}(-1)^n\frac{(1/2)^{2n+1}}{2n + 1} = \arctan\frac{1}{2} \approx 0.4636.$

53. The series in Exercise 50 converges to its sum at a slower rate because its terms approach 0 at a much slower rate.

55. Because the first series is the derivative of the second series, the second series converges for $|x + 1| < 4$ (and perhaps at the endpoints, $x = 3$ and $x = -5$.)

57. Using a graphing utility, you obtain the following partial sums for the left hand side. Note that $1/\pi = 0.3183098862$.

$n = 0$: $S_0 \approx 0.3183098784$

$n = 1$: $S_1 = 0.3183098862$

Section 9.10 Taylor and Maclaurin Series

1. The Taylor series for $f(x)$ converges to $f(x)$ if and only if $R_n(x) \to 0$ as $n \to \infty$.

3. Multiply and divide power series as if they were polynomials.

5. For $c = 0$, you have:

$$f(x) = e^{2x}$$
$$f^{(n)}(x) = 2^n e^{2x} \Rightarrow f^{(n)}(0) = 2^n$$
$$e^{2x} = 1 + 2x + \frac{4x^2}{2!} + \frac{8x^3}{3!} + \frac{16x^4}{4!} + \cdots = \sum_{n=0}^{\infty} \frac{(2x)^n}{n!}.$$

7. For $c = \pi/4$, you have:

$$f(x) = \cos(x) \qquad f\left(\frac{\pi}{4}\right) = \frac{\sqrt{2}}{2}$$
$$f'(x) = -\sin(x) \qquad f'\left(\frac{\pi}{4}\right) = -\frac{\sqrt{2}}{2}$$
$$f''(x) = -\cos(x) \qquad f''\left(\frac{\pi}{4}\right) = -\frac{\sqrt{2}}{2}$$
$$f'''(x) = \sin(x) \qquad f'''\left(\frac{\pi}{4}\right) = \frac{\sqrt{2}}{2}$$
$$f^{(4)}(x) = \cos(x) \qquad f^{(4)}\left(\frac{\pi}{4}\right) = \frac{\sqrt{2}}{2}$$

and so on. Therefore, you have:

$$\cos x = \sum_{n=0}^{\infty} \frac{f^{(n)}(\pi/4)\left[x - (\pi/4)\right]^n}{n!}$$
$$= \frac{\sqrt{2}}{2}\left[1 - \left(x - \frac{\pi}{4}\right) - \frac{\left[x - (\pi/4)\right]^2}{2!} + \frac{\left[x - (\pi/4)\right]^3}{3!} + \frac{\left[x - (\pi/4)\right]^4}{4!} - \cdots\right]$$
$$= \frac{\sqrt{2}}{2}\sum_{n=0}^{\infty} \frac{(-1)^{n(n+1)/2}\left[x - (\pi/4)\right]^n}{n!}.$$

[**Note:** $(-1)^{n(n+1)/2} = 1, -1, -1, 1, 1, -1, -1, 1, \ldots$]

9. For $c = 1$, you have

$$f(x) = \frac{1}{x} = x^{-1} \qquad f(1) = 1$$
$$f'(x) = -x^{-2} \qquad f'(1) = -1$$
$$f''(x) = 2x^{-3} \qquad f''(1) = 2$$
$$f'''(x) = -6x^{-4} \qquad f'''(1) = -6$$

and so on. Therefore, you have

$$\frac{1}{x} = \sum_{n=0}^{\infty} \frac{f^{(n)}(1)(x - 1)^n}{n!}$$
$$= 1 - (x - 1) + \frac{2(x - 1)^2}{2!} - \frac{6(x - 1)^3}{3!} + \cdots$$
$$= 1 - (x - 1) + (x - 1)^2 - (x - 1)^3 + \cdots$$
$$= \sum_{n=0}^{\infty} (-1)^n (x - 1)^n$$

11. For $c = 1$, you have,

$$\begin{aligned} f(x) &= \ln x & f(1) &= 0 \\ f'(x) &= \frac{1}{x} & f'(1) &= 1 \\ f''(x) &= -\frac{1}{x^2} & f''(1) &= -1 \\ f'''(x) &= \frac{2}{x^3} & f'''(1) &= 2 \\ f^{(4)}(x) &= -\frac{6}{x^4} & f^{(4)}(1) &= -6 \\ f^{(5)}(x) &= \frac{24}{x^5} & f^{(5)}(1) &= 24 \end{aligned}$$

and so on. Therefore, you have:

$$\begin{aligned} \ln x &= \sum_{n=0}^{\infty} \frac{f^{(n)}(1)(x-1)^n}{n!} \\ &= 0 + (x-1) - \frac{(x-1)^2}{2!} + \frac{2(x-1)^3}{3!} - \frac{6(x-1)^4}{4!} + \frac{24(x-1)^5}{5!} - \cdots \\ &= (x-1) - \frac{(x-1)^2}{2} + \frac{(x-1)^3}{3} - \frac{(x-1)^4}{4} + \frac{(x-1)^5}{5} - \cdots \\ &= \sum_{n=0}^{\infty} (-1)^n \frac{(x-1)^{n+1}}{n+1}. \end{aligned}$$

13. For $c = 0$, you have

$$\begin{aligned} f(x) &= \sin 3x & f(0) &= 0 \\ f'(x) &= 3\cos 3x & f'(0) &= 3 \\ f''(x) &= -9\sin 3x & f''(0) &= 0 \\ f'''(x) &= -27\cos 3x & f'''(0) &= -27 \\ f^{(4)}(x) &= 81\sin 3x & f^{(4)}(0) &= 0 \end{aligned}$$

and so on. Therefore you have

$$\sin 3x = \sum_{n=0}^{\infty} \frac{f^{(n)}(0)x^n}{n!} = 0 + 3x + 0 - \frac{27x^3}{3!} + 0 + \cdots = \sum_{n=0}^{\infty} \frac{(-1)^n (3x)^{2n+1}}{(2n+1)!}$$

15. For $c = 0$, you have:

$$\begin{aligned} f(x) &= \sec(x) & f(0) &= 1 \\ f'(x) &= \sec(x)\tan(x) & f'(0) &= 0 \\ f''(x) &= \sec^3(x) + \sec(x)\tan^2(x) & f''(0) &= 1 \\ f'''(x) &= 5\sec^3(x)\tan(x) + \sec(x)\tan^3(x) & f'''(0) &= 0 \\ f^{(4)}(x) &= 5\sec^5(x) + 18\sec^3(x)\tan^2(x) + \sec(x)\tan^4(x) & f^{(4)}(0) &= 5 \end{aligned}$$

$$\sec(x) = \sum_{n=0}^{\infty} \frac{f^{(n)}(0)x^n}{n!} = 1 + \frac{x^2}{2!} + \frac{5x^4}{4!} + \cdots.$$

17. The Maclaurin series for $f(x) = \cos x$ is $\sum_{n=0}^{\infty} \frac{(-1)^n x^{2n}}{(2n)!}$.

Because $f^{(n+1)}(x) = \pm\sin x$ or $\pm\cos x$, you have $\left|f^{(n+1)}(z)\right| \le 1$ for all z. So by Taylor's Theorem,

$$0 \le \left|R_n(x)\right| = \left|\frac{f^{(n+1)}(z)}{(n+1)!}x^{n+1}\right| \le \frac{|x|^{n+1}}{(n+1)!}.$$

Because $\lim_{n\to\infty} \frac{|x|^{n+1}}{(n+1)!} = 0$, it follows that $R_n(x) \to 0$ as $n \to \infty$. So, the Maclaurin series for $\cos x$ converges to $\cos x$ for all x.

19. The Maclaurin series for $f(x) = \sinh x$ is $\sum_{n=0}^{\infty} \frac{x^{2n+1}}{(2n+1)!}$.

$f^{(n+1)}(x) = \sinh x$ (or $\cosh x$). For fixed x,

$$0 \le \left|R_n(x)\right| = \left|\frac{f^{(n+1)}(z)}{(n+1)!}x^{n+1}\right| = \left|\frac{\sinh(z)}{(n+1)!}x^{n+1}\right| \to 0 \text{ as } n \to \infty.$$

(The argument is the same if $f^{(n+1)}(x) = \cosh x$). So, the Maclaurin series for $\sinh x$ converges to $\sinh x$ for all x.

21. Because $(1+x)^{-k} = 1 - kx + \frac{k(k+1)x^2}{2!} - \frac{k(k+1)(k+2)x^3}{3!} + \cdots$, you have

$$\begin{aligned}\left[1 + (-x)\right]^{-1/2} &= 1 + \left(\frac{1}{2}\right)x + \frac{(1/2)(3/2)x^2}{2!} + \frac{(1/2)(3/2)(5/2)x^3}{3!} + \cdots \\ &= 1 + \frac{x}{2} + \frac{(1)(3)x^2}{2^2\,2!} + \frac{(1)(3)(5)x^3}{2^3\,3!} + \cdots \\ &= 1 + \sum_{n=1}^{\infty} \frac{1\cdot3\cdot5\cdots(2n-1)x^n}{2^n\,n!}.\end{aligned}$$

23. Because $(1+x)^{-k} = 1 - kx + \frac{k(k+1)x^2}{2!} - \frac{k(k+1)(k+2)x^3}{3!} + \cdots$ you have

$$\begin{aligned}\left[1 + \left(-x^2\right)\right]^{-1/2} &= 1 + \frac{1}{2}x^2 + \frac{(1/2)(3/2)}{2!}x^4 + \frac{(1/2)(3/2)(5/2)x^6}{3!} + \cdots \\ &= 1 + \frac{1}{2}x^2 + \frac{(1)(3)}{2^2\,2!}x^4 + \frac{(1)(3)(5)}{2^3\,3!}x^6 + \cdots \\ &= 1 + \sum_{n=1}^{\infty} \frac{1\cdot3\cdot5\cdots(2n-1)}{2^n\,n!}x^{2n}\end{aligned}$$

25.
$$\begin{aligned}(1+x)^{1/4} &= 1 + \frac{1}{4}x + \frac{(1/4)(-3/4)}{2!}x^2 + \frac{(1/4)(-3/4)(-7/4)}{3!}x^3 + \cdots \\ &= 1 + \frac{1}{4}x - \frac{3}{4^2 2!}x^2 + \frac{3\cdot7}{4^3 3!}x^3 - \frac{3\cdot7\cdot11}{4^4 4!}x^4 + \cdots \\ &= 1 + \frac{1}{4}x + \sum_{n=2}^{\infty} \frac{(-1)^{n+1}3\cdot7\cdot11\cdots(4n-5)}{4^n\,n!}x^n\end{aligned}$$

27. $e^x = \sum_{n=0}^{\infty} \frac{x^n}{n!} = 1 + x + \frac{x^2}{2!} + \frac{x^3}{3!} + \frac{x^4}{4!} + \frac{x^5}{5!} + \cdots$

$e^{x^2/2} = \sum_{n=0}^{\infty} \frac{(x^2/2)^n}{n!} = \sum_{n=0}^{\infty} \frac{x^{2n}}{2^n n!} = 1 + \frac{x^2}{2} + \frac{x^4}{2^2 2!} + \frac{x^6}{2^3 3!} + \frac{x^8}{2^4 4!} + \cdots$

29. $\ln x = \sum_{n=1}^{\infty} (-1)^{n-1} \frac{(x-1)^n}{n}, \; 0 < x \le 2$

$\ln(x+1) = \sum_{n=1}^{\infty} \frac{(-1)^{n-1} x^n}{n}, \; -1 < x \le 1$

31. $\cos x = \sum_{n=0}^{\infty} \frac{(-1)^n x^{2n}}{(2n)!} = 1 - \frac{x^2}{2!} + \frac{x^4}{4!} - \frac{x^6}{6!} + \cdots$

$\cos 4x = \sum_{n=0}^{\infty} \frac{(-1)^n (4x)^{2n}}{(2n)!} = \sum_{n=0}^{\infty} \frac{(-1)^n 4^{2n} x^{2n}}{(2n)!}$

$= 1 - \frac{16x^2}{2!} + \frac{256x^4}{4!} - \cdots$

33. $\arctan x = \sum_{n=0}^{\infty} \frac{(-1)^n x^{2n+1}}{2n+1}$

$\arctan 5x = \sum_{n=0}^{\infty} \frac{(-1)^n (5x)^{2n+1}}{2n+1}$

35. $\cos x = \sum_{n=0}^{\infty} \frac{(-1)^n x^{2n}}{(2n)!} = 1 - \frac{x^2}{2!} + \frac{x^4}{4!} - \cdots$

$\cos x^{3/2} = \sum_{n=0}^{\infty} \frac{(-1)^n (x^{3/2})^{2n}}{(2n)!}$

$= \sum_{n=0}^{\infty} \frac{(-1)^n x^{3n}}{(2n)!}$

$= 1 - \frac{x^3}{2!} + \frac{x^6}{4!} - \cdots$

37. $e^x = 1 + x + \frac{x^2}{2!} + \frac{x^3}{3!} + \frac{x^4}{4!} + \frac{x^5}{5!} + \cdots$

$e^{-x} = 1 - x + \frac{x^2}{2!} - \frac{x^3}{3!} + \frac{x^4}{4!} - \frac{x^5}{5!} + \cdots$

$e^x - e^{-x} = 2x + \frac{2x^3}{3!} + \frac{2x^5}{5!} + \frac{2x^7}{7!} + \cdots$

$\sinh x = \frac{1}{2}(e^x - e^{-x})$

$= x + \frac{x^3}{3!} + \frac{x^5}{5!} + \frac{x^7}{7!} + \cdots = \sum_{n=0}^{\infty} \frac{x^{2n+1}}{(2n+1)!}$

39. $\cos^2(x) = \frac{1}{2}[1 + \cos(2x)]$

$= \frac{1}{2}\left[1 + 1 - \frac{(2x)^2}{2!} + \frac{(2x)^4}{4!} - \frac{(2x)^6}{6!} - \cdots\right] = \frac{1}{2}\left[1 + \sum_{n=0}^{\infty} \frac{(-1)^n (2x)^{2n}}{(2n)!}\right]$

41. $e^{ix} = 1 + ix + \frac{(ix)^2}{2!} + \frac{(ix)^3}{3!} + \frac{(ix)^4}{4!} + \cdots = 1 + ix - \frac{x^2}{2!} - \frac{ix^3}{3!} + \frac{x^4}{4!} + \frac{ix^5}{5!} - \frac{x^6}{6!} - \cdots$

$e^{-ix} = 1 - ix + \frac{(-ix)^2}{2!} + \frac{(-ix)^3}{3!} + \frac{(-ix)^4}{4!} + \cdots = 1 - ix - \frac{x^2}{2!} + \frac{ix^3}{3!} + \frac{x^4}{4!} - \frac{ix^5}{5!} - \frac{x^6}{6!} + \cdots$

$e^{ix} - e^{-ix} = 2ix - \frac{2ix^3}{3!} + \frac{2ix^5}{5!} - \frac{2ix^7}{7!} + \cdots$

$\frac{e^{ix} - e^{-ix}}{2i} = x - \frac{x^3}{3!} + \frac{x^5}{5!} - \frac{x^7}{7!} + \cdots = \sum_{n=0}^{\infty} \frac{(-1)^n x^{2n+1}}{(2n+1)!} = \sin(x)$

43. $x \sin x = x\left(x - \frac{x^3}{3!} + \frac{x^5}{5!} - \cdots\right) = x^2 - \frac{x^4}{3!} + \frac{x^6}{5!} - \cdots = \sum_{n=0}^{\infty} \frac{(-1)^n x^{2n+2}}{(2n+1)!}$

45. $\frac{\sin x}{x} = \frac{x - (x^3/3!) + (x^5/5!) - \cdots}{x} = 1 - \frac{x^2}{2!} + \frac{x^4}{4!} - \cdots = \sum_{n=0}^{\infty} \frac{(-1)^n x^{2n}}{(2n+1)!}, x \ne 0 = 1, x = 0$

47. $f(x) = e^x \sin x$

$$= \left(1 + x + \frac{x^2}{2} + \frac{x^3}{6} + \frac{x^4}{24} + \cdots\right)\left(x - \frac{x^3}{6} + \frac{x^5}{120} - \cdots\right)$$

$$= x + x^2 + \left(\frac{x^3}{2} - \frac{x^3}{6}\right) + \left(\frac{x^4}{6} - \frac{x^4}{6}\right) + \left(\frac{x^5}{120} - \frac{x^5}{12} + \frac{x^5}{24}\right) + \cdots$$

$$= x + x^2 + \frac{x^3}{3} - \frac{x^5}{30} + \cdots$$

$$P_5(x) = x + x^2 + \frac{x^3}{3} - \frac{x^5}{30}$$

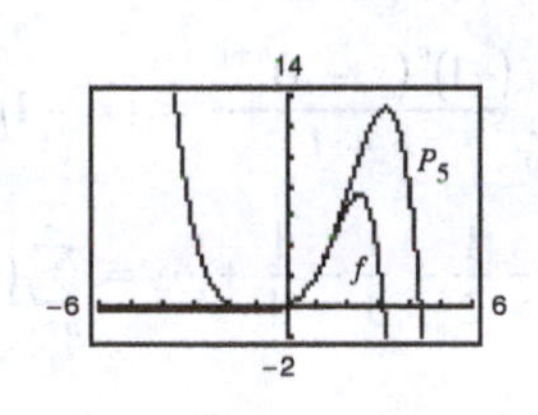

49. $h(x) = \cos x \ln(1 + x)$

$$= \left(1 - \frac{x^2}{2} + \frac{x^4}{24} + \cdots\right)\left(x - \frac{x^2}{2} + \frac{x^3}{3} - \frac{x^4}{4} + \frac{x^5}{5} - \cdots\right)$$

$$= x - \frac{x^2}{2} + \left(\frac{x^3}{3} - \frac{x^3}{2}\right) + \left(\frac{x^4}{4} - \frac{x^4}{4}\right) + \left(\frac{x^5}{5} - \frac{x^5}{6} + \frac{x^5}{24}\right) + \cdots$$

$$= x - \frac{x^2}{2} - \frac{x^3}{6} + \frac{3x^5}{40} + \cdots$$

$$P_5(x) = x - \frac{x^2}{2} - \frac{x^3}{6} + \frac{3x^5}{40}$$

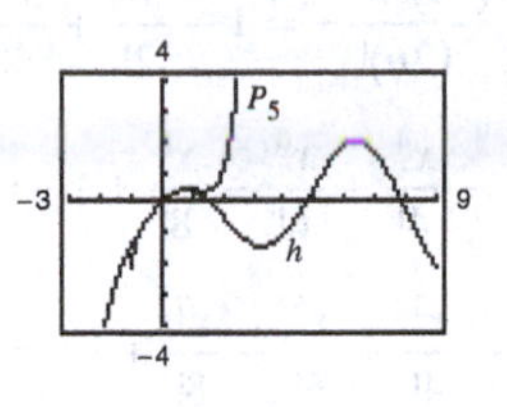

51. $g(x) = \dfrac{\sin x}{1 + x}$. Divide the series for $\sin x$ by $(1 + x)$.

$$\begin{array}{r l}
 & x - x^2 + \frac{5x^3}{6} - \frac{5x^4}{6} + \\
1 + x \,\big)\!\! & x + 0x^2 - \frac{x^3}{6} + 0x^4 + \frac{x^5}{120} + \cdots \\
 & \underline{x + x^2} \\
 & -x^2 - \frac{x^3}{6} \\
 & \underline{-x^2 - x^3} \\
 & \frac{5x^3}{6} + 0x^4 \\
 & \underline{\frac{5x^3}{6} + \frac{5x^4}{6}} \\
 & -\frac{5x^4}{6} + \frac{x^5}{120} \\
 & \underline{-\frac{5x^4}{6} - \frac{5x^5}{6}} \\
 & \vdots
\end{array}$$

$$g(x) = x - x^2 + \frac{5x^3}{6} - \frac{5x^4}{6} + \cdots$$

$$P_4(x) = x - x^2 + \frac{5x^3}{6} - \frac{5x^4}{6}$$

53. $$\int_0^x \left(e^{-t^2} - 1\right) dt = \int_0^x \left[\left(\sum_{n=0}^{\infty} \frac{(-1)^n t^{2n}}{n!}\right) - 1\right] dt$$

$$= \int_0^x \left[\sum_{n=0}^{\infty} \frac{(-1)^{n+1} t^{2n+2}}{(n+1)!}\right] dt$$

$$= \left[\sum_{n=0}^{\infty} \frac{(-1)^{n+1} t^{2n+3}}{(2n+3)(n+1)!}\right]_0^x$$

$$= \sum_{n=0}^{\infty} \frac{(-1)^{n+1} x^{2n+3}}{(2n+3)(n+1)!}$$

55. Because $\ln x = \sum_{n=0}^{\infty} \frac{(-1)^n (x-1)^{n+1}}{n+1} = (x-1) - \frac{(x-1)^2}{2} + \frac{(x-1)^3}{3} - \frac{(x-1)^4}{4} + \cdots, \quad (0 < x \le 2)$

you have $\ln 2 = 1 - \frac{1}{2} + \frac{1}{3} - \frac{1}{4} + \cdots = \sum_{n=1}^{\infty} (-1)^{n+1} \frac{1}{n} \approx 0.6931.$ (10,001 terms)

57. Because $e^x = \sum_{n=0}^{\infty} \frac{x^n}{n!} = 1 + x + \frac{x^2}{2!} + \frac{x^3}{3!} + \cdots$, you have $e^2 = 1 + 2 + \frac{2^2}{2!} + \frac{2^3}{3!} + \cdots = \sum_{n=0}^{\infty} \frac{2^n}{n!} \approx 7.3891.$ (12 terms)

59. Because

$$\cos x = \sum_{n=0}^{\infty} \frac{(-1)^n x^{2n}}{(2n)!} = 1 - \frac{x^2}{2!} + \frac{x^4}{4!} - \frac{x^6}{6!} + \frac{x^8}{8!} - \cdots$$

$$1 - \cos x = \frac{x^2}{2!} - \frac{x^4}{4!} + \frac{x^6}{6!} - \frac{x^8}{8!} + \cdots = \sum_{n=0}^{\infty} \frac{(-1)^n x^{2n+2}}{(2n+2)!}$$

$$\frac{1 - \cos}{x} = \frac{x}{2!} - \frac{x^3}{4!} + \frac{x^5}{6!} - \frac{x^7}{8!} + \cdots = \sum_{n=0}^{\infty} \frac{(-1)^n x^{2n+1}}{(2n+2)!}$$

you have $\lim_{x \to 0} \frac{1 - \cos x}{x} = \lim_{x \to 0} \sum_{n=0}^{\infty} \frac{(-1) x^{2n+1}}{(2n+2)!} = 0.$

61. Because $e^x = 1 + x + \frac{x^2}{2!} + \frac{x^3}{3!} + \cdots$

$$e^x - 1 = x + \frac{x^2}{2!} + \frac{x^3}{3!} + \cdots \sum_{n=0}^{\infty} \frac{x^{n+1}}{(n+1)!}$$

and $\frac{e^x - 1}{x} = 1 + \frac{x}{2!} + \frac{x^2}{3!} + \cdots \sum_{n=0}^{\infty} \frac{x^n}{(n+1)!}$

you have $\lim_{x \to 0} \frac{e^x - 1}{x} = \lim_{x \to 0} \sum \frac{x^n}{(n+1)!} = 1.$

63.
$$\int_0^1 e^{-x^3}\,dx = \int_0^1 \left[\sum_{n=0}^{\infty} \frac{(-x^3)^n}{n!}\right] dx$$

$$= \int_0^1 \left[\sum_{n=0}^{\infty} \frac{(-1)^n x^{3n}}{n!}\right] dx$$

$$= \left[\sum_{n=0}^{\infty} \frac{(-1)^n x^{3n+1}}{(3n+1)n!}\right]_0^1$$

$$= \sum_{n=0}^{\infty} \frac{(-1)^n}{(3n+1)n!}$$

$$= 1 - \frac{1}{4} + \frac{1}{14} - \cdots + (-1)^n \frac{1}{(3n+1)n!} + \cdots$$

Because $\frac{1}{[3(6)+1]6!} < 0.0001$, you need 6 terms.

$$\int_0^1 e^{-x^2}\,dx \approx \sum_{n=0}^{5} \frac{(-1)^n}{(3n+1)n!} \approx 0.8075$$

65. $\int_0^1 \frac{\sin x}{x}\,dx = \int_0^1 \left[\sum_{n=0}^{\infty} \frac{(-1)^n x^{2n}}{(2n+1)!}\right] dx = \left[\sum_{n=0}^{\infty} \frac{(-1)^n x^{2n+1}}{(2n+1)(2n+1)!}\right]_0^1 = \sum_{n=0}^{\infty} \frac{(-1)^n}{(2n+1)(2n+1)!}$

Because $1/(7 \cdot 7!) < 0.0001$, you need three terms:

$\int_0^1 \frac{\sin x}{x}\,dx = 1 - \frac{1}{3 \cdot 3!} + \frac{1}{5 \cdot 5!} - \cdots \approx 0.9461.$ (using three nonzero terms)

Note: You are using $\lim_{x \to 0^+} \frac{\sin x}{x} = 1.$

67. $\int_0^{1/2} \frac{\arctan x}{x}\,dx = \int_0^{1/2}\left(1 - \frac{x^2}{3} + \frac{x^4}{5} - \frac{x^6}{7} + \cdots\right)dx = \left[x - \frac{x^3}{3^2} + \frac{x^5}{5^2} - \frac{x^7}{7^2} + \cdots\right]_0^{1/2}$

Because $1/(9^2 2^9) < 0.0001$, you need four terms.

$$\int_0^{1/2} \frac{\arctan x}{x}\,dx \approx \left(\frac{1}{2} - \frac{1}{3^2 2^3} + \frac{1}{5^2 2^5} - \frac{1}{7^2 2^7}\right) \approx 0.4872$$

Note: You are using $\lim_{x\to 0^+} \frac{\arctan x}{x} = 1$.

69. $\int_{0.1}^{0.3} \sqrt{1 + x^3}\,dx = \int_{0.1}^{0.3}\left(1 + \frac{x^3}{2} - \frac{x^6}{8} + \frac{x^9}{16} - \frac{5x^{12}}{128} + \cdots\right)dx = \left[x + \frac{x^4}{8} - \frac{x^7}{56} + \frac{x^{10}}{160} - \frac{5x^{13}}{1664} + \cdots\right]_{0.1}^{0.3}$

Because $\frac{1}{56}\left(0.3^7 - 0.1^7\right) < 0.0001$, you need two terms.

$$\int_{0.1}^{0.3} \sqrt{1 + x^3}\,dx = \left[(0.3 - 0.1) + \frac{1}{8}\left(0.3^4 - 0.1^4\right)\right] \approx 0.201$$

71. $\int_0^{\pi/2} \sqrt{x}\cos x\,dx = \int_0^{\pi/2}\left[\sum_{n=0}^{\infty} \frac{(-1)^n x^{(4n+1)/2}}{(2n)!}\right]dx = \left[\sum_{n=0}^{\infty} \frac{(-1)^n x^{(4n+3)/2}}{\left(\frac{4n+3}{2}\right)(2n)!}\right]_0^{\pi/2} = \left[\sum_{n=0}^{\infty} \frac{(-1)^n 2x^{(4n+3)/2}}{(4n+3)(2n)!}\right]_0^{\pi/2}$

Because $2(\pi/2)^{23/2}/(23 \cdot 10!) < 0.0001$, you need five terms.

$$\int_0^1 \sqrt{x}\cos x\,dx = 2\left[\frac{(\pi/2)^{3/2}}{3} - \frac{(\pi/2)^{7/2}}{14} + \frac{(\pi/2)^{11/2}}{264} - \frac{(\pi/2)^{15/2}}{10{,}800} + \frac{(\pi/2)^{19/2}}{766{,}080}\right] \approx 0.7040$$

73. From Exercise 27, you have

$$\frac{1}{\sqrt{2\pi}}\int_0^1 e^{-x^2/2}\,dx = \frac{1}{\sqrt{2\pi}}\int_0^1 \sum_{n=0}^{\infty} \frac{(-1)^n x^{2n}}{2^n n!}\,dx = \frac{1}{\sqrt{2\pi}}\left[\sum_{n=0}^{\infty} \frac{(-1)^n x^{2n+1}}{2^n n!(2n+1)}\right]_0^1 = \frac{1}{\sqrt{2\pi}}\sum_{n=0}^{\infty} \frac{(-1)^n}{2^n n!(2n+1)}$$

$$\approx \frac{1}{\sqrt{2\pi}}\left(1 - \frac{1}{2 \cdot 1 \cdot 3} + \frac{1}{2^2 \cdot 2! \cdot 5} - \frac{1}{2^3 \cdot 3! \cdot 7}\right) \approx 0.3412.$$

75. Method 1: Use the half-angle formula $\cos^2 x = \frac{1}{2}(1 + \cos 2x)$.

$$\frac{1}{2} + \frac{1}{2}\left(1 - 2x^2 + \frac{2}{3}x^4\right) = 1 - x^2 + \frac{1}{3}x^4$$

Method 2: Square the series for $\cos x$.

$$\left(1 - \frac{x^2}{2} + \frac{x^4}{4!} - \cdots\right)\left(1 - \frac{x^2}{2} + \frac{x^4}{4!} - \cdots\right) = 1 - x^2 + \left(\frac{x^4}{24} + \frac{x^4}{4} + \frac{x^4}{24}\right) + \cdots = 1 - x^2 + \frac{1}{3}x^4$$

Method 3: Use the definition of Maclaurin series.

$f(x) = \cos^2 x$	$f(0) = 1$
$f'(x) = -2\sin x\cos x$	$f'(0) = 0$
$f''(x) = 2 - 4\cos^2 x$	$f''(0) = -2$
$f'''(x) = 8\sin x\cos x$	$f'''(0) = 0$
$f^{(4)}(x) = 16\cos^2 x - 8$	$f^{(4)}(0) = 8$

$$1 - \frac{2}{2}x^2 + \frac{8}{4!}x^4 = 1 - x^2 + \frac{1}{3}x^4$$

77. Because $\sin x = \sum_{n=0}^{\infty} \frac{(-1)^n x^{2n+1}}{(2n+1)!}$

$$\sum_{n=0}^{\infty} \frac{(-1)^n (x+3)^{2n+1}}{2^2(2n+1)!} = \frac{1}{4}\sin(x+3).$$

79. $y = \left(\tan\theta - \frac{g}{kv_0\cos\theta}\right)x - \frac{g}{k^2}\ln\left(1 - \frac{kx}{v_0\cos\theta}\right)$

$$= (\tan\theta)x - \frac{gx}{kv_0\cos\theta} - \frac{g}{k^2}\left[-\frac{kx}{v_0\cos\theta} - \frac{1}{2}\left(\frac{kx}{v_0\cos\theta}\right)^2 - \frac{1}{3}\left(\frac{kx}{v_0\cos\theta}\right)^3 - \frac{1}{4}\left(\frac{kx}{v_0\cos\theta}\right)^4 - \cdots\right]$$

$$= (\tan\theta)x - \frac{gx}{kv_0\cos\theta} + \frac{gx}{kv_0\cos\theta} + \frac{gx^2}{2v_0^2\cos^2\theta} + \frac{gkx^3}{3v_0^3\cos^3\theta} + \frac{gk^2x^4}{4v_0^4\cos^4\theta} + \cdots$$

$$= (\tan\theta)x + \frac{gx^2}{2v_0^2\cos^2\theta} + \frac{kgx^3}{3v_0^3\cos^3\theta} + \frac{k^2gx^4}{4v_0^4\cos^4\theta} + \cdots$$

81. $f(x) = \begin{cases} e^{-1/x^2}, & x \neq 0 \\ 0, & x = 0 \end{cases}$

(a)

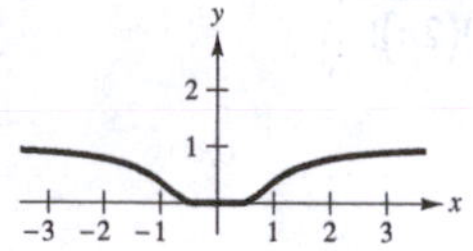

(b) $f'(0) = \lim_{x\to 0} \frac{f(x) - f(0)}{x - 0} = \lim_{x\to 0} \frac{e^{-1/x^2} - 0}{x}$

Let $y = \lim_{x\to 0} \frac{e^{-1/x^2}}{x}$. Then

$$\ln y = \lim_{x\to 0} \ln\left(\frac{e^{-1/x^2}}{x}\right) = \lim_{x\to 0^+}\left[-\frac{1}{x^2} - \ln x\right] = \lim_{x\to 0^+}\left[\frac{-1 - x^2\ln x}{x^2}\right] = -\infty.$$

So, $y = e^{-\infty} = 0$ and you have $f'(0) = 0$.

(c) $\sum_{n=0}^{\infty} \frac{f^{(n)}(0)}{n!}x^n = f(0) + \frac{f'(0)x}{1!} + \frac{f''(0)x^2}{2!} + \cdots = 0 \neq f(x)$ This series converges to f at $x = 0$ only.

83. By the Ratio Test: $\lim_{n\to\infty}\left|\frac{x^{n+1}}{(n+1)!}\cdot\frac{n!}{x^n}\right| = \lim_{n\to\infty}\frac{|x|}{n+1} = 0$ which shows that $\sum_{n=0}^{\infty}\frac{x^n}{n!}$ converges for all x.

85. $\binom{6}{3} = \frac{6\cdot 5\cdot 4}{3!} = \frac{120}{6} = 20$

87. $\binom{-0.8}{5} = \frac{-0.8(-0.8-1)(-0.8-2)(-0.8-3)(-0.8-4)}{5!} = \frac{-73.54368}{120} = -0.612864$

89. $(1+x)^k = \sum_{n=0}^{\infty}\binom{k}{n}x^n$

Example: $(1+x)^2 = \sum_{n=0}^{\infty}\binom{2}{n}x^n = 1 + 2x + x^2$

91. Assume $e = p/q$ is rational. Let $N > q$ and form the following.

$$e - \left[1 + 1 + \frac{1}{2!} + \cdots + \frac{1}{N!}\right] = \frac{1}{(N+1)!} + \frac{1}{(N+2)!} + \cdots$$

Set $a = N!\left[e - \left(1 + 1 + \cdots + \frac{1}{N!}\right)\right]$, a positive integer. But,

$$a = N!\left[\frac{1}{(N+1)!} + \frac{1}{(N+2)!} + \cdots\right] = \frac{1}{N+1} + \frac{1}{(N+1)(N+2)} + \cdots < \frac{1}{N+1} + \frac{1}{(N+1)^2} + \cdots$$

$$= \frac{1}{N+1}\left[1 + \frac{1}{N+1} + \frac{1}{(N+1)^2} + \cdots\right] = \frac{1}{N+1}\left[\frac{1}{1 - \left(\frac{1}{N+1}\right)}\right] = \frac{1}{N}, \text{ a contradiction.}$$

93. Assume the interval is $[-1, 1]$. Let $x \in [-1, 1]$,

$$f(1) = f(x) + (1 - x)f'(x) + \tfrac{1}{2}(1 - x)^2 f''(c),\ c \in (x, 1)$$

$$f(-1) = f(x) + (-1 - x)f'(x) + \tfrac{1}{2}(-1 - x)^2 f''(d),\ d \in (-1, x).$$

So, $f(1) - f(-1) = 2f'(x) + \tfrac{1}{2}(1 - x)^2 f''(c) - \tfrac{1}{2}(1 + x)^2 f''(d)$

$$2f'(x) = f(1) - f(-1) - \tfrac{1}{2}(1 - x)^2 f''(c) + \tfrac{1}{2}(1 + x)^2 f''(d).$$

Because $|f(x)| \le 1$ and $|f''(x)| \le 1$,

$$2|f'(x)| \le |f(1)| + |f(-1)| + \tfrac{1}{2}(1 - x)^2|f''(c)| + \tfrac{1}{2}(1 + x)^2|f''(d)| \le 1 + 1 + \tfrac{1}{2}(1 - x^2) + \tfrac{1}{2}(1 + x)^2 = 3 + x^2 \le 4.$$

So, $|f'(x)| \le 2$.

Note: Let $f(x) = \tfrac{1}{2}(x + 1)^2 - 1$. Then $|f'(x)| \le 1$, $|f''(x)| = 1$ and $f'(1) = 2$.

Review Exercises for Chapter 9

1. $a_n = 6^{n-2}$

$a_1 = 6^1 - 2 = 4$

$a_2 = 6^2 - 2 = 34$

$a_3 = 6^3 - 2 = 214$

$a_4 = 6^4 - 2 = 1294$

$a_5 = 6^5 - 2 = 7774$

3. $a_n = \left(-\tfrac{1}{4}\right)^n$

$a_1 = \left(-\tfrac{1}{4}\right)^1 = -\tfrac{1}{4}$

$a_2 = \left(-\tfrac{1}{4}\right)^2 = \tfrac{1}{16}$

$a_3 = \left(-\tfrac{1}{4}\right)^3 = -\tfrac{1}{64}$

$a_4 = \left(-\tfrac{1}{4}\right)^4 = \tfrac{1}{256}$

$a_5 = \left(-\tfrac{1}{4}\right)^5 = -\tfrac{1}{1024}$

5. $a_n = 4 + \frac{2}{n}$: 6, 5, 4.67, …

Matches (a).

6. $a_n = 4 - \frac{n}{2}$: 3.5, 3, …

Matches (c).

7. $a_n = 10(0.3)^{n-1}$: 10, 3, …

Matches (d).

8. $a_n = 6\left(-\tfrac{2}{3}\right)^{n-1}$: 6, −4, …

Matches (b).

9. $a_n = \dfrac{5n+2}{n}$

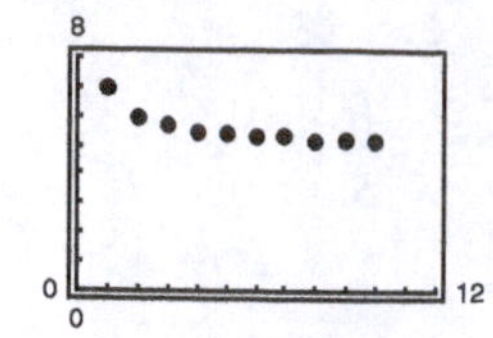

The sequence seems to converge to 5.

$$\lim_{n\to\infty} a_n = \lim_{n\to\infty} \frac{5n+2}{n} = \lim_{n\to\infty}\left(5 + \frac{2}{n}\right) = 5$$

11. $\displaystyle\lim_{n\to\infty} \frac{1}{\sqrt{n}} = 0$

Converges

13. $\displaystyle\lim_{n\to\infty}\left[\left(\frac{2}{5}\right)^n + 5\right] = 0 + 5 = 5$

Converges

15. $a_n = \dfrac{(4n)!}{(4n-1)!} = \dfrac{(4n)(4n-1)!}{(4n-1)!} = 4n$

$\displaystyle\lim_{n\to\infty} a_n = \infty$

Diverges

17. $\displaystyle\lim_{n\to\infty} \frac{e^{2n}}{\ln n} = \lim_{n\to\infty} \frac{2e^{2n}}{1/n} = \infty$

Diverges

19. $a_n = 5n - 2$

21. $a_n = \dfrac{1}{n! + 1}$

23. $a_n = 3 - \dfrac{1}{2n} < 3 - \dfrac{1}{2n+2} = a_{n+1}$

Monotonic; $|a_n| < 3$, bounded

25. (a) $A_n = 8000\left(1 + \dfrac{0.05}{4}\right)^n, \quad n = 1, 2, 3, \ldots$

$$A_1 = 8000\left(1 + \frac{0.05}{4}\right)^1 = \$8100.00$$

$A_2 = \$8201.25$

$A_3 = \$8303.77$

$A_4 = \$8407.56$

$A_5 = \$8512.66$

$A_6 = \$8619.07$

$A_7 = \$8726.80$

$A_8 = \$8835.89$

(b) $A_{40} = \$13{,}148.96$

27. $S_1 = 3$

$S_2 = 3 + \dfrac{3}{2} = \dfrac{9}{2} = 4.5$

$S_3 = 3 + \dfrac{3}{2} + 1 = \dfrac{11}{2} = 5.5$

$S_4 = 3 + \dfrac{3}{2} + 1 + \dfrac{3}{4} = \dfrac{25}{4} = 6.25$

$S_5 = 3 + \dfrac{3}{2} + 1 + \dfrac{3}{4} + \dfrac{3}{5} = \dfrac{137}{20} = 6.85$

29. (a)

n	5	10	15	20	25
S_n	13.2	113.3	873.8	6648.5	50,500.3

The series diverges $\left(\text{geometric } r = \frac{3}{2} > 1\right)$.

(b)

31. $\displaystyle\sum_{n=0}^{\infty}\left(\frac{2}{5}\right)^n = \frac{1}{1 - (2/5)} = \frac{5}{3}$ (Geometric series)

33. $$\sum_{n=0}^{\infty}\left[(0.4)^n + (0.9)^n\right] = \sum_{n=0}^{\infty}(0.4)^n + \sum_{n=0}^{\infty}(0.9)^n = \frac{1}{1-0.4} + \frac{1}{1-0.9} = \frac{5}{3} + 10 = \frac{35}{3}$$

35. (a) $0.\overline{09} = 0.09 + 0.0009 + 0.000009 + \cdots = 0.09(1 + 0.01 + 0.0001 + \cdots) = \displaystyle\sum_{n=0}^{\infty}(0.09)(0.01)^n$

(b) $0.\overline{09} = \dfrac{0.09}{1 - 0.01} = \dfrac{1}{11}$

37. $\sum_{n=0}^{\infty} (1.67)^n$

Geometric series with $r = 1.67 > 1$

Diverges by Theorem 9.6

39. $\sum_{n=0}^{\infty} \frac{2n+1}{3n+2}$

$\lim_{n\to\infty} \frac{2n+1}{3n+2} = \frac{2}{3} \neq 0$

Diverges by Theorem 9.9

41. Year 1: 9600

Year 2: $9600 + 0.92(9600)$

Year 3: $9600 + 0.92(9600) + (0.92)^2(9600)$

Year n: $9600 + \cdots + (0.92)^{n-1}(9600)$

This finite geometric series sum to

$\left(\frac{9600}{1-0.92}\right)(1 - 0.92^n) = 120{,}000(1 - 0.92^n),\ n > 0.$

43. $\sum_{n=1}^{\infty} \frac{2}{6n+1}$

Let $f(x) = \frac{2}{6x+1},\ f'(x) = \frac{-12}{(6x+1)^2} < 0$ for $x \geq 1$

f is positive, continuous, and decreasing for $x \geq 1$.

$\int_1^{\infty} \frac{2}{6x+1}\,dx = \left[\frac{1}{3}\ln(6x+1)\right]_1^{\infty} = \infty$, diverges.

So, the series diverges by Theorem 9.10.

45. $\sum_{n=1}^{\infty} \frac{1}{n^{5/2}}$ is a p-series with $p = \frac{5}{2} > 1$.

So, the series converges.

47. $\sum_{n=1}^{\infty} \left(\frac{1}{n^2} - \frac{1}{n}\right) = \sum_{n=1}^{\infty} \frac{1}{n^2} - \sum_{n=1}^{\infty} \frac{1}{n}$

Because the second series is a divergent p-series while the first series is a convergent p-series, the difference diverges.

49. $\sum_{n=2}^{\infty} \frac{1}{\sqrt[3]{n} - 1}$

$\frac{1}{\sqrt[3]{n} - 1} > \frac{1}{\sqrt[3]{n}}$

Therefore, the series diverges by comparison with the divergent p-series

$\sum_{n=2}^{\infty} \frac{1}{\sqrt[3]{n}} = \sum_{n=2}^{\infty} \frac{1}{n^{1/3}}.$

51. $\sum_{n=1}^{\infty} \frac{1}{\sqrt{n^3 + 2n}}$

$\lim_{n\to\infty} \frac{1/\sqrt{n^3+2n}}{1/(n^{3/2})} = \lim_{n\to\infty} \frac{n^{3/2}}{\sqrt{n^3+2n}} = 1$

By a limit comparison test with the convergent p-series $\sum_{n=1}^{\infty} \frac{1}{n^{3/2}}$, the series converges.

53. $\sum_{n=1}^{\infty} \frac{1 \cdot 3 \cdot 5 \cdots (2n-1)}{2 \cdot 4 \cdot 6 \cdots (2n)}$

$a_n = \frac{1 \cdot 3 \cdot 5 \cdots (2n-1)}{2 \cdot 4 \cdot 6 \cdots (2n)} = \left(\frac{3}{2} \cdot \frac{5}{4} \cdots \frac{2n-1}{2n-2}\right)\frac{1}{2n} > \frac{1}{2n}$

Because $\sum_{n=1}^{\infty} \frac{1}{2n} = \frac{1}{2}\sum_{n=1}^{\infty} \frac{1}{n}$ diverges (harmonic series), so does the original series.

55. $\sum_{n=1}^{\infty} \frac{(-1)^n}{n^5}$ converges by the Alternating Series Test.

$\lim_{n\to\infty} \frac{1}{n^5} = 0$ and $a_{n+1} = \frac{1}{(n+1)^5} < \frac{1}{n^5} = a_n.$

57. $\sum_{n=2}^{\infty} \frac{(-1)^n - n}{n^2 - 3}$ converges by the Alternating Series Test.

$\lim_{n\to\infty} \frac{n}{n^2 - 3} = 0$ and if

$f(x) = \frac{n}{n^2 - 3},\ f'(x) = \frac{-(n^2+3)}{(n^2-3)^2} < 0 \Rightarrow$ terms are decreasing. So, $a_{n+1} < a_n$.

59. $\sum_{n=1}^{\infty} \frac{(-1)^{n+1}\sqrt{n}}{\sqrt[4]{n} + 2}$

$\lim_{n\to\infty} \frac{n^{1/2}}{n^{1/4}} = \lim_{n\to\infty} n^{1/4} = \infty$

Diverges by the nth-Term Test

61. $|R_N| \leq a_{N+1} = \frac{1}{(N+1)^4}$

For an error less than 0.0001, N must satisfy the inequality

$\frac{1}{(N+1)^4} < 0.0001$

$(N+1)^4 > 10{,}000$

$N > \sqrt[4]{10{,}000} - 1 = 9.$

So, use 10 terms.

63. $\sum_{n=1}^{\infty}\left(\frac{3n-1}{2n+5}\right)^n$

$$\lim_{n\to\infty}\sqrt[n]{\left(\frac{3n-1}{2n+5}\right)^n} = \lim_{n\to\infty}\left(\frac{3n-1}{2n+5}\right) = \frac{3}{2} > 1$$

Diverges by Root Test

65. $\sum_{n=1}^{\infty}\frac{2^n}{n^3}$

$$\lim_{n\to\infty}\left|\frac{a_{n+1}}{a_n}\right| = \lim_{n\to\infty}\left|\frac{2^{n+1}}{(n+1)^3}\cdot\frac{n^3}{2^n}\right| = \lim_{n\to\infty}\frac{2n^3}{(n+1)^3} = 2$$

By the Ratio Test, the series diverges.

67. $\sum_{n=1}^{\infty}\frac{n}{e^{n^2}}$

$$\lim_{n\to\infty}\left|\frac{a_{n+1}}{a_n}\right| = \lim_{n\to\infty}\left|\frac{n+1}{e^{(n+1)^2}}\cdot\frac{e^{n^2}}{n}\right| = \lim_{n\to\infty}\left|\frac{e^{n^2}(n+1)}{e^{n^2+2n+1}n}\right| = \lim_{n\to\infty}\left(\frac{1}{e^{2n+1}}\right)\left(\frac{n+1}{n}\right) = (0)(1) = 0 < 1$$

By the Ratio Test, the series converges.

69. (a) Ratio Test: $\lim_{n\to\infty}\left|\frac{a_{n+1}}{a_n}\right| = \lim_{n\to\infty}\frac{(n+1)(3/5)^{n+1}}{n(3/5)^n} = \lim_{n\to\infty}\left(\frac{n+1}{n}\right)\left(\frac{3}{5}\right) = \frac{3}{5} < 1$, converges

(b)

n	5	10	15	20	25
S_n	2.8752	3.6366	3.7377	3.7488	3.7499

(c)

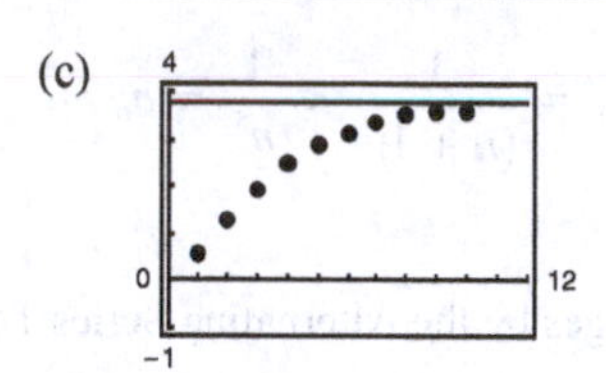

(d) The sum is approximately 3.75.

71. $\sum_{n=1}^{\infty}\frac{4}{n^{2\pi}}$ is a convergent p-series $(p = 2\pi)$.

73. $\sum_{n=1}^{\infty}\frac{5n^3+6}{7n^3+2n}$ diverges by the nth-Term Test for Divergence:

$$\lim_{n\to\infty}\frac{5n^3+6}{7n^3+2n} = \frac{5}{7} \neq 0.$$

75. $\sum_{n=1}^{\infty}\frac{10^n}{4+9^n}$ diverges by a limit comparison with $\sum_{n=1}^{\infty}\left(\frac{10}{9}\right)^n$, a divergent geometric series.

77.

$f(x) = e^{-2x}$, $\quad f(0) = 1$

$f'(x) = -2e^{-2x}$, $\quad f'(0) = -2$

$f''(x) = 4e^{-2x}$, $\quad f''(0) = 4$

$f'''(x) = -8e^{-2x}$, $\quad f'''(0) = -8$

$$P_3(x) = f(0) + f'(0)x + \frac{f''(0)}{2!}x^2 + \frac{f'''(0)}{3!}x^3$$

$$= 1 - 2x + 2x^2 - \frac{4}{3}x^3$$

79.

$f(x) = \frac{1}{x^3} = x^{-3}$ $\quad f(1) = 1$

$f'(x) = -\frac{3}{x^4}$ $\quad f'(1) = -3$

$f''(x) = \frac{12}{x^5}$ $\quad f''(1) = 12$

$f'''(x) = -\frac{60}{x^6}$ $\quad f'''(1) = -60$

$$P_3(x) = 1 - 3(x-1) + \frac{12}{2}(x-1)^2 - \frac{60}{3!}(x-1)^3$$

$$= 1 - 3(x-1) + 6(x-1)^2 - 10(x-1)^3$$

81. $f(x) = \cos x$

$\left|f^{(n+1)}(x)\right| \leq 1$ for all x and all n.

$$\left|R_n(x)\right| = \left|\frac{f^{(n+1)}(z)\,x^{n+1}}{(n+1)!}\right| \leq \frac{(0.75)^{n+1}}{(n+1)!} < 0.001$$

By trial and error, $n = 5$. (3 terms)

83. $\sum_{n=0}^{\infty}\left(\frac{x}{10}\right)^n$

Geometric series which converges only if $|x/10| < 1$ or $-10 < x < 10$.

85. $\sum_{n=0}^{\infty}\frac{(-1)^n(x-2)^n}{(n+1)^2}$

$$\lim_{n\to\infty}\left|\frac{u_{n+1}}{u_n}\right| = \lim_{n\to\infty}\left|\frac{(-1)^{n+1}(x-2)^{n+1}}{(n+2)^2}\cdot\frac{(n+1)^2}{(-1)^n(x-2)^n}\right| = |x-2|$$

$R = 1$

Center: 2

Because the series converges when $x = 1$ and when $x = 3$, the interval of convergence is $[1, 3]$.

87. $\sum_{n=0}^{\infty} n!(x-2)^n$

$$\lim_{n\to\infty}\left|\frac{u_{n+1}}{u_n}\right| = \lim_{n\to\infty}\left|\frac{(n+1)!(x-2)^{n+1}}{n!(x-2)^n}\right| = \infty$$

which implies that the series converges only at the center $x = 2$.

89. (a) $f(x) = \sum_{n=0}^{\infty}\left(\frac{x}{5}\right)^n, (-5, 5)$ (Geometric)

(b) $f'(x) = \sum_{n=1}^{\infty}\frac{n}{5}\left(\frac{x}{5}\right)^{n-1}, (-5, 5)$

(c) $f''(x) = \sum_{n=2}^{\infty}\frac{n(n-1)}{25}\left(\frac{x}{5}\right)^{n-2}, (-5, 5)$

(d) $\int f(x)\,dx = \sum_{n=0}^{\infty}\frac{5}{n+1}\left(\frac{x}{5}\right)^{n+1}, (-5, 5)$

$$\left[\sum_{n=0}^{\infty}\frac{5}{n+1}\left(\frac{-5}{5}\right)^{n+1} = \sum_{n=0}^{\infty}\frac{(-1)^{n+1}5}{n+1}, \quad \text{converges}\right]$$

91.

$$y = \sum_{n=0}^{\infty}(-1)^n\frac{x^{2n}}{4^n(n!)^2}$$

$$y' = \sum_{n=1}^{\infty}\frac{(-1)^n(2n)x^{2n-1}}{4^n(n!)^2} = \sum_{n=0}^{\infty}\frac{(-1)^{n+1}(2n+2)x^{2n+1}}{4^{n+1}[(n+1)!]^2}$$

$$y'' = \sum_{n=0}^{\infty}\frac{(-1)^{n+1}(2n+2)(2n+1)x^{2n}}{4^{n+1}[(n+1)!]^2}$$

$$\begin{aligned}
x^2y'' + xy' + x^2y &= \sum_{n=0}^{\infty}\frac{(-1)^{n+1}(2n+2)(2n+1)x^{2n+2}}{4^{n+1}[(n+1)!]^2} + \sum_{n=0}^{\infty}\frac{(-1)^{n+1}(2n+2)x^{2n+2}}{4^{n+1}[(n+1)!]^2} + \sum_{n=0}^{\infty}(-1)^n\frac{x^{2n+2}}{4^n(n!)^2}\\
&= \sum_{n=0}^{\infty}\left[(-1)^{n+1}\frac{(2n+2)(2n+1)}{4^{n+1}[(n+1)!]^2} + \frac{(-1)^{n+1}(2n+2)}{4^{n+1}[(n+1)!]^2} + \frac{(-1)^n}{4^n(n!)^2}\right]x^{2n+2}\\
&= \sum_{n=0}^{\infty}\left[\frac{(-1)^{n+1}(2n+2)(2n+1+1)}{4^{n+1}[(n+1)!]^2} + (-1)^n\frac{1}{4^n(n!)^2}\right]x^{2n+2}\\
&= \sum_{n=0}^{\infty}\left[\frac{(-1)^{n+1}4(n+1)^2}{4^{n+1}[(n+1)!]^2} + (-1)^n\frac{1}{4^n(n!)^2}\right]x^{2n+2}\\
&= \sum_{n=0}^{\infty}\left[\frac{(-1)^{n+1}1}{4^n(n!)^2} + (-1)^n\frac{1}{4^n(n!)^2}\right]x^{2n+2} = 0
\end{aligned}$$

93. $\frac{2}{3-x} = \frac{2/3}{1-(x/3)} = \frac{a}{1-r}$

$$\sum_{n=0}^{\infty}\frac{2}{3}\left(\frac{x}{3}\right)^n = \sum_{n=0}^{\infty}\frac{2x^n}{3^{n+1}}$$

95. $\dfrac{6}{4-x} = \dfrac{6}{3-(x-1)} = \dfrac{2}{1-\left(\dfrac{x-1}{3}\right)} = \dfrac{a}{1-r} = \displaystyle\sum_{n=0}^{\infty} 2\left(\dfrac{x-1}{3}\right)^n = 2\sum_{n=0}^{\infty} \dfrac{(x-1)^n}{3^n}$

Interval of convergence: $\left|\dfrac{x-1}{3}\right| < 1 \Rightarrow |x-1| < 3 \Rightarrow (-2, 4)$

97. $\ln x = \displaystyle\sum_{n=1}^{\infty} (-1)^{n+1} \dfrac{(x-1)^n}{n}, \quad 0 < x \le 2$

$\ln\left(\dfrac{5}{4}\right) = \displaystyle\sum_{n=1}^{\infty} (-1)^{n+1} \left(\dfrac{(5/4)-1}{n}\right)^n = \sum_{n=1}^{\infty} (-1)^{n+1} \dfrac{1}{4^n n} \approx 0.2231$

99. $e^x = \displaystyle\sum_{n=0}^{\infty} \dfrac{x^n}{n!}, \quad -\infty < x < \infty$

$e^{1/2} = \displaystyle\sum_{n=0}^{\infty} \dfrac{(1/2)^n}{n!} = \sum_{n=0}^{\infty} \dfrac{1}{2^n n!} \approx 1.6487$

101. $\cos x = \displaystyle\sum_{n=0}^{\infty} (-1)^n \dfrac{x^{2n}}{(2n)!}, \quad -\infty < x < \infty$

$\cos\left(\dfrac{2}{3}\right) = \displaystyle\sum_{n=0}^{\infty} (-1)^n \dfrac{2^{2n}}{3^{2n}(2n)!} = 0.7859$

103. $f(x) = \sin x$

$f'(x) = \cos x$

$f''(x) = -\sin x$

$f'''(x) = -\cos x, \cdots$

$\sin(x) = \displaystyle\sum_{n=0}^{\infty} \dfrac{f^{(n)}(x)\left[x - (3\pi/4)\right]^n}{n!}$

$= \dfrac{\sqrt{2}}{2} - \dfrac{\sqrt{2}}{2}\left(x - \dfrac{3\pi}{4}\right) - \dfrac{\sqrt{2}}{2 \cdot 2!}\left(x - \dfrac{3\pi}{4}\right)^2 + \cdots = \dfrac{\sqrt{2}}{2}\displaystyle\sum_{n=0}^{\infty} \dfrac{(-1)^{n(n+1)/2}\left[x - (3\pi/4)\right]^n}{n!}$

105. $3^x = \left(e^{\ln(3)}\right)^x = e^{x\ln(3)}$ and because $e^x = \displaystyle\sum_{n=0}^{\infty} \dfrac{x^n}{n!}$, you have

$3^x = \displaystyle\sum_{n=0}^{\infty} \dfrac{(x \ln 3)^n}{n!} = 1 + x\ln 3 + \dfrac{x^2[\ln 3]^2}{2!} + \dfrac{x^3[\ln 3]^3}{3!} + \dfrac{x^4[\ln 3]^4}{4!} + \cdots.$

107. $f(x) = \dfrac{1}{x}$

$f'(x) = -\dfrac{1}{x^2}$

$f''(x) = \dfrac{2}{x^3}$

$f'''(x) = -\dfrac{6}{x^4}, \cdots$

$\dfrac{1}{x} = \displaystyle\sum_{n=0}^{\infty} \dfrac{f^{(n)}(-1)(x+1)^n}{n!} = \sum_{n=0}^{\infty} \dfrac{-n!(x+1)^n}{n!} = -\sum_{n=0}^{\infty} (x+1)^n, -2 < x < 0$

109. $(1+x)^k = 1 + kx + \dfrac{k(k-1)x^2}{2!} + \dfrac{k(k-1)(k-2)x^3}{3!} + \cdots$

$(1+x)^{1/5} = 1 + \dfrac{x}{5} + \dfrac{(1/5)(-4/5)x^2}{2!} + \dfrac{1/5(-4/5)(-9/5)x^3}{3!} + \cdots$

$= 1 + \dfrac{1}{5}x - \dfrac{1 \cdot 4x^2}{5^2 2!} + \dfrac{1 \cdot 4 \cdot 9x^3}{5^3 3!} - \cdots = 1 + \dfrac{x}{5} + \displaystyle\sum_{n=2}^{\infty} \dfrac{(-1)^{n+1} 4 \cdot 9 \cdot 14 \cdots (5n-6)x^n}{5^n n!} = 1 + \dfrac{x}{5} - \dfrac{2}{25}x^2 + \dfrac{6}{125}x^3 - \cdots$

111. (a) $f(x) = e^{2x}$ $\quad f(0) = 1$

$f'(x) = 2e^{2x}$ $\quad f'(0) = 2$

$f''(x) = 4e^{2x}$ $\quad f''(0) = 4$

$f'''(x) = 8e^{2x}$ $\quad f'''(0) = 8$

$$P(x) = 1 + 2x + \frac{4x^2}{2!} + \frac{8x^3}{3!} = 1 + 2x + 2x^2 + \frac{4}{3}x^3$$

(b) $e^x = \sum_{n=0}^{\infty} \frac{x^n}{n!}, e^{2x} = \sum_{n=0}^{\infty} \frac{(2x)^n}{n!}$

$$P(x) = 1 + 2x + 2x^2 + \frac{4}{3}x^3$$

(c) $e^x \cdot e^x = \left(1 + x + \frac{x^2}{2!} + \cdots\right)\left(1 + x + \frac{x^2}{2!} + \cdots\right)$

$$P(x) = 1 + 2x + 2x^2 + \frac{4}{3}x^3$$

113. $e^x = \sum_{n=0}^{\infty} \frac{x^n}{n!}$

$$e^{6x} = \sum_{n=0}^{\infty} \frac{(6x)^n}{n!}$$

115. $\sin x = \sum_{n=0}^{\infty} \frac{(-1)^n x^{2n+1}}{(2n+1)!}$

$$\sin 5x = \sum_{n=0}^{\infty} \frac{(-1)^n (5x)^{2n+1}}{(2n+1)!}$$

117. $\cos x^3 = 1 - \frac{x^6}{2} + \frac{x^{12}}{24} - \cdots$

$$\int_0^{0.5} \cos x^3 \, dx = \left[x - \frac{x^7}{14} + \frac{x^{13}}{312} - \cdots\right]_0^{.5}$$

$$= 0.5 - 0.000558 + 0.00000039 - \cdots$$

Using the first term,

$$\int_0^{0.5} \cos x^3 \, dx \approx 0.5 \text{ accurate to } 0.01$$

(Exact $\approx$ 0.49944)

Problem Solving for Chapter 9

1. (a) $1\left(\frac{1}{3}\right) + 2\left(\frac{1}{9}\right) + 4\left(\frac{1}{27}\right) + \cdots = \sum_{n=0}^{\infty} \frac{1}{3}\left(\frac{2}{3}\right)^n = \frac{1/3}{1 - (2/3)} = 1$

(b) $0, \frac{1}{3}, \frac{2}{3}, 1$, etc.

(c) $\lim_{n \to \infty} C_n = 1 - \sum_{n=0}^{\infty} \frac{1}{3}\left(\frac{2}{3}\right)^n = 1 - 1 = 0$

3. Let $S = \sum_{n=1}^{\infty} \frac{1}{(2n-1)^2} = \frac{1}{1^2} + \frac{1}{3^2} + \frac{1}{5^2} + \cdots$.

Then

$$\frac{\pi^2}{6} = \frac{1}{1^2} + \frac{1}{2^2} + \frac{1}{3^2} + \frac{1}{4^2} + \cdots = S + \frac{1}{2^2} + \frac{1}{4^2} + \cdots = S + \frac{1}{2^2}\left[1 + \frac{1}{2^2} + \frac{1}{3^2} + \cdots\right] = S + \frac{1}{2^2}\left(\frac{\pi^2}{6}\right).$$

So, $S = \frac{\pi^2}{6} - \frac{1}{4}\frac{\pi^2}{6} = \frac{\pi^2}{6}\left(\frac{3}{4}\right) = \frac{\pi^2}{8}$.

5. (a) Position the three blocks as indicated in the figure.

The bottom block extends 1/6 over the edge of the table, the middle block extends 1/4 over the edge of the bottom block, and the top block extends 1/2 over the edge of the middle block.

The centers of gravity are located at

bottom block: $\frac{1}{6} - \frac{1}{2} = -\frac{1}{3}$

middle block: $\frac{1}{6} + \frac{1}{4} - \frac{1}{2} = -\frac{1}{12}$

top block: $\frac{1}{6} + \frac{1}{4} + \frac{1}{2} - \frac{1}{2} = \frac{5}{12}$.

The center of gravity of the top 2 blocks is

$\left(-\frac{1}{12} + \frac{5}{12}\right)\Big/2 = \frac{1}{6}$, which lies over the bottom block. The center of gravity of the 3 blocks is $\left(-\frac{1}{3} - \frac{1}{12} + \frac{5}{12}\right)\Big/3 = 0$

which lies over the table. So, the far edge of the top block lies $\frac{1}{6} + \frac{1}{4} + \frac{1}{2} = \frac{11}{12}$ beyond the edge of the table.

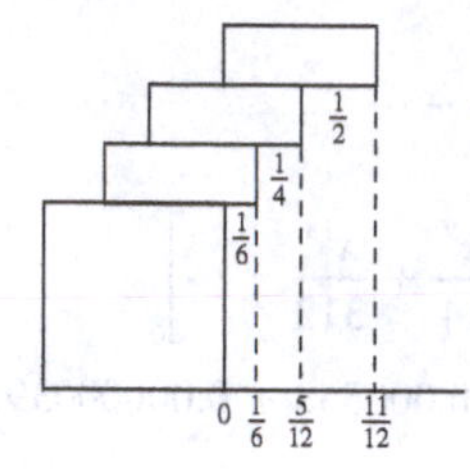

(b) Yes. If there are n blocks, then the edge of the top block lies $\sum_{i=1}^{n} \frac{1}{2i}$ from the edge of the table. Using 4 blocks,

$$\sum_{i=1}^{4} \frac{1}{2i} = \frac{1}{2} + \frac{1}{4} + \frac{1}{6} + \frac{1}{8} = \frac{25}{24}$$

which shows that the top block extends beyond the table.

(c) The blocks can extend any distance beyond the table because the series diverges:

$$\sum_{i=1}^{\infty} \frac{1}{2i} = \frac{1}{2}\sum_{i=1}^{\infty} \frac{1}{i} = \infty.$$

7. (a)
$$e^x = 1 + x + \frac{x^2}{2!} + \cdots = \sum_{n=0}^{\infty} \frac{x^n}{n!}$$

$$xe^x = \sum_{n=0}^{\infty} \frac{x^{n+1}}{n!}$$

$$\int xe^x\,dx = xe^x - e^x + C = \sum_{n=0}^{\infty} \frac{x^{n+2}}{(n+2)n!}$$

Letting $x = 0$, you have $C = 1$. Letting $x = 1$,

$$e - e + 1 = \sum_{n=0}^{\infty} \frac{1}{(n+2)n!} = \frac{1}{2} + \sum_{n=1}^{\infty} \frac{1}{(n+2)n!}.$$

So, $\sum_{n=1}^{\infty} \frac{1}{(n+2)n!} = \frac{1}{2}$.

(b) Differentiating, $xe^x + e^x = \sum_{n=0}^{\infty} \frac{(n+1)x^n}{n!}$.

Letting $x = 1$, $2e = \sum_{n=0}^{\infty} \frac{n+1}{n!} \approx 5.4366$.

9. $a - \frac{b}{2} + \frac{a}{3} - \frac{b}{4} + \cdots = \sum_{n=1}^{\infty} \frac{(-1)^{n+1}(a+b) + (a-b)}{2n}$

If $a = b$, $\sum_{n=1}^{\infty} \frac{(-1)^{n+1}(2a)}{2n} = a\sum_{n=1}^{\infty} \frac{(-1)^{n+1}}{n}$ converges conditionally.

If $a \neq b$, $\sum_{n=1}^{\infty} \frac{(-1)^{n+1}(a+b)}{2n} + \sum_{n=1}^{\infty} \frac{a-b}{2n}$ diverges.

No values of a and b give absolute convergence. $a = b$ implies conditional convergence.

11. Let $b_n = a_n r^n$.

$$(b_n)^{1/n} = (a_n r^n)^{1/n} = a_n^{1/n} \cdot r \to Lr \text{ as } n \to \infty.$$

$$Lr < \frac{1}{r}r = 1.$$

By the Root Test, $\sum b_n$ converges $\Rightarrow \sum a_n r^n$ converges.

13. (a) $\dfrac{1}{0.99} = \dfrac{1}{1 - 0.01} = \displaystyle\sum_{n=0}^{\infty} (0.01)^n$

$= 1 + 0.01 + (0.01)^2 + \cdots$

$= 1.010101\cdots$

(b) $\dfrac{1}{0.98} = \dfrac{1}{1 - 0.02} = \displaystyle\sum_{n=0}^{\infty} (0.02)^n$

$= 1 + 0.02 + (0.02)^2 + \cdots$

$= 1 + 0.02 + 0.0004 + \cdots$

$= 1.0204081632\cdots$

15. (a) Height $= 2\left[1 + \dfrac{1}{\sqrt{2}} + \dfrac{1}{\sqrt{3}} + \cdots\right]$

$= 2\displaystyle\sum_{n=1}^{\infty} \dfrac{1}{n^{1/2}} = \infty \left(p\text{-series, } p = \dfrac{1}{2} < 1\right)$

(b) $S = 4\pi\left[1 + \dfrac{1}{2} + \dfrac{1}{3} + \cdots\right] = 4\pi\displaystyle\sum_{n=1}^{\infty} \dfrac{1}{n} = \infty$

(c) $W = \dfrac{4}{3}\pi\left[1 + \dfrac{1}{2^{3/2}} + \dfrac{1}{3^{3/2}} + \cdots\right]$

$= \dfrac{4}{3}\pi\displaystyle\sum_{n=1}^{\infty} \dfrac{1}{n^{3/2}}$ converges.

CHAPTER 10
Conics, Parametric Equations, and Polar Coordinates

CHAPTER 10
Conics, Parametric Equations, and Polar Coordinates

Section 10.1 Conics and Calculus

1. A parabola is the set of all points that are equidistant from a fixed line (directrix) and a fixed point (focus), not on the line. An ellipse is the set of all points the sum of whose distances from two fixed points (foci) is constant. A hyperbola is the set of all points for which the absolute value of the difference between the distances from two fixed points (foci) is constant.

3. (a) $0 < e < 1$

(b) As e approaches 1, the ellipse becomes more elongated (flattened).

5. $y^2 = 4x$ Parabola

Vertex: $(0, 0)$

$p = 1 > 0$

Opens to the right

Matches (a).

6. $(x + 4)^2 = -2(y - 2)$ Parabola

Vertex: $(-4, 2)$

Opens downward

Matches (c).

7. $\dfrac{y^2}{16} - \dfrac{x^2}{1} = 1$ Hyperbola

Vertices: $(0, \pm 4)$

Matches (c).

8. $\dfrac{(x - 2)^2}{16} + \dfrac{(y + 1)^2}{4} = 1$ Ellipse

Center: $(2, -1)$

Matches (b).

9. $\dfrac{x^2}{4} + \dfrac{y^2}{9} = 1$ Ellipse

Center: $(0, 0)$

Vertices: $(0, \pm 3)$

Matches (f).

10. $\dfrac{(x - 2)^2}{9} - \dfrac{y^2}{4} = 1$ Hyperbola

Vertices: $(5, 0), (-1, 0)$

Matches (d).

11. $(x + 5) + (y - 3)^2 = 0$

$$(y - 3)^2 = -(x + 5) = 4\left(-\tfrac{1}{4}\right)(x + 5)$$

Vertex: $(-5, 3)$

Focus: $\left(-\tfrac{21}{4}, 3\right)$

Directrix: $x = -\tfrac{19}{4}$

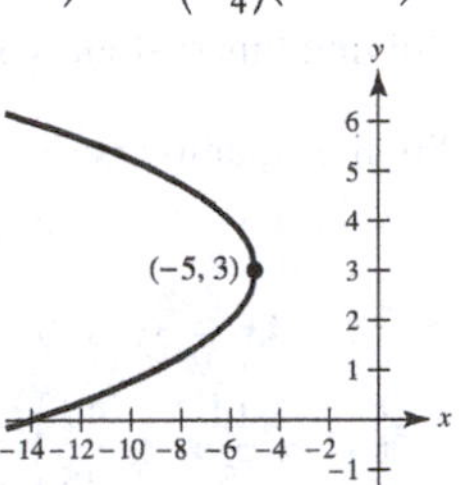

13. $y^2 - 4y - 4x = 0$

$$y^2 - 4y + 4 = 4x + 4$$
$$(y - 2)^2 = 4(1)(x + 1)$$

Vertex: $(-1, 2)$

Focus: $(0, 2)$

Directrix: $x = -2$

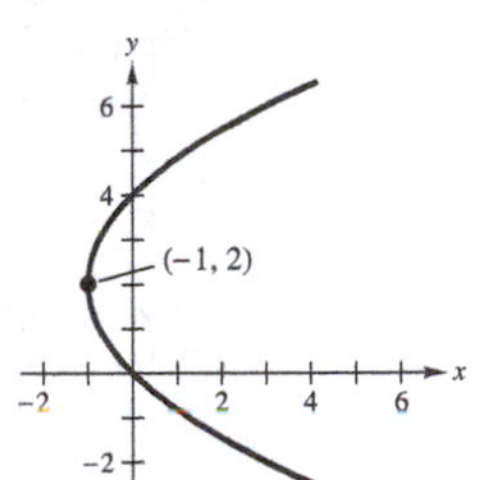

15. $x^2 + 4x + 4y - 4 = 0$

$$x^2 + 4x + 4 = -4y + 4 + 4$$
$$(x + 2)^2 = 4(-1)(y - 2)$$

Vertex: $(-2, 2)$

Focus: $(-2, 1)$

Directrix: $y = 3$

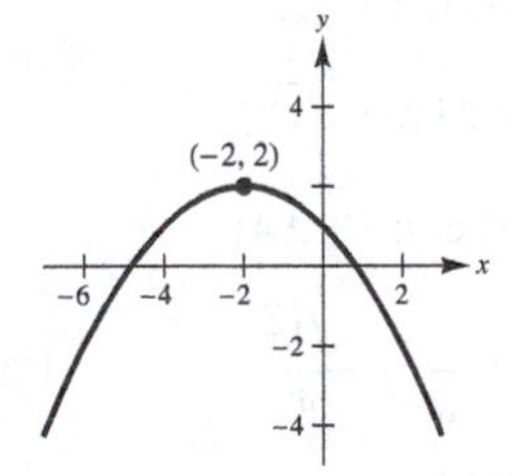

17. Horizontal axis, $p = -2$

$$(y - 4)^2 = 4(-2)(x - 5)$$

19. Vertical axis, $p = 8$

$$(x - 0)^2 = 4(8)(y - 5)$$
$$x^2 = 4(8)(y - 5)$$

21. Vertical axis

$(x - 1)^2 = 4p(y + 1)$

$(-1, -4)$ on parabola: $(-1 - 1)^2 = 4p(-4 + 1)$

$\Rightarrow \quad 1 = p(-3)$

$\Rightarrow \quad p = -\frac{1}{3}$

$(x - 1)^2 = 4\left(-\frac{1}{3}\right)(y + 1)$

23. Because the axis of the parabola is vertical, the form of the equation is $y = ax^2 + bx + c$. Now, substituting the values of the given coordinates into this equation, you obtain

$3 = c, 4 = 9a + 3b + c, 11 = 16a + 4b + c.$

Solving this system, you have $a = \frac{5}{3}, b = -\frac{14}{3}, c = 3.$

So, the equation is

$$y = \frac{5}{3}x^2 - \frac{14}{3}x + 3$$

$$5x^2 - 14x - 3y + 9 = 0$$

$$5\left(x^2 - \frac{14}{5}x + \frac{49}{25}\right) = 3y - 9 + \frac{49}{5}$$

$$5\left(x - \frac{7}{5}\right)^2 = 3y + \frac{4}{5}$$

$$\left(x - \frac{7}{5}\right)^2 = \frac{3}{5}\left(y + \frac{4}{15}\right)$$

$$\left(x - \frac{7}{5}\right)^2 = 4\left(\frac{3}{20}\right)\left(y + \frac{4}{15}\right)$$

25. $16x^2 + y^2 = 16$

$$x^2 + \frac{y^2}{16} = 1$$

$a^2 = 16, b^2 = 1, c^2 = 16 - 1 = 15$

Center: $(0, 0)$

Foci: $\left(0, \pm\sqrt{15}\right)$

Vertices: $(0, \pm 4)$

$e = \frac{c}{a} = \frac{\sqrt{15}}{4}$

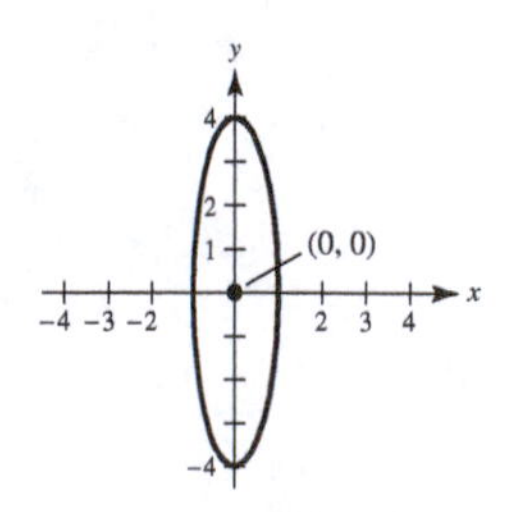

27. $\frac{(x - 3)^2}{16} + \frac{(y - 1)^2}{25} = 1$

$a^2 = 25, b^2 = 16, c^2 = 25 - 16 = 9$

Center: $(3, 1)$

Foci: $(3, 1 + 3) = (3, 4), (3, 1 - 3) = (3, -2)$

Vertices: $(3, 6), (3, -4)$

$e = \frac{c}{a} = \frac{3}{5}$

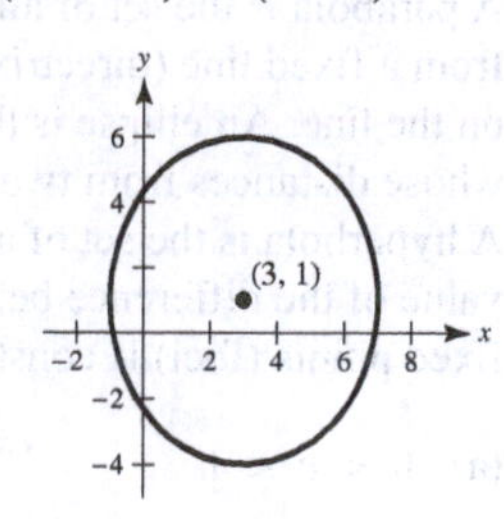

29. $9x^2 + 4y^2 + 36x - 24y + 36 = 0$

$$9\left(x^2 + 4x + 4\right) + 4\left(y^2 - 6y + 9\right) = -36 + 36 + 36 = 36$$

$$\frac{(x + 2)^2}{4} + \frac{(y - 3)^2}{9} = 1$$

$a^2 = 9, b^2 = 4, c^2 = 5$

Center: $(-2, 3)$

Foci: $\left(-2, 3 \pm \sqrt{5}\right)$

Vertices: $(-2, 6), (-2, 0)$

$e = \frac{c}{a} = \frac{\sqrt{5}}{3}$

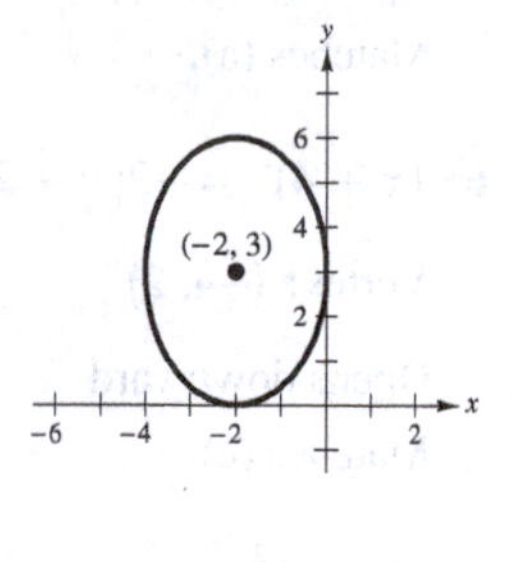

31. Center: $(0, 0)$

Focus: $(5, 0)$

Vertex: $(6, 0)$

Horizontal major axis

$a = 6, c = 5 \Rightarrow b = \sqrt{a^2 - c^2} = \sqrt{11}$

$$\frac{x^2}{36} + \frac{y^2}{11} = 1$$

33. Vertices: $(3, 1), (3, 9)$

Minor axis length: 6

Vertical major axis

Center: $(3, 5)$

$a = 4, b = 3$

$$\frac{(x - 3)^2}{9} + \frac{(y - 5)^2}{16} = 1$$

35. Center: $(0, 0)$

Horizontal major axis

Points on ellipse: $(3, 1), (4, 0)$

Because the major axis is horizontal,

$$\left(\frac{x^2}{a^2}\right) + \left(\frac{y^2}{b^2}\right) = 1.$$

Substituting the values of the coordinates of the given points into this equation, you have

$$\left(\frac{9}{a^2}\right) + \left(\frac{1}{b^2}\right) = 1, \text{ and } \frac{16}{a^2} = 1.$$

The solution to this system is $a^2 = 16, b^2 = \frac{16}{7}$.

So,

$$\frac{x^2}{16} + \frac{y^2}{16/7} = 1, \frac{x^2}{16} + \frac{7y^2}{16} = 1.$$

37. $\frac{x^2}{25} - \frac{y^2}{16} = 1$

$a = 5, b = 4, c = \sqrt{25 + 16} = \sqrt{41}$

Center: $(0, 0)$

Vertices: $(\pm 5, 0)$

Foci: $\left(\pm\sqrt{41}, 0\right)$

$e = \frac{c}{a} = \frac{\sqrt{41}}{5}$

Asymptotes: $y = \pm\frac{b}{a}x = \pm\frac{4}{5}x$

39.

$$9x^2 - y^2 - 36x - 6y + 18 = 0$$

$$9(x^2 - 4x + 4) - (y^2 + 6y + 9) = -18 + 36 - 9$$

$$\frac{(x-2)^2}{1} - \frac{(y+3)^2}{9} = 1$$

$a = 1, b = 3, c = \sqrt{10}$

Center: $(2, -3)$

Vertices: $(1, -3), (3, -3)$

Foci: $\left(2 \pm \sqrt{10}, -3\right)$

$e = \frac{c}{a} = \frac{\sqrt{10}}{1} = \sqrt{10}$

Asymptotes: $y = -3 \pm 3(x - 2)$

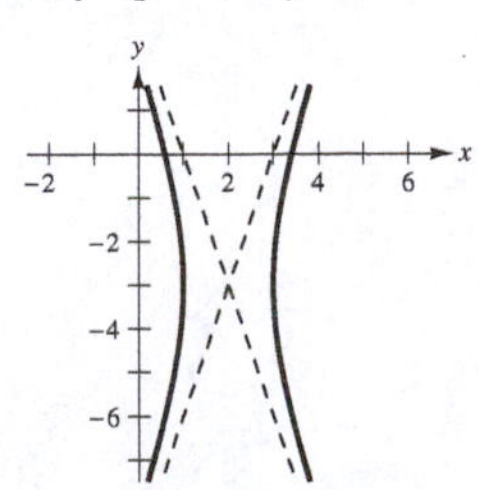

41. Vertices: $(\pm 1, 0)$

Asymptotes: $y = \pm 5x$

Horizontal transverse axis

Center: $(0, 0)$

$a = 1, \frac{b}{a} = 5 \Rightarrow b = 5$

$$\frac{x^2}{1} - \frac{y^2}{25} = 1$$

43. Vertices: $(2, \pm 3)$

Point on graph: $(0, 5)$

Vertical transverse axis

Center: $(2, 0)$

$a = 3$

So, the equation is of the form

$$\frac{y^2}{9} - \frac{(x-2)^2}{b^2} = 1.$$

Substituting the coordinates of the point $(0, 5)$, you have

$$\frac{25}{9} - \frac{4}{b^2} = 1 \quad \text{or} \quad b^2 = \frac{9}{4}.$$

So, the equation is $\frac{y^2}{9} - \frac{(x-2)^2}{9/4} = 1.$

45. Center: $(0, 0)$

Vertex: $(0, 2)$

Focus: $(0, 4)$

Vertical transverse axis

$a = 2, c = 4, b^2 = c^2 - a^2 = 12$

$$\frac{y^2}{4} - \frac{x^2}{12} = 1$$

47. Vertices: $(0, 2), (6, 2)$

Asymptotes: $y = \frac{2}{3}x, y = 4 - \frac{2}{3}x$

Horizontal transverse axis

Center: $(3, 2)$

$a = 3$

Slopes of asymptotes: $\pm\frac{b}{a} = \pm\frac{2}{3}$

So, $b = 2$. Therefore,

$$\frac{(x-3)^2}{9} - \frac{(y-2)^2}{4} = 1.$$

49. (a) $\frac{x^2}{9} - y^2 = 1, \frac{2x}{9} - 2yy' = 0, \frac{x}{9y} = y'$

At $x = 6$: $y = \pm\sqrt{3}, y' = \frac{\pm 6}{9\sqrt{3}} = \frac{\pm 2\sqrt{3}}{9}$

At $(6, \sqrt{3})$: $y - \sqrt{3} = \frac{2\sqrt{3}}{9}(x - 6)$

or $2x - 3\sqrt{3}y - 3 = 0$

At $(6, -\sqrt{3})$: $y + \sqrt{3} = \frac{-2\sqrt{3}}{9}(x - 6)$

or $2x + 3\sqrt{3}y - 3 = 0$

(b) From part (a) you know that the slopes of the normal lines must be $\mp 9/(2\sqrt{3})$.

At $(6, \sqrt{3})$: $y - \sqrt{3} = -\frac{9}{2\sqrt{3}}(x - 6)$

or $9x + 2\sqrt{3}y - 60 = 0$

At $(6, -\sqrt{3})$: $y + \sqrt{3} = \frac{9}{2\sqrt{3}}(x - 6)$

or $9x - 2\sqrt{3}y - 60 = 0$

51. $25x^2 - 10x - 200y - 119 = 0$

$$25\left(x^2 - \tfrac{2}{5}x + \tfrac{1}{25}\right) = 200y + 119 + 1$$

$$25\left(x - \tfrac{1}{5}\right)^2 = 200(y + 1)$$

Parabola

53.
$$3(x - 1)^2 = 6 + 2(y + 1)^2$$

$$3(x - 1)^2 - 2(y + 1)^2 = 6$$

$$\frac{(x - 1)^2}{2} - \frac{(y + 1)^2}{3} = 1$$

Hyperbola

55.
$$9x^2 + 9y^2 - 36x + 6y + 34 = 0$$

$$9\left(x^2 - 4x + 4\right) + 9\left(y^2 + \tfrac{2}{3}y + \tfrac{1}{9}\right) = -34 + 36 + 1$$

$$9(x - 2)^2 + 9\left(y + \tfrac{1}{3}\right)^2 = 3$$

Circle (Ellipse)

57. $9x^2 + 4y^2 - 36x - 24y - 36 = 0$

(a) $9\left(x^2 - 4x + 4\right) + 4\left(y^2 - 6y + 9\right) = 36 + 36 + 36$

$9(x - 2)^2 + 4(y - 3)^2 = 108$

$\frac{(x - 2)^2}{12} + \frac{(y - 3)^2}{27} = 1$

Ellipse

(b) $9x^2 - 4y^2 - 36x - 24y - 36 = 0$

$9\left(x^2 - 4x + 4\right) - 4\left(y^2 + 6y + 9\right) = 36 + 36 - 36$

$\frac{(x - 2)^2}{4} - \frac{(y + 3)^2}{9} = 1$

Hyperbola

(c) $4x^2 + 4y^2 - 36x - 24y - 36 = 0$

$4\left(x^2 - 9x + \frac{81}{4}\right) + 4\left(y^2 - 6y + 9\right) = 36 + 81 + 36$

$\left(x - \frac{9}{2}\right)^2 + (y - 3)^2 = \frac{153}{4}$

Circle

(d) *Sample answer:* Eliminate the y^2-term

59. From Example 6, the circumference of an ellipse is

$$C = 4\int_0^{\pi/2} \sqrt{a^2 - (a^2 - b^2)\sin^2\theta}\, d\theta$$

Because $0 \le \sin^2\theta \le 1$,

$$4\int_0^{2\pi} \sqrt{a^2 - (a^2 - b^2)(1)}\, d\theta \le C \le 4\int_0^{\pi/2} \sqrt{a^2 - (a^2 - b^2)(0)}\, d\theta$$

$$4\int_0^{2\pi} b\, d\theta \le C \le 4\int_0^{\pi/2} a\, d\theta$$

$$2\pi b \le C \le 2\pi a.$$

But, $b < a$, so $2\pi b < C < 2\pi a$.

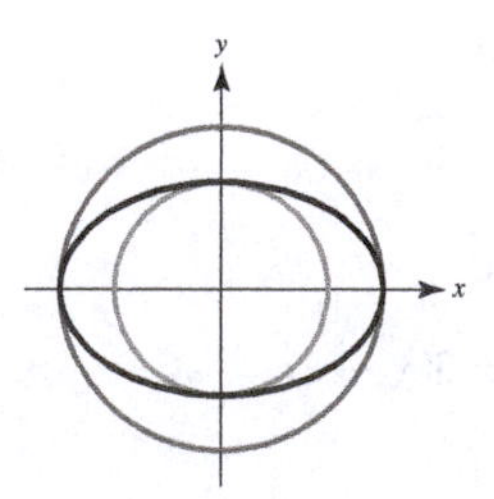

61. Assume that the vertex is at the origin.

$$x^2 = 4py$$
$$(3)^2 = 4p(1)$$
$$\frac{9}{4} = p$$

The pipe is located $\frac{9}{4}$ meters from the vertex.

63. (a) Without loss of generality, place the coordinate system so that the equation of the parabola is $x^2 = 4py$ and, so,

$$y' = \left(\frac{1}{2p}\right)x.$$

So, for distinct tangent lines, the slopes are unequal and the lines intersect.

(b) $x^2 - 4x - 4y = 0$

$$2x - 4 - 4\frac{dy}{dx} = 0$$
$$\frac{dy}{dx} = \frac{1}{2}x - 1$$

At $(0, 0)$, the slope is -1: $y = -x$. At $(6, 3)$, the slope is 2: $y = 2x - 9$. Solving for x,

$$-x = 2x - 9$$
$$-3x = -9$$
$$x = 3$$
$$y = -3.$$

Point of intersection: $(3, -3)$

65. Assume that $y = ax^2$.

$$20 = a(60)^2 \Rightarrow a = \frac{2}{360} = \frac{1}{180} \Rightarrow y = \frac{1}{180}x^2$$

67. Parabola

Vertex: $(0, 4)$

$$x^2 = 4p(y - 4)$$
$$4^2 = 4p(0 - 4)$$
$$p = -1$$
$$x^2 = -4(y - 4)$$
$$y = 4 - \frac{x^2}{4}$$

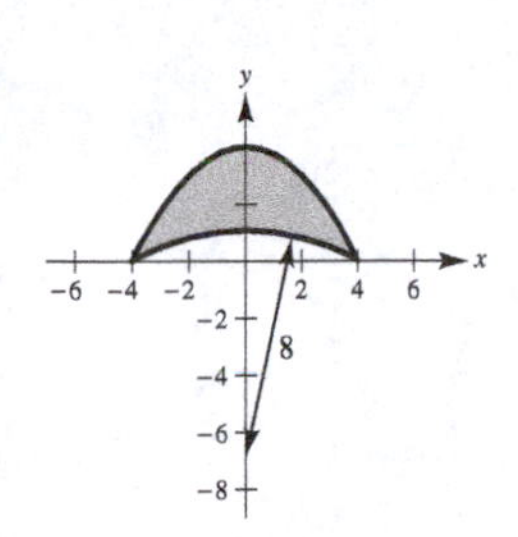

Circle

Center: $(0, k)$

Radius: 8

$$x^2 + (y - k)^2 = 64$$
$$4^2 + (0 - k)^2 = 64$$
$$k^2 = 48$$
$$k = -4\sqrt{3} \quad \text{(Center is on the negative } y\text{-axis.)}$$
$$x^2 + \left(y + 4\sqrt{3}\right)^2 = 64$$
$$y = -4\sqrt{3} \pm \sqrt{64 - x^2}$$

Because the y-value is positive when $x = 0$, we have $y = -4\sqrt{3} + \sqrt{64 - x^2}$.

$$A = 2\int_0^4 \left[\left(4 - \frac{x^2}{4}\right) - \left(-4\sqrt{3} + \sqrt{64 - x^2}\right)\right] dx$$
$$= 2\left[4x - \frac{x^3}{12} + 4\sqrt{3}x - \frac{1}{2}\left(x\sqrt{64 - x^2} + 64 \arcsin\frac{x}{8}\right)\right]_0^4$$
$$= 2\left(16 - \frac{64}{12} + 16\sqrt{3} - 2\sqrt{48} - 32 \arcsin\frac{1}{2}\right) = \frac{16\left(4 + 3\sqrt{3} - 2\pi\right)}{3} \approx 15.536 \text{ square feet}$$

69. $e = \dfrac{c}{a}$

$0.0167 = \dfrac{c}{149{,}598{,}000}$

$c \approx 2{,}498{,}286.6$

Least distance: $a - c = 147{,}099{,}713.4$ km

Greatest distance: $a + c = 152{,}096{,}286.6$ km

71. $e = \dfrac{A - P}{A + P}$

$= \dfrac{(1563 + 4000) - (220 + 4000)}{(1563 + 4000) + (220 + 4000)}$

$= \dfrac{1343}{9783} \approx 0.1373$

73. $e = \dfrac{A - P}{A + P} = \dfrac{35.29 - 0.59}{35.29 + 0.59} = 0.9671$

75. (a) $A = 4\int_0^2 \frac{1}{2}\sqrt{4 - x^2}\,dx = \left[x\sqrt{4 - x^2} + 4\arcsin\left(\frac{x}{2}\right)\right]_0^2 = 2\pi \quad [\text{or, } A = \pi ab = \pi(2)(1) = 2\pi]$

(b) **Disk:** $V = 2\pi\int_0^2 \frac{1}{4}(4 - x^2)\,dx = \frac{1}{2}\pi\left[4x - \frac{1}{3}x^3\right]_0^2 = \frac{8\pi}{3}$

$y = \frac{1}{2}\sqrt{4 - x^2}$

$y' = \dfrac{-x}{2\sqrt{4 - x^2}}$

$\sqrt{1 + (y')^2} = \sqrt{1 + \dfrac{x^2}{16 - 4x^2}} = \sqrt{\dfrac{16 - 3x^2}{4y}}$

$S = 2(2\pi)\int_0^2 y\left(\dfrac{\sqrt{16 - 3x^2}}{4y}\right)dx = \pi\int_0^2 \sqrt{16 - 3x^2}\,dx$

$= \dfrac{\pi}{2\sqrt{3}}\left[\sqrt{3}x\sqrt{16 - 3x^2} + 16\arcsin\left(\dfrac{\sqrt{3}x}{4}\right)\right]_0^2$

$= \dfrac{2\pi}{9}\left(9 + 4\sqrt{3}\pi\right) \approx 21.48$

(c) **Shell:** $V = 2\pi\int_0^2 x\sqrt{4 - x^2}\,dx = -\pi\int_0^2 -2x(4 - x^2)^{1/2}\,dx = -\frac{2\pi}{3}\left[(4 - x^2)^{3/2}\right]_0^2 = \frac{16\pi}{3}$

$x = 2\sqrt{1 - y^2}$

$x' = \dfrac{-2y}{\sqrt{1 - y^2}}$

$\sqrt{1 + (x')^2} = \sqrt{1 + \dfrac{4y^2}{1 - y^2}} = \dfrac{\sqrt{1 + 3y^2}}{\sqrt{1 - y^2}}$

$S = 2(2\pi)\int_0^1 2\sqrt{1 - y^2}\dfrac{\sqrt{1 + 3y^2}}{\sqrt{1 - y^2}}\,dy = 8\pi\int_0^1 \sqrt{1 + 3y^2}\,dy$

$= \dfrac{8\pi}{2\sqrt{3}}\left[\sqrt{3}y\sqrt{1 + 3y^2} + \ln\left|\sqrt{3}y + \sqrt{1 + 3y^2}\right|\right]_0^1$

$= \dfrac{4\pi}{3}\left|6 + \sqrt{3}\ln\left(2 + \sqrt{3}\right)\right| \approx 34.69$

77. From Example 5,

$$C = 4a\int_0^{\pi/2} \sqrt{1 - e^2 \sin^2 \theta}\, d\theta$$

For $\dfrac{x^2}{25} + \dfrac{y^2}{49} = 1$, you have $a = 7, b = 5, c = \sqrt{49 - 25} = 2\sqrt{6}, e = \dfrac{c}{a} = \dfrac{2\sqrt{6}}{7}$.

$$C = 4(7)\int_0^{\pi/2} \sqrt{1 - \frac{24}{49}\sin^2 \theta}\, d\theta \approx 28(1.3558) \approx 37.96$$

79. Area circle $= \pi r^2 = 100\pi$

Area ellipse $= \pi ab = \pi a(10)$

$2(100\pi) = 10\pi a \Rightarrow a = 20$

So, the length of the major axis is $2a = 40$.

81. The transverse axis is horizontal since $(2, 2)$ and $(10, 2)$ are the foci (see definition of hyperbola).

Center: $(6, 2)$

$c = 4, 2a = 6, b^2 = c^2 - a^2 = 7$

So, the equation is $\dfrac{(x - 6)^2}{9} - \dfrac{(y - 2)^2}{7} = 1$.

83. $c = 150, 2a = 0.001(186{,}000), a = 93,$

$b = \sqrt{150^2 - 93^2} = \sqrt{13{,}851}$

$$\frac{x^2}{93^2} - \frac{y^2}{13{,}851} = 1$$

When $y = 75$, you have $x^2 = 93^2\left(1 + \dfrac{75^2}{13{,}851}\right)$

$x \approx 110.3$ mi.

85.

$$\frac{x^2}{a^2} - \frac{y^2}{b^2} = 1$$

$$\frac{2x}{a^2} - \frac{2yy'}{b^2} = 0 \text{ or } y' = \frac{b^2x}{a^2y}$$

$$y - y_0 = \frac{b^2x_0}{a^2y_0}(x - x_0)$$

$$a^2y_0y - a^2{y_0}^2 = b^2x_0x - b^2{x_0}^2$$

$$b^2{x_0}^2 - a^2{y_0}^2 = b^2x_0x - a^2y_0y$$

$$a^2b^2 = b^2x_0x - a^2y_0y$$

$$\frac{x_0x}{a^2} - \frac{y_0y}{b^2} = 1$$

87. False. The parabola is equidistant from the directrix and focus and therefore cannot intersect the directrix.

89. True

91. True

93. Let $\frac{x^2}{a^2} + \frac{y^2}{b^2} = 1$ be the equation of the ellipse with $a > b > 0$. Let $(\pm c, 0)$ be the foci, $c^2 = a^2 - b^2$. Let (u, v) be a point on the tangent line at $P(x, y)$, as indicated in the figure.

$$x^2b^2 + y^2a^2 = a^2b^2$$
$$2xb^2 + 2yy'a^2 = 0$$
$$y' = -\frac{b^2x}{a^2y} \quad \text{Slope at } P(x, y)$$

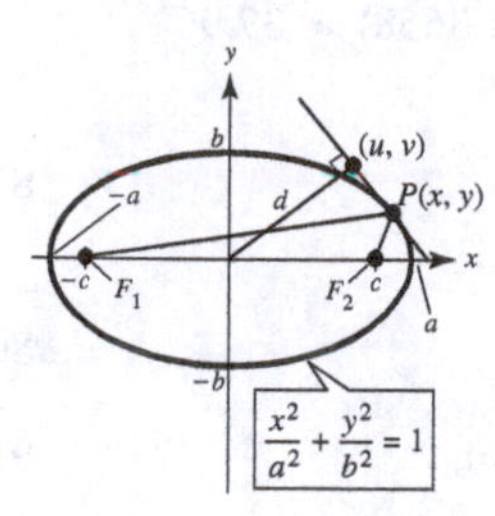

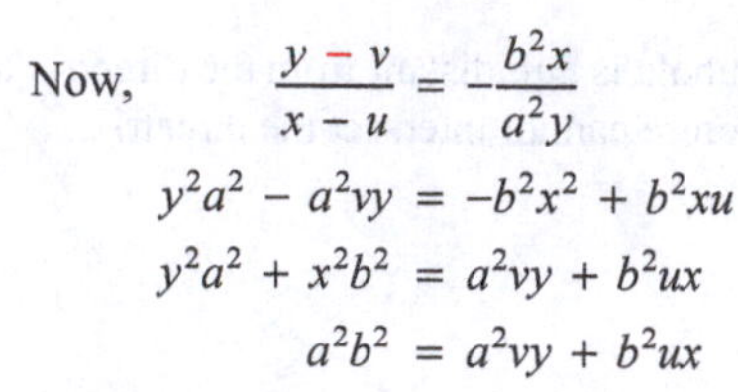

Now, $\frac{y - v}{x - u} = -\frac{b^2x}{a^2y}$

$$y^2a^2 - a^2vy = -b^2x^2 + b^2xu$$
$$y^2a^2 + x^2b^2 = a^2vy + b^2ux$$
$$a^2b^2 = a^2vy + b^2ux$$

Because there is a right angle at (u, v),

$$\frac{v}{u} = \frac{a^2y}{b^2x}$$
$$vb^2x = a^2uy.$$

You have two equations: $a^2vy + b^2ux = a^2b^2$

$$a^2uy - b^2vx = 0.$$

Multiplying the first by v and the second by u, and adding,

$$a^2v^2y + a^2u^2y = a^2b^2v$$
$$y\left[u^2 + v^2\right] = b^2v$$
$$yd^2 = b^2v$$
$$v = \frac{yd^2}{b^2}.$$

Similarly, $u = \frac{xd^2}{a^2}$.

From the figure, $u = d\cos\theta$ and $v = d\sin\theta$. So, $\cos\theta = \frac{xd}{a^2}$ and $\sin\theta = \frac{yd}{b^2}$.

$$\cos^2\theta + \sin^2\theta = \frac{x^2d^2}{a^4} + \frac{y^2d^2}{b^4} = 1$$
$$x^2b^4d^2 + y^2a^4d^2 = a^4b^4$$
$$d^2 = \frac{a^4b^4}{x^2b^4 + y^2a^4}$$

Let $r_1 = PF_1$ and $r_2 = PF_2$, $r_1 + r_2 = 2a$.

$$r_1r_2 = \frac{1}{2}\left[(r_1 + r_2)^2 - r_1^2 - r_2^2\right] = \frac{1}{2}\left[4a^2 - (x + c)^2 - y^2 - (x - c)^2 - y^2\right] = 2a^2 - x^2 - y^2 - c^2 = a^2 + b^2 - x^2 - y^2$$

Finally, $d^2r_1r_2 = \frac{a^4b^4}{x^2b^4 + y^2a^4} \cdot \left[a^2 + b^2 - x^2 - y^2\right]$

$$= \frac{a^4b^4}{b^2(b^2x^2) + a^2(a^2y^2)} \cdot \left[a^2 + b^2 - x^2 - y^2\right]$$
$$= \frac{a^4b^4}{b^2(a^2b^2 - a^2y^2) + a^2(a^2b^2 - b^2x^2)} \cdot \left[a^2 + b^2 - x^2 - y^2\right]$$
$$= \frac{a^4b^4}{a^2b^2\left[a^2 + b^2 - x^2 - y^2\right]} \cdot \left[a^2 + b^2 - x^2 - y^2\right] = a^2b^2, \text{ a constant!}$$

Section 10.2 Plane Curves and Parametric Equations

1. Parametric equations provide position, direction, and speed for a particle at a given time.

3. Different parametric representations can be used to represent various speeds at which objects travel along a given path.

5. $x = 2t - 3$

$y = 3t + 1$

$t = \dfrac{x + 3}{2}$

$y = 3\left(\dfrac{x + 3}{2}\right) + 1 = \dfrac{3}{2}x + \dfrac{11}{2}$

$3x - 2y + 11 = 0$

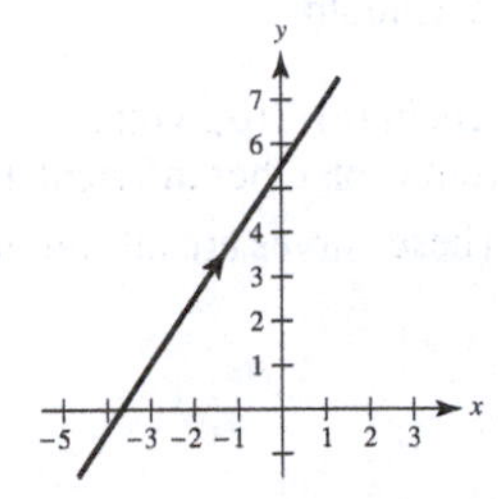

7. $x = t + 1$

$y = t^2$

$y = (x - 1)^2$

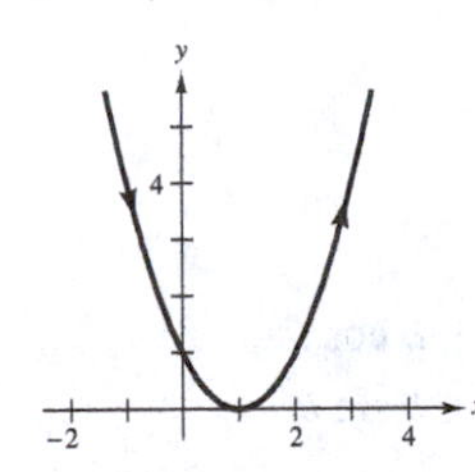

9. $x = t^3$

$y = \frac{1}{2}t^2$

$y = t^3$ implies $t = x^{1/3}$

$y = \frac{1}{2}x^{2/3}$

11. $x = \sqrt{t}$

$y = t - 5$

$x^2 = t$

$y = x^2 - 5, x \geq 0$

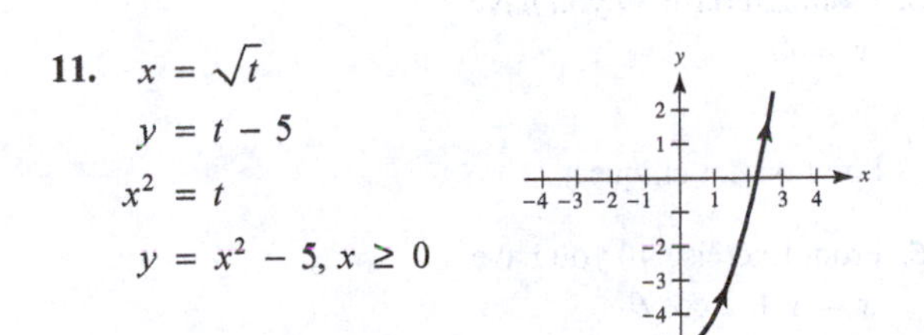

13. $x = t - 3$

$y = \dfrac{t}{t - 3}$

$t = x + 3$

$y = \dfrac{x + 3}{(x + 3) - 3} = 1 + \dfrac{3}{x} = \dfrac{x + 3}{x}$

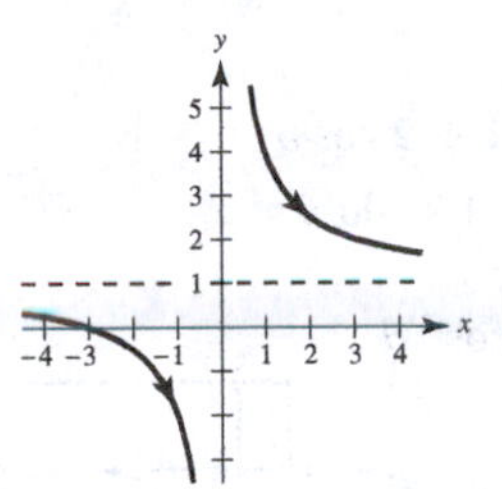

15. $x = 2t$

$y = |t - 2|$

$y = \left|\dfrac{x}{2} - 2\right| = \dfrac{|x - 4|}{2}$

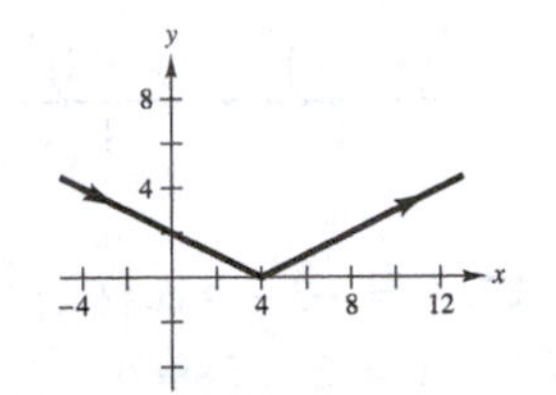

17. $x = e^t, x > 0$

$y = e^{3t} + 1$

$y = x^3 + 1, x > 0$

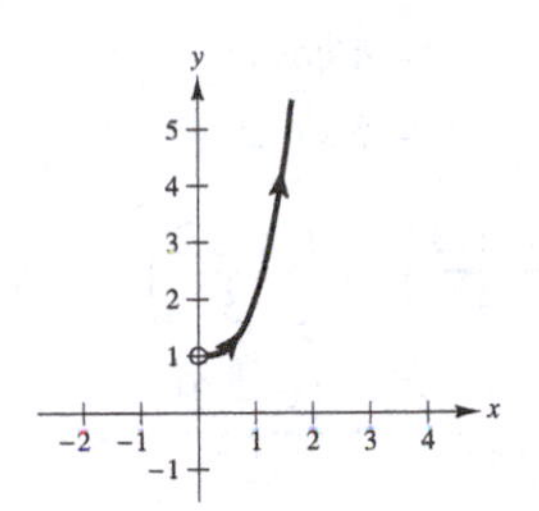

19. $x = \sec\theta$

$y = \cos\theta$

$0 \leq \theta < \dfrac{\pi}{2}, \dfrac{\pi}{2} < \theta \leq \pi$

$xy = 1$

$y = \dfrac{1}{x}$

$|x| \geq 1, |y| \leq 1$

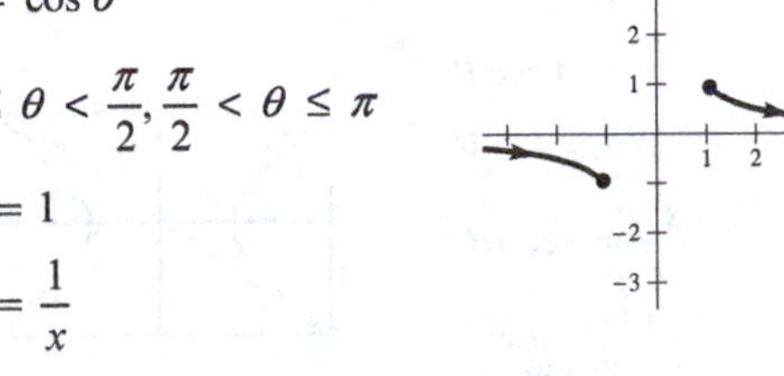

21. $x = 8\cos\theta$

$y = 8\sin\theta$

$x^2 + y^2 = 64\cos^2\theta + 64\sin^2\theta = 64(1) = 64$

Circle

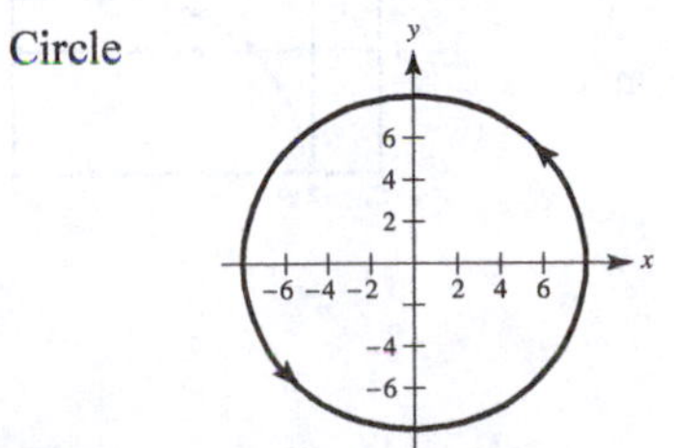

23. $x = 6 \sin 2\theta$

$y = 4 \cos 2\theta$

$$\left(\frac{x}{6}\right)^2 + \left(\frac{y}{4}\right)^2 = \sin^2 2\theta + \cos^2 2\theta = 1$$

$$\frac{x^2}{36} + \frac{y^2}{16} = 1$$

Ellipse

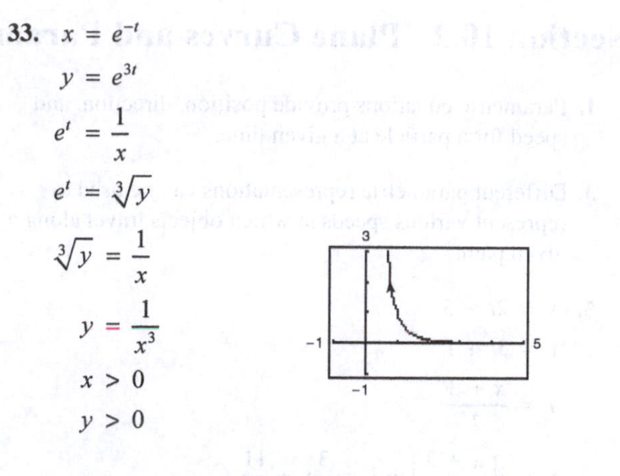

25.

$$x = 4 + 2\cos\theta$$
$$y = -1 + \sin\theta$$
$$\frac{(x-4)^2}{4} = \cos^2\theta$$
$$\frac{(y+1)^2}{1} = \sin^2\theta$$
$$\frac{(x-4)^2}{4} + \frac{(y+1)^2}{1} = 1$$

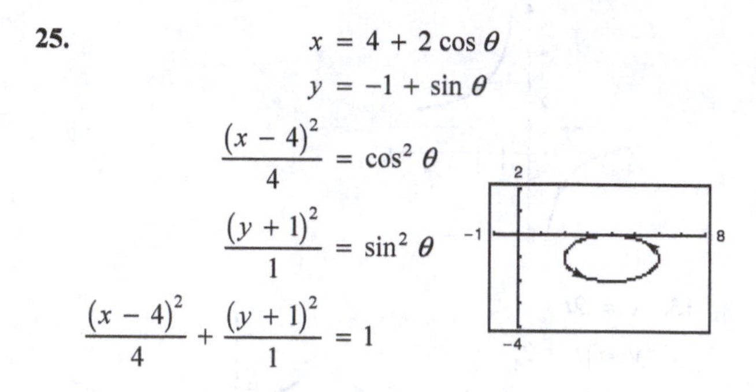

27. $x = -3 + 4\cos\theta$

$y = 2 + 5\sin\theta$

$x + 3 = 4\cos\theta$

$y - 2 = 5\sin\theta$

$$\left(\frac{x+3}{4}\right)^2 + \left(\frac{y-2}{5}\right)^2 = \cos^2\theta + \sin^2\theta = 1$$

$$\frac{(x+3)^2}{16} + \frac{(y-2)^2}{25} = 1$$

Ellipse

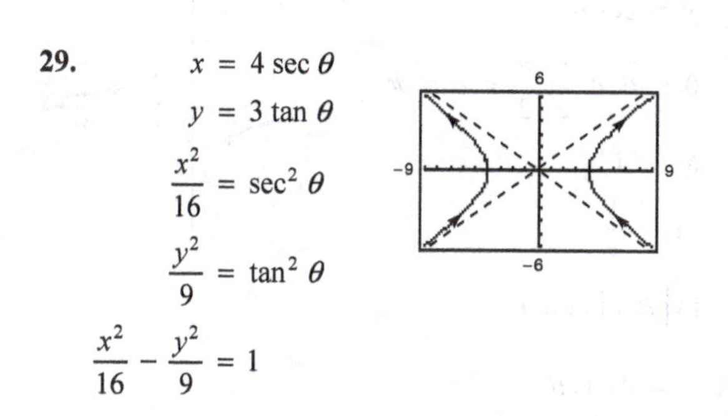

29.

$$x = 4\sec\theta$$
$$y = 3\tan\theta$$
$$\frac{x^2}{16} = \sec^2\theta$$
$$\frac{y^2}{9} = \tan^2\theta$$
$$\frac{x^2}{16} - \frac{y^2}{9} = 1$$

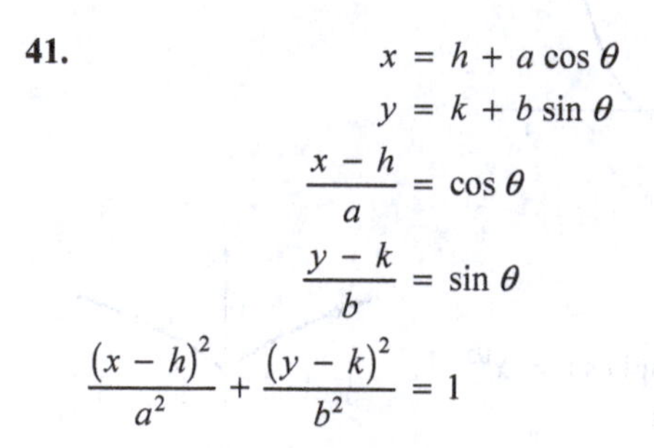

31. $x = t^3$

$y = 3 \ln t$

$y = 3 \ln \sqrt[3]{x} = \ln x$

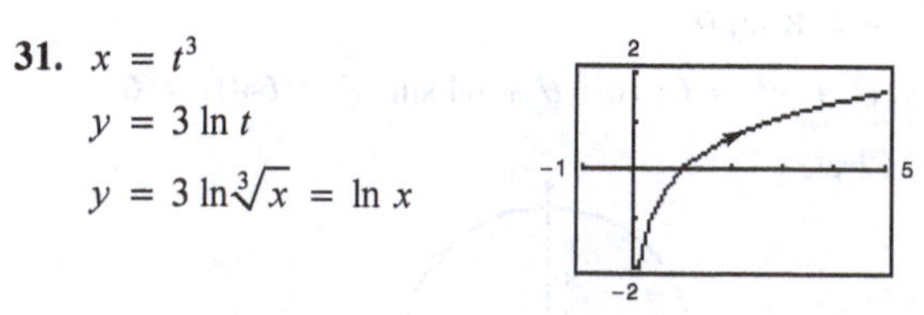

33. $x = e^{-t}$

$y = e^{3t}$

$$e^t = \frac{1}{x}$$
$$e^t = \sqrt[3]{y}$$
$$\sqrt[3]{y} = \frac{1}{x}$$
$$y = \frac{1}{x^3}$$

$x > 0$

$y > 0$

35. Both sets of equations represent the parabola $y = x^2$.

(a) Left-to-right orientation; smooth

(b) Right-to-left orientation; smooth

37. By eliminating the parameters in (a) – (d), you get $y = 2x + 1$. They differ from each other in orientation and in restricted domains. These curves are all smooth except for (b).

39.

$$x = x_1 + t(x_2 - x_1)$$
$$y = y_1 + t(y_2 - y_1)$$
$$\frac{x - x_1}{x_2 - x_1} = t$$
$$y = y_1 + \left(\frac{x - x_1}{x_2 - x_1}\right)(y_2 - y_1)$$
$$y - y_1 = \frac{y_2 - y_1}{x_2 - x_1}(x - x_1)$$
$$y - y_1 = m(x - x_1)$$

41.

$$x = h + a\cos\theta$$
$$y = k + b\sin\theta$$
$$\frac{x-h}{a} = \cos\theta$$
$$\frac{y-k}{b} = \sin\theta$$
$$\frac{(x-h)^2}{a^2} + \frac{(y-k)^2}{b^2} = 1$$

43. From Exercise 39 you have

$x = 4t$

$y = -7t$

Solution not unique

45. From Exercise 40 you have

$x = 1 + 2\cos\theta$

$y = 1 + 2\sin\theta$

Solution not unique

47. From Exercise 41 you have

$a = 5, c = 3, b = \sqrt{a^2 - c^2} = 4$, center: $(2, 0)$

$x = 2 + 5\cos\theta$

$y = 4\sin\theta$

Solution not unique

49. From Exercise 42 you have

$a = 1, c = \sqrt{5}, b = \sqrt{c^2 - a^2} = 2$, center: $(0, 0)$

$x = 2\tan\theta$

$y = \sec\theta$

Note: Transverse axis is vertical, so reverse the roles of x and y.

Solution not unique

51. $y = 6x - 5$

Examples:

$x = t, y = 6t - 5$

$x = t + 1, y = 6t + 1$

53. $y = x^3$

Example

$x = t, \quad y = t^3$

$x = \sqrt[3]{t}, \quad y = t$

$x = \tan t, \quad y = \tan^3 t$

55. $y = 2x - 5$

At $(3, 1)$, $t = 0$: $\quad x = 3 - t$

$y = 2(3 - t) - 5 = -2t + 1$

or, $x = t + 3$

$y = 2t + 1$

57. $y = x^2$

$t = 4$ at $(4, 16)$: $\quad x = t$

$y = t^2$

59. $x = 2(\theta - \sin\theta)$

$y = 2(1 - \cos\theta)$

Not smooth at $\theta = 2n\pi$

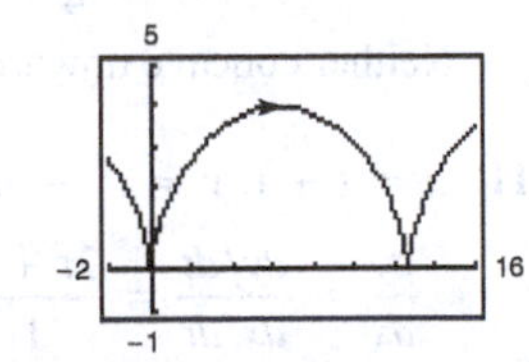

61. $x = \theta - \frac{3}{2}\sin\theta$

$y = 1 - \frac{3}{2}\cos\theta$

Smooth everywhere

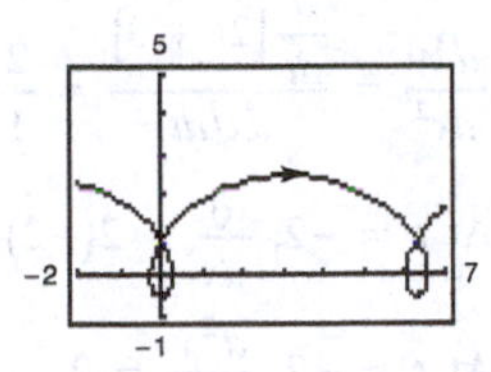

63. $x = 3\cos^3\theta$

$y = 3\sin^3\theta$

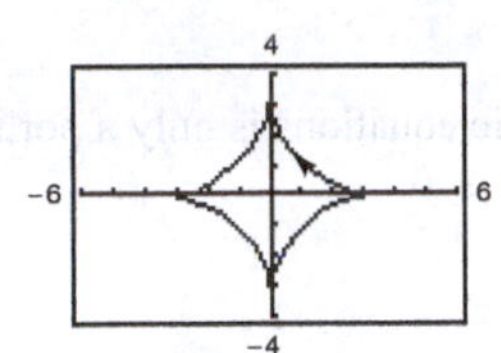

Not smooth at $(x, y) = (\pm 3, 0)$ and $(0, \pm 3)$, or $\theta = \frac{1}{2}n\pi$.

65. $x = 2\cot\theta$

$y = 2\sin^2\theta$

Smooth everywhere

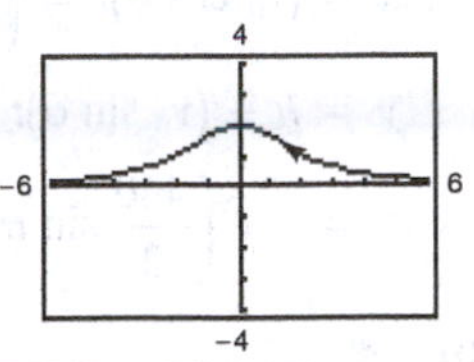

67. For $-1 \le t \le 0$, the orientation is right to left along $y = x^2, 0 \le x \le 1$.

For $0 \le t \le 1$, the orientation is left to right along $y = x^2, 0 \le x \le 1$. There is no definite direction.

69. No. In the interval $0 < \theta < \pi$, $\cos\theta = \cos(-\theta)$ and $\sin^2\theta = \sin^2(-\theta)$. So, the parameter was not changed.

71. Matches (d) because $(4, 0)$ is on the graph.

72. Matches (a) because $(0, 2)$ is on the graph.

73. Matches (b) because $(1, 0)$ is on the graph.

74. Matches (c) because the graph is undefined when $\theta = 0$.

75. When the circle has rolled θ radians, you know that the center is at $(a\theta, a)$.

$\sin\theta = \sin(180° - \theta) = \frac{|AC|}{b} = \frac{|BD|}{b}$ or $|BD| = b\sin\theta$

$\cos\theta = -\cos(180° - \theta) = \frac{|AP|}{-b}$ or $|AP| = -b\cos\theta$

So, $x = a\theta - b\sin\theta$ and $y = a - b\cos\theta$.

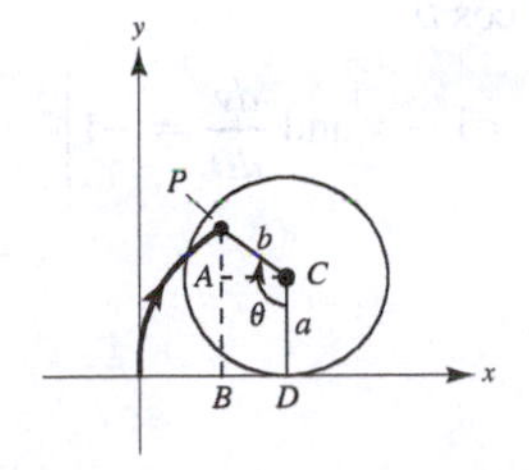

77. False

$x = t^2 \Rightarrow x \geq 0$

$y = t^2 \Rightarrow y \geq 0$

The graph of the parametric equations is only a portion of the line $y = x$ when $x \geq 0$.

79. True. $y = \cos x$

81. (a) $100 \text{ mi/hr} = \frac{(100)(5280)}{3600} = \frac{440}{3} \text{ ft/sec}$

$$x = (v_0 \cos\theta)t = \left(\frac{440}{3}\cos\theta\right)t$$

$$y = h + (v_0 \sin\theta)t - 16t^2$$

$$= 3 + \left(\frac{440}{3}\sin\theta\right)t - 16t^2$$

(b)

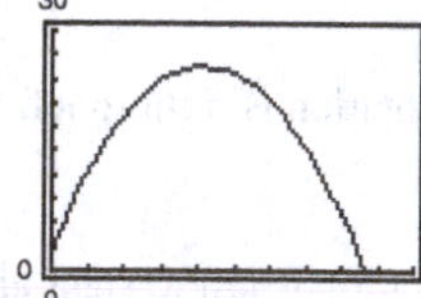

It is not a home run when $x = 400$, $y < 10$.

(c)

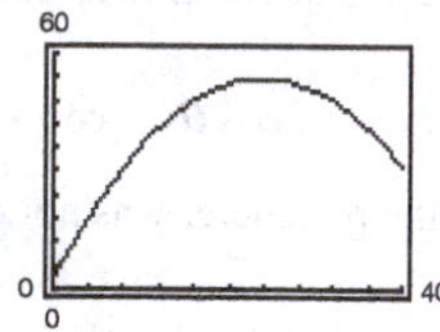

Yes, it's a home run when $x = 400$, $y > 10$.

(d) You need to find the angle θ (and time t) such that

$$x = \left(\frac{440}{3}\cos\theta\right)t = 400$$

$$y = 3 + \left(\frac{440}{3}\sin\theta\right)t - 16t^2 = 10.$$

From the first equation $t = 1200/440\cos\theta$. Substituting into the second equation,

$$10 = 3 + \left(\frac{440}{3}\sin\theta\right)\left(\frac{1200}{440\cos\theta}\right) - 16\left(\frac{1200}{440\cos\theta}\right)^2$$

$$7 = 400\tan\theta - 16\left(\frac{120}{44}\right)^2 \sec^2\theta$$

$$= 400\tan\theta - 16\left(\frac{120}{44}\right)^2 \left(\tan^2\theta + 1\right).$$

You now solve the quadratic for $\tan\theta$:

$$16\left(\frac{120}{44}\right)^2 \tan^2\theta - 400\tan\theta + 7 + 16\left(\frac{120}{44}\right)^2 = 0.$$

$\tan\theta \approx 0.35185 \Rightarrow \theta \approx 19.4°$

Section 10.3 Parametric Equations and Calculus

1. The parametric form of the derivative represents the slope of the tangent line at the point (x, y).

3. If $\frac{dy}{dt} = 0$ and $\frac{dx}{dt} \neq 0$, then there is a horizontal tangent line. If $\frac{dx}{dt} = 0$ and $\frac{dy}{dt} \neq 0$, then there is a vertical tangent line.

5. $\frac{dy}{dx} = \frac{dy/dt}{dx/dt} = \frac{-6}{2t} = -\frac{3}{t}$

7. $\frac{dy}{dx} = \frac{dy/d\theta}{dx/d\theta} = \frac{-2\cos\theta\sin\theta}{2\sin\theta\cos\theta} = -1$

$\left[\text{Note: } x + y = 1 \Rightarrow y = 1 - x \text{ and } \frac{dy}{d\theta} = -1\right]$

9. $x = 4t$, $y = 3t - 2$

$$\frac{dy}{dx} = \frac{dy/dt}{dx/dt} = \frac{3}{4}$$

$$\frac{d^2y}{dx^2} = 0$$

At $t = 3$, slope is $\frac{3}{4}$. (Line)

Neither concave upward nor downward

11. $x = t + 1$, $y = t^2 + 3t$

$$\frac{dy}{dx} = \frac{dy/dt}{dx/dt} = \frac{2t+3}{1} = 2t + 3$$

$$\frac{d^2y}{dx^2} = \frac{\frac{d}{dt}[2t+3]}{dx/dt} = \frac{2}{1} = 2$$

At $t = -2$, $\frac{dy}{dx} = 2(-2) + 3 = -1$.

At $t = -2$, $\frac{d^2y}{dx^2} = 2$.

Concave upward.

13. $x = 4\cos\theta,\ y = 4\sin\theta$

$$\frac{dy}{dx} = \frac{dy/d\theta}{dx/d\theta} = \frac{4\cos\theta}{-4\sin\theta} = \frac{-\cos\theta}{\sin\theta} = -\cot\theta$$

$$\frac{d^2y}{dx^2} = \frac{\frac{d}{d\theta}[-\cot\theta]}{dx/d\theta} = \frac{\csc^2\theta}{-4\sin\theta} = \frac{-1}{4\sin^3\theta} = -\frac{1}{4}\csc^3\theta$$

At $\theta = \frac{\pi}{4}$, $\frac{dy}{dx} = -1$.

At $\theta = \frac{\pi}{4}$, $\frac{d^2y}{dx^2} = \frac{-1}{4\left(\sqrt{2}/2\right)^3} = \frac{-\sqrt{2}}{2}$.

Concave downward

15. $x = 2 + \sec\theta,\ y = 1 + 2\tan\theta$

$$\frac{dy}{dx} = \frac{dy/d\theta}{dx/d\theta} = \frac{2\sec^2\theta}{\sec\theta\tan\theta} = \frac{2\sec\theta}{\tan\theta} = 2\csc\theta$$

$$\frac{d^2y}{dx^2} = \frac{\frac{d}{d\theta}(2\csc\theta)}{dx/d\theta} = \frac{-2\csc\theta\cot\theta}{\sec\theta\tan\theta} = -2\cot^3\theta$$

At $\theta = -\frac{\pi}{3}$, $\frac{dy}{dx} = 2\left(-\frac{2\sqrt{3}}{3}\right) = -\frac{4\sqrt{3}}{3}$.

At $\theta = -\frac{\pi}{3}$, $\frac{d^2y}{dx^2} = -2\left(-\frac{\sqrt{3}}{9}\right) = \frac{2\sqrt{3}}{9}$.

Concave upward

17. $x = \cos^3\theta,\ y = \sin^3\theta$

$$\frac{dy}{dx} = \frac{3\sin^2\theta\cos\theta}{-3\cos^2\theta\sin\theta} = -\tan\theta$$

$$\frac{d^2y}{dx^2} = \frac{-\sec^2\theta}{-3\cos^2\theta\sin\theta} = \frac{1}{3\cos^4\theta\sin\theta} = \frac{\sec^4\theta\csc\theta}{3}$$

At $\theta = 1$, $\frac{dy}{dx} = -\tan\frac{\pi}{4} = -1$.

At $\theta = 1$, $\frac{d^2y}{dx^2} = \frac{\sec^4(\pi/4)\csc(\pi/4)}{3} = \frac{4\sqrt{2}}{3}$.

Concave upward

19. $x = 2\cot\theta,\ y = 2\sin^2\theta$

$$\frac{dy}{dx} = \frac{4\sin\theta\cos\theta}{-2\csc^2\theta} = -2\sin^3\theta\cos\theta$$

At $\left(-\frac{2}{\sqrt{3}}, \frac{3}{2}\right)$, $\theta = \frac{2\pi}{3}$, and $\frac{dy}{dx} = \frac{3\sqrt{3}}{8}$.

Tangent line: $y - \frac{3}{2} = \frac{3\sqrt{3}}{8}\left(x + \frac{2}{\sqrt{3}}\right)$

$3\sqrt{3}x - 8y + 18 = 0$

At $(0, 2)$, $\theta = \frac{\pi}{2}$, and $\frac{dy}{dx} = 0$.

Tangent line: $y - 2 = 0$

At $\left(2\sqrt{3}, \frac{1}{2}\right)$, $\theta = \frac{\pi}{6}$, and $\frac{dy}{dx} = -\frac{\sqrt{3}}{8}$.

Tangent line: $y - \frac{1}{2} = -\frac{\sqrt{3}}{8}\left(x - 2\sqrt{3}\right)$

$\sqrt{3}x + 8y - 10 = 0$

21. $x = t^2 - 4$

$y = t^2 - 2t$

$$\frac{dy}{dx} = \frac{dy/dt}{dx/dt} = \frac{2t - 2}{2t}$$

At $(0, 0)$, $t = 2$, $\frac{dy}{dx} = \frac{1}{2}$.

Tangent line: $y = \frac{1}{2}x$

$2y - x = 0$

At $(-3, -1)$, $t = 1$, $\frac{dy}{dx} = 0$.

Tangent line: $y = -1$

$y + 1 = 0$

At $(-3, 3)$, $t = -1$, $\frac{dy}{dx} = 2$.

Tangent line: $y - 3 = 2(x + 3)$

$2x - y + 9 = 0$

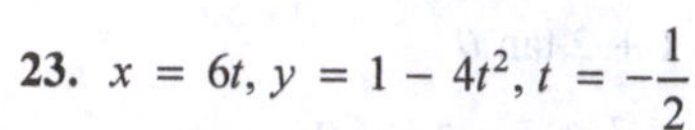

23. $x = 6t,\ y = 1 - 4t^2,\ t = -\frac{1}{2}$

(a), (d)

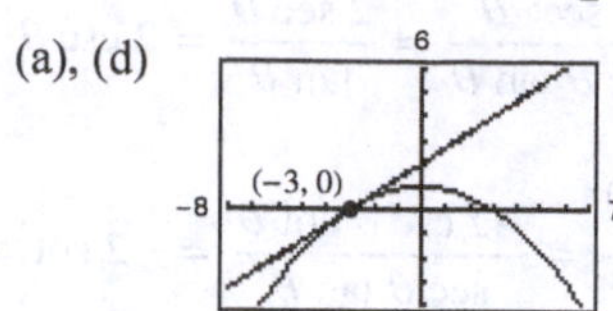

(b) At $t = -\frac{1}{2}$, $(x, y) = (-3, 0)$, and

$$\frac{dx}{dt} = 6,\ \frac{dy}{dt} = 4,\ \frac{dy}{dx} = \frac{2}{3}.$$

(c) $\frac{dy}{dx} = \frac{2}{3}$ at $(-3, 0)$.

$$y - 0 = \frac{2}{3}(x + 3)$$

$$y = \frac{2}{3}x + 2$$

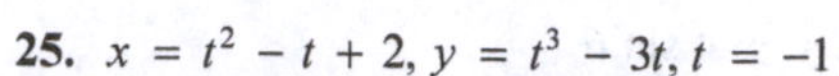

25. $x = t^2 - t + 2,\ y = t^3 - 3t,\ t = -1$

(a), (d)

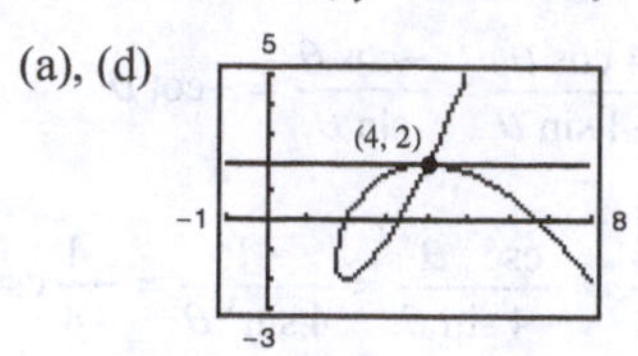

(b) At $t = -1$, $(x, y) = (4, 2)$, and

$$\frac{dx}{dt} = -3,\ \frac{dy}{dt} = 0,\ \frac{dy}{dx} = 0.$$

(c) $\frac{dy}{dx} = 0$ at $(4, 2)$,

$$y - 2 = 0(x - 4)$$

$$y = 2$$

27. $x = 2\sin 2t,\ y = 3\sin t$ crosses itself at the origin, $(x, y) = (0, 0)$.

At this point, $t = 0$ or $t = \pi$.

$$\frac{dy}{dx} = \frac{3\cos t}{4\cos 2t}$$

At $t = 0$: $\frac{dy}{dx} = \frac{3}{4}$ and $y = \frac{3}{4}x$. Tangent Line

At $t = \pi$, $\frac{dy}{dx} = -\frac{3}{4}$ and $y = -\frac{3}{4}x$. Tangent Line

29. $x = t^2 - t,\ y = t^3 - 3t - 1$ crosses itself at the point $(x, y) = (2, 1)$.

At this point, $t = -1$ or $t = 2$.

$$\frac{dy}{dx} = \frac{3t^2 - 3}{2t - 1}$$

At $t = -1$, $\frac{dy}{dx} = 0$ and $y = 1$. Tangent Line

At $t = 2$, $\frac{dy}{dt} = \frac{9}{3} = 3$ and $y - 1 = 3(x - 2)$ or $y = 3x - 5$.

Tangent Line

31. $x = \cos\theta + \theta\sin\theta,\ y = \sin\theta - \theta\cos\theta,\ -2\pi \le \theta \le 2\pi$

$\frac{dy}{d\theta} = \theta\sin\theta = 0$ when

$\theta = 0, \pm\pi, \pm2\pi$.

Also, $\frac{dx}{d\theta} = \theta\cos\theta = 0$ when $\theta = 0, \pm\frac{\pi}{2}, \pm\frac{3\pi}{2}$.

Note that $\theta = 0$ is a cusp.

Horizontal tangents: $\theta = \pm\pi, \pm2\pi \Rightarrow (-1, \pi), (-1, -\pi), (1, -2\pi), (1, 2\pi)$

Vertical tangents: $\theta = \pm\frac{\pi}{2}, \pm\frac{3\pi}{2} \Rightarrow \left(\frac{\pi}{2}, 1\right), \left(\frac{\pi}{2}, -1\right), \left(-\frac{3\pi}{2}, -1\right), \left(-\frac{3\pi}{2}, 1\right)$

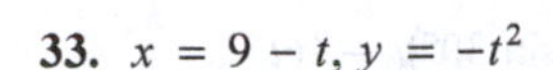

33. $x = 9 - t,\ y = -t^2$

$$\frac{dy}{dt} = -2t = 0 \Rightarrow t = 0$$

$$\frac{dx}{dt} = -1 \neq 0$$

Horizontal tangent at $(9, 0)$

No vertical tangents

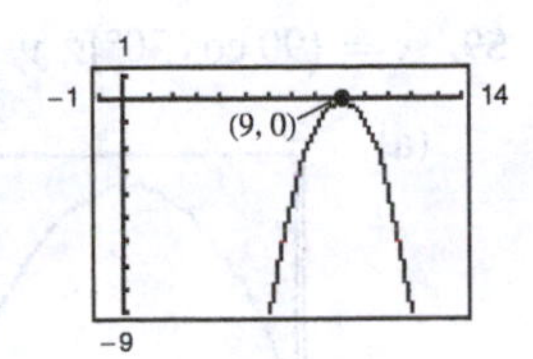

35. $x = t + 4,\ y = t^3 - 12t + 6$

$$\frac{dy}{dt} = 3t^2 - 12 = 3(t + 2)(t - 2) = 0 \Rightarrow t = \pm 2$$

$$\frac{dx}{dt} = 1 \neq 0$$

Horizontal tangents at $(6, -10)$ and $(2, 22)$

No vertical tangents

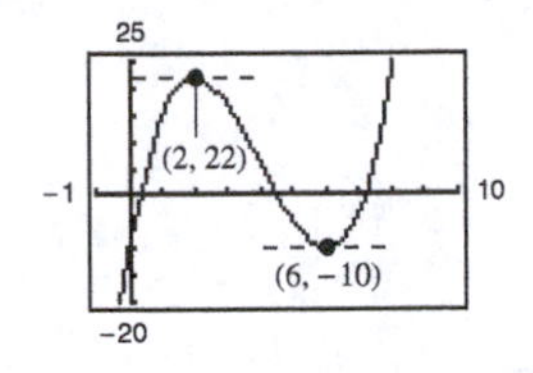

37. $x = 7\cos\theta,\ y = 7\sin\theta$

$$\frac{dy}{dt} = 7\cos\theta = 0 \Rightarrow \theta = \frac{\pi}{2}, \frac{3\pi}{2}$$

$$\frac{dx}{dt} = -7\sin\theta = 0 \Rightarrow \theta = 0, \pi$$

Horizontal tangents at $(0, 7), (0, -7)$

Vertical tangents at $(7, 0), (-7, 0)$

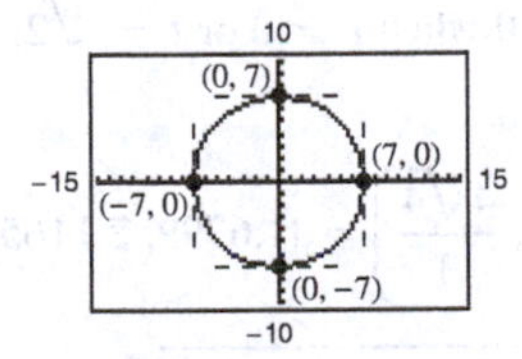

39. $x = 5 + 3\cos\theta,\ y = -2 + \sin\theta$

Horizontal tangents: $\dfrac{dy}{dt} = \cos\theta = 0 \Rightarrow \theta = \dfrac{\pi}{2}, \dfrac{3\pi}{2}$

Points: $(5, -1), (5, -3)$

Vertical tangents: $\dfrac{dx}{dt} = -3\sin\theta = 0 \Rightarrow \theta = 0, \pi$

Points: $(8, -2), (2, -2)$

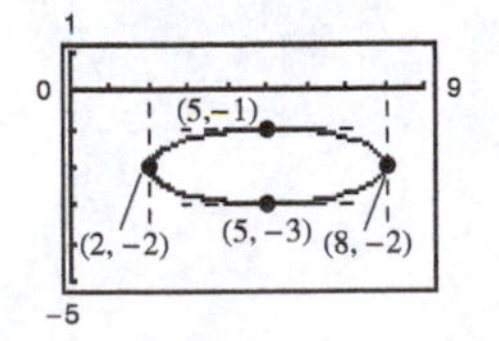

41. $x = \sec\theta,\ y = \tan\theta$

Horizontal tangents: $\dfrac{dy}{d\theta} = \sec^2\theta \neq 0$; None

Vertical tangents: $\dfrac{dx}{d\theta} = \sec\theta\tan\theta = 0$ when

$x = 0, \pi$.

Points: $(1, 0), (-1, 0)$

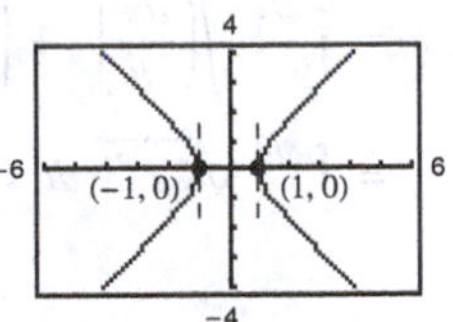

43. $x = 3t^2,\ y = t^3 - t$

$$\frac{dy}{dx} = \frac{dy/dt}{dx/dt} = \frac{3t^2 - 1}{6t} = \frac{t}{2} - \frac{1}{6t}$$

$$\frac{d^2y}{dx^2} = \frac{\dfrac{d}{dt}\left[\dfrac{t}{2} - \dfrac{1}{6t}\right]}{dx/dt} = \frac{\dfrac{1}{2} + \dfrac{1}{6t^2}}{6t} = \frac{3t^2 + 1}{36t^3}$$

Concave upward for $t > 0$

Concave downward for $t < 0$

45. $x = 2t + \ln t,\ y = 2t - \ln t,\ t > 0$

$$\frac{dy}{dx} = \frac{2 - (1/t)}{2 + (1/t)} = \frac{2t - 1}{2t + 1}$$

$$\frac{d^2y}{dx^2} = \left[\frac{(2t + 1)2 - (2t - 1)2}{(2t + 1)^2}\right] \Big/ \left(2 + \frac{1}{t}\right)$$

$$= \frac{4}{(2t + 1)^2} \cdot \frac{t}{2t + 1} = \frac{4t}{(2t + 1)^3}$$

Because $t > 0$, $\dfrac{d^2y}{dx^2} > 0$

Concave upward for $t > 0$

47. $x = \sin t,\ y = \cos t,\ 0 < t < \pi$

$$\frac{dy}{dx} = -\frac{\sin t}{\cos t} = -\tan t$$

$$\frac{d^2y}{dx^2} = -\frac{\sec^2 t}{\cos t} = -\frac{1}{\cos^3 t}$$

Concave upward on $\pi/2 < t < \pi$

Concave downward on $0 < t < \pi/2$

49. $x = 3t + 5,\ y = 7 - 2t,\ -1 \le t \le 3$

$$\frac{dx}{dt} = 3,\ \frac{dy}{dt} = -2$$

$$s = \int_a^b \sqrt{\left(\frac{dx}{dt}\right)^2 + \left(\frac{dy}{dt}\right)^2}\, dt$$

$$= \int_{-1}^{3} \sqrt{9 + 4}\, dt$$

$$\left[\sqrt{13}\, t\right]_{-1}^{3} = 4\sqrt{13} \approx 14.422$$

51. $x = e^{-t}\cos t,\ y = e^{-t}\sin t,\ 0 \le t \le \dfrac{\pi}{2}$

$$\frac{dx}{dt} = -e^{-t}(\sin t + \cos t),\ \frac{dy}{dt} = e^{-t}(\cos t - \sin t)$$

$$s = \int_0^{\pi/2}\sqrt{\left(\frac{dx}{dt}\right)^2 + \left(\frac{dy}{dt}\right)^2}\,dt$$

$$= \int_0^{\pi/2}\sqrt{2e^{-2t}}\,dt = -\sqrt{2}\int_0^{\pi/2} e^{-t}(-1)\,dt$$

$$= \left[-\sqrt{2}e^{-t}\right]_0^{\pi/2}$$

$$= \sqrt{2}\left(1 - e^{-\pi/2}\right) \approx 1.12$$

53. $x = \sqrt{t},\ y = 3t - 1,\ \dfrac{dx}{dt} = \dfrac{1}{2\sqrt{t}},\ \dfrac{dy}{dt} = 3$

$$s = \int_0^1\sqrt{\frac{1}{4t} + 9}\,dt = \frac{1}{2}\int_0^1\frac{\sqrt{1+36t}}{\sqrt{t}}\,dt$$

$$= \frac{1}{6}\int_0^6\sqrt{1+u^2}\,du$$

$$= \frac{1}{12}\left[\ln\left(\sqrt{1+u^2} + u\right) + u\sqrt{1+u^2}\right]_0^6$$

$$= \frac{1}{12}\left[\ln\left(\sqrt{37} + 6\right) + 6\sqrt{37}\right] \approx 3.249$$

$$u = 6\sqrt{t},\ du = \frac{3}{\sqrt{t}}\,dt$$

55. $x = a\cos^3\theta,\ y = a\sin^3\theta,\ \dfrac{dx}{d\theta} = -3a\cos^2\theta\sin\theta,$

$$\frac{dy}{d\theta} = 3a\sin^2\theta\cos\theta$$

$$s = 4\int_0^{\pi/2}\sqrt{9a^2\cos^4\theta\sin^2\theta + 9a^2\sin^4\theta\cos^2\theta}\,d\theta$$

$$= 12a\int_0^{\pi/2}\sin\theta\cos\theta\sqrt{\cos^2\theta + \sin^2\theta}\,d\theta$$

$$= 6a\int_0^{\pi/2}\sin 2\theta\,d\theta = \left[-3a\cos 2\theta\right]_0^{\pi/2} = 6a$$

57. $x = a(\theta - \sin\theta),\ y = a(1 - \cos\theta),$

$$\frac{dx}{d\theta} = a(1 - \cos\theta),\ \frac{dy}{d\theta} = a\sin\theta$$

$$s = 2\int_0^{\pi}\sqrt{a^2(1 - \cos\theta)^2 + a^2\sin^2\theta}\,d\theta$$

$$= 2\sqrt{2}a\int_0^{\pi}\sqrt{1 - \cos\theta}\,d\theta$$

$$= 2\sqrt{2}a\int_0^{\pi}\frac{\sin\theta}{\sqrt{1 + \cos\theta}}\,d\theta$$

$$= \left[-4\sqrt{2}a\sqrt{1 + \cos\theta}\right]_0^{\pi} = 8a$$

59. $x = (90\cos 30°)t,\ y = (90\sin 30°)t - 16t^2$

(a)

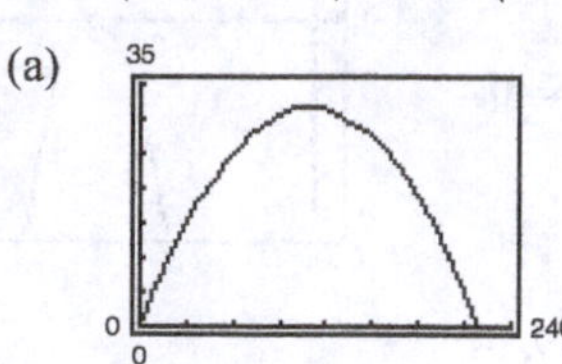

(b) Range: 219.2 ft, $\left(t = \dfrac{45}{16}\right)$

(c) $\dfrac{dx}{dt} = 90\cos 30°,\ \dfrac{dy}{dt} = 90\sin 30° - 32t$

$y = 0$ for $t = \dfrac{45}{16}$.

$$s = \int_0^{45/16}\sqrt{(90\cos 30°)^2 + (90\sin 30° - 32t)^2}\,dt$$

$$\approx 230.8 \text{ ft}$$

61. $x = \dfrac{4t}{1 + t^3},\ y = \dfrac{4t^2}{1 + t^3}$

(a) $x^3 + y^3 = 4xy$

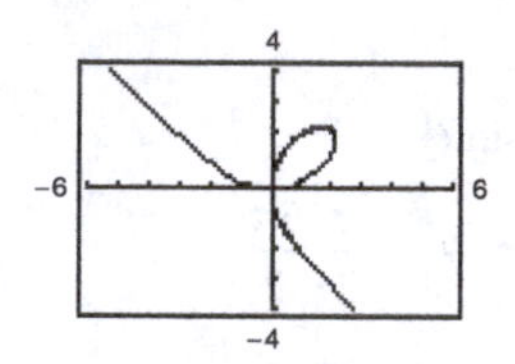

(b) $\dfrac{dy}{dt} = \dfrac{(1 + t^3)(8t) - 4t^2(3t^2)}{(1 + t^3)^2}$

$$= \frac{4t(2 - t^3)}{(1 + t^3)^2} = 0 \text{ when } t = 0 \text{ or } t = \sqrt[3]{2}.$$

Points: $(0, 0), \left(\dfrac{4\sqrt[3]{2}}{3}, \dfrac{4\sqrt[3]{4}}{3}\right) \approx (1.6799, 2.1165)$

(c) $s = 2\displaystyle\int_0^1\sqrt{\left[\frac{4(1 - 2t^3)}{(1 + t^3)^2}\right]^2 + \left[\frac{4t(2 - t^3)}{(1 + t^3)^2}\right]^2}\,dt$

$$= 2\int_0^1\sqrt{\frac{16}{(1 + t^3)^4}\left[t^8 + 4t^6 - 4t^5 - 4t^3 + 4t^2 + 1\right]}\,dt$$

$$= 8\int_0^1\frac{\sqrt{t^8 + 4t^6 - 4t^5 - 4t^3 + 4t^2 + 1}}{(1 + t^3)^2}\,dt \approx 6.557$$

63. $x = 2t, \dfrac{dx}{dt} = 2$

$y = 3t, \dfrac{dy}{dt} = 3$

(a) $S = 2\pi\int_0^3 3t\sqrt{4 + 9}\, dt = 6\sqrt{13}\pi\left[\dfrac{t^2}{2}\right]_0^3 = 6\sqrt{13}\pi\left(\dfrac{9}{2}\right) = 27\sqrt{13}\pi$

(b) $S = 2\pi\int_0^3 2t\sqrt{4 + 9}\, dt = 4\sqrt{13}\pi\left[\dfrac{t^2}{2}\right]_0^3 = 4\sqrt{13}\pi\left(\dfrac{9}{2}\right) = 18\sqrt{13}\pi$

65. $x = 5\cos\theta\ \dfrac{dx}{d\theta} = -5\sin\theta$

$y = 5\sin\theta\ \dfrac{dy}{d\theta} = 5\cos\theta$

$S = 2\pi\int_0^{\pi/2} 5\cos\theta\sqrt{25\sin^2\theta + 25\cos^2\theta}\, d\theta = 10\pi\int_0^{\pi/2} 5\cos\theta\, d\theta = 50\pi\big[\sin\theta\big]_0^{\pi/2} = 50\pi$

[**Note:** This is the surface area of a hemisphere of radius 5]

67. $x = a\cos^3\theta, y = a\sin^3\theta, \dfrac{dx}{d\theta} = -3a\cos^2\theta\sin\theta, \dfrac{dy}{d\theta} = 3a\sin^2\theta\cos\theta$

$S = 4\pi\int_0^{\pi/2} a\sin^3\theta\sqrt{9a^2\cos^4\theta\sin^2\theta + 9a^2\sin^4\theta\cos^2\theta}\, d\theta = 12a^2\pi\int_0^{\pi/2}\sin^4\theta\cos\theta\, d\theta = \dfrac{12\pi a^2}{5}\big[\sin^5\theta\big]_0^{\pi/2} = \dfrac{12}{5}\pi a^2$

69. $x = t^3, \dfrac{dx}{dt} = 3t^2$

$y = t + 2, \dfrac{dy}{dt} = 1$

$S = 2\pi\int_0^2 (t + 2)\sqrt{(3t^2)^2 + 1}\, dt \approx 185.78$

71. $x = \cos^2\theta, \dfrac{dx}{d\theta} = -2\cos\theta\sin\theta$

$y = \cos\theta, \dfrac{dy}{d\theta} = -\sin\theta$

$S = 2\pi\int_0^{\pi/2}\cos\theta\sqrt{4\cos^2\theta\sin^2\theta + \sin^2\theta}\, d\theta$

$= 2\pi\int_0^{\pi/2}\cos\theta\sin\theta\sqrt{4\cos^2\theta + 1}\, d\theta$

$= \dfrac{(5\sqrt{5} - 1)\pi}{6} \approx 5.3304$

73. (a) $x = t - \sin t$
$y = 1 - \cos t$
$0 \le t \le 2\pi$

$x = 2t - \sin(2t)$
$y = 1 - \cos(2t)$
$0 \le t \le \pi$

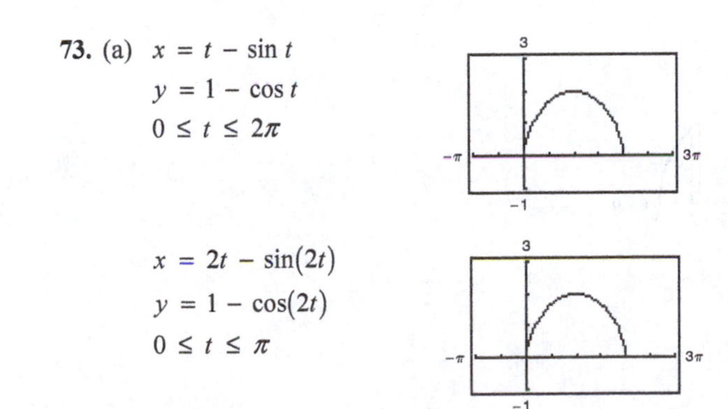

(b) The average speed of the particle on the second path is twice the average speed of a particle on the first path.

(c) $x = \frac{1}{2}t - \sin\left(\frac{1}{2}t\right)$

$y = 1 - \cos\left(\frac{1}{2}t\right)$

The time required for the particle to traverse the same path is $t = 4\pi$.

75. Answers will vary. *Sample answer:*

Let $x = -3t, y = -4t$.

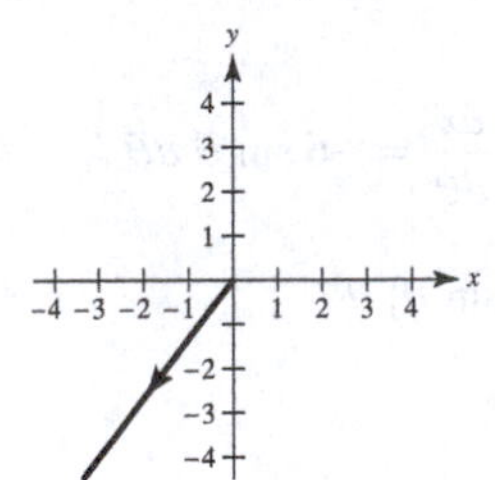

77. Let y be a continuous function of x on $a \le x \le b$. Suppose that $x = f(t), y = g(t)$, and $f(t_1) = a$, $f(t_2) = b$. Then using integration by substitution, $dx = f'(t)\, dt$ and

$\int_a^b y\, dx = \int_{t_1}^{t_2} g(t)f'(t)\, dt.$

79. $x = 2\sin^2\theta$

$y = 2\sin^2\theta\tan\theta$

$\dfrac{dx}{d\theta} = 4\sin\theta\cos\theta$

$$A = \int_0^{\pi/2} 2\sin^2\theta\tan\theta(4\sin\theta\cos\theta)\,d\theta = 8\int_0^{\pi/2}\sin^4\theta\,d\theta = 8\left[\frac{-\sin^3\theta\cos\theta}{4} - \frac{3}{8}\sin\theta\cos\theta + \frac{3}{8}\theta\right]_0^{\pi/2} = \frac{3\pi}{2}$$

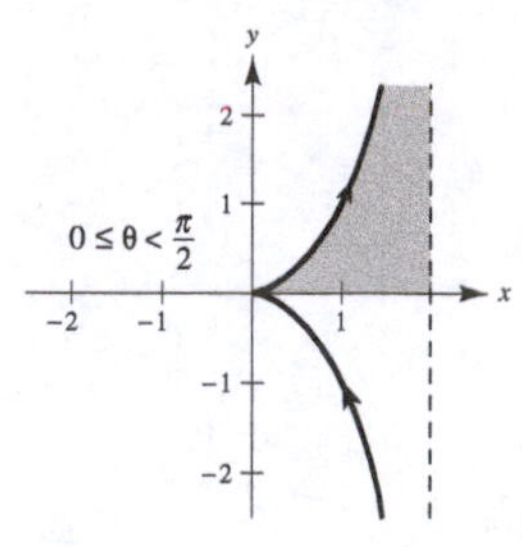

81. πab is area of ellipse (d).

82. $\frac{3}{8}\pi a^2$ is area of asteroid (b).

83. $6\pi a^2$ is area of cardioid (f).

84. $2\pi a^2$ is area of deltoid (c).

85. $\frac{8}{3}ab$ is area of hourglass (a).

86. $2\pi ab$ is area of teardrop (e).

87. $x = \sqrt{t},\ y = 4 - t,\ 0 < t < 4$

$$A = \int_0^2 y\,dx = \int_0^4 (4-t)\frac{1}{2\sqrt{t}}\,dt = \frac{1}{2}\int_0^4\left(4t^{-1/2} - t^{1/2}\right)dt = \left[\frac{1}{2}\left(8\sqrt{t} - \frac{2}{3}t\sqrt{t}\right)\right]_0^4 = \frac{16}{3}$$

$$\bar{x} = \frac{1}{A}\int_0^2 yx\,dx = \frac{3}{16}\int_0^4 (4-t)\sqrt{t}\left(\frac{1}{2\sqrt{t}}\right)dt = \frac{3}{32}\int_0^4 (4-t)\,dt = \left[\frac{3}{32}\left(4t - \frac{t^2}{2}\right)\right]_0^4 = \frac{3}{4}$$

$$\bar{y} = \frac{1}{A}\int_0^2 \frac{y^2}{2}\,dx = \frac{3}{32}\int_0^4 (4-t)^2\frac{1}{2\sqrt{t}}\,dt = \frac{3}{64}\int_0^4\left(16t^{-1/2} - 8t^{1/2} + t^{3/2}\right)dt = \frac{3}{64}\left[32\sqrt{t} - \frac{16}{3}t\sqrt{t} + \frac{2}{5}t^2\sqrt{t}\right]_0^4 = \frac{8}{5}$$

$$(\bar{x}, \bar{y}) = \left(\frac{3}{4}, \frac{8}{5}\right)$$

89. $x = 6\cos\theta,\ y = 6\sin\theta,\ \dfrac{dx}{d\theta} = -6\sin\theta\,d\theta$

$$V = 2\pi\int_{\pi/2}^0 (6\sin\theta)^2(-6\sin\theta)\,d\theta$$

$$= -432\pi\int_{\pi/2}^0 \sin^3\theta\,d\theta$$

$$= -432\pi\int_{\pi/2}^0 \left(1 - \cos^2\theta\right)\sin\theta\,d\theta$$

$$= -432\pi\left[-\cos\theta + \frac{\cos^3\theta}{3}\right]_{\pi/2}^0$$

$$= -432\pi\left(-1 + \frac{1}{3}\right) = 288\pi$$

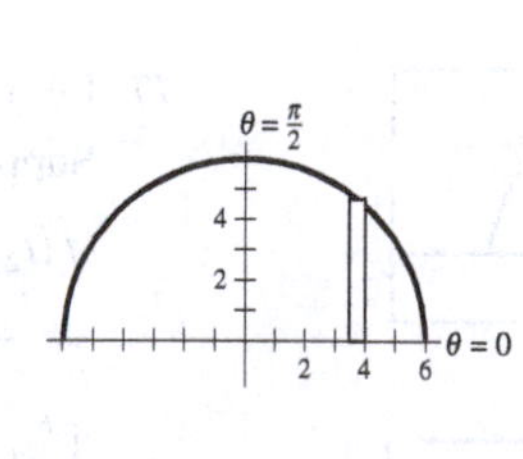

Note: Volume of sphere is $\frac{4}{3}\pi(6^3) = 288\pi$.

91. $x = a(\theta - \sin\theta),\ y = a(1 - \cos\theta)$

(a) $\dfrac{dy}{d\theta} = a\sin\theta,\ \dfrac{dx}{d\theta} = a(1 - \cos\theta)$

$$\frac{dy}{dx} = \frac{a\sin\theta}{a(1-\cos\theta)} = \frac{\sin\theta}{1-\cos\theta}$$

$$\frac{d^2y}{dx^2} = \left[\frac{(1-\cos\theta)\cos\theta - \sin\theta(\sin\theta)}{(1-\cos\theta)^2}\right]\Big/\left[a(1-\cos\theta)\right] = \frac{\cos\theta - 1}{a(1-\cos\theta)^3} = \frac{-1}{a(\cos\theta - 1)^2}$$

(b) At $\theta = \dfrac{\pi}{6}$, $x = a\left(\dfrac{\pi}{6} - \dfrac{1}{2}\right)$, $y = a\left(1 - \dfrac{\sqrt{3}}{2}\right)$, $\dfrac{dy}{dx} = \dfrac{1/2}{1 - \sqrt{3}/2} = 2 + \sqrt{3}.$

Tangent line: $y - a\left(1 - \dfrac{\sqrt{3}}{2}\right) = \left(2 + \sqrt{3}\right)\left(x - a\left(\dfrac{\pi}{6} - \dfrac{1}{2}\right)\right)$

(c) $\dfrac{dy}{dx} = \dfrac{\sin\theta}{1-\cos\theta} = 0 \Rightarrow \sin\theta = 0,\ 1 - \cos\theta \neq 0$

Points of horizontal tangency: $(x, y) = \left(a(2n+1)\pi, 2a\right)$

(d) Concave downward on all open θ-intervals: …, $(-2\pi, 0)$, $(0, 2\pi)$, $(2\pi, 4\pi)$, …

(e) $$s = \int_0^{2\pi} \sqrt{a^2\sin^2\theta + a^2(1-\cos\theta)^2}\, d\theta$$

$$= a\int_0^{2\pi} \sqrt{2 - 2\cos\theta}\, d\theta = a\int_0^{2\pi} \sqrt{4\sin^2\frac{\theta}{2}}\, d\theta = 2a\int_0^{2\pi} \sin\frac{\theta}{2}\, d\theta = \left[-4a\cos\left(\frac{\theta}{2}\right)\right]_0^{2\pi} = 8a$$

93. $x = t + u = r\cos\theta + r\theta\sin\theta = r(\cos\theta + \theta\sin\theta)$

$y = v - w = r\sin\theta - r\theta\cos\theta = r(\sin\theta - \theta\cos\theta)$

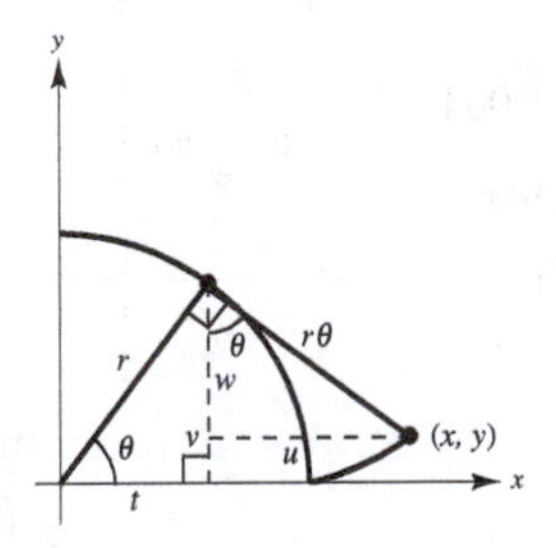

95. (a)

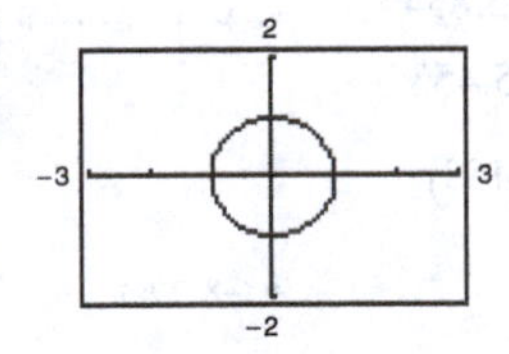

(b) $x = \dfrac{1 - t^2}{1 + t^2},\ y = \dfrac{2t}{1 + t^2},\ -20 \le t \le 20$

The graph (for $-\infty < t < \infty$) is the circle $x^2 + y^2 = 1$, except the point $(-1, 0)$.

Verify: $$x^2 + y^2 = \left(\frac{1-t^2}{1+t^2}\right)^2 + \left(\frac{2t}{1+t^2}\right)^2 = \frac{1 - 2t^2 + t^4 + 4t^2}{(1+t^2)^2} = \frac{(1+t^2)^2}{(1+t^2)^2} = 1$$

(c) As t increases from –20 to 0, the speed increases, and as t increases from 0 to 20, the speed decreases.

97. False. $\dfrac{d^2y}{dx^2} = \dfrac{\dfrac{d}{dt}\left[\dfrac{g'(t)}{f'(t)}\right]}{f'(t)} = \dfrac{f'(t)g''(t) - g'(t)f''(t)}{\left[f'(t)\right]^3}$

99. False. The resulting rectangular equation is a line.

Section 10.4 Polar Coordinates and Polar Graphs

1. r is the directed distance from the origin to the point in the plane. θ is the directed angle, measured counterclockwise from the polar axis to the segment joining the point to the origin.

3. The rectangular coordinate system consists of all points of the form (x, y) where x is the directed distance from the y-axis to the point, and y is the directed distance from the x-axis to the point.

Every point has a unique representation.

The polar coordinate system uses (r, θ) to designate the location of a point.

r is the directed distance to the origin and θ is the angle the point makes with the positive x-axis, measured counterclockwise.

Points do not have a unique polar representation.

5. $\left(8, \dfrac{\pi}{2}\right)$

$x = 8\cos\dfrac{\pi}{2} = 0$

$y = 8\sin\dfrac{\pi}{2} = 8$

$(x, y) = (0, 8)$

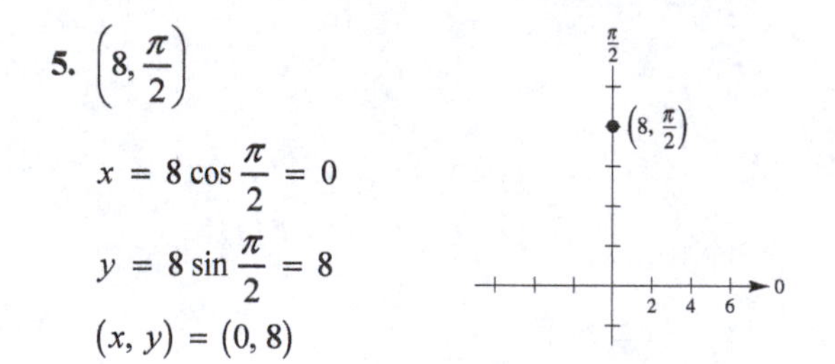

7. $\left(-4, -\dfrac{3\pi}{4}\right)$

$x = -4\cos\left(\dfrac{-3\pi}{4}\right) = -4\left(-\dfrac{\sqrt{2}}{2}\right) = 2\sqrt{2}$

$y = -4\sin\left(\dfrac{-3\pi}{4}\right) = -4\left(-\dfrac{\sqrt{2}}{2}\right) = 2\sqrt{2}$

$(x, y) = \left(2\sqrt{2}, 2\sqrt{2}\right)$

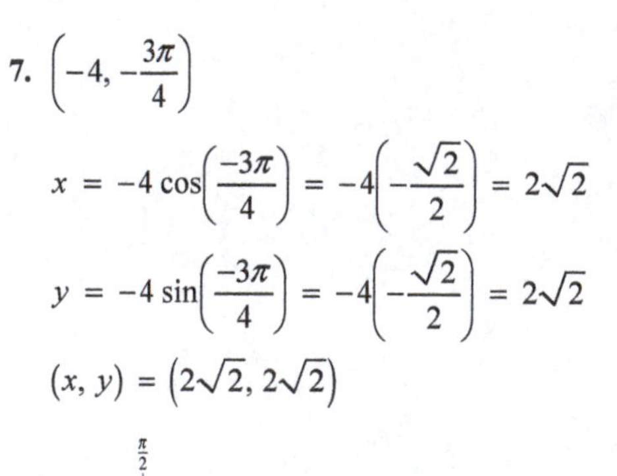

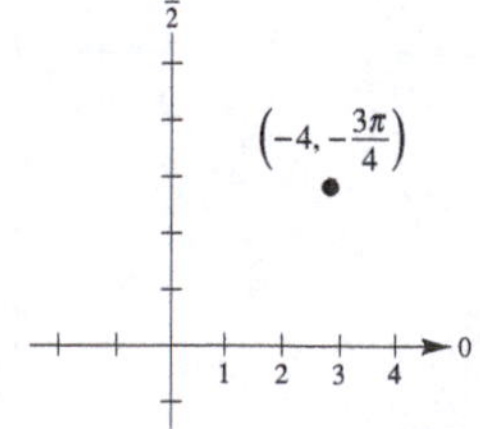

9. $(r, \theta) = \left(7, \dfrac{5\pi}{4}\right)$

$x = 7\cos\dfrac{5\pi}{4} = 7\left(\dfrac{-\sqrt{2}}{2}\right) = -\dfrac{7\sqrt{2}}{2}$

$y = 7\sin\dfrac{5\pi}{4} = 7\left(-\dfrac{\sqrt{2}}{2}\right) = -\dfrac{7\sqrt{2}}{2}$

$(x, y) = \left(-\dfrac{7\sqrt{2}}{2}, -\dfrac{7\sqrt{2}}{2}\right)$

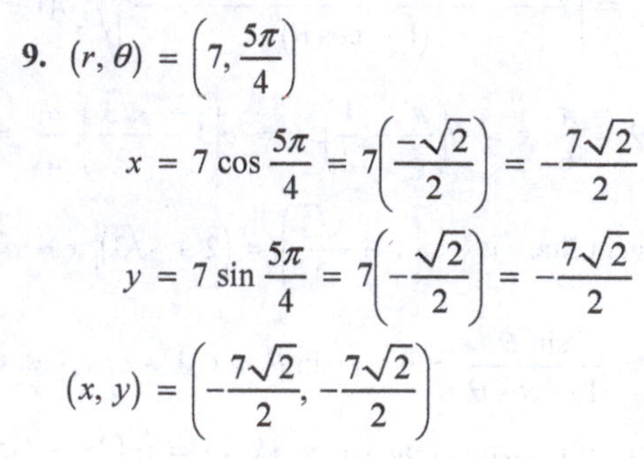

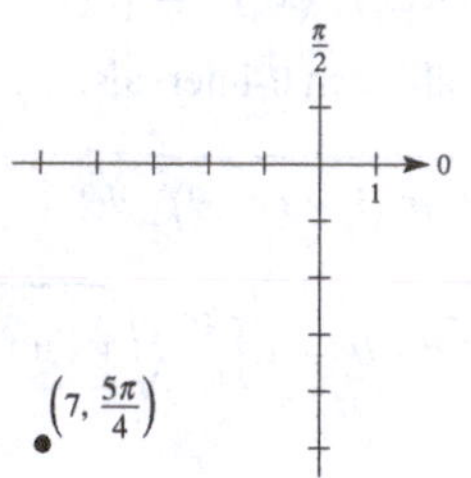

11. $\left(\sqrt{2}, 2.36\right)$

$x = \sqrt{2}\cos(2.36) \approx -1.004$

$y = \sqrt{2}\sin(2.36) \approx 0.996$

$(x, y) \approx (-1.004, 0.996)$

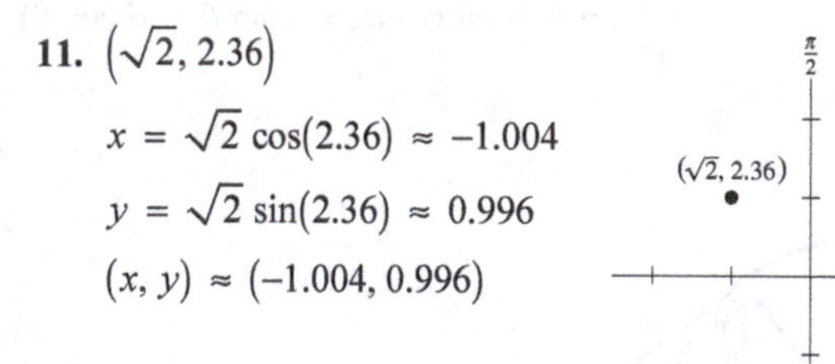

13. $(-8, 0.75)$

$x = -8\cos 0.75 \approx -5.854$

$y = -8\sin 0.75 \approx -5.453$

$(x, y) \approx (-5.854, -5.453)$

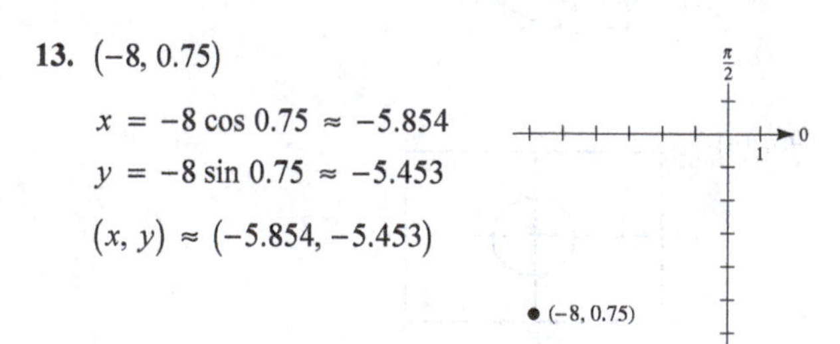

15. $(x, y) = (1, 0)$

$r = \pm 1$

$\tan\theta = 0$

$\theta = 0, \pi$

$(1, 0), (-1, \pi)$

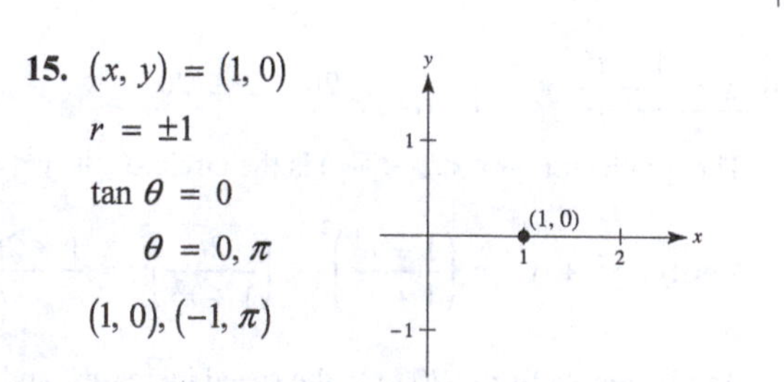

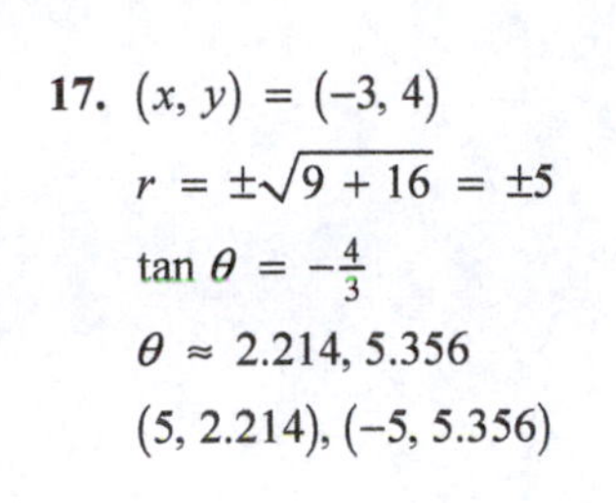

17. $(x, y) = (-3, 4)$

$r = \pm\sqrt{9 + 16} = \pm 5$

$\tan\theta = -\frac{4}{3}$

$\theta \approx 2.214, 5.356$

$(5, 2.214), (-5, 5.356)$

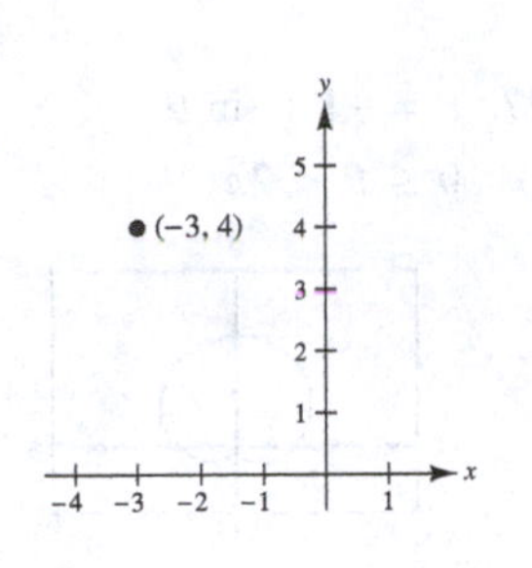

19. $(x, y) = \left(-5, -5\sqrt{3}\right)$

$r = \pm\sqrt{(-5)^2 + \left(-5\sqrt{3}\right)^2} = \pm 10$

$\tan\theta = \dfrac{-5\sqrt{3}}{-5} = \sqrt{3}$

$\theta = \dfrac{\pi}{3}, \dfrac{4\pi}{3}$

$\left(10, \dfrac{4\pi}{3}\right), \left(-10, \dfrac{\pi}{3}\right)$

21. $(x, y) = \left(\sqrt{7}, -\sqrt{7}\right)$

$r = \pm\sqrt{7 + 7} = \pm\sqrt{14}$

$\tan\theta = -\dfrac{\sqrt{7}}{\sqrt{7}} = -1$

$\theta = \dfrac{3\pi}{4}, \dfrac{7\pi}{4}$

$\left(\sqrt{14}, \dfrac{7\pi}{4}\right), \left(-\sqrt{14}, \dfrac{3\pi}{4}\right)$

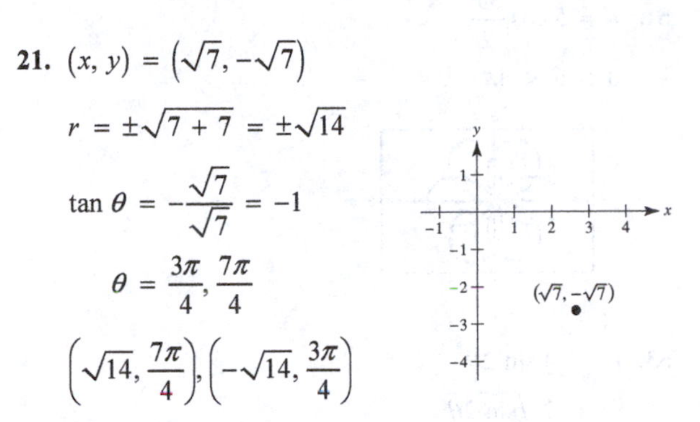

23. $(x, y) = (4, 5)$

$r = \pm\sqrt{4^2 + 5^2} = \pm\sqrt{41}$

$\tan\theta = \dfrac{5}{4}$

$\theta \approx 0.8961, 4.0376$

$\left(\sqrt{41}, 0.8961\right), \left(-\sqrt{41}, 4.0376\right)$

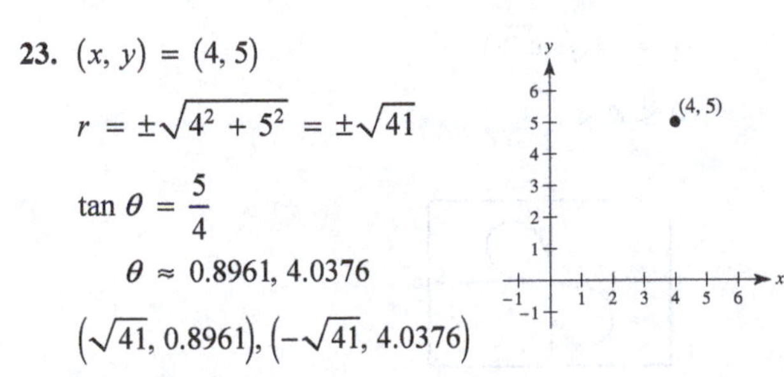

25. $x^2 + y^2 = 9$

$r^2 = 9$

$r = 3$

Circle

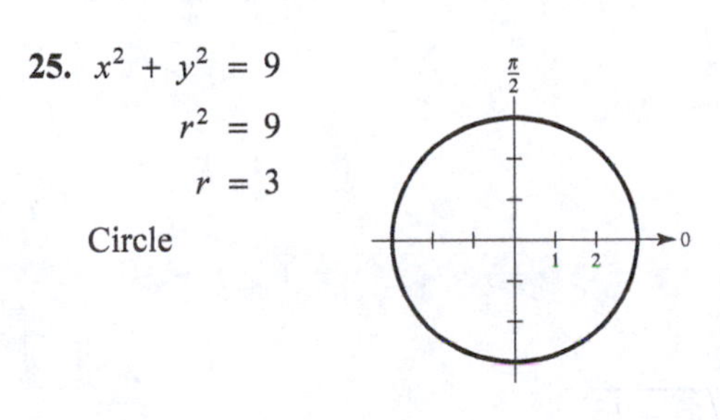

27. $x^2 + y^2 = a^2$

$r = a$

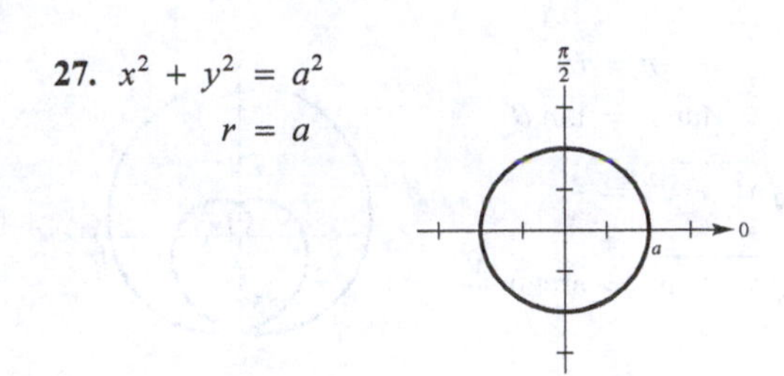

29. $y = 8$

$r\sin\theta = 8$

$r = 8\csc\theta$

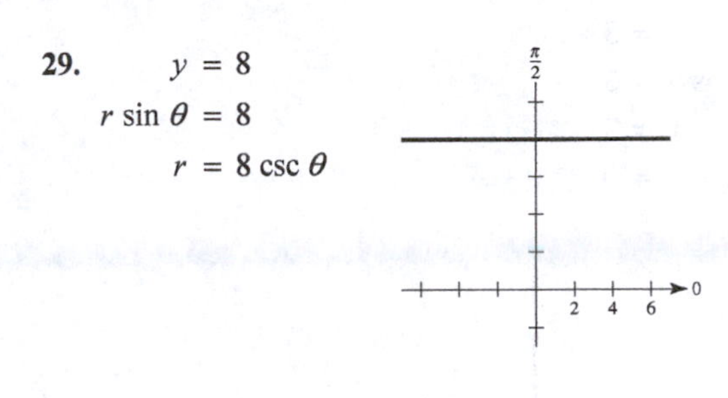

31. $3x - y + 2 = 0$

$3r\cos\theta - r\sin\theta + 2 = 0$

$r(3\cos\theta - \sin\theta) = -2$

$r = \dfrac{-2}{3\cos\theta - \sin\theta}$

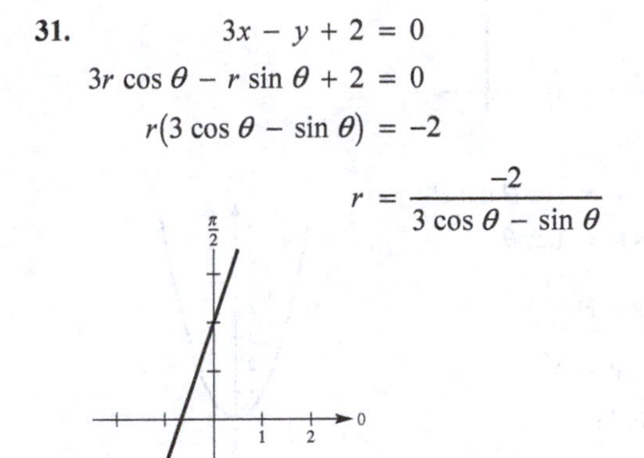

33. $y^2 = 9x$

$r^2\sin^2\theta = 9r\cos\theta$

$r = \dfrac{9\cos\theta}{\sin^2\theta}$

$r = 9\csc^2\theta\cos\theta$

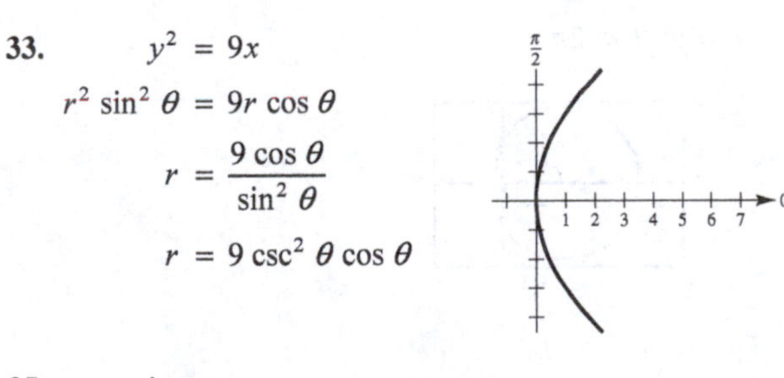

35. $r = 4$

$r^2 = 16$

$x^2 + y^2 = 16$

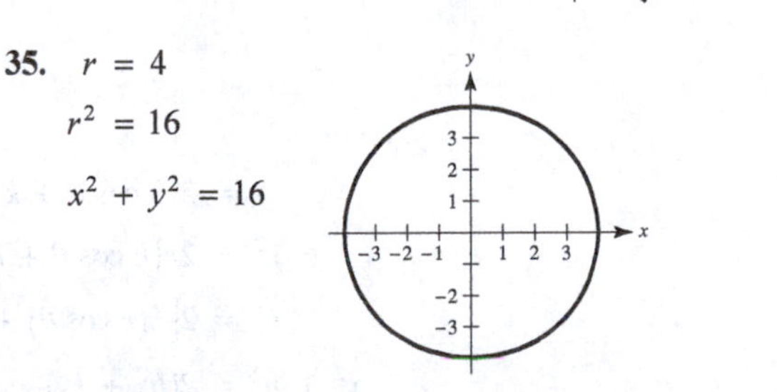

37. $r = 3\sin\theta$

$r^2 = 3r\sin\theta$

$x^2 + y^2 = 3y$

$x^2 + \left(y^2 - 3y + \frac{9}{4}\right) = \frac{9}{4}$

$x^2 + \left(y - \frac{3}{2}\right)^2 = \frac{9}{4}$

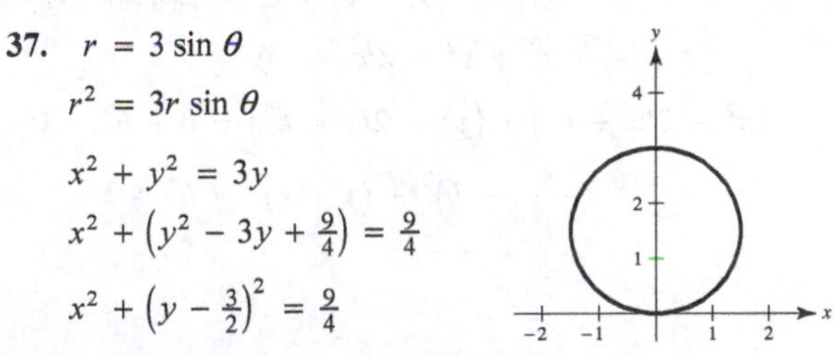

39.
$$\begin{aligned} r &= \theta \\ \tan r &= \tan\theta \\ \tan\sqrt{x^2+y^2} &= \frac{y}{x} \\ \sqrt{x^2+y^2} &= \arctan\frac{y}{x} \end{aligned}$$

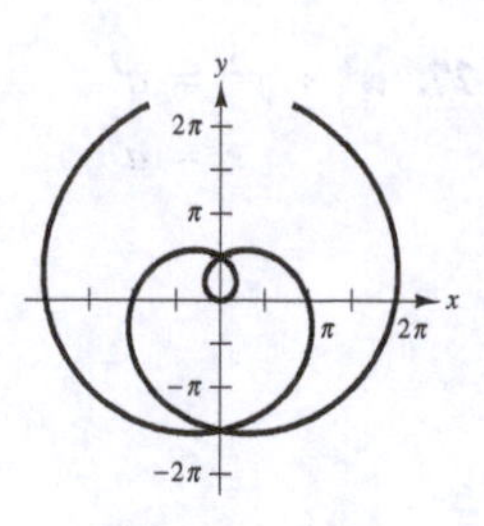

41.
$$\begin{aligned} r &= 3\sec\theta \\ r\cos\theta &= 3 \\ x &= 3 \\ x - 3 &= 0 \end{aligned}$$

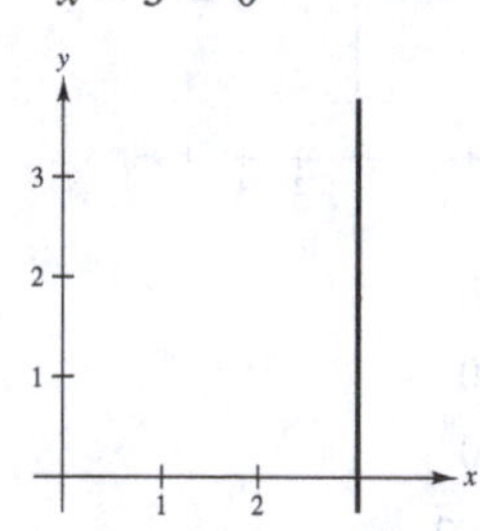

43.
$$\begin{aligned} r &= \sec\theta\tan\theta \\ r\cos\theta &= \tan\theta \\ x &= \frac{y}{x} \\ y &= x^2 \end{aligned}$$

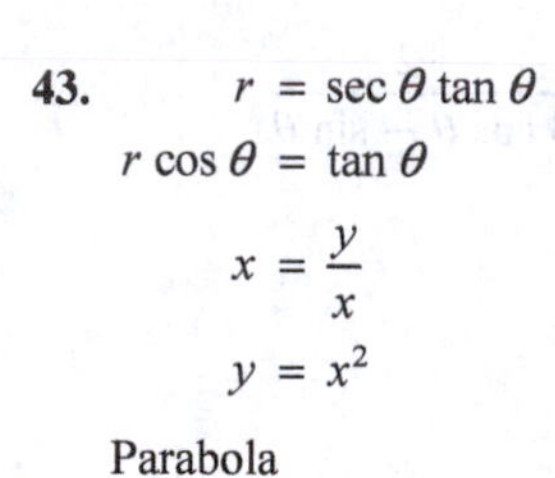

Parabola

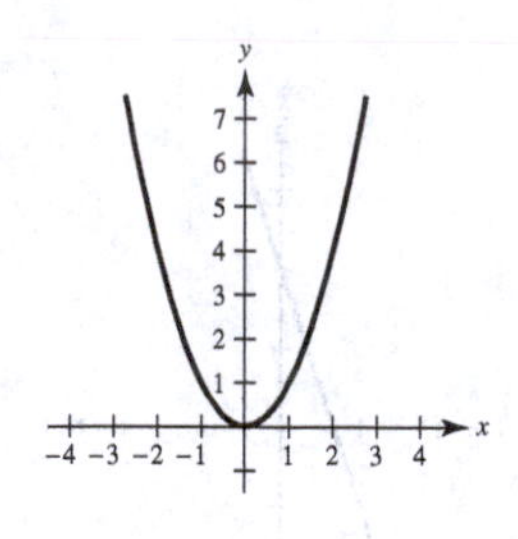

45. $r = 2 - 5\cos\theta$

$0 \le \theta < 2\pi$

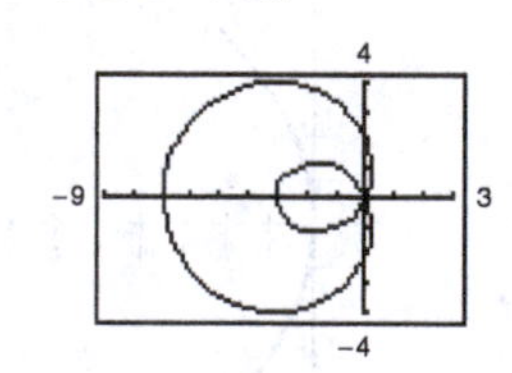

47. $r = -1 + \sin\theta$

$0 \le \theta < 2\pi$

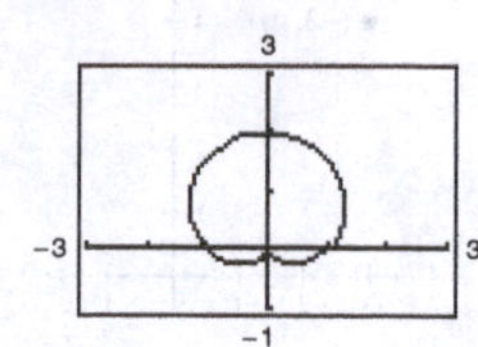

49. $r = \dfrac{2}{1 + \cos\theta}$

$-\pi < \theta < \pi$

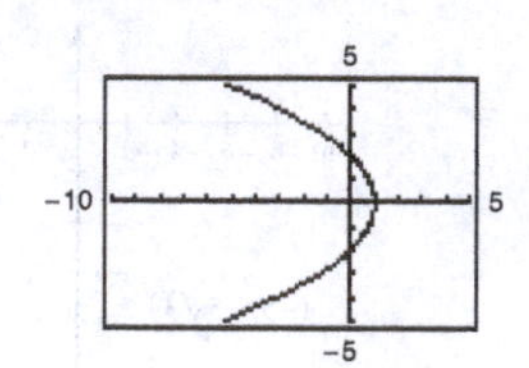

51. $r = 5\cos\dfrac{3\theta}{2}$

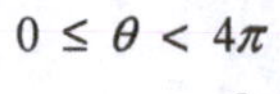

$0 \le \theta < 4\pi$

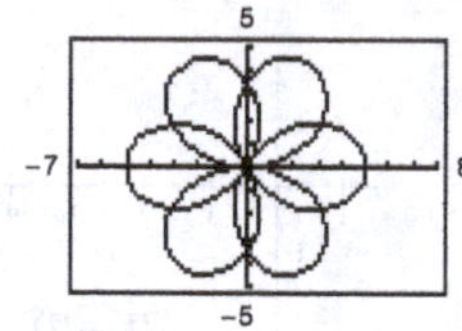

53.
$$\begin{aligned} r^2 &= 4\sin 2\theta \\ r_1 &= 2\sqrt{\sin 2\theta} \\ r_2 &= -2\sqrt{\sin 2\theta} \end{aligned}$$

$0 \le \theta < \dfrac{\pi}{2}$

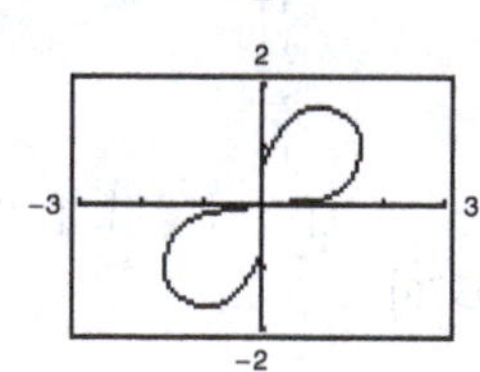

55.
$$\begin{aligned} r &= 2(h\cos\theta + k\sin\theta) \\ r^2 &= 2r(h\cos\theta + k\sin\theta) \\ r^2 &= 2\left[h(r\cos\theta) + k(r\sin\theta)\right] \\ x^2 + y^2 &= 2(hx + ky) \\ x^2 + y^2 - 2hx - 2ky &= 0 \\ \left(x^2 - 2hx + h^2\right) + \left(y^2 - 2ky + k^2\right) &= 0 + h^2 + k^2 \\ (x-h)^2 + (y-k)^2 &= h^2 + k^2 \end{aligned}$$

Radius: $\sqrt{h^2 + k^2}$

Center: (h, k)

57. $r = 4 \sin \theta$

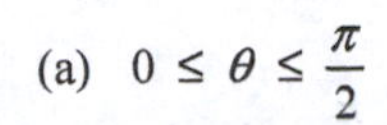

(a) $0 \le \theta \le \dfrac{\pi}{2}$

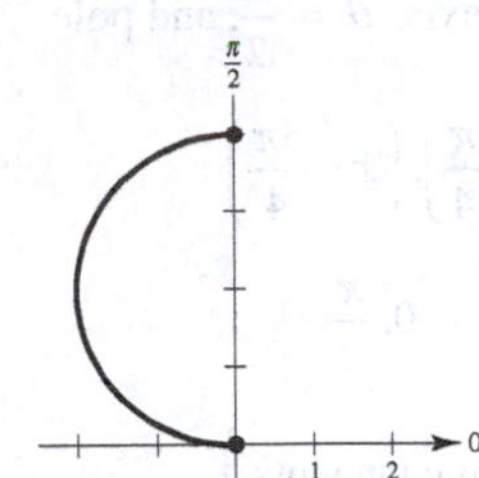

(b) $\dfrac{\pi}{2} \le \theta \le \pi$

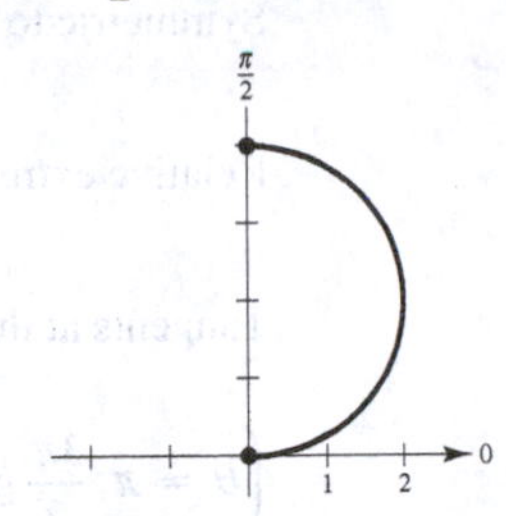

(c) $-\dfrac{\pi}{2} \le \theta \le \dfrac{\pi}{2}$

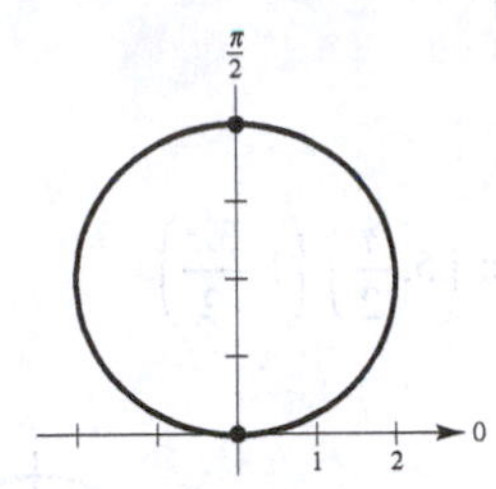

59. $\left(1, \dfrac{5\pi}{6}\right), \left(4, \dfrac{\pi}{3}\right)$

$$d = \sqrt{1^2 + 4^2 - 2(1)(4)\cos\left(\frac{5\pi}{6} - \frac{\pi}{3}\right)}$$

$$= \sqrt{17 - 8\cos\frac{\pi}{2}} = \sqrt{17}$$

61. $(2, 0.5), (7, 1.2)$

$$d = \sqrt{2^2 + 7^2 - 2(2)(7)\cos(0.5 - 1.2)}$$

$$= \sqrt{53 - 28\cos(-0.7)} \approx 5.6$$

63. $r = 2(1 - \sin\theta)$

$$\frac{dy}{dx} = \frac{-2\cos\theta\sin\theta + 2\cos\theta(1 - \sin\theta)}{-2\cos\theta\cos\theta - 2\sin\theta(1 - \sin\theta)}$$

At $(2, 0)$, $\dfrac{dy}{dx} = -1.$

At $\left(3, \dfrac{7\pi}{6}\right)$, $\dfrac{dy}{dx}$ is undefined.

At $\left(4, \dfrac{3\pi}{2}\right)$, $\dfrac{dy}{dx} = 0.$

65. (a), (b) $r = 3(1 - \cos\theta)$

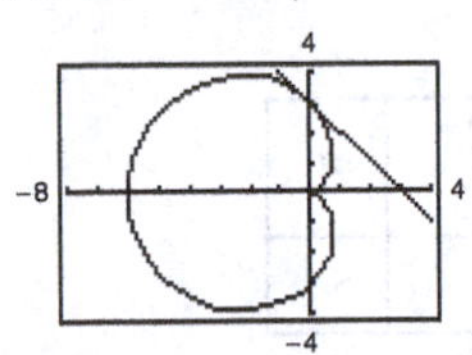

$(r, \theta) = \left(3, \dfrac{\pi}{2}\right) \Rightarrow (x, y) = (0, 3)$

Tangent line: $y - 3 = -1(x - 0)$

$$y = -x + 3$$

(c) At $\theta = \dfrac{\pi}{2}$, $\dfrac{dy}{dx} = -1.0.$

67. (a), (b) $r = 3\sin\theta$

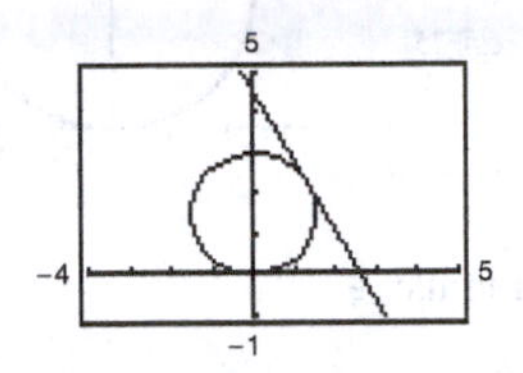

$(r, \theta) = \left(\dfrac{3\sqrt{3}}{2}, \dfrac{\pi}{3}\right) \Rightarrow (x, y) = \left(\dfrac{3\sqrt{3}}{4}, \dfrac{9}{4}\right)$

Tangent line: $y - \dfrac{9}{4} = -\sqrt{3}\left(x - \dfrac{3\sqrt{3}}{4}\right)$

$$y = -\sqrt{3}x + \frac{9}{2}$$

(c) At $\theta = \dfrac{\pi}{3}$, $\dfrac{dy}{dx} = -\sqrt{3} \approx -1.732.$

69. $r = 1 - \sin\theta$

$$\frac{dy}{d\theta} = (1 - \sin\theta)\cos\theta - \cos\theta\sin\theta$$

$$= \cos\theta(1 - 2\sin\theta) = 0$$

$\cos\theta = 0$ or $\sin\theta = \dfrac{1}{2} \Rightarrow \theta = \dfrac{\pi}{2}, \dfrac{3\pi}{2}, \dfrac{\pi}{6}, \dfrac{5\pi}{6}$

Horizontal tangents: $\left(2, \dfrac{3\pi}{2}\right), \left(\dfrac{1}{2}, \dfrac{\pi}{6}\right), \left(\dfrac{1}{2}, \dfrac{5\pi}{6}\right)$

$$\frac{dx}{d\theta} = (-1 + \sin\theta)\sin\theta - \cos\theta\cos\theta$$

$$= -\sin\theta + \sin^2\theta + \sin^2\theta - 1$$

$$= 2\sin^2\theta - \sin\theta - 1$$

$$= (2\sin\theta + 1)(\sin\theta - 1) = 0$$

$\sin\theta = 1$ or $\sin\theta = -\dfrac{1}{2} \Rightarrow \theta = \dfrac{\pi}{2}, \dfrac{7\pi}{6}, \dfrac{11\pi}{6}$

Vertical tangents: $\left(\dfrac{3}{2}, \dfrac{7\pi}{6}\right), \left(\dfrac{3}{2}, \dfrac{11\pi}{6}\right)$

71. $r = 2\csc\theta + 3$

$$\frac{dy}{d\theta} = (2\csc\theta + 3)\cos\theta + (-2\csc\theta\cot\theta)\sin\theta$$
$$= 3\cos\theta = 0$$

$$\theta = \frac{\pi}{2}, \frac{3\pi}{2}$$

Horizontal tangents: $\left(5, \frac{\pi}{2}\right), \left(1, \frac{3\pi}{2}\right)$

73. $r = 5\sin\theta$

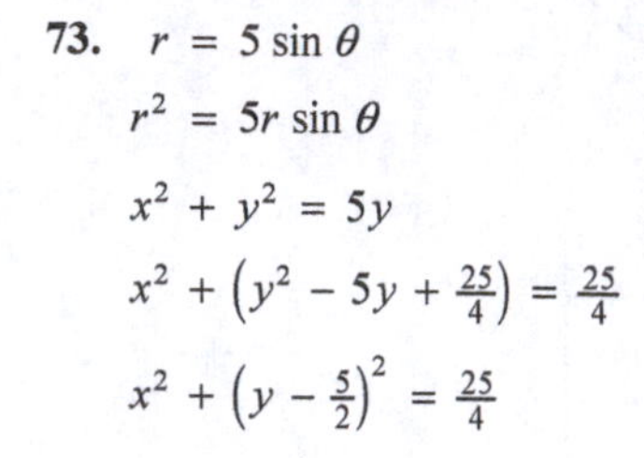

$$r^2 = 5r\sin\theta$$
$$x^2 + y^2 = 5y$$
$$x^2 + \left(y^2 - 5y + \tfrac{25}{4}\right) = \tfrac{25}{4}$$
$$x^2 + \left(y - \tfrac{5}{2}\right)^2 = \tfrac{25}{4}$$

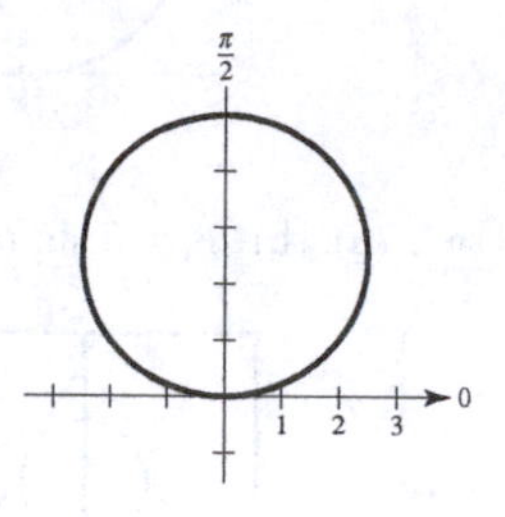

Circle: center: $\left(0, \frac{5}{2}\right)$, radius: $\frac{5}{2}$

Tangent at pole: $\theta = 0$

Note: $f(\theta) = r = 5\sin\theta$

$f(0) = 0, f'(0) \neq 0$

75. $r = 4(1 - \sin\theta)$, Cardioid

Symmetric to y-axis

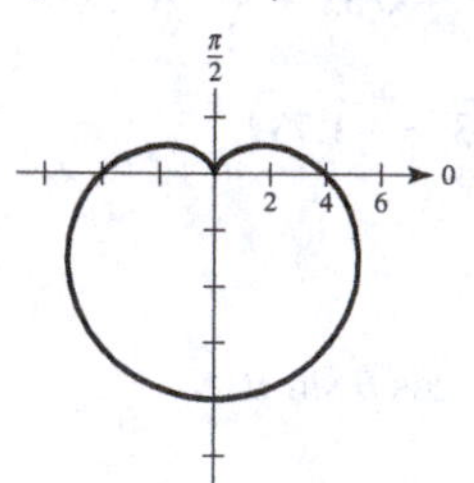

At $\theta = \frac{\pi}{2}, r = 0$

and $\frac{dr}{d\theta} = -4\cos\theta = 0$ at $\theta = \frac{\pi}{2}$.

No tangent at pole

77. $r = 4\cos 3\theta$

Rose curve with three petals.

Tangents at pole: $(r = 0, r' \neq 0)$:

$$\theta = \frac{\pi}{6}, \frac{\pi}{2}, \frac{5\pi}{6}$$

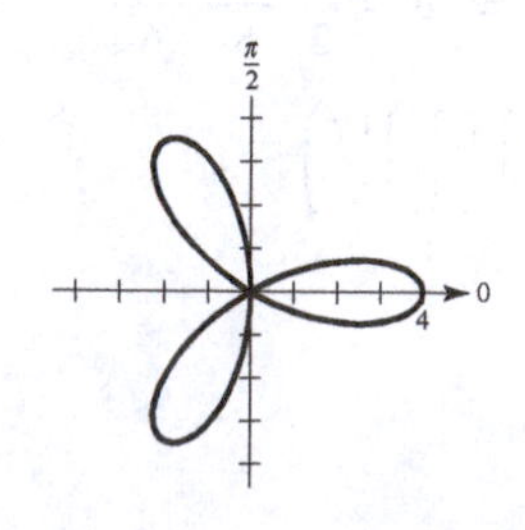

79. $r = 3\sin 2\theta$

Rose curve with four petals

Symmetric to the polar axis, $\theta = \frac{\pi}{2}$, and pole

Relative extrema: $\left(\pm 3, \frac{\pi}{4}\right), \left(\pm 3, \frac{5\pi}{4}\right)$

Tangents at the pole: $\theta = 0, \frac{\pi}{2}$

$\left(\theta = \pi, \frac{3\pi}{2}\right.$ give the same tangents.$\left.\right)$

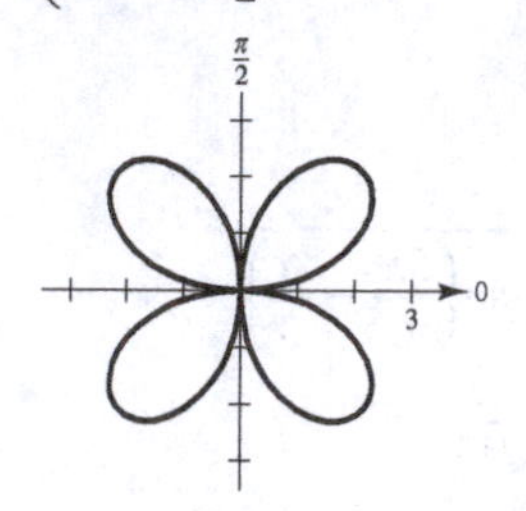

81. $r = 8$

Circle radius 8

$x^2 + y^2 = 64$

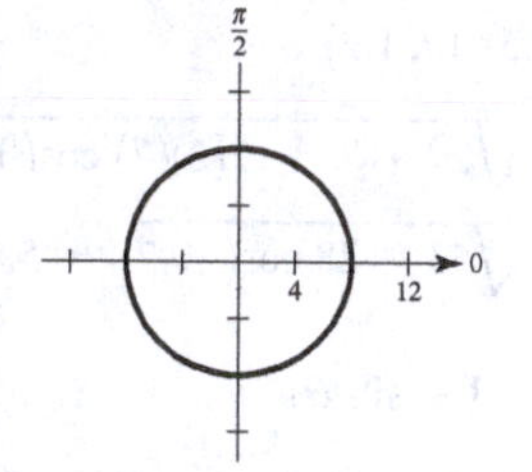

83. $r = 4(1 + \cos\theta)$

Cardioid

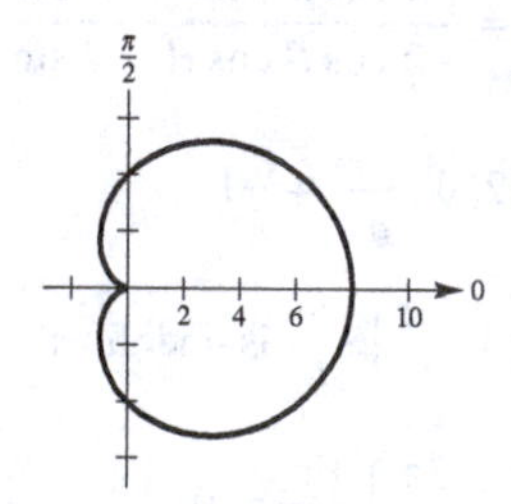

85. $r = 3 - 2\cos\theta$

Limaçon

Symmetric to polar axis

θ	0	$\frac{\pi}{3}$	$\frac{\pi}{2}$	$\frac{2\pi}{3}$	π
r	1	2	3	4	5

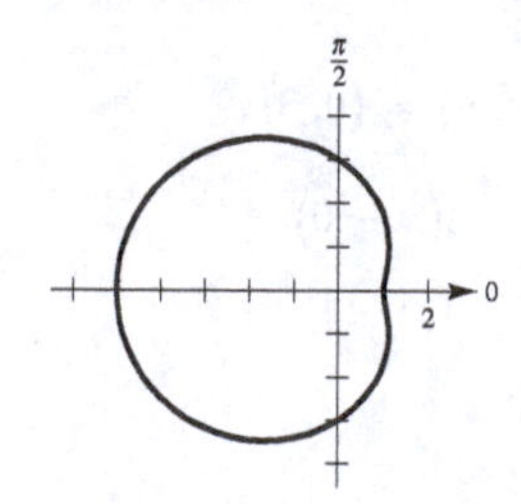

87. $r = -7 \csc \theta$

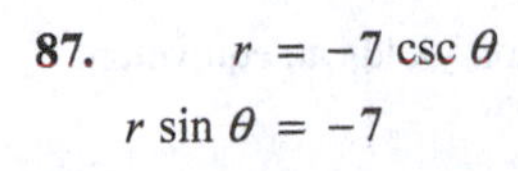

$r \sin \theta = -7$

$y = -7$

Horizontal line

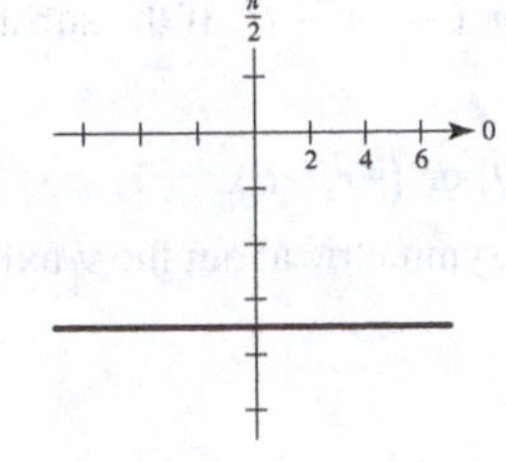

89. $r = 3\theta$

Spiral of Archimedes

Symmetric to $\theta = \dfrac{\pi}{2}$

θ	0	$\frac{\pi}{6}$	$\frac{\pi}{3}$	$\frac{\pi}{2}$	$\frac{2\pi}{3}$	$\frac{5\pi}{6}$	π
r	0	$\frac{\pi}{2}$	π	$\frac{3\pi}{2}$	2π	$\frac{5\pi}{2}$	3π

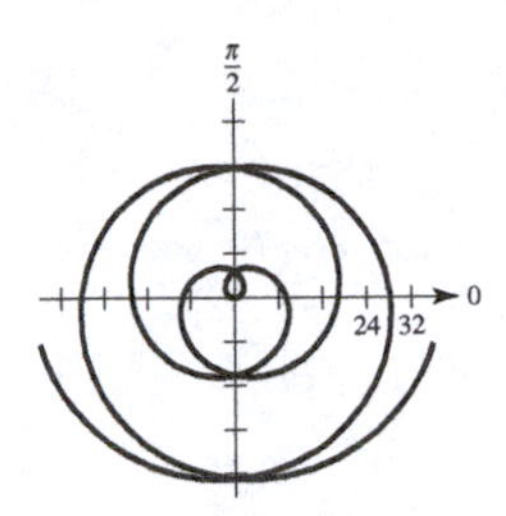

91. $r^2 = 4\cos(2\theta)$

$r = 2\sqrt{\cos 2\theta}, \quad 0 \le \theta \le 2\pi$

Lemniscate

Symmetric to the polar axis, $\theta = \dfrac{\pi}{2}$, and pole

Relative extrema: $(\pm 2, 0)$

θ	0	$\frac{\pi}{6}$	$\frac{\pi}{4}$
r	± 2	$\pm\sqrt{2}$	0

Tangents at the pole: $\theta = \dfrac{\pi}{4}, \dfrac{3\pi}{4}$

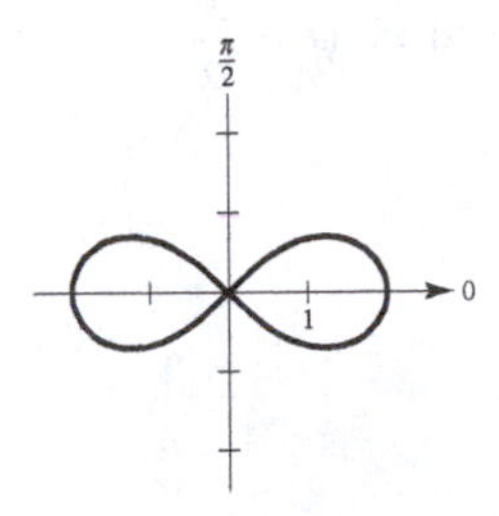

93. Because

$$r = 2 - \sec\theta = 2 - \frac{1}{\cos\theta},$$

the graph has polar axis symmetry and the tangents at the pole are

$$\theta = \frac{\pi}{3}, -\frac{\pi}{3}.$$

Furthermore,

$$r \Rightarrow -\infty \text{ as } \theta \Rightarrow \frac{\pi}{2}^-$$

$$r \Rightarrow \infty \text{ as } \theta \Rightarrow \frac{\pi}{2}^+.$$

Also,

$$r = 2 - \frac{1}{\cos\theta}$$

$$= 2 - \frac{r}{r\cos\theta} = 2 - \frac{r}{x}$$

$$rx = 2x - r$$

$$r = \frac{2x}{1 + x}.$$

So, $r \Rightarrow \pm\infty$ as $x \Rightarrow -1$.

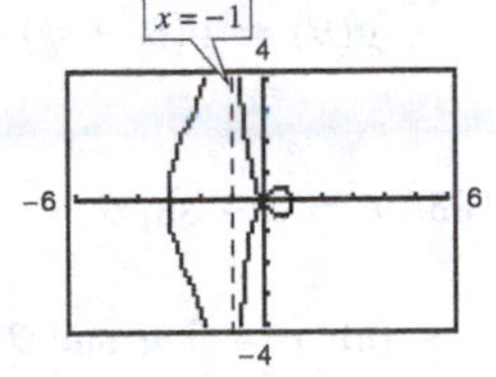

95. $r = \dfrac{2}{\theta}$

Hyperbolic spiral

$r \Rightarrow \infty$ as $\theta \Rightarrow 0$

$$r = \frac{2}{\theta} \Rightarrow \theta = \frac{2}{r} = \frac{2\sin\theta}{r\sin\theta} = \frac{2\sin\theta}{y}$$

$$y = \frac{2\sin\theta}{\theta}$$

$$\lim_{\theta\to 0} \frac{2\sin\theta}{\theta} = \lim_{\theta\to 0} \frac{2\cos\theta}{1} = 2$$

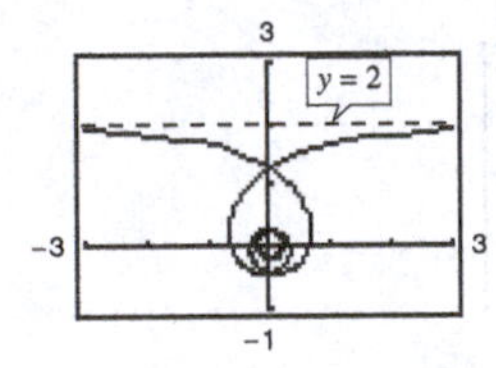

97. The graph is rotated by $\dfrac{\pi}{2}$ (90°).

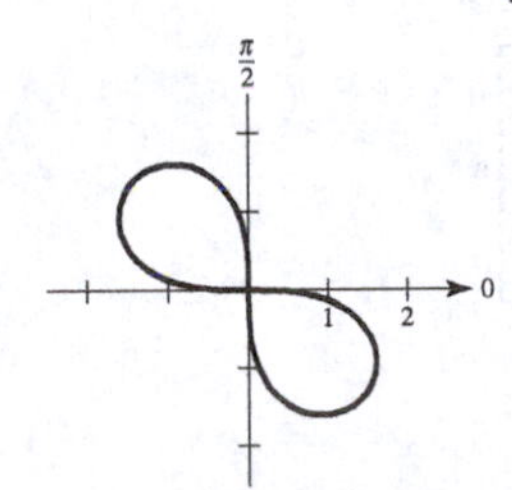

99. (a) To test for symmetry about the x-axis, replace (r, θ) by $(r, -\theta)$ or $(-r, \pi - \theta)$. If the substitution yields an equivalent equation, then the graph is symmetric about the x- axis.

(b) To test for symmetry about the y-axis, replace (r, θ) by $(r, \pi - \theta)$ or $(-r, -\theta)$. If the substitution yields an equivalent equation, then the graph is symmetric about the y-axis.

101. Let the curve $r = f(\theta)$ be rotated by ϕ to form the curve $r = g(\theta)$. If (r_1, θ_1) is a point on $r = f(\theta)$, then $(r_1, \theta_1 + \phi)$ is on $r = g(\theta)$. That is, $g(\theta_1 + \phi) = r_1 = f(\theta_1)$. Letting $\theta = \theta_1 + \phi$, or $\theta_1 = \theta - \phi$, you see that $g(\theta) = g(\theta_1 + \phi) = f(\theta_1) = f(\theta - \phi)$.

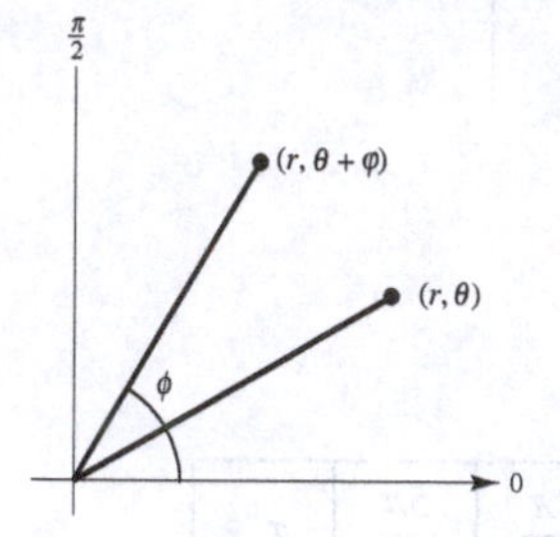

103. $r = 2 - \sin\theta$

(a) $r = 2 - \sin\left(\theta - \dfrac{\pi}{4}\right)$

$= 2 - \dfrac{\sqrt{2}}{2}(\sin\theta - \cos\theta)$

(b) $r = 2 - \sin\left(\theta - \dfrac{\pi}{2}\right) = 2 - (-\cos\theta) = 2 + \cos\theta$

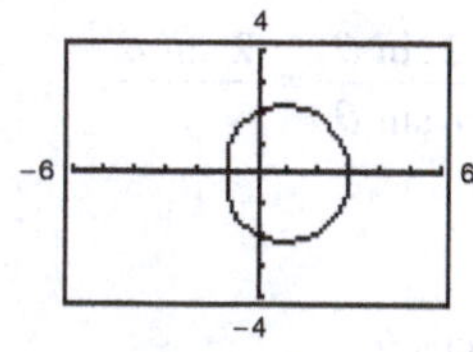

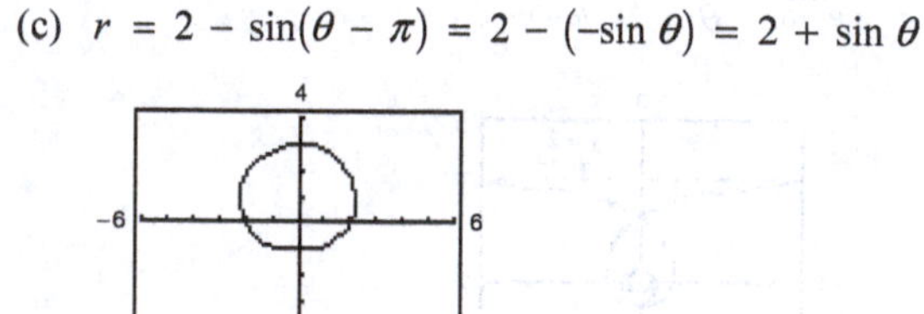

(c) $r = 2 - \sin(\theta - \pi) = 2 - (-\sin\theta) = 2 + \sin\theta$

(d) $r = 2 - \sin\left(\theta - \dfrac{3\pi}{2}\right) = 2 - \cos\theta$

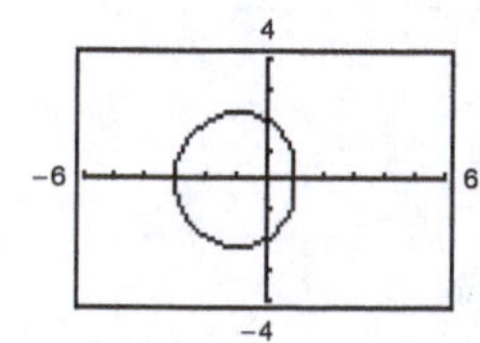

105. (a) $r = 1 - \sin\theta$

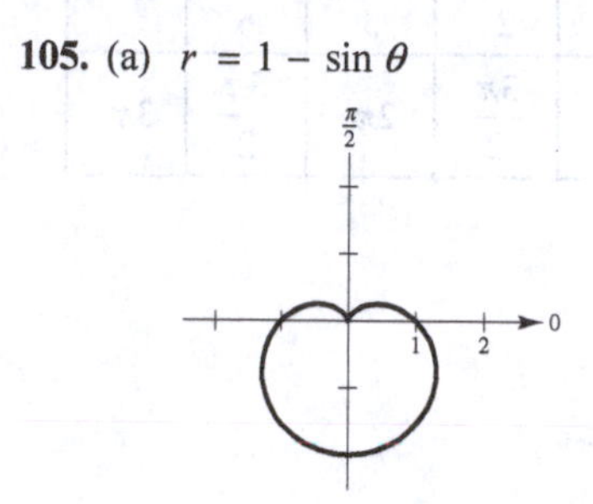

(b) $r = 1 - \sin\left(\theta - \dfrac{\pi}{4}\right)$

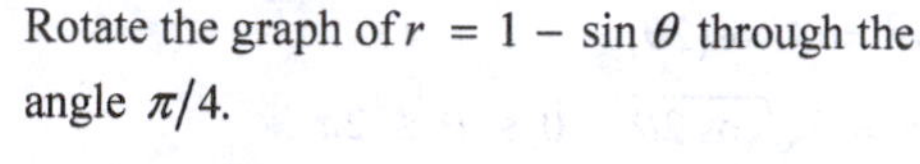

Rotate the graph of $r = 1 - \sin\theta$ through the angle $\pi/4$.

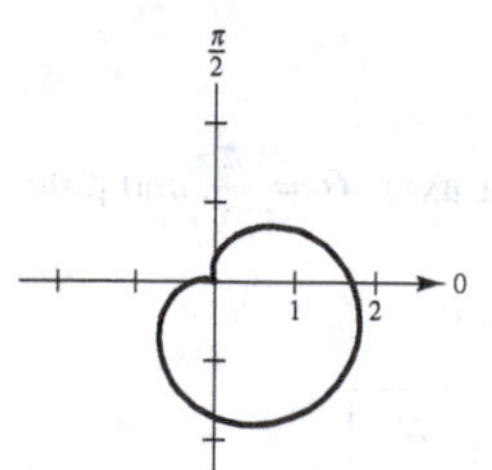

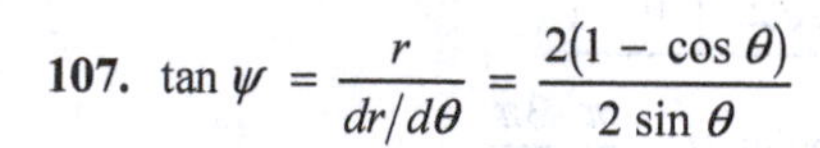

107. $\tan\psi = \dfrac{r}{dr/d\theta} = \dfrac{2(1 - \cos\theta)}{2\sin\theta}$

At $\theta = \pi$, $\tan\psi$ is undefined $\Rightarrow \psi = \dfrac{\pi}{2}$.

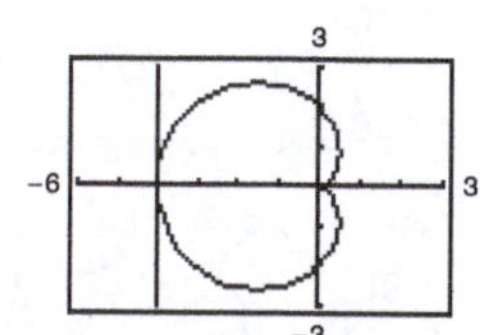

109. $r = 2\cos 3\theta$

$$\tan\psi = \frac{r}{dr/d\theta} = \frac{2\cos 3\theta}{-6\sin 3\theta} = -\frac{1}{3}\cot 3\theta$$

At $\theta = \frac{\pi}{4}$, $\tan\psi = -\frac{1}{3}\cot\left(\frac{3\pi}{4}\right) = \frac{1}{3}$.

$$\psi = \arctan\left(\frac{1}{3}\right) \approx 18.4°$$

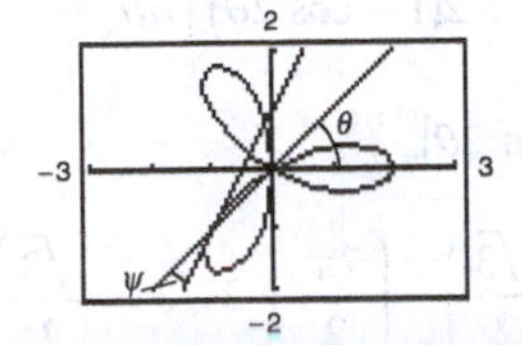

111. $r = \dfrac{6}{1-\cos\theta} = 6(1-\cos\theta)^{-1} \Rightarrow \dfrac{dr}{d\theta} = \dfrac{6\sin\theta}{(1-\cos\theta)^2}$

$$\tan\psi = \frac{r}{\frac{dr}{d\theta}} = \frac{\frac{6}{(1-\cos\theta)}}{\frac{-6\sin\theta}{(1-\cos\theta)^2}} = \frac{1-\cos\theta}{-\sin\theta}$$

At $\theta = \frac{2\pi}{3}$, $\tan\psi = \dfrac{1-\left(-\frac{1}{2}\right)}{-\frac{\sqrt{3}}{2}} = -\sqrt{3}$.

$\psi = \frac{\pi}{3}, (60°)$

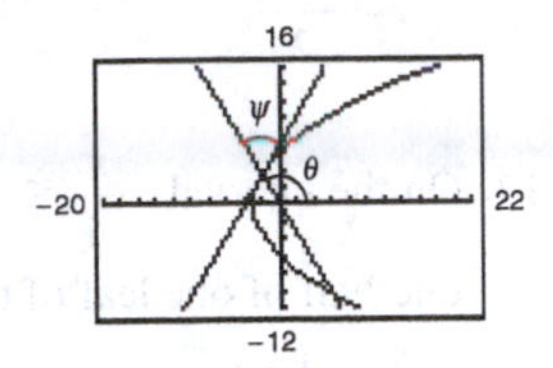

113. True

115. True

Section 10.5 Area and Arc Length in Polar Coordinates

1. You should check to see that f is continuous and either nonnegative or nonpositive on the interval of consideration.

3. $A = \frac{1}{2}\int_\alpha^\beta [f(\theta)]^2\, d\theta$

$$= \frac{1}{2}\int_0^{\pi/2} [4\sin\theta]^2\, d\theta = 8\int_0^{\pi/2} \sin^2\theta\, d\theta$$

5. $A = \frac{1}{2}\int_\alpha^\beta [f(\theta)]^2\, d\theta = \frac{1}{2}\int_{\pi/2}^{3\pi/2} [3 - 2\sin\theta]^2\, d\theta$

7. $A = \frac{1}{2}\int_0^\pi [6\sin\theta]^2\, d\theta$

$$= 18\int_0^\pi \frac{1-\cos 2\theta}{2}\, d\theta = 9\left[\theta - \frac{\sin 2\theta}{2}\right]_0^\pi = 9\pi$$

Note: $r = 6\sin\theta$ is circle of radius 3, $0 \le \theta \le \pi$.

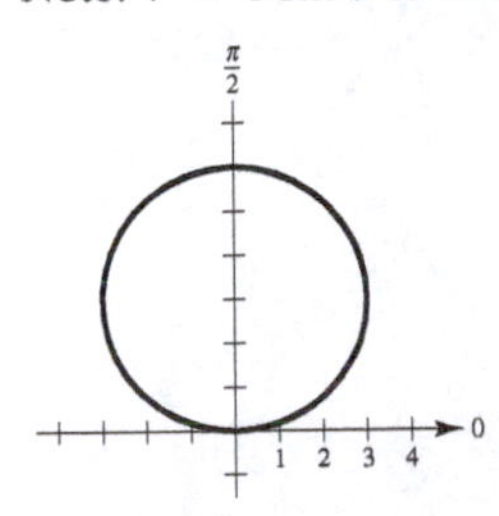

9. $A = 2\left[\frac{1}{2}\int_0^{\pi/6} (2\cos 3\theta)^2\, d\theta\right] = 2\left[\theta + \frac{1}{6}\sin 6\theta\right]_0^{\pi/6} = \frac{\pi}{3}$

11. $A = \frac{1}{2}\int_0^{\pi/4} [\sin 8\theta]^2\, d\theta$

$$= \frac{1}{2}\int_0^{\pi/4} \frac{1-\cos 16\theta}{2}\, d\theta$$

$$= \frac{1}{4}\left[\theta - \frac{\sin 16\theta}{2}\right]_0^{\pi/4}$$

$$= \frac{1}{4}\left[\frac{\pi}{4} - 0\right] = \frac{\pi}{16}$$

13. The region below the polar axis is traced out once for $\pi \le \theta \le 2\pi$.

$$A = \frac{1}{2}\int_\pi^{2\pi} [6 + 5\sin\theta]^2\, d\theta$$

$$= \frac{1}{2}\int_\pi^{2\pi} \left[36 + 60\sin\theta + 25\sin^2\theta\right] d\theta$$

$$= \frac{1}{2}\int_\pi^{2\pi} \left[36 + 60\sin\theta + 25\left(\frac{1-\cos 2\theta}{2}\right)\right] d\theta$$

$$= \frac{1}{2}\left[36\theta - 60\cos\theta + \frac{25\theta}{2} - \frac{25}{4}\sin 2\theta\right]_\pi^{2\pi}$$

$$= \frac{1}{2}\left[(72\pi - 60 + 25\pi) - \left(36\pi + 60 + \frac{25\pi}{2}\right)\right]$$

$$= \frac{97\pi}{4} - 60$$

15. $A = \frac{1}{2}\int_0^{2\pi} [4 + \sin\theta]^2\, d\theta$

$= \frac{1}{2}\int_0^{2\pi} \left[16 + 8\sin\theta + \sin^2\theta\right] d\theta$

$= \frac{1}{2}\int_0^{2\pi} \left[16 + 8\sin\theta + \frac{1 - \cos 2\theta}{2}\right] d\theta$

$= \frac{1}{2}\left[16\theta - 8\cos\theta + \frac{1}{2}\theta - \frac{\sin 2\theta}{4}\right]_0^{2\pi}$

$= \frac{1}{2}\left[(32\pi - 8 + \pi) - (-8)\right]$

$= \frac{33\pi}{2}$

17. On the interval $-\frac{\pi}{4} \le \theta \le 0$, $r = 2\sqrt{\cos 2\theta}$ traces out one-half of one leaf of the lemniscate. So,

$A = 4\frac{1}{2}\int_{-\pi/4}^{0} 4\cos 2\theta\, d\theta$

$= 8\left[\frac{\sin 2\theta}{2}\right]_{-\pi/4}^{0}$

$= 8\left[\frac{1}{2}\right]$

$= 4.$

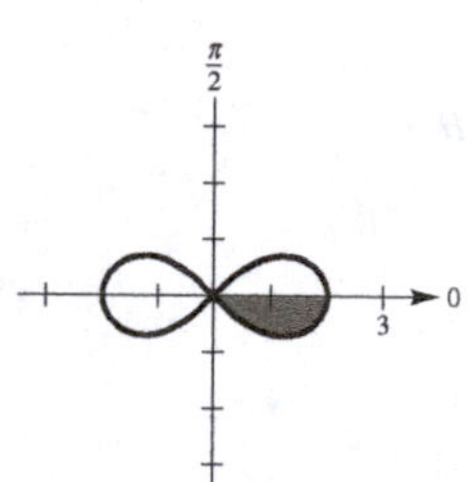

19. $A = \left[2\frac{1}{2}\int_{2\pi/3}^{\pi} (1 + 2\cos\theta)^2\, d\theta\right]$

$= \left[3\theta + 4\sin\theta + \sin 2\theta\right]_{2\pi/3}^{\pi}$

$= \frac{2\pi - 3\sqrt{3}}{2}$

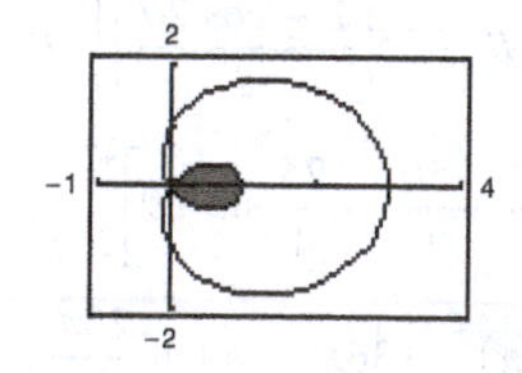

21. The inner loop of $r = 1 + 2\sin\theta$ is traced out on the interval $\frac{7\pi}{6} \le \theta \le \frac{11\pi}{6}$. So,

$A = \frac{1}{2}\int_{7\pi/6}^{11\pi/6} [1 + 2\sin\theta]^2\, d\theta$

$= \frac{1}{2}\int_{7\pi/6}^{11\pi/6} \left[1 + 4\sin\theta + 4\sin^2\theta\right] d\theta$

$= \frac{1}{2}\int_{7\pi/6}^{11\pi/6} \left[1 + 4\sin\theta + 2(1 - \cos 2\theta)\right] d\theta$

$= \frac{1}{2}\left[3\theta - 4\cos\theta - \sin 2\theta\right]_{7\pi/6}^{11\pi/6}$

$= \frac{1}{2}\left[\left(\frac{11\pi}{2} - 2\sqrt{3} + \frac{\sqrt{3}}{2}\right) - \left(\frac{7\pi}{2} + 2\sqrt{3} - \frac{\sqrt{3}}{2}\right)\right]$

$= \frac{1}{2}\left[2\pi - 3\sqrt{3}\right].$

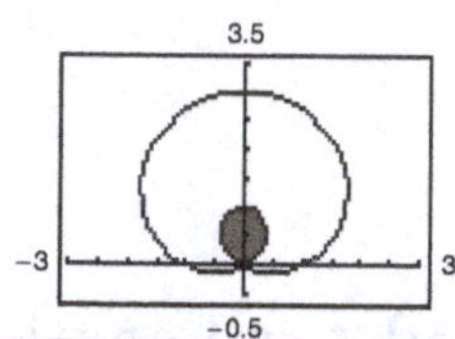

23. The area inside the outer loop is

$2\left[\frac{1}{2}\int_0^{2\pi/3} (1 + 2\cos\theta)^2\, d\theta\right] = \left[3\theta + 4\sin\theta + \sin 2\theta\right]_0^{2\pi/3}$

$= \frac{4\pi + 3\sqrt{3}}{2}.$

From the result of Exercise 17, the area between the loops is

$A = \left(\frac{4\pi + 3\sqrt{3}}{2}\right) - \left(\frac{2\pi - 3\sqrt{3}}{2}\right) = \pi + 3\sqrt{3}.$

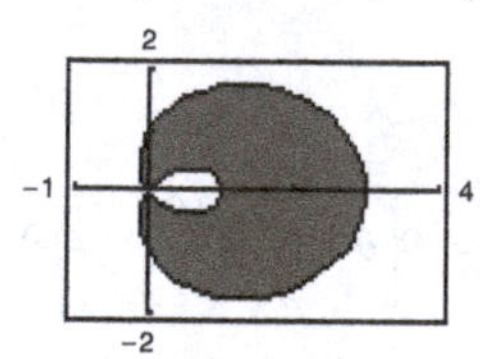

25. The area inside the outer loop is

$$A = 2 \cdot \frac{1}{2}\int_{5\pi/6}^{3\pi/2} [3 - 6\sin\theta]^2\, d\theta$$
$$= \int_{5\pi/6}^{3\pi/2} \left[9 - 36\sin\theta + 36\sin^2\theta\right] d\theta$$
$$= \int_{5\pi/6}^{3\pi/2} \left[9 - 36\sin\theta + 18(1 - \cos 2\theta)\right] d\theta$$
$$= \left[27\theta + 36\cos\theta - 9\sin 2\theta\right]_{5\pi/6}^{3\pi/2} = \left[\frac{81\pi}{2} - \left(\frac{45\pi}{2} - 18\sqrt{3} + \frac{9\sqrt{3}}{2}\right)\right] = 18\pi + \frac{27\sqrt{3}}{2}.$$

The area inside the inner loop is

$$A = 2 \cdot \frac{1}{2}\int_{\pi/6}^{\pi/2} [3 - 6\sin\theta]^2\, d\theta$$
$$= \left[27\theta + 36\cos\theta - 9\sin 2\theta\right]_{\pi/6}^{\pi/2} = \left[\frac{27\pi}{2} - \left(\frac{9\pi}{2} + 18\sqrt{3} - \frac{9\sqrt{3}}{2}\right)\right] = 9\pi - \frac{27\sqrt{3}}{2}.$$

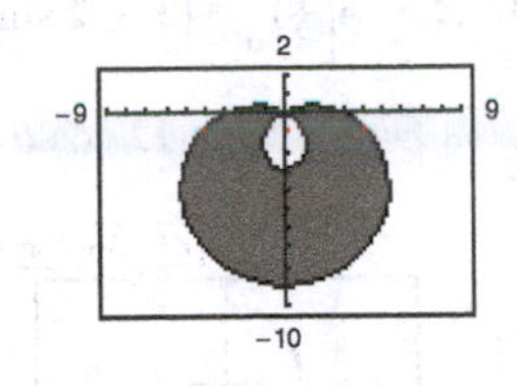

Finally, the area between the loops is $\left[18\pi + \frac{27\sqrt{3}}{2}\right] - \left[9\pi - \frac{27\sqrt{3}}{2}\right] = 9\pi + 27\sqrt{3}.$

27. $r = 1 + \cos\theta$

$r = 1 - \cos\theta$

Solving simultaneously,

$1 + \cos\theta = 1 - \cos\theta$

$2\cos\theta = 0$

$\theta = \frac{\pi}{2}, \frac{3\pi}{2}.$

Replacing r by $-r$ and θ by $\theta + \pi$ in the first equation and solving, $-1 + \cos\theta = 1 - \cos\theta$, $\cos\theta = 1$, $\theta = 0$. Both curves pass through the pole, $(0, \pi)$, and $(0, 0)$, respectively.

Points of intersection: $\left(1, \frac{\pi}{2}\right), \left(1, \frac{3\pi}{2}\right), (0, 0)$

29. $r = 1 + \cos\theta$

$r = 1 - \sin\theta$

Solving simultaneously,

$1 + \cos\theta = 1 - \sin\theta$

$\cos\theta = -\sin\theta$

$\tan\theta = -1$

$\theta = \frac{3\pi}{4}, \frac{7\pi}{4}.$

Replacing r by $-r$ and θ by $\theta + \pi$ in the first equation and solving, $-1 + \cos\theta = 1 - \sin\theta$, $\sin\theta + \cos\theta = 2$, which has no solution. Both curves pass through the pole, $(0, \pi)$, and $(0, \pi/2)$, respectively.

Points of intersection:

$\left(\frac{2 - \sqrt{2}}{2}, \frac{3\pi}{4}\right), \left(\frac{2 + \sqrt{2}}{2}, \frac{7\pi}{4}\right), (0, 0)$

31. $r = 4 - 5\sin\theta$

$r = 3\sin\theta$

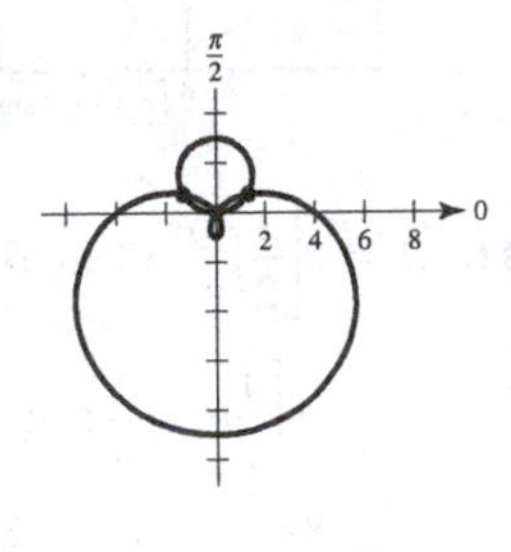

Solving simultaneously,

$4 - 5\sin\theta = 3\sin\theta$

$\sin\theta = \frac{1}{2}$

$\theta = \frac{\pi}{6}, \frac{5\pi}{6}.$

Both curves pass through the pole, $(0, \arcsin 4/5)$, and $(0, 0)$, respectively.

Points of intersection: $\left(\frac{3}{2}, \frac{\pi}{6}\right), \left(\frac{3}{2}, \frac{5\pi}{6}\right), (0, 0)$

33. $r = \frac{\theta}{2}$

$r = 2$

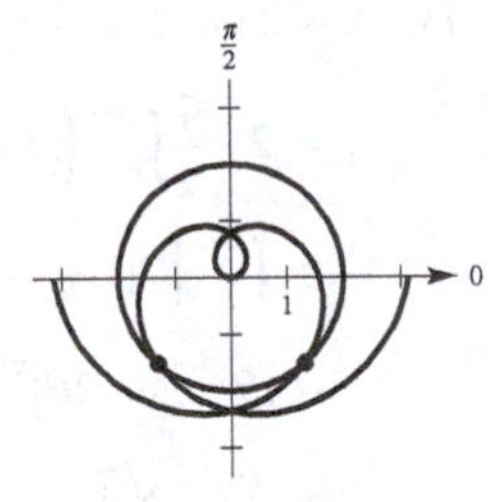

Solving simultaneously, you have

$\theta/2 = 2, \theta = 4.$

Points of intersection:

$(2, 4), (-2, -4)$

35. $r = \cos\theta$

$r = 2 - 3\sin\theta$

Points of intersection:

$(0, 0), (0.935, 0.363),$

$(0.535, -1.006)$

The graphs reach the pole at different times (θ values).

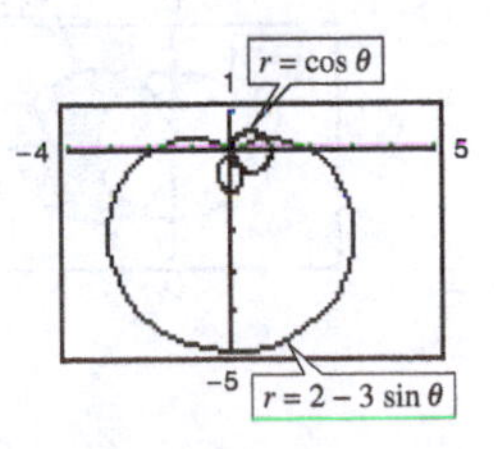

37. The points of intersection for one petal are $(2, \pi/12)$ and $(2, 5\pi/12)$. The area within one petal is

$$A = \frac{1}{2}\int_0^{\pi/12} (4 \sin 2\theta)^2 \, d\theta + \frac{1}{2}\int_{\pi/12}^{5\pi/12} (2)^2 \, d\theta + \frac{1}{2}\int_{5\pi/12}^{\pi/2} (4 \sin 2\theta)^2 \, d\theta$$

$$= 16\int_0^{\pi/12} \sin^2(2\theta) \, d\theta + 2\int_{\pi/12}^{5\pi/12} d\theta \text{ (by symmetry of the petal)}$$

$$= 8\left[\theta - \frac{1}{4} \sin 4\theta\right]_0^{\pi/12} + \left[2\theta\right]_{\pi/12}^{5\pi/12} = \frac{4\pi}{3} - \sqrt{3}.$$

$$\text{Total area} = 4\left(\frac{4\pi}{3} - \sqrt{3}\right) = \frac{16\pi}{3} - 4\sqrt{3} = \frac{4}{3}\left(4\pi - 3\sqrt{3}\right)$$

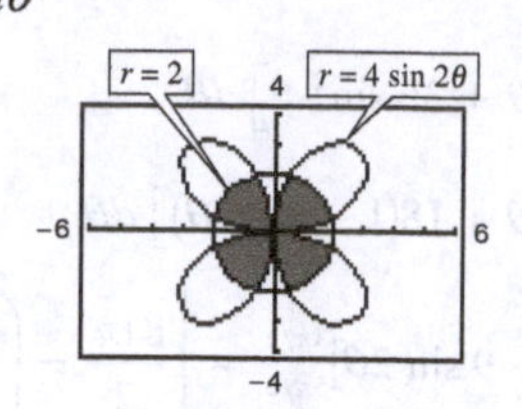

39. $A = 4\left[\frac{1}{2}\int_0^{\pi/2} (3 - 2 \sin \theta)^2 \, d\theta\right]$

$$= 2\left[11\theta + 12 \cos \theta - \sin(2\theta)\right]_0^{\pi/2} = 11\pi - 24$$

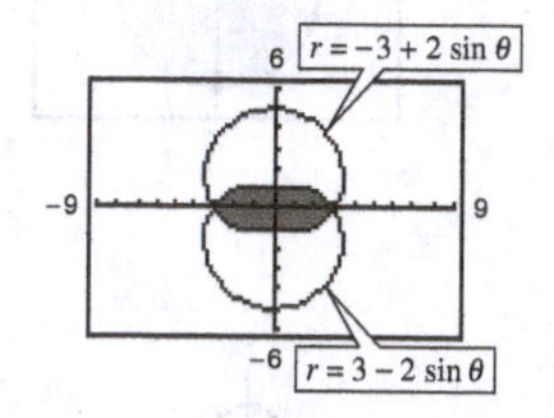

41. $A = 2\left[\frac{1}{2}\int_0^{\pi/6} (4 \sin \theta)^2 \, d\theta + \frac{1}{2}\int_{\pi/6}^{\pi/2} (2)^2 \, d\theta\right]$

$$= 16\left[\frac{1}{2}\theta - \frac{1}{4} \sin(2\theta)\right]_0^{\pi/6} + \left[4\theta\right]_{\pi/6}^{\pi/2}$$

$$= \frac{8\pi}{3} - 2\sqrt{3} = \frac{2}{3}\left(4\pi - 3\sqrt{3}\right)$$

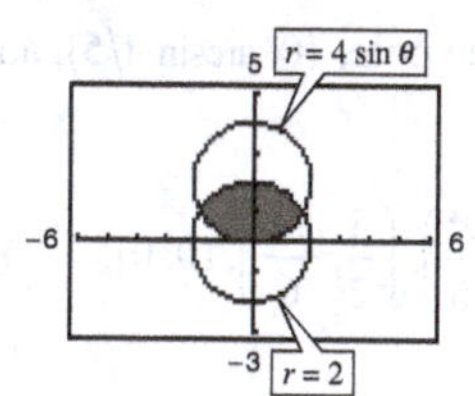

43. $r = 2 \cos \theta = 1 \Rightarrow \theta = \pi/3$

$$A = 2 \cdot \frac{1}{2}\int_0^{\pi/3} \left(\left[2 \cos \theta\right]^2 - 1\right) d\theta$$

$$= \int_0^{\pi/3} \left[2(1 + \cos 2\theta) - 1\right] d\theta$$

$$= \left[\theta + \sin 2\theta\right]_0^{\pi/3}$$

$$= \frac{\pi}{3} + \frac{\sqrt{3}}{2}$$

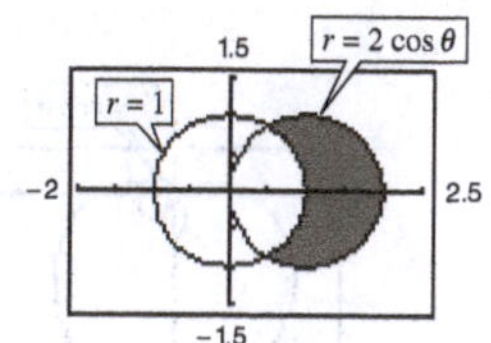

45. $A = 2\left[\frac{1}{2}\int_0^{\pi} \left[a(1 + \cos \theta)\right]^2 \, d\theta\right] - \frac{a^2\pi}{4}$

$$= a^2\left[\frac{3}{2}\theta + 2 \sin \theta + \frac{\sin 2\theta}{4}\right]_0^{\pi} - \frac{a^2\pi}{4}$$

$$= \frac{3a^2\pi}{2} - \frac{a^2\pi}{4} = \frac{5a^2\pi}{4}$$

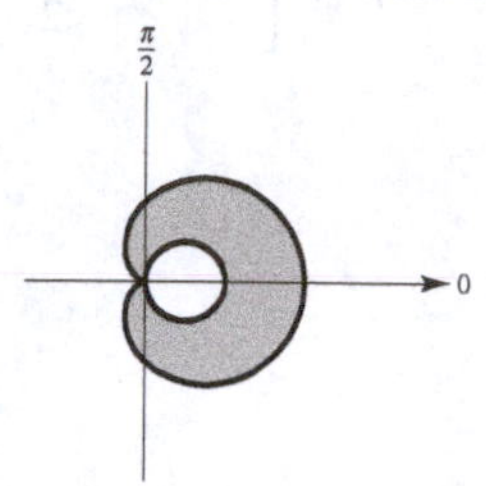

47. $A = \frac{\pi a^2}{8} + \frac{1}{2}\int_{\pi/2}^{\pi} \left[a(1 + \cos \theta)\right]^2 \, d\theta$

$$= \frac{\pi a^2}{8} + \frac{a^2}{2}\int_{\pi/2}^{\pi} \left(\frac{3}{2} + 2 \cos \theta + \frac{\cos 2\theta}{2}\right) d\theta$$

$$= \frac{\pi a^2}{8} + \frac{a^2}{2}\left[\frac{3}{2}\theta + 2 \sin \theta + \frac{\sin 2\theta}{4}\right]_{\pi/2}^{\pi}$$

$$= \frac{\pi a^2}{8} + \frac{a^2}{2}\left[\frac{3\pi}{2} - \frac{3\pi}{4} - 2\right] = \frac{a^2}{2}\left[\pi - 2\right]$$

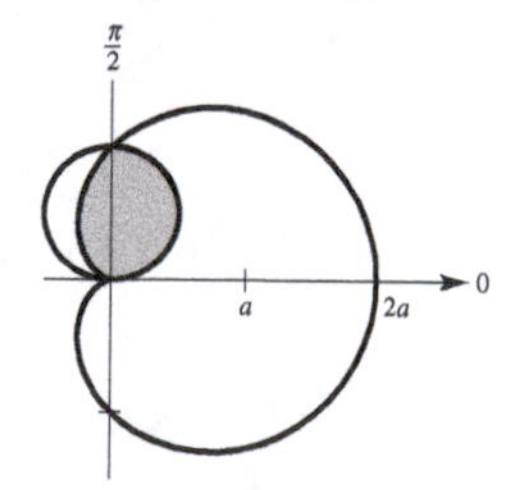

49. (a) $r = a\cos^2\theta$

$r^3 = ar^2\cos^2\theta$

$(x^2 + y^2)^{3/2} = ax^2$

(b)

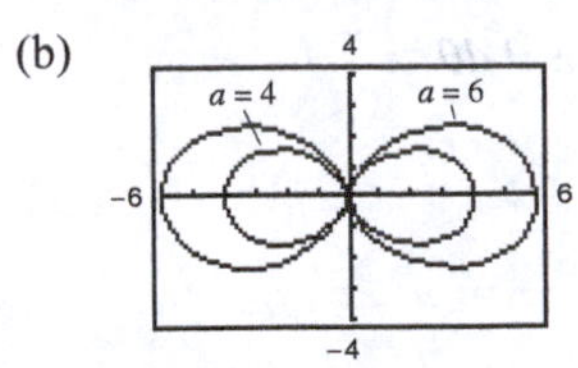

(c) $A = 4\left(\frac{1}{2}\right)\int_0^{\pi/2}\left[(6\cos^2\theta)^2 - (4\cos^2\theta)^2\right]d\theta$

$= 40\int_0^{\pi/2}\cos^4\theta\,d\theta$

$= 10\int_0^{\pi/2}(1 + \cos 2\theta)^2\,d\theta$

$= 10\int_0^{\pi/2}\left(1 + 2\cos 2\theta + \frac{1 - \cos 4\theta}{2}\right)d\theta$

$= 10\left[\frac{3}{2}\theta + \sin 2\theta + \frac{1}{8}\sin 4\theta\right]_0^{\pi/2} = \frac{15\pi}{2}$

51. $r = a\cos(n\theta)$

For $n = 1$:

$r = a\cos\theta$

$A = \pi\left(\frac{a}{2}\right)^2 = \frac{\pi a^2}{4}$

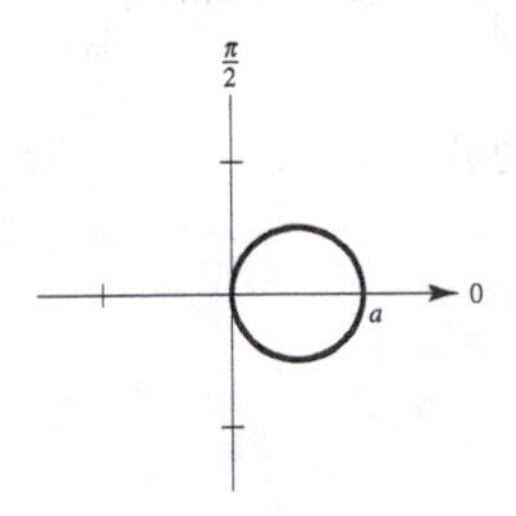

For $n = 2$:

$r = a\cos 2\theta$

$A = 8\left(\frac{1}{2}\right)\int_0^{\pi/4}(a\cos 2\theta)^2\,d\theta = \frac{\pi a^2}{2}$

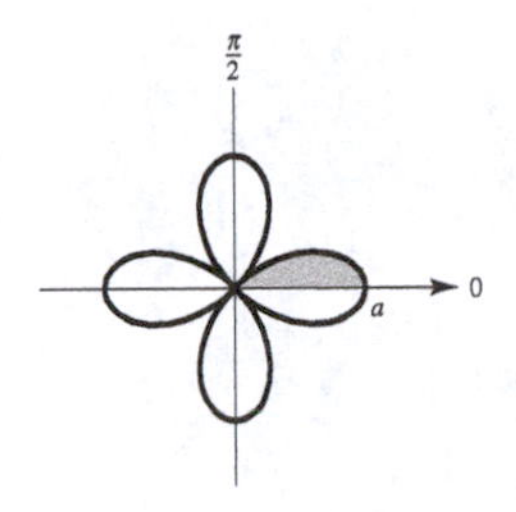

For $n = 3$:

$r = a\cos 3\theta$

$A = 6\left(\frac{1}{2}\right)\int_0^{\pi/6}(a\cos 3\theta)^2\,d\theta = \frac{\pi a^2}{4}$

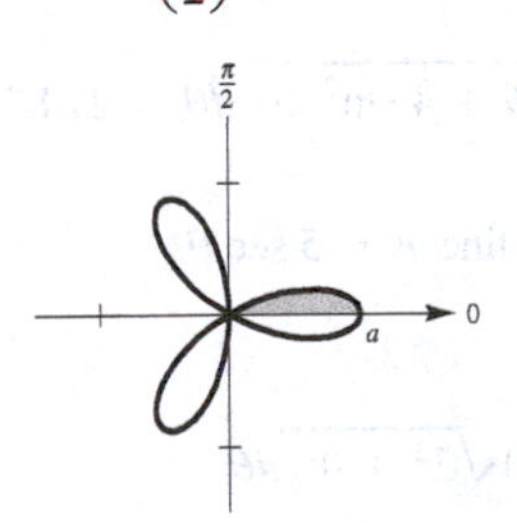

For $n = 4$:

$r = a\cos 4\theta$

$A = 16\left(\frac{1}{2}\right)\int_0^{\pi/8}(a\cos 4\theta)^2\,d\theta = \frac{\pi a^2}{2}$

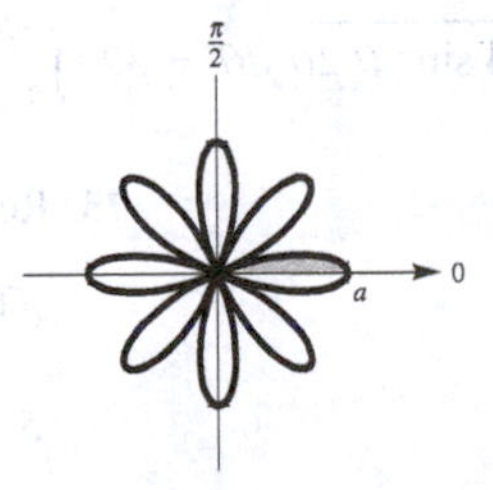

53. $r = 8, r^1 = 0$

$s = \int_0^{\pi/6}\sqrt{8^2 + 0^2}\,d\theta = 8\theta\Big]_0^{\pi/6} = \frac{4\pi}{3}$

(circumference of circle of radius 8)

55. $r = 4\sin\theta$

$r' = 4\cos\theta$

$s = \int_0^{\pi}\sqrt{(4\sin\theta)^2 + (4\cos\theta)^2}\,d\theta$

$= \int_0^{\pi}4\,d\theta = [4\theta]_0^{\pi} = 4\pi$

(circumference of circle of radius 2)

57. $r = 1 + \sin\theta$

$r' = \cos\theta$

$s = 2\int_{\pi/2}^{3\pi/2}\sqrt{(1 + \sin\theta)^2 + (\cos\theta)^2}\,d\theta$

$= 2\sqrt{2}\int_{\pi/2}^{3\pi/2}\sqrt{1 + \sin\theta}\,d\theta$

$= 2\sqrt{2}\int_{\pi/2}^{3\pi/2}\frac{-\cos\theta}{\sqrt{1 - \sin\theta}}\,d\theta$

$= \left[4\sqrt{2}\sqrt{1 - \sin\theta}\right]_{\pi/2}^{3\pi/2}$

$= 4\sqrt{2}(\sqrt{2} - 0) = 8$

59. $r = 2\theta, \left[\theta, \dfrac{\pi}{2}\right]$

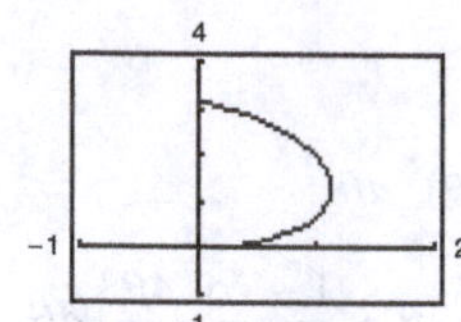

Length ≈ 4.16

61. $r = \dfrac{1}{\theta}, [\pi, 2\pi]$

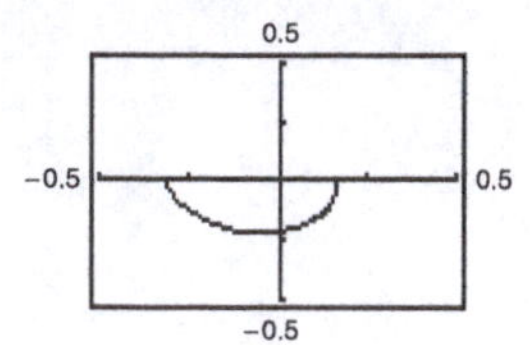

Length ≈ 0.71

63. $r = \sin(3\cos\theta), [0, \pi]$

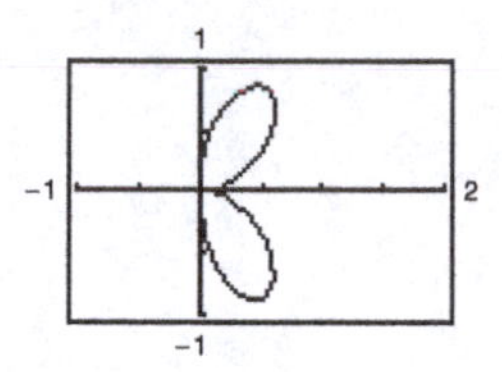

Length ≈ 4.39

65. $r = 6\cos\theta$

$r' = -6\sin\theta$

$$S = 2\pi\int_0^{\pi/2} 6\cos\theta\sin\theta\sqrt{36\cos^2\theta + 36\sin^2\theta}\,d\theta$$

$$= 72\pi\int_0^{\pi/2} \sin\theta\cos\theta\,d\theta$$

$$= \left[36\pi\sin^2\theta\right]_0^{\pi/2}$$

$$= 36\pi$$

67. $r = e^{a\theta}$

$r' = ae^{a\theta}$

$$S = 2\pi\int_0^{\pi/2} e^{a\theta}\cos\theta\sqrt{\left(e^{a\theta}\right)^2 + \left(ae^{a\theta}\right)^2}\,d\theta$$

$$= 2\pi\sqrt{1 + a^2}\int_0^{\pi/2} e^{2a\theta}\cos\theta\,d\theta$$

$$= 2\pi\sqrt{1 + a^2}\left[\frac{e^{2a\theta}}{4a^2 + 1}(2a\cos\theta + \sin\theta)\right]_0^{\pi/2}$$

$$= \frac{2\pi\sqrt{1 + a^2}}{4a^2 + 1}\left(e^{\pi a} - 2a\right)$$

69. $r = 4\cos 2\theta$

$r' = -8\sin 2\theta$

$$S = 2\pi\int_0^{\pi/4} 4\cos 2\theta\sin\theta\sqrt{16\cos^2 2\theta + 64\sin^2\theta\, 2\theta}\,d\theta = 32\pi\int_0^{\pi/4} \cos 2\theta\sin\theta\sqrt{\cos^2 2\theta + 4\sin^2 2\theta}\,d\theta \approx 21.87$$

71. $r = 10\cos\theta, 0 \le \theta < \pi$

Circle of radius 5

Area $= 25\pi$

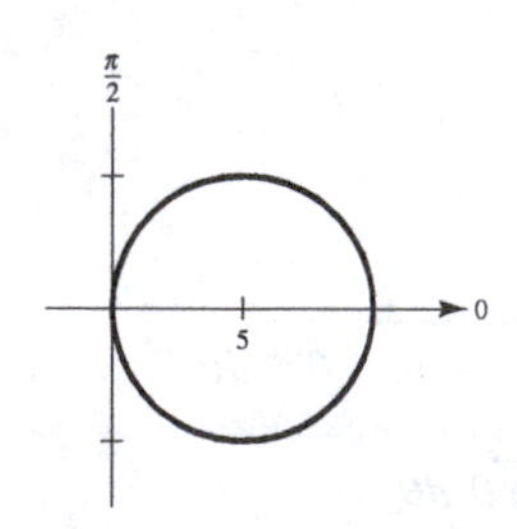

73. Answer will vary. *Sample answer*:

$f(\theta) = \cos^2\theta + 1, g(\theta) = -\frac{3}{2}$

75. Revolve $r = 2$ about the line $r = 5\sec\theta$.

$f(\theta) = 2, f'(\theta) = 0$

$$S = 2\pi\int_0^{2\pi} (5 - 2\cos\theta)\sqrt{2^2 + 0^2}\,d\theta$$

$$= 4\pi\int_0^{2\pi} (5 - 2\cos\theta)\,d\theta$$

$$= 4\pi\left[5\theta - 2\sin\theta\right]_0^{2\pi}$$

$$= 40\pi^2$$

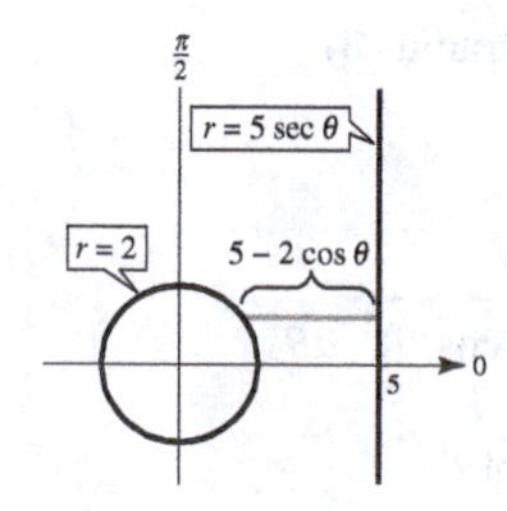

77. $r = 8\cos\theta, 0 \le \theta \le \pi$

(a) $A = \frac{1}{2}\int_0^{\pi} r^2\,d\theta = \frac{1}{2}\int_0^{\pi} 64\cos^2\theta\,d\theta = 32\int_0^{\pi}\frac{1+\cos 2\theta}{2}\,d\theta = 16\left[\theta + \frac{\sin 2\theta}{2}\right]_0^{\pi} = 16\pi$

$\left(\text{Area circle} = \pi r^2 = \pi 4^2 = 16\pi\right)$

(b)

θ	0.2	0.4	0.6	0.8	1.0	1.2	1.4
A	6.32	12.14	17.06	20.80	23.27	24.60	25.08

(c), (d) For $\frac{1}{4}$ of area $(4\pi \approx 12.57)$: 0.42

For $\frac{1}{2}$ of area $(8\pi \approx 25.13)$: $1.57\left(\frac{\pi}{2}\right)$

For $\frac{3}{4}$ of area $(12\pi \approx 37.70)$: 2.73

(e) No, it does not depend on the radius.

79. (a) $r = \theta, \theta \ge 0$

As a increases, the spiral opens more rapidly. If $\theta < 0$, the spiral is reflected about the y-axis.

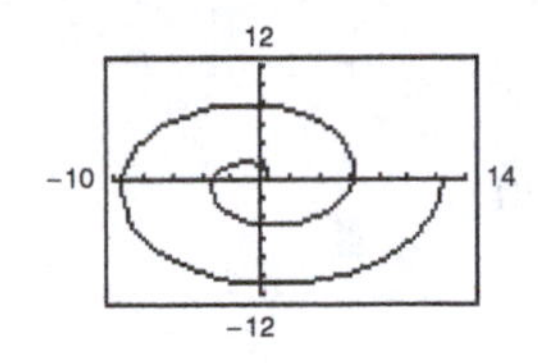

(b) $r = a\theta, \theta \ge 0$, crosses the polar axis for $\theta = n\pi$, n and integer. To see this

$r = a\theta \Rightarrow r\sin\theta = y = a\theta\sin\theta = 0$

for $\theta = n\pi$. The points are $(r, \theta) = (an\pi, n\pi), n = 1, 2, 3, \ldots$.

(c) $f(\theta) = \theta, f'(\theta) = 1$

$$s = \int_0^{2\pi}\sqrt{\theta^2+1}\,d\theta$$

$$= \frac{1}{2}\left[\ln\left(\sqrt{\theta^2+1}+\theta\right)+\theta\sqrt{\theta^2+1}\right]_0^{2\pi}$$

$$= \frac{1}{2}\ln\left(\sqrt{4\pi^2+1}+2\pi\right)+\pi\sqrt{4\pi^2+1} \approx 21.2563$$

(d) $A = \frac{1}{2}\int_{\alpha}^{\beta} r^2\,dr = \frac{1}{2}\int_0^{2\pi}\theta^2\,d\theta = \left[\frac{\theta^3}{6}\right]_0^{2\pi} = \frac{4}{3}\pi^3$

81. The smaller circle has equation $r = a\cos\theta$. The area of the shaded lune is:

$$A = 2\left(\frac{1}{2}\right)\int_0^{\pi/4}\left[(a\cos\theta)^2 - 1\right]d\theta$$

$$= \int_0^{\pi/4}\left[\frac{a^2}{2}(1+\cos 2\theta) - 1\right]d\theta$$

$$= \left[\frac{a^2}{2}\left(\theta + \frac{\sin 2\theta}{2}\right) - \theta\right]_0^{\pi/4}$$

$$= \frac{a^2}{2}\left(\frac{\pi}{4}+\frac{1}{2}\right) - \frac{\pi}{4}$$

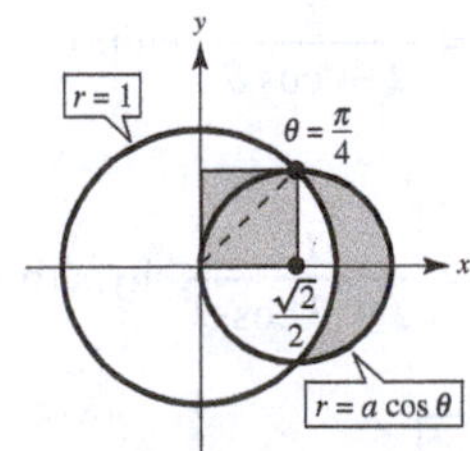

This equals the area of the square, $\left(\frac{\sqrt{2}}{2}\right)^2 = \frac{1}{2}$.

$$\frac{a^2}{2}\left(\frac{\pi}{4}+\frac{1}{2}\right) - \frac{\pi}{4} = \frac{1}{2}$$

$$\pi a^2 + 2a^2 - 2\pi - 4 = 0$$

$$a^2 = \frac{4+2\pi}{2+\pi} = 2$$

$$a = \sqrt{2}$$

Smaller circle: $r = \sqrt{2}\cos\theta$

83. $x = \dfrac{3t}{1 + t^3}, y = \dfrac{3t^2}{1 + t^3}$

(a) $x^3 + y^3 = \dfrac{27(t^3 + t^6)}{(1 + t^3)^3} = \dfrac{27t^3}{(1 + t^3)^2}$

$3xy = \dfrac{27t^3}{(1 + t^3)^2}$

So, $x^3 + y^3 = 3xy$.

$(r\cos\theta)^3 + (r\sin\theta)^3 = 3(r\cos\theta)(r\sin\theta)$

$r = \dfrac{3\cos\theta\sin\theta}{\cos^3\theta + \sin^3\theta}$

(b)

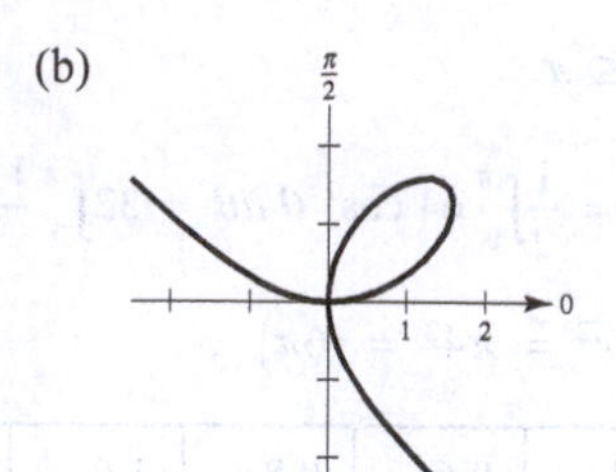

(c) $A = \dfrac{1}{2}\int_0^{\pi/2} r^2\,d\theta = \dfrac{3}{2}$

Section 10.6 Polar Equations of Conics and Kepler's Laws

1. (a) $e = 3$, hyperbola

(b) $e = 1$, parabola

(c) $e = \dfrac{5}{6}$, ellipse

(d) $e = \dfrac{3}{2}$, hyperbola

3. $r = \dfrac{2e}{1 + e\cos\theta}$

(a) $e = 1, r = \dfrac{2}{1 + \cos\theta}$, parabola

(b) $e = 0.5$,

$r = \dfrac{1}{1 + 0.5\cos\theta} = \dfrac{2}{2 + \cos\theta}$, ellipse

(c) $e = 1.5$,

$r = \dfrac{3}{1 + 1.5\cos\theta} = \dfrac{6}{2 + 3\cos\theta}$, hyperbola

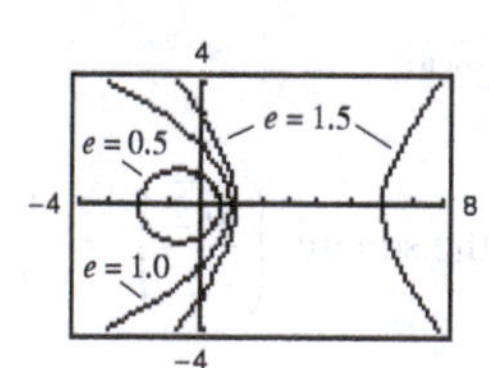

5. $r = \dfrac{4}{1 + e\sin\theta}$

The conic is an ellipse. As $e \to 1^-$, the ellipse becomes more elliptical, and as $e \to 0^+$, it becomes more circular.

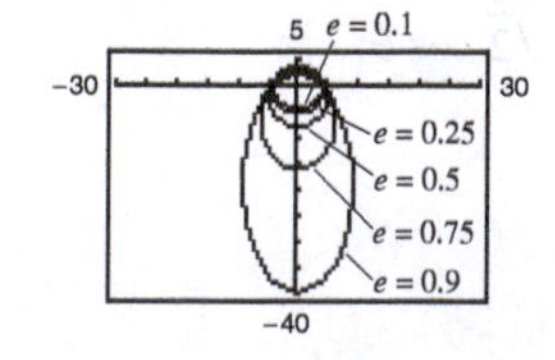

7. Parabola; Matches (c)

8. Ellipse; Matches (f)

9. Hyperbola; Matches (a)

10. Parabola; Matches (e)

11. Ellipse; Matches (b)

12. Hyperbola; Matches (d)

13. $r = \dfrac{1}{1 - \cos\theta}$

Parabola because $e = 1, d = 1$

Distance from pole to directrix: $|d| = 1$

Directrix: $x = -d = -1$

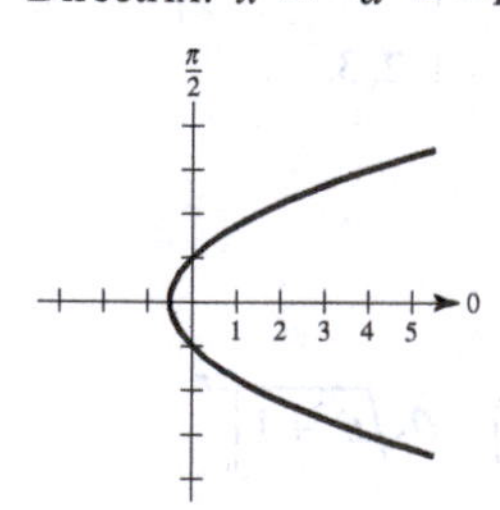

15. $r = \dfrac{7}{4 + 8\sin\theta} = \dfrac{7/4}{1 + 2\sin\theta}$

Hyperbola because $e = 2 > 1, d = \dfrac{7}{8}$

Distance from pole to directrix: $|d| = \dfrac{7}{8}$

Directrix: $y = \dfrac{7}{8}$

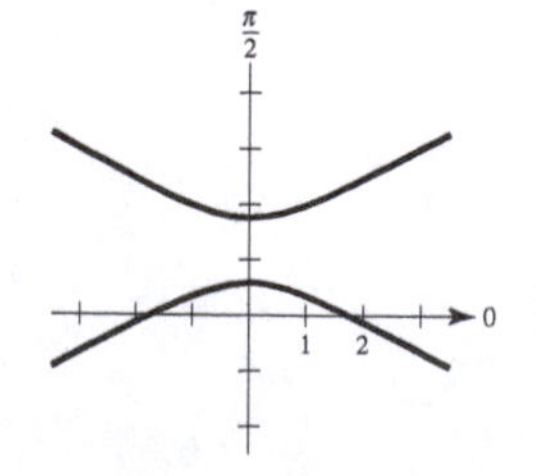

17. $r = \dfrac{6}{-2 + 3\cos\theta} = \dfrac{-3}{1 - 3/2\cos\theta}$

Hyperbola because $e = \dfrac{3}{2} > 0$, $d = -2$

Distance from pole to directrix: $|d| = 2$

Directrix: $x = 2$

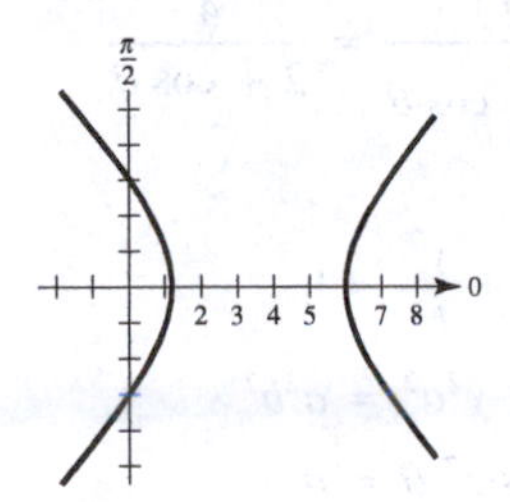

19. $r = \dfrac{6}{2 + \cos\theta} = \dfrac{3}{1 + (1/2)\cos\theta}$

Ellipse because $e = \dfrac{1}{2} < 1$, $d = 6$

Distance from pole to directrix: $|d| = 6$

Directrix: $x = 6$

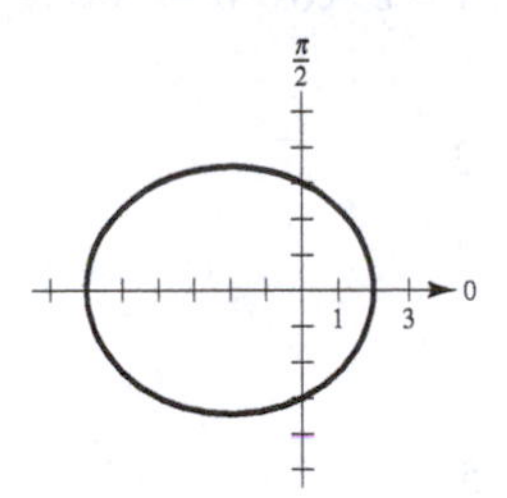

21. $r = \dfrac{300}{-12 + 6\sin\theta} = \dfrac{-25}{1 - \frac{1}{2}\sin\theta}$

Ellipse because $e = \dfrac{1}{2} < 1$, $d = -50$

Distance from pole to directrix: $|d| = 50$

Directrix: $y = -50$

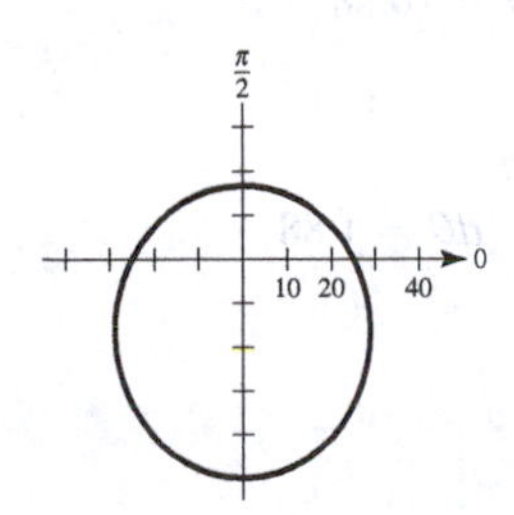

23. $r = \dfrac{3}{-4 + 2\sin\theta} = \dfrac{-\frac{3}{4}}{1 - \frac{1}{2}\sin\theta}$

$e = \dfrac{1}{2}$, Ellipse

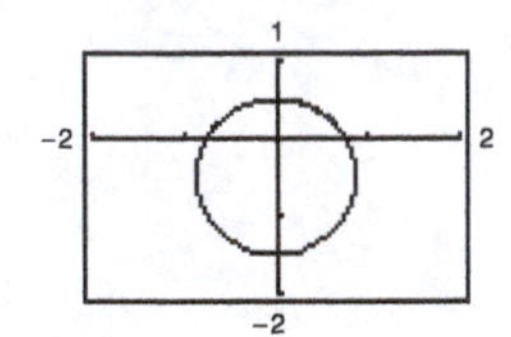

25. $r = \dfrac{-10}{1 - \cos\theta}$

$e = 1$, Parabola

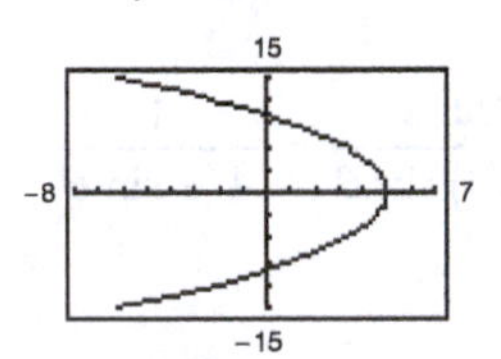

27. $r = \dfrac{4}{1 + \cos\left(\theta - \dfrac{\pi}{3}\right)}$

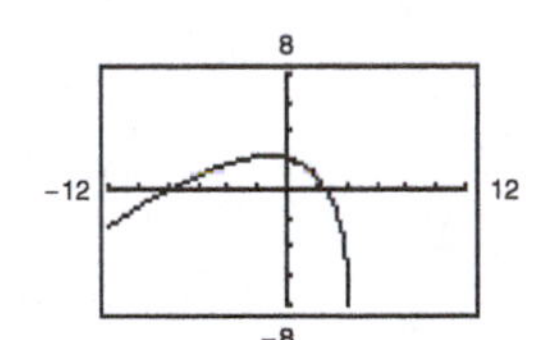

Rotate the graph of $r = \dfrac{4}{1 + \cos\theta}$

$\dfrac{\pi}{3}$ radian counterclockwise.

29. $r = \dfrac{6}{2 + \cos\left(\theta + \dfrac{\pi}{6}\right)}$

Rotate the graph of $r = \dfrac{6}{2 + \cos\theta}$

$\dfrac{\pi}{6}$ radian clockwise.

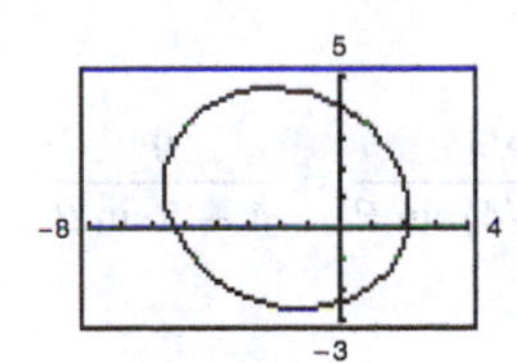

31. Change θ to $\theta + \frac{\pi}{6}$

$$r = \frac{8}{8 + 5\cos\left(\theta + \frac{\pi}{6}\right)}$$

33. Parabola

$e = 1$

$x = -3 \Rightarrow d = 3$

$$r = \frac{ed}{1 - e\cos\theta} = \frac{3}{1 - \cos\theta}$$

35. Ellipse

$e = \frac{1}{4}, y = 1, d = 1$

$$r = \frac{ed}{1 + e\sin\theta} = \frac{1/4}{1 + (1/4)\sin\theta} = \frac{1}{4 + \sin\theta}$$

37. Hyperbola

$e = \frac{4}{3}, x = 2, d = 2$

$$r = \frac{ed}{1 + e\cos\theta} = \frac{(4/3)(2)}{1 + \frac{4}{3}\cos\theta} = \frac{8}{3 + 4\cos\theta}$$

39. Parabola

Vertex: $\left(1, -\frac{\pi}{2}\right)$

$e = 1, d = 2, r = \dfrac{2}{1 - \sin\theta}$

41. Ellipse

Vertices: $(2, 0), (8, \pi)$

$e = \frac{3}{5}, d = \frac{16}{3}$

$$r = \frac{ed}{1 + e\cos\theta} = \frac{16/5}{1 + (3/5)\cos\theta} = \frac{16}{5 + 3\cos\theta}$$

43. Hyperbola

Vertices: $\left(1, \frac{3\pi}{2}\right), \left(9, \frac{3\pi}{2}\right)$

$e = \frac{5}{4}, d = \frac{9}{5}$

$$r = \frac{ed}{1 - e\sin\theta} = \frac{9/4}{1 - (5/4)\sin\theta} = \frac{9}{4 - 5\sin\theta}$$

45. No. The flatness of the ellipse does not depend on the distance between foci.

47. Ellipse, $e = \frac{1}{2}$,

Directrix: $r = 4\sec\theta \Rightarrow x = r\cos\theta = 4$

$$r = \frac{ed}{1 + e\cos\theta} = \frac{\left(\frac{1}{2}\right)4}{1 + \frac{1}{2}\cos\theta} = \frac{4}{2 + \cos\theta}$$

49.

$$\begin{aligned}
\frac{x^2}{a^2} + \frac{y^2}{b^2} &= 1 \\
x^2b^2 + y^2a^2 &= a^2b^2 \\
b^2r^2\cos^2\theta + a^2r^2\sin^2\theta &= a^2b^2 \\
r^2\left[b^2\cos^2\theta + a^2\left(1 - \cos^2\theta\right)\right] &= a^2b^2 \\
r^2\left[a^2 + \cos^2\theta\left(b^2 - a^2\right)\right] &= a^2b^2
\end{aligned}$$

$$\begin{aligned}
r^2 &= \frac{a^2b^2}{a^2 + \left(b^2 - a^2\right)\cos^2\theta} = \frac{a^2b^2}{a^2 - c^2\cos^2\theta} \\
&= \frac{b^2}{1 - (c/a)^2\cos^2\theta} = \frac{b^2}{1 - e^2\cos^2\theta}
\end{aligned}$$

51. $a = 5, c = 4, e = \frac{4}{5}, b = 3$

$$r^2 = \frac{9}{1 - (16/25)\cos^2\theta}$$

53. $a = 3, b = 4, c = 5, e = \frac{5}{3}$

$$r^2 = \frac{-16}{1 - (25/9)\cos^2\theta}$$

55.
$$\begin{aligned}
A &= 2\left[\frac{1}{2}\int_0^{\pi}\left(\frac{3}{2 - \cos\theta}\right)^2 d\theta\right] \\
&= 9\int_0^{\pi}\frac{1}{(2 - \cos\theta)^2}\,d\theta \approx 10.88
\end{aligned}$$

57. $A = \dfrac{1}{2}\displaystyle\int_0^{2\pi}\left(\frac{2}{7 - 6\sin\theta}\right)^2 d\theta \approx 1.88$

59. Vertices: $(123{,}000 + 4000, 0) = (127{,}000, 0)$

$(119 + 4000, \pi) = (4119, \pi)$

$$a = \frac{127{,}000 + 4119}{2} = 65{,}559.5$$

$$c = 65{,}559.5 - 4119 = 61{,}440.5$$

$$e = \frac{c}{a} = \frac{122{,}881}{131{,}119} \approx 0.93717$$

$$r = \frac{ed}{1 - e\cos\theta}$$

$$\theta = 0: r = \frac{ed}{1-e}, \theta = \pi: r = \frac{ed}{1+e}$$

$$2a = 2(65{,}559.5) = \frac{ed}{1-e} + \frac{ed}{1+e}$$

$$131{,}119 = d\left(\frac{e}{1-e} + \frac{e}{1+e}\right) = d\left(\frac{2e}{1-e^2}\right)$$

$$d = \frac{131{,}119(1-e^2)}{2e} \approx 8514.1397$$

$$r = \frac{7979.21}{1 - 0.93717\cos\theta} = \frac{1{,}046{,}226{,}000}{131{,}119 - 122{,}881\cos\theta}$$

When $\theta = 60° = \frac{\pi}{3}$, $r \approx 15{,}015$.

Distance between earth and the satellite is $r - 4000 \approx 11{,}015$ miles.

61. $a = 1.496 \times 10^8$, $e = 0.0167$

$$r = \frac{(1-e^2)a}{1 - e\cos\theta} = \frac{149{,}558{,}278.1}{1 - 0.0167\cos\theta}$$

Perihelion distance: $a(1-e) \approx 147{,}101{,}680$ km

Aphelion distance: $a(1+e) \approx 152{,}098{,}320$ km

63. $a = 4.495 \times 10^9$, $e = 0.0113$

$$r = \frac{(1-e^2)a}{1 - e\cos\theta} = \frac{4{,}494{,}426{,}033}{1 - 0.0113\cos\theta}$$

Perihelion distance: $a(1-e) \approx 4{,}444{,}206{,}500$ km

Aphelion distance: $a(1+e) \approx 4{,}545{,}793{,}500$ km

65. $r = \dfrac{4.498 \times 10^9}{1 - 0.0086\cos\theta}$

(a) $A = \frac{1}{2}\int_0^{\pi/9} r^2\,d\theta \approx 3.591 \times 10^{18}$ km^2

$$165\left[\frac{\frac{1}{2}\int_0^{\pi/2} r^2\,d\theta}{\frac{1}{2}\int_0^{2\pi} r^2\,d\theta}\right] \approx 9.322 \text{ yrs}$$

(b) $\frac{1}{2}\int_\pi^\alpha r^2\,d\theta = 3.591 \times 10^{18}$

By trial and error, $\alpha \approx \pi + 0.361$

$0.361 > \pi/9 \approx 0.349$ because the rays in part (a) are longer than those in part (b)

(c) For part (a),

$$s = \int_0^{\pi/9} \sqrt{r^2 + (dr/d\theta)^2} \approx 1.583 \times 10^9 \text{ km}$$

$$\text{Average per year} = \frac{1.583 \times 10^9}{9.322} \approx 1.698 \times 10^8 \text{ km/yr}$$

For part (b),

$$s = \int_\pi^{\pi+0.361} \sqrt{r^2 + (dr/d\theta)^2}\,d\theta \approx 1.610 \times 10^9 \text{ km}$$

$$\text{Average per year} = \frac{1.610 \times 10^9}{9.322} \approx 1.727 \times 10^8 \text{ km/yr}$$

67. $r_1 = a + c$, $r_0 = a - c$, $r_1 - r_0 = 2c$, $r_1 + r_0 = 2a$

$$e = \frac{c}{a} = \frac{r_1 - r_0}{r_1 + r_0}$$

$$\frac{1+e}{1-e} = \frac{1 + \frac{c}{a}}{1 - \frac{c}{a}} = \frac{a+c}{a-c} = \frac{r_1}{r_0}$$

Review Exercises for Chapter 10

1. $4x^2 + y^2 = 4$

Ellipse

Vertex: $(1, 0)$.

Matches (e)

2. $4x^2 - y^2 = 4$

Hyperbola

Vertex: $(1, 0)$

Matches (c)

3. $y^2 = -4x$

Parabola opening to left.

Matches (b)

4. $y^2 - 4x^2 = 4$

Hyperbola

Vertex: $(0, 2)$

Matches (d)

5. $x^2 + 4y^2 = 4$

Ellipse

Vertex: $(0, 1)$

Matches (a)

6. $x^2 = 4y$

Parabola opening upward.

Matches (f)

7.
$$x^2 + y^2 - 2x - 8y - 8 = 0$$
$$(x^2 - 2x + 1) + (y^2 - 8y + 16) = 8 + 1 + 16$$
$$(x - 1)^2 + (y - 4)^2 = 25$$

Circle

Center: $(1, 4)$

Radius: 1

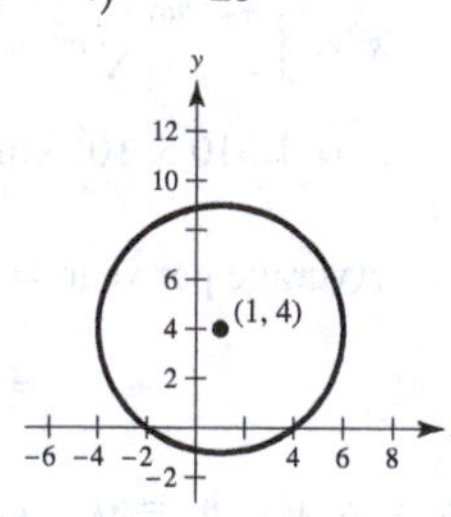

9.
$$3x^2 - 2y^2 + 24x + 12y + 24 = 0$$
$$3(x^2 + 8x + 16) - 2(y^2 - 6y + 9) = -24 + 48 - 18$$
$$\frac{(x + 4)^2}{2} - \frac{(y - 3)^2}{3} = 1$$

Hyperbola

Center: $(-4, 3)$

Vertices: $\left(-4 \pm \sqrt{2}, 3\right)$

Foci: $\left(-4 \pm \sqrt{5}, 3\right)$

Eccentricity: $\dfrac{\sqrt{10}}{2}$

Asymptotes:

$y = 3 \pm \sqrt{\dfrac{3}{2}}(x + 4)$

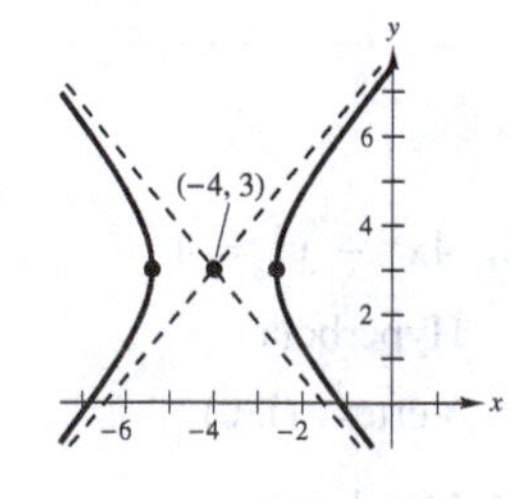

11. $16x^2 + 16y^2 - 16x + 24y - 3 = 0$
$$\left(x^2 - x + \tfrac{1}{4}\right) + \left(y^2 + \tfrac{3}{2}y + \tfrac{9}{16}\right) = \tfrac{3}{16} + \tfrac{1}{4} + \tfrac{9}{16}$$
$$\left(x - \tfrac{1}{2}\right)^2 + \left(y + \tfrac{3}{4}\right)^2 = 1$$

Circle

Center: $\left(\frac{1}{2}, -\frac{3}{4}\right)$

Radius: 1

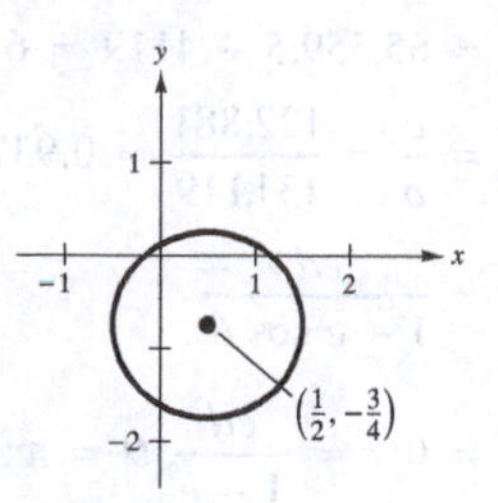

13. $x^2 + 10x - 12y + 13 = 0$
$$x^2 + 10x + 25 = 12y - 13 + 25$$
$$(x + 5)^2 = 12y + 12$$
$$= 12(y + 1)$$
$$= 4(3)(y + 1)$$

Parabola, vertical axis

Vertex: $(-5, -1)$

Directrix: $y = -4$

Focus: $(-5, 2)$

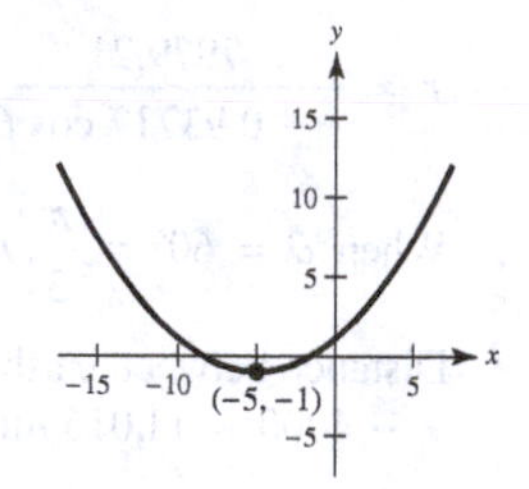

15. Vertex: $(7, 0)$

Directrix: $x = 5$

Parabola opens to the right.

$p = 2$

$(y - 0)^2 = 4(2)(x - 7)$

17. Center: $(0, 1)$

Focus: $(4, 1)$

Vertex: $(6, 1)$

$a = 6, c = 4,$

$b^2 = a^2 - c^2 = 36 - 16 = 20 \Rightarrow b = 2\sqrt{5}$

$$\frac{(x - 0)^2}{6^2} + \frac{(y - 1)^2}{20} = 1$$
$$\frac{x^2}{36} + \frac{(y - 1)^2}{20} = 1$$

19. Vertices: $(3, 1), (3, 7)$

Center: $(3, 4)$

$$\text{Eccentricity} = \frac{2}{3} = \frac{c}{a} \Rightarrow a = 3, c = 2$$

Vertical major axis

$b = \sqrt{9 - 4} = \sqrt{5}$

$$\frac{(x-3)^2}{5} + \frac{(y-4)^2}{9} = 1$$

21. Vertices: $(0, \pm 8) \Rightarrow a = 8$

Center: $(0, 0)$

Vertical transverse axis

Asymptotes:

$$y = \pm 2x \Rightarrow \frac{a}{b} = 2 \Rightarrow \frac{8}{b} = 2 \Rightarrow b = 4$$

$$\frac{y^2}{64} - \frac{x^2}{16} = 1$$

23. Vertices: $(\pm 7, -1)$

Center: $(0, -1)$

Horizontal transverse axis

Foci: $(\pm 9, -1)$

$a = 7, c = 9, b = \sqrt{81 - 49} = \sqrt{32} = 4\sqrt{2}$

$$\frac{x^2}{49} - \frac{(y+1)^2}{32} = 1$$

25. $y = \frac{1}{200}x^2$

(a) $x^2 = 200y$

$x^2 = 4(50)y$

Focus: $(0, 50)$

(b)
$$y = \frac{1}{200}x^2$$
$$y' = \frac{1}{100}x$$
$$\sqrt{1 + (y')^2} = \sqrt{1 + \frac{x^2}{10{,}000}}$$
$$S = 2\pi \int_0^{100} x\sqrt{1 + \frac{x^2}{10{,}000}}\,dx$$
$$\approx 38{,}294.49$$

27. $x = 1 + 8t, y = 3 - 4t$

$$t = \frac{x-1}{8} \Rightarrow y = 3 - 4\left(\frac{x-1}{8}\right) = \frac{7}{2} - \frac{x}{2}$$

$x + 2y - 7 = 0$, Line

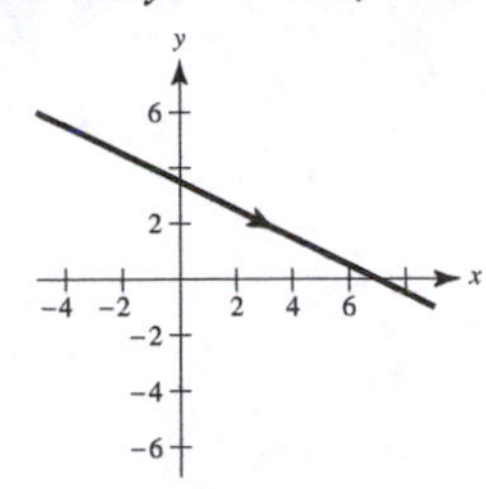

29. $x = \sqrt{t} + 1, y = t - 3$

$x - 1 = \sqrt{t}$

$(x - 1)^2 = t$

$y = (x - 1)^2 - 3, x \geq 1$

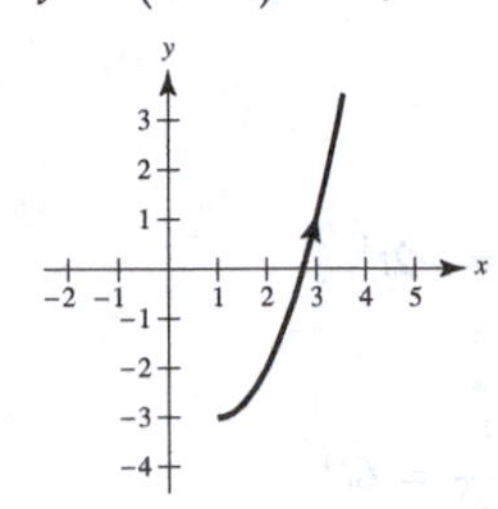

31. $x = 6 \cos \theta, y = 6 \sin \theta$

$$\left(\frac{x}{6}\right)^2 + \left(\frac{y}{6}\right)^2 = 1$$

$x^2 + y^2 = 36$

Circle

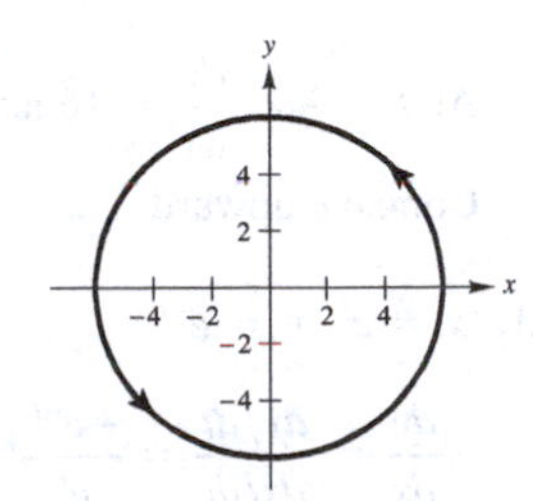

33. $x = 2 + \sec \theta, y = 3 + \tan \theta$

$(x - 2)^2 = \sec^2 \theta = 1 + \tan^2 \theta = 1 + (y - 3)^2$

$(x - 2)^2 - (y - 3)^2 = 1$

Hyperbola

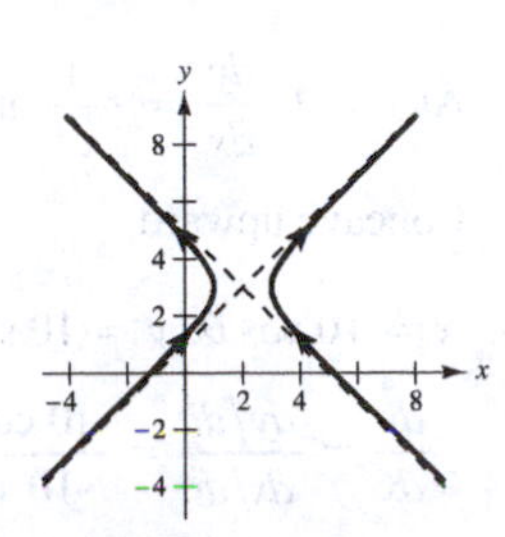

35. $y = 4x + 3$

Examples: $x = t, y = 4t + 3$

$x = t + 1, y = 4(t + 1) + 3 = 4t + 7$

37. $x = \cos 3\theta + 5\cos\theta$

$y = \sin 3\theta + 5\sin\theta$

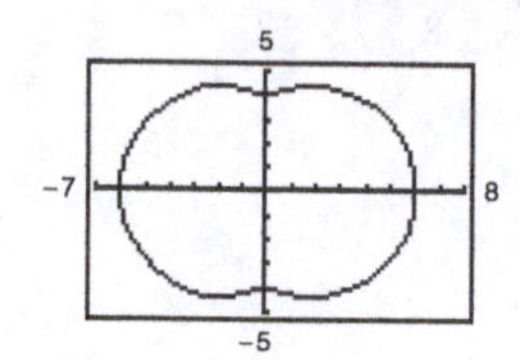

39. $x = 1 + 6t,\ y = 4 - 5t$

$$\frac{dy}{dx} = \frac{dy/dt}{dx/dt} = \frac{-5}{6}$$

$$\frac{d^2y}{dx^2} = 0$$

At $t = 3$, $\frac{dy}{dx} = \frac{-5}{6}$ and $\frac{d^2y}{dx^2} = 0$. (Line)

Neither concave upward nor downward

41. $x = \frac{1}{t},\ y = t^2$

$$\frac{dy}{dx} = \frac{dy/dt}{dx/dt} = \frac{2t}{\left(-1/t^2\right)} = -2t^3$$

$$\frac{d^2y}{dx^2} = \frac{\frac{d}{dt}\left[-2t^3\right]}{dx/dt} = \frac{-6t^2}{\left(-1/t^2\right)} = 6t^4$$

At $t = -2$, $\frac{dy}{dx} = 16$ and $\frac{d^2y}{dx^2} = 96$.

Concave upward

43. $x = e^t,\ y = e^{-t}$

$$\frac{dy}{dx} = \frac{dy/dt}{dx/dt} = \frac{-e^{-t}}{e^t} = -e^{-2t}$$

$$\frac{d^2y}{dx^2} = \frac{\frac{d}{dt}\left(-e^{-2t}\right)}{dx/dt} = \frac{2e^{-2t}}{e^t} = \frac{2}{e^{3t}}$$

At $t = 1$, $\frac{dy}{dx} = -\frac{1}{e^2}$ and $\frac{d^2y}{dx^2} = \frac{2}{e^3}$.

Concave upward

45. $x = 10\cos\theta,\ y = 10\sin\theta$

$$\frac{dy}{dx} = \frac{dy/dt}{dx/dt} = \frac{10\cos\theta}{-10\sin\theta} = -\cot\theta$$

$$\frac{d^2y}{dx^2} = \frac{\frac{d}{d\theta}[-\cot\theta]}{dx/d\theta} = \frac{\csc^2\theta}{-10\sin\theta} = -\frac{1}{10}\csc^3\theta$$

At $\theta = \frac{\pi}{4}$, the slope is $\frac{dy}{dx} = -1$ and $\frac{d^2y}{dx^2} = \frac{-\sqrt{2}}{5}$.

Concave downward

47. $x = \cot\theta,\ y = \sin 2\theta,\ \theta = \frac{\pi}{6}$

(a), (d)

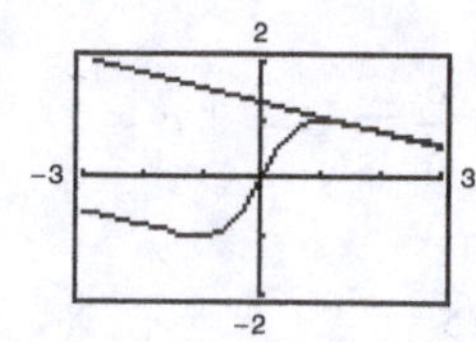

(b) At $\theta = \frac{\pi}{6}$, $\frac{dx}{d\theta} = -4$, $\frac{dy}{d\theta} = 1$, and $\frac{dy}{dx} = -\frac{1}{4}$.

(c) At $\theta = \frac{\pi}{6}$, $(x, y) = \left(\sqrt{3}, \frac{\sqrt{3}}{2}\right)$.

$$y - \frac{\sqrt{3}}{2} = -\frac{1}{4}\left(x - \sqrt{3}\right)$$

$$y = -\frac{1}{4}x + \frac{3\sqrt{3}}{4}$$

49. $x = 5 - t,\ y = 2t^2$

$$\frac{dx}{dt} = -1,\ \frac{dy}{dt} = 4t$$

Horizontal tangent at $t = 0$: $(5, 0)$

No vertical tangents

51. $x = 2 + 2\sin\theta,\ y = 1 + \cos\theta$

$$\frac{dx}{d\theta} = 2\cos\theta,\ \frac{dy}{d\theta} = -\sin\theta$$

$$\frac{dy}{d\theta} = 0 \text{ for } \theta = 0, \pi, 2\pi, \ldots$$

Horizontal tangents: $(x, y) = (2, 2), (2, 0)$

$$\frac{dx}{d\theta} = 0 \text{ for } \theta = \frac{\pi}{2}, \frac{3\pi}{2}, \ldots$$

Vertical tangents: $(x, y) = (4, 1), (0, 1)$

53. $x = t^2 + 1,\ y = 4t^3 + 3,\ 0 \le t \le 2$

$$\frac{dx}{dt} = 2t,\ \frac{dy}{dt} = 12t^2$$

$$s = \int_0^2 \sqrt{(2t)^2 + \left(12t^2\right)^2}\,dt$$

$$= \int_0^2 \sqrt{4t^2 + 144t^4}\,dt$$

$$= \int_0^2 2t\sqrt{1 + 36t^2}\,dt$$

$$= \frac{1}{36}\left[\frac{2}{3}\left(1 + 36t^2\right)^{3/2}\right]_0^2$$

$$= \frac{1}{54}\left[145^{3/2} - 1\right] \approx 32.3154$$

55. $x = 4t,\ y = 3t + 1,\ 0 \le t \le 1$

$$\frac{dx}{dt} = 4,\ \frac{dy}{dt} = 3,\ \sqrt{\left(\frac{dx}{dt}\right)^2 + \left(\frac{dy}{dt}\right)^2} = \sqrt{16 + 9} = 5$$

(a) $S = 2\pi \int_0^1 (3t + 1)\,5\,dt = 10\pi\left[\frac{3t^2}{2} + t\right]_0^1$

$= 10\pi\left[\frac{3}{2} + 1\right] = 25\pi$

(b) $S = 2\pi \int_0^1 4t(5)\,dt = 40\pi\left[\frac{t^2}{2}\right]_0^1 = 20\pi$

57. $x = 3\sin\theta,\ y = 2\cos\theta$

$A = \int_a^b y\,dx = \int_{-\pi/2}^{\pi/2} 2\cos\theta(3\cos\theta)\,d\theta$

$= 6\int_{-\pi/2}^{\pi/2} \frac{1 + \cos 2\theta}{2}\,d\theta$

$= 3\left[\theta + \frac{\sin 2\theta}{2}\right]_{-\pi/2}^{\pi/2}$

$= 3\left[\frac{\pi}{2} + \frac{\pi}{2}\right] = 3\pi$

59. $(r, \theta) = \left(5, \frac{3\pi}{2}\right)$

$x = r\cos\theta = 5\cos\frac{3\pi}{2} = 0$

$y = r\sin\theta = 5\sin\frac{3\pi}{2} = -5$

$(x, y) = (0, -5)$

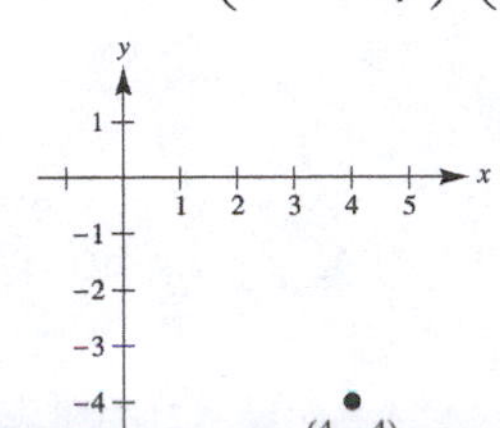

61. $(r, \theta) = \left(\sqrt{7}, 3.25\right)$

$x = r\cos\theta = \sqrt{7}\cos(3.25) \approx -2.6302$

$y = r\sin\theta = \sqrt{7}\sin(3.25) \approx -0.2863$

$(x, y) \approx (-2.6302, -0.2863)$

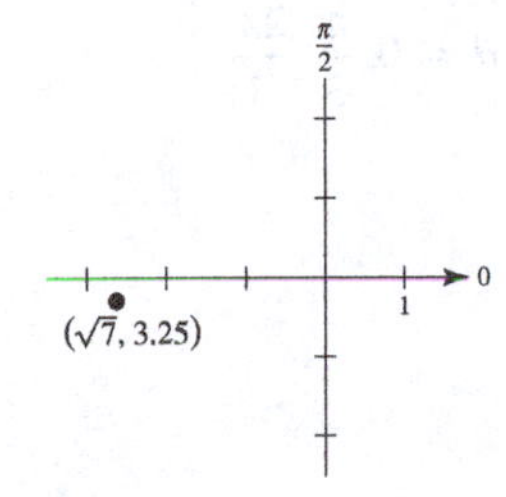

63. $(x, y) = (4, -4)$

$r = \sqrt{4^2 + (-4)^2} = 4\sqrt{2}$

$\theta = \frac{7\pi}{4}$

$(r, \theta) = \left(4\sqrt{2}, \frac{7\pi}{4}\right), \left(-4\sqrt{2}, \frac{3\pi}{4}\right)$

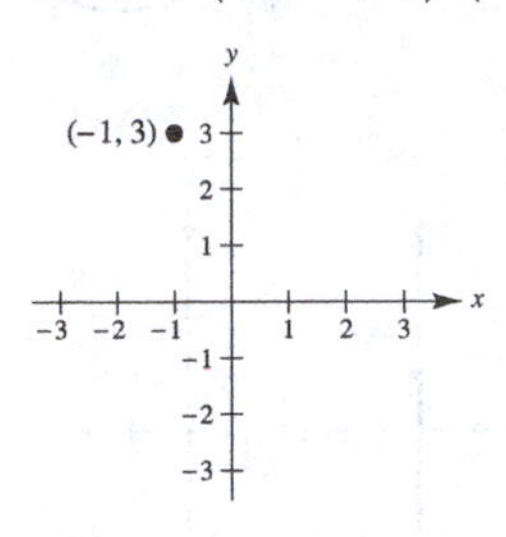

65. $(x, y) = (-1, 3)$

$r = \sqrt{(-1)^2 + 3^2} = \sqrt{10}$

$\theta = \arctan(-3) \approx 1.89(108.43°)$

$(r, \theta) = \left(\sqrt{10}, 1.89\right), \left(-\sqrt{10}, 5.03\right)$

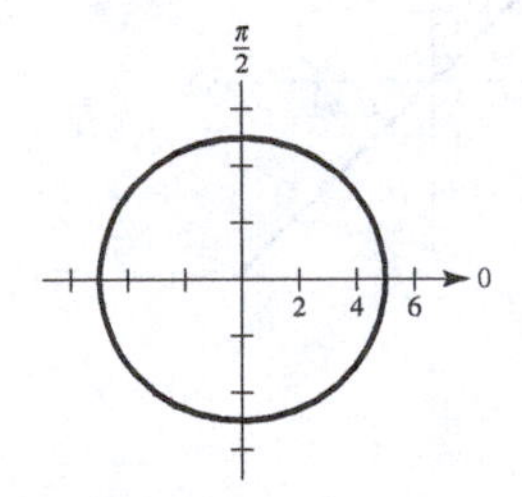

67. $x^2 + y^2 = 25$

$r^2 = 25$

$r = 5$

Circle

69. $y = 9$

$r\sin\theta = 9$

$r = \frac{9}{\sin\theta} = 9\csc\theta$

Horizontal line

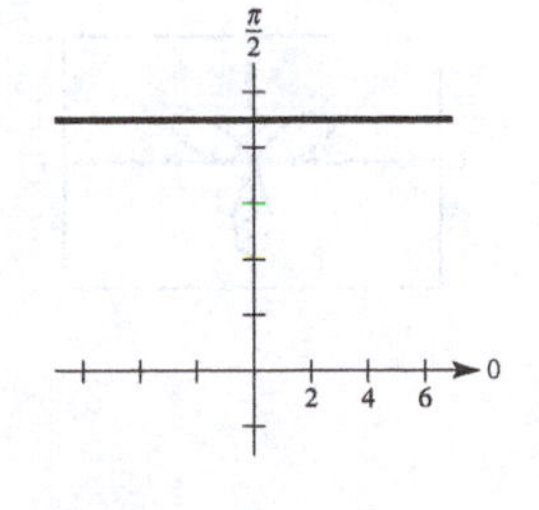

71. $$-x + 4y - 3 = 0$$
$$-r\cos\theta + 4r\sin\theta = 3$$
$$r(4\sin\theta - \cos\theta) = 3$$
$$r = \frac{3}{4\sin\theta - \cos\theta}$$

Line

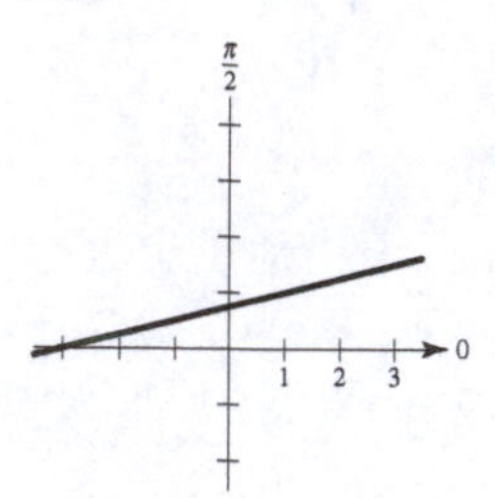

73. $$r = 6\cos\theta$$
$$r^2 = 6r\cos\theta$$
$$x^2 + y^2 = 6x$$
$$x^2 - 6x + 9 + y^2 = 9$$
$$x^2 + y^2 - 6x = 0$$

Circle

75. $$r = -4\sec\theta$$
$$r\cos\theta = -4$$
$$x = -4$$

Vertical line

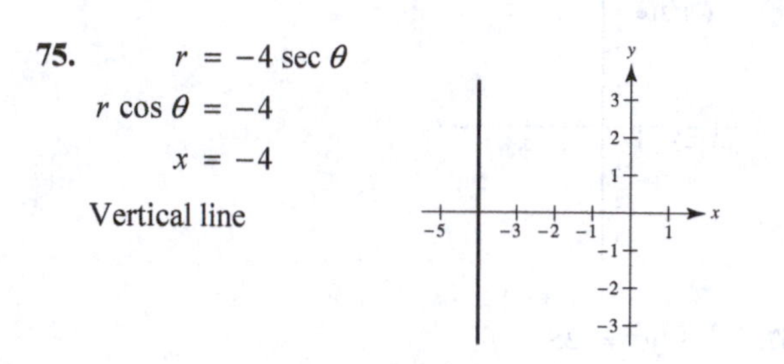

77. $$\theta = \frac{3\pi}{4}$$
$$\tan\theta = -1$$
$$\frac{y}{x} = -1$$
$$y = -x$$

Line

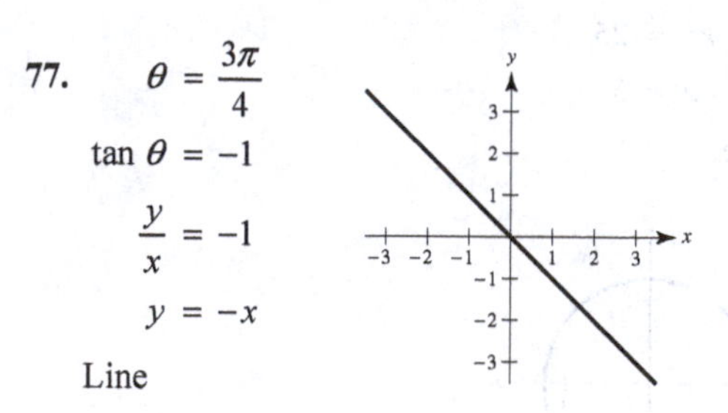

79. $r = \dfrac{3\pi}{2}\sin 3\theta$

$0 \le \theta \le \pi$

81. $r = 4\cos 2\theta\sec\theta$

$0 \le \theta \le \pi$

83. $r = 1 - \cos\theta$, Cardioid

$$\frac{dy}{dx} = \frac{(1 - \cos\theta)\cos\theta + (\sin\theta)\sin\theta}{-(1 - \cos\theta)\sin\theta + (\sin\theta)\cos\theta}$$

Horizontal tangents:

$$\cos\theta - \cos^2\theta + \sin^2\theta = 0$$
$$\cos\theta - \cos^2\theta + \left(1 - \cos^2\theta\right) = 0$$
$$2\cos^2\theta - \cos\theta - 1 = 0$$
$$(2\cos\theta + 1)(\cos\theta - 1) = 0$$
$$\cos\theta = -\frac{1}{2} \Rightarrow \theta = \frac{2\pi}{3}, \frac{4\pi}{3}$$
$$\cos\theta = 1 \Rightarrow \theta = 0$$

Vertical tangents:

$$-\sin\theta + 2\cos\theta\sin\theta = 0$$
$$\sin\theta(2\cos\theta - 1) = 0$$
$$\sin\theta = 0 \Rightarrow \theta = 0, \pi$$
$$\cos\theta = \frac{1}{2} \Rightarrow \theta = \frac{\pi}{3}, \frac{5\pi}{3}$$

Horizontal tangents: $\left(\frac{3}{2}, \frac{2\pi}{3}\right), \left(\frac{3}{2}, \frac{4\pi}{3}\right)$

Vertical tangents: $\left(\frac{1}{2}, \frac{\pi}{3}\right), \left(\frac{1}{2}, \frac{5\pi}{3}\right), (2\pi)$

(There is a cusp at the pole.)

85. $r = 4\sin 3\theta$, Rose curve with three petals

Tangents at the pole: $\sin 3\theta = 0$

$$\theta = 0, \frac{\pi}{3}, \frac{2\pi}{3}$$

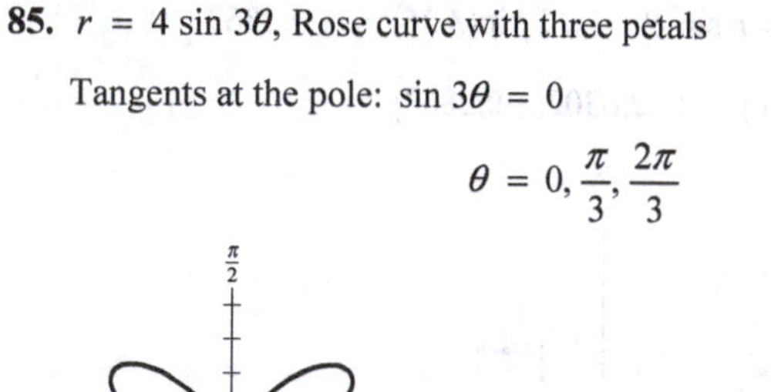

87. $r = 6$, Circle radius 6

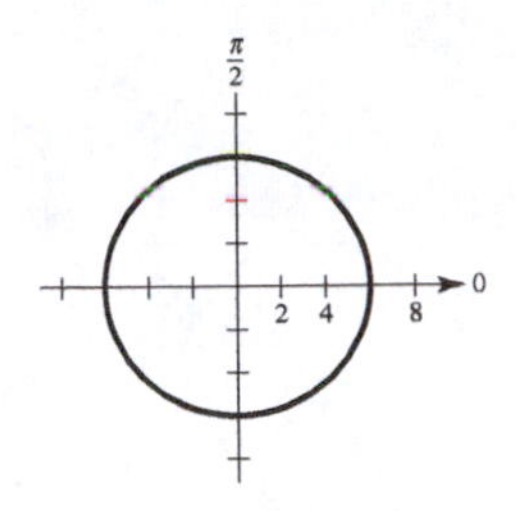

89. $r = -\sec\theta = \dfrac{-1}{\cos\theta}$

$r\cos\theta = -1,\ x = -1$

Vertical line

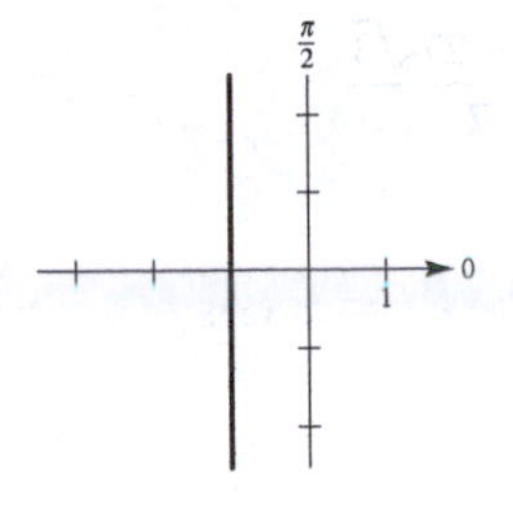

91. $r = 4 - 3\cos\theta$

Limaçon

Symmetric to polar axis

θ	0	$\frac{\pi}{3}$	$\frac{\pi}{2}$	$\frac{2\pi}{3}$	π
r	1	$\frac{5}{2}$	4	$\frac{11}{2}$	7

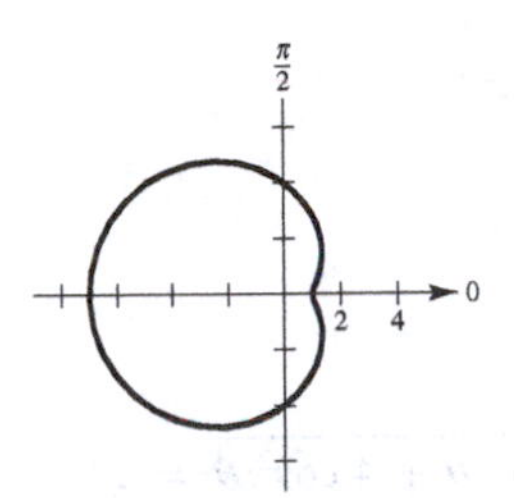

93. $r = 4\theta$

Spiral

Symmetric to $\theta = \dfrac{\pi}{2}$

θ	0	$\frac{\pi}{4}$	$\frac{\pi}{2}$	$\frac{3\pi}{4}$	π	$\frac{3\pi}{2}$	2π
r	0	π	2π	3π	4π	6π	8π

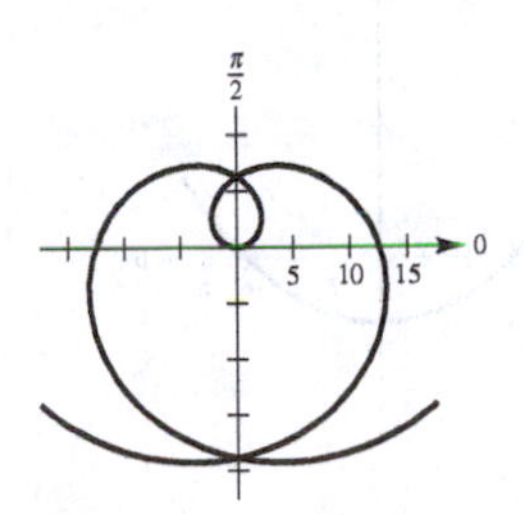

95. $r^2 = 4\sin 2\theta$

$r = 2\sqrt{\sin 2\theta},\ 0 \le \theta \le 2\pi$

Lemniscale

θ	0	$\frac{\pi}{4}$	$\frac{\pi}{2}$	π
r	0	± 2	0	0

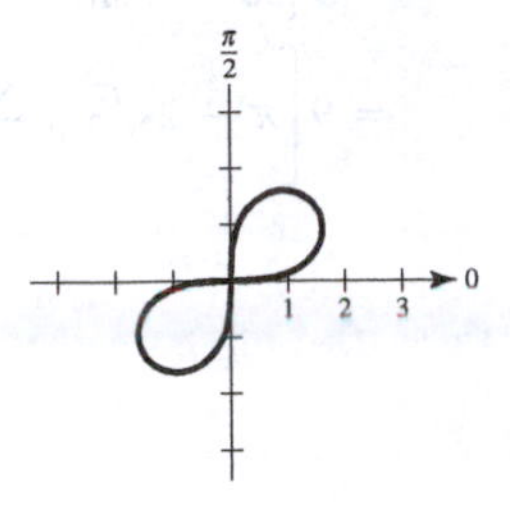

97. $A = 2 \cdot \dfrac{1}{2}\displaystyle\int_0^{\pi/10} [3\cos 5\theta]^2\,d\theta$

$$= \int_0^{\pi/10} 9\left(\frac{1 + \cos(10\theta)}{2}\right) d\theta$$

$$= \frac{9}{2}\left[\theta + \frac{\sin(10\theta)}{2}\right]_0^{\pi/10} = \frac{9}{2}\left[\frac{\pi}{10}\right] = \frac{9\pi}{20}$$

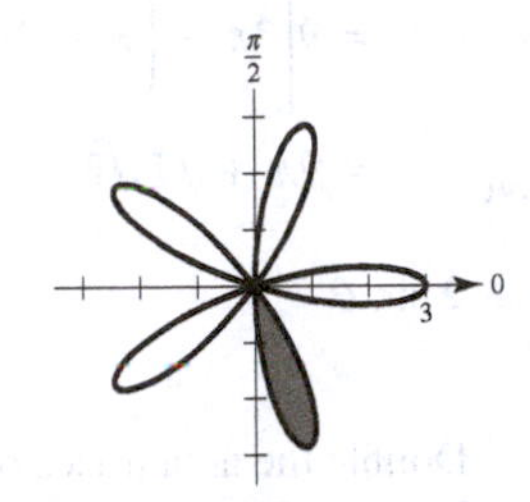

99. $r = 2 + \cos\theta$

$$A = 2\left[\frac{1}{2}\int_0^{\pi} (2 + \cos\theta)^2\,d\theta\right] \approx 14.14,\ \left(\frac{9\pi}{2}\right)$$

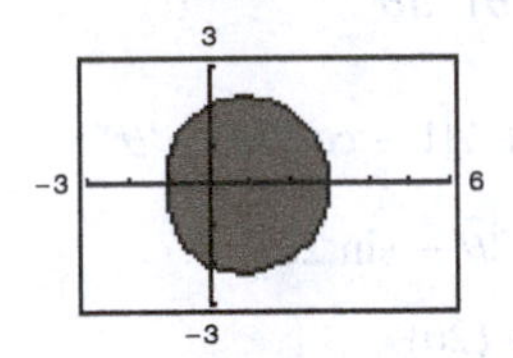

101. $r = 1 - \cos\theta,\ r = 1 + \sin\theta$

$1 - \cos\theta = 1 + \sin\theta$

$\tan\theta = -1 \Rightarrow \theta = \dfrac{3\pi}{4}, \dfrac{7\pi}{4}$

The graphs also intersect at the pole.

Points of intersection:

$$\left(1 + \frac{\sqrt{2}}{2}, \frac{3\pi}{4}\right), \left(1 - \frac{\sqrt{2}}{2}, \frac{7\pi}{4}\right), (0, 0)$$

103. $r = 3 - 6\cos\theta$

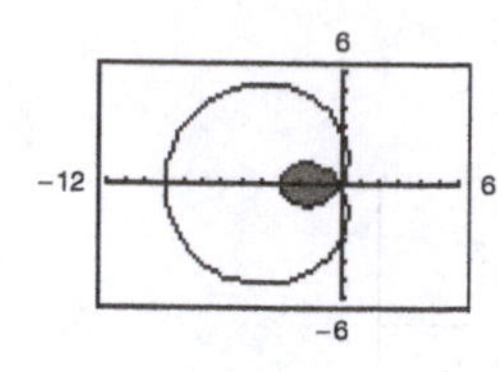

$$A = 2\left[\frac{1}{2}\int_0^{\pi/3}(3 - 6\cos\theta)^2\,d\theta\right]$$
$$= \int_0^{\pi/3}\left[9 - 36\cos\theta + 36\cos^2\theta\right]d\theta$$
$$= 9\int_0^{\pi/3}\left[1 - 4\cos\theta + 2(1 + \cos 2\theta)\right]d\theta$$
$$= 9\left[3\theta - 4\sin\theta + \sin 2\theta\right]_0^{\pi/3}$$
$$= 9\left[\pi - 2\sqrt{3} + \frac{\sqrt{3}}{2}\right] = \frac{18\pi - 27\sqrt{3}}{2}$$

105. $r = 3 - 6\cos\theta$

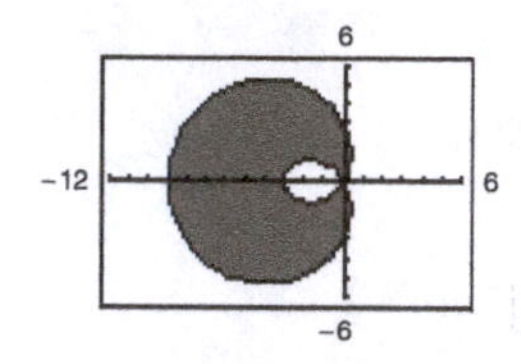

$$A = 2\left[\frac{1}{2}\int_{\pi/3}^{\pi}(3 - 6\cos\theta)^2\,d\theta - \frac{1}{2}\int_0^{\pi/3}(3 - 6\cos\theta)^2\,d\theta\right]$$

From Exercise 103 you have:

$$A = 9\left[3\theta - 4\sin\theta + \sin 2\theta\right]_{\pi/3}^{\pi} - 9\left[3\theta - 4\sin\theta + \sin 2\theta\right]_0^{\pi/3}$$
$$= 9\left[3\pi - \left(\pi - 2\sqrt{3} + \frac{\sqrt{3}}{2}\right)\right] - 9\left[\pi - 2\sqrt{3} + \frac{\sqrt{3}}{2}\right]$$
$$= 9\pi + 27\sqrt{3}$$

107. $r = 5 - 2\sin\theta,\ r = -5 + 2\sin\theta$

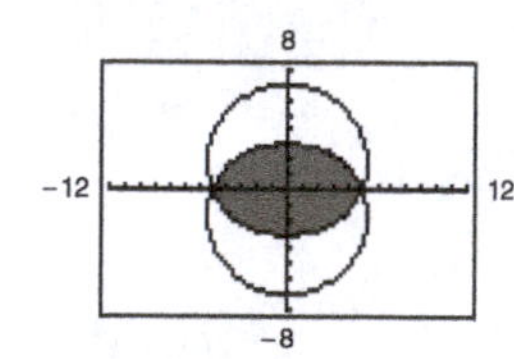

Double the area traced out by $r = 5 - 2\sin\theta$ on the interval $0 \le \theta \le \pi$.

$$A = 2\left(\frac{1}{2}\right)\int_0^{\pi}(5 - 2\sin\theta)^2\,d\theta$$
$$= \int_0^{\pi}\left[25 - 20\sin\theta + 2(1 - \cos 2\theta)\right]d\theta$$
$$= \left[25\theta + 20\cos\theta + 2\theta - \sin 2\theta\right]_0^{\pi}$$
$$= (25\pi - 20 + 2\pi) - (20)$$
$$= 27\pi - 40 \approx 44.823$$

109. $r = 5\cos\theta, \left[\frac{\pi}{2}, \pi\right]$

$$\frac{dr}{d\theta} = -5\sin\theta$$
$$s = \int_{\pi/2}^{\pi}\sqrt{(25\cos^2\theta) + (25\sin^2\theta)}\,d\theta$$
$$= \int_{\pi/2}^{\pi} 5\,d\theta = \left[5\theta\right]_{\pi/2}^{\pi} = \frac{5\pi}{2} \quad \text{(Semicircle)}$$

111. $r = 2\sin\theta$

$$\frac{dr}{d\theta} = 2\cos\theta$$
$$\sqrt{r^2 + \left(\frac{dr}{d\theta}\right)^2} = \sqrt{4\sin^2\theta + 4\cos^2\theta} = 2$$
$$S = 2\pi\int_0^{\pi}(2\sin\theta)\sin\theta(2)\,d\theta$$
$$= 8\pi\int_0^{\pi}\frac{1 - \cos 2\theta}{2}\,d\theta$$
$$= 4\pi\left[\theta - \frac{\sin 2\theta}{2}\right]_0^{\pi}$$
$$= 4\pi^2$$

113. $r = \dfrac{6}{1 - \sin\theta}$

$e = 1$,

Parabola

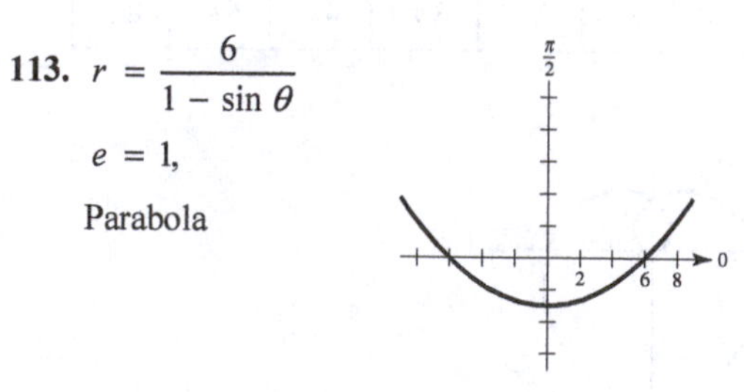

115. $r = \dfrac{6}{3 + 2\cos\theta} = \dfrac{2}{1 + (2/3)\cos\theta}, e = \dfrac{2}{3}$

Ellipse

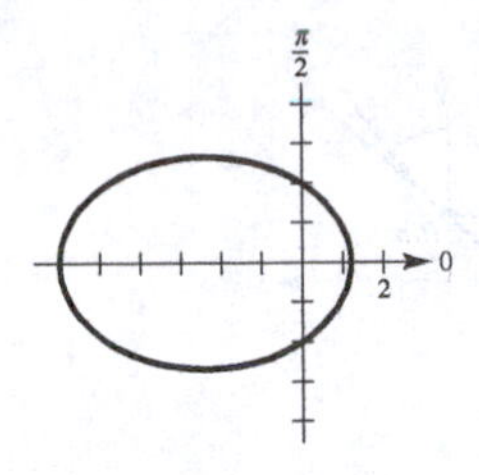

117. $r = \dfrac{4}{2 - 3\sin\theta} = \dfrac{2}{1 - (3/2)\sin\theta}, e = \dfrac{3}{2}$

Hyperbola

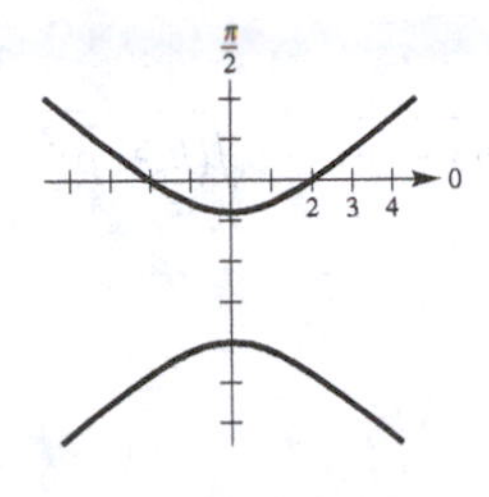

119. Parabola

$e = 1$

$x = 5 \Rightarrow d = 5$

$$r = \frac{ed}{1 + e\cos\theta} = \frac{5}{1 + \cos\theta}$$

121. Hyperbola, $e = 3, y = 3$

$d = 3$

$$r = \frac{ed}{1 + e\sin\theta} = \frac{3(3)}{1 + 3\sin\theta} = \frac{9}{1 + 3\sin\theta}$$

123. Parabola

Vertex: $\left(2, \dfrac{\pi}{2}\right)$

Focus: $(0, 0)$

$e = 1, d = 4$

$$r = \frac{4}{1 + \sin\theta}$$

125. Ellipse

Vertices: $(5, 0), (1, \pi)$

Focus: $(0, 0)$

$a = 3, c = 2, e = \dfrac{2}{3}, d = \dfrac{5}{2}$

$$r = \frac{\left(\frac{2}{3}\right)\left(\frac{5}{2}\right)}{1 - \left(\frac{2}{3}\right)\cos\theta} = \frac{5}{3 - 2\cos\theta}$$

Problem Solving for Chapter 10

1. (a)

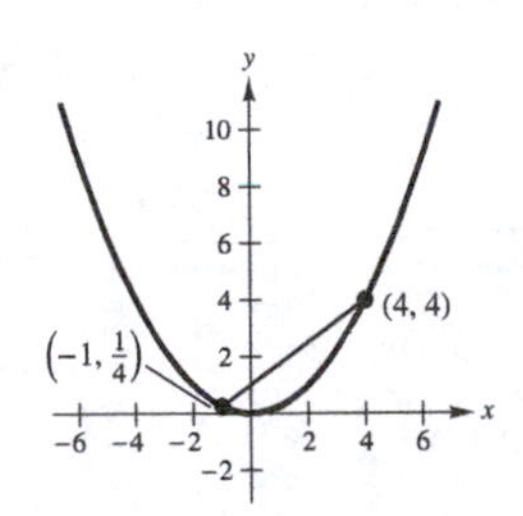

(b) $x^2 = 4y$

$2x = 4y'$

$y' = \dfrac{1}{2}x$

$y - 4 = 2(x - 4) \Rightarrow y = 2x - 4$ Tangent line at $(4, 4)$

$y - \frac{1}{4} = -\frac{1}{2}(x + 1) \Rightarrow y = -\frac{1}{2}x - \frac{1}{4}$ Tangent line at $\left(-1, \frac{1}{4}\right)$

Tangent lines have slopes of 2 and $-\frac{1}{2} \Rightarrow$ perpendicular.

(c) Intersection:

$2x - 4 = -\frac{1}{2}x - \frac{1}{4}$

$8x - 16 = -2x - 1$

$10x = 15$

$x = \frac{3}{2} \Rightarrow \left(\frac{3}{2}, -1\right)$

Point of intersection, $\left(\frac{3}{2}, -1\right)$, is on directrix $y = -1$.

3. Consider $x^2 = 4py$ with focus $F = (0, p)$.

Let $P(a, b)$ be point on parabola.

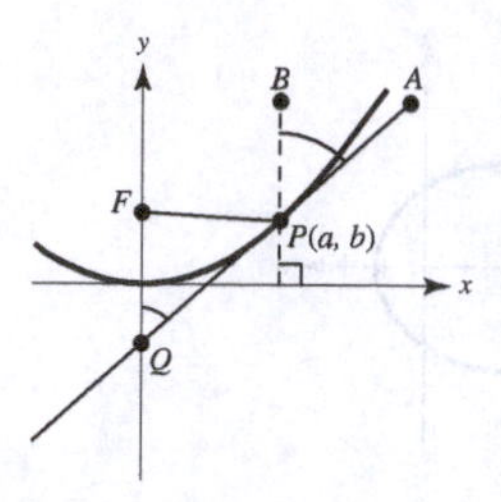

$$2x = 4py' \Rightarrow y' = \frac{x}{2p}$$

$$y - b = \frac{a}{2p}(x - a) \quad \text{Tangent line at } P$$

For $x = 0, y = b + \dfrac{a}{2p}(-a) = b - \dfrac{a^2}{2p} = b - \dfrac{4pb}{2p} = -b.$

So, $Q = (0, -b)$.

ΔFQP is isosceles because

$|FQ| = p + b$

$$|FP| = \sqrt{(a - 0)^2 + (b - p)^2} = \sqrt{a^2 + b^2 - 2bp + p^2} = \sqrt{4pb + b^2 - 2bp + p^2} = \sqrt{(b + p)^2} = b + p.$$

So, $\measuredangle FQP = \measuredangle BPA = \measuredangle FPQ$.

5. (a) $y^2 = \dfrac{t^2(1 - t^2)^2}{(1 + t^2)^2}, x^2 = \dfrac{(1 - t^2)^2}{(1 + t^2)^2}$

$$\frac{1 - x}{1 + x} = \frac{1 - \left(\dfrac{1 - t^2}{1 + t^2}\right)}{1 + \left(\dfrac{1 - t^2}{1 + t^2}\right)} = \frac{2t^2}{2} = t^2$$

So, $y^2 = x^2\left(\dfrac{1 - x}{1 + x}\right)$.

(b)

$$r^2 \sin^2 \theta = r^2 \cos^2 \theta\left(\frac{1 - r\cos\theta}{1 + r\cos\theta}\right)$$

$$\sin^2 \theta(1 + r\cos\theta) = \cos^2 \theta(1 - r\cos\theta)$$

$$r\cos\theta\sin^2\theta + \sin^2\theta = \cos^2\theta - r\cos^3\theta$$

$$r\cos\theta\left(\sin^2\theta + \cos^2\theta\right) = \cos^2\theta - \sin^2\theta$$

$$r\cos\theta = \cos 2\theta$$

$$r = \cos 2\theta \cdot \sec\theta$$

(c)

(d) $r(\theta) = 0$ for $\theta = \dfrac{\pi}{4}, \dfrac{3\pi}{4}$.

So, $y = x$ and $y = -x$ are tangent lines to curve at the origin.

(e) $y'(t) = \dfrac{(1 + t^2)(1 - 3t^2) - (t - t^3)(2t)}{(1 + t^2)^2} = \dfrac{1 - 4t^2 - t^4}{(1 + t^2)^2} = 0$

$$t^4 + 4t^2 - 1 = 0 \Rightarrow t^2 = -2 \pm \sqrt{5} \Rightarrow x = \frac{1 - (-2 \pm \sqrt{5})}{1 + (-2 \pm \sqrt{5})} = \frac{3 \mp \sqrt{5}}{-1 \pm \sqrt{5}} = \frac{3 - \sqrt{5}}{-1 + \sqrt{5}} = \frac{\sqrt{5} - 1}{2}$$

$$\left(\frac{\sqrt{5} - 1}{2}, \pm\frac{\sqrt{5} - 1}{2}\sqrt{-2 + \sqrt{5}}\right)$$

7. (a)

Generated by Mathematica

(b) $(-x, -y) = \left(-\int_0^t \cos\frac{\pi u^2}{2}\,du, -\int_0^t \sin\frac{\pi u^2}{2}\,du\right)$ is on the curve whenever (x, y) is on the curve.

(c) $x'(t) = \cos\frac{\pi t^2}{2}, y'(t) = \sin\frac{\pi t^2}{2},$

$x'(t)^2 + y'(t)^2 = 1$

So, $s = \int_0^a dt = a.$

On $[-\pi, \pi]$, $s = 2\pi.$

9. $r = \dfrac{ab}{a\sin\theta + b\cos\theta}, \quad 0 \le \theta \le \dfrac{\pi}{2}$

$$r(a\sin\theta + b\cos\theta) = ab$$

$$ay + bx = ab$$

$$\frac{y}{b} + \frac{x}{a} = 1$$

Line segment

$\text{Area} = \dfrac{1}{2}ab$

11. Let (r, θ) be on the graph.

$$\sqrt{r^2 + 1 + 2r\cos\theta}\sqrt{r^2 + 1 - 2r\cos\theta} = 1$$

$$\left(r^2 + 1\right)^2 - 4r^2\cos^2\theta = 1$$

$$r^4 + 2r^2 + 1 - 4r^2\cos^2\theta = 1$$

$$r^2\left(r^2 - 4\cos^2\theta + 2\right) = 0$$

$$r^2 = 4\cos^2\theta - 2$$

$$r^2 = 2\left(2\cos^2\theta - 1\right)$$

$$r^2 = 2\cos 2\theta$$

13. If a dog is located at (r, θ) in the first quadrant, then its neighbor is at $\left(r, \theta + \dfrac{\pi}{2}\right)$:

$(x_1, y_1) = (r\cos\theta, r\sin\theta)$ and $(x_2, y_2) = (-r\sin\theta, r\cos\theta)$.

The slope joining these points is $\dfrac{r\cos\theta - r\sin\theta}{-r\sin\theta - r\cos\theta} = \dfrac{\sin\theta - \cos\theta}{\sin\theta + \cos\theta}$ = slope of tangent line at (r, θ).

$$\frac{dy}{dx} = \frac{\frac{dy}{dr}}{\frac{dx}{dr}} = \frac{\frac{dr}{d\theta}\sin\theta + r\cos\theta}{\frac{dr}{d\theta}\cos\theta - r\sin\theta} = \frac{\sin\theta - \cos\theta}{\sin\theta + \cos\theta} \Rightarrow \frac{dr}{d\theta} = -r\frac{dr}{r} = -d\theta \ln r = -\theta + C_1 r = e^{-\theta + C_1} r = Ce^{-\theta}$$

$$r\left(\frac{\pi}{4}\right) = \frac{d}{\sqrt{2}} \Rightarrow r = Ce^{-\pi/4} = \frac{d}{\sqrt{2}} \Rightarrow C = \frac{d}{\sqrt{2}}e^{\pi/4}$$

Finally, $r = \dfrac{d}{\sqrt{2}}e^{((\pi/4) - \theta)}, \theta \ge \dfrac{\pi}{4}.$

15. (a) In ΔOCB, $\cos\theta = \dfrac{2a}{OB} \Rightarrow OB = 2a \cdot \sec\theta.$

In ΔOAC, $\cos\theta = \dfrac{OA}{2a} \Rightarrow OA = 2a \cdot \cos\theta.$

$$r = OP = AB = OB - OA = 2a(\sec\theta - \cos\theta) = 2a\left(\frac{1}{\cos\theta} - \cos\theta\right) = 2a \cdot \frac{\sin^2\theta}{\cos\theta} = 2a \cdot \tan\theta\sin\theta$$

(b) $x = r\cos\theta = (2a\tan\theta\sin\theta)\cos\theta = 2a\sin^2\theta$

$y = r\sin\theta = (2a\tan\theta\sin\theta)\sin\theta = 2a\tan\theta \cdot \sin^2\theta, -\dfrac{\pi}{2} < \theta < \dfrac{\pi}{2}$

Let $t = \tan\theta, -\infty < t < \infty.$

Then $\sin^2\theta = \dfrac{t^2}{1 + t^2}$ and $x = 2a\dfrac{t^2}{1 + t^2}, y = 2a\dfrac{t^3}{1 + t^2}.$

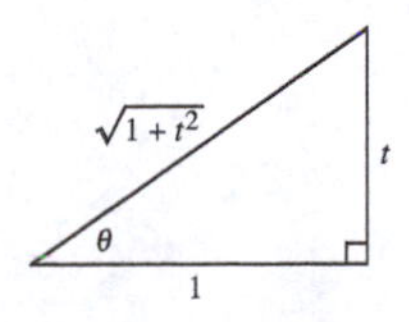

(c)
$$r = 2a \tan\theta \sin\theta$$
$$r\cos\theta = 2a\sin^2\theta$$
$$r^3\cos\theta = 2a\,r^2\sin^2\theta$$
$$\left(x^2 + y^2\right)x = 2ay^2$$
$$y^2 = \frac{x^3}{(2a - x)}$$

17.

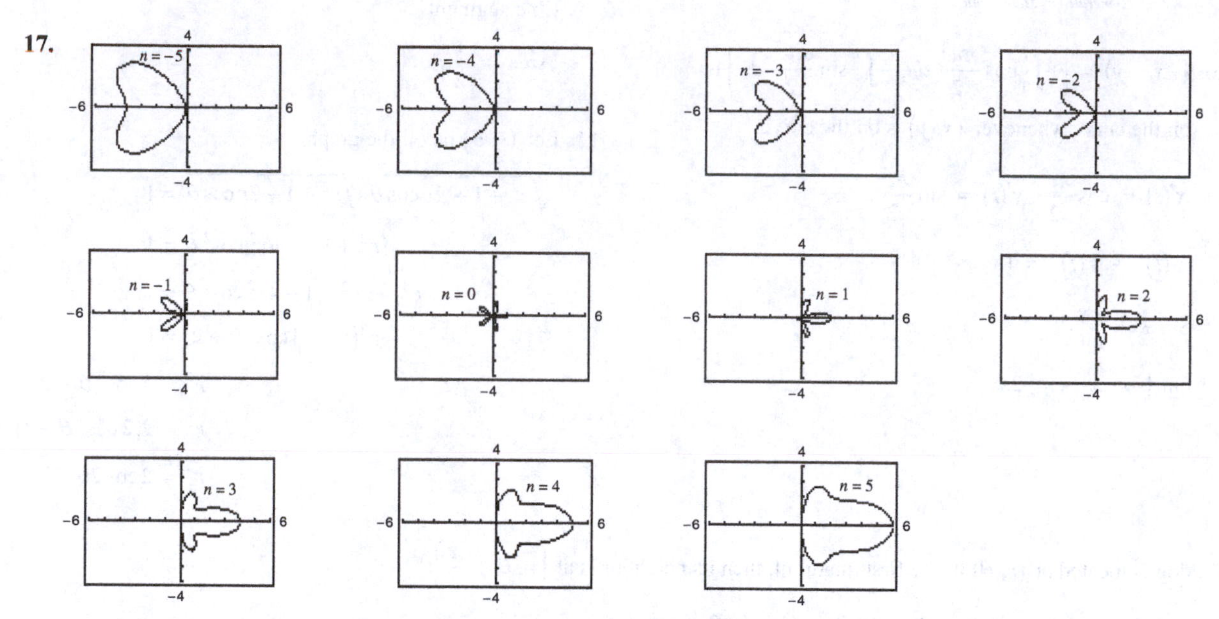

$n = 1, 2, 3, 4, 5$ produce "bells"; $n = -1, -2, -3, -4, -5$ produce "hearts".

Appendix C.1 Real Number and the Real Number Line

1. $0.7 = \frac{7}{10}$

   Rational

3. $\frac{3\pi}{2}$

   Irrational
   (since π is irrational)

5. $4.345\overline{1451}$

   Rational

7. $\sqrt[3]{64} = 4$

   Rational

9. $4\frac{5}{8} = \frac{45}{8}$

   Rational

11. Let $x = 0.36\overline{36}$.

$$100x = 36.36\overline{36}$$
$$-x = -0.36\overline{36}$$
$$99x = 36$$
$$x = \frac{36}{99} = \frac{4}{11}$$

13. Let $x = 0.297\overline{297}$.

$$1{,}000x = 297.297\overline{297}$$
$$-x = -0.297\overline{297}$$
$$999x = 297$$
$$x = \frac{297}{999} = \frac{11}{37}$$

15. Given $a < b$:

    (a) $a + 2 < b + 2$; True

    (b) $5b < 5a$; False

    (c) $5 - a > 5 - b$; True

    (d) $\frac{1}{a} < \frac{1}{b}$; False

    (e) $(a - b)(b - a) > 0$; False

17. x is greater than -3 and less than 3.

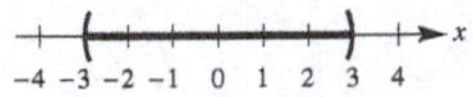

    The interval is bounded.

19. x is less than, or equal to, 5.

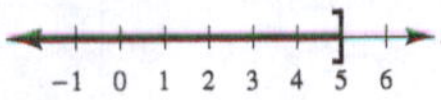

    The interval is unbounded.

21. $y \geq 4, [4, \infty)$

23. $0.03 < r \leq 0.07$,
    $(0.03, 0.07]$

25.
$$2x - 1 \geq 0$$
$$2x \geq 1$$
$$x \geq \frac{1}{2}$$

27.
$$-4 < 2x - 3 < 4$$
$$-1 < 2x < 7$$
$$-\frac{1}{2} < x < \frac{7}{2}$$

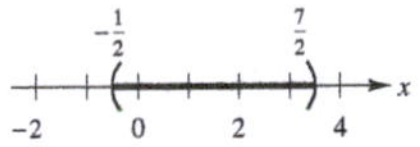

29.
$$\frac{x}{2} + \frac{x}{3} > 5$$
$$3x + 2x > 30$$
$$5x > 30$$
$$x > 6$$

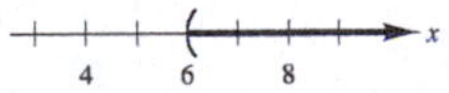

31. $|x| < 1 \Rightarrow -1 < x < 1$

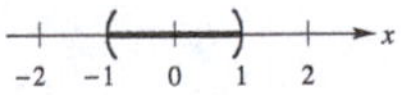

33.
$$\left|\frac{x - 3}{2}\right| \geq 5$$
$$x - 3 \geq 10 \quad \text{or} \quad x - 3 \leq -10$$
$$x \geq 13 \qquad\qquad x \leq -7$$

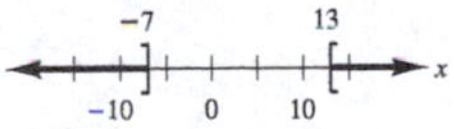

35. $|x - a| < b$

$-b < x - a < b$

$a - b < x < a + b$

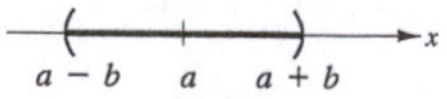

37. $|2x + 1| < 5$

$-5 < 2x + 1 < 5$

$-6 < 2x < 4$

$-3 < x < 2$

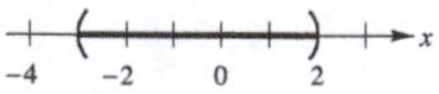

39. $\left|1 - \frac{2x}{3}\right| < 1$

$-1 < 1 - \frac{2x}{3} < 1$

$-2 < -\frac{2x}{3} < 0$

$3 > x > 0$

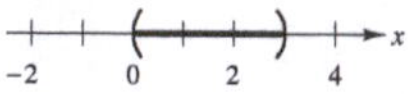

41. $x^2 \le 3 - 2x$

$x^2 + 2x - 3 \le 0$

$(x + 3)(x - 1) \le 0$

Test intervals:

$(-\infty, -3), (-3, 1), (1, \infty)$

Solution: $-3 \le x \le 1$

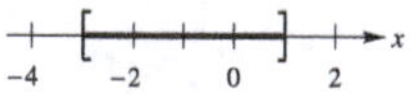

43. $x^2 + x - 1 \le 5$

$x^2 + x - 6 \le 0$

$(x + 3)(x - 2) \le 0$

$x = -3$

$x = 2$

Test intervals: $(-\infty, -3), (-3, 2), (2, \infty)$

Solution: $-3 \le x \le 2$

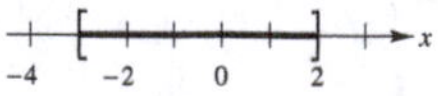

45. $a = -1, b = 3$

Directed distance from a to b: 4

Directed distance from b to a: -4

Distance between a and b: 4

47. (a) $a = 126, b = 75$

Directed distance from a to b: -51

Directed distance from b to a: 51

Distance between a and b: 51

(b) $a = -126, b = -75$

Directed distance from a to b: 51

Directed distance from b to a: -51

Distance between a and b: 51

49. $a = -2, b = 2$

Midpoint: 0

Distance between midpoint and each endpoint: 2

$|x - 0| \le 2$

$|x| \le 2$

51. $a = 0, b = 4$

Midpoint: 2

Distance between midpoint and each endpoint: 2

$|x - 2| > 2$

53. (a) All numbers that are at most 10 units from 12

$|x - 12| \le 10$

(b) All numbers that are at least 10 units from 12

$|x - 12| \ge 10$

55. $a = -1, b = 3$

Midpoint: $\frac{-1 + 3}{2} = 1$

57. (a) $[7, 21]$

Midpoint: 14

(b) $[8.6, 11.4]$

Midpoint: 10

59. $R = 115.95x, C = 95x + 750, R > C$

$115.95x > 95x + 750$

$20.95x > 750$

$x > 35.7995$

$x \ge 36$ units

61. $\left|\dfrac{x-50}{5}\right| \ge 1.645$

$$\frac{x-50}{5} \le -1.645 \quad \text{or} \quad \frac{x-50}{5} \ge 1.645$$
$$x - 50 \le -8.225 \qquad x - 50 \ge 8.225$$
$$x \le 41.775 \qquad x \ge 58.225$$
$$x \le 41 \qquad x \ge 59$$

63. (a) $\pi \approx 3.1415926535$

$$\frac{355}{113} = 3.141592920$$
$$\frac{355}{113} > \pi$$

(b) $\pi \approx 3.1415926535$

$$\frac{22}{7} \approx 3.142857143$$
$$\frac{22}{7} > \pi$$

65. Speed of light: 2.998×10^8 meters per second

Distance traveled in one year = rate × time

$$d = \left(2.998 \times 10^8\right) \times \left(365 \times 24 \times 60 \times 60\right)$$

days × hours × minutes × seconds

$$= \left(2.998 \times 10^8\right) \times \left(3.1536 \times 10^7\right) \approx 9.45 \times 10^{15}$$

This is best estimated by (b).

67. False; 2 is a nonzero integer and the reciprocal of 2 is $\frac{1}{2}$.

69. True

71. True; if $x < 0$, then $|x| = -x = \sqrt{x^2}$.

73. If $a \ge 0$ and $b \ge 0$, then $|ab| = ab = |a||b|$.

If $a < 0$ and $b < 0$, then $|ab| = ab = (-a)(-b) = |a||b|$.

If $a \ge 0$ and $b < 0$, then $|ab| = -ab = a(-b) = |a||b|$.

If $a < 0$ and $b \ge 0$, then $|ab| = -ab = (-a)b = |a||b|$.

75. $\left|\dfrac{a}{b}\right| = \left|a\left(\dfrac{1}{b}\right)\right|$

$$= |a|\left|\frac{1}{b}\right| = |a| \cdot \frac{1}{|b|} = \frac{|a|}{|b|},\ b \ne 0$$

77. $n = 1, \quad |a| = |a|$

$n = 2, \quad |a^2| = |a \cdot a| = |a||a| = |a|^2$

$n = 3, \quad |a^3| = |a^2 \cdot a| = |a^2||a| = |a|^2|a| = |a|^3$

$\vdots$

$$|a^n| = |a^{n-1}a| = |a^{n-1}||a| = |a|^{n-1}|a| = |a|^n$$

79. $|a| \le k \Leftrightarrow \sqrt{a^2} \le k \Leftrightarrow a^2 \le k^2 \Leftrightarrow a^2 - k^2 \le 0 \Leftrightarrow (a + k)(a - k) \le 0 \Leftrightarrow -k \le a \le k,\ k > 0$

81.

$$\left.\begin{aligned} |7-12| &= |-5| = 5 \\ |7| - |12| &= 7 - 12 = -5 \end{aligned}\right\} |7-12| > |7| - |12|$$

$$\left.\begin{aligned} |12-7| &= |5| = 5 \\ |12| - |7| &= 12 - 7 = 5 \end{aligned}\right\} |12-7| = |12| - |7|$$

We know that $|a||b| \ge ab$. Thus, $-2|a||b| \le -2ab$. Since $a^2 = |a|^2$ and $b^2 = |b|^2$, we have

$$|a|^2 + |b|^2 - 2|a||b| \le a^2 + b^2 - 2ab$$
$$0 \le \left(|a| - |b|\right)^2 \le (a-b)^2$$
$$\sqrt{\left(|a| - |b|\right)^2} \le \sqrt{(a-b)^2}$$
$$\big||a| - |b|\big| \le |a - b|.$$

Since $|a| - |b| \le \big||a| - |b|\big|$, we have $|a| - |b| \le |a - b|$. Thus, $|a - b| \ge |a| - |b|$.

Appendix C.2 The Cartesian Plane

1. $d = \sqrt{(4-2)^2 + (5-1)^2}$

$= \sqrt{4+16} = \sqrt{20} = 2\sqrt{5}$

Midpoint: $\left(\dfrac{4+2}{2}, \dfrac{5+1}{2}\right) = (3, 3)$

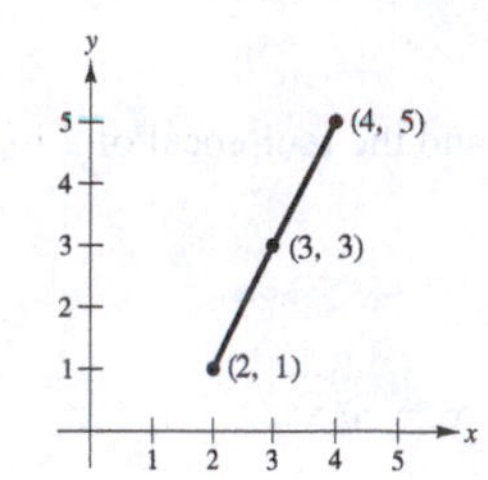

3. $d = \sqrt{\left(\dfrac{1}{2} + \dfrac{3}{2}\right)^2 + (1+5)^2}$

$= \sqrt{4+36} = \sqrt{40} = 2\sqrt{10}$

Midpoint: $\left(\dfrac{(-3/2)+(1/2)}{2}, \dfrac{-5+1}{2}\right) = \left(-\dfrac{1}{2}, -2\right)$

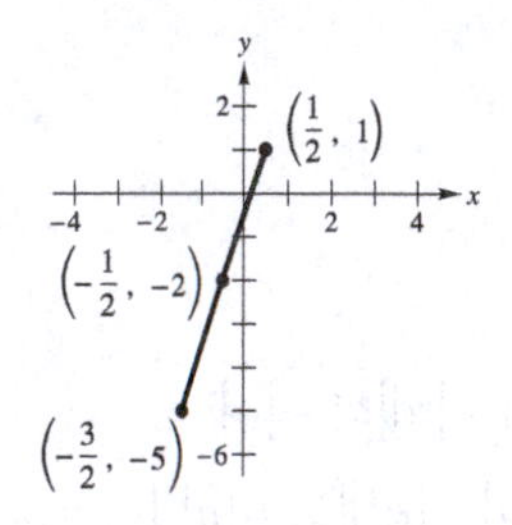

5. $d = \sqrt{(-1-1)^2 + \left(1-\sqrt{3}\right)^2}$

$= \sqrt{4+1-2\sqrt{3}+3} = \sqrt{8-2\sqrt{3}}$

Midpoint: $\left(\dfrac{-1+1}{2}, \dfrac{1+\sqrt{3}}{2}\right) = \left(0, \dfrac{1+\sqrt{3}}{2}\right)$

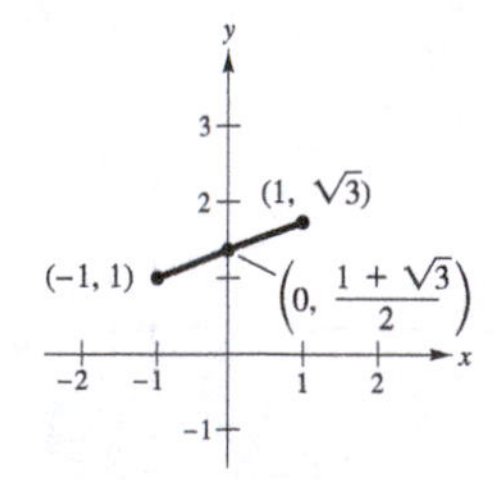

7. $x = -2 \Rightarrow$ quadrants II, III

$y > 0 \Rightarrow$ quadrants I, II

Therefore, quadrant II

9. $xy > 0 \Rightarrow$ quadrants I or III

11. $d_1 = \sqrt{9+36} = \sqrt{45}$

$d_2 = \sqrt{4+1} = \sqrt{5}$

$d_3 = \sqrt{25+25} = \sqrt{50}$

$(d_1)^2 + (d_2)^2 = (d_3)^2$

Right triangle

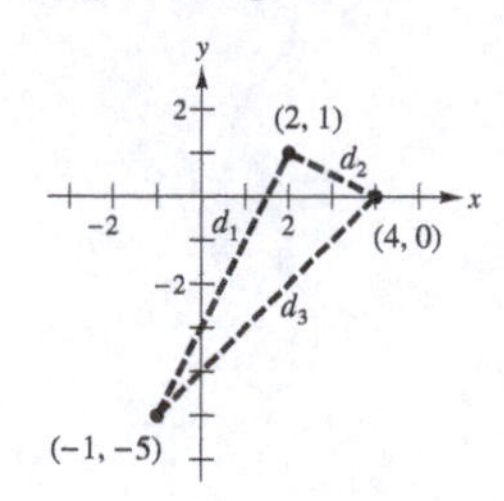

13. $d_1 = d_2 = d_3 = d_4 = \sqrt{5}$

Rhombus

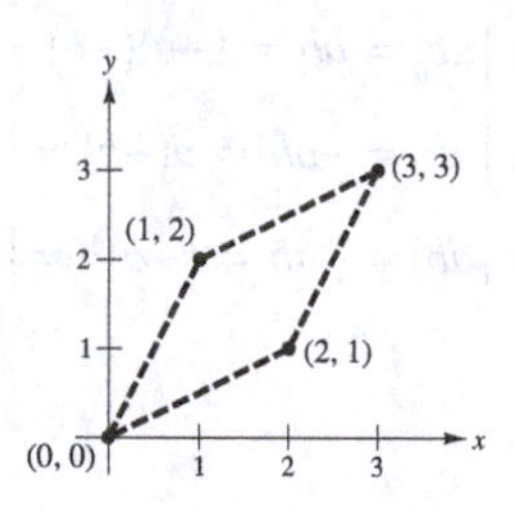

15.

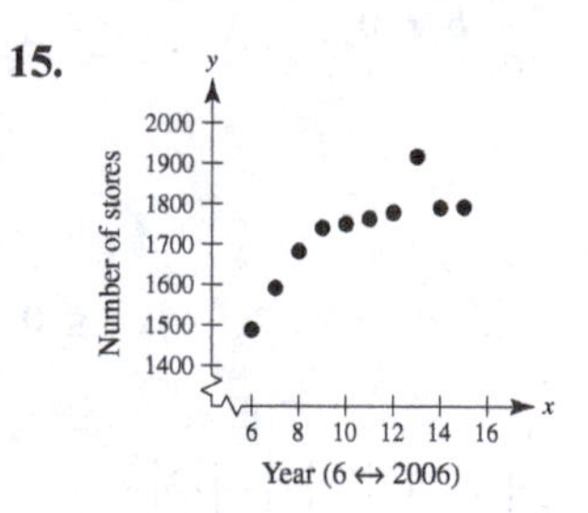

17. $d_1 = \sqrt{4+16} = \sqrt{20} = 2\sqrt{5}$

$d_2 = \sqrt{1+4} = \sqrt{5}$

$d_3 = \sqrt{9+36} = 3\sqrt{5}$

$d_1 + d_2 = d_3$

Collinear

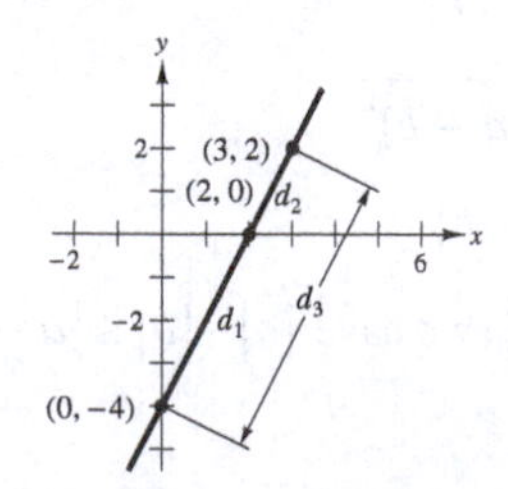

19. $d_1 = \sqrt{1+1} = \sqrt{2}$

$d_2 = \sqrt{9+4} = \sqrt{13}$

$d_3 = \sqrt{16+9} = \sqrt{5}$

$d_1 + d_2 \neq d_3$

Not collinear

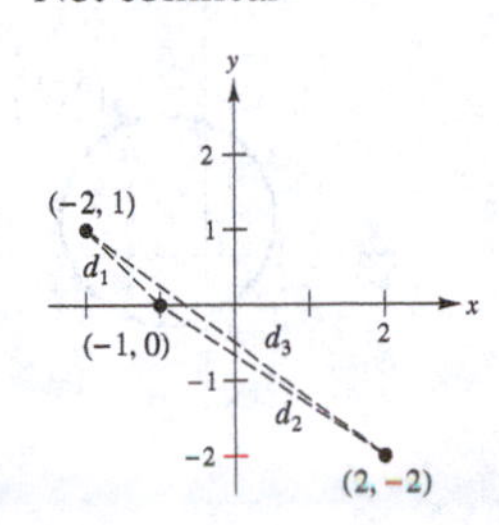

21. $5 = \sqrt{(x-0)^2 + (-4-0)^2}$

$5 = \sqrt{x^2 + 16}$

$25 = x^2 + 16$

$9 = x^2$

$x = \pm 3$

23. $8 = \sqrt{(3-0)^2 + (y-0)^2}$

$8 = \sqrt{9 + y^2}$

$64 = 9 + y^2$

$55 = y^2$

$y = \pm\sqrt{55}$

25. The midpoint of the given line segment is $\left(\frac{x_1 + x_2}{2}, \frac{y_1 + y_2}{2}\right)$.

The midpoint between (x_1, y_1) and $\left(\frac{x_1 + x_2}{2}, \frac{y_1 + y_2}{2}\right)$ is $\left(\frac{x_1 + (x_1 + x_2)/2}{2}, \frac{y_1 + (y_1 + y_2)/2}{2}\right) = \left(\frac{3x_1 + x_2}{4}, \frac{3y_1 + y_2}{4}\right)$.

The midpoint between $\left(\frac{x_1 + x_2}{2}, \frac{y_1 + y_2}{2}\right)$ and (x_2, y_2) $\left(\frac{(x_1 + x_2)/2 + x_2}{2}, \frac{(y_1 + y_2)/2 + y_2}{2}\right) = \left(\frac{x_1 + 3x_2}{4}, \frac{y_1 + 3y_2}{4}\right)$.

Thus, the three points are

$\left(\frac{3x_1 + x_2}{4}, \frac{3y_1 + y_2}{4}\right), \left(\frac{x_1 + x_2}{2}, \frac{y_1 + y_2}{2}\right), \left(\frac{x_1 + 3x_2}{4}, \frac{y_1 + 3y_2}{4}\right)$.

27. Center: $(0, 0)$

Radius: 1

Matches graph (c)

28. Center: $(1, 3)$

Radius: 2

Matches graph (b)

29. Center: $(1, 0)$

Radius: 0

Matches graph (a)

30. Center: $\left(-\frac{1}{2}, \frac{3}{4}\right)$

Radius: $\frac{1}{2}$

Matches graph (d)

31. $(x-0)^2 + (y-0)^2 = (3)^2$

$x^2 + y^2 - 9 = 0$

33. $(x-2)^2 + (y+1)^2 = (4)^2$

$x^2 + y^2 - 4x + 2y - 11 = 0$

35. Radius $= \sqrt{(-1-0)^2 + (2-0)^2} = \sqrt{5}$

$(x+1)^2 + (y-2)^2 = 5$

$x^2 + 2x + 1 + y^2 - 4y + 4 = 5$

$x^2 + y^2 + 2x - 4y = 0$

37. Center = Midpoint = $(3, 2)$

Radius $= \sqrt{10}$

$(x-3)^2 + (y-2)^2 = \left(\sqrt{10}\right)^2$

$x^2 - 6x + 9 + y^2 - 4y + 4 = 10$

$x^2 + y^2 - 6x - 4x + 3 = 0$

39. Place the center of the earth at the origin. Then we have

$x^2 + y^2 = (22{,}000 + 4{,}000)^2$

$x^2 + y^2 = 26{,}000^2$.

41. $$x^2 + y^2 - 2x + 6y + 6 = 0$$
$$(x^2 - 2x + 1) + (y^2 + 6y + 9) = -6 + 1 + 9$$
$$(x - 1)^2 + (y + 3)^2 = 4$$

Center: $(1, -3)$

Radius: 2

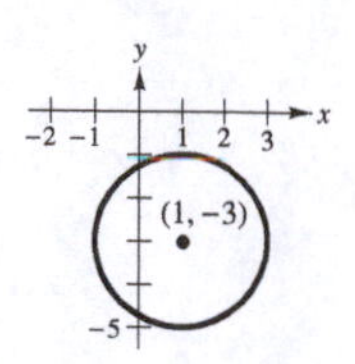

43. $$x^2 + y^2 - 2x + 6y + 10 = 0$$
$$(x^2 - 2x + 1) + (y^2 + 6y + 9) = -10 + 1 + 9$$
$$(x - 1)^2 + (y + 3)^2 = 0$$

Only a point $(1, -3)$

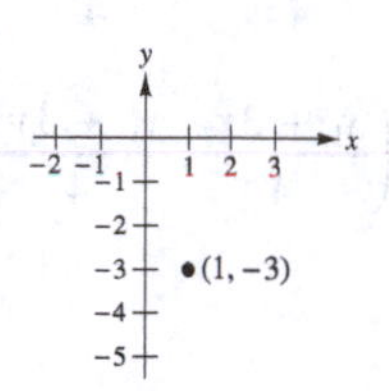

45. $$2x^2 + 2y^2 - 2x - 2y - 3 = 0$$
$$2\left(x^2 - x + \frac{1}{4}\right) + 2\left(y^2 - y + \frac{1}{4}\right) = 3 + \frac{1}{2} + \frac{1}{2}$$
$$\left(x - \frac{1}{2}\right)^2 - \left(y - \frac{1}{2}\right)^2 = 2$$

Center: $\left(\frac{1}{2}, \frac{1}{2}\right)$

Radius: $\sqrt{2}$

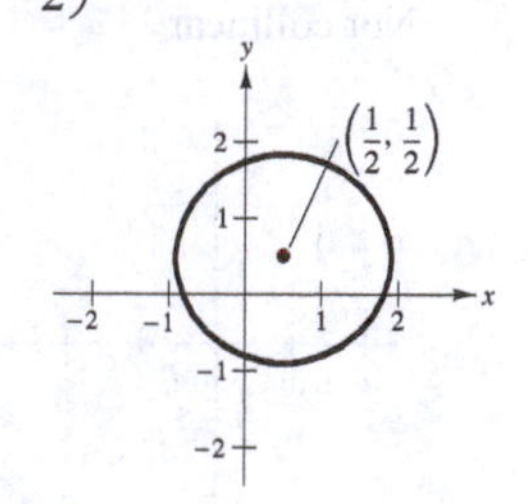

47. $$16x^2 + 16y^2 + 16x + 40y - 7 = 0$$
$$16\left(x^2 + x + \frac{1}{4}\right) + 16\left(y^2 + \frac{5y}{2} + \frac{25}{16}\right) = 7 + 4 + 25$$
$$16\left(x + \frac{1}{2}\right)^2 + 16\left(y + \frac{5}{4}\right)^2 = 36$$
$$\left(x + \frac{1}{2}\right)^2 + \left(y + \frac{5}{4}\right)^2 = \frac{9}{4}$$

Center: $\left(-\frac{1}{2}, -\frac{5}{4}\right)$

Radius: $\frac{3}{2}$

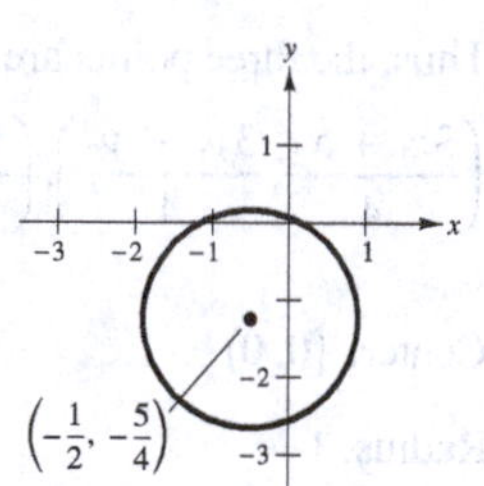

49. (a) $$4x^2 + 4y^2 - 4x + 24y - 63 = 0$$
$$x^2 + y^2 - x + 6y = \frac{63}{4}$$
$$\left(x^2 - x + \frac{1}{4}\right) + (y^2 + 6y + 9) = \frac{63}{4} + \frac{1}{4} + 9$$
$$\left(x - \frac{1}{2}\right)^2 + (y + 3)^2 = 25$$
$$(y + 3)^2 = 25 - \left(x - \frac{1}{2}\right)^2$$
$$y + 3 = \pm\sqrt{25 - \left(x - \frac{1}{2}\right)^2}$$
$$y = -3 \pm \sqrt{25 - \left(x - \frac{1}{2}\right)^2} = \frac{-6 \pm \sqrt{99 + 4x - 4x^2}}{2}$$

(b)

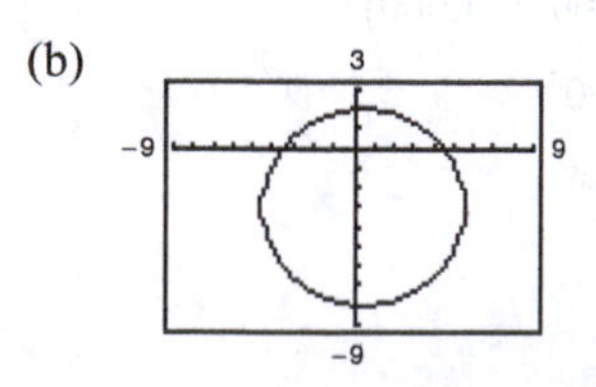

51. $x^2 + y^2 - 4x + 2y + 1 \le 0$

$(x^2 - 4x + 4) + (y^2 + 2y + 1) \le -1 + 4 + 1$

$(x - 2)^2 + (y + 1)^2 \le 4$

Center: $(2, -1)$

Radius: 2

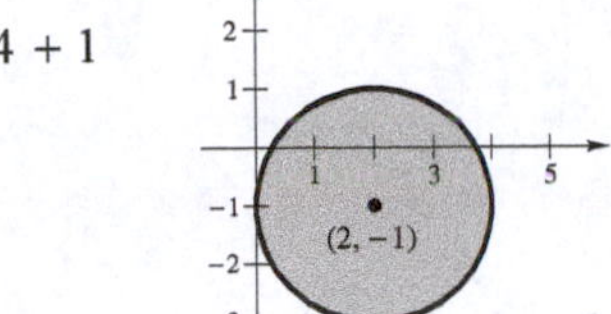

53. The distance between (x_1, y_1) and $\left(\dfrac{2x_1 + x_2}{3}, \dfrac{2y_1 + y_2}{3}\right)$ is

$$d = \sqrt{\left(x_1 - \frac{2x_1 + x_2}{3}\right)^2 + \left(y_1 - \frac{2y_1 + y_2}{3}\right)^2}$$

$$= \sqrt{\left(\frac{x_1 - x_2}{3}\right)^2 + \left(\frac{y_1 - y_2}{3}\right)^2}$$

$$= \sqrt{\frac{1}{9}\left[(x_1 - x_2)^2 + (y_1 - y_2)^2\right]} = \frac{1}{3}\sqrt{(x_1 - x_2)^2 + (y_1 - y_2)^2}$$

which is $\dfrac{1}{3}$ of the distance between (x_1, y_1) and (x_2, y_2).

$$\left(\frac{\left(\frac{2x_1 + x_2}{3}\right) + x_2}{2}, \frac{\left(\frac{2y_1 + y_2}{3}\right) + y_2}{2}\right) = \left(\frac{x_1 + 2x_2}{3}, \frac{y_1 + 2y_2}{3}\right)$$

is the second point of trisection.

55. True; if $ab < 0$ then either a is positive and b is negative (Quadrant IV) or a is negative and b is positive (Quadrant II).

57. True

59. Let one vertex be at $(0, 0)$ and another at $(a, 0)$.

Midpoint of $(0, 0)$ and (d, e) is $\left(\dfrac{d}{2}, \dfrac{e}{2}\right)$.

Midpoint of (b, c) and $(a, 0)$ is $\left(\dfrac{a + b}{2}, \dfrac{c}{2}\right)$.

Midpoint of $(0, 0)$ and $(a, 0)$ is $\left(\dfrac{a}{2}, 0\right)$.

Midpoint of (b, c) and (d, e) is $\left(\dfrac{b + d}{2}, \dfrac{c + e}{2}\right)$.

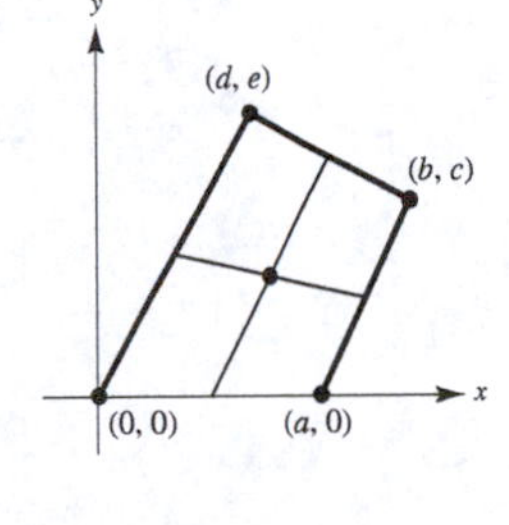

Midpoint of line segment joining $\left(\dfrac{d}{2}, \dfrac{e}{2}\right)$ and $\left(\dfrac{a + b}{2}, \dfrac{c}{2}\right)$ is $\left(\dfrac{a + b + d}{4}, \dfrac{c + e}{4}\right)$.

Midpoint of line segment joining $\left(\dfrac{a}{2}, 0\right)$ and $\left(\dfrac{b + d}{2}, \dfrac{c + e}{2}\right)$ is $\left(\dfrac{a + b + d}{4}, \dfrac{c + e}{4}\right)$.

Therefore the line segments intersect at their midpoints.

61. Let (a, b) be a point on the semicircle of radius r, centered at the origin. We will show that the angle at (a, b) is a right angle by verifying that $d_1^2 + d_2^2 = d_3^2$.

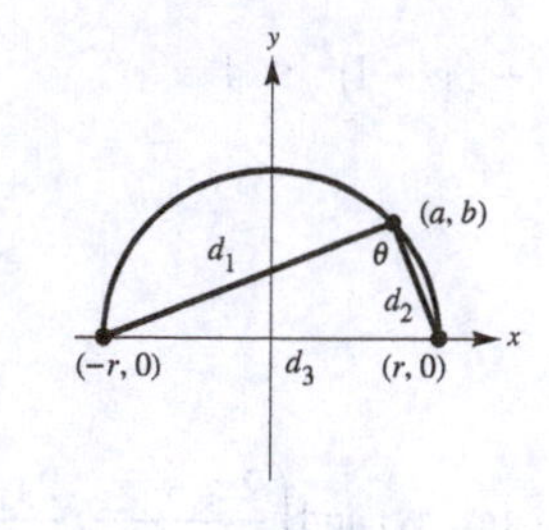

$$d_1^2 = (a + r)^2 + (b - 0)^2$$
$$d_2^2 = (a - r)^2 + (b - 0)^2$$
$$\begin{aligned} d_1^2 + d_2^2 &= \left(a^2 + 2ar + r^2 + b^2\right) + \left(a^2 - 2ar + r^2 + b^2\right) \\ &= 2a^2 + 2b^2 + 2r^2 \\ &= 2\left(a^2 + b^2\right) + 2r^2 \\ &= 2r^2 + 2r^2 \\ &= 4r^2 = (2r)^2 = d_3^2 \end{aligned}$$